Differentiation Rules

$f(t)$	$\dfrac{df}{dt} = \dot{f}(t)$
cu	$c\dot{u}$
$u + v$	$\dot{u} + \dot{v}$
uv	$u\dot{v} + v\dot{u}$
$\dfrac{u}{v}$	$\dfrac{v\dot{u} - u\dot{v}}{v^2}$
$u[v(t)]$	$\dot{u}[v(t)]\dot{v}(t)$

$$\frac{d}{dt} f[x(t), y(t)]$$

$$= \frac{\partial f}{\partial x}\dot{x} + \frac{\partial f}{\partial y}\dot{y}.$$

Differentiation Formulas

$f(t)$	$\dfrac{df}{dt} = \dot{f}(t)$
c	0
t^α	$\alpha t^{\alpha-1}$
e^t	e^t
a^t	$(\ln a)a^t$
$\ln t$	$\dfrac{1}{t}$
$\sin t$	$\cos t$
$\cos t$	$-\sin t$
$\tan t$	$\sec^2 t$
$\cot t$	$-\csc^2 t$
$\sec t$	$\tan t \sec t$
$\csc t$	$-\cot t \csc t$
$\text{arc sin } t$	$\dfrac{1}{\sqrt{1 - t^2}}$
$\text{arc tan } t$	$\dfrac{1}{1 + t^2}$
$\sinh t$	$\cosh t$
$\cosh t$	$\sinh t$
$\tanh t$	$\text{sech}^2\, t$

Polar Coordinates

$$\begin{cases} x = r\cos\theta \\ y = r\sin\theta \end{cases}$$

Spherical Coordinates

$$\begin{cases} x = \rho\sin\phi\cos\theta \\ y = \rho\sin\phi\sin\theta \\ z = \rho\cos\phi \end{cases}$$

Integrals (constant of integration omitted)

1. $$\int u\, dv = uv - \int v\, du$$

2. $$\int \frac{dx}{x} = \ln |$$

3. $$\int \frac{dx}{a^2 + x^2} =$$

4. $$\int \frac{dx}{a^2 - x^2} = \frac{1}{2a}\ln\left|\frac{x + a}{x - a}\right|$$

5. $$\int \frac{dx}{(a^2 + x^2)^n} = \frac{x}{(2n - 2)a^2(a^2 + x^2)^{n-1}} + \frac{2n - 3}{(2n - 2)a^2}\int \frac{dx}{(a^2 + x^2)^{n-1}} \qquad (n > 1)$$

6. $$\int \frac{dx}{\sqrt{a^2 - x^2}} = \text{arc sin}\left(\frac{x}{a}\right)$$

7. $$\int \sqrt{a^2 - x^2}\, dx = \frac{x}{2}\sqrt{a^2 - x^2} + \frac{a^2}{2}\,\text{arc sin}\left(\frac{x}{a}\right)$$

8. $$\int \frac{dx}{\sqrt{x^2 \pm a^2}} = \ln|x + \sqrt{x^2 \pm a^2}|$$

9. $$\int \sqrt{x^2 \pm a^2}\, dx = \frac{x}{2}\sqrt{x^2 \pm a^2} \pm \frac{a^2}{2}\ln|x + \sqrt{x^2 \pm a^2}|$$

(continued inside back cover)

Vectors

$$\mathbf{a}\cdot\mathbf{b} = a_1b_1 + a_2b_2 + a_3b_3$$

$$\mathbf{a}\times\mathbf{b} = \begin{vmatrix} \mathbf{i} & \mathbf{j} & \mathbf{k} \\ a_1 & a_2 & a_3 \\ b_1 & b_2 & b_3 \end{vmatrix}$$

$$\nabla = \mathbf{i}\frac{\partial}{\partial x} + \mathbf{j}\frac{\partial}{\partial y} + \mathbf{k}\frac{\partial}{\partial z}$$

$$\nabla f = \text{grad } f = (f_x, f_y, f_z)$$

$$\nabla\cdot\mathbf{v} = \text{div } \mathbf{v} = u_x + v_y + w_z$$

$$\nabla\times\mathbf{v} = \text{curl } \mathbf{v} = \begin{vmatrix} \mathbf{i} & \mathbf{j} & \mathbf{k} \\ \partial/\partial x & \partial/\partial y & \partial/\partial z \\ u & v & w \end{vmatrix}$$

Calculus

Calculus

Harley Flanders

Robert R. Korfhage

Justin J. Price

PURDUE UNIVERSITY

AP

ACADEMIC PRESS New York and London

ACADEMIC PRESS, INC.
111 Fifth Avenue, New York, New York 10003

United Kingdom Edition published by
ACADEMIC PRESS, INC. (LONDON) LTD.
Berkeley Square House, London W1X 6BA

LIBRARY OF CONGRESS CATALOG CARD NUMBER: 76-86368

AMS (MOS) 1970 Subject Classifications:
26-01, 26A03, 26A06, 26A09

Second Printing, 1971

PRINTED IN THE UNITED STATES OF AMERICA

Preface

Aims of This Book

1. To present calculus and elementary differential equations with a minimum of fuss—through practice, not theory.
2. To stress techniques, applications, and problem solving, rather than definitions, theorems, and proofs.
3. To emphasize numerical aspects such as approximations, order of magnitude, and concrete answers to problems.
4. To organize the topics consistent with the needs of students in their concurrent science and engineering courses.
5. To illustrate the usefulness of computers in applications of calculus.
6. To introduce vector methods and their applications in physical problems.

Why This Approach?

Calculus can be an exciting subject; no other gives so much new scope and power. Yet painful experience has shown that theory and rigor tend to stifle the excitement. The teaching of real variables to freshmen and sophomores has generally been a failure, a great disservice to students, and a source of well-deserved criticism from science and engineering departments.

Our presentation is informal; we reject the practice of writing calculus texts with the style and precision of research papers. Instead of formal definitions, theorems, and proofs, we include intuitive discussions, rules of procedure, and realistic problems. Occasionally we allow ourselves the liberties of circular arguments or slight inconsistencies when expedient. We omit technicalities that almost never occur in practice, rather than clutter the exposition "for the sake of completeness."

The thoughtful student who wants to know more theory will find in Chapter 36 a sketch of some theoretical high points and references to more detailed discussions.

We stress explicit computation, and when appropriate, indicate the value of computers in numerical work. However our book is not a computerized calculus; it is a calculus text that recognizes the increasing importance of computers in all branches of science. (The material on computer applications can be omitted without loss of continuity.)

Organization

This book presupposes reasonable skill in algebraic manipulation, familiarity with the trigonometric functions, and a bit of analytic geometry—graphs of functions and basic facts about straight lines and conic sections. However, some of these topics are reviewed briefly as needed.

The text is divided into three parts, corresponding more or less to a three-semester course. This division and the order of topics are merely guidelines, and can be modified if desired.

Part I presents an elementary introduction to most of the basic material of calculus: derivatives, direction fields, antiderivatives, integrals, volumes by slicing, partial derivatives, low order Taylor polynomials, exponential and trigonometric functions.

Part I is specifically designed for the typical student who needs the basic topics—but not in great depth—in his physics, chemistry, and engineering courses soon after he begins calculus. In particular he needs differentiation and integration of a few standard functions, the most elementary aspects of differential equations, and the concept of partial derivatives.

Part II includes a deepening of the material of Part I and several new topics: inverse functions, interpolation, numerical integration, first and second order differential equations, vectors, double integrals over rectangular regions.

The student should acquire in Parts I and II a working knowledge of the functions he will need in real life—their graphs, rates of growth, orders of magnitude, and interrelations.

Part III completes calculus with harder topics on approximations and several variables: Taylor series, approximate solutions of differential equations, complex functions, double and triple integrals with applications.

The final chapter is a brief introduction to theory.

Order of Topics

Considerable flexibility is possible in the order of topics, particularly after Chapter 19.

The material in Part I will move along briskly; there may be time (for those on a semester system) to include a few chapters from Part II. These can be inserted anywhere after Chapter 7. In Part I three sections are marked [optional].

There is even more leeway for rearrangement of topics in Parts II and III. After 19, several chapters can be omitted, permuted, or postponed. For example, Chapters 21 and 22 on differential equations can very well be left until much later. Chapters 30 (Approximate solutions of differential equations), 31 (Complex numbers), and 35 (Applications of multiple integrals) can be omitted entirely. Furthermore, Parts II and III include many optional sections. Chapter 36 (Calculus theory) may be studied at any time.

A few optional sections and a few exercises involve the use of matrices. These will provide meaningful applications for a previous or concurrent course in linear algebra, but they are not an integral part of the text.

Examples and Exercises

The worked examples are the core of the text. Many times, methods are explained more through choice examples than elaborate discussions. We are always result-oriented and insist on explicit numerical answers.

A number of topics are included as much because they are a source of meaningful, non-contrived problems as they are of interest in themselves.

Physical examples occur in some sections and in many exercises. The science or engineering student is usually impressed by this material, and motivated to learn it.

Students have difficulty with calculus problems for several reasons: (a) inability to perform lengthy algebraic and numerical calculations, (b) too many steps involved (the beginning student often does not know where to start a problem involving several steps), and (c) lack of space perception.

The exercises in Part I are easy and involve few steps. Each exercise set in Parts II and III begins with easy exercises but continues to harder ones, the level of difficulty increasing as the book progresses, so that by Part III the student is solving substantial calculus problems. Altogether, about thirty-one hundred exercises are included.

Those exercise sets containing problems for the computer also contain parallel problems for hand computation.

We recommend use of a slide rule and a book of tables (such as the C. R. C. Standard Mathematical Tables). We include some tables on pages 913–926, and basic formulas in the inside covers.

Illustrations

To solve space problems, the student must be able to make clear drawings. We have purposely restricted the illustrations in this book to simple line drawings, the kind the student can make himself (in fact, we include an optional section on how to do so). We emphasize accuracy in our drawings. For example, a tangent to a sphere which is parallel to the y-axis must actually appear so in a plane projection.

Acknowledgments

It remains to thank our typists, Allene Fritsch, Helen Sutton, Kathy Smith, and Elizabeth Young. Our reviewers, Peter Balise, Ettore Infante,

Paul Mielke, and three unknown reviewers, contributed numerous valuable suggestions for which we are most grateful.

Mistakes in the book are entirely our fault and we shall appreciate corrections.

Lafayette, Indiana

HARLEY FLANDERS
ROBERT R. KORFHAGE
JUSTIN J. PRICE

Contents

PART II

35
36

32. HIGHER PARTIAL DERIVATIVES

33. VECTOR OPERATIONS

34. MULTIPLE INTEGRALS

35. APPLICATIONS OF MULTIPLE INTEGRALS

36. CALCULUS THEORY

MATHEMATICAL TABLES

PART I

1. The Derivative

1. INTRODUCTION

The processes of nature are dynamic. Living matter grows; a planet moves in its orbit; a chemical reaction occurs at a certain rate; a rocket accelerates; a heavy object falls at increasing speed; a quantity of radioactive matter decays; a particle of fluid in a stream flows along its path with varying speed. Differential Calculus is the precise scientific theory that unifies the study of most situations in which there is dynamic change. It is indispensable for your further study of mathematics, physics, chemistry, engineering, modern biology, economics, virtually every exact science, both the theory and the application.

The main objects of study in Differential Calculus are functions. A function is a law which tells how one variable quantity is related to another. Given a particular function, Differential Calculus shows us the precise *rate of change* of the dependent variable relative to change in the independent variable. For example, if the dependent variable x is the distance a particle moves in time t, where t is the independent variable, then $x = x(t)$ is the law of motion of the particle; Calculus tells us the rate of change of distance per unit time, or instantaneous speed. As another example, the pressure P (dependent variable) of a gas confined in a cylinder of variable volume V (independent variable) varies according to the gas law

$$P = \frac{c}{V} \qquad (c \text{ a constant}).$$

From Calculus we learn the rate at which the pressure P changes relative to change in the volume V.

2. SLOPE

Differential Calculus deals with functions whose graphs are smooth curves. The graph of such a function $y = f(x)$ is usually drawn on rectangular graph paper (Fig. 2.1).

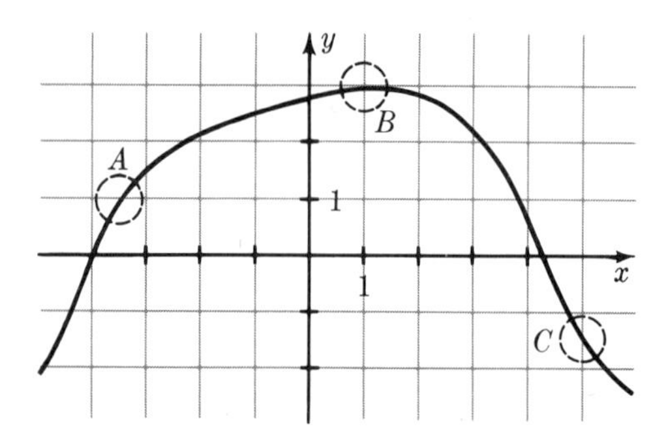

FIG. 2.1

Look at several portions of this graph through a microscope. If the magnification is sufficiently great, small portions appear to be almost straight (Fig. 2.2). Between P and Q in Detail A, the function y increases 0.005 units

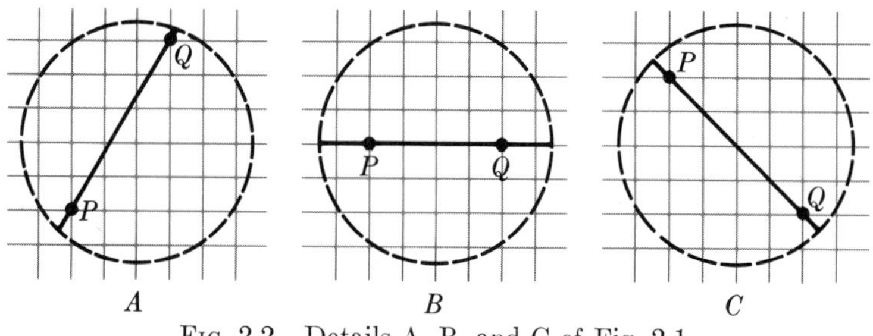

FIG. 2.2 Details A, B, and C of Fig. 2.1.

as x increases 0.003 units. It is reasonable to say that the rate of increase of the function is approximately

$$\frac{0.005}{0.003} = \frac{5}{3}.$$

In Detail B there is no increase in y as x increases 0.004, so the rate of change is approximately 0. Finally, in Detail C the function actually decreases 0.004 as x increases 0.004. In this case the approximate rate of increase is

$$\frac{-0.004}{0.004} = -1.$$

Here is the first idea for finding the rate of growth of a function at a point P. Magnify a small vicinity of P until the curve appears to be almost a straight line. Then define the rate of growth of the function at P to be the rate of change of this line.

In case the graph *is* a straight line, its rate of growth or change is the same at all points. This rate is called the **slope** of the line. It represents the change in y per unit change in x. The slope of a (nonvertical) line is easy to compute. Given any two points (x_0, y_0) and (x_1, y_1) on the line,

$$\text{slope} = \frac{y_1 - y_0}{x_1 - x_0}.$$

This number is the ratio of change in y to change in x. By similar triangles, this ratio depends only on the line, not on the particular pair of points (Fig. 2.3).

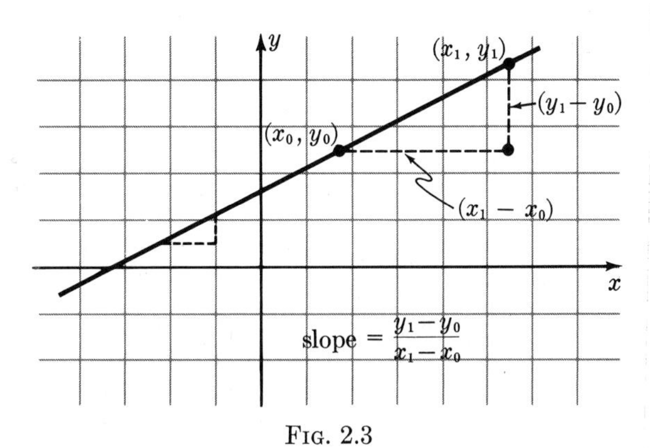

FIG. 2.3

The graph of a linear function

$$f(x) = mx + b$$

is a straight line (Fig. 2.4). Its slope is the coefficient m; let us check this.

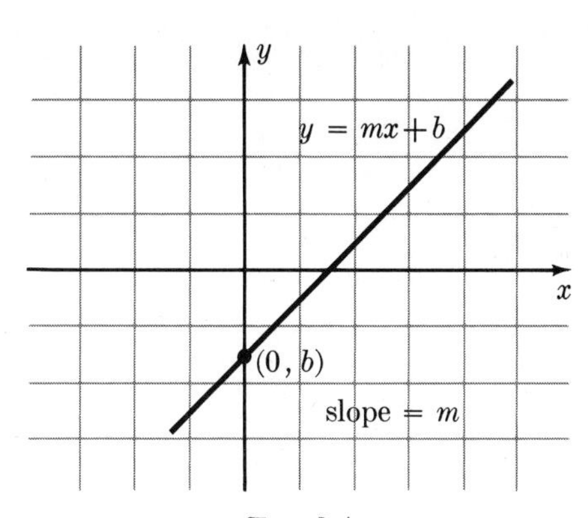

FIG. 2.4

Take any two points (x_0, y_0) and (x_1, y_1) on the graph. Then

$$y_1 = mx_1 + b, \qquad y_0 = mx_0 + b,$$

$$\frac{y_1 - y_0}{x_1 - x_0} = \frac{(mx_1 + b) - (mx_0 + b)}{x_1 - x_0} = \frac{mx_1 - mx_0}{x_1 - x_0} = m.$$

Note the special case $m = 0$. Then the function is

$$f(x) = b,$$

a constant function. Its slope is 0. No amount of change in x can produce any change whatever in y, so naturally the rate of change of y with respect to x is 0. See Fig. 2.5.

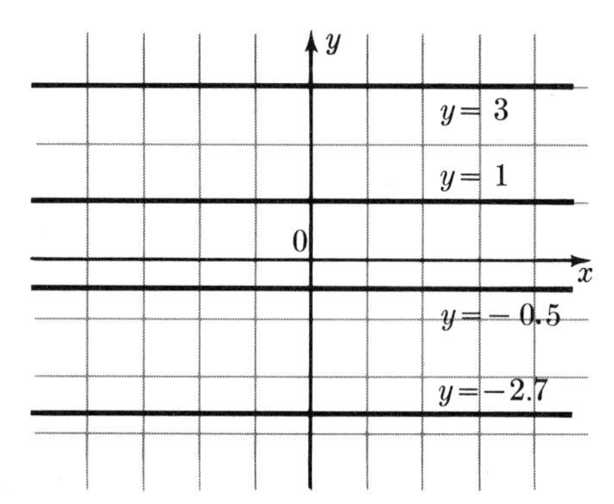

FIG. 2.5 Constant functions.

It is important to have a feeling for the magnitude of slope. Study Fig. 2.6 so you realize how slowly a line of slope $\frac{1}{10}$ grows and how quickly a line of slope 12 grows. Try to imagine slopes of 100 and of 10^{-3}.

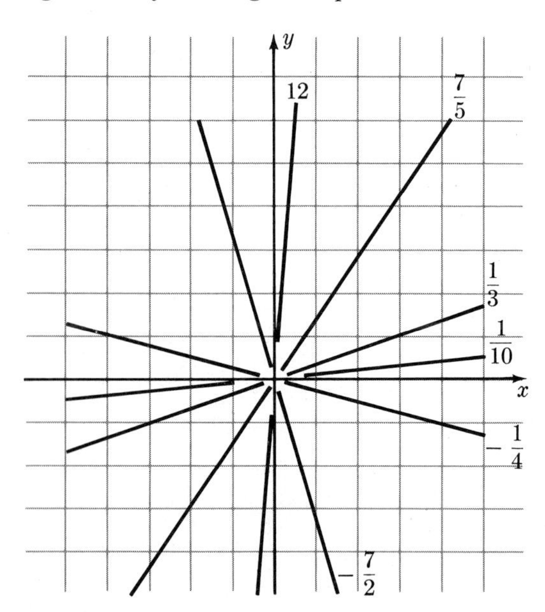

FIG. 2.6 Various slopes.

EXERCISES

Plot the points and find the slope of the line segment joining them:

1. $(1, \frac{1}{4})$, $(4, 1)$
2. $(2, 1)$, $(5, 5)$
3. $(-1, 2)$, $(3, -4)$
4. $(-30, 1)$, $(30, 1)$.

Give the slope of the line:

5. $y = 2x + 3$
6. $y = -7x + 1$
7. $12y - 4x = 3$
8. $\dfrac{y+1}{x-1} = 2$.

Write the equation of the line:

9. through the points $(3, 3)$, $(0, 0)$
10. through the points $(-4, -4)$, $(-1, -1)$
11. through the points $(2, 0)$, $(0, -3)$
12. through the points $(-3, 4)$, $(1, 2)$
13. through the point $(0, 3)$ with slope $m = 1$
14. through the point $(0, 1)$ with slope $m = -7$
15. through the point $(4, 3)$ with slope $m = 5$
16. through the point $(-1, 7)$ with slope $m = -\frac{4}{3}$.
17. A straight line rises 1 inch in 1 mile. What is its slope?
18. A 48-ft-wide highway is sloped for drainage. It is 3 in. higher along the center line than along the edges. What is the slope of the road on each side of the center line?
19. If a small part of the circle $x^2 + y^2 = 1$ near the point $(0, 1)$ is viewed under a microscope, what does it look like?
20. Plot the graph and give the equation of a line that passes through $(0, 3)$ and is parallel to $y = -2x$.

3. THE DERIVATIVE

A function $y = f(x)$ is under scrutiny. Perhaps it arises in a physical situation. Its graph is a smooth curve (Fig. 3.1).

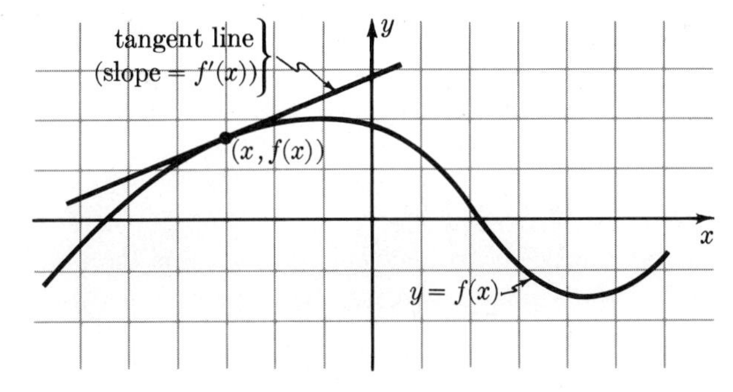

FIG. 3.1

Place a powerful microscope over a point $(x, y) = (x, f(x))$ of the graph. Under high magnification the graph looks nearly straight. Under higher magnification, the graph looks even straighter. Try to imagine the magnification increased indefinitely. The resulting ultimate view is a straight line called the **tangent line** to the graph. Its slope is called the **derivative of** $f(x)$ **at** x and is written

$$f'(x).$$

Thus the tangent line is the straight line through the point $(x, f(x))$ with slope $f'(x)$.

You know one derivative, that of a linear function

$$f(x) = mx + b.$$

Its derivative at each x is

$$f'(x) = m,$$

the slope of the graph.

To find the value of the derivative of any function $y = f(x)$ at $x = a$, you must find the slope of the tangent line to the graph at $(a, f(a))$. See Fig. 3.2.

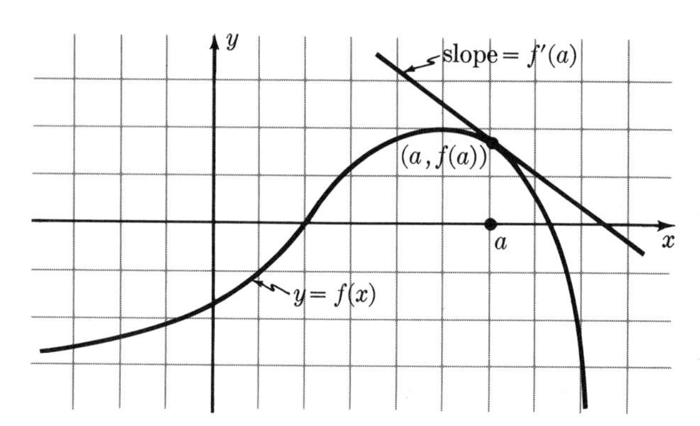

FIG. 3.2

The **slope of a graph** at a point is the slope of its tangent line at the point.

You **differentiate** a function $f(x)$ to form its derivative $f'(x)$. This process of forming the derivative is called **differentiation.** Thus differentiation of $f(x) = 2x - 1$ yields its derivative $f'(x) = 2$; differentiation of $f(x) = -3x$ yields its derivative $f'(x) = -3$.

Fortunately almost every function which occurs in real life can be differentiated. This is important because the derivative turns out to be so useful in applications.

In the following sections you will see the derivative computed in a number of particular cases. In each case the derivative is computed by a sequence of approximations. Each approximation is the slope of a secant to the graph through the point at $x = a$ and a nearby point at $x = a + h$. See Fig. 3.3. As h gets smaller and smaller, the secant is a better and better approximation to the tangent.

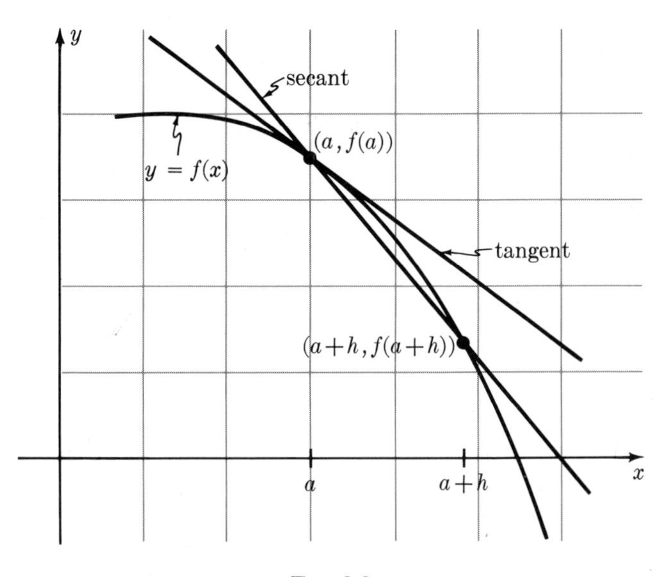

FIG. 3.3

Now please reread this short section several times. It contains the basic idea of Differential Calculus.

4. DERIVATIVE OF $f(x) = x^2$

For a linear function $f(x) = mx + b$, the rate of change of $f(x)$ is constant. For the function $f(x) = x^2$, however, the rate of change is not constant. Graph $y = x^2$. See Fig. 4.1. Near the origin a small change in x produces a small change in y. Farther from the origin, the same change in x causes a larger change in y. For example, as x changes from 0 to 0.1, y increases by $(0.1)^2 - 0^2 = 0.01$, but as x changes from 3 to 3.1, y increases by

$$(3.1)^2 - 3^2 = 0.61.$$

Let us see precisely how the function is changing at various points. First take $x = 0$. The corresponding point of the graph is $(0, 0)$. A nearby point is (h, h^2). The line (secant) through these points has slope

$$\frac{h^2 - 0}{h - 0} = h.$$

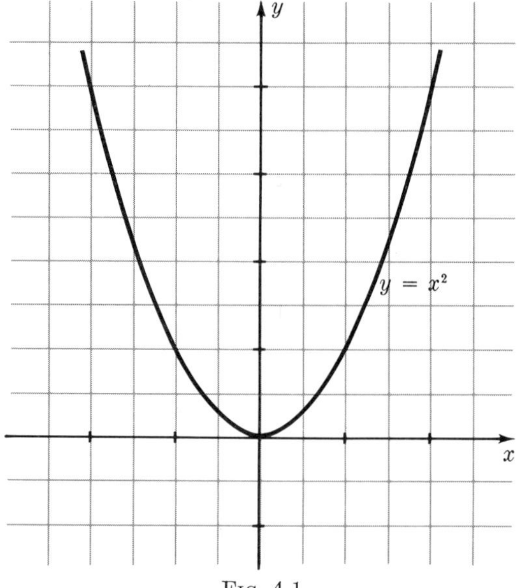

FIG. 4.1

When $h = 0.5$, the slope of the secant is 0.5; when $h = 0.1$, the slope is 0.1. See Fig. 4.2. The smaller h is, the closer the slope of the secant is to 0.

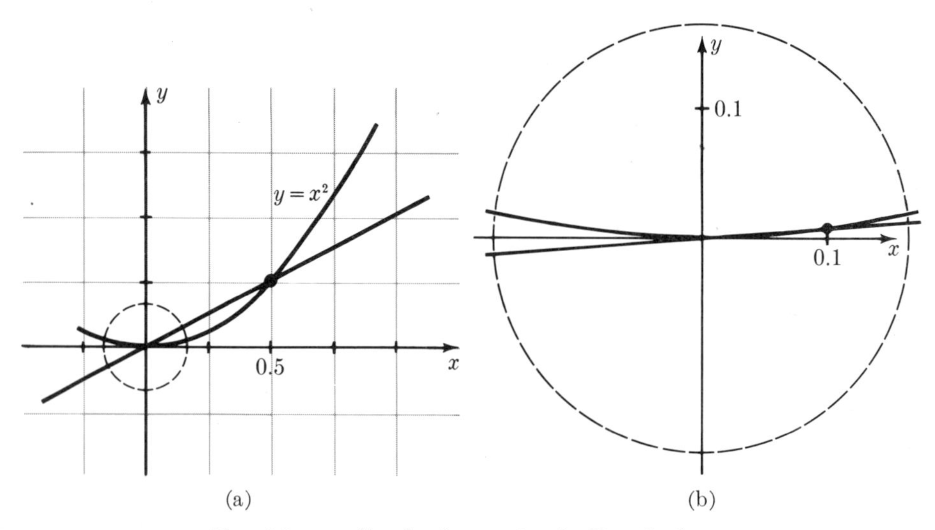

FIG. 4.2 (a) Graph of $y = x^2$. (b) Detail of (a).

Conclusion: the slope of the curve at (0, 0) is 0. Write

$$f(x) = x^2, \qquad f'(0) = 0.$$

Next, try a value different from 0, say $x = a$. Examine the slope of the secant through (a, a^2) and a nearby point $(a + h, (a + h)^2)$:

$$\text{(slope of secant)} = \frac{(a+h)^2 - a^2}{(a+h) - a} = \frac{(a+h)^2 - a^2}{h}.$$

Some values for $a = 1$ are tabulated:

h	1	0.5	0.1	0.001
slope	3	2.5	2.1	2.001

The values of the slope seem to approach the value 2 as h decreases. Try $h = 0.000001$; the new slope is 2.000001, still closer to 2.

Here are similar tables computed for $a = 2$

h	1	0.5	0.1	0.001
slope	5	4.5	4.1	4.001

and $a = 10$:

h	1	0.5	0.1	0.001
slope	21	20.5	20.1	20.001

The data suggest the following conclusions:

a	Slope $= f'(a)$
0	0
1	2
2	4
10	20

In each case the slope is $2a$. See Fig. 4.3. In general,

$$\begin{aligned}\text{(slope of secant)} &= \frac{(a+h)^2 - a^2}{h} \\ &= \frac{(a^2 + 2ah + h^2) - a^2}{h} \\ &= \frac{2ah + h^2}{h} \\ &= 2a + h.\end{aligned}$$

Thus

$$\text{(slope of secant)} = 2a + h.$$

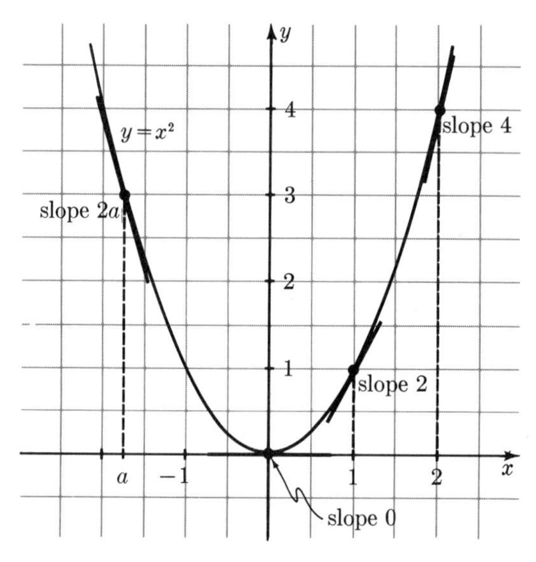

FIG. 4.3

As h becomes smaller and smaller, the slope of the secant more closely approximates $2a$. We conclude that the slope of the tangent line at (a, a^2) is $2a$. This means if $f(x) = x^2$, then $f'(a) = 2a$, or in general, $f'(x) = 2x$.

The derivative of x^2 is $2x$.

Notation

Another notation for the operation of differentiation with respect to x is

$$\frac{d}{dx}\cdot$$

It is applied to a function $y = f(x)$ as follows:

$$\frac{d}{dx}f(x) = f'(x).$$

Also the symbols

$$\frac{df}{dx} \quad \text{and} \quad \frac{dy}{dx}$$

denote $f'(x)$.

Examples.

$$\frac{d}{dx}(mx + b) = m.$$

$$y = x^2, \qquad \frac{dy}{dx} = 2x.$$

$$f(x) = 3x - 1, \qquad \frac{df}{dx} = 3.$$

The notation

$$\frac{dy}{dx}$$

does not indicate the point where the derivative is evaluated. When it is necessary to show the value of the derivative at $x = a$, the following notation is used:

$$\left.\frac{dy}{dx}\right|_{x=a} \quad \text{or simply} \quad \left.\frac{dy}{dx}\right|_{a}.$$

Example.

$$y = x^2, \qquad a = 3.$$

$$\frac{dy}{dx} = 2x,$$

$$\left.\frac{dy}{dx}\right|_{x=3} = 2 \cdot 3 = 6.$$

Similarly,

$$\left.\frac{dy}{dx}\right|_{-2} = 2(-2) = -4.$$

EXERCISES

Find:

1. $\dfrac{dy}{dx}$, for $y = x^2$
2. $\dfrac{dV}{dP}$, where $V = P^2$
3. $\dfrac{dy}{dx}$, for $y = -4(x - 5)$
4. $\dfrac{ds}{dt}$, where $s = -3(4 - 5t)$
5. $\dfrac{d}{dx}(x)$
6. $\dfrac{d}{dx}(x^2)$
7. $\dfrac{df}{dx}$, for $f(x) = 3x + 2$
8. $\dfrac{df}{dx}$, where $f(x) = 12x - 7$
9. $f'(x)$, where $f(x) = 8x$
10. $F'(z)$, for $F(z) = z^2$.

Evaluate:

11. $\dfrac{d}{dx}(x^2)$, at $x = 3, 7, 11$
12. $\dfrac{d}{dP}(P^2)$, at $P = -1, 1, 0$
13. $\dfrac{d}{dx}(13x + 5)$, at $x = 2$
14. $\dfrac{d}{dt}(-13t + v_0)$, for $t = 0$

15. $\left.\dfrac{dy}{dx}\right|_{x=3}$, $y = x^2$

16. $\left.\dfrac{dR}{dI}\right|_{I=-2}$, where $R = I^2$

17. $\left.\dfrac{dv}{dt}\right|_{9}$, $v = -32t + 200$

18. $\left.\dfrac{ds}{dt}\right|_{10}$, $s = t^2$.

Calculate:

19. $f'(-6)$, $f'(12)$, $f'(1)$, where $f(x) = x^2$

20. $G'(-1)$, $G'(0)$, $G'(1)$, if $G(x) = x^2$.

Compute the slope of the secant to $y = x^2$ through (a, a^2) and $(a + h, (a + h)^2)$:

21. for $a = -2$; $h = 1, 0.5, 0.1, 0.001$

22. for $a = 1$; $h = -1, -0.5, -0.1, -0.001$.

23. Find all points on $y = x^2$ where the tangent has slope 6.

24. Find a point on $y = x^2$ where the tangent is parallel to the line $x + 2y + 7 = 0$.

5. DERIVATIVE OF $f(x) = x^3$

Take a point (a, a^3) on the graph (Fig. 5.1) of the function

$$y = f(x) = x^3.$$

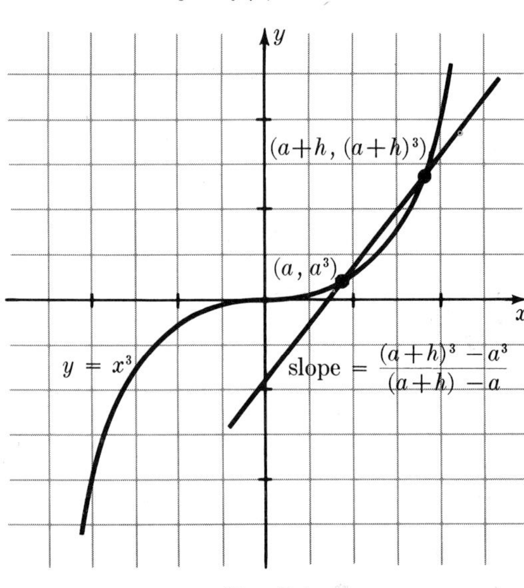

FIG. 5.1

Compute the slope of the secant through (a, a^3) and the nearby point $(a + h, (a + h)^3)$ of the graph:

$$(\text{slope of secant}) = \frac{(a + h)^3 - a^3}{(a + h) - a}.$$

The numerator is

$$(a + h)^3 - a^3 = (a^3 + 3a^2h + 3ah^2 + h^3) - a^3$$
$$= 3a^2h + 3ah^2 + h^3,$$

and the denominator is

$$(a + h) - a = h.$$

Hence

$$(\text{slope of secant}) = \frac{3a^2h + 3ah^2 + h^3}{h} = 3a^2 + 3ah + h^2.$$

Taking h smaller and smaller forces the slope of the secant closer and closer to $3a^2$. Conclusion:

$$(\text{slope of tangent}) = 3a^2.$$

Expressed in derivative notation,

$$\boxed{\frac{d}{dx}(x^3) = 3x^2.}$$

There is another instructive way to do the above computation. This time call the nearby point (x, x^3), but remember that x is close to a. See Fig. 5.2.

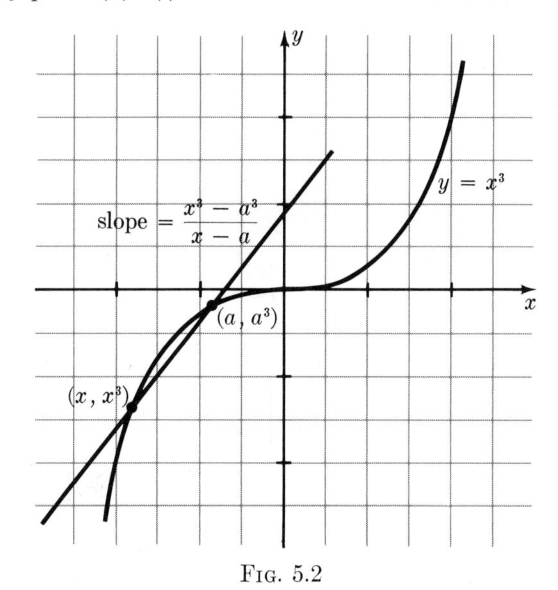

Fig. 5.2

For x near a,

$$(\text{slope of secant}) = \frac{x^3 - a^3}{x - a}.$$

From algebra, $x - a$ is a factor of $x^3 - a^3$:

$$x^3 - a^3 = (x - a)(x^2 + ax + a^2).$$

Hence

$$\text{(slope of secant)} = x^2 + ax + a^2.$$

Now let x approach closer and closer to a. Write

$$x \longrightarrow a$$

to abbreviate this "closer and closer" statement. Then

$$ax \longrightarrow a \cdot a = a^2,$$

$$x^2 = x \cdot x \longrightarrow a \cdot a = a^2,$$

and therefore

$$\text{(slope of secant)} \longrightarrow a^2 + a^2 + a^2 = 3a^2.$$

Conclusion:

$$\frac{d}{dx}(x^3) = 3x^2.$$

Similarly we can derive

$$\frac{d}{dx}(x^4) = 4x^3,$$

$$\frac{d}{dx}(x^5) = 5x^4,$$

and so on. However, we postpone doing so until we actually need these formulas.

EXERCISES

Use differentiation to find:

1. $\dfrac{dy}{dx}$, if $y = x^3$
2. $\dfrac{dF}{dx}$, if $F(x) = x^3$
3. $\dfrac{d}{dx}(4^3)$
4. $\dfrac{d}{dt}(5)\Big|_{t=1}$
5. $\dfrac{dy}{dx}$ at $x = 0, 3, -3,$ if $y = x^3$
6. $\dfrac{ds}{dt}$ for $t = 0, 1, 2, 3$ if $s = t^3$
7. $\dfrac{dG}{dz}\Big|_6$, $G(z) = z^3$
8. $\dfrac{dy}{dx}\Big|_{-4}$, $y = x^3$
9. $V'(4)$, $V'(a)$, if $V(P) = P^3$
10. $s'(5)$, $s'(t_0)$, if $s(t) = t^3$.

Compute the slope of the secant to $y = x^3$ through (a, a^3) and $(a + h, (a + h)^3)$:

11. for $a = 1$; $h = 1, 0.5, 0.1, 0.001$
12. for $a = 1$; $h = -1, -0.5, -0.1, -0.001$.

Compute the slope of the secant to the curve $y = x^3$ through (a, a^3) and (x, x^3):

13. for $a = 2$; $x = 3, 2.5, 2.1, 2.001$
14. for $a = 2$; $x = 1, 1.5, 1.9, 1.99$.
15. Find all points on the curve $y = x^3$ where the slope is 12.

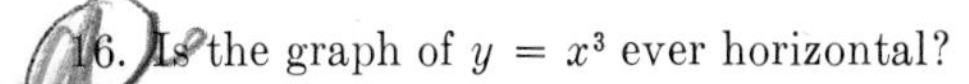

16. Is the graph of $y = x^3$ ever horizontal?
17. Show there do not exist points on the curve $y = x^3$ where the tangent is parallel to the line $x + y + 1 = 0$.
18. Find which of the two curves $y = x^2$ and $y = x^3$ is steeper at $x = \frac{1}{2}$, at $x = 1$, at $x = 2$.
19. Find all positive values of x where $y = x^2$ is steeper than $y = x^3$.
20. Find all positive values of x where $y = x^3$ is steeper than $y = x^2$.

6. DERIVATIVE OF f(x) = 1/x

We seek the derivative of the function

$$f(x) = \frac{1}{x},$$

whose graph is shown in Fig. 6.1.

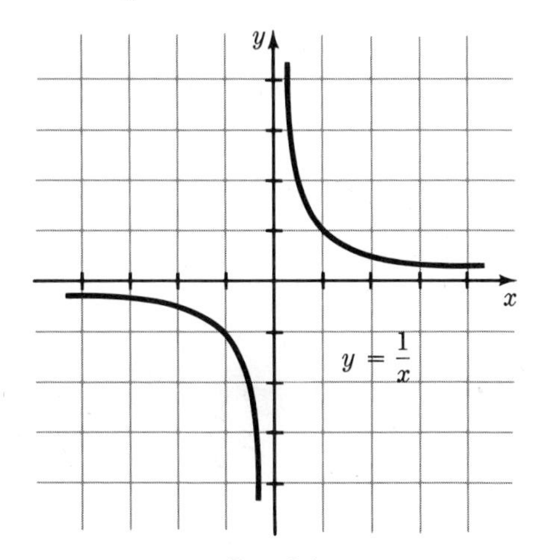

FIG. 6.1

Fix $x_0 \neq 0$. The slope of the secant through $(x_0, f(x_0))$ and a nearby point $(x_0 + h, f(x_0 + h))$ is

$$\text{(slope of secant)} = \frac{f(x_0 + h) - f(x_0)}{(x_0 + h) - x_0}$$

$$= \frac{1}{h}[f(x_0 + h) - f(x_0)]$$

$$= \frac{1}{h}\left[\frac{1}{x_0 + h} - \frac{1}{x_0}\right] = \frac{1}{h}\left[\frac{x_0 - (x_0 + h)}{(x_0 + h)x_0}\right]$$

$$= \frac{1}{h}\left[\frac{-h}{(x_0 + h)x_0}\right] = \frac{-1}{(x_0 + h)x_0}.$$

Let $h \longrightarrow 0$. Then

$$(x_0 + h) \longrightarrow x_0,$$

$$(x_0 + h)x_0 \longrightarrow x_0 \cdot x_0 = x_0^2,$$

$$(\text{slope of secant}) \longrightarrow \frac{-1}{x_0^2}.$$

Conclusion:

$$\boxed{\frac{d}{dx}\left(\frac{1}{x}\right) = \frac{-1}{x^2}.}$$

The derivative is negative since the function decreases as x increases.

EXERCISES

Differentiate:

1. $y = 1/x$
2. $F(t) = 1/t$.

Calculate:

3. $f'(-1)$, $f'(1)$, $f'(a)$, $f'(-a)$, where $f(x) = 1/x$
4. $f'(-\frac{1}{2})$, $f'(\frac{1}{2})$, $f'(2)$, $f'(3)$, if $f(x) = 1/x$
5. $\left.\frac{dy}{dx}\right|_{x=a}$ and $\left.\frac{dy}{dx}\right|_{x=1/a}$, if $y = \frac{1}{x}$
6. $\left.\frac{dV}{dP}\right|_{P=1/4}$ and $\left.\frac{dV}{dP}\right|_{P=4}$, if $V = \frac{1}{P}$
7. $\left.\frac{d}{ds}\left(\frac{1}{s}\right)\right|_{b}$
8. $\left.\frac{d}{dx}\left(\frac{1}{x}\right)\right|_{t}$.
9. Find where the curve $y = 1/x$ has slope $-\frac{1}{2}$.
10. Is the curve $y = 1/x$ ever horizontal?
11. Do the curves $y = 1/x$ and $y = x^3$ ever have the same slope?

Let $f(x) = 1/x$. Find the slope of the secant:

12. through $(1, f(1))$ and $(2, f(2))$
13. through $(10, f(10))$ and $(11, f(11))$
14. through $(100, f(100))$ and $(101, f(101))$.

7. THE TANGENT LINE

Take a point $(a, f(a))$ on the graph of a function $y = f(x)$. The tangent line at this point (Fig. 7.1) has slope $f'(a)$. Its equation is found by the point-slope formula:

$$y - f(a) = f'(a)(x - a),$$

or

$$y = f(a) + f'(a)(x - a).$$

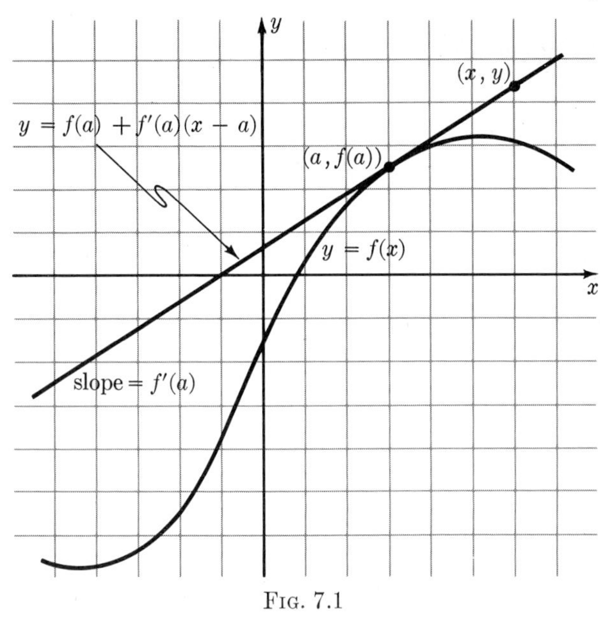

FIG. 7.1

EXAMPLE 7.1

Find the equation of the tangent line to the curve $y = x^2$ at $x = -2$.

Solution: At $x = -2$, $y = (-2)^2 = 4$, and the derivative is

$$\left.\frac{dy}{dx}\right|_{-2} = 2x\Big|_{-2} = 2(-2) = -4.$$

Thus the tangent line passes through $(-2, 4)$ and has slope -4. Its equation is

$$y - 4 = -4(x + 2).$$

Hence

$$\begin{aligned} y &= 4 - 4(x + 2) \\ &= -4x - 4. \end{aligned}$$

Answer: $y = -4x - 4$.

EXAMPLE 7.2

Find where the tangent line to the curve $y = 1/x$ at $x = 3$ meets the x-axis.

Solution: When $x = 3$, $y = \frac{1}{3}$. Set $f(x) = 1/x$. Then

$$f'(x) = -\frac{1}{x^2}, \qquad f'(3) = -\frac{1}{9}.$$

The tangent line has slope $-\frac{1}{9}$ and passes through $(3, \frac{1}{3})$. Its equation is

$$y - \frac{1}{3} = -\frac{1}{9}(x - 3),$$

that is,

$$y = -\frac{1}{9}x + \frac{2}{3}.$$

It crosses the x-axis where $y = 0$, i.e.,

$$-\frac{1}{9}x + \frac{2}{3} = 0.$$

Hence

$$x = 6.$$

Answer: The tangent line meets the x-axis at $(6, 0)$.

The tangents at certain points of a curve may be vertical; at such points the derivative is undefined. However, the equation of the tangent is then simply x = constant.

EXERCISES

Write the equation of the tangent to the curve:

1. $y = x^2$, through $(2, 4)$
2. $y = x^3$, through $(2, 8)$
3. $y = 1/x$, through $(-1, -1)$
4. $y = 1/x$, through $(2, \frac{1}{2})$.

Write the equation of the tangent to the curve:

5. $y = x^2$, at $x = -1$
6. $y = x^3$, at $x = -1$
7. $y = 1/x$, at $x = -3$
8. $y = 1/x$, at $x = 4$.

Find the equation(s) of the tangent(s) to the curve:

9. $y = x^2$, at all points where the slope is 10
10. $y = x^3$, at all points where the slope is 27
11. $y = x^2$, if the tangent crosses the y-axis at $y = -16$
12. $y = x^3$, if the tangent crosses the y-axis at $y = -128$.

Find the equation(s) of the tangent line(s) to the curve and determine where each crosses the coordinate axes if:

13. the curve is $y = x^2$ and the tangent is at the point $(10, 100)$
14. the tangent is determined by the curve $y = x^3$ and the point $(10, 1000)$
15. the curve is defined by the function $F(x) = 1/x$ and the tangent has slope -81
16. the curve is defined by the function $f(x) = 1/x$ and the tangent has slope -6.

Find the area of the triangle bounded by the coordinate axes and the line tangent to $y = 1/x$:

17. at $x = 2$
18. at $x = a$, where $a \neq 0$.

19. Show that the tangents to the parabola $y = x^2$ at $(3, 9)$ and $(-3, 9)$ cross on the y-axis. Where?
20. Exactly one of the lines $y = 3x + b$ is tangent to the parabola $y = x^2$. Which one?
21. Which of the lines $y = -9x + b$ is tangent to the curve $y = 1/x$?

8. RULES FOR DIFFERENTIATING

You know these derivatives:

$$\frac{d}{dx}(mx + b) = m, \qquad \frac{d}{dx}\left(\frac{1}{x}\right) = \frac{-1}{x^2},$$

$$\frac{d}{dx}(x^2) = 2x, \qquad \frac{d}{dx}(x^3) = 3x^2.$$

By learning a few simple rules, you will be able to differentiate all sorts of functions constructed from the ones you already know, for example,

$$\frac{3}{x-1}, \qquad x^3 + x - \frac{1}{x}, \qquad 5x^3 + 2x^2, \; \ldots .$$

First the rules will be stated with examples. Then the rules will be justified.

Sum Rule The derivative of the sum of two functions is the sum of the individual derivatives:

$$\frac{d}{dx}[f(x) + g(x)] = \frac{d}{dx}f(x) + \frac{d}{dx}g(x).$$

EXAMPLE 8.1

Differentiate $x^2 + \dfrac{1}{x}\cdot$

Solution:

$$\frac{d}{dx}\left(x^2 + \frac{1}{x}\right) = \frac{d}{dx}(x^2) + \frac{d}{dx}\left(\frac{1}{x}\right)$$

$$= 2x - \frac{1}{x^2}\cdot$$

Answer: $2x - \dfrac{1}{x^2}\cdot$

EXAMPLE 8.2

Differentiate $x^3 - 8x + 5$.

Solution:

$$\frac{d}{dx}(x^3 - 8x + 5) = \frac{d}{dx}(x^3) + \frac{d}{dx}(-8x + 5) = 3x^2 - 8.$$

Answer: $3x^2 - 8$.

Constant-multiple Rule If a function is multiplied by a constant, its derivative is also multiplied by that constant:

$$\frac{d}{dx}[kf(x)] = kf'(x).$$

EXAMPLE 8.3

Differentiate $8x^3$.

Solution: By the Constant-multiple Rule,

$$\frac{d}{dx}(8x^3) = 8\frac{d}{dx}(x^3) = 8(3x^2).$$

Answer: $24x^2$.

EXAMPLE 8.4

Differentiate $4x^2 - 3x + \frac{5}{x}\cdot$

Solution: First apply the Sum Rule, then the Constant-multiple Rule:

$$\frac{d}{dx}\left(4x^2 - 3x + \frac{5}{x}\right) = \frac{d}{dx}(4x^2) + \frac{d}{dx}(-3x) + \frac{d}{dx}\left(\frac{5}{x}\right)$$

$$= 4\frac{d}{dx}(x^2) - 3\frac{d}{dx}(x) + 5\frac{d}{dx}\left(\frac{1}{x}\right)$$

$$= 4(2x) - 3(1) + 5\left(\frac{-1}{x^2}\right)\cdot$$

Answer: $8x - 3 - \frac{5}{x^2}\cdot$

With a little practice, it becomes easy to do problems like the last one by inspection.

EXAMPLE 8.5

Differentiate $2x^3 + 5x^2 - 7x$.

Answer: $6x^2 + 10x - 7$.

Shifting Rule If $g(x) = f(x + c)$, then

$$\frac{d}{dx} g(x) = \left.\frac{df}{dx}\right|_{x+c} = f'(x + c).$$

EXAMPLE 8.6

Find $\dfrac{d}{dx}\left(\dfrac{1}{x+2}\right)$.

Solution: Set

$$f(x) = \frac{1}{x}$$

and

$$g(x) = f(x + 2) = \frac{1}{x+2}.$$

Then

$$f'(x) = \frac{-1}{x^2};$$

by the Shifting Rule,

$$g'(x) = f'(x + 2) = \frac{-1}{(x+2)^2}.$$

Answer: $\dfrac{-1}{(x+2)^2}$.

The next example shows how the Shifting Rule is used in practice.

EXAMPLE 8.7

Differentiate $(x - 1)^2$.

Solution:

$$\frac{d}{dx}(x^2) = 2x, \qquad \frac{d}{dx}[(x-1)^2] = 2(x-1).$$

CHECK:

$$(x - 1)^2 = x^2 - 2x + 1.$$

By the Sum Rule,

$$\frac{d}{dx}[(x-1)^2] = \frac{d}{dx}(x^2) + \frac{d}{dx}(-2x + 1)$$

$$= 2x - 2 = 2(x - 1).$$

Answer: $2(x - 1)$.

The following example illustrates a technique used over and over in Calculus.

EXAMPLE 8.8

Differentiate $\dfrac{1}{3x-5}$.

Solution: By the Shifting Rule,

$$\frac{d}{dx}\left(\frac{1}{x-c}\right) = \frac{-1}{(x-c)^2}, \qquad c \text{ constant.}$$

Our problem, however, has $3x$ in the denominator, so this formula does not apply. We can *make* it apply by dividing the denominator by 3:

$$\frac{1}{3x-5} = \frac{1}{3}\left(\frac{1}{x-\frac{5}{3}}\right).$$

Apply first the Constant-multiple Rule, then the Shifting Rule:

$$\frac{d}{dx}\left(\frac{1}{3x-5}\right) = \frac{1}{3}\frac{d}{dx}\left(\frac{1}{x-\frac{5}{3}}\right) = \frac{1}{3}\left[\frac{-1}{\left(x-\frac{5}{3}\right)^2}\right]$$

$$= \frac{-1}{3\left(x-\frac{5}{3}\right)^2} = \frac{-3}{(3x-5)^2}.$$

Answer: $\dfrac{-3}{(3x-5)^2}$.

REMARK: If you cannot differentiate an expression, try writing it in an equivalent algebraic form; for example, try factoring, multiplying out factors, combining terms, etc.

Scale Rule If $g(x) = f(kx)$, then

$$\frac{d}{dx}g(x) = k\left.\frac{df}{dx}\right|_{kx} = kf'(kx).$$

EXAMPLE 8.9

Differentiate $(2x)^3$.

Solution: If $f(x) = x^3$, then $f'(x) = 3x^2$. Set $g(x) = f(2x) = (2x)^3$.

By the Scale Rule with $k = 2$,

$$\frac{d}{dx} g(x) = 2f'(2x),$$

$$\frac{d}{dx} (2x)^3 = 2 \cdot 3(2x)^2.$$

Answer: $24x^2$.

REMARK: Compare Example 8.9 with Example 8.3.

EXAMPLE 8.10

Differentiate $\dfrac{1}{3x - 5}$.

Solution: Set

$$f(x) = \frac{1}{x - 5} \qquad \text{and} \qquad g(x) = f(3x) = \frac{1}{3x - 5}.$$

Then

$$f'(x) = \frac{-1}{(x - 5)^2},$$

so by the Scale Rule,

$$\frac{d}{dx}\left(\frac{1}{3x - 5}\right) = \frac{d}{dx} g(x) = 3f'(3x) = 3\left[\frac{-1}{(3x - 5)^2}\right].$$

Answer: $\dfrac{-3}{(3x - 5)^2}$.

REMARK: Compare Example 8.10 with Example 8.8.

The Shifting Rule and the Scale Rule are special cases of a more general rule to be derived in Chapter 15, Section 4:

Chain Rule If $y = f[g(x)]$, then

$$\frac{dy}{dx} = f'[g(x)] \frac{dg}{dx}.$$

begin here

EXAMPLE 8.11

Differentiate $y = (5x - 1)^3 + 4(5x - 1)^2$ by the Chain Rule.

Solution: Set

$$f(x) = x^3 + 4x^2 \qquad \text{and} \qquad g(x) = 5x - 1.$$

Then

$$f'(x) = 3x^2 + 8x \qquad \text{and} \qquad \frac{dg}{dx} = 5.$$

Note that $y = f[g(x)]$. Therefore by the Chain Rule,

$$\begin{aligned}\frac{dy}{dx} &= f'(5x-1)\cdot 5\\ &= [3(5x-1)^2 + 8(5x-1)]\cdot 5\\ &= 5(5x-1)(15x+5).\end{aligned}$$

Answer: $25(5x-1)(3x+1)$.

EXERCISES

Find the derivative:

1. $y = x^3 + \dfrac{1}{x}$
2. $f(x) = x^3 + x^2 + \dfrac{1}{x}$
3. $f(x) = x^2 + 4x + 3$
4. $g(x) = x^3 + x^2 + 5x - 1$
5. $y = \dfrac{1}{x+1}$
6. $g(x) = \dfrac{1}{x-4}$
7. $s(t) = (t-8)^2$
8. $y = (x-5)^3$
9. $y = 3x^2$
10. $y = -\dfrac{3}{x}$.

Calculate the derivative:

11. $y = x^3 - 2x^2$
12. $y = x^3 - \dfrac{8}{x}$
13. $F(x) = 2x^3 - 5x^2 + 2x + 7$
14. $f(x) = \frac{1}{3}x^3 + \frac{1}{2}x^2 + x + 1$
15. $y = \dfrac{1}{2x-1}$
16. $g(x) = \dfrac{3}{5x-2}$
17. $f(x) = (36x-2)^2$
18. $F(x) = (96x-9)^3$
19. $\dfrac{d}{dr}(2\pi r^2 + 2\pi r)$
20. $\dfrac{d}{dr}\left(\dfrac{4}{3}\pi r^3 - 3\pi r^2\right)$.
21. Use the identity $\dfrac{1}{x(x+1)} = \dfrac{1}{x} - \dfrac{1}{x+1}$ to find $\dfrac{d}{dx}\left[\dfrac{1}{x(x+1)}\right]$.
22. Use the identity $\dfrac{x+2}{x+1} = \dfrac{x+1+1}{x+1} = 1 + \dfrac{1}{x+1}$ to find $\dfrac{d}{dx}\left(\dfrac{x+2}{x+1}\right)$.

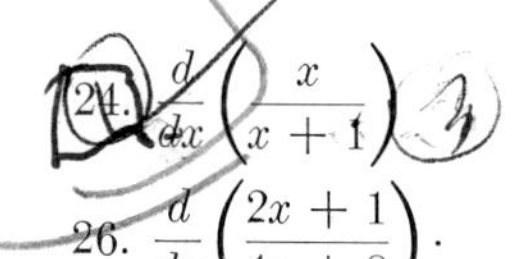

Use the methods of Exs. 21 and 22 to evaluate:

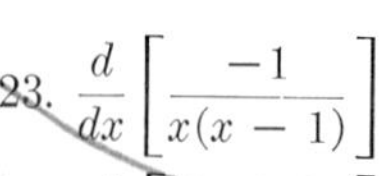

23. $\dfrac{d}{dx}\left[\dfrac{-1}{x(x-1)}\right]$
24. $\dfrac{d}{dx}\left(\dfrac{x}{x+1}\right)$
25. $\dfrac{d}{dx}\left[\dfrac{x+2}{x(x+1)}\right]$
26. $\dfrac{d}{dx}\left(\dfrac{2x+1}{4x+8}\right)$.
27. Find the equation of the tangent to $y = x^3 - 3x + 1$ at $x = 2$.
28. Find the equation of the tangent to $y = 4x^3 - 12x$ at $x = \frac{1}{2}$.
29. The path of an electron in a certain magnetic field is the curve $y = 1 - 2x^2$. When $x = 3$ the electron flies off tangentially. Find the equation of its line of travel.
30. An object, moving from left to right, travels along the path $y = 3x^2$. When $x = 2$ the object escapes the path tangentially. Along what line does it travel? Find its position when $x = 3$.

31. The resistance of a certain resistor as a function of temperature is $R = 40(2 + 0.41T + 0.001T^2)$. Find the rate of change of resistance with respect to temperature (in ohms per degree Kelvin) when the temperature is 15°K.

32. A ball is thrown straight up. Its height at time t is $s = -16t^2 + 96t + 4$. What is the rate of change of height with respect to time (in ft/sec) after 1 sec, after 3 sec, after 5 sec? How high is the ball at these times?

9. REASONS FOR THE RULES

The Sum Rule is true for linear functions. This is easy to see by direct calculation: if

$$f(x) = mx + b \qquad \text{and} \qquad g(x) = nx + c,$$

then $f(x) + g(x) = (m + n)x + (b + c)$. The slope of $f(x)$ is m; the slope of $g(x)$ is n; the slope of $f(x) + g(x)$ is $(m + n)$. Thus

$$\frac{d}{dx}[f(x) + g(x)] = m + n = \frac{d}{dx}[f(x)] + \frac{d}{dx}[g(x)].$$

Consequently, for a pair of linear functions, the slope of the sum function is the sum of the two slopes.

Why is the Sum Rule true for any pair of functions?

Graph two functions $y = f(x)$ and $y = g(x)$, and their sum $y = f(x) + g(x)$. Inspect these graphs near $x = a$. See Fig. 9.1. Under a sufficiently

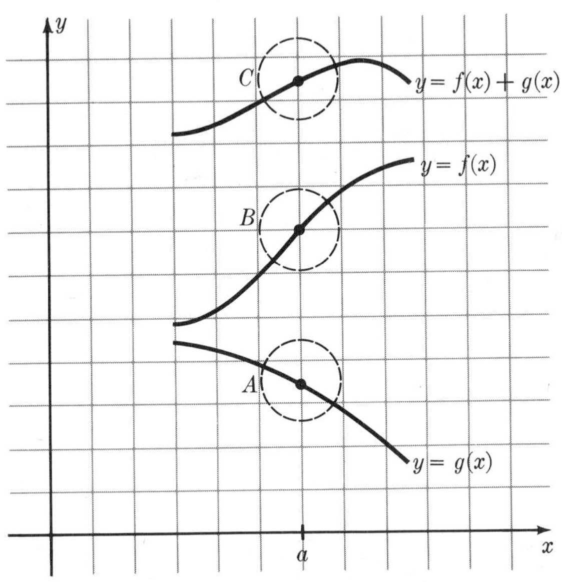

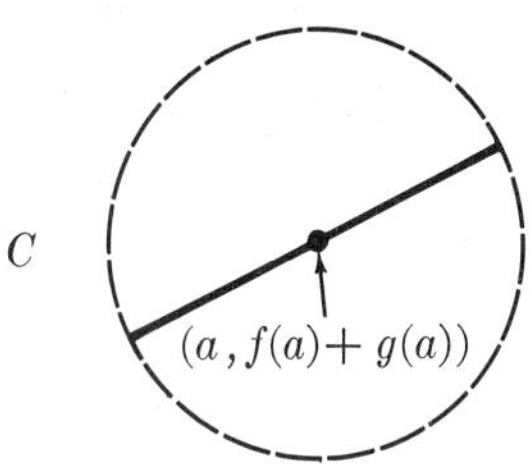

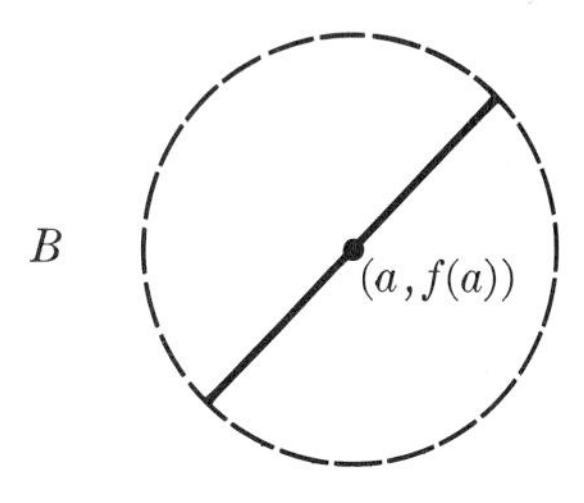

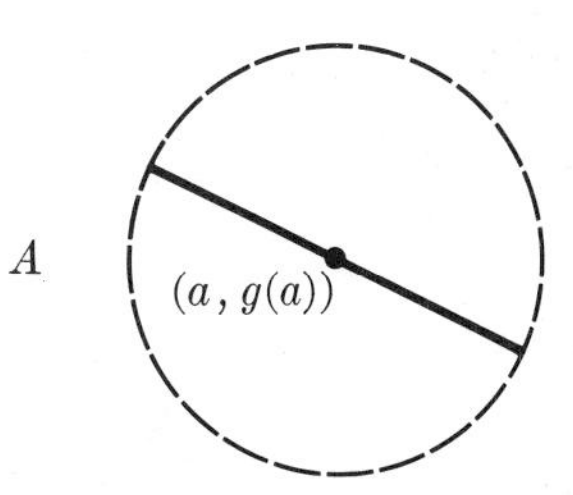

FIG. 9.1 Graphs of $y = g(x)$, of $y = f(x)$, and of $y = f(x) + g(x)$.
A. Detail of $y = g(x)$ with slope $= g'(a)$.
B. Detail of $y = f(x)$ with slope $= f'(a)$.
C. Detail of $y = f(x) + g(x)$ with slope $= f'(a) + g'(a)$.

high-powered microscope, the graphs appear to be straight. Indeed, they are indistinguishable from graphs of linear functions. The function shown in Detail C is the sum of the functions shown in Details A and B. But for all practical purposes these are linear functions, and for linear functions the Sum Rule is true.

The Constant-multiple Rule is true for linear functions. If $g(x) = mx + b$, then $g'(x) = m$. But $kg(x) = (km)x + (kb)$;

$$\frac{d}{dx}[kg(x)] = km = k\frac{d}{dx}[g(x)].$$

A function $f(x)$ looks like a linear function $g(x)$ under a microscope; the constant multiple $kf(x)$ looks like the linear function $kg(x)$. Since the Constant-multiple Rule is true for $g(x)$ it is true for $f(x)$.

The Constant-multiple Rule makes sense if you interpret it this way. If y is changing m times as fast as x, then ky is changing km times as fast as x.

Example.

$$\frac{d}{dx}\left(\frac{1}{x}\right) = \frac{-1}{x^2}, \qquad \frac{d}{dx}\left(\frac{3}{x}\right) = \frac{-3}{x^2}.$$

Figure 9.2 makes this appear plausible.

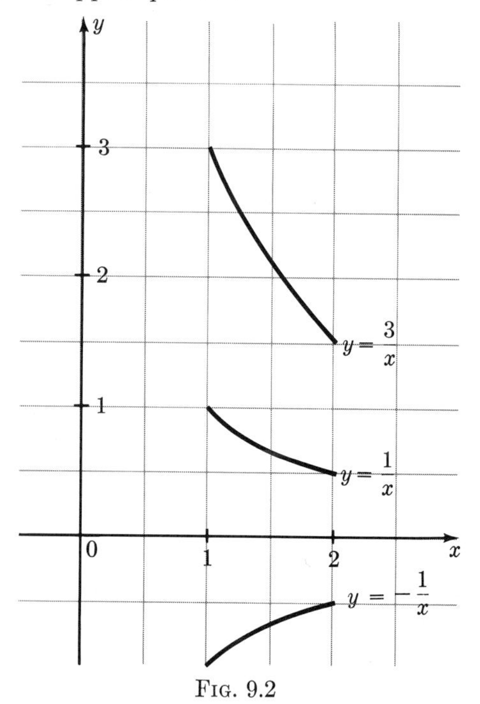

FIG. 9.2

NOTE: k can be negative. On the same figure a portion of the graph of $y = -1/x$ is plotted; it is a reflected image of the graph of $y = 1/x$ in the

x-axis. At each point its slope is

$$\frac{d}{dx}\left(\frac{-1}{x}\right) = -\frac{d}{dx}\left(\frac{1}{x}\right) = \frac{1}{x^2}.$$

The Constant-multiple Rule can be interpreted in terms of a change of units on the y-axis. For example, since one yard equals three feet, a speed of m yd/sec (derivative of distance with respect to time) is the same as a speed of $3m$ ft/sec.

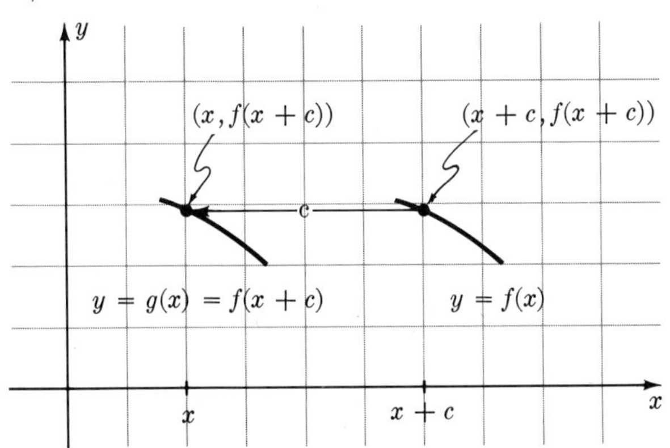

FIG. 9.3

The Shifting Rule is obvious when you look carefully at a graph (Fig. 9.3). The graph of $y = f(x)$ is shifted c units to the left (when c is positive) to form the graph of $y = g(x) = f(x + c)$. The slope of the graph $y = f(x)$ at $x + c$ is the same as the slope of the graph $y = g(x) = f(x + c)$ at x. Examples are shown in Fig. 9.4 and Fig. 9.5.

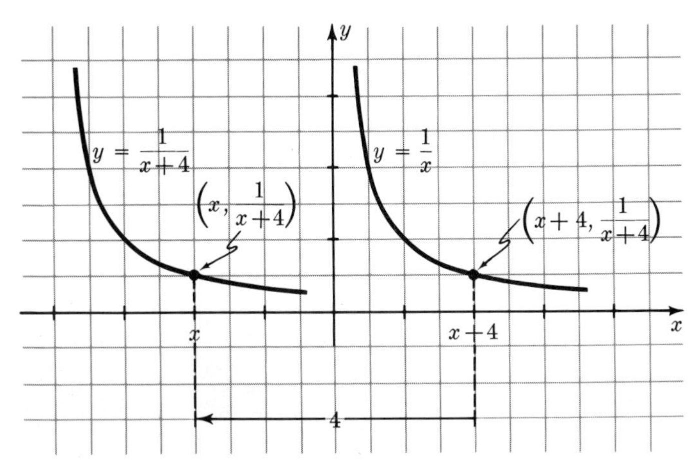

FIG. 9.4 The graph of $y = 1/x$ shifted four units to the left.

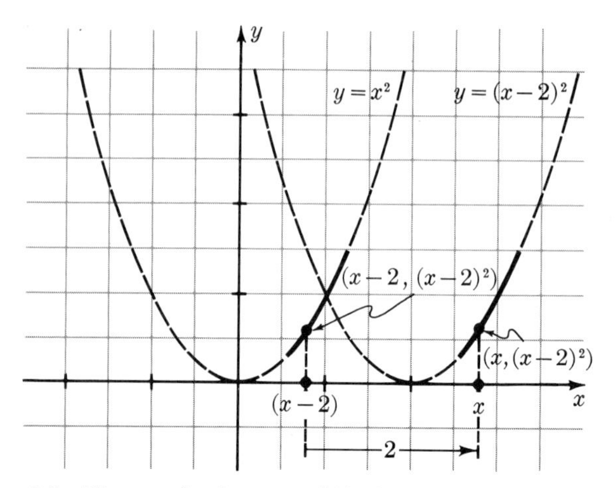

FIG. 9.5 The graph of $y = x^2$ shifted two units to the right.

The Scale Rule for a positive scale factor k is seen by comparing slopes in Fig. 9.6. Look at the curve on the left, the graph of $y = g(x) = f(kx)$. An increase in x from a to $a + h$ produces an increase in y from $g(a) = f(ka)$ to $g(a + h) = f[k(a + h)]$. Precisely the same increase in y is produced on the graph of $y = f(x)$ by increasing x from ka to $k(a + h)$.

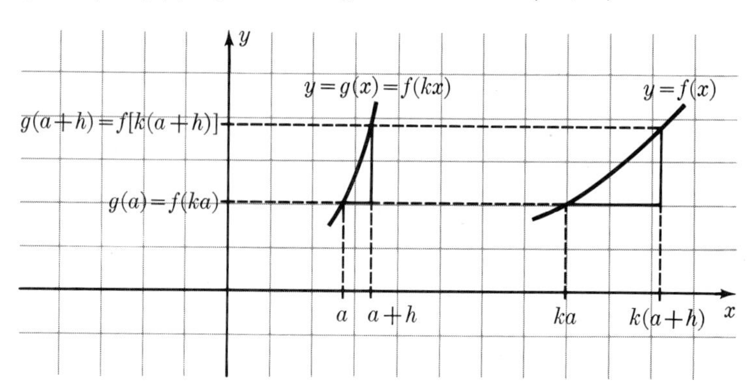

FIG. 9.6

For the curve on the left, the slope at $x = a$ is approximately

$$\frac{g(a + h) - g(a)}{h} = \frac{f[k(a + h)] - f(ka)}{h}.$$

For the curve on the right, the slope at $x = ka$ is approximately

$$\frac{f[k(a + h)] - f(ka)}{kh},$$

which is the preceding expression divided by k. Consequently,

$$\frac{1}{k}\frac{d}{dx}g(x)\bigg|_{x=a} = \frac{1}{k}\frac{d}{dx}f(kx)\bigg|_{x=a} = \frac{d}{dx}f(x)\bigg|_{x=ka}.$$

Hence

$$\frac{d}{dx}g(x) = k\frac{df}{dx}\bigg|_{kx}.$$

The special case $k = -1$ is illustrated in Fig. 9.7. The graph of $y = g(x) = f(-x)$ is the reflection of the graph of $y = f(x)$ in the y-axis. The slope of $g(x) = f(-x)$ at $x = a$ is the negative of the slope of $f(x)$ at $x = -a$.

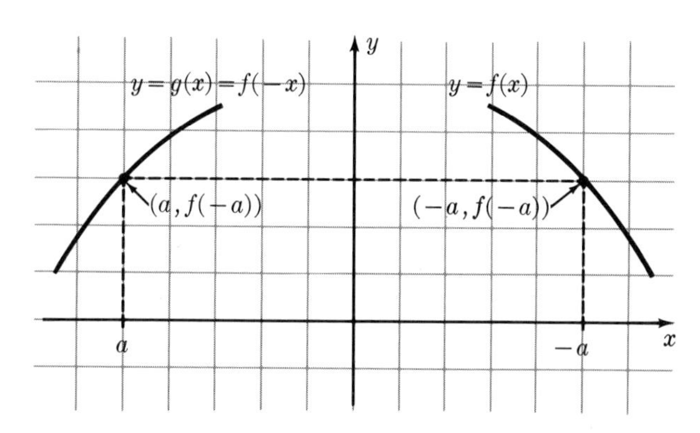

FIG. 9.7

The Scale Rule can be interpreted in terms of a change of units on the x-axis. For example, since one minute equals sixty seconds, a speed of m ft/sec is the same as a speed of $60m$ ft/min.

EXERCISES

Plot the curves. Where more than one equation occurs, plot all curves on the same coordinate axes:

1. $y = \frac{1}{2}x, \quad y = x, \quad y = 2x$
2. $y = x^2, \quad y = -x^2$
3. $y = x^3, \quad y = -x^3$
4. $y = \frac{1}{2}x^2, \quad y = x^2, \quad y = 3x^2$
5. $y = (x - 2)^2$
6. $y = (x + 1)^3$
7. $y = \dfrac{1}{x - 4}$
8. $y = 2x, \quad y = 2(x - 3)$
9. $y = x^2, \quad y = (x - 1)^2, \quad y = -(x - 1)^2, \quad \text{where } x \geq 0$
10. $y = x^3, \quad y = (x + 3)^3, \quad y = -(x + 3)^3, \quad \text{where } x \geq 0$
11. $y = f(x), \quad y = f(-x), \quad \text{if } f(x) = (x - 3)^2 \text{ and } x \geq 3$
12. $y = g(x), \quad y = g(-x), \quad \text{if } g(x) = (x - 1)^3 \text{ and } x \geq 1.$

Plot the curve $y = f(x) + g(x)$:

13. $f(x) = x$ and $g(x) = 1/x$

14. $f(x) = 5x$ and $g(x) = 4/x$.

[*Hint:* First plot the curves $y = f(x)$ and $y = g(x)$ on the coordinate axes, then use these curves to locate points on $y = f(x) + g(x)$.]

10. SECOND DERIVATIVE

Start with a function $y = f(x)$. It has a derivative

$$\frac{dy}{dx} = f'(x),$$

which itself is a function. Since the derivative is a function, it also has a derivative,

$$\frac{d}{dx}[f'(x)].$$

This is called the **second derivative of** f and is written

$$\frac{d^2y}{dx^2} \quad \text{or} \quad f''(x).$$

Example.

$$f(x) = x^3 - 3x.$$

$$f'(x) = 3x^2 - 3,$$

$$f''(x) = \frac{d}{dx}(3x^2 - 3) = 6x.$$

Example.

$$H = 4(T + 2)^3.$$

$$\frac{dH}{dT} = 12(T + 2)^2,$$

$$\frac{d^2H}{dT^2} = 24(T + 2).$$

Example. Let $L = L(t)$ be a function whose first derivative is

$$\frac{dL}{dt} = \frac{1}{t + 1}.$$

Find $\dfrac{d^2L}{dt^2}$.

Solution: $$\frac{d^2L}{dt^2} = \frac{d}{dt}\left(\frac{dL}{dt}\right) = \frac{d}{dt}\left(\frac{1}{t + 1}\right) = \frac{-1}{(t + 1)^2}.$$

EXERCISES

Find the first and second derivatives:

1. $y = 5x^2 - 3$
2. $f(x) = 3x^2 - 2x + 1$
3. $y = 7x^3 + 4x^2 - 3x + 2$
4. $F(x) = \frac{1}{3}x^3 + \frac{1}{2}x^2 + x + 1$
5. $V(P) = P^3 - 7$
6. $s(t) = t^2(1 - t)$
7. $f(x) = (x - 2)^3$
8. $y = (8 - x)^2$
9. $y = (3x)^2$
10. $s(t) = -32t^2 + 96t + 4.$
11. Find $f'(1)$, $f'(3)$, $f''(1)$, $f''(3)$, where $f(x) = 3x^2 - x$.
12. Find $f'(2)$, $f'(3)$, $f''(1)$, $f''(2)$, where $f(x) = 4x^3 - x$.

Evaluate:

13. $\left.\dfrac{d^2}{dx^2}(4x^2 + 3)\right|_{x=2}$
14. $\left.\dfrac{d^2}{dt^2}(5 - 7t^3)\right|_{t=-2}$
15. $\dfrac{d}{dt}\left(\dfrac{dL}{dt}\right)$, where $\dfrac{dL}{dt} = \dfrac{1}{3t - 1}$
16. $F''(2)$, where $\dfrac{dF}{dt} = \dfrac{3}{t} - 7t^2$
17. $\left.\dfrac{d^2V}{dr^2}\right|_{r=3}$, where $V = \frac{4}{3}\pi r^3 - \frac{1}{2}\pi r^2$
18. $\dfrac{dv}{dt}$, where $v = \dfrac{ds}{dt}$ and $s = -\frac{1}{2}gt^2 + v_0t + s_0$
19. $\dfrac{d^2y}{dx^2}$, where $y = -3x + 7$.

Review Exercises

Find all values of x at which the curves have equal slope:

20. $y = -\dfrac{27}{x}$, $\quad y = x^3$
21. $y = 2x^3 - x^2$, $\quad y = x^2 - 4$
22. $y = -3x + 2$, $\quad y = x^3 + 5x^2$
23. $y = \dfrac{1}{2x - 1}$, $\quad y = x^2 - x$.
24. For what values of x are the first and second derivatives of $x^3 - 2x^2 - x$ equal?
25. For what values of x is $x^3 + 6x + 1$ equal to its second derivative?
26. Where does the tangent to the curve $y = kx^2$ at (a, ka^2) meet the x-axis? Assume $k \neq 0$.
27. Show that the tangents to the curve $y = x^2$ at (a, a^2) and at $(a + 1, (a + 1)^2)$ intersect on the curve $y = x^2 - \frac{1}{4}$.
28. For what values of x do $2x^3 + 15x - 1$ and $-3/x$ have the same derivative?
29. Find where the tangent line to $y = 1/(2x - 5)$ at $x = 2$ crosses the line $x = -1$.

2. Curve Sketching

1. USE OF THE DERIVATIVE

The derivative of a function gives the slope of the tangent to its graph. This information is of great help in sketching curves.

> EXAMPLE 1.1
>
> Sketch the curve $y = x^3 - 3x + 1$.

Solution: Examine the derivative:

$$y' = 3x^2 - 3 = 3(x^2 - 1).$$

$$y' > 0 \quad \text{if} \quad x^2 > 1,$$

$$y' < 0 \quad \text{if} \quad x^2 < 1,$$

$$y' = 0 \quad \text{if} \quad x = \pm 1.$$

The curve increases and decreases as indicated below.

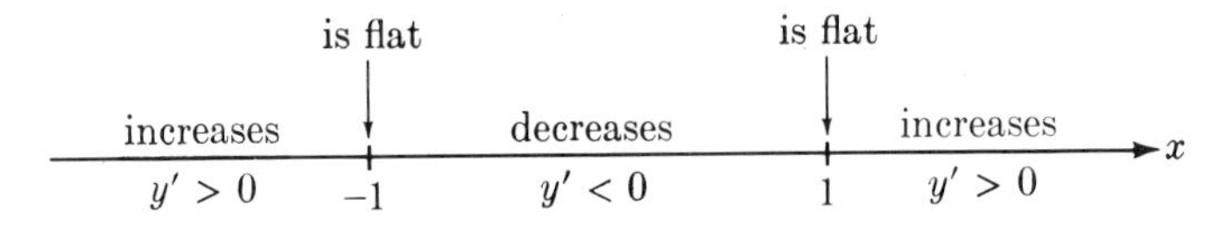

There is a high point on the curve at $x = -1$ and a low point at $x = +1$. Plot the points. At

$$x = -1, \qquad y = (-1)^3 - 3(-1) + 1 = 3;$$

at

$$x = +1, \qquad y = (1)^3 - 3(1) + 1 = -1.$$

Since it is so easy, plot also the point where $x = 0$. At

$$x = 0, \qquad y = (0)^3 - 3(0) + 1 = 1.$$

Now sketch the curve, using the information about its increasing and decreasing sections (Fig. 1.1).

Since $y' = 3(x^2 - 1)$, the slope increases as x increases indefinitely (written $x \longrightarrow \infty$). Similarly, the slope increases as $x = \longrightarrow -\infty$. This is why the curve is progressively steeper to the right and to the left.

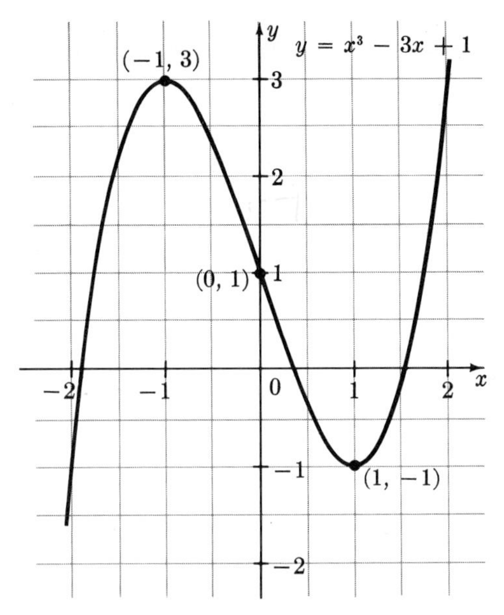

FIG. 1.1

EXAMPLE 1.2

Sketch $y = 0.75 - 0.25x - 0.25x^3$.

Solution:

$$y' = -0.25 - 0.75x^2.$$

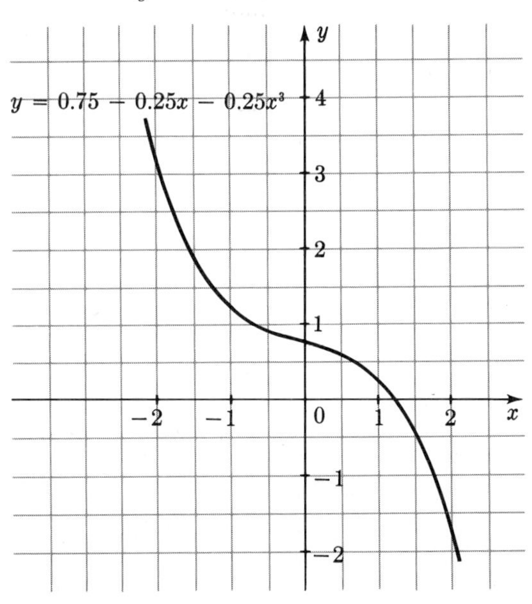

FIG. 1.2

The derivative is always negative. Hence the graph falls steadily from left to right; there are no high points or low points. Notice that as $x \longrightarrow \infty$ or as $x \longrightarrow -\infty$, the slope y' becomes more and more negative. Hence the curve gets steeper and steeper. If you are alert, you will see that the slope is least negative where $x = 0$. There the slope is -0.25. From this information a reasonable sketch of the graph is easily made (Fig. 1.2).

EXERCISES

Sketch the parabola:

1. $y = x^2 - 2x + 1$
2. $y = 4x^2 + 8x + 4$
3. $y = -x^2 + 6x - 9$
4. $y = -x^2 - 4x - 4$
5. $y = x^2 - 2x + 2$
6. $y = x^2 + 4x + 7$
7. $y = -3x^2 - 6x - 1$
8. $y = 12x - 3x^2$.

Sketch the curve:

9. $y = x^3 - 12x$
10. $y = 9x - x^3$
11. $y = 2x^3 - 9x^2 + 12x$
12. $y = x^3 + 9x^2 + 24x$
13. $y = 2x^3 - 9x^2 + 4$
14. $y = -x^3 + 3x^2 + 9x - 1$
15. $y = 16x^3 - 12x^2 + 1$
16. $y = -2x^3 + 15x^2 - 24x - 6$.

Sketch the graph:

17. $f(x) = x^3 + 3x + 2$
18. $f(x) = 1 - 3x - 2x^3$
19. $f(x) = \dfrac{1}{4x + 5}$
20. $f(x) = \dfrac{3}{2x - 1}$
21. $f(x) = 2x - \dfrac{1}{x}$
22. $f(x) = x^2 + \dfrac{1}{4x}$.

2. USE OF THE SECOND DERIVATIVE

The derivative of $f'(x)$ is $f''(x)$. If $f''(x) > 0$, then $f'(x)$ has a positive derivative. Therefore, $f'(x)$ is increasing, which means the slope of the graph $y = f(x)$ is increasing. In Fig. 2.1a, the slope increases from small

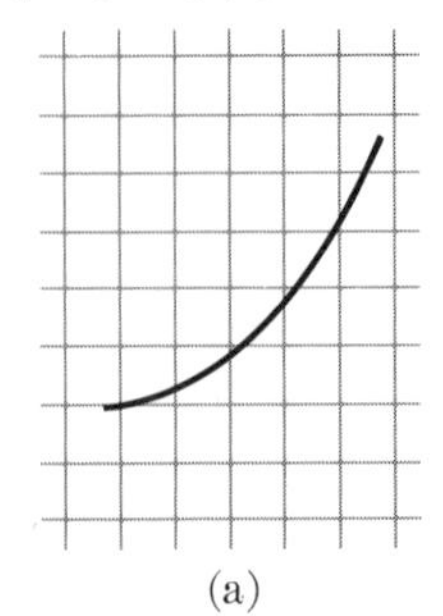

(a)

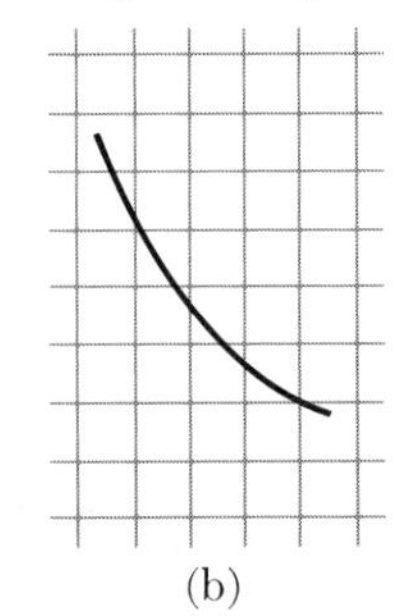

(b)

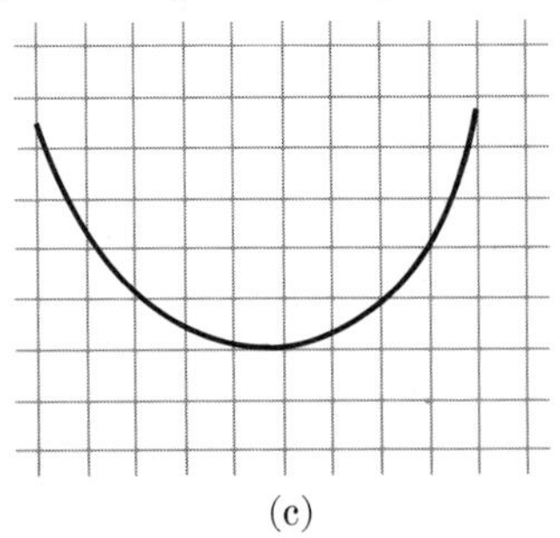

(c)

FIG. 2.1 Concave upwards.

positive values to large positive values. In Fig. 2.1b, the slope *increases* from large negative values to small negative values. (Going from -10 to -2 is an increase.) In Fig. 2.1c, the slope increases from large negative to small negative values, to zero, to small positive, to large positive values. The shape illustrated in Fig. 2.1 is called **concave upwards.**

If $f''(x) < 0$, then $f'(x)$ has a negative derivative. Therefore, $f'(x)$ itself is decreasing. The curve will have one of the shapes in Fig. 2.2. Check

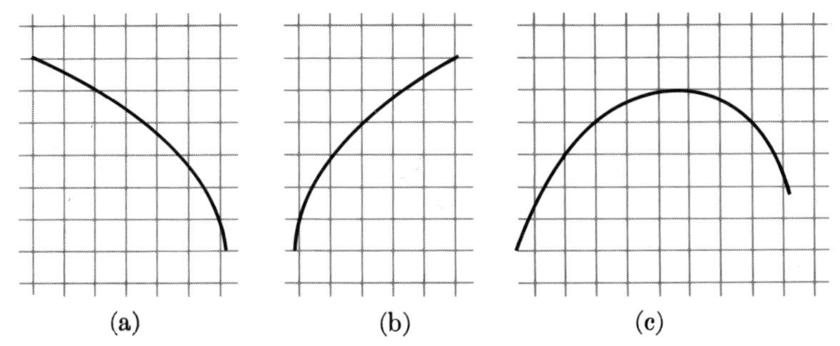

FIG. 2.2 Concave downwards.

for yourself that in each case the slope of the tangent is decreasing. The shape illustrated in Fig. 2.2 is called **concave downwards.**

Knowledge of concavity helps in curve sketching. Look at Example 1.1. That curve was plotted with the information that $y' = 3(x^2 - 1)$. Now take the second derivative, $y'' = 6x$:

$$y'' > 0 \quad \text{if} \quad x > 0,$$

$$y'' < 0 \quad \text{if} \quad x < 0,$$

$$y'' = 0 \quad \text{if} \quad x = 0.$$

The graph cannot look like Fig. 2.3.

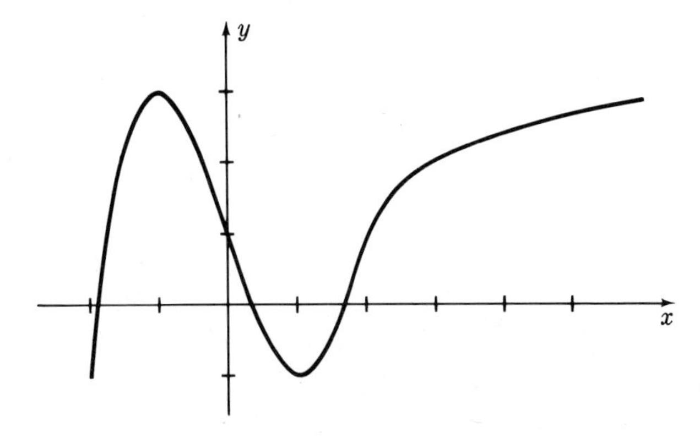

FIG. 2.3

By the sign of y'', the curve is concave upwards when $x > 0$, concave downwards when $x < 0$, and the concavity reverses at $x = 0$. The graph must have the shape shown in Fig. 1.1. It cannot have the shape shown in Fig. 2.3 because the curve sketched there is not always concave upwards when $x > 0$.

Look at Example 1.2. There $y' = -0.25 - 0.75x^2$. The second derivative is $y'' = -1.5x$.

	Second derivative	Concavity
$x < 0$	positive	upwards
$x = 0$	zero	reverses
$x > 0$	negative	downwards

Without this knowledge of the concavity, there might be a temptation to draw the curve as in Fig. 2.4 since you know from the first derivative

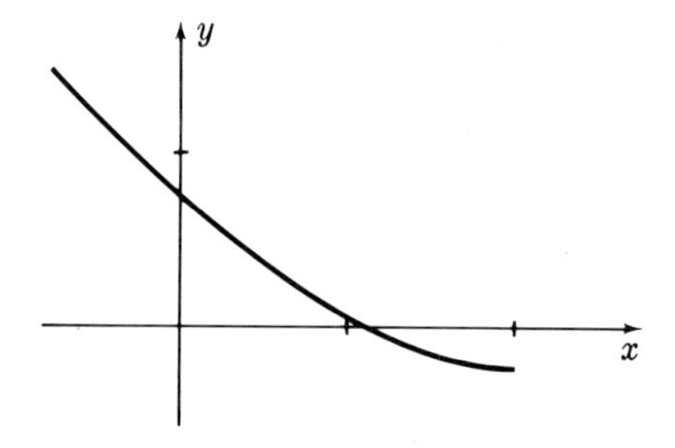

FIG. 2.4

that the graph is always falling. But the sketch is wrong; it shows a curve always concave upwards, whereas the true graph is concave downwards for $x > 0$.

Inflection Points

Correct concavity makes a sketch more accurate. Changes of concavity are especially noteworthy.

A point on a curve at which the concavity reverses is called an **inflection point.**

The graph of $y = f(x)$ has an inflection point where y'' changes from positive to negative, or vice versa.

EXAMPLE 2.1

Sketch $y = (x - 2)^3 + 1$.

Solution: By the Shifting Rule, $y' = 3(x - 2)^2$. Hence $y' > 0$ for all x except $x = 2$. There $y' = 0$.

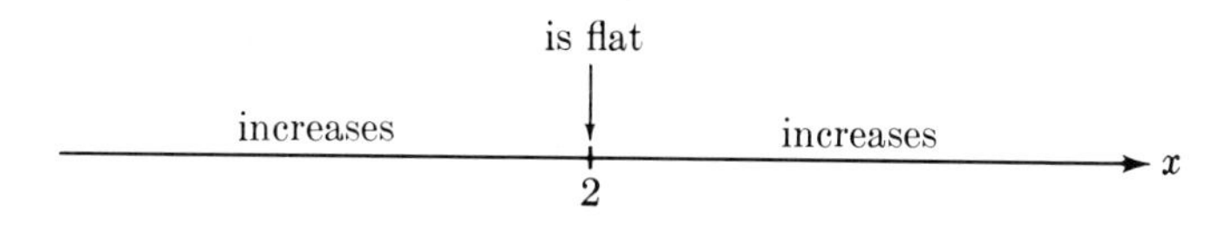

By the Shifting Rule again, $y'' = 3 \cdot 2(x - 2)$.

	Second derivative	Concavity
$x < 2$	negative	downwards
$x = 2$	zero	reverses
$x > 2$	positive	upwards

There is one inflection point: at $x = 2$. There is only one point where the tangent is horizontal: also at $x = 2$. The graph is sketched in Fig. 2.5.

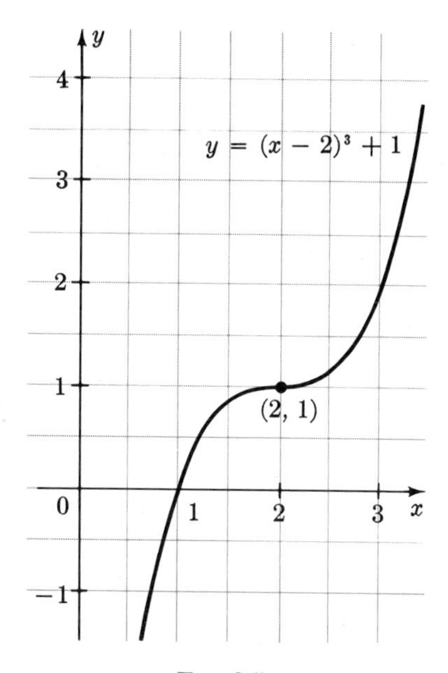

FIG. 2.5

EXERCISES

Find where $y = f(x)$ is concave up and concave down. Locate inflection points:

1. $f(x) = 4(x - 3)^2$
2. $f(x) = -3(x + 7)^2$
3. $f(x) = x^3 + 3x^2 + 3x + 1$
4. $f(x) = 2(x - 2)^3 + 1$
5. $f(x) = (x - 2)^3 + 2x$
6. $f(x) = x^3 + 3x^2 + 7x + 1$
7. $f(x) = -x^3 + 4x - 5$
8. $f(x) = x^3 - 6x^2 + 3x + 1$.

Sketch the curve:

9. $y - 2 = (x - 3)^3$
10. $y + 1 = -4(x + 1)^2$
11. $y = 4(x - 3)^3 + 2(x - 3)^2$
12. $y = (x + 1)^3 + 2$
13. $y = x^3 - x$
14. $y = 1 - 9x - x^3$
15. $y = x^3 - 2x^2 + x + 2$
16. $y = 2 + x - x^3$.

3. HINTS FOR SKETCHING CURVES

(1) Get as much "free" information as you can by inspection.

(a) If easily done, find where $f(x)$ is positive, negative, or zero.

(b) Look for symmetry. If $f(x)$ has the special property that $f(-x) = -f(x)$, then the graph $y = f(x)$ is symmetric through the origin as illustrated in Fig. 3.1a.

Examples.

$$y = x + \frac{1}{x}, \qquad y = x^3 + 5x,$$

$$y = \sin x, \qquad y = x^2 \sin 4x.$$

If $f(x) = f(-x)$, the curve is symmetric about the y-axis (Fig. 3.1b).

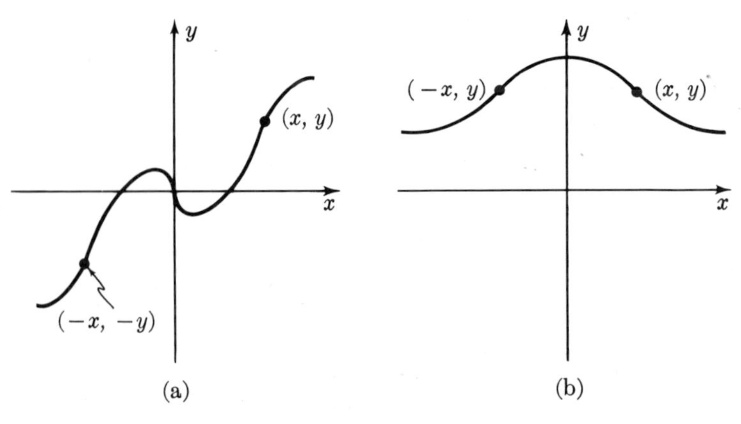

FIG. 3.1

Examples.

$$y = x^2 + \frac{1}{x^2}, \qquad y = 3x^4 + x^2 - 2,$$

$$y = \cos x, \qquad y = x \sin x.$$

(c) Find the behavior of the curve as $x \longrightarrow \infty$ and as $x \longrightarrow -\infty$.

(d) Find values of x, if any, for which the curve is not defined.

Examples.

$$y = x + \frac{1}{x} \quad \text{is not defined for} \quad x = 0,$$

$$y = \sqrt{1 - x} \quad \text{is not defined for} \quad x > 1.$$

(2) Take the derivative. Its sign will tell you where the curve is rising, falling, or flat. Plot all points where the tangent is horizontal, i.e., where $y' = 0$. If you can, find the behavior of y' as $x \longrightarrow \infty$ and as $x \longrightarrow -\infty$.

(3) Take the second derivative. Its sign will indicate the proper concavity. Locate and plot all inflection points.

(4) For more accuracy, plot a few more points. Look for points that are easy to compute. Try $x = 0$, for example, or if not too hard, see where $y = 0$.

These hints are just suggestions; they are not sacred rules. Be flexible; there is no substitute for common sense.

EXAMPLE 3.1

Sketch $y = x + \frac{1}{x}$.

Solution: First notice that there is some important free information. Observe $f(-x) = -f(x)$; therefore, the graph is symmetric through the origin. So, once the curve is sketched for $x > 0$, it can be extended by symmetry to $x < 0$.

Here is some more quick information. If $x > 0$, then $y > 0$. The curve is undefined at $x = 0$, and $y \longrightarrow \infty$ as $x \longrightarrow 0$. If x is large, $y = x + (1/x)$ is slightly larger than x. So as $x \longrightarrow \infty$, the graph is slightly above the line $y = x$.

Combining this information, you expect the graph to be something like Fig. 3.2 for $x > 0$.

To confirm this sketch, take the derivative: $y' = 1 - \frac{1}{x^2}$.

Assuming x is positive,

$$y' < 0 \quad \text{if} \quad x < 1,$$
$$y' = 0 \quad \text{if} \quad x = 1,$$
$$y' > 0 \quad \text{if} \quad x > 1.$$

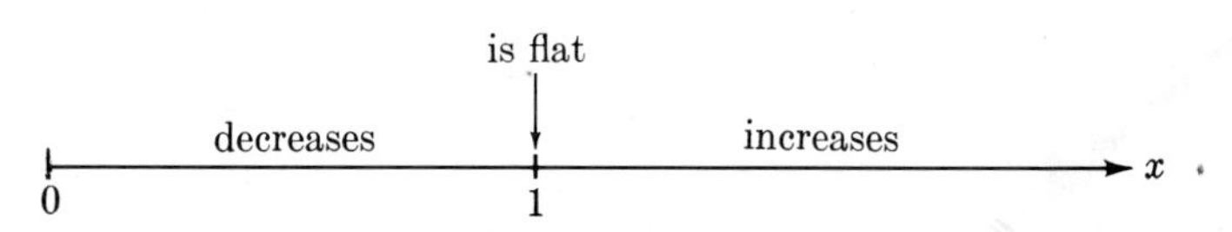

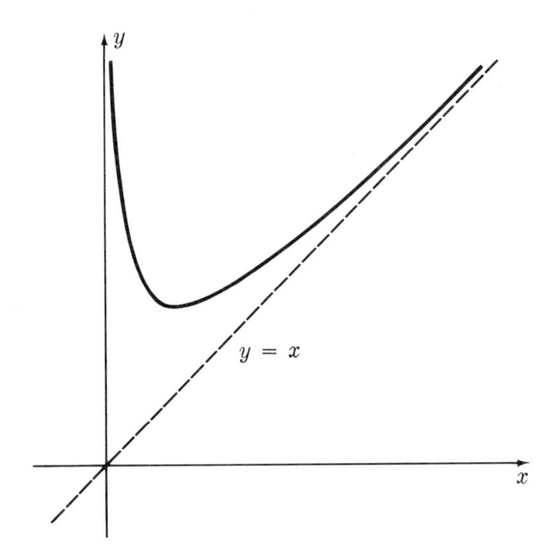

Fig. 3.2

The curve does rise and fall as in Fig. 3.2. It has a minimum point at (1, 2). Now extend the graph by symmetry (Fig. 3.3) to negative values of x.

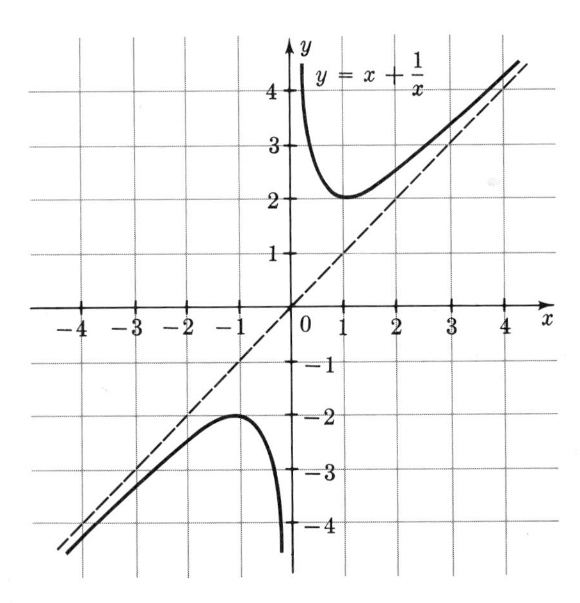

Fig. 3.3

EXERCISES

Sketch the curve $y = f(x)$ for $x \geq 0$, then by symmetry sketch the curve for $x \leq 0$:

1. $f(x) = x^2 + 1$
2. $f(x) = 4 - x^2$
3. $f(x) = x^3 - 3x$
4. $f(x) = 12x - x^3$
5. $f(x) = 3x^3 + x$
6. $f(x) = -12x^3 - 4x$
7. $f(x) = 4x + \dfrac{1}{x}$
8. $f(x) = -x - \dfrac{1}{x}\cdot$

Sketch the curve:

9. $y = x^3 - x$
10. $y = 3x^3 - 2x$
11. $y = x^3 + 8$
12. $y = 3 - x^3$
13. $y = 8x + \dfrac{2}{x}$
14. $y = -x - \dfrac{16}{x}$
15. $y = 1 + \dfrac{1}{x}$
16. $y = 1 - \dfrac{1}{x}\cdot$

Sketch the curve $y = f(x)$:

17. $f(x) = -x^3 + 3x - 4$
18. $f(x) = x^3 - 3x + 5$
19. $f(x) = 1 + 3x + x^3$
20. $f(x) = 4 - x - x^3$
21. $f(x) = x^3 - x^2 - 8x + 4$
22. $f(x) = 7 - 3x^2 - 2x^3$
23. $f(x) = 1 + \dfrac{1}{x+1}$
24. $f(x) = 1 - \dfrac{1}{x-1}\cdot$

Graph the curve:

25. $y = 1 - 6x - 3x^2 - 2x^3$
26. $y = x^3 - 3x^2 + 4x + 1$
27. $y = \dfrac{x}{x+1}$
28. $y = \dfrac{-x}{x+1}$
29. $y = x + \dfrac{1}{x+1}$

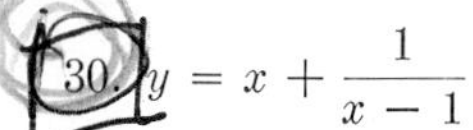

30. $y = x + \dfrac{1}{x-1}$
31. $y = 1 + x + \dfrac{1}{x}$
32. $y = 1 - x - \dfrac{1}{x}\cdot$

3. Maxima and Minima

1. USE OF THE DERIVATIVE

In many problems it is important to know the largest or the smallest value of a function in a certain range.

EXAMPLE 1.1

A ball thrown straight up from the ground reaches a height of $80t - 16t^2$ ft in t sec. How high will it go?

Solution: We know the ball will go up, reach its maximum height, and then come down. So the graph of its height, $y(t) = 80t - 16t^2$, must look

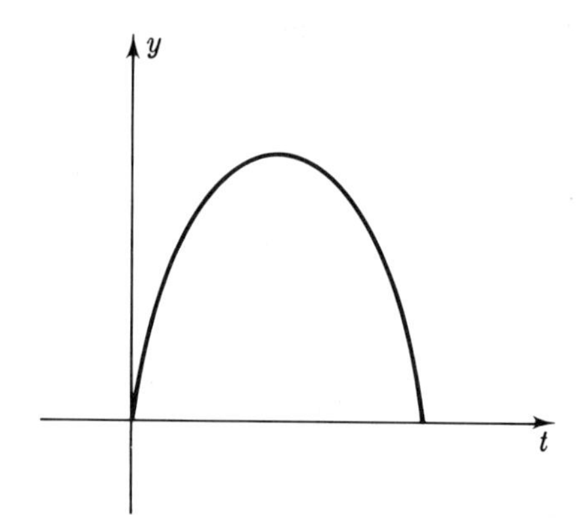

Fig. 1.1

something like Fig. 1.1. The maximum height occurs at the point on the curve where the tangent is horizontal, that is, where $y'(t) = 0$. Since

$$y(t) = 80t - 16t^2,$$

its derivative is

$$y'(t) = 80 - 32t.$$

Hence

$$y'(t) = 0 \qquad \text{when} \quad 80 - 32t = 0,$$

i.e., when

$$t = 2.5 \text{ sec.}$$

At that time the height is

$$y(2.5) = 80(2.5) - 16(2.5)^2 = 100 \text{ ft.}$$

Answer: The ball will reach a maximum height of 100 ft.

EXAMPLE 1.2

What is the largest possible area of a rug whose perimeter is 60 ft?

Solution: Let x be the length and w the width of the rug (Fig. 1.2). Let $A(x) = x(30 - x)$. Since x is a length, $x > 0$; since x is less than half

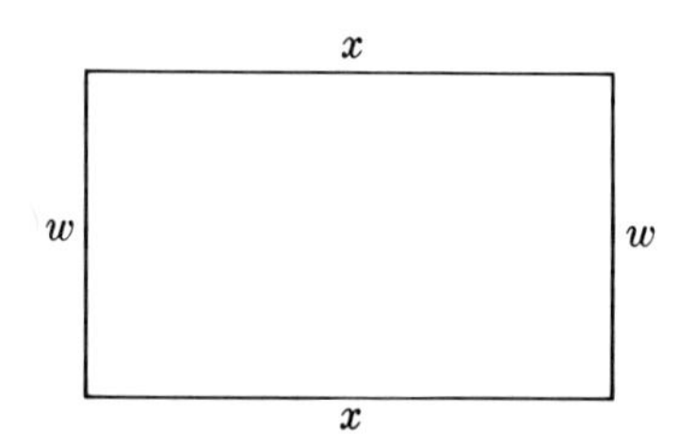

FIG. 1.2

$$\text{perimeter} = 2x + 2w = 60,$$
$$w = 30 - x,$$
$$\text{area} = xw = x(30 - x).$$

the perimeter, $x < 30$. The problem then is to find the largest value of $A(x)$ in the range $0 < x < 30$. The derivative of $A(x)$ is

$$A'(x) = 30 - 2x.$$

Hence,

$$A'(x) > 0 \quad \text{where} \quad x < 15,$$
$$A'(x) = 0 \quad \text{where} \quad x = 15,$$
$$A'(x) < 0 \quad \text{where} \quad x > 15.$$

This information indicates that the graph (Fig. 1.3) has its highest point where $x = 15$, because its shape can be described as follows:

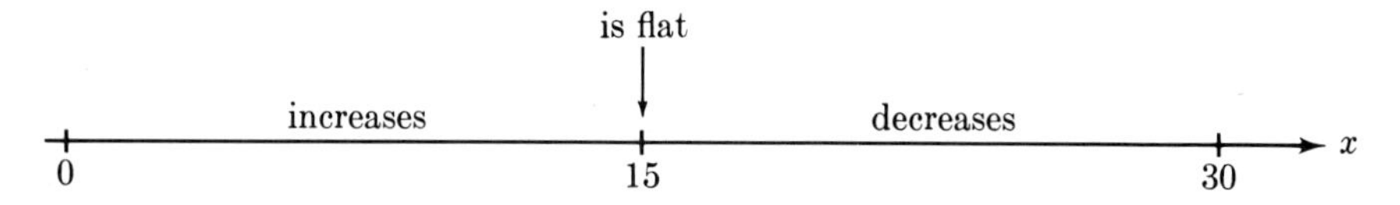

The corresponding area is

$$A(15) = 15(30 - 15) = 225.$$

Answer: 225 ft^2.

Note that $w = 15$ where $x = 15$; the optimal shape is a square. Actually it is unimportant that the perimeter be 60 ft. For any given perimeter, the square always has the largest area. See if you can modify the solution of this example to prove that.

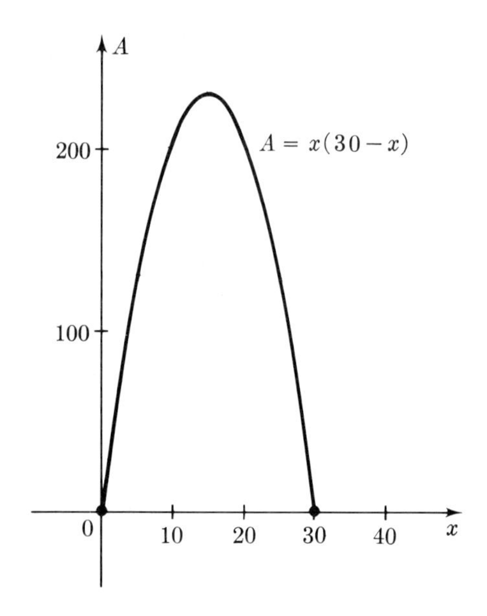

FIG. 1.3 Note that the axes have different scales.

EXERCISES

Find the minimum value of the function:

1. $s(t) = t^2 - 4t + 6$
2. $f(x) = 2x^2 - 9x + 12$
3. $f(x) = 12x + \dfrac{1}{3x}, \quad x > 0.$

Find the maximum value of the function:

4. $s(t) = 4 + 6t - t^2$
5. $A(r) = -3r^2 + 3r + 1$
6. $f(x) = 12x + \dfrac{1}{3x}, \quad x < 0.$
7. A ball thrown straight up reaches a height of $3 + 40t - 16t^2$ ft in t sec. How high will it go?
8. Show that the rectangle of largest possible area, for a given perimeter, is a square.
9. Find the maximum slope of the curve $y = 6x^2 - x^3$.
10. The power output P of a battery is given by $P = EI - RI^2$, where E and R are constants and I is the current. Find the current for which the power output is a maximum.
11. During a typical 8-hr work day the quantity of gravel produced in a plant is $60t + 12t^2 - t^3$ tons, where t represents hours worked. When is the rate of production at a maximum?

12. Find the dimensions of the right circular cone having the greatest volume for a given slant height a.
13. A closed cylindrical can is to be made so that its volume is 52 in^3. Find its dimensions if the total surface is to be a minimum.
14. Find the minimum vertical distance between the curves $y = 27x^3$ and $y = -1/x$ if $x \neq 0$.
15. Find the minimum vertical distance between the curves $y = (x - 3)^2$ and $y = -(x - 1)^2$.

A window of perimeter 16 ft has the form of a rectangle surmounted by a semicircle

16. For what radius of the semicircle is the window area greatest?
17. For what radius of the semicircle is the most light admitted, if the semicircle admits half as much light per unit area as the rectangle admits per unit area?

2. OTHER APPLICATIONS

EXAMPLE 2.1

An open box is constructed by removing a small square from each corner of a tin sheet and then folding up the sides. If the sheet is L in. on each side, what is the largest possible volume of the box?

Solution: Let each cutout square have side x. See Fig. 2.1.

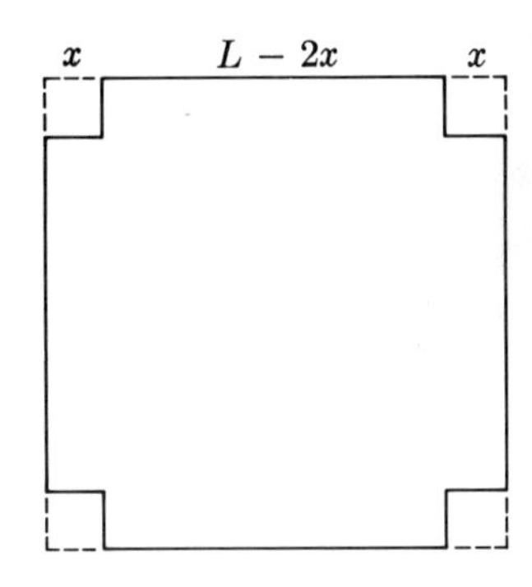

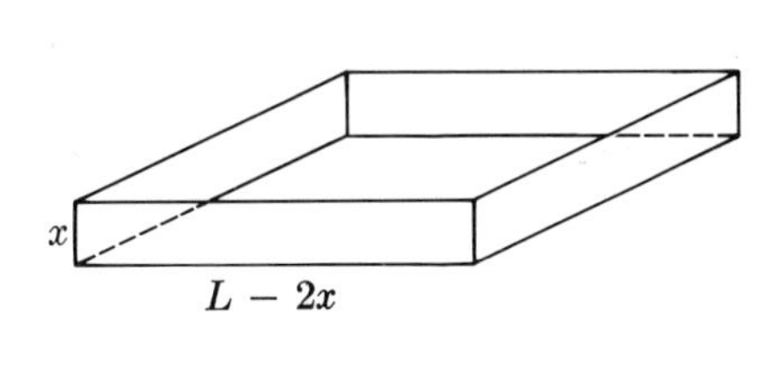

FIG. 2.1

Express the volume of the box as a function of x:

$$\text{Volume} = (\text{area of base}) \cdot (\text{height}).$$

The base is a square of side $L - 2x$, and the height is x. So the volume of the box is

$$\begin{aligned} V(x) = (L - 2x)^2 x &= (L^2 - 4Lx + 4x^2)x \\ &= L^2 x - 4Lx^2 + 4x^3. \end{aligned}$$

By the nature of the problem, x must be positive but less than $L/2$, half the side of the sheet. The problem can now be stated in mathematical terms: Find the largest value of $V(x)$ in the range $0 < x < L/2$.

$$\begin{aligned} V(x) &= L^2x - 4Lx^2 + 4x^3, \\ V'(x) &= L^2 - 8Lx + 12x^2 \\ &= (L - 2x)(L - 6x). \end{aligned}$$

The factor $L - 2x$ is positive because $x < L/2$. Therefore the sign of $V'(x)$ is the same as the sign of $(L - 6x)$:

$$V'(x) > 0 \quad \text{for} \quad x < \frac{L}{6},$$

$$V'(x) = 0 \quad \text{for} \quad x = \frac{L}{6},$$

$$V'(x) < 0 \quad \text{for} \quad x > \frac{L}{6}.$$

The growth of $V = V(x)$ is determined:

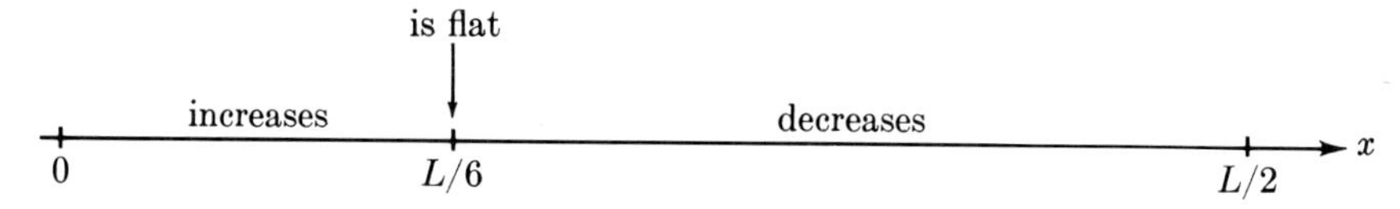

Thus $V(x)$ has its maximum value for $x = L/6$:

$$V\left(\frac{L}{6}\right) = \left(L - \frac{2L}{6}\right)^2 \frac{L}{6} = \left(\frac{2L}{3}\right)^2 \frac{L}{6} = \frac{2L^3}{27}.$$

Answer: $\dfrac{2L^3}{27}$ in^3.

EXAMPLE 2.2

Ship A leaves a port at noon and sails due north at 10 mph. Ship B is 100 mi east of the port at noon and sailing due west at 6 mph. When will the ships be nearest each other?

Solution: Set up axes with the port at the origin and the y-axis pointing north. The relative position of the ships at t hr past noon is shown in Fig. 2.2. The distance between the ships is

$$f(t) = \sqrt{(100 - 6t)^2 + (10t)^2}.$$

There is a technical difficulty here. You do not yet have the tools to compute $f'(t)$ because of that annoying square root. Nevertheless, there is a simple way out. Just *square* $f(t)$. Let

$$g(t) = [f(t)]^2 = (100 - 6t)^2 + (10t)^2.$$

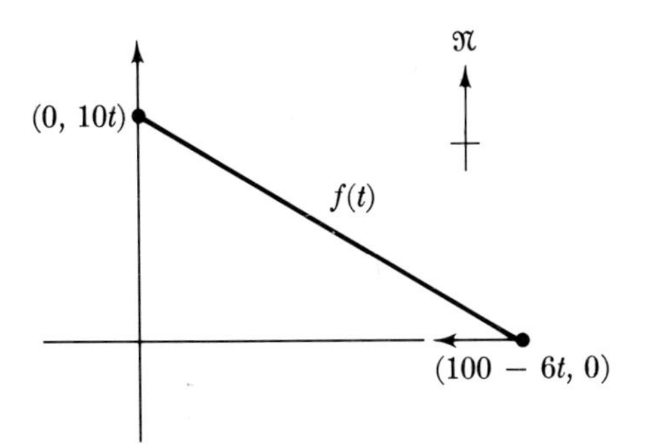

FIG. 2.2

The distance $f(t)$ is smallest precisely when its square $g(t)$ is smallest, and it is easier to minimize $g(t)$ than to minimize $f(t)$.

$$\begin{aligned} g(t) &= (100 - 6t)^2 + (10t)^2 \\ &= 10{,}000 - 1200t + 36t^2 + 100t^2 \\ &= 10{,}000 - 1200t + 136t^2, \\ g'(t) &= -1200 + 272t. \end{aligned}$$

It follows that

$$\begin{aligned} g'(t) &< 0 \qquad \text{when} \quad t < \frac{1200}{272}, \\ g'(t) &= 0 \qquad \text{when} \quad t = \frac{1200}{272}, \\ g'(t) &> 0 \qquad \text{when} \quad t > \frac{1200}{272}. \end{aligned}$$

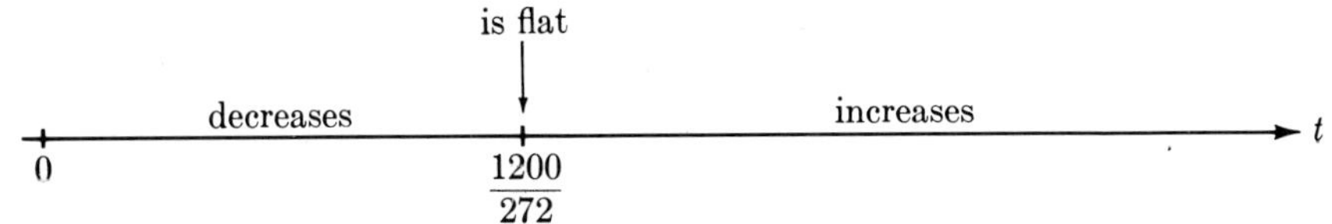

The minimum value of $g(t)$ occurs when

$$t = \frac{1200}{272} \approx 4.41 \text{ hr} \approx 4 \text{ hr, } 25 \text{ min.*}$$

Answer: The ships are closest at approximately 4:25 P.M.

EXAMPLE 2.3

Suppose, in Example 2.2, ship A is already 70 mi north of the port at noon, but otherwise the problem is the same. Now when are the ships closest?

* For $x \approx y$, read, "x is approximately equal to y."

Solution: This time the correct diagram is Fig. 2.3.

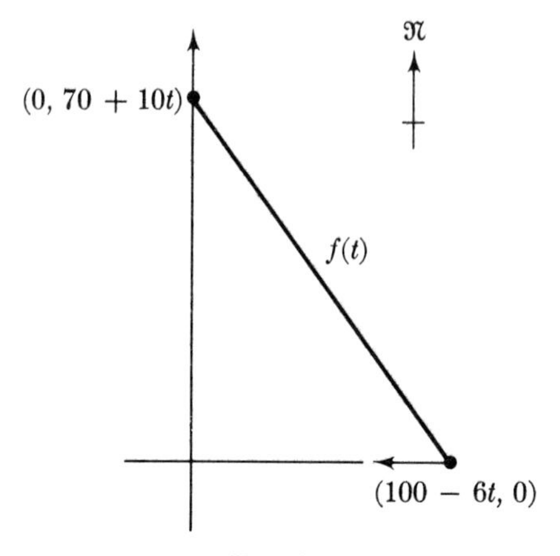

FIG. 2.3

Let

$$g(t) = [f(t)]^2 = (100 - 6t)^2 + (70 + 10t)^2$$
$$= 136t^2 + 200t + 14{,}900,$$
$$g'(t) = 272t + 200.$$

But $g'(t)$ is positive when $t \geq 0$ (at noon and afterwards). That means the distance between the ships is increasing and will continue to increase.

Answer: At noon the ships are as close as they ever will be.

EXAMPLE 2.4

A length of wire 28 in. long is cut into two pieces. One piece is bent into a 3:4:5 right triangle and the other piece is bent into a square. Show that the combined area is at least 18 in².

Solution: This is just a disguised minimum problem: Find the minimum possible combined area and check that it is at least 18 in².

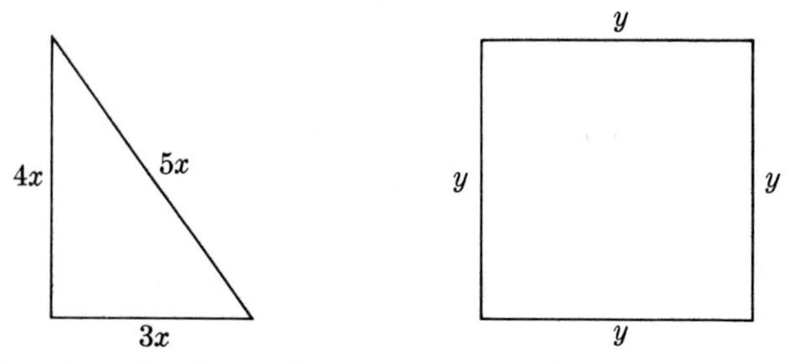

FIG. 2.4 For the triangle: perimeter = $12x$; area = $\frac{1}{2}(3x)(4x) = 6x^2$. For the square: perimeter = $4y$; area = y^2.

To avoid fractions, name the sides of the right triangle $3x$, $4x$, and $5x$. Let y denote the side of the square (Fig. 2.4). The combined perimeter is

$$12x + 4y = 28,$$

hence

$$y = 7 - 3x.$$

The combined area is

$$\begin{aligned} A(x) &= 6x^2 + y^2 \\ &= 6x^2 + (7 - 3x)^2 \\ &= 15x^2 - 42x + 49. \end{aligned}$$

By the nature of the problem, x must be positive. But since the perimeter of the triangle is less than 28 in.,

$$12x < 28 \qquad \text{or} \qquad x < \frac{28}{12} = \frac{7}{3}.$$

So the problem reduces to this: Find the least value of $A(x)$ in the range $0 < x < \frac{7}{3}$. Take the derivative:

$$\begin{aligned} A(x) &= 15x^2 - 42x + 49, \\ A'(x) &= 30x - 42. \end{aligned}$$

Hence

$$\begin{aligned} &A'(x) < 0 \quad \text{for} \quad 30x - 42 < 0, \quad \text{i.e., for} \quad x < \frac{7}{5}, \\ &A'(x) = 0 \quad \text{for} \quad x = \frac{7}{5}, \\ &A'(x) > 0 \quad \text{for} \quad 30x - 42 > 0, \quad \text{i.e., for} \quad x > \frac{7}{5}. \end{aligned}$$

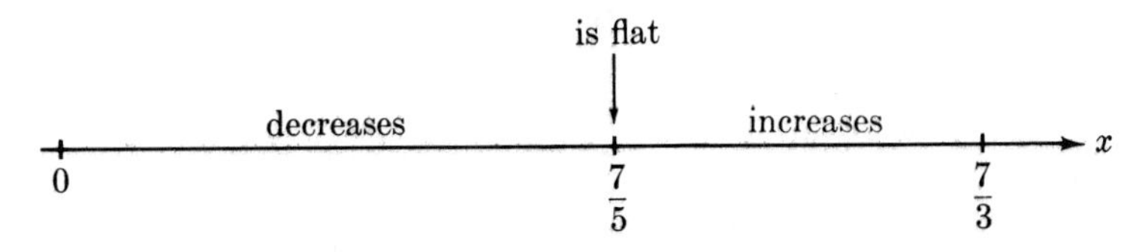

The minimum occurs for $x = \frac{7}{5}$. Since $A(x) = 6x^2 + (7 - 3x)^2$,

$$\begin{aligned} A\left(\frac{7}{5}\right) &= 6\left(\frac{7}{5}\right)^2 + \left(7 - \frac{21}{5}\right)^2 \\ &= \frac{6 \cdot 49}{25} + \left(\frac{14}{5}\right)^2 = \frac{490}{25} = \frac{98}{5} = 19.6. \end{aligned}$$

Answer: The least possible combined area is 19.6 in^2. So the total area always exceeds 18 in^2.

EXAMPLE 2.5

Find the maximum combined area that can be achieved in Example 2.4.

Solution: Recall the diagram

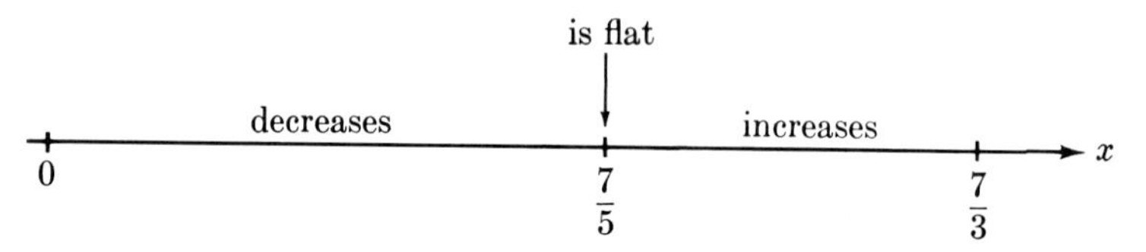

Thus the graph of $y = A(x)$ must have the shape of Fig. 2.5. The maximum

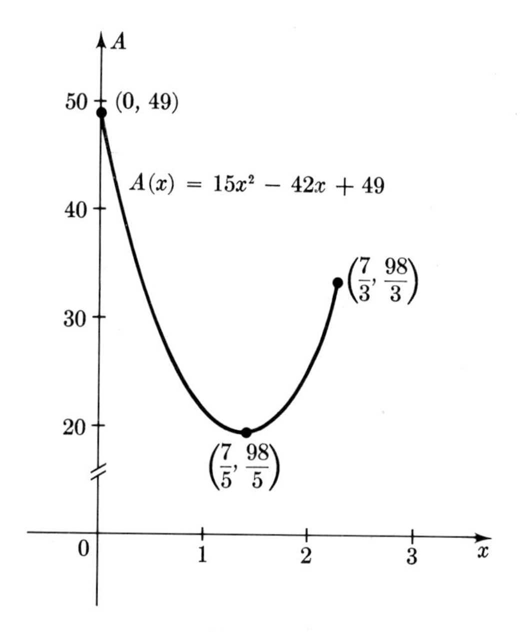

FIG. 2.5

of $A(x)$ in the range $0 \le x \le \frac{7}{3}$ occurs either at $x = 0$ or at $x = \frac{7}{3}$.

$$A(x) = 6x^2 + (7 - 3x)^2.$$

$$A(0) = 7^2 = 49, \qquad A\left(\frac{7}{3}\right) = 6\left(\frac{7}{3}\right)^2 + 0 = \frac{98}{3}.$$

The larger of these numbers is $A(0) = 49$. Therefore the combined area is greatest at $x = 0$. But if $x = 0$, there is no triangle at all, only a square!

Answer: 49 in^2.

An apology to the reader: The maximum area, 49 in^2, can be achieved only if there is no actual cut in the wire. As the example is worded, there is no maximum since the wire must be cut. With a cut, any area less than 49 in^2 is possible but not 49 in^2 itself. Technically speaking our solution is dishonest.

EXERCISES

1. A cylindrical tank (open top) is to hold V gal. How should it be made so as to use the least amount of material?
2. A page is to contain 27 in^2 of print. The margins at the top and bottom are 1.5 in., at the sides 1 in. Find the most economical dimensions of the page.
3. A triangular lot is bounded by two streets intersecting at right angles. The lengths of the street frontage for the lot are a and b ft. Find the dimensions of the largest rectangular building that can be placed on the lot, facing one of the streets.
4. Find the dimensions of the rectangle of maximum area that can be inscribed in the region bounded by the parabola $y = -8x^2 + 16$ and the x-axis.
5. An athletic field of $\frac{1}{4}$-mi perimeter consists of a rectangle with a semicircle at each end. Find the dimensions of the field so that the area of the rectangular portion is the largest possible.
6. As a man starts across a 200-ft bridge, a ship passes directly beneath the center of the bridge. If the ship is moving at the rate of 8 ft/sec and the man at the rate of 5 ft/sec, what is the shortest horizontal distance between them?
7. Find the point on the graph of the equation $y = \sqrt{x}$ nearest to the point $(1, 0)$.
8. Suppose in Ex. 6, the bridge is 50 ft high. Find the shortest *distance* between the man and the ship.
9. Two particles moving in the plane have coordinates $(2t, 8t^3 - 24t + 10)$ and $(2t + 1, 8t^3 + 6t + 1)$ at time t. How close do the particles come to each other?
10. The strength of a rectangular beam is given by $S = xy^2$, where x and y are its cross-sectional dimensions. Find the shape of the strongest beam that can be cut from a circular log of radius r.

Suppose the cost of producing x units is $f(x)$ dollars, and the price of x units is $h(x)$ dollars. Calculate the maximum net revenue possible for a manufacturer if:

11. $f(x) = 7.5x + 400, \qquad h(x) = 10x - 0.0005x^2$
12. $f(x) = 50x + 1200, \qquad h(x) = 55x - 0.001x^2.$
13. The cost per hour in dollars for fuel to operate a certain airliner is given by the equation $F = 0.005v^2$, where v is the speed in mph. If fixed charges amount to \$1200/hr, find the most economical speed for a 1500 mi trip.

3. ON PROBLEM SOLVING

A major part of your time in Calculus and other courses is devoted to solving problems. It is worth your while to develop sound techniques. Here are a few suggestions.

Think. Before plunging into a problem, take a moment to think. Read the problem again. Think about it. What are its essential features? Have you seen a problem like it before? What techniques are needed?

Try to make a rough estimate of the answer. It will help you understand the problem and will serve as a check against unreasonable answers. A car will *not* go 1000 mi in 3 hr; a weight dropped from 10,000 ft will *not* hit the earth at 5 mph; the volume of a tank is *not* -275 gal.

Examine the data. Be sure you understand what is given. Translate the data into mathematical language. Whenever possible, make a clear diagram and label it accurately. Place axes to simplify computations. If you get stuck, check that you are using *all* the data.

Avoid sloppiness.

(a) Avoid sloppiness in language. Mathematics is written in English sentences. A typical mathematical sentence is "$y = 4x + 1$." The equal sign is the verb in this sentence; it means "equals" or "is equal to." The equal sign is not to be used in place of "and," nor as a punctuation mark. *Quantities on opposite sides of an equal sign must be equal.*

Use short simple sentences. Avoid pronouns such as "it" and "which." Give names and use them. Consider the following example.

"To find the minimum of it, differentiate it and set it equal to zero, then solve it which if you substitute it, it is the minimum."

Better: "To find the minimum of $f(x)$, set its derivative $f'(x)$ equal to zero. Let x_0 be the solution of the resulting equation. Then $f(x_0)$ is the minimum value of $f(x)$."

(b) Avoid sloppiness in computation. Do calculations in a sequence of neat, orderly steps. Include all steps except utterly trivial ones. This will help eliminate errors, or at least make errors easier to find. Check any numbers used; be sure that you have not dropped a minus sign or transposed digits.

(c) Avoid sloppiness in units. If you start out measuring in feet, all lengths must be in feet, all areas in square feet, and all volumes in cubic feet. Do not mix feet and acres, seconds and years.

(d) Avoid sloppiness in the answer. Be sure to answer the question that is asked. If the problem asks for the *maximum value* of $f(x)$, the answer is not the *point* where the maximum occurs. If the problem asks for a *formula*, the answer is not a *number*.

EXAMPLE 3.1

Find the minimum of $f(x) = x^2 - 2x + 1$.

Solution 1:

$$2x - 2$$

$$x = 1$$

$$1^2 - 2 \cdot 1 + 1$$

$$0$$

Unbearable. This is just a collection of marks on the paper. There is absolutely no indication of what these marks mean or of what they have to do with the problem. When you write, it is your responsibility to inform the reader what you are doing. Assume he is intelligent, but not a mind reader.

Solution 2:

$$\frac{df}{dx} = 2x - 2 = 0 = 2x = 2 = x = 1$$

$$= f(x) = 1^2 - 2 \cdot 1 + 1 = 0.$$

Poor. The equal sign is badly mauled. This solution contains such enlightening statements as "$0 = 2 = 1$," and it does not explain what the writer is doing.

Solution 3:

$$\frac{df}{dx} = 2x - 2 = 0, \qquad 2x = 2, \qquad x = 1.$$

This is better than Solution 2, but contains two errors. Error 1: The first statement, "$df/dx = 2x - 2 = 0$," muddles two separate steps. First the derivative is computed, then the derivative is equated to zero. Error 2: The solution is incomplete because it does not give what the problem asks for, the minimum value of f. Instead, it gives the point x at which the minimum is assumed.

Solution 4: The derivative of f is

$$f' = 2x - 2.$$

At a minimum, $f' = 0$. Hence

$$2x - 2 = 0, \qquad x = 1$$

The corresponding value of f is

$$f(1) = 1^2 - 2 \cdot 1 + 1 = 0.$$

If $x > 1$, then $f'(x) = 2(x - 1) > 0$, so f is increasing. If $x < 1$, then $f'(x) = 2(x - 1) < 0$, so f is decreasing. Hence f is minimal at $x = 1$, and the minimum value of f is 0.

This solution is absolutely correct, but long. For homework assignments the following is satisfactory (check with your instructor):

Solution 5:

$$f'(x) = 2x - 2.$$

At min, $f' = 0$, $2x - 2 = 0$, $x = 1$. For $x > 1$, $f'(x) = 2(x - 1) > 0$, $f\uparrow$; for $x < 1$, $f'(x) = 2(x - 1) < 0$, $f\downarrow$.

Hence $x = 1$ yields min,

$$f_{\min} = f(1) = 1^2 - 2 \cdot 1 + 1 = 0.$$

The next solution was submitted by a student who took a moment to think.

Solution 6:

$$f(x) = x^2 - 2x + 1 = (x - 1)^2 \geq 0.$$

But

$$f(1) = (1 - 1)^2 = 0.$$

Hence the minimum value of $f(x)$ is 0.

4. Direction Fields

1. TWO BASIC FACTS ABOUT DERIVATIVES

What can be said about a function $f(x)$ whose derivative $f'(x)$ is zero at each x ? Answer: $f(x)$ must be a constant function. To see this, imagine a short horizontal line segment through each point of the plane (Fig. 1.1).

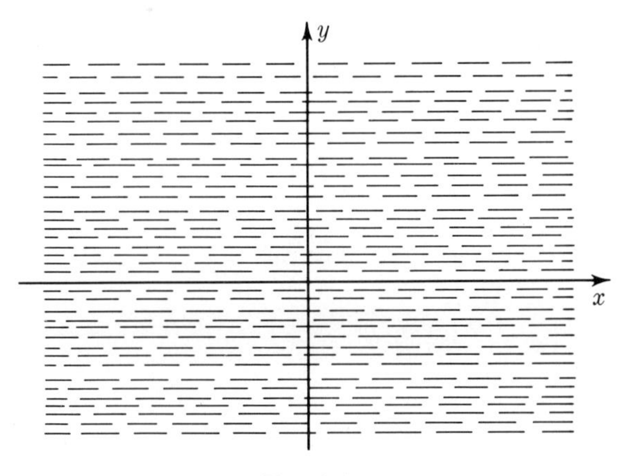

FIG. 1.1

If $f'(x) = 0$, the graph of $y = f(x)$ has slope zero at each point. Hence the direction of the graph at each point is indicated by the line segment at this point (Fig. 1.1). The graph must be a horizontal line.

If $f'(x) = 0$ for each x, then $f(x) = c$, a constant function.

This is reasonable; if the rate of change of $f(x)$ is always zero, $f(x)$ never changes. So $f(x)$ is constant.

Now consider another question: How are two functions $f(x)$ and $g(x)$ related if $f'(x) = g'(x)$ for each x ? The answer is easy if you make one observation: The difference function $[f(x) - g(x)]$ has zero derivative. Indeed $f'(x) = g'(x)$, hence

$$[f(x) - g(x)]' = f'(x) - g'(x) = 0.$$

But a function with derivative zero is constant. Therefore $f(x) - g(x) = c$. See Fig. 1.2.

> Two functions with identical derivatives differ by a constant: If $f'(x) = g'(x)$ for each x, then $f(x) = g(x) + c$.

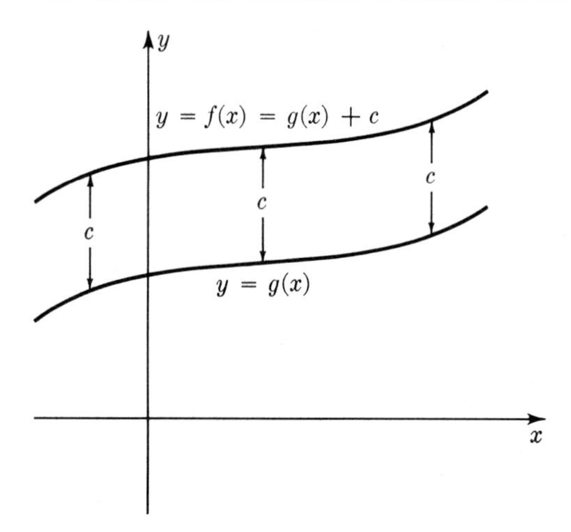

FIG. 1.2

EXERCISES

1. Use the relation $\dfrac{x}{x+1} + \dfrac{1}{x+1} = 1$ to find $\dfrac{d}{dx}\left(\dfrac{x}{x+1}\right)$.

2. Use the method of Ex. 1 to find $\dfrac{d}{dx}\left(\dfrac{x}{x+a}\right)$.

3. Use the relation $\dfrac{2x}{2x+5} + \dfrac{5}{2x+5} = 1$ to find $\dfrac{d}{dx}\left(\dfrac{2x}{2x+5}\right)$.

   [*Hint:* $\dfrac{5}{2x+5} = \dfrac{5}{2}\left(\dfrac{1}{x+\frac{5}{2}}\right)$.]

4. Find: $\dfrac{d}{dx}\left(\dfrac{2x}{3x+4}\right)$.

5. Assume known $\dfrac{d}{dx}\sin^2 x = \sin 2x$. Find $\dfrac{d}{dx}\cos^2 x$.

2. DIRECTION FIELDS

Figure 1.1 is the simplest example of what is called a **direction field.** Figures 2.1 and 2.2 show two familiar examples from nature.

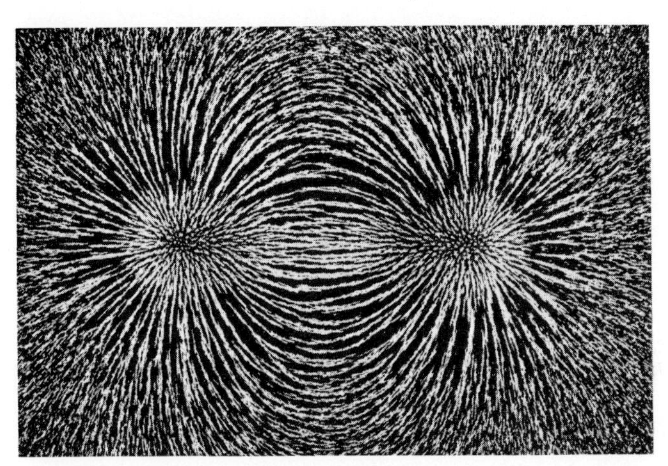

FIG. 2.1 Iron filings arrange themselves in the direction of lines of force around a magnet.

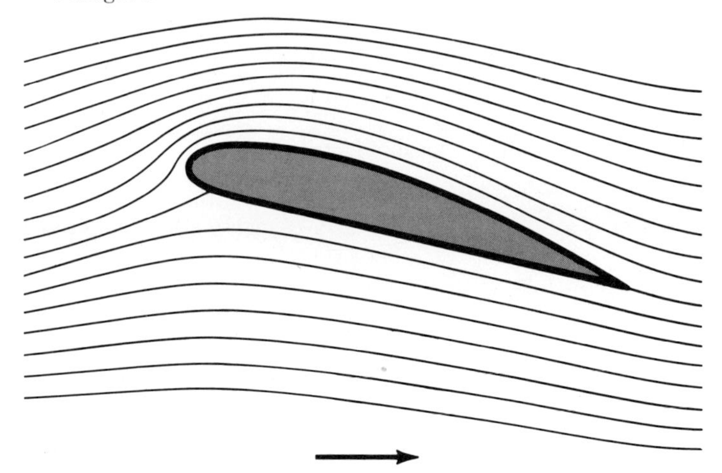

FIG. 2.2 Air flows in streamlines around an airfoil.

In calculus, the derivative of a function indicates the direction of its graph. Given a derivative $f'(x)$, there is a direction field associated with it.

EXAMPLE 2.1

Suppose the derivative of a function is $y' = 0.8$. Describe the function.

Solution: At each point of its graph, the slope is 0.8. Imagine a short line segment of a slope 0.8 at each point of the plane. The direction field in

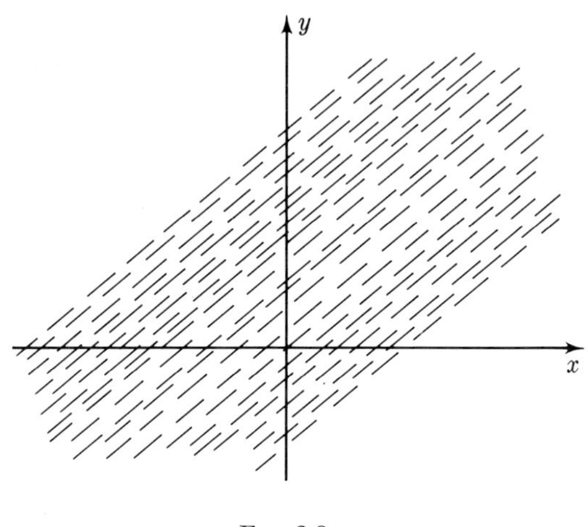

FIG. 2.3

Fig. 2.3 shows that the graph of $y = f(x)$ can only be a straight line of slope 0.8. Now

$$\frac{d}{dx}(0.8x) = 0.8.$$

Hence $y = 0.8x$ is a function whose graph is such a straight line. Any other

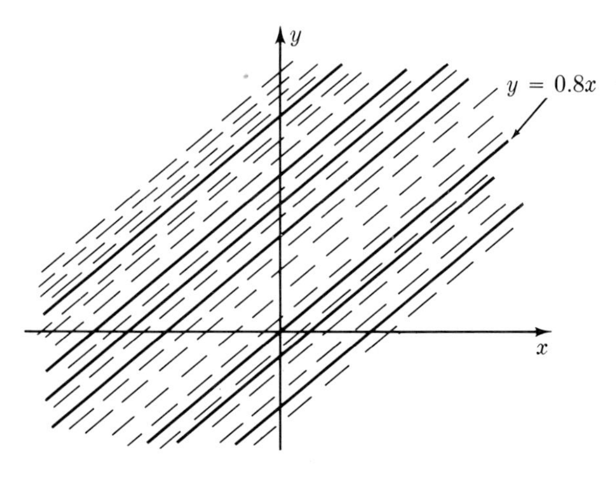

FIG. 2.4

function with the same derivative differs from $0.8x$ by a constant (Fig. 2.4).

Answer: $y = 0.8x + c.$

EXAMPLE 2.2

Describe all functions $y = f(x)$ for which $y' = x$.

Solution: Sketch the associated direction field. At each point (x, y) of the plane, imagine a short line segment whose slope is x. For example, at each point $(0, y)$ there is a segment of slope 0; at each point $(1, y)$ there

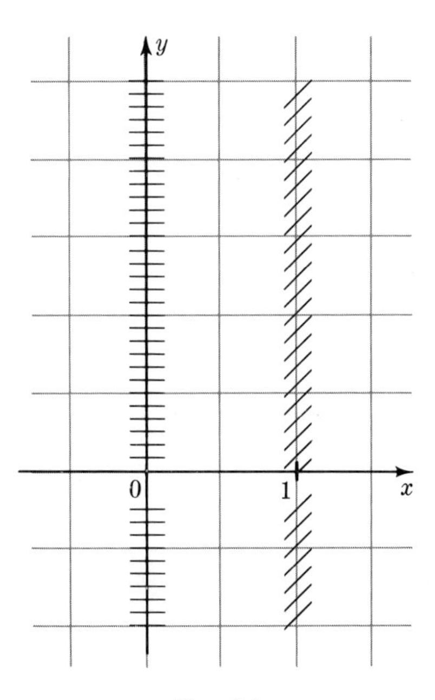

FIG. 2.5

is a segment of slope 1. See Fig. 2.5. Augment this figure by sketching more segments; soon a pattern emerges (Fig. 2.6). It is not hard to find the functions whose graphs are suggested by Fig. 2.6. They all satisfy $y' = x$. Since

$$\frac{d}{dx}\left(\frac{1}{2}x^2\right) = \frac{1}{2}\frac{d}{dx}(x^2) = \frac{1}{2}(2x) = x,$$

one such function is $y = x^2/2$. Any other function with the same derivative differs from this one by a constant (Fig. 2.7).

Answer: $y = \frac{1}{2}x^2 + c.$

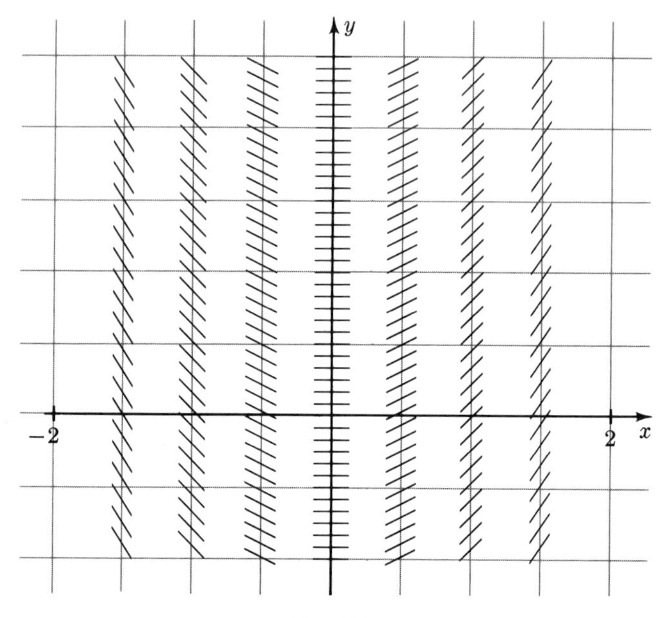

FIG. 2.6

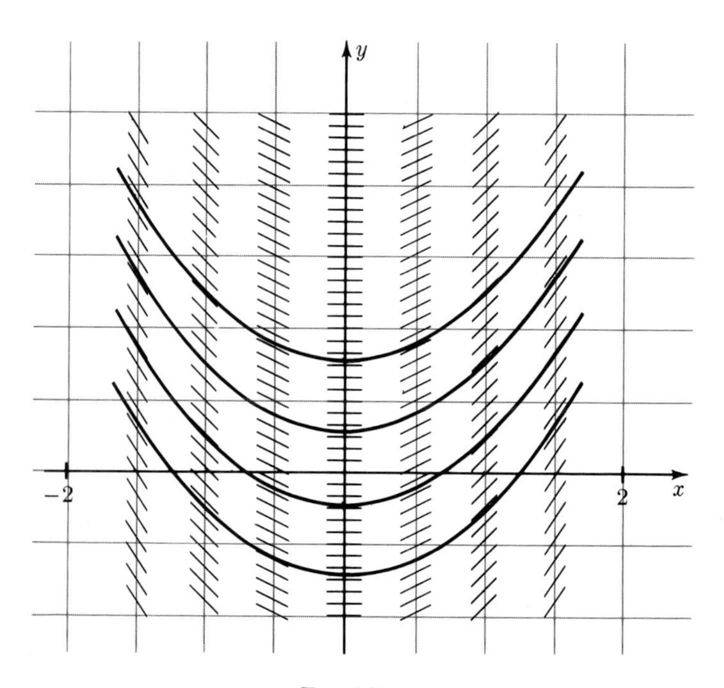

FIG. 2.7

EXAMPLE 2.3

Sketch the direction field associated with $y' = x - 2$. Find all functions $y = f(x)$ satisfying this equation.

Solution: Make a table.

x	$\frac{1}{2}$	1	$\frac{3}{2}$	2	$\frac{5}{2}$	3	$\frac{7}{2}$
$y' = x - 2$	$-\frac{3}{2}$	-1	$-\frac{1}{2}$	0	$\frac{1}{2}$	1	$\frac{3}{2}$

Now sketch the field. At each point $(\frac{1}{2}, y)$ draw a segment of slope $-\frac{3}{2}$; at each point $(1, y)$, a segment of slope -1; etc. Fig. 2.8 results. This figure

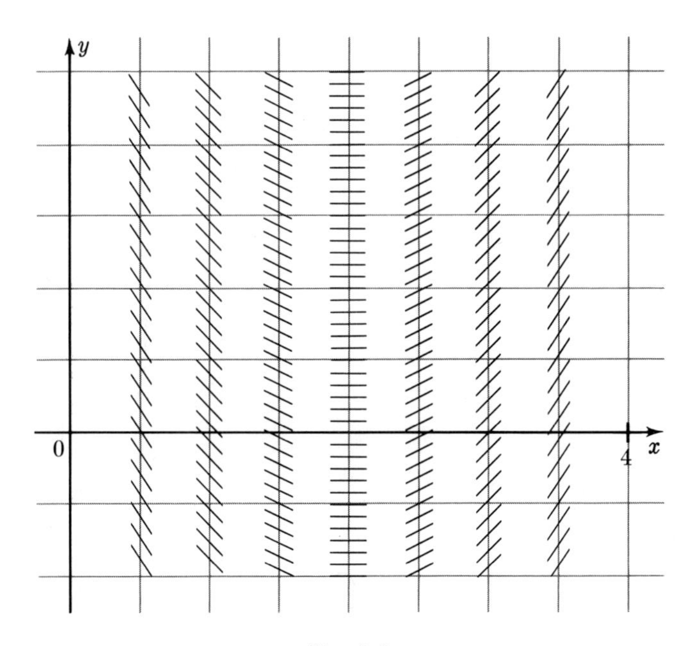

FIG. 2.8

suggests a family of parallel curves, each with a minimum at $x = 2$. See Fig. 2.9. Each curve in Fig. 2.9 is the graph of a function $y = f(x)$ for which $y' = x - 2$. Now

$$\frac{d}{dx}\left(\frac{1}{2}x^2 - 2x\right) = \frac{1}{2}\frac{d}{dx}(x^2) - 2\frac{d}{dx}(x) = x - 2.$$

Hence $y = (x^2/2) - 2x$ is a function with derivative $y' = x - 2$. Any other function with the same derivative differs from this one by a constant.

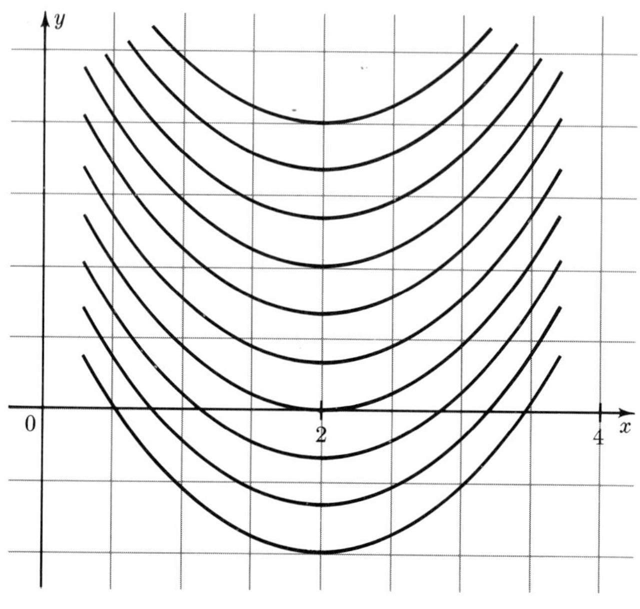

FIG. 2.9

Answer: $y = \frac{1}{2}x^2 - 2x + c.$

(These are the curves in Fig. 2.9.)

EXERCISES

Sketch the direction field and find all functions satisfying the equation:

1. $f'(x) = \frac{1}{2}$
2. $f'(x) = -0.8$
3. $y' = -x$
4. $y' = 2 - x$
5. $f'(x) = x^2$
6. $f'(x) = -x^2 + x$
7. $y' = -1/x^2$
8. $y' = 2/x^2$.

3. ANTIDIFFERENTIATION

Differentiation is the process that leads from a function $f(x)$ to its derivative $f'(x)$. The reverse is called **antidifferentiation**. It is the process of finding a function $f(x)$ whose derivative $f'(x)$ is known. Actually, there is not just one function whose derivative is $f'(x)$, but a family of functions which differ from each other by additive constants.

EXAMPLE 3.1

Find $f(x)$ if $f'(x) = x^2 + \dfrac{2}{x^2}\cdot$

Solution: Since

$$\frac{d}{dx}(x^3) = 3x^2, \qquad \frac{d}{dx}\left(\frac{1}{3}x^3\right) = x^2.$$

Since

$$\frac{d}{dx}\left(\frac{1}{x}\right) = \frac{-1}{x^2}, \qquad \frac{d}{dx}\left(-\frac{2}{x}\right) = \frac{2}{x^2}\cdot$$

Therefore,

$$f(x) = \frac{1}{3}x^3 - \frac{2}{x}$$

is a function whose derivative is

$$x^2 + \frac{2}{x^2}\cdot$$

Any other function with the same derivative differs from

$$\frac{1}{3}x^3 - \frac{2}{x}$$

by a constant.

Answer: $f(x) = \dfrac{1}{3}x^3 - \dfrac{2}{x} + c.$

REMARK: The answer is a family of functions, not a single function.

EXAMPLE 3.2

Find the functions y for which

$$y' = (x + 5)^2 + 3x - 1.$$

Solution: In the last example we found that

$$\frac{d}{dx}\left(\frac{1}{3}x^3\right) = x^2.$$

By the Shifting Rule,

$$\frac{d}{dx}\left[\frac{1}{3}(x + 5)^3\right] = (x + 5)^2.$$

We know also that

$$\frac{d}{dx}\left(\frac{3}{2}x^2\right) = 3x \qquad \text{and} \qquad \frac{d}{dx}(-x) = -1.$$

Therefore,

$$y = \frac{1}{3}(x+5)^3 + \frac{3}{2}x^2 - x$$

is a function whose derivative is $(x+5)^2 + 3x - 1$.

Answer: $y = \frac{1}{3}(x+5)^3 + \frac{3}{2}x^2 - x + c.$

Significance of Additive Constants

Adding a constant c to a function shifts its graph vertically c units. Therefore, antidifferentiation produces a family of parallel curves. For example, in the last section we saw that antidifferentiation of $x - 2$ produced the family of curves shown in Fig. 2.9. Each curve in Fig. 2.9 has an equation of the form

$$y = \frac{1}{2}x^2 - 2x + c.$$

A particular choice of the constant c singles out one curve of the family. In practice, c is usually chosen so that the curve will pass through a given point.

EXAMPLE 3.3

Find the curve $y = f(x)$ passing through the point $(0, 4)$ and such that $y' = x - 2$.

Solution: The curve must be one of the family of curves

$$y = \frac{1}{2}x^2 - 2x + c.$$

To find a suitable value of c, substitute $x = 0$, $y = 4$:

$$4 = \frac{1}{2}(0)^2 - 2(0) + c, \qquad c = 4.$$

Answer: $y = \frac{1}{2}x^2 - 2x + 4.$

EXAMPLE 3.4

Find the curve $y = f(x)$ that satisfies

$$y' = 5x^2 - 3x - \frac{1}{x^2}$$

and passes through the point $(2, -1)$.

Solution: By antidifferentiation,

$$y = \frac{5}{3}x^3 - \frac{3}{2}x^2 + \frac{1}{x} + c.$$

Substitute $x = 2$, $y = -1$:

$$-1 = \frac{5}{3}(2)^3 - \frac{3}{2}(2)^2 + \frac{1}{2} + c$$

$$= \frac{40}{3} - 6 + \frac{1}{2} + c$$

$$c = -\frac{40}{3} + 6 - \frac{1}{2} - 1 = -\frac{53}{6}.$$

Answer: $y = \frac{5}{3}x^3 - \frac{3}{2}x^2 + \frac{1}{x} - \frac{53}{6}.$

EXERCISES

Use antidifferentiation to find $f(x)$ if:

1. $f'(x) = a$
2. $f'(x) = -9$
3. $f'(x) = -x$
4. $f'(x) = 4(x + 3)$
5. $f'(x) = 4x^2 + x$
6. $f'(x) = -3x^2$
7. $f'(x) = -4/x^2$
8. $f'(x) = 5/(x - 1)^2$

Find a function whose derivative is:

9. $x + 1$
10. $(3x - 5)(x + 1)$
11. $3x^2 + 2x + 1$
12. $7x^2 - 5x + 2$
13. $2x^2 - \frac{7}{x^2}$
14. $\left(x + \frac{1}{x}\right)^2$
15. $(x + 4)^2 + \frac{1}{x^2}$
16. $(x - 3)^2 + 2x + \frac{1}{x^2} - 1.$

Find the curve $y = f(x)$ such that:

17. $f'(x) = x - 5$ and $f(1) = 2$
18. $f'(x) = -(x + 3)$ and $f(0) = 4$
19. $\frac{dy}{dx} = 3x^2 - 4x + \frac{1}{x^2}$ and the point (2, 9) is on the curve
20. $\frac{dy}{dx} = (3x - 4)^2 + 5x$ and the point (1, 2) is on the curve [*Hint:* $(3x - 4)^2 = 9(x - \frac{4}{3})^2$].

5. Velocity and Acceleration

1. VELOCITY

During the initial stage of flight, a rocket fired vertically reaches an elevation of $50t^2$ ft above the ground in t sec. How fast is the rocket rising 2 sec after it is fired?

This is a tricky question because it is not really clear what is meant by velocity at an instant. Usually we compute **average velocity** by the formula

$$\text{average velocity} = \frac{\text{displacement}}{\text{time}}$$

applied to a certain time interval. To be more precise, for the time interval between t_1 and t_2,

$$\text{average velocity} = \frac{(\text{position at time } t_2) - (\text{position at time } t_1)}{t_2 - t_1}.$$

This formula does not apply to "instantaneous velocity." Nevertheless, we can compute the average velocity between $t = 2$ and say, $t = 2.01$, or $t = 2.001$, or more generally, $t = 2 + h$.

The average velocity between $t = 2$ and $t = 2 + h$ is

$$\frac{50(2+h)^2 - 50 \cdot 2^2}{(2+h) - 2} = 50 \frac{(2+h)^2 - 2^2}{h}$$

$$= 50 \frac{4h + h^2}{h} = 50(4 + h).$$

The smaller h is, the closer this average velocity is to 200 ft/sec.

The average velocity between $t = t_0$ and $t = t_0 + h$ is

$$\frac{50(t_0+h)^2 - 50t_0^2}{(t_0+h) - t_0} = 50 \frac{(t_0+h)^2 - t_0^2}{h}$$

$$= 50 \frac{2t_0h + h^2}{h} = 50(2t_0 + h).$$

The smaller h is, the stronger we get the message: At $t = t_0$ the instantaneous velocity is $100t_0$ ft/sec.

Here is an important observation. The preceding computation of velocity is exactly the same computation as that needed to find the slope of the curve $y = f(t) = 50t^2$. Thus the velocity of the rocket at time t_0 is numerically the same as the slope of $y = 50t^2$ at $t = t_0$.

This is no accident. The average velocity between $t = t_0$ and $t = t_0 + h$ is

$$\frac{\text{displacement}}{\text{time}} = \frac{f(t_0 + h) - f(t_0)}{h},$$

which is precisely the formula for the slope of the secant between two nearby points on $y = f(t)$. Thus the slope of the secant is like the "average speed" of $y = f(t)$ between t_0 and $t_0 + h$. As the interval gets smaller and smaller, the "average slope" approximates the "instantaneous slope" at t_0, that is, the derivative at t_0.

> If $f(t)$ is the position of a moving body at time t, its **velocity** is defined to be $df(t)/dt$. Its **speed** is defined to be $|df(t)/dt|$.

Notice that df/dt may be negative. This happens, for instance, in the case of a falling body whose position is measured above ground level. Then displacement in any time interval is negative, leading to a negative velocity. Speed ignores the sign of the derivative. It measures only the *rate* of motion, while velocity also takes into account the *direction*.

Notation

The use of a dot to indicate derivative with respect to *time* is common practice. Instead of dy/dt, you often see $\dot{y}(t)$ or just $\dot{y}$.

EXAMPLE 1.1

A ball thrown straight up from the top of the Leaning Tower of Pisa is at a height $180 + 64t - 16t^2$ ft after t sec. Compute:
(a) its velocity after 1 sec,
(b) its maximum height,
(c) its velocity as it hits the ground.

Solution: Let $y(t) = 180 + 64t - 16t^2$. The velocity of the ball after t sec is

$$\dot{y}(t) = 64 - 32t = 32(2 - t).$$

(a) After 1 sec, the velocity is $\dot{y}(1) = 32(2 - 1) = 32$ ft/sec. (Since the velocity is positive, the ball is rising.)

(b) We need the maximum value of $y(t)$. Examine $\dot{y}(t)$:

$$\dot{y}(t) > 0 \quad \text{when} \quad t < 2, \qquad \dot{y}(2) = 0, \qquad \dot{y}(t) < 0 \quad \text{when} \quad t > 2.$$

Therefore the largest value of $y(t)$ is $y(2) = 244$. Since $\dot{y}$ is velocity, the highest point of the trajectory occurs when the velocity is zero. That is because the ball rises at first (positive velocity), then falls (negative velocity). In between is the highest point (zero velocity).

(c) The ball hits the ground when $y(t) = 0$. Solve for t:

$$y(t) = 180 + 64t - 16t^2 = 0,$$

$$4t^2 - 16t - 45 = 0,$$

$$t = \frac{16 \pm \sqrt{16^2 + 4 \cdot 4 \cdot 45}}{8} = 2 \pm \frac{\sqrt{16 + 45}}{2}.$$

There is only one *positive* time t for which $y(t) = 0$; it is

$$t = 2 + \frac{1}{2}\sqrt{61}.$$

At this instant the velocity is

$$\dot{y}\left(2 + \frac{1}{2}\sqrt{61}\right) = 32\left[2 - \left(2 + \frac{1}{2}\sqrt{61}\right)\right]$$

$$= -16\sqrt{61} \quad \text{ft/sec}.$$

Answer: (a) 32 ft/sec. (b) 244 ft. (c) $-16\sqrt{61}$ ft/sec. (The velocity is negative because the ball is falling.)

EXERCISES

1. A projectile shot straight up has height $s = -16t^2 + 980t$ ft after t sec. Compute its average velocity between $t = 2$ and $t = 3$, between $t = 2$ and $t = 2.1$. Compute its instantaneous velocity when $t = 2$.
2. During the initial stages of flight, a rocket reaches an elevation of $50t^2 + 500t$ ft in t sec. What is the average velocity between $t = 2$ and $t = 3$ sec? Find the instantaneous velocity when $t = 2$ and when $t = 3$. What is the average of the two instantaneous velocities?
3. A projectile launched from a plane has elevation $s = -16t^2 + 400t + 8000$ ft after t sec. What is its maximum elevation? Find its vertical velocity after 15 sec, and upon striking the ground.
4. An object projected upward has height $s = -16t^2 + 96t$ ft after t sec. Compute its velocity after 1.5 sec, its maximum height, and the speed with which it strikes the ground.
5. A penny is thrown straight up from the top of a 600-ft tower. After t sec, it is $s = -16t^2 + 24t + 600$ ft above ground. When does the penny begin to descend? What is its speed when 605 ft above ground, while going up, and while coming down?

6. A projectile fired at an angle of 45° with an initial velocity of 860 ft/sec has height $y = 430t\sqrt{2} - 16t^2$, and has a horizontal distance $x = 430t\sqrt{2}$ ft from its point of origin t sec after being fired. When does it strike the ground? How far will it go?
7. A body moves along a horizontal line according to the law $s = t^3 - 9t^2 + 24t$ ft. (a) When is s increasing and when decreasing? (b) When is the velocity increasing and when decreasing? (c) Find the total distance traveled between $t = 0$ and $t = 6$ sec.
8. Solve Ex. 7, assuming that the body moves according to $s = t^3 - 3t^2 - 9t$.

2. ACCELERATION

A falling weight moves faster and faster; its velocity increases; a car with brakes applied moves slower and slower; its velocity decreases. In many applications, it is important to know just how velocity is changing during motion.

> If $v(t)$ is the velocity of a moving object at time t, its **acceleration** is the derivative dv/dt or $\dot{v}(t)$.

Acceleration is the derivative of velocity. It measures the rate of change of velocity during motion. Positive acceleration indicates increasing velocity; negative acceleration, decreasing velocity; zero acceleration, constant velocity.

Remember that velocity itself is a derivative:

$$v(t) = \frac{ds}{dt} = \dot{f}(t),$$

where $s = f(t)$ is the position at time t. Therefore, acceleration is a second derivative, being the derivative of a derivative:

$$\text{acceleration} = \frac{d^2s}{dt^2} = \ddot{f}(t).$$

EXAMPLE 2.1

A ball is $180 + 64t - 16t^2$ ft above the ground at time t sec. Find its acceleration at time t.

Solution: Let $s(t) = 180 + 64t - 16t^2$. The acceleration is

$$\frac{dv}{dt} \quad \text{or} \quad \frac{d^2s}{dt^2}.$$

Differentiate:

$$v(t) = \frac{ds}{dt} = 64 - 32t.$$

Differentiate again:

$$\frac{dv}{dt} = \frac{d^2s}{dt^2} = -32.$$

Answer: -32 ft/sec^2.

REMARK: The minus sign means that velocity is decreasing (from positive to negative to more negative).

EXAMPLE 2.2

A ball is thrown straight up from the top of a hill y_0 ft high with an initial velocity of v_0 ft/sec. Gravity causes a constant negative acceleration, $-g$ ft/sec^2. Find (a) its velocity, (b) its height t sec after release.

Solution: First find a formula for the velocity $v(t)$. Since acceleration is dv/dt, the data is

$$\frac{dv}{dt} = -g.$$

That means $v(t)$ is a function whose derivative is $-g$. In other words, $v(t)$ is an antiderivative of $-g$. One antiderivative is $-gt$; by the discussion in Chapter 4, Section 1, all antiderivatives are of the form $-gt + c$, where c can be any constant. Hence,

$$v(t) = -gt + c.$$

To find the constant c that fits this problem, remember the value of $v(t)$ is given for $t = 0$. Set $t = 0$:

$$v_0 = v(0) = -g \cdot 0 + c.$$

$$c = v_0.$$

Hence

$$v(t) = -gt + v_0,$$

which is the required formula.

Use the same sort of argument to find a formula for $y(t)$, the elevation at time t. Since

$$\frac{dy}{dt} = v(t) = -gt + v_0,$$

$y(t)$ is an antiderivative of $-gt + v_0$. Therefore,

$$y(t) = -\frac{1}{2}gt^2 + v_0t + k,$$

for some appropriate constant k. To find the right value of k, remember that the value of $y(t)$ is given for $t = 0$. Set $t = 0$:

$$y_0 = y(0) = 0 + 0 + k,$$

$$k = y_0.$$

Hence

$$y(t) = -\frac{1}{2}gt^2 + v_0t + y_0.$$

Answer: (a) $v(t) = -gt + v_0$ ft/sec.

(b) $y(t) = -\frac{1}{2}gt^2 + v_0t + y_0$ ft.

In Example 2.2 you solved a **differential equation**. That is an equation involving the derivatives of a function in which the function itself is the unknown. The data of Example 2.2 can be written:

$$\frac{d^2y}{dt^2} = -g, \qquad y(0) = y_0, \quad \dot{y}(0) = v_0.$$

The first equation is the differential equation; the other equations are **initial conditions**. To find $y(t)$, antidifferentiate twice. First you get dy/dt, then $y(t)$ itself. Each antidifferentiation involves a constant to be determined. The two constants are obtained from the two initial conditions, $y(0) = y_0$ and $\dot{y}(0) = v_0$.

EXAMPLE 2.3

An alpha particle enters a linear accelerator. It immediately undergoes a constant acceleration that changes its velocity from 1000 m/sec to 5000 m/sec in 10^{-3} sec. Compute its acceleration. How far does the particle move during this period of 10^{-3} sec?

Solution: For convenience, assume the accelerator lies along the positive x-axis starting at the origin. Also assume the particle enters when $t = 0$, and t sec later reaches position $x(t)$. Then

$$\ddot{x}(t) = a, \qquad \dot{x}(0) = 1000, \qquad x(0) = 0,$$

where a is the unknown constant acceleration. This is the same problem as Example 2.2 with different numbers: a instead of $-g$, $v_0 = 1000$, and $x_0 = 0$.

By exactly the reasoning of Example 2.2, obtain the following formulas:

$$v(t) = at + v_0 = at + 1000,$$

$$x(t) = \frac{1}{2}at^2 + 1000t.$$

Use the first formula to find a. Since $v(10^{-3}) = 5000$,

$$5000 = 10^{-3}a + 1000,$$

$$a = 4 \times 10^6 \quad \text{m/sec}^2.$$

From the second formula,

$$x(10^{-3}) = \frac{1}{2}(4)(10^6)(10^{-3})^2 + (1000)(10^{-3}) = 2 + 1.$$

Answer: $a = 4 \times 10^6$ m/sec^2,
the particle moves 3 m.

EXERCISES

Solve the differential equation:

1. $dy/dx = -16x, \quad y(0) = 12$
2. $dy/dt = 3t^2 + 4, \quad y(1) = -3$
3. $d^2y/dt^2 = -32, \quad y(1) = 48, \quad \dot{y}(0) = 64$
4. $d^2y/dx^2 = 8, \quad y(0) = 2, \quad y'(0) = 1$
5. $d^2y/dt^2 = 2t - 1, \quad y(0) = 1.75, \quad \dot{y}(0) = 0.25$
6. $d^2y/dx^2 = 3 - 4x, \quad y(0) = 2, \quad y'(1) = 6.$
7. A ball is thrown straight up with an initial velocity of 48 ft/sec. Gravity causes a constant negative acceleration, -32 ft/sec^2. How high will the ball go if it is released from a height of 4 ft?
8. An object slides down a 200-ft inclined plane with acceleration 8 ft/sec^2. If the object starts from rest with zero velocity, when does it reach the end of the plane? How fast is it going?
9. Starting from rest, with what constant acceleration must a car proceed to go 75 ft in 5 sec?
10. The makers of a certain automobile advertise that it will accelerate from 0 to 100 mph in 1 min. If the acceleration is constant, how far will the car go in this time?
11. During the initial stages of flight after blast-off, a rocket shot straight up has acceleration $24t$ ft/sec^2. The engine cuts out at $t = 10$ sec, after which only the force of gravity, -32 ft/sec^2, retards its motion. How high will the rocket go?
12. An airplane taking off from a landing field has a run of 3200 ft. If it starts with speed 20 ft/sec, moves with constant acceleration, and makes the run in 40 sec, with what speed does it take off?
13. A subway train starts from rest at a station and accelerates at a rate of 6 ft/sec^2 for 10 sec. It then runs at constant speed for 60 sec, after which it decelerates at 7 ft/sec^2 until it stops at the next station. Find the total distance covered.
14. Gravitation on the moon is 0.165 times that on the earth. If a bullet shot straight up from the earth will rise one mile, how far would it rise if shot on the moon?

6. Circular Functions

1. COSINE AND SINE

There is a most important technicality to understand from the start. In calculus, the unit of angle measurement is the **radian**. It is a very natural unit, taken so the central angle of an arc of length s on a circle of radius r has measure s/r radians. See Fig. 1.1. Since the radian is a ratio of lengths, it is a dimensionless quantity.

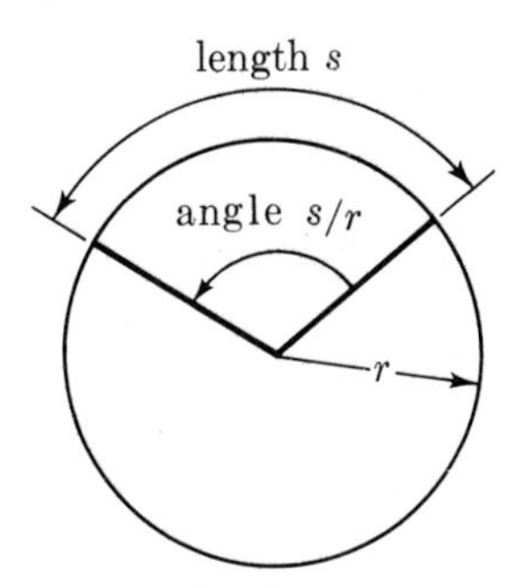

Fig. 1.1

Every single formula in calculus involving angles and trigonometric functions has the angles measured in radians. However, the practical system of degrees, minutes, and seconds (inherited from the ancient Babylonians) dominates scientific measurement. This means that in solving a numerical problem by calculus, you must convert data given in degrees to radians before you start, and convert an answer obtained in radians back to degrees.

Normally one does not write "radian" after a number. The equality

$$\pi \text{ radians} = 180 \text{ degrees}$$

may be written

$$\pi = 180 \text{ degrees} = 180^\circ.$$

The necessary conversion factors are

$$\pi = 180^\circ, \qquad 1 \approx 57.2958^\circ \qquad 0.0174533 \approx 1^\circ.$$

The unit circle

$$x^2 + y^2 = 1$$

has circumference 2π. A particle moving on it with constant speed 1 unit/sec makes a complete circuit in 2π sec. Suppose that at time $t = 0$ the particle is at $(1, 0)$ and moving counterclockwise. Then its position is determined for all future time (and for all past time too). At time t, the particle will be at a point $(x(t), y(t))$, where

$$x^2(t) + y^2(t) = 1.$$

The radius from $(0, 0)$ to $(x(t), y(t))$ will make an angle t (radians) with the positive x-axis (Fig. 1.2).

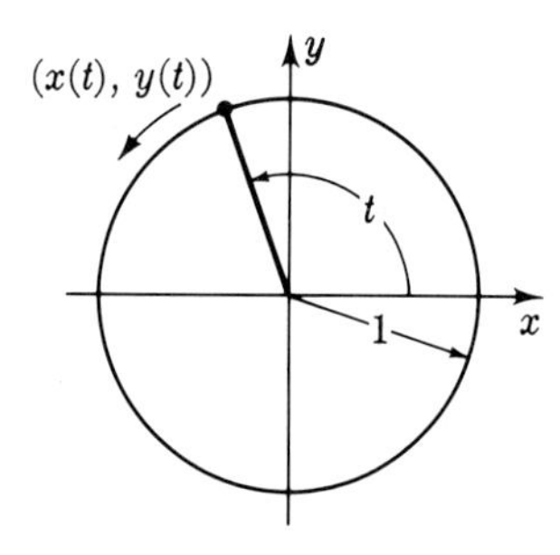

FIG. 1.2

The trigonometric functions **sine** and **cosine** can be defined in terms of this motion:

$$\begin{cases} \cos t = x(t) \\ \sin t = y(t). \end{cases}$$

They are related by

$$\cos^2 t + \sin^2 t = 1.$$

When $t = 0$, the particle is at $(1, 0)$. Hence

$$x(0) = 1, \qquad y(0) = 0,$$

or

$$\cos 0 = 1, \qquad \sin 0 = 0.$$

In any period of 2π sec, a complete circuit is made. Therefore

$$x(t + 2\pi) = x(t), \qquad y(t + 2\pi) = y(t),$$

or

$$\cos(t + 2\pi) = \cos t, \qquad \sin(t + 2\pi) = \sin t.$$

In words, $\cos t$ and $\sin t$ are **periodic functions** with **period** 2π.

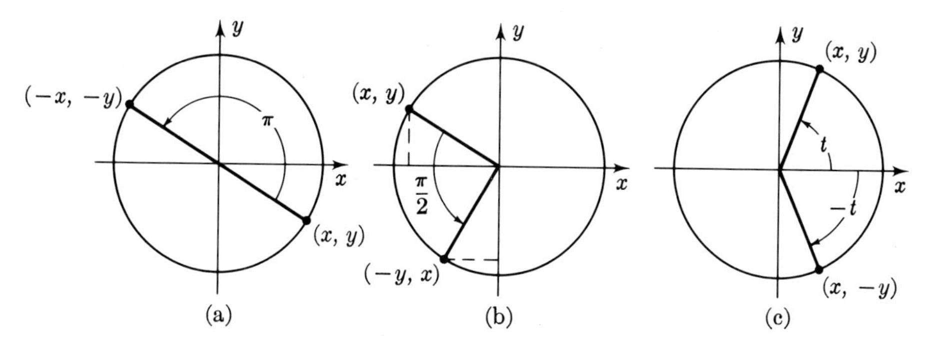

Fig. 1.3

Look at Fig. 1.3. The three drawings illustrate familiar formulas:

$$\text{(a)}\qquad \begin{cases} \cos(t+\pi) = -\cos t \\ \sin(t+\pi) = -\sin t, \end{cases}$$

$$\text{(b)}\qquad \begin{cases} \cos\left(t+\dfrac{\pi}{2}\right) = -\sin t \\ \sin\left(t+\dfrac{\pi}{2}\right) = \cos t, \end{cases}$$

$$\text{(c)}\qquad \begin{cases} \cos(-t) = \cos t \\ \sin(-t) = -\sin t. \end{cases}$$

Each is a special case of the **addition laws:**

$$\begin{cases} \cos(s+t) = \cos s \cos t - \sin s \sin t \\ \sin(s+t) = \sin s \cos t + \cos s \sin t. \end{cases}$$

If you do not remember these important formulas, please accept them now. They will be derived in Chapter 24, Section 6, p. 520.

EXERCISES

Convert to radian measure:

(Example: $30° = \pi/6$, $45° = \pi/4$)

1. $60°$, $150°$, $-240°$, $390°$, $-450°$, $900°$
2. $90°$, $-120°$, $210°$, $420°$, $-330°$, $480°$
3. $45°$, $-180°$, $270°$, $630°$, $-135°$, $495°$
4. $90°$, $135°$, $-315°$, $405°$, $-270°$, $765°$.

Convert to degrees:

5. $\dfrac{\pi}{2}, \quad -\dfrac{2\pi}{3}, \quad \dfrac{5\pi}{3}, \quad 3\pi, \quad -\dfrac{8\pi}{3}, \quad \dfrac{50\pi}{3}$

6. $\pi, \quad -\dfrac{3\pi}{2}, \quad \dfrac{7\pi}{6}, \quad 5\pi, \quad -\dfrac{13\pi}{6}, \quad \dfrac{17\pi}{3}$

7. $\dfrac{\pi}{4}, \quad -\dfrac{7\pi}{4}, \quad \dfrac{5\pi}{4}, \quad -\dfrac{7\pi}{2}, \quad 13\pi, \quad 4\pi$

8. $\dfrac{3\pi}{2}, \quad -\dfrac{3\pi}{4}, \quad \dfrac{9\pi}{4}, \quad -\dfrac{9\pi}{2}, \quad 7\pi, \quad 6\pi.$

Use the addition laws to derive a formula for:

9. $\cos 2t$
10. $\sin 2t$
11. $\sin(s - t)$
12. $\cos(s - t)$.

Compute the coordinates of the point on the unit circle determined by central angle t. See Fig. 1.2:

13. $t = \dfrac{\pi}{4}, \quad -\dfrac{3\pi}{4}, \quad \dfrac{11\pi}{4}$
14. $t = \dfrac{3\pi}{2}, \quad -\dfrac{\pi}{4}, \quad \dfrac{13\pi}{4}$
15. $t = \pi, \quad -\dfrac{2\pi}{3}, \quad \dfrac{13\pi}{6}$
16. $t = \dfrac{\pi}{6}, \quad \dfrac{5\pi}{6}, \quad -\dfrac{2\pi}{3}.$

Find all values of t (in radian measure) between 0 and 2π such that:

17. $\sin t = \sqrt{2}/2$
18. $\sin t = -\frac{1}{2}$
19. $\cos t = -\sqrt{3}/2$
20. $\cos t = \sqrt{2}/2$
21. $\sin(t + \pi) = 1$
22. $\cos(t + \pi) = 1$
23. $\sin t = \cos t.$

Show that:

24. $\cos 3t$ has period $2\pi/3$
25. $\sin 4t - \cos 4t$ has period $\pi/2$
26. $\cos 2\pi t$ has period 1
27. $\cos 2t + 2\cos t$ has period 2π
28. $3 \sin \frac{1}{2}t$ has period 4π.

2. GRAPHS OF sin t AND cos t

First we'll graph $x = \cos t$, paying special attention to the range from 0 to $\pi/2$. We take ten equally spaced points. From tables:

t	0	$\frac{\pi}{20}$	$\frac{2\pi}{20}$	$\frac{3\pi}{20}$	$\frac{4\pi}{20}$	$\frac{5\pi}{20}$	$\frac{6\pi}{20}$	$\frac{7\pi}{20}$	$\frac{8\pi}{20}$	$\frac{9\pi}{20}$	$\frac{\pi}{2}$
$\cos t$	1	0.99	0.95	0.89	0.81	0.71	0.59	0.45	0.31	0.16	0

In terms of degrees, this is a table of $\cos t$ at intervals of 9 degrees, from 0° to 90°.

As t increases from 0 to $\pi/2$, then $\cos t$ decreases from 1 to 0. It decreases very slowly at first: when t moves 10% of the way from 0 to $\pi/2$, then $\cos t$ decreases by only 1%; when t moves 20% of the way, $\cos t$ decreases by 5%. At the other end, $\cos t$ decreases rapidly: 31% of its total drop occurs in the last 20% of the way.

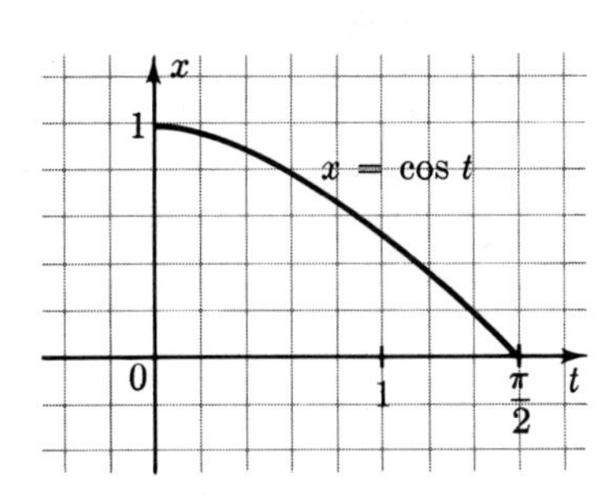

FIG. 2.1

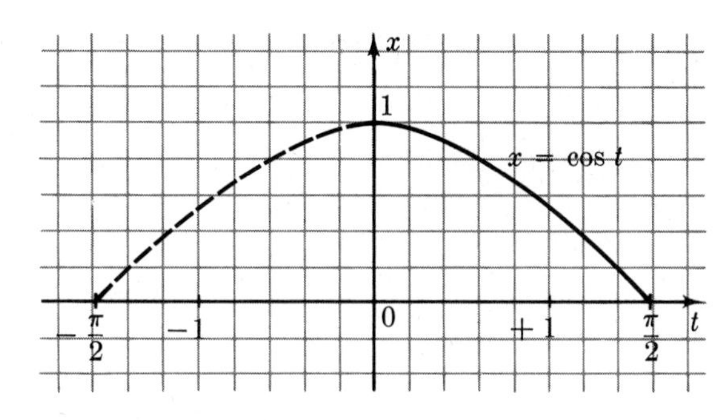

FIG. 2.2

Plotting the points given by the table we obtain Fig. 2.1. Because

$$\cos(-t) = \cos t,$$

we can fill in the graph from $-\pi/2 \le t < 0$ as indicated in Fig. 2.2. Because

$$\cos(t + \pi) = -\cos t,$$

we can fill in the graph for the values $\pi/2 < t \le 3\pi/2$, using what we already have (Fig. 2.3). Finally, because

$$\cos(t + 2\pi) = \cos t,$$

we can sketch the graph as far forwards and backwards as we wish (Fig. 2.4).

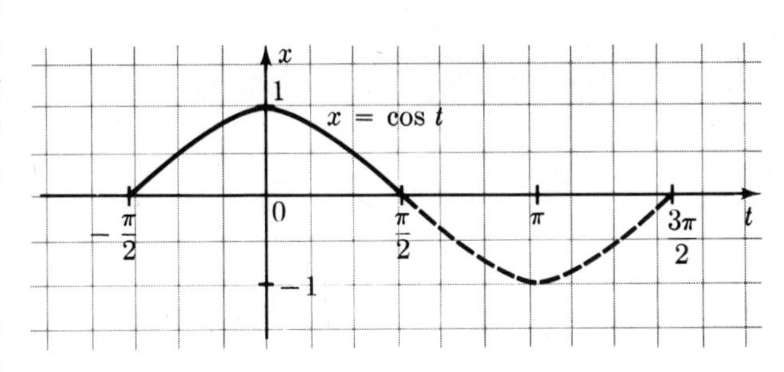

FIG. 2.3

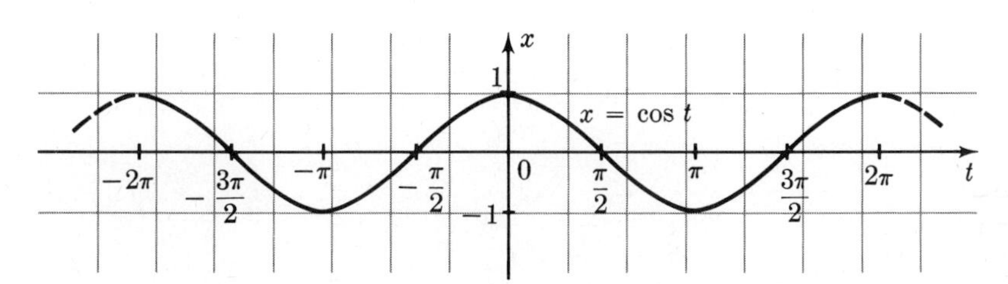

FIG. 2.4

Next, let's graph $x = \sin t$. We could treat this as a separate problem. But since $\sin t$ is related to $\cos t$, we can use the graph of $\cos t$ to find the graph of $\sin t$. Earlier we had

$$\sin\left(t + \frac{\pi}{2}\right) = \cos t,$$

for every value of t. Replace t by $t - \pi/2$:

$$\sin t = \cos\left(t - \frac{\pi}{2}\right).$$

Thus if t is time, the value now of $\sin t$ is the value $\pi/2$ sec ago of $\cos t$. Therefore the graph of $\sin t$ is the graph of $\cos t$ moved bodily $\pi/2$ units to the right (Figs. 2.5 and 2.6).

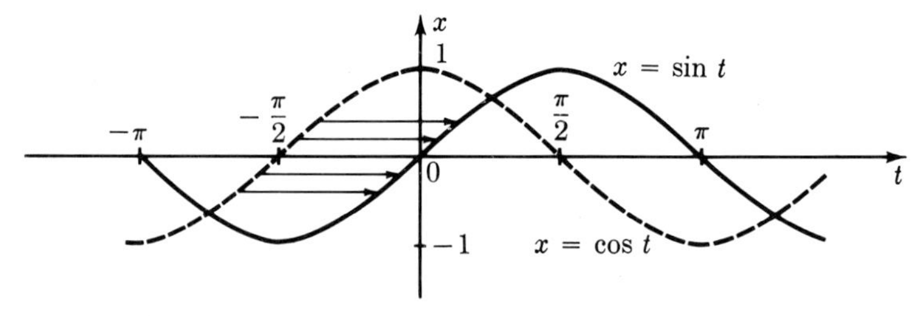

FIG. 2.5

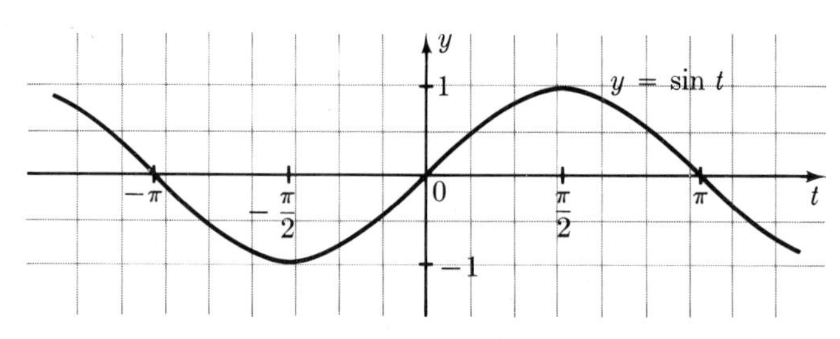

FIG. 2.6

Recall how $\cos t$ and $\sin t$ were defined by the motion of a particle on the unit circle moving with constant speed of 1 unit per second. If the particle

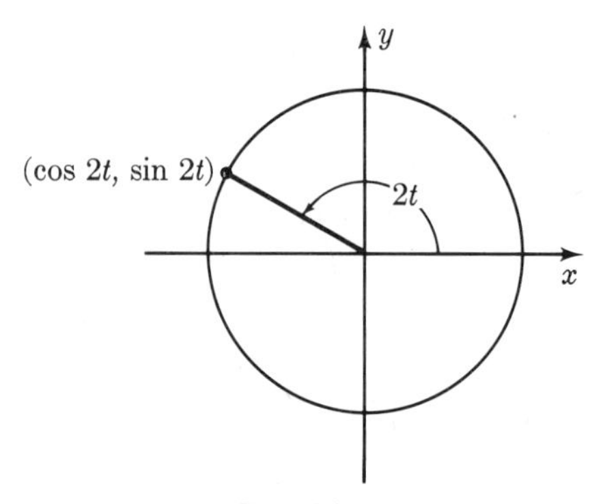

FIG. 2.7

were to move with twice that speed, its position at time t would be $(\cos 2t, \sin 2t)$. See Fig. 2.7. Its x-coordinate $\cos 2t$ would run through the same values as before, only twice as fast. Therefore, the graph of $\cos 2t$ can be plotted by speeding up the graph of $\cos t$ by a factor of 2. See Fig. 2.8.

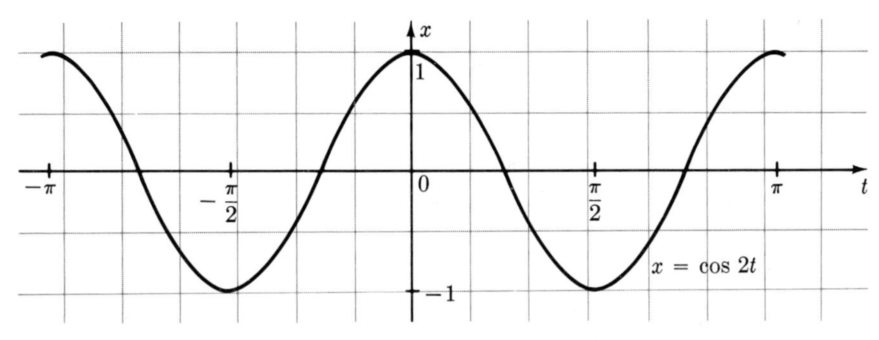

FIG. 2.8

In a similar way, the graph of $\cos 3t$ is the graph of $\cos t$ speeded up by a factor of 3. Analogous statements hold for $\sin 2t$, $\sin 3t$, etc. (Fig. 2.9).

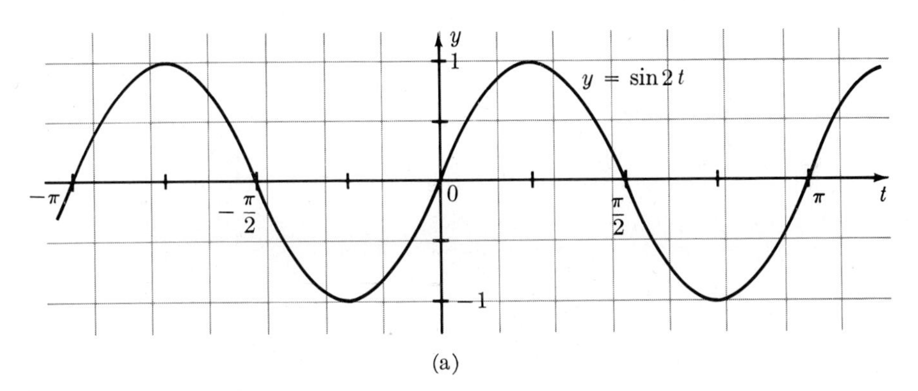

(a)

FIG. 2.9a

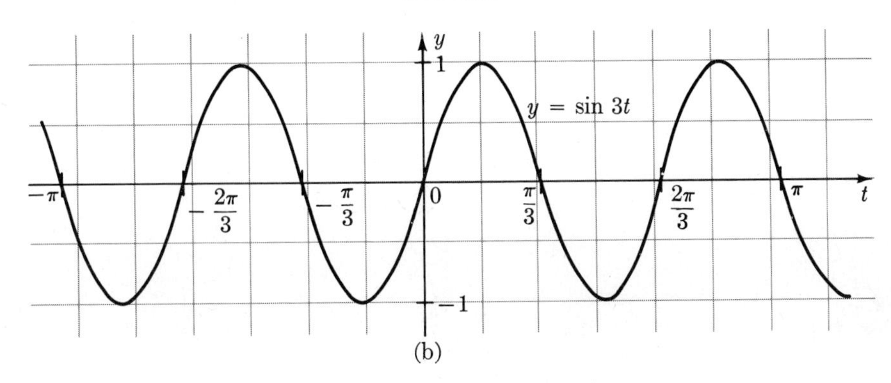

(b)

FIG. 2.9b

Sinusoidal Waves

The graph of $x = \cos t$ trails the graph of $x = \sin t$ by an amount $\pi/2$. See Fig. 2.5. We say that the graphs are **out of phase** by angle $\pi/2$.

More generally, if a point moves uniformly around a circle, its projection on a diameter, when graphed versus time, is a displaced sine curve. Such a graph is called a **sinusoidal wave**.

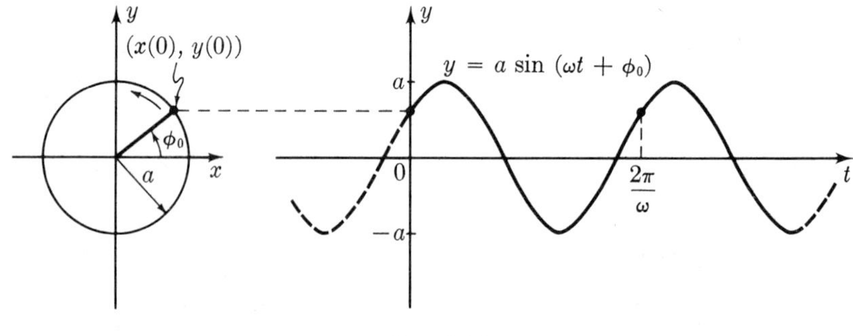

FIG. 2.10

A point rotates on a circle of radius a with uniform angular speed ω. Its y-projection as a function of time is the sinusoidal wave shown in Fig. 2.10. The initial angle ϕ_0 is called the **phase angle**. The equation of the graph is

$$y = a\sin(\omega t + \phi_0).$$

Many natural phenomena are described by sinusoidal waves: vibrations, alternating currents, electromagnetic radiation, sound waves, etc. The parameter ω in these phenomena is the **frequency** of the wave motion and the parameter a is the **amplitude**.

EXERCISES

Sketch the curve:

1. $x = \cos 3t$
2. $x = \sin 4t$
3. $y = \sin \frac{1}{2}t$
4. $y = \cos \frac{1}{2}t$
5. $y = \cos 2\pi t$
6. $y = \sin 2\pi t$
7. $f(t) = \sin(t + \pi)$
8. $f(t) = \cos(t + \pi)$
9. $x = \cos\left(t + \dfrac{\pi}{4}\right)$
10. $x = \sin\left(t + \dfrac{\pi}{4}\right)$
11. $x = 2\sin t$
12. $x = 2\cos t$

13. $y = 2\cos 2t$
14. $y = 2\sin 2t$
15. $x = 1 + \sin t$
16. $x = 1 + \cos t$
17. $f(t) = 1 - \cos t$
18. $f(t) = 1 - \sin t$
19. $x = \sin\left(2t - \dfrac{\pi}{3}\right)$
20. $x = 2\cos\left(3t + \dfrac{\pi}{6}\right)$
21. $y = 3\sin(\pi t + \pi)$
22. $y = 3\sin\left(2\pi t - \dfrac{\pi}{8}\right)$.

3. DERIVATIVES

The two basic formulas are:

$$\frac{d}{dt}(\sin t) = \cos t$$

$$\frac{d}{dt}(\cos t) = -\sin t.$$

You will need them a million times.

Where do they come from? First let us analyze two special cases:

$$\text{(a)} \qquad \left.\frac{d}{dt}(\sin t)\right|_{t=0} = 1,$$

$$\text{(b)} \qquad \left.\frac{d}{dt}(\cos t)\right|_{t=0} = 0.$$

Statement (b) holds because $x = \cos t$ has its maximum at $t = 0$, so its

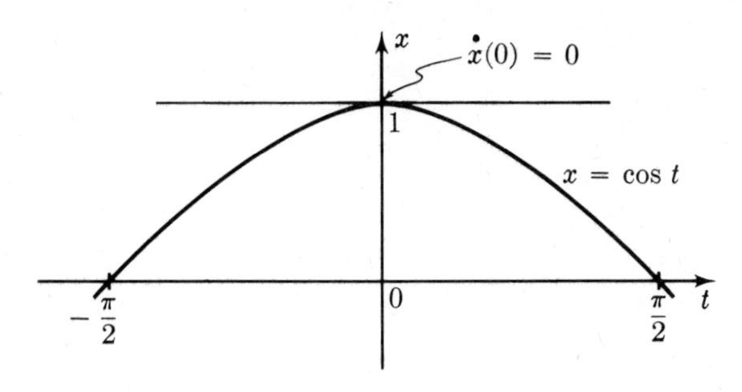

Fig. 3.1

tangent at (0, 1) is horizontal (Fig. 3.1). Statement (a) is plausible because of strong numerical evidence from tables:

t	0.100	0.010	0.001
$\sin t$	0.09983	0.01000	0.00100

If t is small, $\sin t \approx t$. Hence the slope of $x = \sin t$ at the origin must equal the slope of $x = t$ at the origin, namely 1.

From (a) and (b) follow two important facts:

$$\text{(a}'\text{)} \qquad \frac{\sin h}{h} \longrightarrow 1 \quad \text{as} \quad h \longrightarrow 0;$$

$$\text{(b}'\text{)} \qquad \frac{\cos h - 1}{h} \longrightarrow 0 \quad \text{as} \quad h \longrightarrow 0.$$

The quotient in (a′) is just the slope of the secant between (0, 0) and a nearby point $(h, \sin h)$ on the curve $x = \sin t$. Statement (a′) says that the slope of the secant approaches the slope of the tangent at (0, 0). Statement (b′) describes the corresponding situation for $x = \cos t$. (Important. Make sure you understand these points.)

From (a′) and (b′) follow the differentiation formulas for $\sin t$ and $\cos t$. The secant slope for two nearby points of the sine curve is

$$\frac{\sin(t + h) - \sin t}{h}.$$

But

$$\sin(t + h) = \sin t \cos h + \cos t \sin h,$$

therefore

$$\begin{aligned} \frac{\sin(t + h) - \sin t}{h} &= \frac{(\sin t \cos h + \cos t \sin h) - \sin t}{h} \\ &= \frac{\sin t\,(\cos h - 1) + \cos t \sin h}{h} \\ &= \sin t\,\frac{\cos h - 1}{h} + \cos t\,\frac{\sin h}{h}. \end{aligned}$$

This is excellent, because we know

$$\frac{\cos h - 1}{h} \longrightarrow 0 \quad \text{and} \quad \frac{\sin h}{h} \longrightarrow 1 \quad \text{as} \quad h \longrightarrow 0.$$

Hence,

$$\frac{\sin(t + h) - \sin t}{h} \longrightarrow \cos t \quad \text{as} \quad h \longrightarrow 0,$$

so

$$\frac{d}{dt}(\sin t) = \cos t.$$

Similarly, by using the addition law for cosines one establishes the other formula,

$$\frac{d}{dt}(\cos t) = -\sin t.$$

From these formulas and the Scale Rule follow the formulas:

$$\frac{d}{dt}(\sin kt) = k \cos kt$$
$$\frac{d}{dt}(\cos kt) = -k \sin kt.$$

EXERCISES

Find:

1. $\frac{d}{dt}(\sin \pi t)$
2. $\frac{d}{dt}(\cos 3t)$
3. $\frac{d}{dx}(b \cos ax)$
4. $\frac{d}{dx}(b \sin ax)$
5. $\frac{d}{d\theta}\left(1 + \sin \frac{\pi\theta}{2}\right)$
6. $\frac{d}{dt}(3 - \cos 2t)$
7. $\frac{d}{dt}(t + \cos t)$
8. $\frac{d}{dt}(2t - \sin 2t)$.

Differentiate:

9. $x = \sin(t + t_0)$
10. $x = 3 \cos\left(t - \frac{\pi}{4}\right)$
11. $f(t) = t^2 + \cos t$
12. $F(x) = x + \sin(x + 1)$
13. $y = 4x^2 + 3 \sin 4x$
14. $y = x^2 - \pi \cos(\pi x)$
15. $f(t) = \frac{1}{t} - \cos t$
16. $f(t) = 3 \sin 2t - 2 \cos 3t$.

Solve the differential equation:

17. $\frac{dy}{dt} = \cos t, \quad y(\pi/2) = 0$
18. $\frac{dy}{dt} = \sin t, \quad y(0) = 0$
19. $\frac{d^2y}{dx^2} = -\sin x, \quad y(0) = 1, \quad y'(\pi) = -1$
20. $\frac{d^2y}{dt^2} = \cos t + 1, \quad y(0) = -1, \quad \dot{y}(0) = 1.$
21. Prove: $\frac{d}{dt}(\cos t) = -\sin t$. (*Hint:* $\cos(t + h) = ?$)
22. Use $\sin\left(t + \frac{\pi}{2}\right) = \cos t$ to derive the formula for the derivative of the cosine from that of the sine. (*Hint:* Shifting Rule.)

23. If $x(t) = \cos t$ and $y(t) = \sin t$, show that $\dot{x} = -y$, $\dot{y} = x$, $\ddot{x} + x = 0$, $\ddot{y} + y = 0$.
24. If $f(\theta) = a \cos k\theta + b \sin k\theta$, show $\dfrac{d^2}{d\theta^2}[f(\theta)] + k^2 f(\theta) = 0$.
25. Use the second derivative to find where $y = \cos t$ is concave upwards in the interval $0 \le t \le 2\pi$.
26. For what values of t between -2π and 2π does $y = \cos t$ have an inflection point?
27. Find all points on $y = \sin x$ where the slope is $\frac{1}{2}$.
28. The curves $y = \sin kx$ and $y = \cos kx$ intersect at right angles. Find k.

4. APPLICATIONS

EXAMPLE 4.1

A weight hangs 5 ft above the floor from a spring attached to the ceiling. If the weight is pulled down 6 in., then released at time 0, its height at time t will be $y(t) = 5 - \frac{1}{2} \cos \pi t$. Find its velocity and acceleration $\frac{1}{3}$ sec after release.

Solution: The velocity at time t is

$$\dot{y} = \frac{\pi}{2} \sin \pi t,$$

and the acceleration is

$$\ddot{y} = \frac{\pi^2}{2} \cos \pi t.$$

Substitute $t = \frac{1}{3}$ in these formulas.

Answer: velocity $= \dfrac{\pi\sqrt{3}}{4}$ ft/sec.

acceleration $= \dfrac{\pi^2}{4}$ ft/sec^2.

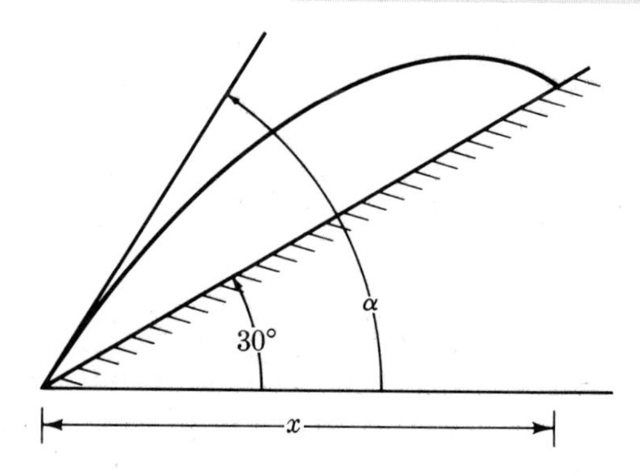

FIG. 4.1

EXAMPLE 4.2

A projectile is fired from the foot of a 30° slope (Fig. 4.1). It hits the hill at a horizontal distance

$$x = \frac{v_0^2}{g}\left[\sin 2\alpha - \frac{1}{\sqrt{3}}(1 + \cos 2\alpha)\right].$$

(Here v_0 is the initial velocity, α is the elevation of the gun, and g is the gravitational constant; air resistance is neglected.) For what angle α does the projectile reach farthest up the slope?

Solution: Find the angle α that maximizes x. This is equivalent to the given problem. Why?

By the nature of the problem, $\pi/6 \leq \alpha \leq \pi/2$. You must maximize

$$x = \frac{v_0^2}{g}\left[\sin 2\alpha - \frac{1}{\sqrt{3}}(1 + \cos 2\alpha)\right]$$

for $\pi/6 \leq \alpha \leq \pi/2$. It is ridiculous to fire at $\pi/6$ or $\pi/2$; the maximum value of x does not occur at either end of the range. Therefore, set the derivative equal to zero:

$$\frac{dx}{d\alpha} = \frac{v_0^2}{g}\left[2 \cos 2\alpha + \frac{2}{\sqrt{3}} \sin 2\alpha\right].$$

Hence $dx/d\alpha = 0$ when

$$\cos 2\alpha = \frac{-1}{\sqrt{3}} \sin 2\alpha, \qquad \text{that is,} \qquad \tan 2\alpha = -\sqrt{3},$$

so $2\alpha = 2\pi/3$ or $2\alpha = 5\pi/3$. Since 2α is certainly not in the fourth quadrant,

$$2\alpha = \frac{2\pi}{3}, \qquad \alpha = \frac{\pi}{3}.$$

Answer: 60°.

EXAMPLE 4.3

Sketch the curve $y = \cos 2x + 2 \cos x$.

Solution: First draw $y = \cos 2x$ and $y = 2 \cos x$ on the same graph (Fig. 4.2). It is possible to add these curves point by point. But the labor can be reduced by plotting

(1) a few points (x, y) easy to compute:

$$x = 0, \quad \frac{\pi}{2}, \quad \pi, \quad \frac{3\pi}{2}, \quad 2\pi;$$

(2) critical points (where the derivative vanishes).

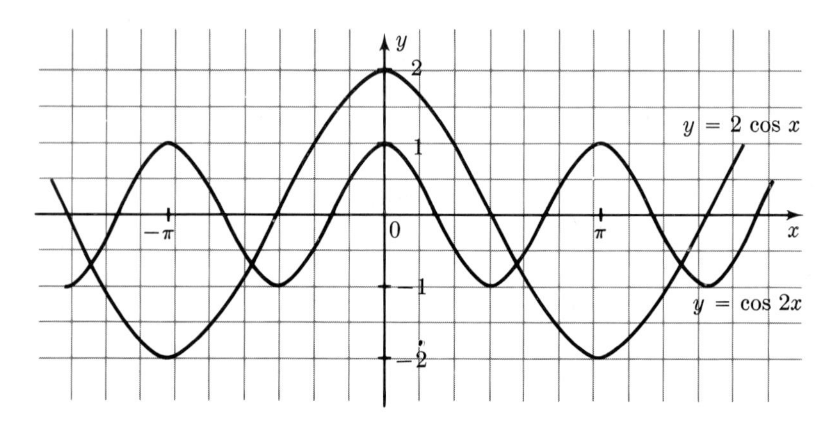

FIG. 4.2

Since the curve is periodic with period 2π, it is enough to sketch it only from 0 to 2π and then extend the sketch periodically. Plot the points in (1). See Fig. 4.3.

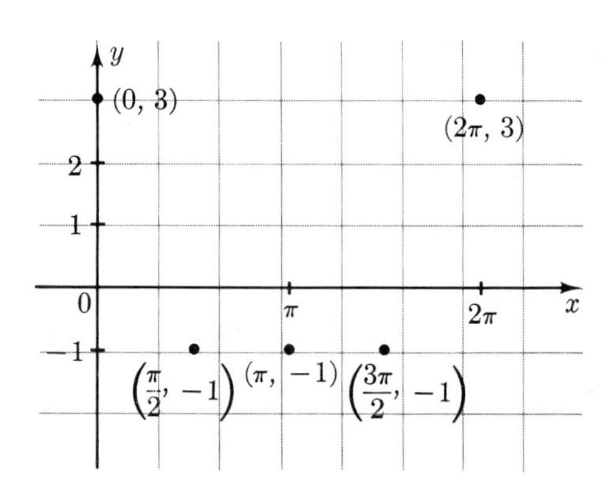

FIG. 4.3

Next, locate the critical points:

$$y = \cos 2x + 2 \cos x.$$

Hence

$$\begin{aligned} \frac{dy}{dx} &= -2 \sin 2x - 2 \sin x \\ &= -2(2 \sin x \cos x + \sin x) \\ &= -2 \sin x\, (2 \cos x + 1). \end{aligned}$$

Therefore $dy/dx = 0$ when either

$$\sin x = 0 \qquad \text{or} \qquad \cos x = -\frac{1}{2}\cdot$$

The corresponding values of x are

$$x = 0, \quad \pi, \quad 2\pi; \qquad x = \frac{2\pi}{3}, \quad \frac{4\pi}{3}.$$

Augment Fig. 4.3 by plotting the critical points and indicating horizontal tangents (Fig. 4.4). These points suggest the sketch in Fig. 4.5.

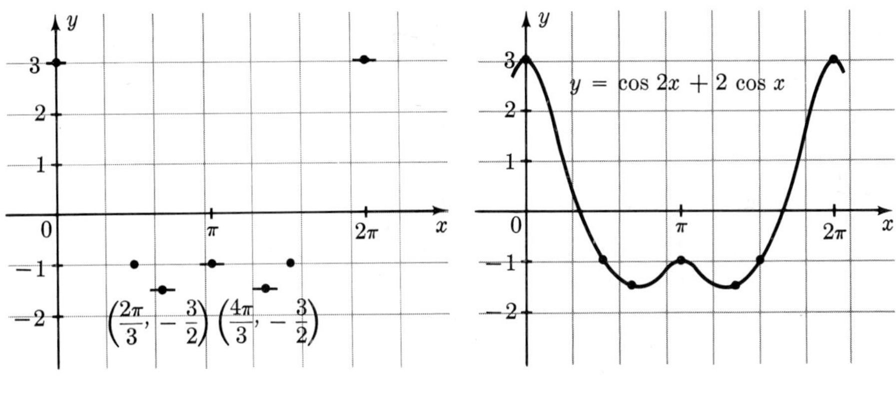

FIG. 4.4 FIG. 4.5

For greater accuracy plot more points, and using the derivative, get the correct slope at various points. Then extend the curve periodically (Fig. 4.6).

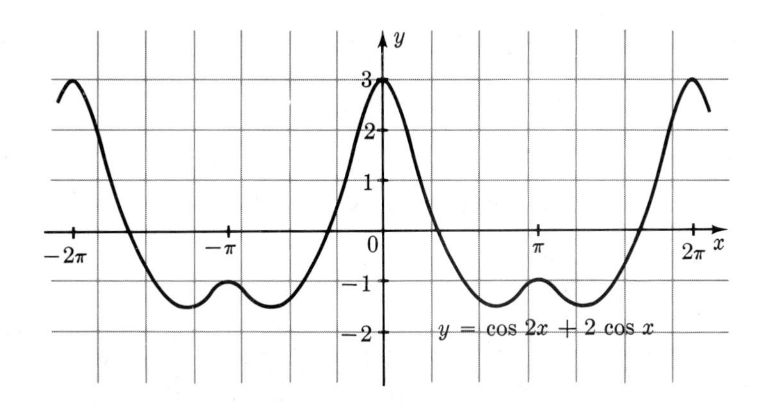

FIG. 4.6

EXERCISES

Sketch the graph of:

1. $y = x + \cos x$
2. $y = x^2 + 2 \cos x$
3. $h(t) = \sin t + \cos t$
4. $F(t) = 4 \sin t + \cos 4t$
5. $y = \sin 2x + 2 \sin x$
6. $y = \frac{1}{2} \sin 2t + \sin t.$

If the position of a point on a line as a function of time is given by $s(t) = A \sin Bt$ [or $s(t) = A \cos Bt$], then the motion is called **simple harmonic motion.** Describe the motion of the point and find its velocity and acceleration:

7. $s(t) = 2 \sin t$
8. $s(t) = 4 \cos 4t$
9. $s(t) = 3 \cos \pi t$
10. $s(t) = 2 \sin \frac{\pi t}{2}.$
11. A spring of length 1 ft is hung vertically. A weight attached to the free end stretches the spring 4 ft. If the weight is displaced 2 ft lower and released, then its distance (measured down from the ceiling) after t sec is $y = 5 + 2 \cos \omega t$, where $\omega^2 = g/4$ and $g = 32.2$ ft/sec². Describe the motion of the weight; give its velocity and acceleration.

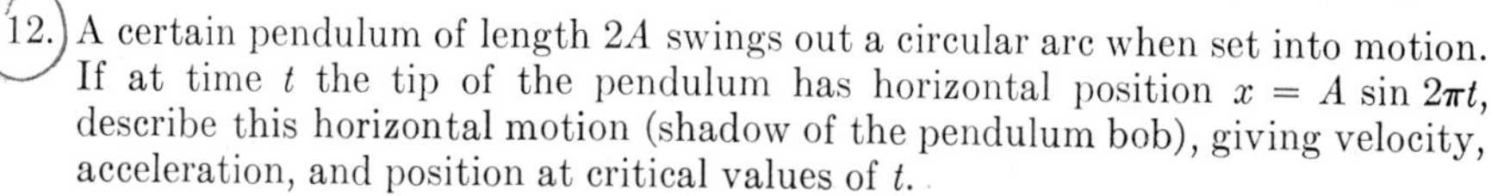

12. A certain pendulum of length $2A$ swings out a circular arc when set into motion. If at time t the tip of the pendulum has horizontal position $x = A \sin 2\pi t$, describe this horizontal motion (shadow of the pendulum bob), giving velocity, acceleration, and position at critical values of t.
13. In Example 4.3 it would have sufficed to sketch the curve only for $0 \leq x \leq \pi$. How could we have known in advance?
14. Suppose in Example 4.2 the hill makes an angle ϕ with the horizontal. Then

$$x = \frac{v_0^2}{g} [\sin 2\alpha - \tan \phi \, (1 + \cos 2\alpha)].$$

Find the maximizing angle α in this case. Interpret geometrically.

15. (cont.) Solve Ex. 14 without calculus in the special case $\phi = 0$.

5. ROUND-OFF ERRORS

In this chapter we have used values of $\sin x$ and $\cos x$ taken from 5-place tables. These values are not exact; they are obtained by rounding-off numbers involving at least 6 decimal places. Even the values in 20-place tables involve small errors.

Whenever we are unable or unwilling to handle numbers with many digits, it is necessary to round them to fewer digits. There are three rules for rounding-off a number, applied regardless of the sign of the number:

(1) If the discarded portion is less than 5000 · · · , drop it.

(2) If the discarded portion is greater than 5000 · · · , drop it and add one to the last digit kept.

(3) If the discarded portion is exactly 5000 · · · , drop it and add zero or one to the last digit kept, making it even.

EXAMPLE 5.1

Round-off 6.14537 to 4, 3, 2, 1, and 0 decimal places.

Answer: 6.1454, 6.145, 6.15, 6.1, 6.

Note that it may be less accurate to round in two steps. For example, we round-off 6.14537 to 6.145, then we round-off 6.145 to 6.14. This is not as accurate as rounding-off 6.14537 directly to 6.15. Similarly, it is better to round-off 6.14537 directly to 6.1, rather than to 6.15 and 6.2 in two steps.

Rounding-off creates uncertainty and error. For example, a 4-place table gives $\sin 1 = 0.8415$ and $\sin 1.8 = 0.9738$. We can be sure only that $\sin 1$ is between 0.84145 and 0.84155, and that $\sin 1.8$ is between 0.97375 and 0.97385. In each case there may be an error as large as 0.00005. According to the tabulated values

$$\sin 1 + \sin 1.8 = 1.8153,$$

but all we really know is that this sum is between 1.81520 and 1.81540.

In computations involving many round-offs the situation may be more critical. Many small errors can accumulate and destroy the accuracy of a computation. For example, in Chapter 10 we shall compute such sums as

$$\sin 0.01 + \sin 0.02 + \sin 0.03 + \cdot \cdot \cdot + \sin 1.00.$$

If 4-place sine tables are used, each term may be in error by as much as 0.00005. If all errors accumulate, the sum may be in error by as much as 0.005, possibly unsatisfactory for the computation. Actually, it is unlikely that the error will be so large. Some of the values in the sine table are overestimates and some are underestimates; therefore they tend to cancel each other. Nevertheless, the degree of precision in the computation is uncertain. In complicated problems, the question of round-off errors can be quite delicate.

EXERCISES

Round to 3 decimal places:

1. 0.4721
2. 0.9436
3. −9.5215
4. 14.0005
5. 0.12345
6. −1.34517
7. 3.442501
8. 4.71399

Round to 3 decimal places before and after performing the indicated calculation. Observe the difference in the results.

9. $0.4126 + 0.3215$
10. $1.7925 + 2.3454$
11. $1.3475 - 0.4934$
12. $4.3244 - 3.1928$
13. $0.3444 + 0.7174$
14. $0.1435 + 0.3216 + 0.4075$
15. $5.5042 - 10(0.2156 + 0.3347)$
16. $0.3127/(0.4136 - 0.4135)$

Calculate the answer to the nearest integer, rounding before and after the calculation. Compare the results.

17. (4.6) (3.5)
18. (4.4) (3.2)
19. (4.4) (3.5)
20. 3.5/2.4
21. 2.4/1.5
22. 2.6/0.6

7. The Exponential Function

1. INTRODUCTION

Let us review some properties of exponents. Begin with powers of 2:

$$2^0 = 1, \quad 2^1 = 2, \quad 2^2 = 2 \cdot 2 = 4, \quad 2^3 = 2 \cdot 2 \cdot 2 = 8,$$

$$2^4 = 2 \cdot 2 \cdot 2 \cdot 2 = 16, \qquad 2^{10} = 1024, \qquad 2^{12} = 4096.$$

Some negative powers:

$$2^{-1} = \frac{1}{2^1} = \frac{1}{2}, \qquad 2^{-3} = \frac{1}{2^3} = \frac{1}{8}, \qquad 2^{-6} = \frac{1}{2^6} = \frac{1}{64}.$$

There are also fractional powers. For example, $2^{1/2}$ is that number whose square is 2:

$$2^{1/2} = \sqrt{2} = 1.41421 \cdots .$$

$2^{5/3}$ is that number whose third power is 2^5:

$$(2^{5/3})^3 = 2^5 = 32,$$

$$2^{5/3} = (32)^{1/3} = \sqrt[3]{32} = 3.17480 \cdots .$$

One more, $2^{-1/2}$ is the reciprocal of $2^{1/2}$:

$$2^{-1/2} = \frac{1}{2^{1/2}} = \frac{1}{1.41421 \cdots} = 0.707106 \cdots .$$

These are all particular values of the function

$$f(x) = 2^x,$$

called the exponential function (with base 2). See Fig. 1.1. There is nothing special about the base 2. For each positive number a, there is an exponential function $f(x) = a^x$. See Fig. 1.2.

Numerical values of a^x are computed by logarithms, using the formula

$$\log (a^x) = x \log a.$$

In the examples that follow, 4-place tables of logs to the base 10 are used.

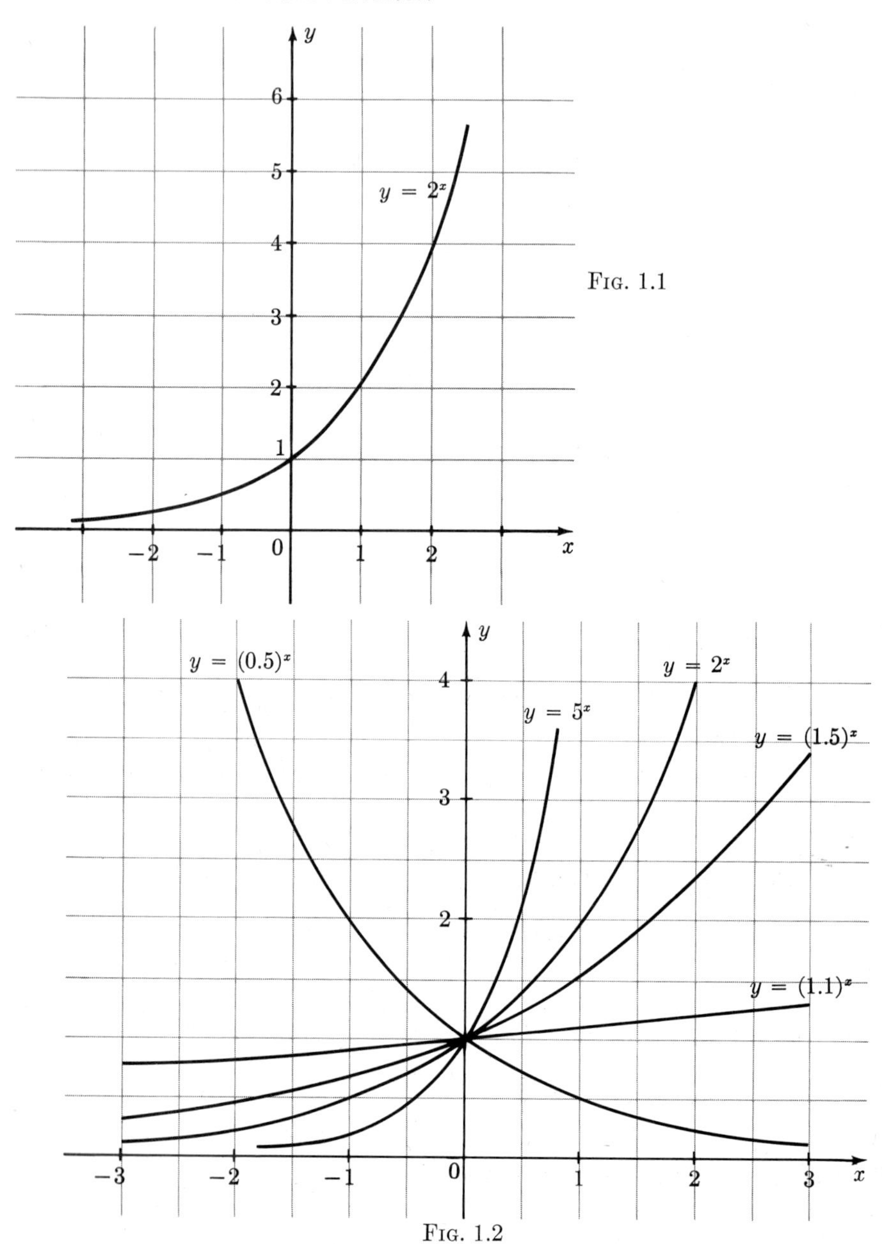

FIG. 1.1

FIG. 1.2

EXAMPLE 1.1

Find $(3.7)^{2.11}$ to four significant figures.*

Solution: Let $x = (3.7)^{2.11}$.

---

* In numerals involving a decimal point, the number of **significant digits** or **figures** is the number of digits from the left-most non-zero digit to the right-most digit. In numerals written without a decimal point, we shall follow the same rule, although this is not universally accepted practice.

$$\log x = (2.11)(\log 3.7)$$
$$\approx (2.11)(0.5682) \approx 1.1989.$$

Answer: $(3.7)^{2.11} \approx 15.81.$

EXAMPLE 1.2

Find $(0.782)^{-3.8}$ to four significant figures.

Solution: Let $x = (0.782)^{-3.8}$.

$$\log x = (-3.8)(\log 0.782)$$
$$\approx (-3.8)(0.8932 - 1) = (-3.8)(-0.1068) \approx 0.4058.$$

Answer: $(0.782)^{-3.8} \approx 2.546.$

REMARK: A crude approximation serves to check whether your answer is in the right ball park.

$$(0.782)^{-3.8} \approx (0.8)^{-4} = [(0.8)^2]^{-2} \approx (0.6)^{-2} = \frac{1}{0.36} \approx 2.8.$$

Laws of Exponents

The laws of exponents, taught in algebra, are three basic identities:

$$(1) \qquad a^x \cdot a^y = a^{x+y}.$$
$$(2) \qquad a^x \cdot b^x = (ab)^x.$$
$$(3) \qquad (a^x)^y = a^{xy}.$$

These important identities will be quite useful later. Indeed, the third one was already used in the computation of $2^{5/3}$ above.

EXERCISES

Evaluate a^x:

1. $a = \frac{1}{2}$; $x = 0, -1, 2, -3$
2. $a = 2$; $x = -2, 0, 4, -1$
3. $a = 3$; $x = 3, -2, 4, 1$
4. $a = \frac{2}{3}$; $x = -1, 0, 2, -3$
5. $a = 27$; $x = \frac{1}{3}, -\frac{1}{3}, \frac{2}{3}, -\frac{4}{3}$
6. $a = \frac{1}{81}$; $x = -\frac{1}{4}, 0, \frac{3}{4}, -\frac{3}{2}$.

Use tables to evaluate to four significant figures:

7. $(2.4)^{4.15}$, $(0.614)^{-5.7}$
8. $(10)^{3.52}$, $(3.142)^{-2.7}$
9. $(2)^{1.91}$, $(15.5)^{-0.9}$
10. $(3.2)^{0.47}$, $(0.008)^{-4.7}$.

Simplify:

11. $9^{-3/2}(\frac{1}{3})^{-3}$, $25^3(25)^{-3/2}$, $4^3(\frac{1}{8})^3$
12. $8^{2/3}(4)^{-1/2}$, $27^{1/3}(27)^{-2/3}$, $2^{-7/2}(8^{9/2})$
13. $16^{-3/4}(\frac{1}{8})^{-5/3}$, $(343^2)^{-1/3}$, $[(4^2)(4^{-1/2})]^{2/3}$
14. $27^{4/3}(3)^{-1}$, $(0.0625^{-1/3})^{3/4}$, $(16^{-1/4})^2(16^{1/4})2^5$.

Use the laws of exponents to prove:

15. $\dfrac{a^x}{a^y} = a^{x-y}$.

16. $\dfrac{a^x}{b^x} = \left(\dfrac{a}{b}\right)^x$.

Find all values of x for which the inequality is satisfied:

17. $3^x \geq 3$

18. $(\frac{1}{2})^x < \frac{1}{2}$

19. $\dfrac{1}{2\sqrt{2}} < 2^x < 4096$

20. $4^x > 2^{x+3}$.

2. DERIVATIVES

To find the slope of the curve $y = a^x$, choose a value x and a nearby value $x + h$. The slope of the corresponding secant is

$$\frac{a^{x+h} - a^x}{h} = \frac{a^x a^h - a^x}{h} = \frac{a^h - 1}{h} \cdot a^x.$$

Here is a curious observation. The quantity

$$\frac{a^h - 1}{h}$$

is the slope of the secant corresponding to 0 and h. See Fig. 2.1. Therefore,

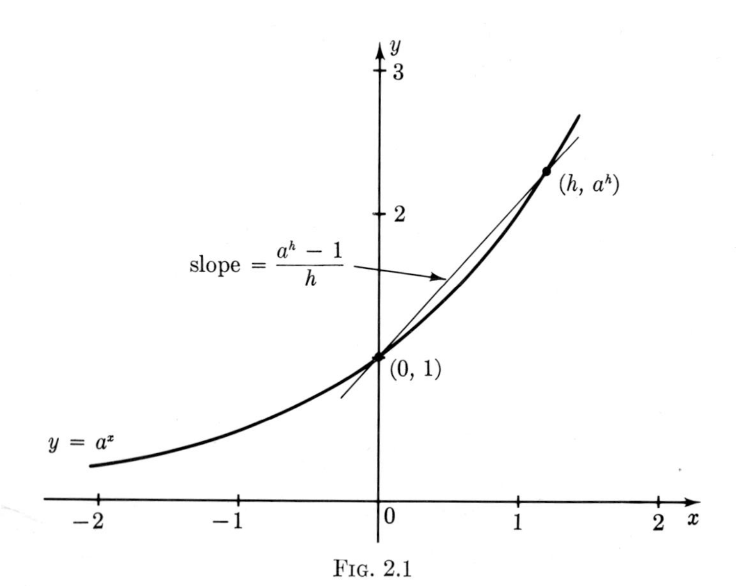

FIG. 2.1

for small values of h,

$$\frac{a^h - 1}{h} \approx \text{(slope of tangent at } x = 0).$$

As $h \longrightarrow 0$, this quantity approaches the derivative of $y = a^x$ at $x = 0$:

$$\frac{a^h - 1}{h} \longrightarrow \frac{d}{dx}(a^x)\Big|_{x=0}.$$

Return to the secant corresponding to x and $x + h$. Its slope is

$$\frac{a^h - 1}{h} \cdot a^x$$

Let h become smaller and smaller. This slope gets closer and closer to the derivative of $y = a^x$. Conclusion:

$$\frac{d}{dx}(a^x) = ka^x, \qquad \text{where} \quad k = \frac{d}{dx}(a^x)\Big|_{x=0}.$$

EXAMPLE 2.1

Show that $\frac{d}{dx}(2^x)\Big|_{x=0} \approx 0.7.$

Solution: Compute

$$\frac{2^h - 1}{h},$$

for $h = 0.01$ and $h = -0.01$.

$h = 0.01$:

$$\log 2^h = (0.01)(\log 2)$$
$$\approx (0.01)(0.30103) \approx 0.00301.$$
$$2^h \approx 1.007.$$
$$\frac{2^h - 1}{h} \approx \frac{0.007}{0.01} = 0.7.$$

$h = -0.01$:

$$\log 2^h = (-0.01)(\log 2)$$
$$\approx -0.00301 = 0.99699 - 1.$$
$$2^h \approx 0.9931.$$
$$\frac{2^h - 1}{h} \approx \frac{0.9931 - 1}{-0.01} = \frac{-0.0069}{-0.01} = 0.69 \approx 0.7.$$

EXAMPLE 2.2

Show that $\dfrac{d}{dx}(3^x)\Big|_{x=0} \approx 1.1.$

Solution: Compute

$$\frac{3^h - 1}{h}$$

for $h = 0.01$ and $h = -0.01$.

$h = 0.01$:

$$\log 3^h \approx (0.01)(0.47712) \approx 0.00477.$$

$$3^h \approx 1.011.$$

$$\frac{3^h - 1}{h} \approx \frac{0.011}{0.01} = 1.1.$$

$h = -0.01$:

$$\log 3^h \approx (-0.01)(0.47712) \approx -0.00477 = 0.99523 - 1.$$

$$3^h \approx 0.9891.$$

$$\frac{3^h - 1}{h} \approx \frac{-0.0109}{-0.01} = 1.09 \approx 1.1.$$

The Number e

In the two last examples we obtained

$$\frac{d}{dx}(2^x)\Big|_{x=0} \approx 0.7, \qquad \frac{d}{dx}(3^x)\Big|_{x=0} \approx 1.1.$$

These results imply

$$\frac{d}{dx}(2^x) \approx (0.7)2^x, \qquad \frac{d}{dx}(3^x) \approx (1.1)3^x.$$

Since 0.7 is less than 1 and 1.1 is greater than 1, they suggest that somewhere between 2 and 3 there is a number a such that

$$\frac{d}{dx}(a^x)\Big|_{x=0} = 1 \qquad \text{or} \qquad \frac{d}{dx}(a^x) = a^x.$$

For that number a, the function a^x is its own derivative! More refined calculations show

$$\frac{d}{dx}(2.7)^x\Big|_{x=0} \approx 0.99325,$$

$$\frac{d}{dx}(2.8)^x \bigg|_{x=0} \approx 1.02962.$$

Narrowing the gap again,

$$\frac{d}{dx}(2.71)^x \bigg|_{x=0} \approx 0.99695,$$

$$\frac{d}{dx}(2.72)^x \bigg|_{x=0} \approx 1.00063.$$

Since 1.00000 is slightly more than $\frac{8}{10}$ of the way between 0.99695 and 1.00063, the desired number is probably slightly more than 2.718.

Conclusion: There is a number, called e, approximately 2.718, such that

$$\frac{d}{dx}(e^x) \bigg|_{x=0} = 1,$$

$$\frac{d}{dx}(e^x) = e^x.$$

The number e has been computed with great accuracy. To 15 places it is

$$e \approx 2.71828\ 18284\ 59045.$$

The number e, like the number π, is a fundamental constant of nature, independent of units of measurement. Its central role will become more and more clear as we proceed. Right now we have the striking property that the exponential function e^x reproduces itself under differentiation:

$$\frac{d}{dx}(e^x) = e^x.$$

The function e^x is of such importance it is called *the* exponential function.

3. EXPONENTIAL FUNCTIONS

The exponential function $y = e^x$ is its own derivative; it satisfies the differential equation

$$\frac{dy}{dx} = y.$$

So does $y = ce^x$ where c is a constant:

$$\frac{dy}{dx} = \frac{d}{dx}(ce^x) = c\frac{d}{dx}(e^x) = ce^x = y.$$

Are there any other functions $y(x)$ satisfying the differential equation? To decide, plot the direction field associated with the differential equation.

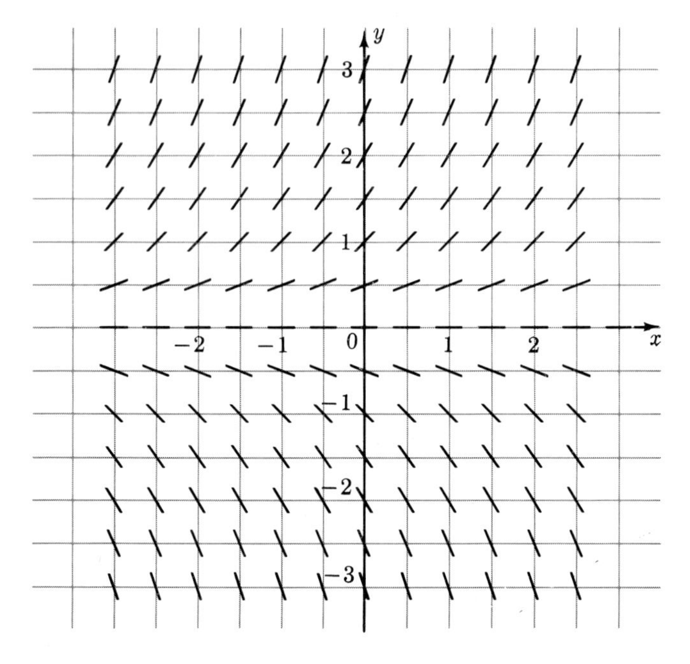

FIG. 3.1 Direction field $\frac{dy}{dx} = y$.

At each point (x, y) it has a short segment of slope y. See Fig. 3.1. A function $y(x)$ satisfies the differential equation

$$\frac{dy}{dx} = y$$

if the direction of its graph at each point coincides with this direction field. Such a function is called a **solution** of the differential equation.

Two facts are apparent from Fig. 3.1.

(1) The graph of each solution crosses the y-axis.

(2) Through each point of the y-axis passes the graph of exactly one solution.

The function ce^x is a solution, and its graph crosses the y-axis at $(0, c)$. By (2), it is the only solution whose graph passes through $(0, c)$.

Now consider any solution. By (1), its graph crosses the y-axis at a point $(0, c)$. Therefore it is the function ce^x.

Conclusion:

> Each solution of the differential equation
>
> $$\frac{dy}{dx} = y$$
>
> is a function of the form
>
> $$y = ce^x$$
>
> for some constant c.

The Function $y = e^{kx}$

Consider the function

$$y = e^{kx},$$

where k is a positive constant. It is obtained from $y = e^x$ by a change of scale (Fig. 3.2).

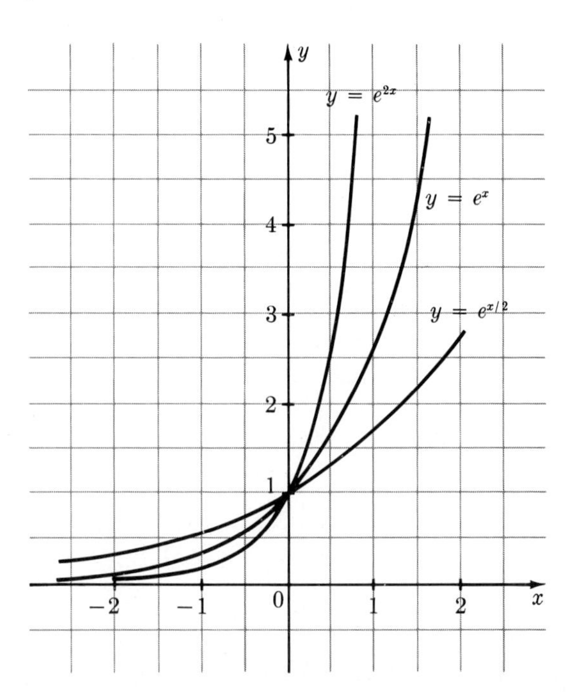

FIG. 3.2

Each function

$$y(x) = e^{kx}, \qquad k > 0,$$

possesses certain basic growth properties:

(1) $y(x) > 0$.
(2) $y(x) \longrightarrow \infty$ as $x \longrightarrow \infty$.
(3) $y(x) \longrightarrow 0$ as $x \longrightarrow -\infty$.
(4) The graph of $y = e^{kx}$ is always rising.

Now consider

$$y(x) = e^{-kx}.$$

Since

$$e^{-kx} = e^{k(-x)},$$

the graph (Fig. 3.3) of $y = e^{-kx}$ is the reflection in the y-axis of the graph of $y = e^{kx}$.

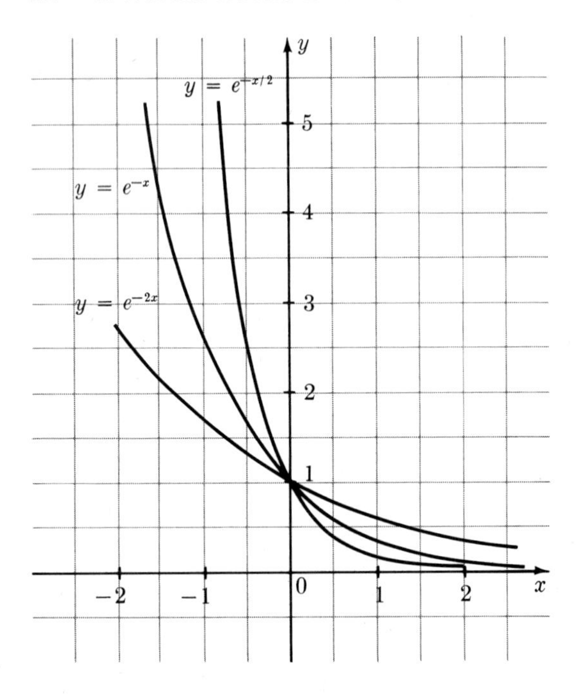

FIG. 3.3

Each function

$$y(x) = e^{-kx}, \qquad k > 0,$$

possesses these growth properties:

(1′) $y(x) > 0$.
(2′) $y(x) \longrightarrow 0$ as $x \longrightarrow \infty$.
(3′) $y(x) \longrightarrow \infty$ as $x \longrightarrow -\infty$.
(4′) The graph of $y = e^{-kx}$ is always falling.

Finally, consider $y = e^{kx}$ with $k = 0$:

$$y = e^{0\cdot x} = e^0 = 1.$$

Derivative of e^{kx}

We find the derivative of $y(x) = e^{kx}$ (for k positive, negative, or zero) by use of the Scale Rule:

$$\frac{d}{dx}[f(kx)] = kf'(kx).$$

Applied to $f(x) = e^x$, this yields the important formula

$$\boxed{\frac{d}{dx}[e^{kx}] = ke^{kx}.}$$

Thus, the derivative of e^{kx} is k times the function itself. Stated differently, $y = e^{kx}$ satisfies the differential equation

$$\frac{dy}{dx} = ky.$$

By practically the same argument as above, we conclude:

> Each solution of the differential equation
>
> $$\frac{dy}{dx} = ky$$
>
> is a function of the form
>
> $$y = ce^{kx}$$
>
> for some constant c.

Values of e^x

The graph of $y = e^x$ is always rising (Fig. 3.4). What is more, it rises arbitrarily high for x large ($e^x \longrightarrow \infty$ as $x \longrightarrow \infty$) and becomes arbitrarily close to 0 for x large negative ($e^x \longrightarrow 0$ as $x \longrightarrow -\infty$). Consequently e^x takes on each positive value once.

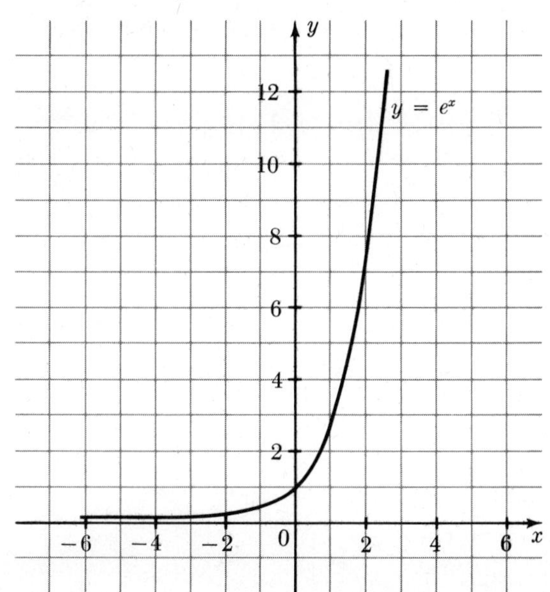

FIG. 3.4

> If $a > 0$, there is one number k for which
>
> $$e^k = a.$$

To find k, take logarithms:

$$k \log_{10} e = \log_{10} a,$$

$$k = \frac{\log_{10} a}{\log_{10} e}.$$

The denominator $\log_{10} e$ is used so often in computations it has a special name M. To ten places,

$$M = \log_{10} e \approx 0.43429\ 44819.$$

Other frequently used constants are

$$\frac{1}{M} \approx 2.30259, \qquad \log_{10} M \approx 0.63778 - 1.$$

Derivative of a^x

This chapter began with the exponential function $y = a^x$ to the base a. If $a > 0$, then

$$a = e^k, \qquad \text{where} \quad k = \frac{\log_{10} a}{M}.$$

By the laws of exponents,

$$a^x = (e^k)^x = e^{kx}.$$

In other words, each exponential function $y = a^x$ for $a > 0$ can be treated like $y = e^{kx}$ for an appropriate k. Thus

$$\frac{d}{dx}(a^x) = \frac{d}{dx}(e^{kx}) = ke^{kx} = ka^x.$$

Therefore

$$\frac{d}{dx}(a^x) = ka^x, \qquad \text{where} \quad k = \frac{\log_{10} a}{M}.$$

REMARK: Logarithms to any positive base b (different from 1) make sense. Recall

$$t = \log_b a \qquad \text{means} \qquad b^t = a.$$

In particular

$$k = \log_e a \qquad \text{means} \qquad e^k = a;$$

therefore the formula for the derivative reads

$$\boxed{\frac{d}{dx}(a^x) = (\log_e a)a^x.}$$

NOTE: If $a = e$, then $\log_e e = 1$ and the formula is particularly simple,

$$\frac{d}{dx}(e^x) = e^x.$$

Tables of **natural logarithms** ($\log_e$) are readily available.

EXERCISES

1. Does $y = e^x + c$ satisfy the differential equation $dy/dx = y$?

Use graph paper to make a large accurate drawing of the direction field in Fig. 3.1 for $-3 \leq x \leq 3$ and $-4 \leq y \leq 4$. Sketch the curve that:

2. passes through $(2, 3)$
3. passes through $(0, -\frac{1}{2})$.
4. Make another large graph for the range $-1 \leq x \leq 1$ and $0 \leq y \leq 3$. Obtain an approximation of the numerical value of e as follows: Approximate the curve through $(0, 1)$ by a "curve" consisting of line segments. When $x = 1$ the height of your curve should be about e. Why?

Differentiate:

5. $f(x) = e^{3x}$
6. $f(x) = e^{-x}$
7. $y = e^{2x+1}$
8. $y = e^{x-1} - e^{-3x}$
9. $y = \dfrac{e^x + e^{-x}}{2}$
10. $y = \dfrac{e^x - e^{-x}}{2}$
11. $f(t) = e^{3t} - 4e^{2t}$
12. $s(t) = \dfrac{e^{-2t}}{2} - e^t$
13. $y = 3e^{4x+1}$
14. $y = (e^{3x})^2$
15. $f(x) = \dfrac{e^{2x} - e^{-x}}{e^x}$
16. $f(x) = e^x(e^{-2x} + e^{-x})$.

Differentiate:

17. $f(x) = 10^x$
18. $f(x) = 10^{-4x}$
19. $y = 5^{x-1}$
20. $y = 5^{2x-1}$
21. $f(x) = 10^{4x-1}$
22. $f(x) = 10^{2-3x}$
23. $y = (10^x + 10^{-x})^2$
24. $y = 3(10^{-4x}) - 4(10^{3-x})$.

On the same graph sketch:

25. $e^x, \quad e^{2x}, \quad e^{x/2}, \quad e^{x/5}$
26. $e^x, \quad e^{-x}, \quad e^{-2x}, \quad e^{-x/3}$.
27. Plot carefully the direction field $dy/dx = \frac{1}{2}y$ for $-2 \leq x \leq 2$, $0 \leq y \leq 3$. Use it to sketch the curve $y = e^{x/2}$.
28. Plot carefully the direction field of $dy/dx = -y$ in the range $-1 \leq x \leq 1$, $0 \leq y \leq 3$. Use it to sketch the curve $y = e^{-x}$.

Sketch the curve:

29. $y = e^{x-1}$
30. $y = e^{x+1}$
31. $y = 1 - e^{-x}$
32. $y = 4e^x - 1$
33. $y = \dfrac{e^x + e^{-x}}{2}$
34. $y = \dfrac{e^x - e^{-x}}{2}$.

Find a function y for which:

35. $y' = e^{2x}$
36. $y' = e^{-2x} + 1$
37. $y' = e^x, \quad y(0) = 2$
38. $y' = e^x - e^{-x}, \quad y(0) = 0$.

39. Verify that $y = 4e^{x} - 2e^{-x}$ satisfies the differential equation $y'' = y$.
40. Verify that $y = ae^{kx} - be^{-kx}$ satisfies the differential equation $y'' = k^2 y$.
41. Verify that $y = ae^{3x} + be^{2x}$ satisfies $y'' - 5y' + 6y = 0$.
42. Verify that $y = \dfrac{e^{x} + e^{-x}}{2}$ satisfies $y' + y = e^{x}$.
43. Find the 58-th derivative of the function in Ex. 42.

4. APPLICATIONS

The exponential function $y = e^{kx}$ satisfies the differential equation

$$\frac{dy}{dx} = ky.$$

Hence e^{kx} grows or decreases at a rate proportional to its own size, depending on whether k is positive or negative.

Bacteria Growth

A colony of bacteria with unlimited food and no enemies grows at a rate proportional to its size. If $N = N(t)$ is the number of bacteria at time t, then

$$\frac{dN}{dt} = kN, \qquad k > 0.$$

Therefore

$$N(t) = Ae^{kt},$$

where A is a constant. If at $t = 0$, the number is N_0, a known quantity, then

$$N_0 = N(0) = Ae^0 = A.$$

So $A = N_0$ is the appropriate constant for this problem. We conclude

$$N(t) = N_0 e^{kt}.$$

EXAMPLE 4.1

There are 10^5 bacteria at the start of an experiment and 3×10^7 after 24 hours. What is the growth law?

Solution: Denote by $N(t)$ the number of bacteria in t hours. Then

$$N'(t) = kN(t).$$

Therefore

$$N(t) = N_0 e^{kt} = 10^5 e^{kt}.$$

To find k, use the information that $N(24) = 3 \times 10^7$:

$$3 \times 10^7 = N(24) = 10^5 e^{24k},$$

$$e^{24k} = 3 \times 10^2 = 300.$$

Taking logs,

$$(24k) \log_{10} e = \log_{10}(300),$$

$$24k = \frac{\log_{10}(300)}{\log_{10} e} \approx 5.7,$$

$$k \approx 0.24.$$

Answer: $N(t) \approx 10^5 e^{0.24t}$, t measured in hours. (Alternately, $N(t) = 10^5(300)^{t/24}$.)

REMARK: Although $N(t)$ jumps by 1 for each bacterium counted, it is treated like a smoothly growing function. Because $N(t)$ is very large, this simplification of the problem leads to highly accurate results.

Radioactive Decay

A radioactive element decays at a rate proportional to the amount present. Its **half-life** is the time in which a given quantity decays to one-half of its original mass.

EXAMPLE 4.2

Carbon-14, ^{14}C, has a half-life of 5668 years. Find its decay law.

Solution: Let $M(t)$ be the mass of ^{14}C at time t, measured in years. Then

$$\frac{dM}{dt} = -\lambda M,$$

where the **decay constant** λ is positive. The equation is written this way because dM/dt is negative. The solution is

$$M(t) = M_0 e^{-\lambda t},$$

where $M_0 = M(0)$, the initial mass. To evaluate λ, use the data:

$$\frac{M_0}{2} = M(5668) = M_0 e^{-5668\lambda},$$

$$e^{-5668\lambda} = \frac{1}{2}.$$

Taking logs,

$$-5668\lambda \log_{10} e = -\log_{10} 2,$$

$$\lambda = \frac{\log_{10} 2}{5668 \log_{10} e} \approx 0.000122.$$

Answer: $M(t) \approx M_0 e^{-(0.000122)t}$. (Alternately, $M(t) = M_0 a^{-t}$ where $a = 2^{1/5668}$.)

Compound interest

EXAMPLE 4.3

$1000 is deposited in a bank giving 5% annual interest compounded daily. Estimate its value in 10 years.

Solution: There is an exact expression for the value:

$$1000\left(1+\frac{0.05}{365}\right)^{10\cdot 365}.$$

Let us get a quick approximation. Suppose interest were compounded not daily, but continuously. Then the value would be growing at a rate proportional to itself. So if $A(t)$ is the amount at time t (in years),

$$\frac{dA}{dt}=0.05A(t).$$

Solving,

$$A(t)=1000e^{0.05t},$$

$$A(10)=1000e^{(0.05)(10)}.$$

Answer: $1648.72.

Here is a table showing the actual and the approximated values after 10 years.

Compounded quarterly	Compounded monthly	Compounded daily	Compounded continuously
$1643.62	$1647.01	$1648.66	$1648.72

EXERCISES

1. Thorium X has a half-life of 3.64 days. Find its decay law. How long will it take for $\frac{2}{3}$ of a quantity to disintegrate?
2. Two pounds of a certain radioactive substance loses $\frac{1}{5}$ of its original mass in 3 days. At what rate is the substance decaying after 4 days?
3. Money compounded continuously will double in a year at what annual rate of interest?
4. How long will it take a sum of money compounded continuously at 7.5% per annum to show a 50% return?
5. A colony of bacteria has a population of 3×10^6 initially, and 9×10^6 two hours later. What is the growth law? How long does it take the colony to double?
6. Assume that population grows at a rate proportional to the population itself. In 1950 the US population was 151 million, in 1960 it was 178 million. Make a prediction for the year 2000.
7. Under ideal conditions the rate of change of pressure above sea level is proportional to the pressure. If the barometer reads 30 in. at sea level, and 25 in. at 4000 ft, find the barometric pressure at 20,000 ft.

8. In a certain calculus course, it was found that the number of students dropping out each day was proportional to the number still enrolled. If 2000 started out and 10% dropped after 28 days, estimate the number left after 12 weeks.

9. A 5-lb sample of radioactive material contains 2 lb of Radium F which has a half-life of 138.3 days and 3 lb of Thorium X which has a half-life of 3.64 days. When will the sample contain equal amounts of Radium F and Thorium X?

10. A salt in solution decomposes into other substances at a rate proportional to the amount still unchanged. If 10 lb of a salt reduces to 5 lb in $\frac{1}{2}$ hr, how much is left after 1 hr?

8. Integration

1. INTRODUCTION

The great Greek mathematician Archimedes used ingenious methods to compute the area bounded by a parabola and a chord (Fig. 1.1).

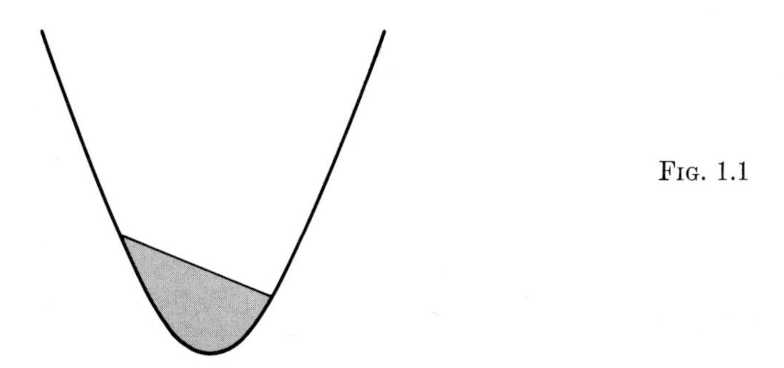

Fig. 1.1

In this chapter we shall develop tools of Calculus which make the solution of this and similar problems routine. These tools are used not only in such area problems, but also in a wide range of scientific and technical applications.

2. AREA UNDER A CURVE

The basic problem: Compute the area of the region in Fig. 2.1. The

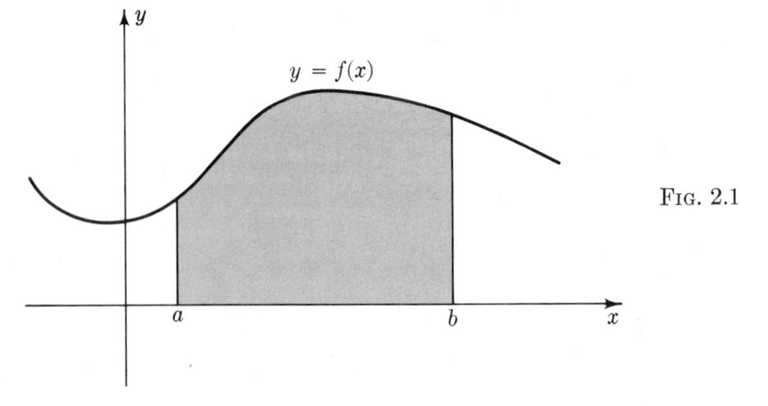

Fig. 2.1

region is bounded by the graph of a positive function $y = f(x)$, the x-axis, and the lines $x = a$ and $x = b$.

Archimedes solved this problem for a few special curves, each by an ingenious special method. In the next few pages you will learn a simple method of solving the problem *in general*, something beyond the reach of the most advanced mathematicians for about 2000 years after Archimedes.

The breakthrough came with the idea of changing the problem from a static one to a dynamic one. Instead of computing the area between two fixed lines, $x = a$ and $x = b$, compute it between a fixed line and a second *moving* line. See Fig. 2.2. Denote by $A(x)$ the area of the region shown in Fig. 2.2. When $x = a$, the area $A(a)$ is zero. As x moves to the right, $A(x)$

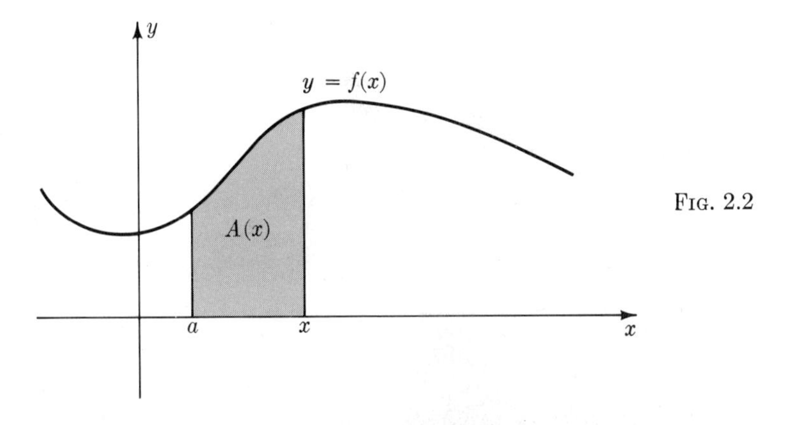

FIG. 2.2

increases. Furthermore, Fig. 2.2 suggests something about the rate of increase of $A(x)$. Where the curve is high $A(x)$ increases rapidly; where the curve is low $A(x)$ increases slowly. Apparently there is a relation between the rate of increase of $A(x)$ and the function $f(x)$.

Let us investigate the derivative of $A(x)$. For this we need to know the approximate value of

$$\frac{A(x + h) - A(x)}{h}$$

for small values of h. The numerator can be interpreted geometrically: It is the area under the curve between x and $x + h$. See Fig. 2.3. When h is small, the shaded region in Fig. 2.3 is approximately a rectangle of base h and height $f(x)$. See Fig. 2.4. Therefore,

$$A(x + h) - A(x) \approx h \cdot f(x).$$

Hence,

$$\frac{A(x + h) - A(x)}{h} \approx f(x),$$

and this approximation improves as $h \longrightarrow 0$. We conclude that $A'(x) = f(x)$,

or

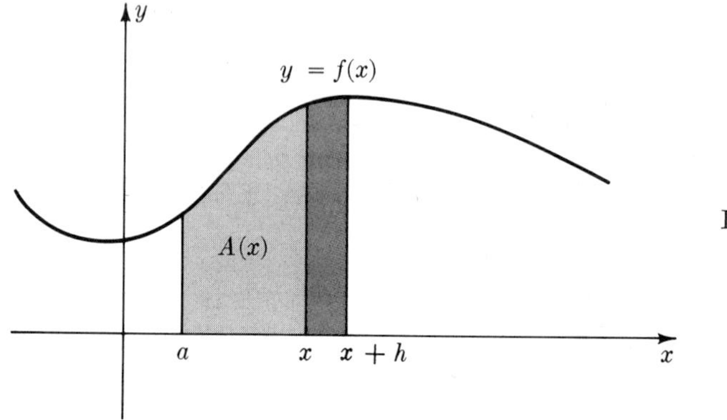

FIG. 2.3

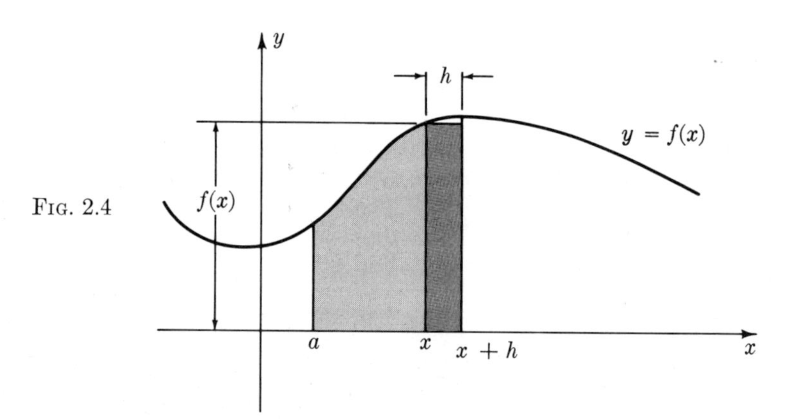

FIG. 2.4

Thus the area under a curve $y = f(x)$ is described by an antiderivative of $f(x)$. In theory, the problem is nearly solved: the desired area $A(x)$ is an antiderivative of $f(x)$. But that does not give a formula for $A(x)$. In practice, you must beg, borrow, or steal an antiderivative. Suppose you are able to find one, $F(x)$. Since any two antiderivatives differ by a constant,

$$A(x) = F(x) + C$$

for some constant C. But you know $A(a) = 0$. Setting $x = a$, you find

$$0 = A(a) = F(a) + C,$$

$$C = -F(a).$$

Therefore,

$$A(x) = F(x) - F(a).$$

This is a formula for the area under $y = f(x)$ between a and x. In particular, the area between a and b is $A(b) = F(b) - F(a)$. This is the breakthrough, the area problem is solved by antidifferentiation!

Area Rule If $y = f(x) \geq 0$ for $a \leq x \leq b$, then the area bounded by its graph, the x-axis, and the lines $x = a$ and $x = b$ is

$$F(b) - F(a),$$

where $F(x)$ is any antiderivative of $f(x)$.

The argument above shows that it makes no difference which antiderivative is used in applying the Area Rule; the value $F(b) - F(a)$ is the same for each antiderivative $F(x)$ of $f(x)$.

Here is a direct verification of that fact. If $F_1(x)$ and $F_2(x)$ are any antiderivatives of $f(x)$, then $F_2(x) = F_1(x) + C$ for some constant C. Hence

$$\begin{aligned} F_2(b) - F_2(a) &= [F_1(b) + C] - [F_1(a) + C] \\ &= F_1(b) - F_1(a). \end{aligned}$$

Thus the constant drops out; the value $F(b) - F(a)$ is the same for each antiderivative.

Notation

$F(b) - F(a)$ is often denoted by $F(x) \Big|_a^b$.

Study this section carefully. It contains the basic idea of Integral Calculus.

3. APPLICATIONS OF THE AREA RULE

EXAMPLE 3.1

Compute the area under the graph of $y = x + 1$ between $x = 1$ and $x = 3$. See Fig. 3.1.

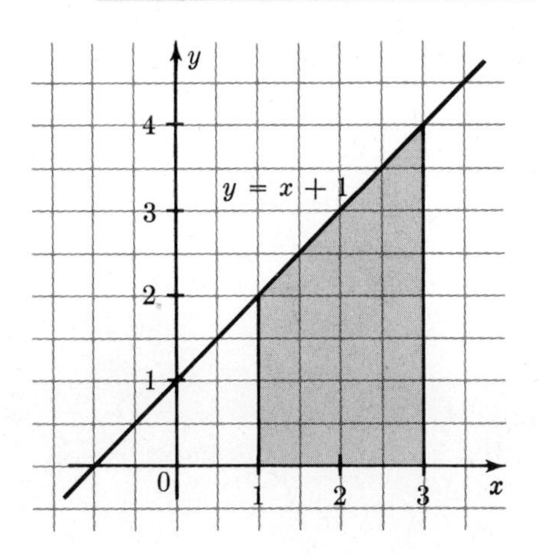

FIG. 3.1

Solution:

$$\text{Area} = F(3) - F(1),$$

where $F(x)$ is any antiderivative of $(x + 1)$. We know that

$$\frac{d}{dx}(x^2) = 2x,$$

hence

$$\frac{d}{dx}\left(\frac{1}{2}x^2\right) = x.$$

We also know that

$$\frac{d}{dx}(x) = 1.$$

Therefore, an antiderivative of $(x + 1)$ is $F(x) = \frac{1}{2}x^2 + x$, so

$$F(3) - F(1) = \left(\frac{1}{2}\cdot 3^2 + 3\right) - \left(\frac{1}{2}\cdot 1^2 + 1\right) = 6.$$

Answer: 6 square units.

CHECK: Since the region is a trapezoid, we can check our answer. The trapezoid has base 2 and legs 2 and 4.

$$\text{Area} = 2\cdot\frac{2 + 4}{2} = 6 \text{ square units.}$$

EXAMPLE 3.2

Compute the area bounded by the parabola $y = 4 - x^2$ and x-axis (Fig. 3.2).

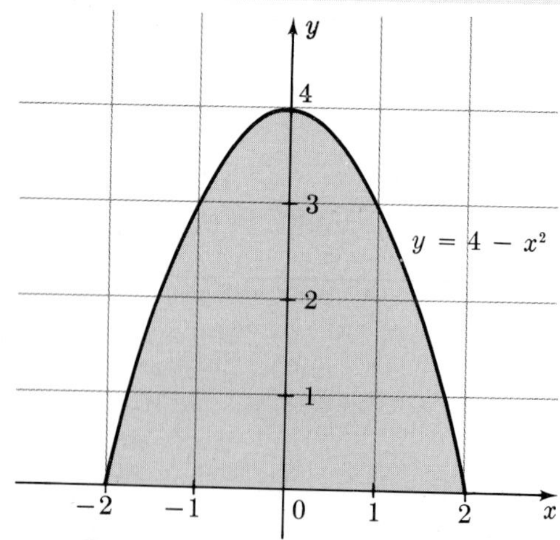

FIG. 3.2

Solution: The parabola intersects the x-axis at $x = 2$ and $x = -2$. Hence

$$\text{Area} = F(2) - F(-2),$$

where $F(x)$ is any antiderivative of $4 - x^2$. Take $F(x) = 4x - \frac{1}{3}x^3$. Then

$$F(2) - F(-2) = \left(4x - \frac{x^3}{3}\right)\bigg|_{-2}^{2} = \left(8 - \frac{8}{3}\right) - \left(-8 + \frac{8}{3}\right) = \frac{32}{3}.$$

Answer: $\frac{32}{3}$ square units.

EXAMPLE 3.3

Compute the area under the curve $y = 1/x^2$ between $x = 3$ and $x = 5$. See Fig. 3.3.

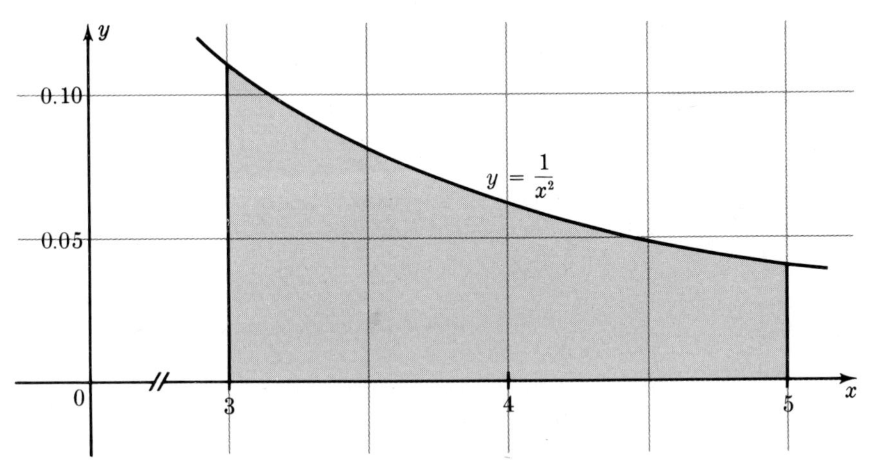

FIG. 3.3 (Note different scales on axes.)

Solution:

$$\text{Area} = F(5) - F(3),$$

where $F(x)$ is any antiderivative of $1/x^2$. One antiderivative is $F(x) = -1/x$, so

$$F(5) - F(3) = \left(-\frac{1}{5}\right) - \left(-\frac{1}{3}\right) = \frac{1}{3} - \frac{1}{5} = \frac{2}{15}.$$

Answer: $\frac{2}{15}$ square units.

EXAMPLE 3.4

Compute the area of a parabolic segment bounded by the curve $y = x^2$ and the line $y = 1$. See Fig. 3.4.

Solution: Compute the area under $y = x^2$ between -1 and 1, and subtract it from the area of the rectangle shown in Fig. 3.4.

$$(\text{Area under } y = x^2) = F(1) - F(-1)$$

$$= \frac{x^3}{3}\bigg|_{-1}^{1} = \frac{(1)^3}{3} - \frac{(-1)^3}{3} = \frac{2}{3}.$$

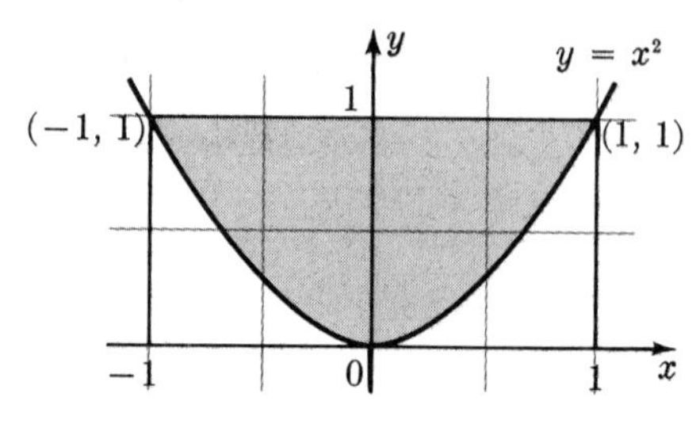

FIG. 3.4

The area of the rectangle is 2.

Answer: $2 - \frac{2}{3} = \frac{4}{3}$ square units.

EXAMPLE 3.5

Find the area enclosed by the ellipse

$$\frac{x^2}{a^2} + \frac{y^2}{b^2} = 1.$$

Solution: The equation may be solved for y if $y \geq 0$:

$$y = b\sqrt{1 - \frac{x^2}{a^2}} = \frac{b}{a}\sqrt{a^2 - x^2}.$$

Take into account the symmetry in the x-axis. Then the area is

$$A = 2\frac{b}{a}[F(a) - F(-a)],$$

where $F(x)$ is any antiderivative of

$$\sqrt{a^2 - x^2}.$$

It looks terribly difficult to find an antiderivative, however, note that $F(a) - F(-a)$ is the area under the curve $y = \sqrt{a^2 - x^2}$ between $x = -a$ and $x = a$. But this curve is a semicircle of radius a, hence its area is $\frac{1}{2}\pi a^2$. Therefore

$$A = \frac{2b}{a}\left(\frac{1}{2}\pi a^2\right) = \pi ab.$$

Answer: πab.

REMARK: If the ellipse is stretched in the y-direction by the factor a/b, it changes to a circle of radius a while its area changes by the factor a/b. See Fig. 3.5. This provides a geometric reason for the answer $\pi ab = (\pi a^2)/(a/b)$.

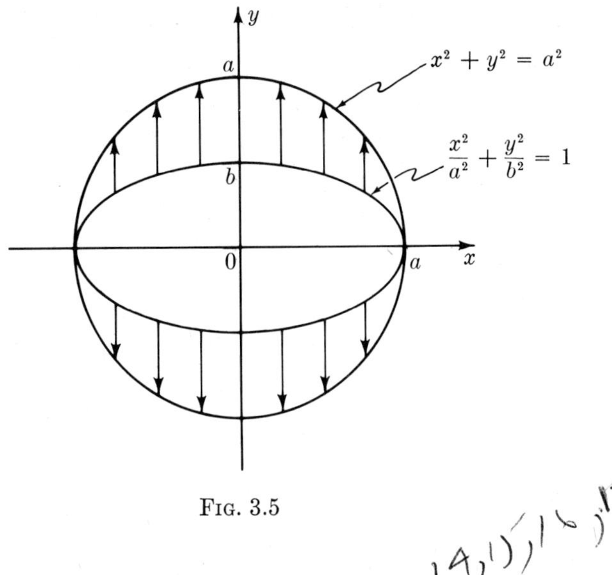

FIG. 3.5

EXERCISES

Compute the area under the graph $y = f(x)$ between $x = a$ and $x = b$:

1. $f(x) = \frac{1}{2}x; \quad a = 1, b = 3$
2. $f(x) = 2 - x; \quad a = 1, b = 2$
3. $f(x) = x + 7; \quad a = -1, b = 2$
4. $f(x) = 3(x - 5); \quad a = 5, b = 7$
5. $y = (x - 1)^2; \quad a = 1, b = 4$
6. $y = 4 - x^2; \quad a = -2, b = 2$
7. $y = 4(x - 3)^2; \quad a = 0, b = 6$
8. $y = 4x - x^2; \quad a = 0, b = 4$
9. $y = \cos x; \quad a = -\frac{\pi}{2}, b = \frac{\pi}{2}$
10. $y = \sin 2x; \quad a = 0, b = \frac{\pi}{2}$
11. $y = \sin x + 4 \cos x; \quad a = 0, b = \frac{\pi}{2}$
12. $y = \frac{1}{x^2} + \sin x; \quad a = \frac{\pi}{2}, b = \pi$
13. $y = e^x; \quad a = 0, b = 1$
14. $y = e^{-x/2}; \quad a = -1, b = 0$
15. $y = e^x - 1; \quad a = 0, b = 4$
16. $y = e^{3-3x}; \quad a = 0, b = 1.$

Find the area bounded by:

17. $y = \frac{1}{2}x^2, \quad y = x$
18. $y = x^2, \quad y = 8 - x^2$
19. $y = e^x$, the line $y = e$, and the y-axis
20. $y = \frac{1}{x^2}$, the line $x = 4$, and the line $y = 4$.

4. THE DEFINITE INTEGRAL

For any function $f(x)$ defined for $a \leq x \leq b$, we write

$$\int_a^b f(x)\,dx = F(b) - F(a),$$

where $F(x)$ is an antiderivative of $f(x)$.

The expression

$$\int_a^b f(x)\,dx$$

is called the **definite integral** of $f(x)$ from a to b.

It is customary also to define

$$\int_b^a f(x)\,dx = F(a) - F(b).$$

It follows that

$$\int_b^a f(x)\,dx = -\int_a^b f(x)\,dx.$$

In terms of the definite integral, the Area Rule of the last section states:

If $f(x) \geq 0$ for $a \leq x \leq b$, then

$$\int_a^b f(x)\,dx$$

is the area of the region bounded by the curve $y = f(x)$, the x-axis, and the lines $x = a$ and $x = b$.

For negative functions this is not true, but the following is:

If $f(x) \leq 0$ for $a \leq x \leq b$, then

$$\int_a^b f(x)\,dx$$

is the negative of the area bounded by the curve $y = f(x)$, the x-axis, $x = a$, and $x = b$.

EXAMPLE 4.1

Show that for $b > 0$,

$$\int_0^b (-x^2)\,dx$$

is the negative of the area bounded by $y = -x^2$, the x-axis, $x = 0$, and $x = b$.

Solution:

$$\int_0^b (-x^2)\,dx = F(b) - F(0),$$

where $F(x)$ is any antiderivative of $-x^2$. Take $F(x) = -\frac{1}{3}x^3$; then

$$\int_0^b (-x^2)\,dx = -\frac{1}{3}b^3 - \left(-\frac{1}{3}\cdot 0^3\right) = -\frac{1}{3}b^3.$$

The region in question is R in Fig. 4.1. Its area is equal to that of region S under the curve $y = x^2$. But the area of S is

$$\int_0^b x^2\,dx = \frac{1}{3}x^3\Big|_0^b = \frac{1}{3}b^3.$$

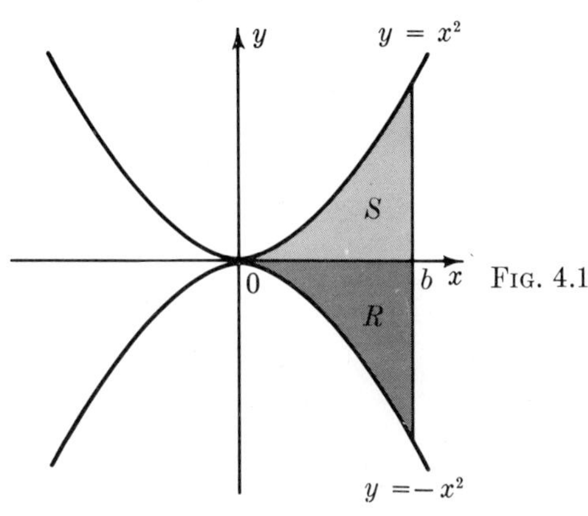

Fig. 4.1

Answer: $\displaystyle\int_0^b (-x^2)\,dx = -\frac{1}{3}b^3$
$= -(\text{area of } R).$

The preceding example is typical for curves below the x-axis. If $f(x) \leq 0$, then $y = -f(x)$ is a curve *above* the axis (Fig. 4.2). The regions R and S

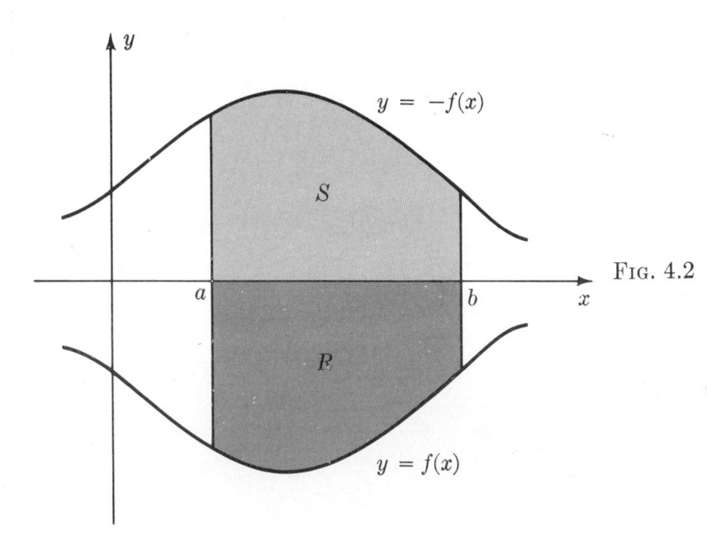

Fig. 4.2

have the same area. We want to show that

$$\int_a^b f(x)\,dx = -(\text{Area of } S).$$

Suppose $F(x)$ is an antiderivative of $f(x)$. Then $-F(x)$ is an antiderivative of $-f(x)$. Hence

$$\begin{aligned}\int_a^b f(x)\,dx = F(b) - F(a) &= -\{[-F(b)] - [-F(a)]\}\\ &= -\int_a^b [-f(x)]\,dx\\ &= -(\text{Area of } S).\end{aligned}$$

This establishes in general the statement preceding Example 4.1.

What can be said about

$$\int_a^c f(x)\,dx,$$

if $f(x)$ is positive for some values of x and negative for others?

To answer this question, we need a simple property of integrals:

> If $a < b < c$, then
>
> $$\int_a^c f(x)\,dx = \int_a^b f(x)\,dx + \int_b^c f(x)\,dx.$$

This property is easily verified:

$$\begin{aligned}\int_a^b f(x)\,dx + \int_b^c f(x)\,dx &= [F(b) - F(a)] + [F(c) - F(b)]\\ &= F(c) - F(a) = \int_a^c f(x)\,dx.\end{aligned}$$

Returning to the question, suppose $f(x) \geq 0$ when $a \leq x \leq b$, but $f(x) \leq 0$ when $b \leq x \leq c$. See Fig. 4.3.

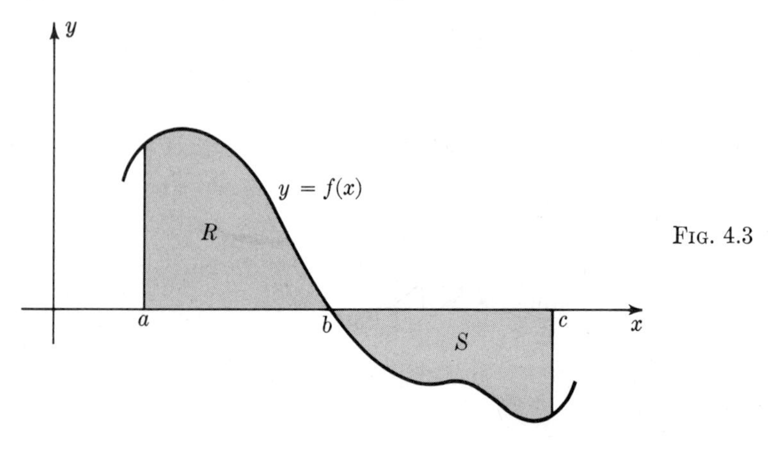

Fig. 4.3

$$\int_a^c f(x)\,dx = \int_a^b f(x)\,dx + \int_b^c f(x)\,dx.$$

The first integral on the right side is positive; it gives the area of R. The second is negative; it gives the negative of the area of S.

In general, if $y = f(x)$ lies partly above the x-axis and partly below, the integral

$$\int_a^c f(x)\,dx$$

adds up the areas, counting those above as positive and those below as negative.

$$\int_a^c f(x)\,dx = [\text{area above the } x\text{-axis bounded by } y = f(x)]$$
$$-[\text{area below the } x\text{-axis bounded by } y = f(x)],$$

all areas taken between $x = a$ and $x = c$.

Alternate statement:

$$\int_a^c f(x)\,dx$$

is the **algebraic area** under $y = f(x)$ from a to c. ("Algebraic" means signed.)

EXAMPLE 4.2

Evaluate the definite integral $\int_0^{2\pi} \sin x\,dx$ and interpret the answer in terms of areas.

Solution:

$$\int_0^{2\pi} \sin x\,dx = F(2\pi) - F(0),$$

where $F(x)$ is any antiderivative of $\sin x$. One antiderivative is $-\cos x$, so

$$\int_0^{2\pi} \sin x\,dx = (-\cos 2\pi) - (-\cos 0) = -1 + 1 = 0.$$

It is certainly not true that there is zero area between $y = \sin x$ and the x-axis. But there is as much area above the axis as below it (Fig. 4.4).

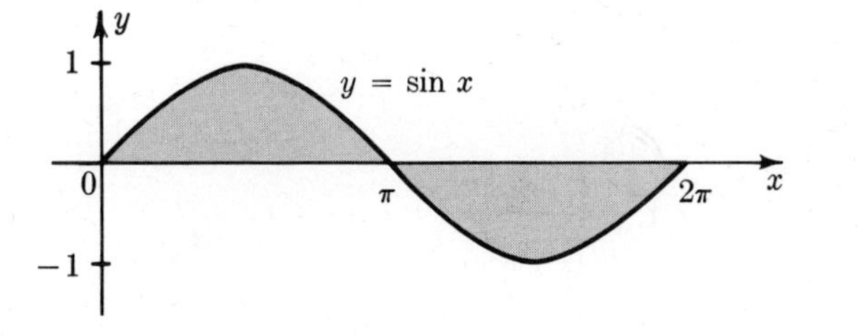

FIG. 4.4

Answer: 0. The areas determined by the curve $y = \sin x$ above and below the x-axis are equal and cancel each other.

EXAMPLE 4.3

One can of paint will cover 50 ft². How much paint is needed to cover the region bounded by $y = \sin x$ and the x-axis between $x = 0$ and $x = 2\pi$? Assume the unit on each axis is one yard.

Solution: If we treat the problem carelessly and integrate $\sin x$ between 0 and 2π, we reach a ridiculous conclusion: it takes no paint at all to cover the region! But remember, the integral

$$\int_a^b f(x)\,dx$$

gives the actual area under $y = f(x)$ only if $f(x) \geq 0$ between a and b.

Look at Fig. 4.4. The two humps of the curve have equal area; we can compute the area of the hump between 0 and π, then double the result.

$$\int_0^\pi \sin x\,dx = (-\cos \pi) - (-\cos 0) = 1 - (-1) = 2.$$

Thus the area of one hump is 2 yd² = 18 ft²; the total area is 36 ft². This will require $\frac{36}{50} = 0.72$ cans of paint.

Answer: 0.72 cans of paint.

EXERCISES

Evaluate:

1. $\displaystyle\int_{-1}^{2} x\,dx$
2. $\displaystyle\int_0^2 (1 - x)\,dx$
3. $\displaystyle\int_0^\pi \sin 2x\,dx$
4. $\displaystyle\int_{\pi/2}^{3\pi/2} \cos 3x\,dx$
5. $\displaystyle\int_{-2}^{-1} \frac{1}{x^2}\,dx$
6. $\displaystyle\int_{-4}^{-2} \left(\frac{1}{x^2} + x\right) dx$
7. $\displaystyle\int_0^{3\pi} (\cos x + \sin x)\,dx$
8. $\displaystyle\int_0^\pi (3 \sin 3x + 2 \cos 2x)\,dx$
9. $\displaystyle\int_{-2}^{2} (3 + 2x - x^2)\,dx$
10. $\displaystyle\int_{-1}^{2} (x - 1)(3x + 1)\,dx$
11. $\displaystyle\int_1^2 \left(\frac{5}{x^2} + e^{1-x}\right) dx$
12. $\displaystyle\int_0^1 (e^{3x} - x^2)\,dx$
13. $\displaystyle\int_a^b (b - x)(x - a)\,dx$
14. $\displaystyle\int_{-a}^{a} (x - a)^2\,dx.$

Compute the geometric area bounded by:

15. $f(x) = \cos 2x$ and the x-axis; $-\frac{\pi}{4} \leq x \leq \frac{\pi}{2}$

16. $f(x) = \sin 4x$ and the x-axis; $0 \leq x \leq \frac{\pi}{2}$

17. $y = x^2 - 5x + 6$ and the x-axis between $x = 0$ and $x = 3$

18. $y = x^2 - 7x + 12$, the x-axis, and the lines $x = 3$ and $x = 5$.

19. Show $\int_a^b + \int_b^c = \int_a^c$ holds even if $b < a$ or $b > c$.

20. Consider the "calculation" $\int_{-1}^{1} \frac{dx}{x^2} = -\frac{1}{x}\Big|_{-1}^{1} = -1 - 1 = -2$. Why is the answer ridiculous? Why is the calculation ridiculous?

5. AREAS AND AVERAGES

The problem of computing the area of a region under the graph of a positive function $y = f(x)$ can be solved by antidifferentiation. Here is another useful approach to the same problem.

Slice the region into a number of thin pieces, each approximately rectangular (Fig. 5.1). The area is approximately the total area of the rec-

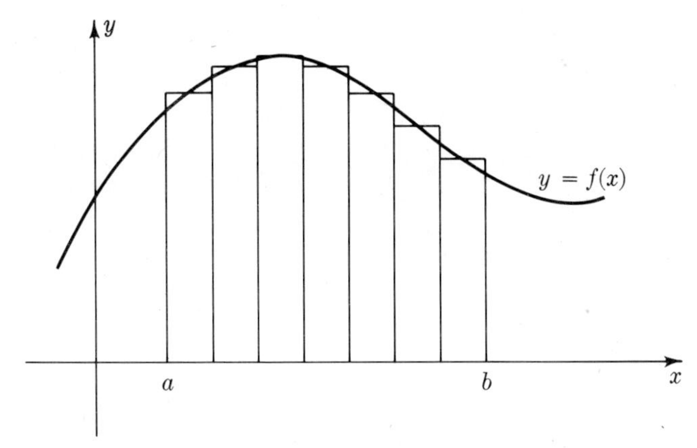

FIG. 5.1

tangles shown in Fig. 5.1. Slicing the region into a larger number of nearly rectangular pieces produces a better estimate of the area (Fig. 5.2).

> The area under a curve can be approximated as closely as desired by the combined areas of rectangles, provided the rectangles are thin enough.

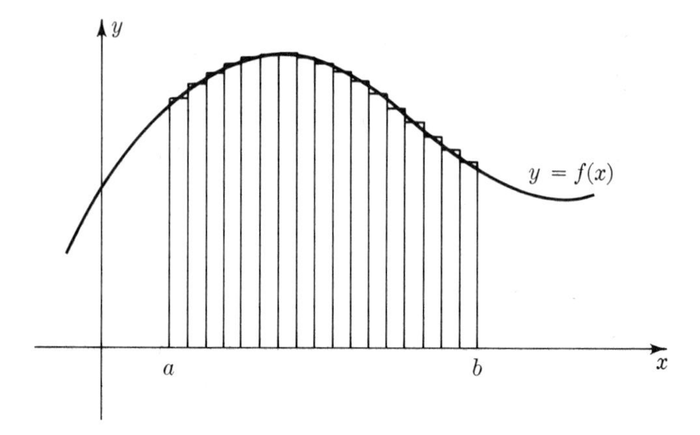

FIG. 5.2

We now translate into mathematical language this process of slicing and approximating by rectangles.

Start by dividing the interval from $x = a$ to $x = b$ into a number of equal parts (subintervals):

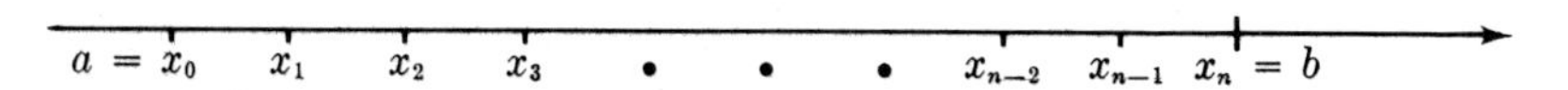

On the first subinterval, which runs from x_0 to x_1, construct a rectangle with that subinterval as its base. For its height choose the value of $f(x)$ at the midpoint $\bar{x}_1$. See Fig. 5.3. The area of this rectangle is $f(\bar{x}_1)\ (x_1 - x_0)$, a good approximation to the area under $y = f(x)$ between x_0 and x_1.

Now repeat the process in the second subinterval. Construct a rectangle whose base is the interval from x_1 to x_2 and whose height is $f(\bar{x}_2)$, where $\bar{x}_2$ is

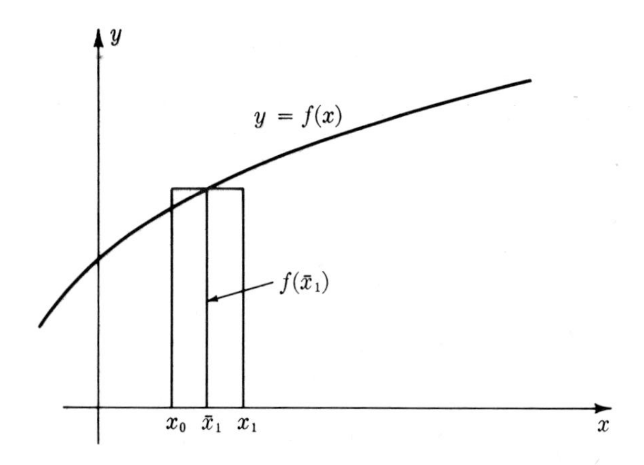

FIG. 5.3

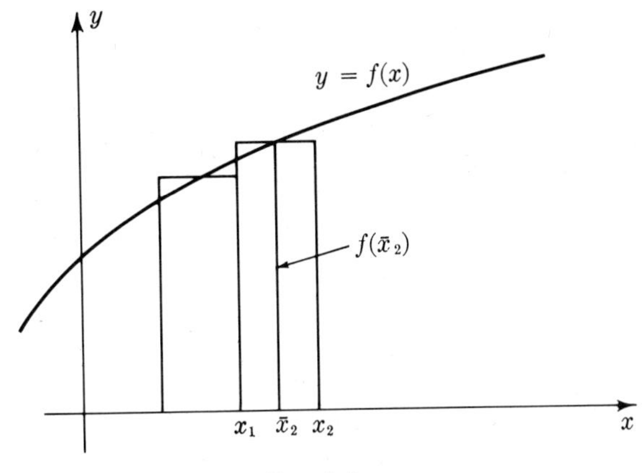

FIG. 5.4

the midpoint (Fig. 5.4). The area of the second rectangle is $f(\bar{x}_2)(x_2 - x_1)$. Continue in this way, constructing a collection of rectangles whose combined areas approximate the area under $y = f(x)$ between $x = a$ and $x = b$. Then

$$\text{Area} = \int_a^b f(x)\,dx \approx \text{(total area of rectangles)}$$
$$= f(\bar{x}_1)(x_1 - x_0) + f(\bar{x}_2)(x_2 - x_1) + \cdots + f(\bar{x}_n)(x_n - x_{n-1}).$$

The quantities $(x_1 - x_0)$, $(x_2 - x_1)$, $(x_3 - x_2)$, $\cdots$ are all equal, being the lengths of the subintervals. Denote their common value $(b - a)/n$ by Δx. With this notation,

$$\int_a^b f(x)\,dx \approx [f(\bar{x}_1) + f(\bar{x}_2) + \cdots + f(\bar{x}_n)]\,\Delta x.$$

We now have a mathematical translation of the statement following Fig. 5.2:

> If $f(x) \geq 0$ for $a \leq x \leq b$, then the sum
>
> $$[f(\bar{x}_1) + f(\bar{x}_2) + \cdots + f(\bar{x}_n)]\,\Delta x$$
>
> is as close as desired to
>
> $$\int_a^b f(x)\,dx,$$
>
> provided $\Delta x = (b - a)/n$ is small enough.

The statement above has a useful interpretation. Divide the interval from a to b into n equal pieces. Then

$$\int_a^b f(x)\,dx \approx [f(\bar{x}_1) + f(\bar{x}_2) + \cdot\cdot\cdot + f(\bar{x}_n)]\,\Delta x$$

$$= [f(\bar{x}_1) + f(\bar{x}_2) + \cdot\cdot\cdot + f(\bar{x}_n)]\,\frac{b-a}{n}$$

$$= \left[\frac{f(\bar{x}_1) + f(\bar{x}_2) + \cdot\cdot\cdot + f(\bar{x}_n)}{n}\right](b-a).$$

The quantity

$$\frac{f(\bar{x}_1) + f(\bar{x}_2) + \cdot\cdot\cdot + f(\bar{x}_n)}{n}$$

is the average of the numbers $f(\bar{x}_1)$, $f(\bar{x}_2)$, $\cdot\cdot\cdot$, $f(\bar{x}_n)$, where the points $\bar{x}_1$, $\bar{x}_2$, $\cdot\cdot\cdot$, $\bar{x}_n$ are evenly distributed through the interval from a to b.

$a \quad \bar{x}_1 \quad \bar{x}_2 \quad \bar{x}_3 \quad \bullet \quad \bullet \quad \bullet \quad \bar{x}_{n-2} \quad \bar{x}_{n-1} \quad \bar{x}_n \quad b$

Thus the integral is approximately $(b - a)$ times the average of a large number of values of the function $f(x)$. The integral is trying to tell us the "average value" of $f(x)$ between a and b.

$$\int_a^b f(x)\,dx = (b-a)\cdot(\text{average value of } f(x) \text{ for } a \le x \le b).$$

This should not be surprising. It says that the area of a region under a curve (Fig. 5.5a) equals the area of a rectangle (Fig. 5.5b) constructed on the same base with height equal to the "average height" of the curve.

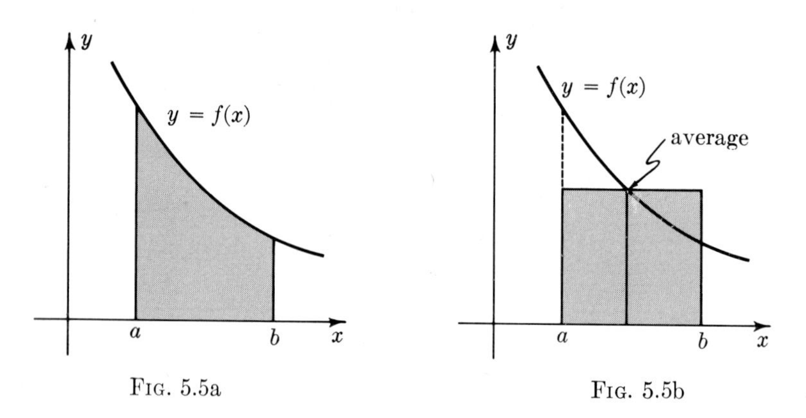

Fig. 5.5a

Fig. 5.5b

EXAMPLE 5.1

The hunger h of a shark of weight x tons is $h = x^2$. What is the average hunger of sharks weighing from 1 to 3 tons?

Solution: By the preceding formula,

$$\int_1^3 x^2\,dx = (3-1)\,(\text{average value of } x^2 \text{ for } 1 \le x \le 3).$$

$$\text{Average value} = \frac{1}{3-1}\int_1^3 x^2\,dx$$

$$= \frac{1}{2}\cdot\frac{x^3}{3}\bigg|_1^3 = \frac{1}{2}\,\frac{3^3-1^3}{3} = \frac{13}{3}.$$

Answer: $\frac{13}{3}$ units of hunger.

Functions Not Necessarily Positive

The sum

$$[f(\bar{x}_1) + f(\bar{x}_2) + \cdots + f(\bar{x}_n)]\,\Delta x$$

can be formed for any function $f(x)$, not only for one with positive values. Suppose for example $f(x) \le 0$ when $a \le x \le b$. On the one hand,

$$\int_a^b f(x)\,dx = -(\text{area of } R),$$

where R is the region shown in Fig. 5.6. On the other hand, if R is approxi-

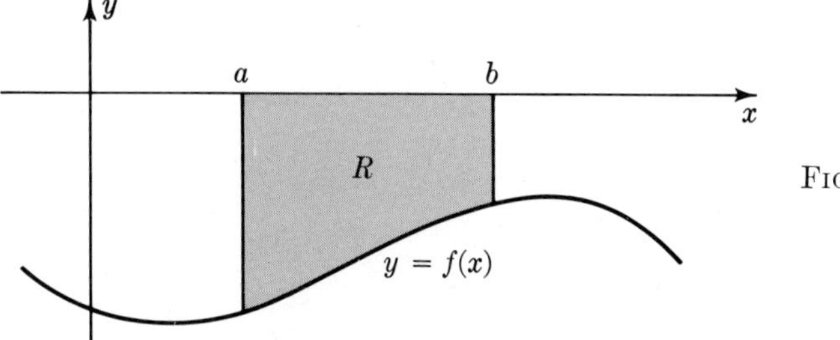

FIG. 5.6

mated by thin rectangles, the area of a typical rectangle is $-f(\bar{x}_i)\,\Delta x$, since $f(\bar{x}_i)$ is negative. Therefore

$$[f(\bar{x}_1) + f(\bar{x}_2) + \cdots + f(\bar{x}_n)]\,\Delta x \approx -(\text{area of } R)$$
$$= \int_a^b f(x)\,dx.$$

In general, the following assertion holds:

> The sum
>
> $$[f(\bar{x}_1) + f(\bar{x}_2) + \cdots + f(\bar{x}_n)]\,\Delta x$$
>
> is as close as desired to
>
> $$\int_a^b f(x)\,dx,$$
>
> provided $\Delta x = (b - a)/n$ is small enough.

We have seen that this statement is true if either $f(x) \ge 0$ or $f(x) \le 0$ throughout the interval from a to b. If $f(x)$ has both positive and negative values, the interval can be divided into subintervals on which either $f(x) \geq 0$ or $f(x) \leq 0$. The statement applies to each subinterval and consequently to the full interval.

Even if $f(x)$ has both positive and negative values, it makes sense to speak of the average value of $f(x)$ between $x = a$ and $x = b$.

> The **average value** of a function $f(x)$ for $a \leq x \leq b$ is
> $$\frac{1}{b-a}\int_a^b f(x)\,dx.$$

EXAMPLE 5.2

Find the average value of $\cos x$ for $0 \leq x \leq \pi$.

Solution: The average value is

$$\frac{1}{\pi - 0}\int_0^\pi \cos x\,dx = \frac{1}{\pi}[F(\pi) - F(0)],$$

where $F(x)$ is any antiderivative of $\cos x$. One antiderivative is $F(x) = \sin x$, hence

$$\text{Average} = \frac{1}{\pi}[\sin \pi - \sin 0] = 0.$$

Answer: 0.

REMARK: The answer seems reasonable from a graph of $y = \cos x$. See Fig. 5.7. The positive and negative values of $\cos x$ cancel each other, so the average value of $\cos x$ is zero.

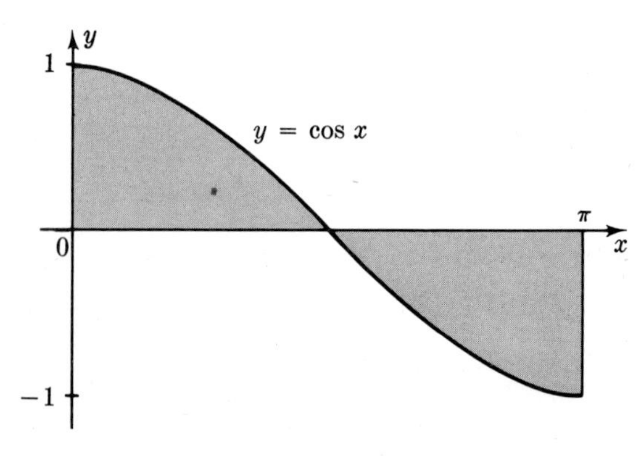

FIG. 5.7

EXERCISES

Find the average value of $y = f(x)$ between $x = a$ and $x = b$:

1. $f(x) = x$; $a = 0, b = 2$
2. $f(x) = x + 1$; $a = 0, b = 2$
3. $f(x) = x^2$; $a = -1, b = 1$
4. $f(x) = 4 - x^2$; $a = -2, b = 2$
5. $f(x) = \sin \pi x$; $a = 0, b = 2$
6. $f(x) = \sin 3\pi x$; $a = 0, b = 2$
7. $f(x) = (x - 4)^2$; $a = 0, b = 4$
8. $f(x) = 1/x^2$; $a = 1, b = 3$.

Approximate the average value of $y = f(x)$ between $x = a$ and $x = b$ by taking the average of the n numbers $f(\bar{x}_1), f(\bar{x}_2), \cdots, f(\bar{x}_n)$. Compare your answer to the average value obtained by integration:

9. $f(x) = x^2$; $a = 0, b = 1, n = 4$
10. $f(x) = x^3$; $a = 0, b = 1, n = 4$
11. $f(x) = \sin x$; $a = 0, b = \frac{\pi}{2}, n = 3$
12. $f(x) = \cos x$; $a = 0, b = \frac{\pi}{2}, n = 3$.
13. If x shares of a certain stock are sold, the market value in dollars per share is $37 + [25 \times 10^5/(x + 500)^2]$. What is the average value per share on sales of 0 to 2000 shares?
14. What is the average area of circles with radius between 1 and 2 ft?
15. The rainfall per day, measured in inches, x days after the beginning of the year, is $0.00002(6511 + 366x - x^2)$. By integration, estimate the average daily rainfall for the first 60 days of the year.
16. A car starting from rest accelerates 11 ft/sec^2. Find its average speed during the first 10 sec.

Compare the average value of $f(x)$ for $a \leq x \leq b$ to the average of the two numbers $f(a)$ and $f(b)$:

17. $f(x) = \sin x$; $a = 0, b = \pi$
18. $f(x) = -x^2 + 3x - 2$; $a = 1, b = 2$
19. $f(x) = e^x$; $a = 0, b = 1$
20. $f(x) = 1/x^2$; $a = -3, b = -1$
21. $f(x) = 3x - 2$; $a = -5, b = 10$
22. $f(x) = \sin x + x$; $a = -2, b = 2$.

9. Applications of Integration

1. INTRODUCTION

The definite integral

$$\int_a^b f(x)\,dx$$

was introduced in Chapter 8 to compute areas. It turns out, however, to be a powerful tool not only in area problems, but in a surprisingly large number of other applications.

Basically the reason is this. The area of a region under the curve $y = f(x)$ can be sliced into a large number of thin pieces, each approximately a rectangle of area $f(\bar{x}_i)\,\Delta x$. The integral "adds up" these pieces. In many other situations, a complicated quantity can be divided into a large number of small parts, each given by $f(\bar{x}_i)\,\Delta x$. The integral "adds up" these parts just as it does for area.

Here are four examples.

(1) AREA. Slice the region from a to b under a curve $y = f(x)$ into thin pieces, each approximately a rectangle of area

$$(\text{height}) \cdot (\text{base}) = f(\bar{x}_i)\,\Delta x.$$

The integral

$$\int_a^b f(x)\,dx$$

"adds up" these areas.

Even the notation is suggestive: $f(x)\,dx$ represents $f(\bar{x}_i)\,\Delta x$, and the symbol

$$\int_a^b f(x)\,dx$$

means "sum" the quantities $f(x)\,dx$ between $x = a$ and $x = b$. The integral sign "$\int$" was originally an "S" for sum.

(2) DISTANCE. If a particle moves to the right along the x-axis with velocity $v(t)$ at time t, how far does it move between $t = a$ and $t = b$?

Divide the time interval into a large number of very short equal time intervals, each of duration Δt. In the i-th short time interval, the velocity is practically constant, so the distance traveled in this short period of time is approximately

$$(\text{velocity}) \cdot (\text{time}) = v(\bar{t}_i)\,\Delta t.$$

The integral

$$\int_a^b v(t)\,dt$$

"adds up" all these little distances and gives the overall distance traveled.

(3) WORK. Suppose at each point of the x-axis there is a force of magnitude $f(x)$ pulling a particle. How much work is done by the force in moving the particle from $x = a$ to $x = b$?

Slice the interval from a to b into a large number of small pieces of length Δx. In the i-th piece the force is nearly constant, so the work it does there is approximately

$$(\text{force}) \cdot (\text{distance}) = f(\bar{x}_i)\,\Delta x.$$

The integral

$$\int_a^b f(x)\,dx$$

"adds up" these little bits of work and gives the total work done.

(4) VOLUME. A base is being shaped on a potter's wheel (Fig. 1.1). For

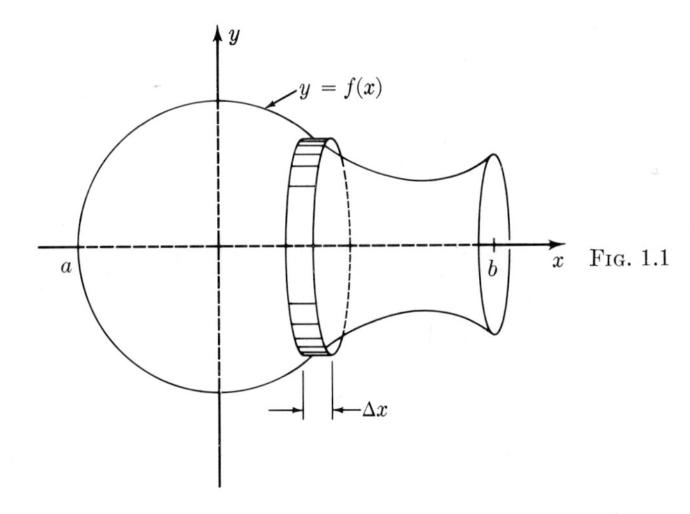

FIG. 1.1

each x between a and b, its cross-section is a circle of radius $f(x)$. What is the volume of the vase?

Slice the vase into thin slabs by cuts perpendicular to the x-axis. Each slab is nearly a cylindrical disk of volume

$$(\text{area of base}) \cdot (\text{thickness}) = [\pi f^2(\bar{x}_i)]\,\Delta x.$$

The integral

$$\int_a^b \pi f^2(x)\,dx$$

"adds up" these small volumes and gives the total volume of the vase.

Summary

The integral "adds up" many small quantities:

AREA = sum of thin rectangles of area $f(\bar{x}_i)\,\Delta x$,
DISTANCE = sum of short distances, $v(\bar{t}_i)\,\Delta t$,
WORK = sum of small amounts of work, $f(\bar{x}_i)\,\Delta x$,
VOLUME = sum of thin cylindrical disks of volume $\pi f^2(\bar{x}_i)\,\Delta x$.

2. AREA

EXAMPLE 2.1

Find the area of the region bounded by two curves $y = f(x)$ and $y = g(x)$, where $f(x) > g(x)$, and the lines $x = a$ and $x = b$. See Fig. 2.1.

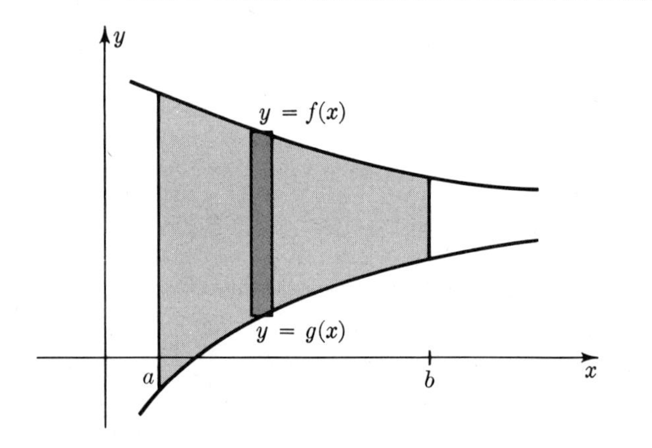

FIG. 2.1

Solution: Think of the region as approximately a large number of thin rectangles. A typical one, shown in Fig. 2.1, has height $[f(\bar{x}_i) - g(\bar{x}_i)]$, width Δx, and area $[f(\bar{x}_i) - g(\bar{x}_i)]\,\Delta x$, which we shall abbreviate as above by $[f(x) - g(x)]\,dx$. The integral

$$\int_a^b [f(x) - g(x)]\,dx$$

"adds up" these areas.

Answer: $\int_a^b [f(x) - g(x)]\,dx.$

REMARK: It is important to know which curve is the upper boundary and which is the lower boundary. If we had reversed them, then the answer would have been

$$\int_a^b [g(x) - f(x)]\,dx,$$

which is the negative of the area. If the curves cross, the upper and lower

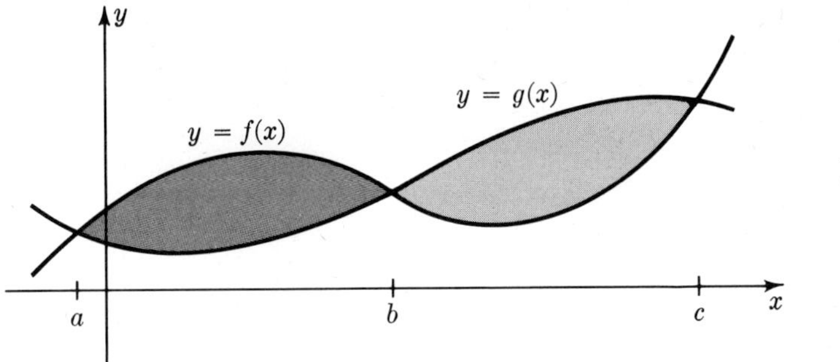

FIG. 2.2

boundaries reverse (Fig. 2.2). The shaded area is computed by

$$\int_a^b [f(x) - g(x)]\,dx + \int_b^c [g(x) - f(x)]\,dx.$$

Under each integral sign, the upper curve is written first. If we compute just

$$\int_a^c [f(x) - g(x)]\,dx,$$

the two areas will be counted with opposite signs and may cancel each other.

EXAMPLE 2.2

Compute the area of the region bounded by the curves $y = e^{x/2}$ and $y = 1/x^2$, and the lines $x = 2$ and $x = 3$.

Solution: Sketch the region (Fig. 2.3). Think of the region between

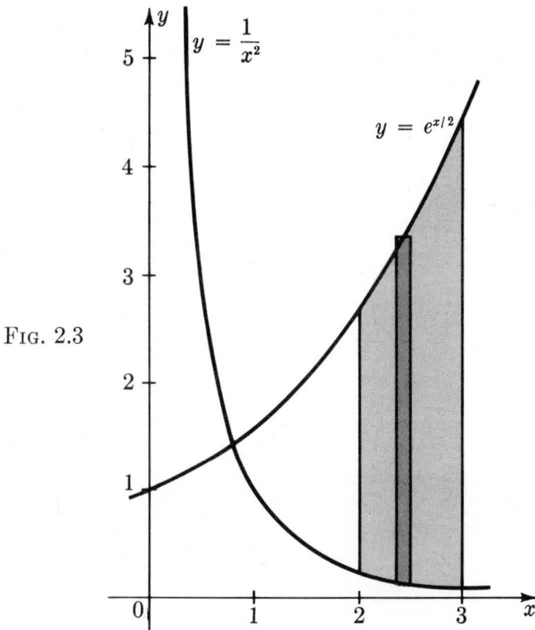

FIG. 2.3

$x = 2$ and $x = 3$ as being composed of rectangular slabs. The area of a typical slab is

$$\left(e^{x/2} - \frac{1}{x^2}\right) dx.$$

Hence

$$\text{Area} = \int_2^3 \left(e^{x/2} - \frac{1}{x^2}\right) dx = F(3) - F(2),$$

where $F(x)$ is an antiderivative of $e^{x/2} - 1/x^2$. Since

$$\frac{d}{dx}(e^{x/2}) = \frac{1}{2}e^{x/2},$$

an antiderivative of $e^{x/2}$ is $2e^{x/2}$. Since

$$\frac{d}{dx}\left(\frac{1}{x}\right) = -\frac{1}{x^2},$$

an antiderivative of $-1/x^2$ is $1/x$. Therefore an antiderivative of $e^{x/2} - 1/x^2$ is $F(x) = 2e^{x/2} + 1/x$. Consequently

$$\begin{aligned}\text{Area} = F(3) - F(2) &= \left(2e^{3/2} + \frac{1}{3}\right) - \left(2e + \frac{1}{2}\right)\\ &= 2(e^{3/2} - e) - \frac{1}{6}\\ &= 2e(e^{1/2} - 1) - \frac{1}{6}.\end{aligned}$$

Answer: $2e(e^{1/2} - 1) - \dfrac{1}{6}$
≈ 3.360 square units.

EXAMPLE 2.3

Compute the area of the region bounded by $y = x^2 - 4$ and $y = \frac{1}{2}x + 1$.

Solution: It is not clear from the statement of the problem what the region is; a graph is indispensable. Make a rough sketch (Fig. 2.4). The region is a parabolic segment. But the figure is not accurate enough; the crucial points P and Q must be known. They can be computed by solving simultaneously

$$\begin{cases} y = x^2 - 4 \\ y = \dfrac{1}{2}x + 1. \end{cases}$$

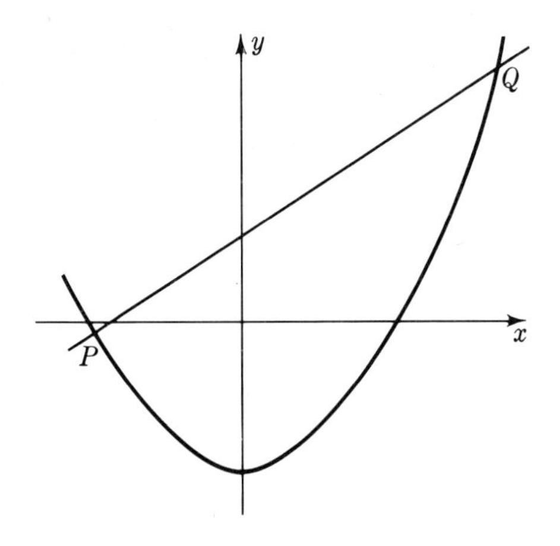

Fig. 2.4

Eliminate y:

$$x^2 - 4 = \frac{1}{2}x + 1,$$

$$2x^2 - x - 10 = 0.$$

Factor:

$$(2x - 5)(x + 2) = 0,$$

$$x = -2, \quad \frac{5}{2}.$$

The corresponding y values are 0 and 9/4. Therefore $P = (-2, 0)$ and $Q = (\frac{5}{2}, \frac{9}{4})$. Now Fig. 2.4 can be replaced by a more accurate sketch (Fig. 2.5).

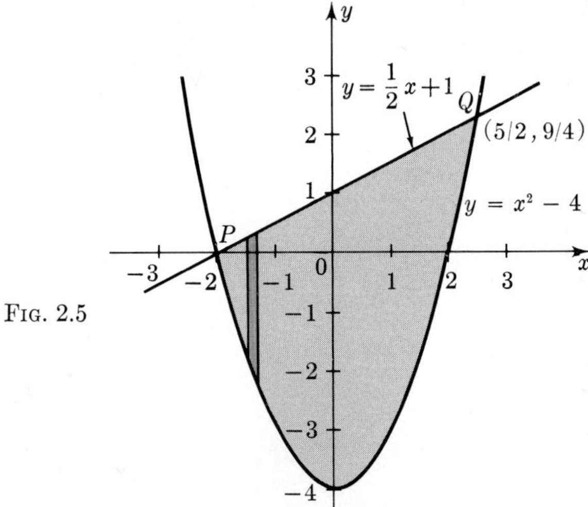

Fig. 2.5

Slice the region between $x = -2$ and $x = \frac{5}{2}$ into thin vertical rectangles. The area of a typical rectangle is

$$\left[\left(\frac{1}{2}x + 1\right) - (x^2 - 4)\right] dx = \left(5 + \frac{1}{2}x - x^2\right) dx.$$

Hence

$$\begin{aligned}
\text{Area} &= \int_{-2}^{5/2} \left(5 + \frac{1}{2}x - x^2\right) dx \\
&= \left(5x + \frac{1}{4}x^2 - \frac{1}{3}x^3\right)\Bigg|_{-2}^{5/2} \\
&= \left[5\left(\frac{5}{2}\right) + \frac{1}{4}\left(\frac{5}{2}\right)^2 - \frac{1}{3}\left(\frac{5}{2}\right)^3\right] - \left[5(-2) + \frac{1}{4}(-2)^2 - \frac{1}{3}(-2)^3\right] \\
&= \frac{729}{48}.
\end{aligned}$$

Answer: $\dfrac{243}{16}$ square units.

EXERCISES

Compute the area of the region bounded by:

1. $y = 8 - x^2$ and $y = -2x$
2. $y = x^2 + 5$ and $y = 6x$
3. $y = 3x^2$ and $y = -3x^2$, and the lines $x = -1$ and $x = 1$
4. $y = 1 - x^2$ and $y = x^2 - 1$
5. $y = x^3 - 5x^2 + 6x$ and $y = x^3$
6. $y = x^2 - 8x$ and $y = x$
7. $y = \cos x$ and $y = \sin x$, between $x = \dfrac{\pi}{4}$ and $x = \dfrac{5\pi}{4}$
8. $y = 2 \sin \dfrac{\pi}{4} x$ and $y = (x - 3)(x - 4)$, between $x = 2$ and $x = 4$
9. $y = e^x$, $y = e^{-x}$, and $y = e^2$
10. $y = \cos x - 1$ and $y = 1 - \cos 2x$, between $x = 0$ and $x = 2\pi$
11. $y = \dfrac{1}{x^2}$, $y = x$, and $y = 8x$ (*Hint:* sketch carefully)
12. $y = \dfrac{1}{x^2}$, $y = 0$, $y = x^2$, and $x = 3$.
13. Find a so that the area bounded by $y = x^2 - a^2$ and $y = a^2 - x^2$ is 9.
14. Find the fraction of the area of one hump of the curve $y = \sin x$ that lies above $y = \frac{1}{2}$.

3. DISTANCE

EXAMPLE 3.1

An α particle enters an accelerator at time $t = 0$. After t sec, its velocity is 10^7t^2 m/sec. How far does the particle move during the first 10^{-2} sec?

Solution: Think of the time interval $t = 0$ to $t = 10^{-2}$ sec divided into tiny subintervals. In a typical subinterval of duration Δt, the velocity is nearly constant, 10^7t^2. The distance moved during this short period is approximately $10^7t^2\,\Delta t$. The integral

$$\int_0^{10^{-2}} 10^7t^2\,dt$$

adds up these tiny distances.

$$(\text{total distance}) = \int_0^{10^{-2}} 10^7t^2\,dt = 10^7 \times \frac{1}{3}t^3\Bigg|_0^{10^{-2}} = \frac{10}{3}\ \text{m}.$$

Answer: $\frac{10}{3}$ m.

EXAMPLE 3.2

A particle moving along the x-axis is at the origin when $t = 0$. At time t its velocity is $\frac{1}{(t+1)^2}$, which is rapidly decreasing. Show that the particle will never pass the point $x = 1$.

Solution: The distance moved between $t = 0$ and $t = b$ is

$$\int_0^b v(t)\,dt = \int_0^b \frac{1}{(t+1)^2}\,dt = F(b) - F(0),$$

where $F(t)$ is an antiderivative of $1/(t+1)^2$. Since

$$\frac{d}{dt}\left(-\frac{1}{t}\right) = \frac{1}{t^2},$$

it follows from the Shifting Rule that

$$\frac{d}{dt}\left(\frac{-1}{t+1}\right) = \frac{1}{(t+1)^2}.$$

Therefore,

$$\int_0^b \frac{1}{(t+1)^2}\,dt = \frac{-1}{t+1}\Bigg|_0^b = \frac{-1}{b+1} - (-1) = 1 - \frac{1}{b+1}.$$

$$\text{Distance moved} = 1 - \frac{1}{b+1}.$$

Answer: At any time $t = b$, no matter how far in the future, the particle will have moved less than one unit. Hence it never passes the point $x = 1$.

Negative Velocity

The formula

$$\text{distance} = \int_a^b v(t)\,dt$$

is valid only when the velocity $v(t)$ is positive during motion. If $v(t) \leq 0$ between $t = a$ and $t = b$, then

$$\int_a^b v(t)\,dt$$

is the negative of the distance traveled. (We hope this reminds you of area bounded by a curve lying below the axis.)

The reason is simple: if $v(t)$ is negative, then the distance traveled during an interval of length Δt is not $v(t)\,\Delta t$, but $|v(t)|\,\Delta t = -v(t)\,\Delta t$. Therefore the integral adds up little quantities $v(t)\,\Delta t$, which are the negatives of little distances. Its value then is the negative of the distance traveled.

If $v(t) \geq 0$ when $a \leq t \leq b$, and $v(t) \leq 0$ when $b \leq t \leq c$, then

$$\int_a^c v(t)\,dt = \int_a^b v(t)\,dt + \int_b^c v(t)\,dt,$$

where the first integral on the right is the distance moved between $t = a$ and $t = b$, and the second is the negative of the distance moved between $t = b$ and $t = c$. For example, if a particle moves to the right between $t = a$ and $t = b$, then moves to the left returning to its original position when $t = c$,

$$\int_a^c v(t)\,dt = 0$$

because

$$\int_a^b v(t)\,dt \qquad \text{and} \qquad \int_b^c v(t)\,dt$$

cancel each other. So

$$\int_a^c v(t)\,dt$$

measures not necessarily distance traveled, but the **displacement**, the difference between the initial position and final position. If $v(t) \geq 0$ throughout the motion, then displacement and distance traveled are the same, but if $v(t)$ changes sign, they are different.

The integral of the velocity

$$\int_a^b v(t)\,dt$$

gives the displacement between time $t = a$ and time $t = b$.

If $v(t) \geq 0$ through the motion, this displacement is the same as the distance traveled. If $v(t) \leq 0$ throughout the motion, it is the negative of the distance traveled.

The preceding assertions are really nothing new. If $F(t)$ denotes the position of the particle at time t, then $F'(t) = v(t)$; in other words $F(t)$ is an antiderivative of $v(t)$. Therefore,

$$\int_a^b v(t)\,dt = F(b) - F(a)$$

$$= (\text{position at time } b) - (\text{position at time } a) = \text{displacement}.$$

EXAMPLE 3.3

A ball is thrown up from the ground at $t = 0$. At time t its velocity is $80 - 32t$ ft/sec. How far does it travel in the first 5 sec?

Solution:

$$80 - 32t > 0 \qquad \text{when} \qquad t < \frac{5}{2},$$

$$80 - 32t < 0 \qquad \text{when} \qquad t > \frac{5}{2}.$$

The distance traveled in the first $\frac{5}{2}$ sec is

$$\int_0^{5/2} (80 - 32t)\,dt = (80t - 16t^2)\Big|_0^{5/2} = 100 \text{ ft}.$$

After $t = \frac{5}{2}$, the velocity is negative. Therefore, the distance traveled in the next $\frac{5}{2}$ sec is

$$-\int_{5/2}^{5} (80 - 32t)\,dt = (-80t + 16t^2)\Big|_{5/2}^{5} = 100 \text{ ft}.$$

Answer: 200 ft.

REMARK: Note in the example that

$$\int_0^5 v(t)\,dt = \int_0^5 (80 - 32t)\,dt = 0.$$

That is because the ball ends up where it started; its net displacement is zero. Actually it moves 100 ft with positive velocity (upwards) and 100 ft with negative velocity (downwards). The integral, however, counts these distances with opposite signs, so they cancel out.

EXERCISES

A body has velocity $v = f(t)$. Find the distance covered between $t = a$ and $t = b$:

1. $f(t) = 2t + 1;\quad a = 0,\ b = 2$
2. $f(t) = -32t;\quad a = 0,\ b = 8$
3. $f(t) = 3 \sin \pi t;\quad a = 0,\ b = 1$
4. $f(t) = 4 \cos \pi t;\quad a = 0,\ b = \dfrac{1}{2}$
5. $f(t) = t + \dfrac{4}{(t+1)^2};\quad a = 0,\ b = 2$
6. $f(t) = t^2 + \dfrac{2}{(3t+9)^2};\quad a = 1,\ b = 4$
7. $f(t) = 3t^2 + 4t + 1;\quad a = 0,\ b = 3$
8. $f(t) = 4t - t^2;\quad a = 0,\ b = 2.$

A body has velocity $v = v(t)$. Find the displacement of the body between $t = a$ and $t = b$:

9. $v(t) = 3t - 1;\quad a = 0,\ t = 1$
10. $v(t) = t - \dfrac{8}{t^2};\quad a = 1,\ b = 4$
11. $v(t) = 4 \sin \dfrac{\pi}{2} t;\quad a = 0,\ b = 3$
12. $v(t) = \cos \dfrac{\pi}{4} t;\quad a = 0,\ b = 8.$

If a body has velocity $v = f(t)$, find the total distance covered and the displacement between $t = a$ and $t = b$:

13. $f(t) = e^{t-2} - 1;\quad a = 1,\ b = 3$
14. $f(t) = t^2 - 3t + 2;\quad a = 0,\ b = 3$
15. $f(t) = \sin t + \cos t;\quad a = 0,\ b = 2\pi$
16. $f(t) = 1 - 2 \cos t;\quad a = 0,\ b = \pi.$

4. WORK

Suppose an object is moved along the x-axis from $x = a$ to $x = b$ by a force of magnitude $f(x)$. Divide the interval into a large number of short pieces, each of length Δx.

$$|\leftarrow \Delta x \rightarrow|$$

$$a = x_0 \quad x_1 \quad x_2 \quad x_3 \quad \bullet \quad \bullet \quad \bullet \quad x_{n-2} \quad x_{n-1} \quad x_n = b$$

The work done moving the object from x_{i-1} to x_i is approximately $f(\bar{x}_i)\,\Delta x$, because in this very short interval the force is nearly constant. These little pieces of work are added up by an integral:

$$\text{Work} = \int_a^b f(x)\, dx.$$

Remark Concerning Units: If force is measured in pounds and distance in feet, then work is measured in foot-pounds. If force is measured in dynes and distance in centimeters, then work is measured in dyne-centimeters.

EXAMPLE 4.1

At each point of the x-axis (marked off in feet) there is a force of $(5x^2 - x + 2)$ pounds pulling an object. Compute the work done in moving it from $x = 1$ to $x = 4$.

Solution:

$$\text{Work} = \int_1^4 (5x^2 - x + 2)\,dx = \left(\frac{5}{3}x^3 - \frac{1}{2}x^2 + 2x\right)\Bigg|_1^4$$

$$= \frac{320}{3} - \frac{19}{6}.$$

Answer: 103.5 ft-lb.

EXAMPLE 4.2

According to Newton's Law of Gravitation, two bodies attract each other with a force proportional to the product of their masses and inversely proportional to the square of the distance between them:

$$\text{force} = k\,\frac{Mm}{x^2},$$

where M and m are the masses, and x is the distance between them. If one of the bodies is fixed at the origin, how much work is done in moving the other from $x = a$ to $x = b$, where a and b are positive? (Assume k, M, and m are known.)

Solution:

$$\text{Work} = \int_a^b \frac{kMm}{x^2}\,dx = kMm\left(-\frac{1}{x}\right)\Bigg|_a^b.$$

Answer: $kMm\left(\frac{1}{a} - \frac{1}{b}\right)$ units of work.

REMARK: When $a < b$ the work is positive because you do work against the gravitational force. But when $a > b$ the free mass moves towards the fixed mass, opposite to your direction of pull. Hence you do negative work. Imagine moving the free mass from a to b and then back to a. The total work is zero. Why?

EXAMPLE 4.3

According to Hooke's Law, when a spring is stretched a short distance, there is a restoring force proportional to the amount of stretching. If 2 ft-lb of work are needed to stretch a certain spring 4 in., how much work is needed to stretch it 1 ft?

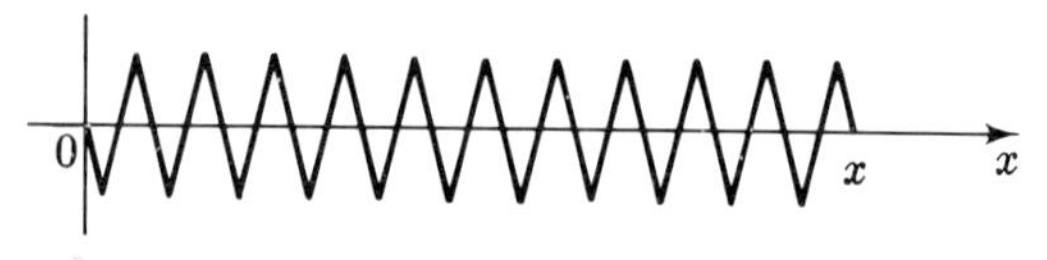

FIG. 4.1

Solution: Imagine the spring placed on the x-axis with its fixed end at the origin. Suppose it is stretched from the origin to x. See Fig. 4.1. By Hooke's Law there is a force of magnitude kx pulling the spring back to the origin. The work required to stretch it from $x = 0$ to $x = b$ is

$$\int_0^b kx\,dx = \frac{k}{2}x^2\Big|_0^b = \frac{kb^2}{2},$$

proportional to the square of the distance. Now 1 ft is 3×4 in., hence the work required to stretch the spring 1 ft is 3^2 times that required to stretch it 4 in.

Answer: $3^2 \times 2 = 18$ ft-lb.

EXERCISES

1. Find the work done by a force $f(x) = 3x + 2$ dynes in moving an object from $x = 1$ cm to $x = 7$ cm.
2. At each point of the x-axis (marked off in feet) there is a force of $(x^2 - 5x + 6)$ lb pushing to the right against an object. Compute the work done in moving it from $x = 1$ to $x = 5$.
3. A 50-ft chain weighing 2 lb/ft is attached to a drum hung from the ceiling. The ceiling is high enough so that the free end of the chain does not touch the floor. How much work is required to wind up the chain around the drum? [*Hint:* When x ft of chain have been wound up, the weight of the chain still unwound is $(100 - 2x)$ lb.]
4. In the previous exercise, suppose that a 200-lb weight is attached to the free end. How much work is required to wind up the chain?
5. The force in pounds required to stretch a certain spring x ft is $F = 8x$. How much work is required to stretch the spring 6 in., 1 ft, 2 ft?
6. A 3-lb force will stretch a spring 0.5 ft. How much work is required to stretch the spring 2 ft?

From Newton's Law of Gravitation, the force of attraction of the earth on an object of weight m lb at a distance x mi from the center of the earth, is $F = \dfrac{(4000)^2\,m}{x^2}$ lb.

7. How much work is required to lift a 1000-lb payload into a 300-mi-high orbit? a 600-mi-high orbit? Assume the radius of the earth is 4000 mi.
8. A 5-lb monkey is attached to the free end of a 20-ft hanging chain which weighs 0.25 lb/ft. The monkey climbs the chain to the top. How much work does he do?

10. Numerical Integration

1. RECTANGULAR APPROXIMATION

In many practical problems, the solution is expressed by a definite integral

$$\int_a^b f(x)\,dx.$$

The value of the integral is $F(b) - F(a)$, where $F(x)$ is an antiderivative of $f(x)$. But suppose you cannot find an antiderivative. The problem is still there, and your boss wants an answer. What do you do? Or suppose $f(x)$ is some physical quantity for which you have only a table of values, for instance velocities measured at intervals of 0.1 sec. Then what?

A practical procedure in problems of this type is to estimate

$$\int_a^b f(x)\,dx$$

if you cannot evaluate it exactly.

Recall that an integral is nearly a sum:

$$\int_a^b f(x)\,dx \approx [f(\bar{x}_1) + f(\bar{x}_2) + \cdots + f(\bar{x}_n)]\,\Delta x.$$

This approximate formula provides a method for estimating the integral. Divide the interval from a to b into n equal pieces of length Δx. Let $\bar{x}_1$, $\bar{x}_2$, $\cdots$ denote their midpoints:

$\bar{x}_1$ $\bar{x}_2$ $\bar{x}_3$ $\bar{x}_n$

a b

Compute $f(\bar{x}_1)$, $f(\bar{x}_2)$, $\cdots$, $f(\bar{x}_n)$. Add these values and multiply by Δx. The result is an approximation to the value of the integral.

Rectangular Rule

$$\int_a^b f(x)\,dx \approx [f(\bar{x}_1) + f(\bar{x}_2) + \cdots + f(\bar{x}_n)]\,\Delta x.$$

The interval from a to b is divided into n equal parts, each of length $\Delta x = (b - a)/n$. The points $\bar{x}_1$, $\bar{x}_2$, $\cdots$ are the midpoints of these parts.

Great accuracy can be achieved by choosing Δx very small, i.e., n very large. Of course the amount of computation increases with n, so a computer may be needed. Often the procedure becomes too laborious for hand computation when n exceeds 15 or 20, but a computer can handle values of n in the thousands.

EXAMPLE 1.1

Estimate $\int_0^4 (x-1)^2\,dx$ and compare to the exact value.

Solution: Divide the interval from 0 to 4 into n equal parts. A reasonable choice of n for hand computation is 16, which means $\Delta x = 0.25$.

0.125 0.625 0.375 3.875

0 1 2 3 4

Tabulate $f(x) = (x-1)^2$ at $\bar{x}_1 = 0.125$, $\bar{x}_2 = 0.375$, $\bar{x}_3 = 0.625$, etc. (Table 1.1).

TABLE 1.1

x	$f(x)$	x	$f(x)$	x	$f(x)$
0.125	0.765625	1.375	0.140625	2.625	2.640625
0.375	0.390625	1.625	0.390625	2.875	3.515625
0.625	0.140625	1.875	0.765625	3.125	4.515625
0.875	0.015625	2.125	1.265625	3.375	5.640625
1.125	0.015625	2.375	1.890625	3.625	6.890625
				3.875	8.265625

$$\int_0^4 (x-1)^2\,dx \approx [f(\bar{x}_1) + f(\bar{x}_2) + \cdots + f(\bar{x}_{16})]\,\Delta x = 9.31250.$$

The exact value of the integral is

$$\int_0^4 (x-1)^2\,dx = \left.\frac{(x-1)^3}{3}\right|_0^4 = \frac{28}{3} = 9.33333\cdots.$$

Notice that carrying 6 decimals in your work does *not* automatically guarantee 6-place accuracy. The answer, however, accurate to 1 decimal, is not bad considering the crude choice $\Delta x = 0.25$. A better estimate is obtained if $\Delta x = 0.1$. That requires more work since then $n = 40$, but still the computation can be done by hand. Here is the result:

$$\int_0^4 (x-1)^2\,dx \approx [f(0.05) + f(0.15) + \cdots + f(3.95)](0.1) = 9.33000,$$

which is closer.

For greater accuracy, use a computer, taking $n = 400$. Result:

$$\int_0^4 (x-1)^2\,dx \approx [f(0.005) + f(0.015) + \cdots + f(3.995)](0.01) \approx 9.33328,$$

correct to within 0.00006.

Answer: $\Delta x = 0.25$: 9.31250; $\Delta x = 0.1$: 9.33000;
$\Delta x = 0.01$: 9.33328; exact: $9\frac{1}{3} = 9.33333 \cdots$.

REMARK: Beware of false accuracy! The computer may give a 10-digit answer, regardless of the inaccuracy of the input data.

EXAMPLE 1.2

Estimate the area under the curve $y = \dfrac{1}{x}$ between $x = 1$ and $x = 3$.

Carry 4 significant digits in your work.

Solution: The area is given by the integral

$$\int_1^3 \frac{1}{x}\,dx.$$

To estimate it, divide the interval $1 \leq x \leq 3$ into 20 subintervals, each of length 0.1. Tabulate $f(x) = 1/x$ at the midpoints of these subintervals (Table 1.2).

TABLE 1.2

x	$f(x)$	x	$f(x)$	x	$f(x)$
1.05	0.9524	1.75	0.5714	2.45	0.4082
1.15	0.8696	1.85	0.5405	2.55	0.3922
1.25	0.8000	1.95	0.5128	2.65	0.3774
1.35	0.7407	2.05	0.4878	2.75	0.3636
1.45	0.6897	2.15	0.4651	2.85	0.3509
1.55	0.6452	2.25	0.4444	2.95	0.3390
1.65	0.6061	2.35	0.4255		

$$\int_1^3 \frac{1}{x}\,dx \approx [f(1.05) + f(1.15) + \cdots + f(2.95)](0.1) \approx 1.098.$$

For greater accuracy, use a computer, taking $n = 200$. Result:

$$\int_1^3 \frac{1}{x}\,dx \approx [f(1.005) + f(1.015) + \cdots + f(2.995)](0.01) \approx 1.098608.$$

Answer: $n = 20$: 1.098; $n = 200$: 1.098608.

More refined calculations show that to 6 significant digits the value is 1.09861.

EXERCISES

Estimate the integral by rectangular approximation, taking the suggested values of n. In each case compute the exact value of the integral and compare it to your estimate.

By hand carrying 4 places, $n = 4$ and $n = 10$:

1. $\int_1^2 (x^2 + x)\, dx$
2. $\int_0^4 (x + 1)^2\, dx$
3. $\int_0^2 e^x\, dx$
4. $\int_0^2 e^{-x}\, dx$;

by computer, $n = 10$ and $n = 100$:

5. $\int_1^2 \frac{1}{x^2}\, dx$
6. $\int_2^3 \left(x + \frac{1}{x^2}\right) dx$;

by computer, $n = 20$ and $n = 200$:

7. $\int_0^2 (1 + \sin x)\, dx$
8. $\int_0^2 3 \cos x\, dx$.

Estimate the integral by rectangular approximation, taking the suggested values of Δx, and carrying 4 places in your work:

9. $\int_{10}^{12} \log_{10} x\, dx;\quad \Delta x = 0.5,\ \Delta x = 0.2$
10. $\int_{10}^{12} x \log_{10} x\, dx;\quad \Delta x = 0.5,\ \Delta x = 0.2$
11. $\int_0^2 \frac{1}{x + 1}\, dx;\quad \Delta x = 0.2,\ \Delta x = 0.01$
12. $\int_0^2 \frac{1}{x^2 + 1}\, dx;\quad \Delta x = 0.2,\ \Delta x = 0.01$
13. $\int_0^2 x \sin x\, dx;\quad \Delta x = 0.2,\ \Delta x = 0.01$
14. $\int_0^2 x \cos x\, dx;\quad \Delta x = 0.2,\ \Delta x = 0.01$
15. $\int_2^4 \left(x + \frac{1}{x}\right) dx;\quad \Delta x = 0.2,\ \Delta x = 0.01$
16. $\int_1^5 \sqrt{x}\, dx;\quad \Delta x = 0.25,\ \Delta x = 0.01$.

2. TRAPEZOIDAL APPROXIMATION

The numerical integration technique of the last section was based on the approximation of areas by rectangles (Fig. 2.1a). Another useful way to approximate areas is by trapezoids (Fig. 2.1b).

Examine a typical trapezoid (Fig. 2.2). It has base Δx and sides $f(x_{i-1})$ and $f(x_i)$. Hence its area is

$$\frac{\Delta x}{2}[f(x_{i-1}) + f(x_i)].$$

Estimate the integral

$$\int_a^b f(x)\, dx$$

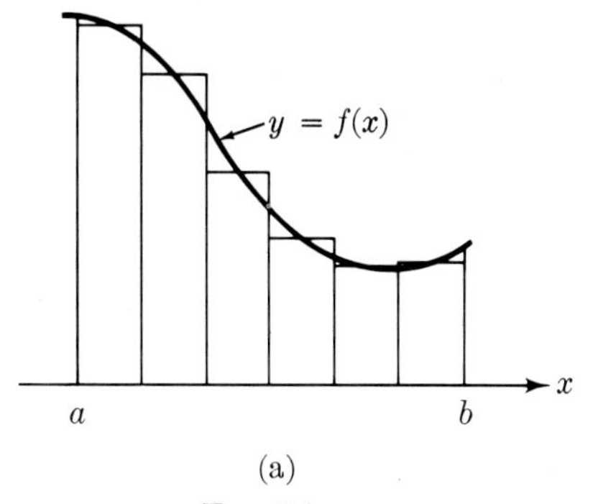

(a)
FIG. 2.1a

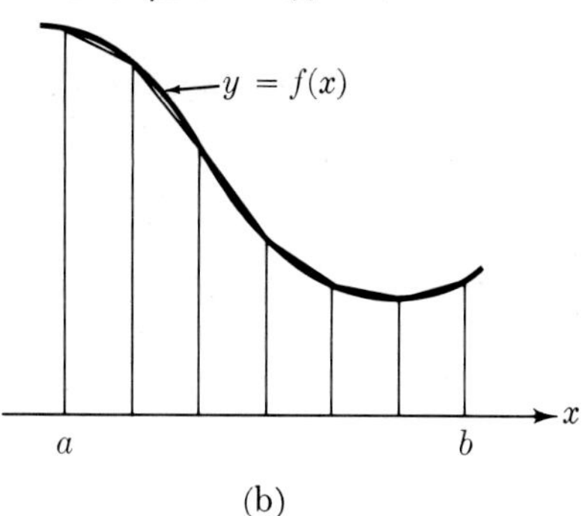

(b)
FIG. 2.1b

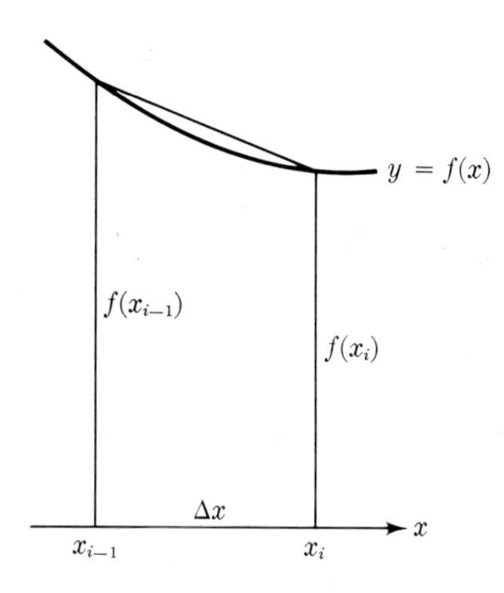

FIG. 2.2

by the combined areas of the trapezoids in Fig. 2.1b:

(total area of trapezoids)

$$= \frac{\Delta x}{2}[f(x_0) + f(x_1)] + \frac{\Delta x}{2}[f(x_1) + f(x_2)] + \cdots + \frac{\Delta x}{2}[f(x_{n-1}) + f(x_n)]$$

$$= \frac{\Delta x}{2}\{[f(x_0) + f(x_1)] + [f(x_1) + f(x_2)] + \cdots + [f(x_{n-1}) + f(x_n)]\}$$

$$= \frac{\Delta x}{2}\{f(x_0) + 2f(x_1) + 2f(x_2) + \cdots + 2f(x_{n-1}) + f(x_n)\}.$$

Trapezoidal Rule

$$\int_a^b f(x)\,dx \approx \frac{\Delta x}{2}\{f(x_0) + 2f(x_1) + 2f(x_2) + \cdots + 2f(x_{n-1}) + f(x_n)\},$$

where $x_0, x_1, \cdots, x_n$ are equally spaced points dividing the interval from $a = x_0$ to $b = x_n$ into n equal parts, each of length $\Delta x = (b - a)/n$.

Each quantity $f(x_i)$, except the first and the last, occurs with coefficient 2. That is because each vertical line in Fig. 2.1b, except the first and the last, is a side of two trapezoids.

Like rectangular approximation, trapezoidal approximation is a simple procedure well-adapted to hand and computer use. It can be made as accurate as desired by choosing Δx very small, i.e., n very large. It has two advantages over the rectangular method, however.

First, the arithmetic is usually easier in hand computation. For example, suppose you estimate

$$\int_0^1 f(x)\,dx$$

with $\Delta x = 0.1$. You must compute $f(0)$, $f(0.1)$, $f(0.2)$, $\cdots$ when using trapezoids; you must compute $f(0.05)$, $f(0.15)$, $f(0.25)$, $\cdots$ when using rectangles. The latter computations generally involve more arithmetic.

Second, and more important, the trapezoid method is superior when the function is given by a table. For example, if $f(0)$, $f(0.1)$, $f(0.2)$, $\cdots$, $f(1.0)$ are given, this data is adequate for trapezoidal approximation to

$$\int_0^1 f(x)\,dx$$

with $\Delta x = 0.1$. For rectangular approximation with this data, you must take $\Delta x = 0.2$ and use only $f(0.1)$, $f(0.3)$, $f(0.5)$, $f(0.7)$, $f(0.9)$. This approximation is inferior since it uses a cruder Δx and only part of the data.

We illustrate use of the Trapezoidal Rule by some examples.

EXAMPLE 2.1

Use the Trapezoidal Rule to estimate $\int_0^1 (x+1)(x-1)^2\,dx$.

Solution: For hand computation take $\Delta x = 0.1$, $n = 10$. Tabulate $f(x) = (x+1)(x-1)^2$ at $x = 0, 0.1, 0.2, \cdots, 1.0$. See Table 2.1.

TABLE 2.1

x	$f(x)$	x	$f(x)$
0.0	1.000	0.6	0.256
0.1	0.891	0.7	0.153
0.2	0.768	0.8	0.072
0.3	0.637	0.9	0.019
0.4	0.504	1.0	0.000
0.5	0.375		

$$\int_0^1 f(x)\,dx \approx \frac{(0.1)}{2}\,[f(0) + 2f(0.1) + 2f(0.2) + \cdots + 2f(0.9) + f(1.0)]$$

$$= 0.4175.$$

For greater accuracy, use a computer, taking $n = 100$:

$$\int_0^1 f(x)\,dx$$

$$\approx \frac{(0.01)}{2}[f(0) + 2f(0.01) + 2f(0.02) + \cdots + 2f(0.99) + f(1.00)]$$

$$\approx 0.416675.$$

Answer: $\Delta x = 0.1: 0.4175$; $\Delta x = 0.01: 0.416675$; exact: $\frac{5}{12} = 0.416666\cdots$.

EXAMPLE 2.2

Use the Trapezoidal Rule to estimate

$$\int_{0.5}^{1.0} \tan x\,dx.$$

Solution: For hand computation choose $\Delta x = 0.1$, $n = 5$, and use a table of tangents of angles in radians:

$$\int_{0.5}^{1.0} \tan x\,dx$$

$$\approx \frac{(0.1)}{2}[\tan(0.5) + 2\tan(0.6) + \cdots + 2\tan(0.9) + \tan(1.0)]$$

$$\approx 0.48681.$$

For greater accuracy, use a computer, taking $n = 50$:

$$\int_{0.5}^{1.0} \tan x\,dx$$

$$\approx \frac{(0.01)}{2}[\tan(0.50) + 2\tan(0.51) + \cdots + 2\tan(0.99) + \tan(1.00)]$$

$$\approx 0.48506.$$

(To 5 places, the actual value is 0.48504.)

Answer: $n = 5: 0.48681$; $n = 50: 0.48506$.

EXERCISES

Carry 4 significant digits in hand computations.

Estimate the integral by trapezoidal approximation, taking the suggested values of n. In each case compare the values with the exact value obtained by integrating:

1. $\int_0^4 (x-1)^2\,dx$; $n = 4$, $n = 10$

2. $\int_1^2 (x^2 + x)\,dx$; $n = 4$, $n = 10$

3. $\int_0^1 \sin x\, dx; \quad n = 10,\ n = 50$

4. $\int_0^1 (\sin x + \cos x)\, dx; \quad n = 10,\ n = 100.$

Estimate the integral by trapezoidal approximation, taking the suggested values of Δx:

5. $\int_0^{\pi/2} \sin^2 x\, dx; \quad \Delta x = \frac{\pi}{12},\ \Delta x = \frac{\pi}{180}$

6. $\int_0^{\pi/2} \cos^2 x\, dx; \quad \Delta x = \frac{\pi}{12},\ \Delta x = \frac{\pi}{180}$

7. $\int_0^1 (x-1)^2(x+1)^2\, dx; \quad \Delta x = 0.1,\ \Delta x = 0.01$

8. $\int_1^{1.5} (x^3 + x)\, dx; \quad \Delta x = 0.1,\ \Delta x = 0.005$

9. $\int_1^2 x\sqrt{x}\, dx; \quad \Delta x = 0.1,\ \Delta x = 0.01$

10. $\int_{10}^{12} x \log_{10} x\, dx; \quad \Delta x = 0.2,\ \Delta x = 0.01$

11. $\int_1^{1.5} (1 + \tan x)\, dx; \quad \Delta x = 0.1,\ \Delta x = 0.005$

12. $\int_0^{\pi/2} \sin\frac{x}{2}\, dx; \quad \Delta x = \frac{\pi}{12},\ \Delta x = \frac{\pi}{180}.$

Estimate the integral by rectangular and trapezoidal approximation. Compare the estimates to the exact value obtained by integrating. Use $n = 10$:

13. $\int_1^2 x^2\, dx$

14. $\int_{0.5}^{1.0} \frac{dx}{x^2}$

15. $\int_0^1 \sin(x+1)\, dx$

16. $\int_0^1 \cos x\, dx.$

17. An automobile starting from rest accelerates for 15 sec. Velocity readings (in feet per second) taken at 1-sec intervals are: 0, 0.5, 2.1, 4.7, 8.3, 13.0, 18.7, 25.5, 33.3, 43.1, 53.0, 63.9, 75.9, 88.9, 102.9, 118.0. Estimate, using trapezoidal approximation, the distance the car traveled during this time.

18. Soundings in feet are taken across a 90-ft river at 5-ft intervals, resulting in the readings: 0, 1.0, 2.5, 5.0, 8.0, 10.5, 11.5, 12.0, 13.0, 12.5, 13.5, 16.0, 16.0, 14.0, 10.5, 9.0, 6.5, 4.0, 0. If the average current at this point in the river is 5 ft/sec, estimate to the nearest million cubic feet the daily flow of water. Use trapezoidal approximation.

3. FLOWCHARTS [optional]

The computations involved in numerical integration are repetitive and boring. In doing them by hand, people become careless and error-prone.

A computer, however, is not subject to boredom and carelessness, and is especially suited for repetitive computations. It can provide speed and accuracy far beyond the capabilities of humans.

Standard programming languages such as Fortran, Algol, and PL/1 simplify the writing of programs for numerical integration and other tasks. Subroutines for common functions, $\sin x$, $\cos x$, e^x, $\sqrt{x}$, and others, are available in these languages and further simplify the work of the applied mathematician, engineer, or scientist.

Rather than use one of these languages, we shall describe computer routines by **flowcharts**. A flowchart is a diagram showing a step-by-step procedure for solving a problem; it can easily be translated into one of the programming languages.

Since flowcharting conventions vary, it is important to understand those we shall use (Fig. 3.1). Boxes represent the steps in a computation, and

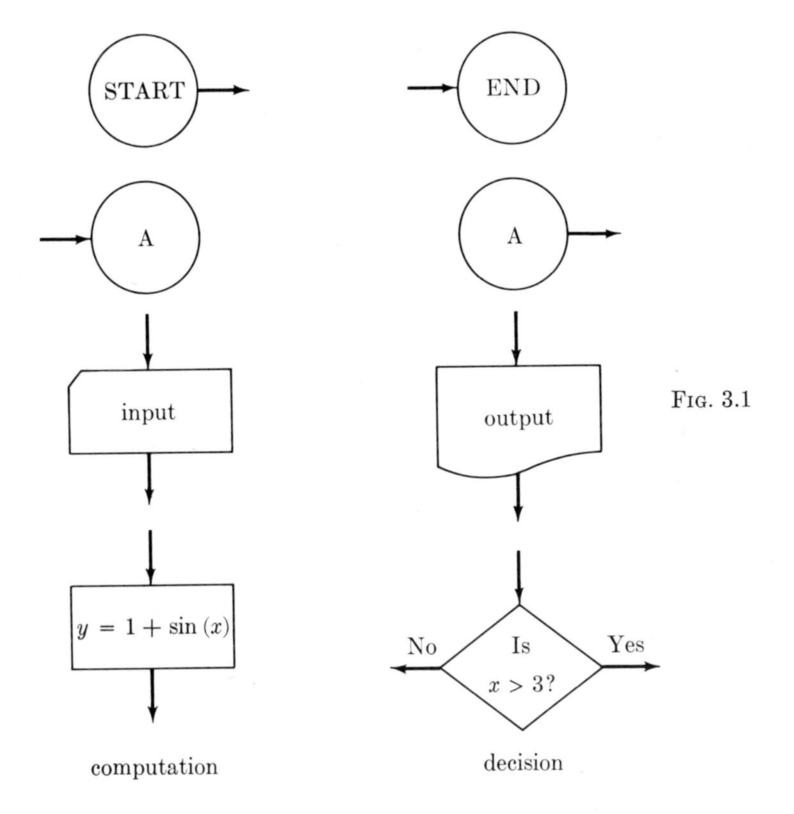

FIG. 3.1

arrows show the sequence of steps. The shape of each box indicates the nature of the step it represents. Circles label the beginning and end, and connections between parts of a long program. Rectangles represent computations. Boxes shaped like punched cards show the input data; boxes shaped like torn-off pieces of paper show the output. Diamonds indicate decision points, where the direction depends on the answer to some question.

The use of these conventions is illustrated in a flowchart (Fig. 3.2) for the trapezoidal approximation of Example 2.1. The statements "$f = 2f$,"

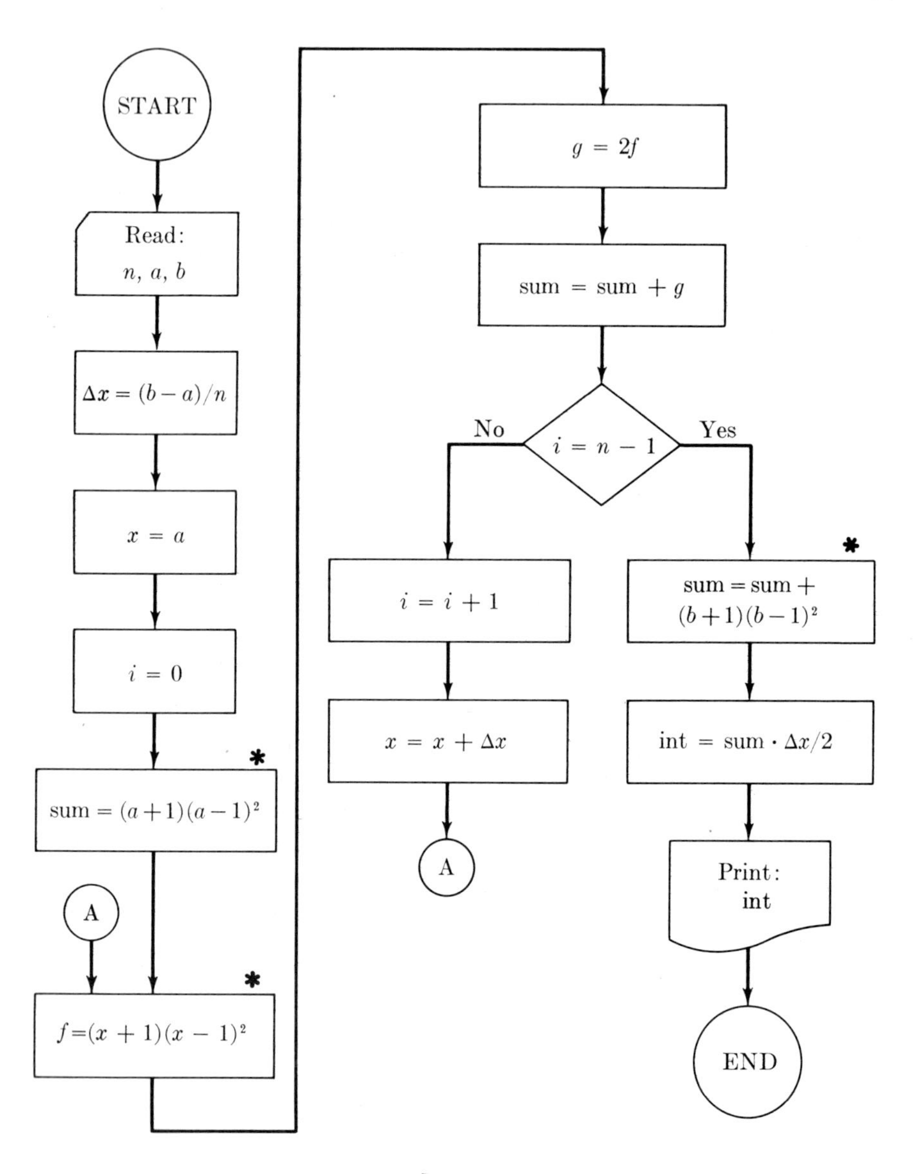

FIG. 3.2

"sum = sum + f," "$i = i + 1$," and "$x = x + \Delta x$" may appear foolish. But the equal sign in boxes is *not* the ordinary mathematical equal sign; instead, it is an abbreviated instruction. For instance, "$f = 2f$" means "calculate a new value for f, namely twice the current value"; and "$i = i + 1$"

means "increase the value of i by 1." (Sometimes a replacement arrow is used to avoid confusion: $f \leftarrow 2f$, $i \leftarrow i + 1$, and so forth.)

In Fig. 3.2, the boxes marked * are the heart of the program. They contain the computation of the values $f(x_0)$, $f(x_1)$, $\cdots$, $f(x_n)$; the rest of the program puts these values together according to the Trapezoidal Rule. Consequently, the flowchart describes the trapezoidal approximation to

$$\int_a^b f(x)\,dx$$

for whichever function $f(x)$ appears in the starred boxes.

Study the flowchart in Fig. 3.2. Be sure you understand each step.

EXERCISES

Construct a flowchart for

1. $y = ax + b$; input a, b, and x
2. x, if $y = ax + b$; input a, b, and y
3. the roots of $ax^2 + bx + c = 0$; input a, b, and c.
4. Construct a flowchart for the Rectangular Rule.
5. Construct a flowchart for this variant of the Rectangular Rule: instead of $f(\bar{x}_i)$, use the value $f(x_i)$ of $f(x)$ at the right-hand endpoint of the subinterval.

11. Space Geometry

1. THREE-DIMENSIONAL FIGURES [optional]

Geometrical and physical problems of the real world are generally three-dimensional. You will benefit by learning how to draw a number of familiar three-dimensional shapes. This skill is indispensable in formulating and solving problems in calculus and its applications to science.

Drawings in perspective are nice, but require more skill than is usually needed for problem solving. In most cases, a shadow projection from a distant light source is adequate. Stand far from an object. Imagine a piece of glass between you and the object, near the object. The light rays from the object to you make a picture on the glass (Fig. 1.1). In such a picture, parallel lines in space project onto parallel lines in the plane of the glass.

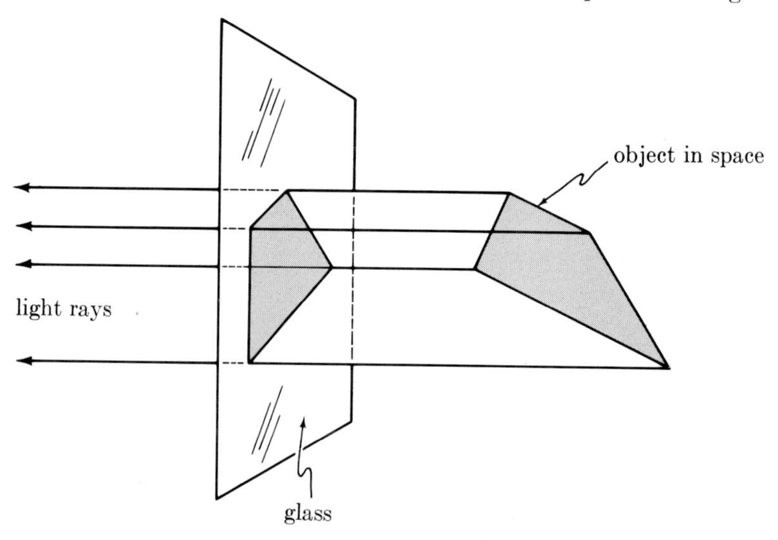

Fig. 1.1

A satisfactory drawing of a solid object must show quite clearly that the object is three-dimensional, and must give you a good feeling for its shape. By projecting in a badly chosen direction, you can lose part of the figure. For example, a cube projected squarely onto the paper yields a square

(Fig. 1.2). This picture completely fails to convey the impression of a cube. Drawings of a cube are most satisfactory if the direction of projection is not parallel to any face. Figure 1.3 is the result of such an oblique projection of a cube whose front side is parallel to the paper.

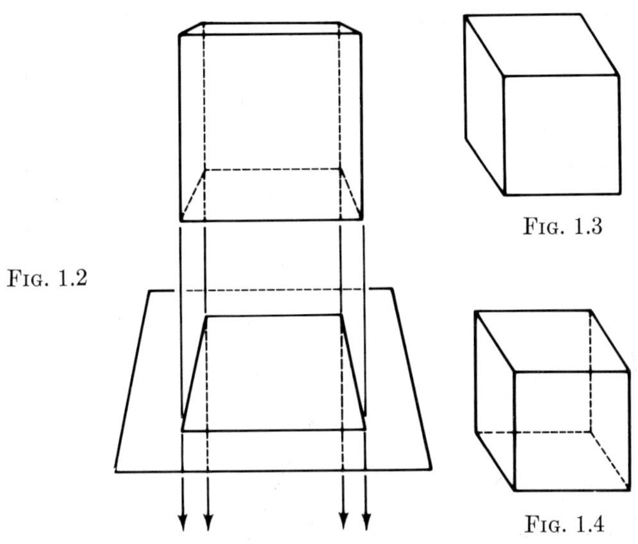

FIG. 1.2

FIG. 1.3

FIG. 1.4

Actually it is very helpful in "seeing" what space object is drawn on the paper if the hidden lines are shown as dotted lines. Add these hidden lines to Fig. 1.3 to obtain Fig. 1.4. Other oblique projections of cubes are shown in Fig. 1.5.

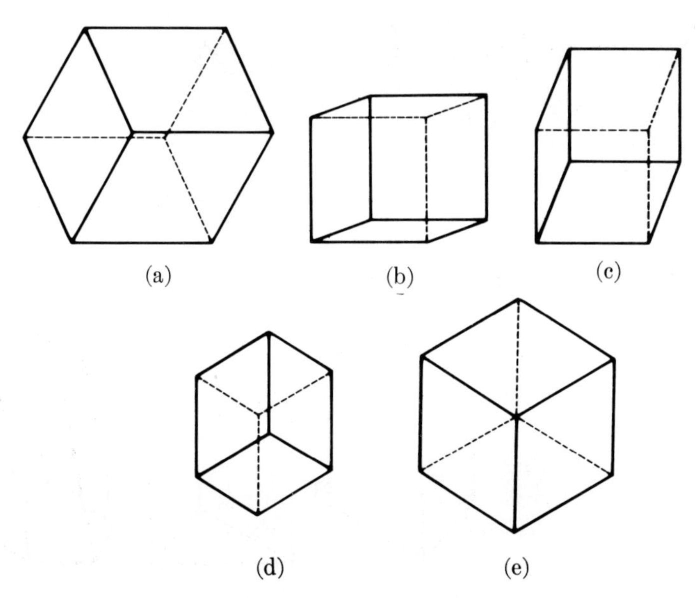

FIG. 1.5 Cubes viewed (a) from slightly above; (b) from slightly below; (c) from well below; (e) from along the main diagonal.

It may require several attempts before you draw a good figure. For example, it is not always obvious at first what is in front and what is behind. The first attempt at a tetrahedron might result in Fig. 1.6. You soon realize that in this position you cannot really see all six edges; one is hidden. Hence there are the two possibilities shown in Fig. 1.7, either a correct figure.

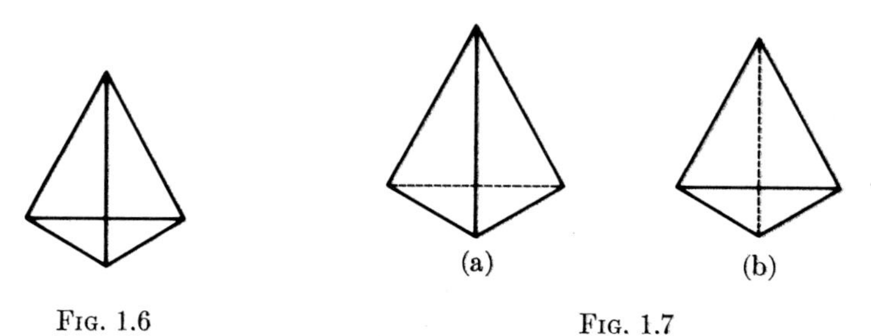

FIG. 1.6

FIG. 1.7

More drawings of tetrahedra are in Fig. 1.8.

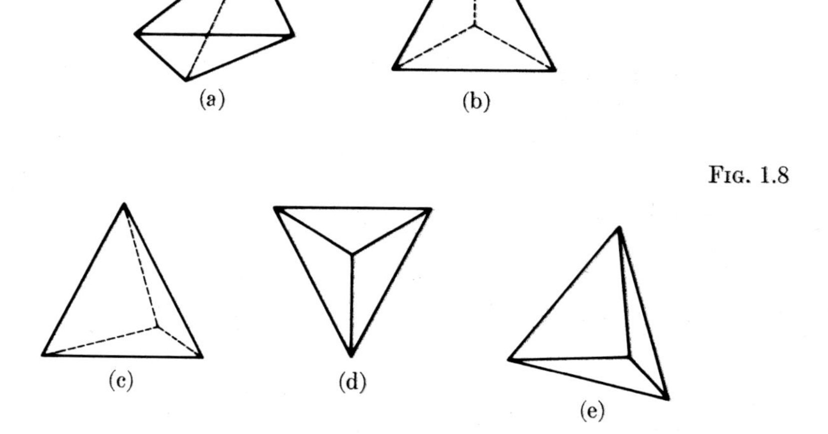

FIG. 1.8

For practice, try a few drawings of prisms (Fig. 1.9).

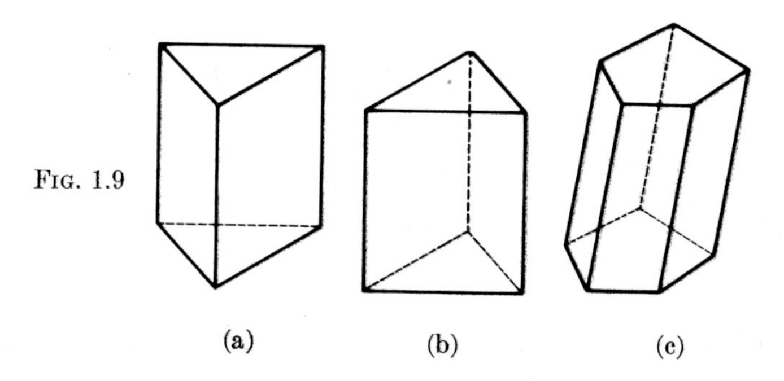

FIG. 1.9

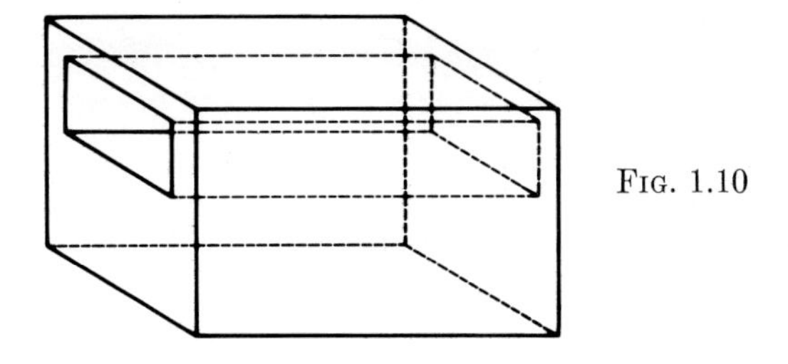

FIG. 1.10

The next drawing, a rectangular solid with a rectangular hole, is more complex (Fig. 1.10). The same drawing with shading added is shown in

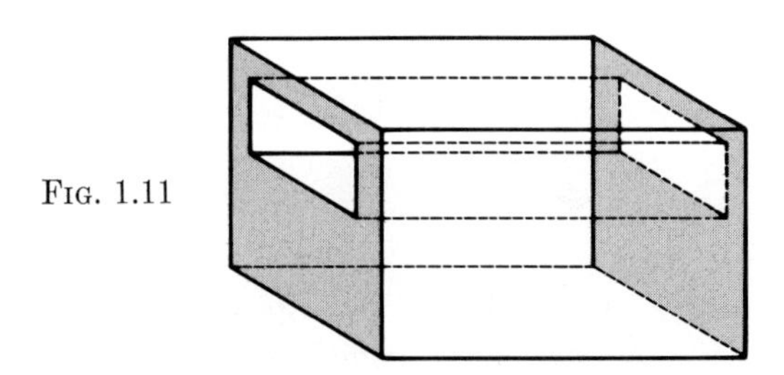

FIG. 1.11

Fig. 1.11. Shading is sometimes used to emphasize spatial features. Variations on the same theme are drawn in Fig. 1.12.

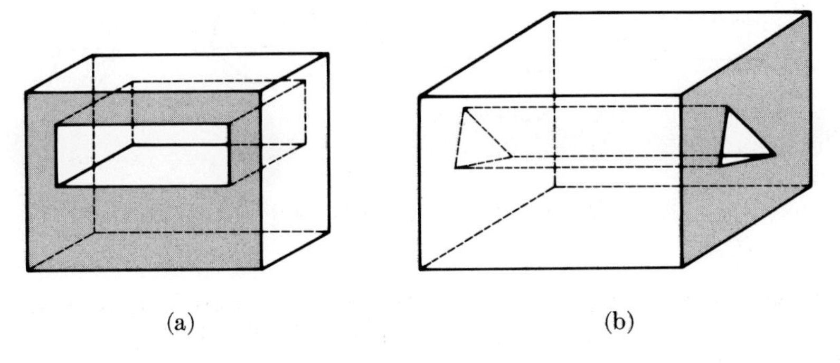

(a) (b)

FIG. 1.12

Important Rules of Drawing

1. Parallel edges of a solid must be drawn parallel.
2. Two parallel edges of equal length of a solid must be drawn of equal length.

Curved figures present problems; it is often quite difficult to get a really accurate representation of a curved solid. However, an adequate sketch is possible with practice.

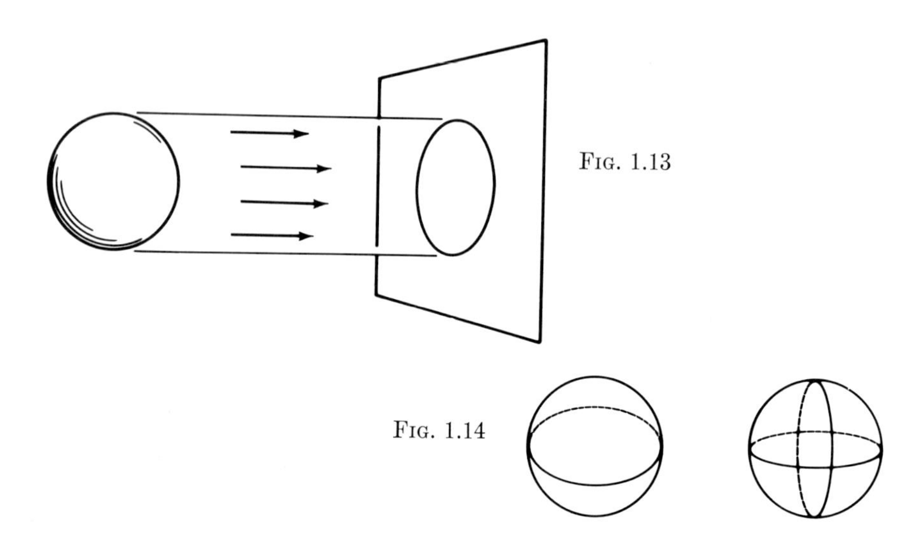

FIG. 1.13

FIG. 1.14

No matter how you project a sphere, you get a circle or ellipse (Fig. 1.13). A few curves added on the surface make the projection look more sphere-like (Fig. 1.14). Various portions of a sphere are drawn in Fig. 1.15.

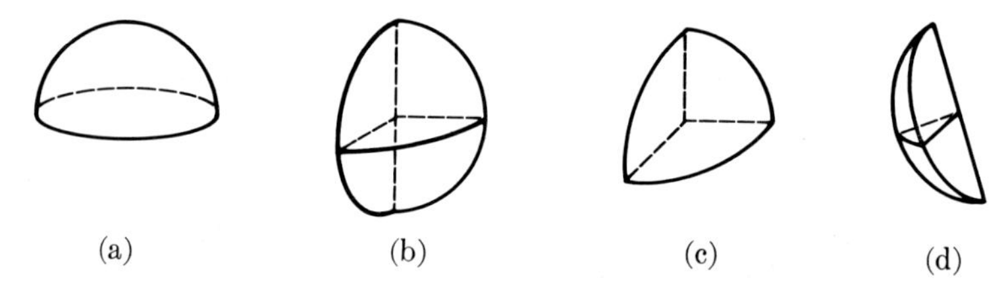

FIG. 1.15 (a) Hemisphere. (b) Quadrant. (c) Octant. (d) Lune.

Drawings of a cylinder and a pipe are shown in Fig. 1.16. The torus (donut) is harder to draw (Fig. 1.17). A cylinder sliced obliquely is shown in

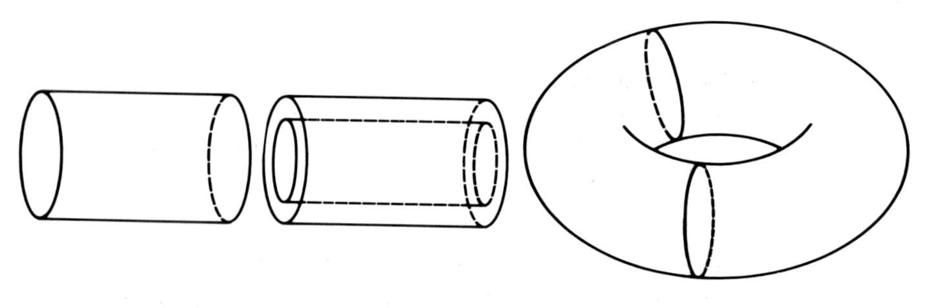

FIG. 1.16

FIG. 1.17

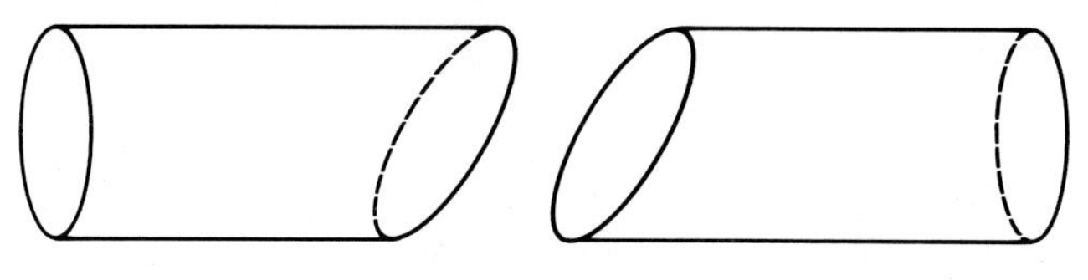

FIG. 1.18

Fig. 1.18. Use of a center line (– — – — – —) for cylinders, cones, etc. is good. Also showing a few cross sections (thin lines, shading) is often quite helpful. We illustrate these techniques with a sliced cylinder and with a cone (Fig. 1.19).

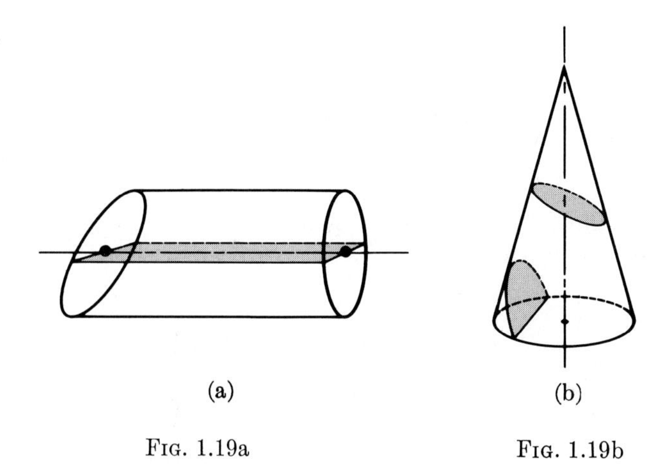

(a) (b)

FIG. 1.19a FIG. 1.19b

EXERCISES

Make a three-dimensional sketch of:

1. a sphere, a right circular cone, and a torus
2. a right pyramid, a triangular prism, and a cylinder
3. a cube with a plane passing through two diagonals
4. a triangular prism cut by a plane at an oblique angle
5. a wedge cut from a triangular prism
6. a pentagonal prism with a cylinder passing through its length
7. a cube surmounted by a triangular prism
8. two cylinders of equal radius, intersecting at right angles
9. the solid common to two cylinders of equal radius intersecting at right angles
10. a sphere with a cylinder passing through its center.

2. GRAPHS

A systematic study of figures in space requires the introduction of coordinates. We set up x-, y-, and z-axes as follows. The plane of this page is the y, z-plane as indicated in Fig. 2.1. The x-axis is conceived perpendicular

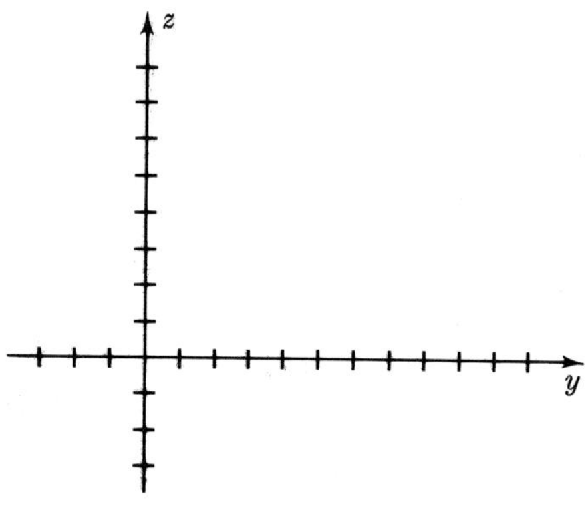

FIG. 2.1

to the page, straight out towards you. By taking an oblique projection from a direction where x, y, and z are all positive, one obtains Fig. 2.2a.

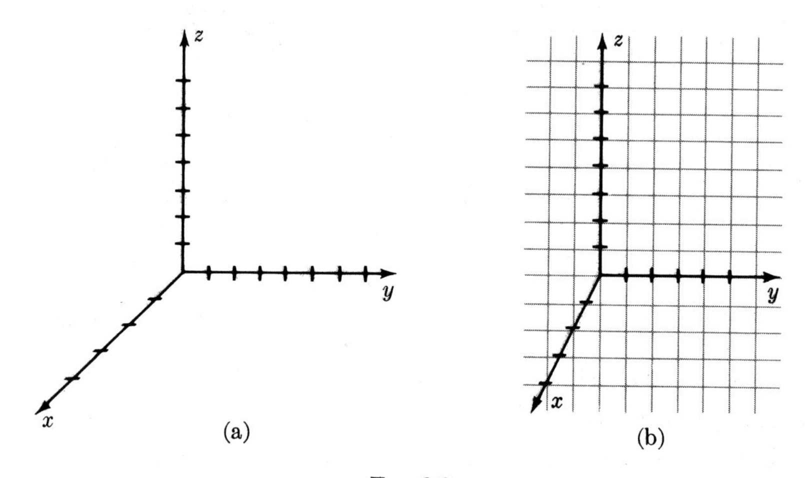

FIG. 2.2

Rectangular graph paper is useful for space graphs. After you fix scales on the perpendicular y- and z-axes, draw the x-axis from the origin towards the lower left with slope about 2 and use the horizontal rulings to fix its scale (Fig. 2.2b).

A point (x, y, z) is located by marking its projection $(x, y, 0)$ in the x, y-plane and going up or down the corresponding amount z. (From the habit of living in the x, y-plane for so long, we think of the z-direction as "up.")

EXAMPLE 2.1

Locate (3, 2, 4) and (1, 3, −2).

Solution: See Fig. 2.3.

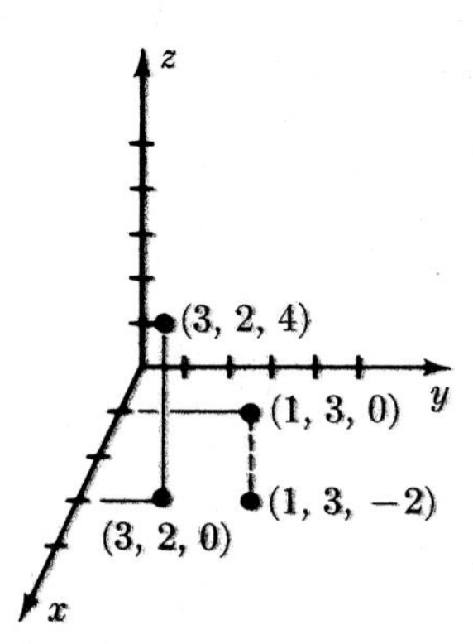

FIG. 2.3

The portion of space where x, y, and z are positive is called the **first octant**. (No one numbers the other seven octants.) Sometimes part of a figure which is not in the first octant is shown; dotted lines indicate it is behind the coordinate planes (Fig. 2.4). The angle at which the x-axis is

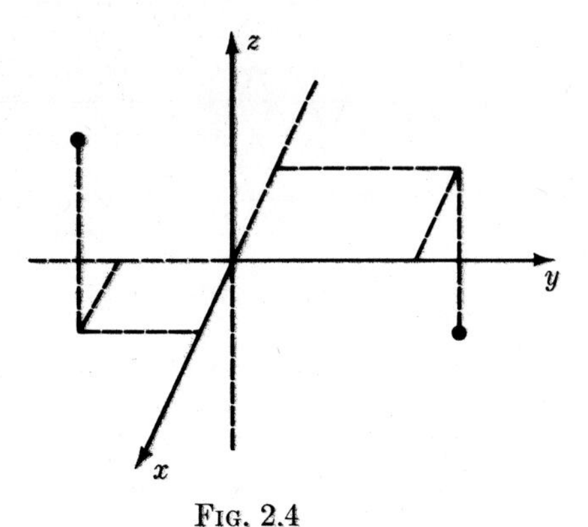

FIG. 2.4

drawn in the y, z-plane is up to you. Choose it so that your drawing is as uncluttered as possible. Actually it is perfectly alright to take a projection into other than the y, z-plane, so that the y- and z-axes are not drawn perpendicular (Fig. 2.5).

There is a type of graph paper, called triangular coordinate paper, which represents the three axes at mutual angles of $2\pi/3$, and has the advantage of rulings in each direction (Fig. 2.6).

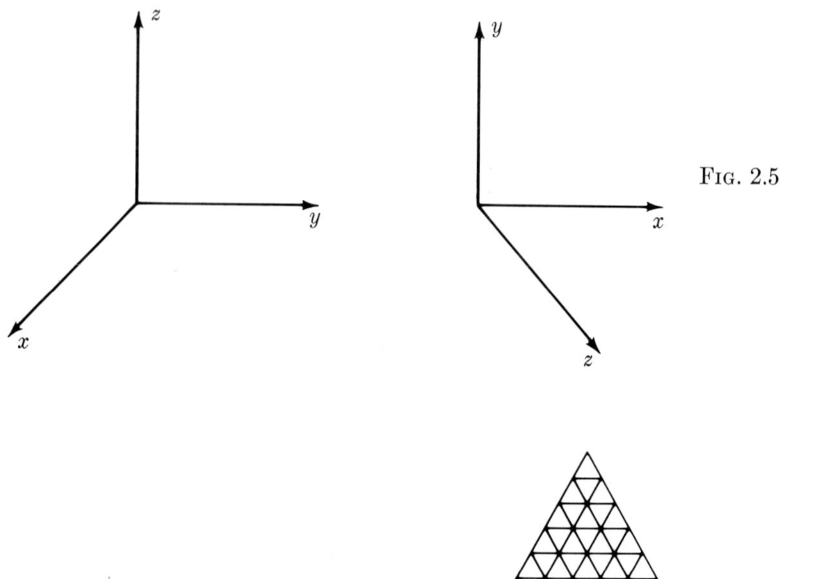

FIG. 2.5

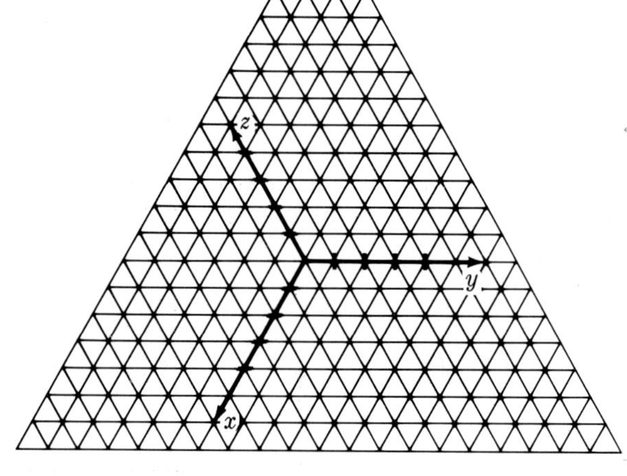

FIG. 2.6

Distance Formula

The distance D between two points whose coordinates are

$$(x_0, y_0, z_0) \qquad \text{and} \qquad (x_1, y_1, z_1)$$

is given by the Distance Formula:

$$D^2 = (x_1 - x_0)^2 + (y_1 - y_0)^2 + (z_1 - z_0)^2.$$

This formula follows from two applications of the Pythagorean Theorem (Fig. 2.7). First

$$E^2 = (y_1 - y_0)^2 + (z_1 - z_0)^2,$$

then

$$\begin{aligned} D^2 &= (x_1 - x_0)^2 + E^2 \\ &= (x_1 - x_0)^2 + (y_1 - y_0)^2 + (z_1 - z_0)^2. \end{aligned}$$

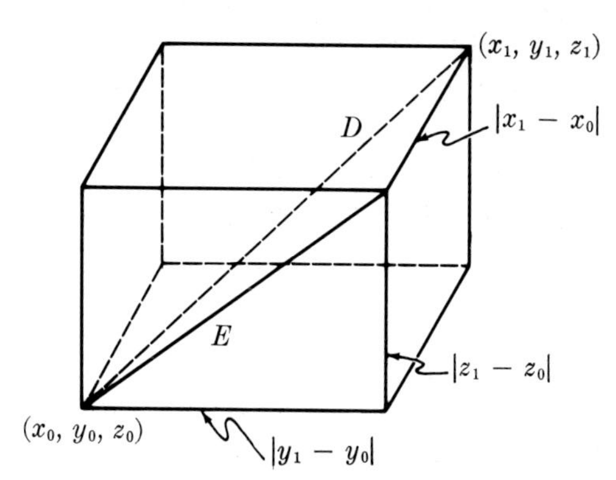

FIG. 2.7

EXERCISES

Draw a graph, locating the points:

1. (1, 2, 3), (1, 3, 4)
2. (2, 4, 1), (2, −4, 1)
3. (−1, 2, 1), (2, 2, −1)
4. (1, −3, 3), (3, 2, −2).

Locate the points and compute the distance between them:

5. (0, 0, 0), (4, 1, 3)
6. (4, 5, 3), (1, 3, 2)
7. (1, 1, 1), (−2, 0, 2)
8. (−3, −2, 4), (−1, 3, −2).
9. Draw the perpendicular from the point (2, 4, 3) to each coordinate axis and find the length of each perpendicular segment.

Draw the parallelepiped with edges parallel to the axes and locate the vertices. The ends of a diagonal are the points:

10. (0, 0, 0), (2, 3, 1)
11. (4, 2, 3), (1, 1, 1).

Are the points vertices of a right triangle? If so, find its area:

12. (0, 0, 0), (2, 3, 0), (5, 1, 1)
13. (3, 1, 0), (4, 2, 4), (3, 4, 1)
14. (−1, 2, −6), (3, −2, 4), (9, −5, 1)
15. (3, 2, 5), (−1, 4, −7), (−5, 8, −5).

Do the points lie on a straight line?

16. (0, 0, 0), (1, 3, 2), (2, 6, 4)
17. (0, 0, 0), (−1, 3, −4), (2, −5, 8)
18. (1, 1, 1), (0, 1, 2), (−1, −3, −5)
19. (1, −1, −2), (−1, 2, 3), (3, −4, −7)
20. (1, 1, 0), (1, 0, 1), (0, 1, 1).

Mark the points A, B, C on a graph; they determine a plane Π. Then mark the points D, E; they determine a line λ. Construct with a straight edge the point where λ intersects Π:

21. $A = (0, 0, 0)$, $B = (1, 2, 0)$, $C = (0, 0, 1)$, $D = (1, 0, 0)$, $E = (0, 1, 0)$
22. A, B, C, D as in Ex. 21, $E = (0, 1, 1)$
23. A, B, C as in Ex. 21, $D = (2, 1, 0)$, $E = (0, 3, 1)$
24. $A = (1, 0, 0)$, $B = (0, 2, 0)$, $C = (0, 0, 3)$, $D = (0, 0, 2)$, $E = (2, 2, 0)$
25. A, B, C, E as in Ex. 24, $D = (0, 1, 1)$.

Mark the points $A_1, B_1, C_1; A_2, B_2, C_2$ on a graph. They determine planes Π_1 and Π_2. Construct with a straight edge their line of intersection:

26. $A_1 = (0, 0, 0), \quad B_1 = (0, 0, 1), \quad C_1 = (1, 2, 0); \quad A_2 = A_1, \quad B_2 = (0, 2, 0), \quad C_2 = (1, 1, 2)$
27. A_1, B_1, C_1 as in Ex. 26; $A_2 = (1, 0, 0), B_2 = B_1, C_2 = (0, 1, 2)$
28. $A_1 = (1, 0, 0), \quad B_1 = (0, 1, 0), \quad C_1 = (0, 0, 2); \quad A_2 = (2, 3, 0), \quad B_2 = (-1, 3, 2), \quad C_2 = (0, 0, 1).$

3. PLANES

In the plane, the graph of a linear equation $ax + by = c$ is a straight line. In space, the graph of a linear equation

$$ax + by + cz = k \qquad (a, b, c \quad \text{not all } 0)$$

is a plane, not a line. The three simplest cases are shown in Fig. 3.1. In

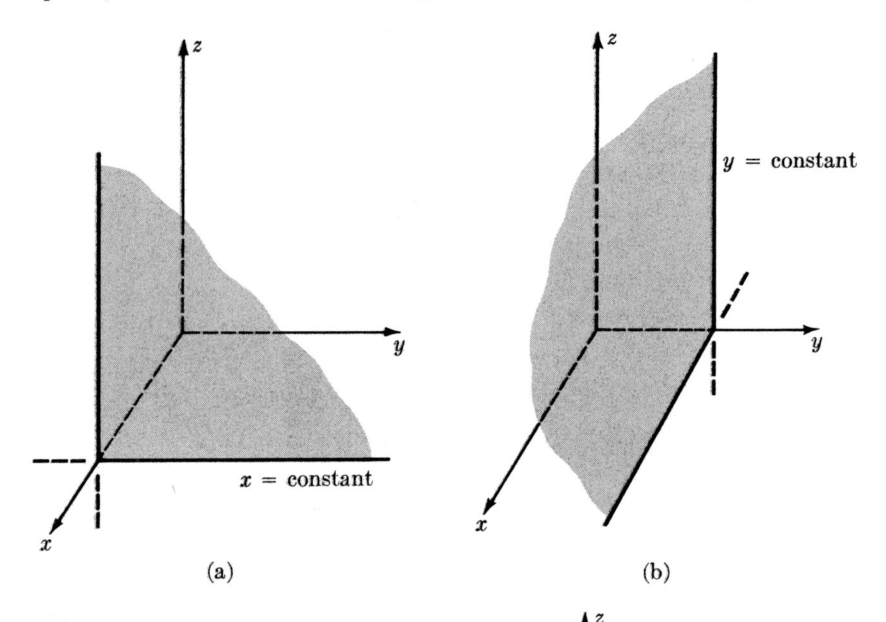

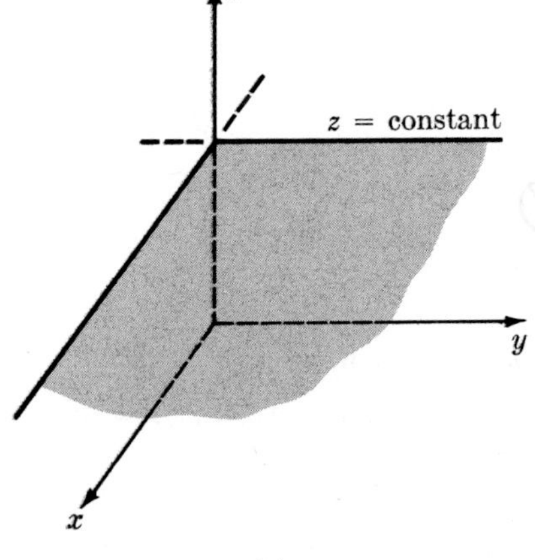

FIG. 3.1

each of these cases two of the coefficients a, b, c are zero. If only one of the coefficients is zero, then the plane is parallel to the corresponding axis (Fig. 3.2).

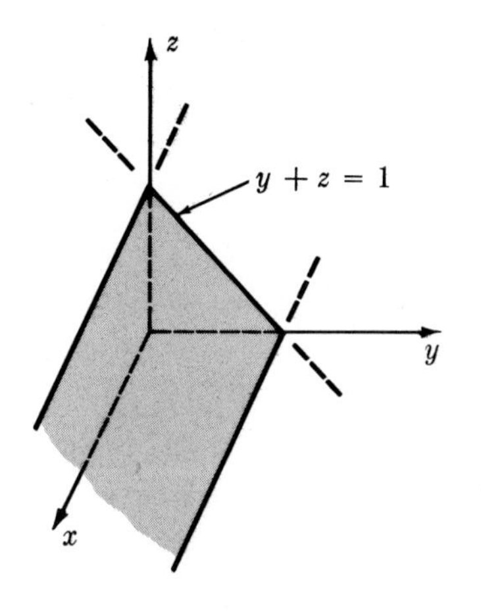

FIG. 3.2a

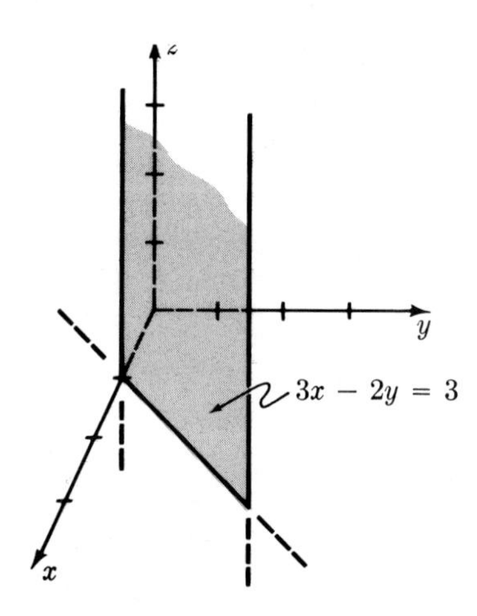

Fig. 3.2b

In the most general case, a, b, c are all different from zero. Often one can locate the plane by finding where it meets each of the three axes. For example, the point where

$$ax + by + cz = k$$

meets the x-axis is obtained by setting $y = z = 0$, and solving the resulting equation for x. The point is $(k/a, 0, 0)$.

EXAMPLE 3.1

Plot $x + y + z = 1$.

Solution: The points where the plane meets the three coordinate axes are

$$(1, 0, 0), \quad (0, 1, 0), \quad (0, 0, 1).$$

The portion in the first octant is shown in Fig. 3.3.

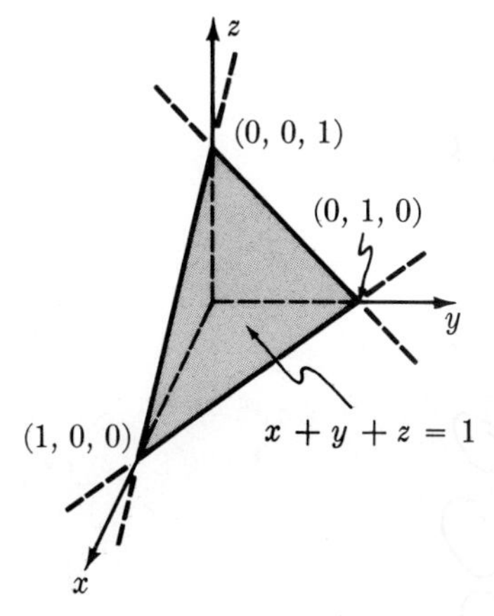

FIG. 3.3

This method fails if $k = 0$, that is, if $ax + by + cz = 0$, for then the plane passes through the origin. In this case however, you can locate the plane by finding its lines of intersection with two of the coordinate planes.

EXAMPLE 3.2

Plot $x - y - z = 0$.

Solution: This plane meets the coordinate plane $z = 0$ in the line $x - y = 0$, and meets the coordinate plane $y = 0$ in the line $x - z = 0$. The portion in the first octant is shown in Fig. 3.4.

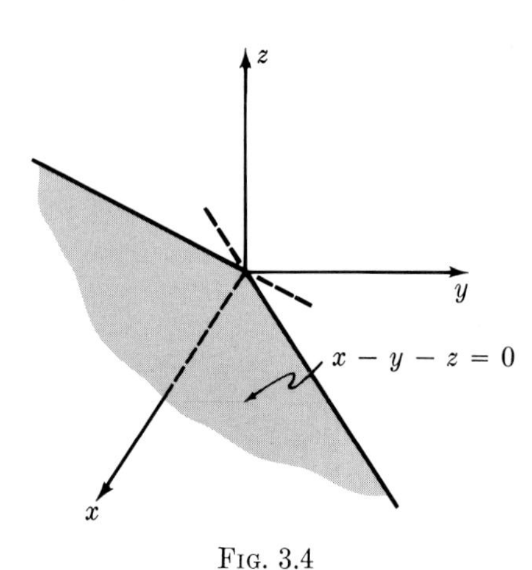

FIG. 3.4

EXERCISES

Sketch the plane:

1. $z = 3$
2. $y = -2$
3. $3y + 2x = 6$
4. $2x - z = 2$
5. $3y + 2x = 0$
6. $2x - z = 0$
7. $x + 2y + 3z = 6$
8. $4x - y + 2z = 4$
9. $y - x + 2z = 2$
10. $3x + 2y - z = 4$
11. $2x - y - z = 0$
12. $6z - 2x + y = 0$.
13. Find the equation of the plane perpendicular to the y-axis and passing through $(-1, -3, 8)$.
14. Find the equation of the plane which contains the whole x-axis and also passes through $(3, -1, 2)$.
15. Find the equation of the plane through $(0, 0, 0)$, $(2, 4, 3)$, and $(-1, -1, -1)$.
16. Find the equation of the plane through $(1, 0, 0)$, $(0, 2, 0)$, and $(0, 0, 3)$.

4. SURFACES

We shall graph some common curved surfaces. The first of these is the right circular cylinder.

EXAMPLE 4.1

Plot the set of points (x, y, z) satisfying $x^2 + y^2 = 1$.

Solution: This surface meets the x, y-plane in the unit circle (Fig. 4.1a). Given a point (x_0, y_0) on this circle, each of the points (x_0, y_0, z), for *any* value of z, is on the surface we seek.

Answer: The surface is the right circular cylinder whose base is the unit circle in the x, y-plane (Fig. 4.1b).

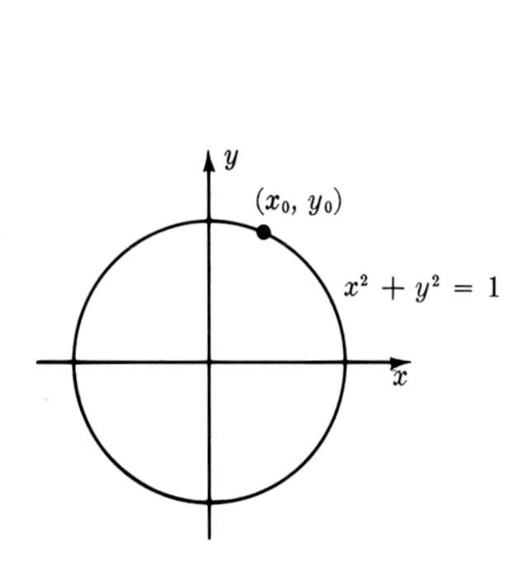

FIG. 4.1a

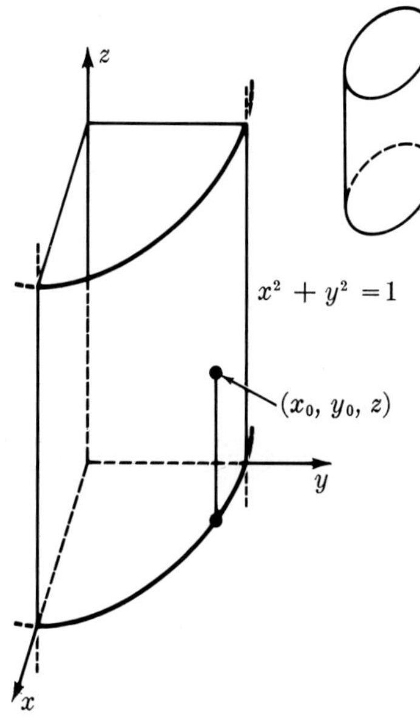

FIG. 4.1b

REMARK: Example 4.1 illustrates an important point. If an equation has one variable missing, then the graph in space is a cylinder (not necessarily right circular) parallel to the axis of the missing variable. For example, the graph in space of $z = y^2$ is obtained by drawing the parabola $z = y^2$ in the y, z-plane, then moving this parabola parallel to the x-axis. The result is a (right parabolic) cylinder whose generators are parallel to the x-axis, and whose cross section by each plane $x = a$ is a copy of the parabola.

EXAMPLE 4.2

Find an equation for a right circular cone.

Solution: Begin by choosing the coordinate system skillfully. Place the origin at the apex of the cone, and the y-axis along the axis of the cone (Fig. 4.2a). Each section by a plane $y = y_0$ is a circle. The radius of this circle is proportional to y. Let a denote the constant of proportionality

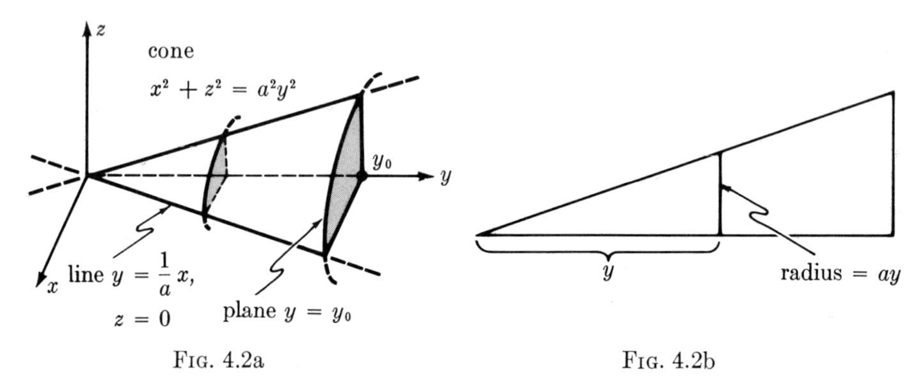

FIG. 4.2a FIG. 4.2b

(Fig. 4.2b). Then

$$x^2 + z^2 = a^2y^2.$$

Note that the cone meets the plane $y = 1$ in the circle $x^2 + z^2 = a^2$ of radius a. This gives one interpretation of the constant a. The cone meets the x, y-plane in the pair of lines $x = \pm ay$ of slope $\pm 1/a$; this gives another interpretation of a.

Answer: $x^2 + z^2 = a^2y^2$.

EXAMPLE 4.3

Find the locus (graph) of $x^2 + y^2 + z^2 = 1$.

Solution: By the Distance Formula, this is the set of points (x, y, z) at distance 1 from $(0, 0, 0)$, i.e., a sphere of radius 1 with center at the origin. One-eighth of it is drawn in Fig. 4.3a and the whole sphere is sketched in Fig. 4.3b.

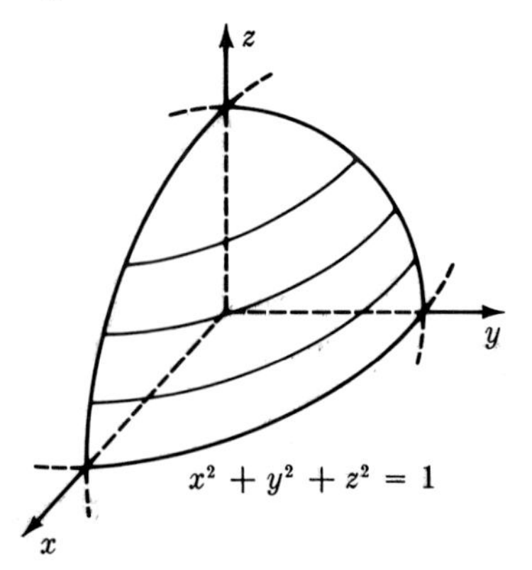

FIG. 4.3a

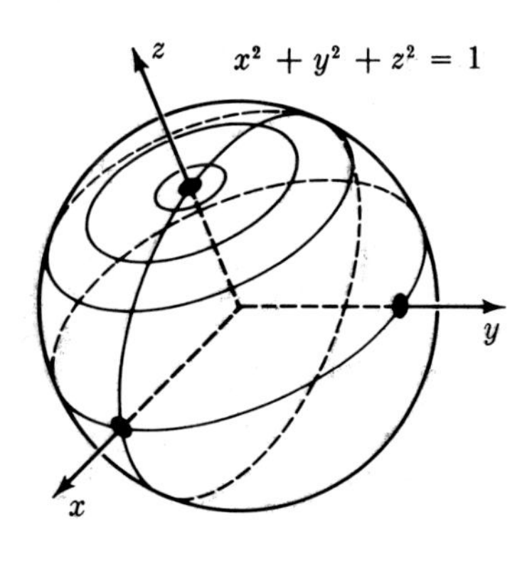

FIG. 4.3b

EXAMPLE 4.4

Plot the locus of $x^2 + y^2 + (z - 1)^2 = 1$.

Solution: By the Distance Formula, this is a sphere of radius 1 with center at (0, 0, 1). It passes through the origin, just touching the x, y-plane at that point. Draw the portion in the first octant (Fig. 4.4).

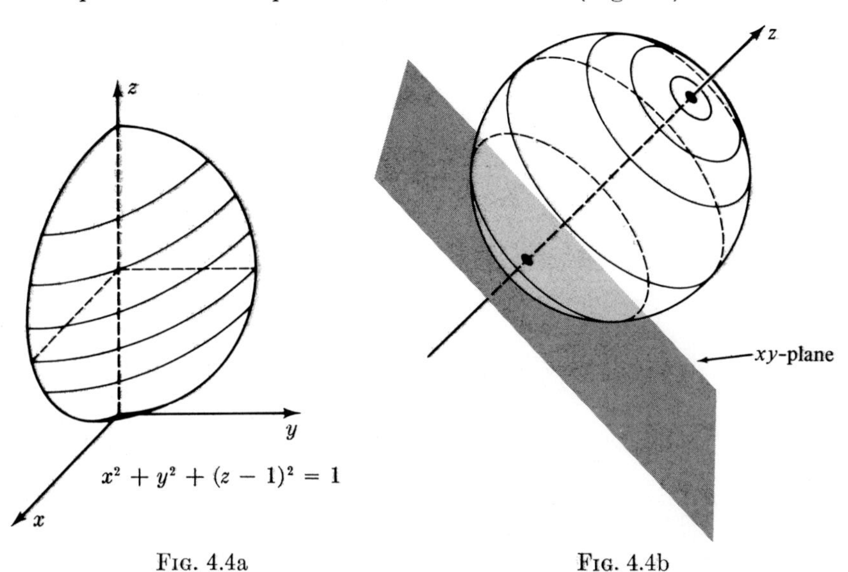

FIG. 4.4a

FIG. 4.4b

EXERCISES

Sketch the surfaces:

1. $x^2 + (z - 1)^2 = 1$
2. $x^2 - 2x + y^2 - 6y + 9 = 0$ [*Hint:* complete the square.]
3. $y = -(x - 4)^2$
4. $z + x^2 - 6x - 9 = 0$
5. $(x - 2)^2 + y^2 + z^2 = 4$
6. $(x + 2)^2 + (y - 1)^2 + z^2 = 1$
7. $x^2 - 6x + y^2 + z^2 + 5 = 0$
8. $x^2 - 2x + y^2 + z^2 = 3$
9. $(x - 1)^2 + z^2 = 9y^2$
10. $x^2 + y^2 - 4y + 4 - 4z^2 = 0.$
11. Find the equation of the sphere with center (1, 2, 3) and radius 4.
12. Find the equation of the sphere with center (1, −1, −1) and which passes through (0, −5, 7).
13. Find the equation of the right circular cylinder which meets the plane $y = -1$ in a circle with center (2, −1, 4) and radius 3.
14. Find the equation of the right circular cone with apex at (3, 0, 0), apex angle $\pi/2$, and axis along the x-axis.
15. Find the equation of the right circular cone with apex (1, 1, 1), with axis parallel to the y-axis, and which passes through the origin.

5. SOLIDS OF REVOLUTION

If a point, curve, or region is revolved about an axis in the same plane, it sweeps out a figure in space. We consider several examples.

(1) The point $(0, b, c)$ in the y, z-plane is revolved about the z-axis. It sweeps out a circle (Fig. 5.1a).

(2) The line $x = 0$, $y = b$ in the y, z-plane is revolved about the z-axis. It sweeps out a cylinder (Fig. 5.1b).

(3) A line $x = 0$, $ay + bz = c$ rotated about the z-axis generates a right circular cone (Fig. 5.1c).

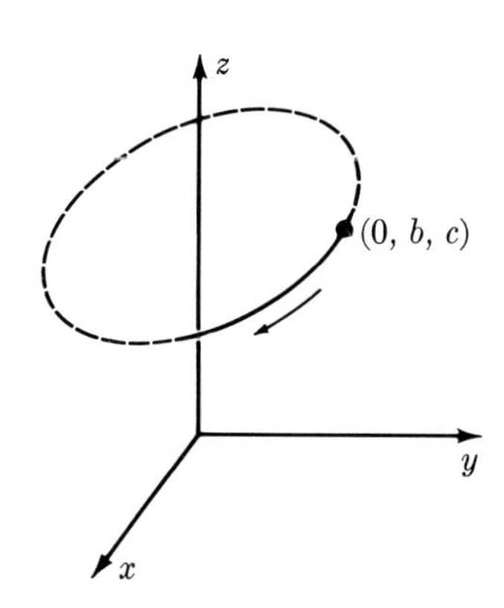

Fig. 5.1a

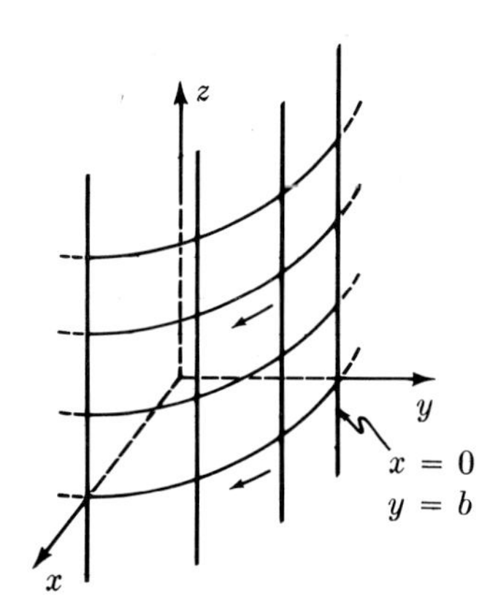

Fig. 5.1b

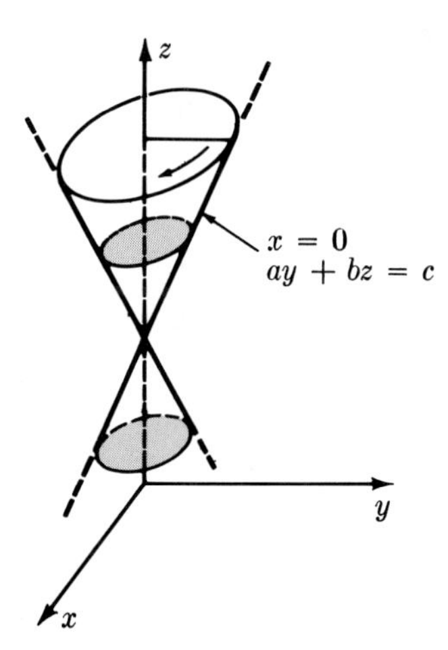

Fig. 5.1c

(4) A circle in the y, z-plane with center on the z-axis sweeps out a sphere when rotated about the z-axis (Fig. 5.2a).

(5) A circle in the y, z-plane, but which does not meet the z-axis, is rotated about the z-axis. It sweeps out a donut-shaped region called a right circular **torus** (Fig. 5.2b).

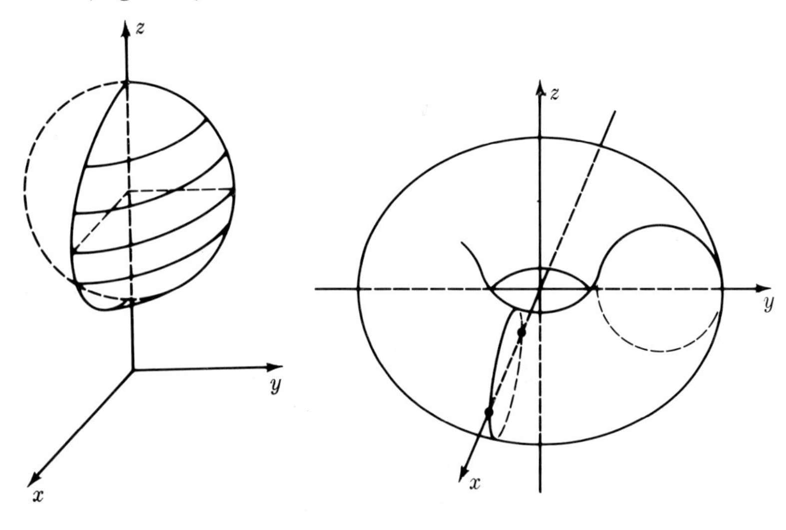

FIG. 5.2a FIG. 5.2b

(6) Consider the region (Fig. 5.3a) in the y, z-plane bounded by the y-axis, the line $y = 1$, and the parabola $z = y^2$. When rotated about the z-axis, it generates the solid illustrated in Fig. 5.3b.

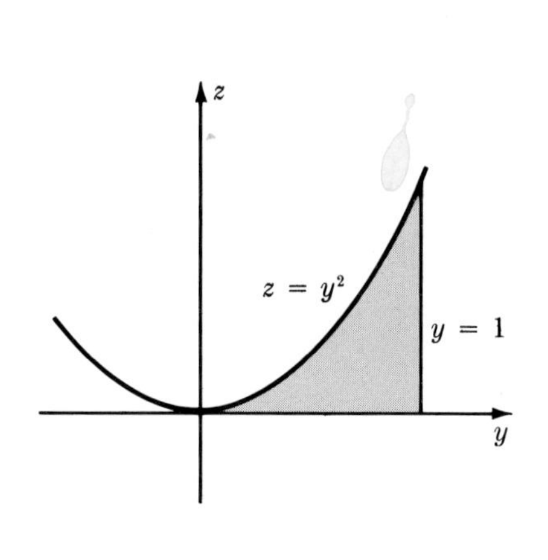

FIG. 5.3a

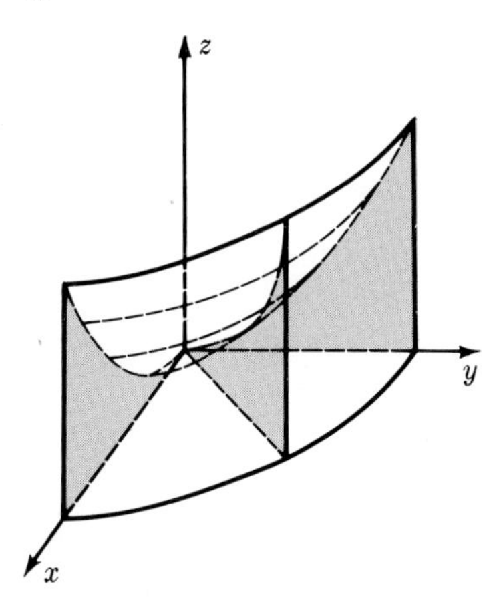

FIG. 5.3b

(7) The region bounded by a triangle in the first quadrant of the x, z-plane is rotated about the x-axis. The result is a ring-shaped solid

(Fig. 5.4). If you consider this solid carefully, you will see that it is bounded by three right circular cones centered on the x-axis.

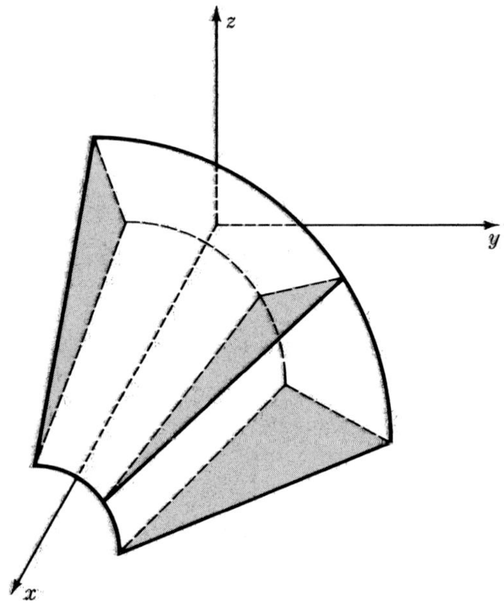

FIG. 5.4

In the next chapter we shall find the volumes of many of these solids.

EXERCISES

Sketch the locus:

1. the point $(a, 0, 0)$ rotated about the x-axis
2. the line $y = 0$, $x = 2$ rotated about the x-axis
3. the line $y = 0$, $z = c$ rotated about the x-axis
4. the line $y = 0$, $z = x$ revolved about the x-axis
5. the line $x = 0$, $z - 2y - 1 = 0$ revolved about the z-axis
6. the circle $x = 0$, $(y - 2)^2 + (z - 2)^2 = 1$ revolved about the z-axis
7. the circle $x = 0$, $(y - 1)^2 + (z - 1)^2 = 1$ rotated about the y-axis
8. the circle $y = 0$, $(x - 1)^2 + (z - 1)^2 = 2$ revolved about the x-axis
9. the parabola $x = 0$, $y = z^2$ rotated about the y-axis
10. the parabola $x = 0$, $y = z^2$ rotated about the z-axis
11. the curve $x = 0$, $yz = 1$ rotated about the y-axis.
12. The region in the y, z-plane bounded by the lines $y = z$ and $y = 2z - 2$ between $y = 2$ and $y = 4$ is rotated about the y-axis. Sketch the solid.
13. Do the same for the region of Ex. 12 rotated about the z-axis.
14. The region in the y, z-plane bounded by the curves $z = e^y$ and $z = 1$ between $y = 0$ and $y = 4$ is rotated about the y-axis. Sketch the solid.

12. Volume

1. INTRODUCTION

The volumes of many solids can be found by integration. The strategy in each case is the same. A solid is sliced into numerous thin pieces, each of which is approximately a familiar shape of known volume. The volume of the i-th piece is $f(\bar{x}_i)\,\Delta x$, the product of its base area $f(\bar{x}_i)$ by its thickness Δx. Addition of all these little pieces leads to the integral

$$\int_a^b f(x)\,dx$$

for the exact volume. [Recall that we abbreviate $f(\bar{x}_i)\,\Delta x$ by $f(x)\,dx$.]

The general plan of attack is executed on a particular solid in four steps: (1) choose a method of slicing the solid; (2) choose a variable x which locates the typical slice, and find the range $a \leq x \leq b$ that applies to the problem; (3) compute the volume $f(x)\,dx$ of the typical slice; (4) find an antiderivative of $f(x)$ and compute

$$\int_a^b f(x)\,dx.$$

Before actually computing volumes, we shall look at two area problems which are solved in the same spirit.

EXAMPLE 1.1

Find the area of a circle of radius r. (Assume the formula $c = 2\pi r$ for circumference.)

Solution: Step 1. Cut the circle into thin concentric rings (Fig. 1.1).

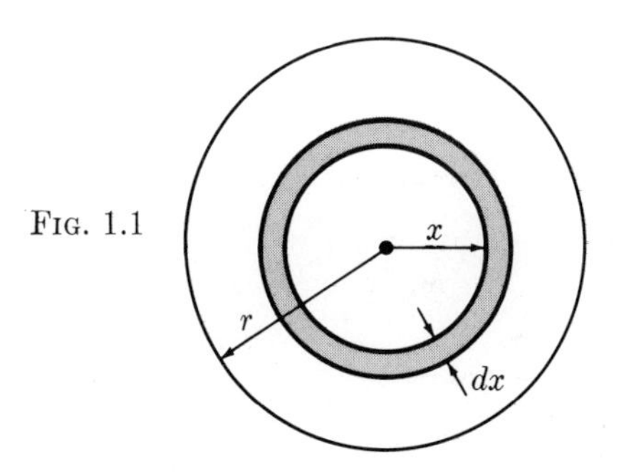

FIG. 1.1

Step 2. Let x denote the radial distance of a ring from the center; $0 \leq x \leq r$.

Step 3. The typical ring has length $2\pi x$ and width dx, hence its area is $2\pi x\, dx$. The area of the circle is

$$\int_0^r 2\pi x\, dx.$$

Step 4. An antiderivative is πx^2, hence

$$\int_0^r 2\pi x\, dx = \pi x^2 \Big|_0^r = \pi r^2.$$

Answer: πr^2.

EXAMPLE 1.2

Find the area of a circular wedge of radius r and central angle α.

Solution: Step 1. Slice the wedge into many thin wedges (Fig. 1.2).

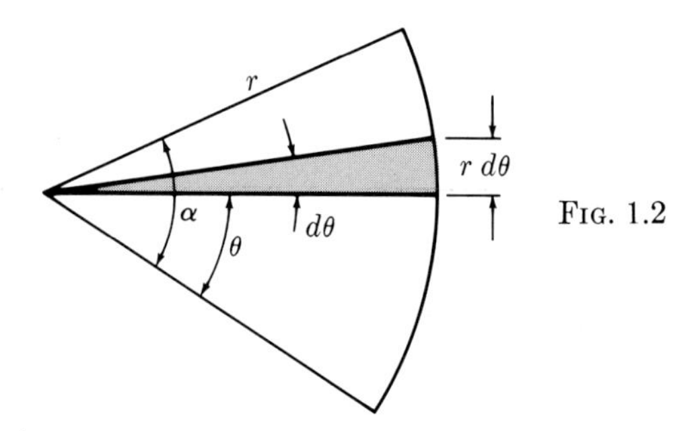

FIG. 1.2

Step 2. Let θ denote the central angle; $0 \leq \theta \leq \alpha$.

Step 3. The thin wedge is almost an isosceles triangle of base $r\, d\theta$, height r, and area

$$\frac{1}{2} r(r\, d\theta) = \frac{1}{2} r^2\, d\theta.$$

The area of the wedge is

$$\int_0^\alpha \frac{1}{2} r^2\, d\theta.$$

Step 4.

$$\int_0^\alpha \frac{1}{2} r^2\, d\theta = \frac{1}{2} r^2 \theta \Big|_0^\alpha = \frac{1}{2} r^2 \alpha.$$

Answer: $\frac{1}{2} r^2 \alpha$.

Note the special case $\alpha = 2\pi$. Then the wedge is the whole pie, of area πr^2, which agrees with the answer in the previous example.

2. VOLUME OF REVOLUTION

A region R in the y, z-plane is revolved about the y-axis (Fig. 2.1). Sup-

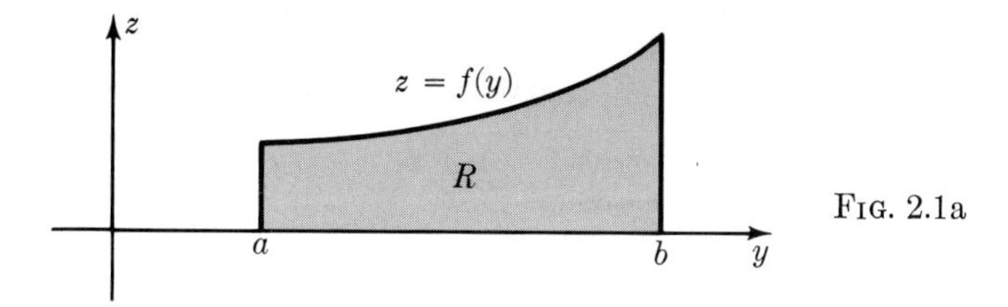

FIG. 2.1a

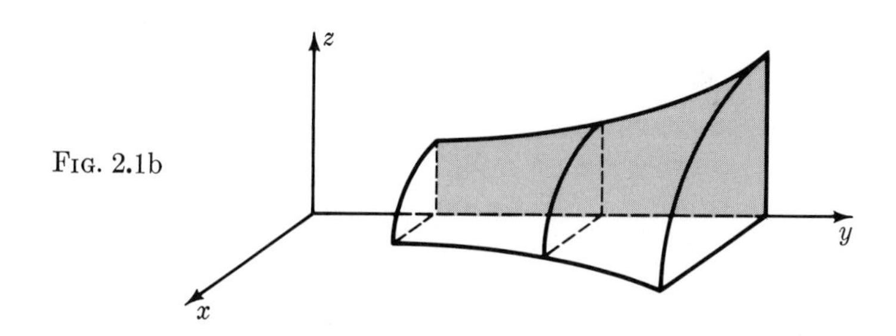

FIG. 2.1b

pose R is the region under a curve $z = f(y)$, where $a \leq y \leq b$. Divide the interval from a to b into many small pieces of width dy. Each of the resulting thin rectangles sweeps out a circular slab (Fig. 2.2). Each slab has radius

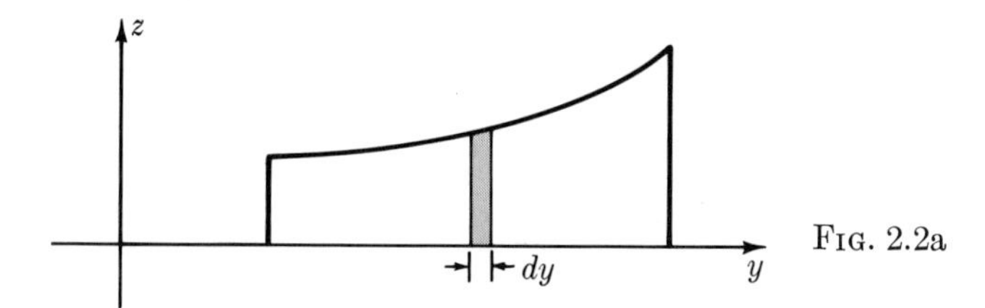

FIG. 2.2a

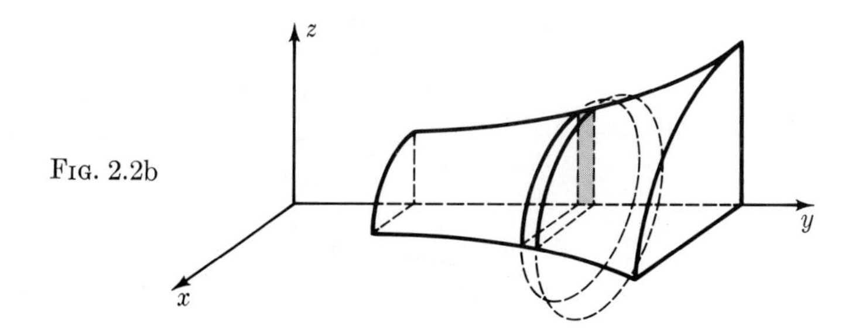

FIG. 2.2b

$f(y)$, base area $\pi f(y)^2$, thickness dy, and volume $\pi f(y)^2\,dy$. Consequently the volume of the solid of revolution is

$$\int_a^b \pi f(y)^2\,dy.$$

EXAMPLE 2.1

Find the volume of a sphere of radius a.

Solution: Step 1. Slice the sphere into parallel slabs by planes.

Step 2. Place the origin at the center of the sphere and choose the y-axis perpendicular to these planes. Thus $-a \le y \le a$. See Fig. 2.3a.

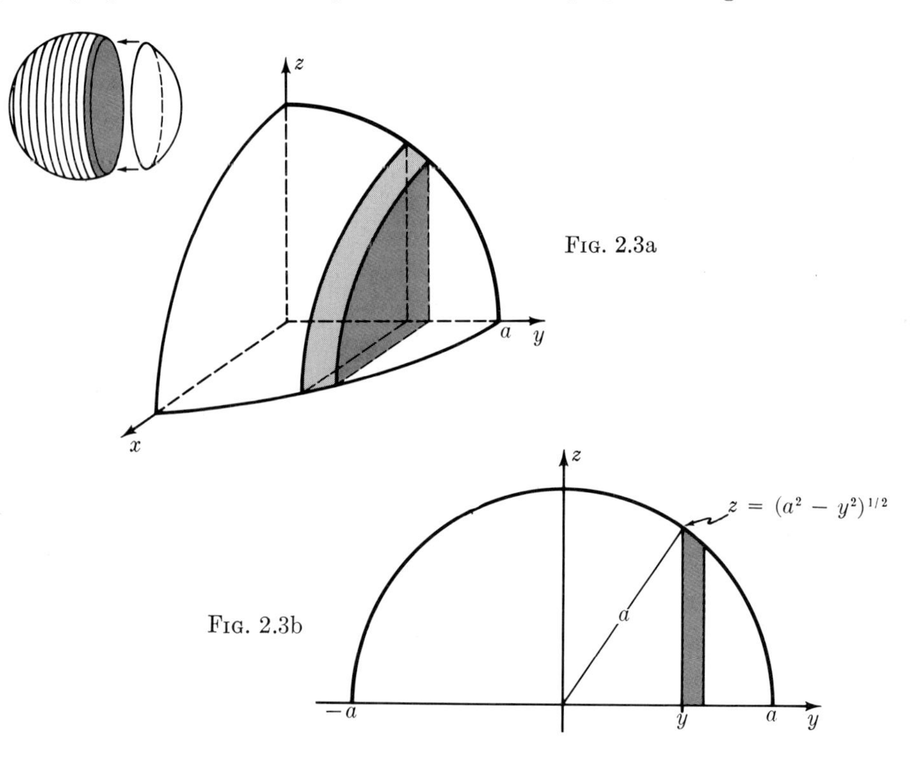

FIG. 2.3a

FIG. 2.3b

Step 3. Each slab has radius $z = \sqrt{a^2 - y^2}$. See Fig. 2.3b. Its thickness is dy, so its volume is $\pi z^2\,dy$. Consequently the volume of the sphere is

$$\int_{-a}^{a} \pi z^2\,dy = \int_{-a}^{a} \pi(a^2 - y^2)\,dy.$$

Step 4. An antiderivative of $(a^2 - y^2)$ is $a^2y - \frac{1}{3}y^3$. Hence

$$\int_{-a}^{a} \pi(a^2 - y^2)\,dy = \pi\left(a^2y - \frac{1}{3}y^3\right)\Bigg|_{-a}^{a}$$

$$= \pi\left(a^3 - \frac{1}{3}a^3\right) - \pi\left(-a^3 + \frac{1}{3}a^3\right) = \frac{4}{3}\pi a^3.$$

Answer: $\frac{4}{3}\pi a^3$.

EXAMPLE 2.2

A right circular cone of height h is constructed over a base of radius a. What is its volume?

Solution: The cone is a solid of revolution, obtained by revolving a right triangle about one leg. Choose the triangle indicated in Fig. 2.4a.

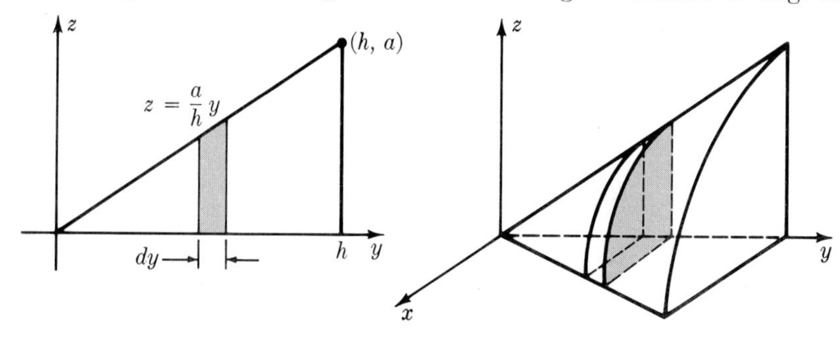

FIG. 2.4a FIG. 2.4b

Rotate this triangle about the y-axis to generate the cone (Fig. 2.4b). Slice the cone into thin slabs, each of width dy. Note (Fig. 2.4a) that the triangle is the region under the curve $z = (a/h)y$, where $0 \le y \le h$. Since each slab has radius $(a/h)y$, its volume is

$$\pi\left(\frac{a}{h}y\right)^2 dy.$$

The volume of the cone is

$$\int_0^h \pi\frac{a^2}{h^2}y^2\,dy = \pi\frac{a^2}{h^2}\left(\frac{1}{3}y^3\right)\Big|_0^h$$

$$= \frac{\pi a^2}{3h^2}h^3.$$

Alternate Solution: Slice the cone into cylindrical shells rather than slabs (Fig. 2.5a), and choose z as the variable. Thus the triangle is sliced into thin strips parallel to the y-axis (Fig. 2.5b). Each thin-walled cylindrical shell has radius z, height $h - (h/a)z$, and thickness dz. Its volume is the product of its three dimensions:

$$(2\pi z)\left(h - \frac{h}{a}z\right)dz.$$

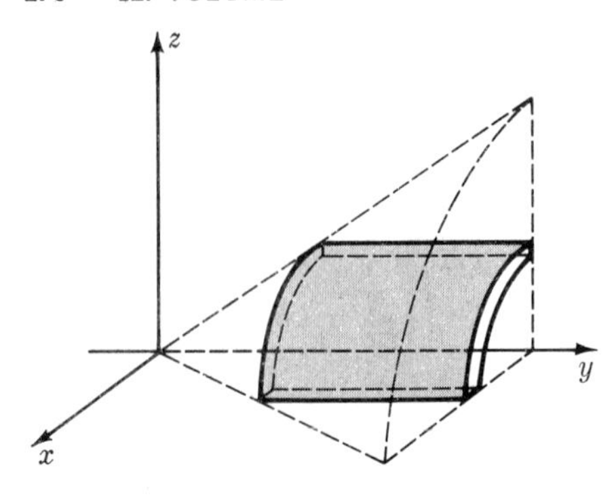

FIG. 2.5a

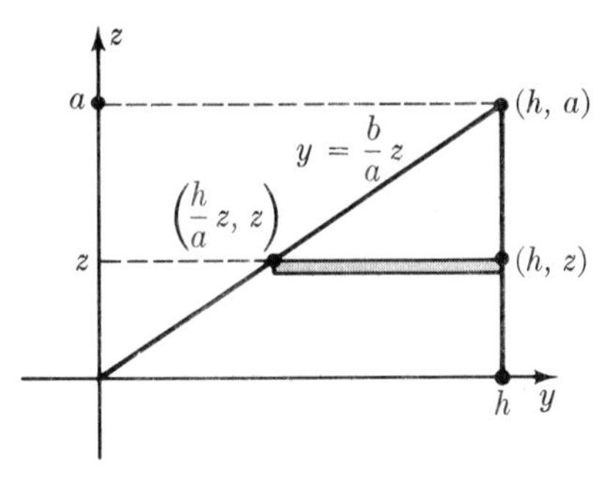

FIG. 2.5b

Hence the volume of the cone is

$$\int_0^a (2\pi z)\left(h - \frac{h}{a}z\right)dz = \int_0^a \left(2\pi hz - \frac{2\pi h}{a}z^2\right)dz$$

$$= \left(\pi hz^2 - \frac{2\pi h}{3a}z^3\right)\Bigg|_0^a = \left(\pi ha^2 - \frac{2\pi}{3}ha^2\right).$$

Answer: $\frac{1}{3}\pi a^2 h.$

EXAMPLE 2.3

The region in the y, z-plane bounded by the parabola $z = y^2$, the line $y = a$, and the y-axis is revolved about the z-axis. (Assume $a > 0$.) What is the resulting volume?

Solution: Slice the plane region by parallels to the y-axis (Fig. 2.6a).

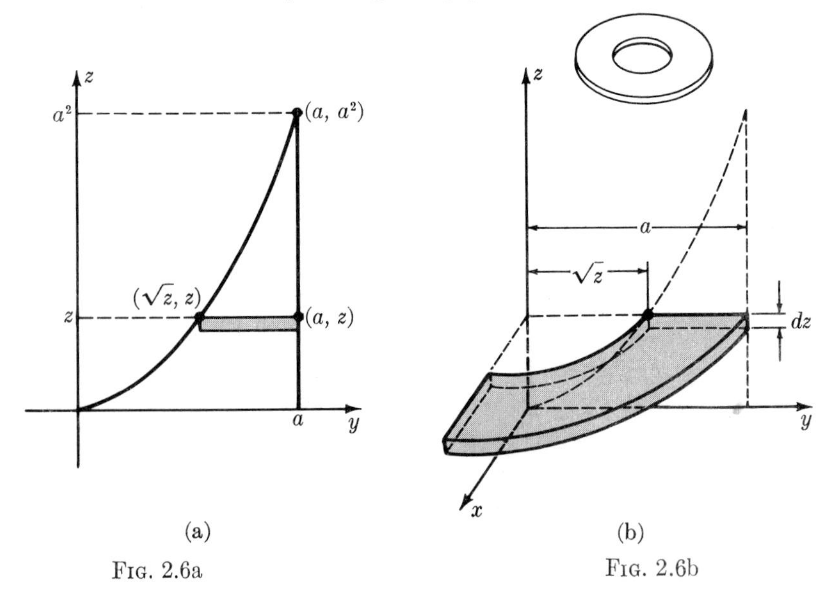

FIG. 2.6a

FIG. 2.6b

When the typical resulting strip is revolved about the z-axis, it sweeps out a thin circular washer (Fig. 2.6b). The base of the washer is the region between two concentric circles. Its area is

$$\pi a^2 - \pi(\sqrt{z})^2 = \pi(a^2 - z),$$

while its thickness is dz. Note that the variable in this solution is z, and $0 \leq z \leq a^2$.

The volume is

$$\int_0^{a^2} \pi(a^2 - z)\,dz = \pi\left(a^2 z - \frac{1}{2}z^2\right)\Big|_0^{a^2}$$
$$= \pi\left(a^4 - \frac{1}{2}a^4\right).$$

Alternate Solution: Proceed as indicated in Fig. 2.7. The region under

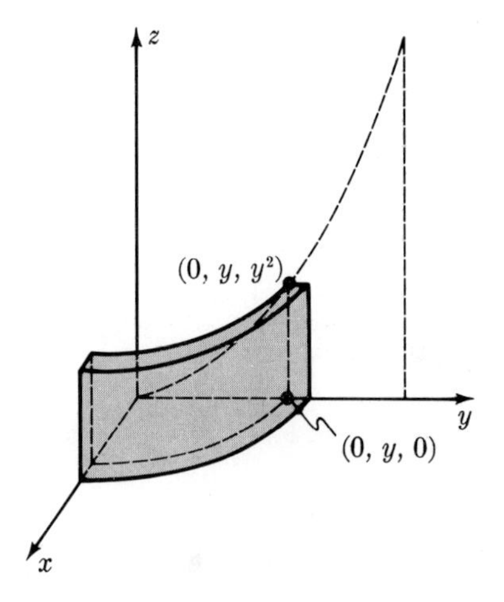

FIG. 2.7

the parabola is split into thin "rectangles" by parallels to the z-axis. Now y is the variable, and the solid of revolution is sliced into thin cylindrical shells. The typical shell has radius y, height y^2, and thickness dy, hence volume $(2\pi y)y^2\,dy$. The volume of the solid is

$$\int_0^a (2\pi y)y^2\,dy = \int_0^a 2\pi y^3\,dy.$$

While we do not yet "officially" know that

$$\frac{d}{dy}(y^4) = 4y^3,$$

it is no secret. Use it to integrate:

$$\int_0^a 2\pi y^3\,dy = 2\pi\left(\frac{1}{4}y^4\right)\Big|_0^a = \frac{\pi}{2}a^4.$$

Answer: $\frac{\pi}{2}a^4$.

2, 5, 9, 12, 14

EXERCISES

The region of the y, z-plane whose boundary curves are given is revolved about the z-axis. Find the volume of the resulting solid of revolution:

1. z-axis, $y = 2z + 3$, $z = 0$, $z = 3$
2. z-axis, $y = \dfrac{3 - z}{2}$, $z = -1$, $z = 1$
3. z-axis, $y = \dfrac{1}{z + 1}$, $z = 0$, $z = 4$
4. z-axis, $y = \dfrac{-3}{z + 4}$, $z = -1$, $z = 1$
5. z-axis, $y = 3\sqrt{z}$, $z = 0$, $z = 4$.
6. The region of the x, y-plane bounded by $y = 16x^2$, $y = 2$, and $y = 4$ is revolved about the y-axis. Find the volume of the resulting solid.
7. The region of the x, y-plane bounded by the x-axis, $y = x + 1$, $x = 1$, and $x = 4$ is revolved about the line $y = -3$. Find the volume of the resulting solid.
8. The region of the x, y-plane bounded by the lines $x = 1$, $y = x$, and $x = -2y + 6$ is revolved about the line $y = -2$. Find the volume of the resulting solid.
9. The region of the x, z-plane bounded by $z = -1$, $z = e^{2x}$, $x = 0$, and $x = 2$ is revolved about the line $z = -1$. Find the volume of the resulting solid.
10. The region of the x, z-plane bounded by the z-axis, $y^2 = \sin z$, $z = 0$, and $z = \pi$ is revolved about the z-axis. Find the volume of the resulting solid.
11. Find the volume of a frustum of a right circular cone with lower radius R_0, upper radius R_1, and height h.
12. Find the volume of a sphere of radius a, with a hole of radius h drilled through its center.
13. A plane at distance h from the center of a sphere of radius a cuts off a spherical cap of height $a - h$. Find its volume.
14. Find the volume of the solid formed by revolving the triangle in the x, y-plane with vertices (1, 1) (0, 2), and (2, 2) about the x-axis.
15. A circular hole is cut on center vertically through a sphere, leaving a ring of height h. Calculate the volume of the ring.
16. A circle of radius r_0 is revolved about a line in the same plane at distance R_0 from the center of the circle. Assume $R_0 > r_0$. Show that the resulting torus has volume $2\pi^2 R_0 r_0^2$. (*Hint:* Use washers; then identify the difficult integral which results as the area of a circle.)
17. The rectangle $-1 \leq x \leq 1$, $-2 \leq y \leq 2$, $z = 0$ moves upwards, its center always on the z-axis, and rotates counterclockwise at a uniform rate as it rises. Suppose it has turned 90° when it reaches $z = 1$. Find the volume swept out.

3. OTHER VOLUMES

The volumes of certain figures other than solids of revolution can also be found with the tools at our disposal.

EXAMPLE 3.1

A cone of height h has an irregular base of area A. Find the volume of the cone.

Solution: Let x denote distance measured from the apex toward the plane of the base (Fig. 3.1). The typical cross-section of the cone by a plane

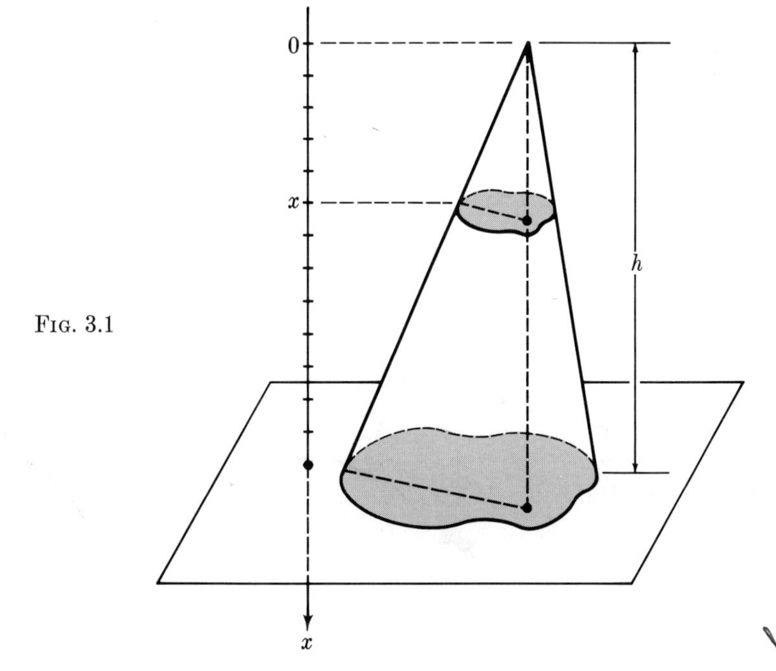

FIG. 3.1

parallel to the base, distance x from the apex, is a plane region similar to the base. The linear dimensions in this cross-section are proportional to x, so its area is proportional to x^2.

Let $f(x)$ denote this area. Then $f(x) = cx^2$. But $f(h) = A$, hence $ch^2 = A$. Therefore $c = A/h^2$ and

$$f(x) = \frac{A}{h^2} x^2.$$

Slice the cone into slabs by planes parallel to the base. A typical slab has base area $f(x)$, thickness dx, and volume

$$f(x)\,dx = \frac{A}{h^2} x^2\,dx.$$

Hence the volume of the cone is

$$\int_0^h \frac{A}{h^2} x^2 \, dx = \frac{A}{h^2}\left(\frac{1}{3} x^3\right)\Big|_0^h = \frac{1}{3} Ah.$$

Answer: $\frac{1}{3} Ah.$

Prismoidal Formula

A **prismoid** is a region in space bounded by two parallel planes and one or more surfaces joining the planes (Fig. 3.2). There is a nice formula for

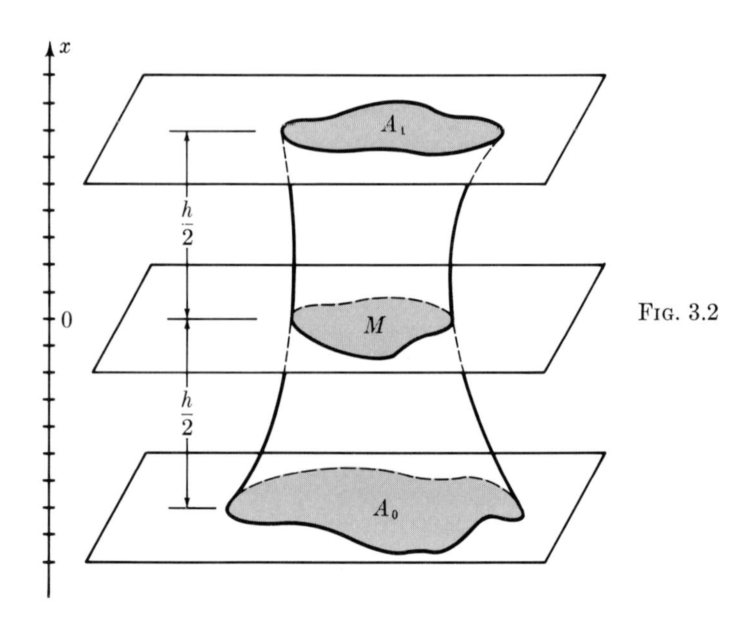

FIG. 3.2

the volume V of a prismoid whose cross-sectional area $f(x)$ is a *quadratic* function of the distance x, measured along an axis perpendicular to the base planes. The formula is

$$V = \frac{h}{6}(A_0 + 4M + A_1),$$

where h is the distance between base planes, A_0 and A_1 are the areas of the bases, and M is the cross-sectional area halfway between the bases.

To derive the formula, choose $x = 0$ at the midsection. Then x runs from $-h/2$ to $h/2$. The hypothesis is

$$f(x) = a + bx + cx^2.$$

Thus

$$A_0 = f\left(-\frac{h}{2}\right) = a - \frac{bh}{2} + \frac{ch^2}{4},$$

$$M = f(0) = a,$$

$$A_1 = f\left(\frac{h}{2}\right) = a + \frac{bh}{2} + \frac{ch^2}{4}.$$

Therefore

$$\frac{h}{6}(A_0 + 4M + A_1) = \frac{h}{6}\left(6a + \frac{ch^2}{2}\right).$$

But the volume, found by slicing, is

$$\begin{aligned}
V &= \int_{-h/2}^{h/2} f(x)\,dx \\
&= \int_{-h/2}^{h/2} (a + bx + cx^2)\,dx \\
&= \left(ax + \frac{1}{2}bx^2 + \frac{1}{3}cx^3\right)\Bigg|_{-h/2}^{h/2} \\
&= \left(\frac{ah}{2} + \frac{bh^2}{8} + \frac{ch^3}{24}\right) - \left(\frac{-ah}{2} + \frac{bh^2}{8} - \frac{ch^3}{24}\right) \\
&= ah + \frac{1}{12}ch^3 \\
&= \frac{h}{6}\left(6a + \frac{ch^2}{2}\right) \\
&= \frac{h}{6}(A_0 + 4M + A_1).
\end{aligned}$$

EXAMPLE 3.2

Find the volume of the tent in Fig. 3.3a.

Solution: Measure x from the top down. The cross-section at position x is a rectangle (Fig. 3.3b). Its shorter dimension is proportional to x and

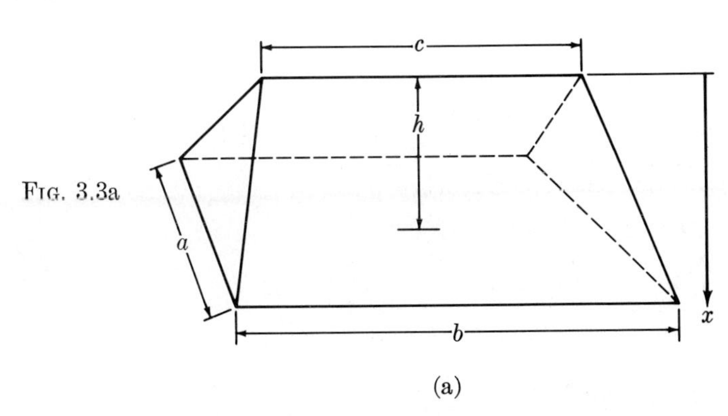

FIG. 3.3a

(a)

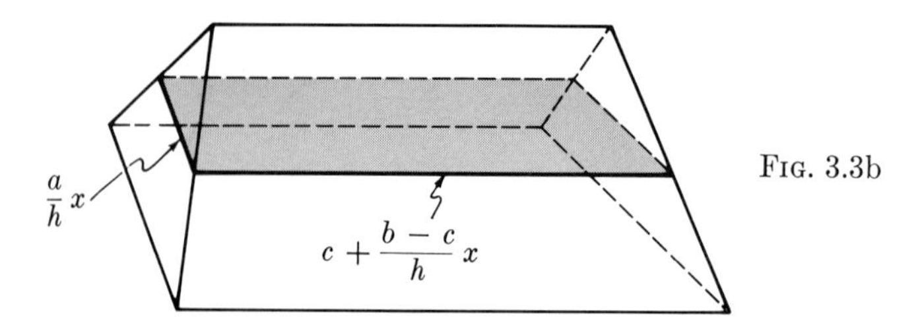

Fig. 3.3b

equal to a at $x = h$, hence it is $(a/h)x$. The longer dimension is a linear function of x, equal to c at $x = 0$ and to b at $x = h$, hence it is $c + [(b - c)/h]x$. The product of these dimensions,

$$\left(\frac{a}{h}x\right)\left(c + \frac{b-c}{h}x\right),$$

is the cross-sectional area. It is a quadratic function of x so the Prismoidal Formula applies:

$$A_0 = 0,$$

$$M = \left(\frac{a}{h}\cdot\frac{h}{2}\right)\left(c + \frac{b-c}{h}\cdot\frac{h}{2}\right) = \frac{a(b+c)}{4},$$

$$A_1 = ab.$$

Hence

$$V = \frac{h}{6}(A_0 + 4M + A_1)$$

$$= \frac{h}{6}\left(0 + 4\frac{a(b+c)}{4} + ab\right) = \frac{h}{6}(2ab + ac).$$

Answer: $\frac{ha}{6}(2b + c)$.

Check: When $c = 0$, the solid is a pyramid; when $c = b$, it is a right prism. In each case the formula above gives the correct volume.

EXAMPLE 3.3

Find the volume of a sphere of radius a by the Prismoidal Formula.

Solution: The cross-sectional area is a quadratic function, hence the formula applies with

$$h = 2a, \qquad A_0 = A_1 = 0, \qquad M = \pi a^2.$$

Conclusion:

$$V = \frac{2a}{6}(0 + 4\pi a^2 + 0).$$

Answer: $\frac{4}{3}\pi a^3$.

Volume of a Sphere

We close this chapter with a surprise. We can use the (known) volume of a sphere to find its surface area.

EXAMPLE 3.4

Find the surface area A of a sphere of radius a.

Solution: Find the volume of the sphere by slicing it into concentric spherical shells (Fig. 3.4). Let $f(x)$ be the surface area of the typical shell

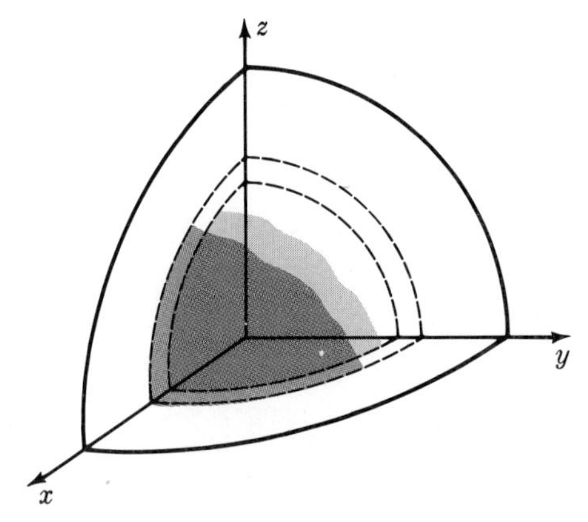

Fig. 3.4

at distance x from the center $(0 \leq x \leq a)$. Then $f(x)$ is proportional to x^2 and $f(a) = A$, hence

$$f(x) = \frac{A}{a^2} x^2.$$

The volume of the shell is

$$f(x)\,dx = \frac{A}{a^2} x^2\,dx.$$

Hence

$$V = \int_0^a \frac{A}{a^2} x^2\,dx = \frac{A}{a^2}\left(\frac{1}{3} x^3\right)\Bigg|_0^a = \frac{Aa}{3}.$$

But

$$V = \frac{4}{3} \pi a^3,$$

therefore

$$\frac{4}{3}\pi a^3 = \frac{Aa}{3}.$$

Answer: $A = 4\pi a^2$.

EXERCISES

Use the Prismoidal Formula to find the volume obtained by revolving the given region of the x, y-plane about the indicated line:

1. bounded by $x = 0, \quad x + y = 2, \quad y = 1$; about the y-axis
2. bounded by $y = 0, \quad y = 3x + 1, \quad x = 0, \quad x = 3$; about the x-axis
3. bounded by $y = 2x^2, \quad y = 2, \quad y = 4$; about the y-axis
4. bounded by $y = x^2 - 2x + 2, \quad y = 1, \quad y = 5$; about $x = 1$
5. bounded by $x^2 + (y - 2)^2 = 16, \quad y = 0, \quad y = 4$; about the y-axis
6. bounded by $(x - 4)^2 + y^2 = 25, \quad x = 2, \quad x = 6$; about the x-axis.

In Ex. 7–10, the given region of the x, y-plane is revolved about the indicated line. Find the resulting volume provided the Prismoidal Formula applies:

7. bounded by the x-axis, $y = x^2, \quad x = 1, \quad x = 4$; about the x-axis
8. bounded by $y = \dfrac{1}{x}, \quad y = 1, \quad y = 3, \quad x = 0$; about the y-axis
9. bounded by $y = 0, \quad y = \sqrt{x}, \quad x = 1, \quad x = 4$; about the x-axis
10. bounded by $x = 0, \quad y = \sqrt{x}, \quad y = 1, \quad y = 4$; about the y-axis.
11. Find the volume of the lower half (by height) of a cone of height 16 cm whose base is a semicircle of radius 4 cm surmounted by a square, one side of which is the diameter of the semicircle.
12. Find the volume of a pyramid of height 8 in., whose base is a 2×4 in. rectangle. What is the volume of the lower half (by height)?
13. An exponential horn is bounded by the 6 surfaces $x = e^{az}$, $x = -e^{az}$, $y = e^{az}$, $y = -e^{az}$, $z = 0$, $z = b$. Find its volume.

13. Partial Derivatives

1. SEVERAL VARIABLES

A geometrical or physical quantity may depend on *several* variables. Here are some examples.

(1) The speed v of sound in an ideal gas is

$$v = \sqrt{\gamma \frac{p}{D}},$$

where D is the density of the gas, p is the pressure, and γ is a constant characteristic of the gas.

(2) A certain wave is formed on a vibrating string. The vertical displacement y at time t of a point of the string initially at $(x, 0)$ is given by

$$y = A \sin kx \sin \omega t,$$

where A, k, and ω are constants (Fig. 1.1).

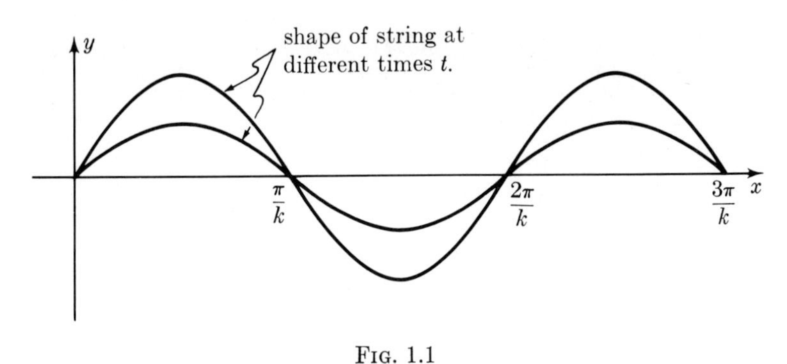

Fig. 1.1

Fig. 1.2

(3) The area of a parallelogram, given by

$$A = sb \sin \alpha,$$

depends on the three variables s, b, and α. See Fig. 1.2.

(4) The area A of a triangle with sides a, b, c is given by the formula of Heron:

$$A^2 = s(s - a)(s - b)(s - c),$$

where s is the semiperimeter $\frac{1}{2}(a + b + c)$. Thus A is a function of the three variables a, b, c.

EXERCISES

1. Express side b of a triangle as a function of side a and angles α and β (Law of Sines).
2. Express side c of a triangle as a function of sides a and b and the included angle γ (Law of Cosines).
3. Express the inradius (radius of inscribed circle) of a triangle as a function of the 3 sides a, b, and c.
4. Express the area of a rectangle as a function of its semiperimeter s and diagonal d.

2. GRAPHS

Suppose a function $z = f(x, y)$ is defined for all points (x, y) in some region of the plane. Its **graph** is the surface in space consisting of all points

$$(x, y, f(x, y)),$$

where (x, y) is in the region of definition. Figure 2.1 illustrates the graph of a function defined for all points (x, y) in a circular disk.

To get an idea of the shape of the surface, draw several sections by planes perpendicular to the x-axis or the y-axis.

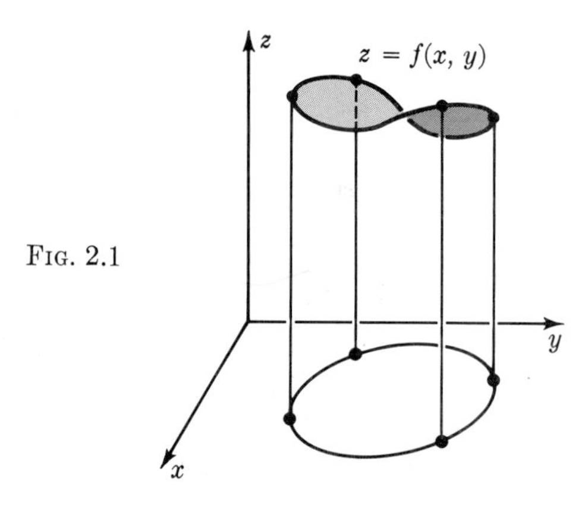

FIG. 2.1

EXAMPLE 2.1

Graph the function $z = f(x, y) = 1 - x^2$.

Solution: The function $f(x, y)$ is independent of y. Its graph is a cylinder with generators parallel to the y-axis. To see this, first graph the parabola $z = 1 - x^2$ in the x, z-plane (Fig. 2.2a). If (a, c) is any point on this parabola and b is any value of y whatsoever, then (a, b, c) is on the graph of $z = f(x, y)$.

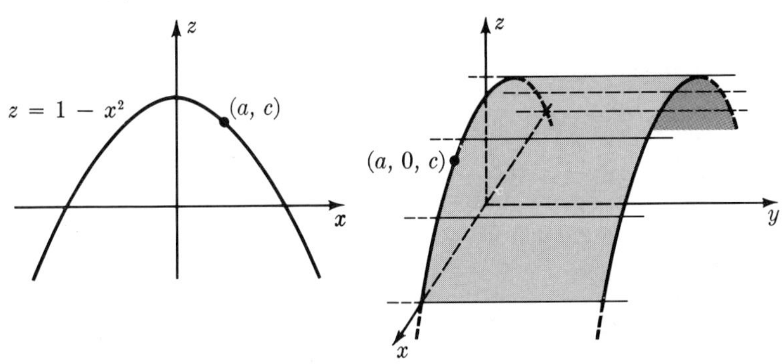

FIG. 2.2a

FIG. 2.2b

Answer: The graph is a parabolic cylinder with generators parallel to the y-axis. See Fig. 2.2b.

EXAMPLE 2.2

Graph the function $z = x^2 + y$.

Solution: Each cross-section by a plane $x = x_0$ is a straight line $z = y + x_0^2$ of slope 1. The surface meets the x, z-plane in the parabola $z = x^2$. See Fig. 2.3a. The figure does not yet convey the shape of the surface, so

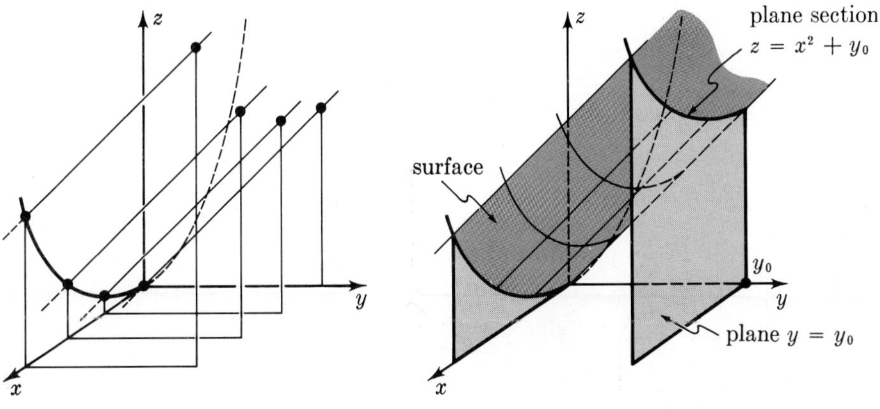

FIG. 2.3a

FIG. 2.3b

look at cross-sections by planes $y = y_0$. See Fig. 2.3b. Each is a parabola $z = x^2 + y_0$. Now the picture is clearer.

> *Answer:* The surface is a cylinder oblique to the x, z-plane; it intersects the x, z-plane in the parabola $z = x^2$.

EXAMPLE 2.3

Graph the function $z = x^2 + y^2$.

Solution: Each cross-section by a plane perpendicular to the x-axis is a parabola $z = x_0^2 + y^2$. This is shown in Fig. 2.4a. But from this figure, it is hard to visualize the surface. You learn more studying the cross-sections of the graph by planes $z = z_0$ perpendicular to the z-axis. Each cross-section is a circle $x^2 + y^2 = z_0$.

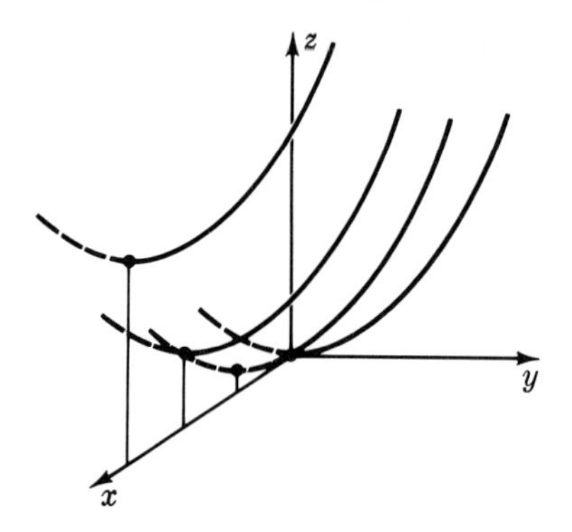

FIG. 2.4a

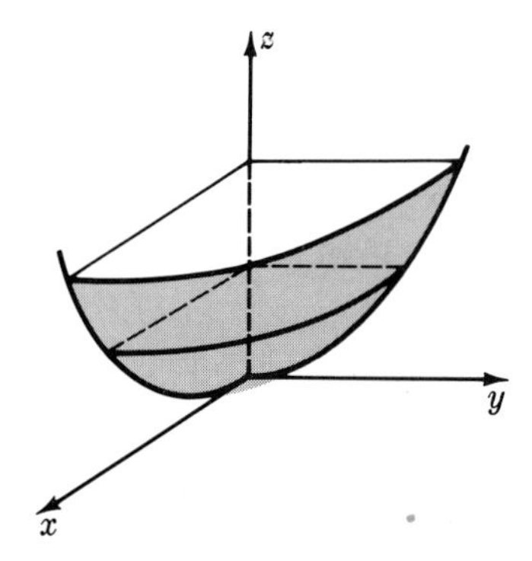

FIG. 2.4b

> *Answer:* The surface is a paraboloid of revolution (Fig. 2.4b).

EXERCISES

Sketch the graphs:

1. $z = f(x, y) = 1 - 2x$
2. $z = f(x, y) = 3 + y$
3. $z = x + y$
4. $z = 3x + 2y$
5. $z = (x - 2)^2$
6. $z = (y + 1)^3$
7. $z = x^3 + y$
8. $z = x^2 - y^2$
9. $z = x^2 + x - y$
10. $z = x^3 + y^3$
11. $z = xy$
12. $z = x^2 + xy + y^2$.
13. Graph the function $z = \sqrt{9 - x^2 - y^2}$ for all points (x, y) in the circle of radius 3 and center $(0, 0)$.

14. Graph the function $z = x^4 + y^4$ for all points (x, y) in the square with vertices $(0, 0)$, $(1, 0)$, $(1, 1)$, and $(0, 1)$.
15. Find the curves formed when planes parallel to the x, z-plane intersect the graph of $z = 2x^2 + 3y^2$.
16. Find all points common to both the plane $x = 3$ and the graph of $z = x^2y$.

3. PARTIAL DERIVATIVES

Take a function of several variables, say

$$w = f(x, y, z).$$

Suppose x and z are held fixed but y is allowed to vary. Then the rate of change of w with respect to y is

$$\frac{dw}{dy} \qquad (x \text{ and } z \text{ fixed}).$$

There is nothing whatever new for us in this situation, except we need a notation which tells us "x and z are fixed." It is customary to call this derivative the **partial derivative** or simply **partial,** and to write

$$\frac{\partial w}{\partial y}$$

instead of dw/dy. This new symbol is read "partial of w with respect to y." (The "∂" is a curly "d.")

Hold y and z fixed. Then the rate of change of w with respect to x is $\partial w/\partial x$. Similarly $\partial w/\partial z$ is the rate of change of w with respect to z when x and y are fixed.

Summarizing: if $w = f(x, y, z)$, then each of the partials

$$\frac{\partial w}{\partial x}, \quad \frac{\partial w}{\partial y}, \quad \frac{\partial w}{\partial z}$$

is the derivative of w with respect to the variable in question, taken while all other variables are held fixed.

EXAMPLE 3.1

Let $z = f(x, y) = xy^2$. Find

$$\frac{\partial z}{\partial x} \quad \text{for} \quad y = 3,$$

$$\frac{\partial z}{\partial y} \quad \text{for} \quad x = -4,$$

$$\frac{\partial z}{\partial x} \quad \text{and} \quad \frac{\partial z}{\partial y} \quad \text{in general.}$$

Solution: If $y = 3$, then $z = 9x$, and

$$\left.\frac{\partial z}{\partial x}\right|_{y=3} = \frac{d}{dx}(9x) = 9.$$

Likewise, if $x = -4$, then $z = -4y^2$, and

$$\left.\frac{\partial z}{\partial y}\right|_{x=-4} = \frac{d}{dy}(-4y^2) = -8y.$$

In general, to compute $\partial z/\partial x$ just differentiate as usual, pretending y is a constant:

$$\frac{\partial z}{\partial x} = \frac{\partial(xy^2)}{\partial x} = y^2 \frac{d}{dx}(x) = y^2.$$

To compute $\partial z/\partial y$, differentiate, pretending x is a constant:

$$\frac{\partial z}{\partial y} = \frac{\partial(xy^2)}{\partial y} = x \frac{d}{dy}(y^2) = 2xy.$$

Answer: $\left.\frac{\partial z}{\partial x}\right|_{y=3} = 9, \qquad \left.\frac{\partial z}{\partial y}\right|_{x=-4} = -8y,$

$$\frac{\partial z}{\partial x} = y^2, \qquad \frac{\partial z}{\partial y} = 2xy.$$

We consider two further examples.

(1) The gas law for a fixed mass of n moles of an ideal gas is

$$P = nR\frac{T}{V},$$

where R is the universal gas constant. Thus P is a function of the two variables T and V:

$$\frac{\partial P}{\partial T} = nR\frac{1}{V},$$

$$\frac{\partial P}{\partial V} = -nR\frac{T}{V^2}.$$

(2) The area A of a parallelogram of base b, slant height s, and angle α is $A = sb \sin \alpha$. The partial derivatives are

$$\frac{\partial A}{\partial s} = b \sin \alpha,$$

$$\frac{\partial A}{\partial b} = s \sin \alpha,$$

$$\frac{\partial A}{\partial \alpha} = sb \cos \alpha.$$

Geometric Interpretation

The graph of $z = f(x, y)$ is a surface in three dimensions. A plane $x = x_0$ cuts the graph in a plane curve $x = x_0$, $z = f(x_0, y)$. See Fig. 3.1a. If this curve is projected straight back onto the y, z-plane, the graph of the function $z = f(x_0, y)$ is obtained (Fig. 3.1b). The partial derivative

$$\frac{\partial f}{\partial y}(x_0, y)$$

is the slope of this graph.

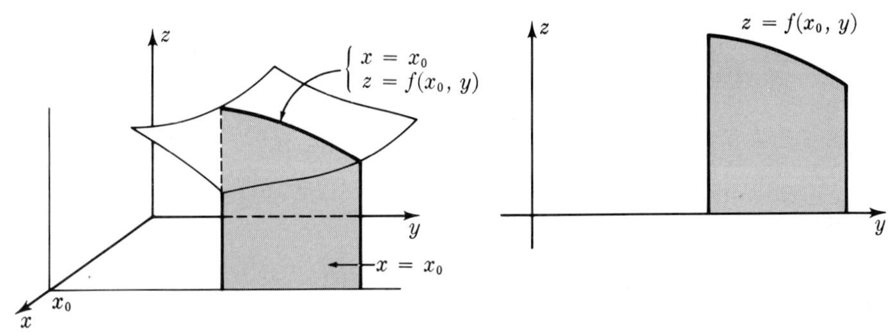

FIG. 3.1a FIG. 3.1b

For example, suppose the graph of the function $z = x + y^2$ is sliced by the plane $x = x_0$. See Fig. 3.2a. The resulting curve is the parabola

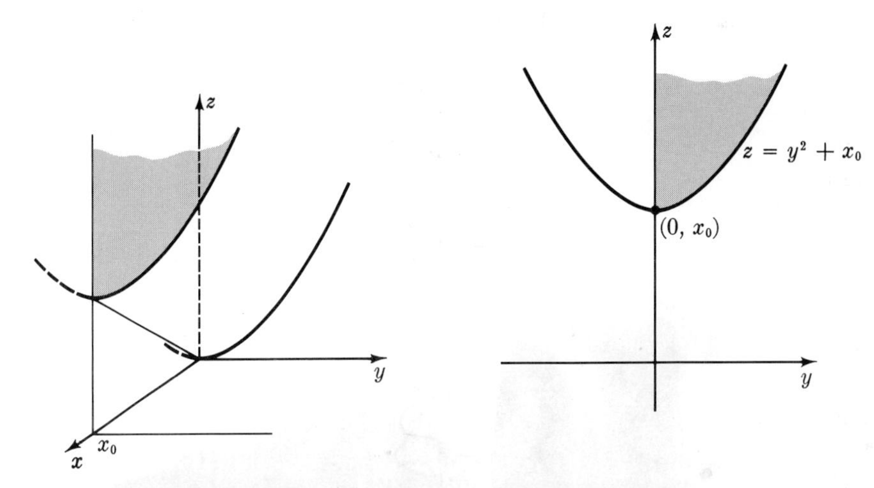

FIG. 3.2a FIG. 3.2b

$x = x_0$, $z = y^2 + x_0$. If this is projected onto the y, z-plane, a parabola is obtained (Fig. 3.2b). Its slope is

$$\frac{\partial z}{\partial y} = 2y.$$

Notation

Unfortunately there are several different notations for partial derivatives in common use. Become familiar with them; they come up again and again in applications.

Suppose

$$w = f(x, y, z).$$

Common notations for $\partial w/\partial x$ are:

$$f_x, \qquad f_x(x, y, z),$$
$$w_x, \qquad w_x(x, y, z),$$
$$D_x f.$$

For example, if

$$w = f(x, y, z) = x^3y^2 \sin z,$$

then

$$f_x = 3x^2y^2 \sin z, \qquad w_y = 2x^3y \sin z, \qquad D_z f = x^3y^2 \cos z.$$

EXERCISES

Find $\dfrac{\partial z}{\partial x}$ and $\dfrac{\partial z}{\partial y}$:

1. $z = f(x, y) = x + 2y$
2. $z = f(x, y) = 3x + 4y$
3. $z = 3xy$
4. $z = x^2y$
5. $z = \dfrac{2x^2}{y+1}$
6. $z = x^2y + xy^2$
7. $z = x \sin y$
8. $z = y^2 \cos x$
9. $z = \sin 2x + \cos 3y$
10. $z = \sin x - \cos y$
11. $z = \sin 2xy$
12. $z = \cos(2x + y)$
13. $z = \dfrac{x}{y} + \dfrac{y}{x}$
14. $z = \dfrac{x^3}{y}$
15. $z = xe^y$
16. $z = \dfrac{1}{x + 2y + 5}$
17. $z = e^{xy}$
18. $z = -3e^{x+y}$
19. $z = e^{2x} \sin y$
20. $z = e^{-y} \cos x$.
21. Let $z = x^2y$. Find $\partial z/\partial x$ for $y = 2$, and $\partial z/\partial y$ for $x = -1$.
22. Let $z = y^2/x$. Find z_x for $y = 3$.
23. Let $w = xy^2z^3$. Find w_x for $y = 2$ and $z = 2$; find w_y for $x = 1$ and $z = 0$; and find w_z for $x = y$.
24. Let $w = xy - xz - yz$. Find $\dfrac{\partial w}{\partial x} + \dfrac{\partial w}{\partial y} + \dfrac{\partial w}{\partial z}$.

4. APPLICATIONS

Suppose

$$z = z(x, y)$$

is a function of two variables defined for certain values of x and y, say

$$x_0 < x < x_1,$$

$$y_0 < y < y_1.$$

Suppose z takes on its minimum value at $(x, y) = (a, b)$. By holding y fixed at $y = b$, the function z becomes a function of x alone with minimum at $x = a$. Hence

$$\frac{\partial z}{\partial x}(a, b) = 0.$$

Similarly

$$\frac{\partial z}{\partial y}(a, b) = 0.$$

These two relations are often enough to locate the points where a function takes on its minimum (or maximum) value. See Fig. 4.1.

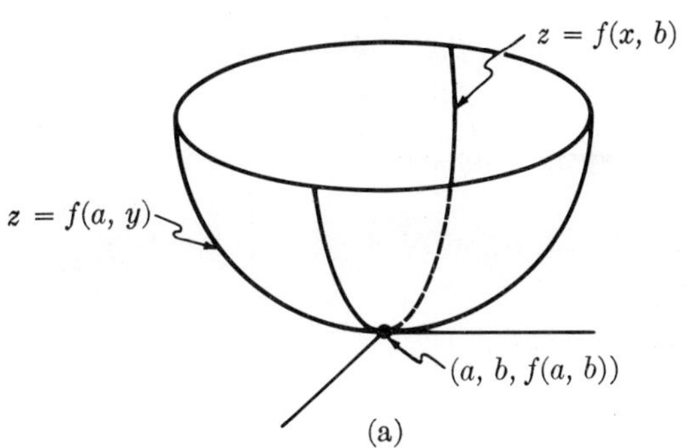

FIG. 4.1 (a) Minimum. (b) Maximum. (c) Neither. In each case $\dfrac{\partial f}{\partial x} = \dfrac{\partial f}{\partial y} = 0.$

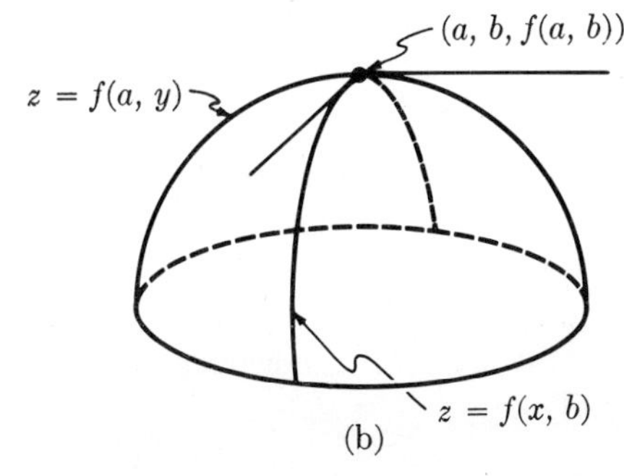

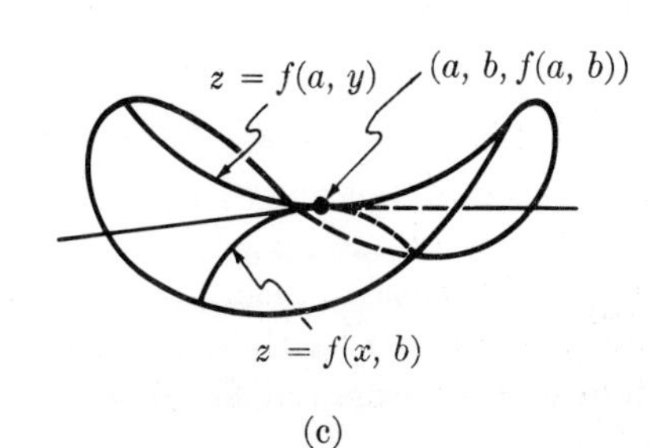

EXAMPLE 4.1

Find the minimum of $z = x^2 - xy + y^2 + 3x$.

Discussion: Is the question reasonable? For a fixed value of y, say $y = b$, the resulting function of x,

$$z(x, b) = x^2 + (3 - b)x + b^2,$$

is a quadratic polynomial whose graph is a parabola turned upward; it has a minimum. Similarly for each fixed $x = a$, the function of y,

$$z(a, y) = y^2 - ay + (a^2 + 3a),$$

has a minimum.

This is fairly good experimental evidence that the function $z(x, y)$ ought to have at least one minimum, probably no maximum.

Solution: The function is defined for all values of x and y. Begin by finding all points (x, y) at which both

$$\frac{\partial z}{\partial x} = 0 \quad \text{and} \quad \frac{\partial z}{\partial y} = 0.$$

Now

$$\frac{\partial z}{\partial x} = \frac{\partial}{\partial x}(x^2 - xy + y^2 + 3x) = 2x - y + 3,$$

$$\frac{\partial z}{\partial y} = \frac{\partial}{\partial y}(x^2 - xy + y^2 + 3x) = -x + 2y,$$

so the conditions are

$$\begin{cases} 2x - y + 3 = 0 \\ -x + 2y = 0. \end{cases}$$

Solve:

$$x = -2, \qquad y = -1.$$

The corresponding value of z is

$$z(-2, -1) = (-2)^2 - (-2)(-1) + (-1)^2 + 3(-2)$$
$$= 4 - 2 + 1 - 6 = -3.$$

Is this value a maximum, a minimum, or neither? (Recall that the vanishing of the derivative of a function $f(x)$ does not guarantee a maximum or minimum, e.g., $f(x) = x^3$ at $x = 0$.)

In this case you can prove that the value $z(-2, -1) = -3$ is the minimum value of z by these algebraic steps. First set up a u, v-coordinate system with its origin at $(x, y) = (-2, -1)$. Take

$$x = u - 2, \qquad y = v - 1,$$

Then

$$z = (u - 2)^2 - (u - 2)(v - 1) + (v - 1)^2 + 3(u - 2)$$
$$= u^2 - uv + v^2 - 3.$$

Next, complete the square:

$$z = \left(u - \frac{v}{2}\right)^2 - \frac{v^2}{4} + v^2 - 3$$

$$= \left(u - \frac{v}{2}\right)^2 + \frac{3}{4}v^2 - 3.$$

Since squares are nonnegative, $z(x, y) \geq -3$.

Answer: -3.

In general, to find possible maximum and minimum values of a function, locate points where all its partial derivatives are zero. Whether a particular one of these actually yields a maximum or a minimum may not be easy to determine. (Later we shall study a second derivative test which sometimes helps.)

EXAMPLE 4.2

Find the rectangular solid of maximum volume whose total edge length is a given constant.

Solution: As drawn in Fig. 4.2, the total length of the 12 edges is

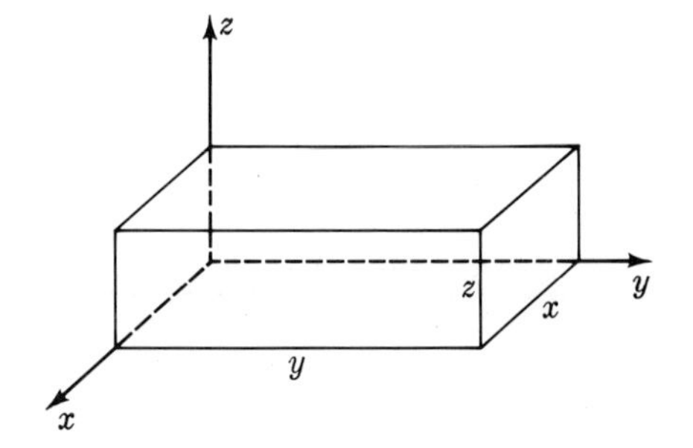

Fig. 4.2

$4x + 4y + 4z$. Thus $4x + 4y + 4z = 4k$, where k is a constant, so

$$x + y + z = k.$$

The volume is

$$V = xyz = xy(k - x - y)$$

$$= kxy - x^2y - xy^2.$$

The conditions for a maximum,

$$\frac{\partial V}{\partial x} = 0, \qquad \frac{\partial V}{\partial y} = 0,$$

are

$$\begin{cases} ky - 2xy - y^2 = 0 \\ kx - x^2 - 2xy = 0. \end{cases}$$

The nature of the geometric problem requires $x > 0$ and $y > 0$. Thus we may cancel y from the first equation and x from the second:

$$\begin{cases} 2x + y = k \\ x + 2y = k. \end{cases}$$

This pair of simultaneous linear equations has the unique solution

$$x = \frac{k}{3}, \qquad y = \frac{k}{3}.$$

Hence also $z = k/3$; the solid is a cube.

Answer: A cube.

EXAMPLE 4.3

What is the largest possible volume, and what are the dimensions of an open rectangular aquarium constructed from 12 ft² of Plexiglas? Ignore the thickness of the plastic.

Solution: See Fig. 4.3. The volume is $V = xyz$. The total surface area

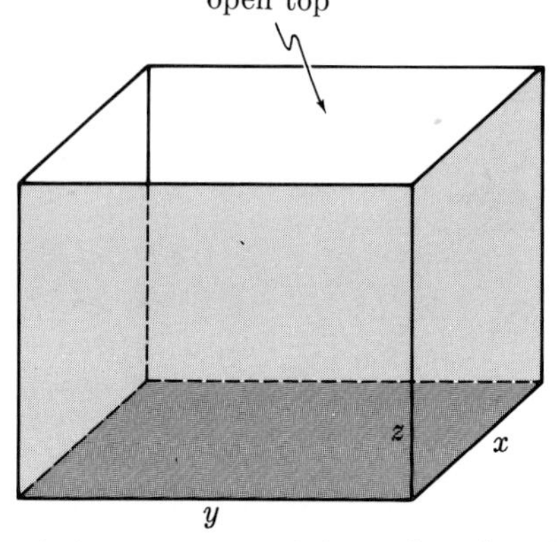

FIG. 4.3

of the bottom and four sides is

$$xy + 2yz + 2zx = 12.$$

Solve for z, then substitute into the formula for V:

$$z = \frac{12 - xy}{2(x + y)},$$

$$V = \frac{(12 - xy)xy}{2(x + y)}.$$

We have not discussed differentiating functions this complicated. Our choices are (a) abandoning the problem because the first way we set it up failed, or (b) trying something else.

Let us analyze what we did. We chose the side lengths as variables because volume and surface area can be expressed in terms of these side lengths. Is this the only way? Perhaps we can use the face areas as variables. They are

$$u = yz, \qquad v = zx, \qquad w = xy.$$

The total surface area, in terms of u, v, w, is

$$w + 2u + 2v = 12.$$

Can we express the volume in terms of u, v, and w? It is unnecessary to solve the system

$$yz = u, \qquad zx = v, \qquad xy = w$$

for x, y, z in terms of u, v, w, and then substitute the resulting expressions in $V = xyz$. We can express V directly in u, v, w:

$$V^2 = (xyz)^2 = x^2y^2z^2 = (yz)(zx)(xy) = uvw.$$

The maximum value of V corresponds to the maximum value of V^2. Now we have formulated a reasonable problem: Maximize uvw subject to the constraint $2u + 2v + w = 12$.

Set

$$\begin{aligned} g(u, v) &= uv(12 - 2u - 2v) \\ &= 12uv - 2u^2v - 2uv^2. \end{aligned}$$

Take partials:

$$\begin{aligned} g_u &= 12v - 4uv - 2v^2, \\ g_v &= 12u - 2u^2 - 4uv. \end{aligned}$$

These partials are 0 where the equations

$$\begin{cases} 12v - 4uv - 2v^2 = 0 \\ 12u - 2u^2 - 4uv = 0 \end{cases}$$

are satisfied. Cancel the factor v in the first and u in the second ($v = 0$ or $u = 0$ doesn't hold water). The equations reduce to

$$\begin{cases} 2u + v = 6 \\ u + 2v = 6. \end{cases}$$

Hence $u = v = 2$. It follows that $w = 4$ and $V^2 = uvw = 16$. The answer, however, should be in terms of x, y, and z. Recall

$$yz = u = 2, \qquad zx = v = 2, \qquad xy = w = 4.$$

Hence $x = y$ and $xy = x^2 = w = 4$. Therefore $x = y = 2$ and $z = 1$.

Answer: The maximum volume is 4 ft³, achieved by a tank of base 2 ft × 2 ft and height 1 ft.

EXERCISES

Find the possible maximum and minimum values:

1. $z = 4 - 2x^2 - y^2$
2. $z = x^2 + y^2 - 1$
3. $z = (x-2)^2 + (y+3)^2$
4. $z = (x-1)^2 + y^2 + 3$
5. $z = x^2 - 2xy + 2y^2 + 4$
6. $z = xy - x^2 - 2y^2 + x + 2y$
7. $z = x - y^2 - x^3$
8. $z = 3x + 12y - x^3 - y^3$.
9. Find the rectangular solid of greatest volume whose total surface area is 24 ft².
10. Find the dimensions of an open-top rectangular box of given volume V, if the surface area is to be a minimum.
11. Find the dimensions of the cheapest open-top rectangular box of given volume V, if the base material costs 3 times as much per ft² as the side material.

5. PARTIAL DIFFERENTIAL EQUATIONS [optional]

An equation which involves a function of several variables and its partial derivatives is called a **partial differential equation.**

EXAMPLE 5.1

Find all functions $z = f(x, y)$ which satisfy the partial differential equation

$$\frac{\partial z}{\partial x} = 0.$$

Solution: Let f be such a function. The equation means that for a constant y_0, the function

$$f(x, y_0)$$

of one variable x has derivative zero, hence is constant:

$$f(x, y_0) = c_0.$$

If another value y_1 is chosen, the function $f(x, y_1)$ is also constant:

$$f(x, y_1) = c_1.$$

There need be no relation between c_0 and c_1. Each choice of y leads to *some* constant c. If y changes, c changes. In other words, c is a function of y.

It is confusing to use "c" for a function, so write $c = g(y)$.

Answer: All functions $f(x, y) = g(y)$ which depend on y alone.

REMARK: A function $f(x, y) = g(y)$ which does not involve x is called **independent of** x.

EXAMPLE 5.2

A function $w = w(x, y, z)$ satisfies the system of partial differential equations

$$\begin{cases} \dfrac{\partial w}{\partial x} = 0 \\ \dfrac{\partial w}{\partial y} = 0. \end{cases}$$

Find all such functions w.

Solution: If w satisfies the system, then it satisfies each of the equations. The first says that w is independent of x and the second that it is independent of y.

Answer: All functions $w(x, y, z) = h(z)$ which depend on z alone.

EXAMPLE 5.3

Show that the function $z = (x - y)^2$ satisfies the partial differential equation

$$\frac{\partial z}{\partial x} + \frac{\partial z}{\partial y} = 0.$$

Solution:

$$z = x^2 - 2xy + y^2,$$

$$\frac{\partial z}{\partial x} = 2x - 2y, \qquad \frac{\partial z}{\partial y} = -2x + 2y,$$

$$\frac{\partial z}{\partial x} + \frac{\partial z}{\partial y} = (2x - 2y) + (-2x + 2y) = 0.$$

EXAMPLE 5.4

Let $z = x + g(y)$, where g is any function of one variable. Show that z satisfies the partial differential equation

$$\frac{\partial z}{\partial x} = 1.$$

Solution:

$$\frac{\partial z}{\partial x} = \frac{\partial (x)}{\partial x} + \frac{\partial}{\partial x}[g(y)] = 1 + 0 = 1.$$

EXAMPLE 5.5

Find all functions $z = f(x, y)$ such that

$$\frac{\partial z}{\partial x} = 2x.$$

Solution: One such function is x^2. If $f(x, y)$ is any such function, then

$$\frac{\partial}{\partial x}[f(x, y) - x^2] = 2x - 2x = 0.$$

Hence

$$f(x, y) - x^2$$

is independent of x; it is a function $g(y)$ of y alone.

Answer: $f(x, y) = x^2 + g(y)$.

EXAMPLE 5.6

Let $z = e^x g(y)$, where $g(y)$ is a function of y alone. Show that

$$\frac{\partial z}{\partial x} = z.$$

Solution: The variable y is fixed when $\partial z/\partial x$ is computed. Hence $g(y)$ acts like a constant:

$$\frac{\partial z}{\partial x} = \frac{\partial}{\partial x}[e^x g(y)] = g(y)\frac{\partial}{\partial x}[e^x]$$

$$= g(y)e^x = z.$$

EXAMPLE 5.7

Let $z = \dfrac{-1}{x + g(y)}\cdot$ Show that $\dfrac{\partial z}{\partial x} = z^2$.

Solution: When differentiated with respect to x, the function $g(y)$ acts like a constant. Hence

$$\frac{\partial z}{\partial x} = \frac{\partial}{\partial x}\left[\frac{-1}{x + g(y)}\right] = \frac{1}{[x + g(y)]^2} = z^2.$$

EXERCISES

Find all functions $z = z(x, y)$ which satisfy the partial differential equation(s):

1. $\dfrac{\partial z}{\partial y} = 0$

2. $\dfrac{\partial z}{\partial x} = 0, \quad \dfrac{\partial z}{\partial y} = 1$

3. $\dfrac{\partial z}{\partial x} = 2x, \quad \dfrac{\partial z}{\partial y} = 3$

4. $\dfrac{\partial z}{\partial x} = -1, \quad \dfrac{\partial z}{\partial y} = 3y^2$

5. $\dfrac{\partial z}{\partial x} = \dfrac{1}{x^2}$

6. $\dfrac{\partial z}{\partial x} = -\dfrac{1}{x^2}, \quad \dfrac{\partial z}{\partial y} = -\dfrac{1}{y^2}$

7. $\dfrac{\partial z}{\partial x} = \sin x, \quad \dfrac{\partial z}{\partial y} = \cos y$

8. $\dfrac{\partial z}{\partial x} = 2e^{2x}, \quad \dfrac{\partial z}{\partial y} = y^2.$

9. Show that $z = (3x - y)^2$ satisfies the partial differential equation $\dfrac{\partial z}{\partial x} + 3\dfrac{\partial z}{\partial y} = 0.$

10. Show that $z = f(x) + y^2$ satisfies $\dfrac{\partial z}{\partial y} = 2y.$

11. Let $z = x^2 - y^2$. Show that $\left(\dfrac{\partial z}{\partial x} + \dfrac{\partial z}{\partial y}\right)\left(\dfrac{\partial z}{\partial x} - \dfrac{\partial z}{\partial y}\right) = 4z.$

12. Find all functions $w = f(x, y, z)$ which satisfy $w_x = 2x$, $w_y = 0$, and $w_z = 1$.

13. Find all functions $w = f(x, y, z)$ which satisfy $\dfrac{\partial w}{\partial x} = a$, $\dfrac{\partial w}{\partial y} = b$, and $\dfrac{\partial w}{\partial z} = c$, where a, b, and c are constants.

14. Find a function $z = f(x, y)$ which satisfies $\dfrac{\partial z}{\partial x} = 2xy$ and $\dfrac{\partial z}{\partial y} = x^2$.

15. Find a function $z = f(x, y)$ which satisfies $\dfrac{\partial z}{\partial x} = e^x \sin y$ and $\dfrac{\partial z}{\partial y} = e^x \cos y$.

16. Find a function $z = f(x, y)$ which satisfies $\partial z/\partial x = e^x y$ and $\partial z/\partial y = e^x + y$.

17. Find a function $z = f(x, y)$ which satisfies $\partial z/\partial x = x^2 y + \cos y$ and $\partial z/\partial y = \frac{1}{3}x^3 - x \sin y + y + 1$.

18. Find constants a and b so that the function $z = x^3 y^2$ satisfies the differential equation $ax\dfrac{\partial z}{\partial x} + by\dfrac{\partial z}{\partial y} = 0.$

14. Approximation Methods

1. INTRODUCTION

One of the fundamental uses of Differential Calculus is the approximation of a curve by its tangent. In this chapter we shall study such straight line approximations, their applications, and their generalizations: approximations by quadratic and cubic curves.

2. APPROXIMATING A CURVE BY ITS TANGENT

Under a microscope a smooth curve appears nearly straight (Fig. 2.1). Its tangent at $(a, f(a))$ is quite close to the curve, at least in a small

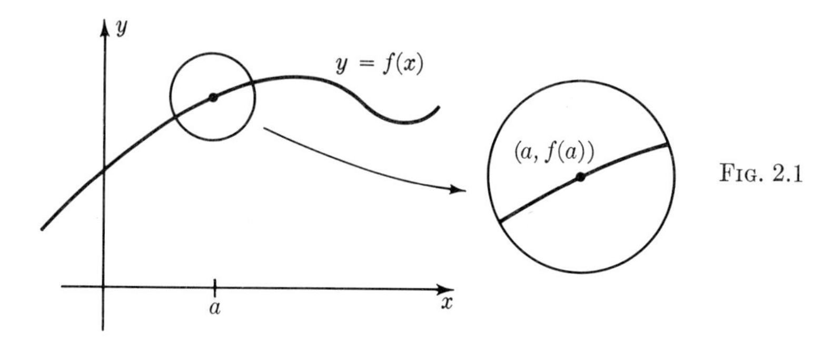

FIG. 2.1

neighborhood of that point (Fig. 2.2). The slope of the tangent line is $f'(a)$. By the point-slope formula, the equation of the tangent line is:

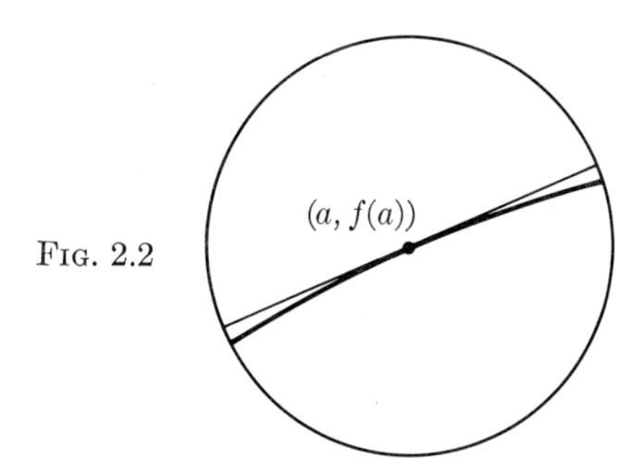

FIG. 2.2

$$y - f(a) = f'(a)(x - a) \qquad \text{or} \qquad y = f(a) + f'(a)(x - a).$$

Since the tangent is close to the curve, it is reasonable to expect that the linear function $f(a) + f'(a)(x - a)$ is a good approximation to the value of $f(x)$, at least for x near a. A good linear approximation is often useful, especially if $f(x)$ is hard to compute.

EXAMPLE 2.1

Find the equation of the line tangent to $y = 4 - (x - 2)^2$ at $(1, 3)$. How closely does it approximate the curve?

Solution: First check the consistency of the data. Is the point $(1, 3)$ really on the curve? Yes, because $3 = 4 - (1 - 2)^2$. Next,

$$y = 4 - (x - 2)^2,$$

$$y' = -2(x - 2).$$

The tangent at $(1, 3)$ has slope $y'(1) = 2$. Its equation is

$$y - 3 = 2(x - 1) \qquad \text{or} \qquad y = 2x + 1.$$

Sketch the curve and its tangent (Fig. 2.3). To see how closely the line

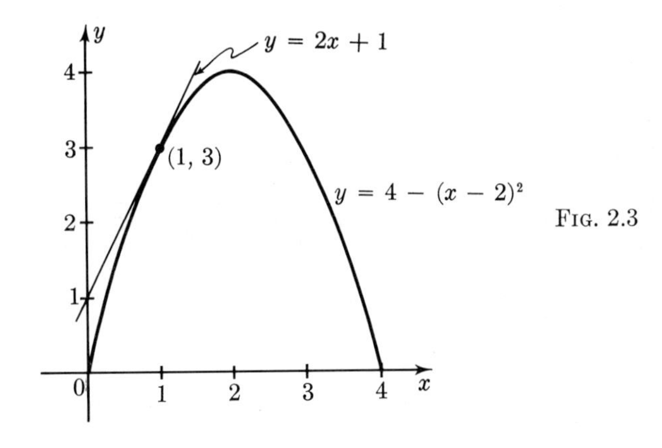

FIG. 2.3

approximates the curve, compute the difference

$$\begin{aligned} y(\text{curve}) - y(\text{line}) &= [4 - (x - 2)^2] - [2x + 1] \\ &= -x^2 + 2x - 1 \\ &= -(x - 1)^2. \end{aligned}$$

Call this difference $e(x)$, the error in approximating the curve by its tangent. Because of the exponent 2, the error is very small when x is near 1. For example, when $x = 1.1$, the curve and the line are 0.01 units apart; when $x = 1.01$, they are 0.0001 units apart.

> *Answer:* $y = 2x + 1$, very close to the curve near $(1, 3)$ since the error $e(x) = -(x - 1)^2$.

The result of this example can be interpreted numerically: $2x + 1$ is a good approximation to $4 - (x - 2)^2$ for values of x near 1. For example, if $0.99 \le x \le 1.01$, the approximation is correct to within 0.0001. When x is within 2 decimal places of 1, then $2x + 1$ is within 4 decimal places of $4 - (x - 2)^2$. This suggests that the error $e(x)$ not only approaches zero as $x \longrightarrow 1$, but also $e(x) \longrightarrow 0$ faster than $x \longrightarrow 1$. To see why, look at the ratio of $e(x)$ to $x - 1$:

$$\frac{e(x)}{x-1} = \frac{-(x-1)^2}{x-1} = -(x-1) \longrightarrow 0 \qquad \text{as} \quad x \longrightarrow 1.$$

Both quantities are small when x is near 1, but $e(x)$ is much smaller.

EXAMPLE 2.2

Find the equation of the line tangent to $y = (x + 1)(x - 2)^2$ at $(-1, 0)$. How closely does it approximate the curve?

Solution: The data is consistent since

$$0 = (-1 + 1)(-1 - 2)^2.$$

Multiply out:

$$y = (x + 1)(x - 2)^2 = x^3 - 3x^2 + 4.$$

Hence

$$y' = 3x^2 - 6x,$$

$$y'(-1) = 9.$$

The equation of the tangent is

$$y - 0 = 9(x + 1).$$

Sketch the curve and its tangent (Fig. 2.4). The error is

$$\begin{aligned} e(x) &= y(\text{curve}) - y(\text{line}) \\ &= (x+1)(x-2)^2 - 9(x+1) \\ &= (x+1)[(x-2)^2 - 9] \\ &= (x+1)(x^2 - 4x - 5) \\ &= (x+1)^2(x-5). \end{aligned}$$

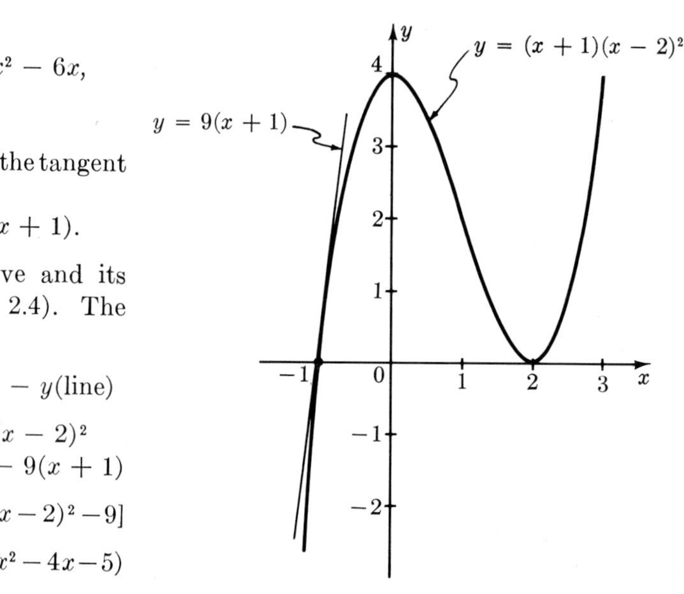

FIG. 2.4

The error is very small when x is near -1 because of the factor $(x + 1)^2$.

The other factor, $(x - 5)$, does no harm; it is approximately -6 when x is near -1.

> *Answer:* $y = 9(x + 1)$. The tangent is very close to the curve near $(-1, 0)$ since $e(x) \approx -6(x + 1)^2$.

REMARK: This example shows

$$(x + 1)(x - 2)^2 \approx 9(x + 1) \qquad \text{for} \quad x \approx -1.$$

More precisely,

$$(x + 1)(x - 2)^2 = 9(x + 1) + e(x),$$

where

$$e(x) = (x + 1)^2(x - 5).$$

Just as $e(x)/(x - 1) \longrightarrow 0$ in the last example, here

$$\frac{e(x)}{x + 1} = (x + 1)(x - 5) \longrightarrow 0 \qquad \text{as} \quad x \longrightarrow -1.$$

These examples illustrate a general principle:

> The equation of the tangent line to $y = f(x)$ at $(a, f(a))$ is
>
> $$y = f(a) + f'(a)(x - a).$$
>
> Let
>
> $$e(x) = f(x) - [f(a) + f'(a)(x - a)]$$
>
> denote the error in approximating the graph by the tangent. Then
>
> $$\frac{e(x)}{x - a} \longrightarrow 0 \qquad \text{as} \quad x \longrightarrow a.$$

The reason is simple:

$$\begin{aligned} e(x) &= f(x) - [f(a) + f'(a)(x - a)] \\ &= [f(x) - f(a)] - f'(a)(x - a), \end{aligned}$$

$$\frac{e(x)}{x - a} = \frac{f(x) - f(a)}{x - a} - f'(a).$$

But as $x \longrightarrow a$,

$$\frac{f(x) - f(a)}{x - a} \longrightarrow f'(a).$$

Therefore,

$$\frac{e(x)}{x - a} \longrightarrow f'(a) - f'(a) = 0 \qquad \text{as} \quad x \longrightarrow a.$$

EXERCISES

Find the equation of the line tangent to the curve at the specified point; also find the error made in approximating the curve by its tangent:

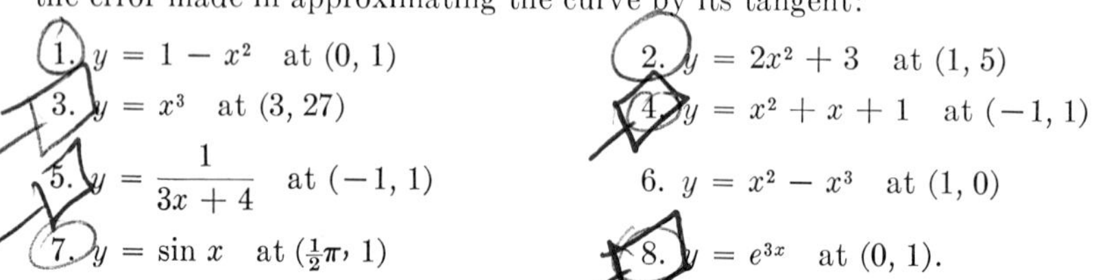

1. $y = 1 - x^2$ at $(0, 1)$
2. $y = 2x^2 + 3$ at $(1, 5)$
3. $y = x^3$ at $(3, 27)$
4. $y = x^2 + x + 1$ at $(-1, 1)$
5. $y = \dfrac{1}{3x + 4}$ at $(-1, 1)$
6. $y = x^2 - x^3$ at $(1, 0)$
7. $y = \sin x$ at $(\frac{1}{2}\pi, 1)$
8. $y = e^{3x}$ at $(0, 1)$.
9. In Ex. 3, let $e(x)$ be the error made in approximating the curve by its tangent. Find the ratio $\dfrac{e(x)}{x - 3}$.
10. Suppose that the population of a country is given by $P = 10^7 e^{t/10}$, where t is the time in years, initially 0. If the growth is assumed linear, how large is the error after 10 years?

Justify the approximation near $x = 0$:

11. $\sin x \approx x$
12. $e^x \approx 1 + x$
13. $\dfrac{1}{1 - x} \approx 1 + x$
14. $\cos x \approx 1$.

3. ROOT FINDING

One of the most common problems in all branches of mathematics is that of solving difficult equations, for example:

$$x^5 - x - 3 = 0, \qquad \cos x = x, \qquad e^x = 4x^2.$$

There are no formulas for exact solutions of these equations. The best we can hope for are close numerical approximations to solutions.

It is customary to write equations so that the right side is zero, for example:

$$x^5 - x - 3 = 0, \qquad \cos x - x = 0, \qquad e^x - 4x^2 = 0.$$

Each equation is then of the form

$$f(x) = 0.$$

A value a for which $f(a) = 0$ is called a **root** or **solution** of the equation $f(x) = 0$. It is also called a **zero** of the function $f(x)$.

Bisection Method

The roots of $f(x) = 0$ are precisely those values of x for which the graph $y = f(x)$ intersects the x-axis. This point of view suggests a "brute force" method of approximating a root if the graph crosses the x-axis. Find two numbers x_0 and x_1 such that $f(x_0) < 0$ and $f(x_1) > 0$. The graph $y = f(x)$ must cross the x-axis between x_0 and x_1 (Fig. 3.1). Let x_2 be the midpoint

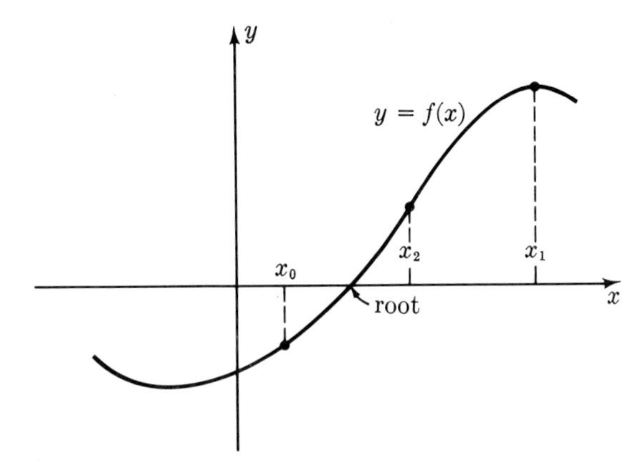

FIG. 3.1

of the segment from x_0 to x_1. Compute $f(x_2)$. If $f(x_2) > 0$, there is a root between x_0 and x_2 (as in Fig. 3.1); if $f(x_2) < 0$, there is a root between x_1 and x_2. In either case, let x_3 be the midpoint of the segment in which the root lies, and repeat the process.

This procedure is called **bisection.** At each step the root is "trapped" in an interval one half the length of the preceding interval. Therefore, after a sufficient number of repetitions, the root is trapped in as small an interval as desired. Thus, any degree of accuracy can be attained by a sufficient amount of work.

Newton–Raphson Method

Another scheme for finding roots is the **Newton–Raphson Method** (also known simply as **Newton's Method**). It is generally more efficient than the Bisection Method because it uses more information about the function.

The idea of the Newton–Raphson Method is simple. You want to know where the graph $y = f(x)$ crosses the x-axis. From a sketch of the graph, or by trial and error, make a guess, x_0. At the point $(x_0, f(x_0))$ of the graph draw the tangent line (Fig. 3.2). The tangent line crosses the x-axis at a point $(x_1, 0)$. If the guess x_0 is close to a root, Fig. 3.2 suggests that x_1 is even closer.

To compute x_1, write the equation of the tangent line:

$$y - f(x_0) = f'(x_0)\,(x - x_0).$$

Since $(x_1, 0)$ is on this line,

$$0 - f(x_0) = f'(x_0)\,(x_1 - x_0).$$

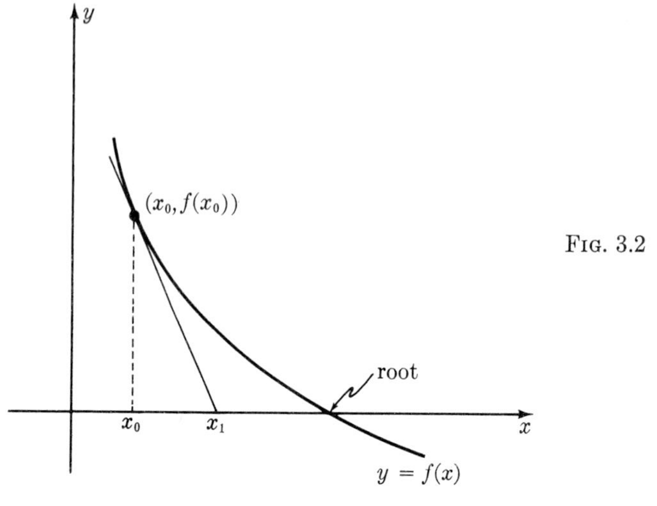

FIG. 3.2

Solve for x_1:

$$x_1 = x_0 - \frac{f(x_0)}{f'(x_0)}.$$

Now repeat the process with x_1 as your new guess (Fig. 3.3). Construct

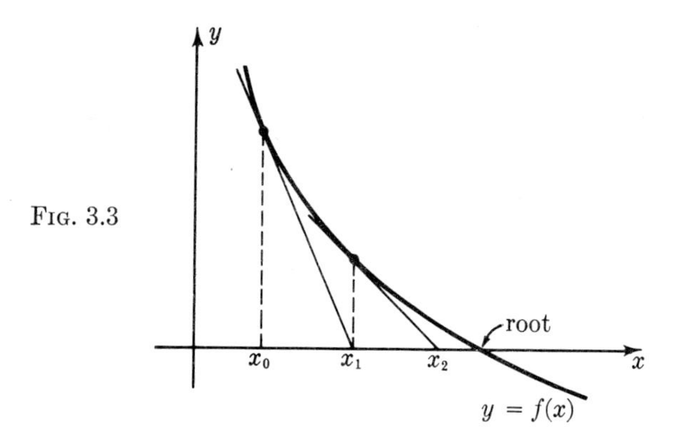

FIG. 3.3

the tangent at $(x_1, f(x_1))$. It crosses the x-axis at x_2, where

$$x_2 = x_1 - \frac{f(x_1)}{f'(x_1)}.$$

Repeat the procedure, obtaining a sequence of approximations $x_0, x_1, x_2, \cdots$, where

$$x_{i+1} = x_i - \frac{f(x_i)}{f'(x_i)}.$$

A typical sequence is sketched in Fig. 3.4.

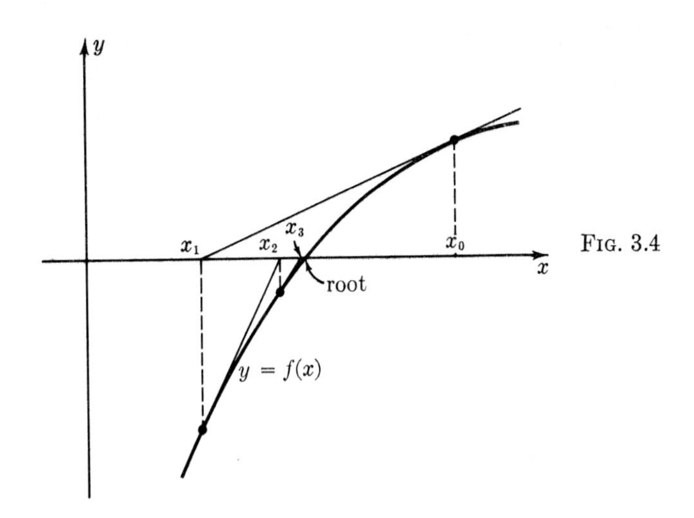

FIG. 3.4

EXAMPLE 3.1

Calculate $\sqrt{2}$. More precisely, use the Newton–Raphson Method to estimate the positive zero of $f(x) = x^2 - 2$. Choose $x_0 = 1$, and obtain the approximations x_1, x_2, and x_3.

Solution:

$$x_{i+1} = x_i - \frac{f(x_i)}{f'(x_i)}.$$

Since $f(x) = x^2 - 2$,

$$x_{i+1} = x_i - \frac{x_i^2 - 2}{2x_i} = \frac{x_i^2 + 2}{2x_i},$$

$$x_{i+1} = \frac{x_i}{2} + \frac{1}{x_i}.$$

Starting with $x_0 = 1$,

$$x_1 = \frac{x_0}{2} + \frac{1}{x_0} = \frac{1}{2} + 1 = \frac{3}{2},$$

$$x_2 = \frac{x_1}{2} + \frac{1}{x_1} = \frac{3}{4} + \frac{2}{3} = \frac{17}{12},$$

$$x_3 = \frac{x_2}{2} + \frac{1}{x_2} = \frac{17}{24} + \frac{12}{17} = \frac{577}{408}.$$

Answer: $x_1 = \frac{3}{2}$, $x_2 = \frac{17}{12}$, $x_3 = \frac{577}{408}$.

Note that

$$x_1 = 1.5, \qquad x_2 = 1.41666 \cdots, \qquad x_3 = 1.4142156 \cdots.$$

From a 6-place table,

$$\sqrt{2} = 1.414214 \cdots.$$

Thus x_3 is correct to within 0.000002.

REMARK: The formula for x_{i+1} in this example may be rewritten

$$x_{i+1} = \frac{1}{2}\left(x_i + \frac{2}{x_i}\right),$$

showing that x_{i+1} is the average of x_i and $2/x_i$. Likewise, approximating $\sqrt{n}$ for any $n > 0$ by the Newton–Raphson Method leads to a similar sequence of estimates:

$$x_{i+1} = \frac{1}{2}\left(x_i + \frac{n}{x_i}\right).$$

Each x_{i+1} is the average of x_i and n/x_i.

EXAMPLE 3.2

Estimate the smallest solution of the equation

$$x^3 - 6x^2 + 9x - 1 = 0.$$

Solution: Let $f(x) = x^3 - 6x^2 + 9x - 1$. Sketch $y = f(x)$ to get a rough idea of where zeros occur.

$$\begin{aligned} f(x) &= x^3 - 6x^2 + 9x - 1, \\ f'(x) &= 3x^2 - 12x + 9 = 3(x^2 - 4x + 3) \\ &= 3(x-1)(x-3). \end{aligned}$$

The curve is flat at $x = 1$ and $x = 3$. Use this information also to plot a few points, then sketch the graph (Fig. 3.5). The figure shows three zeros,

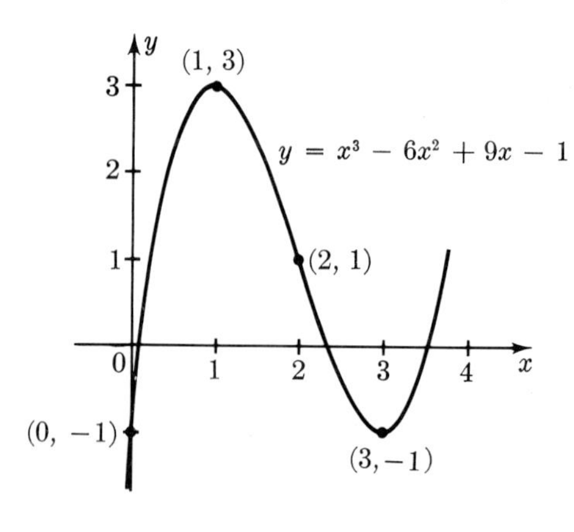

FIG. 3.5

the smallest of which is barely positive. As a first guess try $x_0 = 0.1$. Then use the formula:

$$x_{i+1} = x_i - \frac{f(x_i)}{f'(x_i)}$$

$$= x_i - \frac{x_i^3 - 6x_i^2 + 9x_i - 1}{3(x_i - 1)(x_i - 3)}.$$

Thus

$$x_1 = x_0 - \frac{x_0^3 - 6x_0^2 + 9x_0 - 1}{3(x_0 - 1)(x_0 - 3)}$$

$$= 0.1 - \frac{(0.1)^3 - 6(0.1)^2 + 9(0.1) - 1}{3(-0.9)(-2.9)}$$

$$= 0.1 + \frac{0.159}{7.83} \approx 0.120.$$

Repetition of the process yields:

$$x_2 \approx 0.1206,$$

$$x_3 \approx 0.12061.$$

Since x_2 and x_3 agree to 4 decimal places, it appears that 0.12061 is accurate to at least 4 places.

As a check, compute $f(x_3)$. If x_3 is close to a zero of $f(x)$, then $f(x_3)$ should be close to 0. In fact

$$f(x_3) \approx -0.000036.$$

Answer: 0.12061.

(Starting with the estimates 0 and $\frac{1}{2}$ and using the Bisection Method, it takes 10 iterations to achieve 4-place accuracy.)

REMARK: The notation can be simplified to avoid writing the subscript i repeatedly. This has two advantages: (1) you reduce the possibility of error by writing less and (2) you may wish to manipulate the expression

$$x_i - \frac{f(x_i)}{f'(x_i)}$$

into a form requiring fewer computational steps. Simplify the notation by setting

$$g(x) = x - \frac{f(x)}{f'(x)}.$$

Then

$$x_{i+1} = g(x_i).$$

In the example above, $f(x) = x^3 - 6x^2 + 9x - 1$,

$$g(x) = x - \frac{f(x)}{f'(x)}$$

$$= x - \frac{x^3 - 6x^2 + 9x - 1}{3x^2 - 12x + 9}$$

$$= x - \frac{x(x-3)^2 - 1}{3(x-1)(x-3)}.$$

This form is economical for computation because each computed value of $(x - 3)$ is used twice.

EXAMPLE 3.3

Estimate the solutions of $\cos x = x$.

Solution: On the same graph plot $y = \cos x$ and $y = x$. See Fig. 3.6.

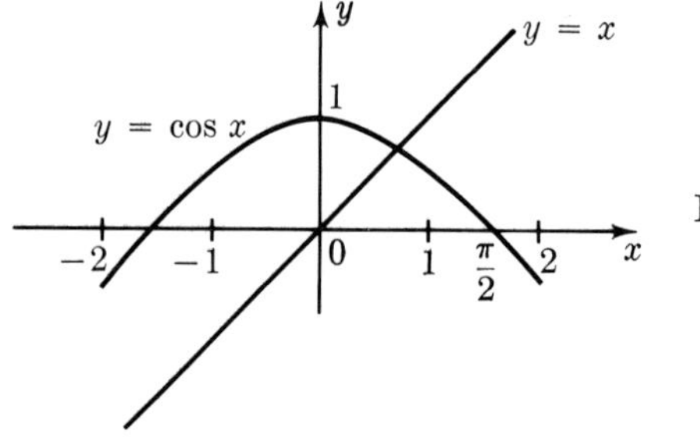

FIG. 3.6

The figure shows that $\cos x = x$ for only one value of x, roughly midway between 0 and $\pi/2 \approx 1.58$. A reasonable first guess is $x_0 = 0.8$. Apply the Newton–Raphson Method to $f(x) = \cos x - x$:

$$x_{i+1} = g(x_i),$$

where

$$g(x) = x - \frac{f(x)}{f'(x)}$$

$$= x - \frac{\cos x - x}{-\sin x - 1}$$

$$= \frac{x \sin x + \cos x}{1 + \sin x}.$$

Let the computer try this one. Results:

$$x_0 = 0.8$$

$$x_1 = 0.739853 \qquad x_3 = 0.739085$$

$$x_2 = 0.739085 \qquad x_4 = 0.739085.$$

The precision is amazing; just two iterations yield 6-place accuracy.

Just for fun, have the computer try some other values of x_0.

Results:

$x_0 = 0$	$x_0 = 0.7$	$x_0 = 1$
$x_1 = 1.000000$	$x_1 = 0.739436$	$x_1 = 0.750364$
$x_2 = 0.750364$	$x_2 = 0.739085$	$x_2 = 0.739113$
$x_3 = 0.739113$	$x_3 = 0.739085$	$x_3 = 0.739085$
$x_4 = 0.739085$	$x_4 = 0.739085$	$x_4 = 0.739085$

Even the crude initial guesses, $x_0 = 0$ and $x_0 = 1$, yield 6-place accuracy after just 3 or 4 iterations.

Answer: 0.739085.

Generally, the Newton–Raphson Method is an effective procedure for finding roots. There are some cases, however, when the method fails. For example, if $f'(x)$ is zero or very small near a root, there may be trouble because of the denominator in the formula

$$x_{i+1} = x_i - \frac{f(x_i)}{f'(x_i)}.$$

In practice, the method usually works provided x_0 is close enough to the root. Still it is comforting that there is the bisection scheme to fall back on.

Computer Approximations

The approximation methods discussed in this chapter are well-adapted for computer use.

The program for the Bisection Method is straightforward (Fig. 3.7). For our flowchart, we assume that $f(x)$ has a zero between a and b, where $a < b$, and that $f(a) < 0 < f(b)$. The zero of $f(x)$ is always within the interval BOT < 0 < TOP. The process terminates when the length of this interval is less than TEST.

For the Newton–Raphson Method, we need $f(x)$ and $f'(x)$, also an initial guess x_0. (There are several computer programs which can formally compute the derivative of a function. For example, such a program would accept "$3x^2 - 2x + 1$" as input, and produce "$6x - 2$" as output. However, since these programs are not available on all computers, the programmer must usually differentiate $f(x)$ himself.) Our flowchart is simplified since we assume that the method will not fail because of a zero derivative or any other reason (Fig. 3.8).

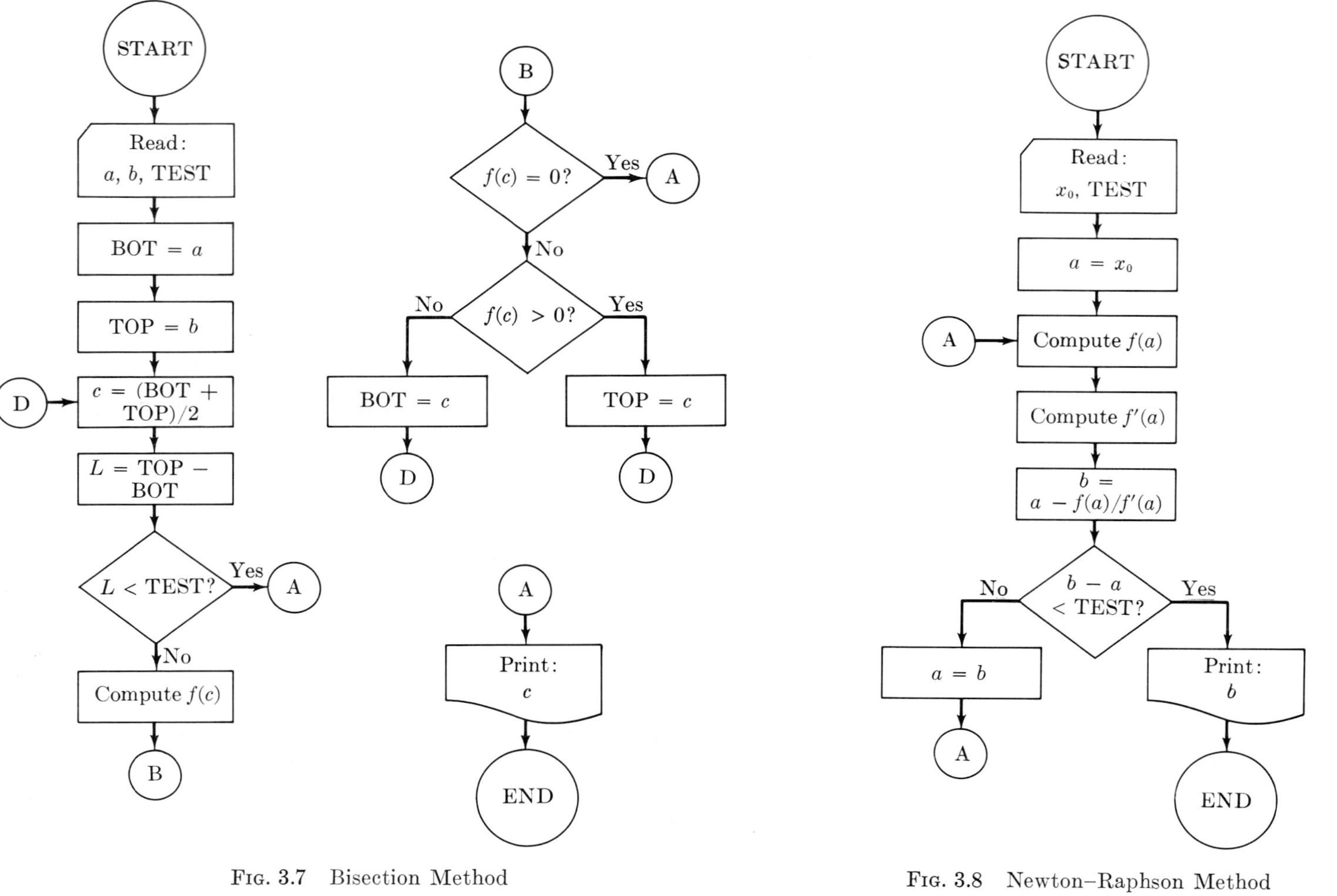

FIG. 3.7 Bisection Method

FIG. 3.8 Newton–Raphson Method

EXERCISES

Estimate to 2 places the root (or roots), using the Bisection Method:

1. $x^3 = 5$
2. $3x^3 = 7x + 3$
3. $x^3 + x^2 - x + 1 = 0$
4. $x^2 = 10.$

Estimate to 3 places the largest root, using Newton's Method:

5. $x^2 = 10$
6. $x^3 + x^2 - 4 = 0$
7. $x^3 - 2x^2 - x + 3 = 0$
8. $x^3 - 3x + 1 = 0$
9. $e^x = 2 \cos x$
10. $e^x = 3x$
11. $x = 2 \sin x$
12. $x = \cos 2x.$
13. Draw a flowchart for the Bisection Method, assuming only that the signs of $f(a)$ and $f(b)$ are different.
14. There are two main reasons for the failure of the Newton–Raphson Method: (1) the derivative is zero at some point near the root of $f(x) = 0$; (2) due to rounding within the computer, the estimates cycle: $x_i = x_{i+2} = x_{i+4} = \cdots$, and $x_{i+1} = x_{i+3} = x_{i+5} = \cdots$. Modify the flowchart of Fig. 3.8 to abort the computation if either of these events occurs.

4. TAYLOR POLYNOMIALS

Previously we approximated a function $f(x)$ near $x = a$ by a linear function $p_1(x) = f(a) + f'(a)(x - a)$. In geometric language, we approximated a curve by its tangent (Fig. 4.1).

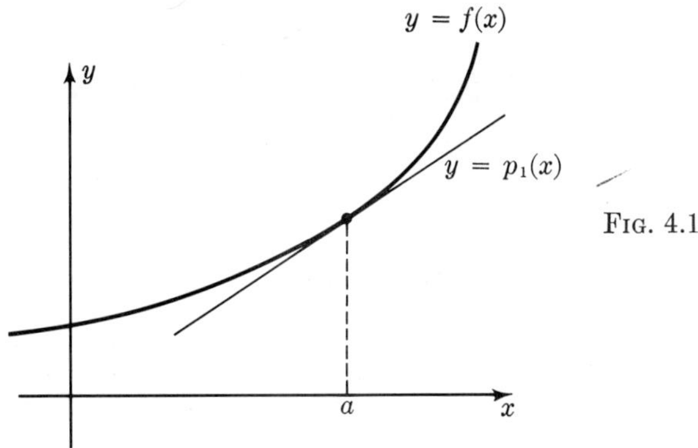

FIG. 4.1

Near $x = a$, the function $p_1(x)$ is a good approximation to $f(x)$ because $p_1(a) = f(a)$ and $p_1'(a) = f'(a)$; both graphs pass through the same point with the same slope. Farther from $x = a$, however, $p_1(x)$ may not be at all close to $f(x)$ since the graph of $y = f(x)$ may bend away from its tangent. The polynomial $p_1(x)$ is the **first degree Taylor polynomial** of $f(x)$ at $x = a$.

For greater accuracy, we shall approximate $f(x)$ by a function whose graph passes through $(a, f(a))$ with the same slope as $y = f(x)$, and which also curves in the same direction as $y = f(x)$. Now the curving of a graph is due to change in its slope. The slope is $f'(x)$, and the rate of change of the slope is the second derivative $f''(x)$. Therefore, we seek a function $p_2(x)$ such that

$$p_2(a) = f(a), \qquad p_2'(a) = f'(a), \qquad \text{and} \qquad p_2''(a) = f''(a).$$

Let us try a quadratic polynomial,

$$p_2(x) = A + B(x - a) + C(x - a)^2.$$

Then

$$p_2(a) = A, \qquad p_2'(a) = B, \qquad p_2''(a) = 2C.$$

(Verify these statements.) Since we want

$$p_2(a) = f(a), \qquad p_2'(a) = f'(a), \qquad p_2''(a) = f''(a),$$

there is no choice but

$$A = f(a), \qquad B = f'(a), \qquad C = \frac{1}{2}f''(a).$$

The polynomial

$$p_2(x) = f(a) + f'(a)\,(x - a) + \frac{1}{2}f''(a)\,(x - a)^2$$

agrees with $f(x)$ at $x = a$, and its derivative and second derivative agree with those of $f(x)$ at $x = a$. The polynomial $p_2(x)$ is the **second degree Taylor polynomial** of $f(x)$ at $x = a$.

The first two terms of $p_2(x)$ are $f(a) + f'(a)\,(x - a)$, which is $p_1(x)$. Thus $p_2(x)$ consists of the tangential linear approximation to $f(x)$ plus another term, $\frac{1}{2}f''(a)\,(x - a)^2$, which (we hope) corrects some of the error in linear approximation.

EXAMPLE 4.1

Approximate e^x near $x = 0$ by its first and second degree Taylor polynomials $p_1(x)$ and $p_2(x)$. Discuss their accuracy.

Solution: Use the formula

$$p_2(x) = f(a) + f'(a)\,(x - a) + \frac{1}{2}f''(a)\,(x - a)^2,$$

where $f(x) = e^x$ and $a = 0$. In this case $f(x) = f'(x) = f''(x) = e^x$, so $f(0) = f'(0) = f''(0) = 1$. Therefore,

$$p_2(x) = 1 + x + \frac{1}{2}x^2.$$

There is no need to compute $p_1(x)$ since it is the linear part of $p_2(x)$:

$$p_1(x) = 1 + x.$$

Table 4.1 compares e^x with $p_1(x)$ and $p_2(x)$ for various values of x.

TABLE 4.1

Small values of x				Larger values of x			
x	e^x	$p_1(x)$	$p_2(x)$	x	e^x	$p_1(x)$	$p_2(x)$
−0.4	0.6703	0.6000	0.6800	−2.0	0.1353	−1.0000	1.0000
−0.3	0.7408	0.7000	0.7450	−1.5	0.2231	−0.5000	0.6250
−0.2	0.8187	0.8000	0.8200	−1.0	0.3679	0.0000	0.5000
−0.1	0.9048	0.9000	0.9050	−0.5	0.6065	0.5000	0.6250
0.0	1.0000	1.0000	1.0000	0.0	1.0000	1.0000	1.0000
0.1	1.1052	1.1000	1.1050	0.5	1.6487	1.5000	1.6250
0.2	1.2214	1.2000	1.2200	1.0	2.7183	2.0000	2.5000
0.3	1.3499	1.3000	1.3450	1.5	4.4817	2.5000	3.6250
0.4	1.4918	1.4000	1.4800	2.0	7.3891	3.0000	5.0000

From the table we see that $p_1(x)$ and $p_2(x)$ are good estimates of e^x for x near 0, but that $p_2(x)$ is considerably better than $p_1(x)$. Both estimates become poor as x moves away from 0, but $p_2(x)$ stays accurate in a wider range because its graph is curved like that of e^x near $x = 0$. See Fig. 4.2.

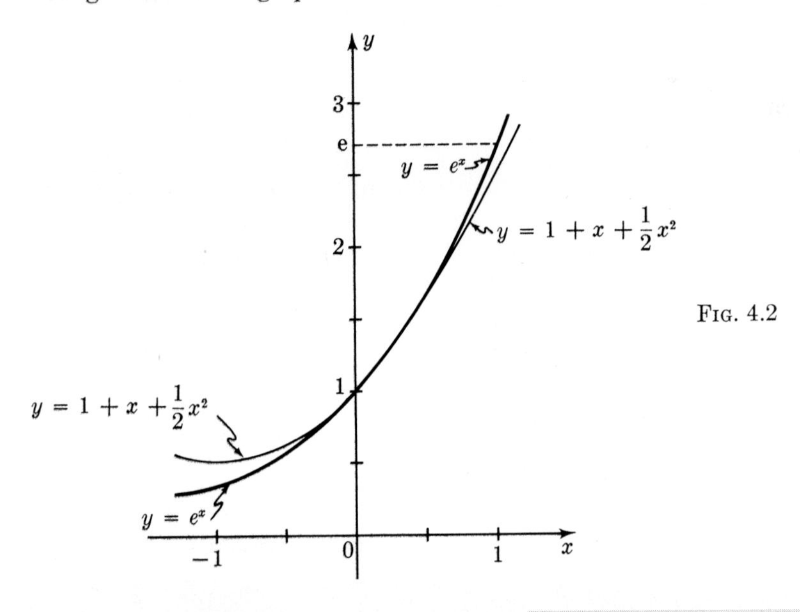

FIG. 4.2

EXAMPLE 4.2

Approximate $1/x$ near $x = 1$ by its first and second degree Taylor polynomials. Discuss their accuracy.

Solution: Use the formula

$$p_2(x) = f(a) + f'(a)(x - a) + \frac{1}{2} f''(a)(x - a)^2,$$

where $f(x) = 1/x$ and $a = 1$. Then $f'(x) = -1/x^2$ and $f''(x) = 2/x^3$ (a fact we shall prove later). Then $f(1) = 1$, $f'(1) = -1$, and $f''(1) = 2$. Hence

$$p_2(x) = 1 - (x - 1) + (x - 1)^2.$$

The polynomial $p_1(x)$ is the linear part of $p_2(x)$:

$$p_1(x) = 1 - (x - 1).$$

Since we shall be dealing with numbers near 1, it is convenient to leave $p_1(x)$ and $p_2(x)$ in terms of $(x - 1)$.

Table 4.2 compares $1/x$ with $p_1(x)$ and $p_2(x)$.

TABLE 4.2

x near 1				Other values of x			
x	$1/x$	$p_1(x)$	$p_2(x)$	x	$1/x$	$p_1(x)$	$p_2(x)$
0.85	1.1765	1.1500	1.1725	0.25	4.0000	1.7500	2.3125
0.90	1.1111	1.1000	1.1100	0.50	2.0000	1.5000	1.7500
0.95	1.0526	1.0500	1.0525	0.75	1.3333	1.2500	1.3125
1.00	1.0000	1.0000	1.0000	1.00	1.0000	1.0000	1.0000
1.05	0.9524	0.9500	0.9525	1.25	0.8000	0.7500	0.8125
1.10	0.9091	0.9000	0.9100	1.50	0.6667	0.5000	0.7500
1.15	0.8696	0.8500	0.8725	1.75	0.5714	0.2500	0.8125

From the table, we see that $p_1(x)$ is a good approximation to $1/x$ provided x is near 1, but that $p_2(x)$ is much better. Both estimates become poor as x moves away from 1, but $p_2(x)$ is accurate in a wider range.

Cubic Approximation

The polynomial $p_2(x)$ was introduced in order to improve the tangential approximation $p_1(x)$. By the same token, we may try for greater precision by using a cubic polynomial

$$p_3(x) = A + B(x - a) + C(x - a)^2 + D(x - a)^3,$$

so chosen that

$$p_3(a) = f(a), \qquad p_3'(a) = f'(a), \qquad p_3''(a) = f''(a), \qquad p_3'''(a) = f'''(a).$$

Here $f'''(x)$ is the third derivative of $f(x)$, that is $f'''(x) = [f''(x)]'$. Differentiating $p_3(x)$ three times, we find

$$p_3(a) = A, \qquad p_3'(a) = B, \qquad p_3''(a) = 2C, \qquad p_3'''(a) = 6D.$$

(Verify these facts.) We are forced to choose

$$A = f(a), \qquad B = f'(a), \qquad C = \frac{1}{2}f''(a), \qquad D = \frac{1}{6}f'''(a).$$

The polynomial

$$p_3(x) = f(a) + f'(a)(x-a) + \frac{1}{2}f''(a)(x-a)^2 + \frac{1}{6}f'''(a)(x-a)^3$$

agrees with $f(x)$ at $x = a$, and its first, second and third derivatives agree with the corresponding derivatives of $f(x)$ at $x = a$. The polynomial $p_3(x)$ is the **third degree Taylor polynomial** of $f(x)$ at $x = a$.

The first two terms of $p_3(x)$ give $p_1(x)$, and the first three terms give $p_2(x)$. Thus $p_3(x)$ consists of $p_2(x)$ plus a term $\frac{1}{6}f'''(a)(x-a)^3$, which (we hope) corrects some of the error made in approximating $f(x)$ by $p_2(x)$.

EXAMPLE 4.3

Find $p_3(x)$, the third degree Taylor polynomial of e^x at $x = 0$. Compare its accuracy with that of $p_2(x)$.

Solution:

$$p_3(x) = p_2(x) + \frac{1}{6}f'''(a)(x-a)^3 = p_2(x) + \frac{1}{6}f'''(0)x^3.$$

But $f(x) = e^x$, so $f'''(x) = e^x$ and $f'''(0) = 1$. From Example 4.1,

$$p_2(x) = 1 + x + \frac{1}{2}x^2.$$

Conclusion:

$$p_3(x) = 1 + x + \frac{1}{2}x^2 + \frac{1}{6}x^3.$$

Table 4.3 shows $p_2(x)$ and $p_3(x)$ for several values of x.

TABLE 4.3

x	e^x	$p_2(x)$	$p_3(x)$
0.1	1.1052	1.1050	1.1052
0.2	1.2214	1.2200	1.2213
0.5	1.6487	1.6250	1.6458
1.0	2.7183	2.5000	2.6667
2.0	7.3891	5.0000	6.3333

The table reveals $p_3(x)$ to be more accurate than $p_2(x)$. Indeed, $p_3(x)$ is extremely precise for small values of x, and is more accurate for larger values of x than is $p_2(x)$.

EXAMPLE 4.4

Approximate $\sin x$ by its third degree Taylor polynomial at $x = 0$. Discuss the accuracy of this approximation.

Solution: Use the formula

$$p_3(x) = f(0) + f'(0)(x - 0) + \frac{1}{2} f''(0)(x - 0)^2 + \frac{1}{6} f'''(0)(x - 0)^3,$$

where $f(x) = \sin x$. Successive derivatives are

$$f'(x) = \cos x, \qquad f''(x) = -\sin x, \qquad f'''(x) = -\cos x;$$

$$f'(0) = 1, \qquad f''(0) = 0, \qquad f'''(0) = -1.$$

Hence,

$$p_3(x) = 0 + (x - 0) + 0 - \frac{1}{6}(x - 0)^3$$

$$= x - \frac{1}{6} x^3.$$

In Chapter 5 we saw that $\sin x \approx x$ if x is small. Now we have the approximation

$$\sin x \approx x - \frac{1}{6} x^3,$$

which is more precise (Table 4.4).

TABLE 4.4

Small values of x			Larger values of x		
x	$\sin x$	$x - \frac{1}{6}x^3$	x	$\sin x$	$x - \frac{1}{6}x^3$
0.00	0.00000	0.00000	0.30	0.29552	0.29550
0.05	0.04998	0.04998	0.50	0.47943	0.47917
0.10	0.09983	0.09983	0.70	0.64422	0.64283
0.15	0.14944	0.14944	0.90	0.78333	0.77850
0.20	0.19867	0.19867	1.10	0.89121	0.87817
0.25	0.24740	0.24740	1.30	0.96356	0.93383

The approximation $\sin x \approx x - \frac{1}{6}x^3$ is extremely accurate. It gives 5-place accuracy for angles at least up to 0.25 radians ($\approx 14.3°$), and 3-place accuracy at least up to 0.5 radians ($\approx 28.6°$). Even the very simple formula

$\sin x \approx x$ is fairly accurate for small angles; it is correct to 3 places for angles up to 0.14 radians ($\approx 8°$).

Estimates Using Taylor Polynomials

Here are some common approximation formulas, all involving third degree Taylor polynomials at $x = 0$:

$$e^x \approx 1 + x + \frac{1}{2}x^2 + \frac{1}{6}x^3,$$

$$\sin x \approx x - \frac{1}{6}x^3,$$

$$\cos x \approx 1 - \frac{1}{2}x^2.$$

The approximation for $\cos x$ is actually $p_3(x)$ although no cubic term appears. That is because the third derivative of $\cos x$ is zero at $x = 0$.

EXERCISES

Sketch the graphs (on the same set of axes) of $y = f(x)$, of $p_1(x) = f(a) + f'(a)(x - a)$, and of $p_2(x) = f(a) + f'(a)(x - a) + \frac{1}{2}f''(a)(x - a)^2$:

1. $y = \sin 2x$; near $a = 0$
2. $y = \frac{1}{2}(e^x + e^{-x})$; near $a = 0$
3. $y = (x - 2)^3 + 3(x - 2)^2 - 4(x - 2) + 1$; near $a = 2$
4. $y = \sin x$; near $a = \frac{\pi}{4}$
5. $y = \cos x + 2\cos\frac{x}{2} + 3\cos\frac{x}{3}$; near $a = 0$
6. $y = (x - 1)^2 + e^x$; near $a = -1$.

Compute the third degree Taylor polynomial at the point specified:

7. $f(x) = e^{-x}$; $x = 0$
8. $f(x) = e^x + \sin x$; $x = 0$
9. $f(x) = e^x + \sin x + \cos x$; $x = 0$
10. $f(x) = 2x^3 - x^2 + 5x - 1$; $x = 1$
11. $f(x) = \cos(20x)$; $x = \frac{\pi}{40}$
12. $f(x) = e^x + e^{-x} + \cos x$; $x = 0$
13. $f(x) = e^{2x}$; $x = 0$.
14. Extend Table 4.3 to the corresponding negative values of x.
15. Extend Table 4.4 to the corresponding negative values of x.
16. Make the table for $\cos x$ corresponding to Table 4.4.

PART II

15. Techniques of Differentiation

1. REVIEW OF DIFFERENTIATION

The derivative is a powerful tool of Calculus, useful in a wide variety of applications. As yet we are limited in its use because there are many functions we cannot differentiate. For example:

$$\frac{1}{x^3}, \qquad x^2e^x, \qquad \tan x, \qquad \frac{x}{x^2+1}, \qquad e^{-x^2}, \qquad \frac{\sin x}{x}.$$

Soon we shall be able to differentiate virtually every function that arises in practice. In this chapter we shall develop three important rules for differentiation that enlarge enormously the class of functions we can handle.

First let us review the derivatives we know.

POWERS OF x:

$$\frac{d}{dx}(x) = 1, \qquad \frac{d}{dx}(x^2) = 2x, \qquad \frac{d}{dx}(x^3) = 3x^2;$$

$$\frac{d}{dx}(c) = 0, \qquad \frac{d}{dx}\left(\frac{1}{x}\right) = -\frac{1}{x^2}.$$

TRIGONOMETRIC FUNCTIONS:

$$\frac{d}{dx}(\sin x) = \cos x, \qquad \frac{d}{dx}(\cos x) = -\sin x.$$

EXPONENTIAL FUNCTION:

$$\frac{d}{dx}(e^x) = e^x.$$

SUM RULE:

$$\frac{d}{dx}[f(x) + g(x)] = \frac{d}{dx}[f(x)] + \frac{d}{dx}[g(x)].$$

CONSTANT MULTIPLE RULE:

$$\frac{d}{dx}[kf(x)] = kf'(x).$$

SHIFTING RULE:

If

$$g(x) = f(x + c),$$

then

$$\frac{d}{dx}g(x) = \frac{df}{dx}\bigg|_{x+c} = f'(x + c).$$

SCALE RULE:

If

$$g(x) = f(kx),$$

then

$$\frac{d}{dx}g(x) = k\frac{df}{dx}\bigg|_{kx} = kf'(kx).$$

EXAMPLE 1.1

Differentiate $y = (x - 6)^2 + 3e^{2x}$.

Solution:

$$\frac{d}{dx}(x^2) = 2x,$$

$$\frac{d}{dx}(x - 6)^2 = 2(x - 6). \qquad \text{(Shifting Rule)}$$

$$\frac{d}{dx}(e^x) = e^x,$$

$$\frac{d}{dx}(e^{2x}) = 2e^{2x}, \qquad \text{(Scale Rule)}$$

$$\frac{d}{dx}(3e^{2x}) = 3 \cdot 2e^{2x}. \qquad \text{(Constant Multiple Rule)}$$

The Sum Rule allows us to combine these derivatives.

Answer: $2(x - 6) + 6e^{2x}$.

EXAMPLE 1.2

Differentiate $y = 3 \sin 5x + \dfrac{2}{x}\cdot$

Solution: Since

$$\frac{d}{dx}(\sin x) = \cos x,$$

$$\frac{d}{dx}(3 \sin 5x) = 3(5 \cos 5x).$$

Since

$$\frac{d}{dx}\left(\frac{1}{x}\right) = -\frac{1}{x^2},$$

$$\frac{d}{dx}\left(\frac{2}{x}\right) = -\frac{2}{x^2}.$$

The Sum Rule allows us to combine these derivatives.

Answer: $15 \cos 5x - \dfrac{2}{x^2}$.

EXERCISES

Differentiate:

1. $f(x) = 2x^3 - x^2$
2. $f(x) = -x^3 + 4x$
3. $f(x) = 2e^{3x}$
4. $f(x) = \sin(-10x)$
5. $f(x) = \dfrac{3}{x+2}$
6. $f(x) = (x-1)^3 + \dfrac{3}{x+4}$
7. $f(x) = 5e^{-x} + \cos 4x$
8. $f(x) = \sin(x-1) + \cos(x-1)$
9. $f(x) = 7e^{x+4} + e^{5x}$
10. $f(x) = A \cos hx + B \cos kx$
11. $f(x) = \sin^2 x + \cos^2 x$
12. $f(x) = \frac{1}{2}(e^{3x} - e^{-3x})$.

2. DERIVATIVE OF A PRODUCT

We still cannot differentiate $x \sin x$ although we know the derivatives of the factors x and $\sin x$. We need a rule for differentiating products.

A natural first guess is that the derivative of $u(x) \cdot v(x)$ is $u'(x) \cdot v'(x)$. However, this is wrong. Try $u(x) = x$ and $v(x) = x^2$. Then $u(x) \cdot v(x) = x^3$, but $u'(x) \cdot v'(x) = 1 \cdot 2x = 2x$, which is not the derivative of x^3. Worse yet, try $u(x) = 1$, a constant function, $v(x)$ any function. Then $u(x) \cdot v(x) = v(x)$, but $u'(x) \cdot v'(x) = 0 \cdot v'(x) = 0$. This says the derivative of every function is zero—ridiculous. Here is the correct rule.

Product Rule If $u = u(x)$ and $v = v(x)$, then

$$\frac{d}{dx}(uv) = u\frac{dv}{dx} + v\frac{du}{dx}.$$

Expressed briefly,

$$(uv)' = uv' + vu'.$$

This rule will be justified in Section 6.

EXAMPLE 2.1

Differentiate $y = x \sin x$.

Solution: Use the Product Rule:

$$\frac{d}{dx}(x \sin x) = x\frac{d}{dx}(\sin x) + (\sin x)\frac{d}{dx}(x)$$

$$= x \cos x + \sin x.$$

Answer: $x \cos x + \sin x$.

EXAMPLE 2.2

Differentiate $y = (x + 1)^3 e^{5x}$.

Solution: By the Product Rule,

$$\frac{d}{dx}[(x + 1)^3 e^{5x}] = (x + 1)^3 \frac{d}{dx}(e^{5x}) + e^{5x}\frac{d}{dx}[(x + 1)^3]$$

$$= (x + 1)^3 \cdot 5e^{5x} + e^{5x} \cdot 3(x + 1)^2$$

$$= e^{5x}(x + 1)^2[5(x + 1) + 3].$$

Answer: $e^{5x}(x + 1)^2(5x + 8)$.

Differentiating Powers

Here is an important application of the Product Rule. Given $u(x)$ and its derivative $u'(x)$, what are the derivatives of $u^2(x)$, $u^3(x)$, etc? Apply the Product Rule to $u(x) \cdot u(x)$:

$$(u^2)' = (uu)' = uu' + uu'$$

$$= 2uu'.$$

Use this information to compute the derivative of u^3:

$$(u^3)' = (u^2u)' = u^2u' + u(u^2)'$$
$$= u^2u' + u(2uu')$$
$$= 3u^2u'.$$

Use this result to compute the derivative of u^4:

$$(u^4)' = (u^3u)' = u^3u' + u(u^3)'$$
$$= u^3u' + u(3u^2u')$$
$$= 4u^3u'.$$

Summarizing:

$$\frac{d}{dx}(u^2) = 2u\frac{du}{dx},$$
$$\frac{d}{dx}(u^3) = 3u^2\frac{du}{dx},$$
$$\frac{d}{dx}(u^4) = 4u^3\frac{du}{dx}.$$

By now the pattern should be clear.

For each positive integer n,

$$\frac{d}{dx}(u^n) = nu^{n-1}\frac{du}{dx}.$$

In particular,

$$\frac{d}{dx}(x^n) = nx^{n-1}.$$

Thus the derivatives of x^3, x^8, x^{20}, x^{50} are respectively $3x^2$, $8x^7$, $20x^{19}$ and $50x^{49}$.

EXAMPLE 2.3

Differentiate $y = (x^2 + 1)^5$.

Solution:

$$\frac{d}{dx}(u^5) = 5u^4\frac{du}{dx}.$$

Take $u = x^2 + 1$. Then $du/dx = 2x$,

$$\frac{d}{dx}[(x^2 + 1)^5] = 5(x^2 + 1)^4 \cdot 2x.$$

Answer: $10x(x^2 + 1)^4$.

REMARK: This example can be done in a clumsy way by using the Binomial Theorem:

$$y = (x^2+1)^5 = x^{10} + 5x^8 + 10x^6 + 10x^4 + 5x^2 + 1,$$

$$\frac{dy}{dx} = 10x^9 + 40x^7 + 60x^5 + 40x^3 + 10x.$$

This method is tedious, and the answer is in a form less suitable for computation than the previous answer.

EXAMPLE 2.4

Differentiate $y = \left(e^{2x} + x^3 + \dfrac{1}{x}\right)^4$.

Solution:

$$\frac{d}{dx}(u^4) = 4u^3\frac{du}{dx}.$$

Take $u = e^{2x} + x^3 + 1/x$. Then $du/dx = 2e^{2x} + 3x^2 - 1/x^2$,

$$\frac{d}{dx}\left[\left(e^{2x} + x^3 + \frac{1}{x}\right)^4\right] = 4\left(e^{2x} + x^3 + \frac{1}{x}\right)^3\left(2e^{2x} + 3x^2 - \frac{1}{x^2}\right).$$

Answer: $4\left(e^{2x} + x^3 + \dfrac{1}{x}\right)^3\left(2e^{2x} + 3x^2 - \dfrac{1}{x^2}\right)$.

Square Roots

Suppose $u(x) > 0$. Then its positive square root $y = \sqrt{u(x)}$ makes sense. To compute dy/dx, first square y:

$$y^2 = u.$$

Then differentiate both sides:

$$2y\frac{dy}{dx} = \frac{du}{dx},$$

$$\frac{dy}{dx} = \frac{1}{2y}\frac{du}{dx} = \frac{1}{2\sqrt{u}}\frac{du}{dx}.$$

If $u(x) > 0$, then

$$\frac{d}{dx}(\sqrt{u}) = \frac{1}{2\sqrt{u}}\frac{du}{dx}.$$

EXAMPLE 2.5

Compute dy/dx:

$$\text{(a)}\quad y = \sqrt{x},$$

$$\text{(b)}\quad y = \sqrt{x^3 + 2},$$

$$\text{(c)}\quad y = \sqrt{x \cos x}.$$

Solution: Use the formula

$$\frac{d}{dx}(\sqrt{u}) = \frac{1}{2\sqrt{u}}\frac{du}{dx}.$$

(a) Take $u = x$. Then

$$\frac{d}{dx}(\sqrt{x}) = \frac{1}{2\sqrt{x}}.$$

(b) Take $u = x^3 + 2$. Then $du/dx = 3x^2$,

$$\frac{d}{dx}(\sqrt{x^3 + 2}) = \frac{1}{2\sqrt{x^3 + 2}}(3x^2).$$

(c) Take $u = x \cos x$. Then

$$\frac{du}{dx} = x\frac{d}{dx}(\cos x) + \cos x\frac{d}{dx}(x) = -x \sin x + \cos x,$$

$$\frac{d}{dx}(\sqrt{x \cos x}) = \frac{1}{2\sqrt{x \cos x}}(-x \sin x + \cos x).$$

Answer: (a) $\dfrac{1}{2\sqrt{x}}$,

(b) $\dfrac{3x^2}{2\sqrt{x^3 + 2}}$,

(c) $\dfrac{\cos x - x \sin x}{2\sqrt{x \cos x}}$.

EXERCISES

Differentiate:

1. $y = x^2 e^x$
2. $y = x^2 \sin x$
3. $y = (x - 1)^4 e^{-3x}$
4. $y = (x^2 + 1)^2 \cos x$
5. $y = \sin x \cos x$
6. $y = e^{2x} \sin 3x$
7. $y = (\sin 2x + 1)^3$
8. $y = (x^5 e^x + 2x)^7$

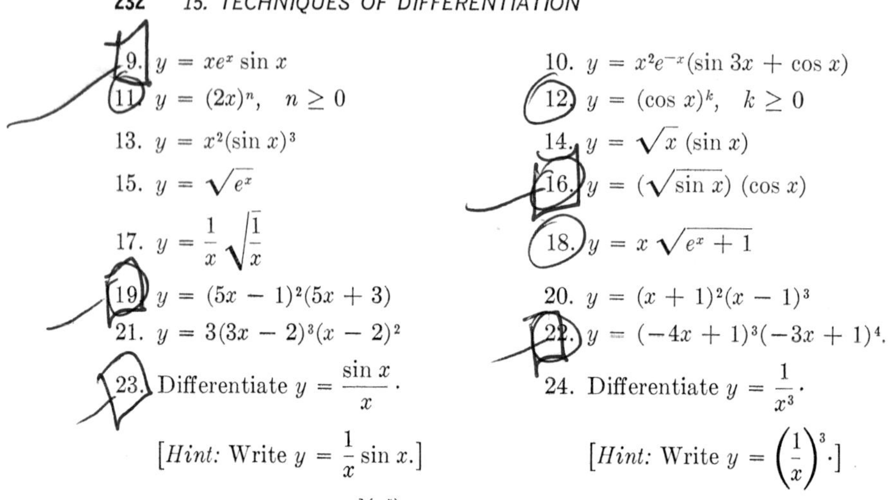

9. $y = xe^x \sin x$
10. $y = x^2e^{-x}(\sin 3x + \cos x)$
11. $y = (2x)^n, \quad n \geq 0$
12. $y = (\cos x)^k, \quad k \geq 0$
13. $y = x^2(\sin x)^3$
14. $y = \sqrt{x}\,(\sin x)$
15. $y = \sqrt{e^x}$
16. $y = (\sqrt{\sin x})\,(\cos x)$
17. $y = \dfrac{1}{x}\sqrt{\dfrac{1}{x}}$
18. $y = x\sqrt{e^x + 1}$
19. $y = (5x - 1)^2(5x + 3)$
20. $y = (x + 1)^2(x - 1)^3$
21. $y = 3(3x - 2)^3(x - 2)^2$
22. $y = (-4x + 1)^3(-3x + 1)^4.$
23. Differentiate $y = \dfrac{\sin x}{x}\cdot$

    [*Hint:* Write $y = \dfrac{1}{x}\sin x.$]
24. Differentiate $y = \dfrac{1}{x^3}\cdot$

    [*Hint:* Write $y = \left(\dfrac{1}{x}\right)^3\cdot$]

25. Verify the formula $\dfrac{d(x^5)}{dx} = 5x^4$ by differentiating the product x^4x; also by differentiating x^3x^2.

3. DERIVATIVE OF A QUOTIENT

We now discuss the rule for differentiating the quotient of two functions. The natural guess

$$\frac{d}{dx}\left[\frac{u(x)}{v(x)}\right] = \frac{du/dx}{dv/dx}$$

is again wrong. (Try to find examples showing this cannot be correct; almost any pair of functions will do.) Here is the correct formula:

Quotient Rule If $u = u(x)$ and $v = v(x) \neq 0$, then

$$\frac{d}{dx}\left(\frac{u}{v}\right) = \frac{v\dfrac{du}{dx} - u\dfrac{dv}{dx}}{v^2}.$$

Expressed briefly,

$$\left(\frac{u}{v}\right)' = \frac{vu' - uv'}{v^2}.$$

This rule will be justified in Section 6.

EXAMPLE 3.1

Differentiate $y = \dfrac{x}{x^2 + 1}\cdot$

Solution: Use the Quotient Rule with $u(x) = x$ and $v(x) = x^2 + 1$:

$$\frac{d}{dx}\left(\frac{x}{x^2+1}\right) = \frac{(x^2+1)\dfrac{d}{dx}(x) - x\dfrac{d}{dx}(x^2+1)}{(x^2+1)^2}$$

$$= \frac{(x^2+1) - x(2x)}{(x^2+1)^2} = \frac{1-x^2}{(x^2+1)^2}.$$

Answer: $\dfrac{1-x^2}{(x^2+1)^2}$.

EXAMPLE 3.2

Differentiate $y = \tan x$.

Solution:

$$\tan x = \frac{\sin x}{\cos x}.$$

Use the Quotient Rule with $u(x) = \sin x$ and $v(x) = \cos x$:

$$\frac{d}{dx}\left(\frac{\sin x}{\cos x}\right) = \frac{\cos x\dfrac{d}{dx}(\sin x) - \sin x\dfrac{d}{dx}(\cos x)}{\cos^2 x}$$

$$= \frac{(\cos x)(\cos x) - (\sin x)(-\sin x)}{\cos^2 x}.$$

But $\cos^2 x + \sin^2 x = 1$.

Answer: $\dfrac{1}{\cos^2 x} = \sec^2 x$.

Derivatives of Powers

As a special case of the Quotient Rule, take $u(x) = 1$. The result is the formula for the derivative of a reciprocal:

$$\frac{d}{dx}\left(\frac{1}{v}\right) = \frac{-1}{v^2}\frac{dv}{dx}.$$

In particular, let $v(x) = u^n(x)$, where n is any positive integer. Since $d(u^n)/dx = nu^{n-1}\,du/dx$,

$$\frac{d}{dx}\left(\frac{1}{u^n}\right) = \frac{-1}{(u^n)^2}\left(nu^{n-1}\frac{du}{dx}\right)$$

$$= \frac{-nu^{n-1}}{u^{2n}}\frac{du}{dx}.$$

Hence

$$\frac{d}{dx}\left(\frac{1}{u^n}\right) = \frac{-n}{u^{n+1}}\frac{du}{dx}.$$

This formula may be written with negative exponents:

$$\frac{d}{dx}(u^{-n}) = -nu^{-n-1}\frac{du}{dx}.$$

Thus you differentiate a negative power of $u(x)$ the same way you differentiate a positive power: bring down the exponent, lower the power by one, then multiply by du/dx. The two formulas can be combined into one.

Power Rule If p is any integer, positive, negative, or zero, then

$$\frac{d}{dx}(u^p) = pu^{p-1}\frac{du}{dx}.$$

In particular,

$$\frac{d}{dx}(x^p) = px^{p-1}.$$

Actually the case $p = 0$ has not been discussed. But in that case the formula holds for a very simple reason. What reason?

EXAMPLE 3.3

Differentiate $y = \dfrac{1}{(x^2 + 5x + 1)^3}$.

Solution: Use the Power Rule with $u = x^2 + 5x + 1$ and $p = -3$:

$$\frac{d}{dx}(u^{-3}) = -3u^{-4}\frac{du}{dx}$$

$$= -3(x^2 + 5x + 1)^{-4}\frac{d}{dx}(x^2 + 5x + 1).$$

Answer: $\dfrac{-3(2x + 5)}{(x^2 + 5x + 1)^4}$.

EXAMPLE 3.4

Differentiate $y = \dfrac{(\sin 2x)^3}{x^5}$.

Solution: By the Quotient Rule,

$$\frac{dy}{dx} = \frac{x^5 \dfrac{d}{dx}[(\sin 2x)^3] - (\sin 2x)^3 \dfrac{d}{dx}(x^5)}{(x^5)^2}.$$

Compute the derivatives of $(\sin 2x)^3$ and x^5:

$$\frac{d}{dx}[(\sin 2x)^3] = 3(\sin 2x)^2 \frac{d}{dx}(\sin 2x) = 3(\sin 2x)^2(\cos 2x) \cdot 2;$$

$$\frac{d}{dx}(x^5) = 5x^4.$$

Substitute these expressions:

$$\frac{dy}{dx} = \frac{x^5[6(\sin 2x)^2 \cos 2x] - 5x^4(\sin 2x)^3}{x^{10}}$$

$$= \frac{(\sin 2x)^2(6x \cos 2x - 5 \sin 2x)}{x^6}.$$

Alternate Solution: Write $y = x^{-5}(\sin 2x)^3$ and use the Product Rule:

$$\frac{dy}{dx} = x^{-5}\frac{d}{dx}[(\sin 2x)^3] + (\sin 2x)^3 \frac{d}{dx}(x^{-5})$$

$$= x^{-5}[6(\sin 2x)^2 \cos 2x] + (\sin 2x)^3(-5x^{-6})$$

$$= x^{-6}(\sin 2x)^2(6x \cos 2x - 5 \sin 2x).$$

Answer: $\dfrac{(\sin 2x)^2(6x \cos 2x - 5 \sin 2x)}{x^6}.$

EXERCISES

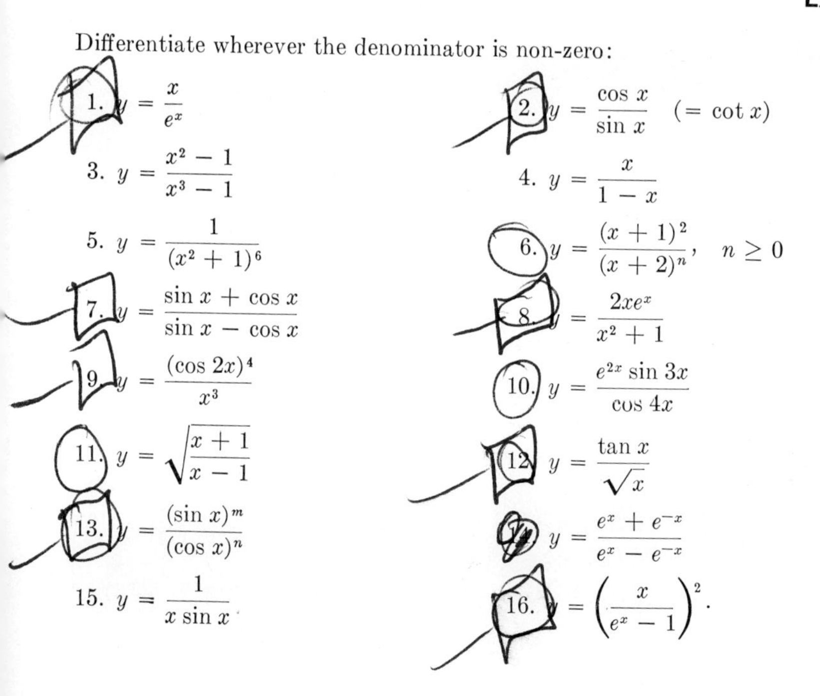

Differentiate wherever the denominator is non-zero:

1. $y = \dfrac{x}{e^x}$

2. $y = \dfrac{\cos x}{\sin x}$ $(= \cot x)$

3. $y = \dfrac{x^2 - 1}{x^3 - 1}$

4. $y = \dfrac{x}{1 - x}$

5. $y = \dfrac{1}{(x^2 + 1)^6}$

6. $y = \dfrac{(x + 1)^2}{(x + 2)^n}, \quad n \geq 0$

7. $y = \dfrac{\sin x + \cos x}{\sin x - \cos x}$

8. $y = \dfrac{2xe^x}{x^2 + 1}$

9. $y = \dfrac{(\cos 2x)^4}{x^3}$

10. $y = \dfrac{e^{2x} \sin 3x}{\cos 4x}$

11. $y = \sqrt{\dfrac{x + 1}{x - 1}}$

12. $y = \dfrac{\tan x}{\sqrt{x}}$

13. $y = \dfrac{(\sin x)^m}{(\cos x)^n}$

14. $y = \dfrac{e^x + e^{-x}}{e^x - e^{-x}}$

15. $y = \dfrac{1}{x \sin x}$

16. $y = \left(\dfrac{x}{e^x - 1}\right)^2.$

4. THE CHAIN RULE

Suppose y is a function of x, and x is a function of t. Then indirectly y is a function of t.

Here are three examples of such **composite functions.**

(1) If $y = \cos(t^2)$, then $y = \cos x$, where $x = t^2$.

(2) When a pebble is dropped into a smooth pond, a circular ripple begins to expand. Its area is πr^2, where r is a function of t. The area is indirectly a function $A(t)$ of time,

$$A(t) = \pi[r(t)]^2.$$

(3) If $y = (x^3 + x^2 + x + 1)^4$, then $y = u^4$, where $u = x^3 + x^2 + x + 1$. In this way, y is the composite of two simpler functions, $f(u) = u^4$ and $u(x) = x^3 + x^2 + x + 1$.

Here is the rule for differentiating composite functions:

> **Chain Rule** If $y = y(x)$ and $x = x(t)$, then
>
> $$\frac{dy}{dt} = \frac{dy}{dx}\frac{dx}{dt}.$$

This rule will be justified in Section 6.

Basically, the Chain Rule says this: if y is changing p times as fast as x and x is changing q times as fast as t, then y is changing pq times as fast as t.

EXAMPLE 4.1

Differentiate $y = \cos(t^2)$.

Solution: Write $y = \cos x$, where $x = t^2$. By the Chain Rule,

$$\frac{dy}{dt} = \frac{dy}{dx}\frac{dx}{dt} = \frac{d}{dx}(\cos x)\frac{d}{dt}(t^2) = (-\sin x)(2t),$$

where $x = t^2$.

Answer: $-2t \sin(t^2)$.

EXAMPLE 4.2

Differentiate $y = e^{x^2-3x}$.

Solution: Write $y = e^u$, where $u = x^2 - 3x$. By the Chain Rule,

$$\frac{dy}{dx} = \frac{dy}{du}\frac{du}{dx} = e^u(2x - 3) = e^{x^2-3x}(2x - 3).$$

Answer: $e^{x^2-3x}(2x - 3)$.

EXAMPLE 4.3

Differentiate $y = (x^3 + x^2 + x + 1)^4$.

Solution: Write $y = u^4$, where $u = x^3 + x^2 + x + 1$. By the Chain Rule,

$$\frac{dy}{dx} = \frac{dy}{du}\frac{du}{dx} = 4u^3(3x^2 + 2x + 1).$$

Answer: $4(x^3 + x^2 + x + 1)^3(3x^2 + 2x + 1)$.

We have already met some special instances of the Chain Rule.

(1) The Shifting Rule:

$$\frac{d}{dx}[f(x + c)] = f'(x + c).$$

Write $y = f(u)$, where $u = x + c$. Then

$$\frac{dy}{dx} = \frac{dy}{du}\frac{du}{dx} = f'(u)\frac{d}{dx}(x + c) = f'(u) = f'(x + c).$$

(2) The Scale Rule:

$$\frac{d}{dx}[f(kx)] = kf'(kx).$$

Write $y = f(u)$, where $u = kx$. Then

$$\frac{dy}{dx} = \frac{dy}{du}\frac{du}{dx} = f'(u)\frac{d}{dx}(kx) = f'(u)k = f'(kx)k.$$

(3) The Power Rule:

$$\frac{d}{dx}[u^p(x)] = pu^{p-1}\frac{du}{dx}.$$

Write $y = u^p$, where $u = u(x)$. Then

$$\frac{dy}{dx} = \frac{dy}{du}\frac{du}{dx} = pu^{p-1}\frac{du}{dx}.$$

EXERCISES

Differentiate:

1. $y = \sin(x^3)$
2. $y = \cos(e^x)$
3. $y = e^{\sin 2x}$
4. $y = \sqrt{2 + \cos(x/5)}$
5. $y = (x^3 + x - 3)^5$
6. $y = \sqrt{x^2 + 7}$
7. $y = e^{ax^2+bx+c}$
8. $y = (\sin\sqrt{x})^2$
9. $y = e^{e^x}$
10. $y = [\cos(3 - x^2)]^3$

11. $y = [(x^3 + 2x)^5 + x^2]^2$

12. $y = \{[(x + 1)^2 + 1]^2 + 1\}^2$

13. $y = e^{x^2} + e^{-x^2}$

14. $y = \sqrt{1 + \sqrt{2x}}$.

5. LIMITS

In order to justify the rules stated in the preceding sections, we must recall the meaning of a derivative.

Given a function $f(x)$, consider the value of

$$\frac{f(a + h) - f(a)}{h},$$

where h is near zero. If there is a number A such that

$$\frac{f(a + h) - f(a)}{h} \approx A,$$

and if this approximation can be made as close as desired by taking h close enough to zero, we say that A is the derivative of $f(x)$ at $x = a$, and write $f'(a) = A$. It is customary to summarize by the notation

$$\lim_{h \to 0} \frac{f(a + h) - f(a)}{h} = f'(a).$$

The left side of this equation is read "the limit of $[f(a + h) - f(a)]/h$ as h approaches zero." Alternative notation is:

$$\frac{f(a + h) - f(a)}{h} \longrightarrow f'(a) \qquad \text{as} \quad h \longrightarrow 0.$$

For example, suppose $f(x) = x^2$ and $a = 3$. Then

$$\frac{f(3 + h) - f(3)}{h} = \frac{(3 + h)^2 - 3^2}{h} = 6 + h.$$

Thus, the closer h is to zero, the more precise is the approximation

$$\frac{f(3 + h) - f(3)}{h} \approx 6.$$

Therefore, the derivative $f'(3)$ is

$$\lim_{h \to 0} \frac{f(3 + h) - f(3)}{h} = 6.$$

EXERCISES

Find the limit:

1. $\lim_{h \to 0} \dfrac{(2 + h)^3 - 8}{h}$

2. $\lim_{h \to 1} \dfrac{h^2 - 1}{h - 1}$

3. $\lim_{h \to 0} \frac{(x+h)^4 - x^4}{h}$

4. $\lim_{x \to 2} (x^3 + 1)$

5. $\lim_{x \to 0} \frac{(1+2x)^2 - 1}{x}$

6. $\lim_{x \to 0} \frac{(1+\frac{1}{2}x)^2 - 1}{x}$

7. $\lim_{h \to 0} \frac{h}{(h+1)^3 - 1}$

8. $\lim_{x \to 0} \frac{1}{(x-1)^3}$

9. $\lim_{x \to 0} \frac{(x+1)^2}{(x-1)^3}$

10. $\lim_{k \to 0} \frac{(2-k)^5 - 32}{k}$.

11. Find $\lim_{x \to 0} \frac{\sqrt{2+x} - \sqrt{2}}{x}$. (*Hint:* "Rationalize" the numerator, i.e., multiply and divide by $\sqrt{2+x} + \sqrt{2}$.)

12. Find $\lim_{x \to 0} \frac{(2+x)^{1/3} - (2)^{1/3}}{x}$. (*Hint:* Proceed as in Ex. 11, using the identity $y^3 - z^3 = (y-z)(y^2 + yz + z^2)$.)

13. Now find $\lim_{x \to 0} \frac{(2+x)^{1/4} - (2)^{1/4}}{x}$.

14. Let $h \to 0$ in the relation

$$\frac{d}{dx}\left[\frac{1}{(x+h)x}\right] = \frac{1}{h}\frac{d}{dx}\left[\frac{1}{x} - \frac{1}{x+h}\right] = \frac{1}{h}\left[-\frac{1}{x^2} + \frac{1}{(x+h)^2}\right]$$

to deduce again that $\frac{d}{dx}\left(\frac{1}{x^2}\right) = \frac{-2}{x^3}$.

6. JUSTIFICATION OF THE RULES

The basic tool needed in this section is the approximation formula,

$$f(x) \approx f(a) + f'(a)(x-a) \qquad \text{for} \quad x \approx a,$$

studied in Chapter 14. It is convenient to write the formula with x replaced by $a + h$:

$$f(a+h) \approx f(a) + f'(a)h \qquad \text{for} \quad h \approx 0.$$

To express this approximation more precisely, write

$$f(a+h) = f(a) + f'(a)h + e(h),$$

where $e(h)$ denotes the error. It is much smaller than h when h is small:

$$\frac{e(h)}{h} = \left[\frac{f(a+h) - f(a)}{h} - f'(a)\right] \longrightarrow 0.$$

The rules for differentiation presented in this chapter can be justified by careful use of this estimate. In order to give the spirit of the justifications without being burdened with technical details, we shall use instead the simpler formula

$$f(a+h) \approx f(a) + f'(a)h.$$

THE PRODUCT RULE

Suppose $f(x) = u(x)v(x)$. The derivative $f'(a)$ is the limit as $h \longrightarrow 0$ of

$$\frac{f(a+h) - f(a)}{h} = \frac{u(a+h)v(a+h) - u(a)v(a)}{h}.$$

We shall use the abbreviations

$$u = u(a), \qquad v = v(a), \qquad u' = u'(a), \qquad v' = v'(a).$$

The numerator of the fraction on the right can be approximated by means of the formulas

$$u(a+h) \approx u + u'h, \qquad v(a+h) \approx v + v'h.$$

Thus

$$\begin{aligned} u(a+h)v(a+h) - u(a)v(a) &\approx (u + u'h)(v + v'h) - uv \\ &= (uv' + vu')h + u'v'h^2. \end{aligned}$$

It follows that

$$\frac{f(a+h) - f(a)}{h} \approx (uv' + vu') + u'v'h.$$

The second term on the right has limit zero as $h \longrightarrow 0$. Therefore

$$f'(a) = \lim_{h\to 0} \frac{f(a+h) - f(a)}{h} = uv' + vu'.$$

In other words,

$$\frac{d}{dx}(uv) = uv' + vu'.$$

THE QUOTIENT RULE

Suppose $f(x) = u(x)/v(x)$ and $v(a) \neq 0$. The derivative $f'(a)$ is the limit as $h \longrightarrow 0$ of

$$\begin{aligned} \frac{f(a+h) - f(a)}{h} &= \frac{1}{h}\left[\frac{u(a+h)}{v(a+h)} - \frac{u(a)}{v(a)}\right] \\ &= \frac{1}{h}\left[\frac{u(a+h)v(a) - u(a)v(a+h)}{v(a+h)v(a)}\right]. \end{aligned}$$

By the approximations

$$u(a+h) \approx u + u'h, \qquad v(a+h) \approx v + v'h,$$

the numerator of the last fraction is approximately

$$(u + u'h)v - u(v + v'h) = (vu' - uv')h,$$

and the denominator is approximately

$$(v + v'h)v = v^2 + v'vh.$$

(If h is very small, then $v^2 + v'vh$ is very close to v^2, hence is non-zero.) Thus

$$\frac{f(a+h) - f(a)}{h} \approx \frac{1}{h}\left[\frac{(vu' - uv')h}{v^2 + v'vh}\right] = \frac{vu' - uv'}{v^2 + v'vh}.$$

The term $v'vh \longrightarrow 0$ as $h \longrightarrow 0$. Therefore

$$f'(a) = \lim_{h\to 0}\left[\frac{f(a+h) - f(a)}{h}\right]$$

$$= \lim_{h\to 0}\left[\frac{vu' - uv'}{v^2 + v'vh}\right] = \frac{vu' - uv'}{v^2}.$$

In other words,

$$\frac{d}{dx}\left(\frac{u}{v}\right) = \frac{vu' - uv'}{v^2}.$$

THE CHAIN RULE

Suppose $y = y(x)$ where $x = x(t)$. Then dy/dt at $t = a$ is the limit as $h \longrightarrow 0$ of

$$\frac{y[x(a+h)] - y[x(a)]}{h}.$$

We shall use the abbreviations

$$x_0 = x(a), \qquad y_0 = y[x(a)], \qquad \frac{dx}{dt} = \left.\frac{dx}{dt}\right|_{t=a}, \qquad \frac{dy}{dx} = \left.\frac{dy}{dx}\right|_{x=x(a)}.$$

We shall estimate $y[x(a+h)]$ by two linear approximations. First, since $x(a+h) \approx x_0 + (dx/dt)h$, we have

$$y[x(a+h)] \approx y[x_0 + \frac{dx}{dt}h].$$

Second, since $y(b+k) \approx y(b) + y'(b)k$, we substitute $b = x_0$ and $k = (dx/dt)h$ to obtain

$$y[x_0 + \frac{dx}{dt}h] \approx y_0 + \frac{dy}{dx}\frac{dx}{dt}h.$$

Hence

$$y[x(a+h)] \approx y_0 + \frac{dy}{dx}\frac{dx}{dt}h,$$

and consequently

$$\frac{y[x(a+h)] - y[x(a)]}{h} \approx \frac{y_0 + \frac{dy}{dx}\frac{dx}{dt}h - y_0}{h} = \frac{dy}{dx}\frac{dx}{dt}.$$

Therefore

$$\frac{dy}{dt} = \lim_{h\to 0} \frac{y[x(a+h)] - y[x(a)]}{h} = \frac{dy}{dx}\frac{dx}{dt}\cdot$$

7. REVIEW

This is a good time to review the formulas introduced in this chapter. Memorize them; you will need them over and over again.

$$\frac{d}{dx}(uv) = u\frac{dv}{dx} + v\frac{du}{dx}\cdot$$ (PRODUCT RULE)

$$\frac{d}{dx}\left(\frac{u}{v}\right) = \frac{v\dfrac{du}{dx} - u\dfrac{dv}{dx}}{v^2}\cdot$$ (QUOTIENT RULE)

For each integer p,

$$\frac{d}{dx}(u^p) = pu^{p-1}\frac{du}{dx}\cdot$$ (POWER RULE)

$$\frac{d}{dx}(\sqrt{u}) = \frac{1}{2\sqrt{u}}\frac{du}{dx}\cdot$$

If y is a function of x, and x is a function of t, then

$$\frac{dy}{dt} = \frac{dy}{dx}\frac{dx}{dt}\cdot$$ (CHAIN RULE)

Be especially careful with the Chain Rule. Approximately 86% of all mistakes in differentiation can be traced to misuse of the Chain Rule. For example:

Typical mistake	Correct answer
$\frac{d}{dx}[(5x)^3] = 3(5x)^2$	$3(5x)^2 \cdot 5$
$\frac{d}{dx}(x^2+x)^5 = 5(x^2+x)^4$	$5(x^2+x)^4(2x+1)$
$\frac{d}{dx}(\sin 4x)^2 = 2\sin 4x \cos 4x$	$8 \sin 4x \cos 4x$
$\frac{d}{dx}(e^{3x}) = e^{3x}$	$3e^{3x}$
$\frac{d}{dx}(\sqrt{1-5x}) = \frac{1}{2\sqrt{1-5x}}$	$\frac{-5}{2\sqrt{1-5x}}\cdot$

Also be careful with the Quotient Rule. Avoid using $uv' - vu'$ as the numerator in place of $vu' - uv'$. To remember which, it is helpful to recall that $u/v = u \cdot v^{-1}$. Thus

$$\frac{d}{dx}\left(\frac{u}{v}\right) = \frac{d}{dx}(uv^{-1}) = v^{-1}\frac{du}{dx} + u\frac{d}{dx}(v^{-1})$$

$$= v^{-1}\frac{du}{dx} - uv^{-2}\frac{dv}{dx} = \frac{vu' - uv'}{v^2}.$$

To test your skill, here is a collection of miscellaneous differentiation problems that can be done with the techniques we have developed.

EXERCISES

Differentiate:

1. $y = \dfrac{2x+1}{2x-1}$
2. $y = (1 + 2\sqrt{x})^3$
3. $y = e^{1/x}$
4. $y = \sqrt{\dfrac{x}{x+1}}$
5. $y = e^{ax}\cos bx$
6. $y = \left(\dfrac{2}{x^2} + \dfrac{3}{x^5}\right)^{-2}$
7. $y = \dfrac{x}{\sqrt{1-x^2}}$
8. $y = x^3e^{-x}$
9. $y = \sin\sqrt{\dfrac{x}{6}}$
10. $y = \dfrac{1}{1+2x+3x^2}$
11. $y = x^2\sqrt{x^2-a^2}$
12. $y = e^{xe^x}$
13. $y = \sin^3 kx$
14. $y = \left(\dfrac{x+1}{x+3}\right)^2$
15. $y = x\sin\dfrac{1}{x}$
16. $y = \dfrac{e^{ax}}{\sqrt{bx}}$
17. $y = \dfrac{4-x}{\sqrt{8x-x^2}}$
18. $y = 3x\cos(x^2)$
19. $y = \dfrac{\sin 2x}{x^3}$
20. $y = \dfrac{1}{(2\sin x)^5}$
21. $y = x^2e^{-3/x}$
22. $y = \sqrt{x + \sqrt{x}}$
23. $y = \cos^4 x - \sin^4 x$
24. $y = \dfrac{1}{1+e^{-x}}$
25. $y = \left(\dfrac{x}{e^{2x}+1}\right)^3$
26. $y = \sqrt{x}/\cos(2\sqrt{x})$
27. $y = xe^{-ax}\sin bx$
28. $y = \dfrac{1-\sqrt{3x}}{1+\sqrt{3x}}$

29. $y = \left(1 + \cos^2 \dfrac{x}{2}\right)^3$

30. $y = \dfrac{x}{(x^3 + 1)^2}$

31. $y = \dfrac{e^x - e^{-x}}{e^x + e^{-x}}$

32. $y = \sqrt{x}\, e^{-\sqrt{x}}$.

33. Show that the derivative of $e^x p(x)$, where $p(x)$ is a polynomial, is a function of the same form.

34. Show that the derivative of $p(x)/q(x)$, where $p(x)$ and $q(x)$ are polynomials, is a function of the same form.

16. Applications of Derivatives

1. RATE PROBLEMS

If two physical quantities are related and both are changing, then their rates of change are also related. Quite a number of physical problems involve this idea.

EXAMPLE 1.1

A large spherical balloon is inflated by a pump that injects 10 ft^3/sec of helium. At the instant when the balloon contains 972π ft^3 of gas, how fast is its radius increasing?

Solution: Denote the radius and volume of the balloon at time t by $r(t)$ and $V(t)$. The derivative $dV/dt = 10$ ft^3/sec is given. The derivative dr/dt is required at a specific time.

The formula for the volume V of a sphere of radius r is $V = \frac{4}{3}\pi r^3$. Hence

$$V(t) = \frac{4}{3}\pi[r(t)]^3.$$

To find a relation between dr/dt and dV/dt, differentiate using the Chain Rule:

$$\frac{dV}{dt} = \frac{dV}{dr}\frac{dr}{dt} = \frac{4}{3}\pi \cdot 3r^2 \cdot \frac{dr}{dt} = 4\pi r^2\frac{dr}{dt}.$$

Solve for dr/dt:

$$\frac{dr}{dt} = \frac{1}{4\pi r^2}\frac{dV}{dt} = \frac{10}{4\pi r^2}.$$

This formula tells the rate of change of the radius at any instant, in terms of the radius. At the instant in question, the volume is 972π ft^3, so the radius can be found by solving

$$\frac{4}{3}\pi r^3 = 972\pi.$$

Thus

$$r^3 = \frac{3}{4\pi}\cdot 972\pi = 729, \qquad r = 9.$$

But when $r = 9$,

$$\frac{dr}{dt} = \frac{10}{4\pi \cdot 9^2} = \frac{10}{324\pi}.$$

Answer: $\dfrac{10}{324\pi} \approx 0.00982$ ft/sec.

EXAMPLE 1.2

A point P moves with increasing speed around a circle of radius 10 ft, starting at Q when $t = 0$. See Fig. 1.1. When $\frac{1}{3}$ of the way around the circle, its speed is 9 ft/sec. At that instant, how fast is the length PQ increasing?

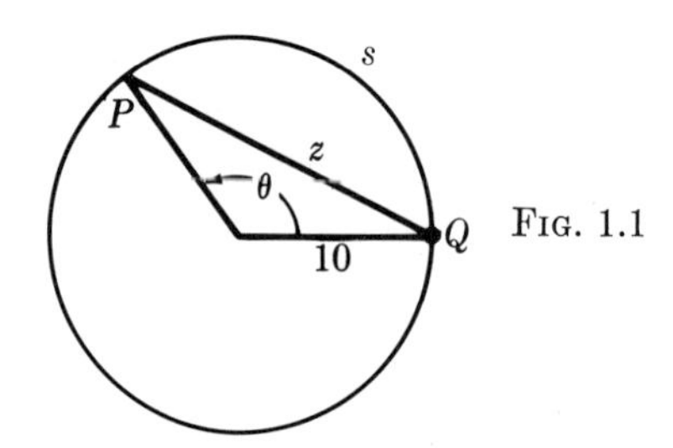

FIG. 1.1

Solution: At time t, let z be the length of PQ, let s be the arc length, and let θ be the central angle (Fig. 1.1). Given: $ds/dt = 9$ ft/sec when $\theta = 2\pi/3$, that is, when $s = 10 \cdot 2\pi/3$ ft. Find dz/dt at that instant.

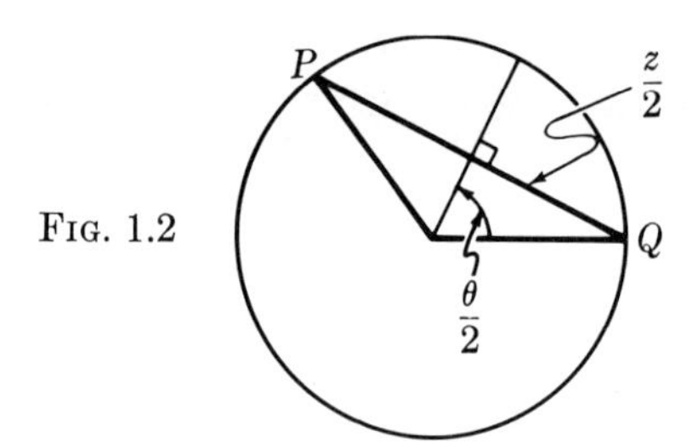

FIG. 1.2

To do so, obtain a relation between z and s. From Fig. 1.2,

$$\frac{z}{2} = 10 \sin \frac{\theta}{2},$$

But from the previous figure, $s = 10\theta$, hence

$$z = 20 \sin \frac{\theta}{2} = 20 \sin \frac{s}{20}.$$

This is the desired relation; now differentiate using the Chain Rule:

$$\frac{dz}{dt} = \left(20 \cos \frac{s}{20}\right) \frac{1}{20} \frac{ds}{dt}.$$

At the instant in question,

$$\cos\frac{s}{20} = \cos\frac{\theta}{2} = \cos\frac{\pi}{3} = \frac{1}{2} \qquad \text{and} \qquad \frac{ds}{dt} = 9.$$

Hence

$$\frac{dz}{dt} = 20 \cdot \frac{1}{2} \cdot \frac{1}{20} \cdot 9 = \frac{9}{2}.$$

Answer: 4.5 ft/sec.

EXAMPLE 1.3

A rectangular tank has a sliding panel S that divides it into two adjustable tanks of width 3 ft. See Fig. 1.3. Water is poured into the left compartment at the rate of 5 ft³/min. At the same time S is moved to the right at the rate of 3 ft/min. When the left compartment is 10 ft long it contains 70 ft³ of water. Is the water level rising or falling? How fast?

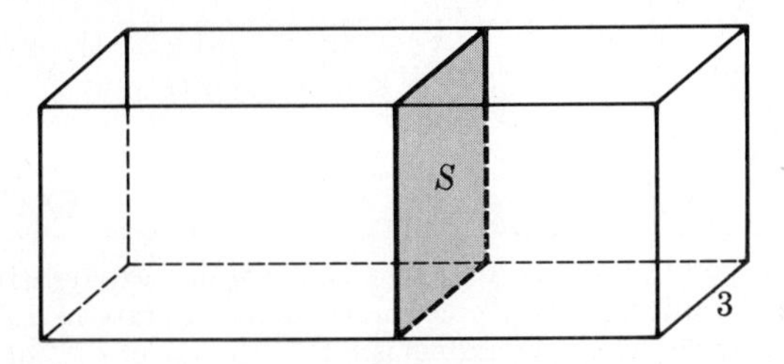

FIG. 1.3

Solution: Let x be the length of the left compartment; let y and V be the depth and the volume of the water in the left compartment. Then x, y, and V are all functions of time. Given:

$$\frac{dx}{dt} = 3 \text{ ft/min}, \qquad \frac{dV}{dt} = 5 \text{ ft}^3\text{/min}.$$

Compute: dy/dt when $x = 10$ and $V = 70$.

To do so, find a relation between x, y, and V. At any instant

$$V = 3xy.$$

Differentiate with respect to t:

$$\frac{dV}{dt} = 3x\frac{dy}{dt} + 3\frac{dx}{dt}y.$$

Substitute the data at the instant in question:

$$5 = 3 \cdot 10 \cdot \frac{dy}{dt} + 3 \cdot 3 \cdot y,$$

$$\frac{dy}{dt} = \frac{5 - 9y}{30}.$$

At the given instant, $V = 70$ and $x = 10$, hence $y = 7/3$. Therefore

$$\frac{dy}{dt} = \frac{5 - 9(7/3)}{30} = -\frac{16}{30}.$$

The water level is falling at the rate of 8/15 ft/min.

Alternate Solution: Instead of differentiating the equation $V = 3xy$, solve for y first, then differentiate:

$$y = \frac{1}{3}\frac{V}{x},$$

$$\frac{dy}{dt} = \frac{1}{3}\frac{x\dfrac{dV}{dt} - V\dfrac{dx}{dt}}{x^2}.$$

At the given instant,

$$\frac{dy}{dt} = \frac{1}{3}\frac{10 \cdot 5 - 70 \cdot 3}{10^2} = -\frac{16}{30}.$$

Answer: Falling at the rate of 8/15 ft/min.

EXERCISES

1. A stone thrown into a pond produces a circular ripple which expands from the point of impact. If the radius of the ripple increases at the rate of 1.5 ft/sec, how fast is the area growing when the radius is 8 ft?
2. Water flows into an inverted conical tank at the rate of 27 ft³/min. When the depth of the water is 2 ft, how fast is the level rising? Assume the height of the tank is 4 ft and the radius at the top is 1 ft.
3. A 6-ft man walks away from a 12-ft lamp post at the rate of 4 ft/sec. How fast is his shadow lengthening when he is 21 ft from the post?
4. Two cars leave an intersection. One travels north at 30 mph, the other east at 40 mph. How fast is the distance between them increasing at the end of 1 min? 5 min?
5. A tetrahedron has vertices at $(0, 0, 0)$, $(x, 0, 0)$, $(0, y, 0)$ and $(0, 0, z)$. Suppose x, y, and z are increasing with time according to the formulas $x = 2t$, $y = t^3$, and $z = 3t^2$. When $t = 2$, how fast is the volume of the tetrahedron increasing? Measure lengths in cm and time in sec.
6. A point moves from the origin along the curve $y = x^3 - 3x^2$ so that its x-coordinate increases 3 ft/sec. Find the rate at which the distance from the point to the origin is increasing when the point is at $(1, -2)$.
7. If the volume of an expanding cube is increasing at the rate of 4 ft³/min, how fast is its surface area increasing when the surface area is 24 ft²?
8. If a chord sweeps (without turning) across a circle of radius 10 ft at the rate of 6 ft/sec, how fast is the length of the chord decreasing when it is $\frac{3}{4}$ of the way across?
9. An observer is standing $\frac{1}{4}$ mi from a railroad track as a train goes by at 80 mph. When the front of the train is 2 mi away from the nearest point on the track to the observer, at what speed is it moving away from him?

10. A lighthouse beacon 2 mi from a straight beach revolves once every 48 sec. How fast is the spot of light moving when it is 5 mi down the beach from the point nearest to the lighthouse? (*Hint:* $d(\tan\theta)/d\theta = \sec^2\theta$.)
11. A 15-ft ladder leans against a vertical wall. If the top slides downward at the rate of 2 ft/sec, find the speed of the lower end when it is 12 ft from the wall.
12. An elevated train on a track 30 ft above the ground crosses a (perpendicular) street at the rate of 50 ft/sec at the instant that an automobile, approaching at the rate of 30 ft/sec, is 40 ft up the street. Find how fast the train and the automobile are separating 2 sec later.
13. A conical water tank is mounted with its vertex down. The angle of the vertex is 60°. Water is pumped into the tank at a rate of 5 ft³/min, and water evaporates from the tank at a rate (ft³/min) equal to 0.01 times the exposed surface area (ft²). At what depth, if any, will the water stop rising?
14. Thread of radius 10^{-3} m is being wound on a ball at the rate of 2 m/sec. Assume that the ball is a perfect sphere at each instant, and consists entirely of thread with no empty space. Find the rate of increase of the radius when the radius is 0.08 m.
15. Two concentric circles are expanding, the outer radius at the rate of 2 ft/sec and the inner one at 5 ft/sec. At a certain instant, the outer radius is 10 ft and the inner radius is 3 ft. At this instant, is the area of the ring between the two circles increasing or decreasing? How fast?
16. A point P moves to the right along the x-axis at the rate of 4 cm/sec. Let R_P denote the region under the curve $y = e^{-x^2}$ between $x = 0$ and $x = P$. Find the rate of change of the area of R_P when P is at $x = 3$.
17. A barber pole is a 3-ft cylinder of radius 4 in. The red spiral stripe on its surface makes exactly 1.5 turns from bottom to top. As the pole rotates at the rate of 40 rpm, how fast does the stripe appear to move vertically?

2. MAXIMA AND MINIMA

Before beginning this section, reread Chapter 3.

Here is the basic problem: find the maximum value or the minimum value of $f(x)$ in an interval $a \leq x \leq b$.

In theory, it is easy to locate the maximum or the minimum:

> The maximum of a smooth function $f(x)$ in the interval $a \leq x \leq b$ occurs either at a value of x where $f'(x) = 0$, or at one of the end points, a or b. The same is true for the minimum of $f(x)$.

This statement is easy to see graphically (Fig. 2.1). At points where

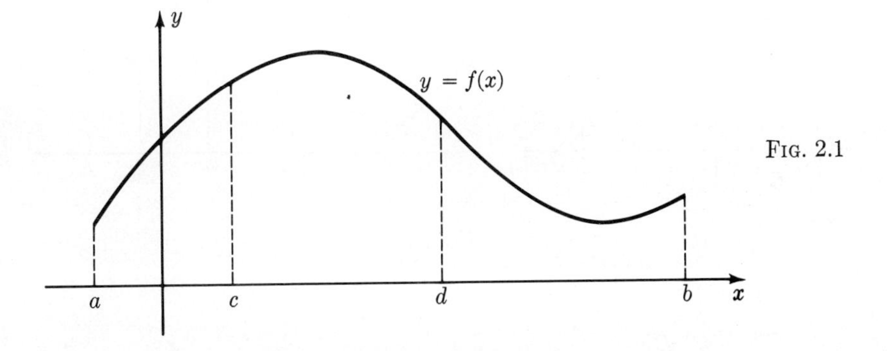

FIG. 2.1

$f'(x) > 0$, the graph is rising ($x = c$ for example). Neither the maximum nor the minimum can occur at such a point because the graph is higher to the right and lower to the left. At points where $f'(x) < 0$, the graph is falling ($x = d$ for example), and for a similar reason neither the maximum nor the minimum can occur there. Hence if $a < x_0 < b$ and $f(x_0)$ is the maximum or the minimum value of $f(x)$, then $f'(x_0) = 0$.

This argument does not apply at the end points a and b, however. (Why not?) The maximum or the minimum may occur at one of the end points without the derivative vanishing there (Fig. 2.2).

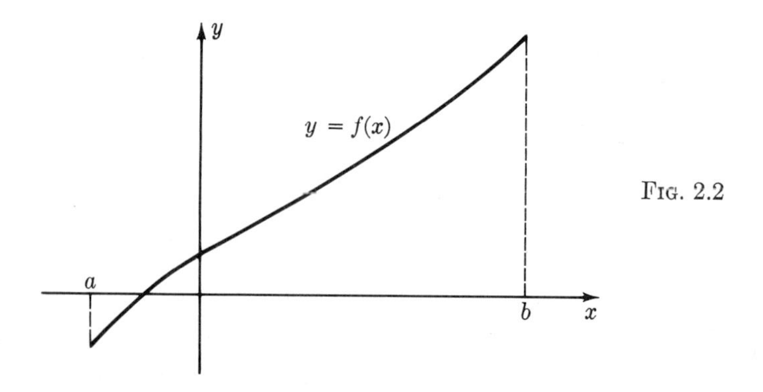

FIG. 2.2

This discussion suggests a procedure for solving the basic problem.

> To find the maximum and minimum of $f(x)$ in the interval $a \leq x \leq b$, locate all points x where $f'(x) = 0$. Call these $x_1, x_2, \cdots, x_n$. The maximum is the largest of the numbers $f(a), f(x_1), f(x_2), \cdots, f(x_n), f(b)$. The minimum is the smallest of these.

In practice, this procedure may involve unnecessary work. The derivative of the function graphed in Fig. 2.3 is zero at x_1, x_2, x_3, x_4, x_5, and x_6.

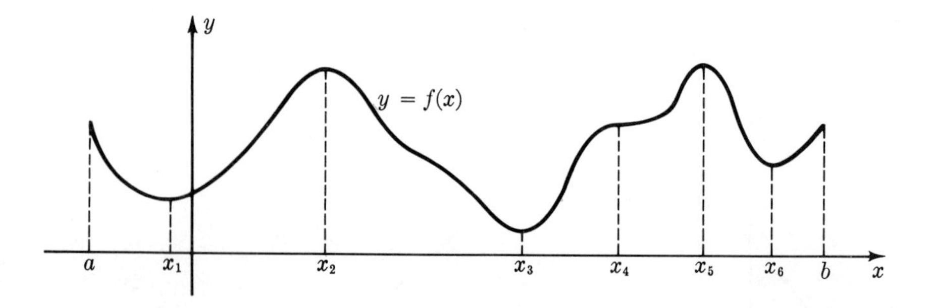

FIG. 2.3

Certainly the maximum does not occur at x_1, x_3, x_4, or x_6. We need a way to eliminate such points quickly.

What can the graph of $y = f(x)$ look like near a point $x = c$ where the derivative is zero? There are only four possible shapes (Fig. 2.4). The figures show a maximum in case (a) and a minimum in case (b). The remaining two cases are neither maxima nor minima.

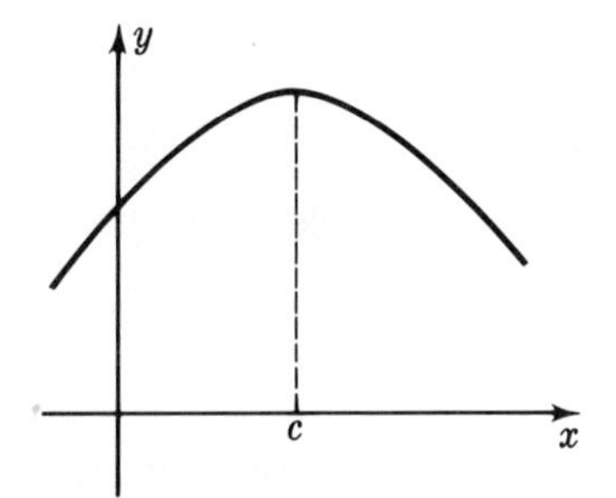

FIG. 2.4a

$f'(x) > 0$	c	$f'(x) < 0$
increases	is flat	decreases

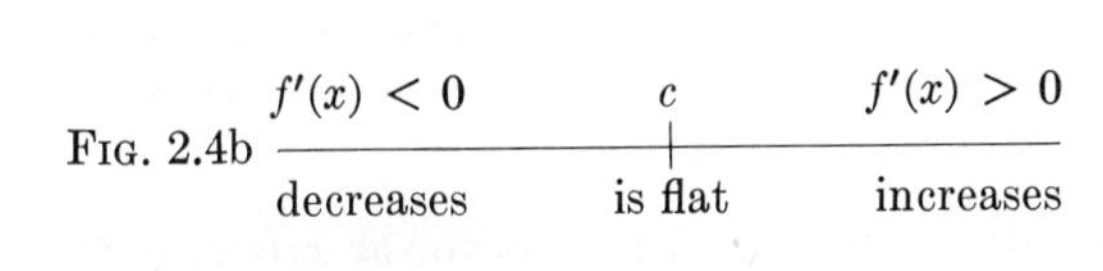

FIG. 2.4b

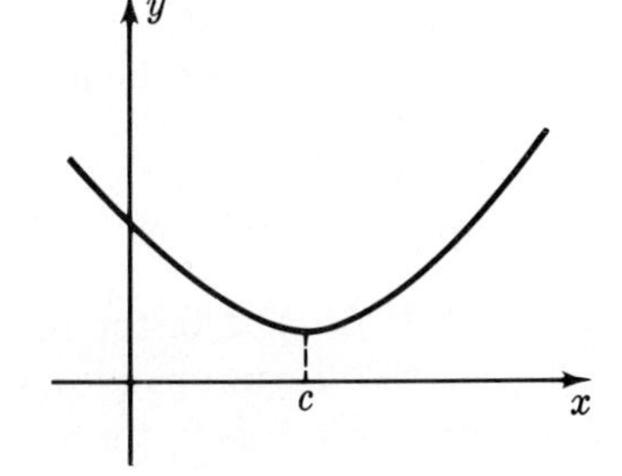

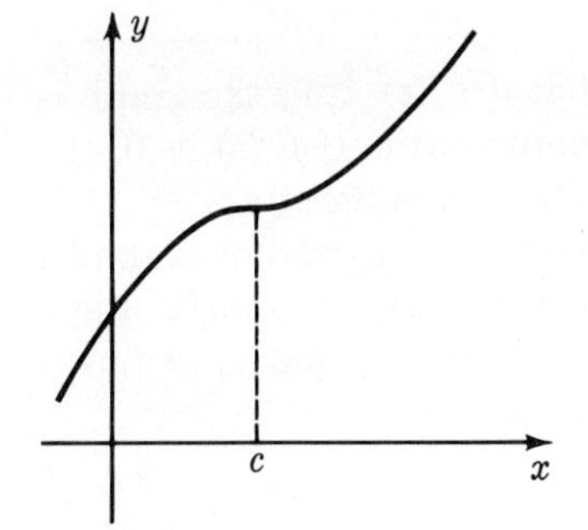

FIG. 2.4c

$f'(x) > 0$	c	$f'(x) > 0$
increases	is flat	increases

FIG. 2.4d

$f'(x) < 0$	c	$f'(x) < 0$
decreases	is flat	decreases

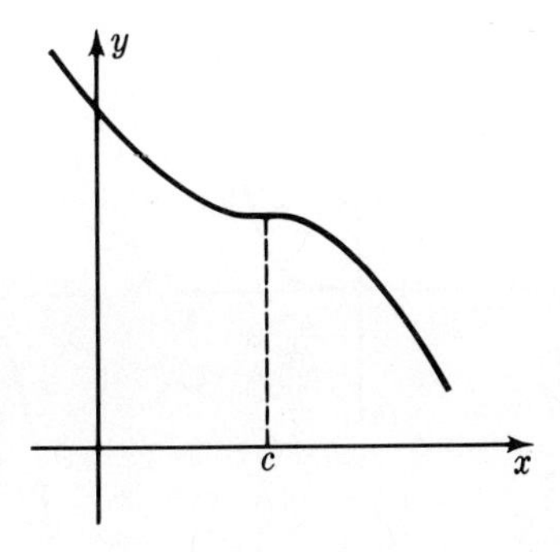

The maximum of $f(x)$ in the interval $a \leq x \leq b$ occurs either at $x = a$, or at $x = b$, or at an interior point $x = c$ where

(1) $f'(c) = 0$, and
(2) $f'(x)$ changes from positive to negative as x increases through c. In other words, $f'(x) > 0$ for $x < c$ and $f'(x) < 0$ for $x > c$.

The minimum occurs either at $x = a$, or at $x = b$, or at an interior point $x = c$ where

(1) $f'(c) = 0$, and
(2) $f'(x)$ changes from negative to positive as x increases through c.

(These statements presuppose that $f'(x) = 0$ only at a finite set of points in the interval $a \leq x \leq b$.)

Sometimes the second derivative can be used to determine how the sign of $f'(x)$ behaves near $x = c$. Suppose $f'(c) = 0$. If $f''(c) > 0$, the derivative $f'(x)$ is increasing at c, hence $f'(x)$ increases from negative to zero to positive. If $f''(c) < 0$, then $f'(x)$ decreases from positive to zero to negative.

Suppose $f'(c) = 0$.

(i) If $f''(c) < 0$, the maximum value of $f(x)$ may occur at $x = c$, but the minimum cannot.
(ii) If $f''(c) > 0$, the minimum value of $f(x)$ may occur at $x = c$, but the maximum cannot.

This makes good sense geometrically. Recall that if $f''(c) < 0$, the graph is concave downwards and so cannot have its minimum at c. If $f''(c) > 0$, the graph is concave upwards and so cannot have its maximum at c.

REMARK: The case $f'(c) = 0$ *and* $f''(c) = 0$ is inconclusive. For example, the functions $f_4(x) = x^4$, $f_5(x) = x^5$, and $f_6(x) = -x^6$ all satisfy these conditions at $c = 0$. Yet $f_4(x)$ has a minimum at $x = 0$, and $f_6(x)$ has a maximum at $x = 0$, whereas $f_5(x)$ has neither (Fig. 2.5).

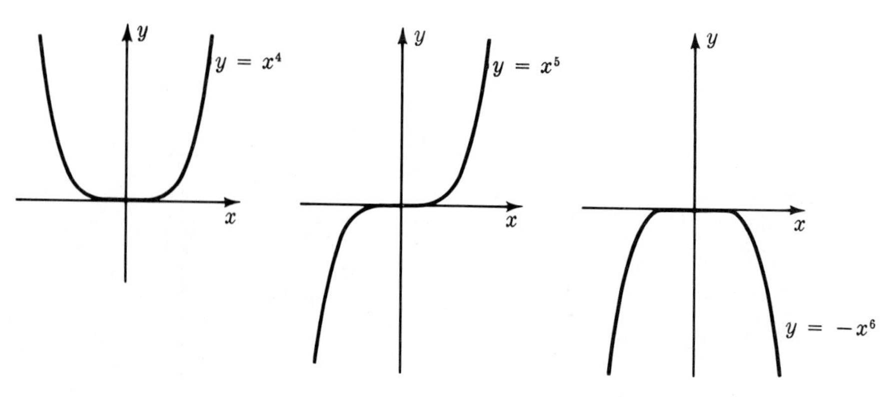

FIG. 2.5a FIG. 2.5b FIG. 2.5c

EXAMPLE 2.1

Find the maximum value of the function $f(x) = \frac{1}{2}x - \sin x$ in the range $0 \le x \le 4\pi$.

Solution: The maximum occurs either at 0 or 4π, or at a zero of the derivative,

$$f'(x) = \frac{1}{2} - \cos x.$$

Thus

$$f'(x) = 0 \qquad \text{if} \quad \cos x = \frac{1}{2}.$$

In the range $0 \le x \le 4\pi$, zeros of the derivative occur at

$$x = \frac{\pi}{3}, \quad \frac{5\pi}{3}, \quad \frac{7\pi}{3}, \quad \frac{11\pi}{3}.$$

Hence, there are six possible places where the maximum can occur. To eliminate some of them, consider the second derivative, $f''(x) = \sin x$:

$$f''\left(\frac{\pi}{3}\right) = f''\left(\frac{7\pi}{3}\right) = \sin\frac{\pi}{3} > 0,$$

$$f''\left(\frac{5\pi}{3}\right) = f''\left(\frac{11\pi}{3}\right) = \sin\frac{5\pi}{3} < 0.$$

The values $5\pi/3$ and $11\pi/3$, for which $f''(x)$ is negative, are candidates for the maximum. The others, $\pi/3$ and $7\pi/3$, are candidates for the minimum; eliminate them. Hence the maximum of $f(x)$ occurs at one of the points

$$0, \quad \frac{5\pi}{3}, \quad \frac{11\pi}{3}, \quad 4\pi.$$

Evaluate $f(x) = \frac{1}{2}x - \sin x$ at each of these points:

$$f(0) = 0,$$

$$f\left(\frac{5\pi}{3}\right) = \frac{5\pi}{6} - \sin\frac{5\pi}{3} = \frac{5\pi}{6} + \frac{\sqrt{3}}{2},$$

$$f\left(\frac{11\pi}{3}\right) = \frac{11\pi}{6} - \sin\frac{11\pi}{3} = \frac{11\pi}{6} + \frac{\sqrt{3}}{2},$$

$$f(4\pi) = 2\pi = \frac{11\pi}{6} + \frac{\pi}{6}.$$

The largest of these is $f(11\pi/3)$.

Alternate Solution: Sketch $y = \frac{1}{2}x - \sin x$ by graphically subtracting the curve $y = \sin x$ from the line $y = \frac{1}{2}x$. See Fig. 2.6. From the figure it is

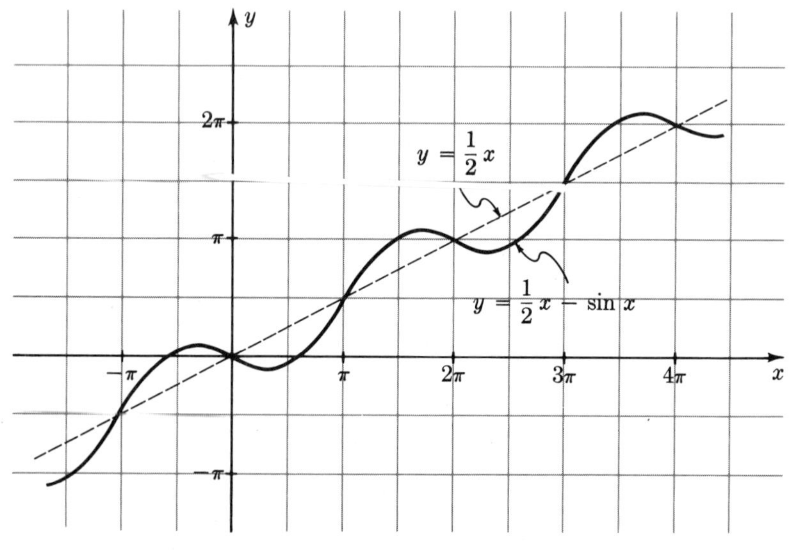

Fig. 2.6

evident that the maximum occurs in the interval $3\pi \leq x \leq 4\pi$. Hence only $11\pi/3$ and 4π are serious candidates.

Answer: $\dfrac{11\pi}{6} + \dfrac{\sqrt{3}}{2} \approx 6.6256.$

Absence of End Points

Some problems may require the maximum value of $f(x)$ for all x or only for $x \geq 0$. In the first case there are no end points; in the second case, there is a left end point but no right end point. Generally, such problems are worked by the techniques of this section, except that it is not necessary to consider end points if there are none.

EXAMPLE 2.2

Find the smallest value of $f(x) = x^4 - 108x$.

Solution: In this problem x is unrestricted; there are no end points. Differentiate:

$$f'(x) = 4x^3 - 108 = 4(x^3 - 27).$$

Thus

$$f'(x) = 0 \qquad \text{for} \quad x = 3.$$

It is easy to see that $f(x)$ has a minimum at $x = 3$. Either observe the sign of $f'(x)$ near $x = 3$, or compute the second derivative $f''(x) = 12x^2$, positive for $x = 3$.

Answer: $f(3) = -243$.

Note that there is no maximum value of $f(x)$ in this problem. In general, a function $f(x)$ need not have a maximum or a minimum unless x is restricted to an interval with two end points. For example, take $f(x) = 1/(1 + x^2)$, where x is unrestricted. This function has maximum value 1, but no minimum value (it takes all values between 0 and 1, excluding zero). As another example, take $f(x) = 1/x$. For $x > 0$, this function has neither a maximum nor a minimum.

Remarks on Finding Maxima and Minima

In most maximum and minimum problems there are only one or two zeros of the derivative to consider, and possibly two end points. Often you can rule out the end points by physical considerations. Then you have to decide which zero of the derivative gives the maximum or the minimum. If it is easy to compute the second derivative, do so. If not, or if the second derivative is zero, try observing the sign of the derivative near the point in question. Better yet, graph the function if that is easy. Be flexible.

EXERCISES

Find the maximum and minimum in the interval indicated:

1. $f(x) = \sin 4\pi x;\quad 0 < x < \frac{1}{2}$
2. $f(x) = 2x^3 - 3x^2 - 12x + 1;\quad -2 \le x \le 3$
3. $f(x) = e^{x^2-2x};\quad 0 \le x \le 2$
4. $f(x) = \sin 2x - \cos 2x;\quad 0 < x < \pi$
5. $f(x) = \dfrac{x}{x^2 - x + 1};\quad x$ unrestricted
6. $f(x) = \dfrac{1}{(x-1)(2-x)};\quad 1 < x < 2$
7. $f(x) = \dfrac{x}{3} + \dfrac{1}{x};\quad 1 \le x \le 2$
8. $f(x) = x^3 e^{-x};\quad 0 \le x$
9. $f(x) = \sin x \sin 2x;\quad 0 \le x \le \dfrac{\pi}{2}$
10. $f(x) = e^{-x} \sin x;\quad 0 \le x \le \pi$
11. $f(x) = \dfrac{1}{x} - \dfrac{1}{x^2} - \dfrac{1}{x^3};\quad 0 < x$
12. $f(x) = x - x^4;\quad 0 \le x \le 1$
13. $f(x) = x - \dfrac{3}{4} \sin x;\quad 0 \le x \le 4\pi$.
14. Approximate to 2 places the point x where $(\sin x)/x$ assumes its minimum value in the interval $\pi \le x \le 2\pi$.
15. Show that $e^x/x \ge e$ for all $x \ge 0$ and that $e^x/x = e$ only for $x = 1$. (*Hint:* Find the minimum of e^x/x.)

3. APPLICATIONS OF MAXIMA AND MINIMA

EXAMPLE 3.1

Compute the volume of the largest right circular cone inscribed in a sphere of radius R.

Solution: The volume V of a cone is

$$V = \frac{1}{3}\pi r^2 h,$$

where r is the radius of its base and h is its height. If a cone is inscribed in a sphere of radius R, there ought to be a relation between r, h, and R. Make a careful drawing of a cross-section (Fig. 3.1).

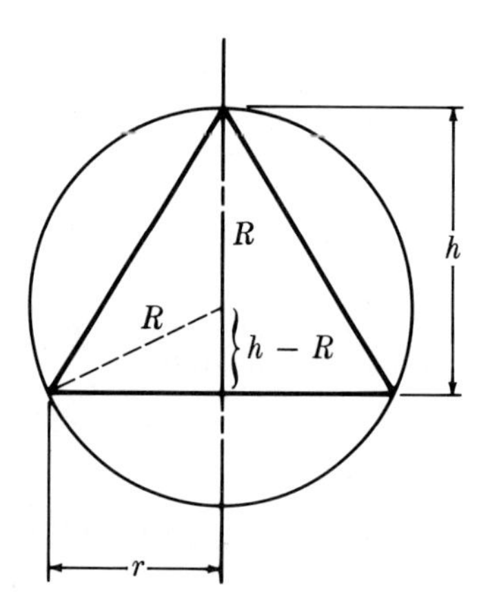

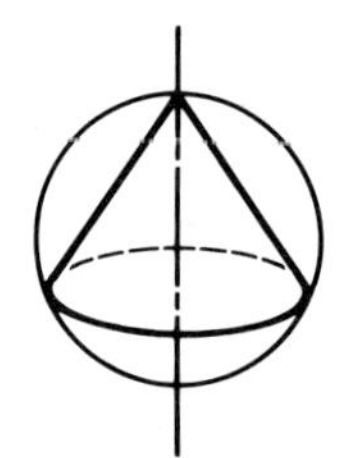

FIG. 3.1

From the drawing,

$$r^2 + (h - R)^2 = R^2,$$

$$r^2 = R^2 - (h - R)^2 = 2Rh - h^2.$$

Substitute:

$$V = \frac{1}{3}\pi r^2 h = \frac{1}{3}\pi(2Rh - h^2)h = \frac{\pi}{3}(2Rh^2 - h^3).$$

By the physical nature of the problem, $0 < h < 2R$. Thus you must maximize

$$V(h) = \frac{\pi}{3}(2Rh^2 - h^3)$$

in the interval $0 < h < 2R$.

There are no end points, hence the maximum occurs at a zero of the derivative:

$$\frac{dV}{dh} = \frac{\pi}{3}(4Rh - 3h^2) = \frac{\pi}{3}h(4R - 3h).$$

Therefore

$$\frac{dV}{dh} = 0 \qquad \text{for} \quad h = 0 \qquad \text{or} \qquad h = \frac{4}{3}R.$$

But $h = 0$ is excluded; the maximum must occur at $4R/3$. Since

$$V(h) = \frac{\pi}{3}h^2(2R - h),$$

$$V\left(\frac{4R}{3}\right) = \frac{\pi}{3}\left(\frac{4R}{3}\right)^2\left(\frac{2R}{3}\right) = \frac{32\pi R^3}{81}.$$

Answer: $\dfrac{32\pi R^3}{81}$.

REMARK: The answer has the correct form; a volume should be a cubic expression. Since the sphere has volume $\frac{4}{3}\pi R^3$, it follows easily that the volume of the largest cone that can be inscribed in a sphere is $\frac{8}{27}$ the volume of the sphere.

EXAMPLE 3.2

A 5-ft fence stands 4 ft from the wall of a house. How long is the shortest ladder that can reach from the ground outside the fence to the wall?

Solution: First, draw a diagram (Fig. 3.2). Now take a moment to

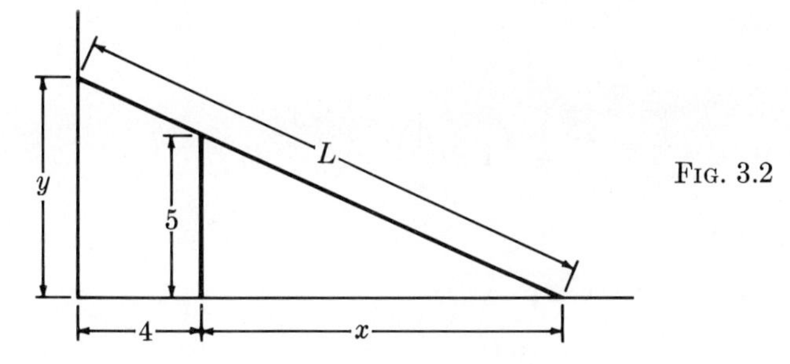

FIG. 3.2

think. If x is very small and positive, the ladder will be nearly vertical, certainly longer than is necessary. If x is large, the ladder will be nearly horizontal, again too long. The best choice of x seems to be somewhere around 5 or 6, surely between 2 and 10. In fact, as x increases starting near 0, it seems that L should decrease, reach a minimum, then increase thereafter.

To start the computation, note that

$$L^2 = (x + 4)^2 + y^2.$$

There is a relation between x and y: by similar triangles,

$$\frac{y}{x + 4} = \frac{5}{x},$$

$$y = \frac{5(x+4)}{x}.$$

Hence,

$$L^2 = (x+4)^2 + \frac{25(x+4)^2}{x^2} = (x+4)^2\left(1 + \frac{25}{x^2}\right).$$

Rather than take the square root, minimize L^2. The range of x is all positive values; there are no end points in this problem. Differentiate L^2:

$$\begin{aligned}\frac{d}{dx}(L^2) &= 2(x+4)\left(1+\frac{25}{x^2}\right) + (x+4)^2\left(\frac{-50}{x^3}\right)\\ &= 2(x+4)\left[1 + \frac{25}{x^2} - \frac{25(x+4)}{x^3}\right]\\ &= 2(x+4)\left[\frac{x^3 + 25x - 25(x+4)}{x^3}\right].\end{aligned}$$

Thus

$$\frac{d}{dx}(L^2) = \frac{2(x+4)(x^3-100)}{x^3}.$$

There is only one positive value of x for which the derivative is zero: $x = \sqrt[3]{100}$. The derivative is negative for $x < \sqrt[3]{100}$, positive for $x > \sqrt[3]{100}$. Thus our physical intuition was correct: L^2 decreases, reaches a minimum near $x = 5$, then increases.

From the formula for L^2,

$$L = (x+4)\sqrt{1+\frac{25}{x^2}} = \left(1+\frac{4}{x}\right)\sqrt{x^2+25}.$$

The minimum value of L is

$$L(\sqrt[3]{100}) = \left(1 + \frac{4}{\sqrt[3]{100}}\right)\sqrt{100^{2/3}+25}.$$

Answer: $\left(1 + \frac{4}{\sqrt[3]{100}}\right)\sqrt{100^{2/3}+25} \approx 12.7$ ft.

EXAMPLE 3.3

The illumination of an object by a light source is directly proportional to the strength of the source and inversely proportional to the square of the distance between the source and the object. Two light bulbs, one 5 times as strong as the other, are 1 yd apart. At what point on the line between the bulbs should a screen be placed so that the illumination it receives is minimal?

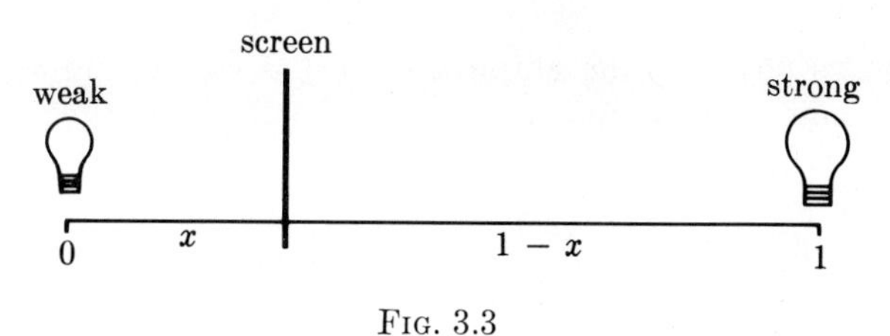

FIG. 3.3

Solution: First draw a diagram (Fig. 3.3). Apparently the screen should be closer to the weaker source; $x < \frac{1}{2}$. Even though one bulb is 5 times as strong as the other, the screen cannot be too close to the weaker bulb because of the inverse square rule. A reasonable guess: x is around 0.3 or 0.4. The illumination from the weaker bulb is

$$I_1 = \frac{k}{x^2},$$

where the constant k depends on the units of measurement. The illumination from the stronger bulb is

$$I_2 = \frac{5k}{(1-x)^2}.$$

The problem is to minimize

$$I = \frac{k}{x^2} + \frac{5k}{(1-x)^2},$$

for $0 < x < 1$. There are no end points in this problem since I is defined neither at $x = 0$ nor at $x = 1$. Differentiate:

$$\frac{dI}{dx} = -\frac{2k}{x^3} + \frac{10k}{(1-x)^3}.$$

This derivative is 0 for

$$\frac{2k}{x^3} = \frac{10k}{(1-x)^3}, \qquad \text{i.e.,} \quad 5x^3 = (1-x)^3.$$

Take cube roots:

$$(\sqrt[3]{5})x = 1 - x,$$

$$x = \frac{1}{1 + \sqrt[3]{5}} \approx 0.369.$$

The physics of the problem suggests this x gives a minimum. As a quick check, take the second derivative:

$$\frac{d^2I}{dx^2} = \frac{6k}{x^4} + \frac{30k}{(1-x)^4},$$

which is always positive (the graph $y = I(x)$ is concave upwards). Hence x does give a minimum, so the guess was not bad. (Note that the minimum does *not* occur where $I_1 = I_2$.)

Answer: $\dfrac{1}{1 + \sqrt[3]{5}} \approx 0.369$ yd from the weaker bulb.

EXAMPLE 3.4

Light travels between two points along the path that requires the least time. In different substances (water, air, glass, etc.) light travels at different speeds. Assume the upper half of the x, y-plane is a substance in which the speed of light is v_1 and the lower half is another substance in which the speed of light is v_2. Describe the path of a light ray traveling between two points in opposite halves of the plane.

Solution: Draw a diagram. Let the two points be $(0, a)$ and $(b, -c)$. See Fig. 3.4. A ray will travel from $(0, a)$ along a straight line to some point $(x, 0)$ and then along another straight line to $(b, -c)$. A value x must be found so that the time of travel is a minimum. Obviously $0 \le x \le b$.

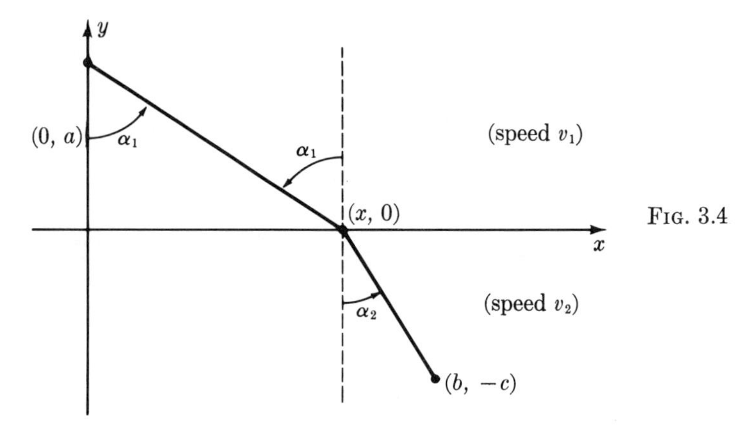

FIG. 3.4

The time required for a ray to travel from $(0, a)$ to $(x, 0)$ is

$$t_1 = \frac{\text{distance}}{\text{speed}} = \frac{\sqrt{x^2 + a^2}}{v_1}.$$

The time required from $(x, 0)$ to $(b, -c)$ is

$$t_2 = \frac{\sqrt{(b - x)^2 + c^2}}{v_2}.$$

Hence you must minimize

$$t = \frac{\sqrt{x^2 + a^2}}{v_1} + \frac{\sqrt{(b - x)^2 + c^2}}{v_2}$$

in the interval $0 \le x \le b$. It is plausible physically that the minimum will not occur at either end point.

Compute dt/dx:

$$\frac{dt}{dx} = \frac{x}{v_1\sqrt{x^2 + a^2}} - \frac{b - x}{v_2\sqrt{(b - x)^2 + c^2}}.$$

But from Fig. 3.4,

$$\frac{x}{\sqrt{x^2 + a^2}} = \sin \alpha_1, \qquad \frac{b - x}{\sqrt{(b - x)^2 + c^2}} = \sin \alpha_2.$$

Hence

$$\frac{dt}{dx} = \frac{\sin \alpha_1}{v_1} - \frac{\sin \alpha_2}{v_2}.$$

The derivative is zero if x is chosen to satisfy

$$\frac{\sin \alpha_1}{v_1} = \frac{\sin \alpha_2}{v_2}.$$

This equation is known as **Snell's Law of Refraction.** To see that it describes the path of least time, note that dt/dx is the difference of two terms. As x increases from 0 to b, the first term, $(\sin \alpha_1)/v_1$, increases steadily starting with 0. The second term, $(\sin \alpha_2)/v_2$, decreases steadily from some positive value to 0. Consequently, dt/dx starts negative at $x = 0$ and steadily increases to a positive value at $x = b$. Therefore, the minimum t occurs at the only x for which $dt/dx = 0$.

Answer: The path is the broken line (Fig. 3.4) for which $\dfrac{\sin \alpha_1}{v_1} = \dfrac{\sin \alpha_2}{v_2}$.

EXAMPLE 3.5

A swampy region shares a long straight border with a region of farm land. A telephone cable is to be constructed connecting two locations, one in each region. Its cost is d_1 dollars per mile in the swampy region and d_2 dollars per mile in the farm land. What path should the cable take for its cost to be least?

Solution: Make a diagram with the x-axis as the border of the two regions and with the given points at $(0, a)$ and $(b, -c)$. See Fig. 3.5. The most economical path must be some broken line as shown. The cost of the cable from $(0, a)$ to $(x, 0)$ is

$$(\text{cost per mile})(\text{distance}) = d_1\sqrt{x^2 + a^2} = \frac{\sqrt{x^2 + a^2}}{1/d_1},$$

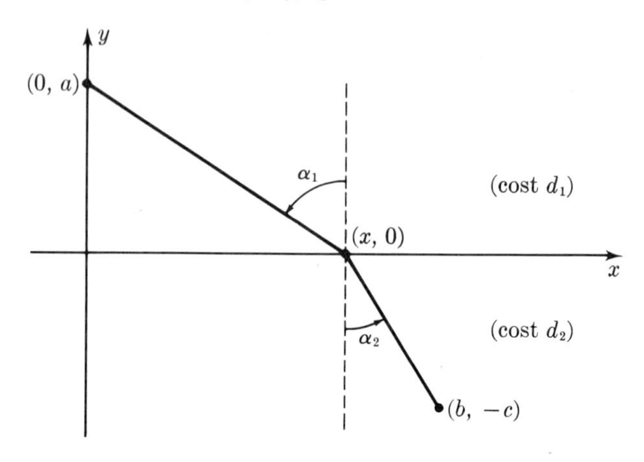

FIG. 3.5

and the cost from $(x, 0)$ to $(b, -c)$ is

$$\frac{\sqrt{(b-x)^2+c^2}}{1/d_2}.$$

Therefore you must minimize the total cost,

$$\frac{\sqrt{x^2+a^2}}{1/d_1}+\frac{\sqrt{(b-x)^2+c^2}}{1/d_2}.$$

But this is precisely the same problem as in the last example!

> *Answer:* The path is that shown in Fig. 3.5 with $\dfrac{\sin\alpha_1}{1/d_1}=\dfrac{\sin\alpha_2}{1/d_2}$.

EXERCISES

1. A man with 640 ft of fencing wishes to enclose a rectangular area and divide it into 5 pens with fences parallel to the short end of the rectangle. What dimensions of the enclosure make its area a maximum?
2. What points on the curve $xy^2 = 1$ are nearest the origin?
3. An open rectangular box has volume 15 ft^3. The length of its base is 3 times its width. Materials for the sides and base cost 60¢ and 40¢ per ft^2, respectively. Find the dimensions of the cheapest such box.
4. Find the dimensions of the rectangle of largest area that can be inscribed in an equilateral triangle of side s, if one side of the rectangle lies on the base of the triangle.
5. Find the dimensions of the rectangle of largest area that can be inscribed in a right triangle with legs of length a and b, if two sides of the rectangle lie along the legs of the triangle.

6. Given n numbers $a_1, a_2, \cdots, a_n$, show that $(x - a_1)^2 + (x - a_2)^2 + \cdots + (x - a_n)^2$ is least when x is the average of the numbers.

7. Of all lines of negative slope through the point (a, b) in the first quadrant, find the one that cuts from the first quadrant a triangle of least area.

8. A wire 30 in. long is cut into two parts, one of which is bent into a circle, and the other into a square. How should the wire be cut so that the sum of the areas of the circle and the square is a minimum?

9. What is the maximum volume of the cylinder generated by rotating a rectangle of perimeter 48 in. about one of its sides?

10. If the equal legs of an isosceles triangle are L ft, how long should the base be to maximize the area of the triangle?

11. Find the line tangent to the curve $y = 4 - x^2$ at a point of the first quadrant that cuts from the first quadrant a triangle of minimum area.

12. The strength of a beam of fixed length and rectangular cross-section is proportional to the width and to the square of the depth of the cross-section. Find the proportions of the beam of greatest strength that can be cut from a circular log.

13. A railroad will run a special train if at least 200 people subscribe. The fare will be \$8 per person if 200 people go, but will decrease 1¢ for each additional person who goes. (For example, if 250 people go, the fare will be \$7.50.) What number of passengers will bring the railroad maximum revenue?

14. Find the two positive numbers x and y for which $x + y = 1$, such that x^3y^4 is maximum.

15. Two posts, 8 ft and 12 ft high, stand 15 ft apart. They are to be stayed by wires attached to a single stake at ground level, and running to the tops of the posts. Where should the stake be placed to use the least amount of wire?

16. One corner of a page of width a is folded over and just reaches the opposite side. See Fig. 3.6. Find x such that the length L of the crease is a minimum.

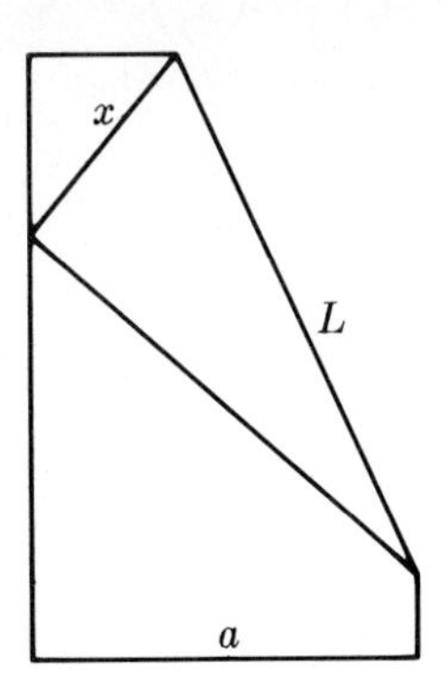

FIG. 3.6

17. A man in a rowboat 3 mi off a long straight shore wants to reach a point 5 mi up the shore. If he can row 2 mph and walk 4 mph, describe his fastest route.

18. Suppose in Ex. 17 the boat has a motor. How fast must the boat be able to go so that the fastest route is entirely by boat?

19. An isosceles triangle is circumscribed about a circle of radius r. Find its altitude if its perimeter is to be as small as possible.

20. Show that $e^x \geq x + 1$ for all values of x.

21. Archimedes proved that the area of the largest triangle inscribed in a given parabolic segment is $\frac{3}{4}$ of the area of the segment (Fig. 1.1, Chapter 8). Verify this for the parabola $y = x^2$.

22. The energy of a diatomic molecule is

$$U = \frac{a}{x^{12}} - \frac{b}{x^6},$$

where a and b are positive constants and x is the distance between the atoms. Find the **dissociation energy,** the maximum of $-U$.

The object of the next two examples is to prove an important inequality: if $a_1, a_2, \cdots, a_n$ are any positive numbers, then

$$\sqrt[n]{a_1 a_2 \cdots a_n} \leq \frac{a_1 + a_2 + \cdots + a_n}{n}.$$

In words, the **geometric mean** of a set of numbers does not exceed the **arithmetic mean** (average). We abbreviate the inequality by the notation

$$G_n \leq A_n.$$

23. Show that the maximum value of the ratio

$$\frac{\sqrt[n+1]{a_1 a_2 \cdots a_n x}}{\frac{1}{n+1}(a_1 + a_2 + \cdots + a_n + x)}$$

occurs for $x = A_n$, and compute the maximum. Conclude that

$$\frac{G_{n+1}}{A_{n+1}} \leq \left(\frac{G_n}{A_n}\right)^{n/(n+1)}.$$

24. By repeated application of Ex. 23, show that

$$\frac{G_n}{A_n} \leq \left(\frac{G_1}{A_1}\right)^{1/(n+1)} = 1,$$

and therefore that

$$G_n \leq A_n.$$

Explain why

$$G_n = A_n \quad \text{if and only if} \quad a_1 = a_2 = \cdots = a_n.$$

17. Inverse Functions, Logarithms

1. INVERSE FUNCTIONS

A function f assigns a number $y = f(x)$ to each number x. In this chapter we are concerned with the question: does the equation $y = f(x)$ determine a number x for each number y? This question can be tricky. For example, suppose $y = \cos x$. If $y = 3$, no value of x is determined since the equation $3 = \cos x$ is never satisfied. But if $y = 1$, infinitely many values of x are determined since the equation $1 = \cos x$ holds for $x = 0$, $\pm 2\pi$, $\pm 4\pi$, etc.

Life is most pleasant when $y = f(x)$ determines a unique x for each y. Let us consider some examples.

Suppose $f(x) = 2x - 5$. The equation $y = f(x) = 2x - 5$ can be solved for x in terms of y:

$$x = \frac{1}{2}(y + 5).$$

For each y, there is a unique x given by this formula.

Geometrically, the situation is simple. The graph of $y = 2x - 5$ is a straight line crossing each horizontal line once (Fig. 1.1). Given y, the graph

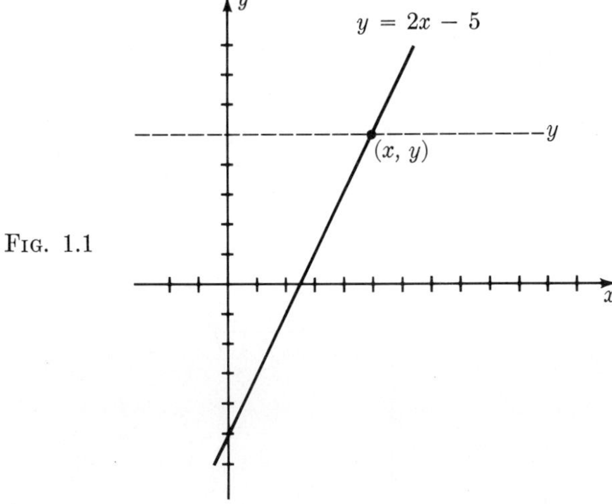

FIG. 1.1

crosses the level y at a point (x, y) where $x = \frac{1}{2}(y + 5)$. Thus each y determines a unique x. In other words, x is a function of y. We write $x = f^{-1}(y)$ and read "f inverse of y," where

$$f^{-1}(y) = \frac{1}{2}(y + 5).$$

The line in Fig. 1.1 is the graph of both $y = f(x)$ and $x = f^{-1}(y)$. This is because the statements

$$y = f(x) \qquad \text{and} \qquad x = f^{-1}(y)$$

are equivalent.

Consider another example, $f(x) = x^3$. The curve $y = x^3$ is always rising; it crosses each horizontal line once (Fig. 1.2). Therefore the equation $y = f(x)$ determines one number x for each number y. We write

$$x = f^{-1}(y) = \sqrt[3]{y}.$$

Each point on the curve can be regarded as

$$(x, x^3) \qquad \text{or as} \qquad (\sqrt[3]{y}, y).$$

It is important to notice that

$$y = x^3 \qquad \text{and} \qquad x = \sqrt[3]{y}$$

are equivalent statements.

These examples illustrate an important situation: if the graph of $f(x)$ is always increasing or always decreasing, then the equation $y = f(x)$ defines a unique x for each y. Thus, associated with $f(x)$ is another function $x = f^{-1}(y)$, called the **inverse function of** $f(x)$.

If the equation $y = f(x)$ can be solved for x in terms of y, then $f^{-1}(y)$ is expressed by an explicit formula. Often, however, this is not possible. For example, $f(x) = x^5 + x$ is an increasing function and so has an inverse

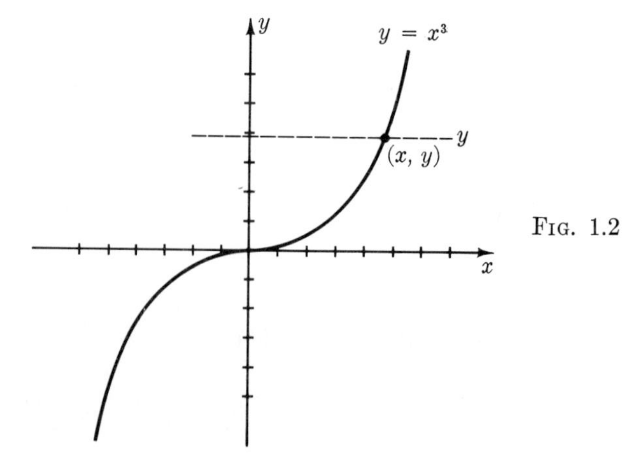

FIG. 1.2

function $f^{-1}(y)$. Yet we cannot solve $x^5 + x = y$ for x; the inverse function exists but there is no formula for it.

Recall that the statements

$$y = f(x) \qquad \text{and} \qquad x = f^{-1}(y)$$

are equivalent. It follows that

$$f^{-1}[f(x)] = f^{-1}[y] = x,$$

$$f[f^{-1}(y)] = f[x] = y.$$

Thus the actions of f and f^{-1} neutralize each other. For example, if $f(x) = 2x - 5$, then $f^{-1}(y) = \frac{1}{2}(y + 5)$, so

$$f^{-1}[f(x)] = f^{-1}[2x - 5] = \frac{1}{2}([2x - 5] + 5) = x.$$

Also,

$$f[f^{-1}(y)] = f\left[\frac{1}{2}(y + 5)\right] = 2\left[\frac{1}{2}(y + 5)\right] - 5 = y.$$

As another example, if $f(x) = x^3$, then $f^{-1}(y) = \sqrt[3]{y}$, so

$$f^{-1}[f(x)] = f^{-1}(x^3) = \sqrt[3]{x^3} = x.$$

Also,

$$f[f^{-1}(y)] = f(\sqrt[3]{y}) = (\sqrt[3]{y})^3 = y.$$

One more example. Imagine a group of men all with different names. For each name x, there is a man $X = f(x)$ with that name. For each man X, there is his name $x = f^{-1}(X)$. For instance,

$$\begin{aligned} f^{-1}[f(\text{john})] &= \text{the name of the man } f(\text{john}) \\ &= \text{the name of the man whose name is john} \\ &= \text{the name of John} = \text{john}. \end{aligned}$$

Thus $f^{-1}[f(x)] = x$. Conversely,

$$\begin{aligned} f[f^{-1}(\text{George})] &= \text{the man whose name is } f^{-1}(\text{George}) \\ &= \text{the man whose name is george} \\ &= \text{George}. \end{aligned}$$

Thus $f[f^{-1}(X)] = X$.

Warning: Do not confuse the inverse function $f^{-1}(x)$ with the reciprocal $1/f(x)$.

EXERCISES

Find the inverse function:

1. $y = 3x - 7$
2. $y = -2x + 5$
3. $y = 4x - 1$
4. $y = 10x + 10$

5. $y = \dfrac{2x - 7}{x + 4} \quad (x \neq -4)$

6. $y = \dfrac{x + 1}{x - 1} \quad (x \neq 1)$

7. $y = \dfrac{x + 2}{x + 3} \quad (x \neq -3)$

8. $y = \dfrac{3x + 4}{2x + 3} \quad \left(x \neq -\dfrac{3}{2}\right)$

9. $y = -\dfrac{1}{x} \quad (x \neq 0)$

10. $y = \sqrt{x + 3} \quad (x > -3)$

11. $y = \sqrt{2x - 8} \quad (x > 4)$

12. $y = \dfrac{\sqrt{x} + 1}{\sqrt{x} - 1} \quad (x > 1).$

13. Let $f(x) = \dfrac{9}{x + 7} - 7$. Show $f^{-1}(x) = f(x)$.

14. Suppose $h(x) = h^{-1}(x)$ and $f(x) = g\{h[g^{-1}(x)]\}$. Show that $f^{-1}(x) = f(x)$.

15. Let $f(x) = 4x - 1$. Plot $y = f(x)$ and $y = f^{-1}(x)$ on the same graph. Verify that the two curves are mirror images in the line $y = x$.

16. Let $f(x) = ax + b$ where $a \neq 0$. Show that the graph of $y = f^{-1}(x)$ is a straight line of slope $1/a$.

17. Let $f(x) = x^3$. Sketch $y = f^{-1}(x)$.

18. Let $f(x) = x^3 + x$. Sketch $y = f^{-1}(x)$.

19. Let $f(x) = x^3 + x$ and let $g(x) = f^{-1}(x)$. Show that $g(x)^3 = -g(x) + x$.

20. Let f and g be the functions in Ex. 19. Compute $g(-2)$, $g(0)$, $g(2)$, $g(10)$, $g(30)$, $g(4^3 + 4)$.

21. Let $f(x) = \int_0^x t^2\,dt$. Find $\int_0^{f^{-1}(x)} t^2\,dt$. Explain.

2. GRAPHS AND DERIVATIVES

Suppose we wish to study an inverse function $x = f^{-1}(y)$. It seems more natural to consider x as the independent variable and y as the dependent variable, so let us study instead $y = f^{-1}(x)$. We have interchanged x and y; whenever the point (x, y) is on the graph of $x = f^{-1}(y)$, then the point (y, x) is on the graph of $y = f^{-1}(x)$, and conversely.

A simple relation exists between the points (x, y) and (y, x): they are mirror images of each other across the line $y = x$. See Fig. 2.1. Since the graph of $y = f^{-1}(x)$ is identical to that of $x = f(y)$, it is the reflection of the graph of $y = f(x)$.

> The graph of $y = f^{-1}(x)$ is the reflection of the graph of $y = f(x)$ across the line $y = x$.

As an example, take $f(x) = 2x - 5$. Then $f^{-1}(y) = \frac{1}{2}(y + 5)$, or $f^{-1}(x) = \frac{1}{2}(x + 5)$. See Fig. 2.2.

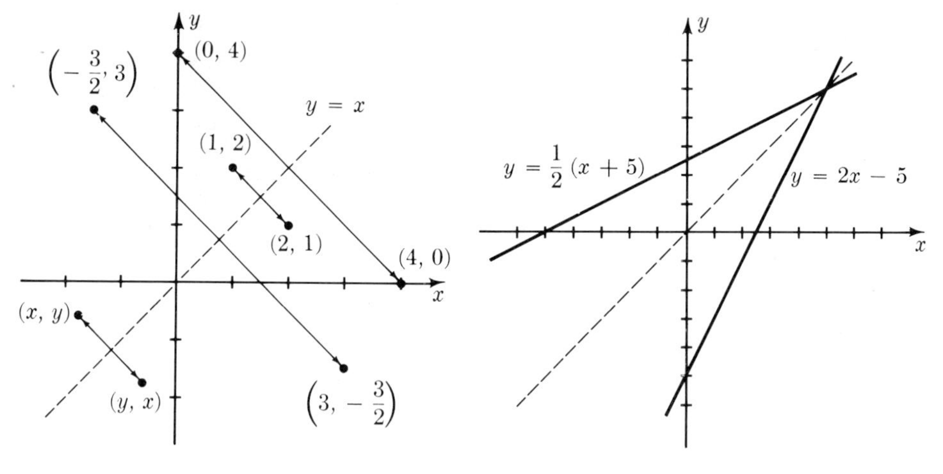

Fig. 2.1

Fig. 2.2

The lines in Fig. 2.2 are reflections of each other across the line $y = x$. Notice that their slopes, 2 and $\frac{1}{2}$, are reciprocals. The same is true of any pair of lines which are reflections of each other in $y = x$. If one line is $y = mx + b$, its reflection is $x = my + b$, that is, $y = (x/m) - (b/m)$. The slopes are m and $1/m$, respectively.

As another example of reflection, take $f(x) = x^3$. Then $f^{-1}(x) = \sqrt[3]{x}$. The graphs are shown in Fig. 2.3.

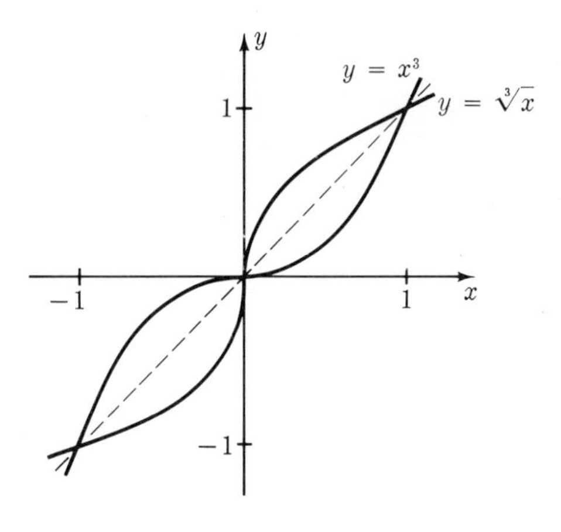

Fig. 2.3

Derivatives of Inverse Functions

The derivative of an inverse function can be expressed in terms of the derivative of the direct function by means of the following rule.

If $g(x)$ is the inverse function of $f(x)$, then

$$\frac{d}{dy}[g(y)]\bigg|_{y=f(x)} = \frac{1}{\dfrac{d}{dx}[f(x)]},$$

provided $f'(x) \neq 0$. Briefly,

$$\frac{dx}{dy} = \frac{1}{\dfrac{dy}{dx}}.$$

The rule makes good sense numerically. If y is changing 5 times as fast as x, then x is changing $\frac{1}{5}$ as fast as y.

The rule makes good sense geometrically, too. The graphs $y = f(x)$ and $y = g(x)$ are reflections of each other (Fig. 2.4). At corresponding points

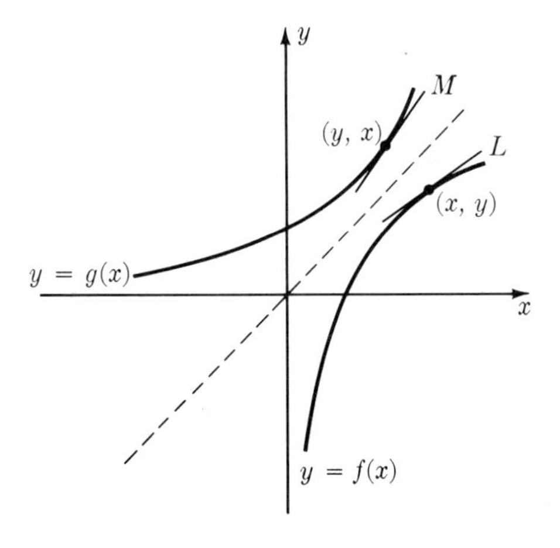

FIG. 2.4

(x, y) and (y, x), the tangent lines L and M also are reflections of each other. Hence their slopes are reciprocals:

$$\text{slope of } M = \frac{1}{\text{slope of } L}.$$

But the slopes of L and M are $f'(x)$ and $g'(y)$, respectively. Therefore

$$g'(y) = \frac{1}{f'(x)}.$$

If L is horizontal, then M is vertical, so the slope of M is not defined. In such a case $g(x)$ does not have a derivative at y.

Applications

EXAMPLE 2.1

Compute the derivative of $x^{1/3}$.

Solution: The function $g(x) = x^{1/3}$ is the inverse of $f(x) = x^3$. Apply the rule with $y = f(x) = x^3$:

$$g'(y) = \frac{1}{f'(x)} = \frac{1}{3x^2}.$$

But $x = y^{1/3}$, therefore

$$g'(y) = \frac{1}{3y^{2/3}}, \qquad \text{that is,} \quad g'(x) = \frac{1}{3x^{2/3}}.$$

Answer: For $x \neq 0$, $\dfrac{d}{dx}(x^{1/3}) = \dfrac{1}{3x^{2/3}}.$

REMARK: Since $y = x^3$ has a horizontal tangent at the origin, $y = x^{1/3}$ has a vertical tangent there. Consequently $x^{1/3}$ has no derivative at $x = 0$, but otherwise the derivative is $1/(3x^{2/3})$.

EXAMPLE 2.2

Compute the derivative of $x^{1/p}$, where p is an odd positive integer.

Solution: Use the same technique as in the last example. The curve $y = f(x) = x^p$, like the curve $y = x^3$, increases steadily. The inverse function, $g(x) = x^{1/p}$, is differentiated according to the formula

$$g'(y) = \frac{1}{f'(x)},$$

where $y = f(x) = x^p$, that is, $x = y^{1/p}$. Since $f'(x) = px^{p-1}$,

$$\begin{aligned} g'(y) = \frac{1}{px^{p-1}} &= \frac{1}{p(y^{1/p})^{p-1}} \\ &= \frac{1}{py^{1-1/p}} \\ &= \frac{1}{p}\, y^{(1/p)-1}. \end{aligned}$$

Answer: If p is odd,

$$\frac{d}{dx}(x^{1/p}) = \frac{1}{p}\, x^{(1/p)-1}, \qquad x \neq 0.$$

EXAMPLE 2.3

Compute the derivative of $x^{1/p}$, where p is an even positive integer.

Solution: Compare the graphs of $y = x^p$ for p odd and for p even (Fig. 2.5). If p is even, the graph crosses no horizontal line below the x-axis,

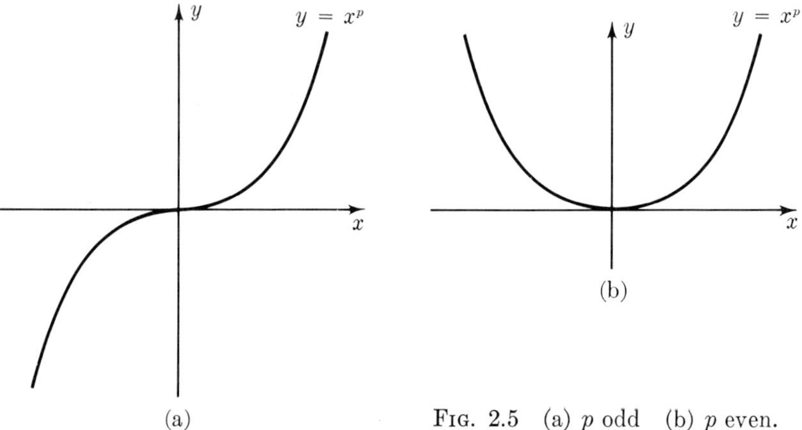

FIG. 2.5 (a) p odd (b) p even.

but crosses each horizontal line above the x-axis twice. Therefore $f^{-1}(y)$ has no meaning if $y < 0$, and two possible meanings if $y > 0$. This ambiguity is avoided by agreeing to consider $y = x^p$ only for $x \geq 0$, and then defining the inverse function $x^{1/p}$. See Fig. 2.6. With this agreement, for p even,

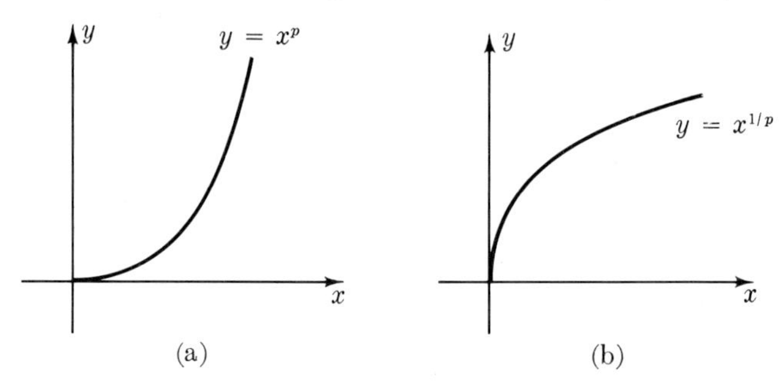

FIG. 2.6 p even.

$x^{1/p}$ is defined only for $x \geq 0$, and has non-negative values. For example, we agree that $x^{1/2}$ means the positive square root of x.

Now that $x^{1/p}$ is defined, the computation of its derivative is precisely the same as in the last example.

Answer: If p is even,

$$\frac{d}{dx}(x^{1/p}) = \frac{1}{p}\, x^{(1/p)-1}, \quad x > 0.$$

A general rule for differentiating fractional powers follows from the last two examples.

Differentiation of Fractional Powers Let $\dfrac{p}{q}$ be a fraction, $q > 0$. Then

$$\frac{d}{dx}(x^{p/q}) = \frac{p}{q}\,x^{(p/q)-1}.$$

If q is even, the rule holds for $x > 0$; if q is odd, it holds for all $x \neq 0$.

This rule says you differentiate fractional powers just as you do integral powers: bring down the power and lower the exponent by one.

The rule is easy to verify. Write

$$x^{p/q} = (x^{1/q})^p,$$

and use the Chain Rule:

$$\begin{aligned}\frac{d}{dx}(x^{p/q}) &= p(x^{1/q})^{p-1}\cdot\frac{d}{dx}(x^{1/q})\\ &= px^{(p-1)/q}\cdot\frac{1}{q}\,x^{(1/q)-1} = \frac{p}{q}\,x^{(p/q)-1}.\end{aligned}$$

This derivation is valid provided $x^{1/q}$ can be differentiated. It can be if $x > 0$ for q even, or if $x \neq 0$ for q odd.

REMARKS: If $p/q \geq 1$, the formula is valid also at $x = 0$, although a special argument is needed to verify it. This will be done in Ex. 16.

Note that $x^{1/3}$ is not the same as $x^{2/6}$. The first function is defined for all x, the second only for $x \geq 0$.

EXAMPLE 2.4

Compute dy/dx if

(a) $y = \sqrt{x}, \quad x > 0;$

(b) $y = \sqrt[3]{x^2 + 1};$

(c) $y = (\sin x)^{-4/3}, \quad \sin x \neq 0.$

Solution:

$$\frac{d}{dx}(x^{p/q}) = \frac{p}{q}\,x^{(p/q)-1}.$$

(a) $y = \sqrt{x} = x^{1/2};$

$$\frac{dy}{dx} = \frac{1}{2}\,x^{(1/2)-1} = \frac{1}{2}\,x^{-1/2} = \frac{1}{2\sqrt{x}}, \qquad x > 0.$$

(b) Use the Chain Rule. Set $y = u^{1/3}$, where $u = x^2 + 1$:

$$\frac{dy}{dx} = \frac{dy}{du}\frac{du}{dx} = \frac{1}{3}\,u^{-2/3}(2x) = \frac{1}{3(x^2+1)^{2/3}}\cdot 2x.$$

(c) Use the Chain Rule. Set $y = u^{-4/3}$, where $u = \sin x$:

$$\frac{dy}{dx} = \frac{dy}{du}\frac{du}{dx} = -\frac{4}{3}u^{-7/3}\cos x = -\frac{4}{3}(\sin x)^{-7/3}\cos x, \qquad \sin x \neq 0.$$

Answer:

(a) $\dfrac{1}{2\sqrt{x}}, \quad x > 0;$

(b) $\dfrac{2x}{3(x^2+1)^{2/3}};$

(c) $\dfrac{-4\cos x}{3(\sin x)^{7/3}}, \quad \sin x \neq 0.$

EXERCISES

Differentiate:

1. $y = 3x^{1/5}$
2. $y = 2x^{1/2}$
3. $y = (10x)^{1/4}$
4. $y = (3x - 1)^{-5/3}$
5. $y = (x^2 - x + 4)^{7/6}$
6. $y = x + (3/\sqrt{x})$
7. $y = \dfrac{(e^x + 1)^{3/2}}{(e^x - 1)^{1/2}}$
8. $y = (3\cos x)^{3/4}$
9. $y = e^{2\sqrt{x}}$
10. $y = (\sin x + \cos x)^{-2/3}$
11. $y = (\sin x)^{p/8}$
12. $y = (x^2 + 1)^{-p/8}$.
13. Let $f(x) = ax^7$, $a > 0$. Find $\dfrac{d[f^{-1}(x)]}{dx}$.
14. Let $f^{-1}(x) = \dfrac{1}{2}\sqrt[3]{x}$. Find $\dfrac{d[f(x)]}{dx}$.
15. Let $f^{-1}(x) = x^{-5}$. Find $\dfrac{d[f(x)]}{dx}$.
16. Find $\dfrac{d}{dx}(x^{4/3})\Big|_{x=0}$ by use of the *definition*

$$\frac{d}{dx}f(x)\Big|_{x=0} = \lim_{x\to 0}\frac{f(x) - f(0)}{x - 0}.$$

Generalize to $x^{p/q}$ for $\dfrac{p}{q} > 1$.

3. NATURAL LOGARITHMS

In this section we study the inverse of the increasing function $y = e^x$. The inverse is denoted by

$$y = \ln x,$$

and is called the **natural logarithm** function. Figure 3.1 shows the graph of $y = \ln x$ as the reflection of the graph of $y = e^x$ across the line $y = x$.

FIG. 3.1

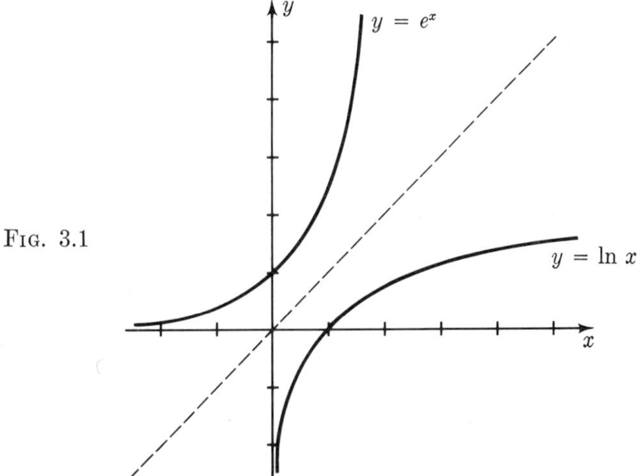

Certain basic properties are immediate from Fig. 3.1.

$\ln x$ is defined only for $x > 0$,
$\ln x$ is an increasing function,
$\ln x < 0 \quad \text{for} \quad 0 < x < 1,$
$\ln 1 = 0,$
$\ln x > 0 \quad \text{for} \quad x > 1.$

Because $\ln x$ and e^x are inverse functions, two further statements are automatic:

$$y = e^x \text{ is equivalent to } x = \ln y,$$

$$\ln(e^x) = x \text{ and } e^{\ln y} = y.$$

Be sure that you can derive the second statement from the first.

Since $\ln e^x = x$, you can make $\ln x$ as large as desired by choosing large values of x. For example

$$\ln e^{10} = 10, \qquad \ln e^{100} = 100, \qquad \ln e^{1000} = 1000, \cdots.$$

Therefore the graph of $y = \ln x$ increases beyond all bounds as x increases (Fig. 3.1). Nevertheless the *rate* of increase is agonizingly slow. Not until $x = e^{10} \approx 22{,}000$ does the curve reach the level $y = 10$; not until $x = e^{100} \approx (2.7)10^{43}$ does it reach the level $y = 100$. Yet it does reach every level eventually:

$$\boxed{\ln x \longrightarrow \infty \qquad \text{as} \quad x \longrightarrow \infty.}$$

Algebraic Properties of ln x

Let us justify the name natural *logarithm* by showing that $\ln x$ behaves the way a logarithm should.

The exponential function e^x has some important algebraic properties:

$$\begin{array}{|lc|}\hline (1) & e^a e^b = e^{a+b}, \\ (2) & e^{-a} = \dfrac{1}{e^a}, \\ (3) & \dfrac{e^a}{e^b} = e^{a-b}, \\ (4) & (e^a)^b = e^{ab}. \\ \hline \end{array}$$

Somehow, these properties ought to rub off onto the inverse function $\ln x$. They do. To see this, set $e^a = x$ and $e^b = y$; that means $a = \ln x$ and $b = \ln y$. Then each of the four statements above can be translated into a statement about $\ln x$ and $\ln y$. For example, property (1) can be translated as follows:

$$e^a e^b = e^{a+b},$$

$$xy = e^{(\ln x + \ln y)}.$$

Hence

$$\ln(xy) = \ln x + \ln y.$$

In a similar way, (2), (3), and (4) can be translated. Here are the results:

$$\begin{array}{|lc|}\hline (1') & \ln(xy) = \ln x + \ln y, \\ (2') & \ln\left(\dfrac{1}{x}\right) = -\ln x, \\ (3') & \ln\left(\dfrac{x}{y}\right) = \ln x - \ln y, \\ (4') & \ln x^b = b \ln x. \\ \hline \end{array}$$

You can make $\ln x$ as large negative as desired by choosing x small. Indeed, as $x \longrightarrow 0$ through positive values, $1/x \longrightarrow \infty$, hence $\ln(1/x) \longrightarrow \infty$.

By (2′),

$$\ln x = -\ln\left(\frac{1}{x}\right) \longrightarrow -\infty.$$

Thus

$$\boxed{\ln x \longrightarrow -\infty \qquad \text{as} \quad x \longrightarrow 0+.}$$

Relation of Natural Logs to Common Logs

Natural logarithms are logarithms to the base e. There is a simple relation between them and logarithms to the base 10. In Chapter 7 it was shown that if $e^k = a$, then

$$k = \frac{\log_{10} a}{\log_{10} e}.$$

But $k = \ln a$. Therefore

$$\ln a = \frac{1}{M} \log_{10} a, \qquad \text{where} \quad M = \log_{10} e \approx 0.43429.$$

Thus natural logs are proportional to common logs, and can be computed from a table of common logs by this formula. Of course tables of natural logs are available.

EXERCISES

Simplify:

1. $\ln e^{a+2}$
2. $e^{\ln x^2}$
3. $e^{-\ln x}$
4. $e^{2 \ln x + 5 \ln y}$
5. $\ln \sqrt{e}$
6. $\ln (1/e^{3/2})$
7. $(\log_2 3)(\log_3 2)$
8. $(\ln 10)(\log_{10} 3)(\log_3 e)$.

Use a common log table to compute to 5 places:

9. $\ln 20$
10. $\ln (0.5)$.

Use a natural log table to compute to 5 significant digits:

11. $\ln \dfrac{(4.71)(2.13)}{(5.09)}$
12. $(\sqrt{9.32})(\sqrt[3]{1.47})$.
13. Given $\ln 2 \approx 0.69$ and $\ln 3 \approx 1.10$. Compute $\ln 12$, $\ln \frac{2}{9}$, $\ln 256$, and $\ln \sqrt{6}$ to 2 places.
14. Show that the estimate $2 < \ln 10 < 3$ follows from the estimate $2.5 < e < 3$.
15. Find the integer n for which $e^n < 1000 < e^{n+1}$.
16. Find the integer n for which $10^n < e^{100} < 10^{n+1}$.

4. DERIVATIVE OF ln x

The natural logarithm function inherits some useful algebraic properties from e^x. It also inherits a simple derivative:

$$\frac{d}{dx}(\ln x) = \frac{1}{x}, \qquad x > 0.$$

This follows easily from the rule for differentiating inverse functions. Let $f(x) = e^x$ and $g(x) = \ln x$, the inverse function. Then

$$g'(y) = \frac{1}{f'(x)}, \qquad \text{where} \quad y = f(x) = e^x.$$

Since $f'(x) = e^x$,

$$g'(y) = \frac{1}{e^x} = \frac{1}{y} \qquad \text{or} \qquad \frac{d}{dy}(\ln y) = \frac{1}{y}.$$

Thus

$$\frac{d}{dx}(\ln x) = \frac{1}{x}.$$

EXAMPLE 4.1

Find the largest value of $\dfrac{\ln x}{x}$.

Solution: Let $y = (\ln x)/x$. This function is defined for all $x > 0$. Differentiate:

$$\frac{dy}{dx} = \frac{x\dfrac{d}{dx}(\ln x) - (\ln x)\dfrac{d}{dx}(x)}{x^2}$$

$$= \frac{x \cdot \dfrac{1}{x} - \ln x}{x^2} = \frac{1 - \ln x}{x^2}.$$

It follows that

$$\frac{dy}{dx} > 0 \qquad \text{if} \quad \ln x < 1,$$

$$\frac{dy}{dx} = 0 \qquad \text{if} \quad \ln x = 1,$$

$$\frac{dy}{dx} < 0 \qquad \text{if} \quad \ln x > 1.$$

Consequently y is largest when $\ln x = 1$, or $x = e$.

Answer: $\dfrac{\ln e}{e} = \dfrac{1}{e}.$

EXAMPLE 4.2

Which is larger, e^π or π^e?

Solution: Take natural logarithms; the larger number has the larger logarithm:

$$\ln e^\pi = \pi, \qquad \ln \pi^e = e \ln \pi.$$

Now, is $\pi > e \ln \pi$ or is $\pi < e \ln \pi$; expressed differently, is

$$\frac{1}{e} > \frac{\ln \pi}{\pi} \qquad \text{or is} \qquad \frac{1}{e} < \frac{\ln \pi}{\pi}?$$

In the last example it was shown that

$$\frac{1}{e} > \frac{\ln x}{x}$$

for all $x > 0$ except $x = e$. In particular

$$\frac{1}{e} > \frac{\ln \pi}{\pi}.$$

Therefore $\pi > e \ln \pi$, which means

$$\ln e^\pi > \ln \pi^e.$$

Answer: $e^\pi > \pi^e$.

The Derivative of ln f(x)

If $f(x)$ has positive values, then $\ln f(x)$ is defined. The derivative of this function is computed by the Chain Rule. Write $y = \ln u$, where $u = f(x)$. Then

$$\frac{dy}{dx} = \frac{dy}{du}\frac{du}{dx} = \left(\frac{d}{du} \ln u\right)\left(\frac{d}{dx} f(x)\right)$$

$$= \frac{1}{u} f'(x) = \frac{1}{f(x)} f'(x).$$

Hence

$$\frac{d}{dx} [\ln f(x)] = \frac{f'(x)}{f(x)},$$

or

$$\boxed{\frac{d}{dx} (\ln u) = \frac{u'}{u}.}$$

EXAMPLE 4.3

Compute the derivative of $y = \ln(x^2 + 1)$.

Solution:

$$\frac{dy}{dx} = \frac{u'}{u} = \frac{(x^2+1)'}{(x^2+1)} = \frac{2x}{x^2+1}.$$

Answer: $\dfrac{2x}{x^2+1}$.

EXAMPLE 4.4

Compute the derivative of $y = \ln\left(\dfrac{1+2x}{1-2x}\right)$, where $-\dfrac{1}{2} < x < \dfrac{1}{2}$.

Solution:

$$\frac{dy}{dx} = \frac{\dfrac{d}{dx}\left(\dfrac{1+2x}{1-2x}\right)}{\left(\dfrac{1+2x}{1-2x}\right)}.$$

Differentiate the quotient in the numerator:

$$\frac{dy}{dx} = \frac{\left(\dfrac{(1-2x)2-(1+2x)(-2)}{(1-2x)^2}\right)}{\left(\dfrac{1+2x}{1-2x}\right)}$$

$$= \frac{\left(\dfrac{4}{(1-2x)^2}\right)}{\left(\dfrac{1+2x}{1-2x}\right)} = \frac{4}{(1-2x)(1+2x)}.$$

Alternate Solution: Use the rules of logarithms before differentiating:

$$y = \ln\left(\frac{1+2x}{1-2x}\right)$$

$$= \ln(1+2x) - \ln(1-2x).$$

Now differentiate:

$$\frac{dy}{dx} = \frac{(1+2x)'}{(1+2x)} - \frac{(1-2x)'}{(1-2x)}$$

$$= \frac{2}{(1+2x)} - \frac{-2}{(1-2x)} = \frac{4}{(1-2x)(1+2x)}.$$

Answer: $\dfrac{4}{(1-2x)(1+2x)}$.

EXERCISES

Differentiate:

1. $y = \ln 5x$
2. $y = 3 \ln 4x$
3. $y = 2 \ln(x^2)$
4. $y = \ln(x^4)$
5. $y = \ln\left(\dfrac{1}{x}\right)$
6. $y = \ln(x^2 + x)$
7. $y = \ln(\sin x)$
8. $y = \ln(\cos x)$
9. $y = \ln(e^x)$
10. $y = \ln\left(\dfrac{x^2+1}{x^2+3}\right)$
11. $y = \ln(\ln x)$
12. $y = \sqrt{\ln x}$
13. $y = \dfrac{\ln(x^2+1)}{x}$
14. $y = x \ln x - x$
15. $y = (\ln x)^2$
16. $y = \ln(x^2 + x + 10)$
17. $y = \ln[\ln(\ln x)]$
18. $y = \ln\left(\dfrac{x^2}{1+x^2}\right)$
19. $y = \dfrac{1}{\ln x}$
20. $y = \dfrac{\ln(x+2)}{\ln(x+1)}$.
21. Graph $y = \dfrac{\ln x}{x}$.
22. Show that $f(x) = \dfrac{\ln x}{x}$ steadily decreases for $x > e$. Find $\dfrac{d[f^{-1}(x)]}{dx}$.
23. (cont.) Let $e \le a < b$. Show that $b^a < a^b$.
24. Graph $y = \dfrac{\ln x}{x^2}$ and find the maximum of y.
25. Compute the second derivative of $\ln x$. What information does it give about the shape of the curve $y = \ln x$?

5. LOGARITHMIC DIFFERENTIATION

Differentiation of products, quotients, and powers can often be simplified by taking natural logs before differentiating. Be careful, however, to take logs of positive functions only.

EXAMPLE 5.1

Compute the derivative of $y = \sqrt[3]{\dfrac{x^2+3}{x^4+1}}$.

Solution: Take the natural log of y and simplify, using rules of logarithms:

$$\ln y = \frac{1}{3}\ln\left(\frac{x^2+3}{x^4+1}\right)$$

$$= \frac{1}{3}[\ln(x^2+3) - \ln(x^4+1)].$$

This is legitimate since both $(x^2 + 3)$ and $(x^4 + 1)$ are positive functions. Now differentiate:

$$\frac{y'}{y} = \frac{1}{3}\left(\frac{2x}{x^2+3} - \frac{4x^3}{x^4+1}\right),$$

$$y' = \frac{y}{3}\left(\frac{2x}{x^2+3} - \frac{4x^3}{x^4+1}\right).$$

On the right side, replace y by its expression in x.

Answer: $\frac{1}{3}\left(\sqrt[3]{\frac{x^2+3}{x^4+1}}\right)\left(\frac{2x}{x^2+3} - \frac{4x^3}{x^4+1}\right).$

EXAMPLE 5.2

Compute the derivative of $y = \frac{\sqrt{x}\, e^{2x}}{(5x+1)^2}$ for $x > 0$.

Solution: Take logs and differentiate:

$$\ln y = \frac{1}{2}\ln x + 2x - 2\ln(5x+1),$$

$$\frac{y'}{y} = \frac{1}{2x} + 2 - 2\left(\frac{5}{5x+1}\right),$$

$$y' = y\left(\frac{1}{2x} + 2 - \frac{10}{5x+1}\right).$$

Answer: $\frac{\sqrt{x}\, e^{2x}}{(5x+1)^2}\left(\frac{1}{2x} + 2 - \frac{10}{5x+1}\right).$

The last two examples can be done without logs. Sometimes, however, it is essential to take logs before differentiating.

EXAMPLE 5.3

Compute the derivative of $y = x^x$, $x > 0$.

Solution: Take logs:

$$\ln y = x \ln x.$$

Now differentiate:

$$\frac{y'}{y} = x\frac{d}{dx}(\ln x) + \ln x \frac{d}{dx}(x)$$

$$= x \cdot \frac{1}{x} + \ln x = 1 + \ln x,$$

$$y' = y(1 + \ln x).$$

Answer: $x^x(1 + \ln x)$.

EXERCISES

Differentiate where valid by logarithmic differentiation:

1. $y = (x^3 + 2)^{1/2}$
2. $y = \left(\dfrac{x+1}{(x+2)^4}\right)^{1/3}$
3. $y = \left(\dfrac{x^2+4}{x+7}\right)^6$
4. $y = \dfrac{(x+1)(x+2)}{(x+4)(x+5)}$
5. $y = (2x+3)^{1/3}e^{-x^2}$
6. $y = \dfrac{x^2e^{3x^3}}{(x+3)^2}$
7. $y = x^{x-1}$
8. $y = (x+2)^{x+3}$
9. $y = x^{1/x}$
10. $y = 2^{2x}$
11. $y = 3^{\ln x}$
12. $y = \left(1 + \dfrac{1}{x}\right)^x$
13. $y = 10^x$
14. $y = x^{x^x}$
15. $y = \dfrac{e^x(x^3-1)}{\sqrt{2x+1}}$
16. $y = \sqrt{\dfrac{x^2-1}{x^2+1}}$.

6. ANTIDERIVATIVE OF 1/x

The basic formula

$$\frac{d}{dx}(\ln x) = \frac{1}{x}$$

supplies an antiderivative of $1/x$ for $x > 0$: the antiderivative is $\ln x$.

If $x < 0$, then $\ln x$ has no meaning but $\ln(-x)$ does, because $(-x)$ is positive. By the Chain Rule (or Scale Rule),

$$\frac{d}{dx}[\ln(-x)] = -\frac{1}{-x} = \frac{1}{x} \qquad \text{for} \quad x < 0.$$

An antiderivative of $1/x$ is

$$\ln x, \qquad \text{for} \quad x > 0,$$

$$\ln(-x), \qquad \text{for} \quad x < 0.$$

Stated briefly, an antiderivative of $1/x$ is

$$\ln |x|, \qquad \text{if} \quad x \neq 0.$$

We now have an antiderivative of x^{-1}. Previously we had an antiderivative for any integral power x^n except x^{-1}:

$$x^n = \frac{d}{dx}\left(\frac{x^{n+1}}{n+1}\right), \qquad n \neq -1.$$

Now that gap is filled.

ln x as an Area

Using our new antiderivative, we see that for $x > 0$,

$$\int_1^x \frac{1}{t}\,dt = \ln t \Big|_1^x = \ln x - \ln 1 = \ln x.$$

Thus $\ln x$ can be expressed as an integral:

$$\ln x = \int_1^x \frac{1}{t}\,dt, \qquad x > 0.$$

This suggests that $\ln x$ can be pictured either as an area (if $x > 1$) or as the negative of an area (if $x < 1$). See Fig. 6.1.

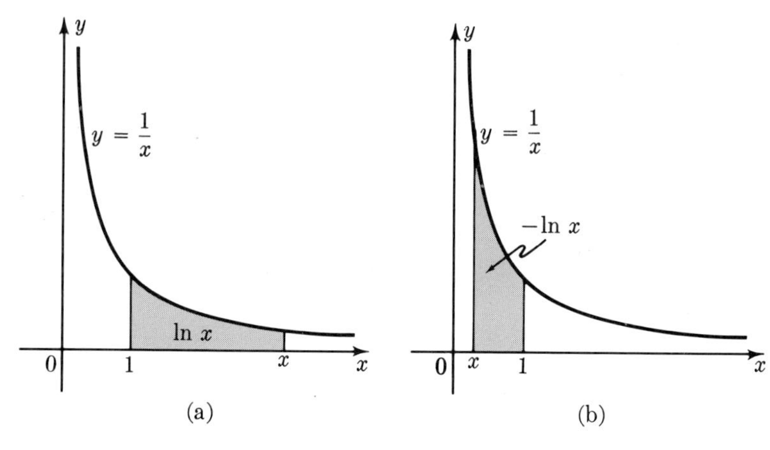

FIG. 6.1a FIG. 6.1b

EXERCISES

Find:

1. $\int_{1/5}^{1} \frac{dt}{t}$

2. $\int_1^{e^x} \frac{dt}{t}$

3. $\int_0^3 \frac{dt}{t+4}$

4. $\int_1^2 \frac{dt}{t+1}$

5. $\displaystyle\int_3^6 \frac{dx}{2x-5}$

6. $\displaystyle\int_2^3 \frac{dx}{x-1}$

7. $\displaystyle\int_0^1 \frac{dt}{2-t}$

8. $\displaystyle\int_1^2 \frac{dt}{3-t}$

9. $\displaystyle\int_0^1 \frac{dt}{4-3t}$

10. $\displaystyle\int_1^3 \frac{dt}{7-2t}\cdot$

11. Show that $\displaystyle\int_a^b \frac{dx}{x} = \int_{1/b}^{1/a} \frac{dx}{x}$ if $0 < a < b$. Interpret in terms of area.

12. Find the area under $y = \dfrac{1}{3x+4}$ from $x = 0$ to $x = 7$.

13. Find the area under $y = \dfrac{1}{4x-7}$ from $x = 2$ to $x = 5$.

14. Find $\displaystyle\int_2^3 \frac{du}{u^2-1}\cdot$ [*Hint:* $\dfrac{1}{u^2-1} = \dfrac{1}{2}\left(\dfrac{1}{u-1} - \dfrac{1}{u+1}\right)$.]

15. Find $\displaystyle\int_3^4 \frac{dx}{x^2-4}\cdot$ (See Ex. 14.)

16. Compute $\displaystyle\int_3^5 \frac{x\,dx}{x^2-4}\cdot$

17. Compute $\displaystyle\int_1^5 \frac{x^3-2x^2+1}{x}\,dx.$

18. Express $\ln \frac{3}{2}$ as an integral and estimate it to 3 places using the Trapezoidal Rule with $\Delta x = 0.1$.

19. Show that the area under the curve $y = 1/x$ between $x = 2$ and $x = 4$ is the same as the area between $x = 50$ and $x = 100$. Find a general principle.

20. Show that $\displaystyle\int_1^{t^2} \frac{1}{x}\,dx = 2\int_1^t \frac{1}{x}\,dx.$

21. Approximate $\displaystyle\int_1^N \frac{1}{x}\,dx$ by the Trapezoidal Rule using $\Delta x = 1$. Deduce that

$$1 + \frac{1}{2} + \frac{1}{3} + \cdots + \frac{1}{N} \approx \ln N + \frac{1}{2} + \frac{1}{2N}\cdot$$

Show that

$$1 + \frac{1}{2} + \frac{1}{3} + \cdots + \frac{1}{10{,}000} \approx 9.71.$$

22. (cont.) Deduce that

$$\frac{1}{N+1} + \frac{1}{N+2} + \frac{1}{N+3} + \cdots + \frac{1}{2N} \approx \ln 2.$$

7. RATE OF GROWTH OF ln x

The function $\ln x$ increases very slowly as $x \longrightarrow \infty$. To see just how slowly, we shall compare the rate of growth of $\ln x$ with that of other functions.

First of all, $\ln x$ grows much slower than x:

$$\frac{\ln x}{x} \longrightarrow 0 \quad \text{as} \quad x \longrightarrow \infty.$$

To verify this, we study the function $y = (\ln x)/x$. Its derivative is

$$\frac{dy}{dx} = \frac{1 - \ln x}{x^2},$$

which is negative when $\ln x > 1$, that is, when $x > e$. So starting at $x = e$, the curve

$$y = \frac{\ln x}{x}$$

decreases. We must show it decreases toward *zero* as x increases.

Note that

$$\frac{\ln 10}{10} \approx \frac{2.303}{10} < 0.24.$$

Since the curve is decreasing,

$$\frac{\ln x}{x} < 0.24 \quad \text{for all} \quad x > 10.$$

Similarly,

$$\frac{\ln 100}{100} = \frac{2 \ln 10}{100} < \frac{2(2.4)}{100} = 0.048.$$

Hence,

$$\frac{\ln x}{x} < 0.048 \quad \text{for all} \quad x > 100.$$

Once more, this time with x *large:*

$$\frac{\ln 10^{100}}{10^{100}} = \frac{100 \ln 10}{10^{100}} < \frac{100(2.4)}{10^{100}} = 2.4 \times 10^{-98}.$$

Consequently,

$$\frac{\ln x}{x} < 2.4 \times 10^{-98} \quad \text{for all} \quad x > 10^{100}.$$

Thus, x overwhelms $\ln x$ as x increases. Even though both functions become large, x becomes so much larger than $\ln x$ that the ratio $(\ln x)/x$ is as small as you please when x is large enough. Consequently,

$$\frac{\ln x}{x} \longrightarrow 0 \quad \text{as} \quad x \longrightarrow \infty.$$

Perhaps we have been unrealistic in comparing $\ln x$ with x. Perhaps we should compare $\ln x$ with a smaller function, for example, with $\sqrt{x}$, which is much smaller than x. How does $\ln x$ compare with $\sqrt{x}$ or with $\sqrt[3]{x}$, which is smaller yet? The answer is that any positive power of x overwhelms $\ln x$.

> If p is any positive number, then
> $$\frac{\ln x}{x^p} \longrightarrow 0 \qquad \text{as} \quad x \longrightarrow \infty.$$

This follows from our study of $(\ln x)/x$. Let $x^p = y$. Then $x = y^{1/p}$ and

$$\frac{\ln x}{x^p} = \frac{\ln y^{1/p}}{y} = \frac{(1/p)\ln y}{y} = \frac{1}{p}\frac{\ln y}{y}.$$

As $x \longrightarrow \infty$, $y \longrightarrow \infty$. Hence $(\ln y)/y \longrightarrow 0$. Therefore

$$\frac{\ln x}{x^p} \longrightarrow 0 \qquad \text{as} \quad x \longrightarrow \infty.$$

As a rough check, let us estimate

$$\frac{\ln x}{\sqrt[3]{x}}$$

for some large values of x.

x	10^3	10^6	10^9	10^{30}	10^{300}
$\dfrac{\ln x}{\sqrt[3]{x}}$ (approx.)	0.69	0.14	0.021	6.9×10^{-9}	6.9×10^{-98}

Thus $\sqrt[3]{x}$ overwhelms $\ln x$ as $x \longrightarrow \infty$.

Not only is $\ln x$ small compared to x, but it is so small that any positive power $(\ln x)^p$ is small compared to x:

> If p is any positive number, then
> $$\frac{(\ln x)^p}{x} \longrightarrow 0 \qquad \text{as} \quad x \longrightarrow \infty.$$

This is just a variant of the statement that $(\ln x)/x^p \longrightarrow 0$ as $x \longrightarrow \infty$, with p replaced by $1/p$. In fact, since $1/p$ is positive,

$$\frac{\ln x}{x^{1/p}} \longrightarrow 0 \qquad \text{as} \quad x \longrightarrow \infty.$$

Therefore

$$\left(\frac{\ln x}{x^{1/p}}\right)^p \longrightarrow 0 \qquad \text{as} \quad x \longrightarrow \infty,$$

that is,

$$\frac{(\ln x)^p}{x} \longrightarrow 0 \quad \text{as} \quad x \longrightarrow \infty.$$

Thus x overwhelms not only $\ln x$, but also any positive power of $\ln x$.

As a check, let us estimate

$$\frac{(\ln x)^5}{x}$$

for some large values of x.

x	10	10^2	10^3	10^4	10^5	10^6	10^{10}	10^{20}	10^{100}
$\frac{(\ln x)^5}{x}$ (approx.)	6.5	20.7	15.7	6.6	2.0	0.50	6.5×10^{-4}	2.1×10^{-12}	6.5×10^{-89}

Thus x eventually overwhelms $(\ln x)^5$.

EXERCISES

Find:

1. $\lim_{x\to\infty} \frac{x}{10^6 \ln x}$
2. $\lim_{x\to\infty} \frac{(x+1)\ln x}{x^2}$
3. $\lim_{x\to\infty} \frac{(\ln x)^3}{x^{1/4}}$
4. $\lim_{x\to\infty} \frac{\log_{10} x}{x}$
5. $\lim_{x\to\infty} \frac{x^2+1}{x(\ln x)^2}$
6. $\lim_{x\to 0} [x \ln x]$ (*Hint:* Let $x = 1/y$.)
7. $\lim_{x\to 0} [x(\ln x)^{20}]$
8. $\lim_{x\to\infty} \frac{x \ln x}{x^p + 1}$ $(p > 1)$
9. $\lim_{x\to\infty} \frac{3(\ln x)^2 + 1}{x}$
10. $\lim_{x\to\infty} \frac{x^2(\ln x)^3}{x^3 \ln x + 1}$.
11. Let $p > 0$ and $q > 0$. Prove $\lim_{x\to\infty} \frac{(\ln x)^p}{x^q} = 0$.
12. For $p > 0$ and $q > 0$, find the maximum value of $f(x) = \frac{(\ln x)^p}{x^q}$, and the value of x at which it occurs.
13. (cont.) Estimate the order of magnitude of this x for $p = 10^5$ and $q = 10^{-5}$.
14. For what values of x can you be sure that $\frac{\ln x}{x} < 10^{-10}$?
15. For what values of x can you be sure that $\frac{(\ln x)^2}{x} < 10^{-10}$?
16. Graph $y = x^x$ for $x > 0$. Find its minimum.
17. Graph $y = x^{1/x}$ for $x > 0$. Find its maximum.

8. RATE OF GROWTH OF e^x

The graphs of $\ln x$ and e^x are mirror images (Fig. 3.1). We know

$$\ln x \longrightarrow \infty \qquad \text{as} \quad x \longrightarrow \infty,$$

but the rate of growth is extremely slow. Since $\ln x$ increases very slowly, e^x ought to increase very rapidly. It does; we can show that as $x \longrightarrow \infty$, e^x eventually overwhelms x^p for any p.

> If p is any positive number, then
>
> $$\frac{e^x}{x^p} \longrightarrow \infty \qquad \text{as} \quad x \longrightarrow \infty.$$

A number is large if its natural logarithm is large. Observe

$$\ln\left(\frac{e^x}{x^p}\right) = \ln e^x - \ln x^p = x - p \ln x.$$

We know that $\ln x$ is much smaller than x. To emphasize this, we write

$$\ln\left(\frac{e^x}{x^p}\right) = x\left(1 - \frac{p \ln x}{x}\right).$$

Since

$$\frac{\ln x}{x} \longrightarrow 0 \qquad \text{as} \quad x \longrightarrow \infty,$$

the second term in the parentheses is less than $\frac{1}{2}$ once x is large enough. Hence,

$$\ln\left(\frac{e^x}{x^p}\right) > \frac{1}{2} x$$

for x large. Consequently,

$$\ln\left(\frac{e^x}{x^p}\right) \longrightarrow \infty \qquad \text{as} \quad x \longrightarrow \infty,$$

which implies

$$\frac{e^x}{x^p} \longrightarrow \infty \qquad \text{as} \quad x \longrightarrow \infty.$$

As an example, we compute

$$\frac{e^x}{x^{10}}$$

for some values of x.

x	5	10	20	30	40	50	100	1000
$\frac{e^x}{x^{10}} \approx$	1.5×10^{-5}	2×10^{-6}	5×10^{-5}	0.018	22.4	5.3×10^4	2.7×10^{23}	2.0×10^{404}

After a slow start, e^x overwhelms x^{10}.

We repeat the two basic facts concerning rate of growth of the logarithm and exponential functions.

If p is any positive number, then

$$\begin{cases} \dfrac{\ln x}{x^p} \longrightarrow 0 \\ \\ \dfrac{e^x}{x^p} \longrightarrow \infty \end{cases} \quad \text{as} \quad x \longrightarrow \infty.$$

EXERCISES

Find:

1. $\lim\limits_{x \to \infty} \dfrac{x^4}{e^x}$

2. $\lim\limits_{x \to \infty} \dfrac{x^7 e^x}{e^{2x} + 3}$

3. $\lim\limits_{x \to \infty} \dfrac{e^{x+1} - e^x}{e^x - x}$

4. $\lim\limits_{x \to \infty} \dfrac{1 + \ln x}{e^{2x}}$.

5. Show that $x^{-10} e^{\sqrt[3]{x}}$ increases towards ∞ as $x \longrightarrow \infty$.
6. Let $f(x) = e^x/x^p$, where $p > 0$. Show that $f'(x) \longrightarrow \infty$ as $x \longrightarrow \infty$.
7. Graph $y = e^x/x$.
8. Find the smallest value of e^x/x^{10}.
9. According to our results $e^x/x^{100} \longrightarrow \infty$ as $x \longrightarrow \infty$. Yet when $x = 100$, its value is $\approx 2.7 \times 10^{-157}$. Is that a contradiction?
10. Verify that $e^7 \approx 10^3$. This is useful in making rough approximations. Use it to get a quick estimate of e^{23}. Find a similar relation between a power of 2 and a power of 10.
11. Arrange these functions in order of their size as $x \longrightarrow \infty$: e^x, e^{10x}, e^{x^2}, e^{e^x}, x^3, 10^x.
12. Find some functions that increase to infinity as $x \longrightarrow \infty$, but much slower than $\ln x$.

18. Trigonometric Functions

1. BASIC PROPERTIES

In this chapter, it is assumed you are familiar with the functions $\sin x$ and $\cos x$. It will help to review Chapter 6.

Recall the definitions of the other trigonometric functions:

$$\tan x = \frac{\sin x}{\cos x}, \qquad \sec x = \frac{1}{\cos x},$$

$$\cot x = \frac{\cos x}{\sin x}, \qquad \csc x = \frac{1}{\sin x}.$$

(NOTE: The cotangent function is sometimes written $\operatorname{ctn} x$.)

Here are some basic properties of these functions.

(1) *Periods.* The functions $\sin x$ and $\cos x$ are periodic with period 2π:

$$\sin(x + 2\pi) = \sin x, \qquad \cos(x + 2\pi) = \cos x.$$

The other four trigonometric functions are defined in terms of $\sin x$ and $\cos x$. They inherit the period 2π. For example,

$$\sec(x + 2\pi) = \frac{1}{\cos(x + 2\pi)} = \frac{1}{\cos x} = \sec x.$$

About $\tan x$ and $\cot x$, more can be said. Recall that

$$\sin(x + \pi) = -\sin x, \qquad \cos(x + \pi) = -\cos x.$$

Therefore,

$$\tan(x + \pi) = \frac{\sin(x + \pi)}{\cos(x + \pi)} = \frac{-\sin x}{-\cos x} = \tan x.$$

Thus $\tan x$ has period π and the same is true of $\cot x$.

The functions $\sin x$, $\cos x$, $\sec x$, and $\csc x$ have period 2π. The functions $\tan x$ and $\cot x$ have period π.

(2) *Parity.* Recall that $\sin(-x) = -\sin x$ and $\cos(-x) = \cos x$. It follows that

$$\tan(-x) = -\tan x, \qquad \sec(-x) = \sec x,$$

$$\cot(-x) = -\cot x, \qquad \csc(-x) = -\csc x.$$

A function $f(x)$ is an **even function** if $f(-x) = f(x)$. A function $f(x)$ is an **odd function** if $f(-x) = -f(x)$.

> The functions $\cos x$ and $\sec x$ are even functions. The functions $\sin x$, $\tan x$, $\cot x$, and $\csc x$ are odd functions.

Do not assume that every function is either odd or even. For example, $x^2 + 2x$ and $\sin x + \cos x$ are neither odd nor even.

The graph of an even function is symmetric about the y-axis; examples are $y = x^2$ and $y = \cos x$. The graph of an odd function is symmetric through the origin; examples are $y = x^3$ and $y = \sin x$. See Fig. 1.1.

The evenness or oddness of a function is called its **parity.**

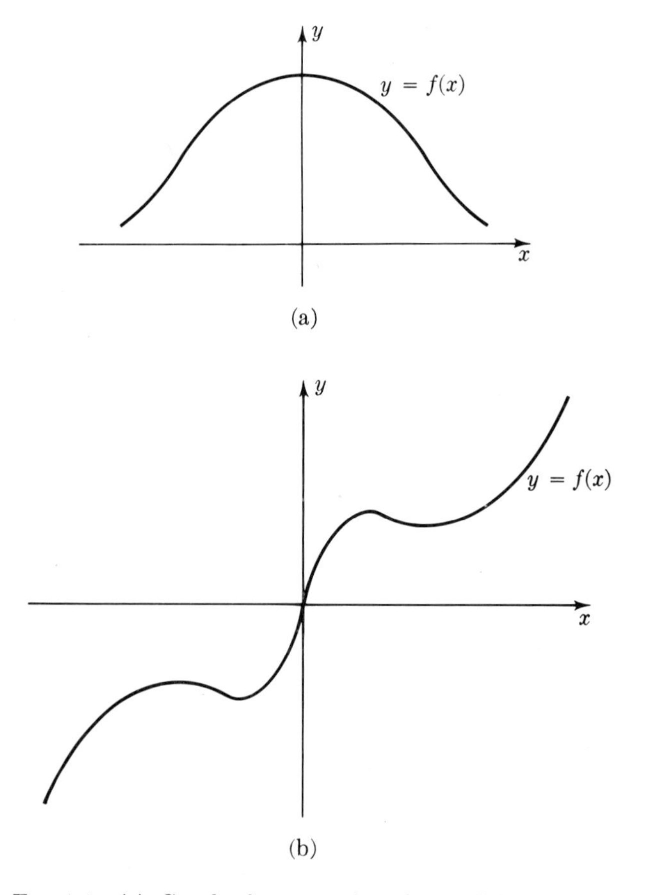

FIG. 1.1 (a) Graph of an even function: $f(x) = f(-x)$.
(b) Graph of an odd function: $f(x) = -f(-x)$.

(3) *Domain of definition.* The functions $\sin x$ and $\cos x$ are defined for all x. Because of zeros in the denominators, $\tan x$ and $\sec x$ are not defined where $\cos x = 0$: at $\pm\pi/2$, $\pm 3\pi/2$, $\pm 5\pi/2$, etc. Similarly, $\cot x$ and $\csc x$ are not defined where $\sin x = 0$: at 0, $\pm\pi$, $\pm 2\pi$, etc.

Graphs of Trigonometric Functions

Let us graph $y = \tan x$. We know that $\tan x$ is an odd function and has period π. It suffices to sketch the curve from 0 to $\pi/2$; the graph can than be extended by oddness and periodicity.

Since

$$\tan x = \frac{\sin x}{\cos x},$$

$\tan x \geq 0$ for $0 \leq x < \pi/2$. Furthermore, as x increases from 0 toward $\pi/2$, the numerator increases starting at 0, while the denominator decreases from 1 to 0. Therefore as x increases starting at 0, the function $\tan x$ increases starting at 0; as $x \longrightarrow \pi/2$, then $\tan x \longrightarrow \infty$. Near $x = 0$ however, $\sin x \approx x$ and $\cos x \approx 1$. Therefore $\tan x \approx x$; the graph has slope 1 at the

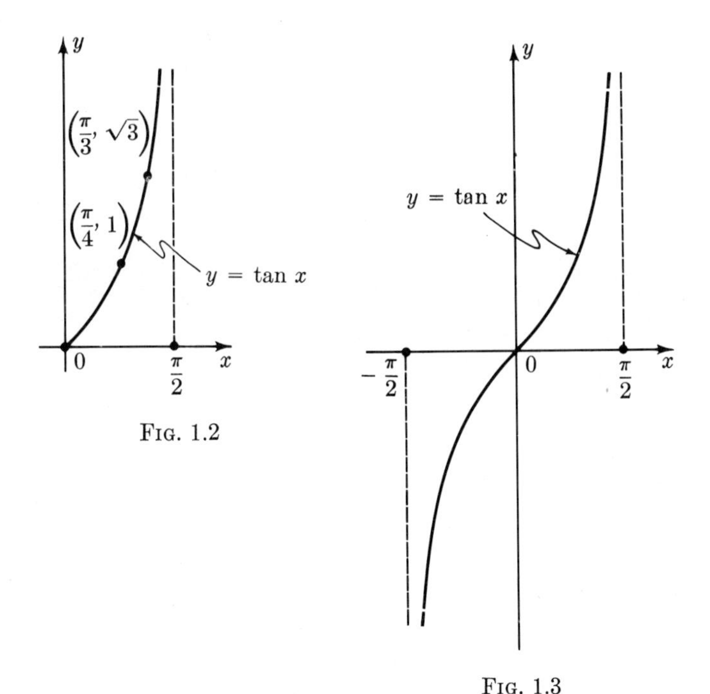

Fig. 1.2

Fig. 1.3

origin (Fig. 1.2). By the oddness of $\tan x$, we extend the graph from 0 to $-\pi/2$. See Fig. 1.3. Since $\tan x$ has period π, the curve in Fig. 1.3 can be extended indefinitely (Fig. 1.4).

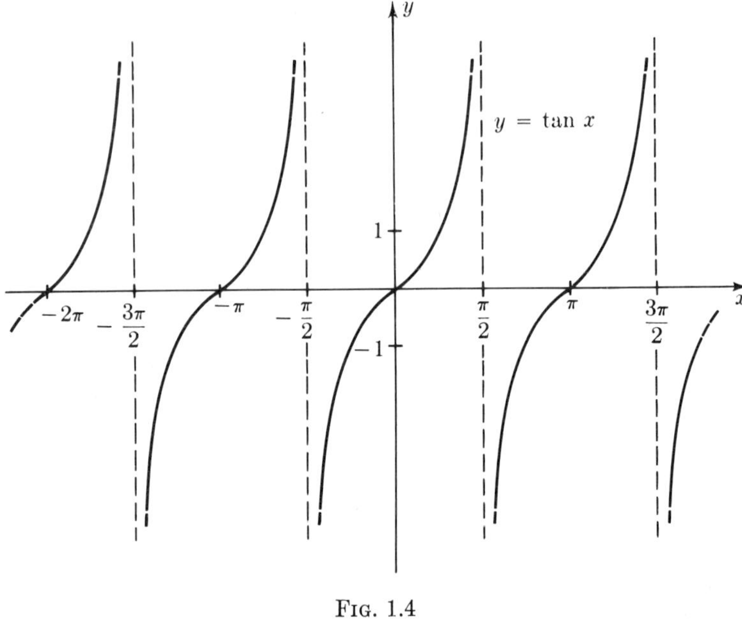

Fig. 1.4

By a similar argument, we obtain the graph of cot x. See Fig. 1.5. In each period of length π, both tan x and cot x take every value once.

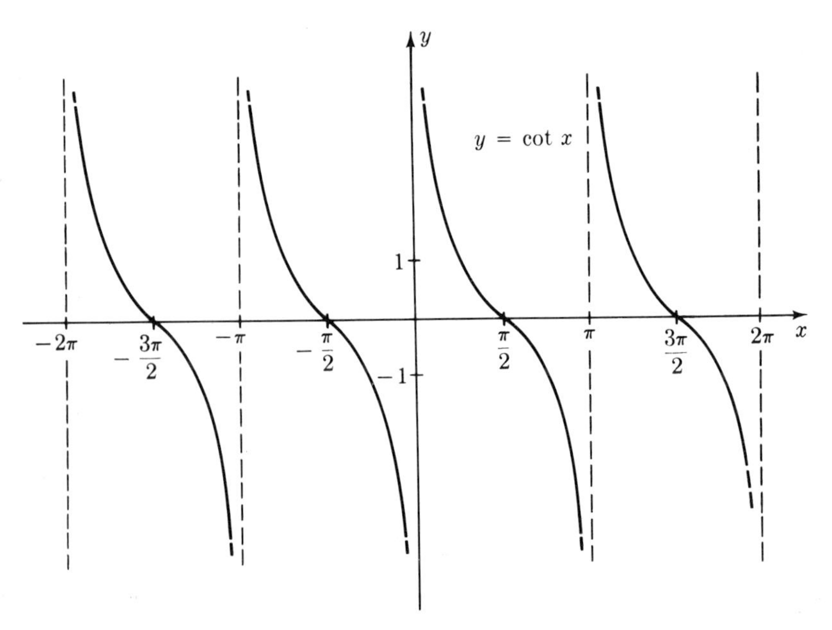

Fig. 1.5

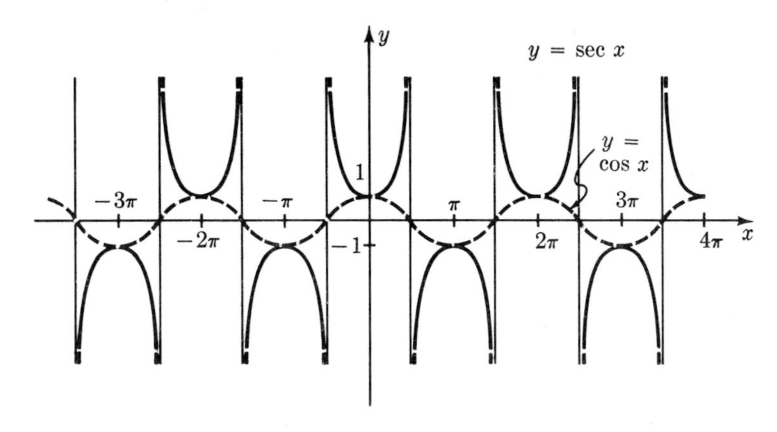

Fig. 1.6

The graphs of $\sec x$ and $\csc x$ are sketched by taking reciprocals of $\cos x$ and $\sin x$. See Figs. 1.6 and 1.7. Both $\sec x$ and $\csc x$ take all values except those between -1 and $+1$; in other words $|\sec x| \geq 1$ and $|\csc x| \geq 1$.

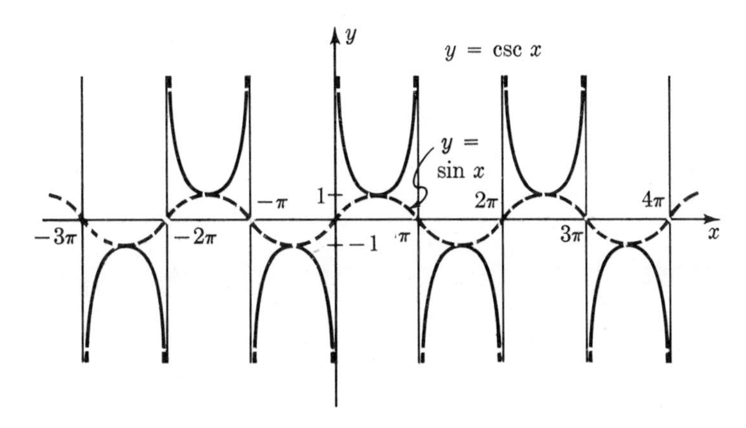

Fig. 1.7

EXERCISES

Find the least period:

1. $\sin 2x$
2. $\tan 3x$
3. $\sin 2\pi x$
4. $\cot x + \csc x$
5. $3 \sec \pi x + \cos \pi x$
6. $3 \cos 2\pi x - 5 \sin 2\pi x$
7. $(\sin x)^{-2}$
8. $\sqrt{1 + \cos \dfrac{x}{2}}$

9. $\sin x \sin 3x$

10. $\tan x + \tan\left(x - \frac{\pi}{6}\right)$

11. $\sin 3x + \cos 4x$

12. $\sin x + \cos \frac{x}{3}$.

Find the parity:

13. $\sin 2x$

14. $\cot 3x$

15. $\left(x + \frac{1}{x}\right)\sin x$

16. $x \sin^3 x + 2\cos x$

17. $e^x - e^{-x}$

18. $\tan(x^2)$

19. $x^2 + x$

20. $\frac{\sin x}{x}$.

21. There are some simple rules of parity in the multiplication of integers:

$$\text{even} \cdot \text{even} = \text{even}, \qquad \text{even} \cdot \text{odd} = \text{even}, \qquad \text{odd} \cdot \text{odd} = \text{odd}.$$

Do the same rules hold in the multiplication of functions?

22. If $f(x)$ is an odd function, show that $f(0) = 0$.

Graph:

23. $y = \tan 3x$

24. $y = \cot 4x$

25. $y = 2\sec 2\pi x$

26. $y = \csc\left(x - \frac{\pi}{4}\right)$

27. $y = \sec x - \csc x$

28. $y = \tan 2x + \cot 2x$

29. $y = \sec^2 x$

30. $y = \sec^3 x$.

2. IDENTITIES

The most common identity relating trigonometric functions is

$$\sin^2 x + \cos^2 x = 1.$$

Dividing both sides first by $\cos^2 x$ and then by $\sin^2 x$, we obtain two identities:

$$\tan^2 x + 1 = \sec^2 x,$$

$$\cot^2 x + 1 = \csc^2 x.$$

These formulas are helpful in expressing one trigonometric function in terms of another. For example,

$$\sin x = \pm\sqrt{1 - \cos^2 x},$$

$$\sec x = \pm\sqrt{\tan^2 x + 1},$$

$$\cot x = \pm\sqrt{\csc^2 x - 1} = \pm\frac{\sqrt{1 - \sin^2 x}}{\sin x}.$$

In each case, some additional information is needed in order to choose the correct sign.

EXAMPLE 2.1

Express $\sin \alpha$ and $\cos \alpha$ in terms of $\tan \alpha$ for $0 < \alpha < \frac{\pi}{2}$.

Solution:

$$\cos \alpha = \frac{1}{\sec \alpha} = \frac{1}{\pm\sqrt{\tan^2 \alpha + 1}},$$

$$\sin \alpha = \tan \alpha \cdot \cos \alpha = \frac{\tan \alpha}{\pm\sqrt{\tan^2 \alpha + 1}}.$$

Since $\sin \alpha$, $\cos \alpha$, and $\tan \alpha$ are positive for $0 < \alpha < \pi/2$, choose the positive square root.

Answer: For $0 < \alpha < \frac{\pi}{2}$,

$$\sin \alpha = \frac{\tan \alpha}{\sqrt{\tan^2 \alpha + 1}};$$

$$\cos \alpha = \frac{1}{\sqrt{\tan^2 \alpha + 1}}.$$

Addition Formulas

The next most important identities are the **addition formulas:**

$$\sin(x + y) = \sin x \cos y + \cos x \sin y,$$

$$\cos(x + y) = \cos x \cos y - \sin x \sin y.$$

(These are probably familiar from Trigonometry; a derivation is given in Chapter 24, Section 6, p. 520.)

A number of related identities are derived from the addition formulas. If y is replaced by $-y$, there follow easily

$$\sin(x - y) = \sin x \cos y - \cos x \sin y,$$

$$\cos(x - y) = \cos x \cos y + \sin x \sin y.$$

If $y = x$, the addition formulas yield the **double angle formulas:**

$$\sin 2x = 2 \sin x \cos x,$$

$$\cos 2x = \cos^2 x - \sin^2 x.$$

The latter formula has alternate forms, obtained on replacing $\cos^2 x$ by $1 - \sin^2 x$, or $\sin^2 x$ by $1 - \cos^2 x$:

$$\cos 2x = 1 - 2\sin^2 x,$$

$$\cos 2x = 2\cos^2 x - 1.$$

From these follow the identities

$$\sin^2 x = \frac{1}{2}(1 - \cos 2x),$$

$$\cos^2 x = \frac{1}{2}(1 + \cos 2x).$$

EXAMPLE 2.2

Find an antiderivative of $\cos^2 x$.

Solution: According to the preceding identity,

$$\cos^2 x = \frac{1}{2}(1 + \cos 2x).$$

An antiderivative of $\frac{1}{2}$ is $\frac{1}{2}x$; an antiderivative of $\frac{1}{2}\cos 2x$ is $\frac{1}{4}\sin 2x$.

Answer: $\frac{1}{2}x + \frac{1}{4}\sin 2x.$

EXAMPLE 2.3

Compute the derivative of $y = \sin x \cos x$.

Solution: By the Product Rule,

$$\begin{aligned}\frac{dy}{dx} &= \sin x \frac{d}{dx}(\cos x) + \cos x \frac{d}{dx}(\sin x)\\ &= (\sin x)(-\sin x) + \cos x(\cos x)\\ &= \cos^2 x - \sin^2 x.\end{aligned}$$

Alternate Solution: By the double angle formula, $\sin x \cos x = \frac{1}{2}\sin 2x$. By the Chain Rule,

$$\frac{dy}{dx} = \frac{d}{dx}\left(\frac{1}{2}\sin 2x\right) = \frac{1}{2}(\cos 2x)\cdot 2 = \cos 2x.$$

Note that the two answers agree: $\cos 2x = \cos^2 x - \sin^2 x$.

Answer: $\cos 2x.$

EXERCISES

1. Suppose $0 < x < \pi/2$. Express $\sin x$ in terms of
   (a) $\cot x$ (b) $\sec x$.

2. Express $\cos(x - \pi/3)$ in terms of $\sin x$ and $\cos x$.
3. Express $\cot^2 x$ in terms of $\cos^2 x$.
4. Express $\sin 3x$ in terms of $\sin x$. [*Hint:* $\sin 3x = \sin(2x + x)$.]
5. Express $\cos^4 x$ in terms of $\cos 2x$.
6. Show that $\tan 2x = \dfrac{2 \tan x}{1 - \tan^2 x}$.
7. Using two different identities, show that

$$\sin 15° = \tfrac{1}{4}(\sqrt{6} - \sqrt{2}) = \tfrac{1}{2}\sqrt{2 - \sqrt{3}}.$$

8. Compute the derivative of $\cos^2 x - \sin^2 x$ in two different ways.
9. Show that $\tan x/2 = \pm\sqrt{\dfrac{1 - \cos x}{1 + \cos x}}$, where the sign is + if $0 < x < \pi$ and − if $-\pi < x < 0$.
10. Prove that $\cos x \cos y = \tfrac{1}{2}[\cos(x + y) + \cos(x - y)]$.
11. Find similar formulas for $\sin x \cos y$ and $\sin x \sin y$.
12. Prove that

$$\cos x \cos 2x \cos 4x \cos 8x$$
$$= \frac{1}{8}(\cos x + \cos 3x + \cos 5x + \cdots + \cos 13x + \cos 15x).$$

3. DERIVATIVES

The basic differentiation formulas are these:

$$\frac{d}{dx}(\sin x) = \cos x \qquad \frac{d}{dx}(\cot x) = -\csc^2 x$$

$$\frac{d}{dx}(\cos x) = -\sin x \qquad \frac{d}{dx}(\sec x) = \sec x \tan x$$

$$\frac{d}{dx}(\tan x) = \sec^2 x \qquad \frac{d}{dx}(\csc x) = -\csc x \cot x.$$

The first two formulas are known; the rest follow easily. For example,

$$\frac{d}{dx}(\cot x) = \frac{d}{dx}\left(\frac{\cos x}{\sin x}\right) = \frac{-\sin^2 x - \cos^2 x}{\sin^2 x} = \frac{-1}{\sin^2 x} = -\csc^2 x;$$

$$\frac{d}{dx}(\csc x) = \frac{d}{dx}\left(\frac{1}{\sin x}\right) = \frac{-\cos x}{\sin^2 x} = -\csc x \cot x.$$

We look next at some useful relations between the trigonometric functions and their derivatives. If $y = \sin x$, then

$$\frac{dy}{dx} = \cos x = \pm\sqrt{1 - \sin^2 x} = \pm\sqrt{1 - y^2}.$$

For $-\pi/2 \leq x \leq \pi/2$, we choose the positive square root since $\cos x \geq 0$. Thus $y = \sin x$ satisfies the differential equation

$$\frac{dy}{dx} = \sqrt{1 - y^2}, \qquad -\frac{\pi}{2} \leq x \leq \frac{\pi}{2}.$$

By similar reasoning, we see that $y = \tan x$ satisfies the differential equation

$$\frac{dy}{dx} = 1 + y^2.$$

Indeed,

$$\frac{d}{dx}(\tan x) = \sec^2 x = 1 + \tan^2 x.$$

Similarly, $y = \sec x$ satisfies

$$\frac{dy}{dx} = \begin{cases} y\sqrt{y^2 - 1}, & 0 \leq x < \dfrac{\pi}{2}, \\ -y\sqrt{y^2 - 1}, & -\dfrac{\pi}{2} < x \leq 0. \end{cases}$$

EXERCISES

Differentiate:

1. $\sin \dfrac{x}{3}$
2. $\tan 10x$
3. $\cos^3 x$
4. $\sec^2 x$
5. $x \cot 2x$
6. $\sec \dfrac{1}{x}$
7. $\sqrt{2 - \csc^2 x}$
8. $\sin 3x + \cos 4x$
9. $\left(1 + \cos^5 \dfrac{x}{3}\right)^2$
10. $e^{-x} \csc x$
11. $\dfrac{1 + \sin x}{1 - \sin x}$
12. $\dfrac{1}{3 + \cos^2 5x}$
13. $\sec^3(2x^2)$
14. $e^{\sin 3x}$.
15. Find the 42nd derivative of $\sin x$.

16. Verify the formulas given in the text for the derivatives of $\tan x$ and $\sec x$.
17. Express $d(\csc x)/dx$ in terms of $\csc x$ for $0 < x < \pi/2$; thus find a differential equation which $\csc x$ satisfies.
18. Find a differential equation satisfied by $f(x) = 5 \sin 2x - \cos 2x$. (*Hint:* Differentiate twice.)

Use trigonometric identities to verify the formula:

19. $\dfrac{d}{dx}(\sec^4 x - \tan^4 x) = 4 \sec^2 x \tan x$
20. $\dfrac{d}{dx}(x - \sin x) = 2 \sin^2 \dfrac{x}{2}$
21. $\dfrac{d}{dx}(x + \sec x - \tan x) = \dfrac{\sin x}{1 + \sin x}$.
22. Verify that $\sin^2 x$ and $-\cos^2 x$ have identical derivatives. Does that mean that $\sin^2 x = -\cos^2 x$; if not, what does it mean?

4. APPLICATIONS

EXAMPLE 4.1

A point P moves counterclockwise at a constant speed around a circle of radius 50 ft, making 1 rpm. The tangent at P crosses the line $\overline{OA}$ at a point T. See Fig. 4.1. Compute the speed of T when $\theta = \pi/4$.

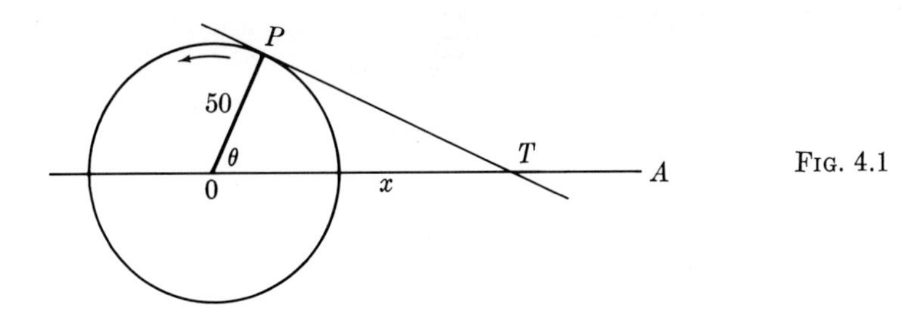

FIG. 4.1

Solution: We are given $d\theta/dt = 2\pi$ rad/min. We are asked to find dx/dt where $x = \overline{0T}$. A relation between θ and x is seen from the right triangle $0PT$, namely, $x = 50 \sec \theta$. Differentiating with respect to t:

$$\frac{dx}{dt} = 50 \sec \theta \tan \theta \frac{d\theta}{dt}.$$

When $\theta = \pi/4$, $\sec \theta = \sqrt{2}$ and $\tan \theta = 1$. Hence $dx/dt = 50\sqrt{2} \cdot 2\pi$.

Answer: $100\pi\sqrt{2}$ ft/min.

EXAMPLE 4.2

A lighthouse stands 2 mi off a long straight shore, opposite a point P. Its light rotates counterclockwise at the constant rate of 1 rpm. How fast is the beam moving along the shore as it passes a point 3 mi to the right of P?

Solution: Set up axes with P at the origin and x-axis along the shore (Fig. 4.2a). The beam hits the shore at x. The rate of change of the angle θ is given:

$$\dot{\theta} = 2\pi \text{ rad/min}.$$

The problem: compute the derivative $\dot{x}$ at the instant when $x = 3$.

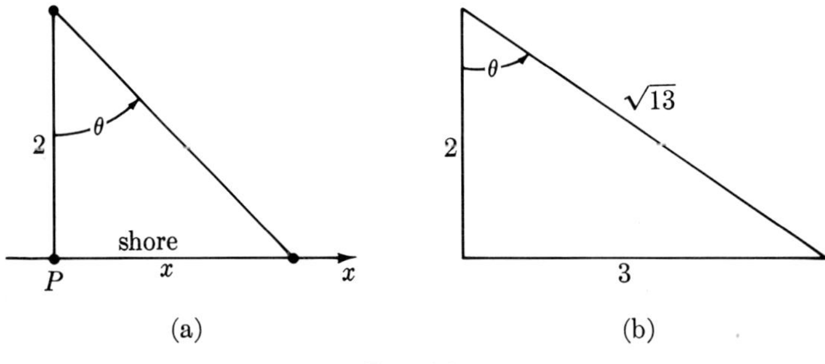

FIG. 4.2

There is a relation between x and θ suggested by the figure:

$$\frac{x}{2} = \tan\theta \qquad \text{or} \qquad x = 2\tan\theta.$$

Now compute $\dot{x}$ by the Chain Rule:

$$\dot{x} = \frac{dx}{d\theta}\dot{\theta} = (2\sec^2\theta)(2\pi) = 4\pi\sec^2\theta.$$

At the instant in question, $x = 3$. From Fig. 4.2b, $\sec\theta = \frac{1}{2}\sqrt{13}$. Hence

$$\dot{x} = 4\pi\left(\frac{13}{4}\right) = 13\pi.$$

Answer: 13π mi/min.

EXAMPLE 4.3

A 5-ft fence stands 4 ft from a house. What is the angle of inclination of the shortest ladder that can stand outside the fence and lean against the wall?

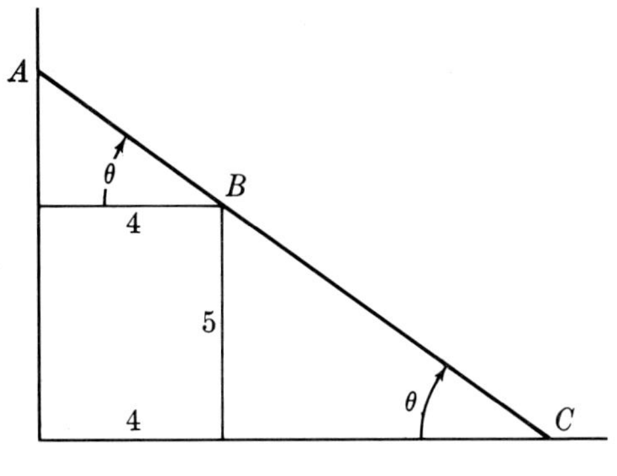

FIG. 4.3

Solution: Draw a diagram (Fig. 4.3). If θ is near 0 or near $\pi/2$, the ladder will be long. The best angle θ is probably near $\pi/4$.

Let L be the length of the ladder,

$$L = \overline{AB} + \overline{BC} = 4 \sec \theta + 5 \csc \theta.$$

The problem: find the angle θ in the range $0 < \theta < \pi/2$ that minimizes $L = L(\theta)$. Now

$$\begin{aligned} \frac{dL}{d\theta} &= 4 \sec \theta \tan \theta - 5 \csc \theta \cot \theta \\ &= 4 \frac{1}{\cos \theta} \frac{\sin \theta}{\cos \theta} - 5 \frac{1}{\sin \theta} \frac{\cos \theta}{\sin \theta} \\ &= \frac{4 \sin^3 \theta - 5 \cos^3 \theta}{\sin^2 \theta \cos^2 \theta}. \end{aligned}$$

As θ increases, the numerator changes from negative to zero to positive. Hence, L is minimized for just one value of θ:

$$\frac{dL}{d\theta} = 0 \quad \text{if} \quad 4 \sin^3 \theta = 5 \cos^3 \theta, \qquad \text{that is,} \quad \tan^3 \theta = \tfrac{5}{4}.$$

> *Answer:* The angle θ whose tangent is $\sqrt[3]{\frac{5}{4}} \approx 1.077$. From tables, $\theta \approx 47°8'$.

(A similar problem was discussed in Example 3.2, Chapter 16, p. 257.)

> EXAMPLE 4.4
>
> A man is in a rowboat 1 mi off shore. His home is 5 mi farther along the shore. If he can walk twice as fast as he can row, what is his quickest way home?

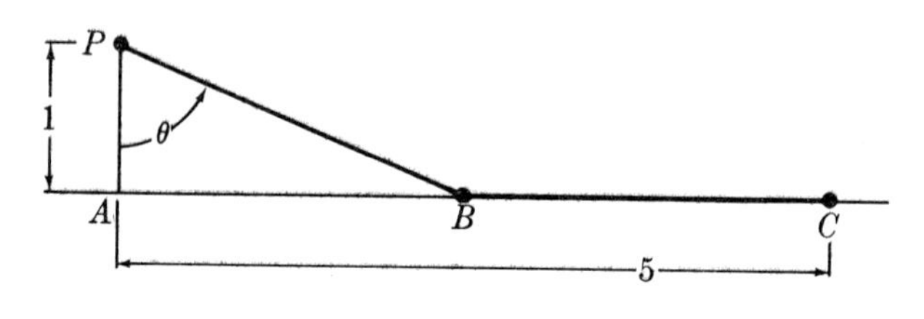

FIG. 4.4

Solution: Draw a diagram (Fig. 4.4). He should row to a point B, then walk to his home at C. Express everything in terms of the angle θ:

$$\overline{PB} = \sec\theta, \qquad \overline{BC} = 5 - \overline{AB} = 5 - \tan\theta.$$

Let s be his rowing speed and $2s$ his walking speed. The time required to reach home is

$$t = \frac{\overline{PB}}{s} + \frac{\overline{BC}}{2s} = \frac{\sec\theta}{s} + \frac{5 - \tan\theta}{2s}.$$

Since B must be between A and C, angle θ is at least 0 and at most the angle whose tangent is 5.

$$\frac{dt}{d\theta} = \frac{\sec\theta\tan\theta}{s} - \frac{\sec^2\theta}{2s}$$

$$= \frac{\sin\theta}{s\cos^2\theta} - \frac{1}{2s\cos^2\theta} = \frac{2\sin\theta - 1}{2s\cos^2\theta}.$$

The derivative is 0 if $\sin\theta = \frac{1}{2}$, that is $\theta = \pi/6$. Its sign changes from minus to plus as θ increases through $\pi/6$. Hence t has its minimum at $\theta = \pi/6$. Notice that $\pi/6$ falls within the required range of θ because $\tan\pi/6 < 5$.

Answer: The man should row to shore at an angle $\theta = \pi/6$ then walk the rest of the way (Fig. 4.4).

EXAMPLE 4.5

Describe the isosceles triangle of smallest area that circumscribes a circle of radius r.

Solution: Draw a few pictures (Fig. 4.5). If the triangle is too thin (Fig. 4.5a) or too wide (Fig. 4.5b), its area is unnecessarily large. The most efficient choice seems to be in between, perhaps equilateral (Fig. 4.5c).

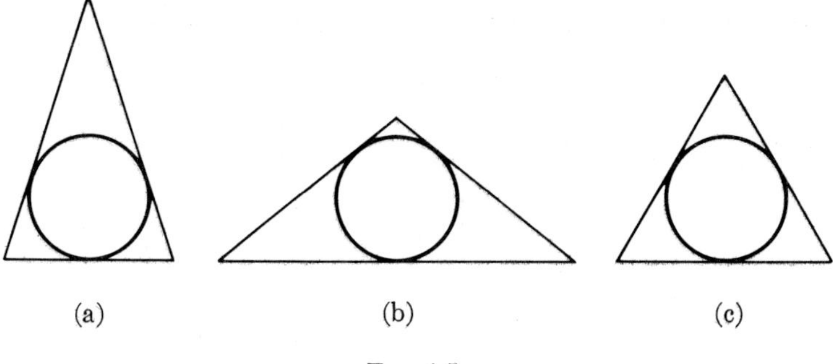

Fig. 4.5

Let 2θ be the apex angle, and express the area of the triangle in terms of θ. See Fig. 4.6.

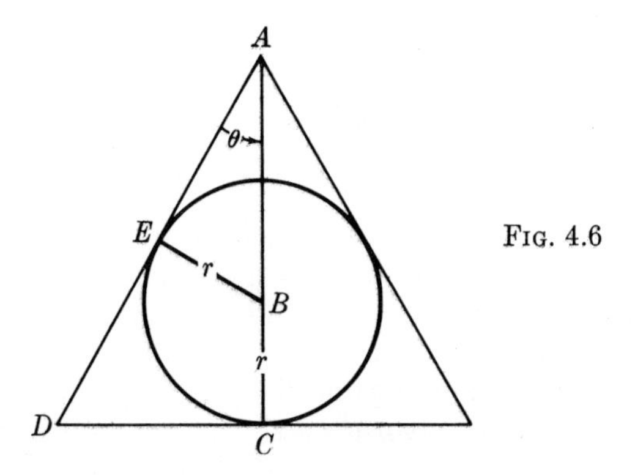

Fig. 4.6

$$\text{Area} = \frac{1}{2}\,\text{base} \cdot \text{height} = \overline{CD} \cdot \overline{AC}.$$

Now $\overline{AC} = \overline{AB} + \overline{BC} = r \csc \theta + r$. From triangle ACD, observe that $\overline{CD} = \overline{AC} \tan \theta = (r \csc \theta + r) \tan \theta$. Therefore

$$\begin{aligned} \text{Area} = A(\theta) &= (r \csc \theta + r)^2 \tan \theta \\ &= r^2(\csc \theta + 1)^2 \tan \theta. \end{aligned}$$

The function $A(\theta)$ is to be minimized. The range of θ is $0 < \theta < \pi/2$; there are no end points to consider.

$$\begin{aligned} \frac{dA}{d\theta} &= r^2 \left[(\csc \theta + 1)^2 \frac{d}{d\theta} (\tan \theta) + \tan \theta \frac{d}{d\theta} (\csc \theta + 1)^2 \right] \\ &= r^2[(\csc \theta + 1)^2 \sec^2 \theta + \tan \theta \cdot 2(\csc \theta + 1)\,(-\csc \theta \cot \theta)] \\ &= r^2(\csc \theta + 1)[(\csc \theta + 1) \sec^2 \theta - 2 \csc \theta]. \end{aligned}$$

In the range $0 < \theta < \pi/2$, $\csc\theta + 1$ is positive. Hence, $dA/d\theta = 0$ if

$$(\csc\theta + 1)\sec^2\theta - 2\csc\theta = 0.$$

Multiply through by $\sin\theta\cos^2\theta$:

$$\begin{aligned} 1 + \sin\theta - 2\cos^2\theta &= 0, \\ 2\sin^2\theta + \sin\theta - 1 &= 0, \\ (2\sin\theta - 1)(\sin\theta + 1) &= 0. \end{aligned}$$

Therefore

$$\sin\theta = \frac{1}{2} \qquad \text{or} \qquad \sin\theta = -1.$$

Since $0 < \theta < \pi/2$, the case $\sin\theta = -1$ is irrelevant. Hence, the minimum occurs for $\sin\theta = \frac{1}{2}$, i.e., for $\theta = \pi/6$. It follows that the desired triangle is equilateral. Its side is

$$2\overline{CD} = 2r\left(\csc\frac{\pi}{6} + 1\right)\tan\frac{\pi}{6} = 2r(2+1)\frac{1}{\sqrt{3}} = 2r\sqrt{3}.$$

Answer: An equilateral triangle of side $2r\sqrt{3}$.

REMARK: For physical reasons we were so confident that $A(\theta)$ had a *minimum* at $\pi/6$ (not a maximum or neither) we did not check. Actually the check is easy. Rather than setting the derivative equal to zero immediately, we can express it in the form

$$\frac{dA}{d\theta} = r^2(\csc\theta + 1)\frac{(2\sin\theta - 1)(\sin\theta + 1)}{\cos^2\theta\sin\theta}.$$

All factors are positive throughout the interval $0 < \theta < \pi/2$ except $(2\sin\theta - 1)$. Hence the sign of $dA/d\theta$ is the same as the sign of $(2\sin\theta - 1)$, which changes from minus to plus as θ increases through $\pi/6$. Thus $A(\theta)$ definitely has a minimum at $\pi/6$.

EXERCISES

1. Sketch the curve $y = 2\sin x + \cos 2x$ for $0 \le x \le \pi/2$. Indicate the maximum and minimum points.
2. A point P moves at the rate of 3 rpm counterclockwise around the circle of radius 50 ft centered at the origin. Determine how fast the distance between P and the point $(50\sqrt{2}, 0)$ is increasing when $\theta = 3\pi/4$. See Fig. 4.7.

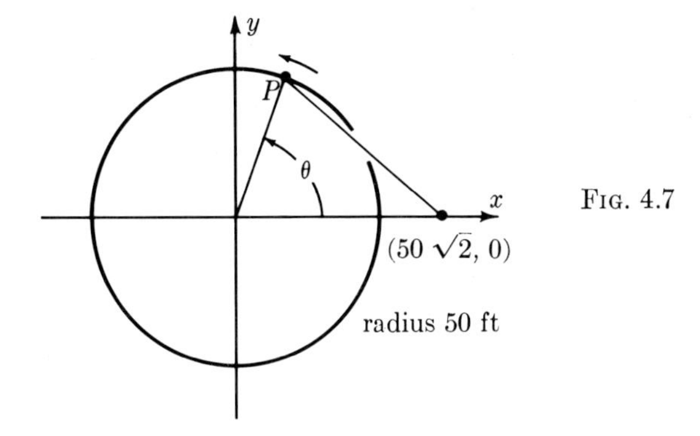

FIG. 4.7

3. The equal sides of an isosceles triangle are 10 in. If the vertex angle is increasing 1°/sec, how fast is the area of the triangle changing when the vertex angle is 45°? When is the area greatest?

4. A point P_1 moves around a circle of radius 3 ft at a steady rate of 2 rpm. A point P_2 moves in the same sense around a concentric circle of radius 5 ft at a steady rate of 1 rpm. If the points are 2 ft apart at $t = 0$, when is the first time their separation will be maximal?

5. A sector is cut from a circular piece of paper. The remaining paper is formed into a cone by joining together the edges of the sector without overlap. Find the sector angle that maximizes the volume of the cone.

6. An 8-ft ladder leans against a 4-ft fence. What is the largest horizontal distance the ladder can reach beyond the fence?
(*Hint:* Use the angle between the ladder and the ground as the variable.)

7. A weight hangs at the end of an 8-ft rope rigged up by the pulley system shown in Fig. 4.8. The pulleys A and B are stationary and 2 ft apart on the same level. The pulley at C is moveable. The weight hangs as far below the level of A and B as possible. How far is that? (Ignore the small pulley diameters.)

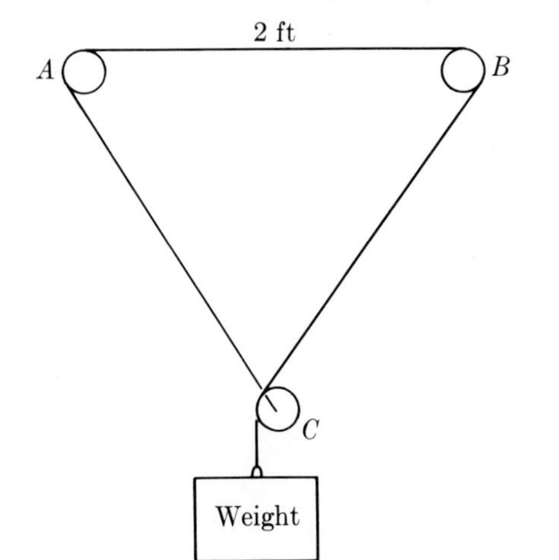

FIG. 4.8

8. A low-flying jet passes 450 ft directly over an observer on the ground. Shortly its angle of elevation is 30° and decreasing at the rate of 20°/sec. Compute the plane's speed.

9. Suppose Fig. 4.9 represents an electric clock of radius 3 in. and that Q is the tip of its second hand. How fast is the point P moving at 20 sec past 3:47 p.m.?

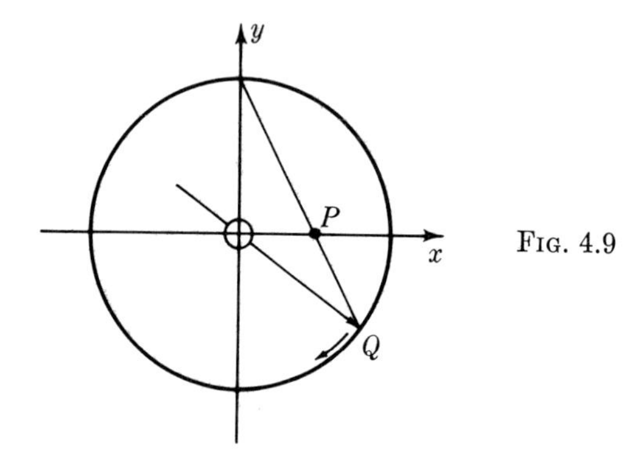

FIG. 4.9

10. Find the length of the longest pole that can be moved horizontally around the corner shown in Fig. 4.10.
(*Hint:* For each angle θ between 0 and 90°, the ladder must be no longer than the diagonal line in Fig. 4.10.)

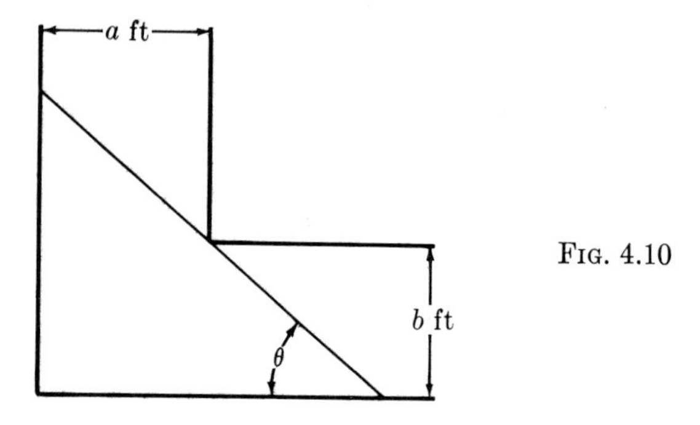

FIG. 4.10

11. Suppose in Ex. 10, the pole can be tilted so its ends touch the floor and the ceiling, which is c ft from the floor. Now how long can the pole be?

5. INVERSE TRIGONOMETRIC FUNCTIONS

Inverse of sin *x*

The sine of an angle does not completely determine the angle. For example, if $\sin x = \frac{1}{2}\sqrt{2}$, then x may be $\pi/4$, or $3\pi/4$, or $9\pi/4$, or $11\pi/4$, etc. This is clear from the graph of $\sin x$. See Fig. 5.1. The curve $y = \sin x$

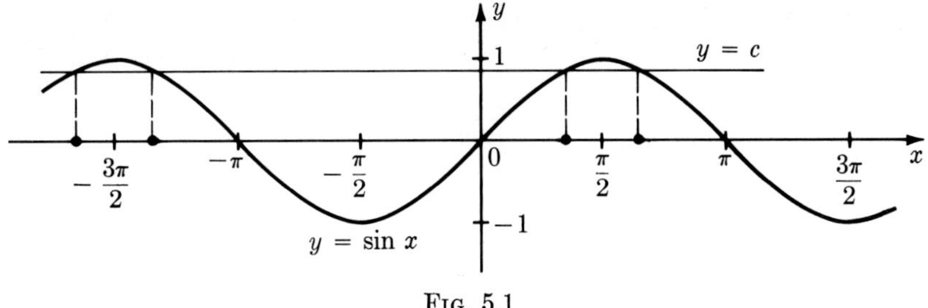

Fig. 5.1

intersects each horizontal line $y = c$, where $-1 \leq c \leq 1$, infinitely often; there are infinitely many values of x for which $\sin x = c$.

If, however, x is restricted to the interval $-\pi/2 \leq x \leq \pi/2$, then the equation $\sin x = c$ has a unique solution for each value of c in the interval $-1 \leq c \leq 1$. That is because the graph of the restricted function $y = \sin x$ is increasing (Fig. 5.2). There exists an inverse function, called arc sin:

$$y = \sin x \qquad \text{is equivalent to} \qquad x = \text{arc sin } y.$$

Thus arc sin y is the unique angle in the range from $-\pi/2$ to $\pi/2$ whose sine is y.

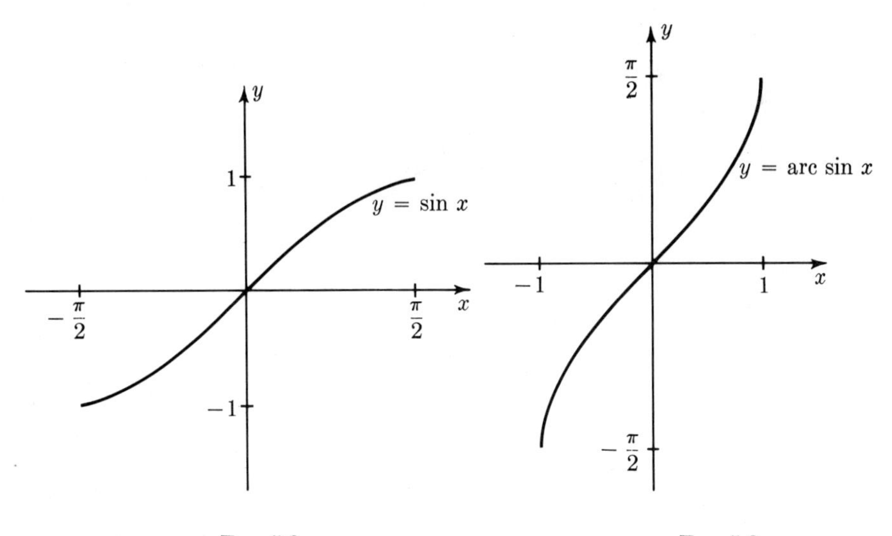

Fig. 5.2

Fig. 5.3

The graph of the function arc sin x is the reflection of the curve in Fig. 5.2 across the line $y = x$. See Fig. 5.3. The curve $y = \text{arc sin } x$ has vertical tangents at $(1, \pi/2)$ and $(-1, -\pi/2)$, reflecting the horizontal tangents of $y = \sin x$ at $(\pi/2, 1)$ and $(-\pi/2, -1)$.

> The function arc sin x is the inverse of sin x. It is defined for $-1 \le x \le 1$, and its values range from $-\pi/2$ to $\pi/2$.

Examples:

$$\text{arc sin } 1 = \frac{\pi}{2} \qquad \text{arc sin}(-1) = -\frac{\pi}{2}$$

$$\text{arc sin } \frac{1}{2} = \frac{\pi}{6} \qquad \text{arc sin}(\sin x) = x, \qquad -\frac{\pi}{2} \le x \le \frac{\pi}{2}$$

$$\text{arc sin } 0 = 0 \qquad \sin(\text{arc sin } x) = x$$

$$\text{arc sin}\left(-\frac{1}{\sqrt{2}}\right) = -\frac{\pi}{4} \qquad \text{arc sin}(-x) = -\text{ arc sin } x.$$

Inverse of cos x

Just as for sin x, the inverse of cos x can be defined only if x is suitably restricted. We agree to confine x to the interval $0 \le x \le \pi$. See Fig. 5.4.

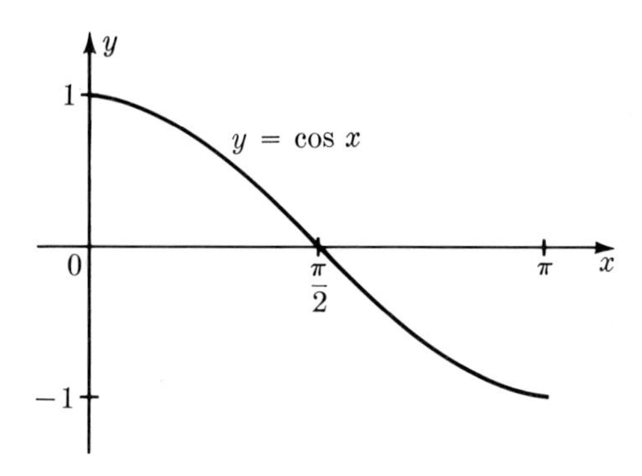

FIG. 5.4

In this interval the cosine function is decreasing; therefore the equation $\cos x = c$ has a unique solution for each c in the interval $-1 \le c \le 1$. The inverse function is called arc cos:

$$y = \cos x \qquad \text{is equivalent to} \qquad x = \text{arc cos } y.$$

Thus arc cos y is the unique angle in the range from 0 to π whose cosine is y.

The graph of the function arc cos x is the reflection of the curve in Fig. 5.4 across the line $y = x$. See Fig. 5.5.

> The function arc cos x is the inverse of cos x. It is defined for $-1 \le x \le 1$, and its values range from 0 to π.

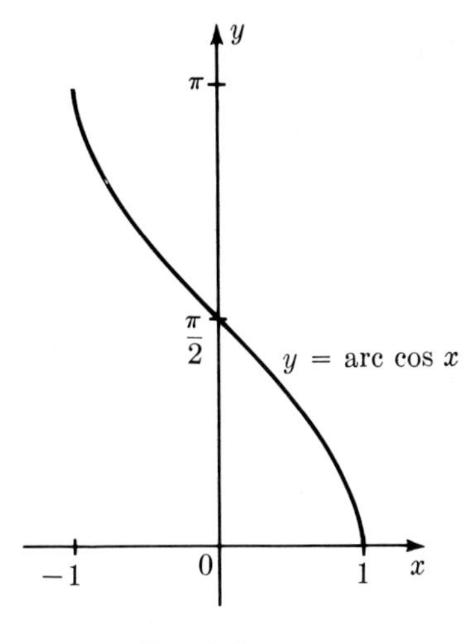

FIG. 5.5

Examples:

$$\arc\cos 1 = 0 \qquad \arc\cos(-1) = \pi$$

$$\arc\cos\frac{1}{2} = \frac{\pi}{3} \qquad \arc\cos(\cos x) = x, \qquad 0 \le x \le \pi$$

$$\arc\cos 0 = \frac{\pi}{2} \qquad \cos(\arc\cos x) = x$$

$$\arc\cos\left(-\frac{1}{\sqrt{2}}\right) = \frac{3\pi}{4} \qquad \arc\cos(-x) = \pi - \arc\cos x.$$

To understand the last example, look at Fig. 5.5. The curve is symmetric with respect to the point $(0, \pi/2)$. Hence $\pi/2$ is the average of arc cos x and arc cos $(-x)$, i.e.,

$$\frac{\pi}{2} = \frac{1}{2}[\arc\cos x + \arc\cos(-x)],$$

from which $\arc\cos(-x) = \pi - \arc\cos x$. For example,

$$\arc\cos\left(-\frac{1}{\sqrt{2}}\right) = \pi - \arc\cos\left(\frac{1}{\sqrt{2}}\right), \qquad \text{i.e.} \quad \frac{3\pi}{4} = \pi - \frac{\pi}{4}.$$

There is a relation between the functions arc sin x and arc cos x:

$$\boxed{\arc\sin x + \arc\cos x = \frac{\pi}{2}.}$$

To understand this relation, draw arc sin x and arc cos x on the same graph (Fig. 5.6). Then translate arc cos x downward $\pi/2$ units; the result is the graph of $(\arc\cos x) - \pi/2$. Clearly this curve is the reflection of the curve

$y = \arcsin x$ in the x-axis. Therefore

$$\left(\arccos x - \frac{\pi}{2}\right) + \arcsin x = 0.$$

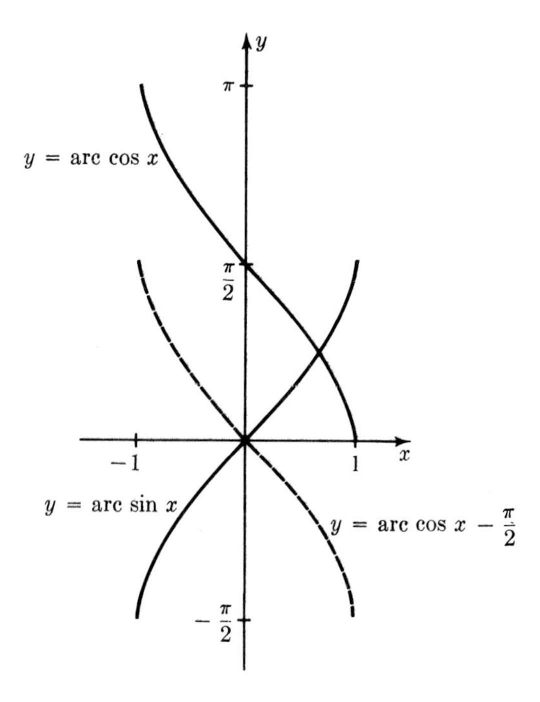

FIG. 5.6

Inverse of tan x

Recall the graph of $\tan x$ (Fig. 5.7). Consider only one branch of the graph, the one for which $-\pi/2 < x < \pi/2$. Then the equation $\tan x = c$ has a unique solution for each number c. There exists an inverse function called arc tan:

$$y = \tan x \qquad \text{is equivalent to} \qquad x = \arctan y.$$

Thus $\arctan y$ is the unique angle between $-\pi/2$ and $\pi/2$ whose tangent is y.

The graph of the function $\arctan x$ is the reflection across the line $y = x$ of one branch of the curve $y = \tan x$. See Fig. 5.8.

> The function $\arctan x$ is the inverse of $\tan x$. It is defined for all x, and takes values between $-\pi/2$ and $\pi/2$.
>
> As $x \longrightarrow \infty$, $\quad \arctan x \longrightarrow \dfrac{\pi}{2}$;
>
> as $x \longrightarrow -\infty$, $\quad \arctan x \longrightarrow -\dfrac{\pi}{2}$.

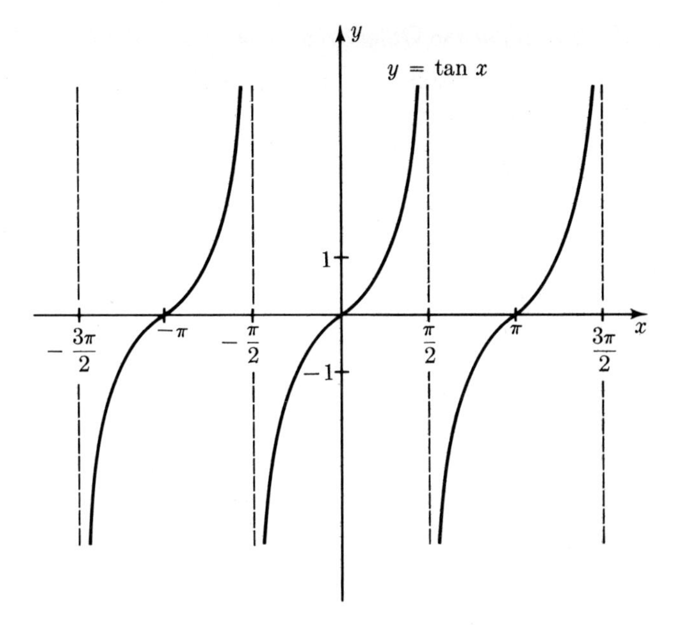

FIG. 5.7

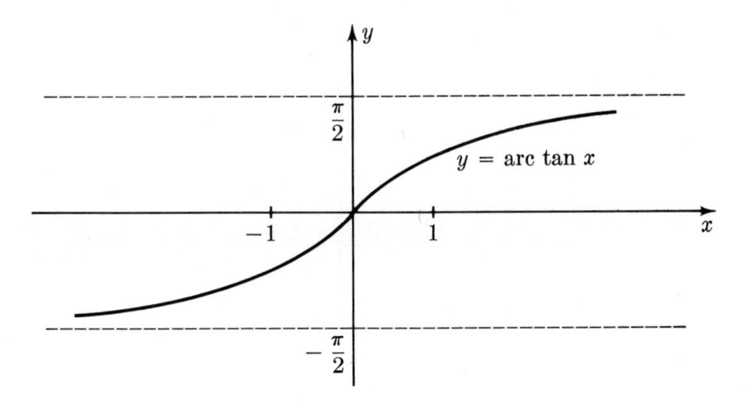

FIG. 5.8

Examples:

$$\text{arc tan } 0 = 0 \qquad \text{arc tan}(\tan x) = x, \qquad -\frac{\pi}{2} < x < \frac{\pi}{2}$$

$$\text{arc tan } 1 = \frac{\pi}{4} \qquad \text{arc tan } (100) \approx 1.5608$$

$$\text{arc tan}(-\sqrt{3}) = -\frac{\pi}{3} \qquad \text{arc tan } (3437.7) \approx 89^\circ\, 59' \approx 1.5705$$

(*note:* $\pi/2 \approx 1.5708$).

$$\text{arc tan}(-x) = -\text{ arc tan } x$$

Inverses of the Other Trigonometric Functions

Inverses of the functions cot x, sec x, and csc x can be defined in a similar manner. Rather than discuss them in detail, we shall show their graphs (Figs. 5.9–5.11) and list a few basic relations.

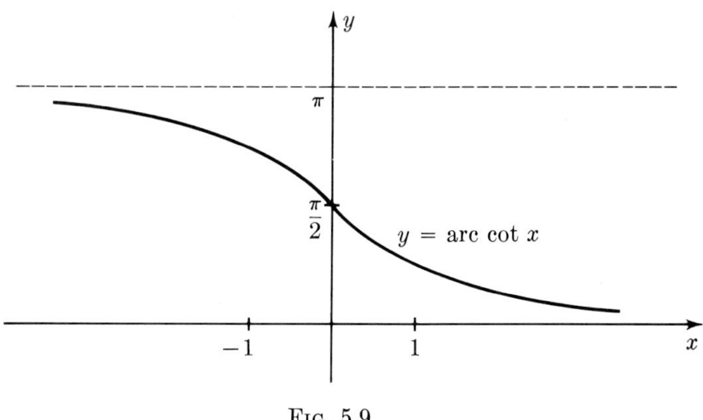

FIG. 5.9

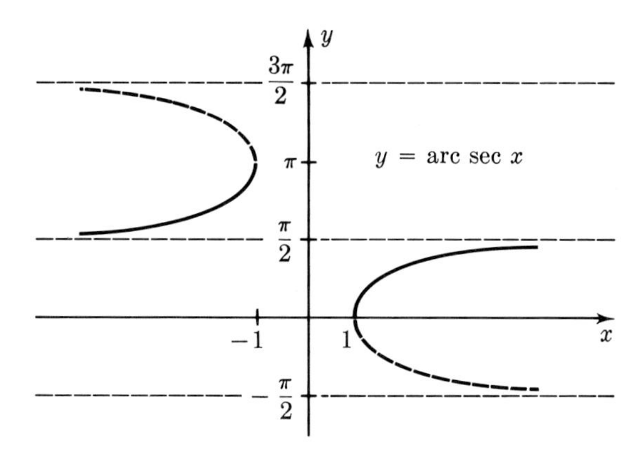

FIG. 5.10 $0 \leq \text{arc sec } x < \pi/2$ if $x \geq 1$; $\pi/2 < \text{arc sec } x \leq \pi$ if $x \leq -1$.

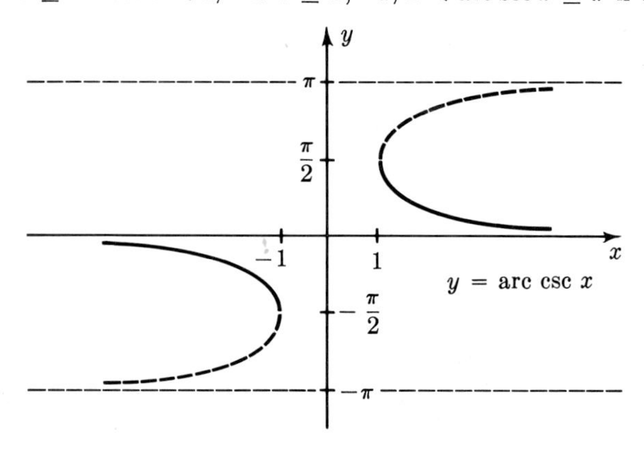

FIG. 5.11 $0 < \text{arc csc } x \leq \pi/2$ if $x \geq 1$; $-\pi/2 \leq \text{arc csc } x < 0$ if $x \leq -1$.

The following relations hold:

$$\arctan x + \operatorname{arc}\cot x = \frac{\pi}{2} \qquad \operatorname{arc}\sec x = \arccos\left(\frac{1}{x}\right)$$

$$\operatorname{arc}\sec x + \operatorname{arc}\csc x = \frac{\pi}{2} \qquad \operatorname{arc}\csc x = \arcsin\left(\frac{1}{x}\right).$$

Notation

The following alternate notation is very common for inverse trigonometric functions:

$$\arcsin x = \sin^{-1} x, \qquad \arctan x = \tan^{-1} x, \qquad \text{etc.}$$

Warning: Do not confuse

$$\sin^{-1} x \qquad \text{with} \qquad \frac{1}{\sin x}.$$

The notation $\sin^{-1} x$ is a bit awkward because we do write $\sin^n x$ for $(\sin x)^n$ when $n > 0$.

EXERCISES

Evaluate:

1. $\arcsin\left(\frac{1}{\sqrt{2}}\right)$
2. $\arccos\left(-\frac{1}{\sqrt{2}}\right)$
3. $\arctan \sqrt{3}$
4. $\operatorname{arc}\sec 1$
5. $\arcsin\frac{1}{2} - \arcsin\left(-\frac{\sqrt{3}}{2}\right)$
6. $\cos[\arccos(0.35)]$
7. $\cos\left(\arcsin\frac{3}{5}\right)$
8. $\arctan\left(\tan\frac{\pi}{4}\right)$
9. $\arctan(0.62) + \operatorname{arc}\cot(0.62)$
10. $\arccos\left(\frac{1}{2}\right) - \arccos\left(-\frac{1}{2}\right).$
11. Show that $2 \arccos x = \arccos(2x^2 - 1)$ if $0 < x < 1$.
    (*Hint:* Use a double-angle formula.)
12. Show that $\arctan a + \arctan b = \arctan\frac{a + b}{1 - ab}$ if $|a| < 1$ and $|b| < 1$.
13. Show that $\arctan\frac{1}{2} + \arctan\frac{1}{3} = \frac{\pi}{4}.$
14. Show that $\arctan x + \arctan\frac{1}{x} = \frac{\pi}{2} \qquad (x > 0).$
15. Explain how the graphs in Figs. 5.9, 5.10, and 5.11 are obtained.

6. DERIVATIVES OF INVERSE FUNCTIONS

The derivatives of the inverse trigonometric functions are:

$$\frac{d}{dx}(\text{arc sin } x) = \frac{1}{\sqrt{1-x^2}} \qquad \frac{d}{dx}(\text{arc tan } x) = \frac{1}{1+x^2}$$

$$\frac{d}{dx}(\text{arc cos } x) = \frac{-1}{\sqrt{1-x^2}} \qquad \frac{d}{dx}(\text{arc cot } x) = \frac{-1}{1+x^2}$$

$$\frac{d}{dx}(\text{arc sec } x) = \begin{cases} \dfrac{1}{x\sqrt{x^2-1}}, & x > 1 \\ \dfrac{-1}{x\sqrt{x^2-1}}, & x < -1 \end{cases}$$

$$\frac{d}{dx}(\text{arc csc } x) = \begin{cases} \dfrac{-1}{x\sqrt{x^2-1}}, & x > 1 \\ \dfrac{1}{x\sqrt{x^2-1}}, & x < -1. \end{cases}$$

We shall justify the first three formulas; these are by far the most important.

First consider arc sin x. If $y = \text{arc sin } x$, then $x = \sin y$. By the rule for differentiation of inverse functions,

$$\frac{dy}{dx} = \frac{1}{\dfrac{dx}{dy}} = \frac{1}{\dfrac{d}{dy}(\sin y)} = \frac{1}{\cos y}, \qquad \cos y \neq 0.$$

But

$$\cos y = \pm\sqrt{1-\sin^2 y} = \pm\sqrt{1-x^2}.$$

However, $\cos y > 0$ since y is between $-\pi/2$ and $\pi/2$ (by definition of arc sin x). Hence, the positive square root is the correct one. Therefore

$$\frac{d}{dx}(\text{arc sin } x) = \frac{1}{\cos y} = \frac{1}{\sqrt{1-x^2}}, \qquad -1 < x < 1.$$

This formula is not valid at ± 1 because there the denominator is zero. In fact, the function $y = \text{arc sin } x$ has no derivative at $x = +1$ and $x = -1$; its graph has vertical tangents at these values of x.

The formula for the derivative of arc cos x follows immediately. Since

$$\text{arc cos } x = \frac{\pi}{2} - \text{arc sin } x,$$

$$\frac{d}{dx}(\text{arc cos } x) = -\frac{d}{dx}(\text{arc sin } x) = \frac{-1}{\sqrt{1-x^2}}.$$

Next, consider arc tan x. If $y = \text{arc tan } x$, then $x = \tan y$.

$$\frac{dy}{dx} = \frac{1}{\frac{dx}{dy}} = \frac{1}{\sec^2 y} = \frac{1}{1 + \tan^2 y} = \frac{1}{1 + x^2}.$$

Verification of the remaining differentiation formulas is left as an exercise.

EXERCISES

Differentiate:

1. $\text{arc sin } \frac{x}{3}$
2. $\text{arc cos } 2x$
3. $\text{arc tan}(x^2)$
4. $x \text{ arc sin}(2x + 1)$
5. $(\text{arc sin } 3x)^2$
6. $\text{arc tan } \sqrt{x}$
7. $\text{arc cot } \frac{1}{x}$
8. $\text{arc sin } \frac{x}{x + 3}$
9. $\text{arc tan } \frac{x - 1}{x + 1}$
10. $\text{arc tan } \frac{1}{x} + \text{arc cot } x$
11. $2x \text{ arc tan } 2x - \ln \sqrt{1 + 4x^2}$
12. $x \text{ arc cot } x + \ln \sqrt{1 + x^2}$.
13. Verify the formulas given in the text for the derivatives of arc cot x, arc sec x, and arc csc x.

7. APPLICATIONS OF INVERSE TRIGONOMETRIC FUNCTIONS

EXAMPLE 7.1

Find the area under the curves

$$\text{(a)} \quad y = \frac{1}{\sqrt{1 - x^2}}, \quad \text{from } 0 \text{ to } \frac{1}{2},$$

$$\text{(b)} \quad y = \frac{1}{1 + x^2}, \quad \text{from } 0 \text{ to } 1.$$

Solution: From the differentiation formulas for inverse trigonometric functions, we see that

$$\frac{d}{dx}(\text{arc sin } x) = \frac{1}{\sqrt{1 - x^2}}, \qquad -1 < x < 1,$$

$$\frac{d}{dx}(\text{arc tan } x) = \frac{1}{1 + x^2}.$$

Thus,

$$\text{arc sin } x \quad \text{is an antiderivative of} \quad \frac{1}{\sqrt{1-x^2}},$$

$$\text{arc tan } x \quad \text{is an antiderivative of} \quad \frac{1}{1+x^2}.$$

Therefore

$$\int_0^{1/2} \frac{dx}{\sqrt{1-x^2}} = \text{arc sin } x \Big|_0^{1/2} = \frac{\pi}{6} - 0,$$

and

$$\int_0^1 \frac{dx}{1+x^2} = \text{arc tan } x \Big|_0^1 = \frac{\pi}{4} - 0.$$

Answer: (a) $\frac{\pi}{6}$; (b) $\frac{\pi}{4}$.

EXAMPLE 7.2

Show that $\int_0^z \frac{dx}{1+x^2} < \frac{\pi}{2}$ no matter how large z is.

Solution:

$$\int_0^z \frac{dx}{1+x^2} = \text{arc tan } x \Big|_0^z = \text{arc tan } z.$$

By definition, all values of arc tan z are between $-\pi/2$ and $\pi/2$, so the integral is less than $\pi/2$.

REMARK: As $z \longrightarrow \infty$, arc tan $z \longrightarrow \pi/2$. For this reason we write

$$\int_0^\infty \frac{dx}{1+x^2} = \frac{\pi}{2}.$$

In geometric terms, the area under the curve $y = 1/(1+x^2)$ between 0 and z is close to $\pi/2$ when z is large (Fig. 7.1). Furthermore, the larger

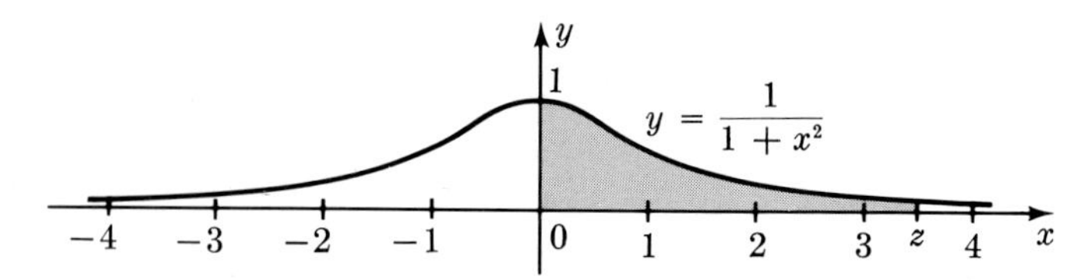

FIG. 7.1 Area of shaded region is nearly $\pi/2$.

z is, the closer the area is to $\pi/2$. Integrals of this type will be studied in Chapter 29.

EXAMPLE 7.3

The circle shown in Fig. 7.2 has radius a ft. As the point P moves to the right at the rate of 7 ft/sec, how fast does the arc BQ increase?

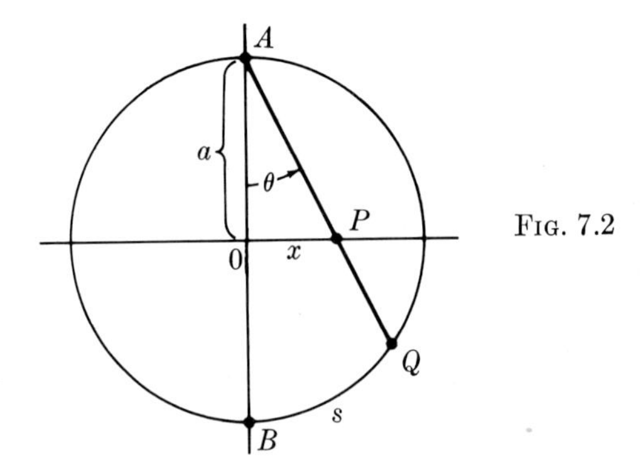

FIG. 7.2

Solution: Express the arc length s in terms of the angle θ. From plane geometry,

$$s = 2a\theta,$$

where θ is measured in radians. From the triangle AOP,

$$\theta = \arc \tan \frac{x}{a}\cdot$$

Hence,

$$s = 2a \arc \tan \frac{x}{a}\cdot$$

Differentiate with respect to time:

$$\begin{aligned} \dot{s} &= 2a \frac{d}{dt}\left(\arc \tan \frac{x}{a}\right) \\ &= 2a \frac{1}{1 + \left(\frac{x}{a}\right)^2}\left(\frac{\dot{x}}{a}\right) = \frac{2a^2}{x^2 + a^2} \cdot 7. \end{aligned}$$

Answer: $\dfrac{14a^2}{x^2 + a^2}$ ft/sec.

EXAMPLE 7.4

The Statue of Liberty is 150 ft tall and stands on a 150-ft pedestal. How far from the base should you stand so you can photograph the statue with largest possible angle? Assume camera level is 5 ft.

Solution: Draw a diagram, labelling the various distances and angles as indicated (Fig. 7.3). The problem is to choose x in such a way that the

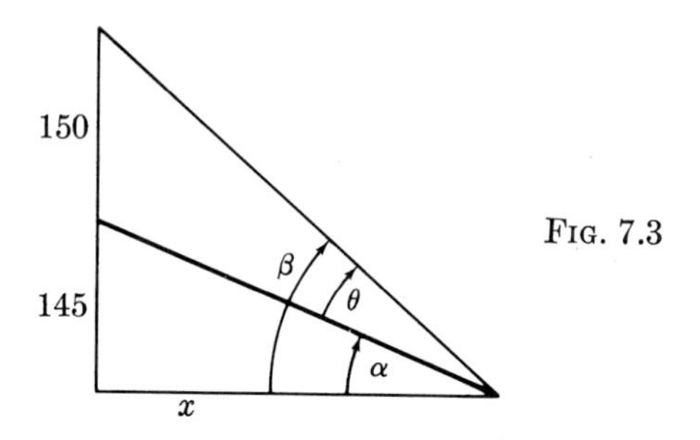

FIG. 7.3

angle θ is greatest. If x is very small or very large, θ will be small. A reasonable guess at the optimal value of x is 200–300 ft.

Express θ as a function of x. From Fig. 7.3,

$$\theta = \beta - \alpha,$$

$$\cot \alpha = \frac{x}{145}, \qquad \cot \beta = \frac{x}{295}.$$

Hence

$$\theta = \text{arc cot}\,\frac{x}{295} - \text{arc cot}\,\frac{x}{145}.$$

This is the function of x to be maximized. The range of x is $x > 0$; there are no end points.

Differentiate:

$$\frac{d\theta}{dx} = \frac{-\dfrac{1}{295}}{1 + \left(\dfrac{x}{295}\right)^2} + \frac{\dfrac{1}{145}}{1 + \left(\dfrac{x}{145}\right)^2} = \frac{145}{(145)^2 + x^2} - \frac{295}{(295)^2 + x^2}.$$

Therefore

$$\frac{d\theta}{dx} = 0 \quad \text{if} \quad \frac{145}{(145)^2 + x^2} = \frac{295}{(295)^2 + x^2}.$$

Solve for x:

$$\begin{aligned}
(145)\,(295)^2 + 145x^2 &= (295)\,(145)^2 + 295x^2, \\
x^2(295 - 145) &= (145)\,(295)\,(295 - 145), \\
x^2 &= (145)\,(295).
\end{aligned}$$

The only positive root of this equation is

$$x = \sqrt{(145)(295)} = 5\sqrt{1711} \approx 5(41.36) = 206.8.$$

Answer: Approximately 206.8 ft.

EXERCISES

1. A large picture hangs on the wall with its top T ft above eye level, and its bottom B ft above eye level. How far should you stand from the wall so that the angle at your eye subtended by the picture is as large as possible?
2. A balloon is released from eye level and rises 10 ft/sec. According to an observer 100 ft from the point of release, how fast is the balloon's elevation angle increasing 4 sec later?
3. Compute $\displaystyle\int_0^{\sqrt{3}/2} \frac{dx}{\sqrt{1-x^2}}$.
4. Show that $\displaystyle\int_0^1 \frac{dx}{1+x^2} = 3\int_1^{\sqrt{3}} \frac{dx}{1+x^2}$.
5. Suppose the circle shown in Fig. 7.4 has radius 10 ft. Suppose also that x moves to the right in such a way that the length of the chord $\overline{AB}$ increases at the rate of 1 in./sec. How fast is the angle θ increasing when $\overline{AB}$ is 15 ft?

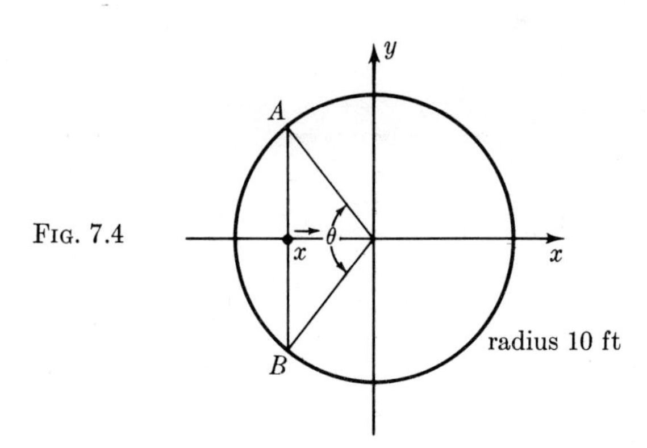

FIG. 7.4

8. HYPERBOLIC FUNCTIONS

The hyperbolic functions are certain combinations of exponential functions, with properties similar to those of the trigonometric functions. They are useful in solving differential equations and in evaluating integrals.

The three basic hyperbolic functions are the **hyperbolic sine,** the **hyperbolic cosine,** and the **hyperbolic tangent:**

$$\sinh x = \frac{e^x - e^{-x}}{2}, \qquad \cosh x = \frac{e^x + e^{-x}}{2},$$

$$\tanh x = \frac{\sinh x}{\cosh x} = \frac{e^x - e^{-x}}{e^x + e^{-x}}.$$

These functions are defined for all values of x. Their graphs are shown in

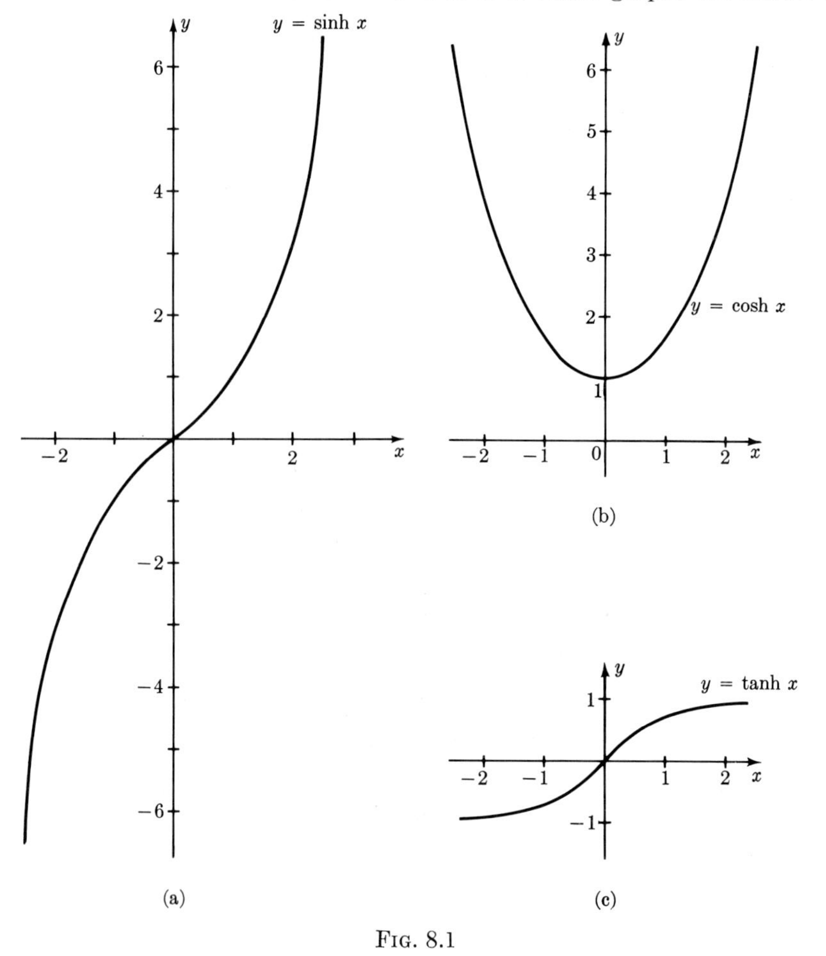

FIG. 8.1

Fig. 8.1. The less useful **hyperbolic cotangent, hyperbolic secant,** and **hyperbolic cosecant** are defined by

$$\coth x = \frac{1}{\tanh x}, \qquad \operatorname{sech} x = \frac{1}{\cosh x}, \qquad \operatorname{csch} x = \frac{1}{\sinh x}.$$

There are numerical tables for all of the hyperbolic functions.

From the definitions,

$$\sinh(-x) = -\sinh x, \qquad \cosh(-x) = \cosh x, \qquad \tanh(-x) = -\tanh x.$$

Thus $\sinh x$ and $\tanh x$ are odd functions, while $\cosh x$ is an even function.

Next,

$$\cosh^2 x = \left(\frac{e^x + e^{-x}}{2}\right)^2 = \frac{1}{4}(e^{2x} + 2 + e^{-2x}),$$

$$\sinh^2 x = \left(\frac{e^x - e^{-x}}{2}\right)^2 = \frac{1}{4}(e^{2x} - 2 + e^{-2x}).$$

Hence

$$\boxed{\cosh^2 x - \sinh^2 x = 1.}$$

It follows easily that

$$\tanh^2 x + \operatorname{sech}^2 x = 1, \qquad \coth^2 x - \operatorname{csch}^2 x = 1.$$

Note the similarity to trigonometric identities, except for sign. Virtually every identity involving trigonometric functions has an analogue involving hyperbolic functions. (But you must be careful with signs!) For example,

$$\cosh(u + v) = \cosh u \cosh v + \sinh u \sinh v,$$

$$\tanh 2x = \frac{2 \tanh x}{1 + \tanh^2 x}.$$

Each such identity can be proved by replacing the hyperbolic functions which occur with the corresponding combinations of exponential functions.

Inverse Hyperbolic Functions

The function $\sinh x$ increases steadily, and takes on each real value. Hence $\sinh x$ has an inverse function, written

$$\sinh^{-1} x \qquad \text{or} \qquad \arg\sinh x,$$

such that

$$\text{if} \quad x = \sinh y, \qquad \text{then} \quad y = \sinh^{-1} x.$$

Warning: Do not confuse

$$\sinh^{-1} x \qquad \text{with} \qquad \frac{1}{\sinh x}.$$

The graph of $y = \sinh^{-1} x$ is shown in Fig. 8.2. Since $\sinh x$ is given in terms of the exponential function, it stands to reason that $\sinh^{-1} x$ can be

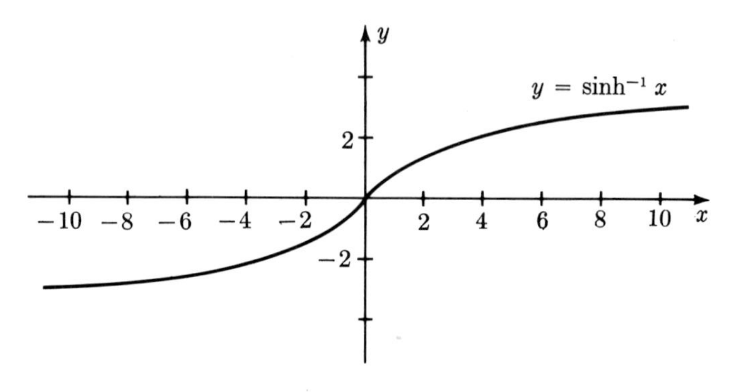

FIG. 8.2

expressed in terms of the logarithm. Indeed, if

$$y = \sinh^{-1} x,$$

then

$$x = \sinh y = \frac{e^y - e^{-y}}{2}.$$

Hence

$$e^y - 2x - e^{-y} = 0,$$

$$e^{2y} - 2xe^y - 1 = 0.$$

This is a quadratic for e^y; solve it:

$$e^y = x \pm \sqrt{x^2 + 1}.$$

Since $e^y > 0$, the plus sign must be chosen,

$$e^y = x + \sqrt{x^2 + 1},$$

$$y = \ln(x + \sqrt{x^2 + 1}).$$

The final result is

$$\boxed{\sinh^{-1} x = \ln(x + \sqrt{x^2 + 1}).}$$

The inverse hyperbolic cosine is defined only for $x \geq 1$. By a similar argument,

$$\boxed{\cosh^{-1} x = \ln (x + \sqrt{x^2 - 1}), \qquad x \geq 1.}$$

The inverse hyperbolic tangent is defined for $-1 < x < 1$ and can be expressed by the formula

$$\boxed{\tanh^{-1} x = \frac{1}{2} \ln \left(\frac{1 + x}{1 - x}\right), \qquad -1 < x < 1.}$$

Graphs of $\cosh^{-1} x$ and $\tanh^{-1} x$ are shown in Fig. 8.3.

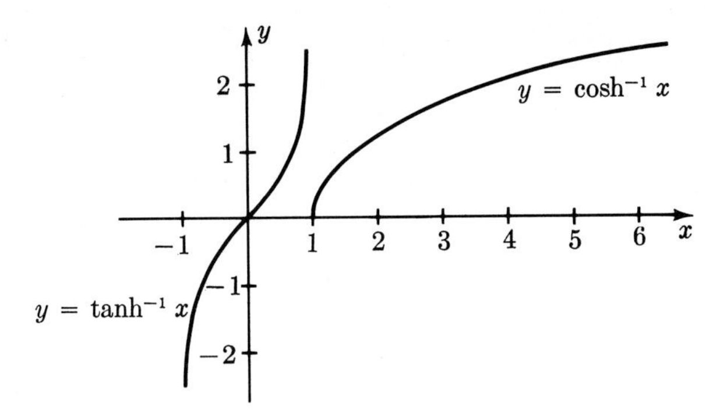

FIG. 8.3

Derivatives

Derivatives of the hyperbolic functions are easily obtained. For example,

$$\frac{d}{dx}(\sinh x) = \frac{d}{dx}\left(\frac{e^x - e^{-x}}{2}\right) = \frac{e^x + e^{-x}}{2} = \cosh x.$$

Similarly,

$$\begin{array}{ll} \dfrac{d}{dx}(\sinh x) = \cosh x, & \dfrac{d}{dx}(\coth x) = -\operatorname{csch}^2 x, \\ \dfrac{d}{dx}(\cosh x) = \sinh x, & \dfrac{d}{dx}(\operatorname{sech} x) = -\operatorname{sech} x \tanh x, \\ \dfrac{d}{dx}(\tanh x) = \operatorname{sech}^2 x, & \dfrac{d}{dx}(\operatorname{csch} x) = -\operatorname{csch} x \coth x. \end{array}$$

Derivatives of the inverse functions are found in the usual way. For example, let $y = \sinh^{-1} x$. Then

$$x = \sinh y, \qquad \frac{dx}{dy} = \cosh y = \sqrt{1 + \sinh^2 y} = \sqrt{1 + x^2},$$

$$\frac{d}{dx}(\sinh^{-1} x) = \frac{1}{dx/dy} = \frac{1}{\sqrt{1 + x^2}}.$$

In this way one obtains

$$\begin{aligned} \frac{d}{dx}(\sinh^{-1} x) &= \frac{1}{\sqrt{1 + x^2}}, \\ \frac{d}{dx}(\cosh^{-1} x) &= \frac{1}{\sqrt{x^2 - 1}}, \\ \frac{d}{dx}(\tanh^{-1} x) &= \frac{1}{1 - x^2}. \end{aligned}$$

EXERCISES

1. Express $7e^{2x} + 4e^{-2x}$ in terms of $\sinh 2x$ and $\cosh 2x$.
2. Prove $\sinh(x + y) = \sinh x \cosh y + \cosh x \sinh y$.
3. Express $\cosh 2x$ in terms of
   (a) $\cosh x$, (b) $\sinh x$.
4. Verify that $(\cosh x + \sinh x)^n = \cosh nx + \sinh nx$.
5. Verify $\tanh^{-1} x = \dfrac{1}{2} \ln \left(\dfrac{1 + x}{1 - x}\right)$ for $-1 < x < 1$.
6. Find the 77-th derivative of $\sinh x$.

Differentiate:

7. $\sinh 5x$
8. $\cosh \sqrt{x}$
9. $\cosh^2 3x - \sinh^2 3x$
10. $(\tanh x)^3$
11. $\dfrac{1}{3} e^{2x} (2 \cosh x - \sinh x)$
12. $1 + x \sinh^2 x$.
13. Show that $y = a \sinh cx + b \cosh cx$ satisfies the differential equation $d^2y/dx^2 = c^2y$.
14. Describe the behavior as $x \to \infty$:
    (a) $\tanh x$
    (b) $\dfrac{\sinh x}{e^x}$
    (c) $\operatorname{sech} x$
    (d) $\dfrac{\cosh x}{\cosh 2x}$.
15. Prove
    (a) $\dfrac{d}{dx} (\cosh^{-1} x) = \dfrac{1}{\sqrt{x^2 - 1}}$
    (b) $\dfrac{d}{dx} (\tanh^{-1} x) = \dfrac{1}{1 - x^2}$.
16. If the coordinates of a moving point are $x = a \cosh t$ and $y = b \sinh t$, where $a > 0$ and $b > 0$, describe its path.
    (*Hint:* Eliminate t.)

19. Techniques of Integration

1. INDEFINITE INTEGRALS

We have developed rules for differentiation of most functions that arise in practice. The reverse process, antidifferentiation, is much harder. There is no systematic procedure for antidifferentiation, rather, a few important techniques and a large bag of miscellaneous tricks. What is worse, there are functions which are not derivatives of any common function, for example

$$e^{-x^2}, \qquad \frac{\sin x}{x}, \qquad \frac{1}{\sqrt{(1 - x^2)\,(1 - k^2x^2)}}.$$

In this chapter, we discuss some basic techniques of antidifferentiation, a few of the more common tricks, and the use of integral tables.

First, some notation. The symbol

$$\int f(x)\,dx,$$

called the **indefinite integral** of $f(x)$, denotes the most general antiderivative of $f(x)$. For example,

$$\int x^2\,dx = \frac{x^3}{3} + C,$$

$$\int \cos x\,dx = \sin x + C,$$

$$\int \frac{dx}{1 + x^2} = \arctan x + C.$$

To each differentiation formula, there corresponds an indefinite integral formula. For instance, to

$$\frac{d}{dx}(\tan x) = \sec^2 x$$

there corresponds

$$\int \sec^2 x\,dx = \tan x + C.$$

us list some integration formulas we know.

$\int x^n\,dx = \dfrac{x^{n+1}}{n+1} + C \quad (n \neq -1)$	$\int \dfrac{dx}{x} = \ln\lvert x\rvert + C$
$\int e^x\,dx = e^x + C$	$\int \sec x \tan x\,dx = \sec x + C$
$\int \cos x\,dx = \sin x + C$	$\int \csc x \cot x\,dx = -\csc x + C$
$\int \sin x\,dx = -\cos x + C$	$\int \dfrac{dx}{\sqrt{1-x^2}} = \arcsin x + C \quad (\lvert x\rvert < 1)$
$\int \sec^2 x\,dx = \tan x + C$	$\int \dfrac{dx}{1+x^2} = \arctan x + C$
$\int \csc^2 x\,dx = -\cot x + C$	$\int \dfrac{dx}{x\sqrt{x^2-1}} = \operatorname{arc\,sec} x + C \quad (x > 1)$

2. DIFFERENTIALS, SUBSTITUTIONS

We start with a technique that adapts known integration formulas to new situations.

Suppose we want an antiderivative for $\sin^2 x$. Since

$$\int x^2\,dx = \frac{x^3}{3} + C,$$

it is natural to suspect that

$$\int \sin^2 x\,dx = \frac{\sin^3 x}{3} + C.$$

But this is wrong; indeed,

$$\frac{d}{dx}\left[\frac{\sin^3 x}{3}\right] = \sin^2 x \cos x.$$

Therefore a correct formula is

$$\int \sin^2 x \cos x\,dx = \frac{\sin^3 x}{3} + C,$$

which is fine, but it does not help us find an antiderivative of $\sin^2 x$. That problem must await a discussion of trigonometric identities in Section 6.

Similarly,

$$\int (x^2+1)^3\,dx \neq \frac{(x^2+1)^4}{4} + C.$$

In fact,

$$\frac{d}{dx}\left[\frac{(x^2+1)^4}{4}\right] = (x^2+1)^3\cdot 2x, \quad \text{not} \quad (x^2+1)^3.$$

Therefore a correct formula is

$$\int (x^2+1)^3\cdot 2x\,dx = \frac{(x^2+1)^4}{4} + C.$$

But that does not solve the given problem. There is no short cut to finding

$$\int (x^2+1)^3\,dx.$$

We must multiply out and integrate each term:

$$\int (x^2+1)^3\,dx = \int (x^6+3x^4+3x^2+1)\,dx$$

$$= \frac{x^7}{7} + \frac{3x^5}{5} + x^3 + x + C.$$

These examples show that the plausible guess

$$\int [u(x)]^n\,dx = \frac{1}{n+1}[u(x)]^{n+1} + C$$

is wrong. The correct formula, a consequence of the Chain Rule, is

$$\int u^n \frac{du}{dx}\,dx = \frac{u^{n+1}}{n+1} + C, \qquad n \neq -1.$$

Along with u^n, there must be the factor du/dx.

For a similar reason, the formula

$$\int e^x\,dx = e^x + C$$

does not imply

$$\int e^{u(x)}\,dx = e^{u(x)} + C.$$

The correct formula is

$$\int e^{u(x)}\frac{du}{dx}\,dx = e^{u(x)} + C.$$

Along with $e^{u(x)}$, there must be the factor du/dx. For example,

$$\int e^{x^2}\cdot 2x\,dx = e^{x^2} + C.$$

Here $u = x^2$ and $du/dx = 2x$. If the factor $2x$ were missing, we could not integrate.

> Suppose $F(x)$ is an antiderivative of $f(x)$. Then
>
> $$\int f[u(x)] \frac{du}{dx}\, dx = F[u(x)] + C.$$

The preceding assertion is just a restatement of the Chain Rule:

$$\frac{d}{dx}[F(u)] = \frac{dF}{du}\frac{du}{dx} = f(u)\frac{du}{dx}\cdot$$

For example, if $f(x) = \sec^2 x$ and $F(x) = \tan x$, then

$$\int \sec^2 u \frac{du}{dx}\, dx = \tan u + C.$$

Each of the formulas listed in the last section gives rise to a more general formula in this way. For instance, from

$$\int \cos x\, dx = \sin x + C$$

follows

$$\int \cos u \frac{du}{dx}\, dx = \sin u + C;$$

from

$$\int \frac{dx}{1 + x^2} = \arctan x + C$$

follows

$$\int \frac{1}{1 + u^2}\frac{du}{dx}\, dx = \arctan u + C.$$

Typical applications of these two new formulas are:

$$\int \cos(3x + 1) \cdot 3\, dx = \sin(3x + 1) + C,$$

$$\int \cos(\sqrt{x}) \frac{1}{2\sqrt{x}}\, dx = \sin(\sqrt{x}) + C,$$

and

$$\int \frac{1}{1 + e^{2x}} e^x\, dx = \arctan(e^x) + C,$$

$$\int \frac{1}{1 + x^4} \cdot 2x\, dx = \arctan(x^2) + C.$$

Differentials

The indefinite integral symbol

$$\int f(x)\,dx$$

denotes the general antiderivative of $f(x)$. Why write the "differential" dx? Because we can introduce a notation that simplifies integration formulas.

For each function $u = u(x)$, we introduce the formal "differential"

$$du = u'(x)\,dx.$$

With this notation we replace

$$\int f[u(x)]\frac{du}{dx}\,dx$$

by

$$\int f(u)\,du.$$

Not only is the notation simpler, but also it reminds us of the important factor du/dx.

Properties of differentials which will be used repeatedly:

$$\boxed{\begin{aligned} d(u+v) &= du + dv; \\ d(cu) &= c\,du, \qquad c \quad \text{a constant;} \\ df(u) &= f'(u)\,du. \end{aligned}}$$

The last property follows from the Chain Rule:

$$df(u) = \frac{d}{dx}[f(u)]\,dx = f'(u)\frac{du}{dx}\,dx = f'(u)\,du.$$

EXAMPLE 2.1

Evaluate $\int (e^x + x)^2(e^x + 1)\,dx$.

Solution: Let $u = e^x + x$. Then

$$du = \frac{du}{dx}\,dx = (e^x + 1)\,dx.$$

Hence,

$$\int (e^x + x)^2(e^x + 1)\,dx = \int u^2\,du$$

$$= \frac{u^3}{3} + C = \frac{(e^x + x)^3}{3} + C.$$

Answer: $\dfrac{(e^x + x)^3}{3} + C.$

EXAMPLE 2.2

Evaluate $\displaystyle\int \frac{\arctan x}{1+x^2}\,dx.$

Solution: Notice that

$$\frac{d}{dx}(\arctan x) = \frac{1}{1+x^2}.$$

If $u = \arctan x$, then

$$du = \frac{du}{dx}\,dx = \frac{1}{1+x^2}\,dx.$$

Hence

$$\int \frac{\arctan x}{1+x^2}\,dx = \int u\,du = \frac{u^2}{2} + C = \frac{(\arctan x)^2}{2} + C.$$

Answer: $\displaystyle\frac{(\arctan x)^2}{2} + C.$

EXAMPLE 2.3

Find $\displaystyle\int \frac{\ln x}{x}\,dx.$

Solution: Set $u = \ln x$, then $du = dx/x$, so

$$\int \frac{\ln x}{x}\,dx = \int u\,du = \frac{u^2}{2} + C = \frac{(\ln x)^2}{2} + C.$$

Answer: $\displaystyle\frac{(\ln x)^2}{2} + C.$

EXERCISES

Find the indefinite integral:

1. $\displaystyle\int 2\cos 2x\,dx$
2. $\displaystyle\int \frac{2x}{1+x^2}\,dx$
3. $\displaystyle\int 2xe^{x^2}\,dx$
4. $\displaystyle\int \frac{\ln^2 x}{x}\,dx$
5. $\displaystyle\int \frac{-2x}{\sqrt{1-x^2}}\,dx$
6. $\displaystyle\int \sin x\cos x\,dx$

7. $\int \sin^3 x \cos x \, dx$

8. $\int (3x^2 + 1)(x^3 + x)^5 \, dx$

9. $\int \frac{2x}{(1 + x^2)^3} \, dx$

10. $\int \frac{3x^2}{4 + x^3} \, dx$

11. $\int \frac{-\sin x}{\cos^2 x} \, dx$

12. $\int (1 + \sin x)^3 \cos x \, dx$

13. $\int 5 \sec^2 5x \, dx$

14. $\int \sec^4 x \tan x \, dx$

15. $\int \frac{e^x + 2x}{e^x + x^2 + 1} \, dx$

16. $\int \sqrt{x - 2} \, dx$

17. $\int 8x \sqrt{4x^2 + 1} \, dx$

18. $\int \frac{e^x - e^{-x}}{e^x + e^{-x}} \, dx$

19. $\int \frac{e^{\sqrt{x}}}{2\sqrt{x}} \, dx$

20. $\int \frac{dx}{x \ln x}.$

3. SUMS, CONSTANT FACTORS

Here are two simple integral formulas:

$$\int [f(x) + g(x)] \, dx = \int f(x) \, dx + \int g(x) \, dx,$$

$$\int cf(x) \, dx = c \int f(x) \, dx.$$

The first formula splits an integration problem into two parts. The second allows constant factors to slide across the integral sign. Both are derived from simple properties of differentiation. How?

Warning: Although constants can slide across the integral sign, variables cannot:

$$x \int x^2 \, dx \neq \int x^3 \, dx.$$

EXAMPLE 3.1

Evaluate $\int \left(e^x + \frac{1}{x^3}\right) dx.$

Solution:

$$\int \left(e^x + \frac{1}{x^3}\right) dx = \int e^x \, dx + \int x^{-3} \, dx.$$

Both integrals on the right are known:

$$\int e^x\,dx = e^x + C,$$

$$\int x^{-3}\,dx = \frac{x^{-2}}{-2} + C' = -\frac{1}{2x^2} + C'.$$

Answer: $e^x - \dfrac{1}{2x^2} + C + C'.$

REMARK: Since both C and C' are arbitrary constants, so is $C + C'$. Therefore, the answer can just as well be written as $e^x - 1/2x^2 + C$.

EXAMPLE 3.2

Evaluate $\displaystyle\int xe^{x^2}\,dx.$

Solution: Set $u = x^2$, then $du = 2x\,dx$. Hence $x\,dx = \frac{1}{2}\,du$, so

$$\int xe^{x^2}\,dx = \int e^{x^2}x\,dx = \int e^u \cdot \frac{1}{2}\,du$$

$$= \frac{1}{2}\int e^u\,du = \frac{1}{2}e^u + C = \frac{1}{2}e^{x^2} + C.$$

Here is a slightly different way to look at the problem. If $u = x^2$, then $du = 2x\,dx$. The integral contains $x\,dx$, which is not quite du; it lacks a factor of 2. Supply this factor by multiplying and dividing by 2:

$$\int xe^{x^2}\,dx = \frac{2}{2}\int xe^{x^2}\,dx$$

$$= \frac{1}{2}\int e^{x^2}\cdot 2x\,dx = \frac{1}{2}\int e^u\,du.$$

Answer: $\dfrac{1}{2}e^{x^2} + C.$

EXAMPLE 3.3

Evaluate $\displaystyle\int \frac{dx}{\sqrt{3x+5}}.$

Solution: Set $u = 3x + 5$. Then $du = 3\,dx$ and

$$\int \frac{dx}{\sqrt{3x+5}} = \int \frac{\frac{1}{3}\,du}{\sqrt{u}} = \frac{1}{3}\int \frac{du}{\sqrt{u}}$$

$$= \frac{1}{3}\int u^{-1/2}\,du = \frac{1}{3}(2u^{1/2}) + C = \frac{2}{3}\sqrt{3x+5} + C.$$

Alternate Solution: Set $u^2 = 3x + 5$. Then $2u\,du = 3\,dx$ and

$$\int \frac{dx}{\sqrt{3x+5}} = \frac{2}{3}\int \frac{u\,du}{u} = \frac{2}{3}u + C$$

$$= \frac{2}{3}\sqrt{3x+5} + C.$$

Answer: $\frac{2}{3}\sqrt{3x+5} + C.$

EXAMPLE 3.4

Find $\displaystyle\int \frac{4x-5}{x^2+1}\,dx.$

Solution:

$$\int \frac{4x-5}{x^2+1}\,dx = 4\int \frac{x\,dx}{x^2+1} - 5\int \frac{dx}{x^2+1}.$$

In the first integral on the right, the numerator is nearly the differential of the denominator. Set $u = x^2 + 1$, so $du = 2x\,dx$, and throw in the needed factor 2:

$$4\int \frac{x\,dx}{x^2+1} = 2\int \frac{2x\,dx}{x^2+1} = 2\int \frac{du}{u}$$

$$= 2\ln|u| + C_1 = 2\ln(x^2+1) + C_1 = \ln(x^2+1)^2 + C_1.$$

The second integral is arc tan $x + C_2$.

Answer: $\ln(x^2+1)^2 - 5 \text{ arc tan } x + C.$

EXAMPLE 3.5

Find $\displaystyle\int e^{-x^2}\,dx.$

DISCUSSION: The only formula that might apply is

$$\int e^u\,du = e^u + C,$$

with $u = -x^2$ and $du = -2x\,dx$. Unfortunately, a factor of x is lacking and there is nothing we can do about it. There is no simple function whose derivative is e^{-x^2}.

EXERCISES

Find:

1. $\int (3x+1)^5\,dx$
2. $\int (x^3+\sin x)\,dx$
3. $\int x(x^2+1)^3\,dx$
4. $\int (e^x+3x)(e^x+3)\,dx$
5. $\int \cos 3x\,dx$
6. $\int \sin(x+2)\,dx$
7. $\int (\sin ax+\cos ax)\,dx$
8. $\int x\sin(x^2)\,dx$
9. $\int e^{3x}\,dx$
10. $\int \frac{1}{2}(e^{2x}+e^{-2x})\,dx$
11. $\int \frac{e^{2\sqrt{x}}}{\sqrt{x}}\,dx$
12. $\int x^2e^{-x^3}\,dx$
13. $\int \frac{dx}{\sqrt{1+5x}}$
14. $\int \frac{\cos x}{(1-\sin x)^2}\,dx$
15. $\int \frac{x^3}{1+x^4}\,dx$
16. $\int \frac{3x-4}{1+x^2}\,dx$
17. $\int \frac{\ln(3x+1)}{3x+1}\,dx$
18. $\int (x+2)\sqrt[3]{x^2+4x+1}\,dx$
19. $\int \frac{dx}{(5-3x)^2}$
20. $\int \sin 5x\cos^2 5x\,dx$
21. $\int \tan^3\frac{x}{2}\sec^2\frac{x}{2}\,dx$
22. $\int \tan^2 3x\,dx.$

4. OTHER SUBSTITUTIONS

Frequently an integral can be simplified by an appropriate substitution.

EXAMPLE 4.1

Evaluate $\int x\sqrt{x+1}\,dx.$

Solution: Set $u^2=x+1$. Then $x=u^2-1$, so $\sqrt{x+1}=u$, and $dx=2u\,du$. Thus

$$\int x\sqrt{x+1}\,dx=\int (u^2-1)\cdot u\cdot 2u\,du=2\int (u^4-u^2)\,du$$

$$=2\left(\frac{u^5}{5}-\frac{u^3}{3}\right)+C=\frac{2u^3}{15}(3u^2-5)+C,$$

where $u = \sqrt{x+1}$. Hence

$$\int x\sqrt{x+1}\,dx = \frac{2(x+1)^{3/2}}{15}[3(x+1)-5] + C.$$

Alternate Solution: Set $u = x + 1$. Then

$$\int x\sqrt{x+1}\,dx = \int (u-1)\sqrt{u}\,du = \int (u^{3/2} - u^{1/2})\,du$$

$$= 2\left(\frac{u^{5/2}}{5} - \frac{u^{3/2}}{3}\right) + C = \frac{2(x+1)^{3/2}}{15}(3x-2) + C$$

as before.

Answer: $\dfrac{2}{15}(x+1)^{3/2}(3x-2) + C.$

EXAMPLE 4.2

Find $\displaystyle\int \frac{x\,dx}{(x-a)^3}$.

Solution: Set $u = x - a$. Then $x = u + a$ and $dx = du$:

$$\int \frac{x\,dx}{(x-a)^3} = \int \frac{(u+a)\,du}{u^3} = \int \left(\frac{1}{u^2} + \frac{a}{u^3}\right) du$$

$$= -\frac{1}{u} - \frac{a}{2u^2} + C = -\frac{1}{x-a} - \frac{a}{2(x-a)^2} + C.$$

Answer: $\dfrac{-2x+a}{2(x-a)^2} + C.$

EXAMPLE 4.3

Evaluate $\displaystyle\int \frac{dx}{1+\sqrt[3]{x}}$.

Solution: Substitute $x = u^3$ and $dx = 3u^2\,du$:

$$\int \frac{dx}{1+\sqrt[3]{x}} = 3\int \frac{u^2\,du}{1+u}.$$

By long division,

$$\frac{u^2}{1+u} = u - 1 + \frac{1}{1+u}.$$

Hence

$$\int \frac{dx}{1+\sqrt[3]{x}} = 3\int \left(u - 1 + \frac{1}{1+u}\right) du = 3\left(\frac{u^2}{2} - u + \ln|1+u|\right) + C$$

$$= 3\left(\frac{1}{2}x^{2/3} - x^{1/3} + \ln|1 + x^{1/3}|\right) + C.$$

Answer:

$$\frac{3}{2}x^{2/3} - 3x^{1/3} + \ln|1 + x^{1/3}|^3 + C.$$

EXAMPLE 4.4

Find $\displaystyle\int \frac{dx}{a^2 + x^2}$.

Solution: We already have the formula

$$\int \frac{dx}{1 + x^2} = \arctan x + C.$$

The given integral is so like this one, we try to change it into this form. We set $x = ay$; then $dx = a\,dy$:

$$\int \frac{dx}{a^2 + x^2} = \int \frac{a\,dy}{a^2 + (ay)^2} = \frac{1}{a}\int \frac{dy}{1 + y^2} = \frac{1}{a}\arctan y + C.$$

Answer: $\displaystyle\frac{1}{a}\arctan\frac{x}{a} + C.$

EXERCISES

Find:

1. $\displaystyle\int x\sqrt{x+3}\,dx$
2. $\displaystyle\int x^2\sqrt{x+3}\,dx$
3. $\displaystyle\int \frac{x}{\sqrt{2x+5}}\,dx$
4. $\displaystyle\int (x^2 + x + 1)\sqrt{x+1}\,dx$
5. $\displaystyle\int \frac{x^2\,dx}{(x-1)^3}$
6. $\displaystyle\int \frac{x^3 - 5}{(x+2)^2}\,dx$
7. $\displaystyle\int \frac{dx}{1+\sqrt{x}}$
8. $\displaystyle\int \frac{x}{1+\sqrt{x}}\,dx$
9. $\displaystyle\int \frac{dx}{x+\sqrt[4]{x}}$
10. $\displaystyle\int \frac{\sqrt{x}+1}{x+3}\,dx$

11. $\displaystyle\int \frac{e^{2x}}{1+e^{4x}}\,dx$

12. $\displaystyle\int e^{2x}\sqrt{1+e^{x}}\,dx$

13. $\displaystyle\int \frac{dx}{1+(5x+2)^2}$

14. $\displaystyle\int \frac{dx}{\sqrt{1-4x^2}}$

15. $\displaystyle\int \frac{dx}{\sqrt{a^2-x^2}},\quad a>0$

16. $\displaystyle\int \frac{x+1}{a^2+b^2x^2}\,dx$

17. $\displaystyle\int \frac{x^2\,dx}{1+3x^2}$ (long division)

18. $\displaystyle\int e^{\cos x}\sin x\,dx$

19. $\displaystyle\int \frac{x^4}{2+x^2}\,dx$

20. $\displaystyle\int \frac{x^3\,dx}{\sqrt{a^2-x^2}}\quad (u^2=a^2-x^2)$

21. $\displaystyle\int \frac{x^3}{x^2+1}\,dx\quad (u=x^2)$

22. $\displaystyle\int x^3\sqrt{x^2+1}\,dx\quad (u^2=x^2+1).$

5. EVALUATION OF DEFINITE INTEGRALS

Indefinite integrals are studied mainly for the purpose of evaluating definite integrals. Recall that

$$\int_a^b f(x)\,dx = F(b)-F(a),$$

where $F(x)$ is an antiderivative of $f(x)$.

We have seen examples in which an indefinite integral

$$\int f(x)\,dx$$

is converted by a substitution $u=u(x)$ into an indefinite integral

$$\int g(u)\,du.$$

The latter integral is evaluated in terms of u, and then u is replaced by its expression in x. For computation of definite integrals, however, it is not necessary to convert back to x.

> Suppose $\int f(x)\,dx$ is converted into $\int g(u)\,du$ by the substitution $u=u(x)$. Then
>
> $$\int_a^b f(x)\,dx = \int_c^d g(u)\,du,$$
>
> where $c=u(a)$ and $d=u(b)$.

Thus once the integral is changed into an integral in u, the computation can be done entirely in terms of u, *provided* the limits of the integral are changed correctly. This rule will be justified shortly.

EXAMPLE 5.1

Compute $\int_0^4 \sqrt{x^2+9}\cdot 2x\,dx.$

Solution: (Old way.) First evaluate the indefinite integral

$$\int \sqrt{x^2+9}\cdot 2x\,dx.$$

Make the substitution $u = x^2 + 9$, $du = 2x\,dx$:

$$\int \sqrt{x^2+9}\cdot 2x\,dx = \int u^{1/2}\,du = \frac{u^{3/2}}{3/2} + C.$$

Now change back to x:

$$\int \sqrt{x^2+9}\cdot 2x\,dx = \frac{2}{3}(x^2+9)^{3/2} + C.$$

Therefore,

$$\int_0^4 \sqrt{x^2+9}\cdot 2x\,dx = \frac{2}{3}(x^2+9)^{3/2}\Big|_0^4 = \frac{2}{3}(5^3-3^3) = \frac{196}{3}.$$

Solution: (New way.) Again substitute $u = x^2 + 9$. Note that

$$u = 25 \quad \text{for} \quad x = 4 \qquad \text{and} \qquad u = 9 \quad \text{for} \quad x = 0.$$

Therefore,

$$\int_0^4 \sqrt{x^2+9}\cdot 2x\,dx = \int_9^{25} u^{1/2}\,du = \frac{2}{3}u^{3/2}\Big|_9^{25} = \frac{2}{3}(5^3-3^3).$$

Alternate Solution: (Avoiding fractional exponents.) Set $u^2 = x^2 + 9$. Then $2u\,du = 2x\,dx$. Now

$$u = 5 \quad \text{for} \quad x = 4 \qquad \text{and} \qquad u = 3 \quad \text{for} \quad x = 0.$$

Therefore,

$$\int_0^4 \sqrt{x^2+9}\cdot 2x\,dx = \int_3^5 u\cdot 2u\,du = \frac{2}{3}u^3\Big|_3^5 = \frac{2}{3}(5^3-3^3).$$

Answer: $\frac{196}{3}$.

EXAMPLE 5.2

Evaluate $\displaystyle\int_{1/2}^{1/\sqrt{2}} \frac{\arcsin x}{\sqrt{1-x^2}}\,dx.$

Solution: Substitute

$$u = \arcsin x, \qquad du = \frac{dx}{\sqrt{1-x^2}},$$

and note that

$$u\left(\frac{1}{2}\right) = \frac{\pi}{6}, \qquad u\left(\frac{1}{\sqrt{2}}\right) = \frac{\pi}{4}.$$

Therefore,

$$\int_{1/2}^{1/\sqrt{2}} \frac{\arcsin x}{\sqrt{1-x^2}}\,dx = \int_{\pi/6}^{\pi/4} u\,du = \frac{u^2}{2}\bigg|_{\pi/6}^{\pi/4} = \frac{1}{2}\left(\frac{\pi^2}{16} - \frac{\pi^2}{36}\right).$$

Answer: $\dfrac{5\pi^2}{288}$.

Justification of Change of Limits

Suppose that a substitution $u = u(x)$ converts

$$f(x)\,dx \qquad \text{into} \qquad g(u)\,du.$$

That means

$$\int f(x)\,dx = \int g[u(x)]\frac{du}{dx}\,dx.$$

In other words

$$f(x) = g[u(x)]\frac{du}{dx}.$$

Let $G(u)$ be an antiderivative of $g(u)$. Then $F(x) = G[u(x)]$ is an antiderivative of $f(x)$:

$$\frac{dF}{dx} = \frac{dG}{du}\frac{du}{dx} = g(u)\frac{du}{dx} = f(x).$$

Therefore,

$$\int_a^b f(x)\,dx = F(b) - F(a);$$

$$\int_c^d g(u)\,du = G(d) - G(c).$$

But if $u(a) = c$ and $u(b) = d$, then

$$F(b) = G[u(b)] = G(d),$$
$$F(a) = G[u(a)] = G(c).$$

Consequently,

$$\int_a^b f(x)\,dx = \int_c^d g(u)\,du.$$

EXERCISES

Compute the definite integral by making an appropriate substitution and changing the limits of integration:

1. $\displaystyle\int_0^2 x^3\sqrt{x^4+9}\,dx$

2. $\displaystyle\int_0^5 x\sqrt{x+4}\,dx$

3. $\displaystyle\int_{-1}^3 xe^{x^2}\,dx$

4. $\displaystyle\int_1^e \frac{\ln x}{x}\,dx$

5. $\displaystyle\int_4^5 \frac{x}{(x-2)^3}\,dx$

6. $\displaystyle\int_1^2 \frac{\sqrt{x^2-1}}{x}\,dx$

7. $\displaystyle\int_{\pi/6}^{\pi/2} \frac{\cos x}{\sqrt{\sin x}}\,dx$

8. $\displaystyle\int_0^{12} \frac{x^2}{\sqrt{2x+1}}\,dx$

9. $\displaystyle\int_0^3 \frac{dx}{1+\sqrt{1+x}}$

10. $\displaystyle\int_{-\ln 2}^{-(1/2)\ln 2} \frac{e^x\,dx}{\sqrt{1-e^{2x}}}$

11. $\displaystyle\int_0^1 \frac{x^3\,dx}{\sqrt{4-x^2}}$

12. $\displaystyle\int_0^2 \frac{dx}{(x+2)\sqrt{x+1}}.$

6. USE OF IDENTITIES

The function under an integral sign is called the **integrand.** Keep in mind the possibility of simplifying the integrand by algebraic manipulation. Such tactics as factoring, combining fractions, and using trigonometric identities may convert a function into an equivalent form which is easier to integrate.

EXAMPLE 6.1

Evaluate $\displaystyle\int \sqrt{\frac{1+x}{1-x}}\,dx.$

Solution: Simplify the integrand by writing

$$\sqrt{\frac{1+x}{1-x}} = \sqrt{\frac{1+x}{1-x}} \cdot \frac{\sqrt{1+x}}{\sqrt{1+x}} = \frac{1+x}{\sqrt{1-x^2}}.$$

Then

$$\int \sqrt{\frac{1+x}{1-x}}\,dx = \int \frac{dx}{\sqrt{1-x^2}} + \int \frac{x\,dx}{\sqrt{1-x^2}}.$$

Both integrals on the right can be evaluated easily.

Answer: $\arcsin x - \sqrt{1-x^2} + C.$

Trigonometric Identities

EXAMPLE 6.2

Find $\int \cos^3 x\,dx.$

Solution: Convert the integrand into powers of $\sin x$, reserving a factor of $\cos x$ for the differential:

$$\cos^3 x = \cos^2 x \cos x = (1 - \sin^2 x) \cos x.$$

Hence

$$\int \cos^3 x\,dx = \int \cos x\,dx - \int \sin^2 x \cos x\,dx$$

$$= \sin x - \frac{\sin^3 x}{3} + C.$$

Answer: $\sin x - \frac{\sin^3 x}{3} + C.$

EXAMPLE 6.3

Find $\int \cos^3 x \sin^2 x\,dx.$

Solution:

$$\cos^3 x \sin^2 x = \cos^2 x \sin^2 x \cos x = (1 - \sin^2 x) \sin^2 x \cos x.$$

$$\int \cos^3 x \sin^2 x\,dx = \int \sin^2 x \cos x\,dx - \int \sin^4 x \cos x\,dx$$

$$= \frac{\sin^3 x}{3} - \frac{\sin^5 x}{5} + C.$$

Answer: $\frac{\sin^3 x}{3} - \frac{\sin^5 x}{5} + C.$

EXAMPLE 6.4

Find $\int \tan x\,dx.$

Solution:

$$\int \tan x\,dx = \int \frac{\sin x}{\cos x}\,dx = \int \frac{-du}{u},$$

where $u = \cos x$.

Answer: $-\ln|\cos x| + C = \ln|\sec x| + C.$

EXAMPLE 6.5

Find $\int \tan^2 x\,dx.$

Solution:

$$\int \tan^2 x\,dx = \int (\sec^2 x - 1)\,dx = \int \sec^2 x\,dx - \int dx.$$

Answer: $\tan x - x + C.$

EXAMPLE 6.6

Find $\int \tan^3 x\,dx.$

Solution:

$$\int \tan^3 x\,dx = \int \tan x\,(\sec^2 x - 1)\,dx$$

$$= \int \tan x \sec^2 x\,dx - \int \tan x\,dx.$$

The first integral is of the form

$$\int u\,du,$$

where $u = \tan x$; the second integral was done in Example 6.4.

Answer: $\dfrac{\tan^2 x}{2} + \ln|\cos x| + C.$

EXAMPLE 6.7

Find $\int \sin^2 x\,dx.$

Solution: Use the identity $\sin^2 x = \frac{1}{2}(1 - \cos 2x)$:

$$\int \sin^2 x\,dx = \frac{1}{2}\int dx - \frac{1}{2}\int \cos 2x\,dx.$$

Answer: $\dfrac{x}{2} - \dfrac{\sin 2x}{4} + C.$

Completing the Square

We have already seen the formulas

$$\int \frac{dx}{a^2 + x^2} = \frac{1}{a} \arctan \frac{x}{a} + C,$$

$$\int \frac{dx}{\sqrt{a^2 - x^2}} = \arcsin \frac{x}{a} + C, \qquad |x| \leq a.$$

Other integrals of this type are:

$$\int \frac{dx}{a^2 - x^2} = \frac{1}{2a} \ln \left| \frac{a + x}{a - x} \right| + C,$$

$$\int \frac{dx}{\sqrt{x^2 - a^2}} = \ln |x + \sqrt{x^2 - a^2}| + C, \qquad |x| \geq a,$$

$$\int \frac{dx}{\sqrt{x^2 + a^2}} = \ln |x + \sqrt{x^2 + a^2}| + C.$$

In later sections we show how these formulas are found. (Their correctness may be checked immediately by differentiating.)

The preceding formulas are useful when the integrand involves a quadratic polynomial or the square root of a quadratic polynomial. The basic trick is completing the square.

EXAMPLE 6.8

Evaluate $\displaystyle\int \frac{dx}{x^2 - 10x + 29}$.

Solution: Complete the square:

$$x^2 - 10x + 29 = x^2 - 10x + 25 + 4 = (x - 5)^2 + 2^2.$$

$$\int \frac{dx}{x^2 - 10x + 29} = \int \frac{dx}{2^2 + (x - 5)^2} = \int \frac{du}{a^2 + u^2}$$

$$= \frac{1}{a} \arctan \frac{u}{a} + C,$$

where $u = x - 5$ and $a = 2$.

Answer: $\displaystyle\frac{1}{2} \arctan \frac{x - 5}{2} + C.$

EXAMPLE 6.9

Evaluate $\displaystyle\int \frac{dx}{\sqrt{3 - x - x^2}}$.

Solution: Complete the square:

$$3 - x - x^2 = 3 - (x^2 + x) = 3 - \left(x^2 + x + \frac{1}{4}\right) + \frac{1}{4}$$

$$= \frac{13}{4} - \left(x + \frac{1}{2}\right)^2.$$

$$\int \frac{dx}{\sqrt{3 - x - x^2}} = \int \frac{dx}{\sqrt{\dfrac{13}{4} - \left(x + \dfrac{1}{2}\right)^2}}$$

$$= \int \frac{du}{\sqrt{a^2 - u^2}} = \text{arc sin}\,\frac{u}{a} + C,$$

where $u = x + \frac{1}{2} = (2x + 1)/2$ and $a = \sqrt{\frac{13}{4}} = \sqrt{13}/2$

Answer: $\text{arc sin}\left(\dfrac{2x + 1}{\sqrt{13}}\right) + C.$

EXAMPLE 6.10

Evaluate $\displaystyle\int \frac{dx}{\sqrt{5x^2 - 2x}}$.

Solution: Complete the square:

$$5x^2 - 2x = 5\left(x^2 - \frac{2}{5}x + \frac{1}{25} - \frac{1}{25}\right) = 5\left[\left(x - \frac{1}{5}\right)^2 - \frac{1}{25}\right].$$

Therefore

$$\int \frac{dx}{\sqrt{5x^2 - 2x}} = \int \frac{dx}{\sqrt{5}\sqrt{\left(x - \dfrac{1}{5}\right)^2 - \dfrac{1}{25}}}$$

$$= \frac{1}{\sqrt{5}} \int \frac{du}{\sqrt{u^2 - a^2}} = \frac{1}{\sqrt{5}} \ln |u + \sqrt{u^2 - a^2}| + C,$$

where $u = x - \frac{1}{5}$ and $a = \frac{1}{5}$. But

$$u + \sqrt{u^2 - a^2} = \frac{1}{5}(5x - 1 + \sqrt{25x^2 - 10x}),$$

hence

$$\int \frac{dx}{\sqrt{5x^2 - 2x}} = \frac{1}{\sqrt{5}} \left(\ln |5x - 1 + \sqrt{25x^2 - 10x}| - \ln 5 \right) + C.$$

Absorb $(\ln 5)/\sqrt{5}$ into C.

Answer: $\frac{1}{\sqrt{5}} \ln |5x - 1 + \sqrt{25x^2 - 10x}| + C.$

EXERCISES

Find:

1. $\int (\sin^2 3x + \cos^2 3x)\, dx$
2. $\int \cos x \csc x\, dx$
3. $\int \cos^3 x \sin^4 x\, dx$
4. $\int (\cos x - \sin x)^2\, dx$
5. $\int \sin^3 x \cos^2 x\, dx$
6. $\int \sin^3 ax\, dx$
7. $\int \cos^5 3x\, dx$
8. $\int \tan^2 x\, dx$
9. $\int \cos^4 x\, dx$
10. $\int \sin^2 \frac{x}{3} \cos^2 \frac{x}{3}\, dx$
11. $\int \tan^4 x\, dx$
12. $\int \tan^5 x\, dx$
13. $\int \sec^n kx \tan kx\, dx$
14. $\int \tan^2 x \sec^4 x\, dx$
15. $\int \frac{dx}{1 - \sin x}$
16. $\int \sin x \sqrt{\frac{\sec x + 1}{\sec x - 1}}\, dx.$

Evaluate the definite integrals:

17. $\int_0^{\pi/2} \sin 2x\, dx$
18. $\int_0^1 \cos^2 \pi x\, dx$
19. $\int_0^{2\pi} \cos 3x \cos 4x\, dx$ $\quad [(\cos A)(\cos B) = \frac{1}{2}(\cos ? + \cos ?).]$
20. $\int_0^{2\pi} \sin x \cos 3x\, dx.$

Find:

21. $\int \frac{dx}{x^2 + 2x + 5}$
22. $\int \frac{dx}{2x^2 + x + 6}$

23. $\displaystyle\int \frac{dx}{\sqrt{6x - x^2}}$

24. $\displaystyle\int \frac{3x + 10}{\sqrt{x^2 + 2x + 5}}\, dx$

25. $\displaystyle\int \frac{x\, dx}{\sqrt{4x - x^2}}$

26. $\displaystyle\int \frac{x^2\, dx}{x^2 - 4x + 9}$ (long division)

27. $\displaystyle\int \frac{x\, dx}{\sqrt{3x^4 - 4x^2 + 1}}$

28. $\displaystyle\int \frac{2x\, dx}{1 - x^2 - x^4}$

29. $\displaystyle\int \frac{dx}{bx - ax^2}$ $a > 0,\ b > 0$

30. $\displaystyle\int \frac{dx}{a^2x^2 + x}$.

7. PARTIAL FRACTIONS

Any fraction of the form

$$\frac{cx + d}{(x - a)(x - b)}$$

can be split into the sum of two simpler fractions

$$\frac{A}{x - a} + \frac{B}{x - b}.$$

This decomposition into **partial fractions** simplifies integration since each term is easy to integrate.

EXAMPLE 7.1

Decompose $\dfrac{2x + 1}{(x - 3)(x - 4)}$ into partial fractions.

Solution: Write

$$\frac{2x + 1}{(x - 3)(x - 4)} = \frac{A}{x - 3} + \frac{B}{x - 4},$$

where A and B are constants to be determined. Multiply through by $(x - 3)(x - 4)$:

$$\begin{aligned} 2x + 1 &= A(x - 4) + B(x - 3) \\ &= (A + B)x - (4A + 3B). \end{aligned}$$

The coefficients of x on both sides of this identity must be equal. Hence

$$A + B = 2.$$

Also, the constants on both sides must be equal. Hence,

$$-4A - 3B = 1.$$

The unknowns A and B must satisfy these two equations simultaneously.

Solve: $A = -7$, $B = 9$. Therefore,

$$\frac{2x+1}{(x-3)(x-4)} = \frac{-7}{x-3} + \frac{9}{x-4}.$$

Alternate Solution: There is a different way to compute A and B. Return to the equation

$$2x + 1 = A(x-4) + B(x-3).$$

This must hold for every value of x, in particular for $x = 3$ and $x = 4$:

$$x = 3:\quad 6 + 1 = A(3-4) + 0,$$
$$A = -7;$$
$$x = 4:\quad 8 + 1 = 0 + B(4-3),$$
$$B = 9.$$

Answer: $\dfrac{2x+1}{(x-3)(x-4)} = \dfrac{-7}{x-3} + \dfrac{9}{x-4}.$

EXAMPLE 7.2

Evaluate $\displaystyle\int \frac{dx}{a^2 - x^2}$.

Solution: Write

$$\frac{1}{a^2 - x^2} = \frac{1}{(a-x)(a+x)}$$
$$= \frac{A}{a-x} + \frac{B}{a+x}.$$

Multiply through by $(a-x)(a+x)$:

$$1 = A(a+x) + B(a-x).$$

Set $x = a$ to obtain $A = 1/2a$; then set $x = -a$ to obtain $B = 1/2a$. Hence

$$\frac{1}{a^2 - x^2} = \frac{1}{2a}\left(\frac{1}{a-x} + \frac{1}{a+x}\right).$$

Therefore,

$$\int \frac{dx}{a^2 - x^2} = \frac{1}{2a}\left(\int \frac{dx}{a-x} + \int \frac{dx}{a+x}\right)$$
$$= \frac{1}{2a}(-\ln|a-x| + \ln|a+x|) + C.$$

Answer: $\dfrac{1}{2a}\ln\left|\dfrac{a+x}{a-x}\right| + C.$

Rational Functions

A **rational function** is the quotient of two polynomials. To integrate a rational function $p(x)/q(x)$, use partial fractions.

If degree $[p(x)] \geq$ degree $[q(x)]$, first divide $p(x)$ by $q(x)$:

$$\frac{p(x)}{q(x)} = r(x) + \frac{s(x)}{q(x)}.$$

Here $r(x)$ is a polynomial and $s(x)$ is a polynomial whose degree is less than that of $q(x)$.

EXAMPLE 7.3

Evaluate $\displaystyle\int \frac{x^3 + 4}{x^2 + x}\,dx.$

Solution: Divide $x^3 + 4$ by $x^2 + x$:

$$\frac{x^3 + 4}{x^2 + x} = x - 1 + \frac{x + 4}{x^2 + x}.$$

Hence

$$\int \frac{x^3 + 4}{x^2 + x}\,dx = \int (x - 1)\,dx + \int \frac{x + 4}{x^2 + x}\,dx$$

$$= \frac{x^2}{2} - x + \int \frac{x + 4}{x^2 + x}\,dx.$$

The problem is now reduced to evaluating the last integral. Write

$$\frac{x + 4}{x^2 + x} = \frac{x + 4}{x(x + 1)} = \frac{A}{x} + \frac{B}{x + 1}.$$

Multiply by $x(x + 1)$:

$$x + 4 = A(x + 1) + Bx.$$

Set $x = 0$ and $x = -1$ to obtain $A = 4$ and $B = -3$. Thus

$$\frac{x + 4}{x^2 + x} = \frac{4}{x} - \frac{3}{x + 1},$$

$$\int \frac{x + 4}{x^2 + x}\,dx = 4\int \frac{dx}{x} - 3\int \frac{dx}{x + 1}$$

$$= 4 \ln |x| - 3 \ln |x + 1| + C.$$

Answer: $\displaystyle\frac{x^2}{2} - x + \ln \left| \frac{x^4}{(x + 1)^3} \right| + C.$

Partial fractions are useful in the integration of rational functions $p(x)/q(x)$ provided the denominator can be completely factored into linear and quadratic factors. In practice, this is hard to do for polynomials of degree 3 or more, except in special cases.

Assume the degree of $q(x)$ exceeds that of $p(x)$, and that $q(x)$ is factored into linear and quadratic factors. Then for each factor $(x - a)$ there is a term:

$$\frac{A}{x-a}.$$

If $(x - a)^2$ occurs, there are two terms:

$$\frac{A_1}{x-a} + \frac{A_2}{(x-a)^2}.$$

If $(x - a)^3$ occurs, there are three terms:

$$\frac{A_1}{x-a} + \frac{A_2}{(x-a)^2} + \frac{A_3}{(x-a)^3}.$$

For each quadratic factor $x^2 + ax + b$ there is a term:

$$\frac{Ax+B}{x^2+ax+b}.$$

If $(x^2 + ax + b)^2$ occurs, there are two terms:

$$\frac{Ax+B}{x^2+ax+b} + \frac{Cx+D}{(x^2+ax+b)^2},$$

and so on. For instance:

$$\frac{1}{(x-a)(x-b)(x-c)} = \frac{A}{x-a} + \frac{B}{x-b} + \frac{C}{x-c},$$

$$\frac{1}{(x-a)^2(x-b)} = \frac{A}{x-a} + \frac{B}{(x-a)^2} + \frac{C}{(x-b)},$$

$$\frac{1}{(x-a)(x^2+bx+c)} = \frac{A}{x-a} + \frac{Bx+C}{x^2+bx+c},$$

$$\frac{1}{(x-a)(x^2+b^2)^2} = \frac{A}{x-a} + \frac{Bx+C}{x^2+b^2} + \frac{Dx+E}{(x^2+b^2)^2},$$

$$\frac{1}{x^4-1} = \frac{1}{(x-1)(x+1)(x^2+1)} = \frac{A}{x-1} + \frac{B}{x+1} + \frac{Cx+D}{x^2+1}.$$

EXAMPLE 7.4

Evaluate $\displaystyle\int \frac{dx}{x^4-1}.$

Solution: Write

$$\frac{1}{x^4 - 1} = \frac{A}{x - 1} + \frac{B}{x + 1} + \frac{Cx + D}{x^2 + 1}.$$

Multiply through by $(x - 1)(x + 1)(x^2 + 1)$:

$$1 = A(x + 1)(x^2 + 1) + B(x - 1)(x^2 + 1) + Cx(x - 1)(x + 1) + D(x - 1)(x + 1).$$

Set $x = 1$ and $x = -1$ to obtain $A = -B = \frac{1}{4}$. Set $x = 0$ to obtain $1 = A - B - D = \frac{1}{4} + \frac{1}{4} - D$. Hence $D = -\frac{1}{2}$. Choose any other value of x to find C. Try $x = 2$, for example:

$$1 = 15A + 5B + 6C + 3D = \frac{15}{4} - \frac{5}{4} + 6C - \frac{3}{2},$$

hence $C = 0$. Therefore

$$\frac{1}{x^4 - 1} = \frac{1}{4}\left(\frac{1}{x - 1}\right) - \frac{1}{4}\left(\frac{1}{x + 1}\right) - \frac{1}{2}\left(\frac{1}{x^2 + 1}\right),$$

$$\int \frac{dx}{x^4 - 1} = \frac{1}{4}\int \frac{dx}{x - 1} - \frac{1}{4}\int \frac{dx}{x + 1} - \frac{1}{2}\int \frac{dx}{x^2 + 1}$$

$$= \frac{1}{4}\ln|x - 1| - \frac{1}{4}\ln|x + 1| - \frac{1}{2}\arctan x + C.$$

Answer: $\ln\left|\frac{x - 1}{x + 1}\right|^{1/4} - \frac{1}{2}\arctan x + C.$

EXAMPLE 7.5

Evaluate $\displaystyle\int \frac{2x + 5}{(x - 1)(x + 3)^2}\,dx.$

Solution: Write

$$\frac{2x + 5}{(x - 1)(x + 3)^2} = \frac{A}{x - 1} + \frac{B}{x + 3} + \frac{C}{(x + 3)^2}.$$

Multiply through by $(x - 1)(x + 3)^2$:

$$2x + 5 = A(x + 3)^2 + B(x - 1)(x + 3) + C(x - 1).$$

Set $x = 1$ to obtain $A = \frac{7}{16}$; set $x = -3$ to obtain $C = \frac{1}{4}$. Choose any other value of x to find B, for example, $x = 0$:

$$5 = 9A - 3B - C = \frac{63}{16} - 3B - \frac{1}{4},$$

from which $B = -\frac{7}{16}$. Therefore,

$$\frac{2x+5}{(x-1)(x+3)^2} = \frac{7}{16}\left(\frac{1}{x-1}\right) - \frac{7}{16}\left(\frac{1}{x+3}\right) + \frac{1}{4}\left(\frac{1}{(x+3)^2}\right),$$

$$\int \frac{2x+5}{(x-1)(x+3)^2}\,dx = \frac{7}{16}\int \frac{dx}{x-1} - \frac{7}{16}\int \frac{dx}{x+3} + \frac{1}{4}\int \frac{dx}{(x+3)^2}.$$

Answer: $\frac{7}{16}\ln\left|\frac{x-1}{x+3}\right| - \frac{1}{4(x+3)} + C.$

EXERCISES

Decompose into partial fractions:

1. $\frac{1}{(x+1)(x-1)}$
2. $\frac{x}{(x+2)(x+3)}$
3. $\frac{x^2}{(x+1)(x-2)}$
4. $\frac{1}{(x+1)(x+2)(x+3)}$
5. $\frac{x}{(x+1)(x+2)(x+3)}$
6. $\frac{1}{(x+1)(x^2+4)}$
7. $\frac{x^4}{(x^2+1)^2}$
8. $\frac{x^3-1}{x(x^2+1)}$
9. $\frac{x+1}{(x-1)(x^2+4)}$
10. $\frac{1}{x(x+1)^2}$.

Find:

11. $\int \frac{dx}{x^2-3x+2}$
12. $\int \frac{dx}{(x-2)(x+4)}$
13. $\int \frac{x+3}{x^2+x}\,dx$
14. $\int \frac{x^2+1}{x^2-5x+6}\,dx$
15. $\int \frac{2x+3}{x^3+x}\,dx$
16. $\int \frac{x\,dx}{(x+1)^2(x-3)}$
17. $\int \frac{dx}{(x-2)^2(x^2+9)}$
18. $\int \frac{dx}{3x^2-13x+4}$
19. $\int \frac{x^4\,dx}{x^3-1}$
20. $\int \frac{x\,dx}{x^4-1}$
21. $\int \frac{x^3\,dx}{x^2+3x+2}$
22. $\int \frac{dx}{(x-1)(x-2)(x-3)}$
23. $\int \frac{x^2+x+1}{(x-3)(x^2+2x+2)}\,dx$
24. $\int \frac{dx}{x(x-3)^2}$
25. $\int \frac{\cos\theta\,d\theta}{\sin^2\theta+7\sin\theta+10}$
26. $\int \frac{dx}{\sqrt{x}\,(1+\sqrt{x})\,(2+\sqrt{x})}$.

8. TRIGONOMETRIC SUBSTITUTIONS

We have not yet integrated $\sec x$ and $\csc x$. The integral of $\sec x$ is done by a trick:

$$\sec x = \frac{\sec x\,(\sec x + \tan x)}{(\sec x + \tan x)}$$

$$= \frac{\sec x \tan x + \sec^2 x}{\sec x + \tan x}$$

$$= \frac{1}{\sec x + \tan x}\,\frac{d}{dx}(\sec x + \tan x).$$

Hence

$$\int \sec x\,dx = \int \frac{d\,(\sec x + \tan x)}{\sec x + \tan x} = \ln|\sec x + \tan x| + C.$$

In a similar manner, we may derive the formula

$$\int \csc x\,dx = -\ln|\csc x + \cot x| + C.$$

Integrals involving $a^2 - x^2$ or $a^2 + x^2$ are often simplified by trigonometric substitutions. The substitution $x = a \sin\theta$ changes $a^2 - x^2$ into $a^2 \cos^2\theta$; the substitution $x = a\tan\theta$ changes $a^2 + x^2$ into $a^2\sec^2\theta$.

EXAMPLE 8.1

Evaluate $\displaystyle\int \frac{dx}{x^2\sqrt{4 - x^2}}$.

Solution: Set $x = 2\sin\theta$. Then

$$\int \frac{dx}{x^2\sqrt{4 - x^2}} = \int \frac{2\cos\theta\,d\theta}{(2\sin\theta)^2\sqrt{4 - 4\sin^2\theta}}$$

$$= \int \frac{2\cos\theta\,d\theta}{4\sin^2\theta \cdot 2\cos\theta} = \frac{1}{4}\int \frac{d\theta}{\sin^2\theta}.$$

Hence

$$\int \frac{dx}{x^2\sqrt{4 - x^2}} = \frac{1}{4}\int \csc^2\theta\,d\theta = -\frac{1}{4}\cot\theta + C.$$

As a final step, express $\cot\theta$ in terms of x. This can be done quickly by drawing a right triangle (Fig. 8.1) that shows $x = 2\sin\theta$. It follows that

$$\cot\theta = \frac{\sqrt{4 - x^2}}{x}.$$

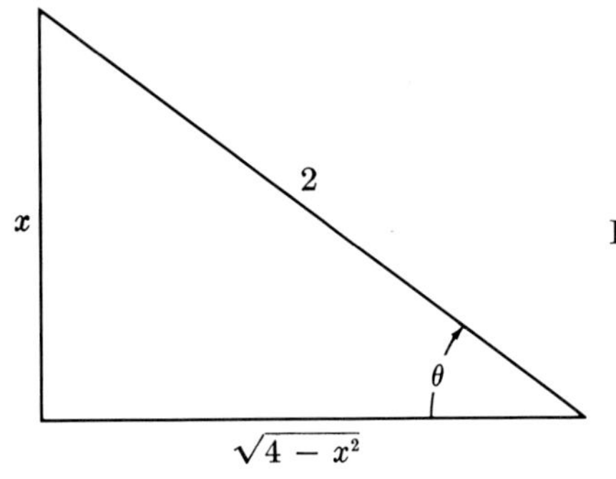

Fig. 8.1

Answer: $\quad -\dfrac{\sqrt{4 - x^2}}{4x} + C.$

EXAMPLE 8.2

Evaluate $\displaystyle\int \frac{dx}{\sqrt{a^2 + x^2}}$.

Solution: Set $x = a \tan \theta$. Then

$$\int \frac{dx}{\sqrt{a^2 + x^2}} = \int \frac{a \sec^2 \theta \, d\theta}{\sqrt{a^2(1 + \tan^2 \theta)}} = \int \frac{a \sec^2 \theta \, d\theta}{a \sec \theta}$$

$$= \int \sec \theta \, d\theta = \ln |\sec \theta + \tan \theta| + C.$$

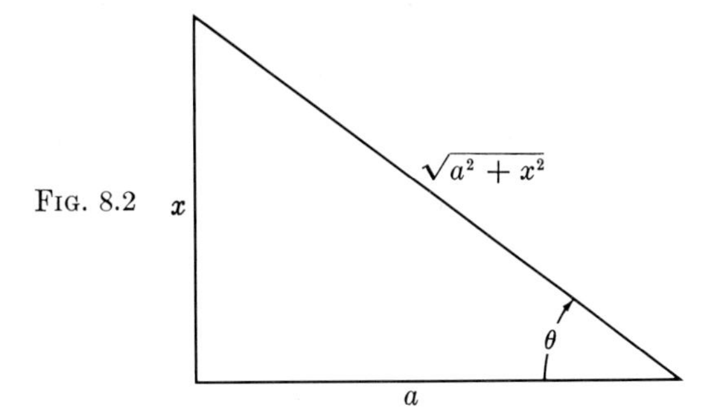

Fig. 8.2

Draw a right triangle (Fig. 8.2) showing $x = a \tan \theta$, and read off $\sec \theta$. Thus

$$\int \frac{dx}{\sqrt{a^2 + x^2}} = \ln \left| \frac{\sqrt{a^2 + x^2}}{a} + \frac{x}{a} \right| + C$$

$$= \ln |\sqrt{a^2 + x^2} + x| - \ln |a| + C.$$

Absorb $-\ln|a|$ into C.

Answer: $\ln|\sqrt{x^2+a^2}+x|+C.$

EXAMPLE 8.3

Compute the definite integral $\displaystyle\int_0^1 \frac{dx}{(1+x^2)^2}\cdot$

Solution: Substitute $x = \tan\theta$. Then $\theta = 0$ for $x = 0$, and $\theta = \pi/4$ for $x = 1$. Hence

$$\int_0^1 \frac{dx}{(1+x^2)^2} = \int_0^{\pi/4} \frac{\sec^2\theta\, d\theta}{(1+\tan^2\theta)^2}$$

$$= \int_0^{\pi/4} \frac{d\theta}{\sec^2\theta} = \int_0^{\pi/4} \cos^2\theta\, d\theta.$$

Use the identity

$$\cos^2\theta = \frac{1}{2}(1+\cos 2\theta):$$

$$\int_0^1 \frac{dx}{(1+x^2)^2} = \int_0^{\pi/4}\left(\frac{1}{2}+\frac{1}{2}\cos 2\theta\right)d\theta = \left(\frac{1}{2}\theta + \frac{1}{4}\sin 2\theta\right)\Bigg|_0^{\pi/4}$$

$$= \left(\frac{1}{2}\cdot\frac{\pi}{4}+\frac{1}{4}\sin\frac{\pi}{2}\right) - \left(\frac{1}{2}\cdot 0 + \frac{1}{4}\sin 0\right)$$

$$= \left(\frac{\pi}{8}+\frac{1}{4}\right) - 0.$$

Answer: $\dfrac{\pi}{8}+\dfrac{1}{4}\cdot$

EXAMPLE 8.4

Evaluate the indefinite integral $\displaystyle\int \frac{dx}{(1+x^2)^2}\cdot$

Solution: From the solution of the last example,

$$\int \frac{dx}{(1+x^2)^2} = \frac{\theta}{2} + \frac{\sin 2\theta}{4} + C, \qquad x = \tan\theta.$$

It remains to express this function of θ in terms of x. Draw a right triangle showing $x = \tan\theta$. See Fig. 8.3. Thus

$$\theta = \arctan x,$$

$$\sin 2\theta = 2\sin\theta\cos\theta = 2\left(\frac{x}{\sqrt{1+x^2}}\right)\left(\frac{1}{\sqrt{1+x^2}}\right)\cdot$$

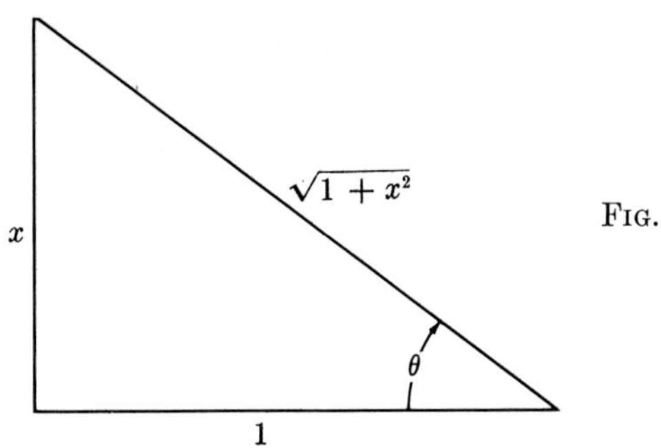

FIG. 8.3

Answer: $\dfrac{1}{2}\arctan x + \dfrac{x}{2(1+x^2)} + C.$

EXERCISES

Use a trigonometric substitution:

1. $\displaystyle\int \frac{dx}{\sqrt{1+x^2}}$
2. $\displaystyle\int x^3\sqrt{1+x^2}\,dx$
3. $\displaystyle\int \frac{x^3\,dx}{\sqrt{4-x^2}}$
4. $\displaystyle\int \frac{x^2\,dx}{\sqrt{a^2-x^2}}$.
5. $\displaystyle\int \frac{dx}{x^2\sqrt{16-x^2}}$
6. $\displaystyle\int \frac{dx}{x\sqrt{x^2+1}}$
7. $\displaystyle\int \frac{\sqrt{x^2+a^2}}{x^2}\,dx$
8. $\displaystyle\int \frac{1+x^2}{1-x^2}\,dx$
9. $\displaystyle\int \frac{x^2\,dx}{(1+x^2)^2}$
10. $\displaystyle\int \frac{\sqrt{x^2-9}}{x}\,dx$

In integrals involving the expression $x^2 + a^2$, the substitution $x = a\sinh\theta$ is often useful. Recall that $\cosh^2 x - \sinh^2 x = 1$ and $\sinh^{-1} x = \ln(x + \sqrt{x^2+1})$. Use this substitution:

11. $\displaystyle\int \frac{dx}{\sqrt{x^2+a^2}}$
12. $\displaystyle\int \sqrt{x^2+a^2}\,dx$
13. $\displaystyle\int \frac{x^2\,dx}{\sqrt{x^2+a^2}}$
14. $\displaystyle\int \frac{\sqrt{x^2+a^2}}{x^2}\,dx.$

9. INTEGRATION BY PARTS

In this section, we discuss a technique that converts an integration problem into a different integration problem, hopefully an easier one. The technique depends on the Product Rule:

$$\frac{d}{dx}(uv) = u\frac{dv}{dx} + v\frac{du}{dx},$$

where u and v are functions of x.

This rule can be expressed in terms of differentials. Multiply through formally by the symbol dx:

$$\frac{d}{dx}(uv)\,dx = u\frac{dv}{dx}\,dx + v\frac{du}{dx}\,dx.$$

The term on the left is $d(uv)$. On the right we have

$$\frac{dv}{dx}\,dx = dv, \qquad \frac{du}{dx}\,dx = du.$$

Product Rule in Differential Form

$$d(uv) = u\,dv + v\,du.$$

Rearranging terms in this rule, we have

$$u\,dv = d(uv) - v\,du,$$

and consequently

$$\int u\,dv = \int d(uv) - \int v\,du.$$

But

$$\int d(uv) = uv + C.$$

(Why?) Thus we obtain the formula for integration by parts:

Integration by Parts

$$\int u\,dv = uv - \int v\,du.$$

(The constant is absorbed into the second integral.)

This formula converts the problem of integrating $u\,dv$ into that of integrating $v\,du$. With luck, the second integration is easier. There is no guarantee, however, that it need be.

EXAMPLE 9.1

Evaluate $\int x \cos x\,dx$.

Solution: Interpret the integral as

$$\int x\,d(\sin x).$$

Apply the formula for integration by parts with

$$u = x, \qquad du = dx; \qquad dv = \cos x \, dx, \qquad v = \sin x,$$

to obtain

$$\int x \cos x \, dx = uv - \int v \, du$$

$$= x \sin x - \int \sin x \, dx.$$

This is a bit of luck; the integral on the right is easy. Conclusion:

$$\int x \cos x \, dx = x \sin x + \cos x + C.$$

Answer: $x \sin x + \cos x + C.$

EXAMPLE 9.2

Evaluate $\int xe^x \, dx.$

Solution: Use integration by parts with

$$u = x, \qquad du = dx; \qquad dv = e^x \, dx, \qquad v = e^x.$$

Therefore

$$\int xe^x \, dx = xe^x - \int e^x \, dx$$

$$= xe^x - e^x + C.$$

Answer: $e^x(x - 1) + C.$

REMARK: It is possible to choose u and v differently:

$$\int xe^x \, dx = \int u \, dv,$$

where

$$u = e^x, \qquad du = e^x \, dx; \qquad dv = x \, dx, \qquad v = \frac{x^2}{2}.$$

Then integration by parts yields

$$\int xe^x \, dx = e^x \cdot \frac{x^2}{2} - \int \frac{x^2}{2} e^x \, dx,$$

which is true but does not help; the integral on the right is harder than the given one. Thus, it may be crucial how you factor the integrand into u and dv. It is also possible that no choice of u and dv helps.

Here are some further examples of integration by parts.

EXAMPLE 9.3

Evaluate $\int \arcsin x\, dx$.

Solution: Use integration by parts with

$$u = \arcsin x, \qquad du = \frac{dx}{\sqrt{1-x^2}}; \qquad dv = dx, \qquad v = x.$$

Therefore

$$\int \arcsin x\, dx = uv - \int v\, du = x \arcsin x - \int \frac{x\, dx}{\sqrt{1-x^2}}.$$

In the integral on the right, substitute $y^2 = 1 - x^2$, $y\, dy = -x\, dx$:

$$-\int \frac{x\, dx}{\sqrt{1-x^2}} = \int \frac{y\, dy}{y} = \int dy = y + C = \sqrt{1-x^2} + C.$$

Answer: $x \arcsin x + \sqrt{1-x^2} + C$.

NOTE: "y" was used in the substitution step because "u" was used in the first step. Always take care not to confuse variables.

EXAMPLE 9.4

Compute $\int_1^2 \ln x\, dx$.

Solution: Use integration by parts with

$$u = \ln x, \qquad du = \frac{dx}{x}; \qquad dv = dx, \qquad v = x.$$

Therefore

$$\int_1^2 \ln x\, dx = uv\Big|_1^2 - \int_1^2 v\, du = x \ln x\Big|_1^2 - \int_1^2 x \cdot \frac{dx}{x}$$

$$= (2 \ln 2 - 1 \ln 1) - x\Big|_1^2.$$

Answer: $2 \ln 2 - 1$.

Repeated Integration by Parts

Some problems require two or more integrations by parts.

EXAMPLE 9.5

Evaluate $\int x (\ln x)^2\, dx$.

Solution: Integrate by parts with

$$u = (\ln x)^2, \qquad du = \frac{2 \ln x \, dx}{x}; \qquad dv = x\,dx, \qquad v = \frac{x^2}{2}.$$

Therefore

$$\int x (\ln x)^2 \, dx = \frac{x^2 (\ln x)^2}{2} - \int \frac{x^2}{2} \cdot \frac{2 \ln x}{x} \, dx$$

$$= \frac{x^2 (\ln x)^2}{2} - \int x \ln x \, dx.$$

The problem now is to evaluate

$$\int x \ln x \, dx,$$

which is similar to the original integral except that $\ln x$ appears only to the first power. Therefore another integration by parts should reduce the integral to

$$\int x \, dx.$$

Try it. Integrate by parts again with

$$u = \ln x, \qquad du = \frac{dx}{x}; \qquad dv = x\,dx, \qquad v = \frac{x^2}{2}.$$

Therefore

$$\int x \ln x \, dx = \frac{x^2 \ln x}{2} - \int \frac{x^2}{2} \cdot \frac{dx}{x}$$

$$= \frac{x^2 \ln x}{2} - \frac{1}{2} \int x \, dx$$

$$= \frac{x^2 \ln x}{2} - \frac{x^2}{4} + C.$$

Combine the results:

$$\int x(\ln x)^2 \, dx = \frac{x^2(\ln x)^2}{2} - \left(\frac{x^2 \ln x}{2} - \frac{x^2}{4} + C\right).$$

Answer: $\dfrac{x^2}{4} [2(\ln x)^2 - 2 \ln x + 1] - C.$

EXAMPLE 9.6

Evaluate $\int x^3 e^x \, dx.$

Solution: Integrate by parts three times:

$$\int x^3e^x\,dx = x^3e^x - 3\int x^2e^x\,dx,$$

$$\int x^2e^x\,dx = x^2e^x - 2\int xe^x\,dx,$$

$$\int xe^x\,dx = xe^x - \int e^x\,dx = xe^x - e^x + C.$$

Combine the results:

$$\int x^3e^x\,dx = x^3e^x - 3x^2e^x + 6xe^x - 6e^x + C.$$

Answer: $e^x(x^3 - 3x^2 + 6x - 6) + C.$

EXAMPLE 9.7

Evaluate $\int e^x \cos x\,dx.$

Solution: Denote the integral by J. Integrate by parts with

$$u = e^x, \qquad du = e^x\,dx; \qquad dv = \cos x\,dx, \qquad v = \sin x.$$

Therefore

$$J = e^x \sin x - \int e^x \sin x\,dx.$$

Integrate by parts again with

$$u = e^x, \qquad du = e^x\,dx; \qquad dv = \sin x\,dx, \qquad v = -\cos x.$$

Therefore

$$\int e^x \sin x\,dx = -e^x \cos x + \int e^x \cos x\,dx.$$

The integral on the right is J again! Hence

$$\int e^x \sin x\,dx = -e^x \cos x + J.$$

Have we gone in a circle? No, because substitution of this expression in the first equation for J yields

$$J = e^x \sin x + e^x \cos x - J.$$

The minus sign on the right saves us from disaster. Solve for J:

$$J = \frac{1}{2}(e^x \sin x + e^x \cos x).$$

Don't forget the constant of integration.

Answer: $\frac{1}{2}e^x(\sin x + \cos x) + C.$

EXERCISES

Find:

1. $\int x \sin x\, dx$
2. $\int x^2 \cos 3x\, dx$
3. $\int xe^{2x}\, dx$
4. $\int x \sec^2 x\, dx$
5. $\int \sqrt{x} \ln x\, dx$
6. $\int \text{arc cos } x\, dx$
7. $\int \text{arc tan } x\, dx$
8. $\int \ln(x^2 + 1)\, dx$
9. $\int x \text{ arc tan } x\, dx$
10. $\int x^3 e^{x^2}\, dx$
11. $\int e^{2x} \sin 3x\, dx$
12. $\int x^2 e^{-x}\, dx$
13. $\int x \cosh x\, dx$
14. $\int x^2 \sinh 3x\, dx$
15. $\int x^2 \cos ax\, dx$
16. $\int \sec^3 x\, dx.$

Compute:

17. $\int_\pi^{2\pi} x \cos x\, dx$
18. $\int_1^e (\ln x)^2\, dx$
19. $\int_1^2 x^2 \ln x\, dx$
20. $\int_0^{1/2} \text{arc sin } x\, dx$
21. $\int_0^1 e^x(x + 3)^2\, dx$
22. $\int_0^{\pi/2} e^{2x} \sin x\, dx.$

10. REDUCTION FORMULAS

The integral

$$\int x^2 e^x\, dx$$

requires two integrations by parts. Each integration lowers the power of x by one until x disappears. In the same way

$$\int x^3 e^x\, dx$$

requires three integrations by parts, and

$$\int x^4 e^x\, dx$$

four integrations by parts. It is convenient to have a **reduction formula,** a formula that reduces

$$\int x^n e^x\,dx \qquad \text{to} \qquad \int x^{n-1}e^x\,dx.$$

Repeated use of such a formula reduces

$$\int x^n e^x\,dx \qquad \text{to} \qquad \int e^x\,dx.$$

EXAMPLE 10.1

Derive a reduction formula for $\int x^n e^x\,dx$.

Solution: Integrate by parts with

$$u = x^n, \qquad du = nx^{n-1}\,dx; \qquad dv = e^x\,dx, \qquad v = e^x:$$

$$\int x^n e^x\,dx = x^n e^x - \int e^x \cdot nx^{n-1}\,dx.$$

Answer: $\int x^n e^x\,dx = x^n e^x - n\int x^{n-1}e^x\,dx.$

REMARK: For abbreviation, write

$$J_n = \int x^n e^x\,dx.$$

Then the reduction formula is

$$J_n = x^n e^x - nJ_{n-1}.$$

EXAMPLE 10.2

Evaluate $\int x^5 e^x\,dx$.

Solution: Use the reduction formula just derived to compute J_5. By the reduction formula with $n = 5$,

$$J_5 = x^5 e^x - 5J_4.$$

By the reduction formula with $n = 4$,

$$J_4 = x^4 e^x - 4J_3.$$

Thus

$$\begin{aligned} J_5 &= x^5 e^x - 5(x^4 e^x - 4J_3) \\ &= x^5 e^x - 5x^4 e^x + 20J_3. \end{aligned}$$

But by repeated use of the reduction formula,

$$\begin{aligned} J_3 &= x^3e^x - 3J_2 \\ &= x^3e^x - 3(x^2e^x - 2J_1) \\ &= x^3e^x - 3x^2e^x + 6J_1 \\ &= x^3e^x - 3x^2e^x + 6(xe^x - J_0). \end{aligned}$$

The integral J_0 is easy:

$$J_0 = \int x^0e^x\,dx = e^x + C.$$

Hence

$$J_3 = e^x(x^3 - 3x^2 + 6x - 6) + C,$$

and consequently

$$J_5 = x^5e^x - 5x^4e^x + 20e^x(x^3 - 3x^2 + 6x - 6) + C.$$

Answer:
$e^x(x^5 - 5x^4 + 20x^3 - 60x^2 + 120x - 120) + C.$

Question: Study the polynomial in the answer. How does each term follow from the preceding term? Can you write down the value of

$$\int x^6e^x\,dx$$

by inspection?

EXAMPLE 10.3

Derive a reduction formula for $\int (\cos x)^n\,dx$.

Solution: Write

$$J_n = \int (\cos x)^n\,dx = \int (\cos x)^{n-1}\cos x\,dx,$$

and integrate by parts with

$$u = (\cos x)^{n-1}, \qquad du = -(n-1)(\cos x)^{n-2}\sin x\,dx;$$

$$dv = \cos x\,dx, \qquad v = \sin x:$$

$$\begin{aligned} \int (\cos x)^n\,dx &= (\cos x)^{n-1}\sin x + \int (n-1)(\cos x)^{n-2}\sin^2 x\,dx \\ &= (\cos x)^{n-1}\sin x + (n-1)\int (\cos x)^{n-2}(1 - \cos^2 x)\,dx. \end{aligned}$$

Therefore,

$$J_n = (\cos x)^{n-1}\sin x + (n-1)J_{n-2} - (n-1)J_n.$$

Combine the terms in J_n:

$$nJ_n = (\cos x)^{n-1} \sin x + (n-1)J_{n-2}.$$

Now divide by n.

Answer: If $J_n = \int (\cos x)^n \, dx$, then

$$J_n = \frac{(\cos x)^{n-1} \sin x}{n} + \frac{n-1}{n} J_{n-2}.$$

REMARK: This reduction formula lowers the power of $\cos x$ by two. Therefore, repeated application will ultimately reduce J_n to J_0 or J_1, according as n is even or odd. But both of these are easy:

$$J_0 = \int (\cos x)^0 \, dx = \int dx = x + C,$$

$$J_1 = \int \cos x \, dx = \sin x + C.$$

EXAMPLE 10.4

Compute $\int_0^{\pi/2} (\cos x)^6 \, dx$.

Solution: Set

$$I_n = \int_0^{\pi/2} (\cos x)^n \, dx.$$

Then by the reduction formula of the last example,

$$I_n = \left.\frac{(\cos x)^{n-1} \sin x}{n}\right|_0^{\pi/2} + \frac{n-1}{n} \int_0^{\pi/2} (\cos x)^{n-2} \, dx.$$

Hence,

$$I_n = 0 + \frac{n-1}{n} I_{n-2}.$$

Apply this formula with $n = 6$, then repeat with $n = 4$ and $n = 2$:

$$I_6 = \frac{5}{6} I_4 = \frac{5}{6}\left(\frac{3}{4} I_2\right) = \frac{5}{6} \cdot \frac{3}{4}\left(\frac{1}{2} I_0\right).$$

Therefore,

$$\int_0^{\pi/2} (\cos x)^6 \, dx = \frac{5 \cdot 3 \cdot 1}{6 \cdot 4 \cdot 2} \int_0^{\pi/2} dx = \frac{5 \cdot 3 \cdot 1}{6 \cdot 4 \cdot 2} \cdot \frac{\pi}{2}.$$

Answer: $\frac{5\pi}{32}$.

EXERCISES

Derive the reduction formula:

1. $\int (\ln x)^n\, dx = x (\ln x)^n - n \int (\ln x)^{n-1}\, dx, \quad n \neq -1$

2. $\int (\sin x)^n\, dx = -\frac{1}{n} (\sin x)^{n-1} \cos x + \frac{n-1}{n} \int (\sin x)^{n-2}\, dx$

3. $\int x^n \sin x\, dx = -x^n \cos x + n \int x^{n-1} \cos x\, dx$

4. $\int (\tan x)^n\, dx = \frac{(\tan x)^{n-1}}{n-1} - \int (\tan x)^{n-2}\, dx, \quad n \neq -1$

5. $\int \frac{dx}{(1+x^2)^n} = \frac{1}{2n-2} \frac{x}{(1+x^2)^{n-1}} + \frac{2n-3}{2n-2} \int \frac{dx}{(1+x^2)^{n-1}}, \quad n \neq 1.$

Using an appropriate reduction formula, compute the definite integral:

6. $\int_1^2 (\ln x)^2\, dx$

7. $\int_0^{\pi/2} \sin^7 x\, dx$

8. $\int_0^{\pi/2} \sin^8 x\, dx$

9. $\int_0^{\pi/4} \tan^{10} x\, dx$

10. $\int_0^1 \frac{dx}{(1+x^2)^3}$

11. $\int_{\pi/2}^{\pi} x^4 \sin x\, dx.$

12. Prove $\int_a^b (x-a)^m (b-x)^n\, dx = \frac{m!n!}{(m+n+1)!} (b-a)^{m+n+1}.$

11. TRICKS WITH DIFFERENTIALS [optional]

Integrals which involve $\sqrt{x^2 \pm a^2}$ can be handled by certain formal tricks. Examples are

$$\int \frac{dx}{\sqrt{x^2 \pm a^2}} \quad \text{and} \quad \int \sqrt{x^2 \pm a^2}\, dx.$$

Set

$$y^2 = x^2 \pm a^2.$$

Then

$$y\, dy = x\, dx, \qquad \frac{dx}{y} = \frac{dy}{x}.$$

In elementary ratio and proportion it is shown that if

$$\frac{a}{b} = \frac{c}{d}, \quad \text{then} \quad \frac{a}{b} = \frac{a+c}{b+d}.$$

Hence

$$\frac{dx}{y} = \frac{dx + dy}{x + y} = \frac{d(x + y)}{(x + y)}.$$

Conclusion:

$$\int \frac{dx}{y} = \int \frac{d(x + y)}{(x + y)} = \ln |x + y| + C,$$

$$\int \frac{dx}{\sqrt{x^2 \pm a^2}} = \ln |x + \sqrt{x^2 \pm a^2}| + C.$$

(This was previously obtained by trigonometric substitutions.)

Next,

$$y\,dx = \frac{y^2\,dx}{y} = \frac{x^2 \pm a^2}{y}\,dx$$

$$= \frac{x^2\,dx}{y} \pm a^2 \frac{dx}{y} = x\,dy \pm a^2 \frac{dx}{y}.$$

But also $y\,dx = d(xy) - x\,dy$. Add:

$$2y\,dx = d(xy) \pm a^2 \frac{dx}{y}.$$

Integrate, using the first integral:

$$\int y\,dx = \frac{1}{2}\,xy \pm \frac{a^2}{2} \ln |x + y| + C,$$

$$\int \sqrt{x^2 \pm a^2}\,dx = \frac{1}{2}\,x\sqrt{x^2 \pm a^2} + \frac{a^2}{2} \ln |x + \sqrt{x^2 \pm a^2}| + C.$$

Similarly one can evaluate

$$\int \frac{\sqrt{x^2 \pm a^2}}{x}\,dx, \qquad \int \sqrt{a^2 - x^2}\,dx, \qquad \text{etc.}$$

EXERCISES

Find in the order given:

1. $\displaystyle\int \frac{\sqrt{x^2 \pm a^2}}{x}\,dx$

2. $\displaystyle\int \frac{\sqrt{x^2 \pm a^2}\,dx}{x^2}$

3. $\displaystyle\int \frac{\sqrt{x^2 \pm a^2}\,dx}{x^3}$

4. $\displaystyle\int \frac{dx}{x\sqrt{x^2 \pm a^2}}$

5. $\displaystyle\int \frac{dx}{x^2\sqrt{x^2 \pm a^2}}$

6. $\displaystyle\int \frac{\sqrt{x^2 \pm a^2}\,dx}{x^4}$

7. $\displaystyle\int x^2\sqrt{x^2 \pm a^2}\,dx.$

Find in the order given:

8. $\displaystyle\int \sqrt{a^2 - x^2}\,dx$

9. $\displaystyle\int \frac{dx}{x\sqrt{a^2 - x^2}}$

10. $\displaystyle\int \frac{\sqrt{a^2 - x^2}}{x}\,dx$

11. $\displaystyle\int \frac{x^2\,dx}{\sqrt{a^2 - x^2}}$

12. $\displaystyle\int \frac{\sqrt{a^2 - x^2}}{x^2}\,dx$

13. $\displaystyle\int \frac{dx}{x^2\sqrt{a^2 - x^2}}.$

12. INTEGRAL TABLES

Inside the two covers of this book there is a short table of indefinite integrals. Much longer tables are available, for example those in the C.R.C. Standard Mathematical Tables and in Pierce's Table of Integrals.

A busy scientist does not want to bother with various tricks each time he encounters an integral; he uses integral tables. Not only do they save time, but they help eliminate errors.

We suggest that you get one of the more complete integral tables and spend some time browsing through it. Become familiar with the type of integral you can expect to find there.

Not every integral is listed in a table, but many can be transformed into integrals which are listed.

EXAMPLE 12.1

Use integral tables to evaluate $\displaystyle\int x^3\sqrt{3 - 4x^2}\,dx.$

Solution: Most tables include a section on integrals involving $\sqrt{a^2 - x^2}$. Formula 177 of the C.R.C. tables states

$$\int x^3\sqrt{a^2 - x^2}\,dx = -\left(\frac{1}{5}x^2 + \frac{2}{15}a^2\right)(a^2 - x^2)^{3/2}.$$

This is very close to what is wanted, except that $\sqrt{3 - 4x^2}$ appears instead of $\sqrt{a^2 - x^2}$. There are two simple ways of modifying the integrand: write either

$$\sqrt{3 - 4x^2} = \sqrt{4\left(\frac{3}{4} - x^2\right)} = 2\sqrt{\frac{3}{4} - x^2},$$

or

$$\sqrt{3 - 4x^2} = \sqrt{3 - (2x)^2} = \sqrt{3 - u^2}, \qquad \text{where} \quad u = 2x.$$

Use the first method with $a^2 = \frac{3}{4}$:

$$\int x^3 \sqrt{3 - 4x^2}\, dx = 2 \int x^3 \sqrt{\frac{3}{4} - x^2}\, dx$$

$$= -2\left(\frac{1}{5}x^2 + \frac{2}{15}\cdot\frac{3}{4}\right)\left(\frac{3}{4} - x^2\right)^{3/2}$$

Use the second method with $u = 2x$ and $a^2 = 3$:

$$\int x^3 \sqrt{3 - 4x^2}\, dx = \int \left(\frac{u}{2}\right)^3 \sqrt{3 - u^2} \cdot \frac{1}{2}\, du$$

$$= \frac{1}{16} \int u^3 \sqrt{3 - u^2}\, du$$

$$= -\frac{1}{16}\left(\frac{1}{5}u^2 + \frac{2}{15}\cdot 3\right)(3 - u^2)^{3/2}$$

$$= -\frac{1}{16}\left(\frac{1}{5}\cdot 4x^2 + \frac{2}{5}\right)(3 - 4x^2)^{3/2}.$$

A little algebra shows that both answers agree.

Answer: $-\frac{1}{40}(2x^2 + 1)(3 - 4x^2)^{3/2} + C.$

EXAMPLE 12.2

Use integral tables to evaluate $\int e^{2x}(\sin x)^3\, dx.$

Solution: In the C.R.C. integral tables under EXPONENTIAL FORMS is Formula 416:

$$\int e^{ax}(\sin bx)^n\, dx$$

$$= \frac{1}{a^2 + n^2b^2}\left[(a \sin bx - nb \cos bx)e^{ax}(\sin bx)^{n-1} + n(n-1)b^2 \int e^{ax}(\sin bx)^{n-2}\, dx\right].$$

This is a reduction formula which lowers the power of $\sin bx$ by two. Apply it with $a = 2$, $b = 1$, $n = 3$:

$$\int e^{2x}(\sin x)^3\, dx$$

$$= \frac{1}{4 + 9}\left[(2 \sin x - 3 \cos x)e^{2x}(\sin x)^2 + 6 \int e^{2x}(\sin x)\, dx\right].$$

The integral on the right is given by Formula 407 with $a = 2$, $b = 1$. Its value is

$$\frac{e^{2x}(2\sin x - \cos x)}{5}.$$

Substitute this into the preceding equation to obtain the answer.

Answer: $$\frac{e^{2x}}{13}\left[(2\sin x - 3\cos x)\sin^2 x + \frac{6}{5}(2\sin x - \cos x)\right] + C.$$

Integral tables use abbreviations for common expressions. For instance, one section of the C.R.C. tables contains formulas involving X and $\sqrt{X}$, where $X = a + bx + cx^2$.

EXAMPLE 12.3

Use integral tables to evaluate

$$\int \frac{\sqrt{5x^2 + 2x + 3}}{x}\,dx.$$

Solution: Apply Formula 218 of the C.R.C. tables:

$$\int \frac{\sqrt{X}}{x}\,dx = \sqrt{X} + \frac{b}{2}\int \frac{dx}{\sqrt{X}} + a\int \frac{dx}{x\sqrt{X}}.$$

The integrals on the right are given by Formulas 193 and 214:

$$\int \frac{dx}{\sqrt{X}} = \frac{1}{\sqrt{c}}\ln\left(\sqrt{X} + x\sqrt{c} + \frac{b}{2\sqrt{c}}\right), \qquad c > 0,$$

$$\int \frac{dx}{x\sqrt{X}} = -\frac{1}{\sqrt{a}}\ln\left(\frac{\sqrt{X} + \sqrt{a}}{x} + \frac{b}{2\sqrt{a}}\right), \qquad a > 0.$$

In this case $X = 3 + 2x + 5x^2$; set $a = 3$, $b = 2$, $c = 5$, to obtain

$$\int \frac{\sqrt{5x^2 + 2x + 3}}{x}\,dx = \sqrt{5x^2 + 2x + 3} + \frac{1}{\sqrt{5}}\ln\left(\sqrt{5x^2 + 2x + 3} + x\sqrt{5} + \frac{1}{\sqrt{5}}\right) - \sqrt{3}\ln\left(\frac{\sqrt{5x^2 + 2x + 3} + \sqrt{3}}{x} + \frac{1}{\sqrt{3}}\right) + C.$$

The C.R.C. tables use $\ln(\cdot\;\cdot\;\cdot)$; absolute values are understood.

Answer:
$$\sqrt{X} + \frac{1}{\sqrt{5}} \ln\left|\sqrt{X} + x\sqrt{5} + \frac{1}{\sqrt{5}}\right| - \sqrt{3}\ln\left|\frac{\sqrt{X}+\sqrt{3}}{x} + \frac{1}{\sqrt{3}}\right| + C,$$
where $X = 5x^2 + 2x + 3$.

EXERCISES

Find, using tables:

1. $\displaystyle\int e^{-2x} \sin 5x \, dx$
2. $\displaystyle\int \sqrt{4 - x^2} \, dx$
3. $\displaystyle\int x^2 \sqrt{1 - 4x^2} \, dx$
4. $\displaystyle\int \frac{x^2 \, dx}{\sqrt{4 - 3x^2}}$
5. $\displaystyle\int \frac{x^2 \, dx}{2 + 5x^2}$
6. $\displaystyle\int (4 - x^2)^{5/2} \, dx$
(trigonometric substitution)
7. $\displaystyle\int x^3 \sin \frac{x}{2} \, dx$
8. $\displaystyle\int (\ln x)^4 \, dx$
9. $\displaystyle\int \frac{x^2 - 6x - 2}{x\sqrt{10x^2 + 7}} \, dx$
10. $\displaystyle\int (x + 3)^2 \sqrt{x^2 + 2x + 5} \, dx.$

Compute:

11. $\displaystyle\int_0^1 \frac{x^4 + 2x^2 - 3}{x^4 + 2x^2 + 1} \, dx$
12. $\displaystyle\int_0^1 \frac{x^2 \, dx}{1 + 3x^2}$
13. $\displaystyle\int_0^\pi e^{3x} \cos^6 x \, dx$
14. $\displaystyle\int_0^{2\pi} \cos^2 x \sin^8 x \, dx$
15. $\displaystyle\int_0^1 \frac{dx}{(1 + 3x)^2(2 + 5x)}$
16. $\displaystyle\int_1^2 (x \ln x)^3 \, dx.$

20. Interpolation and Numerical Integration

1. INTERPOLATION

We begin this chapter by considering an important scientific technique: fitting a curve to data.

Suppose in an accurate experiment, a quantity x is measured at 11 time readings. For example,

t	0	0.1	0.2	$\cdots$	0.8	0.9	1.0
x	2.783	3.142	4.003	$\cdots$	2.001	1.833	1.801

In order to find a mathematical relation between these readings, we seek a function which fits the data, that is, a function $x(t)$ defined for all values of t in the interval $0 \leq t \leq 1$, and satisfying the conditions $x(0) = 2.783$, $x(0.1) = 3.142, \cdots, x(1.0) = 1.801$. In other words, we want a function whose graph passes through 11 given points. Finding such a function is called **interpolation.**

The simple case of linear interpolation is familiar. For example, from a 5-place table,

$$\log 3.1920 \approx 0.50406 \qquad \text{and} \qquad \log 3.1930 \approx 0.50420.$$

To estimate $\log 3.1927$, add $\frac{7}{10}$ of the difference $(0.50420 - 0.50406)$ to 0.50406:

$$\log 3.1927 \approx 0.50416.$$

This is equivalent to finding the linear function $f(x)$ that fits the two points $(3.1920, 0.50406)$ and $(3.1930, 0.50420)$, and writing

$$\log 3.1927 \approx f(3.1927).$$

As is easily computed,

$$f(x) = 0.50406 + (0.14)(x - 3.1920).$$

Hence

$$\log 3.1927 \approx 0.50406 + (0.14)(3.1927 - 3.1920) \approx 0.50416.$$

In this example the data consist of only two points and we can fit a straight line graph (linear polynomial). In case the data consist of several points, we seek the polynomial of least degree that fits.

Polynomial Interpolation

Here is the basic problem. Given $n + 1$ points

$$(x_0, y_0), \quad (x_1, y_1), \quad (x_2, x_2), \quad \cdots, \quad (x_n, y_n),$$

where

$$x_0 < x_1 < x_2 < \cdots < x_n,$$

find a polynomial $y = p(x)$ of least degree whose graph passes through the given points, i.e.,

$$p(x_0) = y_0, \quad p(x_1) = y_1, \quad \cdots, \quad p(x_n) = y_n.$$

Let us start with three points. Suppose we wish to interpolate

$$(x_0, y_0), \quad (x_1, y_1), \quad (x_2, y_2), \qquad x_0 < x_1 < x_2,$$

by a polynomial of least degree. A first degree polynomial generally won't do since its graph is a straight line. So we try a quadratic polynomial

$$y = A + Bx + Cx^2$$

and hope to choose its three coefficients so that its graph passes through the three given points.

EXAMPLE 1.1

Fit a quadratic to the points (0, 1), (1, 2), (2, 4).

Solution: Set $y = A + Bx + Cx^2$. Substitute $x = 0$, $y = 1$; $x = 1$, $y = 2$; $x = 2$, $y = 4$:

$$\begin{cases} A \qquad\qquad\quad = 1 \\ A + \ B + \ C = 2 \\ A + 2B + 4C = 4. \end{cases}$$

The solution of these simultaneous linear equations is $A = 1$, $B = C = \frac{1}{2}$.

Answer: $1 + \dfrac{1}{2}x + \dfrac{1}{2}x^2.$

This example illustrates the **method of undetermined coefficients.** The next example demonstrates the method for more than three points.

EXAMPLE 1.2

Fit a cubic to the four points $(-1, \frac{1}{2})$, (0, 1), (1, 2), (2, 4).

Solution: Set

$$p(x) = A + Bx + Cx^2 + Dx^3.$$

The data yields four equations:

$$\begin{cases} A - B + C - D = \frac{1}{2} \\ A = 1 \\ A + B + C + D = 2 \\ A + 2B + 4C + 8D = 4. \end{cases}$$

After some labor one finds the solution:

$$A = 1, \qquad B = \tfrac{2}{3}, \qquad C = \tfrac{1}{4}, \qquad D = \tfrac{1}{12}.$$

Answer: $1 + \frac{2}{3}x + \frac{1}{4}x^2 + \frac{1}{12}x^3$.

Fitting polynomials is important in approximation problems. For example, we may need many values of $f(x)$, a function difficult to compute. Then it is convenient to have a polynomial (easy to compute) which approximates $f(x)$. A basic method of finding one is fitting a polynomial to several points on the graph of $y = f(x)$.

EXAMPLE 1.3

Approximate $y = 2^x$, using the values at $x = 0, 1, 2$.

Solution: By Example 1.1, the quadratic polynomial $p(x) = 1 + \frac{1}{2}x + \frac{1}{2}x^2$ passes through the three points $(0, 2^0)$, $(1, 2^1)$, and $(2, 2^2)$.

REMARK: The approximation is quite good for $0 \le x \le 2$. One can show (not easily) that the error $\epsilon(x)$ in

$$2^x = 1 + \frac{1}{2}x + \frac{1}{2}x^2 + \epsilon(x)$$

satisfies

$$|\epsilon(x)| \le 0.05 \qquad \text{for} \quad -0.2 \le x \le 2.1,$$

$$|\epsilon(x)| \le 0.29 \qquad \text{for} \quad -0.5 \le x \le 2.5.$$

Figure 1.1 shows the error in the interval $0 \le x \le 2$.

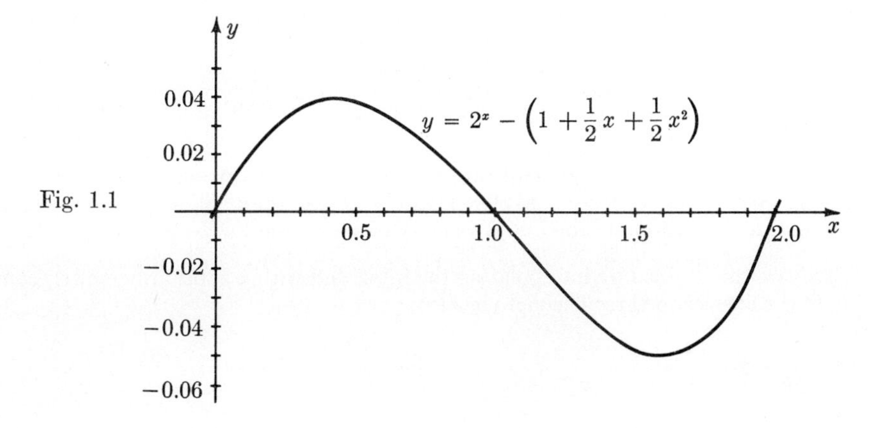

Fig. 1.1

EXERCISES

Fit a quadratic polynomial to the points:

1. $(0, 2), (2, 0), (3, 1)$
2. $(-2, 4), (0, -1), (1, 2)$
3. $(-2, -3), (2, 0), (5, 1)$
4. $(-4, 2), (-3, 3), (3, 9)$
5. $(-4, 1), (-1, 2), (2, 3)$
6. $(1, -2), (2, 1), (10, 1)$
7. $(2, 4), (3, 1), (4, -1)$
8. $(0, 7), (2, 3), (3, 1)$
9. $(-4, 4), (0, 2), (2, 1)$
10. $(-4, 2), (-3, 5), (3, 2)$.

Fit a cubic polynomial to the points:

11. $(-1, 1), (0, 2), (1, 4), (2, 6)$
12. $(-2, -3), (1, -1), (2, 0), (5, 4)$
13. $(-4, 4), (0, 0), (2, 2), (4, -4)$
14. $(1, 5), (2, 4), (3, 1), (4, -1)$
15. $(-2, -8), (0, -4), (2, 0), (3, 2)$
16. $(-2, 0), (-1, 0), (1, 2), (2, 2)$
17. $(-3, 1), (-2, 3), (-1, 0), (0, -1)$
18. $(-1, -2), (1, 0), (3, 2), (6, 5)$
19. $(-3, 4), (2, 3), (5, 0), (7, -3)$.

Fit a quadratic polynomial to the function, using the values at $x = 0, 1, 2$:

20. $y = \dfrac{6}{1 + x}$
21. $y = e^x$
22. $y = x^3$
23. $y = \dfrac{x}{1 + x}$.

Fit a cubic polynomial to the function, using the values at $x = 0, 1, 2, 3$:

24. $y = 2^x$
25. $y = x^4 - x$
26. $y = \dfrac{6}{1 + x}$
27. $y = \dfrac{x^2}{1 + x}$
28. $y = \ln(x + 1)$.
29. During a three hour period, the distance from the tide line to a fixed mark is measured:

t	0	1	2	3	hr
s	20	10	8	15	ft

Estimate to 2 decimals the minimum distance from the tide line to the mark during this period.

30. A rocket is fired from zero altitude at time $t = 0$. Its altitude is 200 ft at $t = 5$ sec, 600 ft at $t = 10$ sec, and 3000 ft at $t = 15$ sec. Fit a polynomial to the data, and from it estimate the initial velocity of the rocket.

Estimate the integral by interpolating the integrand with a cubic polynomial; compute the answer to three decimal places:

31. $\displaystyle\int_0^3 x \sin^2 \frac{\pi x}{4}\, dx$
32. $\displaystyle\int_1^4 (\ln x)^2\, dx$.

2. LAGRANGE INTERPOLATION

We state the fundamental fact about polynomial fitting.

Given $n + 1$ points

$$(x_0, y_0), \quad (x_1, y_1), \quad \cdots, \quad (x_n, y_n),$$

with $x_0 < x_1 < \cdots < x_n$, there is precisely one polynomial of degree at most n that passes through these points.

This statement contains two important assertions. First, there is *some* polynomial of degree at most n which fits the given points. Second, this polynomial is *unique*, it is the only one. For us, the first assertion is the more important one. We shall verify it by actually writing down a polynomial that fills the bill. Fortunately we can avoid the method of undetermined coefficients, which becomes complicated for more than three or four points; a neater method is available.

Given $n + 1$ points as above, consider the polynomial

$$(x - x_1)(x - x_2) \cdots (x - x_n).$$

Its degree is n and it has zeros at $x = x_1, x = x_2, \cdots, x = x_n$. At $x = x_0$ its value is

$$(x_0 - x_1)(x_0 - x_2) \cdots (x_0 - x_n) \neq 0.$$

Consequently, the polynomial

$$p_0(x) = \frac{(x - x_1)(x - x_2) \cdots (x - x_n)}{(x_0 - x_1)(x_0 - x_2) \cdots (x_0 - x_n)}$$

has degree n, has value 1 at $x = x_0$, and has zeros at $x_1, x_2, \cdots, x_n$. Thus $p_0(x)$ solves the interpolation problem in the special case $y_0 = 1$, $y_1 = 0, \cdots, y_n = 0$.

Similarly, for each $j = 0, 1, \cdots, n$, set

$$p_j(x) = \frac{(x - x_0)(x - x_1) \cdots (x - x_{j-1})(x - x_{j+1}) \cdots (x - x_n)}{(x_j - x_0)(x_j - x_1) \cdots (x_j - x_{j-1})(x_j - x_{j+1}) \cdots (x_j - x_n)}.$$

(There is no factor $x - x_j$ in the numerator and no factor $x_j - x_j$ in the denominator.) Then $p_j(x)$ has degree n and

$$p_j(x_j) = 1, \qquad \text{but} \qquad p_j(x_i) = 0 \quad \text{if} \quad i \neq j.$$

Thus $p_j(x)$ solves the interpolation problem in the special case

$$y_0 = 0, \quad y_1 = 0, \quad \cdots, \quad y_{j-1} = 0, \quad y_j = 1, \quad y_{j+1} = 0, \quad \cdots, \quad y_n = 0.$$

Finally, set

$$p(x) = y_0p_0(x) + y_1p_1(x) + \cdots + y_np_n(x).$$

This is it! Check:

$$p(x_0) = y_0 \cdot 1 + y_1 \cdot 0 + \cdots + y_n \cdot 0 = y_0,$$

$$p(x_1) = y_0 \cdot 0 + y_1 \cdot 1 + \cdots + y_n \cdot 0 = y_1,$$

$$\cdots\cdots\cdots\cdots$$

$$p(x_n) = y_0 \cdot 0 + y_1 \cdot 0 + \cdots + y_n \cdot 1 = y_n.$$

Since $p(x)$ is the sum of n-th degree polynomials, $p(x)$ has degree at most n (some terms may drop out). The formula for $p(x)$ is the **Lagrange interpolation formula.**

Lagrange Interpolation Formula The polynomial of degree at most n that fits $n + 1$ points

$$(x_0, y_0), \quad (x_1, y_1), \quad \cdots, \quad (x_n, y_n), \qquad x_0 < x_1 < \cdots < x_n,$$

is

$$p(x) = y_0p_0(x) + y_1p_1(x) + \cdots + y_np_n(x),$$

where

$$p_j(x) = \frac{(x - x_0) \cdots (x - x_{j-1})(x - x_{j+1}) \cdots (x - x_n)}{(x_j - x_0) \cdots (x_j - x_{j-1})(x_j - x_{j+1}) \cdots (x_j - x_n)}$$

for $j = 0, 1, \cdots, n$.

EXAMPLE 2.1

Fit a quadratic to (0, 1), (1, 2), and (2, 4).

Solution: Use the Lagrange interpolation formula:

$$p_0(x) = \frac{(x-1)(x-2)}{(0-1)(0-2)} = \frac{1}{2}(x^2 - 3x + 2),$$

$$p_1(x) = \frac{(x-0)(x-2)}{(1-0)(1-2)} = -x^2 + 2x,$$

$$p_2(x) = \frac{(x-0)(x-1)}{(2-0)(2-1)} = \frac{1}{2}(x^2 - x).$$

The desired polynomial is

$$p(x) = 1 \cdot p_0(x) + 2 \cdot p_1(x) + 4 \cdot p_2(x)$$

$$= \frac{1}{2}(x^2 - 3x + 2) + 2(-x^2 + 2x) + \frac{4}{2}(x^2 - x)$$

$$= \frac{1}{2}x^2 + \frac{1}{2}x + 1.$$

Answer: $\frac{1}{2}x^2 + \frac{1}{2}x + 1.$

Lagrange's formula shows that there *is* a polynomial of degree at most n through $n + 1$ given points $(x_0, y_0), \cdots, (x_n, y_n)$. Now we must verify that there is *only one.*

Suppose there were two different ones, $p(x)$ and $q(x)$. Their difference $r(x)$ would be a non-zero polynomial of degree at most n. But

$$r(x_0) = p(x_0) - q(x_0) = 0, \cdots, r(x_n) = p(x_n) - q(x_n) = 0,$$

so $r(x)$ would have $n + 1$ *different* zeros. However, it is shown in algebra that this is impossible for a non-zero polynomial of degree *less than* $n + 1$. Consequently, there cannot be two different polynomials of degree less than $n + 1$ which fit $n + 1$ given points.

EXERCISES

Use Lagrange interpolation to fit a polynomial to the data:

1. $(-3, 4), (-2, 0), (0, -2), (1, 1)$
2. $(-2, -2), (-1, 2), (0, 3), (1, 0), (2, 1)$
3. $(0, 8), (2, 4), (4, 2), (6, 1)$
4. $(-3, -1), (-1, 0), (1, 0), (3, 1).$
5. $(-2, -5), (0, 2), (2, 0), (4, -2), (6, 3)$
6. $(-1, 0), (1, 0), (2, 2), (5, 0), (8, -1), (10, 0)$

Use Lagrange interpolation to fit the function at the given points:

7. $y = \sin \frac{\pi x}{2}, \quad x = 0, 1, 2, 3, 4$
8. $y = e^{-x} \cos \frac{\pi x}{2}, \quad x = 0, 1, 2, 3$
9. $y = (1 + \sin \pi x)\left(\cos \frac{2\pi x}{3}\right), \quad x = \frac{1}{2}, \frac{3}{4}, 1$
10. $y = x^2(x + 1)^2, \quad x = -1, 0, 1, 2.$

Estimate the integral by fitting an interpolating polynomial to the integrand; give a 2-decimal answer:

11. $\int_{-1/2}^{1} x^3 \cos \pi x \, dx$, 4-point fit
12. $\int_{1}^{3} \frac{dx}{x\sqrt{1 + x}}$, 3-point fit
13. $\int_{0}^{4} e^{-x^2} \, dx$, 3-point fit.

3. SIMPSON'S RULE

Interpolation is used to derive numerical integration formulas. We know such a formula, the Trapezoidal Rule (Chapter 10):

$$\int_a^b f(x)\,dx \approx \frac{h}{2}[f(x_0) + 2f(x_1) + \cdots + 2f(x_{n-1}) + f(x_n)],$$

where $h = (b - a)/n$, and

$$x_0 = a, \quad x_1 = a + h, \quad x_2 = a + 2h, \quad \cdots, \quad x_{n-1} = a + (n-1)h, \quad x_n = b.$$

A convenient abbreviation is

$$\int_a^b f(x)\,dx \approx \frac{h}{2}[f_0 + 2f_1 + 2f_2 + \cdots + 2f_{n-1} + f_n],$$

where $f_i = f(x_i)$.

In this chapter, we derive several numerical integration formulas. In order to compare their usefulness, we test each of them on the integral

$$\int_0^2 e^{-x^2}\,dx.$$

Its exact value is not known, but from the U.S. Department of Commerce Table of Normal Distributions,

$$\int_0^2 e^{-x^2}\,dx \approx 0.88208\ 10350\ 6.$$

We shall carry our calculations to six significant figures since the C.R.C. table of the exponential function gives six places.

In Table 3.1 are listed the results of applying the Trapezoidal Rule to the above integral, using several values of n.

TABLE 3.1. TRAPEZOIDAL APPROXIMATIONS

n	$\int_0^2 e^{-x^2}\,dx$	n	$\int_0^2 e^{-x^2}\,dx$
2	0.877037	8	0.881704
4	0.880618	10	0.881837
6	0.881415	20	0.882020

Simpson's Rule

The idea behind numerical integration formulas is simple: in order to estimate

$$\int_a^b f(x)\,dx,$$

approximate $f(x)$ by a polynomial $p(x)$, then estimate the given integral by

$$\int_a^b p(x)\, dx.$$

Each choice of $p(x)$ gives rise to a different numerical integration formula.

Of particular importance is the following formula, obtained by using a *quadratic* $p(x)$.

Simpson's Three-Point Rule

$$\int_a^b f(x)\, dx \approx \frac{h}{3}\left[f(a) + 4f\left(\frac{a+b}{2}\right) + f(b)\right],$$

where $h = \frac{1}{2}(b - a)$.

The formula can be abbreviated

$$\int_a^b f(x)\, dx \approx \frac{h}{3}[f(x_0) + 4f(x_1) + f(x_2)],$$

where $x_0 = a, \qquad x_1 = \frac{a+b}{2}, \qquad x_2 = b.$

We obtain the rule by the following steps. First we approximate $f(x)$ by the quadratic $p(x)$ that fits the three points

$$(x_0, f(x_0)), \qquad (x_1, f(x_1)), \qquad (x_2, f(x_2)).$$

Then we show that

$$\int_a^b p(x)\, dx = \frac{h}{3}[f(x_0) + 4f(x_1) + f(x_2)].$$

Since $p(x_0) = f(x_0)$, $p(x_1) = f(x_1)$, and $p(x_2) = f(x_2)$, what we actually must do is show that the equation

$$\int_a^b p(x)\, dx = \frac{h}{3}[p(x_0) + 4p(x_1) + p(x_2)]$$

holds. In fact, this equation holds for each quadratic as we now show.

If $p(x) = A + Bx + Cx^2$, then

$$\begin{aligned}\int_a^b p(x)\, dx &= \left(Ax + \frac{1}{2}Bx^2 + \frac{1}{3}Cx^3\right)\Big|_a^b \\ &= A(b - a) + \frac{1}{2}B(b^2 - a^2) + \frac{1}{3}C(b^3 - a^3) \\ &= (b - a)\left[A + \frac{1}{2}B(b + a) + \frac{1}{3}C(b^2 + ba + a^2)\right].\end{aligned}$$

Now $h = \frac{1}{2}(b - a)$, $x_0 = a$, $x_1 = \frac{1}{2}(a + b)$, and $x_2 = b$. Consequently,

$$\frac{h}{3}[p(x_0) + 4p(x_1) + p(x_2)]$$

$$= \frac{(b-a)}{6}\left\{[A + Ba + Ca^2] + 4\left[A + B\left(\frac{a+b}{2}\right) + C\left(\frac{a+b}{2}\right)^2\right] + [A + Bb + Cb^2]\right\}$$

$$= \left(\frac{b-a}{6}\right)[6A + 3B(a+b) + 2C(a^2 + ab + b^2)]$$

$$= (b-a)\left[A + \frac{1}{2}B(a+b) + \frac{1}{3}C(a^2 + ab + b^2)\right].$$

This is equal to

$$\int_a^b p(x)\,dx,$$

which we wanted to show.

REMARK 1: If $p(x)$ is the cross-sectional area of a prismoid, then the formula

$$\int_a^b p(x)\,dx = \frac{h}{3}[p(x_0) + 4p(x_1) + p(x_2)]$$

is the Prismoidal Rule of Chapter 12.

REMARK 2: Surprisingly, this formula is correct not only for quadratics, but also for cubics.

> Simpson's Three-Point Rule is exact for polynomials of degree three or less. In other words,
>
> $$\int_a^b f(x)\,dx = \frac{h}{3}[f(x_0) + 4f(x_1) + f(x_2)]$$
>
> if $f(x)$ is a polynomial of degree three or less.

The proof for cubics is similar to the proof for quadratics; we leave it as an exercise.

Does the Three-Point Rule provide a *good* approximation to

$$\int_a^b f(x)\,dx\,?$$

That depends on the graph of $f(x)$. See Fig. 3.1.

The same interpolating quadratic $p(x)$ appears in Fig. 3.1a and Fig. 3.1b, since the three points which determine $p(x)$ are the same. Yet the

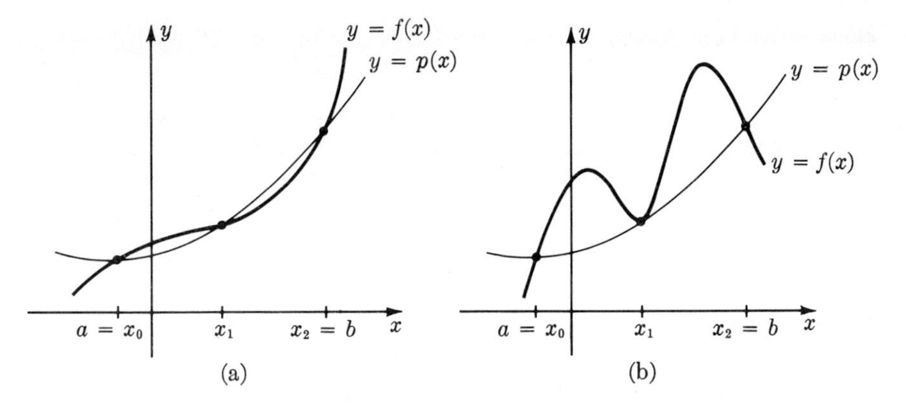

Fig. 3.1

approximation

$$\int_a^b f(x)\,dx \approx \int_a^b p(x)\,dx$$

is good in Fig. 3.1a and poor in Fig. 3.1b.

In the first case, $p(x)$ approximates $f(x)$ closely. Furthermore,

$$\int_{x_0}^{x_1} p(x)\,dx < \int_{x_0}^{x_1} f(x)\,dx,$$

whereas

$$\int_{x_1}^{x_2} p(x)\,dx > \int_{x_1}^{x_2} f(x)\,dx,$$

so the errors tend to cancel.

In Fig. 3.1b, however, $p(x)$ is a poor approximation to $f(x)$. To make matters worse, the errors do not cancel, they accumulate. The trouble is that the points x_0, x_1, x_2 are too widely spaced. It is much better to apply the Three-Point Rule *twice*, from x_0 to x_1 and from x_1 to x_2 (Fig. 3.2).

For greater accuracy divide the interval $a \leq x \leq b$ into an even number of equal pieces, and apply Simpson's Three-Point Rule to each pair of

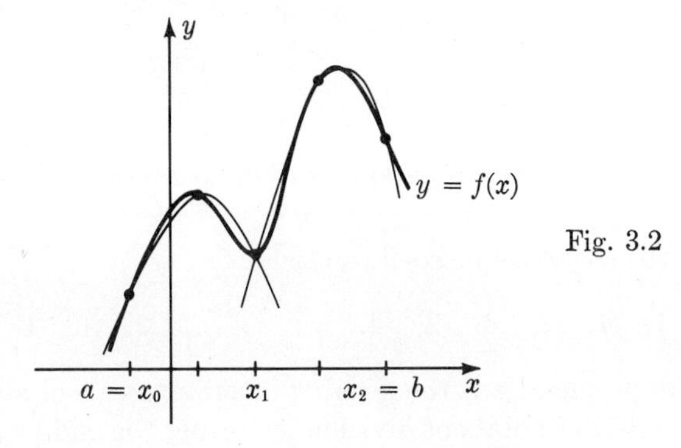

Fig. 3.2

consecutive subintervals:

$$\begin{array}{ccccccccc} & h & h & & & & h & h & \\ a = x_0 & x_1 & x_2 & x_3 & x_4 \cdots\cdots & x_{2n-2} & x_{2n-1} & x_{2n} = b. \end{array}$$

Set

$$h = \frac{b-a}{2n},$$

the length of each subinterval. Write

$$\int_a^b f(x)\,dx = \int_{x_0}^{x_2} + \int_{x_2}^{x_4} + \int_{x_4}^{x_6} + \cdots + \int_{x_{2n-2}}^{x_{2n}}.$$

Three-point approximation yields

$$\int_{x_0}^{x_2} f(x)\,dx \approx \frac{h}{3}[f_0 + 4f_1 + f_2],$$

$$\int_{x_2}^{x_4} f(x)\,dx \approx \frac{h}{3}[f_2 + 4f_3 + f_4],$$

$$\int_{x_4}^{x_6} f(x)\,dx \approx \frac{h}{3}[f_4 + 4f_5 + f_6],$$

$$\cdots\cdots\cdots\cdots$$

$$\int_{x_{2n-2}}^{x_{2n}} f(x)\,dx \approx \frac{h}{3}[f_{2n-2} + 4f_{2n-1} + f_{2n}].$$

Add these; the result is Simpson's Rule.

Simpson's Rule

$$\int_a^b f(x)\,dx \approx$$

$$\frac{h}{3}[f_0 + 4f_1 + 2f_2 + 4f_3 + 2f_4 + 4f_5 + \cdots + 2f_{2n-2} + 4f_{2n-1} + f_{2n}],$$

where $h = (b-a)/2n$ and $f_0, f_1, \cdots, f_{2n}$ are the values of f at the successive points of division of $a \leq x \leq b$ into $2n$ equal parts. The rule is exact whenever $f(x)$ is a polynomial of degree 3 or less.

Note that the sequence of coefficients is

$$1,\ 4,\ 2,\ 4,\ 2,\ 4,\ \cdots,\ 4,\ 2,\ 4,\ 1.$$

In Simpson's approximation, the number of subintervals of $a \leq x \leq b$ is $2n$ (even). The number of points of division, counting the end points, is $2n+1$ (odd). These odd numbers are used to describe various versions of Simpson's

Rule. Thus a 7-point Simpson approximation refers to the division shown here:

$$a = x_0 \quad x_1 \quad x_2 \quad x_3 \quad x_4 \quad x_5 \quad x_6 = b.$$

In this case $n = 3$, $2n = 6$, $2n + 1 = 7$, and the rule states

$$\int_a^b f(x)\,dx \approx \frac{h}{3}[f_0 + 4f_1 + 2f_2 + 4f_3 + 2f_4 + 4f_5 + f_6].$$

We apply several versions of Simpson's Rule to our test integral

$$\int_0^2 e^{-x^2}\,dx.$$

See Table 3.2.

TABLE 3.2. $(2n + 1)$-POINT SIMPSON APPROXIMATIONS

$2n + 1$	$\int_0^2 e^{-x^2}\,dx$	$2n + 1$	$\int_0^2 e^{-x^2}\,dx$
3	0.829944	9	0.882066
5	0.881812	11	0.882073
7	0.882031	21	0.882081

Tables 3.1 and 3.2 show that Simpson's Rule is much more precise than the Trapezoidal Rule. For instance, with 5 points of subdivision, the errors are approximately 0.001463 (Trapezoidal) versus 0.000269 (Simpson); with 11 points of subdivision, 0.000244 (Trapezoidal) versus 0.000008 (Simpson).

EXERCISES

Use Simpson's Three-Point Rule to estimate the integral; give a 2-decimal answer:

1. $\int_3^5 x^3\,dx$
2. $\int_{-1}^1 \cos x\,dx$
3. $\int_0^2 xe^x\,dx$
4. $\int_1^5 \frac{dx}{x}$
5. $\int_{\pi/2}^{5\pi/2} \frac{\sin x}{x}\,dx$
6. $\int_1^2 e^{-x^2}\,dx$
7. $\int_{-1}^1 e^x \tan x\,dx$
8. $\int_2^4 \frac{dx}{x^2 + x + 1}.$

Use Simpson's Rule with 3, 5, and 9 points to estimate the integral. Give a 4-decimal answer (the different results show that you do *not* have 4-decimal accuracy):

9. $\int_0^4 x^2\sqrt{x + 1}\,dx$
10. $\int_0^8 \frac{dx}{x^3 + x + 1}$

11. $\int_2^6 x \ln x \, dx$

12. $\int_{0.6}^{1.4} \ln \tan x \, dx$ (*Hint:* Express ln in terms of $\log_{10}$.)

13. $\int_{-2}^{2} x\sqrt{9 - x^3} \, dx$

14. $\int_0^2 \sin x \, dx.$

With a computer, use Simpson's Rule to estimate the integral; give a 6-decimal answer:

15. $\int_0^4 e^{-x^2} \, dx,$ 101 points

16. $\int_0^{\pi/2} \frac{dx}{\sqrt{1 - \frac{1}{2}\sin^2 x}},$ 21 points

17. $\int_1^3 \frac{\sin x}{x} \, dx,$ 101 points

18. $\int_0^{10} \frac{dx}{\sqrt{1 + x^3}},$ 101 points.

19. Verify that Simpson's Three-Point Rule is exact for cubics.

20. Show that Simpson's Three-Point Rule is exact for $\int_{-h}^{h} x^5 \, dx.$

21. (cont.) Does Ex. 20 mean Simpson's Three-Point Rule is exact for 5-th degree polynomials?

22. Draw a flowchart for Simpson's Rule with $2n + 1$ points.

4. NEWTON–COTES RULES [optional]

The Trapezoidal Rule is based on linear interpolation, Simpson's Rule on quadratic interpolation. Rules based on interpolation of degree three or more are known as Newton–Cotes Rules.

The 4-point Newton–Cotes Rule is based on division of the interval $a \leq x \leq b$ into three equal parts:

$$a = x_0 \qquad x_1 \qquad x_2 \qquad x_3 = b.$$

The rule asserts that

$$\int_a^b f(x) \, dx \approx \int_a^b p_3(x) \, dx,$$

where $p_3(x)$ is the polynomial of degree at most three that agrees with $f(x)$ at x_0, x_1, x_2, x_3.

We need a formula for the integral on the right. We could find $p_3(x)$ explicitly by Lagrange interpolation and then integrate it. A quicker method, however, is this.

The answer will be of the form

$$\int_a^b p_3(x) \, dx = h[c_0 f_0 + c_1 f_1 + c_2 f_2 + c_3 f_3],$$

where c_0, c_1, c_2, c_3 are constants and $h = (b - a)/3$. (Why?) By symmetry (reverse the x-axis), we guess $c_0 = c_3$ and $c_1 = c_2$;

$$\int_a^b p_3(x)\,dx = h[c_0(f_0 + f_3) + c_1(f_1 + f_2)].$$

We find the constants c_0 and c_1 by examining special cases of this formula. First we take $x_0 = -3$, $x_1 = -1$, $x_2 = 1$, $x_3 = 3$ and $h = 2$:

$$\int_{-3}^3 p_3(x)\,dx = 2\left\{c_0[f(-3) + f(3)] + c_1[f(-1) + f(1)]\right\}.$$

Next we choose $p_3(x) = 1$, obtaining

$$6 = 2(2c_0 + 2c_1).$$

Last we choose $p_3(x) = x^2$:

$$18 = 2(18c_0 + 2c_1).$$

From these equations $c_0 = \frac{3}{8}$ and $c_1 = \frac{9}{8}$. The final result is the Three-Eighths Rule.

Three-Eighths Rule

$$\int_a^b f(x)\,dx \approx \frac{3h}{8}[f_0 + 3f_1 + 3f_2 + f_3],$$

where $h = \frac{1}{3}(b - a)$.

Just as Simpson's Three-Point Rule was extended to a more general rule, the Three-Eighths Rule can also be extended. For example, divide the interval into 6 subintervals:

$a = x_0 \quad x_1 \quad x_2 \quad x_3 \quad x_4 \quad x_5 \quad x_6 = b.$

Write

$$\int_a^b f(x)\,dx = \int_{x_0}^{x_3} + \int_{x_3}^{x_6}.$$

By the Three-Eighths Rule,

$$\int_{x_0}^{x_3} f(x)\,dx \approx \frac{3h}{8}[f_0 + 3f_1 + 3f_2 + f_3],$$

$$\int_{x_3}^{x_6} f(x)\,dx \approx \frac{3h}{8}[f_3 + 3f_4 + 3f_5 + f_6].$$

Add; the result is a 7-point rule:

$$\int_a^b f(x)\,dx \approx \frac{3h}{8}[f_0 + 3f_1 + 3f_2 + 2f_3 + 3f_4 + 3f_5 + f_6],$$

where $h = \frac{1}{6}(b - a)$.

Both the Three-Eighths Rule and Simpson's Rule are exact for cubic polynomials, but not for fourth degree polynomials. The two rules provide approximately the same degree of accuracy.

Interpolation with polynomials of degree 4 and 5 leads to the 5-point and 6-point Newton–Cotes Rules, etc. The coefficients that occur are tedious to compute; we simply state some of the more useful results.

Higher Order Newton–Cotes Rules

5-point:

$$\int_a^b f(x)\,dx \approx \frac{2h}{45}\,[7f_0 + 32f_1 + 12f_2 + 32f_3 + 7f_4],$$

where $h = \frac{1}{4}(b - a)$;

7-point:

$$\int_a^b f(x)\,dx \approx \frac{h}{140}\,[41f_0 + 216f_1 + 27f_2 + 272f_3 + 27f_4 + 216f_5 + 41f_6],$$

where $h = \frac{1}{6}(b - a)$;

9-point:

$$\int_a^b f(x)\,dx \approx \frac{4h}{14{,}175}\,[989f_0 + 5888f_1 - 928f_2 + 10{,}496f_3 - 4540f_4 + 10{,}496f_5 - 928f_6 + 5888f_7 + 989f_8],$$

where $h = \frac{1}{8}(b - a)$.

Repeated application of these rules produces further rules. For example, the 5-point rule applied twice produces a nine-point rule.

Nine-point 5-point Rule:

$$\int_a^b f(x)\,dx \approx \frac{2h}{45}\,[7f_0 + 32f_1 + 12f_2 + 32f_3 + 14f_4 + 32f_5 + 12f_6 + 32f_7 + 7f_8],$$

where $h = \frac{1}{8}(b - a)$.

There are many other approximate integration rules based on rather complicated arguments. One stands out because of the simplicity of its coefficients; it is about as accurate as the 5-point rule.

Weddle's Rule

$$\int_a^b f(x)\,dx \approx \frac{3h}{10}\,[f_0 + 5f_1 + f_2 + 6f_3 + f_4 + 5f_5 + f_6],$$

where $h = \frac{1}{6}(b - a)$.

Applied to our test integral, these rules yield the estimates in Table 4.1.

TABLE 4.1. NEWTON–COTES APPROXIMATIONS

Rule	$\int_0^2 e^{-x^2}\,dx$
$\frac{3}{8}$-Rule	0.862224
5-point	0.885270
7-point	0.881916
Seven-point $\frac{3}{8}$-rule	0.881925
9-point	0.882087
Nine-point 5-point	0.882082
Weddle	0.882086

Comparing the results of Tables 3.1, 3.2, and 4.1 is instructive. (Recall that the true value of the integral is 0.8820810 to 7 places.) Most accurate is the Simpson 21-point approximation. Next is the 5-point rule applied to 9 division points.

The error in the various rules is discussed in Section 7.

EXERCISES

Estimate the integral, using the 5-point Simpson's Rule and the 5-point Newton–Cotes Rule; give a 2-decimal answer:

1. $\displaystyle\int_1^5 \frac{x^4+1}{x^5+1}\,dx$
2. $\displaystyle\int_{-4}^0 \sqrt{1+x+4x^2}\,dx$
3. $\displaystyle\int_0^4 \frac{e^x}{x+1}\,dx$
4. $\displaystyle\int_0^8 x^2e^{-x}\,dx$
5. $\displaystyle\int_{-2}^2 e^x\cos\frac{\pi x}{2}\,dx$
6. $\displaystyle\int_0^1 \ln\cos x\,dx.$

Estimate the integral using the 7-point Three-Eighths-Rule, the 7-point Newton–Cotes Rule, and Weddle's Rule; give a 4-decimal answer:

7. $\displaystyle\int_1^7 \frac{x+1}{x^3+x+1}\,dx$
8. $\displaystyle\int_{-3}^3 \ln(x+5)\,dx$
9. $\displaystyle\int_0^{1.5} \frac{\sin x}{x+1}\,dx$
10. $\displaystyle\int_{-1.5}^{1.5} \sqrt{x^3+4}\,dx$
11. $\displaystyle\int_1^7 e^{-x}\ln x\,dx$
12. $\displaystyle\int_0^3 x\sin\frac{x}{2}\,dx.$

Values of a function are given at several points. Integrate the function numerically over the given range of points, using one of the rules of this section:

13. $(0, 0), (1, 4), (2, -1), (3, 4)$
14. $(-1, 4), (2, 7), (5, 4), (8, 0)$
15. $(-2, 3), (-1, 1), (0, 7), (1, 5), (2, -1)$
16. $(0, 3), (1, -1), (2, 0), (3, 1), (4, 5)$
17. $(0, 7), (1, 5), (2, 4), (3, 1), (4, -2), (5, -3), (6, -4)$
18. $(-3, -5), (-2, -1), (-1, 1), (0, 1), (1, 1), (2, 4), (3, 3)$.

In Ex. 19–23, use a computer to estimate $\int_1^{25} f(x)\,dx$ by Simpson's Rule, and by each of the rules of this section. Use $x = 1, 2, 3, \cdots, 25$:

19. $f(x) = e^{-x}\dfrac{x^3+1}{x^2+1}$

20. $f(x) = \dfrac{\sin x}{x}$

21. $f(x) = x \ln x$

22. $f(x) = \sqrt{1+e^x}$

23. only the tabulated values of $f(x)$ are known:

x	$f(x)$	x	$f(x)$	x	$f(x)$	x	$f(x)$	x	$f(x)$
1	1	6	8	11	11	16	25	21	0
2	1	7	13	12	14	17	11	22	5
3	2	8	21	13	25	18	10	23	5
4	3	9	8	14	13	19	21	24	10
5	5	10	3	15	12	20	5	25	15

24. Carry out the details in the derivation of the 4-point Three-Eighths Rule.
25. Verify that the Three-Eighths Rule is exact for cubics in any interval $a \le x \le b$, not only $-3 \le x \le 3$ as discussed on p. 387.
26. Use Lagrange Interpolation to derive the Three-Eighths Rule. (Reread the first three paragraphs of the section.)
27. Draw a flowchart for the 5-point Newton–Cotes Rule.
28. Draw a flowchart for Weddle's Rule.

5. SYMMETRIES IN DEFINITE INTEGRALS

The rest of this chapter is devoted to certain techniques for evaluating and estimating definite integrals, and to discussion of the errors in various numerical integration formulas.

In this section, we discuss a few labor-saving methods for evaluating definite integrals.

Sometimes an integral

$$\int_a^b f(x)\,dx$$

can be simplified because the integrand $f(x)$ has certain symmetry. We look at several cases.

Even Functions

> If $f(x)$ is an even function, i.e.,
> $$f(x) = f(-x),$$
> then
> $$\int_{-a}^{a} f(x)\,dx = 2\int_{0}^{a} f(x)\,dx.$$

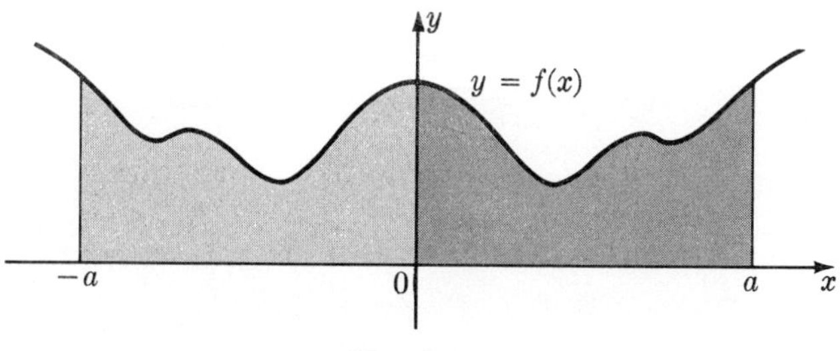

FIG. 5.1

Look at a graph to see why this is so (Fig. 5.1). It can also be seen by a substitution:

$$\int_{-a}^{0} f(x)\,dx = \int_{a}^{0} f(-u)\,(-du) = \int_{a}^{0} f(u)\,(-du) = \int_{0}^{a} f(u)\,du.$$

(Justify each of these steps.) Hence,

$$\int_{-a}^{a} f(x)\,dx = \int_{-a}^{0} f(x)\,dx + \int_{0}^{a} f(x)\,dx = 2\int_{0}^{a} f(x)\,dx.$$

EXAMPLE 5.1

Find $\displaystyle\int_{-\pi/2}^{\pi/2} \cos\frac{x}{3}\,dx.$

Solution: The integrand is an even function:

$$\cos\left(\frac{-x}{3}\right) = \cos\left(\frac{x}{3}\right).$$

Therefore,

$$\int_{-\pi/2}^{\pi/2} \cos\frac{x}{3}\,dx = 2\int_{0}^{\pi/2} \cos\frac{x}{3}\,dx = 6\sin\frac{x}{3}\bigg|_{0}^{\pi/2} = 6\sin\frac{\pi}{6} = 3.$$

Answer: 3.

Symmetry is useful in approximate integration also.

EXAMPLE 5.2

Estimate $\int_{-\pi/2}^{\pi/2} \cos\frac{x}{3}\,dx$, by Simpson's Rule with 3 points.

Solution:

$$\int_{-\pi/2}^{\pi/2} \cos\frac{x}{3}\,dx \approx \frac{\pi}{6}\left(\cos\frac{-\pi}{6} + 4\cos 0 + \cos\frac{\pi}{6}\right)$$

$$= \frac{\pi}{6}\left(\frac{\sqrt{3}}{2} + 4 + \frac{\sqrt{3}}{2}\right)$$

$$= \frac{\pi}{6}(4 + \sqrt{3}) \approx 3.0013.$$

(The exact value is 3.)

Alternate Solution: First observe $\cos(x/3)$ is even. Hence

$$\int_{-\pi/2}^{\pi/2} \cos\frac{x}{3}\,dx = 2\int_{0}^{\pi/2} \cos\frac{x}{3}\,dx.$$

Now use Simpson's Rule on the interval $0 \leq x \leq \pi/2$:

$$\int_{-\pi/2}^{\pi/2} \cos\frac{x}{3}\,dx \approx 2\left(\frac{1}{3}\right)\left(\frac{\pi}{4}\right)\left(\cos 0 + 4\cos\frac{\pi}{12} + \cos\frac{\pi}{6}\right)$$

$$\approx \frac{\pi}{6}[1 + 4(0.96593) + (0.86603)]$$

$$\approx \frac{\pi}{6}(5.72975) \approx 3.0000.$$

(There is a cumulative round-off error here. Using 7-place tables we get 3.00007.)

Answer: 3.0000.

REMARK: In the second solution we exploited the evenness of $\cos(x/3)$. That enabled us to apply Simpson's Rule on a smaller interval and obtain greater accuracy.

Sometimes a function may be "even" with respect to an axis other than the y-axis. For example,

$$\int_{4}^{6} (x - 5)^2\,dx = 2\int_{5}^{6} (x - 5)^2\,dx$$

because the function $(x - 5)^2$ is symmetric about the line $x = 5$. See Fig. 5.2.

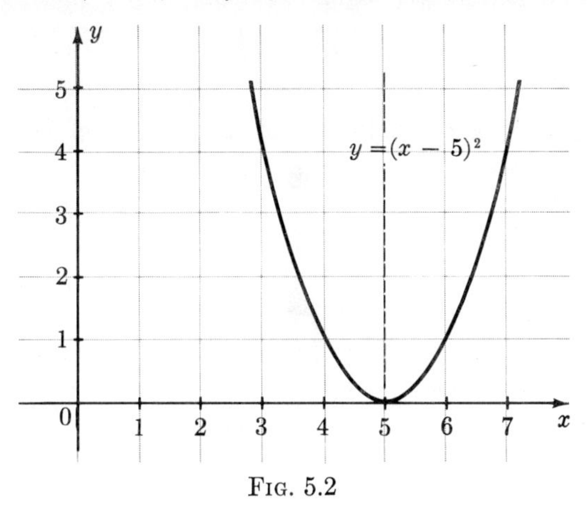

FIG. 5.2

Odd Functions

> If $f(x)$ is an odd function, i.e.,
> $$f(-x) = -f(x),$$
> then
> $$\int_{-a}^{a} f(x)\,dx = 0.$$

This can be seen from a graph (Fig. 5.3). It also can be derived by a substitution:

$$\int_{-a}^{a} f(x)\,dx = \int_{a}^{-a} f(-x)\,(-dx) = \int_{a}^{-a} [-f(x)]\,[-dx]$$
$$= \int_{a}^{-a} f(x)\,dx = -\int_{-a}^{a} f(x)\,dx.$$

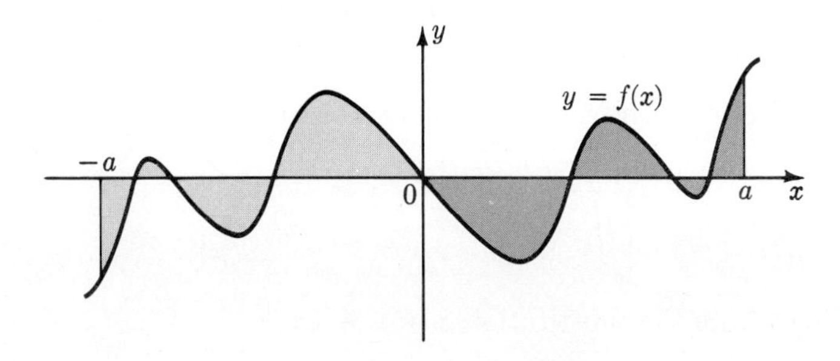

FIG. 5.3

(Justify each of these steps.) Hence,

$$2\int_{-a}^{a} f(x)\,dx = 0, \qquad \int_{-a}^{a} f(x)\,dx = 0.$$

This property of odd functions can save a lot of computation. For example,

$$\int_{-\pi/4}^{\pi/4} \sin^9 x\,dx = 0, \qquad \int_{-2}^{2} \frac{x^3\,dx}{x^4 + x^2 + 1} = 0,$$

by inspection.

The graph of an odd function is symmetric with respect to the origin. A function may be symmetric with respect to some other point of the x-axis.

EXAMPLE 5.3

Find $\displaystyle\int_0^6 x(x-1)(x-2)(x-3)(x-4)(x-5)(x-6)\,dx.$

Solution: The point of symmetry is (3, 0). To exploit this make the change of variable $x = u + 3$. Then

$$x(x-1)\cdots(x-6) = g(u),$$

where

$$g(u) = (u+3)(u+2)(u+1)u(u-1)(u-2)(u-3).$$

Thus

$$\int_0^6 x(x-1)\cdots(x-6)\,dx = \int_{-3}^{3} g(u)\,du.$$

But $g(-u) = -g(u)$, so the integral is 0.

Answer: 0.

Periodic Functions

Suppose $f(x)$ is periodic of period 2π. That is,

$$f(x + 2\pi) = f(x)$$

for each value of x. The first relation we notice is

$$\boxed{\int_a^b f(x)\,dx = \int_{a+2\pi}^{b+2\pi} f(x)\,dx.}$$

This follows from the substitution $x = u + 2\pi$:

$$\int_{a+2\pi}^{b+2\pi} f(x)\,dx = \int_a^b f(u + 2\pi)\,du = \int_a^b f(u)\,du.$$

It can also be seen from a graph (Fig. 5.4).

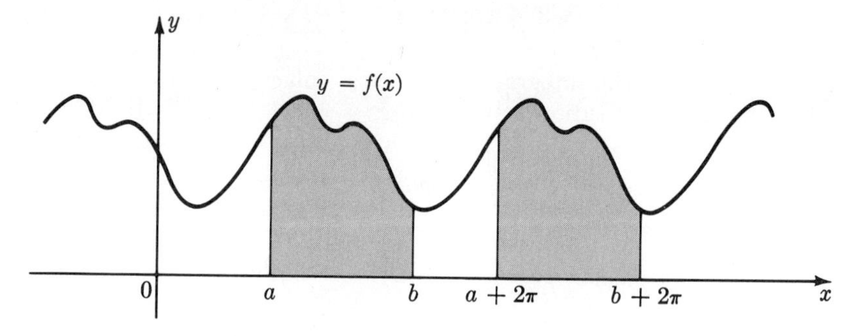

FIG. 5.4

The next relation says that the integral can be evaluated over a full period by evaluating it over *any* full period:

$$\int_a^{a+2\pi} f(x)\,dx = \int_0^{2\pi} f(x)\,dx.$$

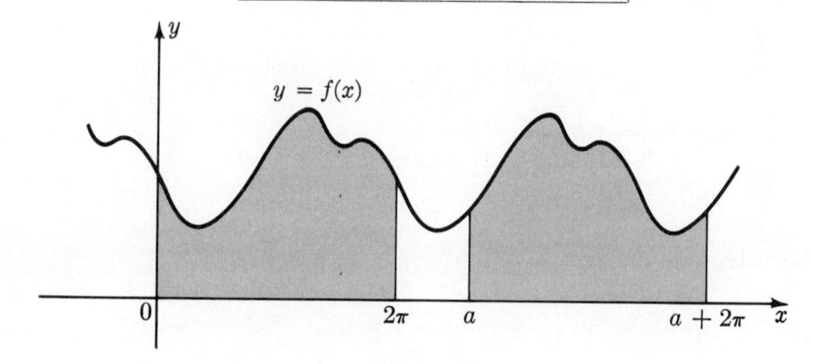

FIG. 5.5

This again is clear from a graph (Fig. 5.5). It can also be justified by shorthand:

$$\begin{aligned}\int_a^{a+2\pi} &= \int_0^{a+2\pi} - \int_0^a \\ &= \int_0^{2\pi} + \int_{2\pi}^{a+2\pi} - \int_0^a \\ &= \int_0^{2\pi} + \int_0^a - \int_0^a \\ &= \int_0^{2\pi}.\end{aligned}$$

EXAMPLE 5.4

Show that $\int_0^{2\pi} \sqrt{1+\sin^3 x}\, dx = \int_0^{2\pi} \sqrt{1+\cos^3 x}\, dx.$

Solution: In the integral on the left, replace $\sin x$ by $\cos(\pi/2 - x)$. Then make the substitution $u = \pi/2 - x$:

$$\int_0^{2\pi} \sqrt{1+\sin^3 x}\, dx = \int_0^{2\pi} \sqrt{1+\cos^3\left(\frac{\pi}{2} - x\right)}\, dx$$

$$= \int_{\pi/2}^{-3\pi/2} \sqrt{1+\cos^3 u}\,(-du)$$

$$= \int_{-3\pi/2}^{\pi/2} \sqrt{1+\cos^3 u}\, du = \int_0^{2\pi} \sqrt{1+\cos^3 u}\, du.$$

The next example uses the same idea.

EXAMPLE 5.5

Suppose n is a positive integer and $f(t)$ is any function defined for $-1 \le t \le 1$. Show that

$$\int_0^{2\pi} f(\sin nx)\, dx = \int_0^{2\pi} f(\cos nx)\, dx.$$

Solution: Substitute

$$x = \frac{\pi}{2n} - u$$

into the integral on the left. Then

$$\sin nx = \sin\left(\frac{\pi}{2} - nx\right) = \cos nu, \qquad dx = -du,$$

$$\int_0^{2\pi} f(\sin nx)\, dx = -\int_{\pi/2n}^{(\pi/2n)-2\pi} f(\cos nu)\, du$$

$$= \int_{(\pi/2n)-2\pi}^{\pi/2n} f(\cos nu)\, du = \int_0^{2\pi} f(\cos nu)\, du.$$

We have used two facts:

(i) $f(\cos nu)$ has period 2π,

(ii) $\int_a^{a+2\pi} = \int_0^{2\pi}$, where $a = \dfrac{\pi}{2n} - 2\pi$.

EXAMPLE 5.6

Compute $\int_0^{2\pi} \sin^2(100x)\, dx.$

Solution: Set

$$A = \int_0^{2\pi} \cos^2(100x)\,dx \qquad \text{and} \qquad B = \int_0^{2\pi} \sin^2(100x)\,dx.$$

By the last example, $A = B$. On the other hand,

$$\begin{aligned} A + B &= \int_0^{2\pi} \cos^2(100x)\,dx + \int_0^{2\pi} \sin^2(100x)\,dx \\ &= \int_0^{2\pi} [\cos^2(100x) + \sin^2(100x)]\,dx \\ &= \int_0^{2\pi} 1 \cdot dx = 2\pi. \end{aligned}$$

Hence, $A = \pi$.

Alternate Solution: Use the identity

$$\sin^2\theta = \frac{1}{2} + \frac{1}{2}\cos 2\theta.$$

$$\int_0^{2\pi} \sin^2(100x)\,dx = \frac{1}{2}\int_0^{2\pi} dx + \frac{1}{2}\int_0^{2\pi} \cos(200x)\,dx.$$

The first integral on the right is π. The second is zero because $\cos 200x$ makes a whole number of complete cycles from 0 to 2π.

Answer: π.

A function $f(x)$ is called **periodic with period** p if

$$f(x + p) = f(x).$$

Each statement about functions of period 2π can be modified to a statement about functions of any period p.

EXERCISES

Determine if the function is even, odd, or neither:

1. $y = e^{-x^2}$
2. $y = x \sin x$
3. $y = x^3 - 5x$
4. $y = x \cos x$
5. $y = \ln(x^2)$
6. $y = \ln(x^3)$
7. $y = \sin x + \cos x$
8. $y = e^{x^2} \cos x$.

Find a line $x = k$ of even or odd symmetry:

9. $y = (x^2 - 1)(x + 3)$
10. $y = (x^2 + 2x)(x + 3)(x + 5)$

11. $y = \sin(x + 1)$
12. $y = e^{-(x+1)^2}$
13. $y = \sin x + \cos x$
14. $y = \dfrac{1}{x^2 + x + 1}$
15. $y = \sqrt{1 + x^2}$
16. $y = e^{\sin x}$.

Reduce the given integral by symmetry to one over a shorter interval; do not evaluate:

17. $\displaystyle\int_{-1}^{2} (x^3 - 5x)\, dx$
18. $\displaystyle\int_{-4}^{4} \sin(x^2)\, dx$
19. $\displaystyle\int_{-2}^{3} \frac{2x - 1}{x^2 - x + 1}\, dx$
20. $\displaystyle\int_{-4}^{0} \sqrt{x^2 + 4x + 6}\, dx$.

Reduce the given integral by symmetry and periodicity to one over a shorter interval; do not evaluate:

21. $\displaystyle\int_{0}^{3\pi} (\sin x + \cos x)\, dx$
22. $\displaystyle\int_{0}^{\pi} (\sin x + \cos 2x)\, dx$
23. $\displaystyle\int_{0}^{2\pi} \sin^2 x\, dx$
24. $\displaystyle\int_{0}^{3\pi} \frac{\sin x}{2 + \cos x}\, dx$
25. $\displaystyle\int_{-100\pi}^{100\pi} \sin\left(\frac{x}{12} - 4\right) dx$
26. $\displaystyle\int_{1}^{4} \sin(\pi x + 3)\, dx$.

(*Hint:* Use the addition formula for the sine.)

6. ESTIMATES OF INTEGRALS

In applications, important quantities may occur as definite integrals which are difficult or impossible to compute. In such cases, one can use numerical integration formulas. But if only a rough estimate is needed, there are some simple quick techniques worth knowing. Here are a few of the most basic ones.

Inequalities for Definite Integrals

(1) If $f(x) \geq 0$ for $a \leq x \leq b$, then $\displaystyle\int_a^b f(x)\, dx \geq 0$.

(2) If $f(x) \leq g(x)$ for $a \leq x \leq b$, then $\displaystyle\int_a^b f(x)\, dx \leq \int_a^b g(x)\, dx$.

(3) If $m \leq f(x) \leq M$ for $a \leq x \leq b$, then

$$m(b - a) \leq \int_a^b f(x)\, dx \leq M(b - a).$$

(4)
$$\left| \int_a^b f(x)\, dx \right| \leq \int_a^b |f(x)|\, dx \qquad (a < b).$$

(5) If $|f(x)| \leq M$, and either $g(x) \geq 0$ for all x or $g(x) \leq 0$ for all x, where $a \leq x \leq b$, then $\displaystyle\left| \int_a^b f(x) g(x)\, dx \right| \leq M \left| \int_a^b g(x)\, dx \right|$.

COMMENTS: Inequality (1) goes back to the definition of the definite integral of a nonnegative function as an area. In fact, with area in mind a bit more can be said:

> If $f(x)$ is a smooth function and $f(x) \geq 0$ for $a \leq x \leq b$, then actually
> $$\int_a^b f(x)\,dx > 0$$
> unless $f(x) = 0$ for all x.

Inequality (2) follows from (1). If $f(x) \leq g(x)$, then $[g(x) - f(x)] \geq 0$. From (1),

$$\int_a^b [g(x) - f(x)]\,dx \geq 0.$$

Hence

$$\int_a^b g(x)\,dx - \int_a^b f(x)\,dx \geq 0,$$

which implies (2). Notice that the sharpened version of (1) implies a sharpened version of (2):

> If $f(x) \leq g(x)$ for $a \leq x \leq b$, then actually
> $$\int_a^b f(x)\,dx < \int_a^b g(x)\,dx$$
> unless $f(x) = g(x)$ for all x.

Inequality (3) follows from (2). Let $g(x)$ be the constant function $g(x) = M$. If $f(x) \leq M$, then

$$\int_a^b f(x)\,dx \leq \int_a^b g(x)\,dx = \int_a^b M\,dx = M(b - a).$$

The other half of (3) is proved in a similar way.

Inequality (3) has a simple geometric meaning for positive functions (Fig. 6.1). The area of the region under $y = f(x)$ between a and b is larger than that of a rectangle, base $b - a$, height m, and smaller than that of a rectangle, base $b - a$, height M.

Inequality (4) follows from (2), and from simple properties of absolute values:

$$|a| = +a \quad \text{or} \quad -a; \qquad a \leq |a|, \qquad -a \leq |a|.$$

Thus

$$\left|\int_a^b f(x)\,dx\right| = +\int_a^b f(x)\,dx \qquad \text{or} \qquad -\int_a^b f(x)\,dx.$$

But $f(x) \leq |f(x)|$ and $-f(x) \leq |f(x)|$. Hence, from (2),

$$\left|\int_a^b f(x)\,dx\right| = \int_a^b \pm f(x)\,dx \leq \int_a^b |f(x)|\,dx.$$

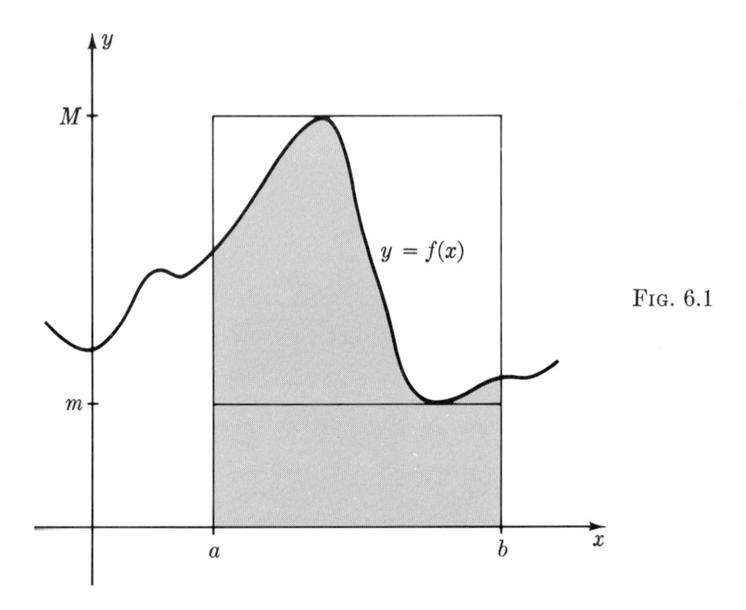

FIG. 6.1

This inequality also has a geometric meaning. The integral

$$\int_a^b f(x)\,dx$$

is an algebraic sum of areas (Fig. 6.2a). But the integral

$$\int_a^b |f(x)|\,dx$$

counts all areas positive; it allows no cancellation (Fig. 6.2b). Therefore its value is greater than or equal to the absolute value of

$$\int f(x)\,dx.$$

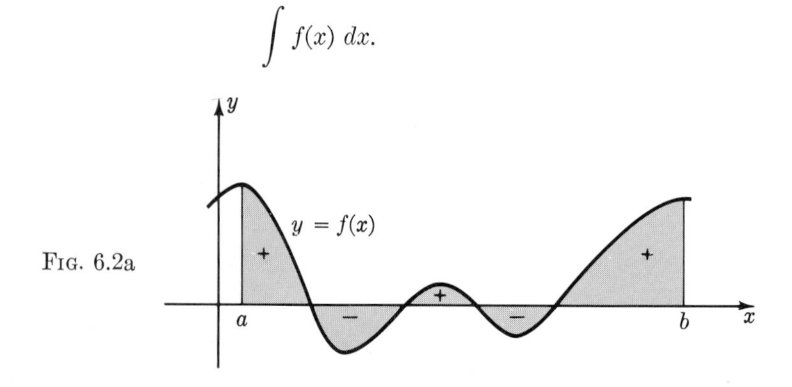

FIG. 6.2a

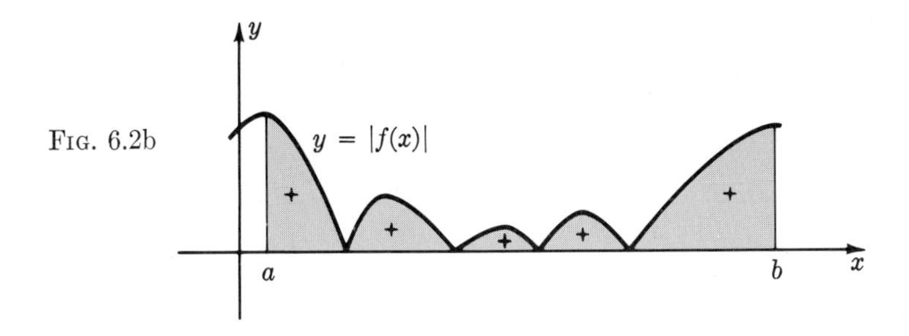

FIG. 6.2b

The integrals are unequal when there is cancellation, equal otherwise:

$$\left| \int_a^b f(x)\,dx \right| < \int_a^b |f(x)|\,dx$$

if $f(x)$ changes sign in the interval $a \le x \le b$. Otherwise,

$$\left| \int_a^b f(x)\,dx \right| = \int_a^b |f(x)|\,dx.$$

Inequality (5) follows from (2) and (4): If $|f(x)| \le M$, and $g(x)$ does not change sign, then

$$\begin{aligned} \left| \int_a^b f(x)g(x)\,dx \right| &\le \int_a^b |f(x)| \cdot |g(x)|\,dx \\ &\le \int_a^b M|g(x)|\,dx = M \left| \int_a^b g(x)\,dx \right|. \end{aligned}$$

Final comment: In each of the five integral inequalities, it is assumed that $a < b$. If $a > b$, the rule

$$\int_a^b f(x)\,dx = -\int_a^b f(x)\,dx$$

converts the integral into one whose upper limit is larger than its lower limit. Using this rule, we see that the inequalities are reversed in (1), (2), (3), and (4) if $a > b$. Inequality (5) is valid in all cases.

Applications

EXAMPLE 6.1

Show that $\displaystyle\int_1^3 \frac{dx}{1 + x^4} < \frac{1}{3}$.

Solution: Since

$$\frac{1}{1 + x^4} < \frac{1}{x^4},$$

it follows from inequality (2) that

$$\int_1^3 \frac{dx}{1+x^4} < \int_1^3 \frac{dx}{x^4} = -\frac{1}{3x^3}\Big|_1^3 = \frac{1}{3} - \frac{1}{81} < \frac{1}{3}.$$

Remark: This example illustrates an important technique: replacing the integrand $f(x)$ by a slightly larger one, $g(x)$, which is easily integrated. Generally, this technique yields better results than does use of inequality (3). For instance, all we can deduce from (3) is

$$\int_1^3 \frac{dx}{1+x^4} < 1.$$

(Here $a = 1$, $b = 3$, and $M = \frac{1}{2}$, the largest value of $1/(1+x^4)$ in the interval $1 \le x \le 3$.) Nevertheless, (3) is effective for very quick estimates, especially when great accuracy is not needed.

EXAMPLE 6.2

Estimate $\displaystyle\int_6^8 \frac{dx}{x^3 + x + \sin 2x}$.

Solution: Find bounds for the integrand. Since $-1 \le \sin 2x \le 1$ and $x^3 + x$ is increasing,

$$6^3 + 6 - 1 < x^3 + x + \sin 2x < 8^3 + 8 + 1$$

in the range $6 \le x \le 8$. Take reciprocals (and reverse inequalities):

$$\frac{1}{521} < \frac{1}{x^3 + x + \sin 2x} < \frac{1}{221}.$$

Use inequality (3) with $a = 6$, $b = 8$, $m = \frac{1}{521}$, $M = \frac{1}{221}$:

$$\frac{2}{521} < \int_6^8 \frac{dx}{x^3 + x + \sin 2x} < \frac{2}{221}.$$

Answer:

$$0.0038 < \int_6^8 \frac{dx}{x^3 + x + \sin 2x} < 0.0091.$$

EXAMPLE 6.3

Show that $\displaystyle\left|\int_0^{2\pi} \frac{\sin x\, dx}{1+x^2}\right| < \frac{\pi}{2}$.

Solution: Regard the integral as

$$\int_0^{2\pi} f(x)g(x)\, dx, \qquad f(x) = \sin x, \qquad g(x) = \frac{1}{1+x^2}.$$

Since $|f(x)| \leq 1$ and $g(x) > 0$, use inequality (5):

$$\left| \int_0^{2\pi} \frac{\sin x \, dx}{1+x^2} \right| \leq \int_0^{2\pi} \frac{dx}{1+x^2} = \arctan x \Big|_0^{2\pi} = \arctan 2\pi.$$

But $\arctan x < \pi/2$ for all values of x, hence

$$\left| \int_0^{2\pi} \frac{\sin x \, dx}{1+x^2} \right| < \frac{\pi}{2}.$$

EXAMPLE 6.4

Show that $\int_0^3 e^{-x^2} \, dx$ is a good approximation for $\int_0^{100} e^{-x^2} \, dx$. How good?

Solution: The error in this approximation is

$$\int_0^{100} e^{-x^2} \, dx - \int_0^3 e^{-x^2} \, dx = \int_3^{100} e^{-x^2} \, dx.$$

Since e^{-x^2} is extremely small even for moderate values of x, the integral on the right is small. Estimate the integral by (3). Because e^{-x^2} is a decreasing function, its largest value in the interval $3 \leq x \leq 100$ occurs at the left end point; this value is e^{-9}. By (3),

$$\int_3^{100} e^{-x^2} \, dx < (100 - 3)e^{-9} < 97(0.000124) < 0.0124.$$

A better estimate can be obtained using (2). Note that $x^2 \geq 3x$ when $x \geq 3$. In this range therefore, $e^{-x^2} \leq e^{-3x}$ (which can be integrated). Hence

$$\int_3^{100} e^{-x^2} \, dx \leq \int_3^{100} e^{-3x} \, dx = \frac{e^{-3x}}{-3} \Big|_3^{100} = \frac{e^{-9}}{3} - \frac{e^{-300}}{3} < \frac{e^{-9}}{3} < 0.000042.$$

Answer:

$$\int_0^{100} e^{-x^2} \, dx - \int_0^3 e^{-x^2} \, dx < 4.2 \times 10^{-5}.$$

REMARK: This example illustrates an important labor-saving device. If you intend to estimate

$$\int_0^{100} e^{-x^2} \, dx$$

by use of a numerical integration formula, you can save a great deal of work by applying the formula to

$$\int_0^3 e^{-x^2} \, dx.$$

By ignoring the rest of the integral, you introduce an error less than 0.000042. If that is not precise enough, you might apply numerical integration to

$$\int_0^4 e^{-x^2}\,dx.$$

Then by the same argument as in the example, the error you introduce by ignoring the rest of the integral will be less than $e^{-16}/4 < 3 \times 10^{-8}$.

EXERCISES

In the following exercises you are asked to find bounds for certain integrals. There are generally several ways to estimate each integral. Try to obtain the bound given or to improve on it. If you cannot, at least find *some* bound.

Show:

1. $\dfrac{\pi}{2} < \displaystyle\int_0^{\pi/2} \frac{d\theta}{\sqrt{1 - k^2 \sin^2 \theta}} < \frac{\pi}{2\sqrt{1 - k^2}} \quad (0 < k^2 < 1)$

2. $\displaystyle\int_0^1 x \ln(1 + x)\,dx < \frac{\ln 2}{2}$

3. $\displaystyle\int_1^2 \frac{dx}{x^3 + 3x + 1} < \frac{1}{5}$

4. $\displaystyle\int_1^3 \frac{x\,dx}{(1 + x)^3} < \frac{1}{4}$

5. $\displaystyle\int_0^{100} e^{-x} \sin^2 x\,dx < 1$

6. $\displaystyle\int_0^{\pi/3} \sin 2\theta \cos \theta\,d\theta < \frac{\sqrt{3}}{2}$

7. $\displaystyle\int_5^{10} \frac{x^2\,dx}{3 + 2x} < \frac{75}{4}$

8. $\displaystyle\int_0^1 \frac{dx}{4 + x^3} > \ln \frac{5}{4}\cdot$

Without evaluating the integral, prove:

9. $\dfrac{3}{10} < \displaystyle\int_1^4 \frac{dx}{x^2 + x + 1} < \frac{3}{4}$

10. $3\sqrt{23} < \displaystyle\int_2^5 \sqrt{3x^3 - 1}\,dx < 10\sqrt{15} - \frac{8}{5}\sqrt{6}$

11. $3 \ln \dfrac{5}{3} < \displaystyle\int_3^5 \frac{\sqrt{3 + 2x}}{x}\,dx < \sqrt{13} \ln \frac{5}{3}$

12. $\dfrac{8}{\sqrt{13}} < \displaystyle\int_3^5 \frac{x\,dx}{\sqrt{3 + 2x}} < \frac{8}{3}$

13. $2 < \displaystyle\int_0^4 \frac{dx}{1 + \sin^2 x} < 4$

14. $\dfrac{99\pi}{400} < \displaystyle\int_1^{100} \frac{\arctan x}{x^2}\,dx < \frac{99\pi}{200}$

15. $\dfrac{609(\ln 2)^2}{4} < \displaystyle\int_2^5 x^3 (\ln x)^2\,dx < \frac{609(\ln 5)^2}{4}$

16. $(1 - e^{-1}) \ln 10 < \displaystyle\int_1^{10} \frac{1 - e^{-x}}{x}\,dx < \ln 10.$

17. Estimate how closely $\int_0^{10} e^{-x} \sin^2 x\, dx$ approximates $\int_0^{100} e^{-x} \sin^2 x\, dx$.

18. Find b such that $\int_b^1 \frac{\sin x}{x}\, dx$ approximates $\int_0^1 \frac{\sin x}{x}\, dx$ to within 5×10^{-5}. (Recall $\sin x < x$ for $x > 0$.)

7. ERRORS IN NUMERICAL INTEGRATION [optional]

It is important to have some idea of the error made when using a numerical integration formula. Often the error is expressed as an integral which can be estimated by techniques of the last section. We shall illustrate by a discussion of the error in the Trapezoidal Rule.

First we prove a preliminary result:

> If $f(a) = 0$ and $f(b) = 0$, then
> $$\int_a^b f(x)\, dx = -\frac{1}{2}\int_a^b (x - a)(b - x) f''(x)\, dx.$$

We derive this formula by two applications of integration by parts; in each case the "uv" part drops out.

$$\begin{aligned}\int_a^b [(x - a)(b - x)] f''(x)\, dx &= -\int_a^b [(b - x) - (x - a)] f'(x)\, dx \\ &= +\int_a^b (-2) f(x)\, dx = -2\int_a^b f(x)\, dx.\end{aligned}$$

Now follows another preliminary result:

> Suppose $f(a) = 0$ and $f(b) = 0$, and suppose
> $$|f''(x)| \leq M, \qquad a \leq x \leq b.$$
> Then
> $$\left|\int_a^b f(x)\, dx\right| \leq \frac{M}{12}(b - a)^3.$$

Indeed, from the preceding result,

$$\left|\int_a^b f(x)\, dx\right| = \frac{1}{2}\left|\int_a^b (x - a)(b - x) f''(x)\, dx\right|.$$

But $(x - a)(b - x) \geq 0$ between a and b, so we can use inequality (5) of the last section:

$$\left|\int_a^b f(x)\, dx\right| \leq \frac{1}{2} M \int_a^b (x - a)(b - x)\, dx.$$

A routine computation shows that the integral on the right is $\frac{1}{6}(b-a)^3$. Conclusion:

$$\left|\int_a^b f(x)\,dx\right| \leq \frac{M}{12}(b-a)^3.$$

We are now ready for the basic error estimate in the Trapezoidal Rule.

Two-Point Trapezoidal Rule with Error Suppose $|f''(x)| \leq M$ for $a \leq x \leq b$. Set $h = b - a$. Then

$$\int_a^b f(x)\,dx = \frac{h}{2}[f(a) + f(b)] + \epsilon,$$

where

$$|\epsilon| \leq \frac{Mh^3}{12}.$$

The trapezoidal approximation was obtained by fitting a *linear* function $p(x)$ to the points $(a, f(a))$ and $(b, f(b))$, then writing

$$\int_a^b f(x)\,dx \approx \int_a^b p(x)\,dx = \frac{h}{2}[f(a) + f(b)].$$

The error to be estimated is therefore

$$\epsilon = \int_a^b f(x)\,dx - \int_a^b p(x)\,dx = \int_a^b [f(x) - p(x)]\,dx.$$

But look at the function $g(x) = f(x) - p(x)$. Note two of its properties:

(i) $g(a) = g(b) = 0,$

(ii) $g''(x) = f''(x).$

(Verify these.) From (ii), $|g''(x)| \leq M$. Now the preceding result applies to $g(x)$:

$$\left|\int_a^b g(x)\,dx\right| \leq \frac{M}{12}(b-a)^3,$$

that is,

$$|\epsilon| = \left|\int_a^b [f(x) - p(x)]\,dx\right| \leq \frac{M}{12}h^3.$$

This is the stated error estimate.

Now we estimate the error in the n-point trapezoidal approximation.

Trapezoidal Rule with Error Divide the interval $a \le x \le b$ into n equal parts by inserting points:

$$a = x_0 < x_1 < x_2 < \cdots < x_n = b.$$

Set

$$h = \frac{b - a}{n}.$$

Suppose $|f''(x)| \le M$ for all x. Then

$$\int_a^b f(x)\,dx = \frac{h}{2}[f_0 + 2f_1 + 2f_2 + \cdots + 2f_{n-1} + f_n] + \epsilon,$$

where

$$|\epsilon| \le \frac{M(b - a)}{12}h^2.$$

Write

$$\int_a^b f(x)\,dx = \int_{x_0}^{x_1} + \int_{x_1}^{x_2} + \cdots + \int_{x_{n-1}}^{x_n},$$

and apply the two-point estimate to each integral on the right. The error in each case is at most $Mh^3/12$, so the total error is at most n times this:

$$|\epsilon| \le n\frac{Mh^3}{12} = \frac{M}{12}(nh)h^2 = \frac{M(b - a)}{12}h^2.$$

EXAMPLE 7.1

Suppose you estimate ln 2, approximating $\int_1^2 \frac{dx}{x}$ by the Trapezoidal Rule with $h = 0.1$. Find a bound for the error.

Solution:

$$|\text{error}| \le \frac{M(b - a)}{12}h^2,$$

where $a = 1$, $b = 2$, $h = 0.1$, and M is a bound for the second derivative of $f(x) = 1/x$. But $f''(x) = 2/x^3$, hence $|f''(x)| \le 2$ for $1 \le x \le 2$. Therefore

$$|\text{error}| \le \frac{2(2 - 1)}{12}(0.1)^2 = \frac{1}{6}(0.01).$$

Answer: $|\text{error}| < 0.0017$.

In some cases the actual error is much smaller than the trapezoidal estimate indicates. For example, if $f(x)$ has a point of inflection (change of concavity) in the interval, the trapezoidal errors tend to cancel (Fig. 7.1a).

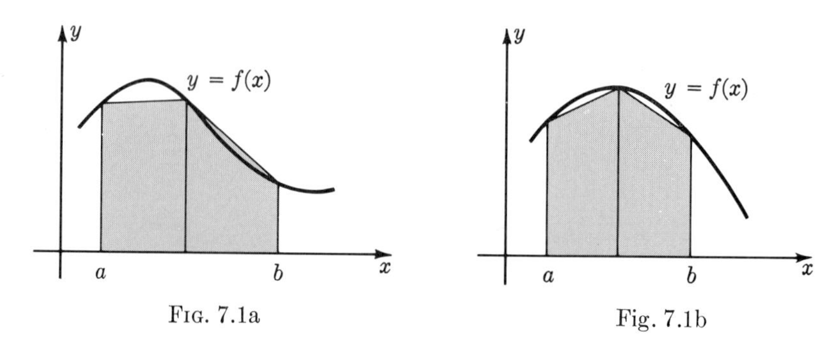

FIG. 7.1a

Fig. 7.1b

In Fig. 7.1b, however, there is no cancellation: all errors have the same sign. Since the estimate $M(b - a)h^2/12$ was made assuming all errors accumulate, the trapezoidal error estimate exceeds the actual error more in Fig. 7.1a than in Fig. 7.1b.

Other Error Estimates

By similar (although much harder) methods one finds the following estimates for the errors in other numerical integration methods.

THREE-POINT SIMPSON'S RULE

$$\int_a^b f(x)\,dx = \frac{h}{3}[f_0 + 4f_1 + f_2] + \epsilon,$$

$$|\epsilon| \le \frac{Mh^5}{90}, \qquad \text{where} \quad |f^{(4)}(x)| \le M.$$

THREE-EIGHTHS RULE

$$\int_a^b f(x)\,dx = \frac{3h}{8}[f_0 + 3f_1 + 3f_2 + f_3] + \epsilon,$$

$$|\epsilon| \le \frac{3Mh^5}{80}, \qquad \text{where} \quad |f^{(4)}(x)| \le M.$$

5-POINT NEWTON–COTES

$$\int_a^b f(x)\,dx = \frac{2h}{45}[7f_0 + 32f_1 + 12f_2 + 32f_3 + 7f_4] + \epsilon,$$

$$|\epsilon| \le \frac{8Mh^7}{945}, \qquad \text{where} \quad |f^{(6)}(x)| \le M.$$

7-POINT NEWTON–COTES

$$\int_a^b f(x)\,dx = \frac{h}{140}[41f_0 + 216f_1 + 27f_2 + 272f_3 + 27f_4 + 216f_5 + 41f_6] + \epsilon,$$

$$|\epsilon| \le \frac{9Mh^9}{1400}, \qquad \text{where} \quad |f^{(8)}(x)| \le M.$$

9-POINT NEWTON–COTES

$$\int_a^b f(x)\,dx = \frac{4h}{14175}[989f_0 + 5888f_1 - 928f_2 + 10496f_3 - 4540f_4$$
$$+ 10496f_5 - 928f_6 + 5888f_7 + 989f_8] + \epsilon,$$

$$|\epsilon| \le \frac{2368Mh^{11}}{467775}, \qquad \text{where} \quad |f^{(10)}(x)| \le M.$$

WEDDLE'S RULE

$$\int_a^b f(x)\,dx = \frac{3h}{10}[f_0 + 5f_1 + f_2 + 6f_3 + f_4 + 5f_5 + f_6] + \epsilon,$$

$$|\epsilon| \le \frac{h^7}{1400}(10M + 9h^2N), \qquad \text{where} \quad |f^{(6)}(x)| \le M, \qquad |f^{(8)}(x)| \le N.$$

EXAMPLE 7.2

Suppose you estimate ln 2 as in Example 7.1, but use Simpson's Rule with $h = 0.1$. Find a bound for the error.

Solution: Divide the interval $1 \le x \le 2$ into 5 equal parts. Write

$$\ln 2 = \int_1^2 \frac{dx}{x} = \int_{1.0}^{1.2} + \int_{1.2}^{1.4} + \int_{1.4}^{1.6} + \int_{1.6}^{1.8} + \int_{1.8}^{2.0},$$

and apply the three-point Simpson's Rule to each integral on the right. In each case the error is at most

$$\frac{M(0.1)^5}{90},$$

where M is a bound for the fourth derivative of $f(x) = 1/x$. Since

$$f^{(4)}(x) = \frac{2 \cdot 3 \cdot 4}{x^5} \qquad \text{and} \qquad 1 \le x \le 2,$$

take $M = 2 \cdot 3 \cdot 4 = 24$. Thus each of the 5 errors is at most

$$\frac{24(0.1)^5}{90},$$

and the combined error is at most 5 times this much.

Answer: $|\text{error}| < 1.4 \times 10^{-5}$.

REMARK: We were wasteful using the same value of M in each of the 5 integrals. For a sharper estimate, we can use

$$M = \frac{24}{1^5}, \quad \frac{24}{(1.2)^5}, \quad \frac{24}{(1.4)^5}, \quad \frac{24}{(1.6)^5}, \quad \frac{24}{(1.8)^5},$$

respectively. Then we find

$$|\text{error}| < 4.7 \times 10^{-6}.$$

EXERCISES

Estimate the error in using the Trapezoidal Rule, Simpson's Rule, and the Three-Eighths Rule to approximate the integral:

1. $\int_0^1 \sin x\,dx, \quad h = \frac{1}{6}$
2. $\int_{\pi/4}^{\pi/2} x \sin x\,dx, \quad h = \pi/24$
3. $\int_1^3 \ln x\,dx, \quad h = \frac{1}{3}$
4. $\int_0^1 e^{-x}\,dx, \quad h = \frac{1}{6}$.

Estimate the error in using the 7-point Newton–Cotes Rule and Weddle's Rule to approximate the integral:

5. $\int_0^2 \cos x\,dx, \quad h = \frac{1}{3}$
6. $\int_1^4 \ln x\,dx, \quad h = \frac{1}{2}$
7. $\int_{-2}^1 x^2 e^x\,dx, \quad h = \frac{1}{2}$
8. $\int_0^{\pi/2} \sin x \cos 2x\,dx, \quad h = \pi/12$.
9. Suppose you are to estimate $\int_0^6 x^2 e^{-x}\,dx$, using $h = 0.25$. Find one of the methods of this chapter which will guarantee an error of at most 5×10^{-5}.
10, Write a program to carry out the estimation chosen in Ex. 9.
11. Show that multiplying the number of points by a factor of 10 in the Trapezoidal Rule improves the error estimate by a factor of 100.
12. Suppose $f(-1) = f(0) = f(1) = 0$. Show (integration by parts) that

$$\int_{-1}^1 f(x)\,dx = -\frac{1}{6}\left[\int_{-1}^0 x(x+1)^2 f'''(x)\,dx + \int_0^1 x(x-1)^2 f'''(x)\,dx\right].$$

13. (cont.) Show further that

$$72 \int_{-1}^1 f(x)\,dx = \int_{-1}^0 (x+1)^3(3x-1) f^{(4)}(x)\,dx + \int_0^1 (x-1)^3(3x+1) f^{(4)}(x)\,dx.$$

14. (cont.) Conclude that if $f(-1) = f(0) = f(1) = 0$ and $f^{(4)}(x) = 0$ for $-1 < x < 1$, then $\int_{-1}^1 f(x)\,dx = 0$.

15. (cont.) Suppose $f(-1) = f(0) = f(1) = 0$ and $|f^{(4)}(x)| \leq M$ for $-1 \leq x \leq 1$. Use Ex. 13 to prove that $\left| \int_{-1}^{1} f(x)\,dx \right| \leq \dfrac{M}{90}$.

16. (cont.) Deduce the error estimate in the three-point Simpson's Rule.

17. Suppose Simpson's Rule is used with 5, 9, or $2n + 1$ points. Find an error estimate.

18. Do the same for the $(3n + 1)$-point Three-Eighths Rule.

19. Do the same for the $(4n + 1)$-point 5-point Newton–Cotes Rule.

20. Consider the estimate in the second box on p. 405. Stretching the interval $a \leq x \leq b$ by a factor c appears to divide the left-hand side of the inequality by c and the right-hand side by c^3, or does it?

21. First Order Differential Equations

1. INTRODUCTION

In Chapter 4 we discussed differential equations of the type

$$\frac{dy}{dx} = f(x).$$

Each solution is an antiderivative of $f(x)$. In Chapter 7 we discussed the differential equation

$$\frac{dy}{dx} = y,$$

each solution of which is a function $y = ce^x$.

These differential equations are instances of the most general **first order differential equation,**

$$\frac{dy}{dx} = q(x, y),$$

where $q(x, y)$ is a function of two variables. "First order" refers to the presence of the first derivative only, not the second or third, etc.

Recall that a differential equation $dy/dx = q(x, y)$ defines a **direction field**. At each point (x, y) where $q(x, y)$ is defined, imagine a short line segment of slope $q(x, y)$, as in Fig. 1.1. A **solution** of the differential equation is a function $y = y(x)$ whose graph has slope matching that of the direction field (Fig. 1.2).

The most important problem in this subject is the **initial-value problem:**

> **Initial-Value Problem** Find the solution $y = y(x)$ of the differential equation
>
> $$\frac{dy}{dx} = q(x, y)$$
>
> whose graph passes through a given point (a, b), that is, which satisfies the **initial condition**
>
> $$y(a) = b.$$

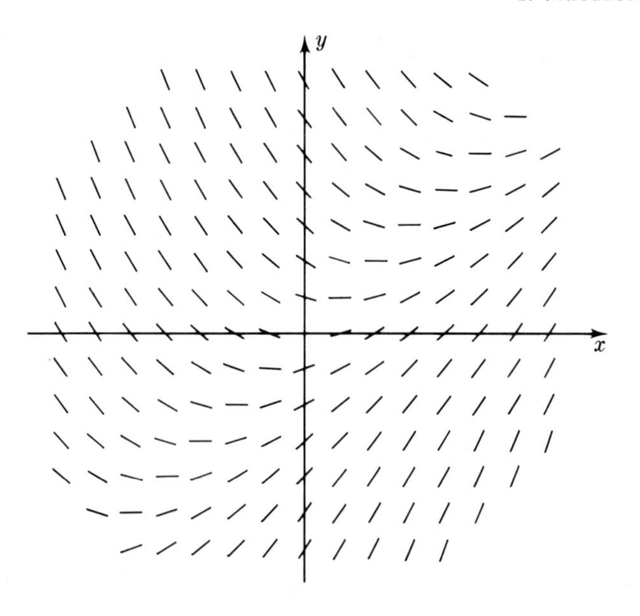

FIG. 1.1

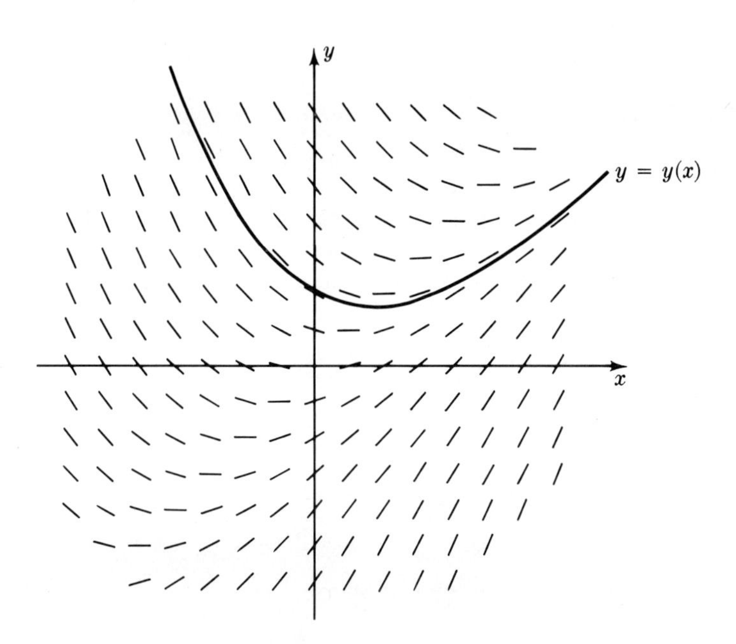

FIG. 1.2

Figure 1.3 shows two solutions of

$$\frac{dy}{dx} = q(x, y),$$

one satisfying the initial condition $y(0) = 0$, the other satisfying the initial condition $y(1) = -1$.

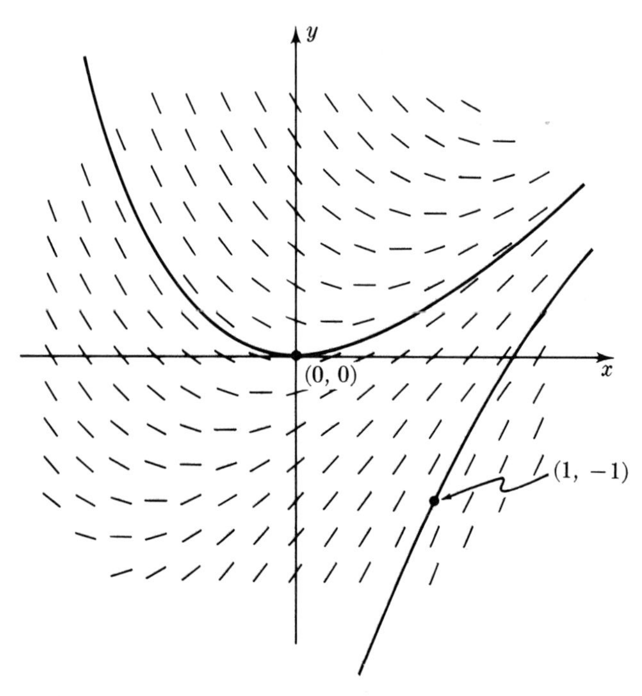

FIG. 1.3

Usually we first find the general solution to the differential equation, ignoring the initial condition. Somewhere along the line we are forced to integrate, thus introducing a constant of integration. We then select the constant so that the initial condition is satisfied.

EXAMPLE 1.1

Find a solution of the initial-value problem

$$\begin{cases} \dfrac{dy}{dx} = x^2 \\ y(-3) = 1. \end{cases}$$

Solution: Antidifferentiate:

$$y(x) = \frac{1}{3}x^3 + c.$$

This is the general solution of the differential equation. Now select the constant by substituting the initial condition:

$$1 = \frac{1}{3}(-3)^3 + c, \qquad c = 10.$$

Answer: $y(x) = \frac{1}{3}x^3 + 10.$

EXERCISES

Solve the initial-value problem:

1. $\frac{dy}{dx} = x^3, \quad y(0) = 0$
2. $\frac{dy}{dx} = x^2 + x - 5, \quad y(0) = -2$
3. $\frac{dy}{dx} = \frac{3}{x}, \quad y(1) = 1$
4. $\frac{dy}{dx} = \cos 2x, \quad y(\pi) = 0$
5. $\frac{dy}{dx} = e^{-x}, \quad y(0) = 5$
6. $\frac{dy}{dx} = xe^{x^2}, \quad y(0) = 1$
7. $\frac{dy}{dx} = y, \quad y(1) = 10$
8. $\frac{dy}{dx} = y, \quad y(1) = 0$
9. $\frac{dy}{dx} = x + \sin x, \quad y(0) = 0$
10. $\frac{dy}{dx} = \sqrt{x+1}, \quad y(0) = 2$
11. $\frac{dy}{dx} = \frac{x}{\sqrt{x^2+1}}, \quad y(0) = 0$
12. $\frac{dy}{dx} = xe^x, \quad y(1) = 5$
13. $\frac{dy}{dx} = (1 + x^2)(1 + x^3), \quad y(0) = 0.$
14. Sketch the direction field of $dy/dx = y^2$. Draw the solutions passing through $(0, 1)$; through $(0, -1)$. Rewrite the equation as $dx/dy = y^{-2}$ and solve.
15. Solve $xy' + y = 2x$.
    [*Hint:* Compute $(xy)'$.]
16. Solve $y' + y = e^{-x}\sin x$.
    [*Hint:* Compute $(e^x y)'$.]
17. Solve $x + yy' = x^2 + y^2$.
    [*Hint:* Compute the derivative of $\ln(x^2 + y^2)$.]
18. Solve $xy' - y = x^2 \sin x$.
    [Find the trick yourself this time.]

2. SEPARATION OF VARIABLES

The general first order differential equation is

$$\frac{dy}{dx} = q(x, y).$$

In this section, we study the special case in which $q(x, y) = f(x)g(y)$, the product of a function of x alone by a function of y alone. Examples are

$$\frac{dy}{dx} = x^2y^3, \qquad \frac{dy}{dx} = e^x \cos y, \qquad \frac{dy}{dx} = \frac{y}{\sqrt{1 + x^2}}.$$

Not included in this special case are

$$\frac{dy}{dx} = x + y, \qquad \frac{dy}{dx} = \sin(xy), \qquad \frac{dy}{dx} = \sqrt{1 + x^2 + y^2},$$

since in each of these equations the right-hand side is not of the form $f(x)g(y)$.

Solve

$$\frac{dy}{dx} = f(x)g(y)$$

by first "separating the variables," putting all the "y" on one side and all the "x" on the other. Write

$$\frac{dy}{g(y)} = f(x)\,dx.$$

Then integrate both sides (if you can):

$$G(y) = F(x) + c,$$

where $G(y)$ is an antiderivative of $1/g(y)$ and $F(x)$ is an antiderivative of $f(x)$. If this equation can be solved for y in terms of x, the resulting function is a solution of the differential equation. (This technique will be justified after several examples.)

Note that in separating the variables you must assume that $g(y) \neq 0$. If $g(y) = 0$ for $y = y_0$, it is simple to check that the constant function $y(x) = y_0$ is a solution of the equation.

EXAMPLE 2.1

Solve $\dfrac{dy}{dx} = xy.$

Solution: Obviously $y(x) = 0$ is a solution. Now assume $y(x) \neq 0$ and separate variables:

$$\frac{dy}{y} = x\,dx, \qquad \int \frac{dy}{y} = \int x\,dx,$$

$$\ln|y| = \frac{1}{2}x^2 + c,$$

$$|y| = ke^{x^2/2} \qquad (k = e^c),$$

where the constant $k = e^c$ is positive. If $y \geq 0$, then $|y| = y$; if $y < 0$, then $|y| = -y$. Hence

$$y = \begin{cases} ke^{x^2/2} & \text{if} \quad y \geq 0 \\ -ke^{x^2/2} & \text{if} \quad y < 0. \end{cases}$$

This is equivalent to

$$y = ae^{x^2/2},$$

where a is any non-zero constant.

Answer: $y = ae^{x^2/2}$, a any constant. (Note that the solution $y(x) = 0$ is included in this answer.)

Check:

$$\frac{dy}{dx} = \frac{d}{dx}(ae^{x^2/2}) = axe^{x^2/2} = xy.$$

EXAMPLE 2.2

Solve $\dfrac{dy}{dx} = \dfrac{2x(y-1)}{x^2+1}$.

Solution: Obviously $y(x) = 1$ is a solution. Now assume $y(x) \neq 1$ and separate variables:

$$\frac{dy}{y-1} = \frac{2x\,dx}{x^2+1}, \qquad \int \frac{dy}{y-1} = \int \frac{2x\,dx}{x^2+1},$$

$$\ln|y-1| = \ln(x^2+1) + c,$$

$$|y-1| = e^c(x^2+1) = k(x^2+1).$$

Therefore

$$y - 1 = \pm k(x^2+1),$$

where k is a positive constant. The answer may be written

$$y = 1 + a(x^2+1),$$

where a is an arbitrary constant. This includes the special solution $y(x) = 1$.

Answer: $y = 1 + a(x^2+1)$.

Check:

$$\frac{dy}{dx} = 2ax = 2x\,\frac{a(x^2+1)}{x^2+1} = \frac{2x(y-1)}{x^2+1}.$$

EXAMPLE 2.3

Solve $\dfrac{dy}{dx} = e^{x-y}$.

Solution:

$$\frac{dy}{dx} = e^x e^{-y}, \qquad e^y\, dy = e^x\, dx, \qquad e^y = e^x + c.$$

Answer: $y = \ln(e^x + c), \qquad e^x + c > 0.$

Check:

$$\frac{dy}{dx} = \frac{1}{e^x + c}\frac{d}{dx}(e^x + c) = \frac{e^x}{e^x + c} = \frac{e^x}{e^y} = e^{x-y}.$$

These examples show that the method works. Why does it work? Suppose $y = y(x)$ satisfies

$$\frac{dy}{dx} = f(x)g(y).$$

Let $G(y)$ be any antiderivative of $1/g(y)$. Note that $G(y)$ is indirectly a function of x. By the Chain Rule,

$$\frac{dG}{dx} = \frac{dG}{dy}\frac{dy}{dx} = \frac{1}{g(y)} f(x)g(y) = f(x).$$

Hence $G(y)$ is an antiderivative of $f(x)$. Therefore, if $F(x)$ is *any* antiderivative of $f(x)$,

$$G(y) = F(x) + c.$$

If you can solve this equation for y as a function of x, you have the general solution. If not, then at least you have a relation between x and y, often adequate for applications.

EXAMPLE 2.4

Solve the initial-value problem $\dfrac{dy}{dx} = \dfrac{2x}{5y^4 - 1}, \quad y(1) = 0.$

Solution:

$$(5y^4 - 1)\, dy = 2x\, dx, \qquad y^5 - y = x^2 + c.$$

We cannot solve this fifth degree equation for y; it is hopeless. Still we can substitute the initial data:

$$0 = 1 + c, \qquad c = -1;$$
$$y^5 - y = x^2 - 1.$$

This relation between x and y is as far as we shall get with the problem. But it is adequate for a graph and can be used to calculate y (given x) to any degree of accuracy.

Answer: $y^5 - y = x^2 - 1.$

EXAMPLE 2.5

Solve the initial-value problem $\frac{dy}{dx} = x^2y^2, \quad y(0) = b.$

Solution: Obviously $y(x) = 0$ is the solution if $b = 0$. Assume $y \neq 0$ and separate variables:

$$\frac{dy}{y^2} = x^2\,dx, \qquad -\frac{1}{y} = \frac{x^3}{3} + c.$$

Substitute the initial data:

$$-\frac{1}{b} = 0 + c = c.$$

Hence

$$-\frac{1}{y} = \frac{x^3}{3} - \frac{1}{b} = \frac{bx^3 - 3}{3b}.$$

Answer: $y = \frac{3b}{3 - bx^3}\cdot$ (The answer includes the solution $y(x) = 0$.)

EXAMPLE 2.6

Find all curves with the following geometric property: the slope of the curve at each point P is twice the slope of the line through P and the origin.

Solution: Translate the geometrical property into an equation. Suppose the graph of $y(x)$ is such a curve and $P = (x, y)$ a point on it. The slope at P is dy/dx. The slope of the line through P and $(0, 0)$ is y/x. Hence, the geometrical property is expressed by the differential equation

$$\frac{dy}{dx} = 2\frac{y}{x}\cdot$$

Obviously $y(x) = 0$ is a solution. Now assume $y(x) \neq 0$ and separate variables:

$$\frac{dy}{y} = 2\frac{dx}{x}, \qquad \ln|y| = 2\ln|x| + c = \ln x^2 + c.$$

Take exponentials, setting $k = e^c$:

$$|y| = kx^2.$$

As in Example 2.1, it follows that $y = ax^2$, where a is any constant.

Answer: The curves are the parabolas $y = ax^2$ and the line $y = 0$. See Fig. 2.1.

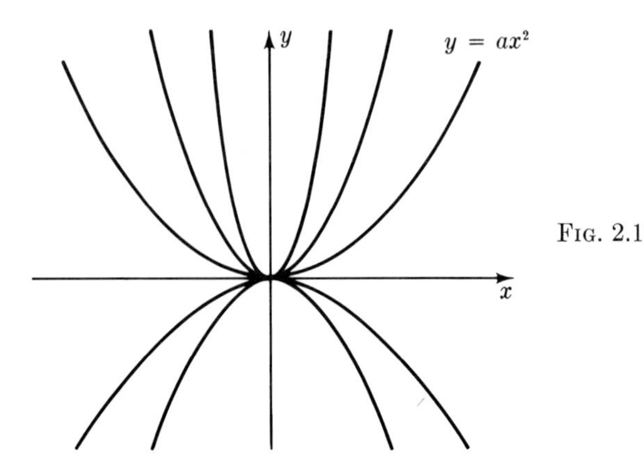

FIG. 2.1

EXAMPLE 2.7

Find all curves whose slope at each point P is the reciprocal of the slope of the line through P and the origin.

Solution: The problem calls for all functions $y(x)$ satisfying

$$\frac{dy}{dx} = \frac{x}{y}.$$

Separate variables:

$$y\,dy = x\,dx, \qquad \frac{1}{2}y^2 = \frac{1}{2}x^2 + c.$$

Answer: The curves are the rectangular hyperbolas $y^2 - x^2 = c$. See Fig. 2.2.

Note the special case $c = 0$, $y = \pm x$.

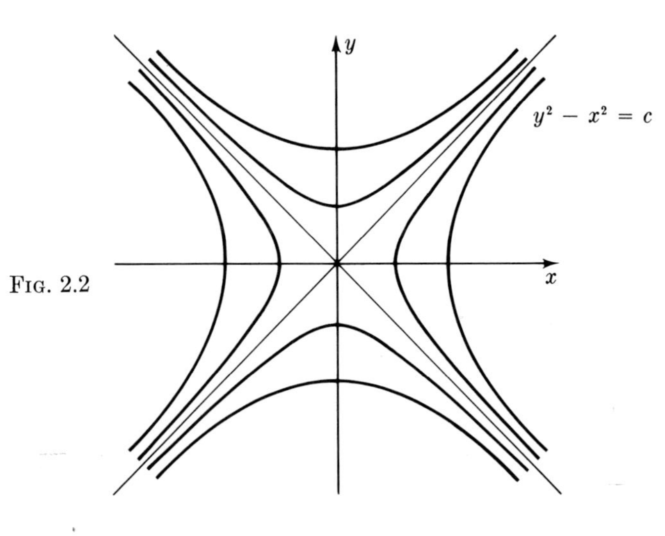

FIG. 2.2

EXERCISES

Solve:

1. $\dfrac{dy}{dx} = \dfrac{x^2}{y}$
2. $\dfrac{dy}{dx} = \dfrac{1}{xy}$
3. $\dfrac{dy}{dx} = \sqrt{\dfrac{y}{x}}$
4. $\dfrac{dy}{dx} = \dfrac{x+1}{xy}$
5. $\dfrac{dy}{dx} = xe^y$
6. $\dfrac{dy}{dx} = \left(\dfrac{1+y}{1+x}\right)^2.$

Show that the substitution $y = ux$ changes the differential equation into an equation for $u = u(x)$ whose variables separate, and solve:

7. $\dfrac{dy}{dx} = \dfrac{x^2 + y^2}{2x^2}$
8. $\dfrac{dy}{dx} = \dfrac{x+y}{x-y}.$

Solve the initial-value problem:

9. $\dfrac{dy}{dx} = -\dfrac{y}{2x}, \quad y(1) = 3$
10. $\dfrac{dy}{dx} = 2y \cot x, \quad y\left(\dfrac{\pi}{2}\right) = 5$
11. $\dfrac{dy}{dx} = \dfrac{(y-1)(y-2)}{x}, \quad y(4) = 0$
12. $\dfrac{dy}{dx} = y^3 \sin x, \quad y(0) = \dfrac{1}{2}.$

Solve the given differential equation; interpret the equation and its solution geometrically:

13. $\dfrac{dy}{dx} = -\dfrac{y}{x}$
14. $\dfrac{dy}{dx} = -\dfrac{x}{y}$
15. $\dfrac{dy}{dx} = \dfrac{y}{x}.$

3. LINEAR DIFFERENTIAL EQUATIONS

A differential equation of the form

$$\frac{dy}{dx} + p(x)y = q(x)$$

is called a **first order linear differential equation.**

The term "linear" can be described in the following way. Imagine a "black box" or "processor" that converts a function $y(x)$ into a function $w(x)$, as in Fig. 3.1.

$y(x)$ —input→ [black box] —output→ $w(x)$.

Fig. 3.1

The black box is called **linear** if from a linear combination* of inputs, it produces the same linear combination of outputs (Fig. 3.2):

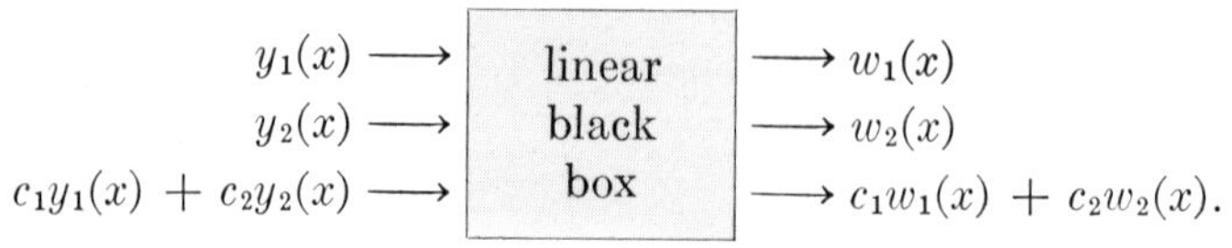

$$\begin{array}{rcl} y_1(x) \longrightarrow & \text{linear} & \longrightarrow w_1(x) \\ y_2(x) \longrightarrow & \text{black} & \longrightarrow w_2(x) \\ c_1y_1(x) + c_2y_2(x) \longrightarrow & \text{box} & \longrightarrow c_1w_1(x) + c_2w_2(x). \end{array}$$

FIG. 3.2

In particular, the black box that converts y into $dy/dx + p(x)y$ is linear: when the input is $c_1y_1 + c_2y_2$, the output is

$$\frac{d}{dx}(c_1y_1 + c_2y_2) + p(x)\,(c_1y_1 + c_2y_2)$$
$$= c_1\left(\frac{dy_1}{dx} + p(x)y_1\right) + c_2\left(\frac{dy_2}{dx} + p(x)y_2\right)$$
$$= c_1(\text{output}_1) + c_2(\text{output}_2).$$

This is why the differential equation

$$\frac{dy}{dx} + p(x)y = q(x)$$

is called linear.

Not every black box is linear, however. For instance, consider the black box in Fig. 3.3.

$$y \longrightarrow \boxed{\text{black box}} \longrightarrow \sqrt{\frac{dy}{dx}}.$$

FIG. 3.3

This box is not linear since the output of a sum is not the sum of the outputs (Fig. 3.4).

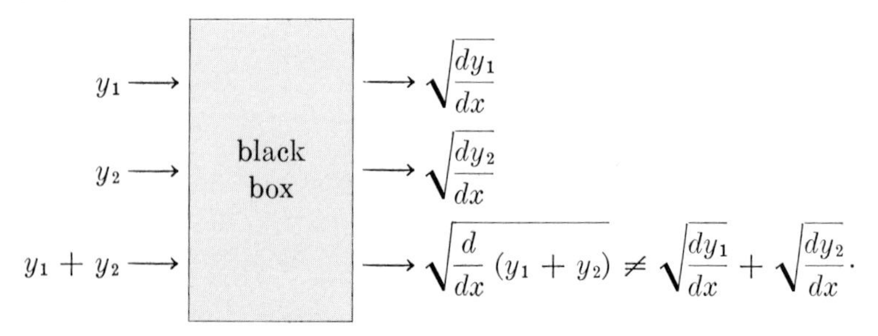

FIG. 3.4

* An expression $aX + bY + cZ + \cdots$, where $a, b, c, \cdots$ are numbers, is called a **linear combination** of $X, Y, Z, \cdots$.

Figure 3.5 shows another non-linear process.

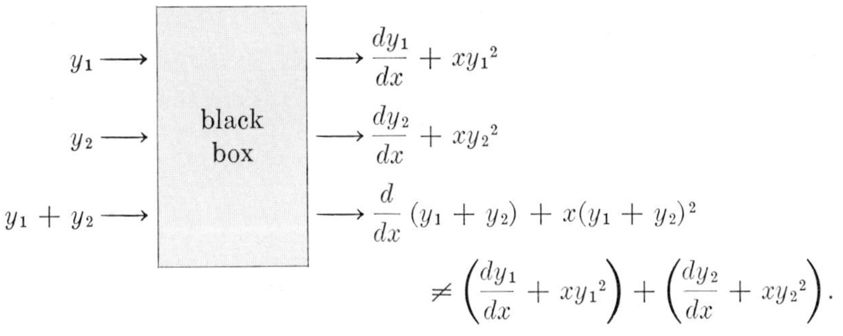

$$y_1 \longrightarrow \boxed{\text{black box}} \longrightarrow \frac{dy_1}{dx} + xy_1^2$$

$$y_2 \longrightarrow \qquad \longrightarrow \frac{dy_2}{dx} + xy_2^2$$

$$y_1 + y_2 \longrightarrow \qquad \longrightarrow \frac{d}{dx}(y_1 + y_2) + x(y_1 + y_2)^2$$

$$\neq \left(\frac{dy_1}{dx} + xy_1^2\right) + \left(\frac{dy_2}{dx} + xy_2^2\right).$$

FIG. 3.5

A Property of Linear Processes

An extremely useful property of linear black boxes is shown in Fig. 3.6.

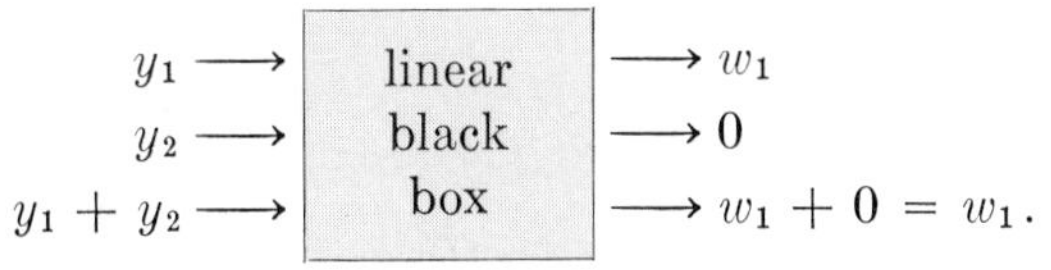

$$y_1 \longrightarrow \boxed{\text{linear black box}} \longrightarrow w_1$$

$$y_2 \longrightarrow \qquad \longrightarrow 0$$

$$y_1 + y_2 \longrightarrow \qquad \longrightarrow w_1 + 0 = w_1.$$

FIG. 3.6

If the output of y_2 is 0, then the output of $y_1 + y_2$ is the output of y_1. Hence output is unchanged when input is augmented by a function which is converted to zero.

This simple fact is the basis for a systematic study of linear differential equations. We return to the process that really interests us, the one that converts $y(x)$ into $dy/dx + p(x)y$. For brevity, we write

$$\frac{dy}{dx} + p(x)y = Ly,$$

indicating that Ly is the output corresponding to the input y. The differential equation in question is

$$\frac{dy}{dx} + p(x)y = q(x),$$

or

$$Ly = q.$$

A **solution** is an input $y(x)$ that results in the given output $q(x)$.

Suppose we can find a solution y_1. That means $Ly_1 = q$. If z is any function for which $Lz = 0$, then $L(y_1 + z) = q$. Thus $y_1 + z$ is also a solution. Furthermore, *every* solution must be of the form $y_1 + z$. To see why, assume y_1 and y_2 are solutions:

$$Ly_1 = q, \qquad Ly_2 = q.$$

Consider the difference $y_1 - y_2$. Because L is a *linear* process,

$$L(y_1 - y_2) = Ly_1 - Ly_2 = q - q = 0.$$

Therefore, when the input is $y_1 - y_2$, the output is 0; this says the difference of two solutions is one of the functions z.

Any two solutions of the differential equation

$$\frac{dy}{dx} + p(x)y = q(x)$$

differ by a solution of the **homogeneous equation**

$$\frac{dy}{dx} + p(x)y = 0.$$

Therefore, the general solution of

$$\frac{dy}{dx} + p(x)y = q(x)$$

is

$$u(x) + z(x),$$

where $u(x)$ is any solution of the equation and $z(x)$ is the general solution of the associated homogeneous equation.

Because of this analysis, the differential equation (often called the **non-homogeneous equation**)

$$\frac{dy}{dx} + p(x)y = q(x)$$

is solved in three steps:

(1) Find the general solution of the associated homogeneous equation

$$\frac{dy}{dx} + p(x)y = 0.$$

(2) Find any solution of

$$\frac{dy}{dx} + p(x)y = q(x).$$

(3) Add the results.

In the next two sections we shall treat steps (1) and (2) separately.

EXERCISES

Suppose the input to a black box is $y(x)$ and the output is one of the following. Determine in each case whether the black box is linear or non-linear:

1. $\int_0^1 y(x)\,dx$
2. $y(x) + x$
3. $\dfrac{d^2y}{dx^2} + 6\dfrac{dy}{dx} + y$
4. $3y(x)$
5. $\ln |y(x)|$
6. $y(x+1)$.
7. What is the general solution of the differential equation $dy/dx = x^2$? Is your answer a particular solution plus the general solution of the associated homogeneous equation?
8. A big black box consists of two linear black boxes in series:

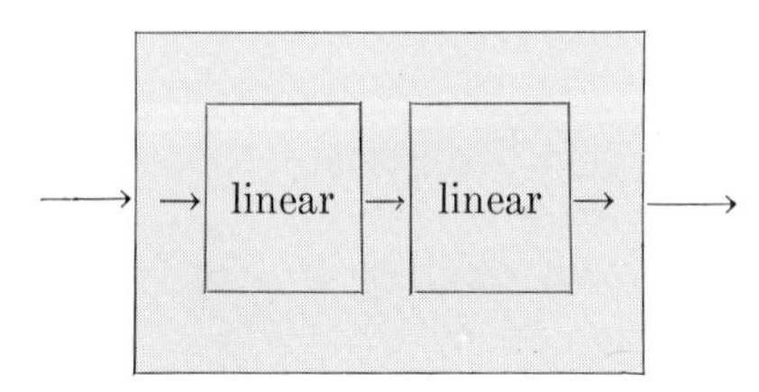

Is the big box linear?

4. HOMOGENEOUS EQUATIONS

Let us solve the homogeneous equation

$$\frac{dy}{dx} + p(x)y = 0.$$

Obviously $y(x) = 0$ is a solution. Assume $y(x) \not\equiv 0$ and separate variables:

$$\frac{dy}{y} = -p(x)\,dx, \qquad \ln |y| = -P(x) + c,$$

where $P(x)$ is any antiderivative of $p(x)$. Now exponentiate both sides, setting $k = e^c$:

$$|y(x)| = ke^{-P(x)}.$$

Hence,

$$y(x) = \begin{cases} ke^{-P(x)} & \text{if } \; y \geq 0 \\ -ke^{-P(x)} & \text{if } \; y < 0. \end{cases}$$

These facts are summarized in the following statement.

The general solution of

$$\frac{dy}{dx} + p(x)y = 0$$

is

$$y(x) = ae^{-P(x)},$$

where $P(x)$ is any antiderivative of $p(x)$ and a is a constant.

EXAMPLE 4.1

Find the general solution of $\dfrac{dy}{dx} + (\sin x)y = 0.$

Solution: The equation is of the form

$$\frac{dy}{dx} + p(x)y = 0,$$

where $p(x) = \sin x$. An antiderivative of $p(x)$ is $P(x) = -\cos x$. By the formula, the general solution is

$$y(x) = ae^{-P(x)} = ae^{\cos x}.$$

Answer: $y(x) = ae^{\cos x}.$

EXAMPLE 4.2

Solve the initial-value problem $y' + 3y = 0, \quad y(1) = 4.$

Solution: In this case $p(x) = 3$. An antiderivative is $P(x) = 3x$ and the general solution is

$$y(x) = ae^{-3x}.$$

To choose the correct constant, substitute $x = 1$, $y = 4$:

$$4 = ae^{-3}, \qquad a = 4e^{3}.$$

Answer: $y(x) = 4e^{3(1-x)}.$

EXAMPLE 4.3

Solve the initial-value problem $\dfrac{dy}{dx} + \dfrac{1}{x}y = 0, \quad y(1) = 2.$

Solution: Since $p(x) = 1/x$ is not defined at $x = 0$, the solution may behave badly near $x = 0$. Therefore consider only $x > 0$.

An antiderivative of $1/x$ is $\ln x$; the general solution is

$$y(x) = ae^{-\ln x} = \frac{a}{x}$$

(which $\longrightarrow \infty$ as $x \longrightarrow 0$). To choose the correct constant, substitute $x = 1$, $y = 2$:

$$2 = \frac{a}{1}.$$

Answer: $y(x) = \frac{2}{x}, \quad x > 0.$

EXERCISES

Solve:

1. $\frac{dy}{dx} + 3\frac{y}{x} = 0$
2. $L\frac{di}{dt} + Ri = 0, \quad L, R$ constants
3. $\frac{dy}{d\theta} + y \tan \theta = 0$
4. $\frac{1}{2x}\frac{dy}{dx} + \frac{y}{x^2 + 1} = 0$
5. $x\frac{dy}{dx} + (1 + x)y = 0.$

Solve the initial-value problem:

6. $\frac{dy}{dx} + xy = 0, \quad y(0) = -1$
7. $(1 - x^2)\frac{dy}{dx} + xy = 0, \quad y(0) = 3$
8. $\frac{dy}{dx} + y\sqrt{x} = 0, \quad y(0) = 2$
9. $\frac{dy}{dx} + \frac{2y}{x} = 0, \quad y(2) = \frac{1}{5}.$
10. Solve $y'' - y' = 0$ by substituting $v = y'$.

5. NON-HOMOGENEOUS EQUATIONS

We now consider the non-homogeneous equation

$$\frac{dy}{dx} + p(x)y = q(x).$$

By our analysis of the problem, we need only one particular solution of this equation, to be added to the general solution of the associated homogeneous equation. We shall discuss two methods for finding a particular solution.

METHOD 1. GUESSING.

This works particularly well when

(a) $p(x)$ is a constant, and

(b) $q(x)$ and all of its derivatives can be expressed in terms of a few functions. For example, a quadratic polynomial and its derivatives can be expressed in terms of 1, x, and x^2. The function $\sin x$ and its derivatives can be expressed in terms of $\sin x$ and $\cos x$. Each of the functions

$$e^x, \qquad xe^x, \qquad x^2e^x, \qquad x\cos x, \qquad e^x \sin x$$

has a similar property; it and all its derivatives involve only a few functions. However, $1/x$ is not of this type; each of its derivatives involves a different function.

EXAMPLE 5.1

Find a solution of $\dfrac{dy}{dx} - y = x^2$.

Solution: The right side of the equation is a quadratic polynomial. Guess a quadratic polynomial

$$y(x) = Ax^2 + Bx + C,$$

and try to determine suitable coefficients A, B, C. Substitute $y(x)$ into the differential equation:

$$(2Ax + B) - (Ax^2 + Bx + C) = x^2,$$

$$-Ax^2 + (2A - B)x + (B - C) = x^2.$$

Equate coefficients:

$$-A = 1, \qquad 2A - B = 0, \qquad B - C = 0,$$

from which

$$A = -1, \qquad B = -2, \qquad C = -2.$$

Answer: $y(x) = -(x^2 + 2x + 2)$ is a solution.

EXAMPLE 5.2

Find a solution of $\dfrac{dy}{dx} - y = e^{3x}$.

Solution: Try $y = Ae^{3x}$. The equation becomes

$$3Ae^{3x} - Ae^{3x} = e^{3x}.$$

Hence,

$$2A = 1, \qquad A = \frac{1}{2}.$$

Answer: $y(x) = \dfrac{1}{2}e^{3x}$ is a solution.

EXAMPLE 5.3

Find a solution of $\dfrac{dy}{dx} - y = xe^{3x}$.

Solution: Notice that xe^{3x} and all its derivatives involve only the functions xe^{3x} and e^{3x}. Try

$$y(x) = Axe^{3x} + Be^{3x}.$$

The differential equation becomes

$$(Ae^{3x} + 3Axe^{3x} + 3Be^{3x}) - (Axe^{3x} + Be^{3x}) = xe^{3x},$$

$$2Axe^{3x} + (A + 2B)e^{3x} = xe^{3x}.$$

Thus, $2A = 1$, $A + 2B = 0$, from which $A = \frac{1}{2}$, $B = -\frac{1}{4}$.

Answer: $y(x) = \dfrac{1}{2}xe^{3x} - \dfrac{1}{4}e^{3x}$

is a solution.

EXAMPLE 5.4

Find a solution of $\dfrac{dy}{dx} - y = \sin x$.

Solution: All derivatives of $\sin x$ involve only $\sin x$ and $\cos x$. Try

$$y(x) = A\cos x + B\sin x.$$

The differential equation becomes

$$(-A\sin x + B\cos x) - (A\cos x + B\sin x) = \sin x,$$

$$-(A + B)\sin x + (B - A)\cos x = \sin x.$$

Therefore

$$-(A + B) = 1, \qquad B - A = 0.$$

Thus $A = B = -\frac{1}{2}$.

Answer: $y(x) = -\dfrac{1}{2}\cos x - \dfrac{1}{2}\sin x$

is a solution.

There is a special situation in which the method must be modified. Consider for instance the differential equation

$$\frac{dy}{dx} - y = e^{x}.$$

Try $y(x) = Ae^{x}$:

$$\frac{d}{dx}(Ae^{x}) - Ae^{x} = e^{x},$$

$$Ae^{x} - Ae^{x} = e^{x}, \qquad 0 = e^{x},$$

which is impossible. The guess fails because e^x is a solution of the homogeneous equation

$$\frac{dy}{dx} - y = 0.$$

Consequently, substituting $y = Ae^x$ makes the left side zero; there is no hope of equating it to the right side. The function xe^x is a solution, however, as is easily checked. This is an example of a general situation:

> If $q(x)$ is a solution of the homogeneous equation
>
> $$\frac{dy}{dx} + p(x)y = 0,$$
>
> then $y(x) = xq(x)$ is a solution of the non-homogeneous equation
>
> $$\frac{dy}{dx} + p(x)y = q(x).$$

This fact soon will seem more natural when we discuss Method 2. (See Example 5.8.) It is easily verified:

$$\begin{aligned}\frac{dy}{dx} + p(x)y &= \frac{d}{dx}[xq(x)] + p(x)[xq(x)]\\ &= q(x) + x\left(\frac{dq}{dx} + p(x)q(x)\right)\\ &= q(x) + x \cdot 0 = q(x).\end{aligned}$$

EXAMPLE 5.5

Find a solution of $\dfrac{dy}{dx} - \dfrac{1}{x}y = x.$

Solution: Notice that $y(x) = x$ is a solution of the homogeneous equation

$$\frac{dy}{dx} - \frac{1}{x}y = 0.$$

But x occurs also on the right-hand side of the differential equation. Therefore, according to the rule, a solution is $x \cdot x = x^2$. Check it!

Answer: $y(x) = x^2$ is a solution.

Summary

Here is a list of standard guesses for solving

$$\frac{dy}{dx} + p(x)y = q(x).$$

$q(x)$	Guess
a, constant	A, constant
$ax + b$	$Ax + B$
$ax^2 + bx + c$	$Ax^2 + Bx + C$
e^x	Ae^x
$\sin x, \quad \cos x$	$A \cos x + B \sin x$
$e^x(ax + b)$	$e^x(Ax + B)$
$e^x(ax^2 + bx + c)$	$e^x(Ax^2 + Bx + C)$
$(ax + b) \sin x$	$(Ax + B) \cos x + (Cx + D) \sin x$
$e^x \sin x, \quad e^x \cos x$	$e^x(A \cos x + B \sin x)$

METHOD 2.

Begin with any non-zero solution $u(x)$ of the homogeneous equation

$$\frac{du}{dx} + p(x)u = 0.$$

Look for a solution $y(x)$ of the non-homogeneous equation in the form

$$y(x) = u(x) \cdot v(x).$$

The unknown function is now $v(x)$. Substitute $y = uv$ and $y' = uv' + vu'$ into the differential equation $y' + py = q$:

$$uv' + vu' + puv = q,$$

$$uv' + v(u' + pu) = q.$$

But the expression in parentheses is 0. Hence,

$$uv' = q.$$

Since u is not 0,

$$\frac{dv}{dx} = \frac{q(x)}{u(x)}.$$

Antidifferentiate to find $v(x)$. (This may be difficult or impossible, a disadvantage of the method.)

If $V(x)$ is an antiderivative, then $V(x)u(x)$ is a solution of the differential equation. If you keep the constant of integration, then you get $[V(x) + c]u(x)$, which is the general solution of the differential equation. (Why?)

EXAMPLE 5.6

Find the general solution of $\dfrac{dy}{dx} - \dfrac{y}{x} = x^3 + 1.$

Solution: The associated homogeneous equation

$$\frac{du}{dx} - \frac{u}{x} = 0$$

has a solution $u(x) = x$. Set $y = xv$ and substitute:

$$x\frac{dv}{dx} + v - \frac{xv}{x} = x^3 + 1,$$

$$x\frac{dv}{dx} = x^3 + 1, \qquad \frac{dv}{dx} = x^2 + \frac{1}{x}\cdot$$

Antidifferentiate:

$$v = \frac{1}{3}x^3 + \ln|x| + c.$$

The general solution is $xv(x)$.

Answer: $y(x) = \frac{1}{3}x^4 + x\ln|x| + cx.$

EXAMPLE 5.7

Solve $\dfrac{dy}{dx} + 2xy = x.$

Solution: Solve the associated homogeneous equation

$$\frac{du}{dx} + 2xu = 0:$$

$$\frac{du}{u} = -2x\,dx, \qquad \ln|u| = -x^2 + c.$$

One solution is

$$u(x) = e^{-x^2}.$$

Now set

$$y = e^{-x^2}v,$$

so

$$y' = -2xe^{-x^2}v + e^{-x^2}v'.$$

Substitute into the differential equation:

$$-2xe^{-x^2}v + e^{-x^2}v' + 2xe^{-x^2}v = x,$$

$$v' = xe^{x^2}.$$

Antidifferentiate:

$$v(x) = \frac{1}{2}e^{x^2} + c.$$

Hence the general solution is

$$y = e^{-x^2}v = \frac{1}{2} + ce^{-x^2}.$$

Hindsight: We should have guessed the particular solution $y = \frac{1}{2}$.

Alternate Solution: Separate variables:

$$\frac{dy}{dx} = x(1 - 2y), \qquad \frac{dy}{1 - 2y} = x\,dx.$$

$$-\frac{1}{2}\ln|1 - 2y| = \frac{1}{2}x^2 - \frac{1}{2}a,$$

$$\ln|1 - 2y| = a - x^2,$$

$$|1 - 2y| = 2ke^{-x^2} \qquad (2k = e^a).$$

It follows that

$$1 - 2y = \pm 2ke^{-x^2},$$

$$y = \frac{1}{2} + ce^{-x^2}, \qquad c \text{ any constant.}$$

Answer: $y(x) = \dfrac{1}{2} + ce^{-x^2}.$

Next we derive an assertion made in the discussion of Method 1 (rather than just pulling it out of a hat).

EXAMPLE 5.8

Find a solution of $\dfrac{dy}{dx} + p(x)y = q(x)$ if $q(x)$ is a solution of the associated homogeneous equation.

Solution: Try a solution $y(x) = v(x) \cdot q(x)$.

$$\frac{dy}{dx} + py = \frac{d}{dx}(vq) + pvq = q\frac{dv}{dx} + v\frac{dq}{dx} + pvq$$

$$= q\frac{dv}{dx} + v\left(\frac{dq}{dx} + pq\right) = q\frac{dv}{dx} + 0.$$

Hence, the differential equation becomes

$$q\frac{dv}{dx} = q, \qquad \frac{dv}{dx} = 1, \qquad v(x) = x.$$

Answer: $y(x) = xq(x)$ is a solution.

EXERCISES

Find a particular solution:

1. $\dfrac{dy}{dx} + 2y = x$

2. $\dfrac{dy}{dx} - y = 3x - 2$

3. $\dfrac{dy}{dx} + 4y = e^x$

4. $\dfrac{dy}{dx} - y = xe^x$

5. $\dfrac{dy}{dx} + y = x^2e^x$

6. $\dfrac{dy}{dx} + 2y = \cos x$

7. $2\dfrac{dy}{dx} + 5y = 3 \sin x$

8. $6\dfrac{dy}{dx} + y = e^{3x} - x - 1$

9. $\dfrac{dy}{dx} + 2y = e^{-x} \cos x$

10. $L\dfrac{di}{dt} + Ri = e^{-kt}$

11. $L\dfrac{di}{dt} + Ri = E \cos t$

12. $\dfrac{dy}{d\theta} + ay = \cos 2\theta$

13. $\dfrac{dy}{dx} - y = x^5$

14. $\dfrac{dy}{dx} + 2y = e^{-2x}$

15. $\dfrac{dy}{dx} - y = e^x$

Use Method 2 of the text to find a particular solution:

16. $\dfrac{dy}{dx} + xy = x$

17. $\dfrac{dy}{dx} - \dfrac{2y}{x} = x^2 + 1$

18. $\dfrac{dy}{dx} - \dfrac{y}{x} = \dfrac{1}{x^2}$

19. $\dfrac{dy}{dx} - \dfrac{2xy}{x^2 + 1} = x.$

Find the general solution:

20. $\dfrac{dy}{dx} + 3y = 0$

21. $\dfrac{dy}{dx} + 3y = 1$

22. $\dfrac{dy}{dx} - 2y = 3x^2 + 2x + 1$

23. $\dfrac{dy}{dx} + y = xe^{2x} + 1$

24. $\dfrac{dy}{dx} - 2y = x^2e^x$

25. $L\dfrac{di}{dt} + Ri = E \cos \omega t$

26. $L\dfrac{di}{dt} + Ri = Ee^{kt}$

27. $\dfrac{dy}{dx} + 2xy = 2x$

28. $\dfrac{dy}{dx} + \dfrac{2xy}{x^2 + 1} - x$

29. $\dfrac{dy}{dx} + 3y = e^{-3x}.$

30. Show that the **Bernoulli Equation** $\dfrac{dy}{dx} + p(x)y = q(x)y^n$ can be reduced to a linear equation by the substitution $z = y^{1-n}$.

31. (cont.) Solve $y' + xy = \sqrt{y}$; find a formula for the answer, but do not try to evaluate it.

32. Suppose $q(x)$ satisfies $|q(x)| \leq M$ and that $y(x)$ is a solution of the initial-value problem $\dfrac{dy}{dx} + y = q(x), \quad y(0) = 0$. Show that $|y(x)| \leq M$ for $x \geq 0$.

33. Sketch the direction field of $y' = x + y$. By inspection of the field find a particular solution of $y' - y = x$.

6. APPLICATIONS

We now consider some applications of the preceding material. In each case we must first interpret the geometrical or physical data in terms of a differential equation.

EXAMPLE 6.1

Find all curves $y = y(x)$ with the property that the tangent line at each point (x, y) on the curve meets the x-axis at $(x + 3, 0)$.

Solution: Sketch the data (Fig. 6.1). The slope of the tangent line is dy/dx on the one hand and $[y - 0]/[x - (x + 3)] = -y/3$ on the other.

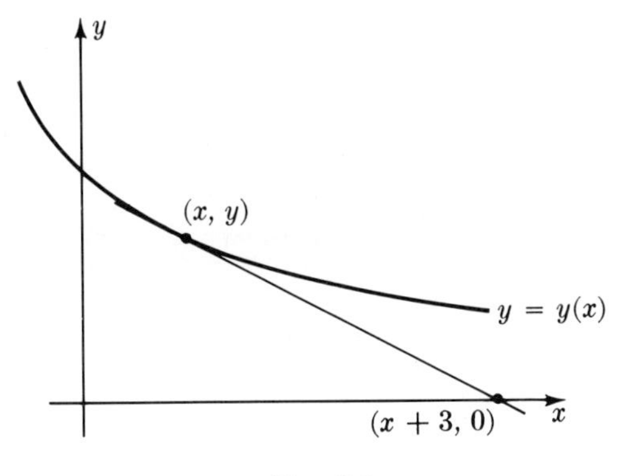

FIG. 6.1

Hence the differential equation is

$$\frac{dy}{dx} = -\frac{1}{3}y.$$

Answer: $y = ke^{-x/3}$.

EXAMPLE 6.2

Find all curves which intersect each of the rectangular hyperbolas $xy = c$ at right angles.

Solution: The family of hyperbolas is shown in Fig. 6.2. First construct the direction field which at each point (x, y) is tangent to the hyperbola $xy = c$ through the point.

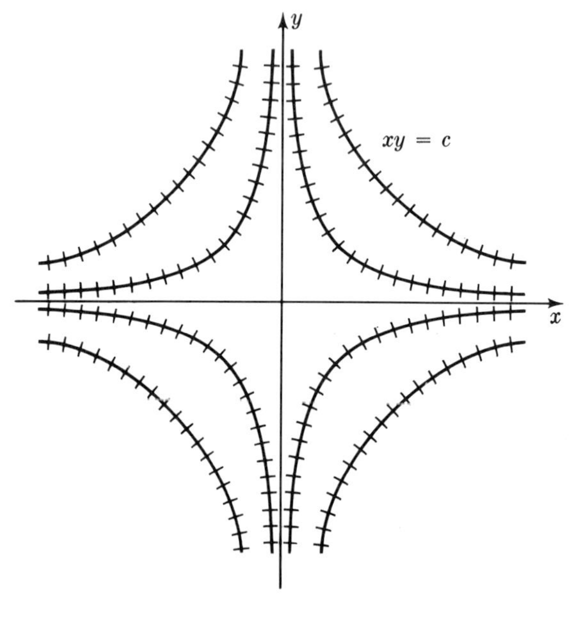

FIG. 6.2

Differentiate:

$$y + xy' = 0.$$

Hence

$$y' = -\frac{y}{x}.$$

The direction field perpendicular to the hyperbola through (x, y) is determined by the negative reciprocal of y'. In other words, the differential equation of the direction field is

$$\frac{dy}{dx} = \frac{x}{y}.$$

Separate variables:

$$y\,dy = x\,dx, \qquad y^2 = x^2 - k.$$

Answer: The perpendicular curves (Fig. 6.3) are the rectangular hyperbolas $x^2 - y^2 = k$.

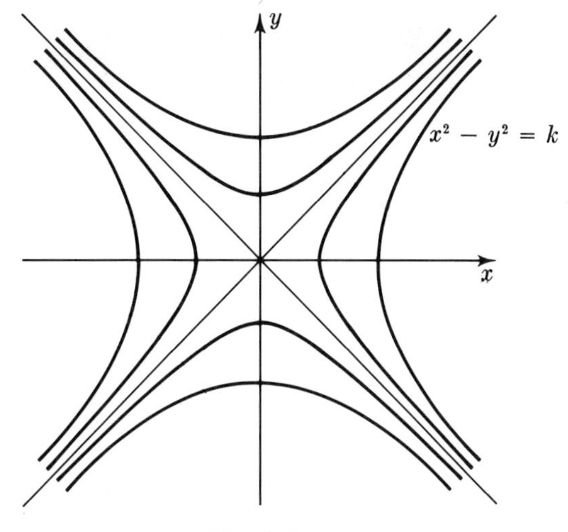

FIG. 6.3

EXAMPLE 6.3

A spherical mothball evaporates at a rate proportional to its surface area. If half of it evaporates in 3 weeks, when will the mothball disappear?

Solution: Let $r = r(t)$ denote the radius at time t in weeks, with $r(0) = r_0$. The volume at time t is $V(t) = \frac{4}{3}\pi r^3 = \frac{4}{3}\pi[r(t)]^3$, and the surface area is $A(t) = 4\pi r^2 = 4\pi[r(t)]^2$.

The data are

(i) $\dfrac{dV}{dt} = -kA,$

(ii) $V(3) = \dfrac{1}{2} V(0) = \dfrac{1}{2} \cdot \dfrac{4}{3} \pi r_0{}^3.$

By (i),

$$\frac{d}{dt}\left(\frac{4}{3}\pi r^3\right) = -4k\pi r^2,$$

that is,

$$4\pi r^2 \frac{dr}{dt} = -4k\pi r^2,$$

from which

$$\frac{dr}{dt} = -k.$$

Hence

$$r = -kt + c.$$

Evaluate the constant c by substituting the initial condition, $r = r_0$ when $t = 0$:

$$r_0 = c.$$

Hence

$$r = r_0 - kt.$$

To find k, use condition (ii):

$$\frac{4}{3}\pi[r(3)]^3 = \frac{1}{2}\cdot\frac{4}{3}\pi r_0{}^3, \quad \text{hence} \quad r(3) = \frac{r_0}{\sqrt[3]{2}}\cdot$$

But $r(3) = r_0 - 3k$, consequently

$$\frac{r_0}{\sqrt[3]{2}} = r_0 - 3k, \qquad k = \frac{r_0}{3}\left(1 - \frac{1}{\sqrt[3]{2}}\right)\cdot$$

Therefore, the formula for r is

$$r = r_0 - kt = r_0\left[1 - \frac{t}{3}\left(1 - \frac{1}{\sqrt[3]{2}}\right)\right]\cdot$$

From this it follows that the mothball disappears ($r = 0$) when

$$t = \frac{3}{\left(1 - \frac{1}{\sqrt[3]{2}}\right)} \approx 14.5.$$

Answer: The mothball disappears after approximately 14.5 weeks.

EXAMPLE 6.4

Let $T = T(t)$ denote the average temperature of a small piece of hot metal in a cooling bath of fixed temperature 50°. According to Newton's Law of Cooling, the metal cools at a rate proportional to the difference $T - 50$. Suppose the metal cools from 250° to 150° in 30 sec. How long would it take to cool from 450° to 60° ?

Solution: In mathematical terms, the Law of Cooling is a linear differential equation:

$$\frac{dT}{dt} = -k(T - 50),$$

that is,

$$\frac{dT}{dt} + kT = 50k.$$

The associated homogeneous equation

$$\frac{dT}{dt} + kT = 0,$$

has the general solution

$$T = ce^{-kt}.$$

It is easy to guess the particular solution $T = 50$. Conclusion: the general solution is

$$T = 50 + ce^{-kt}.$$

Setting $t = 0$, we find

$$T(0) = 50 + c, \qquad c = T(0) - 50.$$

Hence

$$T = 50 + [T(0) - 50]e^{-kt}.$$

We must determine the constant k. By hypothesis, if $T(0) = 250$, then $T(30) = 150$:

$$T(30) = 150 = 50 + [250 - 50]e^{-30k},$$

$$100 = 200e^{-30k}, \qquad k = -\frac{1}{30}\ln\frac{1}{2} = \frac{1}{30}\ln 2.$$

Therefore,

$$T = 50 + [T(0) - 50]e^{-kt}, \qquad \text{where} \quad k = \frac{1}{30}\ln 2.$$

In this formula we substitute $T = 60$ and $T(0) = 450$, then solve for t:

$$60 = 50 + [450 - 50]e^{-kt}, \qquad \frac{10}{400} = e^{-kt},$$

$$t = -\frac{1}{k}\ln\frac{1}{40} = \frac{1}{k}\ln 40 = \frac{30\ln 40}{\ln 2}.$$

Answer: $\dfrac{30\ln 40}{\ln 2} \approx 160$ sec.

EXAMPLE 6.5

Water flows from an open tank through a small hole of area A ft² at the base. When the depth in the tank is x ft, the rate of flow is $(0.60)A\sqrt{2gx}$, where $g = 32.17$ ft/sec². How long does it take to empty a full spherical tank of diameter 10 ft through a circular hole of diameter 2 in. at the bottom? (Assume a small hole at the top admits air.)

Solution: Let V denote the volume when the depth is x. Then

$$\frac{dV}{dt} = -k\sqrt{x}, \qquad \text{where} \quad k = (0.60)A\sqrt{2g}.$$

By the Chain Rule,

$$\frac{dV}{dx}\frac{dx}{dt} = -k\sqrt{x}.$$

Finding dV/dx is a problem in volume by slicing. Draw a cross section of the tank (Fig. 6.4). The change in volume is $dV \approx \pi y^2\,dx$. Hence

$$\frac{dV}{dx} = \pi y^2 = \pi[5^2 - (x-5)^2] = \pi(10x - x^2).$$

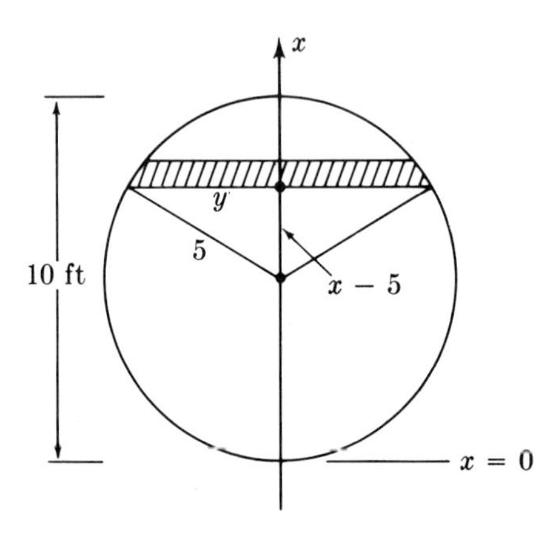

FIG. 6.4

The differential equation is

$$\pi(10x - x^2)\frac{dx}{dt} = -k\sqrt{x},$$

$$(10x^{1/2} - x^{3/2})\frac{dx}{dt} = \frac{-k}{\pi}, \qquad x(0) = 10.$$

Separate variables and integrate:

$$\frac{20}{3}x^{3/2} - \frac{2}{5}x^{5/2} = -\frac{k}{\pi}t + c.$$

Substitute the initial condition, $x = 10$ when $t = 0$:

$$\left(\frac{200}{3} - \frac{200}{5}\right)\sqrt{10} = c, \qquad c = \frac{80}{3}\sqrt{10}.$$

Solve for t when $x = 0$:

$$0 = -\frac{k}{\pi}t + c, \qquad t = \frac{\pi c}{k},$$

where $k = (0.60)A\sqrt{2g}$, and A is the area of a circle of 2 in. diameter,

$$A = \pi\left(\frac{1}{12}\right)^2 \text{ ft}^2.$$

Thus

$$k = (0.60)A\sqrt{2g} = \frac{\pi\sqrt{2g}}{240}.$$

The answer to the problem is

$$t = \frac{\pi c}{k} = \frac{\dfrac{80\pi\sqrt{10}}{3}}{\dfrac{\pi\sqrt{2g}}{240}} = 6400\sqrt{\frac{5}{g}} \quad \text{sec.}$$

Answer: $6400\sqrt{\dfrac{5}{g}} \approx 2520$ sec.

EXAMPLE 6.6

When a uniform steel rod of length L and cross-sectional area 1 is subjected to a tension T, it stretches to length $L + D$, where

$$\frac{D}{L} = kT, \qquad k \quad \text{a constant.}$$

(This is Hooke's Law, valid if T does not exceed a certain bound.)

Suppose the rod is suspended vertically by one end. How much does it stretch? Assume the weight of the rod is W.

Solution: First imagine the rod lying on a horizontal surface. Calibrate the rod by measuring distances from one end. In this way each point of the rod is identified with a number x between 0 and L.

Now suppose the rod is hung by the "zero" end. Let $y(x)$ denote the distance from the top to the point designated by x. Since the rod is stretched due to its own weight, $y(x) > x$ for $x > 0$. The amount the rod stretches is $y(L) - L$.

Consider a short portion of the rod from x to $x + h$. See Fig. 6.5. It is under tension $T(x)$, approximately the weight of the portion of the rod between the points marked x and L, i.e.,

$$T(x) \approx W\frac{L - x}{L}.$$

How much is this portion stretched by the tension $T(x)$? Its original length is h. When stretched, its ends are at $y(x)$ and $y(x + h)$, so its new length is $y(x + h) - y(x)$. The increase in length is $y(x + h) - y(x) - h$. By Hooke's Law,

$$\frac{y(x + h) - y(x) - h}{h} \approx kW\frac{L - x}{L}.$$

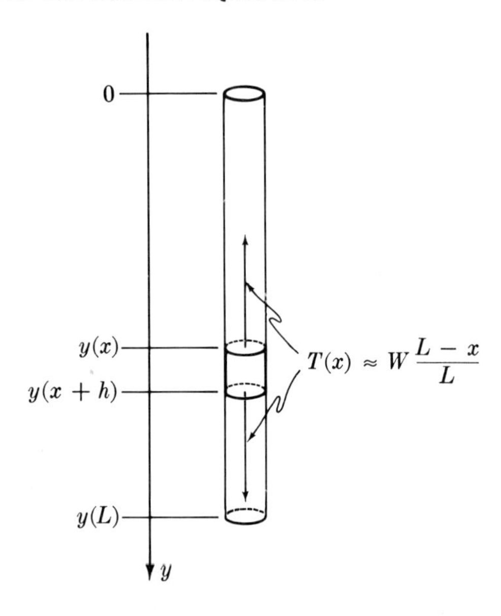

FIG. 6.5

Let $h \longrightarrow 0$:

$$\frac{dy}{dx} - 1 = kW\frac{L-x}{L}.$$

This is a differential equation for the position $y(x)$ of the point x. Solve it subject to the initial condition $y(0) = 0$. First integrate both sides:

$$y - x = -\frac{kW}{2}\frac{(L-x)^2}{L} + c.$$

By the initial condition,

$$0 = -\frac{kWL}{2} + c,$$

hence

$$y - x = \frac{1}{2}kWL - \frac{1}{2}\frac{kW}{L}(L-x)^2.$$

Set $x = L$:

$$y(L) - L = \frac{1}{2}kWL.$$

Answer: The hanging rod stretches $\frac{1}{2}kWL$ units of length.

EXERCISES

1. Find all curves which intersect each of the ellipses $x^2 + 3y^2 = c$ at right angles. Sketch the curves. (See Example 6.2.)
2. Find all curves which intersect each of the curves $y = ce^{-x^2}$ at right angles. Sketch them.
3. The **half-life** of a radioactive substance is the length of time in which a quantity of the substance is reduced to half its original mass. The rate of decay is proportional to the undecayed mass. If the half-life of a substance is 100 years, how long will it take a quantity of the substance to decay to $\frac{1}{10}$ its original mass?
4. (cont.) If $\frac{1}{4}$ of a radioactive substance decays in 2 hr, what is its half-life?
5. The population of a city is now 110,000. Ten years ago it was 100,000. If the rate of growth is proportional to the population, predict the population 20 years from now.
6. When quenched in oil at 50°C, a piece of hot metal cools from 250°C to 150°C in 30 sec. At what constant oil temperature will the metal cool from 250°C to 150°C in 45 sec? (See Example 6.4.)
7. (cont.) The same piece of metal is quenched in oil whose temperature is 50°C, but rising 0.5°C/sec. How long will it take for the metal to cool from 250°C to 150°C?
8. When a cool object is placed in hot gas it heats up at a rate proportional to the difference in temperature between the gas and the object. A cold metal bar at 0°C is placed in a tank of gas at constant temperature 200°C. In 40 sec its temperature rises to 50°C. How long would it take to heat up from 20°C to 100°C?
9. (cont.) How hot should the gas be so that the metal bar heats up from 0°C to 50°C in 20 sec?
10. Two substances, U and V, combine chemically to form a substance X; one gram of U combines with 2 gm of V to form 3 gm of X. Suppose 50 gm of U are allowed to react with 100 gm of V. According to the Law of Mass Action, the chemicals combine at a rate proportional to the product of the untransformed masses. Denote by $x(t)$ the mass of X at time t. (The reaction starts at $t = 0$.) Show that $x(t)$ satisfies the initial-value problem
$$\frac{dx}{dt} = k\left(50 - \frac{x}{3}\right)\left(100 - \frac{2x}{3}\right), \qquad x(0) = 0.$$
If 75 gm of X are formed in the first second, find a formula for $x(t)$.
11. (cont.) Suppose that 50 gm of U are allowed to react with 150 gm of V. (Not all of V will be transformed.) Set up an initial-value problem for $x(t)$ and solve. Verify that $x(t) \longrightarrow 150$ as $t \longrightarrow \infty$.
12. A cylindrical tank of diameter 10 ft contains water to a depth of 4 ft. How long will it take the tank to empty through a 2-in. hole in the bottom? (See Example 6.5.)
13. (cont.) Suppose the tank in Ex. 12 is an inverted paraboloid of revolution containing water to a depth of 4 ft, and that the diameter of the tank at the 4-ft level is 10 ft. How long will it take the tank to empty through a 2-in. hole in the bottom?
14. A falling body of mass m is subject to a downward force mg, where g is the gravitational constant. Due to air resistance, it is subject also to a retarding (up-

ward) force proportional to its velocity. Verify that Newton's Law, $F = ma$, asserts in this case that

$$\frac{dv}{dt} + \frac{k}{m} v = g.$$

Solve, and show that the velocity does not increase indefinitely but approaches a limiting value. (Assume an initial velocity v_0.)

15. A thermometer is plunged into a hot liquid. A few seconds later it records T_0°C. Five seconds later it records T_1°C. Five seconds later still it records T_2°C. Show that the temperature of the liquid is

$$\frac{T_1^2 - T_0 T_2}{-T_0 + 2T_1 - T_2}.$$

(Assume the reading T changes at a rate proportional to the difference between T and the actual temperature.)

16. In Example 6.6, is the amount of stretching of the top half of the rod half the total stretching?

17. Suppose the rod of Example 6.6 is a long right circular cone. Suspend it vertically by its base. How much does it stretch?

22. Second Order Linear Equations

1. INTRODUCTION

In this chapter we discuss second order linear differential equations with constant coefficients:

$$\frac{d^2x}{dt^2} + p\frac{dx}{dt} + qx = r(t),$$

where p and q are constants. Equations of this type have many physical applications, particularly to elastic or electric phenomena.

First order linear equations are solved by finding one particular solution and adding it to the general solution of the associated homogeneous equation. The same method applies to second order linear equations for exactly the same reason. Regard the left-hand side of the equation as a "black box"; an input $x(t)$ leads to an output $\ddot{x} + p\dot{x} + qx$. The sum $x + y$ of two inputs leads to the sum of the corresponding outputs:

$$(x + y)^{\cdot\cdot} + p(x + y)^{\cdot} + q(x + y) = (\ddot{x} + p\dot{x} + qx) + (\ddot{y} + p\dot{y} + qy).$$

A constant multiple ax of an input leads to the same constant multiple of the output:

$$(ax)^{\cdot\cdot} + p(ax)^{\cdot} + q(ax) = a(\ddot{x} + p\dot{x} + qx).$$

Thus the "black box" is linear.

> The general solution of
>
> $$\frac{d^2x}{dt^2} + p\frac{dx}{dt} + qx = r(t)$$
>
> is
>
> $$x(t) + z(t),$$
>
> where $x(t)$ is any solution of the differential equation and $z(t)$ is the general solution of the homogeneous equation
>
> $$\frac{d^2x}{dt^2} + p\frac{dx}{dt} + qx = 0.$$

Therefore, as in the last chapter, we first treat homogeneous equations, then look for particular solutions of non-homogeneous equations.

2. HOMOGENEOUS EQUATIONS

In this section we study homogeneous equations

$$\ddot{x} + p\dot{x} + qx = 0,$$

where p and q are constants. There are three important special cases, each with $p = 0$, namely $q = 0$, $q = -k^2$, and $q = k^2$.

CASE 1: $\ddot{x} = 0$. This is easy. The solution is

$$x = at + b.$$

CASE 2: $\ddot{x} - k^2x = 0$. Here, the second derivative of x is proportional to x itself, which reminds us of the exponential function. It is not hard to guess two solutions: $x = e^{kt}$ and $x = e^{-kt}$. By linearity,

$$x = ae^{kt} + be^{-kt}$$

is also a solution. In Section 4 we shall prove this is the general solution.

CASE 3: $\ddot{x} + k^2x = 0$. Again it is easy to guess two solutions: $x = \cos kt$ and $x = \sin kt$. In Section 4 we shall prove the general solution is

$$x = a \cos kt + b \sin kt.$$

Summary

Differential equation	General solution
$\ddot{x} = 0$	$x = at + b$
$\ddot{x} - k^2x = 0$	$x = ae^{kt} + be^{-kt}$
$\ddot{x} + k^2x = 0$	$x = a \cos kt + b \sin kt$

These special cases are important because each homogeneous equation

$$\ddot{x} + p\dot{x} + qx = 0$$

can be reduced to one of them. The trick is analogous to reducing a quadratic equation $X^2 + pX + q = 0$ to the form $Y^2 \pm k^2 = 0$ by completing the square. Set

$$x(t) = e^{ht}y(t),$$

where $y(t)$ is a new unknown function and h is a constant. Compute derivatives by the Product Rule:

$$\begin{aligned} \dot{x} &= he^{ht}y + e^{ht}\dot{y} \\ &= e^{ht}(\dot{y} + hy); \\ \ddot{x} &= he^{ht}(\dot{y} + hy) - e^{ht}(\ddot{y} + h\dot{y}) \\ &= e^{ht}(\ddot{y} + 2h\dot{y} + h^2y). \end{aligned}$$

Substitute these expressions into the differential equation:

$$e^{ht}(\ddot{y} + 2h\dot{y} + h^2y) + pe^{ht}(\dot{y} + hy) + qe^{ht}y = 0.$$

Cancel e^{ht} and group terms:

$$\ddot{y} + (2h + p)\dot{y} + (h^2 + ph + q)y = 0.$$

Now choose $h = -p/2$ to make the term in $\dot{y}$ drop out. The coefficient of y is then

$$h^2 + ph + q = \frac{p^2}{4} - \frac{p^2}{2} + q = \frac{-p^2 + 4q}{4},$$

and the differential equation becomes

$$\ddot{y} - \frac{p^2 - 4q}{4}y = 0.$$

This is one of the three special cases, depending on whether $p^2 - 4q$ is zero, positive, or negative.

EXAMPLE 2.1

Solve $\ddot{x} + 6\dot{x} + 9x = 0$.

Solution: Here

$$p = 6, \qquad q = 9, \qquad h = -\frac{p}{2} = -3, \qquad \frac{p^2 - 4q}{4} = 0.$$

The substitution $x = e^{-3t}y$ transforms the differential equation into

$$\ddot{y} = 0,$$

from which

$$y = at + b.$$

Answer: $x = e^{-3t}(at + b)$, where a and b are constants.

EXAMPLE 2.2

Solve $\ddot{x} - 3\dot{x} + 2x = 0$.

Solution: Here

$$p = -3, \qquad q = 2, \qquad h = -\frac{p}{2} = \frac{3}{2}, \qquad \frac{p^2 - 4q}{4} = \frac{1}{4} = \left(\frac{1}{2}\right)^2.$$

The substitution $x = e^{3t/2}y$ transforms the differential equation into

$$\ddot{y} - \left(\frac{1}{2}\right)^2 y = 0,$$

one of the standard forms. Conclusion:

$$y = ae^{t/2} + be^{-t/2}.$$

Answer: $x = e^{3t/2}(ae^{t/2} + be^{-t/2}) = ae^{2t} + be^{t}.$

EXAMPLE 2.3

Solve $\ddot{x} + 2\dot{x} + 5x = 0$.

Solution: In this example

$$p = 2, \qquad q = 5, \qquad h = -\frac{p}{2} = -1, \qquad \frac{p^2 - 4q}{4} = -4 = -2^2.$$

Set $x = e^{-t}y$. Then

$$\ddot{y} + 2^2 y = 0,$$

$$y = a\cos 2t + b\sin 2t.$$

Answer: $x = e^{-t}(a\cos 2t + b\sin 2t).$

Formula for Solving $\ddot{x} + p\dot{x} + qx = 0$

When you set

$$x = e^{-(p/2)t}y,$$

the differential equation

$$\ddot{x} + p\dot{x} + q = 0$$

is transformed into

$$\ddot{y} - \frac{\Delta}{4}y = 0, \qquad \text{where} \quad \Delta = p^2 - 4q.$$

Its solution depends on the sign of Δ:

CASE 1: $\Delta = 0$. Then

$$x(t) = e^{-(p/2)t}(at + b).$$

CASE 2: $\Delta > 0$. Write $\Delta = \sigma^2$. Then

$$x(t) = e^{-(p/2)t}(ae^{(\sigma/2)t} + be^{-(\sigma/2)t}) = ae^{[(-p+\sigma)/2]t} + be^{[(-p-\sigma)/2]t}.$$

CASE 3: $\Delta < 0$. Write $\Delta = -\sigma^2$. Then

$$x(t) = e^{-(p/2)t}\left[a \cos\left(\frac{\sigma}{2} t\right) + b \sin\left(\frac{\sigma}{2} t\right)\right].$$

Now compare these three cases with the three cases describing the roots of the quadratic equation

$$X^2 + pX + q = 0.$$

By the quadratic formula, the roots are

$$\alpha = -\frac{p}{2} + \frac{1}{2}\sqrt{\Delta}, \qquad \beta = -\frac{p}{2} - \frac{1}{2}\sqrt{\Delta}.$$

CASE 1: $\Delta = 0$. The roots are real and equal:

$$\alpha = \beta = -\frac{p}{2}.$$

CASE 2: $\Delta > 0$. Write $\Delta = \sigma^2$. The roots are real and distinct:

$$\alpha = \frac{-p + \sigma}{2}, \qquad \beta = \frac{-p - \sigma}{2}.$$

CASE 3: $\Delta < 0$. Write $\Delta = -\sigma^2$. The roots are complex:

$$\alpha = \frac{-p + \sigma i}{2}, \qquad \beta = \frac{-p - \sigma i}{2}, \qquad i = \sqrt{-1}.$$

To solve the differential equation

$$\ddot{x} + p\dot{x} + qx = 0,$$

find the roots α and β of the quadratic equation

$$X^2 + pX + q = 0.$$

1. If $\alpha = \beta$, then the solution is

$$x = e^{\alpha t}(at + b).$$

2. If α and β are real and distinct, then the solution is

$$x = ae^{\alpha t} + be^{\beta t}.$$

3. If α and β are complex,

$$\alpha = h + ki, \qquad \beta = h - ki,$$

then the solution is

$$x = e^{ht}(a \cos kt + b \sin kt).$$

EXAMPLE 2.4

Solve each equation by the preceding rule:
(a) $\ddot{x} + 6\dot{x} + 9x = 0$,
(b) $\ddot{x} - 3\dot{x} + 2x = 0$,
(c) $\ddot{x} + 2\dot{x} + 5x = 0$.

Solution:

(a) $X^2 + 6X + 9 = 0, \qquad (X + 3)^2 = 0, \quad \alpha = \beta = -3,$

$x = e^{-3t}(at + b).$

(b) $X^2 - 3X + 2 = 0, \quad (X - 2)(X - 1) = 0,$

$\alpha = 2, \quad \beta = 1, \quad x = ae^{2t} + be^{t}.$

(c) $X^2 + 2X + 5 = 0,$

$$\alpha = \frac{1}{2}(-2 + \sqrt{-16}) = -1 + 2i,$$

$$\beta = \frac{1}{2}(-2 - \sqrt{-16}) = -1 - 2i,$$

$x = e^{-t}(a \cos 2t + b \sin 2t).$

Answer:
(a) $x = e^{-3t}(at + b)$,
(b) $x = ae^{2t} + be^{t}$,
(c) $x = e^{-t}(a \cos 2t + b \sin 2t)$.

Leibniz's Formula

In addition to the formula $(uv)^{\cdot} = \dot{u}v + u\dot{v}$ for the first derivative of a product, it is convenient to know the formula

$$(uv)^{\cdot\cdot} = \ddot{u}v + 2\dot{u}\dot{v} + u\ddot{v}$$

for the second derivative. The general formula, called **Leibniz's Formula,** involves binomial coefficients:

$$\frac{d^n}{dt^n}(uv) = \sum_{k=0}^{n} \binom{n}{k} \frac{d^{n-k}u}{dt^{n-k}} \frac{d^k v}{dt^k}.$$

EXERCISES

Find the general solution:

1. $\ddot{x} - 6\dot{x} + 5x = 0$
2. $\ddot{x} + 7\dot{x} + 10x = 0$
3. $\dfrac{d^2r}{d\theta^2} + 4r = 0$
4. $\dfrac{d^2y}{dx^2} + 6\dfrac{dy}{dx} + 13y = 0$
5. $2\dfrac{d^2y}{dx^2} - \dfrac{dy}{dx} + y = 0$
6. $\ddot{x} - 8\dot{x} + 16x = 0$

7. $\ddot{x} + 6\dot{x} = 0$

8. $4\dfrac{d^2y}{dx^2} + 4\dfrac{dy}{dx} + y = 0$

9. $\dfrac{d^2y}{dx^2} + 5\dfrac{dy}{dx} + 4y = 0$

10. $2\dfrac{d^2x}{dt^2} - 5\dfrac{dx}{dt} - 3x = 0$

11. $\dfrac{\ddot{x}}{a^3} - 4\dfrac{\dot{x}}{a} + 4ax = 0$

12. $\ddot{x} + \dot{x} + x = 0.$

Find the most general solution, subject to the indicated condition:

13. $\dfrac{d^2y}{dx^2} + 9y = 0, \quad y(0) = 0$

14. $\dfrac{d^2y}{dx^2} + 4y = 0, \quad y'(0) = 0$

15. $\ddot{x} - 4\dot{x} + 4x = 0, \quad x(0) = 1$

16. $\ddot{x} + 6\dot{x} + 8 = 0, \quad y(0) = 0$

17. $\dfrac{d^2r}{d\theta^2} - 2\dfrac{dr}{d\theta} + 2r = 0, \quad r(\pi) = 1$

18. $\dfrac{d^2r}{d\theta^2} - 2\dfrac{dr}{d\theta} + 2r = 0, \quad r'(\pi) = 1$

19. $\dfrac{d^2r}{d\theta^2} + r = 0, \quad r\left(\dfrac{\pi}{6}\right) = 0$

20. $\dfrac{d^2r}{d\theta^2} - 5\dfrac{dr}{d\theta} + 6r = 0, \quad r'(0) = 3.$

21. If a and b are positive constants, show that each solution of $\dfrac{d^2y}{dx^2} + a\dfrac{dy}{dx} + by = 0$ tends to zero as $x \longrightarrow \infty$.

22. Let $u(t)$ and $v(t)$ be solutions of $\ddot{x} + P(t)\dot{x} + Q(t)x = 0$. Set $w = \dot{u}v - u\dot{v}$. Show that $\dot{w} + Pw = 0$. Solve for w. (w is called the **Wronskian** of u and v.)

3. PARTICULAR SOLUTIONS

We return to the equation

$$\ddot{x} + p\dot{x} + qx = r(t).$$

The problem is to find any one solution. There are several ingenious methods for doing this, some quite complicated. We consider only the easiest one: guessing. This works just as in the last chapter.

EXAMPLE 3.1

Find a solution of $\ddot{x} + 3\dot{x} - x = t^2 - 1$.

Solution: Try $x = A + Bt + Ct^2$. Then $\dot{x} = B + 2Ct$ and $\ddot{x} = 2C$, so substitution in the equation yields

$$2C + 3(B + 2Ct) - (A + Bt + Ct^2) = t^2 - 1.$$

Equate coefficients:

$$\begin{aligned} t^2: &\quad -C = 1, \\ t: &\quad -B + 6C = 0, \\ 1: &\quad -A + 3B + 2C = -1. \end{aligned}$$

Therefore $C = -1$, $B = -6$, $A = -19$.

Answer: $x = -19 - 6t - t^2$ is a solution.

EXAMPLE 3.2

Find a solution of $\ddot{x} + 3\dot{x} - x = e^{-4t}$.

Solution: Try $x = Ae^{-4t}$. Then $\dot{x} = -4Ae^{-4t}$ and $\ddot{x} = 16Ae^{-4t}$. Substitute, and cancel e^{-4t}:

$$16A - 12A - A = 1, \quad 3A = 1.$$

Answer: $x = \frac{1}{3}e^{-4t}$ is a solution.

EXAMPLE 3.3

Find a solution of $\ddot{x} + 3\dot{x} - x = \cos t$.

Solution: Try $x = A\cos t + B\sin t$. Then $\dot{x} = -A\sin t + B\cos t$ and $\ddot{x} = -A\cos t - B\sin t$. Substitute:

$$(-A\cos t - B\sin t) + 3(-A\sin t + B\cos t) - (A\cos t + B\sin t) = \cos t.$$

Equate coefficients of $\cos t$ and $\sin t$:

$$\begin{cases} -2A + 3B = 1 \\ -3A - 2B = 0. \end{cases}$$

The solution of this system is $A = -\frac{2}{13}$ and $B = \frac{3}{13}$.

Answer: $x = \frac{1}{13}(-2\cos t + 3\sin t)$ is a solution.

EXAMPLE 3.4

Find the general solution of $\ddot{x} - x = t$.

Solution: The quadratic equation $X^2 - 1 = 0$ has roots 1 and -1. Hence the solution of the homogeneous equation $\ddot{x} - x = 0$ is $x = ae^t + be^{-t}$. We guess the particular solution $x = -t$.

Answer: $x(t) = ae^t + be^{-t} - t$, where a and b are constants.

Initial-Value Problems

The general solution of the differential equation

$$\ddot{x} + p\dot{x} + qx = r(t)$$

involves two arbitrary constants. By choosing them suitably you can usually find a solution satisfying two additional conditions. Most important is the case of initial conditions:

Find a solution $x = x(t)$ of the differential equation

$$\ddot{x} + p\dot{x} + qx = r(t)$$

such that

$$x(t_0) = a, \qquad \dot{x}(t_0) = b.$$

EXAMPLE 3.5

Solve the initial-value problem $\ddot{x} - x = t, \quad x(0) = 0, \quad \dot{x}(0) = 1.$

Solution: By the last example, the general solution of the differential equation is

$$x = ae^{t} + be^{-t} - t.$$

Hence

$$\dot{x} = ae^{t} - be^{-t} - 1.$$

Substitute the initial conditions:

$$\begin{cases} a + b = 0 \\ a - b - 1 = 1. \end{cases}$$

This system has the solution $a = 1$ and $b = -1$.

Answer: $x = e^{t} - e^{-t} - t.$

EXERCISES

Find a particular solution:

1. $\ddot{x} + 3\dot{x} = 1$
2. $2\ddot{x} + 3\dot{x} = 5$
3. $\ddot{x} - 4x = 2t + 1$
4. $\ddot{x} + x = t^2$
5. $\ddot{x} + 3\dot{x} - x = t^2 + 4t + 6$
6. $\ddot{x} + \dot{x} + x = t^3$
7. $\ddot{x} + \dot{x} + 2x = 2e^{3t}$
8. $\ddot{x} + x = \cosh t$
9. $3\dfrac{d^2y}{dx^2} + \dfrac{dy}{dx} - y = e^{-2x} + x$
10. $2\dfrac{d^2y}{dx^2} - 3y = \sin x$
11. $\dfrac{d^2y}{dx^2} + \dfrac{dy}{dx} - 3y = 4\sin 2x$
12. $\dfrac{d^2y}{dx^2} + 2\dfrac{dy}{dx} + y = 1 + x + \cos 3x$
13. $\dfrac{d^2y}{dx^2} + \dfrac{dy}{dx} + 2y = e^{x}\cos x$
14. $2\dfrac{d^2y}{dx^2} - y = x\cos x$
15. $\dfrac{d^2y}{dx^2} + y = e^{x} + e^{2x} + e^{3x} + e^{4x} + e^{5x}$
16. $\dfrac{d^2y}{dx^2} - \dfrac{dy}{dx} - y = xe^{3x}.$

Find the general solution:

17. $\ddot{x} + x = t^2$
18. $\ddot{x} - 7\dot{x} + 10x = 3e^{t}$

19. $\ddot{x} + \dot{x} - 6x = te^{-t}$

20. $\ddot{x} + \dot{x} - 5x = 2t + 3$

21. $\ddot{x} + 3\dot{x} = \cosh 2t$

22. $3\dfrac{d^2y}{dx^2} - \dfrac{dy}{dx} + y = x^2 + 5$

23. $3\dfrac{d^2i}{dt^2} + 4\dfrac{di}{dt} + 2i = 10\cos t$

24. $\dfrac{d^2r}{d\theta^2} - 2r = \sin\theta + \cos\theta.$

Solve the initial-value problem:

25. $\ddot{x} + x = 2t - 5, \quad x(0) = 0, \quad \dot{x}(0) = 1$

26. $\ddot{x} + 3x = e^t, \quad x(0) = 0, \quad \dot{x}(0) = 1$

27. $\ddot{x} - 4x = \sin 2t, \quad x(0) = 1, \quad \dot{x}(0) = 0$

28. $\ddot{x} - 2\dot{x} - 15x = 1, \quad x(1) = 0, \quad \dot{x}(1) = 0$

29. $4\ddot{x} - 7\dot{x} + 3x = e^{2t}, \quad x(0) = -1, \quad \dot{x}(0) = 2$

30. $\ddot{x} + 2\dot{x} + x = t^2, \quad x(0) = 0, \quad \dot{x}(0) = 5.$

31. The usual method for finding a particular solution of $\ddot{x} - 4x = e^{2t}$ fails. Why? Try $x = Ate^{2t}$.

32. (cont.) Find a solution of $\ddot{x} + x = \sin t$ by guessing.

33. Let z be a solution of the homogeneous equation $\ddot{x} + p\dot{x} + qx = 0$. Substitute $x(t) = z(t)u(t)$ in the equation $\ddot{x} + p\dot{x} + qx = r(t)$. Show that the resulting equation for u can be reduced to a first order equation by setting $w = \dot{u}$.

34. (cont.) Apply this technique to find a particular solution of $\ddot{x} + x = \sin t$.

35. Show that the equation $\ddot{x} + k^2x = r(t)$ has the particular solution

$$x(t) = \frac{1}{k}\int_0^t \sin k(t - s)r(s)\,ds.$$

You may presuppose the formula

$$\frac{d}{dt}\left(\int_0^t F(s, t)\,ds\right) = F(t, t) + \int_0^t \frac{\partial F}{\partial s}\,ds.$$

4. JUSTIFICATIONS [optional]

In this section we justify two assertions made in Section 2:

1. The general solution of $\ddot{x} - k^2x = 0$ is $x = ae^{kt} + be^{-kt}$.
2. The general solution of $\ddot{x} + k^2x = 0$ is $x = a\cos kt + b\sin kt$.

First, suppose $x(t)$ is any solution of $\ddot{x} - k^2x = 0$. We must find constants a and b such that

$$x(t) = ae^{kt} + be^{-kt}.$$

Where will these constants come from? Idea: if $x(t)$ *is* of this form, we should be able to express a and b in terms of x. Let us try. If

$$x = ae^{kt} + be^{-kt},$$

then

$$\dot{x} = ake^{kt} - bke^{-kt}.$$

Multiply the first equation by k; then add the two equations:

$$kx + \dot{x} = 2ake^{kt},$$

$$2ak = e^{-kt}(kx + \dot{x}).$$

Similarly

$$2bk = e^{kt}(kx - \dot{x}).$$

In other words, *if* $x = ae^{kt} + be^{-kt}$, then a and b are given by these equations.

Now work backwards. Suppose $x(t)$ is any solution of $\ddot{x} - k^2x = 0$. Define two functions

$$u(t) = e^{-kt}(kx + \dot{x}),$$

$$v(t) = e^{kt}(kx - \dot{x}),$$

which ought to be constants. Compute the derivative of $u(t)$:

$$\begin{aligned} \dot{u} &= -ke^{-kt}(kx + \dot{x}) + e^{-kt}(k\dot{x} + \ddot{x}) \\ &= e^{-kt}[(\ddot{x} - k^2x) + (-k\dot{x} + k\dot{x})] \\ &= 0, \end{aligned}$$

since $\ddot{x} - k^2x = 0$. Similarly $\dot{v} = 0$. Thus u and v *are* constants; they may be written

$$u = 2ak, \qquad v = 2bk.$$

Therefore if x is any solution of $\ddot{x} - k^2x = 0$, then

$$\begin{cases} e^{-kt}(kx + \dot{x}) = 2ak \\ e^{kt}(kx - \dot{x}) = 2bk, \end{cases}$$

that is,

$$\begin{cases} kx + \dot{x} = 2ake^{kt} \\ kx - \dot{x} = 2bke^{-kt}, \end{cases}$$

where a and b are constants. Add, then cancel $2k$:

$$x = ae^{kt} + be^{-kt};$$

x *has* the asserted form.

Next suppose $x(t)$ is any solution of $\ddot{x} + k^2x = 0$. Problem: find constants a and b such that

$$x = a \cos kt + b \sin kt.$$

Set

$$u(t) = kx \cos kt - \dot{x} \sin kt,$$

$$v(t) = kx \sin kt + \dot{x} \cos kt.$$

As in the previous case, these functions ought to be constants. Differentiate:

$$\begin{aligned} \dot{u} &= (k\dot{x} \cos kt - k^2x \sin kt) - (\ddot{x} \sin kt + k\dot{x} \cos kt) \\ &= -(\ddot{x} + k^2x) \sin kt = 0, \end{aligned}$$

similarly $\dot{v} = 0$. Thus u and v *are* constants. Write

$$u = ak, \qquad v = bk.$$

Therefore, if x is any solution of $\ddot{x} + k^2x = 0$, then

$$\begin{cases} kx \cos kt - \dot{x} \sin kt = ak \\ kx \sin kt + \dot{x} \cos kt = bk, \end{cases}$$

where a and b are constants. Multiply the first equation by $\cos kt$, the second by $\sin kt$, then add. The result is

$$x(t) = a \cos kt + b \sin kt.$$

5. APPLICATIONS

In applications, the expression

$$a \cos kt + b \sin kt$$

occurs frequently. There is a useful equivalent expression for it. Assume a and b are not both zero. If c denotes the positive square root of $a^2 + b^2$, then

$$\left(-\frac{a}{c}\right)^2 + \left(\frac{b}{c}\right)^2 = 1.$$

Thus the point $(b/c, -a/c)$ lies on the unit circle (Fig. 5.1). Hence there is an angle kt_0, determined up to an integer multiple of 2π, such that

$$\sin kt_0 = -\frac{a}{c}, \qquad \cos kt_0 = \frac{b}{c}.$$

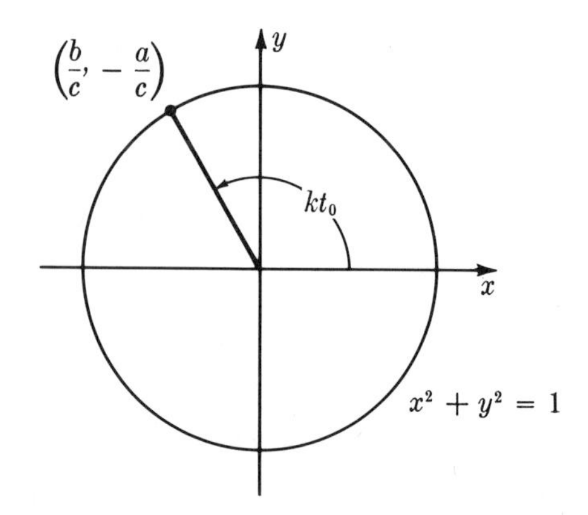

FIG. 5.1

By the addition law for the sine,

$$\begin{aligned} a\cos kt + b\sin kt &= c\left(\frac{a}{c}\cos kt + \frac{b}{c}\sin kt\right) \\ &= c(-\sin kt_0 \cos kt + \cos kt_0 \sin kt) \\ &= c\sin k(t - t_0). \end{aligned}$$

Thus the expression $a\cos kt + b\sin kt$ is nothing but a sine function (sinusoidal wave) in disguise.

[Alternatively, t_0 may be defined by $a/c = \cos kt_0$, $b/c = \sin kt_0$, from which $a\cos kt + b\sin kt = c\cos k(t - t_0)$.]

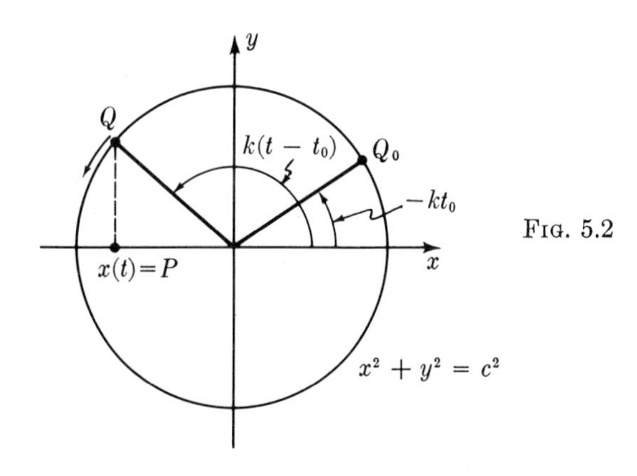

FIG. 5.2

Simple Harmonic Motion

Imagine a point Q moving counterclockwise about the circle $x^2 + y^2 = c^2$. See Fig. 5.2. Starting at Q_0 when $t = 0$, the point Q moves with constant speed, making k revolutions in 2π sec. Its projection P on the x-axis has coordinate

$$x(t) = c\cos k(t - t_0).$$

As Q traverses its circle, P oscillates between $(c, 0)$ and $(-c, 0)$, making one complete oscillation in $2\pi/k$ sec. The motion of P is called **simple harmonic motion** with **amplitude** c, **period** $2\pi/k$, and **time lag** t_0 (**phase angle** $-kt_0$).

Note that the speed of P, unlike that of Q, is not constant. Why? When is it greatest?

REMARK: Since $\cos\alpha = \sin(\pi/2 - \alpha)$, harmonic motion may also be described in terms of a sine:

$$x(t) = c\cos k(t - t_0) = c\cos k(t_0 - t) = c\sin\left[\frac{\pi}{2} - k(t_0 - t)\right].$$

Conclusion:

$$x(t) = c\sin k(t - t_1), \qquad \text{where} \quad t_1 = t_0 - \frac{\pi}{2k}.$$

EXAMPLE 5.1 (Simple harmonic motion)

A particle moves along the x-axis. The only force acting on it is a force directed towards the origin and proportional to the displacement of the particle from the origin (a spring). Describe the motion. (Assume the spring has negligible length when unloaded.)

Solution: Call the mass m. The force is x times a negative constant, $-mk^2$, with $k > 0$. According to Newton's Law of Motion,

$$\text{mass} \times \text{acceleration} = \text{force}:$$

$$m\ddot{x} = -mk^2x, \qquad \ddot{x} + k^2x = 0.$$

The solution is

$$x = a \cos kt + b \sin kt = c \cos k(t - t_0),$$

where the constants c and t_0 depend on the initial position and velocity. Hence the motion is simple harmonic.

Answer: $x = c \cos k(t - t_0)$.

EXAMPLE 5.2

Solve the same problem
(a) with initial conditions $x(0) = 0, \quad \dot{x}(0) = v_0 > 0$;
(b) with initial conditions $x(0) = x_0 > 0, \quad \dot{x}(0) = 0$.

Solution: By Example 5.1, the general solution is $x = c \cos k(t - t_0)$, where c and k are positive. Now determine c and t_0.

(a) Since $\dot{x} = -ck \sin k(t - t_0)$, the initial conditions are:

$$0 = c \cos(-kt_0), \qquad v_0 = -ck \sin(-kt_0),$$

that is,

$$\cos kt_0 = 0, \qquad \sin kt_0 = \frac{v_0}{ck}\cdot$$

From the first equation, $kt_0 = \pi/2$ or $kt_0 = -\pi/2$; from the second, $\sin kt_0 > 0$. Hence the only possibility is $kt_0 = \pi/2$, that is, $t_0 = \pi/2k$. It follows that $v_0 = ck$, so $c = v_0/k$. Therefore the desired solution is

$$x = \frac{v_0}{k} \cos\left(kt - \frac{\pi}{2}\right) = \frac{v_0}{k} \sin kt.$$

(b) In this case the initial conditions are:

$$x_0 = c \cos(-kt_0), \qquad 0 = -ck \sin(-kt_0),$$

that is,

$$\cos kt_0 = \frac{x_0}{c}, \qquad \sin kt_0 = 0.$$

From the second equation, $kt_0 = 0$ or π; from the first, $\cos kt_0 > 0$. Hence

$kt_0 = 0$. It follows that $c = x_0$, and the desired solution is

$$x = x_0 \cos kt.$$

Answer: (a) $x = \dfrac{v_0}{k} \sin kt;$

(b) $x = x_0 \cos kt.$

REMARK: As this example shows, it is simpler to describe harmonic motion by the sine form when the initial position is $x_0 = 0$ and by the cosine form when the initial velocity is $v_0 = 0$.

EXAMPLE 5.3

A 16-lb weight hangs at rest on the end of an 8-ft spring attached to a ceiling. When the spring is stretched, it exerts a restoring force proportional to displacement from equilibrium position. Suppose the weight is pulled down 6 in. and released. Describe its motion.

Solution: Let x denote the distance from the ceiling in feet (Fig. 5.3). The force due to the spring is $F(x) = -cx$. When the spring is at rest, $F = -16$ and $x = 8$. Hence $c = 2$ and $F(x) = -2x$ lb.

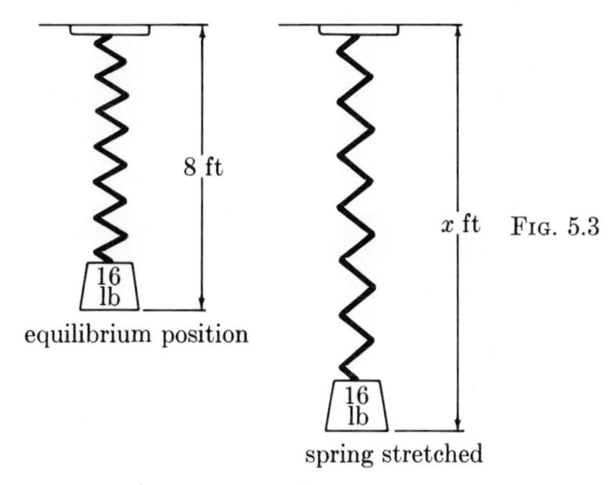

FIG. 5.3

When the weight is at position x, the forces acting on it are its weight, 16 lb, and the spring, $-2x$ lb. Since the units of force and length are pounds and feet, mass must be in slugs: $m = 16/g = \frac{16}{32}$ slugs. Thus mass $\times$ acceleration = force,

$$\frac{16}{32}\ddot{x} = 16 - 2x, \qquad \ddot{x} + 4x = 32.$$

$$\text{Initial conditions:} \quad x(0) = 8.5, \qquad \dot{x}(0) = 0.$$

The associated homogeneous equation has solution

$$x = c \cos 2(t - t_0),$$

and it is easy to guess the particular solution, $x = 8$. Consequently,

$$x = c \cos 2(t - t_0) + 8$$

is the general solution. It describes simple harmonic motion centered about $x = 8$ rather than $x = 0$. That is reasonable since you expect the weight to oscillate about its equilibrium point at $x = 8$.

The initial conditions are

$$x(0) = 8.5, \qquad \dot{x}(0) = 0.$$

From these you can solve for c and t_0. But good scientists are lazy; they refuse to do the same work twice. Use the answer to part (b) of Example 5.2. It describes harmonic motion with 0 initial velocity by $x = x_0 \cos kt$, where x_0 is initial displacement. In this problem, initial displacement is $\frac{1}{2}$. Hence, you can write down the answer.

Answer: $x = \dfrac{1}{2} \cos 2t + 8.$

Other Applications

EXAMPLE 5.4

A particle of mass 1 attached to a spring slides along a straight surface. It is subject to a restoring force proportional to displacement and a retarding force (due to friction) assumed proportional to velocity. Describe the motion of the particle assuming it starts at the origin (equilibrium point) with initial velocity v_0.

Solution: The forces are $-k^2x$ (spring) and $-p\dot{x}$ (friction), where k and p are positive constants. The equation of motion is

$$\text{mass} \times \text{acceleration} = \text{force},$$

$$\ddot{x} = -p\dot{x} - k^2x, \qquad \ddot{x} + p\dot{x} + k^2x = 0.$$

$$\text{Initial conditions:} \quad x(0) = 0, \qquad \dot{x}(0) = v_0.$$

Before solving, think about the problem. If there is no friction, that is if p is zero, then the motion is simple harmonic. If the friction is small, then the motion should be nearly simple harmonic motion, except for a gradual slowing down due to friction. If the friction is large, the motion should be considerably inhibited, in fact there might not be oscillations.

From this physical reasoning, it seems clear that the relative size of the constants k and p is crucial.

To solve the differential equation, examine the roots of the corresponding quadratic equation:

$$X^2 + pX + k^2 = 0.$$

They are

$$\alpha = -\frac{p}{2} + \frac{1}{2}\sqrt{\Delta}, \qquad \beta = -\frac{p}{2} - \frac{1}{2}\sqrt{\Delta}, \qquad \text{where} \quad \Delta = p^2 - 4k^2.$$

The nature of the solutions depends on the sign of Δ. In terms of k and p, the crucial question is whether $p > 2k$, $p < 2k$, or $p = 2k$.

CASE 1: $p < 2k$ (underdamped case: friction small compared to spring force). In this case $\Delta < 0$; set $\Delta = -4q^2$. Then the general solution of the differential equation is

$$x = e^{-(p/2)t}(a \cos qt + b \sin qt).$$

Use the first initial condition, $x(0) = 0$; it implies $a = 0$. Hence

$$x = be^{-(p/2)t} \sin qt.$$

Now use the second initial condition, $\dot{x}(0) = v_0$:

$$\dot{x}(t) = b\left(-\frac{p}{2}e^{-(p/2)t} \sin qt + qe^{-(p/2)t} \cos qt\right),$$

$$v_0 = \dot{x}(0) = bq, \qquad b = \frac{v_0}{q}.$$

Therefore, the solution of the initial-value problem is:

$$x = \frac{v_0}{q} e^{-(p/2)t} \sin qt.$$

This is a damped oscillatory motion (Fig. 5.4). The particle oscillates with constant period $2\pi/q$, but the amplitude $\longrightarrow 0$ as $t \longrightarrow \infty$.

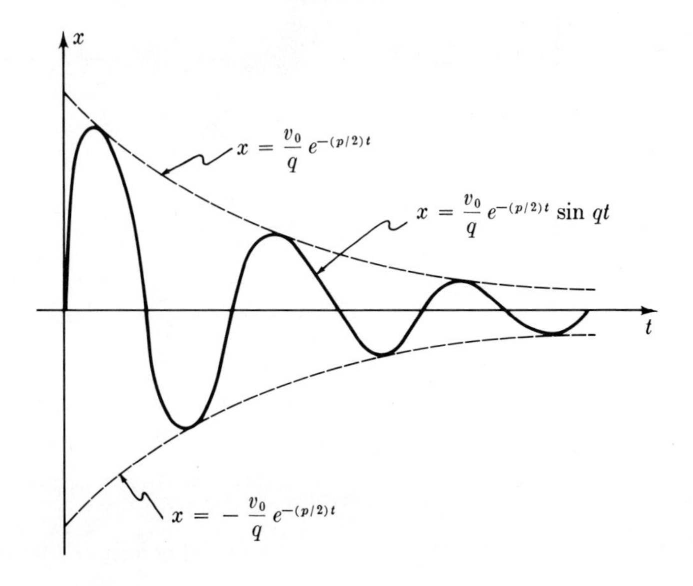

FIG. 5.4

CASE 2: $p > 2k$ (overdamped case: friction large compared to spring force). Now $\Delta > 0$. The roots of the quadratic are

$$\alpha = \frac{1}{2}(-p + \sqrt{\Delta}), \qquad \beta = \frac{1}{2}(-p - \sqrt{\Delta}).$$

Both roots are negative ($\sqrt{\Delta} = \sqrt{p^2 - 4k^2} < p$). Let us call the roots $-r$ and $-s$, where $r > s > 0$. The general solution is

$$x = ae^{-rt} + be^{-st}.$$

Since

$$\dot{x} = -are^{-rt} - bse^{-st},$$

the initial conditions become

$$0 = a + b, \qquad v_0 = -ar - bs.$$

It follows that

$$-a = b = \frac{v_0}{r - s}.$$

Therefore, the solution of the initial-value problem is

$$x = c(e^{-st} - e^{-rt}), \qquad \text{where} \quad c = \frac{v_0}{r - s} > 0.$$

The graph of x as a function of t is shown in Fig. 5.5. The particle moves away from the origin at first, then reverses direction and approaches the origin, never quite reaching it again.

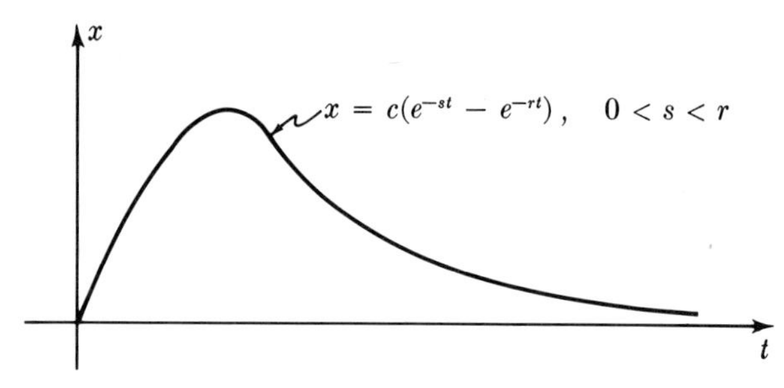

FIG. 5.5

CASE 3: $p = 2k$ (borderline case, critical damping). Now $\Delta = 0$, and the quadratic equation has two equal roots, $\alpha = \beta = -p/2$. The general solution is

$$x = e^{-(p/2)t}(a + bt).$$

Thus $x(0) = a$ and $\dot{x}(0) = b - \frac{1}{2}pa$, so the initial conditions reduce to

$$0 = a, \qquad v_0 = b.$$

Therefore the solution of the initial-value problem is

$$x = v_0te^{-(p/2)t} = v_0te^{-kt}.$$

By solving $\dot{x} = 0$, it is seen that x reaches its maximum value $2v_0/pe = v_0/ke$ when $t = 2/p$. See Fig. 5.6. The particle moves $2v_0/pe$ units from the origin,

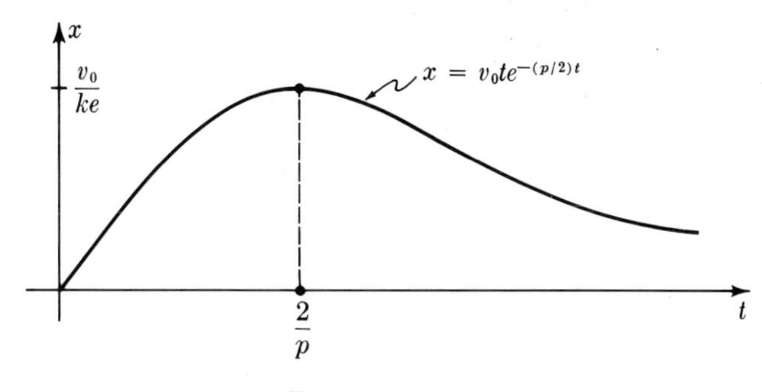

FIG. 5.6

reverses direction and approaches the origin again as $t \longrightarrow \infty$. As is physically plausible, the maximum distance from the origin is proportional to the initial velocity, inversely proportional to the coefficient of friction.

Answer: Let $\Delta = p^2 - 4k^2$.

If $\Delta < 0$, then $x = \dfrac{v_0}{q} e^{-(p/2)t} \sin qt$, where $q = \dfrac{1}{2}\sqrt{-\Delta}$.

If $\Delta > 0$, then $x = \dfrac{v_0}{r - s}(e^{-st} - e^{-rt})$, where

$$r = \frac{1}{2}(p + \sqrt{\Delta}), \qquad s = \frac{1}{2}(p - \sqrt{\Delta}).$$

If $\Delta = 0$, then $x = v_0te^{-(p/2)t} = v_0te^{-kt}$.

The next example provides a simple mechanical model for a variety of physical phenomena.

EXAMPLE 5.5

Suppose in Example 5.4 an external force $F = A \sin \omega t$ is imposed. What is the nature of the motion when t is large? (Describe the solution $x(t)$ as $t \longrightarrow \infty$.) Assume small friction.

Solution: The differential equation is

$$\ddot{x} + p\dot{x} + k^2x = A \sin \omega t.$$

Guess a particular solution of the form

$$x = a \cos \omega t + b \sin \omega t.$$

Substitute; after some computation the differential equation becomes

$$[(k^2 - \omega^2)a + p\omega b] \cos \omega t + [(k^2 - \omega^2)b - p\omega a] \sin \omega t = A \sin \omega t.$$

Equate coefficients of $\cos \omega t$ and $\sin \omega t$:

$$\begin{cases} ra + sb = 0 \\ -sa + rb = A \end{cases} \qquad \text{where} \quad r = k^2 - \omega^2 \quad \text{and} \quad s = p\omega.$$

Solve:

$$a = \frac{-As}{r^2 + s^2}, \qquad b = \frac{Ar}{r^2 + s^2}.$$

Hence, a particular solution is

$$x = \frac{A}{r^2 + s^2}(-s \cos \omega t + r \sin \omega t).$$

By the usual method, convert it to the form

$$x = \frac{A}{\sqrt{r^2 + s^2}} \sin \omega(t - t_0),$$

where

$$\sin \omega t_0 = \frac{s}{\sqrt{r^2 + s^2}} \qquad \text{and} \qquad \cos \omega t_0 = \frac{r}{\sqrt{r^2 + s^2}}.$$

If friction is small, the general solution of the homogeneous equation

$$\ddot{x} + p\dot{x} + k^2 x = 0$$

is

$$x = ce^{-(p/2)t} \sin q(t - t_1), \qquad q = \frac{1}{2}\sqrt{4k^2 - p^2}.$$

(See the answer to Example 5.4, first case.)

Combine results: the general solution of

$$\ddot{x} + p\dot{x} + k^2 x = A \sin \omega t$$

is

$$x = ce^{-(p/2)t} \sin q(t - t_1) + \frac{A}{\sqrt{(k^2 - \omega^2)^2 + (p\omega)^2}} \sin \omega(t - t_0).$$

The constants c and t_1 can be determined from initial conditions, but they are not needed to predict the behavior of x for large values of t. As $t \longrightarrow \infty$, the term $ce^{-(p/2)t} \sin q(t - t_1) \longrightarrow 0$. (It is called a **transient.**) Thus after sufficient time has elapsed,

$$x(t) \approx C \sin \omega(t - t_0);$$

all that is visible is simple harmonic motion. The amplitude of this motion is

$$C = \frac{A}{\sqrt{(k^2 - \omega^2)^2 + (p\omega)^2}}.$$

In case p is small and $\omega \approx k$, the denominator is small, hence the amplitude C is large. This is the phenomenon of **resonance.** It occurs when there is a periodic force with frequency near the "natural frequency" of the system. Destructive vibrations in machinery and vibrations caused by soldiers marching in step on a bridge are results of resonance.

The approximation

$$x(t) \approx C \sin \omega(t - t_0),$$

valid for large values of t, is called the **steady state solution.** Notice that it is independent of the initial conditions; these affect only the constants in the transient term.

Answer: For large t, $x(t) \approx C \sin \omega(t - t_0)$, where

$$C = \frac{A}{\sqrt{(k^2 - \omega^2)^2 + (p\omega)^2}} = \frac{A}{\sqrt{r^2 + s^2}},$$

and

$$\begin{cases} \sin \omega t_0 = \dfrac{s}{\sqrt{r^2 + s^2}} \\ \cos \omega t_0 = \dfrac{r}{\sqrt{r^2 + s^2}} \end{cases} \qquad \begin{cases} r = k^2 - \omega^2 \\ s = p\omega. \end{cases}$$

The standard electric model for this situation is a simple circuit with resistance R, inductance L, capacitance C, current I, and with voltage $E = -A \cos \omega t$. The equation of the circuit is

$$\frac{d^2I}{dt^2} + \frac{R}{L}\frac{dI}{dt} + \frac{1}{LC} I = \frac{1}{L}\frac{dE}{dt} = \frac{A\omega}{L} \sin \omega t.$$

The steady state solution is

$$I(t) \approx \frac{A\omega}{L\sqrt{\left(\dfrac{1}{LC} - \omega^2\right)^2 + \left(\dfrac{R}{L}\omega\right)^2}} \sin \omega(t - t_0).$$

Resonance (tuned circuit) occurs when $\omega^2 = 1/LC$. Then $\sin \omega t_0 = 1$ and $\cos \omega t_0 = 0$. Hence $\omega t_0 = \pi/2$ and

$$I(t) \approx \frac{A}{R} \sin\left(\omega t - \frac{\pi}{2}\right).$$

EXERCISES

Express as $c \cos k(t - t_0)$ and as $c \sin k(t - t_1)$. Use trigonometric tables if necessary:

1. $\cos 3t + \sin 3t$
2. $\sin t - \cos t$

3. $\sqrt{3}\cos 2t - \sin 2t$

4. $\cos t + \sqrt{3}\sin t$

5. $4\sin t - 3\cos t$

6. $\sin\frac{t}{2} + 5\cos\frac{t}{2}\cdot$

The next four problems concern a particle in simple harmonic motion about the origin. Its position at time t is $x(t)$ and its velocity is $v(t)$.

7. Find $x(t)$ if the period is 2 sec, the amplitude is 6, and $x(0) = 3$.
8. Find $x(t)$ if $x(0) = 0$, $v(0) = 5$, and the period is 2 sec.
9. Find the amplitude if $x(0) = 0$, $v(0) = 10$, and $v(3) = 5$.
10. Estimate the amplitude to 3 significant digits if $x(0) = 0$, $v(0) = 10$, and $x(1) = 6$.
11. A 16-lb weight hangs at rest on the end of an 8-ft spring attached to a ceiling. (See Example 5.3.) The weight is pulled down k in. and released. As it passes through its original (equilibrium) position, its speed is 1.5 ft/sec. Find k.
12. (cont.) Suppose the weight slides along a wall. Due to friction there is a retarding force proportional to the velocity. The motion is a damped vibration with period $2\pi/\sqrt{3}$ sec. After how long will the amplitude of the oscillation be reduced to half its original magnitude?
13. Suppose the weight in Example 5.3 is w lb, not 16 lb. What is the period of its motion?
14. A pendulum is made of a small weight at the end of a long wire. Its motion is described by the differential equation

$$\frac{d^2\theta}{dt^2} + \frac{g}{L}\sin\theta = 0,$$

where θ is the angle between the wire and the vertical. If the pendulum swings only through a small arc, then $\sin\theta$ can be approximated by θ, thus simplifying the differential equation. Do so and find the approximate period of the pendulum.

15. A cylindrical buoy floats vertically in the water. Its weight is 100 lb and its diameter is 2 ft. When depressed slightly and released, it oscillates with simple harmonic motion. Find the period of the oscillation.
[*Hint:* This is just a spring problem in disguise. Use Archimedes' Law: A body in water is subjected to an upward buoyant force equal to the weight of the water displaced. Take the density of water to be 62.4 lb/ft^3.]
16. In the overdamped motion of Fig. 5.5, find the maximum value of $x(t)$. Show that $x(t) > 0$ for $t > 0$ and $x(t) < 0$ for $t < 0$.
17. Consider the underdamped motion of Fig. 5.4. Show that the graph crosses the t-axis infinitely often for $t > 0$.
18. A rocket sled is subjected to $6g$ acceleration for 5 sec. After the engine shuts down, the sled undergoes a deceleration (ft/sec^2) equal to 0.05 times its velocity (ft/sec). What is the speed of the sled 10 sec after engine shutdown? How far has it traveled?

The external forces acting on a projectile are gravity and air resistance. At low altitude and low speed, it may be assumed that air resistance is proportional to speed.

If a projectile is shot straight up, it rises to its maximum height and then falls to the ground. Whether it takes longer to rise or to fall, or equal times, is not obvious.

However, as we shall see, in falling there is a terminal speed. Hence a projectile shot up with initial speed faster than the terminal speed necessarily takes longer in falling than in rising. The next five examples show this is *always* so.

19. A projectile is shot straight up with initial velocity v_0. Show that its height satisfies the initial-value problem $\ddot{y} + k\dot{y} = -g, \quad y(0) = 0, \quad \dot{y}(0) = v_0$. Derive the solution

$$y(t) = A(1 - e^{-kt}) - \frac{g}{k}t, \qquad \text{where} \quad A = \frac{g + kv_0}{k^2}.$$

20. (cont.) Show that v approaches a terminal velocity as $t \longrightarrow \infty$. Find it.

21. (cont.) Show that the projectile reaches its maximum height at time

$$t_1 = \frac{1}{k}\ln\left(\frac{g + kv_0}{g}\right).$$

Show that the projectile returns to ground at time $t_2 > 0$, where

$$\left(\frac{g + kv_0}{g}\right)(1 - e^{-kt_2}) = kt_2.$$

22. (cont.) Show that $t_1 \longrightarrow v_0/g$ as $k \longrightarrow 0$ and interpret.
(*Hint:* Express the derivative of $\ln(1 + cx)$ at $x = 0$ as a limit; alternatively use the first order Taylor Approximation.)
Guess what t_2 approaches as $k \longrightarrow 0$.

23. (cont.) Prove $(t_2 - t_1) > t_1$. Begin by showing that $\dot{y}(t_1 + \tau) + \dot{y}(t_1 - \tau) > 0$ for $\tau > 0$. Interpret physically. Then integrate the inequality over the interval $0 \le \tau \le t_1$ to obtain $y(t_1 + \tau) > y(t_1 - \tau)$ for $\tau > 0$. Deduce that $2t_1 < t_2$.

24. (cont.) Now try a concrete problem. Suppose the initial velocity is 500 ft/sec and the maximum height is 3000 ft. Assume $g = 32.2$ ft/sec². Estimate k to 3 significant digits. Then estimate t_1 and t_2 to 2 significant digits.

23. Vectors

1. INTRODUCTION

Vector analysis is a very useful tool for handling problems in more than one dimension. Its main advantages are (1) equations in vector form are independent of choice of coordinate axes; they are well-suited to describe physical situations; (2) each vector equation replaces three ordinary equations; and (3) several frequently occurring procedures can be summarized neatly in vector form.

Let the origin **0** be fixed once and for all. A **vector** in space is a directed line segment which begins at **0**; it is completely determined by its terminal point. Denote vectors by bold-faced letters **x**, **v**, **F**, **r**, etc. (In written work use $\underline{x}$ or $\vec{x}$ instead of **x**.)

A point (x, y, z) in space is often identified with the vector **x** from the origin to the point.

A vector is determined by two quantities, *length* (or *magnitude*) and *direction*. Many physical quantities are vectors: force, velocity, acceleration, electric field intensity, etc.

It is emphasized that the origin **0** is fixed, and that each vector starts at **0**. Vectors are often *drawn* starting at other points, but in computations, they all originate at **0**. For example, if a force **F** is applied at a point **x**, we often draw Fig. 1.1a because it is suggestive. But the correct figure is Fig. 1.1b. One must specify both the force vector **F** (magnitude and direction) and its point of application **x**.

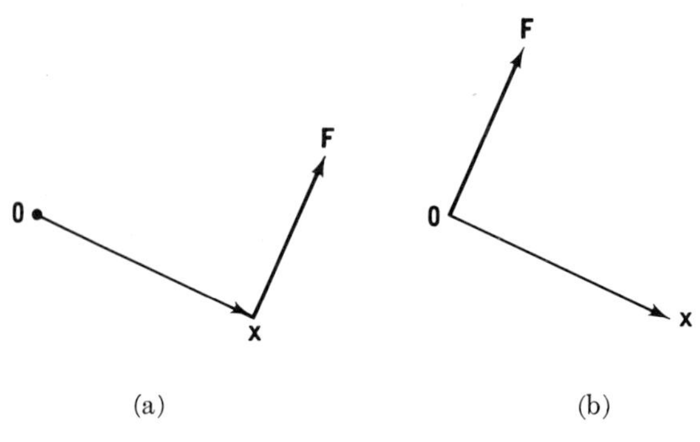

(a) (b)

Fig. 1.1

With respect to coordinate axes, each vector $\mathbf{x}$ has three **components** (coordinates) x, y, and z. We write

$$\mathbf{x} = (x, y, z)$$

to indicate these components. See Fig. 1.2.

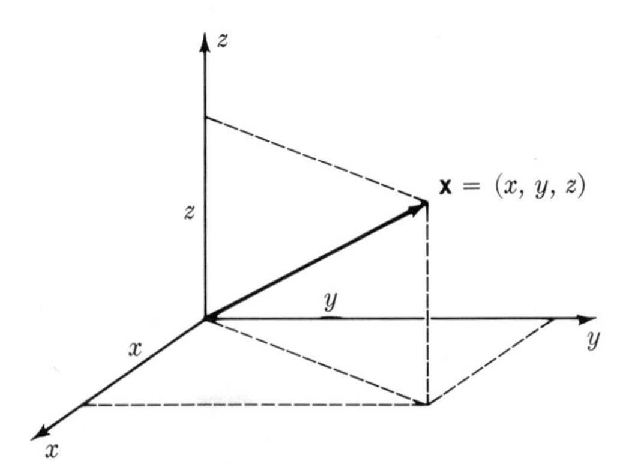

Fig. 1.2

Sometimes it is convenient to index the components, writing $\mathbf{x} = (x_1, x_2, x_3)$ instead of $\mathbf{x} = (x, y, z)$. The zero vector (origin) will be written $\mathbf{0} = (0, 0, 0)$. For this vector alone, direction is undefined.

2. VECTOR ALGEBRA

Addition

The **sum** $\mathbf{v} + \mathbf{w}$ of two vectors is defined by the parallelogram law (Fig. 2.1). The points $\mathbf{0}$, $\mathbf{v}$, $\mathbf{w}$, and $\mathbf{v} + \mathbf{w}$ are vertices of a parallelogram with $\mathbf{v} + \mathbf{w}$ opposite to $\mathbf{0}$.

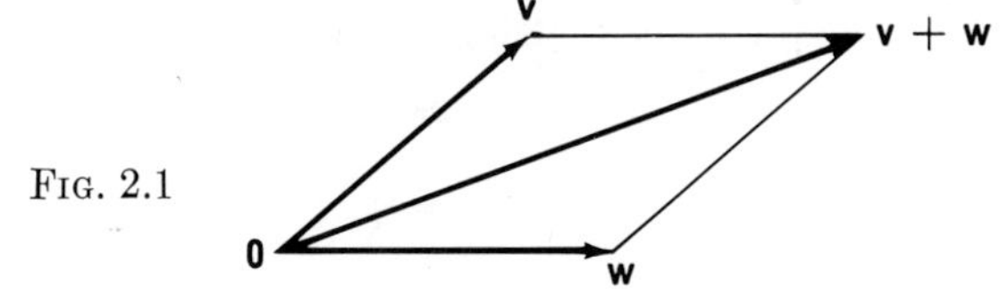

Fig. 2.1

Vectors are added numerically by adding their components:

$$\boxed{(v_1, v_2, v_3) + (w_1, w_2, w_3) = (v_1 + w_1, v_2 + w_2, v_3 + w_3).}$$

For example,

$$(-1, 3, 2) + (1, 1, 4) = (0, 4, 6), \qquad (0, 0, 1) + (-1, 0, 1) = (-1, 0, 2).$$

The projection on each axis of the sum $\mathbf{v} + \mathbf{w}$ is the sum of the corresponding projections of $\mathbf{v}$ and of $\mathbf{w}$. This is shown in Fig. 2.2 for two vectors in the x_1, x_2-plane.

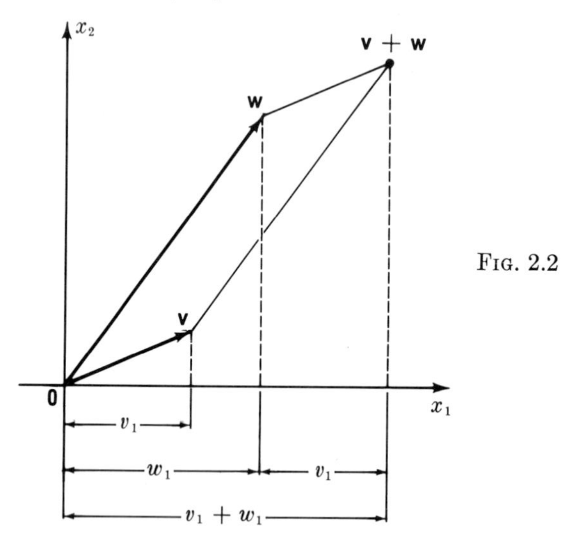

FIG. 2.2

Scalar Multiplication

Next consider **scalar multiplication** of vectors. Let $\mathbf{v}$ be a vector and a be a number (**scalar**). The product $a\mathbf{v}$ is the vector whose length is $|a|$ times the length of $\mathbf{v}$ and which points in the same direction as $\mathbf{v}$ if $a > 0$, in the opposite direction if $a < 0$. If $a = 0$, then $a\mathbf{v} = \mathbf{0}$. In components,

$$a(v_1, v_2, v_3) = (av_1, av_2, av_3).$$

If a particle moving in a certain direction doubles its speed, its velocity vector is doubled; if a horse pulling a cart in a certain direction triples its effort, the force vector triples. Figure 2.3 illustrates multiples of a vector.

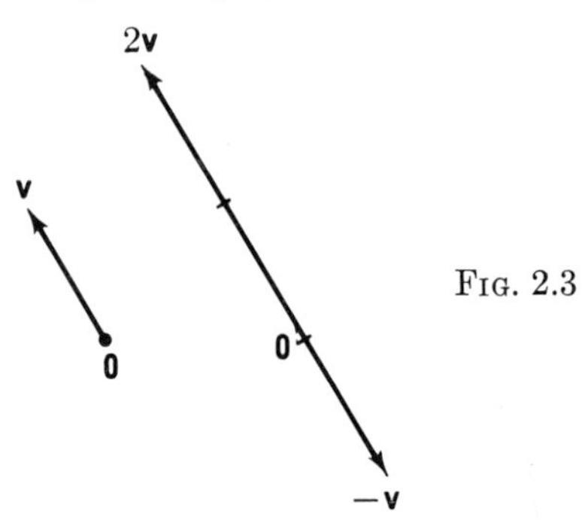

FIG. 2.3

Write $\mathbf{v} - \mathbf{w}$ for $\mathbf{v} + (-\mathbf{w})$. See Fig. 2.4. The segment from the tip of $\mathbf{w}$ to the tip of $\mathbf{v}$ (the dotted line in Fig. 2.4c) has the same length and direction as **v** − **w**. Hence if two points are represented by vectors **v** and **w**, the distance between them is the length of **v** − **w**.

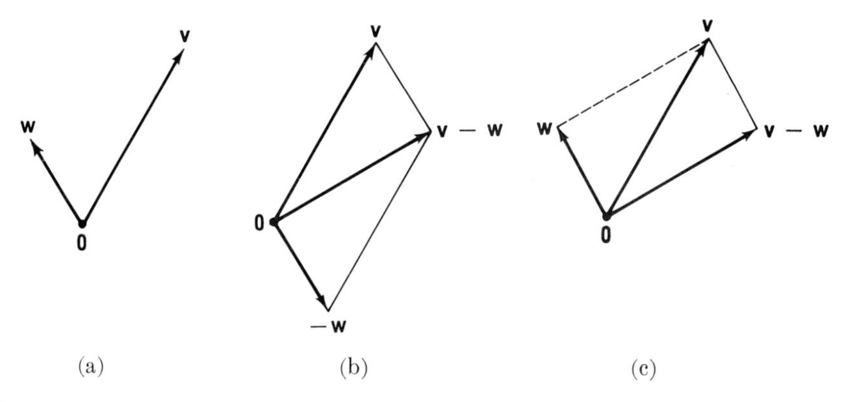

Fig. 2.4

The **length** of a vector **v** is denoted by $|\mathbf{v}|$. It is the distance from **0** to the terminal point of **v**. By the Distance Formula,

$$|\mathbf{v}|^2 = |(v_1, v_2, v_3)|^2 = v_1^2 + v_2^2 + v_3^2.$$

For example, $|(1, 2, 3)|^2 = 1 + 4 + 9$; the vector (1, 2, 3) has length $\sqrt{14}$.

As another example, the distance between two points $\mathbf{x} = (x_1, x_2, x_3)$ and $\mathbf{y} = (y_1, y_2, y_3)$ is the length of the vector $\mathbf{x} - \mathbf{y}$. Since $\mathbf{x} - \mathbf{y} = (x_1 - y_1, x_2 - y_2, x_3 - y_3)$, the distance between the points is

$$|(x_1 - y_1, x_2 - y_2, x_3 - y_3)| = [(x_1 - y_1)^2 + (x_2 - y_2)^2 + (x_3 - y_3)^2]^{1/2}.$$

Dot Product

There is another important vector operation, the **scalar product** or **dot product** of two vectors. Let **v** and **w** be vectors, and let θ be the angle between them (Fig. 2.5a). Define

$$\mathbf{v} \cdot \mathbf{w} = |\mathbf{v}|\,|\mathbf{w}| \cos\theta.$$

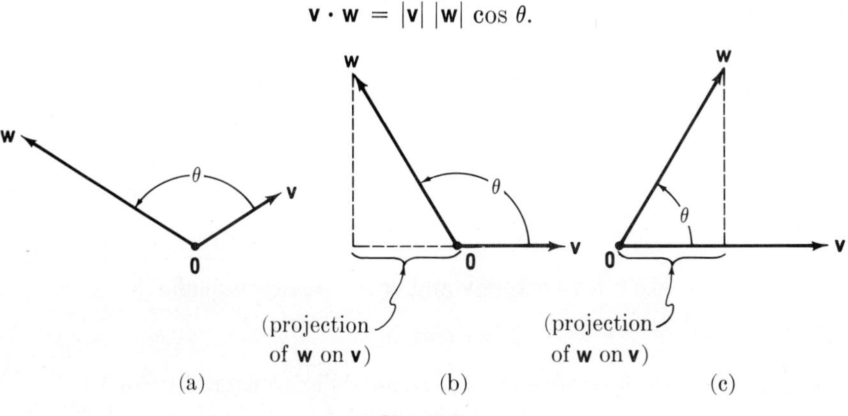

Fig. 2.5

Since $\cos(-\theta) = \cos\theta$, you can measure θ from $\mathbf{v}$ to $\mathbf{w}$ or from $\mathbf{w}$ to $\mathbf{v}$. Note (Fig. 2.5b, c) that $|\mathbf{w}|\cos\theta$ is the (signed) projection of $\mathbf{w}$ on $\mathbf{v}$, hence $\mathbf{v}\cdot\mathbf{w}$ is $|\mathbf{v}|$ times the projection of $\mathbf{w}$ on $\mathbf{v}$.

IMPORTANT: The scalar product of two vectors is a *scalar* (number), not a vector.

The numerical rule for computing dot products is

$$\boxed{\mathbf{v}\cdot\mathbf{w} = v_1w_1 + v_2w_2 + v_3w_3,}$$

an important formula. Let us prove it. See Fig. 2.6. By the Law of Cosines,

$$|\mathbf{v}-\mathbf{w}|^2 = |\mathbf{v}|^2 + |\mathbf{w}|^2 - 2|\mathbf{v}|\,|\mathbf{w}|\cos\theta = |\mathbf{v}|^2 + |\mathbf{w}|^2 - 2\mathbf{v}\cdot\mathbf{w}.$$

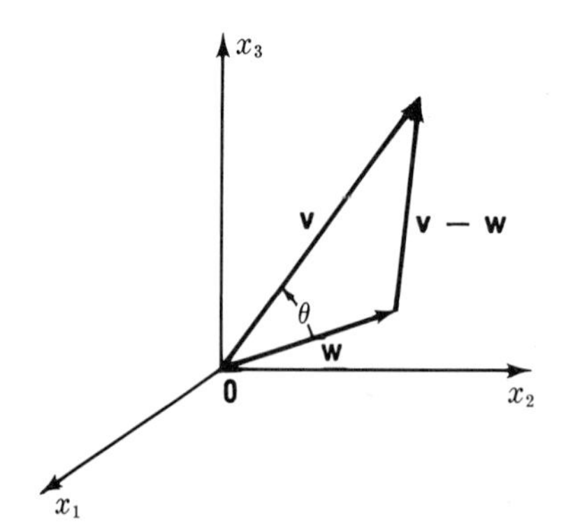

FIG. 2.6

Hence

$$\begin{aligned}\mathbf{v}\cdot\mathbf{w} &= \frac{1}{2}\Big[|\mathbf{v}|^2 + |\mathbf{w}|^2 - |\mathbf{v}-\mathbf{w}|^2\Big]\\ &= \frac{1}{2}\Big[|(v_1, v_2, v_3)|^2 + |(w_1, w_2, w_3)|^2 - |(v_1 - w_1, v_2 - w_2, v_3 - w_3)|^2\Big]\\ &= \frac{1}{2}\Big[(v_1^2 + v_2^2 + v_3^2) + (w_1^2 + w_2^2 + w_3^2)\\ &\qquad - (v_1 - w_1)^2 - (v_2 - w_2)^2 - (v_3 - w_3)^2\Big]\\ &= v_1w_1 + v_2w_2 + v_3w_3.\end{aligned}$$

Two vectors $\mathbf{v}$ and $\mathbf{w}$ are perpendicular if $\theta = \pi/2$, i.e., if $\cos\theta = 0$. This can be expressed very neatly as follows:

> The condition for vectors $\mathbf{v}$ and $\mathbf{w}$ to be perpendicular is
>
> $$\mathbf{v}\cdot\mathbf{w} = 0.$$
>
> (The vector $\mathbf{0}$ is considered perpendicular to every vector.)

For example, $(1, 2, 3)$ and $(-1, -1, 1)$ are perpendicular because

$$(1, 2, 3) \cdot (-1, -1, 1) = -1 - 2 + 3 = 0.$$

There is a connection between lengths and dot products. The dot product of a vector $\mathbf{v}$ with itself is $\mathbf{v} \cdot \mathbf{v} = |\mathbf{v}|^2 \cos 0 = |\mathbf{v}|^2$.

> For any vector $\mathbf{v}$,
>
> $\mathbf{v} \cdot \mathbf{v} = |\mathbf{v}|^2 = v_1^2 + v_2^2 + v_3^2.$

From the dot product can be found the angle θ between any two nonzero vectors $\mathbf{v}$ and $\mathbf{w}$. Indeed,

$$\cos \theta = \frac{\mathbf{v} \cdot \mathbf{w}}{|\mathbf{v}|\,|\mathbf{w}|}.$$

EXAMPLE 2.1

Find the angle between $\mathbf{v} = (1, 2, 1)$ and $\mathbf{w} = (3, -1, 1)$.

Solution:

$$\mathbf{v} \cdot \mathbf{w} = 3 - 2 + 1 = 2,$$

$$|\mathbf{v}|^2 = 1 + 4 + 1 = 6, \qquad |\mathbf{w}|^2 = 9 + 1 + 1 = 11.$$

Hence

$$\cos \theta = \frac{2}{\sqrt{6}\,\sqrt{11}}.$$

Answer: $\operatorname{arc\,cos} \dfrac{2}{\sqrt{66}}$.

EXAMPLE 2.2

The point $(1, 1, 2)$ is joined to the points $(1, -1, -1)$ and $(3, 0, 4)$ by lines L_1 and L_2. What is the angle θ between these lines?

Solution: The vector

$$\mathbf{v} = (1, -1, -1) - (1, 1, 2) = (0, -2, -3)$$

is parallel to L_1 (but starts at $\mathbf{0}$). Likewise

$$\mathbf{w} = (3, 0, 4) - (1, 1, 2) = (2, -1, 2)$$

is parallel to L_2. Hence

$$\cos \theta = \frac{\mathbf{v} \cdot \mathbf{w}}{|\mathbf{v}|\,|\mathbf{w}|} = \frac{0 + 2 - 6}{\sqrt{0 + 4 + 9}\,\sqrt{4 + 1 + 4}} = \frac{-4}{\sqrt{13}\,\sqrt{9}}.$$

Answer: $\theta = \operatorname{arc\,cos} \left(\dfrac{-4}{3\sqrt{13}} \right)$.

Direction Cosines

It is customary to use the notation

$$\mathbf{i} = (1, 0, 0), \qquad \mathbf{j} = (0, 1, 0), \qquad \mathbf{k} = (0, 0, 1)$$

for the three unit-length vectors along the positive coordinate axes (Fig. 2.7). If $\mathbf{v}$ is any vector, then

$$\begin{aligned}\mathbf{v} &= (v_1, v_2, v_3)\\ &= v_1(1, 0, 0) + v_2(0, 1, 0) + v_3(0, 0, 1)\\ &= v_1\mathbf{i} + v_2\mathbf{j} + v_3\mathbf{k}.\end{aligned}$$

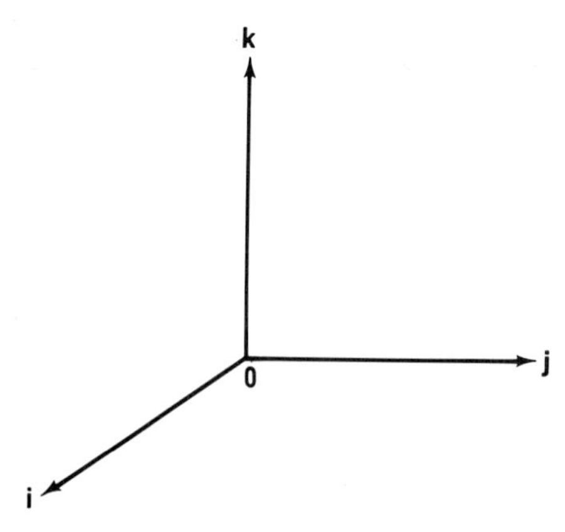

FIG. 2.7

Thus $\mathbf{v}$ is the sum of three vectors $v_1\mathbf{i}$, $v_2\mathbf{j}$, $v_3\mathbf{k}$ which lie along the three coordinate axes. The components v_1, v_2, v_3 can be interpreted as dot products:

$$\mathbf{v} \cdot \mathbf{i} = (v_1, v_2, v_3) \cdot (1, 0, 0) = v_1.$$

Similarly, $v_2 = \mathbf{v} \cdot \mathbf{j}$ and $v_3 = \mathbf{v} \cdot \mathbf{k}$.

Now suppose $\mathbf{u}$ is a **unit vector** i.e., a vector of length one (Fig. 2.8).

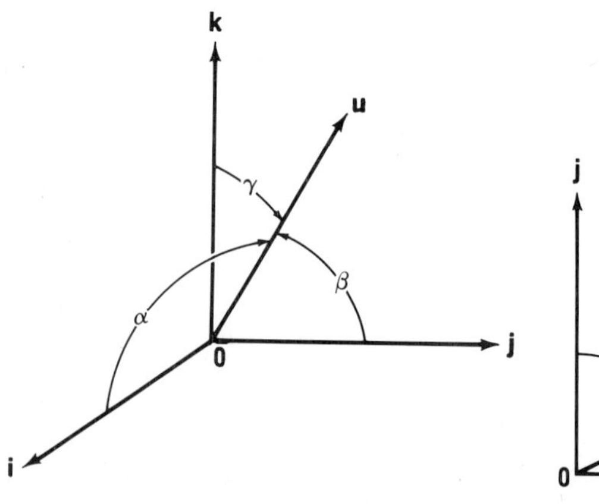

FIG. 2.8

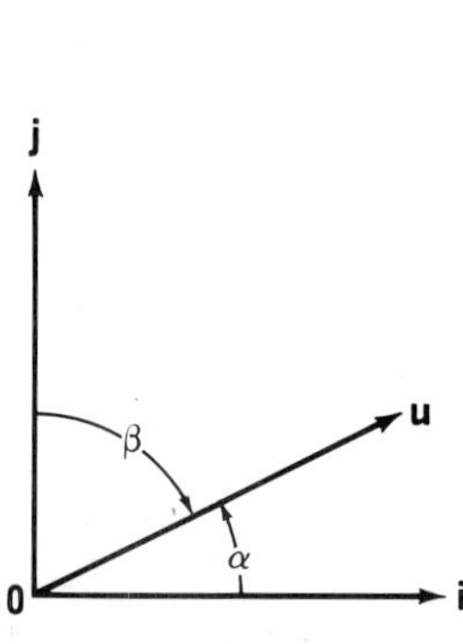

FIG. 2.9

Let α be the angle from $\mathbf{i}$ to $\mathbf{u}$. Define β and γ similarly. Then $\mathbf{u} \cdot \mathbf{i} = \cos\alpha$, $\mathbf{u} \cdot \mathbf{j} = \cos\beta$, and $\mathbf{u} \cdot \mathbf{k} = \cos\gamma$. Hence

$$\begin{aligned} \mathbf{u} &= \cos\alpha\,\mathbf{i} + \cos\beta\,\mathbf{j} + \cos\gamma\,\mathbf{k} \\ &= (\cos\alpha, \cos\beta, \cos\gamma). \end{aligned}$$

Since $|\mathbf{u}| = 1$,

$$\cos^2\alpha + \cos^2\beta + \cos^2\gamma = 1.$$

Unit vectors are direction indicators. Any non-zero vector $\mathbf{v}$ is a positive multiple of a unit vector $\mathbf{u}$ in the same direction as $\mathbf{v}$. In fact $\mathbf{v} = |\mathbf{v}|\mathbf{u}$, so

$$\mathbf{u} = \frac{1}{|\mathbf{v}|}\mathbf{v} \qquad (\mathbf{v} \neq \mathbf{0}).$$

> Each non-zero vector $\mathbf{v}$ can be expressed as
>
> $$\mathbf{v} = |\mathbf{v}|\mathbf{u}, \qquad \mathbf{u} \quad \text{a unit vector,}$$
>
> or as
>
> $$\mathbf{v} = |\mathbf{v}|(\cos\alpha, \cos\beta, \cos\gamma).$$
>
> The numbers $\cos\alpha$, $\cos\beta$, $\cos\gamma$ are called the **direction cosines** of $\mathbf{v}$. They satisfy
>
> $$\cos^2\alpha + \cos^2\beta + \cos^2\gamma = 1.$$

If $\mathbf{u}$ is a unit vector (Fig. 2.9) in the plane of $\mathbf{i}$ and $\mathbf{j}$, then

$$\mathbf{u} = \cos\alpha\,\mathbf{i} + \cos\beta\,\mathbf{j}.$$

Since $\mathbf{u}$ is a unit vector, $\cos^2\alpha + \cos^2\beta = 1$. But, as is seen in Fig. 2.9, $\cos\beta = \sin\alpha$. Therefore, the preceding equation simply says

$$\cos^2\alpha + \sin^2\alpha = 1.$$

Summary

ADDITION OF VECTORS:

$$(v_1, v_2, v_3) + (w_1, w_2, w_3) = (v_1 + w_1, v_2 + w_2, v_3 + w_3).$$

SCALAR MULTIPLICATION:

$$a(v_1, v_2, v_3) = (av_1, av_2, av_3).$$

LENGTH:

$$|\mathbf{v}|^2 = \mathbf{v} \cdot \mathbf{v} = v_1^2 + v_2^2 + v_3^2.$$

DOT PRODUCT:

$$\mathbf{v} \cdot \mathbf{w} = |\mathbf{v}|\,|\mathbf{w}| \cos\theta = v_1w_1 + v_2w_2 + v_3w_3.$$

VECTORS $\mathbf{i}$, $\mathbf{j}$, $\mathbf{k}$:

These are unit vectors in the direction of the positive x-axis, y-axis, z-axis, respectively. If $\mathbf{v} = (v_1, v_2, v_3)$, then $\mathbf{v} = v_1\mathbf{i} + v_2\mathbf{j} + v_3\mathbf{k}$, where $v_1 = \mathbf{v} \cdot \mathbf{i}$, $v_2 = \mathbf{v} \cdot \mathbf{j}$, $v_3 = \mathbf{v} \cdot \mathbf{k}$.

DIRECTION COSINES:

If $\mathbf{u}$ is a unit vector, then $\mathbf{u} = \cos\alpha\,\mathbf{i} + \cos\beta\,\mathbf{j} + \cos\gamma\,\mathbf{k}$, where α, β, γ are the angles to $\mathbf{u}$ from $\mathbf{i}$, $\mathbf{j}$, $\mathbf{k}$, respectively. Furthermore $\cos^2\alpha + \cos^2\beta + \cos^2\gamma = 1$. Any non-zero vector $\mathbf{v}$ can be written as $\mathbf{v} = |\mathbf{v}|\mathbf{u} = |\mathbf{v}|(\cos\alpha, \cos\beta, \cos\gamma)$. The numbers $\cos\alpha$, $\cos\beta$, $\cos\gamma$ are the direction cosines of $\mathbf{v}$.

EXERCISES

Find:

1. $(1, 2, -3) + (4, 0, 7)$
2. $(-1, -1, 0) - (3, 5, 2)$
3. $(8, 2, 1) \cdot (3, 0, 5)$
4. $(-1, -1, -1) \cdot (1, 2, 3)$
5. $(1, 2, 3) - 6(0, 3, -1)$
6. $4[(1, -2, 7) - (1, 1, 1)]$
7. $\frac{1}{3}[(1, -1, -1) + (5, 3, 0) + (-1, -2, 4)]$
8. $(1, 0, 2) \cdot [(1, 4, 1) + (2, 0, -3)]$
9. $|(2, -4, 7)|$
10. $|(1/\sqrt{3})\,(-1, 1, 1)|$
11. $|3\mathbf{i} - \mathbf{j} + \mathbf{k}|$
12. $|(-1, -1, 0) - (3, 5, 2)|$.

Find the angle between the vectors:

13. $(1, 1, -1)$ and $(2, 0, 4)$
14. $(2, 2, 2)$ and $(-2, 2, -2)$.

Find the direction cosines of:

15. $\mathbf{v} = (1, 0, 1)$
16. $\mathbf{v} = (-1, -1, -1)$
17. $\mathbf{v} = (2, 1, -3)$
18. $\mathbf{v} = (4, -7, -4)$.
19. Find the angle between the line joining $(0,0,0)$ to $(1,1,1)$ and the line joining $(1,0,0)$ to $(0,1,0)$.
20. Prove the **Cauchy–Schwarz** inequality:
$$|\mathbf{v} \cdot \mathbf{w}| \le |\mathbf{v}|\,|\mathbf{w}|.$$
When does equality hold?
21. Prove the identity $|\mathbf{v} + \mathbf{w}|^2 - |\mathbf{v} - \mathbf{w}|^2 = 4\,\mathbf{v} \cdot \mathbf{w}$.
22. Show that $\frac{1}{2}(\mathbf{v} + \mathbf{w})$ is the midpoint of the segment from $\mathbf{v}$ to $\mathbf{w}$. (*Hint:* Use the parallelogram law.)
23. (cont.) Use Ex. 22 to show that the segments joining the midpoints of opposite sides of a (skew) quadrilateral bisect each other. [*Hint:* $(\mathbf{u} + \mathbf{v}) + (\mathbf{w} + \mathbf{z}) = (\mathbf{v} + \mathbf{w}) + (\mathbf{u} + \mathbf{z})$.]
24. Show that $\frac{1}{3}(\mathbf{u} + \mathbf{v} + \mathbf{w})$ is the intersection of the medians of the triangle with vertices $\mathbf{u}$, $\mathbf{v}$, and $\mathbf{w}$.

3. VECTOR CALCULUS

Suppose a vector $\mathbf{x}$ depends on time. For example, $\mathbf{x}$ may represent the position of a moving object at time t or the gravitational force on an orbiting satellite at time t. To indicate that $\mathbf{x}$ is a function of time, write

$$\mathbf{x} = \mathbf{x}(t);$$

in components,

$$\mathbf{x}(t) = (x(t), y(t), z(t)).$$

Thus a vector function is a single expression for three ordinary (scalar) functions

$$x = x(t), \qquad y = y(t), \qquad z = z(t).$$

What is the derivative of a vector function? Think of $\mathbf{x} = \mathbf{x}(t)$ as tracing a path in space (Fig. 3.1). For h small, the difference vector

$$\mathbf{x}(t + h) - \mathbf{x}(t)$$

represents the secant from $\mathbf{x}(t)$ to $\mathbf{x}(t + h)$. The difference quotient

$$\frac{\mathbf{x}(t + h) - \mathbf{x}(t)}{h}$$

represents this (short) secant divided by the small number h. The limit as $h \longrightarrow 0$ is called the **derivative** of the vector function:

$$\boxed{\dot{\mathbf{x}}(t) = \frac{d\mathbf{x}}{dt} = \lim_{h \to 0} \frac{\mathbf{x}(t + h) - \mathbf{x}(t)}{h}.}$$

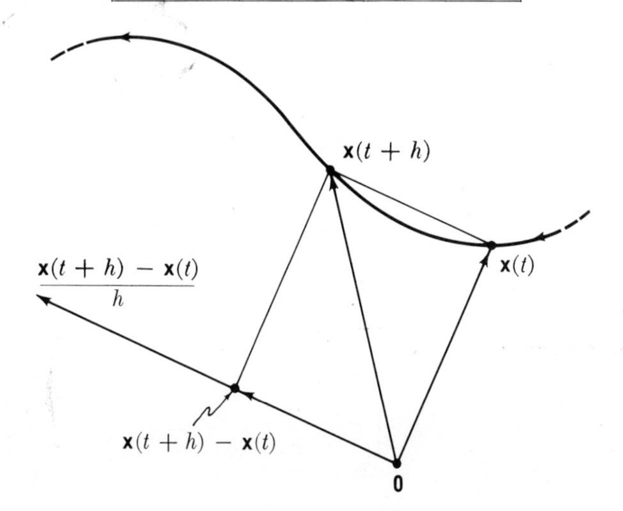

FIG. 3.1

The derivative $\dot{\mathbf{x}}(t)$ is a vector in the direction of the tangent to the curve because the tangent is the limiting position of the secant.

To compute the derivative, express all vectors in components:

$$\frac{\mathbf{x}(t+h)-\mathbf{x}(t)}{h} = \frac{1}{h}[(x(t+h), y(t+h), z(t+h)) - (x(t), y(t), z(t))]$$

$$= \frac{1}{h}(x(t+h)-x(t), y(t+h)-y(t), z(t+h)-z(t))$$

$$= \left(\frac{x(t+h)-x(t)}{h}, \frac{y(t+h)-y(t)}{h}, \frac{z(t+h)-z(t)}{h}\right).$$

It follows that

$$\lim_{h\to 0}\frac{\mathbf{x}(t+h)-\mathbf{x}(t)}{h} = \left(\lim_{h\to 0}\frac{x(t+h)-x(t)}{h}, \cdots, \lim_{h\to 0}\frac{z(t+h)-z(t)}{h}\right)$$

$$= \left(\frac{dx}{dt}, \frac{dy}{dt}, \frac{dz}{dt}\right).$$

The result is

$$\dot{\mathbf{x}} = \frac{d\mathbf{x}}{dt} = \left(\frac{dx}{dt}, \frac{dy}{dt}, \frac{dz}{dt}\right).$$

The derivative of a vector function

$$\mathbf{x}(t) = (x(t), y(t), z(t))$$

is the vector function

$$\frac{d\mathbf{x}}{dt} = \left(\frac{dx}{dt}, \frac{dy}{dt}, \frac{dz}{dt}\right).$$

If a particle moves along a path $\mathbf{x}(t)$, its **velocity** is the vector function

$$\mathbf{v}(t) = \frac{d\mathbf{x}}{dt}.$$

The magnitude $|\mathbf{v}(t)|$ of the velocity is called the **speed.** It is a scalar (numerical) function. The direction of $\mathbf{v}(t)$ is tangential to the path of motion.

EXAMPLE 3.1

The position of a moving particle at time t is (t, t^2, t^3). Find its velocity vector and its speed.

Solution: Let $\mathbf{x}(t) = (t, t^2, t^3)$. Then

$$\mathbf{v}(t) = \dot{\mathbf{x}}(t) = (1, 2t, 3t^2),$$

$$|\mathbf{v}(t)|^2 = 1 + (2t)^2 + (3t^2)^2 = 1 + 4t^2 + 9t^4.$$

Answer: $\mathbf{v}(t) = (1, 2t, 3t^2)$,

speed $= \sqrt{1 + 4t^2 + 9t^4}$.

EXAMPLE 3.2

If $\mathbf{x}(t)$ is a vector function whose derivative is zero, show that $\mathbf{x}(t) = \mathbf{c}_0$, a constant vector.

Solution:

$$\dot{\mathbf{x}}(t) = (\dot{x}_1(t), \dot{x}_2(t), \dot{x}_3(t)) = (0, 0, 0).$$

Hence $\dot{x}_i(t) = 0$, and so $x_i(t) = c_i$ (constant) for $i = 1, 2, 3$. Therefore

$$\mathbf{x}(t) = (c_1, c_2, c_3) = \mathbf{c}_0.$$

REMARK: Physically, this example simply says that an object with zero velocity is standing still.

Differentiation Formulas

The following formulas are essential for differentiating vector functions. Each can be verified by differentiating components.

$$\frac{d}{dt}[f(t)\mathbf{x}(t)] = \dot{f}(t)\mathbf{x}(t) + f(t)\dot{\mathbf{x}}(t),$$

$$\frac{d}{dt}[\mathbf{x}(t) + \mathbf{y}(t)] = \dot{\mathbf{x}}(t) + \dot{\mathbf{y}}(t),$$

$$\frac{d}{dt}[\mathbf{x}(t) \cdot \mathbf{y}(t)] = \dot{\mathbf{x}}(t) \cdot \mathbf{y}(t) + \mathbf{x}(t) \cdot \dot{\mathbf{y}}(t).$$

$$\frac{d}{dt}\mathbf{x}[s(t)] = \frac{d\mathbf{x}}{ds}\frac{ds}{dt} \qquad \text{(Chain Rule)}$$

To establish the first formula, for example, write

$$f(t)\mathbf{x}(t) = (f(t)x_1(t), f(t)x_2(t), f(t)x_3(t)).$$

Then

$$\begin{aligned}\frac{d}{dt}[f(t)\mathbf{x}(t)] &= \left(\frac{d}{dt}[f(t)x_1(t)], \frac{d}{dt}[f(t)x_2(t)], \frac{d}{dt}[f(t)x_3(t)]\right)\\ &= (\dot{f}(t)x_1(t) + f(t)\dot{x}_1(t), \dot{f}(t)x_2(t) + f(t)\dot{x}_2(t), \dot{f}(t)x_3(t) + f(t)\dot{x}_3(t))\\ &= \dot{f}(t)\,(x_1(t), x_2(t), x_3(t)) + f(t)\,(\dot{x}_1(t), \dot{x}_2(t), \dot{x}_3(t))\\ &= \dot{f}(t)\mathbf{x}(t) + f(t)\dot{\mathbf{x}}(t).\end{aligned}$$

EXAMPLE 3.3

Differentiate $t^2\mathbf{x}(t)$, where $\mathbf{x}(t) = (\cos 3t, \sin 3t, t)$.

Solution: Apply the first formula above:

$$\frac{d}{dt}[t^2\mathbf{x}(t)] = 2t\,\mathbf{x}(t) + t^2\dot{\mathbf{x}}(t)$$

$$= 2t(\cos 3t, \sin 3t, t) + t^2(-3\sin 3t, 3\cos 3t, 1).$$

Answer:

$(2t\cos 3t - 3t^2\sin 3t, 2t\sin 3t + 3t^2\cos 3t, 3t^2)$.

EXAMPLE 3.4

Suppose $\mathbf{x}(t)$ is a moving *unit* vector. Show that $\mathbf{x}(t)$ is always perpendicular to its velocity vector $\mathbf{v}(t)$.

Solution: Verify that $\mathbf{x}(t)\cdot\mathbf{v}(t) = 0$:

$$\mathbf{x}\cdot\mathbf{v} = (x_1, x_2, x_3)\cdot\left(\frac{dx_1}{dt}, \frac{dx_2}{dt}, \frac{dx_3}{dt}\right)$$

$$= x_1\frac{dx_1}{dt} + x_2\frac{dx_2}{dt} + x_3\frac{dx_3}{dt} = \frac{1}{2}\frac{d}{dt}[x_1^2 + x_2^2 + x_3^2].$$

But $x_1^2 + x_2^2 + x_3^2 = 1$ for every t, since $\mathbf{x}$ is a unit vector. Hence $\mathbf{x}\cdot\mathbf{v} = 0$.

Alternate Solution:

$$\mathbf{x}(t)\cdot\mathbf{x}(t) = |\mathbf{x}(t)|^2 = 1,$$

$$\frac{d}{dt}[\mathbf{x}(t)\cdot\mathbf{x}(t)] = 0.$$

But by the third differentiation formula on the previous page,

$$\frac{d}{dt}[\mathbf{x}(t)\cdot\mathbf{x}(t)] = \dot{\mathbf{x}}(t)\cdot\mathbf{x}(t) + \mathbf{x}(t)\cdot\dot{\mathbf{x}}(t)$$

$$= 2\mathbf{x}(t)\cdot\dot{\mathbf{x}}(t).$$

Thus $\mathbf{x}(t)\cdot\dot{\mathbf{x}}(t) = 0$, that is, $\mathbf{x}(t)\cdot\mathbf{v}(t) = 0$.

EXERCISES

Differentiate:

1. $\mathbf{x}(t) = (e^t, e^{2t}, e^{3t})$
2. $\mathbf{x}(t) = (t^4, t^5, t^6)$
3. $\mathbf{x}(t) = (t + 1, 3t - 1, 4t)$
4. $\mathbf{x}(t) = (t^2, 0, t^3)$.

Find the velocity and the speed:

5. $\mathbf{x}(t) = (t^2, t^3 + t^4, 1)$
6. $\mathbf{x}(t) = (2t - 1, 3t + 1, -2t + 1)$
7. $\mathbf{x}(t) = (A \cos \omega t, A \sin \omega t, Bt)$
8. $\mathbf{x}(t) = (a_1 t + b_1, a_2 t + b_2, a_3 t + b_3)$.
9. Suppose that $\mathbf{x} = \mathbf{x}(t)$ is a moving point such that $\dot{\mathbf{x}}(t)$ is always perpendicular to $\mathbf{x}(t)$. Show that $\mathbf{x}(t)$ moves on a sphere with center at $\mathbf{0}$. (*Hint:* Differentiate $|\mathbf{x}|^2$.)
10. Suppose $\mathbf{x}(t) \neq \mathbf{0}$. Show that $\dfrac{d}{dt}|\mathbf{x}(t)| = \dfrac{1}{|\mathbf{x}|}\mathbf{x} \cdot \dot{\mathbf{x}}$.
11. Prove the formula $\dfrac{d}{dt}[\mathbf{x}(t) + \mathbf{y}(t)] = \dot{\mathbf{x}}(t) + \dot{\mathbf{y}}(t)$.
12. Prove the formula $\dfrac{d}{dt}[\mathbf{x}(t) \cdot \mathbf{y}(t)] = \dot{\mathbf{x}} \cdot \mathbf{y} + \mathbf{x} \cdot \dot{\mathbf{y}}$.

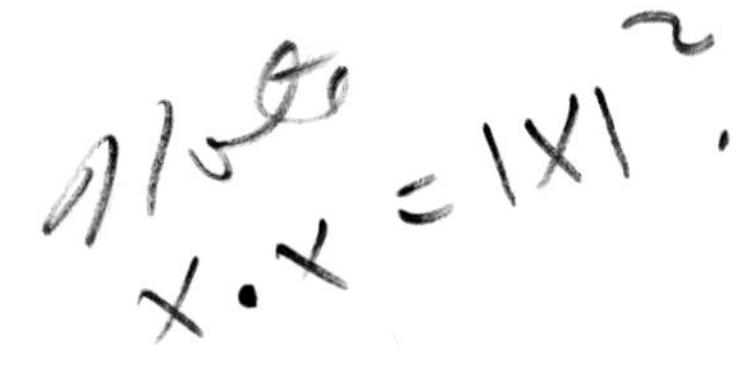

13. Prove the chain rule $\dfrac{d}{dt}\mathbf{x}[s(t)] = \dfrac{d\mathbf{x}}{ds}\dfrac{ds}{dt}$.
14. Suppose $\mathbf{x}(t)$ is a space curve which does not pass through $\mathbf{0}$, and that $\mathbf{x}(t_0)$ is the point of the curve closest to $\mathbf{0}$. Show that $\mathbf{x}(t_0) \cdot \dot{\mathbf{x}}(t_0) = 0$.
15. Suppose that $\mathbf{x}(t)$ and $\mathbf{y}(\tau)$ are two space curves which do not intersect. Suppose the distance $\mathbf{x}(t) - \mathbf{y}(\tau)$ is minimal at $t = t_0$ and $\tau = \tau_0$. Show that the vector $\mathbf{x}(t_0) - \mathbf{y}(\tau_0)$ is perpendicular to the tangents to the two curves at $\mathbf{x}(t_0)$ and $\mathbf{y}(\tau_0)$, respectively.

4. SPACE CURVES

In this section we study the arc lengths and the tangents of curves in the plane and in space.

When dealing with vectors in a plane, assume the plane is the x, y-plane. Then all vectors have the form $\mathbf{x} = (x_1, x_2, 0)$. For simplicity write $\mathbf{x} = (x_1, x_2)$.

Length of a Curve

Let $\mathbf{x} = \mathbf{x}(t)$ represent a curve in space. How long is the part of the curve between the points $\mathbf{x}(t_0)$ and $\mathbf{x}(t_1)$?

This question is attacked through the velocity vector $\dot{\mathbf{x}}(t)$. Recall that $\dot{\mathbf{x}}(t)$ points in the direction of the tangent line to the curve (limit of secant). See Fig. 4.1. The length of the velocity vector is speed, the rate at which

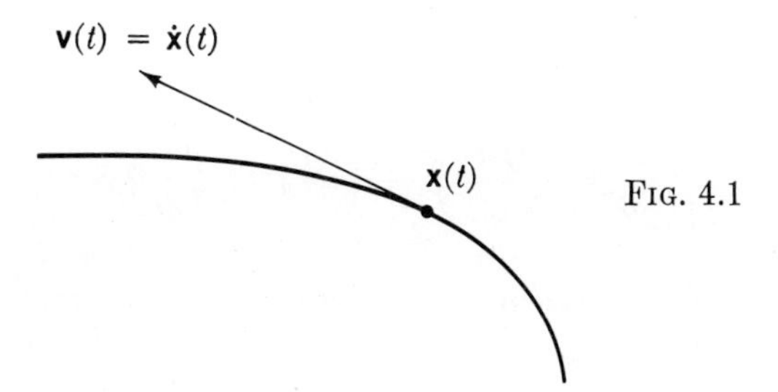

FIG. 4.1

distance along the curve increases with respect to time. If s denotes the distance measured from a fixed point $\mathbf{x}(t_0)$, then

$$\frac{ds}{dt} = |\mathbf{v}(t)| = |\dot{\mathbf{x}}(t)|.$$

Therefore,

$$\left(\frac{ds}{dt}\right)^2 = |\dot{\mathbf{x}}(t)|^2 = \left|\left(\frac{dx}{dt}, \frac{dy}{dt}, \frac{dz}{dt}\right)\right|^2$$
$$= \left(\frac{dx}{dt}\right)^2 + \left(\frac{dy}{dt}\right)^2 + \left(\frac{dz}{dt}\right)^2.$$

In term of differentials,

$$ds = \sqrt{\left(\frac{dx}{dt}\right)^2 + \left(\frac{dy}{dt}\right)^2 + \left(\frac{dz}{dt}\right)^2}\, dt.$$

This formula has a simple geometric interpretation. See Fig. 4.2. The tiny bit of arc length ds corresponds to three "displacements" dx, dy, and dz along the coordinate axes. By the Distance Formula,

$$(ds)^2 = (dx)^2 + (dy)^2 + (dz)^2.$$

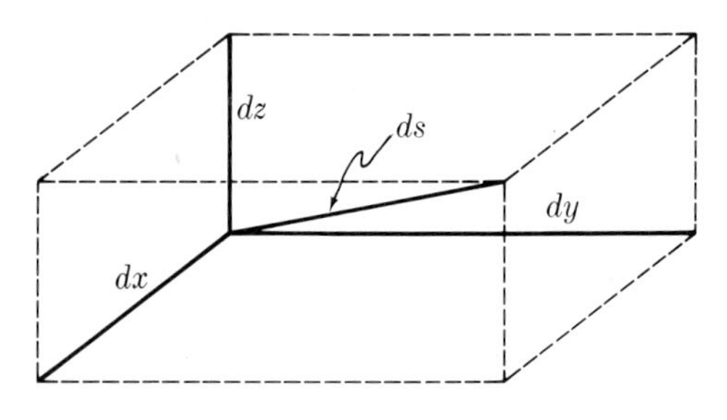

Fig. 4.2

Divide by $(dt)^2$ and take square roots to obtain

$$\frac{ds}{dt} = \sqrt{\left(\frac{dx}{dt}\right)^2 + \left(\frac{dy}{dt}\right)^2 + \left(\frac{dz}{dt}\right)^2}.$$

This is the time derivative of arc length. Integrate it to obtain the arc length itself.

Suppose $\mathbf{x}(t)$ describes a curve in space. Let $s(t)$ denote the length of the curve measured from a fixed initial point. Then

$$\frac{ds}{dt} = \sqrt{\dot{x}^2 + \dot{y}^2 + \dot{z}^2}.$$

The **length** of the curve from $\mathbf{x}(t_0)$ to $\mathbf{x}(t_1)$ is

$$L = \int_{t_0}^{t_1} \sqrt{\dot{x}^2 + \dot{y}^2 + \dot{z}^2}\, dt.$$

For plane curves the formula is slightly simpler because $z = 0$.

Suppose $\mathbf{x}(t) = (x(t), y(t))$ describes a plane curve. The length of the curve from $\mathbf{x}(t_0)$ to $\mathbf{x}(t_1)$ is

$$L = \int_{t_0}^{t_1} \sqrt{\dot{x}^2 + \dot{y}^2}\, dt.$$

If the curve is the graph of a function $y = f(x)$, then its length from $(x_0, f(x_0))$ to $(x_1, f(x_1))$ is

$$L = \int_{x_0}^{x_1} \sqrt{1 + (f')^2}\, dx.$$

The last formula is a special case of the preceding one. Indeed, set $x = t$, $y = f(t)$, where $x_0 \leq t \leq x_1$. Then $\dot{x} = 1$ and $\dot{y} = \dot{f}$, so

$$\frac{ds}{dt} = \sqrt{\dot{x}^2 + \dot{y}^2} = \sqrt{1 + \dot{f}^2} = \sqrt{1 + (f')^2}.$$

The formula for L follows.

EXAMPLE 4.1

Find the length of the parabola $\mathbf{x}(t) = (t, t^2)$, $0 \leq t \leq 1$.

Solution: This plane curve is a parabola because

$$x = t, \qquad y = t^2,$$

hence $y = x^2$. Its length is

$$\int_0^1 \sqrt{\dot{x}^2 + \dot{y}^2}\, dt = \int_0^1 \sqrt{1 + (2t)^2}\, dt.$$

From integral tables,

$$\int_0^1 \sqrt{1 + 4t^2}\, dt = \frac{\sqrt{5}}{2} + \frac{1}{4}\ln(2 + \sqrt{5}).$$

Answer: $\frac{1}{4}[2\sqrt{5} + \ln(2 + \sqrt{5})] \approx 1.479.$

EXAMPLE 4.2

Find the length of the curve $y = \sin x$ for $0 \leq x \leq \pi$.

Solution: The length L is given by

$$L = \int_0^\pi \sqrt{1 + \left(\frac{dy}{dx}\right)^2}\, dx = \int_0^\pi \sqrt{1 + \cos^2 x}\, dx.$$

The exact evaluation of this integral is impossible. It can, however, be approximated by Simpson's Rule.

Answer: $L = \int_0^\pi \sqrt{1 + \cos^2 x}\, dx \approx 3.820.$

EXAMPLE 4.3

Find the length of the curve $\mathbf{x}(t) = (t \cos t, t \sin t, 2t)$, $0 \leq t \leq 4\pi$. Sketch the curve.

Solution: Since

$$x^2 + y^2 = (t \cos t)^2 + (t \sin t)^2 = t^2 = \frac{1}{4} z^2,$$

the curve lies on the right circular cone $z^2 = 4(x^2 + y^2)$. As t increases, z steadily increases, while the projection $(t \cos t, t \sin t)$ of $\mathbf{x}(t)$ on the x, y-plane traces a spiral. Hence the space curve $\mathbf{x} = \mathbf{x}(t)$ is a spiral on the surface of the cone (Fig. 4.3).

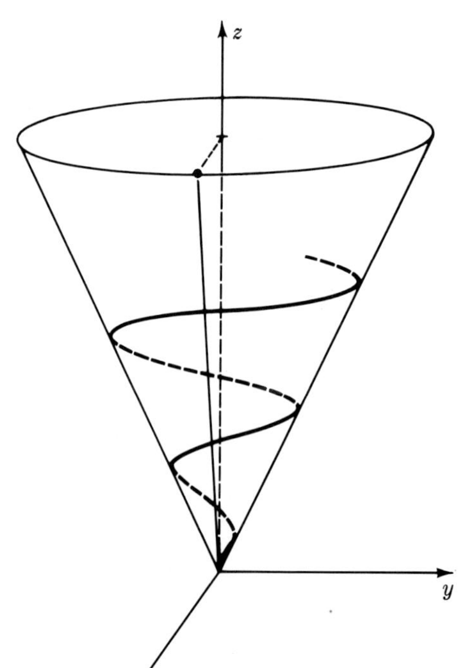

FIG. 4.3

Compute ds/dt:

$$\left(\frac{ds}{dt}\right)^2 = \left(\frac{dx}{dt}\right)^2 + \left(\frac{dy}{dt}\right)^2 + \left(\frac{dz}{dt}\right)^2$$
$$= (\cos t - t\sin t)^2 + (\sin t + t\cos t)^2 + (2)^2$$
$$= 5 + t^2.$$

Hence

$$L = \int_0^{4\pi} \sqrt{5+t^2}\,dt = \frac{1}{2}[t\sqrt{5+t^2} + 5\ln(t + \sqrt{5+t^2})]\Big|_0^{4\pi}.$$

Answer: $L = \dfrac{1}{2}\left[4\pi a + 5\ln(4\pi + a) - \dfrac{5}{2}\ln 5\right],$

where $a = \sqrt{5 + 16\pi^2}$. $L \approx 86.3$.

Unit Tangent Vector

EXAMPLE 4.4

Plot the locus $\mathbf{x}(t) = (t^2, t^3)$.

Solution: The locus is the plane curve described by

$$x = t^2, \qquad y = t^3.$$

Hence

$$x^3 = y^2, \qquad y = \pm x^{3/2}.$$

The curve is defined only for $x \geq 0$. For each positive value of x, there are two values of y. (See Fig. 4.4.)

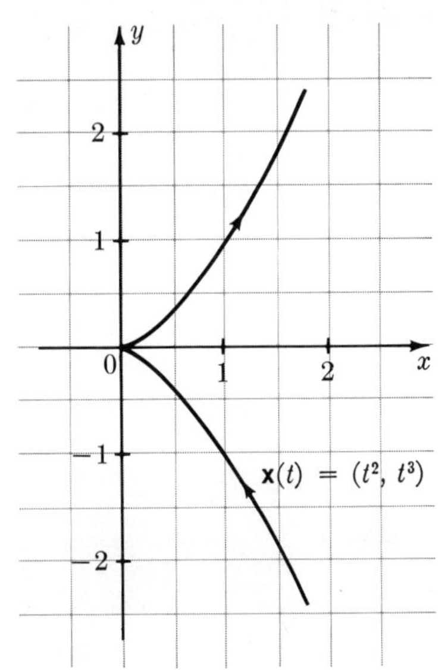

FIG. 4.4

REMARK: The sharp point at the origin is called a **cusp**. At that point, a particle moving along the curve changes direction abruptly. Note that its velocity at the origin is zero since

$$\mathbf{v} = \dot{\mathbf{x}} = (2t, 3t^2), \qquad \mathbf{v}(0) = \mathbf{0}.$$

In fact, an abrupt change in direction can occur only when the velocity vector is zero. Physically, this seems plausible; a moving particle cannot change direction suddenly unless it slows down to an instantaneous stop at the "corner."

To avoid such curves with cusps as shown in Fig. 4.4, we study only curves $\mathbf{x}(t)$ for which the velocity $\dot{\mathbf{x}}(t)$ never equals zero.

Suppose a particle moves along a curve $\mathbf{x}(t)$. Its velocity vector $\mathbf{v}(t) = d\mathbf{x}/dt$ has length ds/dt and is directed along the tangent to the curve; hence

$$\mathbf{v} = \frac{d\mathbf{x}}{dt} = \frac{ds}{dt}\mathbf{T},$$

where $\mathbf{T}$ is a unit vector in the tangential direction. But by the Chain Rule,

$$\mathbf{v} = \frac{d\mathbf{x}}{dt} = \frac{d\mathbf{x}}{ds}\frac{ds}{dt}.$$

Compare these two expressions for $\mathbf{v}$; the result is

$$\frac{ds}{dt}\frac{d\mathbf{x}}{ds} = \frac{ds}{dt}\mathbf{T}.$$

Therefore $d\mathbf{x}/ds = \mathbf{T}$ since $ds/dt \neq 0$ is assumed.

If $\mathbf{x}(t)$ represents a space curve, then

$$\frac{d\mathbf{x}}{ds} = \mathbf{T}$$

is the **unit tangent vector** to the curve. In terms of the velocity vector $\mathbf{v}$,

$$\mathbf{T} = \frac{\mathbf{v}}{|\mathbf{v}|}.$$

(It is assumed $\mathbf{v} \neq \mathbf{0}$.)

EXAMPLE 4.5

Find the unit tangent vector to the curve $\mathbf{x}(t) = (t, t^2, t^3)$ at the point $\mathbf{x}(1) = (1, 1, 1)$.

Solution:

$$\mathbf{T} = \frac{\mathbf{v}}{|\mathbf{v}|},$$

where

$$\mathbf{v} = \dot{\mathbf{x}} = (1, 2t, 3t^2),$$

$$|\mathbf{v}|^2 = 1 + 4t^2 + 9t^4.$$

Hence

$$\mathbf{T} = \frac{1}{\sqrt{1 + 4t^2 + 9t^4}}(1, 2t, 3t^2).$$

Now substitute $t = 1$.

Answer: $\mathbf{T} = \dfrac{1}{\sqrt{14}}(1, 2, 3).$

EXERCISES

1. Find the arc length of $\mathbf{x}(t) = (a_1 t + b_1, a_2 t + b_2, a_3 t + b_3)$ for $0 \le t \le 1$.
2. Find the length of $\mathbf{x}(t) = (t^2, t^3)$ for $0 \le t \le a$.
3. Find the length of $\mathbf{x}(t) = (t, \sin t, \cos t)$ for $0 \le t \le 2\pi$.
4. Set up the length of $y = x^3$ for $-1 \le x \le 1$, but do not evaluate the integral.
5. Set up the length of $y = ax^n$ for $x_0 \le x \le x_1$, but do not evaluate the integral.
6. Set up the length of $\mathbf{x}(t) = (t^m, t^n, t^r)$ for $0 \le t \le b$, but do not evaluate the integral.
7. Find the length of $y = -x^2 + 2x$ for $-1 \le x \le 1$.
8. Carefully plot $\mathbf{x}(t) = (t^2, t^4 + t^5)$ for t near 0.
9. Find the unit tangent $\mathbf{T}$ to the curve $\mathbf{x}(t) = (t, \cos t, \sin t)$ at $t = 0$.
10. Find the unit tangent $\mathbf{T}$ to the curve $\mathbf{x}(t) = (3t - 1, 4t, -2t + 1)$ at any point.
11. Find the unit tangent $\mathbf{T}$ to the curve $\mathbf{x}(t) = (a_1 t + b_1, a_2 t + b_2, a_3 t + b_3)$ at any point.
12. Find the unit tangent $\mathbf{T}$ to the curve $\mathbf{x}(t) = (t \cos t, t \sin t, 2t)$ at any point.

5. CURVATURE

The curvature of a curve is a quantity which tells how fast the direction of the curve is changing relative to arc length.

> The magnitude of the rate of change of the unit tangent $\mathbf{T}$ with respect to arc length is called the **curvature** of a curve, and is denoted by k:
>
> $$k = \left|\frac{d\mathbf{T}}{ds}\right|.$$

Since the length of $\mathbf{T}$ is constant, $\mathbf{T}$ changes in direction only. Thus the curvature k measures its rate of change of direction. The curvature is a geometric quantity; it does not depend on how the curve is parameterized.

EXAMPLE 5.1

A curve has curvature zero. What is the curve?

Solution: A natural guess is a straight line. Let us prove this is so. We are given $k = 0$. Therefore,

$$\left|\frac{d\mathbf{T}}{ds}\right| = 0,$$

hence

$$\frac{d\mathbf{T}}{ds} = \mathbf{0}.$$

Consequently $\mathbf{T}$ is constant,

$$\mathbf{T} = \mathbf{T}_0 = (t_1, t_2, t_3).$$

But $d\mathbf{x}/ds = \mathbf{T}_0$, which means

$$\frac{dx}{ds} = t_1, \qquad \frac{dy}{ds} = t_2, \qquad \frac{dz}{ds} = t_3.$$

Integrating,

$$x = a + t_1 s, \qquad y = b + t_2 s, \qquad z = c + t_3 s.$$

In vector notation,

$$\mathbf{x} = \mathbf{x}_0 + s\mathbf{T}_0.$$

But this equation describes a line: to the vector $\mathbf{x}_0 = (a, b, c)$ are added multiples of a fixed unit vector $\mathbf{T}_0$. By the parallelogram law, a line is traced (Fig. 5.1).

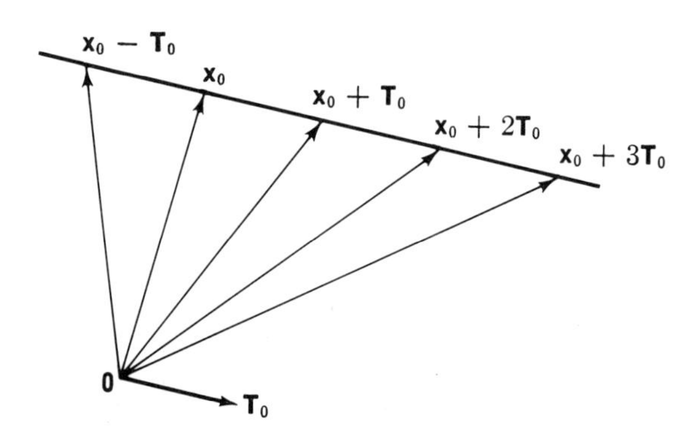

FIG. 5.1

Answer: A straight line.

Here are three formulas for curvature; they will be derived at the end of this section:

If $\mathbf{x} = \mathbf{x}(t)$ is a space curve, then

$$k = \frac{[|\dot{\mathbf{x}}|^2\,|\ddot{\mathbf{x}}|^2 - (\dot{\mathbf{x}} \cdot \ddot{\mathbf{x}})^2]^{1/2}}{|\dot{\mathbf{x}}|^3}.$$

If $\mathbf{x} = (x(t), y(t))$ is a plane curve, then

$$k = \frac{\dot{x}\ddot{y} - \dot{y}\ddot{x}}{(\dot{x}^2 + \dot{y}^2)^{3/2}}.$$

If a plane curve is the graph of a function $y = f(x)$, then

$$k = \frac{|f''(x)|}{[1 + f'(x)^2]^{3/2}}.$$

EXAMPLE 5.2

Find the curvature of a circle of radius a.

Solution: Let the equation of the circle be $x^2 + y^2 = a^2$. Thus

$$y = \pm\sqrt{a^2 - x^2}.$$

(This equation describes either the upper or lower half of the circle depending on whether the positive or negative square root is chosen.) Differentiate:

$$y' = \frac{-x}{\pm\sqrt{a^2 - x^2}} = -\frac{x}{y}.$$

Differentiate again:

$$y'' = -\frac{y - xy'}{y^2} = -\frac{y - x\,(-x/y)}{y^2}$$

$$= -\frac{x^2 + y^2}{y^3} = -\frac{a^2}{y^3},$$

Now

$$1 + y'^2 = 1 + \left(-\frac{x}{y}\right)^2 = \frac{y^2 + x^2}{y^2} = \frac{a^2}{y^2}.$$

Hence by the formula for curvature,

$$k = \frac{|-a^2/y^3|}{(a^2/y^2)^{3/2}} = \frac{a^2}{a^3} = \frac{1}{a}.$$

Alternate Solution: Write

$$\mathbf{x}(t) = (a\cos t, a\sin t).$$

(This describes the circle by its central angle t.) Then

$$\dot{\mathbf{x}} = (-a \sin t, a \cos t),$$

$$|\dot{\mathbf{x}}|^2 = \left(\frac{ds}{dt}\right)^2 = (-a \sin t)^2 + (a \cos t)^2 = a^2,$$

$$|\dot{\mathbf{x}}| = \frac{ds}{dt} = a.$$

Hence

$$\mathbf{T} = \frac{\dot{\mathbf{x}}}{|\dot{\mathbf{x}}|} = (-\sin t, \cos t).$$

Differentiate:

$$\frac{ds}{dt}\frac{d\mathbf{T}}{ds} = \frac{d\mathbf{T}}{dt} = (-\cos t, -\sin t).$$

Take lengths, substituting $a = ds/dt$:

$$a\left|\frac{d\mathbf{T}}{ds}\right| = [(-\cos t)^2 + (-\sin t)^2]^{1/2} = 1,$$

$$k = \left|\frac{d\mathbf{T}}{ds}\right| = \frac{1}{a}.$$

Answer: $k = \dfrac{1}{a}.$

REMARK: The curvature of a circle is the reciprocal of its radius. This is reasonable on two counts. First, the curvature is the same at all points of a circle. Second, it is small for large circles, since the larger the circle the more slowly its direction changes per unit of arc length.

The Unit Normal

The vector $d\mathbf{T}/ds$ has length k, the curvature. Therefore

$$\frac{d\mathbf{T}}{ds} = k\mathbf{N},$$

where $\mathbf{N}$ is a unit vector in the direction of $d\mathbf{T}/ds$. (We assume $k \neq 0$.) Since $\mathbf{T}$ is a unit vector, $\mathbf{T}$ is perpendicular to $d\mathbf{T}/ds$; this was shown in Example 3.4. The vector $\mathbf{N}$ is called the **unit normal vector** to the curve (Fig. 5.2).

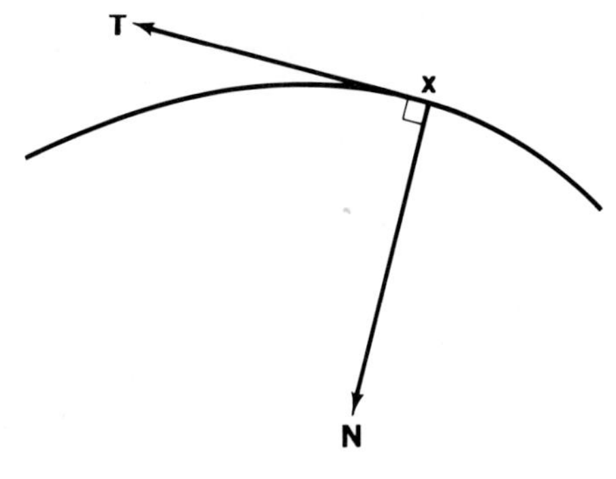

FIG. 5.2

Let us summarize some basic facts concerning curves.

> Let $\mathbf{x}(t)$ represent a curve in space.
>
> $$\frac{d\mathbf{x}}{ds} = \mathbf{T}.$$
>
> $$\frac{d\mathbf{T}}{ds} = k\mathbf{N}, \qquad k = k(s).$$
>
> $$|\mathbf{T}| = |\mathbf{N}| = 1, \qquad \mathbf{T}\cdot\mathbf{N} = 0.$$

The further study of space curves, not pursued here, begins with an analysis of $d\mathbf{N}/ds$. That leads to another quantity, the torsion, which measures how fast the plane of **T** and **N** is turning around the tangent line.

> EXAMPLE 5.3
>
> Compute **T**, **N**, and k for the circular spiral (helix) $\mathbf{x}(t) = (a\cos t, a\sin t, bt)$. Assume $a > 0$ and $b > 0$.

Solution: The projection of $\mathbf{x}(t)$ on the x, y-plane is $(a\cos t, a\sin t, 0)$. As a particle describes the curve $\mathbf{x}(t)$, its projection describes a circle of radius a. The third component of $\mathbf{x}(t)$ is bt; the particle moves upward at a steady rate. Thus, the curve is a spiral; it is circular but steadily rising (Fig. 5.3). Differentiate $\mathbf{x} = (a\cos t, a\sin t, bt)$:

$$\dot{\mathbf{x}} = (-a\sin t, a\cos t, b).$$

Introduce $c > 0$ by $c^2 = a^2 + b^2$. Then

$$|\dot{\mathbf{x}}|^2 = a^2 + b^2 = c^2, \qquad \frac{ds}{dt} = |\dot{\mathbf{x}}| = c,$$

and

$$\mathbf{T} = \frac{1}{c}\dot{\mathbf{x}} = \frac{1}{c}(-a\sin t, a\cos t, b).$$

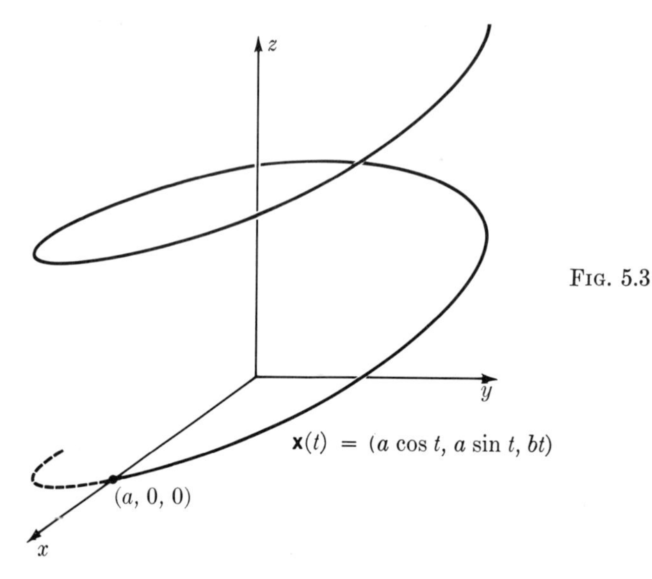

FIG. 5.3

To find k and $\mathbf{N}$, use the relation $k\mathbf{N} = \dfrac{d\mathbf{T}}{ds}$:

$$k\mathbf{N} = \frac{d\mathbf{T}}{ds} = \frac{d\mathbf{T}/dt}{ds/dt} = \frac{1}{c}\frac{d\mathbf{T}}{dt} = \frac{a}{c^2}(-\cos t, -\sin t, 0).$$

Since $k \geq 0$ and $\mathbf{N}$ is a unit vector,

$$k = \frac{a}{c^2}, \qquad \mathbf{N} = (-\cos t, -\sin t, 0).$$

Answer: $k = \dfrac{a}{a^2 + b^2}$,

$$\mathbf{T} = \frac{1}{\sqrt{a^2 + b^2}}(-a \sin t, a \cos t, b),$$

$$\mathbf{N} = (-\cos t, -\sin t, 0).$$

REMARK: If $b = 0$, the spiral degenerates into a circle of radius a and the curvature k reduces to $1/a$, which agrees with Example 5.2.

Derivations

We now derive the three formulas on p. 489. By the Chain Rule,

$$\dot{\mathbf{x}} = \frac{ds}{dt}\frac{d\mathbf{x}}{ds} = \frac{ds}{dt}\mathbf{T}, \qquad \ddot{\mathbf{x}} = \frac{d^2s}{dt^2}\mathbf{T} + \frac{ds}{dt}\frac{d\mathbf{T}}{dt} = \frac{d^2s}{dt^2}\mathbf{T} + \left(\frac{ds}{dt}\right)^2\frac{d\mathbf{T}}{ds}.$$

Hence

$$|\dot{\mathbf{x}}|^2 = \dot{\mathbf{x}} \cdot \dot{\mathbf{x}} = \left(\frac{ds}{dt}\right)^2, \quad \dot{\mathbf{x}} \cdot \ddot{\mathbf{x}} = \frac{ds}{dt}\frac{d^2s}{dt^2},$$

$$|\ddot{\mathbf{x}}|^2 = \ddot{\mathbf{x}} \cdot \ddot{\mathbf{x}} = \left(\frac{d^2s}{dt^2}\right)^2 + \left(\frac{ds}{dt}\right)^4 \left|\frac{d\mathbf{T}}{ds}\right|^2 = \left(\frac{d^2s}{dt^2}\right)^2 + \left(\frac{ds}{dt}\right)^4 k^2.$$

Consequently

$$|\dot{\mathbf{x}}|^2 |\ddot{\mathbf{x}}|^2 - (\dot{\mathbf{x}} \cdot \ddot{\mathbf{x}})^2 = \left(\frac{ds}{dt}\right)^6 k^2 = |\dot{\mathbf{x}}|^6 k^2.$$

The first formula for k follows.

If $\mathbf{x} = (x(t), y(t))$ is a plane curve, then

$$\dot{\mathbf{x}} = (\dot{x}, \dot{y}) \quad \text{and} \quad \ddot{\mathbf{x}} = (\ddot{x}, \ddot{y}),$$

hence

$$\begin{aligned} |\dot{\mathbf{x}}|^2 |\ddot{\mathbf{x}}|^2 - (\dot{\mathbf{x}} \cdot \ddot{\mathbf{x}})^2 &= (\dot{x}^2 + \dot{y}^2)(\ddot{x}^2 + \ddot{y}^2) - (\dot{x}\ddot{x} + \dot{y}\ddot{y})^2 \\ &= (\dot{x}\ddot{y} - \dot{y}\ddot{x})^2, \end{aligned}$$

so the second formula follows.

Finally, if the plane curve is the graph of $y = f(x)$, apply the second formula with $t = x$ and $\mathbf{x} = (t, f(t)) = (x, f(x))$. Then $\dot{x} = 1$, $\ddot{x} = 0$, $\dot{y} = f'(x)$, and $\ddot{y} = f''(x)$, so the third formula follows by direct substitution.

EXERCISES

Find the curvature:

1. $y = x^2$; at $x = 1$
2. $\mathbf{x}(t) = (t^3, t^2)$; at $t = 1$
3. $\mathbf{x}(t) = (t, t^2, t^3)$; at $t = -1$
4. $\mathbf{x}(t) = (a_1t + a_2t^3, b_1t + b_2t^3, c_1t + c_2t^3)$; at $t = 0$.
5. Let $\mathbf{x} = \mathbf{x}(s)$ be a plane curve. Show that $d\mathbf{N}/ds = -k\mathbf{T}$. (*Hint:* Differentiate $\mathbf{T} \cdot \mathbf{N} = 0$ and $\mathbf{N} \cdot \mathbf{N} = 1$.)
6. Find the point of the plane curve $y = x^2$ where k is maximum.
7. Find the point of $y = \sin x$ where k is maximum, $0 < x < \pi$.
8. Find the curvature of $y = x^3$ at $x = 0$ and at $x = 1$.
9. Show that the curvature of a plane curve at an inflection point is zero.
10. Let the tangent line of plane curve intersect the x-axis with angle α. Show that $k = |d\alpha/ds|$. [*Hint:* Write $\mathbf{T} = (\cos\alpha, \sin\alpha)$.]
11. A point moves along the curve $y = e^x$ at the rate of 3 in./sec. How fast is the tangent turning when the point is at $(2, e^2)$?
12. Compute the maximum and minimum curvature of an ellipse with semimajor axis a, semiminor axis b. Check the case $a = b$.
13. From a graph, predict the behavior of the curvature of $y = 1/x$ as $x \longrightarrow 0$ and as $x \longrightarrow \infty$. Verify your prediction.

24. Applications of Vectors

1. VELOCITY AND ACCELERATION

If $\mathbf{x} = \mathbf{x}(t)$ represents the position of a moving particle, its **velocity** is

$$\mathbf{v}(t) = \dot{\mathbf{x}}(t)$$

and its **acceleration** is

$$\mathbf{a}(t) = \dot{\mathbf{v}}(t) = \ddot{\mathbf{x}}(t).$$

These definitions extend those of Chapter 5 for velocity and acceleration in straight line motion. Notice that velocity and acceleration are vectors, each having magnitude and direction. The direction of the velocity is the direction the particle is moving. The direction of the acceleration is the direction the particle is turning. The following example shows that the direction of the acceleration is not necessarily that of the velocity; it may even be perpendicular to the velocity.

EXAMPLE 1.1

The path of a particle moving around the circle $x^2 + y^2 = r^2$ is given by $\mathbf{x}(t) = (r \cos \omega t, r \sin \omega t)$, where ω is a constant. Find its velocity and acceleration vectors.

Solution: Differentiate to find $\mathbf{v}$ and $\mathbf{a}$:

$$\mathbf{v}(t) = \dot{\mathbf{x}}(t) = r\omega(-\sin \omega t, \cos \omega t),$$

$$\mathbf{a}(t) = \dot{\mathbf{v}}(t) = r\omega^2(-\cos \omega t, -\sin \omega t) = -\omega^2\mathbf{x}(t).$$

Answer: $\mathbf{v}(t) = r\omega(-\sin \omega t, \cos \omega t)$,

$\mathbf{a}(t) = r\omega^2(-\cos \omega t, -\sin \omega t)$.

REMARK: The speed, $|\mathbf{v}| = r\omega$, is constant; the motion is uniform circular motion. The velocity $\mathbf{v}(t)$ is perpendicular to the position vector $\mathbf{x}(t)$ since $\mathbf{x}(t) \cdot \mathbf{v}(t) = 0$. This is expected since each tangent to a circle is perpendicular to the corresponding radius. But $\mathbf{a}(t) = -\omega^2\mathbf{x}(t)$, so the acceleration vector $\mathbf{a}(t)$ is directed opposite to the position vector $\mathbf{x}(t)$. See Fig. 1.1. What is the physical meaning of this phenomenon?

Remember that $\mathbf{a}(t)$ is the rate of change of the velocity vector. Observe the velocity vectors at t and an instant later at $t + h$. See Fig. 1.2. The difference $\mathbf{v}(t + h) - \mathbf{v}(t)$ is nearly parallel, but oppositely directed, to $\mathbf{x}(t)$.

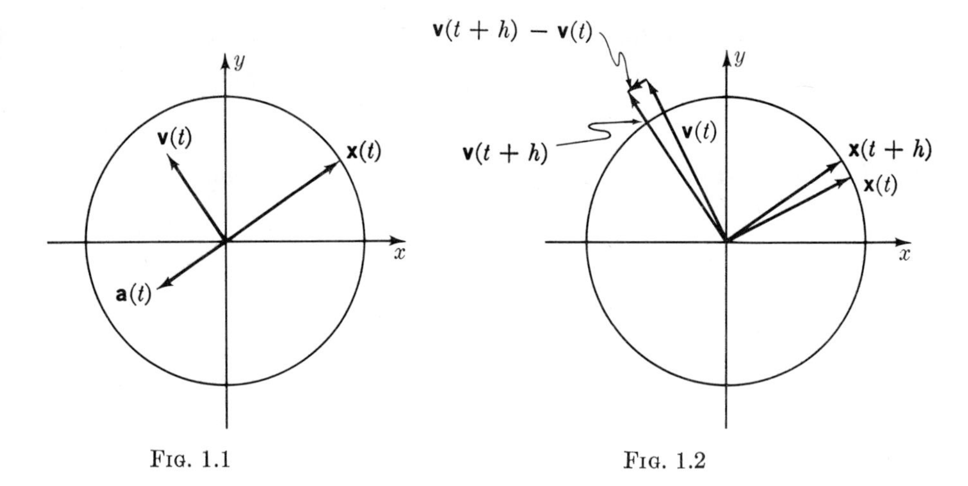

FIG. 1.1 FIG. 1.2

Thus the velocity is changing in a direction opposite to $\mathbf{x}(t)$. It seems reasonable, therefore, that $\mathbf{a}(t) = -c\mathbf{x}(t)$, where $c > 0$.

Newton's Law of Motion

This famous principle states that

$$\text{force} = \text{mass} \times \text{acceleration}.$$

But force and acceleration are vectors, both have magnitude and direction. Thus Newton's Law is a vector equation:

$$\mathbf{F} = m\ddot{\mathbf{x}}.$$

This is equivalent to three scalar equations for the components:

$$F_1 = m\ddot{x}_1, \qquad F_2 = m\ddot{x}_2, \qquad F_3 = m\ddot{x}_3.$$

EXAMPLE 1.2

A particle of mass m is subject to zero force. What is its trajectory?

Solution: By Newton's Law,

$$m\ddot{\mathbf{x}} = \mathbf{0}, \qquad \ddot{\mathbf{x}} = \mathbf{0}.$$

Since $\ddot{\mathbf{x}} = \dot{\mathbf{v}}$,

$$\frac{d}{dt}(\mathbf{v}) = \mathbf{0}.$$

Integrate once; $\mathbf{v}$ is constant:

$$\mathbf{v} = \mathbf{v}_0, \qquad \frac{d\mathbf{x}}{dt} = \mathbf{v}_0.$$

Integrate again:

$$\mathbf{x} = t\mathbf{v}_0 + \mathbf{x}_0.$$

The result is a straight line.

> *Answer:* The trajectory is a straight line, traversed at constant speed.

REMARK: Let us check the second integration in components. The equation

$$\frac{d\mathbf{x}}{dt} = \mathbf{v}_0$$

means

$$\dot{x}_1 = v_{01}, \qquad \dot{x}_2 = v_{02}, \qquad \dot{x}_3 = v_{03},$$

where the v_{0j} are constants. Integrating,

$$x_1 = tv_{01} + x_{01}, \qquad x_2 = tv_{02} + x_{02}, \qquad x_3 = tv_{03} + x_{03}.$$

Written as a vector equation, this is simply $\mathbf{x} = t\mathbf{v}_0 + \mathbf{x}_0$.

> EXAMPLE 1.3
>
> A shell is fired at an angle α with the ground. What is its path? Neglect air resistance.

Solution: Draw a figure, taking the axes as indicated (Fig. 1.3). Let $\mathbf{v}_0$ be the initial velocity vector, so $\mathbf{v}_0 = v_0(\cos\alpha, \sin\alpha)$, where v_0 is the initial speed. Let m denote the mass of the shell. The force of gravity at each point is constant,

$$\mathbf{F} = (0, -mg).$$

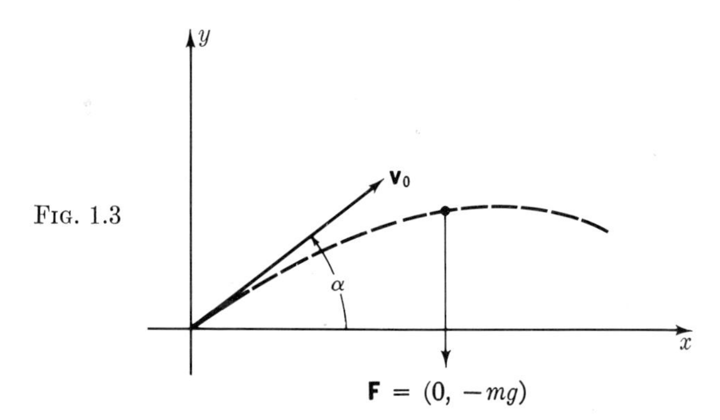

FIG. 1.3

The equation of motion is

$$m\mathbf{a} = \mathbf{F},$$

that is,

$$\frac{d^2\mathbf{x}}{dt^2} = (0, -g).$$

Integrate:

$$\frac{d\mathbf{x}}{dt} = (0, -gt) + \mathbf{v}_0.$$

Integrate again, noting that $\mathbf{x}_0 = \mathbf{0}$ by the choice of axes:

$$\mathbf{x} = (0, -\frac{1}{2}gt^2) + t\mathbf{v}_0.$$

Hence

$$(x(t), y(t)) = (0, -\frac{1}{2}gt^2) + tv_0(\cos\alpha, \sin\alpha)$$

$$= (v_0t\cos\alpha, v_0t\sin\alpha - \frac{1}{2}gt^2).$$

To describe the path, eliminate t:

$$x = v_0t\cos\alpha, \qquad t = \frac{x}{v_0\cos\alpha},$$

$$y = v_0t\sin\alpha - \frac{1}{2}gt^2 = x\tan\alpha - \frac{g}{2v_0^2\cos^2\alpha}x^2.$$

The graph of this quadratic is a parabola.

Answer: $x = (v_0\cos\alpha)t, \quad y = (v_0\sin\alpha)t - \frac{1}{2}gt^2,$
where v_0 is the initial speed. The path is a parabola:

$$y = x\tan\alpha - \frac{g}{2v_0^2\cos^2\alpha}x^2.$$

EXAMPLE 1.4

In Example 1.3, what is the maximum range?

Solution: The shell hits ground when $y = 0$:

$$\left(v_0\sin\alpha - \frac{1}{2}gt\right)t = 0.$$

This equation has two roots. The root $t = 0$ indicates the initial point. We want the other root,

$$t = \frac{2v_0\sin\alpha}{g}.$$

The range is the value of x at this time:

$$x = (v_0\cos\alpha)\left(\frac{2v_0\sin\alpha}{g}\right) = \frac{v_0^2}{g}\sin 2\alpha.$$

Clearly x is maximum when $\sin 2\alpha = 1$, or $\alpha = \pi/4$. The maximum range is v_0^2/g.

Answer: The maximum range is v_0^2/g. It is obtained by firing at 45°.

REMARK: If the initial speed is doubled, the maximum range is quadrupled. Is this reasonable? (How much more gunpowder is necessary?)

Components of Acceleration

The arc length s, the unit tangent $\mathbf{T}$, the unit normal $\mathbf{N}$, and the curvature k are geometric properties of a path. How are the velocity and acceleration of a particle moving on the path related to these quantities? The question is partly answered by the equation

$$\mathbf{v} = \frac{ds}{dt}\mathbf{T},$$

which says that the motion is directed along the tangent with speed ds/dt.

For further information, differentiate $\mathbf{v}$ with respect to time, using the Chain Rule carefully:

$$\mathbf{a} = \dot{\mathbf{v}} = \frac{d^2s}{dt^2}\mathbf{T} + \frac{ds}{dt}\dot{\mathbf{T}}.$$

But

$$\dot{\mathbf{T}} = \frac{d\mathbf{T}}{dt} = \frac{ds}{dt}\frac{d\mathbf{T}}{ds} = \frac{ds}{dt}k\mathbf{N},$$

where k is the curvature. Therefore

$$\mathbf{a} = \frac{d^2s}{dt^2}\mathbf{T} + k\left(\frac{ds}{dt}\right)^2\mathbf{N}.$$

This is an important equation in mechanics. It says that the acceleration is composed of two perpendicular components. The first is a tangential component with magnitude $\ddot{s}$, the rate of change of the speed. The second is a normal component, directed along $\mathbf{N}$ with magnitude $k\dot{s}^2$. It is called the **centripetal acceleration.**

EXAMPLE 1.5

A particle moves along a circle. Find its velocity and acceleration.

Solution: Let r denote the radius and let $\theta = \theta(t)$ denote the central angle at time t. Place the circle in the x, y-plane with center at $\mathbf{0}$. Then the path is given by

$$\mathbf{x}(t) = r(\cos\theta, \sin\theta).$$

Differentiate:

$$\mathbf{v} = \dot{\mathbf{x}} = r\dot{\theta}(-\sin\theta, \cos\theta).$$

It follows that

$$\mathbf{T} = (-\sin\theta, \cos\theta), \qquad \frac{ds}{dt} = r\dot{\theta} = r\omega(t).$$

Here $\omega(t) = \dot{\theta}(t)$ represents the instantaneous angular speed. Thus

$$\mathbf{v} = r\omega(-\sin\theta, \cos\theta) = r\omega\mathbf{T}.$$

Differentiate:

$$\begin{aligned}\mathbf{a} = \dot{\mathbf{v}} &= r\dot{\omega}\mathbf{T} + r\omega\dot{\mathbf{T}} \\ &= r\dot{\omega}(-\sin\theta, \cos\theta) + r\omega^2(-\cos\theta, -\sin\theta) \\ &= r\dot{\omega}\mathbf{T} + r\omega^2\mathbf{N}.\end{aligned}$$

Answer: $\mathbf{v} = r\omega\mathbf{T}$, $\mathbf{a} = r\dot{\omega}\mathbf{T} + r\omega^2\mathbf{N}$, where $\omega = \dot{\theta}$ is the angular speed.

REMARK: When the motion is uniform (ω constant), then $\mathbf{a} = r\omega^2\mathbf{N}$, so the acceleration is all centripetal, perpendicular to the direction of motion. This agrees with the answer to Example 1.1.

EXERCISES

1. A hill makes angle β with the ground (Fig. 1.4). A shell is fired from the base of the hill at angle α with the ground. Show that the x-component of the position where shell strikes the hill is $x = (2v_0^2/g)(\sin\alpha\cos\alpha - \tan\beta\cos^2\alpha)$.

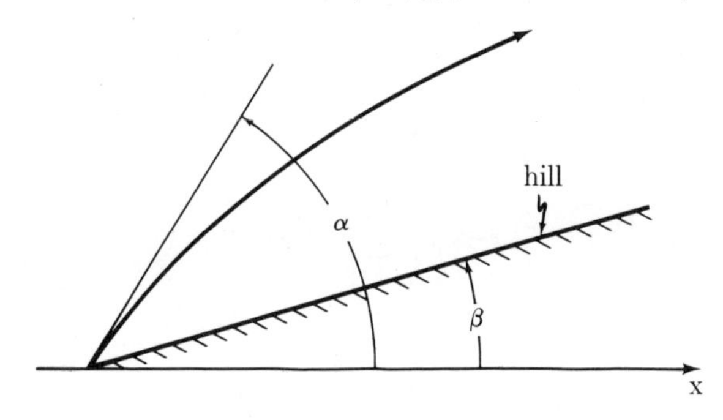

FIG. 1.4

2. (cont.) Find the maximum of x as a function of α. Show that it occurs for

$$\frac{\pi}{2} - \alpha = \alpha - \beta, \quad \text{that is,} \quad \alpha = \frac{1}{2}\left(\frac{\pi}{2} + \beta\right).$$

3. Let $\mathbf{x}(t) = (t, t^2)$. Find $\mathbf{v}(t)$ and $\mathbf{a}(t)$.
4. (cont.) Find the tangential and normal components of $\mathbf{a}$ at $t = 0$ and $t = -1$.
5. A particle moves along the curve $y = \sin x$ with constant speed 1. Find the tangential and normal components of $\mathbf{a}$ at $x = 0$ and at $x = \pi/2$.
6. Find the tangential and normal components of acceleration for $\mathbf{x}(t) = (\cos t^2, \sin t^2)$.
7. Find the tangential and normal components of acceleration for $\mathbf{x}(t) = (a\cos\omega t, a\sin\omega t, bt)$, where ω is constant.

8. A particle moves with constant speed 1 on the surface of the unit sphere $|\mathbf{x}| = 1$. Show that the normal component of the acceleration has magnitude at least 1.

9. A particle moves on the surface $z = x^2 + y^2$ with constant speed 1. At a certain instant t_0 it passes through $\mathbf{0}$. Show that the tangential component of $\mathbf{a}$ is $\mathbf{0}$ and the normal component is $(\ddot{x}(t_0), \ddot{y}(t_0), 2)$. Show also with $\dot{x}\ddot{x} + \dot{y}\ddot{y} = 0$ at t_0.

2. INTEGRALS

Suppose a particle moves along a path $\mathbf{x} = \mathbf{x}(t)$ from $\mathbf{x}(t_0)$ to $\mathbf{x}(t_1)$, acted on by a force $\mathbf{F} = \mathbf{F}(t)$. How much work is done by the force?

From physics we learn that only the component of the force in the direction of motion does work, and that the amount of work done in a small displacement of length ds is

$$dW = F_a\, ds,$$

where F_a is the average component of force in the direction of motion. The total work done by the force can be expressed as a definite integral.

Draw the unit tangent $\mathbf{T}$, the force $\mathbf{F}$, and a small portion of the path of length ds. See Fig. 2.1. Since $\mathbf{T}$ is a unit vector in the direction of motion, the component of the force $\mathbf{F}$ in the direction of motion is the dot product

$$\mathbf{F} \cdot \mathbf{T}.$$

Hence

$$dW = (\mathbf{F} \cdot \mathbf{T})\, ds.$$

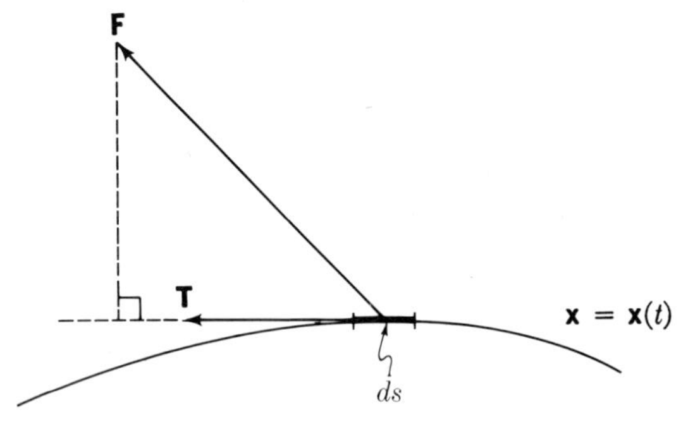

FIG. 2.1

Replace ds by $(ds/dt)\, dt$. This makes good physical sense: The length ds traveled in a short time period dt is given by speed $\times$ time $= (ds/dt)\, dt$. The result is

$$dW = \mathbf{F} \cdot \mathbf{T} \frac{ds}{dt}\, dt.$$

But

$$\mathbf{T} \frac{ds}{dt} = \mathbf{v} = \frac{d\mathbf{x}}{dt},$$

hence

$$dW = (\mathbf{F} \cdot \mathbf{v})\, dt = \left(\mathbf{F} \cdot \frac{d\mathbf{x}}{dt}\right) dt.$$

The total work done is computed by an integral that "adds up" these small bits of work from $\mathbf{x}(t_0)$ to $\mathbf{x}(t_1)$:

$$W = \int_{t_0}^{t_1} (\mathbf{F} \cdot \mathbf{v})\, dt = \int_{t_0}^{t_1} \left(\mathbf{F} \cdot \frac{d\mathbf{x}}{dt}\right) dt.$$

Replace $(d\mathbf{x}/dt)\, dt$ by $d\mathbf{x}$. The result is

$$W = \int_{\mathbf{x}(t_0)}^{\mathbf{x}(t_1)} \mathbf{F} \cdot d\mathbf{x}.$$

This type of integral is called a **line integral.** It arises naturally in connection with work, but has many other practical applications in physics. The evaluation of a line integral involves nothing more than the evaluation of an ordinary integral.

Suppose a particle moves on a curve $\mathbf{x}(t)$ from $\mathbf{x}(t_0)$ to $\mathbf{x}(t_1)$ and is subject to a force $\mathbf{F}(t)$. Then the work done by the force is given by the line integral

$$\int_{t_0}^{t_1} \left(\mathbf{F} \cdot \frac{d\mathbf{x}}{dt}\right) dt = \int_{\mathbf{x}(t_0)}^{\mathbf{x}(t_1)} \mathbf{F} \cdot d\mathbf{x}.$$

In the integral on the right,

$$d\mathbf{x} = \left(\frac{dx}{dt}, \frac{dy}{dt}, \frac{dz}{dt}\right) dt.$$

Let $\mathbf{F}(t) = (F_1(t), F_2(t), F_3(t))$. Then the line integral is evaluated as an ordinary integral:

$$\int_{\mathbf{x}(t_0)}^{\mathbf{x}(t_1)} \mathbf{F} \cdot d\mathbf{x} = \int_{t_0}^{t_1} \left[F_1(t)\frac{dx}{dt} + F_2(t)\frac{dy}{dt} + F_3(t)\frac{dz}{dt}\right] dt.$$

EXAMPLE 2.1

Evaluate the line integral $\int_{\mathbf{x}(0)}^{\mathbf{x}(3)} \mathbf{F} \cdot d\mathbf{x}$, where $\mathbf{F} = (3, -1, 2)$ and $\mathbf{x}(t) = (t, t^2, t^3)$.

Solution:

$$d\mathbf{x} = (1, 2t, 3t^2)\, dt,$$

$$\mathbf{F} \cdot d\mathbf{x} = (3, -1, 2) \cdot (1, 2t, 3t^2)\, dt = (3 - 2t + 6t^2)\, dt.$$

Therefore

$$\int_{\mathbf{x}(0)}^{\mathbf{x}(3)} \mathbf{F} \cdot d\mathbf{x} = \int_0^3 (3 - 2t + 6t^2)\, dt = (3t - t^2 + 2t^3)\Big|_0^3 = (9 - 9 + 54).$$

Answer: 54.

EXAMPLE 2.2

Under the action of a force $\mathbf{F}(t)$, a particle moves on a path $\mathbf{x}(t)$ from $\mathbf{x}(t_0)$ to $\mathbf{x}(t_1)$. Let W denote the work done by the force. From Newton's Law, show that $W = \frac{1}{2}m|\mathbf{v}(t_1)|^2 - \frac{1}{2}m|\mathbf{v}(t_0)|^2$.

Solution:

$$W = \int_{t_0}^{t_1} \mathbf{F} \cdot \dot{\mathbf{x}}\, dt.$$

According to Newton's Law, $\mathbf{F} = m\ddot{\mathbf{x}}$, and so

$$\mathbf{F} \cdot \dot{\mathbf{x}} = m\ddot{\mathbf{x}} \cdot \dot{\mathbf{x}}.$$

But observe that

$$2\ddot{\mathbf{x}} \cdot \dot{\mathbf{x}} = \frac{d}{dt}(\dot{\mathbf{x}} \cdot \dot{\mathbf{x}}).$$

Therefore

$$\mathbf{F} \cdot \dot{\mathbf{x}} = \frac{1}{2} m \frac{d}{dt}(\dot{\mathbf{x}} \cdot \dot{\mathbf{x}}) = \frac{1}{2} m \frac{d}{dt} |\mathbf{v}|^2,$$

so

$$W = \int_{t_0}^{t_1} \mathbf{F} \cdot \dot{\mathbf{x}}\, dt = \int_{t_0}^{t_1} \frac{1}{2} m \frac{d}{dt} |\mathbf{v}|^2\, dt = \frac{1}{2} m|\mathbf{v}(t_1)|^2 - \frac{1}{2} m|\mathbf{v}(t_0)|^2.$$

REMARK: The quantity $\frac{1}{2}m|\mathbf{v}|^2$ is the **kinetic energy** of the particle. The result of this example is the Law of Conservation of Energy: work done equals change in kinetic energy.

Integrals of Vector Functions

Another kind of integral which occurs often is the integral of a vector function.

> Suppose $\mathbf{u}(t) = (u_1(t), u_2(t), u_3(t))$ is defined for $a \le t \le b$. Then
>
> $$\int_a^b \mathbf{u}(t)\, dt = \left(\int_a^b u_1(t)\, dt, \int_a^b u_2(t)\, dt, \int_a^b u_3(t)\, dt\right).$$

Notice that the integral of a vector function is a *vector*, whereas a line integral is a *scalar*.

EXAMPLE 2.3

Let $\mathbf{u}(t) = (1, t - 1, t^2)$. Find $\displaystyle\int_{-2}^{3} \mathbf{u}(t)\, dt$.

Solution:

$$\int_{-2}^{3} \mathbf{u}(t)\, dt = \int_{-2}^{3} (1, t - 1, t^2)\, dt$$

$$= \left(\int_{-2}^{3} dt, \int_{-2}^{3} (t - 1)\, dt, \int_{-2}^{3} t^2\, dt\right) = \left(5, -\frac{5}{2}, \frac{35}{3}\right).$$

Answer: $(5, -\frac{5}{2}, \frac{35}{3})$.

The integral of a vector function $\mathbf{u}(t)$ is particularly easy to evaluate if an antiderivative of $\mathbf{u}(t)$ is known.

If

$$\mathbf{u}(t) = \frac{d}{dt}\mathbf{w}(t),$$

then

$$\int_a^b \mathbf{u}(t)\,dt = \int_a^b \frac{d\mathbf{w}}{dt}\,dt = \mathbf{w}(b) - \mathbf{w}(a).$$

To prove this, simply check the three components; for each of these use our old rule for evaluating an integral. Here is an example:

$$\int_0^2 (2t, 3t^2, 4t^3)\,dt = \int_0^2 \frac{d}{dt}(t^2, t^3, t^4)\,dt = (t^2, t^3, t^4)\Big|_0^2 = (4, 8, 16).$$

The following application shows the importance in physics of vector-valued integrals. Suppose a particle moves under the influence of a force $\mathbf{F}$, so that

$$\mathbf{F} = m\ddot{\mathbf{x}} = m\frac{d\dot{\mathbf{x}}}{dt}.$$

Integrate with respect to t on the interval $t_0 \leq t \leq t_1$:

$$\int_{t_0}^{t_1} \mathbf{F}\,dt = \int_{t_0}^{t_1} m\frac{d\dot{\mathbf{x}}}{dt}\,dt,$$

hence,

$$\int_{t_0}^{t_1} \mathbf{F}\,dt = m\dot{\mathbf{x}}(t_1) - m\dot{\mathbf{x}}(t_0).$$

The quantity $m\dot{\mathbf{x}}$ is the **momentum** of the particle, so the right-hand side is its change in momentum. The left-hand side is called the **impulse** of the force during the time from t_0 to t_1. This equation is a form of the Law of Conservation of Momentum: impulse equals change in momentum.

EXERCISES

Evaluate:

1. $\int_{\mathbf{x}(0)}^{\mathbf{x}(1)} (t, 2t, 3t)\cdot d\mathbf{x};\quad \mathbf{x}(t) = (1, t, t^2)$

2. $\displaystyle\int_{\mathbf{x}(0)}^{\mathbf{x}(\pi)} (\cos t, \sin t) \cdot d\mathbf{x}; \quad \mathbf{x}(t) = (\sin t, -\cos t)$

3. $\displaystyle\int_{\mathbf{x}(0)}^{\mathbf{x}(2\pi)} (0, 0, 3t) \cdot d\mathbf{x}; \quad \mathbf{x}(t) = (\cos t, \sin t, t + 1)$

4. $\displaystyle\int_{\mathbf{x}(-1)}^{\mathbf{x}(1)} (e^t, e^t, e^t) \cdot d\mathbf{x}; \quad \mathbf{x}(t) = (t + 1, t - 1, t)$

5. $\displaystyle\int_{(0,0,0)}^{(1,1,1)} (t, -t^2, t) \cdot d\mathbf{x}$; straight path

6. $\displaystyle\int_{\mathbf{x}(0)}^{\mathbf{x}(2\pi)} \left(\frac{-y}{x^2 + y^2}, \frac{x}{x^2 + y^2}\right) \cdot d\mathbf{x}; \quad \mathbf{x}(t) = (a \cos t, a \sin t)$

7. $\displaystyle\int_{\mathbf{x}(0)}^{\mathbf{x}(1)} x\,dx + y\,dy; \quad \mathbf{x}(t) = (t^2, t^3)$

8. $\displaystyle\int_{(1,1,2)}^{(-1,0,4)} x\,dy + y\,dz + z\,dx$; straight path.

9. Find the work done by the uniform gravitational field $\mathbf{F} = (0, 0, -g)$ in moving a particle from $(0, 0, 1)$ to $(1, 1, 0)$ along a straight path.

10. Find the work done by the central force field $\mathbf{F} = -\dfrac{1}{|\mathbf{x}|^3}\mathbf{x}$ in moving a particle from $(2, 2, 2)$ to $(1, 1, 1)$ along a straight path.

Evaluate:

11. $\displaystyle\int_0^1 (1 + t, 1 + 2t, 1 + 3t)\,dt$

12. $\displaystyle\int_0^{2\pi} (\cos t, \sin t, 1)\,dt$

13. $\displaystyle\int_{-1}^1 (t^3, t^4, t^5)\,dt$

14. $\displaystyle\int_1^4 \left(\frac{1}{t}, \frac{1}{t^2}, \frac{1}{t^3}\right) dt.$

15. The force $\mathbf{F}(t) = (1 - t, 1 - t^2, 1 - t^3)$ acts from $t = 0$ to $t = 2$. Find its impulse.

16. The force $\mathbf{F}(t) = (e^t, e^{2t}, e^{3t})$ acts from $t = -1$ to $t = 0$. Find its impulse.

17. Let $\mathbf{x} = \mathbf{x}(t)$, where $a \le t \le b$, be a plane curve which does not pass through $(0, 0)$. Show that $\dfrac{1}{2}\displaystyle\int_{\mathbf{x}(a)}^{\mathbf{x}(b)} (x\,dy - y\,dx)$ is the area in Fig. 2.2.

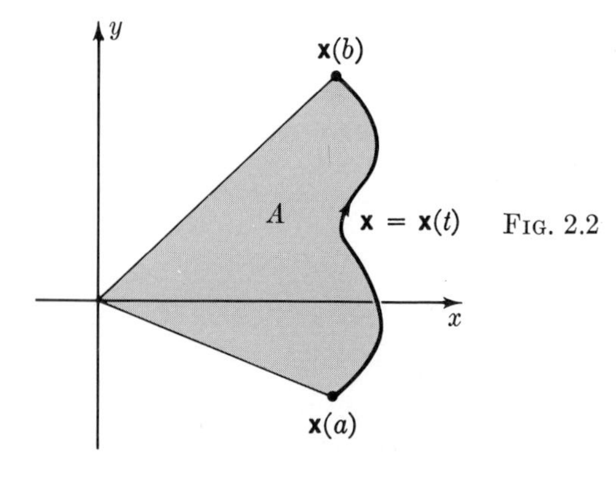

FIG. 2.2

3. MORE ON CURVES

We have been discussing curves given by vector functions $\mathbf{x}(t)$. Any curve defined this way is the image of the one-dimensional t-axis (or a part of it) in three-dimensional space. As t runs along the t-axis, the corresponding point describes a space curve (Fig. 3.1).

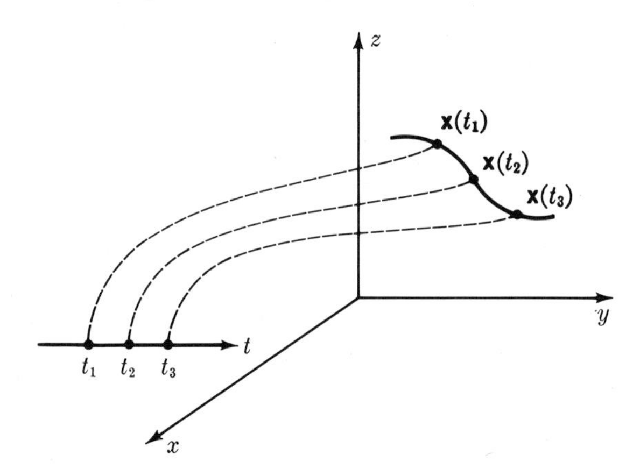

FIG. 3.1

The description of a curve by a vector function $\mathbf{x}(t)$ is called a **parametric representation** of the curve. The variable t is called the **parameter.**

In the plane, there are two other ways of presenting a curve:

(1) as the graph of a function $y = f(x)$,
(2) as the graph of an equation $F(x, y) = 0$.

In case (1), it is easy to parameterize the graph. Just set $x = t$. Then

$$\mathbf{x}(t) = (t, f(t))$$

is a parametric representation of the curve.

In case (2), it is not always easy to parameterize the graph. Here are three examples.

The circle

$$x^2 + y^2 = a^2$$

is easy; parameterize by the central angle θ:

$$(x, y) = (a \cos \theta, a \sin \theta).$$

The similar locus

$$x^4 + y^4 = 1$$

is less obvious:

$$(x, y) = (\pm \sqrt{\cos \theta}, \pm \sqrt{\sin \theta}).$$

The locus

$$x^6 + xy + y^8 = 1$$

is beyond our skill; it is difficult to learn anything at all about a curve whose equation is this complicated.

There are many ways to parameterize a curve. If $\mathbf{x} = \mathbf{x}(t)$ is one parameterization and t is a function of another variable $t = t(u)$, then the composite vector function

$$\mathbf{x} = \mathbf{x}[t(u)]$$

is another parameterization of the same curve (or at least part of it). For example,

$$\mathbf{x}(t) = (t + 1, 3t - 2)$$

describes a straight line in the plane. Let $t = 2u$. Then

$$\mathbf{x}[t(u)] = (2u + 1, 6u - 2)$$

describes the same line. Next, let $t = u^2$. Then

$$\mathbf{x}[t(u)] = (u^2 + 1, 3u^2 - 2)$$

describes part of that line [the part given by $(t + 1, 3t - 2)$ for $t \geq 0$].

Some Special Cases

Certain parameterizations are worth knowing. The ellipse

$$\frac{x^2}{a^2} + \frac{y^2}{b^2} = 1$$

in Fig. 3.2 is parameterized by

$$(x, y) = (a \cos \theta, b \sin \theta).$$

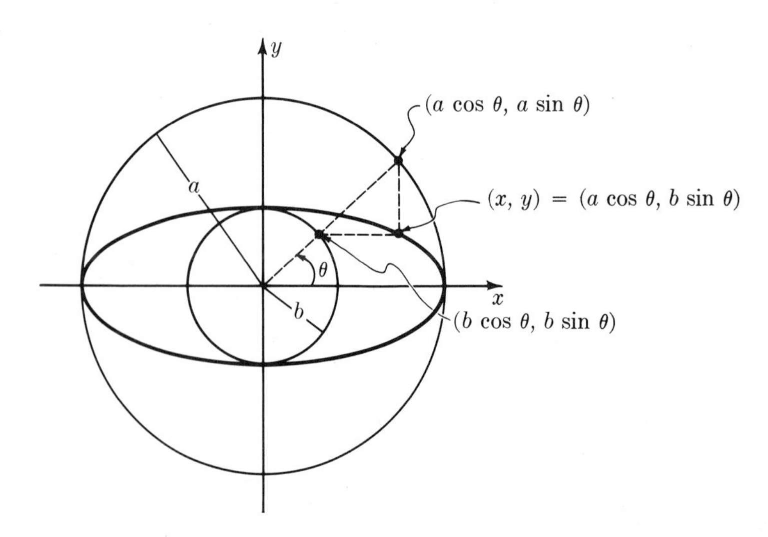

FIG. 3.2

As a check,

$$\frac{(a\cos\theta)^2}{a^2}+\frac{(b\sin\theta)^2}{b^2}=1.$$

The hyperbola

$$\frac{x^2}{a^2}-\frac{y^2}{b^2}=1$$

is parameterized by

$$(x, y) = (a\cosh t, b\sinh t).$$

Recall

$$\cosh t=\frac{e^t+e^{-t}}{2},\qquad \sinh t=\frac{e^t-e^{-t}}{2},$$

$$\cosh^2 t-\sinh^2 t=1.$$

See Ex. 6 below for an interpretation of t.

Besides the parameterization in terms of sine and cosine, there is another useful one for the circle

$$x^2+y^2=1.$$

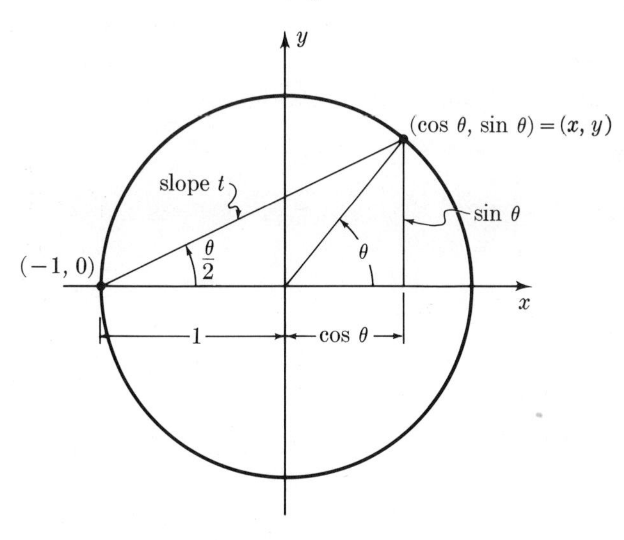

Fig. 3.3

The line (Fig. 3.3) through $(-1, 0)$ of slope t meets the circle in two points, $(-1, 0)$ and $(x, y) = (\cos\theta, \sin\theta)$. The inscribed angle at $(-1, 0)$ is half the central angle, hence

$$t=\tan\frac{\theta}{2}\cdot$$

But from the larger triangle,

$$t = \frac{y}{1 + x} = \frac{\sin \theta}{1 + \cos \theta}.$$

To express x and y in terms of t, solve simultaneously

$$\begin{cases} x^2 + y^2 = 1 \\ y = t(1 + x). \end{cases}$$

Substitute the expression for y into the first equation:

$$x^2 + t^2(1 + x)^2 = 1,$$

$$t^2(1 + x)^2 + (x^2 - 1) = 0.$$

Factor:

$$(x + 1)[t^2(x + 1) + (x - 1)] = 0.$$

You do not want the root $x = -1$ because it corresponds to the known point of intersection $(-1, 0)$. Hence

$$t^2(x + 1) + x - 1 = 0, \qquad (t^2 + 1)x + (t^2 - 1) = 0,$$

$$x = \frac{1 - t^2}{1 + t^2}, \qquad y = t(1 + x) = \frac{2t}{1 + t^2}.$$

The final result is a parameterization of the circle:

$$\begin{cases} x = \dfrac{1 - t^2}{1 + t^2} \\ y = \dfrac{2t}{1 + t^2} \end{cases} \quad \text{where} \quad t = \frac{y}{1 + x};$$

$$\mathbf{x} = \left(\frac{1 - t^2}{1 + t^2}, \frac{2t}{1 + t^2}\right), \qquad t = \frac{y}{1 + x}.$$

In terms of θ, these are precisely the half-angle formulas:

$$\begin{cases} \cos \theta = \dfrac{1 - \tan^2 \dfrac{\theta}{2}}{1 + \tan^2 \dfrac{\theta}{2}} \\ \sin \theta = \dfrac{2 \tan \dfrac{\theta}{2}}{1 + \tan^2 \dfrac{\theta}{2}} \end{cases} \qquad \tan \frac{\theta}{2} = \frac{\sin \theta}{1 + \cos \theta}.$$

Intersection of Surfaces

A space curve may be presented in other than parametric form, for example, as the intersection of two surfaces.

EXAMPLE 3.1

Describe the curve common to the two circular cylinders $x^2 + y^2 = 1$ and $x^2 + z^2 = 1$.

Solution: Draw the portion in the first octant (Fig. 3.4a). The intersection consists of two closed curves which intersect at right angles (Fig. 3.4b) at $(1, 0, 0)$ and $(-1, 0, 0)$. To find the nature of these curves, subtract the two equations:

$$(x^2 + y^2) - (x^2 + z^2) = 1 - 1,$$

$$y^2 - z^2 = 0, \qquad y = \pm z.$$

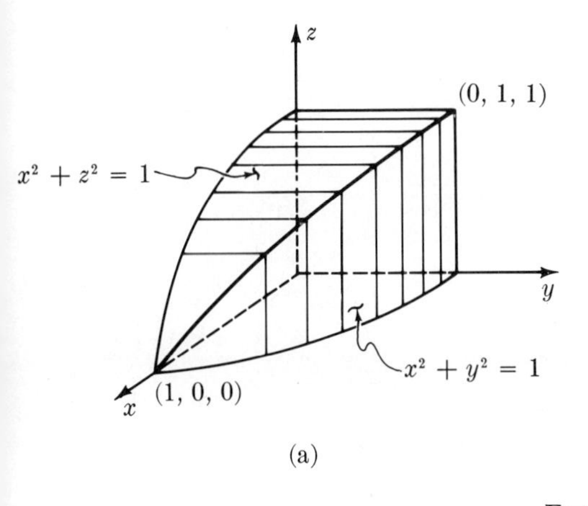

(a)

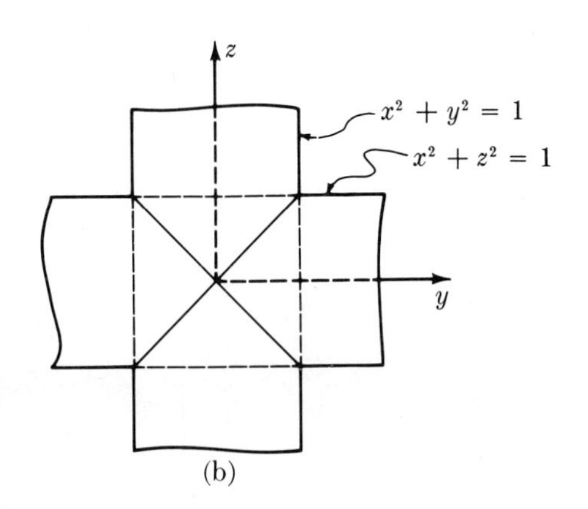

(b)

FIG. 3.4

Thus each point on the intersection lies either on the plane $y = z$ or on the plane $y = -z$. Hence the curves of intersection are two ellipses, the intersections of the cylinders with two perpendicular planes, $y = z$ and $y = -z$.

Answer: Two ellipses.

Note the parameterization

$$\mathbf{x} = (\cos\theta, \sin\theta, \pm\sin\theta)$$

of the pair of ellipses. This is correct because $(x, y) = (\cos\theta, \sin\theta)$ parameterizes $x^2 + y^2 = 1$, and $z = \pm y = \pm\sin\theta$.

EXERCISES

1. Show that $\mathbf{x}(t) = t\mathbf{a} + (1 - t)\mathbf{b}$ parameterizes the line passing through $\mathbf{a}$ and $\mathbf{b}$. Interpret t and $(1 - t)$ as weights at $\mathbf{a}$ and $\mathbf{b}$ with center of gravity $\mathbf{x}$.
2. Show that $\mathbf{x}(t) = (t^2 - 1, t^3 - t^2)$ parameterizes $y^2 = x^3 + x^2$. Graph the curve.
3. A circle of radius a rolls along the x-axis. Initially its center is $(0, a)$. The point on the circle initially at $(0, 0)$ traces a curve called the **cycloid.** Show that $\mathbf{x}(\theta) = a(\theta - \sin\theta, 1 - \cos\theta)$ is a parameterization of the cycloid. Graph the curve.
4. (cont.) Show that the area under one arch of the cycloid is 3 times the area of the circle.
   (*Hint:* Express $y\,dx$ in terms of θ and integrate.)
5. (cont.) Find the arc length of one arch of the cycloid.
6. The hyperbola $\dfrac{x^2}{a^2} - \dfrac{y^2}{b^2} = 1$ is parameterized by $\mathbf{x}(t) = (a\cosh t, b\sinh t)$. Show that t is related to the area A by $t = 2A/ab$. See Fig. 3.5.
   (*Hint:* Use Ex. 17, Section 2.)

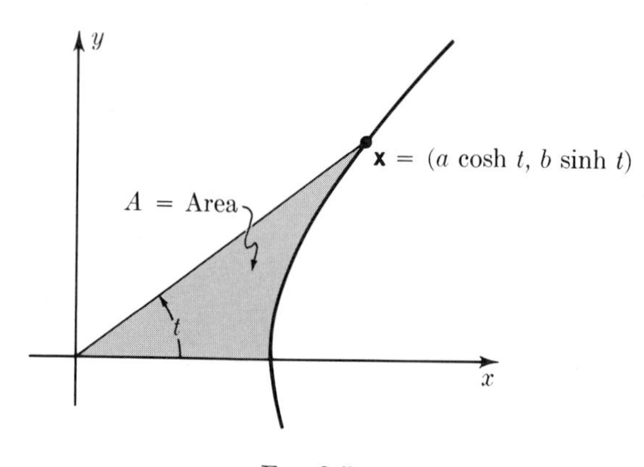

FIG. 3.5

7. Express the curve $\mathbf{x}(t) = (t, t^2, t^3)$ as the intersection of two surfaces.
8. Parameterize the intersection of the two surfaces $z = x^2 - y^2$ and $x^2 + y^2 = 1$.
9. Parameterize the intersection of the surfaces $y^2 - z^2 = 1$ and $x = y^2 + z^2$.
10. Express the helix $\mathbf{x}(t) = (\cos t, \sin t, t)$ as the intersection of two surfaces.
11. Show that the circle $x^2 + y^2 = 1$ lies inside of the curve $x^4 + y^4 = 1$.
    (*Hint:* $\cos^4\theta \le \cos^2\theta$ and $\sin^4\theta \le \sin^2\theta$.)
12. Show that the "witch of Agnesi" $y = 1/(1 + x^2)$ is parameterized by $\mathbf{x} = \mathbf{x}(\theta) = 2(\cot\theta, \sin^2\theta)$. Sketch the curve.
13. Show that the "folium of Descartes" $x^3 + y^3 = 3axy$ is parameterized by $\mathbf{x} = \mathbf{x}(t) = (1 + t^3)^{-1}(3at, 3at^2)$. Sketch the curve.

4. POLAR COORDINATES

The **polar coordinates** of a point $\mathbf{x} \neq \mathbf{0}$ in the plane are the distance $r = |\mathbf{x}|$ of $\mathbf{x}$ from $\mathbf{0}$, and the angle θ from the positive x-axis to the vector $\mathbf{x}$, measured counterclockwise (Fig. 4.1). The angle θ is determined up to a multiple of 2π.

We shall write polar coordinates $\{r, \theta\}$ with curly braces to distinguish them from rectangular coordinates (x, y).

Any value of r is allowed, even $r = 0$ (the origin) and negative r; the point $\{-r, \theta\}$ is the reflection of $\{r, \theta\}$ through the origin (Fig. 4.2) and is identical with $\{r, \theta + \pi\}$. Note that θ is undefined at the origin.

The rectangular coordinates (x, y) of the point with polar coordinates $\{r, \theta\}$ are given (Fig. 4.3) by

$$\begin{cases} x = r\cos\theta \\ y = r\sin\theta. \end{cases}$$

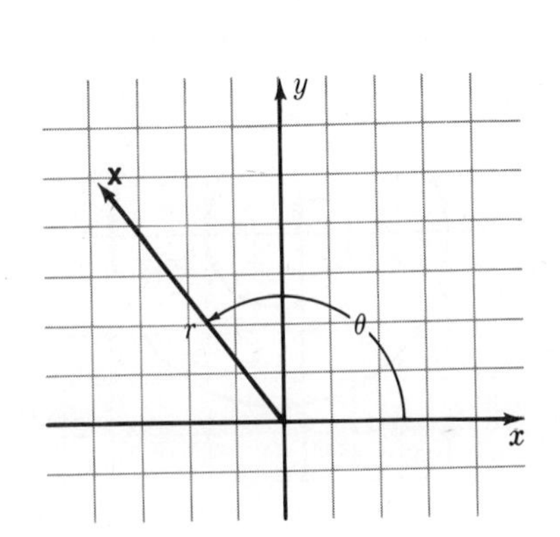

Fig. 4.1

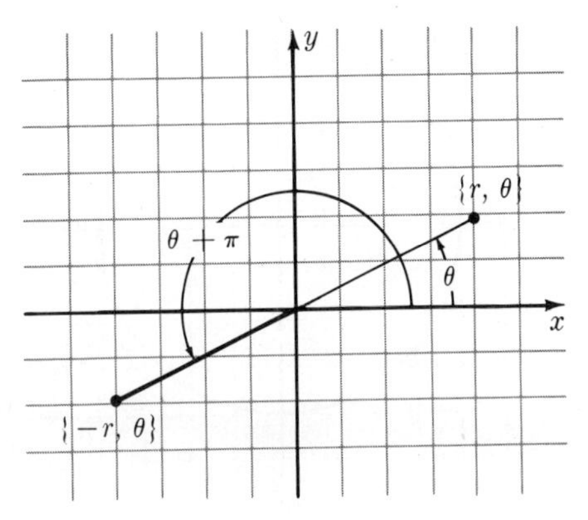

Fig. 4.2

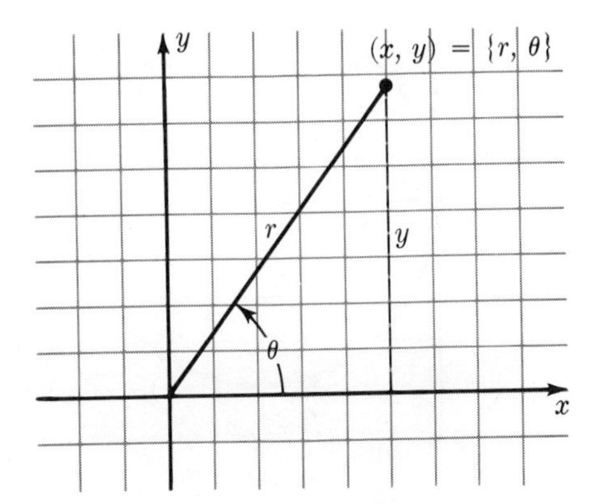

Fig. 4.3

Conversely, given (x, y) we find $\{r, \theta\}$ from Fig. 4.3:

$$r^2 = x^2 + y^2, \qquad \begin{cases} \cos\theta = \dfrac{x}{r} \\ \sin\theta = \dfrac{y}{r}. \end{cases}$$

(The single formula for the angle, $\tan\theta = y/x$, is not adequate to distinguish quadrants. For example, $\tan 0 = \tan\pi = 0$.)

The number r is called the **radius** of the point $\mathbf{x} = \{r, \theta\}$ and θ is called the **polar angle.**

A curve may be presented by a relation between r and θ, often in the form $r = f(\theta)$. Three examples are shown in Fig. 4.4. Note that Fig. 4.4c is

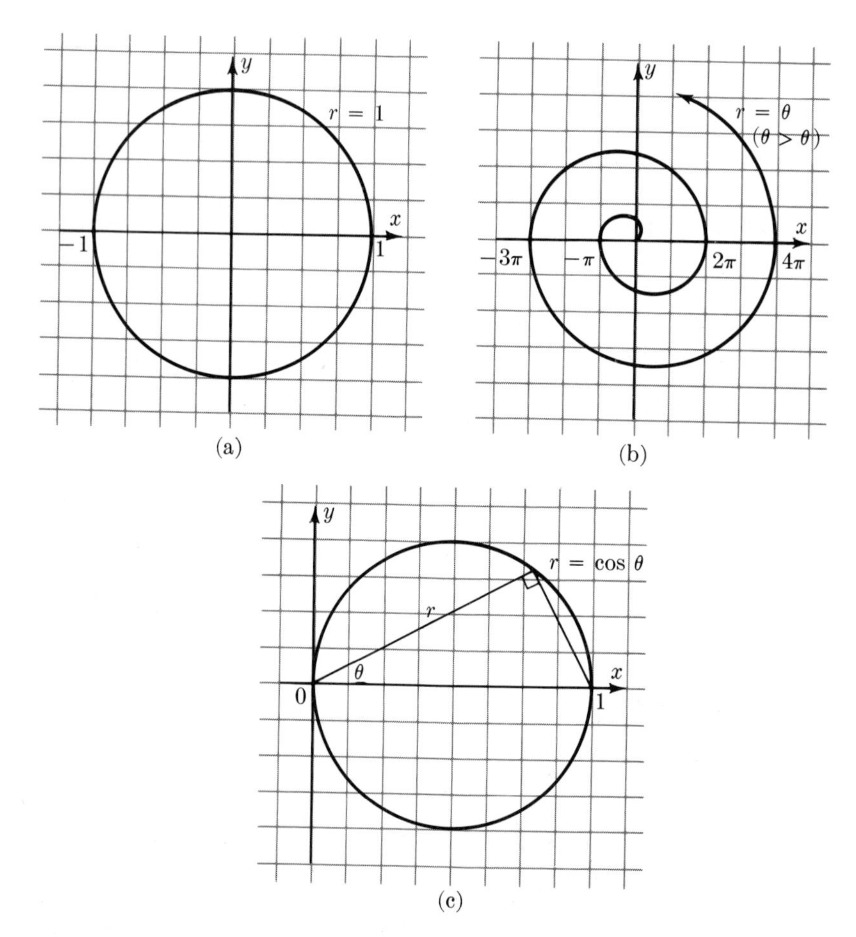

FIG. 4.4

a circle:

$$r = \cos\theta, \qquad r^2 = r\cos\theta, \qquad x^2 + y^2 = x,$$

$$(x^2 - x) + y^2 = 0, \qquad \left(x - \frac{1}{2}\right)^2 + y^2 = \frac{1}{4}.$$

The center is $(\frac{1}{2}, 0)$, the radius is $\frac{1}{2}$.

Length

Suppose a curve is given in the form

$$r = r(t), \qquad \theta = \theta(t), \qquad t_0 \le t \le t_1.$$

What is the length? Write

$$\mathbf{x} = (r\cos\theta, r\sin\theta) = r(\cos\theta, \sin\theta).$$

Differentiate:

$$\begin{aligned}\dot{\mathbf{x}} &= \dot{r}(\cos\theta, \sin\theta) + r\dot{\theta}(-\sin\theta, \cos\theta)\\ &= \dot{r}\mathbf{u} + r\dot{\theta}\mathbf{w},\end{aligned}$$

where

$$\mathbf{u} = (\cos\theta, \sin\theta) \qquad \text{and} \qquad \mathbf{w} = (-\sin\theta, \cos\theta).$$

Now

$$\begin{aligned}\left(\frac{ds}{dt}\right)^2 &= |\dot{\mathbf{x}}|^2 = \dot{\mathbf{x}} \cdot \dot{\mathbf{x}}\\ &= (\dot{r}\mathbf{u} + r\dot{\theta}\mathbf{w}) \cdot (\dot{r}\mathbf{u} + r\dot{\theta}\mathbf{w})\\ &= \dot{r}^2\mathbf{u}\cdot\mathbf{u} + 2r\dot{r}\dot{\theta}\mathbf{u}\cdot\mathbf{w} + r^2\dot{\theta}^2\mathbf{w}\cdot\mathbf{w}.\end{aligned}$$

But $\mathbf{u}\cdot\mathbf{u} = 1$, $\mathbf{w}\cdot\mathbf{w} = 1$, and $\mathbf{u}\cdot\mathbf{w} = 0$. Hence,

$$\left(\frac{ds}{dt}\right)^2 = \dot{r}^2 + r^2\dot{\theta}^2.$$

It follows that

$$\boxed{\text{Length} = \int_{t_0}^{t_1} \sqrt{\dot{r}^2 + r^2\dot{\theta}^2}\, dt.}$$

Figure 4.5 provides an aid to memory. The "right triangle" has sides dr, $r\,d\theta$, and ds, so the Pythagorean Theorem suggests

$$(ds)^2 = (dr)^2 + r^2(d\theta)^2.$$

Suppose a curve is given by $r = r(\theta)$. This is a special case of the previous situation with $\theta = t$ and $r = r(t)$. The length formula specializes to

$$\boxed{L = \int_{\theta_0}^{\theta_1} \sqrt{r^2 + \left(\frac{dr}{d\theta}\right)^2}\, d\theta.}$$

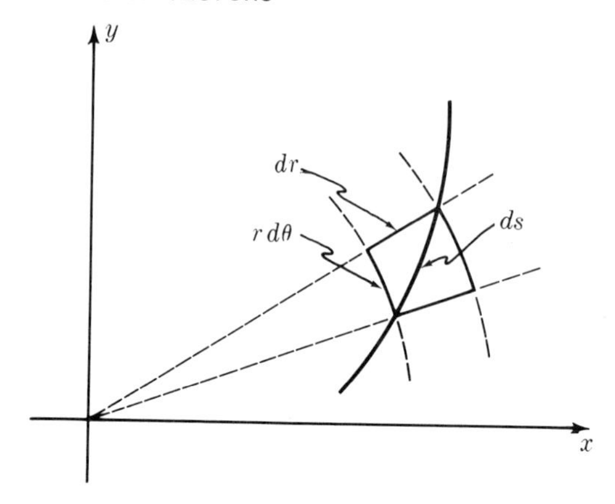

Fig. 4.5

EXAMPLE 4.1

Find the length of the spiral $r = \theta^2$, $0 \leq \theta \leq 2\pi$.

Solution:

$$L = \int_0^{2\pi} \sqrt{r^2 + \left(\frac{dr}{d\theta}\right)^2}\, d\theta = \int_0^{2\pi} \sqrt{\theta^4 + (2\theta)^2}\, d\theta$$

$$= \int_0^{2\pi} \theta\sqrt{\theta^2 + 4}\, d\theta = \frac{1}{3}(\theta^2 + 4)^{3/2}\Big|_0^{2\pi}.$$

Answer: $L = \dfrac{8}{3}(\pi^2 + 1)^{3/2} - \dfrac{8}{3}.$

Area

There are problems which require the area swept out by the segment joining **0** to a moving point on a curve (Fig. 4.6).

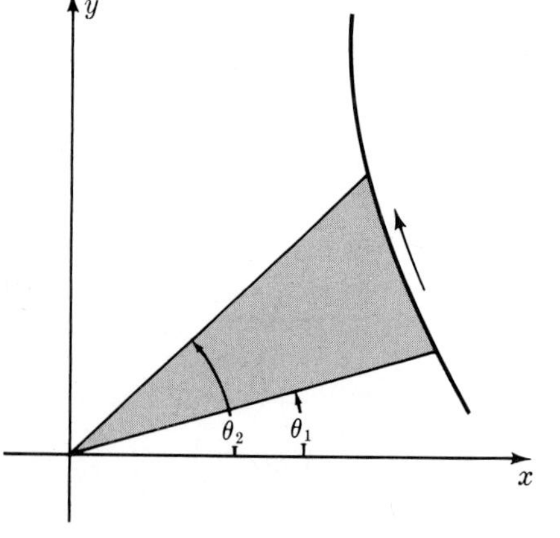

Fig. 4.6

Suppose the curve is given by

$$r = r(t), \qquad \theta = \theta(t), \qquad t_0 \le t \le t_1.$$

In a small time interval dt, a thin triangle of base $r\,d\theta$ and height r (ignoring negligible errors) is swept out (Fig. 4.7). Hence

$$dA = \frac{1}{2} r^2\,d\theta = \frac{1}{2} r^2 \frac{d\theta}{dt}\,dt,$$

$$\boxed{A = \int_{t_0}^{t_1} \frac{1}{2} r^2 \frac{d\theta}{dt}\,dt.}$$

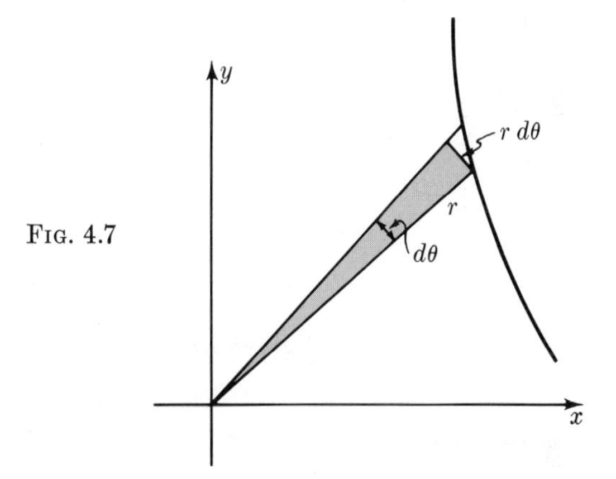

Fig. 4.7

In the special case that the curve is given by $r = r(\theta)$ for $\theta_0 \le \theta \le \theta_1$, choose $t = \theta$. The formula specializes to

$$\boxed{A = \int_{\theta_0}^{\theta_1} \frac{1}{2} r(\theta)^2\,d\theta.}$$

EXAMPLE 4.2

Find the area of the "four-petal rose" $r = a\cos 2\theta$.

Solution: Graph the curve carefully (Fig. 4.8). The portion on which $0 \le \theta \le \pi/2$ is emphasized. Note that $r < 0$ for $\pi/4 < \theta < \pi/2$. Because of symmetry it suffices to find the area of half of one petal. Thus

$$\begin{aligned} A &= 8\int_0^{\pi/4} \frac{1}{2}(a\cos 2\theta)^2\,d\theta \\ &= 4a^2 \int_0^{\pi/4} \cos^2 2\theta\,d\theta = 2a^2 \int_0^{\pi/2} \cos^2 t\,dt. \end{aligned}$$

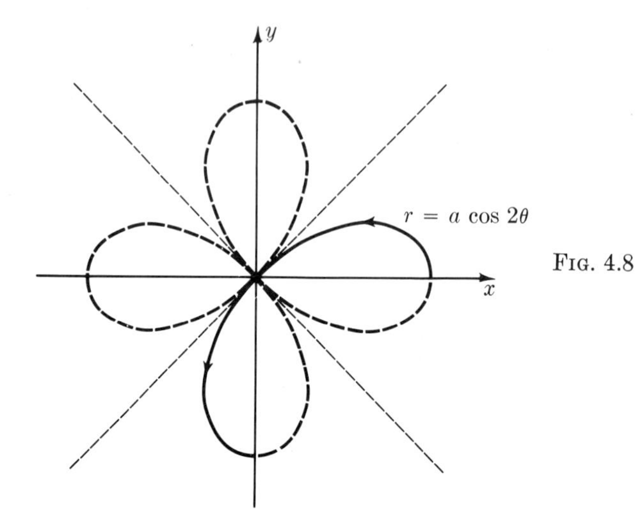

FIG. 4.8

Answer: $A = \dfrac{\pi a^2}{2}$.

Summary

Suppose a curve is given in polar coordinates by

$$r = r(t), \quad \theta = \theta(t), \quad t_0 \le t \le t_1 \qquad r = r(\theta), \quad \theta_0 \le \theta \le \theta_1.$$

ARC LENGTH:

$$L = \int_{t_0}^{t_1} \sqrt{\left(\frac{dr}{dt}\right)^2 + r^2\left(\frac{d\theta}{dt}\right)^2}\, dt \qquad L = \int_{\theta_0}^{\theta_1} \sqrt{r^2 + \left(\frac{dr}{d\theta}\right)^2}\, d\theta.$$

AREA:

$$A = \int_{t_0}^{t_1} \frac{1}{2} r^2 \frac{d\theta}{dt}\, dt \qquad A = \int_{\theta_0}^{\theta_1} \frac{1}{2} r^2(\theta)\, d\theta.$$

EXERCISES

Express in rectangular coordinates:

1. $\{1, \pi/2\}$
2. $\{-1, 3\pi/2\}$
3. $\{2, \pi/4\}$
4. $\{-4, 3\pi/4\}$
5. $\{1, \pi/6\}$
6. $\{2, 5\pi/6\}$.

Express in polar coordinates:

7. $(1, -1)$
8. $(-1, -1)$
9. $(-\sqrt{3}, 1)$
10. $(-\frac{1}{2}, \sqrt{3}/2)$.

11. Find the length of the "spiral of Archimedes" $r = a\theta$ from $\theta = 0$ to $\theta = 1$.
12. Set up an integral for the length of the "four-petal rose" $r = a \cos 2\theta$.
13. Set up an integral for the length of the "three-petal rose" $r = a \cos 3\theta$.
14. Find the length of the "one-petal rose" $r = a \sin \theta$. Precisely what is this curve?
15. Find the area enclosed by the "three-petal rose" $r = a \cos 3\theta$.
16. Find the area enclosed by the "$(2n + 1)$-petal rose" $r = a \cos (2n + 1)\theta$.
17. Find the area enclosed by the "$4n$-petal rose" $r = a \cos 2n\theta$.
18. Find the area enclosed by the curve $r = a \cos^2 2n\theta$.
19. Find the area enclosed by the "cardioid" $r = a(1 - \cos \theta)$.
20. Show that the "cissoid of Diocles" $y^2 = x^3/(a - x)$ can be expressed in polar coordinates by $r = a(\sec \theta - \cos \theta)$; sketch the curve.
21. Find the area enclosed by the figure eight, the "lemniscate of Bernoulli" $r^2 = a^2 \cos 2\theta$. Sketch the curve.
22. Sketch the "strophoid" $r = a \cos 2\theta \sec \theta$. Find the area of the closed loop.
23. Sketch the "limaçon of Pascal" $r = b + a \cos \theta$ in the three cases $0 < a < b$, $0 < a = b$, and $0 < b < a$. In the third case compute the area between the two loops.

5. POLAR VELOCITY AND ACCELERATION [optional]

Let us think of a curve given by

$$r = r(t), \qquad \theta = \theta(t)$$

as the path of a particle. What are its velocity and acceleration vectors?

In the above discussion of length, the perpendicular unit vectors

$$\mathbf{u} = (\cos \theta, \sin \theta) \qquad \text{and} \qquad \mathbf{w} = (-\sin \theta, \cos \theta)$$

were introduced. In using polar coordinates, it is natural to express the velocity and acceleration in terms of these vectors (Fig. 5.1). Note that

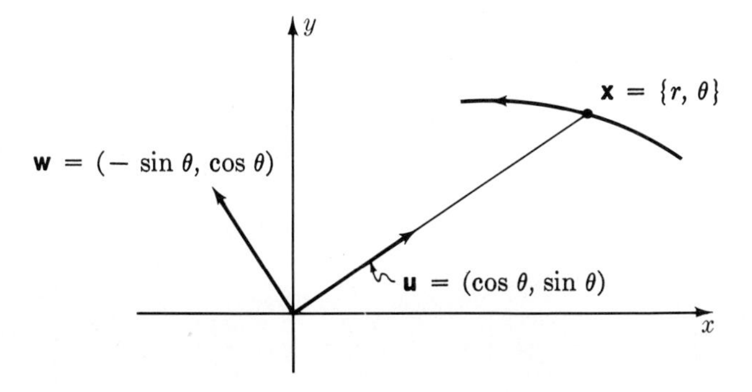

FIG. 5.1

$\mathbf{x} = r\mathbf{u}$ and that

$$\dot{\mathbf{u}} = \dot{\theta}\mathbf{w}, \qquad \dot{\mathbf{w}} = -\dot{\theta}\mathbf{u}.$$

Differentiate $\mathbf{x} = r\mathbf{u}$:

$$\mathbf{v} = \dot{\mathbf{x}} = \dot{r}\mathbf{u} + r\dot{\theta}\mathbf{w}.$$

Differentiate again:

$$\mathbf{a} = \dot{\mathbf{v}} = (\ddot{r}\mathbf{u} + \dot{r}\dot{\theta}\mathbf{w}) + (\dot{r}\dot{\theta}\mathbf{w} + r\ddot{\theta}\mathbf{w} - r\dot{\theta}^2\mathbf{u}).$$

Hence

$$\mathbf{a} = (\ddot{r} - r\dot{\theta}^2)\mathbf{u} + (r\ddot{\theta} + 2\dot{r}\dot{\theta})\mathbf{w}.$$

This is an important formula for motion problems involving a **central force,**

$$\mathbf{F} = f(t)\mathbf{u}.$$

Suppose a unit mass moves under the action of such a central force. Since $\mathbf{a} = \mathbf{F}$, the component of $\mathbf{a}$ in the direction of $\mathbf{w}$ is zero:

$$r\ddot{\theta} + 2\dot{r}\dot{\theta} = 0 \qquad \text{or} \qquad \frac{1}{2}r^2\ddot{\theta} + r\dot{r}\dot{\theta} = 0.$$

This is the same as

$$\frac{d}{dt}\left(\frac{1}{2}r^2\dot{\theta}\right) = 0 \qquad \text{or} \qquad \frac{1}{2}r^2\dot{\theta} = \text{constant}.$$

But

$$\frac{1}{2}r^2\dot{\theta} = \frac{dA}{dt},$$

the rate at which central area is swept out by the curve. It follows that the same area is swept out in equal time anywhere along the path. This is Kepler's Second Planetary Law; it is a case of the Law of Conservation of Angular Momentum.

Summary

Suppose a curve is given in polar coordinates by

$$r = r(t), \qquad \theta = \theta(t).$$

VELOCITY:

$$\mathbf{v} = \frac{dr}{dt}\mathbf{u} + r\frac{d\theta}{dt}\mathbf{w};$$

ACCELERATION:

$$\mathbf{a} = \left[\frac{d^2r}{dt^2} - r\left(\frac{d\theta}{dt}\right)^2\right]\mathbf{u} + \left[r\frac{d^2\theta}{dt^2} + 2\frac{dr}{dt}\frac{d\theta}{dt}\right]\mathbf{w},$$

where $\mathbf{u} = (\cos\theta, \sin\theta)$ and $\mathbf{w} = (-\sin\theta, \cos\theta)$.

EXERCISES

This set of exercises develops Kepler's First and Third Laws of Planetary Motion. Assume a particle of unit mass is moving in a central force field given by the inverse square law:

$$\mathbf{F} = -\frac{1}{r^2}\mathbf{u}.$$

1. Show that the equations of motion are $r^2\dot{\theta} = J$, $\ddot{r} - \dfrac{J^2}{r^3} = -\dfrac{1}{r^2}$, where J is a constant.
2. Show that $\dot{r}^2 + \dfrac{J^2}{r^2} = \dfrac{2}{r} + C$, where C is a constant. This equation is essentially the Law of Conservation of Energy.
   (*Hint:* Multiply the second equation in Ex. 1 by $\dot{r}$ and integrate.)
3. Set $p = \dfrac{1}{r}\cdot$ Show that $\dfrac{\dot{p}^2}{p^4} + J^2p^2 = 2p + C$.
4. Imagine $\theta = \theta(t)$ solved for t as a function of θ and this substituted into $p = p(t)$. Thus p may be considered as a function of θ. Show that
   $J^2\left(\dfrac{dp}{d\theta}\right)^2 = \dfrac{\dot{p}^2}{p^4}$, and conclude that $J^2\left[\left(\dfrac{dp}{d\theta}\right)^2 + p^2\right] = 2p + C$.
5. Show that $\dfrac{d^2p}{d\theta^2} + p = \dfrac{1}{J^2}\cdot$ (*Hint:* Differentiate the previous relation.)
6. Use the methods of Chapter 22 to find the general solution of the preceding differential equation: $p = A\cos\theta + B\sin\theta + \dfrac{1}{J^2}$, where A and B are constants.
7. Show that by a suitable choice of the x-axis, this may be written
   $\dfrac{1}{r} = \dfrac{1}{J^2}(1 - e\cos\theta)$, where e is a constant, the **eccentricity** of the orbit, $e \geq 0$.
8. By passing to rectangular coordinates, show that the orbit is a conic section.
9. Suppose $e = 0$. Show that the orbit is a circle with center at $\mathbf{0}$, and that the speed is constant.
10. Suppose $e = 1$. Show the orbit is a parabola with focus at the origin and opening in the positive x-direction.
11. Suppose $e > 1$. Show that the orbit is a branch of a hyperbola with one focus at the origin.
12. Suppose $0 < e < 1$. Show that the orbit is the ellipse $\dfrac{(x-c)^2}{a^2} + \dfrac{y^2}{b^2} = 1$,
    where $a = \dfrac{J^2}{1-e^2}$, $b = \dfrac{J^2}{\sqrt{1-e^2}}$, and $c = ae$.
    [By Ex. 8–12, each closed orbit is an ellipse (or circle), Kepler's First Law.]
13. (cont.) Show that $a^2 = b^2 + c^2$. Conclude that the foci of the ellipse are $(0, 0)$ and $(2c, 0)$.
14. (cont.) Let T denote the **period** of the orbit, the time necessary for a complete revolution. Show that $\dfrac{J}{2}T = \pi ab$. (*Hint:* Use Kepler's Second Law.)
15. Conclude that $T^2 = 4\pi^2a^3$. This is Kepler's Third Law: The square of the period of a planetary orbit is proportional to the cube of its semimajor axis.

6. ROTATION OF AXES [optional]

Suppose the coordinate axes in the plane are rotated about $\mathbf{0}$ through angle α to a new position. How are the coordinates of a point with respect to this new coordinate system related to the old coordinates?

The unit vector $\mathbf{i}$ rotates to the unit vector $\bar{\mathbf{i}} = (\cos\alpha, \sin\alpha)$ in the old system (Fig. 6.1). Similarly $\mathbf{j}$ rotates to $\bar{\mathbf{j}} = (-\sin\alpha, \cos\alpha)$. If a point $\mathbf{x}$ has coordinates $(\bar{x}, \bar{y})$ in the new system, then

$$\begin{aligned}\mathbf{x} &= \bar{x}\bar{\mathbf{i}} + \bar{y}\bar{\mathbf{j}}\\ &= \bar{x}(\cos\alpha, \sin\alpha) + \bar{y}(-\sin\alpha, \cos\alpha)\\ &= (\bar{x}\cos\alpha - \bar{y}\sin\alpha, \bar{x}\sin\alpha + \bar{y}\cos\alpha).\end{aligned}$$

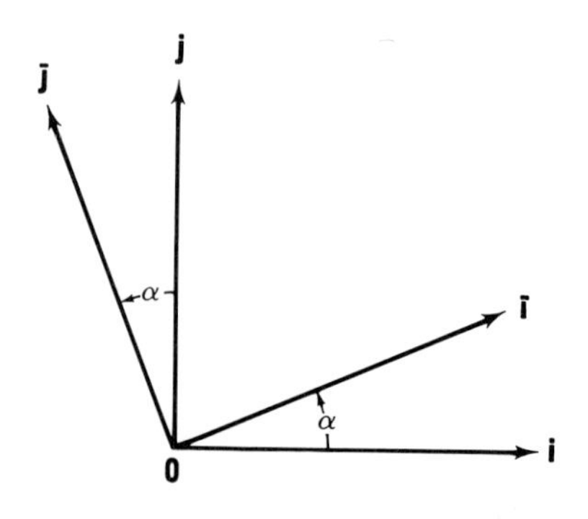

FIG. 6.1

It follows that its old coordinates (x, y) are related to its new ones $(\bar{x}, \bar{y})$ by

$$\boxed{\begin{aligned}x &= \bar{x}\cos\alpha - \bar{y}\sin\alpha\\ y &= \bar{x}\sin\alpha + \bar{y}\cos\alpha.\end{aligned}}$$

This law of rotation of coordinates is quite important. Although the formulas are tricky to remember, it is not hard to visualize Fig. 6.1, and to remember $\mathbf{x} = x\mathbf{i} + y\mathbf{j} = \bar{x}\bar{\mathbf{i}} + \bar{y}\bar{\mathbf{j}}$, from which they follow.

As an application we derive the addition formulas for sine and cosine, unfinished business from Chapter 6. Suppose a unit vector $\mathbf{x}$ makes angle θ with the $\bar{x}$-axis (Fig. 6.2). Then $\bar{x} = \cos\theta$, $\bar{y} = \sin\theta$. But $\mathbf{x}$ makes angle $(\theta + \alpha)$ with the x-axis, hence $x = \cos(\theta + \alpha)$, $y = \sin(\theta + \alpha)$. Substitute this data into the rotation formulas:

$$\begin{cases}\cos(\theta + \alpha) = \cos\theta\cos\alpha - \sin\theta\sin\alpha\\ \sin(\theta + \alpha) = \cos\theta\sin\alpha + \sin\theta\cos\alpha.\end{cases}$$

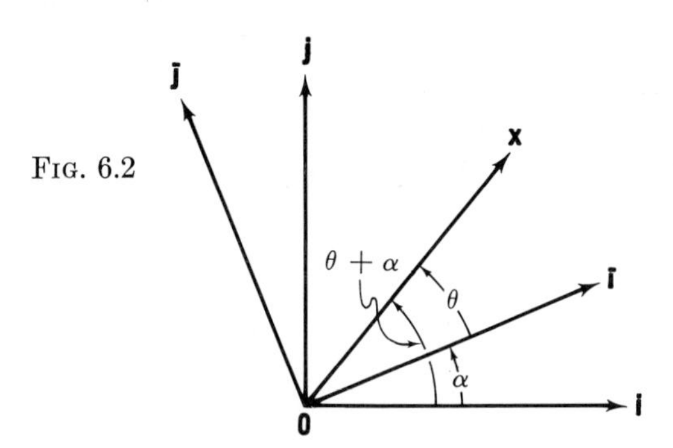

FIG. 6.2

EXERCISES

1. Suppose the frame $\bar{\mathbf{i}}, \bar{\mathbf{j}}$ rotates uniformly with respect to the fixed frame $\mathbf{i}, \mathbf{j}$ so that $\bar{\mathbf{i}} = (\cos \omega t, \sin \omega t)$ and $\bar{\mathbf{j}} = (-\sin \omega t, \cos \omega t)$, where ω is constant. Show that $\dot{\bar{\mathbf{i}}} = \omega \bar{\mathbf{j}}$ and $\dot{\bar{\mathbf{j}}} = -\omega \bar{\mathbf{i}}$.

2. (cont.) Now suppose that a particle is moving. Its path $\mathbf{x}(t)$ may be represented as $\mathbf{x}(t) = x(t)\mathbf{i} + y(t)\mathbf{j}$ in the fixed frame, or as $\mathbf{x}(t) = \bar{x}(t)\,\bar{\mathbf{i}} + \bar{y}(t)\,\bar{\mathbf{j}}$ in the rotating frame. To an observer in the rotating frame, its apparent velocity is $\mathbf{v}_{\text{app}} = \dot{\bar{x}}\,\bar{\mathbf{i}} + \dot{\bar{y}}\,\bar{\mathbf{j}}$. How is this related to the actual velocity?

3. (cont.) Now find the relation between the apparent acceleration $\mathbf{a}_{\text{app}} = \ddot{\bar{x}}\,\bar{\mathbf{i}} + \ddot{\bar{y}}\,\bar{\mathbf{j}}$ and the true acceleration.

4. Find the semimajor and semiminor axes of the rectangular hyperbola $xy = 1$. (*Hint:* Rotate by $\pi/4$.)

5. Rotation of coordinates $x = \bar{x}\cos\alpha - \bar{y}\sin\alpha$, $y = \bar{x}\sin\alpha + \bar{y}\cos\alpha$ takes the equation $Ax^2 + Bxy + Cy^2 = 1$ into an equation of the same form: $\bar{A}\bar{x}^2 + \bar{B}\bar{x}\bar{y} + \bar{C}\bar{y}^2 = 1$. Show that $A + C = \bar{A} + \bar{C}$ and $B^2 - 4AC = \bar{B}^2 - 4\bar{A}\bar{C}$.

6. (cont.) Show that $\bar{B} = 0$ if α is chosen to satisfy $(C - A)\sin 2\alpha + B\cos 2\alpha = 0$.

7. (cont.) Deduce that each graph $Ax^2 + Bxy + Cy^2 = 1$ is a central conic section (possibly degenerate).

Apply the method of Exs. 6 and 7 to the given equation. Describe the graph; if it is an ellipse or hyperbola, find the semimajor and semiminor axes:

8. $x^2 - 4xy + 4y^2 = 1$
9. $x^2 - 3xy + y^2 = 1$
10. $4x^2 + xy - 4y^2 = 1$
11. $x^2 - 2xy + 3y^2 = 1$
12. $x^2 - xy = 1$.

25. Several Variables

1. PLANES

Any pair of non-collinear vectors $\mathbf{v}_1$ and $\mathbf{v}_2$ determines a plane Q through the origin. Each point $\mathbf{x}$ of Q is a linear combination of $\mathbf{v}_1$ and $\mathbf{v}_2$, that is, $\mathbf{x} = a\mathbf{v}_1 + b\mathbf{v}_2$. See Fig. 1.1. Conversely, the set of all such linear combinations fills out the plane Q.

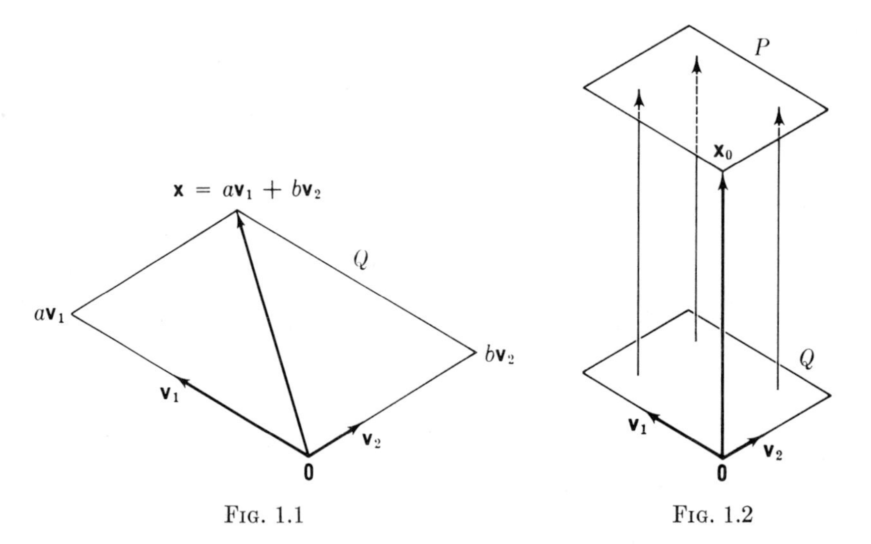

Fig. 1.1

Fig. 1.2

Suppose P is any plane parallel to the plane Q and $\mathbf{x}_0$ is any point of P. Adding $\mathbf{x}_0$ to each point of Q translates the plane Q parallel to itself (Fig. 1.2) so as to coincide with P. Thus each point $\mathbf{x}$ of P satisfies

$$\mathbf{x} = \mathbf{x}_0 + a\mathbf{v}_1 + b\mathbf{v}_2.$$

A plane P may be described in a different way: by a point $\mathbf{x}_0$ on P and a vector $\mathbf{n}$ **normal** (perpendicular) to P. See Fig. 1.3. If $\mathbf{x}$ is any point of P, then the vector $\mathbf{x} - \mathbf{x}_0$ is perpendicular to $\mathbf{n}$:

$$(\mathbf{x} - \mathbf{x}_0) \cdot \mathbf{n} = 0.$$

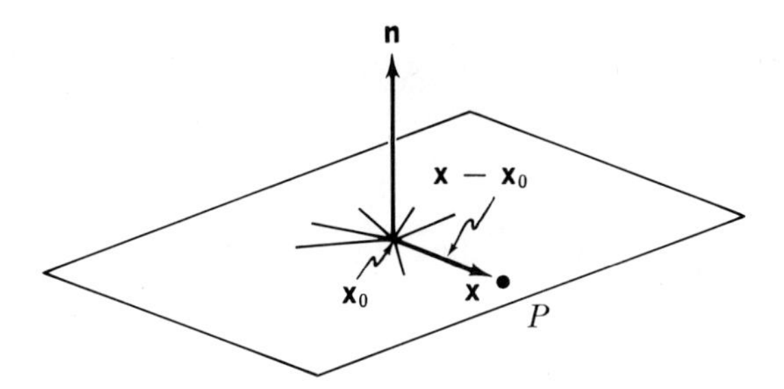

FIG. 1.3

This is a vector equation for P. If $\mathbf{x} = (x, y, z)$ and $\mathbf{x}_0 = (x_0, y_0, z_0)$, and $\mathbf{n} = (a, b, c)$, the vector equation is equivalent to

$$a(x - x_0) + b(y - y_0) + c(z - z_0) = 0.$$

Equivalently,

$$ax + by + cz = d \qquad \text{or} \qquad \mathbf{x} \cdot \mathbf{n} = d,$$

where $d = ax_0 + by_0 + cz_0$. Conversely, each such linear equation represents a plane perpendicular to (a, b, c). (Why?)

EXAMPLE 1.1

Write the equation of the plane through $(1, 2, 3)$ that is perpendicular to the vector $(2, 5, 7)$.

Solution: Just substitute into the formula

$$a(x - x_0) + b(y - y_0) + c(z - z_0) = 0:$$

$$2(x - 1) + 5(y - 2) + 7(z - 3) = 0.$$

Answer: $2x + 5y + 7z = 33.$

EXAMPLE 1.2

Find a unit vector perpendicular to the plane $3x + 4y - 2z = 6$.

Solution: Write the equation of the plane as

$$(x, y, z) \cdot (3, 4, -2) = 6.$$

This is in the form

$$\mathbf{x} \cdot \mathbf{n} = d.$$

Thus $\mathbf{n}$ is normal to the plane. A unit normal vector is

$$\frac{\mathbf{n}}{|\mathbf{n}|} = \frac{(3, 4, -2)}{\sqrt{9 + 16 + 4}}.$$

Answer: $\dfrac{1}{\sqrt{29}}(3, 4, -2).$

EXAMPLE 1.3

Compute the distance from the point (1, 2, 4) to the plane

$$3x + 4y - 2z = 6.$$

Solution: Choose any point $\mathbf{x}_0$ in the plane, say (2, 0, 0). A vector equation for the plane is

$$(\mathbf{x} - \mathbf{x}_0) \cdot \mathbf{n} = 0,$$

where $\mathbf{x}_0 = (2, 0, 0)$ and $\mathbf{n} = (3, 4, -2)$. Let $\mathbf{p} = (1, 2, 4)$. The perpendicular distance from $\mathbf{p}$ to the plane is the projection of the vector from $\mathbf{x}_0$ to $\mathbf{p}$ on the normal vector $\mathbf{n}$. See Fig. 1.4. The length of this projection

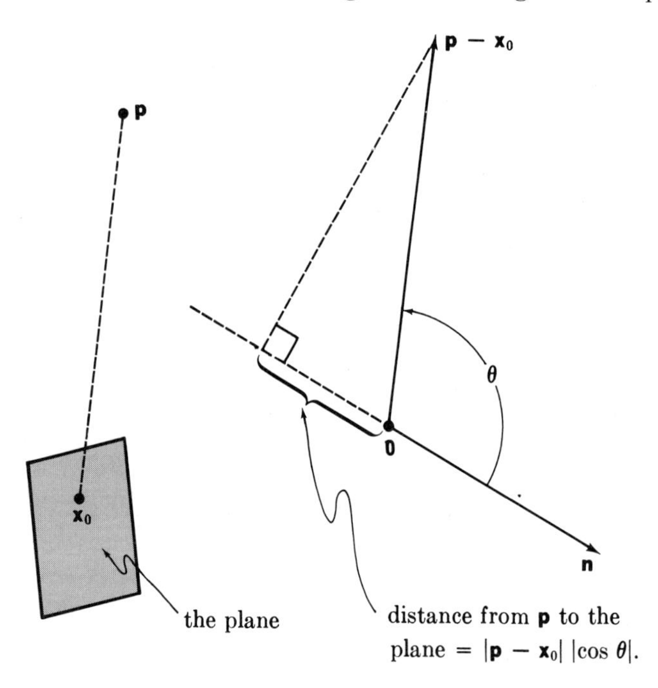

FIG. 1.4

is the absolute value of $|\mathbf{p} - \mathbf{x}_0| \cos \theta$. But

$$|\mathbf{p} - \mathbf{x}_0| \cos \theta = (\mathbf{p} - \mathbf{x}_0) \cdot \frac{\mathbf{n}}{|\mathbf{n}|}$$

$$= (-1, 2, 4) \cdot \frac{1}{\sqrt{29}} (3, 4, -2) = \frac{-3}{\sqrt{29}}.$$

Answer: $\dfrac{3}{\sqrt{29}}$.

EXERCISES

Find the equation of the plane through $\mathbf{x}_0$ and perpendicular to $\mathbf{n}_0$:

1. $\mathbf{x}_0 = (1, 0, 0)$ $\mathbf{n}_0 = (0, 1, 0)$
2. $\mathbf{x}_0 = \mathbf{n}_0 = (1, 1, 1)$
3. $\mathbf{x}_0 = (2, -5, -3)$, $\mathbf{n}_0 = (1, -1, 2)$
4. $\mathbf{x}_0 = (1, 3, 0)$, $\mathbf{n}_0 = (-2, -2, 1)$.

Find a normal $\mathbf{n}$ to the plane:

5. $x - y - z = 0$
6. $x - y - z = 1$
7. $x + 2z = 4$
8. $-x - y + 3z = 4$.
9. Find the equation of the plane through the origin parallel to $x - y + 3z = 1$.
10. Find the distance from $(1, 0, 0)$ to the plane $2x - 2y + 3z = 4$.
11. Find the distance from $(-1, 1, -1)$ to the plane $-3x - y + 4z = 1$.
12. Find a unit normal to the plane through $(1, 0, 0)$, $(0, -2, 0)$, and $(0, 0, 3)$.
13. Find the equation of the plane through $(-1, 0, 0)$, $(0, -1, 0)$, and $(0, 0, -1)$.
14. Find the equation of the plane through $(1, 2, 0)$, $(-1, 0, 3)$, and $(0, 1, 1)$.

2. CHAIN RULE

Before beginning this section, please review Chapter 13, the introduction to functions of several variables and partial derivatives, and please review the Chain Rule in Chapter 15, Section 4, p. 236.

The Chain Rule for functions of one variable gives the derivative of a composite function: if $y = f(x)$ where $x = x(t)$, then

$$\frac{dy}{dt} = \frac{dy}{dx}\frac{dx}{dt}.$$

There is a corresponding chain rule for functions of several variables. Suppose $z = f(x, y)$ where $x = x(t)$ and $y = y(t)$. Thus z is indirectly a function of t. The Chain Rule asserts that

$$\frac{dz}{dt} = \frac{\partial z}{\partial x}\frac{dx}{dt} + \frac{\partial z}{\partial y}\frac{dy}{dt}.$$

This Chain Rule and the general theory of several variables can be better understood by thinking in terms of vectors. A function $z = f(x, y)$ of two variables may be viewed as a function $z = f(\mathbf{x})$ of a single vector variable, $\mathbf{x} = (x, y)$. To each vector $\mathbf{x}$, it assigns a number z. See Fig. 2.1.

In the Chain Rule stated above, the composite function $z(t) = f[x(t), y(t)]$ can be thought of as $z(t) = f[\mathbf{x}(t)]$. If t is time, then $\mathbf{x}(t)$ represents the path of a moving particle in the plane, and the composite function $f[\mathbf{x}(t)]$ assigns a number z to each value of t. The Chain Rule is a formula for the rate of change of $f[\mathbf{x}(t)]$ with respect to t. For instance,

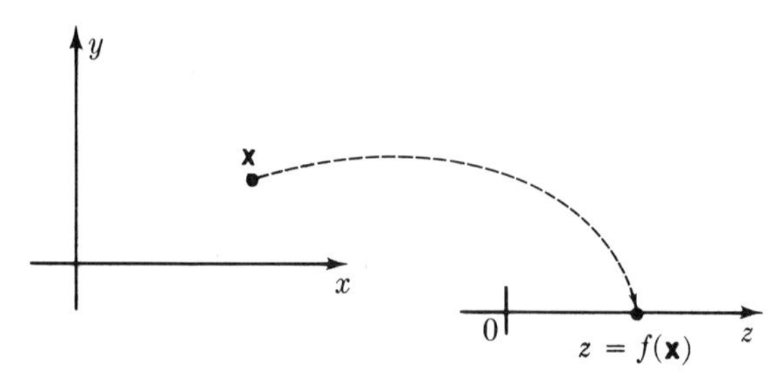

FIG. 2.1

$f(\mathbf{x})$ might be the temperature at position $\mathbf{x}$. Then the Chain Rule tells how fast the temperature is changing as the particle moves along the curve $\mathbf{x}(t)$.

Chain Rule Let $z = f(x, y)$, where $x = x(t)$ and $y = y(t)$. Then

$$\frac{dz}{dt} = \frac{\partial z}{\partial x}\frac{dx}{dt} + \frac{\partial z}{\partial y}\frac{dy}{dt},$$

where $\dfrac{\partial z}{\partial x}$ and $\dfrac{\partial z}{\partial y}$ are evaluated at $(x(t), y(t))$. In briefer notation,

$$\dot{z} = f_x\dot{x} + f_y\dot{y}.$$

In terms of vectors, $z = f[\mathbf{x}(t)]$,

$$\frac{dz}{dt} = f_x[\mathbf{x}(t)]\dot{x}(t) + f_y[\mathbf{x}(t)]\dot{y}(t).$$

Similar rules hold for functions of more than two variables. For instance, if $w = f(x, y, z)$, where $x = x(t)$, $y = y(t)$, $z = z(t)$, then

$$\frac{dw}{dt} = \frac{\partial w}{\partial x}\frac{dx}{dt} + \frac{\partial w}{\partial y}\frac{dy}{dt} + \frac{\partial w}{\partial z}\frac{dz}{dt}.$$

We omit the (difficult) proof of the general Chain Rule, but do derive several special cases at the end of this section.

EXAMPLE 2.1

Let $w = f(x, y, z) = xy^2z^3$, where $x = t\cos t$, $y = e^t$, and $z = \ln(t^2 + 2)$. Compute $\dfrac{dw}{dt}$ at $t = 0$.

Solution: There is a direct but tedious way to do the problem. Write

$$w = (t\cos t)e^{2t}[\ln(t^2 + 2)]^3.$$

Differentiate, then set $t = 0$. That's quite a job!

Use of the Chain Rule is much simpler:

$$\dot{w}(0) = \frac{\partial w}{\partial x}\,\dot{x}(0) + \frac{\partial w}{\partial y}\,\dot{y}(0) + \frac{\partial w}{\partial z}\,\dot{z}(0),$$

where the partial derivatives are evaluated at

$$\mathbf{x}_0 = (x(0),\, y(0),\, z(0)) = (0,\, 1,\, \ln 2).$$

Since $w = xy^2z^3$,

$$\frac{\partial w}{\partial x}\bigg|_{\mathbf{x}_0} = y^2z^3\bigg|_{\mathbf{x}_0} = (\ln 2)^3, \qquad \frac{\partial w}{\partial y}\bigg|_{\mathbf{x}_0} = 2xyz^3\bigg|_{\mathbf{x}_0} = 0,$$

$$\frac{\partial w}{\partial z}\bigg|_{\mathbf{x}_0} = 3xy^2z^2\bigg|_{\mathbf{x}_0} = 0,$$

hence

$$\dot{w}(0) = (\ln 2)^3\dot{x}(0) + 0 + 0.$$

But

$$\dot{x}(0) = (\cos t - t \sin t)\bigg|_0 = 1,$$

therefore $\dot{w}(0) = (\ln 2)^3$.

Answer: $(\ln 2)^3$.

Another Version of the Chain Rule

Suppose $z = f(x, y)$, where this time x and y are functions of *two* variables, $x = x(s, t)$ and $y = y(s, t)$. Then indirectly, z is a function of the variables s and t. There is a chain rule for computing $\partial z/\partial s$ and $\partial z/\partial t$:

Chain Rule If $z = f(x, y)$ is a function of two variables x and y, where $x = x(s, t)$ and $y = y(s, t)$, then

$$\begin{cases} \dfrac{\partial z}{\partial s} = \dfrac{\partial z}{\partial x}\dfrac{\partial x}{\partial s} + \dfrac{\partial z}{\partial y}\dfrac{\partial y}{\partial s} \\[2ex] \dfrac{\partial z}{\partial t} = \dfrac{\partial z}{\partial x}\dfrac{\partial x}{\partial t} + \dfrac{\partial z}{\partial y}\dfrac{\partial y}{\partial t}, \end{cases}$$

where $\dfrac{\partial z}{\partial x}$ and $\dfrac{\partial z}{\partial y}$ are evaluated at $(x(s, t),\, y(s, t))$.

This Chain Rule is a consequence of the previous one. For instance, to compute $\partial z/\partial s$, hold t fixed, making $x(s, t)$ and $y(s, t)$ effectively functions of the one variable s. Then apply the previous Chain Rule.

EXAMPLE 2.2

Let $w = x^2y$, where $x = s^2 + t^2$ and $y = \cos st$. Compute $\dfrac{\partial w}{\partial s}$.

Solution:

$$\begin{aligned}\frac{\partial w}{\partial s} &= \frac{\partial w}{\partial x}\frac{\partial x}{\partial s} + \frac{\partial w}{\partial y}\frac{\partial y}{\partial s}\\ &= (2xy)2s + x^2(-t \sin st)\\ &= 2(s^2+t^2)(\cos st)2s + (s^2+t^2)^2(-t\sin st).\end{aligned}$$

Answer: $(s^2+t^2)[4s \cos st - t(s^2+t^2)\sin st]$.

The next example is important in physical applications.

EXAMPLE 2.3

If $w = f(x, y)$, where $x = r\cos\theta$ and $y = r\sin\theta$, show that

$$\left(\frac{\partial w}{\partial x}\right)^2 + \left(\frac{\partial w}{\partial y}\right)^2 = \left(\frac{\partial w}{\partial r}\right)^2 + \frac{1}{r^2}\left(\frac{\partial w}{\partial \theta}\right)^2.$$

Solution: Use the Chain Rule to compute $\partial w/\partial r$ and $\partial w/\partial\theta$:

$$\frac{\partial w}{\partial r} = \frac{\partial w}{\partial x}\frac{\partial x}{\partial r} + \frac{\partial w}{\partial y}\frac{\partial y}{\partial r} = \frac{\partial w}{\partial x}\cos\theta + \frac{\partial w}{\partial y}\sin\theta;$$

$$\begin{aligned}\frac{\partial w}{\partial \theta} &= \frac{\partial w}{\partial x}\frac{\partial x}{\partial \theta} + \frac{\partial w}{\partial y}\frac{\partial y}{\partial \theta} = \frac{\partial w}{\partial x}(-r\sin\theta) + \frac{\partial w}{\partial y}r\cos\theta\\ &= r\left(-\frac{\partial w}{\partial x}\sin\theta + \frac{\partial w}{\partial y}\cos\theta\right).\end{aligned}$$

From these formulas

$$\left(\frac{\partial w}{\partial r}\right)^2 = \left(\frac{\partial w}{\partial x}\right)^2\cos^2\theta + 2\frac{\partial w}{\partial x}\frac{\partial w}{\partial y}\sin\theta\cos\theta + \left(\frac{\partial w}{\partial y}\right)^2\sin^2\theta,$$

$$\frac{1}{r^2}\left(\frac{\partial w}{\partial \theta}\right)^2 = \left(\frac{\partial w}{\partial x}\right)^2\sin^2\theta - 2\frac{\partial w}{\partial x}\frac{\partial w}{\partial y}\sin\theta\cos\theta + \left(\frac{\partial w}{\partial y}\right)^2\cos^2\theta.$$

Add:

$$\begin{aligned}\left(\frac{\partial w}{\partial r}\right)^2 + \frac{1}{r^2}\left(\frac{\partial w}{\partial \theta}\right)^2 &= \left[\left(\frac{\partial w}{\partial x}\right)^2 + \left(\frac{\partial w}{\partial y}\right)^2\right](\cos^2\theta + \sin^2\theta)\\ &= \left(\frac{\partial w}{\partial x}\right)^2 + \left(\frac{\partial w}{\partial y}\right)^2.\end{aligned}$$

Special Cases

We shall verify the Chain Rule

$$\frac{dz}{dt} = \frac{\partial z}{\partial x}\frac{dx}{dt} + \frac{\partial z}{\partial y}\frac{dy}{dt}$$

for several special cases of the function $z = f(x, y)$.

(1) $z = f(x, y)$ depends only on x. Then $z = f(x)$ and the Chain Rule reduces to

$$\frac{dz}{dt} = \frac{dz}{dx}\frac{dx}{dt},$$

nothing but the old Chain Rule for functions of one variable.

(2) $z = f(x, y) = g(x)h(y)$, the product of a function of x alone by a function of y alone. Then

$$\frac{\partial z}{\partial x} = \frac{dg}{dx}h(y) \qquad \text{and} \qquad \frac{\partial z}{\partial y} = g(x)\frac{dh}{dy},$$

so the Chain Rule reduces to

$$\frac{dz}{dt} = \frac{dg}{dx}h(y)\frac{dx}{dt} + g(x)\frac{dh}{dy}\frac{dy}{dt}.$$

By the one-variable Chain Rule,

$$\frac{dg}{dx}\frac{dx}{dt} = \frac{dg}{dt} \qquad \text{and} \qquad \frac{dh}{dy}\frac{dy}{dt} = \frac{dh}{dt}.$$

Hence

$$\frac{dz}{dt} = \frac{dg}{dt}h(y) + g(x)\frac{dh}{dt},$$

nothing but the Product Rule applied to

$$z(t) = g[x(t)] \cdot h[y(t)].$$

(3) $z = f(x, y) = g(x, y) + h(x, y)$, where the Chain Rule has been proved for $g(x, y)$ and $h(x, y)$. Then the Chain Rule for $f(x, y)$ follows easily.

By combining (2) and (3), the Chain Rule follows for a large class of functions, such as

$$x^2y^3, \qquad Ax^2 + Bxy + Cxy^2 + Dx^2y^2 + Exy^3, \qquad e^x \cos y, \qquad \frac{e^{x+y}}{1 + x^2}.$$

EXERCISES

Find dz/dt by the Chain Rule:

1. $z = e^{xy}$; $x = 3t + 1$, $y = t^2$
2. $z = x/y$; $x = t + 1$, $y = t - 1$
3. $z = x^2 \cos y - x$; $x = t^2$, $y = 1/t$
4. $z = x/y$; $x = \cos t$, $y = 1 + t^2$.

Find dw/dt by the Chain Rule:

5. $w = xyz$; $x = t^2$, $y = t^3$, $z = t^4$
6. $w = e^x \cos(y + z)$; $x = 1/t$, $y = t^2$, $z = -t$

7. $w = e^{-x}y^2 \sin z; \quad x = t,\ y = 2t,\ z = 4t$

8. $w = (e^{-x} \sec z)/y^2; \quad x = t^2,\ y = 1 + t,\ z = t^3.$

Find $\partial z/\partial s$ and $\partial z/\partial t$ by the Chain Rule:

9. $z = x^3/y^2; \quad x = s^2 - t,\ y = 2st$

10. $z = (x + y^2)^4; \quad x = se^t,\ y = se^{-t}$

11. $z = \sqrt{1 + x^2 + y^2}; \quad x = st^2,\ y = 1 + st$

12. $z = e^{x^2y}; \quad x = \dfrac{s}{\sqrt{1 + t^2}},\ y = st.$

13. The radius r and height h of a conical tank increase at rates $\dot{r} = 0.3$ in./hr and $\dot{h} = 0.5$ in./hr. Find the rate of increase $\dot{V}$ of the volume when $r = 6$ ft and $h = 30$ ft.

14. Suppose the Chain Rule has been verified for $f(x, y)$ and $g(x, y)$. Show it is true for the product $f(x, y)g(x, y)$.

15. Suppose the Chain Rule has been verified for $g(x, y)$. Show it is true for the reciprocal $1/g(x, y)$ provided $g(x, y) \neq 0$.

A function $w = f(x, y, z)$ is **homogeneous of degree *n*** if $f(tx, ty, tz) = t^n f(x, y, z)$ for all $t > 0$. The condition of homogeneity can be written vectorially:

$$f(t\mathbf{x}) = t^n f(\mathbf{x}).$$

Show that the function is homogeneous: What degree?

16. $x^2 + yz$

17. $x - y + 2z$

18. $x^3 + y^3 + z^3 - 3xyz$

19. $x^2 e^{-y/z}$

20. $\dfrac{xyz}{x^4 + y^4 + z^4}$

21. $\dfrac{1}{x + y}\cdot$

22. Suppose f and g are homogeneous of degree m and n respectively. Show that fg is homogeneous of degree mn.

23. Let $f(x, y, z)$ be homogeneous of degree n. Show that f_x is homogeneous of degree $n - 1$. (Exception: $n = 0$ and f constant.)

24. Let $f(x, y, z)$ be homogeneous of degree n. Prove **Euler's Relation:** $xf_x + yf_y + zf_z = nf$.
(*Hint:* Differentiate $f(tx, ty, tz) = t^n f(x, y, z)$ with respect to t, using the Chain Rule; then set $t = 1$.)

25. Verify Euler's Relation for the functions in Exs. 18 and 19.

26. The right-hand side of the differential equation

$$\frac{dy}{dx} = \frac{y}{x + y}$$

is homogeneous of degree 0. Make the substitution $y(x) = xu(x)$ and show that the resulting equation for $u(x)$ is

$$x\frac{du}{dx} = \frac{-u^2}{1 + u}\cdot$$

Separate variables and solve.

27. Solve by the method of Ex. 26:

$$\frac{dy}{dx} = \frac{x^2 + y^2}{2xy} + \frac{y}{x}.$$

28. Prove the Chain Rule in the special case of a linear function of x and y, where x and y are linear functions of t.

29. Suppose $F(x, y, z, t)$ is a function of 4 variables. Suppose $x = x(t)$, $y = y(t)$, $z = z(t)$, and $f(t)$ is the composite function

$$f(t) = F[x(t), y(t), z(t), t].$$

Show that

$$\dot{f} = \frac{\partial F}{\partial x}\dot{x} + \frac{\partial F}{\partial y}\dot{y} + \frac{\partial F}{\partial z}\dot{z} + \frac{\partial F}{\partial t}.$$

3. TANGENT PLANE

The graph of

$$z = f(x, y)$$

is a surface. Given a point $\mathbf{x}_0 = (x_0, y_0, z_0)$ on this surface, we are going to describe the plane tangent to the surface at $\mathbf{x}_0$.

> Let $z = f(x, y)$ represent a surface and let $\mathbf{x}_0 = (x_0, y_0, z_0)$ be a fixed point on it. Consider all curves lying on the surface and passing through $\mathbf{x}_0$. Their velocity vectors fill out a plane through the origin. The parallel plane through $\mathbf{x}_0$ is called the **tangent plane** at $\mathbf{x}_0$.

Let us see why this is so. Any curve on the surface (Fig. 3.1) is given by

$$\mathbf{x}(t) = (x(t), y(t), z(t)) = (x(t), y(t), f[x(t), y(t)]).$$

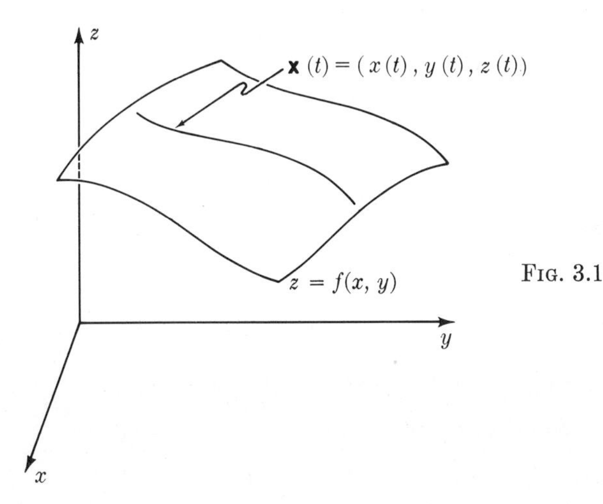

FIG. 3.1

Its velocity vector is found by the Chain Rule:

$$\mathbf{v}(t) = \dot{\mathbf{x}}(t) = (\dot{x}(t), \dot{y}(t), f_x \dot{x} + f_y \dot{y}).$$

Let us assume time is measured so that $\mathbf{x}(0) = \mathbf{x}_0$, that is,

$$x(0) = x_0, \qquad y(0) = y_0, \qquad z(0) = z_0.$$

Then

$$\begin{aligned}\mathbf{v}(0) &= (\dot{x}(0), \dot{y}(0), f_x(x_0, y_0)\dot{x}(0) + f_y(x_0, y_0)\dot{y}(0)) \\ &= \dot{x}(0)(1, 0, f_x(x_0, y_0)) + \dot{y}(0)(0, 1, f_y(x_0, y_0)).\end{aligned}$$

The vectors

$$\mathbf{w}_1 = (1, 0, f_x(x_0, y_0)) \qquad \text{and} \qquad \mathbf{w}_2 = (0, 1, f_y(x_0, y_0))$$

depend only on the function f and the point $\mathbf{x}_0$, not on the particular curve (Fig. 3.2).

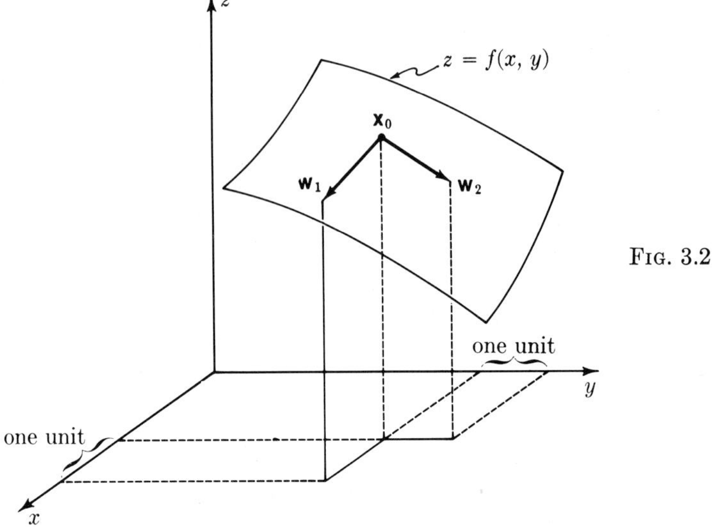

FIG. 3.2

> Suppose a curve lies on the surface $z = f(x, y)$ and passes through
>
> $$\mathbf{x}_0 = (x_0, y_0, z_0).$$
>
> Then its velocity vector has the form
>
> $$\mathbf{v} = a\mathbf{w}_1 + b\mathbf{w}_2,$$
>
> where
>
> $$\mathbf{w}_1 = (1, 0, f_x(x_0, y_0)),$$
> $$\mathbf{w}_2 = (0, 1, f_y(x_0, y_0)),$$
>
> and a and b are constants, $a = \dot{x}(0)$ and $b = \dot{y}(0)$.

Given any pair of numbers a and b, there are curves $\mathbf{x}(t)$ which lie on the surface, pass through $\mathbf{x}_0$ when $t = 0$, and have $\dot{x}(0) = a$ and $\dot{y}(0) = b$. One such is

$$\mathbf{x}(t) = (x_0 + at, y_0 + bt, f(x_0 + at, y_0 + bt)),$$

Its velocity vector is $a\mathbf{w}_1 + b\mathbf{w}_2$. Therefore, all vectors $a\mathbf{w}_1 + b\mathbf{w}_2$ actually occur as velocity vectors. It follows that the velocity vectors fill out a plane through the origin. The parallel plane through $\mathbf{x}_0$ is the tangent plane at $\mathbf{x}_0$. It consists of all points

$$\mathbf{x} = \mathbf{x}_0 + a\mathbf{w}_1 + b\mathbf{w}_2, \qquad a \text{ and } b \quad \text{arbitrary.}$$

EXAMPLE 3.1

Find the tangent plane to $z = x^2 + y$ at $(1, 1, 2)$.

Solution:

$$\mathbf{w}_1 = \left(1, 0, \frac{\partial z}{\partial x}(1, 1)\right) = (1, 0, 2),$$

$$\mathbf{w}_2 = \left(0, 1, \frac{\partial z}{\partial y}(1, 1)\right) = (0, 1, 1).$$

Hence the typical velocity vector is

$$\begin{aligned}\mathbf{v} &= a\mathbf{w}_1 + b\mathbf{w}_2\\ &= a(1, 0, 2) + b(0, 1, 1) = (a, b, 2a + b).\end{aligned}$$

The tangent plane consists of all points

$$(1, 1, 2) + (a, b, 2a + b) = (a + 1, b + 1, 2a + b + 2),$$

where a and b are arbitrary.

Answer: All points $(a + 1, b + 1, 2a + b + 2)$.

REMARK: The typical point on this tangent plane is

$$\mathbf{x} = (x, y, z) = (a + 1, b + 1, 2a + b + 2).$$

Thus

$$a = x - 1, \qquad b = y - 1,$$

and

$$z = 2a + b + 2 = 2(x - 1) + (y - 1) + 2 = 2x + y - 1.$$

Consequently the *equation* of the tangent plane is

$$z = 2x + y - 1.$$

The Normal

The vector

$$\mathbf{n} = (-f_x, -f_y, 1)$$

is perpendicular to both tangent vectors $\mathbf{w}_1$ and $\mathbf{w}_2$:

$$\mathbf{n} \cdot \mathbf{w}_1 = (-f_x, -f_y, 1) \cdot (1, 0, f_x) = -f_x + 0 + f_x = 0,$$

$$\mathbf{n} \cdot \mathbf{w}_2 = (-f_x, -f_y, 1) \cdot (0, 1, f_y) = 0 - f_y + f_y = 0.$$

Consequently $\mathbf{n}$ is perpendicular to each vector $a\mathbf{w}_1 + b\mathbf{w}_2$:

$$\mathbf{n} \cdot (a\mathbf{w}_1 + b\mathbf{w}_2) = a(\mathbf{n} \cdot \mathbf{w}_1) + b(\mathbf{n} \cdot \mathbf{w}_2) = 0.$$

In other words, $\mathbf{n}$ is perpendicular to tangent plane at $\mathbf{x}_0$. Using the formulas of Section 1, we may write the equation of that plane.

The tangent plane to the surface $z = f(x, y)$ at the point $\mathbf{x}_0 = (x_0, y_0, z_0)$ is given by

$$-f_x \cdot (x - x_0) - f_y \cdot (y - y_0) + (z - z_0) = 0,$$

where f_x and f_y are evaluated at (x_0, y_0). The vector form of this equation is

$$(\mathbf{x} - \mathbf{x}_0) \cdot \mathbf{n} = 0, \qquad \mathbf{n} = (-f_x, -f_y, 1).$$

EXAMPLE 3.2

Find an equation for the tangent plane to $z = x^2 + y$ at $(1, 1, 2)$.

Solution: Write $f(x, y) = x^2 + y$.

$$f_x(x, y) = 2x, \qquad f_y(x, y) = 1.$$

At $(1, 1, 2)$,

$$f_x = 2, \qquad f_y = 1.$$

Hence the equation is

$$-2(x - 1) - (y - 1) + (z - 2) = 0.$$

Answer: $2x + y - z = 1$.

The vector $(-f_x, -f_y, 1)$ is perpendicular to the tangent plane and has a positive z-component (it is the upward normal rather than the downward normal). We introduce the unit vector in the same direction.

The **unit normal at $\mathbf{x}_0$ to the surface** $z = f(x, y)$ is the vector

$$\mathbf{N} = \frac{1}{\sqrt{f_x^2 + f_y^2 + 1}} (-f_x, -f_y, 1),$$

where f_x and f_y are evaluated at (x_0, y_0).

EXAMPLE 3.3

Find the unit normal at $(1, 1, 2)$ to the surface $z = x^2 + y$.

Solution:

$$(-f_x, -f_y, 1) = (-2, -1, 1).$$

$$\mathbf{N} = \frac{1}{\sqrt{2^2 + 1^2 + 1^2}}(-2, -2, 1).$$

Answer: $\mathbf{N} = \dfrac{1}{\sqrt{6}}(-2, -1, 1).$

EXAMPLE 3.4

How far is $\mathbf{0}$ from the tangent plane to $z = x^2 + y$ at $(1, 1, 2)$?

Solution: The distance is the length of the projection of $\mathbf{x}_0 = (1, 1, 2)$ on $\mathbf{N}$. See Fig. 3.3.

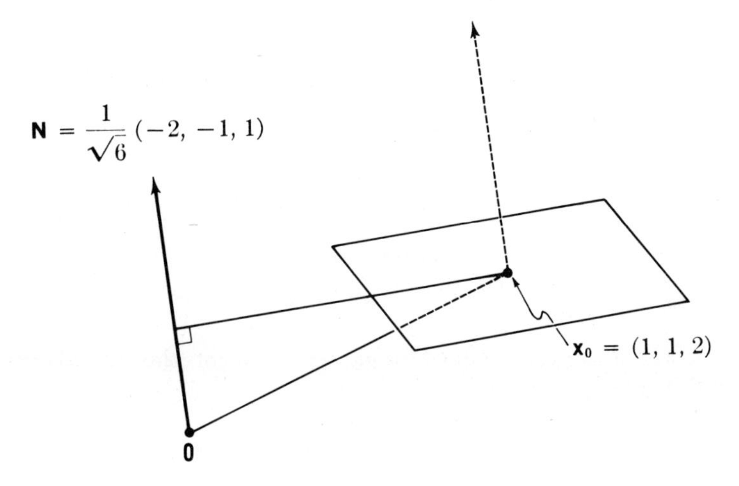

FIG. 3.3

$$\text{Distance} = |\mathbf{x}_0 \cdot \mathbf{N}|$$

$$= \left| (1, 1, 2) \cdot \frac{1}{\sqrt{6}}(-2, -1, 1) \right|$$

$$= \frac{1}{\sqrt{6}}|-2 - 1 + 2| = \frac{1}{\sqrt{6}}.$$

Answer: $\dfrac{1}{\sqrt{6}}.$

EXERCISES

Give the equation $z = ax + by + c$ of the tangent plane to the surface at the indicated point:

1. $z = x^2 - y^2$; $x = 0, y = 0$
2. $z = x^2 - y^2$; $x = 1, y = -1$
3. $z = x^2 + 4y^2$; $x = 2, y = 1$
4. $z = x^2e^y$; $x = -1, y = 2$
5. $z = x^2y + y^3$; $x = -1, y = 2$
6. $z = x \cos y + y \cos x$; $x = 0, y = 0$.

Find the unit normal to the surface at the indicated point:

7. $z = x + x^2y^3 + y$; $x = 0, y = 0$
8. $z = x^3 + y^3$; $x = 1, y = -1$
9. $z = x^2 + xy + y^2$; $x = -1, y = 2$
10. $z = \sqrt{1 - x^2 - y^2}$; $x = \frac{1}{2}, y = \frac{1}{2}$.
11. Show that the tangent plane to the hyperbolic paraboloid $z = x^2 - y^2$ at $(0, 0, 0)$ intersects the surface in a pair of straight lines.
12. (cont.) Show that the conclusion is valid for the tangent plane at *any* point of the surface.
13. (cont.) Show that the property of the tangent planes in Ex. 12 is also valid for the hyperboloid of one sheet $z^2 = x^2 + y^2 - 1$.

4. GRADIENT

The gradient field of a function f on a region is the assignment of a certain vector, called grad f, to each point of that region.

Suppose $z = f(x, y)$ is defined on a region R of the x, y-plane. The **gradient** of f is the vector

$$\text{grad } f = (f_x, f_y).$$

Likewise, if $w = f(x, y, z)$ is defined on a region R of space, the **gradient** of f is the vector

$$\text{grad } f = (f_x, f_y, f_z).$$

The **gradient field** of a function f is the assignment of the vector grad f to each point of the region R.

For example, if

$$f(x, y) = x^2 + y, \qquad \text{grad } f = (2x, 1);$$

if

$$f(x, y, z) = |\mathbf{x}|^2 = x^2 + y^2 + z^2, \qquad \text{grad } f = (2x, 2y, 2z) = 2\mathbf{x}.$$

In this section, we discuss several uses of the gradient. The first of these concerns notation. Certain formulas are simplified if expressed in terms of gradients. An example is the Chain Rule, which asserts

$$\dot{z} = f_x\dot{x} + f_y\dot{y}$$

if $z = f(x, y)$ and $x = x(t)$, $y = y(t)$. But grad $f = (f_x, f_y)$ and $\dot{\mathbf{x}} = (\dot{x}, \dot{y})$. Therefore, in vector notation

$$\dot{z} = (\text{grad } f) \cdot \dot{\mathbf{x}}.$$

Level Curves

Imagine a surface $z = f(x, y)$ above a portion of the x, y-plane. In the plane we can draw a contour map of the surface by indicating curves of constant altitude. These are called **contour lines** or **level curves.** Figure 4.1

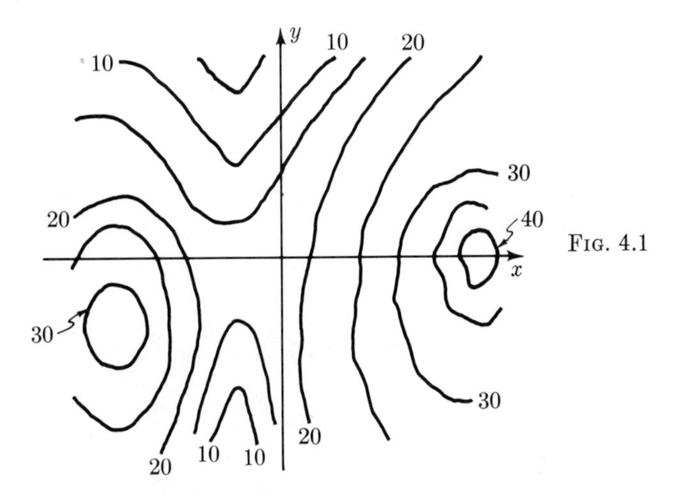

FIG. 4.1

shows a contour map with two hills and a pass between them. Level curves are obtained by slicing the surface $z = f(x, y)$ with planes $z = c$ for various constants c. Each plane $z = c$ intersects the surface in a plane curve (Fig. 4.2). The projection of this curve onto the x, y-plane is the **level curve at level** c. It is the graph of $f(x, y) = c$. Where level curves are close together the surface is steep; where they are far apart it is relatively flat.

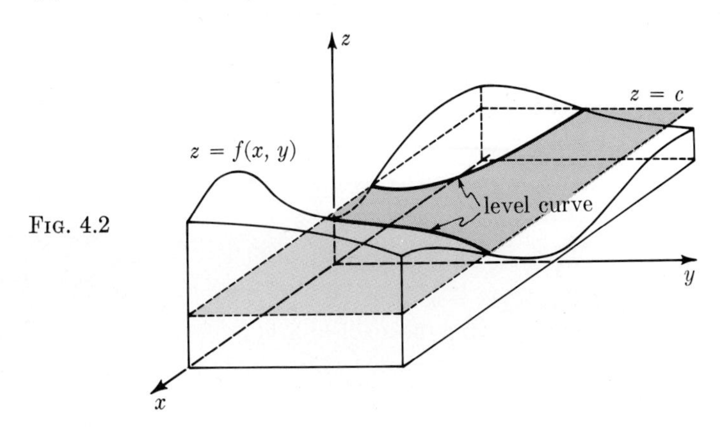

FIG. 4.2

An important relation exists between the gradient field and the level curves of a function.

> The gradient field of f is **orthogonal** (perpendicular) to the level curves of f.

For suppose a particle moves along the level curve $f(x, y) = c$. Let its position at time t be $\mathbf{x}(t) = (x(t), y(t))$. Then

$$f[x(t), y(t)] = c.$$

Therefore the time derivative is zero:

$$f_x\dot{x} + f_y\dot{y} = 0, \qquad (\text{grad } f) \cdot \dot{\mathbf{x}} = 0.$$

Hence grad f is orthogonal to the tangent to the level curve.

Level Surfaces

We cannot graph a function of three variables

$$w = f(x, y, z)$$

(the graph would be four-dimensional). We can, however, learn a good deal about the function by plotting in three-space the **level surfaces**

$$f(x, y, z) = \text{constant}.$$

For example, the level surfaces of

$$f(x, y, z) = x^2 + y^2 + z^2$$

are the spheres

$$x^2 + y^2 + z^2 = c^2$$

centered at the origin.

> The gradient field of f is orthogonal to the level surfaces of f.

The proof of this statement is practically identical to the proof of the analogous statement in two variables.

EXERCISES

Plot the level curves and the gradient field in the region $|x| \leq 3$, $|y| \leq 3$:

1. $z = x - 2y$
2. $z = x^2 - y^2$
3. $z = x^2 + y$
4. $z = x^2 + 4y^2$.

Describe the level surfaces:

5. $w = x + y + z$

6. $w = x^2 + 4y^2 + 9z^2$

7. $w = xyz$

8. $w = x^2 + y^2 - z^2$.

9. For each function in Exs. 5–8, find the gradient field.

10. Suppose $z = f(r, \theta)$ is given in terms of polar coordinates. Show that

$$\operatorname{grad} z = f_r \mathbf{u} + \frac{1}{r} f_\theta \mathbf{w},$$

where $\mathbf{u} = (\cos\theta, \sin\theta)$ and $\mathbf{w} = (-\sin\theta, \cos\theta)$.

11. Find $\operatorname{grad}(r^{-2}\cos 2\theta)$.

5. DIRECTIONAL DERIVATIVE

Given a function $w = f(x, y, z)$ and a point $\mathbf{x}$ in space, we may ask how fast the function is changing at $\mathbf{x}$ in various directions. (A direction is indicated by a unit vector $\mathbf{u}$.)

> The **directional derivative** of $f(x, y, z)$ at a point $\mathbf{x}$ in the direction $\mathbf{u}$ is
>
> $$D_{\mathbf{u}}f(\mathbf{x}) = \frac{d}{dt} f(\mathbf{x} + t\mathbf{u})\Big|_{t=0}.$$

Think of a directional derivative this way. Imagine a particle moving along a straight line with constant velocity $\mathbf{u}$, passing through the point $\mathbf{x}$ when $t = 0$. See Fig. 5.1. To each point $\mathbf{x} + t\mathbf{u}$ of its path is assigned the number

$$w(t) = f(\mathbf{x} + t\mathbf{u}).$$

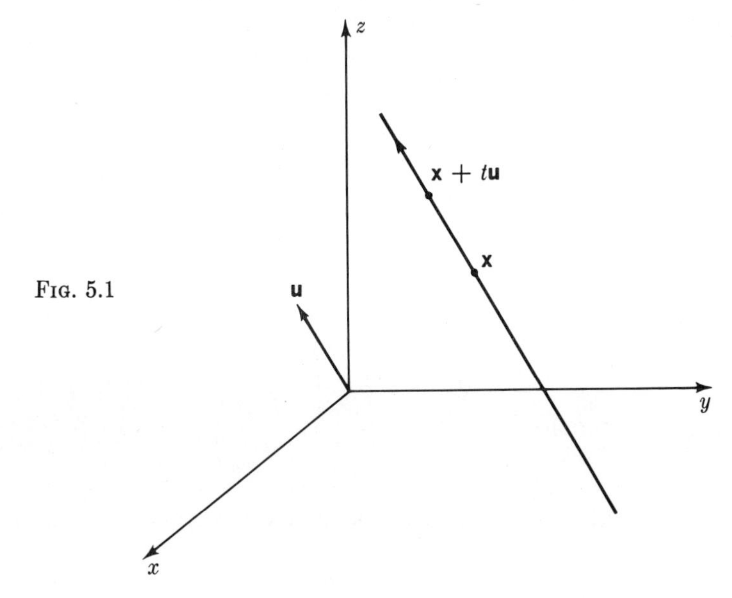

FIG. 5.1

Then

$$D_{\mathbf{u}}f(\mathbf{x}) = w'(0),$$

the rate of change of $w(t)$ as the particle moves through the point $\mathbf{x}$. For example, suppose $f(x, y, z)$ is the steady temperature at each point (x, y, z) of a fluid. Suppose a particle moves with unit speed through a point $\mathbf{x}$ in the direction $\mathbf{u}$. Then $D_{\mathbf{u}}f(\mathbf{x})$ measures the time rate of change of the particle's temperature.

There is a handy formula for directional derivatives. Let

$$\mathbf{p}(t) = \mathbf{x} + t\mathbf{u}.$$

Then $w(t) = f[\mathbf{p}(t)]$, a composite function. By the Chain Rule,

$$\begin{aligned} D_{\mathbf{u}}f(\mathbf{x}) &= \dot{w}(0) = \frac{d}{dt} f[\mathbf{p}(t)]\bigg|_{t=0} \\ &= (\text{grad } f) \cdot \dot{\mathbf{p}}(0) = (\text{grad } f) \cdot \mathbf{u}. \end{aligned}$$

Since $\mathbf{u}$ is a unit vector, (grad f) $\cdot$ $\mathbf{u}$ is simply the projection of grad f on $\mathbf{u}$.

> The derivative of f in the direction $\mathbf{u}$ is the projection of grad f on $\mathbf{u}$:
>
> $$D_{\mathbf{u}}f(\mathbf{x}) = (\text{grad } f) \cdot \mathbf{u}.$$

In particular if $\mathbf{u} = \mathbf{i} = (1, 0, 0)$, then

$$D_{\mathbf{i}}f(\mathbf{x}) = (\text{grad } f) \cdot (1, 0, 0) = \left(\frac{\partial f}{\partial x}, \frac{\partial f}{\partial y}, \frac{\partial f}{\partial z}\right) \cdot (1, 0, 0) = \frac{\partial f}{\partial x}.$$

A similar situation holds for $\mathbf{j} = (0, 1, 0)$ and $\mathbf{k} = (0, 0, 1)$.

> The directional derivatives of $f(x, y, z)$ in the directions $\mathbf{i}$, $\mathbf{j}$, and $\mathbf{k}$ are the partial derivatives:
>
> $$D_{\mathbf{i}}f = \frac{\partial f}{\partial x}, \qquad D_{\mathbf{j}}f = \frac{\partial f}{\partial y}, \qquad D_{\mathbf{k}}f = \frac{\partial f}{\partial z}.$$

EXAMPLE 5.1

Compute the directional derivatives of $f(x, y, z) = xy^2z^3$ at $(3, 2, 1)$, in the direction of the vectors

(a) $(-2, -1, 0)$, (b) $(5, 4, 1)$.

Solution:

$$D_{\mathbf{u}}f(\mathbf{x}) = (\text{grad } f) \cdot \mathbf{u},$$

where $\mathbf{u}$ is a unit vector in the desired direction, and grad f is evaluated at

(3, 2, 1). Since

$$\frac{\partial f}{\partial x} = y^2z^3, \qquad \frac{\partial f}{\partial y} = 2xyz^3, \qquad \text{and} \qquad \frac{\partial f}{\partial z} = 3xy^2z^2,$$

$$\operatorname{grad} f \Big|_{(3,2,1)} = (4, 12, 36).$$

Thus

$$D_{\mathbf{u}}f(\mathbf{x}) = (4, 12, 36) \cdot \mathbf{u}.$$

(a) $\mathbf{u} = \dfrac{1}{\sqrt{5}}(-2, -1, 0)$,

$$D_{\mathbf{u}}f(\mathbf{x}) = (4, 12, 36) \cdot \frac{1}{\sqrt{5}}(-2, -1, 0) = \frac{-20}{\sqrt{5}}.$$

(b) $\mathbf{u} = \dfrac{1}{\sqrt{42}}(5, 4, 1)$,

$$D_{\mathbf{u}}f(\mathbf{x}) = (4, 12, 36) \cdot \frac{1}{\sqrt{42}}(5, 4, 1) = \frac{104}{\sqrt{42}}.$$

Answer: (a) $\dfrac{-20}{\sqrt{5}}$; (b) $\dfrac{104}{\sqrt{42}}$.

Question: In what direction is a given function f increasing fastest? In other words, at a fixed point $\mathbf{x}$ in space, for which unit vector $\mathbf{u}$ is

$$D_{\mathbf{u}}f(\mathbf{x})$$

largest? Now

$$D_{\mathbf{u}}f(\mathbf{x}) = (\operatorname{grad} f) \cdot \mathbf{u} = |\operatorname{grad} f| \cos \theta,$$

where θ is the angle between grad f and $\mathbf{u}$. Therefore, the largest value of $D_{\mathbf{u}}f(\mathbf{x})$ is $|\operatorname{grad} f|$, taken where $\cos \theta = 1$, that is, $\theta = 0$.

The direction of most rapid increase of $f(x, y, z)$ at a point $\mathbf{x}$ is the direction of the gradient. The derivative in that direction is $|\operatorname{grad} f|$.

The direction of most rapid decrease is opposite to the direction of the gradient.

EXAMPLE 5.2

Find the direction of most rapid increase of the function

$$f(x, y, z) = x^2 + yz$$

at (1, 1, 1) and give the rate of increase in this direction.

Solution:

$$\text{grad } f \Big|_{(1,1,1)} = (2x, z, y) \Big|_{(1,1,1)} = (2, 1, 1).$$

The most rapid increase is

$$D_{\mathbf{u}} f = |\text{grad } f| = \sqrt{2^2 + 1^2 + 1^2} = \sqrt{6},$$

where **u** is the direction of grad f:

$$\mathbf{u} = \frac{\text{grad } f}{|\text{grad } f|} = \frac{1}{\sqrt{6}} (2, 1, 1).$$

Answer: $D_{\mathbf{u}} f \Big|_{(1,1,1)} = \sqrt{6}$

for $\mathbf{u} = \dfrac{1}{\sqrt{6}} (2, 1, 1).$

We conclude this section with two examples involving directions of most rapid change of a function.

Consider water running down a hill from a spring at a point P. See Fig. 5.2. The water descends as quickly as possible. What is the path of the stream?

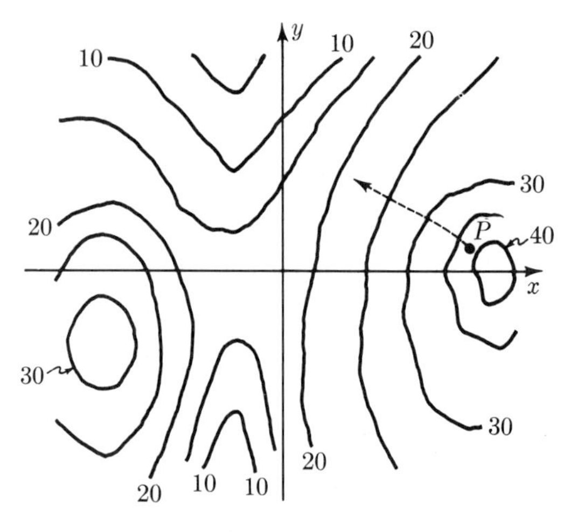

FIG. 5.2

From physics, change in kinetic energy [$\frac{1}{2}$ (speed)^2] equals change in potential energy [height]. Hence the speed of a water particle depends only on how far it has descended (its altitude). Since the speed at a given time does not depend on direction, the particle "chooses" the direction of steepest descent (most rapid change of altitude). Let the hill be represented by the surface $z = f(x, y)$. Then water will flow in the direction of $-$ grad f, that is, perpendicular to the level curves.

Next, consider a function $z = f(x, y)$ and all curves $\mathbf{x}(t) = (x(t), y(t))$ in the x, y-plane which pass through a fixed point $\mathbf{x}_0$ with speed 1. Along which of these is $f(x, y)$ increasing the fastest at $\mathbf{x}_0$?

To find the direction of most rapid increase, write

$$z(t) = f[\mathbf{x}(t)].$$

By the Chain Rule,

$$\dot{z} = (\text{grad } f) \cdot \dot{\mathbf{x}} = (\text{grad } f) \cdot \mathbf{v},$$

where $\mathbf{v}$ is the velocity vector. But $\mathbf{v}$ is a unit vector; hence $\dot{z}$ is the directional derivative in the direction of $\mathbf{v}$ (tangential to the curve $\mathbf{x}(t)$). Thus $\dot{z}$ is greatest for those curves whose tangents at $\mathbf{x}_0$ point in the direction of grad f; such curves are orthogonal to the level curves.

EXERCISES

Find the directional derivative of $f(x, y, z)$ at $\mathbf{x}_0$ in the directions of $\mathbf{v}_1$, $\mathbf{v}_2$, and $\mathbf{v}_3$:

1. $f = x + y + z$; $\mathbf{x}_0 = \mathbf{0}$, $\mathbf{v}_1 = (1, 0, 0)$, $\mathbf{v}_2 = (0, 1, 0)$, $\mathbf{v}_3 = (0, 0, 1)$
2. $f = xy + yz + zx$; $\mathbf{x}_0 = (1, 1, 1)$, $\mathbf{v}_1 = (1, 1, 1)$, $\mathbf{v}_2 = (1, -1, 1)$, $\mathbf{v}_3 = (-1, -1, -1)$
3. $f = xyz$; $\mathbf{x}_0 = (1, -1, 2)$, $\mathbf{v}_1 = (1, 1, 0)$, $\mathbf{v}_2 = (1, 0, 1)$, $\mathbf{v}_3 = (0, 1, 1)$
4. $f = x^2y^2z^2$; $\mathbf{x}_0 = (-1, -1, -1)$, $\mathbf{v}_1 = (1, 2, 3)$, $\mathbf{v}_2 = (1, 1, 0)$, $\mathbf{v}_3 = (3, 2, 1)$.

Find the largest directional derivative of $f(x, y, z)$ at $\mathbf{x}_0$:

5. $f = x^3 + y^2 + z$; $\mathbf{x}_0 = \mathbf{0}$
6. $f = xyz$; $\mathbf{x}_0 = (-1, -1, -1)$
7. $f = x^2 + y^2 + z^2$; $\mathbf{x}_0 = (1, 2, 2)$
8. $f = x^2 - y^2 + 4z^2$; $\mathbf{x}_0 = (-1, -1, 1)$.

6. APPLICATIONS

A **vector field** is the assignment of a vector $\mathbf{F}(\mathbf{x})$ to each point $\mathbf{x}$ of a region R in space. (The gradient of a function is one example.)

Let $\mathbf{F}(\mathbf{x})$ be a vector field on R and suppose $\mathbf{x}(t)$ is a path in R from $\mathbf{x}_0 = \mathbf{x}(t_0)$ to $\mathbf{x}_1 = \mathbf{x}(t_1)$. The line integral

$$\int_{\mathbf{x}_0}^{\mathbf{x}_1} \mathbf{F} \cdot d\mathbf{x}$$

is defined over this path and is computed by the ordinary integral

$$\int_{t_0}^{t_1} \mathbf{F}[\mathbf{x}(t)] \cdot \dot{\mathbf{x}}(t)\, dt.$$

Its value generally depends on the path $\mathbf{x}(t)$ connecting $\mathbf{x}_0$ and $\mathbf{x}_1$.

EXAMPLE 6.1

Let $\mathbf{F}(\mathbf{x})$ be the vector field $\mathbf{F}(\mathbf{x}) = (x, y, x + y + z)$, $\mathbf{x} = (x, y, z)$. Let $\mathbf{x}_0 = (0, 0, 0)$ and $\mathbf{x}_1 = (1, 1, 1)$. Compute the line integral $\int_{\mathbf{x}_0}^{\mathbf{x}_1} \mathbf{F} \cdot d\mathbf{x}$ over the paths

$$\text{(a)} \quad \mathbf{x}(t) = (t, t, t), \qquad \text{(b)} \quad \mathbf{x}(t) = (t, t^2, t^3).$$

Solution: Notice that for both paths, $\mathbf{x}_0 = \mathbf{x}(0)$ and $\mathbf{x}_1 = \mathbf{x}(1)$.

(a) On this path $\mathbf{x} = (t, t, t)$ and $\dot{\mathbf{x}}(t) = (1, 1, 1)$.

$$\int_{\mathbf{x}_0}^{\mathbf{x}_1} \mathbf{F} \cdot d\mathbf{x} = \int_0^1 \mathbf{F}[\mathbf{x}(t)] \cdot \dot{\mathbf{x}}(t)\, dt$$

$$= \int_0^1 (t, t, 3t) \cdot (1, 1, 1)\, dt$$

$$= \int_0^1 5t\, dt = \frac{5}{2}.$$

(b) This time $\mathbf{x} = (t, t^2, t^3)$ and $\dot{\mathbf{x}}(t) = (1, 2t, 3t^2)$.

$$\int_{\mathbf{x}_0}^{\mathbf{x}_1} \mathbf{F} \cdot d\mathbf{x} = \int_0^1 (t, t^2, t + t^2 + t^3) \cdot (1, 2t, 3t^2)\, dt$$

$$= \int_0^1 [t + 2t^3 + 3(t^3 + t^4 + t^5)]\, dt$$

$$= \int_0^1 (t + 5t^3 + 3t^4 + 3t^5)\, dt = \frac{1}{2} + \frac{5}{4} + \frac{3}{5} + \frac{1}{2} = \frac{57}{20}.$$

Answer: (a) $\frac{5}{2}$; (b) $\frac{57}{20}$.

In an important special case, the line integral does not depend on the path $\mathbf{x}(t)$, but only on the initial and terminal points $\mathbf{x}_0$ and $\mathbf{x}_1$:

If the vector field $\mathbf{F}(\mathbf{x})$ is the gradient of some function f,

$$\mathbf{F} = \operatorname{grad} f,$$

then

$$\int_{\mathbf{x}_0}^{\mathbf{x}_1} \mathbf{F} \cdot d\mathbf{x} = f(\mathbf{x}_1) - f(\mathbf{x}_0).$$

Thus the value of the line integral is independent of the path connecting $\mathbf{x}_0$ and $\mathbf{x}_1$.

This assertion is easily verified by means of the Chain Rule,

$$\frac{d}{dt} f[\mathbf{x}(t)] = (\operatorname{grad} f) \cdot \dot{\mathbf{x}}.$$

If $\mathbf{F} = \text{grad } f$, then

$$\int_{\mathbf{x}_0}^{\mathbf{x}_1} \mathbf{F} \cdot d\mathbf{x} = \int_{t_0}^{t_1} (\text{grad } f) \cdot \dot{\mathbf{x}}(t)\, dt$$

$$= \int_{t_0}^{t_1} \frac{d}{dt} f[\mathbf{x}(t)]\, dt = f[\mathbf{x}(t)] \Big|_{t_0}^{t_1} = f(\mathbf{x}_1) - f(\mathbf{x}_0).$$

REMARK: Not every vector field is the gradient of some function. For example, the field $\mathbf{F}(\mathbf{x}) = (x, y, x + y + z)$ is not a gradient;

$$\int_{\mathbf{x}_0}^{\mathbf{x}_1} \mathbf{F} \cdot d\mathbf{x}$$

is *not* independent of the path (see Example 6.1). When a vector field is a gradient will be discussed further in Chapter 33, Section 4, p. 801.

Application to Physics

If $\mathbf{F} = \text{grad } f$ is a force, the net work done by this force in moving a particle from $\mathbf{x}_0$ to $\mathbf{x}_1$ is $f(\mathbf{x}_1) - f(\mathbf{x}_0)$, independent of the path. The function f, which is unique up to an additive constant, is called the **potential** of the force.

An important example is that of a central force subject to the inverse square law, for instance, the electric force $\mathbf{E}$ on a unit charge at $\mathbf{x}$ due to a unit charge of the same sign at the origin. The magnitude of the vector $\mathbf{E}$ is inversely proportional to $|\mathbf{x}|^2$. Its direction is the same as that of $\mathbf{x}$. See Fig. 6.1. The unit vector in the direction of $\mathbf{x}$ is

$$\frac{\mathbf{x}}{|\mathbf{x}|} = \frac{\mathbf{x}}{\rho}, \qquad \rho = |\mathbf{x}|.$$

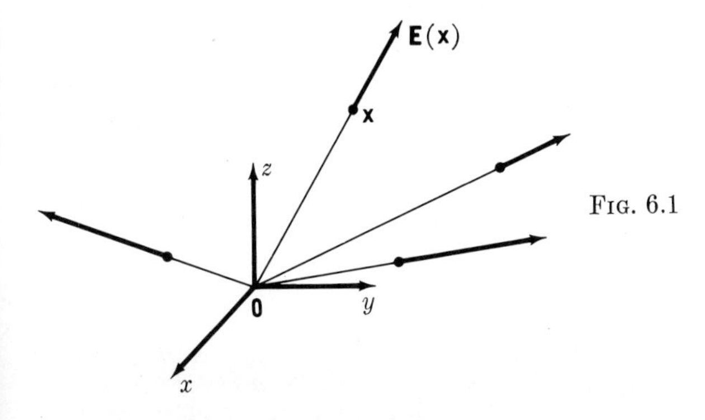

FIG. 6.1

Therefore, expressed in suitable units,

$$\mathbf{E} = \frac{1}{\rho^2} \frac{\mathbf{x}}{\rho} = \frac{\mathbf{x}}{\rho^3}.$$

The force field $\mathbf{E}$ is defined at all points of space except the origin. We shall prove that $\mathbf{E}$ is the gradient of a function, in fact that

$$\mathbf{E} = \text{grad } f, \qquad \text{where} \quad f(x, y, z) = -\frac{1}{\rho} = \frac{-1}{\sqrt{x^2 + y^2 + z^2}}.$$

Let us compute the gradient of $f = -1/\rho$:

$$\frac{\partial f}{\partial x} = \frac{x}{(x^2 + y^2 + z^2)^{3/2}} = \frac{x}{\rho^3},$$

and similarly

$$\frac{\partial f}{\partial y} = \frac{y}{\rho^3}, \qquad \frac{\partial f}{\partial z} = \frac{z}{\rho^3}.$$

Therefore

$$\text{grad } f = \left(\frac{x}{\rho^3}, \frac{y}{\rho^3}, \frac{z}{\rho^3}\right) = \frac{1}{\rho^3}(x, y, z)$$

$$= \frac{\mathbf{x}}{\rho^3} = \mathbf{E}.$$

Since $\mathbf{E} = \text{grad } f$, it follows from our discussion of line integrals that

$$\int_{\mathbf{x}_0}^{\mathbf{x}_1} \mathbf{E} \cdot d\mathbf{x} = f(\mathbf{x}_1) - f(\mathbf{x}_0) = \frac{1}{|\mathbf{x}_0|} - \frac{1}{|\mathbf{x}_1|}.$$

The right-hand side is the **potential difference** or **voltage.** It represents the work done by the electric force when a unit charge moves from $\mathbf{x}_0$ to $\mathbf{x}_1$ *along any path.*

If $\mathbf{x}_1$ is far out, then $1/|\mathbf{x}_1|$ is small, so

$$\int_{\mathbf{x}_0}^{\mathbf{x}_1} \mathbf{E} \cdot d\mathbf{x} \approx \frac{1}{|\mathbf{x}_0|}.$$

As $\mathbf{x}_1$ moves farther out, the approximation improves. In mathematical shorthand, write

$$\int_{\mathbf{x}_0}^{\mathbf{x}_1} \mathbf{E} \cdot d\mathbf{x} \longrightarrow \frac{1}{|\mathbf{x}_0|} \qquad \text{as} \quad |\mathbf{x}_1| \longrightarrow \infty,$$

or

$$\int_{\mathbf{x}_0}^{\infty} \mathbf{E} \cdot d\mathbf{x} = \frac{1}{|\mathbf{x}_0|}.$$

Physical conservation laws are usually derived by identifying an appropriate vector field with the gradient of a function and then evaluating a line integral.

EXERCISES

1. Compute $\int_{(0,1)}^{(0,-1)} (xy, 1 + 2y) \cdot d\mathbf{x}$
   (a) along the straight path,
   (b) along the semicircular path passing through $(-1, 0)$.
2. Let $\mathbf{F} = (3x^2y^2z, 2x^3yz, x^3y^2)$. Show that $\int_{(0,0,0)}^{(1,1,1)} \mathbf{F} \cdot d\mathbf{x}$ is independent of the path, and evaluate it.
3. Let $\mathbf{F} = (x^2 + yz, y^2 + zx, z^2 + xy)$. Show that $\int_{(0,0,0)}^{(a,b,c)} \mathbf{F} \cdot d\mathbf{x}$ is independent of the path, and evaluate it.
4. Let θ denote the polar angle in the plane. Show that grad $\theta = \left(\dfrac{-y}{x^2 + y^2}, \dfrac{x}{x^2 + y^2}\right)$.
5. Find $\int \dfrac{-y\,dx + x\,dy}{x^2 + y^2}$ over the circle $|\mathbf{x}| = a$. Explain why the answer does not have to be 0, and why it is independent of the radius a.
6. Let $\mathbf{F} = \mathbf{x}/|\mathbf{x}|^5$ and suppose $\mathbf{a} \neq \mathbf{0}$. Show that $\int_{\mathbf{a}}^{\infty} \mathbf{F} \cdot d\mathbf{x}$, taken along any path from $\mathbf{a}$ which does not pass through $\mathbf{0}$ and goes out indefinitely, depends only on $a = |\mathbf{a}|$. Evaluate the integral.
7. (cont.) Do the same for $\mathbf{F} = \mathbf{x}/|\mathbf{x}|^n$, for any $n > 2$,
8. (cont.) Show that $\mathbf{F} = \mathbf{x}/|\mathbf{x}|^2$ is a gradient.

7. IMPLICIT FUNCTIONS

Often a function $y = g(x)$ is defined only as the root of an equation

$$f(x, y) = 0,$$

which may be hard or impossible to solve explicitly. In such a case, the equation is said to define an **implicit function** $y = g(x)$. For example, Fig. 7.1 shows part of the graph of

$$y^6 + y + xy - x = 0.$$

Near the origin, this equation defines y as an implicit function of x. (It is hopeless to express y as an *explicit* function of x.)

What is the derivative of an implicit function $y = g(x)$ defined by $f(x, y) = 0$? Substitute $y = g(x)$:

$$f[x, g(x)] = 0.$$

Differentiate with respect to x using the Chain Rule:

$$f_x + f_y \cdot g' = 0, \qquad g' = -\frac{f_x}{f_y}.$$

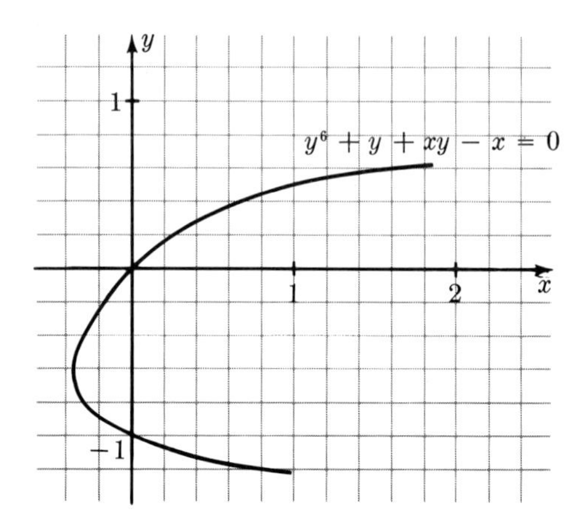

FIG. 7.1

> If y is an implicit function of x defined by
> $$f(x, y) = 0,$$
> then
> $$\frac{dy}{dx} = -\frac{f_x(x, y)}{f_y(x, y)}$$
> at each point (x, y) where $f(x, y) = 0$ and $f_y(x, y) \neq 0$.

Differentiation of implicit functions is called **implicit differentiation.**

EXAMPLE 7.1

Find $\left.\frac{dy}{dx}\right|_{(0,0)}$ where $y = g(x)$ is defined by $y^6 + y + xy - x = 0$.

Solution:

$$f(x, y) = y^6 + y + xy - x,$$

$$f_x = y - 1, \qquad f_y = 6y^5 + 1 + x.$$

Hence

$$\frac{dy}{dx} = -\frac{f_x}{f_y} = -\frac{y - 1}{6y^5 + 1 + x}.$$

At (0, 0),

$$\frac{dy}{dx} = -\frac{-1}{1} = 1.$$

Alternate Solution: Differentiate the equation

$$y^6 + y + xy - x = 0,$$

treating y as a function of x:

$$6y^5 \frac{dy}{dx} + \frac{dy}{dx} + x\frac{dy}{dx} + y - 1 = 0,$$

$$\frac{dy}{dx} = -\frac{y-1}{6y^5 + 1 + x}.$$

Answer: $\left.\frac{dy}{dx}\right|_{(0,0)} = 1.$

REMARK: The technique in the alternate solution is equivalent to use of the rule

$$\frac{dy}{dx} = -\frac{f_x}{f_y}$$

because the rule was derived by that very technique.

EXAMPLE 7.2

Let $y = \sqrt{1 - x^2}$. Compute y' and y'' by differentiating implicitly $x^2 + y^2 - 1 = 0$.

Solution: Differentiate:

$$2x + 2yy' = 0, \qquad y' = -\frac{x}{y}.$$

Differentiate again:

$$y'' = -\frac{y - xy'}{y^2} = -\frac{y - x\left(-\frac{x}{y}\right)}{y^2} = -\frac{y^2 + x^2}{y^3} = -\frac{1}{y^3},$$

since $x^2 + y^2 = 1$.

Answer:

$$y' = -\frac{x}{y} = -\frac{x}{\sqrt{1-x^2}},$$

$$y'' = -\frac{1}{y^3} = \frac{1}{(1-x^2)^{3/2}}.$$

Applications

Implicit differentiation is useful when a function must be maximized or minimized subject to certain restrictions.

EXAMPLE 7.3

A cylindrical container (right circular) is required to have a given volume V. The material on the top and bottom is k times as expensive as the material on the sides. What are the proportions of the most economical container?

Solution: The cost C of the container is proportional to

$$(\text{area of side}) + k(\text{area of top} + \text{area of bottom}).$$

Let r and h denote the radius and height of the container. In the proper units,

$$C = 2\pi rh + k(2\pi r^2),$$

where

$$\pi r^2 h = V, \qquad \text{a constant.}$$

We must minimize C subject to this restriction.

One approach is obvious: solve the last equation for h and substitute into the equation for C. Then C is an explicit function of r which can be minimized.

It is simpler, however, not to make the substitution, but to consider C as a function of r anyway (as if the substitution had been made). Differentiate implicitly:

$$\frac{dC}{dr} = 2\pi\left(r\frac{dh}{dr} + h + 2kr\right).$$

Now differentiate the equation for V with respect to r:

$$2\pi rh + \pi r^2\frac{dh}{dr} = 0, \qquad \frac{dh}{dr} = -\frac{2h}{r}.$$

Substitute this value of dh/dr into the preceding equation:

$$\frac{dC}{dr} = 2\pi\left[r\left(\frac{-2h}{r}\right) + h + 2kr\right] = 2\pi(2kr - h).$$

Hence

$$\frac{dC}{dr} = 0 \qquad \text{when} \quad h = 2kr.$$

It is easily verified that C is minimal for $h = 2kr$. Since $\pi r^2 h$ is constant, h is large if r is small and decreases as r increases. Therefore $(2kr - h)$ increases from negative to positive as r increases. Thus dC/dr satisfies the conditions for C to have a minimum at $h = 2kr$.

Answer: height $= 2k \times$ radius.

REMARK: The special case $k = 1$ is interesting. All parts of the cylinder are equally expensive; the cheapest cylinder is the one with least surface area. Conclusion: Of all cylinders with fixed volume, the one with least surface area is the one whose height is twice its radius.

EXAMPLE 7.4

Find the greatest distance between the origin and a point of the curve $x^4 + y^4 = 1$.

Solution: Draw a graph (Fig. 7.2). Because of symmetry we need consider only $x \geq 0$ and $y \geq 0$. Since the curve lies outside of the circle

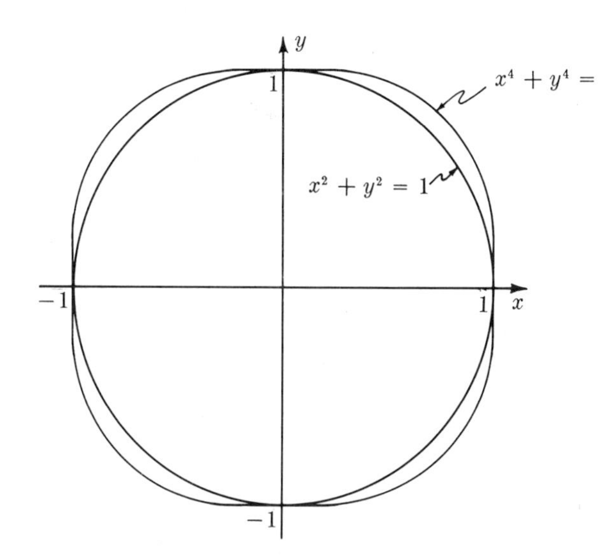

Fig. 7.2

$x^2 + y^2 = 1$, the maximum distance is greater than 1 and occurs at some point (x, y) where $x > 0$ and $y > 0$.

The square of the distance from any point (x, y) to the origin is $x^2 + y^2$. Hence, we must maximize

$$L^2 = x^2 + y^2$$

subject to

$$x^4 + y^4 - 1 = 0.$$

Differentiate both relations with respect to x:

$$\frac{d}{dx}(L^2) = 2x + 2yy', \qquad 4x^3 + 4y^3y' = 0.$$

It follows that

$$\frac{d}{dx}(L^2) = 2x + 2y\left(-\frac{x^3}{y^3}\right) = 2x\left(\frac{y^2 - x^2}{y^2}\right).$$

This derivative vanishes in the first quadrant only for $x = y$. Hence the maximum distance occurs at the point (x, y) of the curve in the first quadrant for which $x = y$. Thus

$$x^4 + y^4 = 1, \qquad x = y,$$

from which

$$x = y = \frac{1}{\sqrt[4]{2}},$$

and

$$L^2 = x^2 + y^2 = \frac{1}{\sqrt{2}} + \frac{1}{\sqrt{2}} = \sqrt{2}$$

Answer: $\sqrt[4]{2}$.

EXERCISES

Find dy/dx:

1. $x + y = x \sin y$
2. $x^2 + y^3 = xy$
3. $e^{xy} = 3xy^2$
4. $x^4 - y^4 = 3x^2y^3$
5. $e^x \sin y = e^y \cos x$
6. $x^3y^3 = x^2 - y^2 + 1$
7. $x^4 + 3y^6 = 1$
8. $x^5 + y^5 = xy + 1$.
9. Find the maximum and minimum values of $f(x, y, z) = x^4 + y^4 + z^4$ on the surface of the unit sphere $x^2 + y^2 + z^2 = 1$.
10. (cont.) Deduce that $\frac{1}{3}(x^2 + y^2 + z^2)^2 \leq x^4 + y^4 + z^4 \leq (x^2 + y^2 + z^2)^2$ for *any* (x, y, z).
11. (cont.) Find the corresponding inequalities for $n \geq 3$ relating $x^n + y^n + z^n$ and $x^2 + y^2 + z^2$. (Assume $x \geq 0$, $y \geq 0$, $z \geq 0$ if n is odd.)
12. (cont.) Find the largest A and the smallest B so that $A(x^4 + y^4 + z^4)^3 \leq (x^6 + y^6 + z^6)^2 \leq B(x^4 + y^4 + z^4)^3$ for all (x, y, z).

8. LEAST SQUARES [optional]

In a certain physical situation it is assumed that a variable y is some linear function of time,

$$y = At + B.$$

An experiment produces readings

$$(t_1, y_1), \quad (t_2, y_2), \quad \cdots, \quad (t_n, y_n),$$

which are plotted in Fig. 8.1. The points are supposed to be collinear, but are not exactly so because of experimental error, round-off error, etc. For which constants A and B does the straight line $y = At + B$ most closely fit the data?

The answer depends on what is meant by "closely." Probably the most popular measure of fit is the **least squares** fit: The line is chosen to minimize the sum of the squares of the vertical deviations from the line (Fig. 8.2).

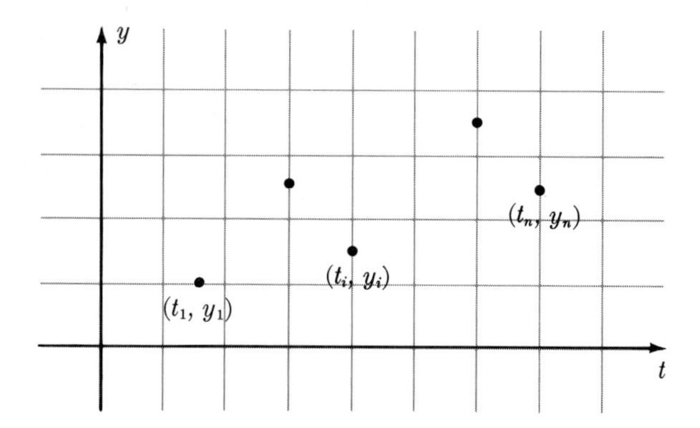

FIG. 8.1

EXAMPLE 8.1

Find the line $y = At + B$ that is the least squares fit to the points $(0, 2)$, $(1, 3)$, $(2, 3)$.

Solution: Write $y(t) = At + B$, and choose A and B to minimize

$$\begin{aligned} f(A, B) &= [y(0) - 2]^2 + [y(1) - 3]^2 + [y(2) - 3]^2 \\ &= (B - 2)^2 + (A + B - 3)^2 + (2A + B - 3)^2. \end{aligned}$$

Necessary conditions for a minimum are

$$\frac{\partial f}{\partial A} = \frac{\partial f}{\partial B} = 0.$$

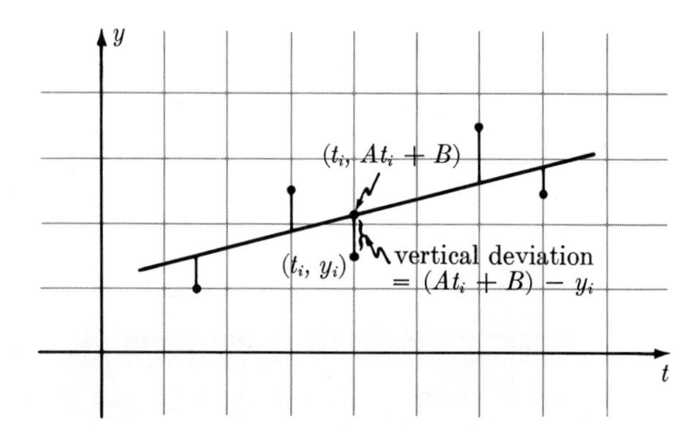

FIG. 8.2

Differentiate:

$$\frac{\partial f}{\partial A} = 2(A + B - 3) + 4(2A + B - 3) = 2(5A + 3B - 9),$$

$$\frac{\partial f}{\partial B} = 2(B - 2) + 2(A + B - 3) + 2(2A + B - 3) = 2(3A + 3B - 8).$$

Set these partial derivatives equal to zero:

$$\begin{cases} 5A + 3B - 9 = 0 \\ 3A + 3B - 8 = 0. \end{cases}$$

The system has a unique solution: $A = \frac{1}{2}$ and $B = \frac{13}{6}$.

Now $f(A, B)$ must have a minimum at $(\frac{1}{2}, \frac{13}{6})$; here is why. If either A or B is large, then $f(A, B)$ is large (by inspection). On the circle $A^2 + B^2 = (1000)^2$, for example, $f(A, B)$ is very large. The minimum of $f(A, B)$ in the region bounded by this circle occurs either on the boundary or at a point where $\partial f/\partial A = \partial f/\partial B = 0$. But the boundary is ruled out, hence the minimum occurs at $(\frac{1}{2}, \frac{13}{6})$, the only point where $\partial f/\partial A = \partial f/\partial B = 0$.

Answer: $y = \dfrac{1}{2}t + \dfrac{13}{6}$. (See Fig. 8.3.)

To solve the general problem of least squares fit, imitate the method used in this example. Given readings

$$(t_1, y_1), \quad (t_2, y_2), \quad \cdots, \quad (t_n, y_n)$$

for n distinct time values, seek a linear function $y = At + B$ which

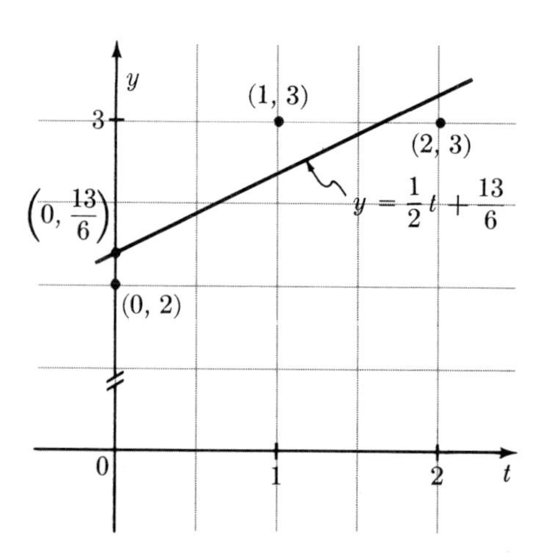

Fig. 8.3

minimizes*

$$f(A, B) = \sum_{i=1}^{n} [(At_i + B) - y_i]^2.$$

Now

$$\frac{\partial f}{\partial A} = 2 \sum_{i=1}^{n} t_i(At_i + B - y_i) = 2\left[A\left(\sum_{i=1}^{n} t_i^2\right) + B\left(\sum_{i=1}^{n} t_i\right) - \left(\sum_{i=1}^{n} t_i y_i\right)\right],$$

$$\frac{\partial f}{\partial B} = 2 \sum_{i=1}^{n} (At_i + B - y_i) = 2\left[A\left(\sum_{i=1}^{n} t_i\right) + nB - \left(\sum_{i=1}^{n} y_i\right)\right].$$

Set these partial derivatives equal to zero to obtain two equations for the two unknowns A and B:

$$\begin{cases} \left(\sum t_i^2\right) A + \left(\sum t_i\right) B = \sum t_i y_i \\ \left(\sum t_i\right) A + nB = \sum y_i. \end{cases}$$

All coefficients in this system of equations are computable from the data. There is a unique solution (A, B), and by the same reasoning as used in the example, the minimum value of f occurs at (A, B).

EXAMPLE 8.2

Find the least squares straight line fit to the data

$(0, 0), (1, 1), (2, 0), (3, 1), (4, 0)$.

Solution: Here $n = 5$ and

$$\sum t_i = 0 + 1 + 2 + 3 + 4 = 10,$$

$$\sum t_i^2 = 0^2 + 1^2 + 2^2 + 3^2 + 4^2 = 30,$$

$$\sum y_i = 0 + 1 + 0 + 1 + 0 = 2,$$

$$\sum t_i y_i = 0 \cdot 0 + 1 \cdot 1 + 2 \cdot 0 + 3 \cdot 1 + 4 \cdot 0 = 4.$$

The system of equations is

$$\begin{cases} 30A + 10B = 4 \\ 10A + 5B = 2. \end{cases}$$

* The Σ (sigma) denotes *summation*. See p. 596 for details on this notation.

Its solution is $A = 0$ and $B = \frac{2}{5}$.

Answer: $y = \dfrac{2}{5}\cdot$ See Fig. 8.4.

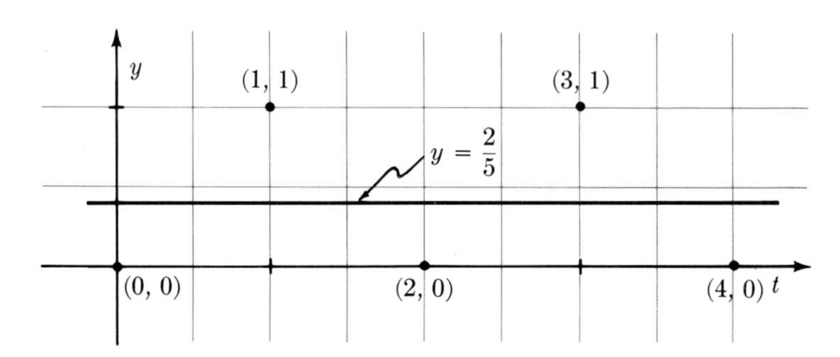

FIG. 8.4

EXAMPLE 8.3

Find the least squares straight line fit to the data $(0, 0), (1, 1), (2, 0), (3, 1)$.

Solution: Here $n = 4$ and

$$\sum t_i = 0 + 1 + 2 + 3 = 6,$$

$$\sum t_i^2 = 0 + 1 + 4 + 9 = 14,$$

$$\sum y_i = 0 + 1 + 0 + 1 = 2,$$

$$\sum t_i y_i = 0 + 1 + 0 + 3 = 4.$$

The system of equations is

$$\begin{cases} 14A + 6B = 4 \\ 6A + 4B = 2. \end{cases}$$

Its solution is $A = \frac{1}{5}$ and $B = \frac{1}{5}$.

Answer: $y = \dfrac{1}{5}t + \dfrac{1}{5}\cdot$

REMARK: It is interesting to compare this case (Fig. 8.5) with that of Fig. 8.4.

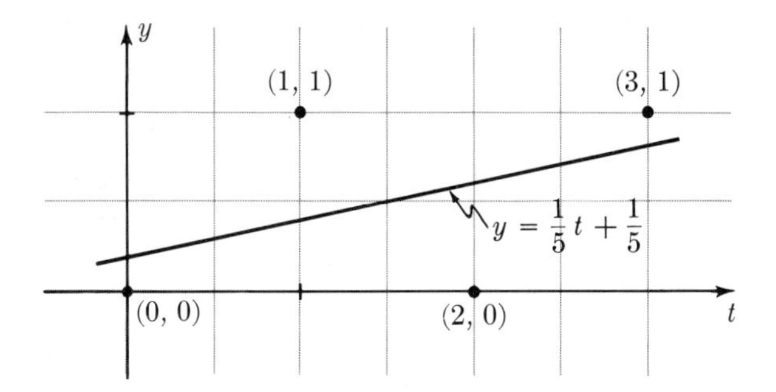

FIG. 8.5

The least squares idea can be used in more complicated situations. For example, one might seek the *quadratic* polynomial $y = At^2 + Bt + C$ that most closely fits the data

$$(t_1, y_1), \quad (t_2, y_2), \quad \cdots, \quad (t_n, y_n).$$

Then A, B, and C must be found so that

$$F(A, B, C) = \sum_{i=1}^{n} [(At_i^2 + Bt_i + C) - y_i]^2$$

is minimized.

Still more complicated is the problem of approximating a *function*, rather than discrete data. For example, suppose $y = t^2$ is to be approximated on the interval $0 \leq t \leq 1$ by a linear function $y = At + B$ in the sense of least squares. Then

$$F(A, B) = \int_0^1 [(At + B) - t^2]^2 \, dt$$

must be minimized.

EXERCISES

Find the least squares straight line fit to the data; plot the data and the line you obtain:

1. $(0, 0), (1, 1), (2, 3)$
2. $(1, 0), (2, -1), (3, -4)$
3. $(0, 0), (\frac{1}{2}, \frac{1}{4}), (1, 1), (\frac{3}{2}, \frac{9}{4}), (2, 4)$
4. $(1, 1), (2, \frac{1}{2}), (3, \frac{1}{3}), (4, \frac{1}{4})$
5. $(1, 5.0), (2, 5.3), (3. 5.4), (4, 5.6)$

6. $(-2, 3.0)$, $(-1, 1.8)$, $(0, 1.1)$, $(1, 0.6)$
7. $(0, 0)$, $(1, -1)$, $(2, -2)$, $(3, -1)$, $(4, 0)$
8. $(-3, 0)$, $(-2, 0)$, $(-1, 0)$, $(0, 1)$, $(1, 2)$, $(2, 2)$, $(3, 2)$
9. $(-1, 6.2)$, $(0, 5.4)$, $(1, 5.5)$, $(2, 5.0)$
10. $(1, -1)$, $(2, 1)$, $(3, -1)$, $(4, 1)$, $\cdots$, $(9, -1)$.
11. Approximate the function $y = t^2$ on the interval $0 \leq t \leq 1$ by a linear function $y = At + B$ in the sense of least squares. In other words, minimize

$$F(A, B) = \int_0^1 [(At + B) - t^2]^2 \, dt.$$

12. (cont.) Do the same for $y = t^3$.
13. (cont.) Do the same for $y = e^t$.
14. Prove that the least squares approximation to a function $y(t)$ on $0 \leq t \leq 1$ by a linear function is $y = 6(2\beta - \alpha)t + 2(2\alpha - 3\beta)$, where

$$\alpha = \int_0^1 y(t) \, dt, \qquad \beta = \int_0^1 t y(t) \, dt.$$

9. HILL CLIMBING [optional]

In theory, to locate the maximum and minimum of a function of several variables, you compute the partial derivatives, set them equal to zero, and solve the resulting system of equations. The maximum and minimum occur at solution points or on the boundary. In practice, however, the system of equations computed from the partials may be difficult or impossible to solve. You are then forced to seek approximate solutions.

Of course the computer can be used to find these approximate solutions. But it is often simpler and more direct to forget the partials and to work with the function itself in a technique called **hill climbing.**

To climb a wooded mountain, you head constantly uphill, with the certainty that you will reach a peak (perhaps not the highest). Similarly, to find a maximum of a function, the computer starts at a point (x_0, y_0), finds a neighboring point (x_1, y_1) with $f(x_1, y_1) > f(x_0, y_0)$, and proceeds "up the hill." If $f(x_0, y_0)$ is greater than the function values $f(x, y)$ at all neighboring points (x, y), then the computer has reached a peak (perhaps not the highest). Minima of a function are located similarly by heading down the hill.

Since the computer cannot examine *all* neighboring points, you must choose which it will examine. The most common choice is a square grid centered at the starting point. Thus starting at (x_0, y_0), take the eight neighbors $(x_0, y_0 \pm h)$, $(x_0 \pm h, y_0)$, $(x_0 \pm h, y_0 \pm h)$, where h is the grid size (Fig. 9.1).

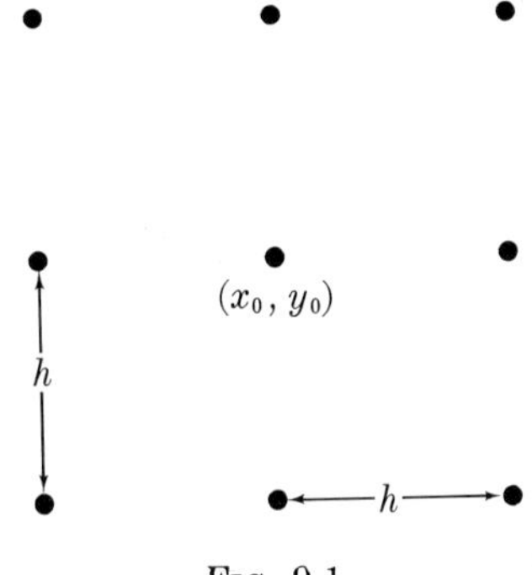

FIG. 9.1

The success of this method depends on the grid size h. If h is too small, the process takes a long time; if h is too large, the desired maximum (or minimum) may be bypassed. Often the h initially chosen is replaced by a smaller one as the maximum of the function is approached.

EXAMPLE 9.1

Find the minimum of $f(x, y) = x^2 + 2y^2 + x$ by hill climbing. Start the process at (0, 0), at (1, 1), and at (−2, −1) with an initial grid size $h = 0.2$.

Solution: Beginning at (0, 0), examine the function values at $(0, \pm 0.2)$, $(\pm 0.2, 0)$, $(\pm 0.2, \pm 0.2)$. Among these the lowest value is $f(-0.2, 0) = -0.16$. Now choose the point $(-0.2, 0)$ and repeat the process from there.

Similarly, at (1, 1), the neighboring points are $(1, 1 \pm 0.2)$, $(1 \pm 0.2, 1)$, and $(1 \pm 0.2, 1 \pm 0.2)$. Among these (0.8, 0.8) has the smallest function value: $f(0.8, 0.8) = 2.72$. Hence choose (0.8, 0.8) as the next point. Table 9.1 shows the points chosen at each step, with $h = 0.2$.

TABLE 9.1. $h = 0.2$.

Step	Points		
0	(0, 0)	(1, 1)	(−2, −1)
1	(−0.2, 0)	(0.8, 0.8)	(−1.8, −0.8)
2	(−0.4, 0)	(0.6, 0.6)	(−1.6, −0.6)
3	(−0.6, 0)	(0.4, 0.4)	(−1.4, −0.4)
4	—	(0.2, 0.2)	(−1.2, −0.2)
5	—	(0, 0)	(−1.0, 0)
6	—	(−0.2, 0)	(−0.8, 0)
7	—	(−0.4, 0)	(−0.6, 0)
8	—	(−0.6, 0)	(−0.4, 0)

From each starting point, the process leads to

$$f(-0.4, 0) = f(-0.6, 0) = -0.24.$$

From any other initial point (x, y) the process leads also to the points $(-0.4, 0)$ and $(-0.6, 0)$.

The next stage of the process is to choose a smaller h and examine the corresponding closer neighbors of $(-0.4, 0)$ and $(-0.6, 0)$. The choice of $h = 0.1$ leads to $f(-0.5, 0) = -0.25$. Smaller values of h, say 0.01 or 0.0001, do not lead to smaller values of $f(x, y)$.

Answer: $f(-0.5, 0) = -0.25$.

EXAMPLE 9.2

By hill climbing, find the maximum and minimum of $f(x, y) = x^2y + x$ in the square $-2 \leq x \leq 2$ and $0 \leq y \leq 4$.

Solution: Since no starting point is given, try a coarse grid $h = 1$ over the entire square to see how the function looks (Fig. 9.2).

y \ x	-2	-1	0	1	2
4	14	3	0	5	18
3	10	2	0	4	14
2	6	1	0	3	10
1	2	0	0	2	6
0	-2	-1	0	1	2

FIG. 9.2

The figure suggests that $f(x, y)$ has its maximum value in the square at $(2, 4)$ and its minimum value at $(-2, 0)$. Starting at these points and using any smaller h, you may confirm this.

Answer: Maximum $f(2, 4) = 18$;
Minimum $f(-2, 0) = -2$.

REMARK: The figure shows the danger of picking the initial point arbitrarily. The maximum is found if you start with a point (x_0, y_0) satisfying $2x_0y_0 \geq -1$ (see Ex. 19); other initial points can lead to the false maximum at $(-2, 4)$. The minimum can be found starting from any initial point (x_0, y_0) in the square; however, small values of h must be used if $|x_0|$ is small so that the search will move off the y-axis.

The shape of the boundary of a region can influence the success of hill climbing. For example if the boundary is circular, few of the rectangular lattice points actually lie on the boundary. This makes it more difficult to find the maximum or minimum. Working in polar coordinates gives better results.

EXAMPLE 9.3

Approximate to 3 decimals the maximum of $f(x, y) = x^2y$ in the disk $x^2 + y^2 \leq 1$.

Solution: The function $f(x, y)$ is negative for $y < 0$ and $x \neq 0$, and positive for $y > 0$ and $x \neq 0$. Also $f(-x, y) = f(x, y)$, hence it suffices to seek the maximum of $f(x, y)$ in the first quadrant portion of the disk. In polar coordinates,

$$f\{r, \theta\} = (r \cos \theta)^2(r \sin \theta) = r^3 \cos^2 \theta \sin \theta,$$

and this must be maximized in the region $0 \leq r \leq 1$, $0 \leq \theta \leq \pi/2$.

Since $f(r, \theta)$ increases as r increases, only points on the unit circle $r = 1$ need be considered. Choose $\Delta\theta = 0.05\pi$ and begin the hill climb at $\{1, 0\}$.

$\{r, \theta\}$	$\{1, 0\}$	$\{1, 0.05\pi\}$	$\{1, 0.10\pi\}$	$\{1, 0.15\pi\}$	$\{1, 0.20\pi\}$	$\{1, 0.25\pi\}$
$f(r, \theta)$	0	0.152	0.299	0.360	0.385	0.353

The maximum occurs near $\{1, 0.20\pi\}$. Try a finer grid:

$$f(1, 0.19\pi) \approx 0.384, \qquad f(1, 0.21\pi) \approx 0.385.$$

Answer: 0.385.
(Exact answer: $2/\sqrt{27} \approx 0.38490$.)

The technique of hill climbing works equally well for functions of arbitrarily many variables. You merely have more neighbors to examine at each point. For example, in three dimensions a point has 26 neighbors.

EXERCISES

Find the maximum and minimum on the square $-1 \leq x, y \leq 1$. Compute by hand starting at $(0, 0)$:

1. $x^2 + y$
2. $(x^2 - 1)(y^2 + 1)$
3. $\dfrac{x}{2 - y}$
4. $\dfrac{x + 2}{y^2 + 1}$
5. $e^x \cos \pi y$
6. xe^{-y}
7. $e^x \ln(3 + y)$
8. $\sqrt{\dfrac{x + 2}{y + 2}}$
9. $\dfrac{2x + 1}{\sqrt{x + y + 3}}$
10. $e^x(2xy - 1)$.

Find the maximum and minimum on the disk $x^2 + y^2 \le 1$. Compute by hand, starting at $(0, 0)$:

11. $x^2 + y$

12. x^3y.

Find the maximum and minimum on the rectangle $0 \le x \le 1$, $0 \le y \le 2\pi$. Compute by hand, starting at $(0, 0)$:

13. $(3x - 2) \sin y \cos \left(y + \frac{\pi}{3}\right)$.

14. $\dfrac{2x^2 - 1}{2 + \sin y}$.

Find the maximum and minimum on the square $0 \le x, y \le 5$. Use a computer:

15. $x \sin x + \sin y$

16. $\sin x \cos y + x^2$

17. $x^2y + 3xy^2 - y^3$

18. $\dfrac{x - y}{\sqrt{x + y^2 + 1}}$.

19. Carefully sketch the level curves of $f(x, y) = x^2y + x$ in the square $-2 \le x \le 2$ and $0 \le y \le 4$. Show, as stated in the solution to Example 9.2, that hill climbing from any initial point (x_0, y_0) satisfying $2x_0y_0 \ge -1$ will lead to the true maximum.

10. DIFFERENTIALS

The Chain Rule may be expressed in terms of differentials. The **differential** of a function $f(x, y, z)$ is the formal expression

$$df = f_x\, dx + f_y\, dy + f_z\, dz.$$

It may happen that x, y, and z are themselves functions of other variables, say u and v. Then f is a (composite) function of u and v, and as such it has a differential

$$df = f_u\, du + f_v\, dv.$$

Is this the same as the previous df? Yes, for by the Chain Rule,

$$\begin{aligned} f_u\, du + f_v\, dv &= (f_x x_u + f_y y_u + f_z z_u)\, du + (f_x x_v + f_y y_v + f_z z_v)\, dv \\ &= f_x(x_u\, du + x_v\, dv) + f_y\,(y_u\, du + y_v\, dv) + f_z\,(z_u\, du + z_v\, dv) \\ &= f_x\, dx + f_y\, dy + f_z\, dz. \end{aligned}$$

Differentials will be used in Chapter 33, Section 7, p. 816.

26. Double Integrals

1. INTRODUCTION

In Chapter 8, we introduced the definite integral in order to compute the area under a curve. Now we introduce a new tool, the double integral, in order to compute the volume under a surface (Fig. 1.1).

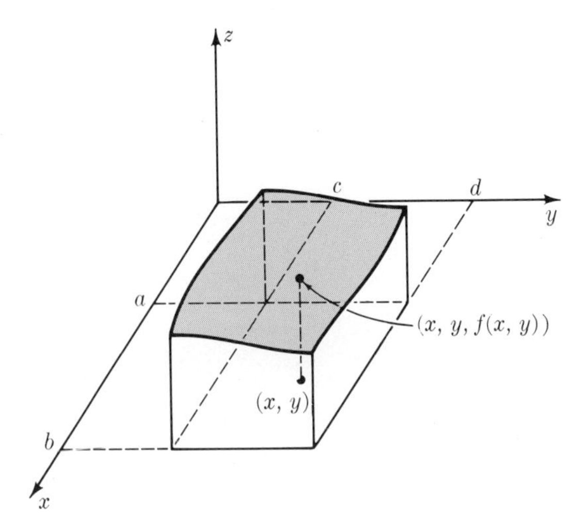

FIG. 1.1

The prismatic solid shown in Fig. 1.1 is bounded above by a surface $z = f(x, y)$ and below by a rectangle in the x, y-plane,

$$a \leq x \leq b, \qquad c \leq y \leq d.$$

Just as the area under a curve can be approximated by thin rectangles, so the volume of this solid can be approximated by thin rectangular blocks.

Partition its base into a large number of equal small rectangles by drawing segments parallel to the x- and the y-axes (Fig. 1.2). This amounts to

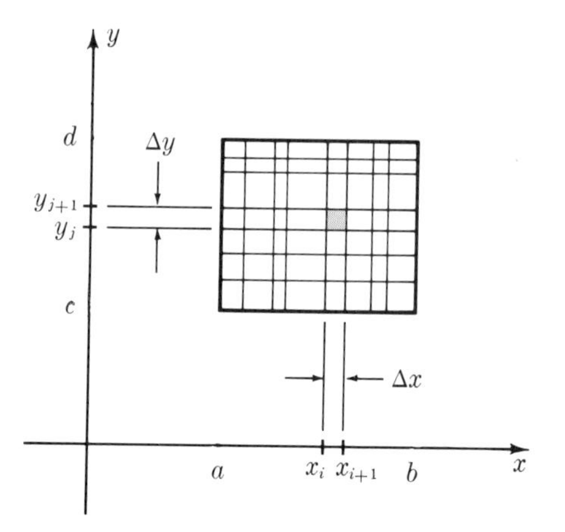

FIG. 1.2

partitioning both the interval $a \leq x \leq b$ and the interval $c \leq y \leq d$:

$$a = x_0 < x_1 < x_2 < \cdots < x_m = b,$$

$$c = y_0 < y_1 < y_2 < \cdots < y_n = d.$$

Write

$$x_{i+1} - x_i = \frac{1}{m}(b - a) = \Delta x, \qquad y_{j+1} - y_j = \frac{1}{n}(d - c) = \Delta y.$$

The part of the solid above the i, j-th small rectangle is approximately a thin rectangular column of height $f(\bar{x}_i, \bar{y}_j)$, where $(\bar{x}_i, \bar{y}_j)$ is the midpoint of the base (Fig. 1.3).

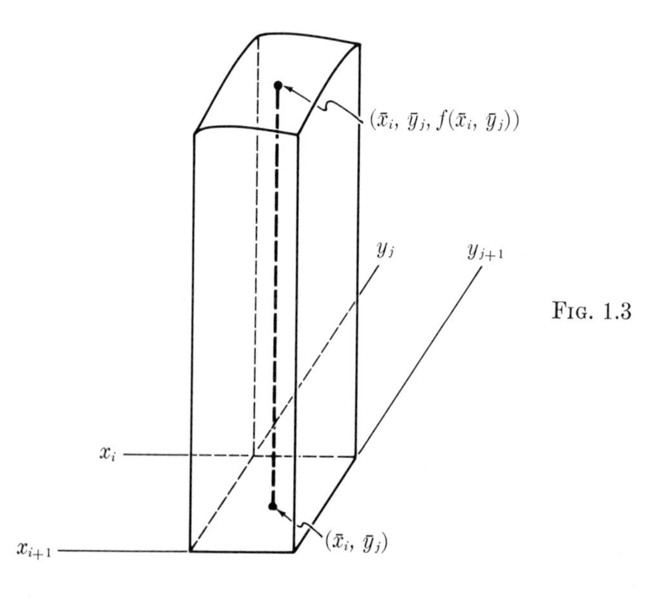

FIG. 1.3

The volume of the column is

$$f(\bar{x}_i, \bar{y}_j)\ \Delta x\ \Delta y.$$

Add* these volumes to approximate the total volume V:

$$V \approx \sum_{\substack{i=1,2,\cdots,m\\ j=1,2,\cdots,n}} f(\bar{x}_i, \bar{y}_j)\ \Delta x\ \Delta y.$$

An alternate notation is

$$V \approx \sum_{i=1}^{m}\sum_{j=1}^{n} f(\bar{x}_i, \bar{y}_j)\ \Delta x\ \Delta y.$$

Now partition the base of the solid into more and more small rectangles (Take m and n larger and larger, hence Δx and Δy smaller and smaller.) The approximation approaches the true volume. Write

$$V = \lim_{\substack{\Delta x \to 0\\ \Delta y \to 0}} \sum\sum f(\bar{x}_i, \bar{y}_j)\ \Delta x\ \Delta y,$$

which suggests the notation

$$V = \iint\limits_{\substack{a \le x \le b\\ c \le y \le d}} f(x, y)\ dx\ dy,$$

or simply

$$V = \iint\limits_{R} f(x, y)\ dx\ dy,$$

where R denotes the rectangular base.

This quantity V is the **double integral** of the function $f(x, y)$. How to evaluate it will be discussed in the following sections. There are, however, several general properties of the double integral worth mentioning here.

First of all, if $f(x, y)$ takes negative values over part of the rectangle, the volume there will be counted negative, just as the integral

$$\int_a^b f(x)\ dx$$

counts area below the x-axis negative.

Next, the integral is additive:

$$\iint\limits_{R} [f(x, y) + g(x, y)]\ dx\ dy = \iint\limits_{R} f(x, y)\ dx\ dy + \iint\limits_{R} g(x, y)\ dx\ dy.$$

* Recall that Σ denotes summation. See p. 596 for details on this notation.

Finally, any constant factor may be brought outside:

$$\iint_R a\, f(x, y)\, dx\, dy = a \iint_R f(x, y)\, dx\, dy.$$

Thus the double integral is a linear operator. (The same is true, of course, of the single integral.)

2. SPECIAL CASES

In certain cases where $f(x, y)$ has a particularly simple form, it is easy to evaluate the double integral

$$\iint_R f(x, y)\, dx\, dy.$$

As before, R denotes the rectangle

$$a \leq x \leq b, \qquad c \leq y \leq d.$$

Suppose first that $f(x, y) = h$, a constant. Then the double integral represents the volume of a rectangular solid of height h. This volume is h times the area of the base:

$$\iint_R h\, dx\, dy = h \times (\text{area } R) = h(b - a)(d - c).$$

Suppose next that $f(x, y)$ is a function of x alone,

$$f(x, y) = g(x).$$

Then each cross-section of the solid by a plane parallel to the z, x-plane is an identical plane region (Fig. 2.1a, b). The volume is the area of this

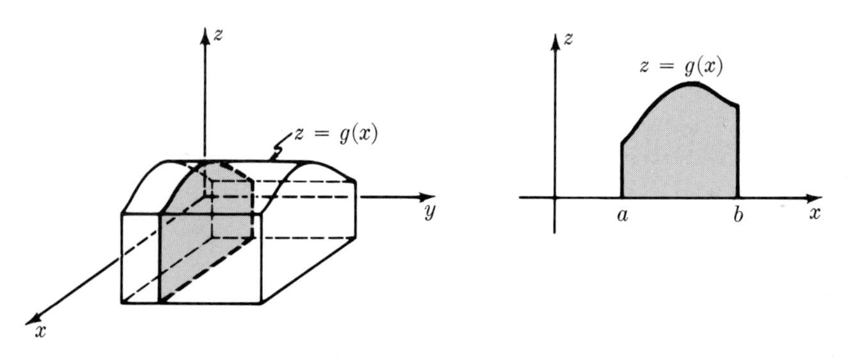

FIG. 2.1a FIG. 2.1b

region times the y-length of the rectangle:

$$\iint_R g(x)\,dx\,dy = \left(\int_a^b g(x)\,dx\right)(d - c).$$

Similarly if $f(x, y) = h(y)$,

$$\iint_R h(y)\,dx\,dy = \left(\int_c^d h(y)\,dy\right)(b - a).$$

These formulas are special cases of the following one:

$$\boxed{\iint_R g(x)\,h(y)\,dx\,dy = \left(\int_a^b g(x)\,dx\right)\left(\int_c^d h(y)\,dy\right).}$$

This formula gives a double integral as the product of two ordinary (single) integrals. It applies to such functions as

$$x^5y^7, \qquad e^{x-y}, \qquad e^x \cos y, \qquad \text{etc.}$$

The formula will be derived after we work some examples.

EXAMPLE 2.1

Find $\displaystyle\iint_{0\le x,\ y\le 1} x^4y^6\,dx\,dy.$

Solution:

$$\iint_{0\le x,\ y\le 1} x^4y^6\,dx\,dy = \left(\int_0^1 x^4\,dx\right)\left(\int_0^1 y^6\,dy\right) = \frac{1}{5}\cdot\frac{1}{7}\cdot$$

Answer: $\dfrac{1}{35}\cdot$

EXAMPLE 2.2

Find $\displaystyle\iint_{\substack{0\le x\le 1\\ \pi/2\le y\le\pi}} e^x \cos y\,dx\,dy.$

Solution: Let R denote the rectangle, $0 \le x \le 1$ and $\pi/2 \le y \le \pi$. Then

$$\iint_R = \left(\int_0^1 e^x\,dx\right)\left(\int_{\pi/2}^{\pi} \cos y\,dy\right) = \left(e^x\Big|_0^1\right)\left(\sin y\Big|_{\pi/2}^{\pi}\right) = (e - 1)(-1).$$

Answer: $1 - e$.

QUESTION: The answer is negative. Why?

EXAMPLE 2.3

Find $\displaystyle\iint\limits_{\substack{0\le x\le 1\\ -2\le y\le -1}} e^{x-y}\,dx\,dy.$

Solution: Since $e^{x-y} = e^x e^{-y}$,

$$\iint\limits_R e^{x-y}\,dx\,dy = \left(\int_0^1 e^x\,dx\right)\left(\int_{-2}^{-1} e^{-y}\,dy\right)$$

$$= \left(e^x\Big|_0^1\right)\left(-e^{-y}\Big|_{-2}^{-1}\right) = (e-1)(e^2-e).$$

Answer: $e(e-1)^2$.

EXAMPLE 2.4

Find $\displaystyle\iint\limits_{\substack{1\le x\le 2\\ -1\le y\le 1}} (x^2y - 3xy^2)\,dx\,dy.$

Solution: Use the linear property of the double integral:

$$\iint\limits_R (x^2y - 3xy^2)\,dx\,dy = \iint\limits_R x^2y\,dx\,dy - 3\iint\limits_R xy^2\,dx\,dy.$$

Evaluate these two integrals separately:

$$\iint\limits_R x^2y\,dx\,dy = \left(\int_1^2 x^2\,dx\right)\left(\int_{-1}^1 y\,dy\right) = 0;$$

$$\iint\limits_R xy^2\,dx\,dy = \left(\int_1^2 x\,dx\right)\left(\int_{-1}^1 y^2\,dy\right) = \frac{3}{2}\cdot\frac{2}{3} = 1.$$

Answer: -3.

Justification of the Formula

$$\iint\limits_{\substack{a\le x\le b\\ c\le y\le d}} g(x)\,h(y)\,dx\,dy = \left(\int_a^b g(x)\,dx\right)\left(\int_c^d h(y)\,dy\right).$$

The volume of a solid (the left-hand side) is evaluated by slicing the solid into slabs, then adding up (Fig. 2.2a). Fix y, then slice the solid by the planes parallel to the z, x-plane at y and at $y + dy$. The result is a slab of thickness dy. The projection (Fig. 2.2b) of this slab on the z, x-plane is the

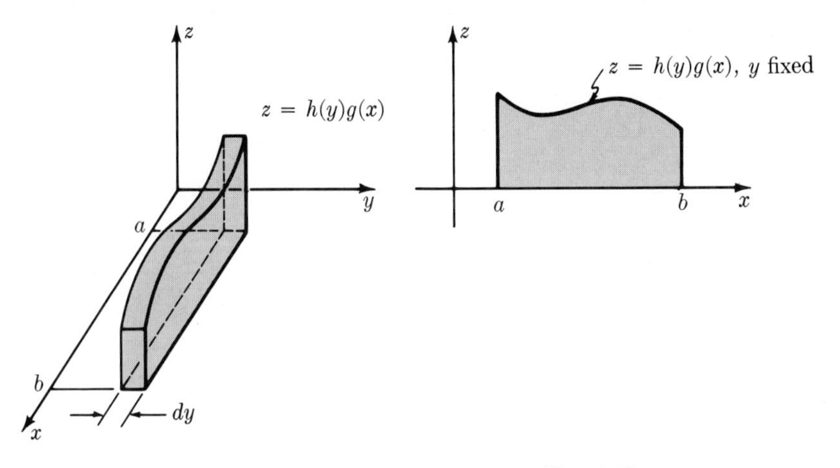

FIG. 2.2a

FIG. 2.2b

region in that plane under the curve $z = z(x) = h(y)g(x)$. (Remember y is fixed!) Hence its area is

$$\int_a^b z(x)\,dx = \int_a^b h(y)\,g(x)\,dx = h(y)\int_a^b g(x)\,dx.$$

The volume of the slab is its area times its thickness:

$$dV = \left(h(y)\int_a^b g(x)\,dx\right)dy = \left(\int_a^b g(x)\,dx\right)h(y)\,dy.$$

On the right-hand side, the first factor is a constant, therefore adding the slabs yields

$$V = \int_c^d\left(\int_a^b g(x)\,dx\right)h(y)\,dy = \left(\int_a^b g(x)\,dx\right)\left(\int_c^d h(y)\,dy\right).$$

This is the desired formula.

EXERCISES

Evaluate:

1. $\iint (3x - 1)\,dx\,dy;\qquad -1 \le x \le 2,\quad 0 \le y \le 5$
2. $\iint e^y\,dx\,dy;\qquad -1 \le x \le 1,\quad 0 \le y \le \ln 2$
3. $\iint x^2y^2\,dx\,dy;\qquad -1 \le x \le 0,\quad 0 \le y \le 1$

4. $\iint x^2y^2\,dx\,dy;\qquad -1 \le x, y \le 1$

5. $\iint x^3y^3\,dx\,dy;\qquad 0 \le x, y \le 1$

6. $\iint (x - y)\,dx\,dy;\qquad 0 \le x, y \le 1$

7. $\iint (x^2 - y^2)\,dx\,dy;\qquad 0 \le x, y \le 1$

8. $\iint (x^{17} - y^{17})\,dx\,dy;\qquad 0 \le x, y \le 1$

9. $\iint (e^{x^2} - e^{y^2})\,dx\,dy;\qquad 0 \le x, y \le 1$

10. $\iint \cos x \cos y\,dx\,dy;\qquad 0 \le x \le \frac{\pi}{4},\quad 0 \le y \le \frac{\pi}{2}$

11. $\iint (x^2 + y^2)\,dx\,dy;\qquad -1 \le x, y \le 1$

12. $\iint (x - y)^2\,dx\,dy;\qquad -1 \le x, y \le 1$

13. $\iint (x - y)^3\,dx\,dy;\qquad -1 \le x, y \le 1$

14. $\iint (x - y)^{75}\,dx\,dy;\qquad -1 \le x, y \le 1$

15. $\iint (x + y)^{93}\,dx\,dy;\qquad -1 \le x, y \le 1$

16. $\iint \frac{x^2}{y^3}\,dx\,dy;\qquad 1 \le x \le 2,\quad 1 \le y \le 4$

17. $\iint \frac{x}{1 + y^2}\,dx\,dy;\qquad 0 \le x \le 2,\quad 0 \le y \le 1$

18. $\iint xy \ln x\,dx\,dy;\qquad 1 \le x \le 4,\quad -1 \le y \le 2$

19. $\iint x \ln(xy)\,dx\,dy;\qquad 2 \le x \le 3,\quad 1 \le y \le 2$

20. $\iint e^{x+y} \cos 2x\,dx\,dy;\qquad 0 \le x \le \pi,\quad 1 \le y \le 2.$

3. ITERATED INTEGRALS

The formula in the last section for integrating products $g(x)h(y)$ is a useful one, but it is inadequate for many functions, such as $f(x, y) = 1/(x + y)$ and $f(x, y) = y \cos(xy)$. In this section we derive the most general method for evaluating double integrals, the method of iterated integration. This method includes the previous rule for products as a special case.

The problem is to compute

$$V = \iint\limits_{\substack{a \leq x \leq b \\ c \leq y \leq d}} f(x, y)\, dx\, dy.$$

Consider the integral as a volume to be found by slicing.

Fix a value of y and slice the region by the corresponding plane parallel to the x, z-plane (Fig. 3.1).

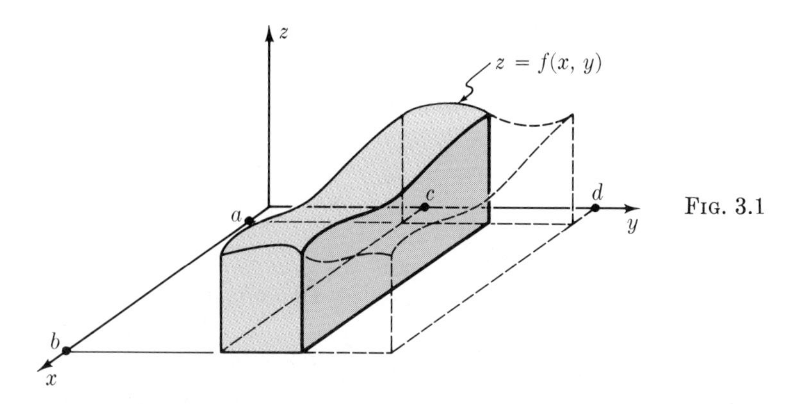

FIG. 3.1

The resulting cross-section has area $A(y)$. Write the volume to the left of this slice as $V(y)$. (Thus $V(c) = 0$, $V(d) =$ the desired volume.) It is fundamental to the whole integration process that

$$\frac{dV}{dy} = A(y).$$

(Please review Chapter 8, Section 2, p. 108, and Chapter 12, Section 1, p. 171.) Hence

$$V = \int_c^d A(y)\, dy.$$

Now $A(y)$, the area of the cross-section (Fig. 3.2), can be expressed as a simple integral. Indeed, $A(y)$ is the area under the curve $z = f(x, y)$ from

FIG. 3.2

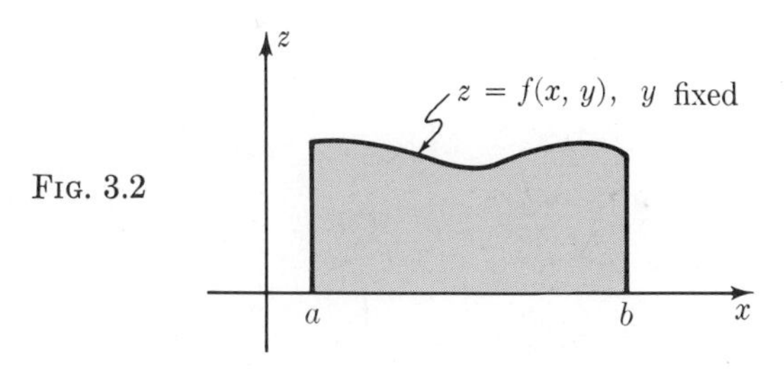

$x = a$ to $x = b$. (Remember y is *fixed* in this process.) Thus

$$A(y) = \int_a^b f(x, y)\,dx,$$

where y is treated as a constant in computing the integral.

The final result is the following pair of formulas.

Iteration Formulas

$$\iint\limits_{\substack{a \le x \le b \\ c \le y \le d}} f(x, y)\,dx\,dy = \int_c^d \left(\int_a^b f(x, y)\,dx \right) dy$$

$$= \int_a^b \left(\int_c^d f(x, y)\,dy \right) dx.$$

(The second form is obtained by reversing the roles of x and y in the argument.)

REMARK: The expressions on the right are **iterated integrals,** i.e., repetitions of simple integrals. They are often written this way:

$$\int_c^d dy \int_a^b f(x, y)\,dx, \qquad \int_a^b dx \int_c^d f(x, y)\,dy.$$

EXAMPLE 3.1

Find $\displaystyle\iint\limits_{\substack{0 \le x \le 1 \\ 1 \le y \le 2}} \frac{dx\,dy}{x + y}$.

Solution:

$$\iint = \int_0^1 \left(\int_1^2 \frac{dy}{x + y} \right) dx.$$

Now for fixed x,

$$\int_1^2 \frac{dy}{x + y} = \ln(x + y) \Big|_{y=1}^{y=2} = \ln(2 + x) - \ln(1 + x).$$

Hence

$$\iint = \int_0^1 [\ln(2 + x) - \ln(1 + x)]\,dx.$$

But

$$\int \ln u\,du = u \ln u - u + C,$$

therefore,

$$\int_0^1 [\ln(2+x) - \ln(1+x)]\,dx$$

$$= \Big[(2+x)\ln(2+x) - (2+x) - (1+x)\ln(1+x) + (1+x)\Big]\Big|_0^1$$

$$= 3\ln 3 - 2\ln 2 - 2\ln 2 = 3\ln 3 - 4\ln 2.$$

Answer: $\ln\dfrac{27}{16}$.

An important feature of the Iteration Formulas above is that the iteration may be done in either order. Sometimes the computation is difficult in one order but relatively easy in the opposite order.

EXAMPLE 3.2

Let R be the region $0 \le x \le 1$ and $0 \le y \le \pi$. Find

$$\iint_R y\cos(xy)\,dx\,dy.$$

Solution: Here is one set-up:

$$\iint_R = \int_0^1 \left(\int_0^\pi y\cos(xy)\,dy\right) dx.$$

The inner integral,

$$\int_0^\pi y\cos(xy)\,dy,$$

while not impossible to integrate (by parts), is messy. The alternate procedure is iteration in the other order:

$$\iint_R = \int_0^\pi \left(\int_0^1 y\cos(xy)\,dx\right) dy.$$

Since y is constant in the inner integration, this can be rewritten as

$$\iint_R = \int_0^\pi y\left(\int_0^1 \cos(xy)\,dx\right) dy.$$

Now

$$\int_0^1 \cos(xy)\,dx = \frac{1}{y}\sin(xy)\Big|_{x=0}^{x=1} = \frac{\sin y}{y}.$$

Hence

$$\iint_R = \int_0^\pi y\frac{\sin y}{y}\,dy = \int_0^\pi \sin y\,dy = 2.$$

Alternate writing of the solution:

$$\iint\limits_R y\cos(xy)\,dx\,dy = \int_0^\pi y\,dy\int_0^1 \cos(xy)\,dx$$

$$= \int_0^\pi y\left(\frac{\sin(xy)}{y}\Bigg|_0^1\right)dy = \int_0^\pi \sin y\,dy = 2.$$

Answer: 2.

EXERCISES

Evaluate:

1. $\iint \frac{dx\,dy}{(x+y)^2}$; $\quad 0 \le x \le 1, \quad 1 \le y \le 2$
2. $\iint \frac{x}{y}\,dx\,dy$; $\quad 0 \le x \le 1, \quad 1 \le y \le 5$
3. $\iint \frac{x^2}{y^2}\,dx\,dy$; $\quad -1 \le x \le 1, \quad 1 \le y \le 3$
4. $\iint e^{x+y}\,dx\,dy$; $\quad -1 \le x, y \le 0$
5. $\iint y^2\sin(xy)\,dx\,dy$; $\quad 0 \le x \le 2\pi, \quad 0 \le y \le 1$
6. $\iint (1-2x)\sin(y^2)\,dx\,dy$; $\quad 0 \le x, y \le 1$
7. $\iint e^y\sin(x/y)\,dx\,dy$; $\quad -\pi/2 \le x \le \pi/2, \quad 1 \le y \le 2$
8. $\iint (1-x+2y)^2\,dx\,dy$; $\quad 3 \le x \le 4, \quad 1 \le y \le 2$
9. $\iint (1+x+y)(3+x-y)\,dx\,dy$; $\quad 2 \le x, y \le 3$
10. $\iint \sin(x+y)\,dx\,dy$; $\quad 0 \le x, y \le \pi/2$
11. $\iint (x+y)^n\,dx\,dy$; $\quad 0 \le x \le 1, \quad 1 \le y \le 2$
12. $\iint (x-y)^n\,dx\,dy$; $\quad 0 \le x, y \le 1.$
13. Suppose $f(-x, -y) = -f(x, y)$. Prove $\iint f(x, y)\,dx\,dy = 0$; $\quad -1 \le x, y \le 1.$
14. Suppose $f(x, -y) = -f(x, y)$. Prove $\iint f(x, y)\,dx\,dy = 0$; $\quad -1 \le x, y \le 1.$

15. Find the constant A that best approximates $f(x, y)$ on the square $0 \leq x, y \leq 1$ in the least squares sense. In other words, minimize

$$\iint [f(x, y) - A]^2 \, dx \, dy; \qquad 0 \leq x, y \leq 1.$$

16. (cont.) Find the least squares linear approximation

$$A + Bx + Cy$$

to $f(x, y) = xy$ on the square $0 \leq x, y \leq 1$.

17. (cont.) Show that the coefficients of the least squares linear approximation $A + Bx + Cy$ to $f(x, y)$ on the square $0 \leq x, y \leq 1$ satisfy

$$\begin{cases} A + \frac{1}{2}B + \frac{1}{2}C = \iint f \, dx \, dy \\ \frac{1}{2}A + \frac{1}{3}B + \frac{1}{4}C = \iint x f \, dx \, dy \\ \frac{1}{2}A + \frac{1}{4}B + \frac{1}{3}C = \iint y f \, dx \, dy. \end{cases}$$

4. APPLICATIONS

In this section we study mass, center of gravity, and volume.

Mass and Density

Suppose a sheet of non-homogeneous material covers the rectangle $a \leq x \leq b$ and $c \leq y \leq d$. See Fig. 4.1. At each point (x, y), let $\rho(x, y)$

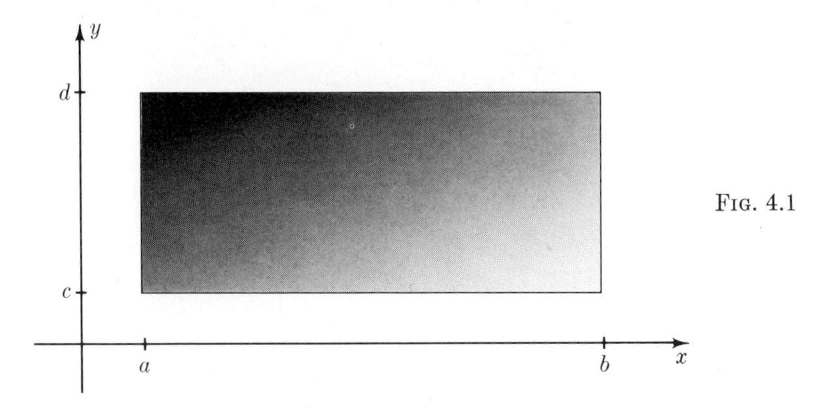

Fig. 4.1

denote the **density** of the material, i.e., the mass per unit area. (Dimensionally, planar density is mass divided by length squared. Common units are gm/cm^2 and lb/ft^2.)

The mass of a small rectangular portion of the sheet (Fig. 4.2) is

$$dM \approx \rho(x, y) \, dx \, dy.$$

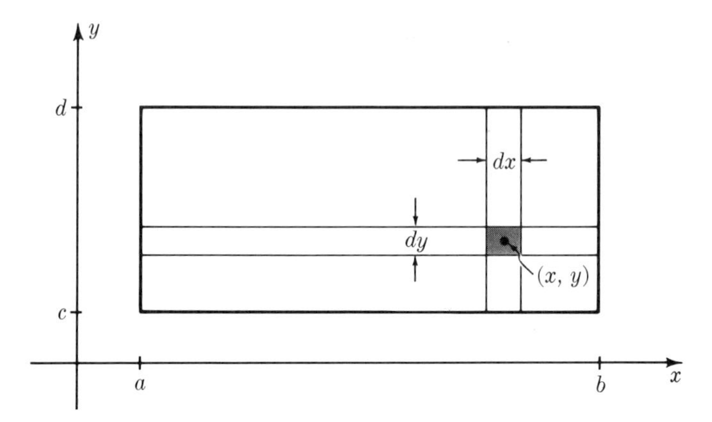

FIG. 4.2

Therefore the total mass of the sheet is

$$M = \iint\limits_{\substack{a \leq x \leq b \\ c \leq y \leq d}} \rho(x, y)\, dx\, dy.$$

EXAMPLE 4.1

The density (lb/ft^2) at each point of a one-foot square of plastic is the product of the four distances of the point from the sides of the square. Find the total mass.

Solution: Take the square in the position $0 \leq x,\ y \leq 1$. Then

$$\rho(x, y) = x(1 - x)y(1 - y),$$

$$M = \iint\limits_{R} \rho(x, y)\, dx\, dy$$

$$= \left(\int_0^1 x(1 - x)\, dx\right)\left(\int_0^1 y(1 - y)\, dy\right) = \frac{1}{6} \cdot \frac{1}{6} = \frac{1}{36}.$$

Answer: $\dfrac{1}{36}$ lb.

Moment and Center of Gravity

Suppose gravity (perpendicular to the plane of the figure) acts on the rectangular sheet of Fig. 4.1. The sheet is to be suspended by a single point so that it balances parallel to the floor. This point of balance is the **center of gravity** of the sheet and is denoted $\bar{\mathbf{x}} = (\bar{x}, \bar{y})$.

The center of gravity $\bar{\mathbf{x}}$ is found in three steps:

(1) Find the mass M.

(2) Find the **moment** of the sheet with respect to the origin. This is the vector

$$\mathbf{m} = \iint_R \rho\,\mathbf{x}\,dx\,dy = \Big(\iint_R \rho\,x\,dx\,dy,\ \iint_R \rho\,y\,dx\,dy\Big)$$

Here R denotes the region the sheet occupies, $a \le x \le b$ and $c \le y \le d$, and $\rho = \rho(x, y)$ is the density.

(3) Divide the moment by the mass to obtain the center of gravity:

$$\bar{\mathbf{x}} = \frac{1}{M}\,\mathbf{m}.$$

This formula will be derived after two examples.

EXAMPLE 4.2

Find the center of gravity of a homogeneous rectangular sheet.

Solution: "Homogeneous" means the density ρ is constant. Take the sheet in the position $0 \le x \le a$ and $0 \le y \le b$. The mass is $M = \rho ab$, and the moment is

$$\begin{aligned}\mathbf{m} = \iint_R \rho\,\mathbf{x}\,dx\,dy = \rho\iint_R \mathbf{x}\,dx\,dy &= \rho\,\Big(\iint_R x\,dx\,dy,\ \iint_R y\,dx\,dy\Big)\\ &= \rho\,\Big(\int_0^a x\,dx\int_0^b dy,\ \int_0^a dx\int_0^b y\,dy\Big)\\ &= \rho\,\Big(\frac{1}{2}a^2b,\ \frac{1}{2}ab^2\Big).\end{aligned}$$

The center of gravity is

$$\bar{\mathbf{x}} = \frac{1}{M}\,\mathbf{m} = \frac{1}{\rho ab}\,\rho\,\Big(\frac{1}{2}a^2b,\ \frac{1}{2}ab^2\Big) = \frac{1}{2}\,(a, b).$$

This is the midpoint (intersection of the diagonals) of the rectangle. (Of course the rectangle balances on its midpoint; no one needs calculus for this, but it is reassuring that the analytic method gives the right answer.)

Answer: The midpoint of the rectangle.

EXAMPLE 4.3

A rectangular sheet over the region $1 \le x \le 2$ and $1 \le y \le 3$ has density $\rho(x, y) = xy$. Find its center of gravity.

Solution: The mass is

$$M = \iint xy\,dx\,dy = \int_1^2 x\,dx \int_1^3 y\,dy = \frac{3}{2}\cdot 4 = 6.$$

The moment is

$$\mathbf{m} = \iint_R xy\,\mathbf{x}\,dx\,dy = \Big(\iint_R x^2y\,dx\,dy,\ \iint_R xy^2\,dx\,dy\Big)$$

$$= \Big(\int_1^2 x^2\,dx \int_1^3 y\,dy,\ \int_1^2 x\,dx \int_1^3 y^2\,dy\Big)$$

$$= \Big(\frac{7}{3}\cdot 4,\ \frac{3}{2}\cdot\frac{26}{3}\Big) = \frac{1}{3}(28, 39).$$

Therefore

$$\bar{\mathbf{x}} = \frac{1}{M}\mathbf{m} = \frac{1}{6}\cdot\frac{1}{3}(28, 39) = \Big(\frac{14}{9}, \frac{13}{6}\Big).$$

Note that $\bar{\mathbf{x}}$ is inside the rectangle, a little northeast of center. Could you have predicted this?

Answer: $\bar{\mathbf{x}} = \Big(\frac{14}{9}, \frac{13}{6}\Big).$

The formula for center of gravity is derived by balancing the rectangular sheet on various knife edges.

Suppose a knife edge passes through $\bar{\mathbf{x}}$ and the sheet balances (Fig. 4.3). Divide the sheet into many small rectangles. The turning moments of these pieces about the knife edge must add up to zero.

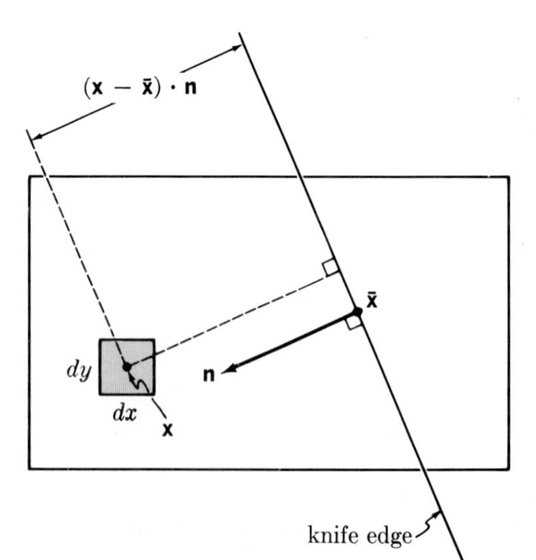

FIG. 4.3

Let $\mathbf{n}$ be a unit vector in the plane of the knife edge perpendicular to the knife edge. A small rectangle with sides dx and dy located at $\mathbf{x}$ has (signed) distance $(\mathbf{x} - \bar{\mathbf{x}}) \cdot \mathbf{n}$ from the knife edge and has mass $\rho\, dx\, dy$. Hence its turning moment is

$$\rho (\mathbf{x} - \bar{\mathbf{x}}) \cdot \mathbf{n}\, dx\, dy.$$

The sum of all such must vanish:

$$\iint_R \rho (\mathbf{x} - \bar{\mathbf{x}}) \cdot \mathbf{n}\, dx\, dy = 0.$$

Since $\bar{\mathbf{x}}$ and $\mathbf{n}$ are constant, this relation may be written

$$\mathbf{n} \cdot \iint_R \rho\, \mathbf{x}\, dx\, dy = (\mathbf{n} \cdot \bar{\mathbf{x}}) \iint_R \rho\, dx\, dy,$$

or

$$\mathbf{n} \cdot \mathbf{m} = M \mathbf{n} \cdot \bar{\mathbf{x}}.$$

This equation of balance is true for each choice of the knife edge (each choice of the unit vector $\mathbf{n}$). Hence,

$$\mathbf{n} \cdot (\mathbf{m} - M\bar{\mathbf{x}}) = \mathbf{0}$$

for each unit vector $\mathbf{n}$. This means the component of $\mathbf{m} - M\bar{\mathbf{x}}$ in each direction is zero. Therefore,

$$\mathbf{m} - M\bar{\mathbf{x}} = \mathbf{0},$$

$$\mathbf{m} = M\bar{\mathbf{x}}.$$

Volume in Polar Coordinates

A surface lies over a region (Fig. 4.4a) of the x, y-plane described in polar coordinates by $r_0 \le r \le r_1$ and $\theta_0 \le \theta \le \theta_1$. The surface is described

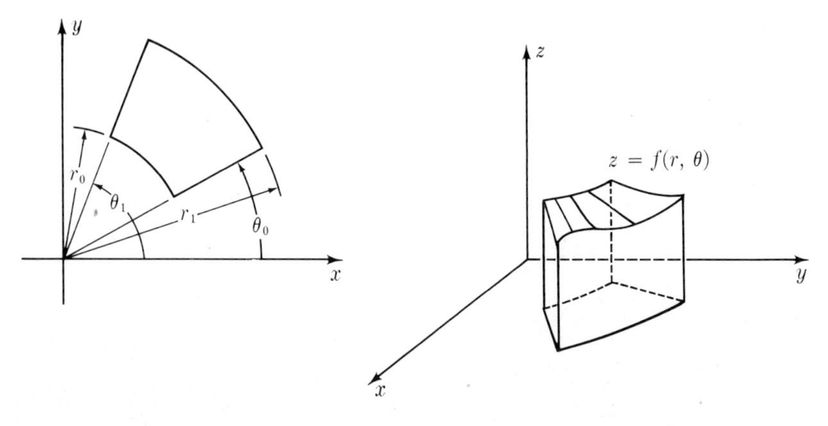

FIG. 4.4a FIG. 4.4b

(Fig. 4.4b) by $z = f(r, \theta)$. What is the volume of the region under this surface and over the r, θ-plane?

Split the interval $r_0 \le r \le r_1$ into small pieces. Do the same for $\theta_0 \le \theta \le \theta_1$. The result is a decomposition of the plane region into many small, almost rectangular, regions (Fig. 4.5). Each has dimensions dr and

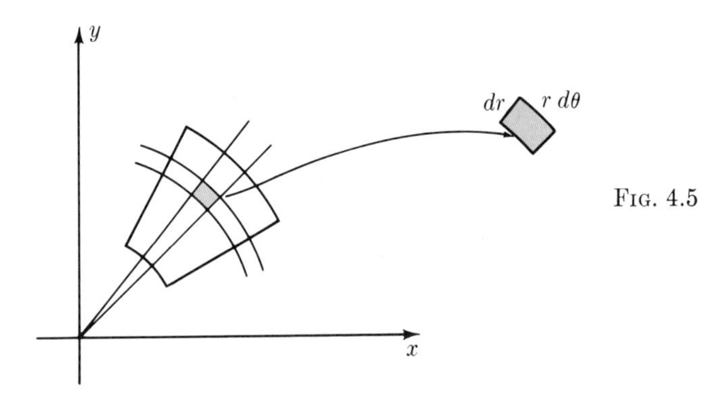

FIG. 4.5

$r\,d\theta$, hence area $dA = r\,dr\,d\theta$. The portion of the solid lying over this elementary "rectangle" has height $z = f(r, \theta)$, hence its volume is $dV = f(r, \theta)\,r\,dr\,d\theta$. See Fig. 4.6. The total volume is

$$V = \iint_R f(r, \theta)\,r\,dr\,d\theta = \int_{\theta_0}^{\theta_1} \left(\int_{r_0}^{r_1} f(r, \theta)\,r\,dr \right) d\theta$$
$$= \int_{r_0}^{r_1} r \left(\int_{\theta_0}^{\theta_1} f(r, \theta)\,d\theta \right) dr.$$

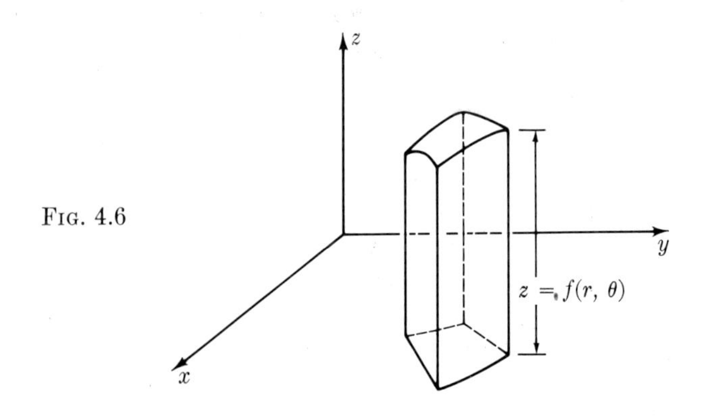

FIG. 4.6

EXAMPLE 4.4

Find the volume under the cone $z = r$, and over the region $0 \le r \le a$ and $0 \le \theta \le \pi/2$.

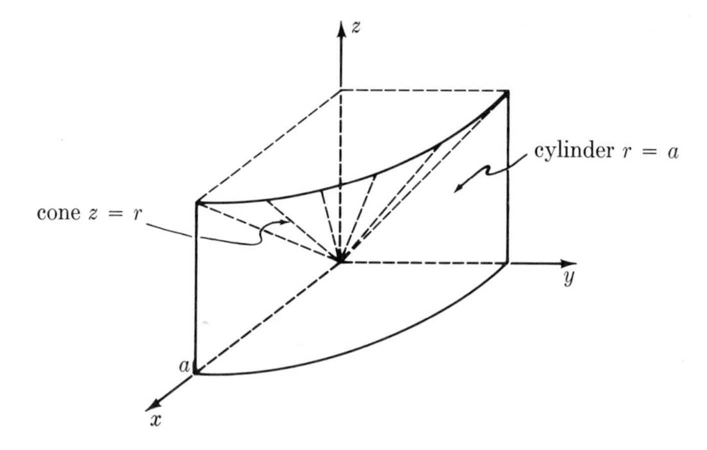

FIG. 4.7

Solution: The solid in question is shown in Fig. 4.7. Its volume is

$$V = \iint z\, r\, dr\, d\theta = \iint r^2\, dr\, d\theta$$

$$= \int_0^a r^2\, dr \int_0^{\pi/2} d\theta = \frac{a^3}{3} \cdot \frac{\pi}{2}.$$

Answer: $V = \dfrac{\pi a^3}{6}.$

Mass and Center of Gravity in Polar Coordinates

The mass and the center of gravity of a non-homogeneous sheet over the region in Fig. 4.5 can be computed by double integrals. Since the element of area is

$$dA = r\, dr\, d\theta,$$

the elementary mass is

$$dM = \rho(r, \theta)\, r\, dr\, d\theta.$$

Note that ρ must be expressed in polar coordinates. Thus

$$\boxed{M = \iint_R \rho\, r\, dr\, d\theta.}$$

To find the moment, express $\mathbf{x}$ in polar coordinates:

$$\mathbf{x} = (x, y) = (r \cos \theta, r \sin \theta).$$

Then $\mathbf{\bar{x}} = \mathbf{m}/M$, where

$$\mathbf{m} = \iint_R \mathbf{x}\, dM = \iint_R \mathbf{x}\, \rho\, r\, dr\, d\theta = \iint_{\substack{r_0 \le r \le r_1 \\ \theta_0 \le \theta \le \theta_1}} (r\cos\theta, r\sin\theta)\, \rho\, r\, dr\, d\theta.$$

EXAMPLE 4.5

Find the center of gravity of a homogeneous sheet in the shape of the semicircle $0 \le r \le a$ and $0 \le \theta \le \pi$.

Solution: Here ρ is constant. The mass is

$$M = \frac{1}{2}\pi a^2 \rho.$$

The moment is

$$\mathbf{m} = \iint_R \mathbf{x}\, \rho\, r\, dr\, d\theta = \iint_R (r\cos\theta, r\sin\theta)\, \rho\, r\, dr\, d\theta$$

$$= \rho \left(\iint_R r^2 \cos\theta\, dr\, d\theta, \iint_R r^2 \sin\theta\, dr\, d\theta \right)$$

$$= \rho \left(\int_0^a r^2\, dr \int_0^\pi \cos\theta\, d\theta, \int_0^a r^2\, dr \int_0^\pi \sin\theta\, d\theta \right) = \rho \left(0, \frac{2}{3} a^3\right).$$

Therefore

$$\mathbf{\bar{x}} = \frac{1}{M}\mathbf{m} = \frac{2}{\pi a^2}\left(0, \frac{2}{3}a^3\right).$$

Answer: $\mathbf{\bar{x}} = \left(0, \dfrac{4a}{3\pi}\right)$.

EXAMPLE 4.6

Find the center of gravity of a sheet with density $\rho = r$ covering the quarter circle $0 \le r \le a$ and $0 \le \theta \le \pi/2$.

Solution: The mass is

$$M = \iint_R \rho\, r\, dr\, d\theta = \iint_R r^2\, dr\, d\theta = \int_0^a r^2\, dr \int_0^{\pi/2} d\theta = \frac{\pi a^3}{6}.$$

The moment is

$$\mathbf{m} = \iint_R \mathbf{x}\,\rho\, r\, dr\, d\theta = \iint_R (r\cos\theta,\, r\sin\theta)\, r^2\, dr\, d\theta$$

$$= \int_0^a r^3\, dr \int_0^{\pi/2} (\cos\theta,\, \sin\theta)\, d\theta = \frac{a^4}{4}(1, 1).$$

Therefore

$$\bar{\mathbf{x}} = \frac{1}{M}\mathbf{m} = \frac{6}{\pi a^3} \cdot \frac{a^4}{4}(1, 1).$$

Answer: $\bar{\mathbf{x}} = \left(\frac{3a}{2\pi}, \frac{3a}{2\pi}\right).$

Could you have predicted that $\bar{\mathbf{x}}$ lies on the line $y = x$ and $|\bar{\mathbf{x}}| > a/2$?

EXERCISES

Find the volume under the surface and over the portion of the x, y-plane indicated. Draw a figure in each case:

1. $z = 2 - (x^2 + y^2)$; $\quad -1 \le x, y \le 1$
2. $z = 1 - xy$; $\quad 0 \le x, y \le 1$
3. $z = x^2 + 4y^2$; $\quad 0 \le x \le 2, \quad 0 \le y \le 1$
4. $z = \sin x \sin y$; $\quad 0 \le x, y \le \pi$
5. $z = x^2y + y^2x$; $\quad 1 \le x \le 2, \quad 2 \le y \le 3$
6. $z = (1 + x^3)y^2$; $\quad -1 \le x, y \le 1.$

A sheet of non-homogeneous material of density ρ gm/cm² covers the indicated rectangle. Find the mass of the sheet, assuming lengths are measured in centimeters:

7. $\rho = 3(1 + x)(1 + y)$; $\quad 0 \le x, y \le 1$
8. $\rho = 1 - 0.2xy$; $\quad 0 \le x \le 1, \quad 1 \le y \le 1.5$
9. $\rho = 3 + 0.1x$; $\quad 2 \le x \le 3, \quad -1 \le y \le 1$
10. $\rho = 4e^{x+y} - 2$; $\quad 0 \le x \le 1, \quad 0 \le y \le 0.5.$

Find the center of gravity of each rectangular sheet, density as given:

11. $\rho = (1 - x)(1 - y) + 1$; $\quad 0 \le x, y \le 1$
12. $\rho = \sin x$; $\quad 0 \le x \le \pi, \quad 0 \le y \le 1$
13. $\rho = \sin x(1 - \sin y)$; $\quad \pi/2 \le x, y \le \pi$
14. $\rho = 10 - e^{x+y}$; $\quad 0 \le x, y \le 1$
15. $\rho = 1 + x^2 + y^2$; $\quad -1 \le x \le 1, \quad 1 \le y \le 4$
16. $\rho = 2 + x^2y^2$; $\quad -1 \le x \le 1, \quad 0 \le y \le 1.$

Find the volume over the region given in polar coordinates and under the indicated surface. Sketch:

17. $z = x^2 + y^2$; $\quad 1 \leq r \leq 2, \quad 0 \leq \theta \leq \pi/2$
18. $z = x$; $\quad 0 \leq r \leq 1, \quad -\pi/2 \leq \theta \leq \pi/2$
19. $z = xy$; $\quad 1 \leq r \leq 2, \quad \pi/4 \leq \theta \leq \pi/2$.
20. Use polar coordinates to find the volume of a hemisphere of radius a.
21. Find the volume of the region bounded by the two paraboloids of revolution $z = x^2 + y^2$ and $z = 4 - 3(x^2 + y^2)$.
22. Find the volume of the lens-shaped region common to the sphere of radius 1 centered at (0, 0, 0) and the sphere of radius 1 centered at (0, 0, 1).
23. A drill of radius b bores on center through a sphere of radius a, where $a > b$. How much material is removed?
24. Find the center of gravity of a sheet with uniform density in the shape of a quarter circle $0 \leq r \leq a$ and $0 \leq \theta \leq \pi/2$.
25. (cont.) Now do it the easy way, using the result of Example 4.5.
26. Find the center of gravity of a homogeneous wedge of pie in the position $0 \leq r \leq a$ and $0 \leq \theta \leq \alpha$.
27. (cont.) Hold the radius a of Ex. 26 fixed, but let $\alpha \longrightarrow 0$. What is the limiting position of the center of gravity?
28. A quarter-circular sheet in the position $0 \leq r \leq a$ and $0 \leq \theta \leq \pi/2$ has density $\rho = a^2 - r^2$. Find its center of gravity.

5. NUMERICAL INTEGRATION [optional]

In this section we discuss one method for approximating double integrals. It is an extension of Simpson's Rule studied in Chapter 20.

Let us recall Simpson's Rule. To approximate an integral

$$\int_a^b f(x)\,dx,$$

we divide the interval $a \leq x \leq b$ into $2m$ equal parts of length h:

$$a = x_0 < x_1 < x_2 < \cdots < x_{2m} = b, \qquad h = \frac{b-a}{2m},$$

and use the formula

$$\int_a^b f(x)\,dx \approx \frac{h}{3}\sum_{i=0}^{2m} B_i\,f(x_i),$$

where the coefficients are 1, 4, 2, 4, 2, 4, 2, $\cdots$, 2, 4, 1.

We extend Simpson's Rule to double integrals in the following way. To approximate

$$\iint_R f(x, y)\,dx\,dy,$$

where R denotes the rectangle $a \leq x \leq b$ and $c \leq y \leq d$, we divide the x-interval into $2m$ parts as before and also divide the y-interval into $2n$ equal parts of length k:

$$c = y_0 < y_1 < y_2 < \cdots < y_{2n} = d, \qquad k = \frac{d - c}{2n}.$$

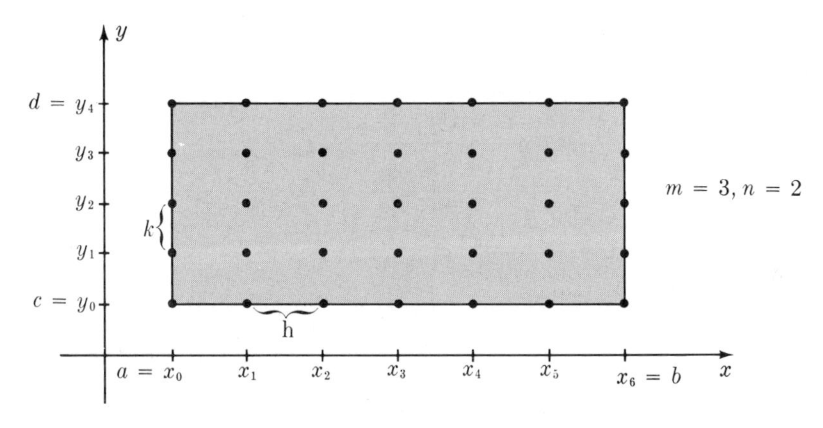

FIG. 5.1

We obtain $(2m + 1)(2n + 1)$ points of the rectangle R shown in Fig. 5.1. The Rule is

$$\iint_R f(x, y)\, dx\, dy \approx \frac{hk}{9} \sum_{i=0}^{2m} \sum_{j=0}^{2n} A_{ij} f(x_i, y_j),$$

where the coefficients A_{ij} are certain products of the coefficients in the ordinary Simpson's Rule. Precisely,

$$A_{ij} = B_i C_j,$$

where $B_0, B_1, \cdots, B_{2m}$ are the coefficients in the ordinary Simpson's Rule

$$\int_a^b p(x)\, dx \approx \frac{h}{3} \sum_{i=0}^{2m} B_i\, p(x_i),$$

and $C_0, C_1, \cdots, C_{2n}$ are the coefficients in the ordinary Simpson's Rule

$$\int_c^d q(y)\, dy \approx \frac{k}{3} \sum_{j=0}^{2n} C_j\, q(y_j).$$

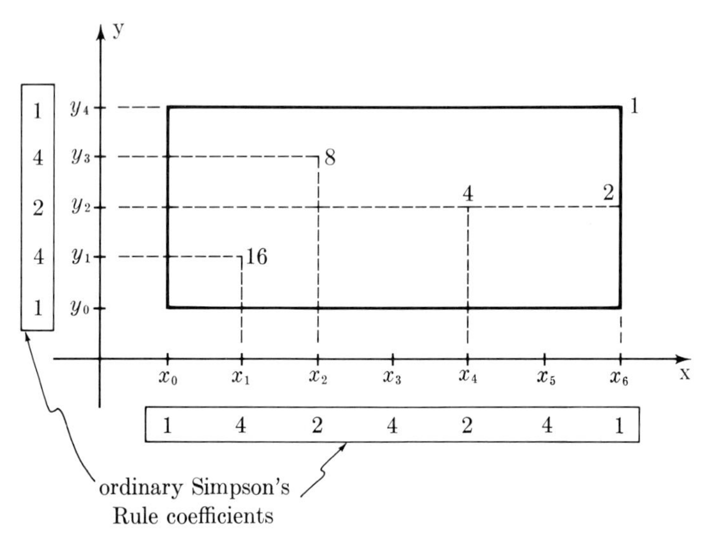

FIG. 5.2

In Fig. 5.2, several of these products are formed. Since B_i and C_j take values 1, 2, and 4, the coefficients A_{ij} take values 1, 2, 4, 8, and 16. The A_{ij} can be written in an array (matrix) corresponding to the points (x_i, y_j) as in Fig. 5.2. For example, if $m = 3$ and $n = 2$, the array is

$$\begin{bmatrix} 1 & 4 & 2 & 4 & 2 & 4 & 1 \\ 4 & 16 & 8 & 16 & 8 & 16 & 4 \\ 2 & 8 & 4 & 8 & 4 & 8 & 2 \\ 4 & 16 & 8 & 16 & 8 & 16 & 4 \\ 1 & 4 & 2 & 4 & 2 & 4 & 1 \end{bmatrix}.$$

EXAMPLE 5.1

Estimate $\displaystyle\iint_{0 \le x, y \le 1} (x + y)^3 \, dx \, dy$ by Simpson's Rule with $m = n = 1$.

Compare the result with the exact answer.

Solution: Here $h = k = \frac{1}{2}$. The array of coefficients is

$$[A_{ij}] = \begin{bmatrix} 1 & 4 & 1 \\ 4 & 16 & 4 \\ 1 & 4 & 1 \end{bmatrix}.$$

Write the values $(x_i + y_j)^3$ in an array:

$$[(x_i + y_j)^3] = \begin{bmatrix} (1+0)^3 & \left(1+\frac{1}{2}\right)^3 & (1+1)^3 \\ \left(\frac{1}{2}+0\right)^3 & \left(\frac{1}{2}+\frac{1}{2}\right)^3 & \left(\frac{1}{2}+1\right)^3 \\ (0+0)^3 & \left(0+\frac{1}{2}\right)^3 & (0+1)^3 \end{bmatrix} = \begin{bmatrix} 1 & \frac{27}{8} & 8 \\ \frac{1}{8} & 1 & \frac{27}{8} \\ 0 & \frac{1}{8} & 1 \end{bmatrix}.$$

Now estimate the integral by

$$\frac{hk}{9}\sum_{i=0}^{2}\sum_{j=0}^{2} A_{ij}(x_i + y_j)^3.$$

To evaluate this sum, multiply corresponding terms of the two arrays, and add the nine products:

$$\frac{1}{9}\left(\frac{1}{2}\right)\left(\frac{1}{2}\right)\left[1\cdot 1 + 4\cdot\frac{27}{8} + 1\cdot 8 + 4\cdot\frac{1}{8}\right.$$
$$\left. + 16\cdot 1 + 4\cdot\frac{27}{8} + 1\cdot 0 + 4\cdot\frac{1}{8} + 1\cdot 1\right]$$
$$= \frac{1}{36}\left[1 + \frac{27}{2} + 8 + \frac{1}{2} + 16 + \frac{27}{2} + \frac{1}{2} + 1\right] = \frac{54}{36} = \frac{3}{2}.$$

The exact value is

$$\iint\limits_{0\le x,y\le 1} (x+y)^3\,dx\,dy = \int_0^1\left(\int_0^1 (x+y)^3\,dx\right)dy$$
$$= \frac{1}{4}\int_0^1 [(y+1)^4 - y^4]\,dy$$
$$= \frac{1}{20}[(y+1)^5 - y^5]\Big|_0^1 = \frac{30}{20} = \frac{3}{2}.$$

Answer: Simpson's Rule gives the estimate $\frac{3}{2}$, which is exact.

REMARK 1: Because Simpson's Rule is exact for cubics, the double integral rule is exact for cubics in two variables. (See exercises below.)

REMARK 2: The array of values $[f(x_i, y_j)]$ is arranged to conform to the layout of points (x_i, y_j) in the plane (Fig. 5.1).

The next example is an integral which cannot be evaluated exactly, only approximated.

EXAMPLE 5.2

Estimate $\displaystyle\iint_{0\le x,y\le\pi/2} \sin(xy)\,dx\,dy$, using $m = n = 1$.

Solution: Here $h = k = \pi/4$, and the array of coefficients is

$$[A_{ij}] = \begin{bmatrix} 1 & 4 & 1 \\ 4 & 16 & 4 \\ 1 & 4 & 1 \end{bmatrix}.$$

The array of values of $\sin(xy)$ is

$$[\sin x_iy_j] = \begin{bmatrix} \sin 0 & \sin\frac{\pi^2}{8} & \sin\frac{\pi^2}{4} \\ \sin 0 & \sin\frac{\pi^2}{16} & \sin\frac{\pi^2}{8} \\ \sin 0 & \sin 0 & \sin 0 \end{bmatrix}.$$

The estimate is

$$\iint_{0\le x,y\le\pi/2} \sin(xy)\,dx\,dy \approx \frac{\pi^2}{144}\left(16\sin\frac{\pi^2}{16} + 8\sin\frac{\pi^2}{8} + \sin\frac{\pi^2}{4}\right).$$

Error Estimate

The error estimate for Simpson's Rule in two variables is analogous to that in one variable. (See Chapter 20, Section 7, p. 405.) For $m = n = 1$ it is

$$|\text{error}| \le \frac{hk}{45}[h^4M + k^4N],$$

where

$$\left|\frac{\partial^4 f}{\partial x^4}\right| \le M \qquad \text{and} \qquad \left|\frac{\partial^4 f}{\partial y^4}\right| \le N.$$

For general m and n it is

$$\frac{(b-a)(d-c)}{180}[h^4M + k^4N].$$

We omit the proof.

EXAMPLE 5.3

Estimate the error in Example 5.2.

Solution:

$$\frac{\partial^4}{\partial x^4}(\sin xy) = y^4\sin xy, \qquad \frac{\partial^4}{\partial y^4}(\sin xy) = x^4\sin xy.$$

But $|\sin xy| \le 1$. Hence in the square $0 \le x,\ y \le \pi/2$,

$$\left|\frac{\partial^4 f}{\partial x^4}\right| = |x^4 \sin xy| \le \left(\frac{\pi}{2}\right)^4, \qquad \left|\frac{\partial^4 f}{\partial y^4}\right| = |y^4 \sin xy| \le \left(\frac{\pi}{2}\right)^4.$$

Apply the error estimate, with $m = n = 1$, $h = k = \pi/4$, and $M = N = (\pi/2)^4$:

$$|\text{error}| \le \frac{1}{45}\left(\frac{\pi}{4}\right)^2\left[2\left(\frac{\pi}{4}\right)^4\left(\frac{\pi}{2}\right)^4\right] = \frac{1}{45}\frac{\pi^{10}}{2^{15}} \approx 0.064.$$

EXERCISES

1. Use tables to complete Example 5.2.
2. Do Example 5.2 using $m = n = 2$, and compare your estimate to that of Ex. 1. Also estimate the error.
3. Write a program for $\displaystyle\iint_{0 \le x, y \le 1} (x + y)^6\, dx\, dy$; $m = n = 4$. Run the program and compare your estimate to the exact value.
4. Write a program for $\displaystyle\iint_R \frac{dx\, dy}{xy}$; $m = 10$, $n = 5$, where R is the rectangle $1 \le x \le 3$ and $4 \le y \le 5$. Run the program and compare your estimate with the exact value.
5. Suppose $f(x, y) = p(x)q(y)$. Show that the double integral Simpson's Rule estimate is just the product of the Simpson's Rule estimate for $\int p(x)\, dx$ by that for $\int q(y)\, dy$.
6. (cont.) Conclude that the rule is exact for polynomials involving only x^3y^3, x^3y^2, x^2y^3, x^3y, x^2y^2, xy^3, and lower degree terms.
7. The analogue of the Trapezoidal Rule is
$$\iint_{0 \le x, y \le 1} f(x, y)\, dx\, dy \approx \frac{1}{4}\left[f(0, 0) + f(0, 1) + f(1, 1) + f(1, 0)\right].$$
Show this rule is exact for polynomials $f(x, y) = A + Bx + Cy + Dxy$.
8. (cont.) Find the corresponding rule for a rectangle $a \le x \le b$, $c \le x \le d$, divided into rectangles of size h by k with $h = (b - a)/m$ and $k = (d - c)/n$.
9. (cont.) Let S denote the unit square $0 \le x,\ y \le 1$. Suppose $f(x, y) = 0$ at its four vertices. Prove that
$$\iint_S f(x, y)\, dx\, dy = -\frac{1}{2}\iint_S y(1 - y) f_{yy}(x, y)\, dx\, dy$$
$$-\frac{1}{4}\int_0^1 x(1 - x)[f_{xx}(x, 1) + f_{xx}(x, 0)]\, dx.$$

10. (cont.) Suppose also that $|f_{xx}| \leq M$ and $|f_{yy}| \leq N$ on S. Prove that

$$\left| \iint_S f(x, y)\, dx\, dy \right| \leq \frac{1}{12} M + \frac{1}{12} N.$$

11. (cont.) Conclude that for any function $f(x, y)$, the error in the trapezoidal estimate (Ex. 7) is at most $(M + N)/12$.
(*Hint:* Use the result of Ex. 7 and interpolation.)

12. (cont.) Suppose $f(x, y)$ is defined on $0 \leq x \leq h$, $0 \leq y \leq k$, and $f = 0$ at the vertices of this rectangle R. Suppose also $|f_{xx}| \leq M$ and $|f_{yy}| \leq N$. Deduce that

$$\left| \iint_R f(x, y)\, dx\, dy \right| \leq \frac{hk}{12} (h^2 M + k^2 N).$$

PART III

27. Taylor Approximations

1. INTRODUCTION

In this chapter, we continue our study of approximation of functions by Taylor polynomials, begun in Chapter 14.

Why approximate functions by polynomials? Because values of polynomials can be computed by addition and multiplication, simple operations well-suited for hand or machine computations.

Suppose, for example, you need a 6-place table of

$$f(x) = e^{3x^2}$$

at 1000 equally spaced values of x between -1 and 1. If possible, find a polynomial $p(x)$ such that

$$e^{3x^2} = p(x) + \epsilon(x),$$

where $|\epsilon(x)|$ is less than 5×10^{-7} for $-1 \leq x \leq 1$. Then program a computer to tabulate the corresponding values of $p(x)$.

We shall discuss methods for finding polynomial approximations and obtain estimates for the errors in such approximations.

2. POLYNOMIALS

We begin with a basic algebraic property of polynomials: every polynomial can be expressed not only in powers of x, but also in powers of $(x - a)$, where a is any number. This form of the polynomial is convenient for computations near $x = a$.

EXAMPLE 2.1

Express $x^2 + x + 2$ in powers of $x - 1$.

Solution: Set $u = x - 1$. Then $x = u + 1$, and

$$x^2 + x + 2 = (u + 1)^2 + (u + 1) + 2$$
$$= (u^2 + 2u + 1) + (u + 1) + 2 = u^2 + 3u + 4.$$

Answer: $(x - 1)^2 + 3(x - 1) + 4.$

EXAMPLE 2.2

Express $x^3 - 6x^2 + 11x - 6$ in powers of $x - 2$.

Solution: Set $u = x - 2$. Then $x = u + 2$, and

$$x^3 - 6x^2 + 11x - 6 = (u + 2)^3 - 6(u + 2)^2 + 11(u + 2) - 6$$
$$= u^3 - u.$$

Answer: $(x - 2)^3 - (x - 2)$.

REMARK: The answer reveals a symmetry about the point (2, 0) not evident in the original expression for the polynomial.

EXAMPLE 2.3

Express x^4 in powers of $x + 1$.

Solution: $x^4 = [(x + 1) - 1]^4$.

Answer: $(x + 1)^4 - 4(x + 1)^3 + 6(x + 1)^2 - 4(x + 1) + 1$.

The method used in these examples is simple. To express

$$p(x) = A_0 + A_1x + A_2x^2 + A_3x^3 + \cdots + A_nx^n$$

in powers of $x - a$, write $u = x - a$. Then substitute $u + a$ for x:

$$p(x) = A_0 + A_1(u + a) + A_2(u + a)^2 + \cdots + A_n(u + a)^n.$$

Expand each of the powers by the Binomial Formula and collect like powers of u. The result is a polynomial in $u = x - a$, as desired.

This method is laborious when the degree of $p(x)$ exceeds three or four. We now discuss a simpler, more systematic method.

Suppose $p(x)$ is a polynomial expressed in powers of $x - a$:

$$p(x) = A_0 + A_1(x - a) + A_2(x - a)^2 + \cdots + A_n(x - a)^n.$$

What are the coefficients $A_0, A_1, A_2 \cdots$ in terms of $p(x)$?

There is an easy way to compute A_0. Just replace x by a. Then all terms on the right vanish except the first:

$$p(a) = A_0.$$

Now modify this trick to compute A_1. Differentiate $p(x)$:

$$p'(x) = A_1 + 2A_2(x - a) + \cdots + nA_n(x - a)^{n-1}.$$

Substitute $x = a$; again all terms vanish except the first:

$$p'(a) = A_1.$$

Differentiate again to find A_2:

$$p''(x) = 2A_2 + 3 \cdot 2A_3(x - a) + \cdots + n(n-1)A_n(x-a)^{n-2}.$$

Substitute $x = a$:

$$p''(a) = 2A_2.$$

Once again:

$$p'''(x) = 3 \cdot 2A_3 + \cdots + n(n-1)(n-2)A_n(x-a)^{n-3},$$

$$p'''(a) = 3 \cdot 2A_3 = 3!\,A_3.$$

Continuing in this way yields

$$p^{(4)}(a) = 4!\,A_4, \quad p^{(5)}(a) = 5!\,A_5, \quad \cdots, \quad p^{(n)}(a) = n!\,A_n.$$

(Here $p^{(k)}$ is the k-th derivative.)

If $p(x)$ is a polynomial of degree n and if a is a number, then

$$p(x) = p(a) + p'(a)(x-a) + \frac{1}{2!}p''(a)(x-a)^2 + \frac{1}{3!}p'''(a)(x-a)^3 + \cdots + \frac{1}{n!}p^{(n)}(a)(x-a)^n.$$

EXAMPLE 2.4

Express $p(x) = x^3 - x^2 + 1$

(a) in powers of $x - \frac{1}{2}$, (b) in powers of $x - 10$.

Solution: Use the preceding formula with $n = 3$. Compute three derivatives:

$$p'(x) = 3x^2 - 2x, \qquad p''(x) = 6x - 2, \qquad p'''(x) = 6.$$

For (a), evaluate at $x = \frac{1}{2}$:

$$p\left(\frac{1}{2}\right) = \frac{7}{8}, \qquad p'\left(\frac{1}{2}\right) = -\frac{1}{4}, \qquad p''\left(\frac{1}{2}\right) = 1, \qquad p'''\left(\frac{1}{2}\right) = 6.$$

By the formula,

$$p(x) = \frac{7}{8} - \frac{1}{4}\left(x - \frac{1}{2}\right) + \frac{1}{2!} \cdot 1\left(x - \frac{1}{2}\right)^2 + \frac{1}{3!} \cdot 6\left(x - \frac{1}{2}\right)^3.$$

For (b), evaluate at $x = 10$:

$$p(10) = 901, \qquad p'(10) = 280, \qquad p''(10) = 58, \qquad p'''(10) = 6.$$

By the formula,

$$p(x) = 901 + 280(x - 10) + \frac{1}{2!} \cdot 58(x-10)^2 + \frac{1}{3!} \cdot 6(x-10)^3.$$

Answer:

(a) $\frac{7}{8} - \frac{1}{4}\left(x - \frac{1}{2}\right) + \frac{1}{2}\left(x - \frac{1}{2}\right)^2 + \left(x - \frac{1}{2}\right)^3$,

(b) $901 + 280(x - 10) + 29(x - 10)^2 + (x - 10)^3$.

The next example illustrates the computational advantages gained by expanding polynomials in powers of $x - a$.

EXAMPLE 2.5

Let $p(x) = x^3 - x^2 + 1$. Compute $p(0.50028)$ to 5 places.

Solution: Use answer (a) of the preceding example:

$$p(0.50028) = p\left(\frac{1}{2} + 0.00028\right)$$

$$= \frac{7}{8} - \frac{1}{4}(0.00028) + \frac{1}{2}(0.00028)^2 + (0.00028)^3.$$

The last two terms on the right are smaller than 10^{-7}. Therefore to 5 places, $p(0.50028)$ agrees with

$$\frac{7}{8} - \frac{1}{4}(0.00028) = 0.87500 - 0.00007.$$

Answer: 0.87493.

Notation

Because we shall write polynomials frequently, it is convenient to use the notation

$$\sum_{i=0}^{n} A_i x^i = A_0 + A_1 x + A_2 x^2 + \cdots + A_n x^n.$$

The sigma $\sum$ denotes "summation" and the notation $\sum_{i=0}^{n}$ means the index i of summation ranges from 0 to n. For example

$$\sum_{i=0}^{3} (2i + 1)x^i = (2 \cdot 0 + 1)x^0 + (2 \cdot 1 + 1)x^1$$
$$+ (2 \cdot 2 + 1)x^2 + (2 \cdot 3 + 1)x^3 = 1 + 3x + 5x^2 + 7x^3,$$

$$\sum_{i=0}^{4} \frac{1}{i+1} x^i = 1 + \frac{1}{2}x + \frac{1}{3}x^2 + \frac{1}{4}x^3 + \frac{1}{5}x^4.$$

The index i need not start at $i = 0$. For example,

$$\sum_{i=3}^{7} i^2 x^i = 9x^3 + 16x^4 + 25x^5 + 36x^6 + 49x^7.$$

The formula for an n-th degree polynomial in powers of $x - a$ now can be abbreviated:

$$p(x) = \sum_{i=0}^{n} \frac{p^{(i)}(a)}{i!} (x - a)^i.$$

Here $p^{(i)}(a)$ denotes the i-th derivative of $p(x)$ evaluated at $x = a$, with the special convention $p^{(0)}(a) = p(a)$. (Also recall the convention $0! = 1$.)

EXERCISES

Expand in powers of $x - a$:

1. $x^2 + 5x + 2;\quad a = 1$
2. $x^3 - 3x^2 + 4x;\quad a = 2$
3. $2x^3 + 5x^2 + 13x + 10;\quad a = -1$
4. $x^4 - 5x^2 + x + 2;\quad a = 2$
5. $2x^4 + 5x^3 + 4x + 16;\quad a = -2$
6. $3x^3 - 2x^2 - 2x + 1;\quad a = 1$
7. $5x^5 + 4x^4 - 3x^3 - 2x^2 + x + 1;\quad a = -1$
8. $x^5 + 2x^4 + 3x^2 + 4x + 5;\quad a = -2$
9. $x^4 - 7x^3 + 5x^2 + 3x - 6;\quad a = 0.$

Evaluate to 4 significant digits:

10. $x^3 - 3x^2 + 2x + 1;\quad x = 1.004$
11. $x^5 + x^4 + x^3 + x^2 + x + 1;\quad x = 1.994$
12. $4x^4 - 3x^2 + 10x + 12;\quad x = -0.9890$
13. $10x^3 + 12x^2 - 6x - 5;\quad x = -3.042.$

3. TAYLOR POLYNOMIALS

Consider this problem: Given a function $f(x)$ and a number a, find a polynomial $p(x)$ which approximates $f(x)$ for values of x near a.

Recall the Taylor polynomials introduced in Chapter 14:

$$p_1(x) = f(a) + f'(a)(x - a),$$

$$p_2(x) = f(a) + f'(a)(x - a) + \frac{1}{2}f''(a)(x - a)^2,$$

$$p_3(x) = f(a) + f'(a)(x - a) + \frac{1}{2}f''(a)(x - a)^2 + \frac{1}{6}f'''(a)(x - a)^3.$$

We showed in several examples that these are good approximations to $f(x)$ near $x = a$, indeed, that each one is a better approximation than the previous one. Now we introduce Taylor polynomials of higher degree.

The n-th **degree Taylor polynomial** of $f(x)$ at $x = a$ is

$$p_n(x) = f(a) + f'(a)(x - a) + \frac{1}{2!}f''(a)(x - a)^2 + \cdots + \frac{1}{n!}f^{(n)}(a)(x - a)^n.$$

When $f(x)$ is itself a polynomial of degree n, then

$$p_n(x) = f(x).$$

This was shown in the last section; $p_n(x)$ is precisely the expression for $f(x)$ in powers of $x - a$. Furthermore, in this case,

$$p_n(x) = p_{n+1}(x) = p_{n+2}(x) = \cdots.$$

(Why?) Thus for an n-th degree polynomial $f(x)$, the n-th degree and all higher Taylor polynomials *equal* $f(x)$.

There is good reason to suspect that any function $f(x)$ is approximated near $x = a$ by its n-th degree Taylor polynomial $p_n(x)$. This is because $p_n(x)$ is constructed so that

$$p_n(a) = f(a), \quad p_n'(a) = f'(a), \quad p_n''(a) = f''(a), \cdots, p_n^{(n)}(a) = f^{(n)}(a).$$

Thus $p_n(x)$ mimics $f(x)$ and its first n derivatives at $x = a$. Therefore near $x = a$, the approximation $p_n(x) \approx f(x)$ should be very accurate. The following formula gives the error in this approximation. It will be justified in Section 6.

Taylor's Formula with Remainder Suppose $f(x)$ has derivatives up to and including $f^{(n+1)}(x)$ near $x = a$. Write

$$f(x) = p_n(x) + r_n(x),$$

where $p_n(x)$ is the n-th degree Taylor polynomial at $x = a$ and $r_n(x)$ is the remainder (or error). Then

$$r_n(x) = \frac{1}{n!}\int_a^x (x - t)^n f^{(n+1)}(t)\, dt.$$

Usually the integral expressing $r_n(x)$ cannot be computed exactly. Nevertheless, the integral can be estimated. Here is one important estimate:

Estimate of Remainder Suppose

$$f(x) = p_n(x) + r_n(x),$$

where $p_n(x)$ is the n-th Taylor polynomial at $x = a$. If

$$|f^{(n+1)}(x)| \leq M$$

in some interval including $x = a$, say $b \leq x \leq c$, then

$$|r_n(x)| \leq \frac{M}{(n+1)!}|x - a|^{n+1} \qquad \text{for} \quad b \leq x \leq c.$$

This assertion is verified as follows. Apply Estimate (5) on p. 398:

$$|r_n(x)| = \frac{1}{n!}\left|\int_a^x (x - t)^n f^{(n+1)}(t)\, dt\right|$$

$$\leq \frac{M}{n!}\left|\int_a^x (x - t)^n\, dt\right| = \frac{M}{(n+1)!}|x - a|^{n+1}.$$

EXAMPLE 3.1

Find the n-th degree Taylor polynomial of $f(x) = e^x$ at $x = 0$. Estimate the error.

Solution:

$$f(x) = e^x, \qquad f'(x) = e^x, \qquad f''(x) = e^x, \quad \cdots;$$

$$f(0) = 1, \qquad f'(0) = 1, \qquad f''(0) = 1, \quad \cdots.$$

Hence

$$p_n(x) = \sum_{i=0}^{n} \frac{f^{(i)}(0)}{i!} x^i = \sum_{i=0}^{n} \frac{x^i}{i!}.$$

The $(n + 1)$-th derivative is $f^{(n+1)}(t) = e^t$. If $x \geq 0$, its largest value for t between 0 and x is e^x. By the remainder estimate with $M = e^x$,

$$|r_n(x)| \leq \frac{e^x}{(n + 1)!} x^{n+1} \qquad \text{for} \quad x \geq 0.$$

If $x \leq 0$, the largest value of $f^{(n+1)}(t)$ between 0 and x is $e^0 = 1$. By the remainder estimate with $M = 1$,

$$|r_n(x)| \leq \frac{|x|^{n+1}}{(n + 1)!}.$$

Answer: $p_n(x) = 1 + x + \dfrac{x^2}{2!} + \dfrac{x^3}{3!} + \cdots + \dfrac{x^n}{n!}$,

$$|r_n(x)| \leq \frac{e^x}{(n + 1)!} x^{n+1} \qquad \text{for} \quad x \geq 0,$$

$$|r_n(x)| \leq \frac{|x|^{n+1}}{(n + 1)!} \qquad \text{for} \quad x \leq 0.$$

EXAMPLE 3.2

Find the n-th degree Taylor polynomial of $f(x) = e^x$ at $x = 1$. Estimate the remainder.

Solution:

$$p_n(x) = \sum_{i=0}^{n} \frac{f^{(i)}(1)}{i!} (x - 1)^i = \sum_{i=0}^{n} \frac{e}{i!} (x - 1)^i.$$

When $|x - 1| \leq c$, the largest value of e^x is e^{1+c}. Thus

$$|r_n(x)| \leq \frac{e^{1+c}}{(n + 1)!} |x - 1|^{n+1}.$$

Alternate Solution: Think!

$$e^x = e^{1+(x-1)} = e \cdot e^{x-1}.$$

Write $u = x - 1$. By the previous example,

$$e^u = \sum_{i=0}^{n} \frac{1}{i!} u^i + r_n(u).$$

Hence,

$$e^x = e \cdot e^u = e \sum_{i=0}^{n} \frac{1}{i!} (x - 1)^i + e r_n(x - 1),$$

where

$$e|r_n(x-1)| \le e\frac{e^c}{(n+1)!}|x-1|^{n+1}.$$

Answer: $$p_n(x) = \sum_{i=0}^{n} \frac{e}{i!}(x-1)^i,$$

$$|r_n(x)| \le \frac{e^{c+1}}{(n+1)!}|x-1|^{n+1} \quad \text{for} \quad |x-1| \le c.$$

EXAMPLE 3.3

Find the n-th degree Taylor polynomial of

$$f(x) = \frac{1}{1-x}$$

at $x = 0$. Estimate the remainder.

Solution: $f(x) = (1-x)^{-1}$,

$$f'(x) = (1-x)^{-2}, \qquad f''(x) = 2(1-x)^{-3}, \qquad f'''(x) = 3!(1-x)^{-4}, \quad \cdots .$$

It follows that $f^{(i)}(0) = i!$,

$$p_n(x) = \sum_{i=0}^{n} \frac{f^{(i)}(0)}{i!}x^i = \sum_{i=0}^{n} x^i.$$

Try to estimate the remainder; fix c between 0 and 1: $0 < c < 1$. Then for $|x| \le c$,

$$|f^{(n+1)}(x)| = \left|\frac{(n+1)!}{(1-x)^{n+2}}\right| \le \frac{(n+1)!}{(1-c)^{n+2}}.$$

Hence for $-c \le x \le c$,

$$|r_n(x)| \le \frac{1}{(n+1)!}\frac{(n+1)!}{(1-c)^{n+2}} \cdot c^{n+1} = \frac{c^{n+1}}{(1-c)^{n+2}} = \frac{1}{1-c}\left(\frac{c}{1-c}\right)^{n+1}.$$

This is a satisfactory error estimate provided $c < \frac{1}{2}$. For then

$$\frac{c}{1-c} < 1, \qquad \text{so} \qquad \left(\frac{c}{1-c}\right)^{n+1} \quad \text{is small.}$$

But if $c > \frac{1}{2}$, then

$$\frac{c}{1-c} > 1, \qquad \text{so} \qquad \left(\frac{c}{1-c}\right)^{n+1} \quad \text{is large;}$$

the estimate is useless.

Fortunately there is another approach. The polynomial

$$p_n(x) = 1 + x + x^2 + \cdots + x^n$$

is the sum of a geometric progression, so by the standard formula

$$p_n(x) = \frac{1 - x^{n+1}}{1 - x}.$$

We can compute the error *exactly:*

$$r_n(x) = \frac{1}{1-x} - p_n(x) = \frac{1}{1-x} - \frac{1 - x^{n+1}}{1-x} = \frac{x^{n+1}}{1-x}.$$

Answer: $p_n(x) = 1 + x + x^2 + \cdots + x^n,$

$$r_n(x) = \frac{x^{n+1}}{1-x} \quad \text{for} \quad x \neq 1.$$

REMARK: If $|x| < 1$, the answer shows that $|r_n(x)|$ is small because $|x|^{n+1}$ is small. On the other hand if $|x| > 1$, then $|r_n(x)|$ is large. For instance,

$$p_n(2) = 1 + 2 + 2^2 + 2^3 + \cdots + 2^n,$$

hardly a good approximation to

$$f(2) = \frac{1}{1-2} = -1.$$

The point is that the Taylor polynomials of a function $f(x)$ at $x = a$ may provide poor approximations to $f(x)$ when x is far from a. How near x must be for good approximations depends on the function $f(x)$.

EXAMPLE 3.4

Find the Taylor polynomials for $\sin x$ at $x = 0$. Estimate the remainders.

Solution: Compute derivatives:

$$f(x) = \sin x, \qquad f'(x) = \cos x, \qquad f''(x) = -\sin x,$$

$$f'''(x) = -\cos x, \qquad f^{(4)}(x) = \sin x, \qquad \cdots,$$

repeating in cycles of four. At $x = 0$, the values are

$$0, \quad 1, \quad 0, \quad -1, \quad 0, \quad 1, \quad 0, \quad -1, \quad 0, \quad \cdots.$$

Hence the n-th degree Taylor polynomial of $\sin x$ is

$$p_n(x) = x - \frac{x^3}{3!} + \frac{x^5}{5!} - \frac{x^7}{7!} + \frac{x^9}{9!} - \cdots,$$

where the last term is $\pm x^n/n!$ if n is odd and $\pm x^{n-1}/(n-1)!$ if n is even. For example,

$$p_3(x) = p_4(x) = x - \frac{x^3}{3!},$$

$$p_5(x) = p_6(x) = x - \frac{x^3}{3!} + \frac{x^5}{5!},$$

$$p_7(x) = p_8(x) = x - \frac{x^3}{3!} + \frac{x^5}{5!} - \frac{x^7}{7!}.$$

Thus $p_{2m-1}(x) = p_{2m}(x)$ and the sign of the last term is plus if m is odd, minus if m is even:

$$p_{2m-1}(x) = p_{2m}(x) = x - \frac{x^3}{3!} + \frac{x^5}{5!} - \cdots + (-1)^{m-1}\frac{x^{2m-1}}{(2m-1)!}$$

$$= \sum_{i=1}^{m} (-1)^{i-1}\frac{x^{2i-1}}{(2i-1)!}.$$

The remainder estimate is easy: $|f^{(n+1)}(x)| \leq 1$ because $f^{(n+1)}(x) = \pm \sin x$ or $\pm \cos x$. Hence

$$|r_n(x)| \leq \frac{|x|^{n+1}}{(n+1)!}.$$

Since $p_{2m-1}(x) = p_{2m}(x)$, if follows that

$$|r_{2m-1}(x)| = |r_{2m}(x)| \leq \frac{|x|^{2m+1}}{(2m+1)!}.$$

Answer: $\sin x = p_{2m}(x) + r_{2m}(x)$,

$$p_{2m}(x) = \sum_{i=1}^{m} (-1)^{i-1}\frac{x^{2i-1}}{(2i-1)!}, \quad |r_{2m}(x)| \leq \frac{|x|^{2m+1}}{(2m+1)!}.$$

[Note: $p_{2m-1}(x) = p_{2m}(x)$.]

REMARK: For the cosine, a similar argument shows

$$\cos x = 1 - \frac{x^2}{2!} + \frac{x^4}{4!} - \frac{x^6}{6!} + \cdots + (-1)^m\frac{x^{2m}}{(2m)!} + r_{2m}(x),$$

$$|r_{2m}(x)| = |r_{2m+1}(x)| \leq \frac{|x|^{2m+2}}{(2m+2)!}.$$

EXAMPLE 3.5

Find the n-th degree Taylor polynomial of $\ln x$ at $x = 5$. Estimate the remainder.

Solution: Let $f(x) = \ln x$. Then

$$f'(x) = \frac{1}{x}, \qquad f''(x) = -\frac{1}{x^2}, \qquad f'''(x) = \frac{1 \cdot 2}{x^3},$$

$$f^{(4)}(x) = -\frac{3!}{x^4}, \quad \cdots, \quad f^{(i)}(x) = (-1)^{i-1}\frac{(i-1)!}{x^i}.$$

The coefficient of $(x-5)^i$ in the Taylor polynomial is

$$\frac{1}{i!}f^{(i)}(5) = (-1)^{i-1} \cdot \frac{(i-1)!}{i!} \cdot \frac{1}{5^i} = (-1)^{i-1}\frac{1}{i \cdot 5^i}.$$

Therefore,

$$p_n(x) = \ln 5 + \frac{1}{1 \cdot 5}(x-5) - \frac{1}{2 \cdot 5^2}(x-5)^2$$

$$+ \frac{1}{3 \cdot 5^3}(x-5)^3 - \cdots + (-1)^{n-1}\frac{1}{n \cdot 5^n}(x-5)^n.$$

For an error estimate, assume $5 - c \leq x \leq 5 + c$, where $0 < c < 5$. Then

$$|r_n(x)| \leq \frac{M}{(n+1)!}|x-5|^{n+1} \leq \frac{M}{(n+1)!}c^{n+1},$$

where M is a bound for $|f^{(n+1)}(x)|$. Now

$$|f^{(n+1)}(x)| = \left| \pm \frac{n!}{x^{n+1}} \right| \leq \frac{n!}{(5-c)^{n+1}} \qquad \text{for} \quad 5 - c \leq x \leq 5 + c.$$

Use the number on the right-hand side as M. The resulting estimate is

$$|r_n(x)| \leq \frac{1}{(n+1)!}\frac{n!}{(5-c)^{n+1}}c^{n+1} = \frac{1}{n+1}\left(\frac{c}{5-c}\right)^{n+1}.$$

If $0 \leq c \leq 2.5$, then

$$0 \leq \frac{c}{5-c} \leq 1,$$

and

$$\frac{1}{n+1}\left(\frac{c}{5-c}\right)^{n+1} \longrightarrow 0 \qquad \text{as} \quad n \longrightarrow \infty.$$

Answer: For $|x - 5| \leq c \leq 2.5$,

$$p_n(x) = \ln 5 + \sum_{i=1}^{n}(-1)^{i-1}\frac{1}{i \cdot 5^i}(x-5)^i,$$

$$|r_n(x)| \leq \frac{1}{n+1}\left(\frac{c}{5-c}\right)^{n+1}.$$

Summary

TAYLOR'S FORMULA WITH REMAINDER:

$$f(x) = p_n(x) + r_n(x),$$

$$p_n(x) = \sum_{i=0}^{n} \frac{f^{(i)}(a)}{i!}(x-a)^i, \qquad r_n(x) = \frac{1}{n!}\int_a^x (x-t)^n f^{(n+1)}(t)\,dt.$$

ESTIMATE OF THE REMAINDER:

If $|f^{(n+1)}(t)| \le M$ for all t between a and x, then

$$|r_n(x)| \le \frac{M}{(n+1)!}|x-a|^{n+1}.$$

SPECIAL FUNCTIONS:

$$e^x = \sum_{i=0}^{n} \frac{x^i}{i!} + r_n(x), \qquad |r_n(x)| \le \begin{cases} \dfrac{e^x x^{n+1}}{(n+1)!} & \text{if } x \ge 0 \\[2ex] \dfrac{|x|^{n+1}}{(n+1)!} & \text{if } x \le 0. \end{cases}$$

$$\frac{1}{1-x} = \sum_{i=0}^{n} x^i + \frac{x^{n+1}}{1-x} \qquad \text{if } x \ne 1.$$

$$\sin x = \sum_{i=1}^{m} \frac{(-1)^{i-1}}{(2i-1)!}x^{2i-1} + r_{2m-1}(x), \qquad |r_{2m-1}(x)| \le \frac{|x|^{2m+1}}{(2m+1)!}.$$

$$\cos x = \sum_{i=0}^{m} \frac{(-1)^i}{(2i)!}x^{2i} + r_{2m}(x), \qquad |r_{2m}(x)| \le \frac{|x|^{2m+2}}{(2m+2)!}.$$

EXERCISES

Find the Taylor polynomials at the given point; estimate the remainder:

1. $f(x) = \sin 2x; \quad x = 0$
2. $f(x) = \sin 2x; \quad x = \pi/2$
3. $f(x) = xe^x; \quad x = 0$
4. $f(x) = xe^x; \quad x = 1$
5. $f(x) = x^2 \ln x; \quad x = 1$
6. $f(x) = x^2 \ln x; \quad x = e$
7. $f(x) = x^2e^{-x}; \quad x = 0$
8. $f(x) = x^2e^{-x}; \quad x = 1$
9. $f(x) = x \sin x; \quad x = 0$
10. $f(x) = x \sin x; \quad x = \pi/2$

11. $f(x) = \sin x + \cos x; \quad x = 0$
12. $f(x) = \cosh x; \quad x = 0$
13. $f(x) = \sinh x + \sin x; \quad x = 0$
14. $f(x) = 1 + e^x + e^{2x}; \quad x = 0.$
15. Let $p(x) = x^4 - 4x^3 + 6x^2 - 3x + 2$. Estimate the error in the range $\frac{3}{4} \leq x \leq \frac{5}{4}$ if this polynomial is approximated by its linear Taylor polynomial about $x = 1$.
16. Compare the Taylor polynomials for $\sin x$ at $x = \pi/2$ with those for $\cos x$ at $x = 0$.
17. Let $p_n(x)$ be the n-th degree Taylor polynomial of $f(x)$ at $x = a$. Verify that $p_n^{(i)}(a) = f^{(i)}(a)$, for $i = 0, 1, 2, \cdots, n$.
18. Let $p_n(x)$ be the n-th degree Taylor polynomial of $f(x)$ at $x = 0$, where $n \geq 2$. Show that the curves $y = f(x)$ and $y = p_n(x)$ have the same curvature at $x = 0$.

4. APPLICATIONS

Taylor's Formula with Remainder provides a practical method for approximating functions. First, it gives a simple procedure for obtaining a polynomial approximation. Second, it supplies an estimate of the error.

EXAMPLE 4.1

Find a polynomial $p(x)$ such that $|e^x - p(x)| < 0.001$

(a) for all x in the interval $-\frac{1}{2} \leq x \leq \frac{1}{2}$,

(b) for all x in the interval $-2 \leq x \leq 2$.

Solution: By the answer to Example 3.1, a logical choice for $p(x)$ is one of the Taylor polynomials

$$p_n(x) = 1 + x + \frac{x^2}{2!} + \frac{x^3}{3!} + \cdots + \frac{x^n}{n!}.$$

We want to choose n so that

$$|e^x - p_n(x)| < 10^{-3}.$$

To minimize computation and round-off error, we prefer n as small as possible.

(a) If $-\frac{1}{2} \leq x \leq \frac{1}{2}$, then

$$|r_n(x)| \leq \frac{e^{1/2}}{(n+1)!}\left(\frac{1}{2}\right)^{n+1}.$$

We choose n so that

$$\frac{e^{1/2}}{(n+1)!}\left(\frac{1}{2}\right)^{n+1} < \frac{1}{1000},$$

that is,

$$\frac{1}{(n+1)!\,2^{n+1}} < \frac{1}{1000\,e^{1/2}} < \frac{1}{1648}.$$

A few trials show

$$\frac{1}{4!2^4} = \frac{1}{384}, \qquad \frac{1}{5!2^5} = \frac{1}{3840}.$$

Hence we take $n + 1 = 5$, that is, $n = 4$. The desired polynomial is

$$p_4(x) = 1 + x + \frac{x^2}{2!} + \frac{x^3}{3!} + \frac{x^4}{4!}.$$

(b) If $-2 \leq x \leq 2$, then

$$|r_n(x)| \leq \frac{e^2}{(n+1)!} \cdot 2^{n+1}.$$

This time we choose n so that

$$\frac{e^2}{(n+1)!} \cdot 2^{n+1} < \frac{1}{1000},$$

that is,

$$\frac{2^{n+1}}{(n+1)!} < \frac{1}{7389}.$$

A few trials show

$$\frac{2^{10}}{10!} = \frac{1024}{3{,}628{,}800} = \frac{4}{14{,}175} \approx \frac{1}{3544},$$

$$\frac{2^{11}}{11!} = \frac{2^{10}}{10!} \cdot \frac{2}{11} \approx \frac{2}{38{,}981} \approx \frac{1}{19{,}500}.$$

Hence we take $n + 1 = 11$, that is, $n = 10$. The desired polynomial is

$$p_{10}(x) = 1 + x + \frac{x^2}{2!} + \frac{x^3}{3!} + \cdots + \frac{x^{10}}{10!}.$$

Answer: (a) $p(x) = \sum_{i=0}^{4} \frac{x^i}{i!}$; (b) $p(x) = \sum_{i=0}^{10} \frac{x^i}{i!}$.

Tables of Sines and Cosines

Suppose we want to construct 5-place tables of $\sin x$ and $\cos x$. A logical method is approximation by Taylor polynomials. These should be of low degree (few terms) to limit the number of arithmetic operations necessary and to prevent accumulation of round-off errors. Note that we need tabulate $\sin x$ and $\cos x$ only for $0 \leq x \leq \pi/4$. (Why?)

Let us concentrate on $\sin x$; a similar discussion applies to $\cos x$. First we consider the third degree Taylor polynomial at $x = 0$,

$$p_3(x) = x - \frac{x^3}{6}.$$

From Section 3,

$$|\sin x - p_3(x)| \leq \frac{x^5}{5!} = \frac{x^5}{120}.$$

For 5-place accuracy, the error must be less than 5×10^{-6}. Therefore we want

$$\frac{x^5}{120} < 5 \times 10^{-6}, \qquad x^5 < 6 \times 10^{-4}.$$

An easy computation with slide rule (or 4-place log tables) shows this is the case if $x < 0.22$ rad $\approx 12.5°$. Thus up to about $12°$, the third degree Taylor polynomial yields 5-place accuracy.

Next we try

$$p_5(x) = x - \frac{x^3}{3!} + \frac{x^5}{5!}.$$

Since

$$|\sin x - p_5(x)| \leq \frac{x^7}{7!},$$

5-place accuracy is obtained if

$$\frac{x^7}{7!} < 5 \times 10^{-6}, \qquad x^7 < 0.0252.$$

Using the C.R.C. table of seventh powers, we find this is the case provided $x < 0.59$ rad $\approx 34°$.

Next we try

$$p_7(x) = x - \frac{x^3}{3!} + \frac{x^5}{5!} - \frac{x^7}{7!}.$$

The error is

$$|\sin x - p_7(x)| \leq \frac{x^9}{9!}.$$

For angles up to $45°$, this error is at most

$$\frac{1}{9!}\left(\frac{\pi}{4}\right)^9 \approx \frac{1}{9!}(0.7854)^9 < \frac{1}{9!}(0.79)^9.$$

The C.R.C. table of ninth powers shows $(79)^9$ is slightly less than 12×10^{16}. Hence $(0.79)^9$ is slightly less than 12×10^{-2}. The C.R.C. tables show also that $1/9! \approx 0.276 \times 10^{-5}$. Therefore

$$|\text{error}| < (12 \times 10^{-2})(0.28 \times 10^{-5}) < 4 \times 10^{-7}.$$

Conclusion: For $0 \le x \le \pi/4$, the Taylor polynomial $p_7(x)$ yields 6-place accuracy in approximating $\sin x$. Furthermore, since $p_7(x)$ involves only four terms, the probability of large accumulated round-off error is low. Thus $p_7(x)$ provides a practical way of constructing a table of sines.

It is not necessary to limit ourselves to Taylor polynomials at $x = 0$. For example, if we are interested only in angles close to 45°, it is better to use Taylor polynomials at $\pi/4$. They provide greater accuracy for the same amount of computation.

The third degree Taylor polynomial of $\sin x$ at $\pi/4$ is

$$p_3(x) = \left(\frac{\sqrt{2}}{2}\right)\left[1 + \left(x - \frac{\pi}{4}\right) - \frac{1}{2}\left(x - \frac{\pi}{4}\right)^2 - \frac{1}{6}\left(x - \frac{\pi}{4}\right)^3\right],$$

and

$$|\sin x - p_3(x)| \le \frac{1}{4!}\left(x - \frac{\pi}{4}\right)^4.$$

If x differs from 45° by at most 0.1 rad $\approx$ 5.7°, then the error is bounded by

$$\frac{1}{4!}(0.1)^4 < 4.2 \times 10^{-6}.$$

Hence for x between 39.3° and 50.7°, $p_3(x)$ yields 5-place accuracy. Between 44° and 46°, the error is bounded by

$$\frac{1}{4!}(0.0175)^4 < 4 \times 10^{-9}, \qquad (1° \approx 0.01745 \text{ rad})$$

and so $p_3(x)$ yields 8-place accuracy.

We see that near $\pi/4$, the Taylor polynomial $p_3(x)$ about $x = \pi/4$ gives the same accuracy as does $p_7(x)$ about $x = 0$. This is typical of Taylor polynomials: for values near $x = a$, the accuracy achieved by an n-th degree Taylor polynomial about $x = 0$ is matched by a lower degree Taylor polynomial about $x = a$. Generally the lower degree polynomial means less computation and less round-off error.

What is not typical is that every other coefficient is zero in the Taylor polynomials of $\sin x$ and $\cos x$ about $x = 0$. For computation, this is excellent; it means relatively little computation yields extraordinary accuracy. For example, the polynomial $p_7(x)$ of $\sin x$ actually involves only four terms, yet provides an approximation to within $|x|^9/9!$. This is why the Taylor polynomials about points other than $x = 0$ are rarely used for $\sin x$ and $\cos x$.

Taylor Polynomials versus Interpolation Polynomials

Both Taylor and interpolation polynomials can be used to approximate a function $f(x)$. An n-th degree interpolation polynomial agrees with $f(x)$ at $n + 1$ points $x_0 < x_1 < \cdots < x_n$, whereas an n-th degree Taylor polynomial agrees with $f(x)$ and its first n derivatives at *one* point $x = a$. As a result, the Taylor polynomial generally approximates $f(x)$ closely near a, but not necessarily over the interval $x_0 \leq x \leq x_n$. The interpolation polynomial, however, approximates $f(x)$ well throughout the interval, but not as well as the Taylor polynomial near $x = a$. For example, consider the function

$$f(x) = 2^x.$$

The interpolation polynomial

$$p(x) = 1 + \frac{1}{2}x + \frac{1}{2}x^2$$

agrees with $f(x)$ at $x = 0$, at $x = 1$, and at $x = 2$. Compare it with the second degree Taylor polynomial of $f(x)$ at $x = 1$,

$$p_2(x) = f(1) + f'(1)(x - 1) + \frac{1}{2}f''(1)(x - 1)^2.$$

Now

$$f(x) = 2^x = e^{x \ln 2},$$

$$f'(x) = (\ln 2)e^{x \ln 2} = (\ln 2)2^x, \qquad f''(x) = (\ln 2)^2 2^x.$$

Hence

$$\begin{aligned} p_2(x) &= 2 + 2(\ln 2)(x - 1) + (\ln 2)^2(x - 1)^2 \\ &\approx 2 + 1.38630(x - 1) + 0.48046(x - 1)^2. \end{aligned}$$

Now tabulate values to 3 places (Table 4.1).

TABLE 4.1

x	2^x	$p(x)$	$p_2(x)$	x	2^x	$p(x)$	$p_2(x)$
0.0	1.000	1.000	1.094	1.2	2.297	2.320	2.296
0.2	1.149	1.120	1.198	1.4	2.639	2.680	2.631
0.4	1.320	1.280	1.341	1.6	3.031	3.080	3.005
0.6	1.516	1.480	1.522	1.8	3.482	3.520	3.416
0.8	1.741	1.720	1.742	2.0	4.000	4.000	3.867
1.0	2.000	2.000	2.000				

The table indicates that from 0.4 to 1.6, the better approximation is $p_2(x)$, but farther from $x = 1$, at 0.0, 0.2, 1.8, and 2.0, the better approximation is $p(x)$.

EXERCISES

1. Find a polynomial $p(x)$ such that $|e^{-x^2} - p(x)| < 0.001$ for all x in the interval $-1 \leq x \leq 1$.
2. What degree Taylor polynomial about $x = 0$ is needed to approximate $f(x) = \cos x$ for $-\pi/4 \leq x \leq \pi/4$ to 5 decimal places?
3. Estimate the error in approximating $f(x) = \ln(1 + x)$ for $-\frac{1}{3} \leq x \leq \frac{1}{3}$ by its 10-th degree Taylor polynomial about $x = 0$.
4. Approximate $f(x) = 1/(1 - x)^2$ for $-\frac{1}{4} \leq x \leq \frac{1}{4}$ to 3 decimal places by a Taylor polynomial about $x = 0$.
5. Find the 4-th degree interpolation polynomial matching $f(x) = \sin x$ at $x = 0$, $\pm\pi/4$, $\pm\pi/2$.
6. With use of a computer, calculate the error at 10 or more points if $\sin x$ is approximated on $-\pi/2 \leq x \leq \pi/2$ by the polynomial in Ex. 5. Do the same for the 5-th degree Taylor polynomial at $x = 0$. From these calculations, estimate the maximum error for each polynomial over this range.
7. Show that $|\sin x - p_9(x)| < 5 \times 10^{-6}$ for $0 \leq x \leq \pi/2$, where $p_9(x)$ is the 9-th degree Taylor polynomial for $\sin x$ at $x = 0$.
8. Find the smallest positive integer n such that for $0 \leq x \leq \frac{1}{3}$,
$$\left| \frac{1}{1 - x} - (1 + x + x^2 + \cdots + x^n) \right| < 5 \times 10^{-6}.$$
9. Compute $\sin(5\pi/8)$ to 5-place accuracy. [Use Taylor polynomials of $\sin x$, but not at $x = 0$.]
10. If $f'(a) = f''(a) = \cdots = f^{(n-1)}(a) = 0$, but $f^{(n)}(a) \neq 0$, show that a reasonable approximation to $f(x)$ near a is $f(x) \approx f(a) + \dfrac{f^n(a)}{n!}(x - a)^n$.
11. (cont.) Suppose n is even; show that $f(x)$ has a maximum at $x = a$ if $f^{(n)}(a) < 0$, and a minimum at $x = a$ if $f^{(n)}(a) > 0$. Suppose n is odd; show that $f(x)$ has neither a maximum nor a minimum at $x = a$.

5. TAYLOR SERIES

Consider once again the Taylor polynomials and remainders of the exponential function at $x = 0$:

$$e^x = \sum_{i=0}^{n} \frac{x^i}{i!} + r_n(x),$$

where

$$|r_n(x)| \leq \frac{g(x)\,|x|^{n+1}}{(n + 1)!}; \qquad \begin{cases} g(x) = 1 & \text{for } x < 0 \\ g(x) = e^x & \text{for } x \geq 0. \end{cases}$$

This estimate shows that no matter what x is, the error is very small if n is large enough. In other words, for fixed x,

$$\frac{g(x)\,|x|^{n+1}}{(n + 1)!} \longrightarrow 0 \qquad \text{as} \quad n \longrightarrow \infty.$$

Here is the reason. The number x is fixed; set $A = |x|$. Pick m so that $m > 10A$. Then

$$\frac{A}{m+1} < \frac{1}{10}, \qquad \frac{A}{m+2} < \frac{1}{10}, \qquad \frac{A}{m+3} < \frac{1}{10}, \qquad \cdots, \qquad \frac{A}{m+k} < \frac{1}{10},$$

for any k. If $n = m + k$, then

$$\frac{|x|^n}{n!} = \frac{A^n}{n!} = \frac{A^m}{m!} \cdot \frac{A}{m+1} \cdot \frac{A}{m+2} \cdots \frac{A}{m+k}$$

$$< \frac{A^m}{m!} \cdot \frac{1}{10} \cdot \frac{1}{10} \cdots = \frac{A^m}{m!} \cdot \frac{1}{10^k}.$$

Now $k \longrightarrow \infty$ as $n \longrightarrow \infty$. Since $g(x) A^m/m!$ is fixed, the right-hand term $\longrightarrow 0$ as $n \longrightarrow \infty$. This means that $r_n(x) \longrightarrow 0$ as $n \longrightarrow \infty$ for each x, hence

$$e^x \quad \text{and the Taylor polynomial} \quad \sum_{i=0}^{n} \frac{x^i}{i!}$$

are as close together as we please, once n is sufficiently large. We introduce the notation

$$\boxed{e^x = \sum_{i=0}^{\infty} \frac{x^i}{i!}}$$

to express precisely this relationship.

Similarly

$$\boxed{\begin{aligned} \sin x &= \sum_{i=1}^{\infty} \frac{(-1)^{i-1}}{(2i-1)!} x^{2i-1}, \\ \cos x &= \sum_{i=0}^{\infty} \frac{(-1)^i}{(2i)!} x^{2i}. \end{aligned}}$$

Such expansions of functions into what look like polynomials of infinite degree, are called **Taylor series**. A familiar one is the infinite geometric series.

$$\boxed{\frac{1}{1-x} = \sum_{i=0}^{\infty} x^i \qquad \text{for} \quad |x| < 1.}$$

EXERCISES

Find Taylor series about the given point:

1. $f(x) = \sin 3x; \quad x = 0$
2. $f(x) = \cos \frac{1}{2}x; \quad x = 0$
3. $f(x) = \sin^2 x; \quad x = 0$
4. $f(x) = \cos^2(2x - 1); \quad x = \frac{1}{2}$
5. $f(x) = e^{-2x}; \quad x = 0$
6. $f(x) = \cosh x; \quad x = 0$
7. $f(x) = e^{3x+2}; \quad x = -\frac{2}{3}$
8. $f(x) = \sin x + 2\cos x; \quad x = 0$
9. $f(x) = \dfrac{1}{1 - 3x}; \quad x = 0$
10. $f(x) = \dfrac{1}{2 + x}; \quad x = 0$
11. $f(x) = \dfrac{1}{x}; \quad x = 2.$

6. DERIVATION OF TAYLOR'S FORMULA

We shall derive Taylor's Formula as stated in Section 3:

$$f(x) = p_n(x) + r_n(x),$$

where

$$p_n(x) = \sum_{i=0}^{n} \frac{f^{(i)}(a)}{i!}(x - a)^i$$

and

$$r_n(x) = \frac{1}{n!}\int_a^x (x - t)^n f^{(n+1)}(t)\, dt.$$

This is a consequence of the following assertion (actually a special case of Taylor's Formula):

> If
>
> $$g(a) = g'(a) = g''(a) = \cdots = g^{(n)}(a) = 0,$$
>
> then
>
> $$g(x) = \frac{1}{n!}\int_a^x (x - t)^n g^{(n+1)}(t)\, dt.$$

If this assertion is correct, how does Taylor's Formula follow? Suppose $f(x)$ is any function and $p_n(x)$ is its n-th degree Taylor polynomial at $x = a$. Set

$$g(x) = f(x) - p_n(x).$$

Now $p_n(x)$ agrees with $f(x)$, and its first n derivatives agree with those of $f(x)$ at $x = a$. Therefore

$$g(a) = g'(a) = g''(a) = \cdots = g^{(n)}(a) = 0.$$

Thus $g(x)$ satisfies the conditions of the assertion, so

$$g(x) = \frac{1}{n!}\int_a^x (x-t)^n g^{(n+1)}(t)\,dt.$$

But $g(x) = f(x) - p_n(x)$ and $g^{(n+1)}(t) = f^{(n+1)}(t)$ because the $(n+1)$-th derivative of the polynomial $p_n(x)$ is zero. Therefore,

$$f(x) - p_n(x) = \frac{1}{n!}\int_a^x (x-t)^n f^{(n+1)}(t)\,dt,$$

which is precisely Taylor's Formula.

Let us return to the assertion and verify it for low values of n. We integrate by parts, noting that a and x are fixed.

CASE $n = 0$:

$$\frac{1}{0!}\int_a^x (x-t)^0 g'(t)\,dt = \int_a^x g'(t)\,dt = g(t)\Big|_a^x = g(x).$$

(Do not forget $g(a) = 0$.)

CASE $n = 1$: To evaluate

$$\frac{1}{1!}\int_a^x (x-t)g''(t)\,dt,$$

set $u(t) = (x-t)$ and $v(t) = g'(t)$. Then

$$\int_a^x (x-t)g''(t)\,dt = \int_a^x u\,dv = u(t)v(t)\Big|_a^x - \int_a^x v(t)u'(t)\,dt.$$

Therefore

$$\int_a^x (x-t)g''(t)\,dt = (x-t)g'(t)\Big|_a^x + \int_a^x g'(t)\,dt = \int_a^x g'(t)\,dt = g(x),$$

by the previous case. (Do not forget $g'(a) = 0$.)

CASE $n = 2$: Set $u(t) = \frac{1}{2}(x-t)^2$ and $v(t) = g''(t)$. Then

$$\frac{1}{2!}\int_a^x (x-t)^2 g'''(t)\,dt = \int_a^x u\,dv$$

$$= u(t)v(t)\Big|_a^x - \int_a^x v(t)u'(t)\,dt = 0 + \int_a^x (x-t)g''(t)\,dt = g(x),$$

by the previous case. Note that $v(a) = 0$ by the hypothesis $g''(a) = 0$.

The general case is handled the same way. One integration by parts, with

$$u(t) = \frac{(x-t)^n}{n!} \qquad \text{and} \qquad v(t) = g^{(n)}(t),$$

reduces

$$\frac{1}{n!}\int_a^x (x-t)^n g^{(n+1)}(t)\,dt \qquad \text{to} \qquad \frac{1}{(n-1)!}\int_a^x (x-t)^{n-1} g^{(n)}(t)\,dt.$$

The latter is $g(x)$ by the previous case.

7. ERROR IN INTERPOLATION [optional]

The error in approximating a function $f(x)$ by an interpolation polynomial of degree n can be estimated in terms of the size of the $(n+1)$-th derivative of f. We shall state an error estimate here, but postpone its proof until Chapter 36, Section 4, p. 903.

Let $p(x)$ be the polynomial of degree n or less that satisfies

$$p(x_0) = f(x_0), \qquad p(x_1) = f(x_1), \quad \cdots, \quad p(x_n) = f(x_n),$$

where

$$a = x_0 < x_1 < x_2 < \cdots < x_n = b.$$

Suppose

$$|f^{(n+1)}(x)| \le M$$

on the interval $a \le x \le b$. Then for each x in the interval,

$$|f(x) - p(x)| \le \frac{M}{(n+1)!}|(x-x_0)(x-x_1)\cdots(x-x_n)|.$$

Note the similarity between this estimate and the corresponding error estimate for the n-th degree Taylor polynomial:

$$|f(x) - p_n(x)| \le \frac{M}{(n+1)!}|x - x_0|^{n+1}.$$

EXAMPLE 7.1

Let $p(x)$ be the interpolation polynomial that agrees with $\ln x$ at $x = 3, 3.25, 3.5, 3.75, 4$. Find an estimate of $|\ln x - p(x)|$ in the interval $3 \le x \le 4$.

Solution: Apply the preceding estimate to $f(x) = \ln x$ with $n = 4$. Then $f^{(5)}(x) = -24/x^5$; hence for $3 \le x \le 4$,

$$|f^{(5)}(x)| \le \frac{24}{3^5} = \frac{8}{81}.$$

Take $M = \frac{8}{81}$; the error estimate is

$$|\ln x - p(x)| \le \frac{1}{5!}\cdot\frac{8}{81}E(x),$$

where

$$E(x) = |(x-3)(x-3\tfrac{1}{4})(x-3\tfrac{1}{2})(x-3\tfrac{3}{4})(x-4)|.$$

If $3 \le x \le 3\frac{1}{4}$, then

$$|x-3| \le \tfrac{1}{4}, \quad |x-3\tfrac{1}{4}| \le \tfrac{1}{4}, \quad |x-3\tfrac{1}{2}| \le \tfrac{1}{2}, \quad |x-3\tfrac{3}{4}| \le \tfrac{3}{4}, \quad |x-4| \le 1,$$

hence

$$E(x) \le \frac{1}{4}\cdot\frac{1}{4}\cdot\frac{1}{2}\cdot\frac{3}{4}\cdot 1 = \frac{3}{128}.$$

Similarly if $3\frac{1}{4} \le x \le 3\frac{1}{2}$, then

$$E(x) \le \frac{1}{2}\cdot\frac{1}{4}\cdot\frac{1}{4}\cdot\frac{1}{2}\cdot\frac{3}{4} = \frac{3}{256} < \frac{3}{128}.$$

Likewise (or by symmetry) $E(x) \le 3/128$ if $3\frac{1}{2} \le x \le 3\frac{3}{4}$, or if $3\frac{3}{4} \le x \le 4$.

Thus $3/128$ is an upper bound for $E(x)$ on the interval $3 \le x \le 4$. Therefore

$$|\ln x - p(x)| \le \frac{1}{5!}\cdot\frac{8}{81}\cdot\frac{3}{128} = \frac{1}{51840} < 2 \times 10^{-5}.$$

Answer: $|\ln x - p(x)| < 2 \times 10^{-5}$ for $3 \le x \le 4$.

EXERCISES

1. Estimate the error in approximating $f(x) = 2^x$ by the interpolation polynomial $p(x) = 1 + \frac{1}{2}x + \frac{1}{2}x^2$ over the range $0 \le x \le 2$. Compare this with the computed error. (See p. 160.)
2. Estimate the error in approximating $f(x) = \sin x$ for $-\pi/2 \le x \le \pi/2$ by the interpolation polynomial matching the function at $x = 0, \pm\pi/4, \pm\pi/2$.
3. Suppose that e^x is to be approximated for $0 \le x \le 2$ by an interpolation polynomial matching the function at $n+1$ evenly spaced points. Find n as small as you can such that the error is less than 0.001 throughout the interval.
4. Find a cubic polynomial $p(x)$ satisfying

$$p(0) = \sin 0 \qquad p(\pi/2) = \sin \pi/2$$
$$p'(0) = \cos 0 \qquad p'(\pi/2) = \cos \pi/2.$$

   This $p(x)$ is a **first order approximation** to $\sin x$ *both* at 0 and at $\pi/2$.
5. (cont.) How well does $p(x)$ approximate $\sin x$ at $x = \pi/60, \pi/6, \pi/4, \pi/3, 29\pi/60, 2\pi/3$?
6. (cont.) Find the cubic polynomial that approximates x^4 to first order both at $x = 0$ and $x = 1$. Compare it graphically to the interpolation cubic for x^4 fitted at $x = 0, \frac{1}{3}, \frac{2}{3}, 1$.

7. Show that the function $E(x)$ in the solution of Example 7.1 takes its maximum between 3 and $3\frac{1}{4}$, not between $3\frac{1}{4}$ and $3\frac{1}{2}$.
[*Hint:* The function $E(x)$ is symmetric in the line $x = 3\frac{1}{2}$ so only the interval $3 < x < 3\frac{1}{2}$ need be considered. If $3\frac{1}{4} < x < 3\frac{1}{2}$, show that $E(x) < E(x - \frac{1}{4})$.]

8. NEWTON INTERPOLATION [optional]

The Lagrange interpolation formula supplies a polynomial $y = p(x)$ of degree at most n to fit data

$$(x_0, y_0), \quad (x_1, y_1), \quad \cdots, \quad (x_n, y_n).$$

In many cases the points $x_0, x_1, x_2, \cdots$ are equally spaced, that is, each interval $x_i \leq x \leq x_{i+1}$ has the same length $(x_{i+1} - x_i) = (x_n - x_0)/n$. In such cases, the **Newton interpolation formula** is usually more convenient than the Lagrange interpolation formula.

The Newton formula is based on the **forward difference table** (Table 8.1).

TABLE 8.1

x_0	y_0				
		Δ_0			
x_1	y_1		$\Delta_0^{(2)}$		
		Δ_1		$\Delta_0^{(3)}$	
x_2	y_2		$\Delta_1^{(2)}$		
		Δ_2		$\Delta_1^{(3)}$	$\cdot\ \cdot\ \cdot$
x_3	y_3		$\Delta_2^{(2)}$	$\cdot$	
		Δ_3	$\cdot$	$\cdot$	
x_4	y_4	$\cdot$	$\cdot$	$\cdot$	
$\cdot$	$\cdot$	$\cdot$	$\cdot$	$\cdot$	
$\cdot$	$\cdot$	$\cdot$	$\cdot$	$\cdot$	
$\cdot$	$\cdot$	$\cdot$	$\cdot$	$\cdot$	$\cdot\ \cdot\ \cdot$
$\cdot$	$\cdot$	$\cdot$	$\cdot$	$\Delta_{n-2}^{(3)}$	
$\cdot$	$\cdot$	$\cdot$	$\Delta_{n-2}^{(2)}$		
$\cdot$	$\cdot$	Δ_{n-1}			
x_n	y_n				

The third column gives the **first forward differences**

$$\Delta_0 = y_1 - y_0, \quad \Delta_1 = y_2 - y_1, \quad \cdots, \quad \Delta_{n-1} = y_n - y_{n-1};$$

the next column gives the **second forward differences**

$$\Delta_0^{(2)} = \Delta_1 - \Delta_0, \quad \Delta_1^{(2)} = \Delta_2 - \Delta_1, \quad \cdots, \quad \Delta_{n-2}^{(2)} = \Delta_{n-1} - \Delta_{n-2},$$

and so on. Each column (after the second) lists the successive differences of the previous column. The table terminates with the single n-th difference $\Delta_0^{(n)}$.

The numbers $y_0, y_1, \cdots, y_n$ may be experimental readings, or they may be the computed values of a function $y_i = f(x_i)$.

EXAMPLE 8.1

Compute the forward difference table through the 4-th differences for $y = x^3 - x$ on the interval $-2 \le x \le 2$, taking x at intervals of 0.5.

Solution: Be careful of the signs. The results are in Table 8.2.

TABLE 8.2

x	y	Δ	$\Delta^{(2)}$	$\Delta^{(3)}$	$\Delta^{(4)}$
−2.0	−6.000				
		4.125			
−1.5	−1.875		−2.250		
		1.875		0.750	
−1.0	0.000		−1.500		0.000
		0.375		0.750	
−0.5	0.375		−0.750		0.000
		−0.375		0.750	
0.0	0.000		0.000		0.000
		−0.375		0.750	
0.5	−0.375		0.750		0.000
		0.375		0.750	
1.0	0.000		1.500		0.000
		1.875		0.750	
1.5	1.875		2.250		
		4.125			
2.0	6.000				

The third differences in Table 8.2 are constant. This fact ought to be related to the function $y = x^3 - x$ being a polynomial of *degree* 3. Actually, the following is true; the proof is left to Exs. 13–17.

> For a polynomial of degree n, the n-th forward differences (equally spaced) are constant.

There is another important observation to be made from Table 8.2. Given the first entry in each column starting with y_0, one can reconstruct the whole body of the table. That is, given

$$y_0 = -6.000, \quad \Delta_0 = 4.125, \quad \Delta_0^{(2)} = -2.250,$$

$$\Delta_0^{(3)} = 0.750, \quad \Delta_0^{(4)} = 0.000, \quad \Delta_0^{(5)} = 0.000 \quad \cdots,$$

one can reconstruct the body of the table and recapture, in particular, the data $y_0, y_1, \cdots, y_8$.

This is the basis for the Newton interpolation formula:

Newton Interpolation Formula Let the interval $a \le x \le b$ be divided into n subintervals of equal length $h = (b - a)/n$:

$$a = x_0 < x_1 < x_2 < \cdots < x_{n-1} < x_n = b.$$

Let data

$$(x_0, y_0), \quad (x_1, y_1), \quad \cdots, \quad (x_n, y_n)$$

be given and construct the corresponding forward difference table. Then the polynomial

$$\begin{aligned} p_n(x) = y_0 &+ \frac{\Delta_0}{h}(x - x_0) + \frac{\Delta_0^{(2)}}{(2!)h^2}(x - x_0)(x - x_1) \\ &+ \frac{\Delta_0^{(3)}}{(3!)h^3}(x - x_0)(x - x_1)(x - x_2) \\ &+ \cdots + \frac{\Delta_0^{(n)}}{(n!)h^n}(x - x_0)(x - x_1) \cdots (x - x_{n-1}) \end{aligned}$$

interpolates the data.

This will be verified for $n = 1$, 2, and 3.

For $n = 1$,

$$p_1(x) = y_0 + \frac{\Delta_0}{h}(x - x_0),$$

$$p_1(x_0) = y_0, \quad p_1(x_1) = y_0 + \Delta_0 = y_1.$$

For $n = 2$,

$$p_2(x) = p_1(x) + \frac{\Delta_0^{(2)}}{(2!)h^2}(x - x_0)(x - x_1),$$

$$p_2(x_0) = p_1(x_0) = y_0, \qquad p_2(x_1) = p_1(x_1) = y_1,$$

$$\begin{aligned} p_2(x_2) &= y_0 + \frac{\Delta_0}{h}(2h) + \frac{\Delta_0^{(2)}}{(2!)h^2}(2h)(h) \\ &= y_0 + 2\Delta_0 + \Delta_0^{(2)} = (y_0 + \Delta_0) + (\Delta_0 + \Delta_0^{(2)}) \\ &= y_1 + \Delta_1 = y_2. \end{aligned}$$

For $n = 3$,

$$p_3(x) = p_2(x) + \frac{\Delta_0^{(3)}}{(3!)h^3}(x - x_0)(x - x_1)(x - x_2).$$

Since the added term is 0 at x_0, x_1, x_2, and since $p_2(x)$ has the correct values, it follows that

$$p_3(x_0) = y_0, \qquad p_3(x_1) = y_1, \qquad p_3(x_2) = y_2.$$

Now

$$\begin{aligned} p_3(x_3) &= y_0 + 3\Delta_0 + 3\Delta_0^{(2)} + \Delta_0^{(3)} \\ &= (y_0 + \Delta_0) + 2(\Delta_0 + \Delta_0^{(2)}) + (\Delta_0^{(2)} + \Delta_0^{(3)}) \\ &= y_1 + 2\Delta_1 + \Delta_1^{(2)} = (y_1 + \Delta_1) + (\Delta_1 + \Delta_1^{(2)}) \\ &= y_2 + \Delta_2 = y_3. \end{aligned}$$

The general formula can be established similarly.

Note two facts:

(a) As each additional point x_n is included in the interpolation, one new term is added; the previous terms are not altered. (This is most convenient; compare with the Lagrange interpolation formula.)

(b) The Newton interpolation polynomial $p_n(x)$ is closely analogous to the Taylor polynomial.

EXAMPLE 8.2

Interpolate the data

x	0.0	0.2	0.4	0.6
y	3.1	2.5	2.0	2.4

Solution: Compute the forward differences (Table 8.3).

TABLE 8.3

x	y	Δ	$\Delta^{(2)}$	$\Delta^{(3)}$
0.0	3.1			
		−0.6		
0.2	2.5		0.1	
		−0.5		0.8
0.4	2.0		0.9	
		0.4		
0.6	2.4			

Now substitute:

$$p_3(x) = 3.1 - \frac{0.6}{0.2}x + \frac{0.1}{2(0.2)^2}x(x - 0.2) + \frac{0.8}{6(0.2)^3}x(x - 0.2)(x - 0.4)$$

$$= 3.1 - 3x + 1.25x(x - 0.2) + \frac{100}{6}x(x - 0.2)(x - 0.4).$$

Answer: $y = 3.1 - (1.91666 \cdots)x - 8.75x^2 + (16.66 \cdots)x^3.$

EXERCISES

Find the interpolating polynomial:

1. (1, 3), (2, 5), (3, 10)
2. (−1, 4), (0, −1), (1, 2)
3. (0, −5), (1, −2) (2, 4), (3, 5)
4. (−1, 4), (0, −1), (1, 2), (2, 3)
5. (−1, 4), (0, −1), (1, 2), (2, −1)
6. (0, 7), (2, 4), (4, −2), (6, 1)
7. (−2, 3), (−1, 0), (0, 0), (1, −2), (2, 1)
8. (3, 5), (4, 3), (5, 6), (6, 2), (7, 7)
9. (−2, 1), (0, 2), (2, 0), (4, 0), (6, −2)
10. (−3, −5), (−2, −8), (−1, −1), (0, 5), (1, 10)
11. (0, 2), (1, 3), (2, 5), (3, 7), (4, 11), (5, 13)
12. (−2, 1), (−1, 2), (0, 3), (1, 5), (2, 8), (3, 13).
13. Set $f_0(x) = 1$ and $f_n(x) = (x - 1)(x - 2) \cdots (x - n)$ for $n > 0$. Show that $\Delta f_n = f_n(x + 1) - f_n(x) = nf_{n-1}(x)$.
14. (cont.) Show that $1, f_1(x), \cdots, f_n(x)$ form a **basis** for polynomials of degree n. That is, if $p(x)$ is a polynomial of degree at most n, then there are unique constants $a_0, a_1, \cdots, a_n$ such that
$$p(x) = a_0f_0(x) + a_1f_1(x) + \cdots + a_nf_n(x).$$
15. (cont.) Show that $\Delta^{(n)}f_n$ is constant.
16. (cont.) Let $p(x)$ have degree at most n. Show that $\Delta^{(n)}p(x)$ is constant.
17. (cont.) Conclude that in any equally spaced forward difference table of an n-th degree polynomial $p(x)$, the n-th differences are constant.

28. Power Series

1. RADIUS OF CONVERGENCE

In this chapter we study power series

$$a_0 + a_1(x - x_0) + a_2(x - x_0)^2 + \cdots + a_n(x - x_0)^n + \cdots$$

and their applications. In most of our examples $x_0 = 0$, but the discussion applies to $x_0 \neq 0$ as well.

Power series serve two important purposes. First, they express known functions in a form particularly suitable for computation. Second, they define functions which are not simple to specify otherwise. Certainly nobody objects to defining a function by a polynomial,

$$f(x) = a_0 + a_1x + a_2x^2 + \cdots + a_nx^n.$$

Then why not define a function by a power series,

$$f(x) = a_0 + a_1x + a_2x^2 + \cdots,$$

providing the meaning of such an *infinite* sum is understood? We shall see, in fact, that in many ways power series resemble polynomials.

Recall the geometric series:

$$\frac{1}{1 - x} = 1 + x + x^2 + \cdots + x^n + \cdots, \qquad |x| < 1.$$

The infinite sum is shorthand for the following precise statement:

$$\frac{1}{1 - x} = 1 + x + x^2 + \cdots + x^n + \epsilon_n(x),$$

where for each x in the interval $-1 < x < 1$, the error $\epsilon_n(x) \longrightarrow 0$ as $n \longrightarrow \infty$. In other words, for each x in the interval $-1 < x < 1$, the approximation

$$\frac{1}{1 - x} \approx 1 + x + x^2 + \cdots + x^n$$

can be made as accurate as desired by taking enough terms. For example, let $x = \frac{1}{2}$:

$$\frac{1}{1 - \dfrac{1}{2}} \approx 1 + \frac{1}{2} + \left(\frac{1}{2}\right)^2 + \cdots + \left(\frac{1}{2}\right)^n.$$

The left side is 2; the right side is a finite geometric series whose sum is

$$\frac{1 - \left(\dfrac{1}{2}\right)^{n+1}}{1 - \dfrac{1}{2}} = 2\left[1 - \left(\frac{1}{2}\right)^{n+1}\right]$$

$$= 2 - \frac{1}{2^n}.$$

Thus the error is 2^{-n}, which can be made as small as desired by taking enough terms (n large enough).

Note that if $|x| \geq 1$, the geometric series

$$1 + x + x^2 + x^3 + \cdots$$

has no meaning; its **partial sums**

$$1 + x + x^2 + \cdots + x^n$$

do not approximate anything (they have no limit as $n \longrightarrow \infty$). We describe this state of affairs by saying that the geometric series **converges** for $|x| < 1$ and **diverges** for $|x| \geq 1$.

Other examples of power series were discussed in the last chapter:

$$e^x = 1 + x + \frac{x^2}{2!} + \frac{x^3}{3!} + \cdots = \sum_{n=0}^{\infty} \frac{x^n}{n!},$$

$$\sin x = x - \frac{x^3}{3!} + \frac{x^5}{5!} - \frac{x^7}{7!} + \cdots = \sum_{n=1}^{\infty} (-1)^{n-1} \frac{x^{2n-1}}{(2n-1)!},$$

$$\cos x = 1 - \frac{x^2}{2!} + \frac{x^4}{4!} - \frac{x^6}{6!} + \cdots = \sum_{n=0}^{\infty} (-1)^n \frac{x^{2n}}{(2n)!}.$$

These series were derived from Taylor's Formula with Remainder. Each converges for all values of x, whereas the geometric series converges only for $|x| < 1$.

We state without proof the fundamental fact concerning convergence and divergence of power series:

Given a power series

$$a_0 + a_1(x - x_0) + a_2(x - x_0)^2 + \cdots + a_n(x - x_0)^n + \cdots,$$

precisely one of the following three cases holds:

(i) The series converges only for $x = x_0$.
(ii) The series converges for all values of x.
(iii) There is a positive number R such that the series converges for each x satisfying $|x - x_0| < R$ and diverges for each x satisfying $|x - x_0| > R$. (See Fig. 1.1.)

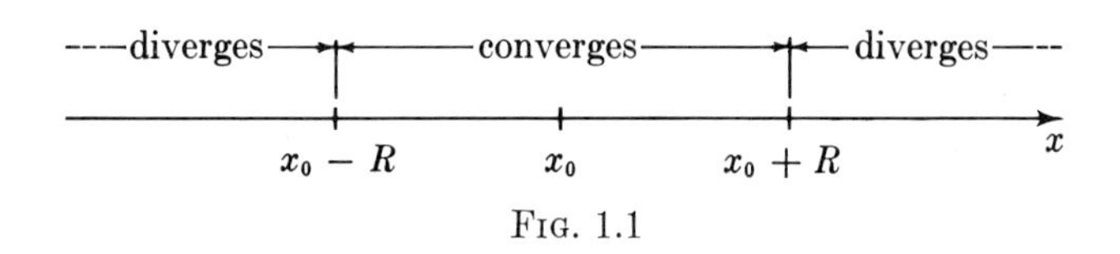

FIG. 1.1

REMARKS: Case (i) is an extreme case, unimportant and uninteresting. It occurs when the coefficients a_n become so large that the power series can converge only if all terms after the first vanish. An example is

$$1 + x + 2^2x^2 + 3^3x^3 + 4^4x^4 + \cdots.$$

For each non-zero x, note that $|n^nx^n| = |nx|^n \longrightarrow \infty$ as $n \longrightarrow \infty$.

Case (ii) is the opposite extreme and occurs when the coefficients $a_0, a_1, a_2, \cdots$ become small very rapidly. The series for e^x is an example. Here the general term is $x^n/n!$, which for each x tends to zero very quickly since the coefficients $a_n = 1/n!$ become small so fast.

Case (iii) lies in between. The coefficients do not increase so rapidly that the series never converges (except for $x = x_0$), nor do they decrease so rapidly that the series always converges. A typical example is the geometric series

$$1 + x + x^2 + \cdots + x^n + \cdots,$$

where each $a_n = 1$. This series converges for $|x| < 1$ and diverges for $|x| \geq 1$, hence $R = 1$.

Note that in Case (iii) nothing is said about the endpoints $x = x_0 + R$ and $x = x_0 - R$. The series may or may not converge at either point.

The number R in Case (iii) is called the

radius of convergence.

By convention, $R = 0$ in Case (i), convergence for $x = x_0$ only, and $R = \infty$ in Case (ii), convergence for all x. (The word "radius" will be clarified in Chapter 31, Section 7, p. 736.)

The interval $x_0 - R < x < x_0 + R$ in Case (iii) is called the

interval of convergence.

(Refer to Fig. 1.1.) By convention, the interval of convergence in Case (i) is the single point x_0; in Case (ii) it is the entire x-axis.

EXAMPLE 1.1

Find the sum of the power series and its radius of convergence R:

(a) $1 + \frac{x}{3} + \frac{x^2}{3^2} + \cdots + \frac{x^n}{3^n} + \cdots,$

(b) $1 + 5x + 5^2x^2 + \cdots + 5^nx^n + \cdots.$

Solution: Each is a geometric series

$$\sum_{n=0}^{\infty} y^n = \frac{1}{1-y}, \qquad |y| < 1.$$

In (a), $y = x/3$; hence the series converges for $|x/3| < 1$ and diverges for $|x/3| \geq 1$, that is, converges for $|x| < 3$ and diverges for $|x| \geq 3$. Thus $R = 3$. In (b), $y = 5x$, which implies convergence for $|x| < \frac{1}{5}$ and divergence for $|x| \geq \frac{1}{5}$.

Answer: (a) $\dfrac{1}{1 - \dfrac{x}{3}} = \dfrac{3}{3-x}, \quad R = 3;$

(b) $\dfrac{1}{1-5x}, \quad R = \dfrac{1}{5}.$

Note that the series with smaller coefficients has the larger radius of convergence.

EXAMPLE 1.2

Find the radius of convergence and the sum of the power series

$$1 - x^2 + \frac{x^4}{2!} - \frac{x^6}{3!} + \frac{x^8}{4!} - \cdots.$$

Solution: The series has the form

$$1 + y + \frac{y^2}{2!} + \frac{y^3}{3!} + \cdots, \qquad \text{where} \quad y = -x^2.$$

Since the series converges to e^y for all values of y, we may replace y by $-x^2$ for any value of x.

Answer: $R = \infty$; the series converges to e^{-x^2} for all values of x.

EXERCISES

Find the sum of the series and its radius of convergence:

1. $1 + (x - 3) + (x - 3)^2 + (x - 3)^3 + \cdots + (x - 3)^n + \cdots$
2. $1 + \left(\frac{x + 10}{2}\right) + \left(\frac{x + 10}{2}\right)^2 + \left(\frac{x + 10}{2}\right)^3 + \cdots + \left(\frac{x + 10}{2}\right)^n + \cdots$
3. $1 - x^2 + x^4 - x^6 + \cdots + (-1)^n x^{2n} + \cdots$
4. $\left(\frac{x}{3}\right)^2 + \left(\frac{x}{3}\right)^3 + \left(\frac{x}{3}\right)^4 + \cdots + \left(\frac{x}{3}\right)^n + \cdots$
5. $1 + \frac{(x + 1)}{1!} + \frac{(x + 1)^2}{2!} + \frac{(x + 1)^3}{3!} + \cdots + \frac{(x + 1)^n}{n!} + \cdots$
6. $1 - \frac{2x}{1!} + \frac{4x^2}{2!} - \frac{8x^3}{3!} + \cdots + \frac{(-2x)^n}{n!} + \cdots$
7. $1 + (5x)^3 + (5x)^6 + (5x)^9 + \cdots + (5x)^{3n} + \cdots$
8. $2x - \frac{8x^3}{3!} + \frac{32x^5}{5!} - \frac{128x^7}{7!} + \cdots + (-1)^{n-1}\frac{(2x)^{2n-1}}{(2n - 1)!} + \cdots$
9. $1 - \frac{(x - 1)^4}{1!} + \frac{(x - 1)^8}{2!} - \frac{(x - 1)^{12}}{3!} + \cdots + (-1)^n \frac{(x - 1)^{4n}}{n!} + \cdots.$

Find the sum of the series and all values of x for which the series converges:

10. $\frac{1}{x} + \frac{1}{x^2} + \frac{1}{x^3} + \cdots + \frac{1}{x^n} + \cdots$
11. $1 + e^x + e^{2x} + e^{3x} + \cdots + e^{nx} + \cdots$
12. $\cos^2 x + \cos^4 x + \cos^6 x + \cdots + \cos^{2n} x + \cdots$
13. $1 - \frac{\sin 3x}{1!} + \frac{\sin^2 3x}{2!} - \frac{\sin^3 3x}{3!} + \cdots + (-1)^n \frac{\sin^n 3x}{n!} + \cdots$
14. $1 + \frac{\ln x}{1!} + \frac{\ln^2 x}{2!} + \frac{\ln^3 x}{3!} + \cdots + \frac{\ln^n x}{n!} + \cdots$
15. $1 + 2\sqrt{x} + 4x + 8x\sqrt{x} + 16x^2 + \cdots + (2\sqrt{x})^n + \cdots$
16. $\ln x + \ln(x^{1/2}) + \ln(x^{1/4}) + \cdots + \ln(x^{1/2^n}) + \cdots$
17. $1 + (x^2 + a^2) + (x^2 + a^2)^2 + (x^2 + a^2)^3 + \cdots$
18. $(x^2 - 1) + \frac{(x^2 - 1)^2}{2} + \frac{(x^2 - 1)^3}{4} + \cdots + \frac{(x^2 - 1)^n}{2^{n-1}} + \cdots$
19. $1 - \frac{x^{2/3}}{2!} + \frac{x^{4/3}}{4!} - \frac{x^{6/3}}{6!} + \cdots + (-1)^n \frac{x^{2n/3}}{(2n)!} + \cdots.$

2. RATIO TEST

For most power series which arise in practice, the radius of convergence can be found by the following criterion. (There is an intuitive justification at the end of the section.)

Ratio Test Suppose the power series

$$a_0 + a_1(x - x_0) + a_2(x - x_0)^2 + \cdots + a_n(x - x_0)^n + \cdots$$

has non-zero coefficients. If

$$\left|\frac{a_n}{a_{n+1}}\right| \longrightarrow R \quad \text{as} \quad n \longrightarrow \infty,$$

where R is 0, positive, or ∞, then R is the radius of convergence.

EXAMPLE 2.1

Find the radius of convergence in each case:

(a) $1 + \dfrac{x}{1} + \dfrac{x^2}{2} + \dfrac{x^3}{3} + \cdots + \dfrac{x^n}{n} + \cdots,$

(b) $(x - 5) - 4(x - 5)^2 + 9(x - 5)^3 - \cdots + (-1)^{n-1}n^2(x - 5)^n + \cdots,$

(c) $1 + \dfrac{x}{2 + 1} + \dfrac{x^2}{2^2 + 2} + \dfrac{x^3}{2^3 + 3} + \cdots + \dfrac{x^n}{2^n + n} + \cdots,$

(d) $1 - x + \dfrac{x^2}{2^2} - \dfrac{x^3}{3^3} + \cdots + (-1)^n \dfrac{x^n}{n^n} + \cdots,$

(e) $\dfrac{x^3}{\sqrt{3}} + \dfrac{x^6}{\sqrt{6}} + \dfrac{x^9}{\sqrt{9}} + \cdots + \dfrac{x^{3n}}{\sqrt{3n}} + \cdots.$

Solution: In each case apply the Ratio Test.

(a) Here $a_n = 1/n$ and

$$\left|\frac{a_n}{a_{n+1}}\right| = \frac{1}{n} \Big/ \frac{1}{n+1} = \frac{n+1}{n}.$$

Hence

$$\left|\frac{a_n}{a_{n+1}}\right| \longrightarrow 1 \quad \text{as} \quad n \longrightarrow \infty,$$

so $R = 1$.

(b) $a_n = (-1)^{n-1}n^2$,

$$\left|\frac{a_n}{a_{n+1}}\right| = \frac{n^2}{(n+1)^2} = \left(\frac{n}{n+1}\right)^2 \longrightarrow 1 \quad \text{as} \quad n \longrightarrow \infty,$$

so $R = 1$.

(c)

$$\left|\frac{a_n}{a_{n+1}}\right| = \frac{1}{2^n + n} \Big/ \frac{1}{2^{n+1} + n + 1} = \frac{2^{n+1} + n + 1}{2^n + n}$$

$$= \frac{2 + (n+1) \cdot 2^{-n}}{1 + n \cdot 2^{-n}} \longrightarrow \frac{2+0}{1+0} = 2$$

as $n \longrightarrow \infty$. Hence $R = 2$.

(d)

$$\left|\frac{a_n}{a_{n+1}}\right| = \frac{1}{n^n} \Big/ \frac{1}{(n+1)^{n+1}}$$

$$= \frac{(n+1)^{n+1}}{n^n} = (n+1)\left(\frac{n+1}{n}\right)^n > n + 1.$$

Hence $|a_n/a_{n+1}| \longrightarrow \infty$ as $n \longrightarrow \infty$, so $R = \infty$; the series converges for all values of x.

(e) The Ratio Test does not apply directly because "two-thirds" of the coefficients in this power series are zero. Nevertheless, the series may be written

$$\frac{y}{\sqrt{3}} + \frac{y^2}{\sqrt{6}} + \frac{y^3}{\sqrt{9}} + \cdots + \frac{y^n}{\sqrt{3n}} + \cdots,$$

where $y = x^3$. The Ratio Test does apply to the series in this form:

$$\frac{\sqrt{3(n+1)}}{\sqrt{3n}} = \sqrt{\frac{n+1}{n}} \longrightarrow 1.$$

Hence the y-series converges for $|y| < 1$ and diverges for $|y| > 1$. Therefore, the original series converges for $|x^3| < 1$ and diverges for $|x^3| > 1$, i.e., for $|x| < 1$ and $|x| > 1$, respectively. Hence $R = 1$.

Answer: (a) 1; (b) 1; (c) 2; (d) ∞; (e) 1.

REMARK: The ratio test does not apply to all series. An example is

$$1 + 2x + x^2 + 2x^3 + x^4 + 2x^5 + \cdots.$$

For this series

$$\frac{a_0}{a_1} = \frac{1}{2}, \qquad \frac{a_1}{a_2} = 2, \qquad \frac{a_2}{a_3} = \frac{1}{2}, \qquad \frac{a_3}{a_4} = 2, \qquad \text{etc.}$$

The ratios are alternately $\frac{1}{2}$ and 2; they do not have a limit and so the ratio test does not apply. Nevertheless, the given series is a perfectly decent one. It is in fact the sum of two geometric series,

$$1 + x^2 + x^4 + \cdots = \frac{1}{1 - x^2},$$

and

$$2x + 2x^3 + 2x^5 + \cdots = \frac{2x}{1 - x^2},$$

both of which converge for $|x| < 1$ and diverge for $|x| \geq 1$. Hence

$$1 + 2x + x^2 + 2x^3 + \cdots = \frac{1}{1 - x^2} + \frac{2x}{1 - x^2} = \frac{1 + 2x}{1 - x^2}, \qquad R = 1.$$

Intuitive Meaning of the Ratio Test

Suppose the power series

$$a_0 + a_1(x - x_0) + a_2(x - x_0)^2 + \cdots + a_n(x - x_0)^n + \cdots$$

has non-zero coefficients and

$$\left|\frac{a_n}{a_{n+1}}\right| \longrightarrow R > 0 \qquad \text{as} \quad n \longrightarrow \infty.$$

It follows that

$$\frac{|a_{n+1}(x - x_0)^{n+1}|}{|a_n(x - x_0)^n|} = \frac{|x - x_0|}{|a_n/a_{n+1}|} \longrightarrow \frac{|x - x_0|}{R},$$

which means that eventually the size of each term is approximately $|x - x_0|/R$ times the size of the preceding term. Thus, the series itself behaves like the geometric series

$$1 + \left(\frac{x - x_0}{R}\right) + \left(\frac{x - x_0}{R}\right)^2 + \cdots,$$

which converges if $|x - x_0|/R < 1$, that is, in the interval $|x - x_0| < R$. It seems quite reasonable that the given series converges in the same interval, hence that its radius of convergence is R.

If

$$\left|\frac{a_n}{a_{n+1}}\right| \longrightarrow \infty,$$

then

$$\frac{|a_{n+1}(x - x_0)^{n+1}|}{|a_n(x - x_0)^n|} = \frac{|x - x_0|}{|a_n/a_{n+1}|} \longrightarrow 0.$$

Therefore the terms of the series decrease more rapidly than those of the geometric series

$$1 + \left(\frac{x - x_0}{R}\right) + \left(\frac{x - x_0}{R}\right)^2 + \cdots,$$

no matter how large R is. Consequently, the series converges in the interval $|x - x_0| < R$ for each and every positive number R. That means it converges for all x.

If

$$\left|\frac{a_n}{a_{n+1}}\right| \longrightarrow 0,$$

then

$$\frac{|a_{n+1}(x - x_0)^{n+1}|}{|a_n(x - x_0)^n|} = \frac{|x - x_0|}{|a_n/a_{n+1}|} \longrightarrow \infty,$$

provided $x \neq x_0$; the size of successive terms *increases*. Therefore, the series converges only for $x = x_0$.

EXERCISES

Find the radius of convergence:

1. $1 + x + 2x^2 + 3x^3 + \cdots + nx^n + \cdots$
2. $(x - 1) + \dfrac{(x-1)^2}{2} + \dfrac{(x-1)^3}{3} + \cdots + \dfrac{(x-1)^n}{n} + \cdots$
3. $x + \dfrac{x^2}{3} + \dfrac{x^3}{5} + \dfrac{x^4}{7} + \cdots + \dfrac{x^n}{2n-1} + \cdots$
4. $x - \dfrac{x^2}{5} + \dfrac{x^3}{9} - \dfrac{x^4}{17} + \cdots + (-1)^{n-1}\dfrac{x^n}{2^n+1} + \cdots$
5. $1 + \dfrac{x}{1 \cdot 2} + \dfrac{x^2}{2 \cdot 4} + \dfrac{x^3}{3 \cdot 8} + \cdots + \dfrac{x^n}{n \cdot 2^n} + \cdots$
6. $\dfrac{(x+1)}{1 \cdot 2 \cdot 3} + \dfrac{(x+1)^2}{2 \cdot 3 \cdot 4} + \dfrac{(x+1)^3}{3 \cdot 4 \cdot 5} + \cdots + \dfrac{(x+1)^n}{n(n+1)(n+2)} + \cdots$
7. $x + \sqrt{2}\,x^2 + \sqrt{3}\,x^3 + \cdots + \sqrt{n}\,x^n + \cdots$
8. $1 - \dfrac{x}{2} + \dfrac{x^2}{2^4} - \dfrac{x^3}{2^9} + \cdots + \dfrac{(-x)^n}{2^{n^2}} + \cdots$
9. $(e - 1)x + (e^2 - 1)x^2 + (e^3 - 1)x^3 + \cdots + (e^n - 1)x^n + \cdots$
10. $-\dfrac{2x}{1^3} + \dfrac{3x^2}{2^3} - \dfrac{4x^3}{3^3} + \cdots + (-1)^n\dfrac{(n+1)x^n}{n^3} + \cdots$
11. $\dfrac{1}{4} + \dfrac{3(x-5)}{4^2} + \dfrac{3^2(x-5)^2}{4^3} + \cdots + \dfrac{3^n(x-5)^n}{4^{n+1}} + \cdots$
12. $1 + x + 2!x^2 + 3!x^3 + \cdots + n!x^n + \cdots$

13. $\dfrac{x^2}{2+\ln 2}+\dfrac{x^3}{3+\ln 3}+\dfrac{x^4}{4+\ln 4}+\cdots+\dfrac{x^n}{n+\ln n}+\cdots$

14. $1+\dfrac{1}{2}x+\dfrac{1\cdot 3}{2\cdot 4}x^2+\dfrac{1\cdot 3\cdot 5}{2\cdot 4\cdot 6}x^3+\cdots+\dfrac{1\cdot 3\cdot 5\cdots(2n-1)}{2\cdot 4\cdot 6\cdots(2n)}x^n+\cdots$

15. $\dfrac{1}{30}+\left(\dfrac{1}{5^2}-\dfrac{1}{6^2}\right)x+\left(\dfrac{1}{5^3}-\dfrac{1}{6^3}\right)x^2+\cdots+\left(\dfrac{1}{5^{n+1}}-\dfrac{1}{6^{n+1}}\right)x^n+\cdots$

16. $1+x^2+x^{10}+x^{12}+x^{20}+x^{22}+\cdots+x^{10n}+x^{10n+2}+\cdots$

17. $4\cdot 5x^4+8\cdot 9x^8+\cdots+4n(4n+1)x^{4n}+\cdots$

18. $1+(1+2)x+(1+2+4)x^2+(1+2+4+8)x^3+\cdots$
$+(1+2+2^2+\cdots+2^n)x^n+\cdots$

19. $x+\dfrac{x^2}{2^2}+\dfrac{x^3}{3^3}+\cdots+\dfrac{x^n}{n^n}+\cdots$

20. $x+\dfrac{x^2}{(2!)^2}+\dfrac{x^3}{(3!)^2}+\cdots+\dfrac{x^n}{(n!)^2}+\cdots.$

3. EXPANSIONS OF FUNCTIONS

For many applications, it is convenient to expand functions in power series. We have already done this for e^x, $\sin x$, and $\cos x$ by computing the coefficient of x^n from the formula

$$a_n = \frac{f^{(n)}(0)}{n!}.$$

Generally, however, computation of higher derivatives is extremely laborious. Try to find the seventh derivative of $\tan x$ or of $1/(1+x^4)$ and you will soon agree.

In this section we describe several techniques for deriving power series without tedious differentiation. They all depend on one basic principle:

> If $f(x)$ can be expressed as a series of powers of $x - x_0$, then this series is unique. In fact the coefficient of $(x-x_0)^n$ must be
>
> $$\frac{f^{(n)}(x_0)}{n!}.$$

Thus if you find a power series for $f(x)$ at x_0 by any method, fair or foul, then you have it! There is no other series for $f(x)$ at x_0. The details of the proof are quite hard, so we shall omit them.

We state the first new technique for obtaining power series, substitution.

In a power series

$$a_0 + a_1x + a_2x^2 + \cdots,$$

the variable x may be replaced by any expression with values in the interval of convergence of the series.

EXAMPLE 3.1

Express $\dfrac{1}{1+x^2}$ as a power series at $x = 0$ and describe its interval of convergence.

Solution: In the geometric series

$$\frac{1}{1-x} = 1 + x + x^2 + x^3 + \cdots,$$

convergent for $|x| < 1$, replace x by $-x^2$. This is valid provided $|-x^2| < 1$, that is, provided $|x| < 1$.

$$\frac{1}{1+x^2} = \frac{1}{1-(-x^2)} = 1 + (-x^2) + (-x^2)^2 + (-x^2)^3 + \cdots$$

$$= 1 - x^2 + x^4 - x^6 + x^8 - \cdots.$$

Answer: $\dfrac{1}{1+x^2} = 1 - x^2 + x^4 - \cdots$ for $|x| < 1$.

EXAMPLE 3.2

Express $\cos\sqrt{x}$ as a power series at $x = 0$.

Solution: Recall

$$\cos x = 1 - \frac{x^2}{2!} + \frac{x^4}{4!} - \cdots, \qquad \text{for all } x.$$

Replace x by $\sqrt{x}$ (assume $x \geq 0$):

$$\cos\sqrt{x} = 1 - \frac{x}{2!} + \frac{x^2}{4!} - \frac{x^3}{6!} + \cdots.$$

Answer: $\cos\sqrt{x} = 1 - \dfrac{x}{2!} + \dfrac{x^2}{4!} - \dfrac{x^3}{6!} + \cdots$ for $x \geq 0$.

EXAMPLE 3.3

Express $\left(1 - \dfrac{1}{2}\sin x\right)^{-1}$ in powers of $\sin x$.

Solution: The geometric series

$$(1 - y)^{-1} = 1 + y + y^2 + y^3 + \cdots$$

converges for $|y| < 1$. Since $|\frac{1}{2}\sin x| < 1$, it is fair game to replace y by $\frac{1}{2}\sin x$:

$$\left(1 - \frac{1}{2}\sin x\right)^{-1} = 1 + \left(\frac{1}{2}\sin x\right) + \left(\frac{1}{2}\sin x\right)^2 + \cdots.$$

Answer:

$$\left(1 - \frac{1}{2}\sin x\right)^{-1} = \sum_{n=0}^{\infty} \frac{\sin^n x}{2^n} \quad \text{for all } x.$$

Now another important technique, stated without proof:

> Two power series may be added, subtracted, multiplied, or divided within their common interval of convergence. (Division by 0 is excluded.)

EXAMPLE 3.4

Express $\cosh x$ in a power series at $x = 0$.

Solution: $\cosh x = \frac{1}{2}(e^x + e^{-x})$. But

$$e^x = 1 + x + \frac{x^2}{2!} + \frac{x^3}{3!} + \cdots \quad \text{for all } x,$$

$$e^{-x} = 1 - x + \frac{x^2}{2!} - \frac{x^3}{3!} + \cdots \quad \text{for all } x.$$

Add these series and divide by 2:

$$\frac{1}{2}(e^x + e^{-x}) = 1 + \frac{x^2}{2!} + \frac{x^4}{4!} + \cdots.$$

Answer: $\cosh x = 1 + \dfrac{x^2}{2!} + \dfrac{x^4}{4!} + \cdots$ for all x.

EXAMPLE 3.5

Compute the terms up to x^6 in the power series of $x^2 e^x \sin 2x$ at $x = 0$.

Solution:

$$x^2 e^x \sin 2x$$

$$= x^2\left(1 + x + \frac{x^2}{2!} + \frac{x^3}{3!} + \cdots\right)\left(2x - \frac{(2x)^3}{3!} + \frac{(2x)^5}{5!} - + \cdots\right).$$

Since only terms involving x^6 and lower powers are required, it suffices to compute the product

$$x^2\left(1 + x + \frac{x^2}{2!} + \frac{x^3}{3!}\right)\left(2x - \frac{(2x)^3}{3!}\right) = x^2\left(1 + x + \frac{x^2}{2} + \frac{x^3}{6}\right)\left(2x - \frac{4x^3}{3}\right)$$

$$= x^2\left[2x + 2x^2 + \left(1 - \frac{4}{3}\right)x^3 + \left(\frac{1}{3} - \frac{4}{3}\right)x^4 + \cdots\right].$$

Answer: For all x,

$$x^2e^x \sin 2x = 2x^3 + 2x^4 - \frac{1}{3}x^5 - x^6 + \cdots.$$

Odd and Even Functions

The power series at $x = 0$ for the odd function $\sin x$ lacks x^2, x^4, x^6, $\cdots$. Likewise the series at $x = 0$ for the even function $\cos x$ lacks x, x^3, x^5, $\cdots$. Each illustrates a useful principle:

If $f(x)$ is an odd function, $f(-x) = -f(x)$, then its power series at $x = 0$ has the form

$$f(x) = a_1x + a_3x^3 + a_5x^5 + \cdots.$$

If $g(x)$ is an even function, $g(-x) = g(x)$, then its power series at $x = 0$ has the form

$$g(x) = a_0 + a_2x^2 + a_4x^4 + \cdots.$$

These statements are easy to remember: an odd (even) function involves only odd (even) powers of x at $x = 0$.

The proofs of the principles depend in turn on another pair of important facts:

The derivative of an odd function is an even function; the derivative of an even function is an odd function.

Reason: Suppose $f(x)$ is odd, $f(-x) = -f(x)$. Differentiate, using the Chain Rule on the left-hand side: $-f'(-x) = -f'(x)$. Hence $f'(-x) = f'(x)$, so $f'(x)$ is even. Likewise, if $g(-x) = g(x)$, then $g'(-x) = -g'(x)$.

Now return to power series of odd and even functions. Suppose $f(x)$ is odd and at $x = 0$ has the power series

$$f(x) = a_0 + a_1x + a_2x^2 + \cdots.$$

Since $f(x)$ is odd, $f(0) = 0$. Hence $a_0 = 0$. The derivative $f'(x)$ is even and its derivative $f''(x)$ is again odd. Hence $f''(0) = 0$. But $a_2 = f''(0)/2!$, so

$a_2 = 0$. Likewise $a_4 = 0$, $a_6 = 0$, $\cdots$, so

$$f(x) = a_1 x + a_3 x^3 + a_5 x^5 + \cdots.$$

The statement about even functions can be proved similarly.

EXAMPLE 3.6

Find the power series for $\tan x$ at $x = 0$ up to terms in x^7.

Solution: The power series for $\tan x$ is obtained by long division: the series for $\sin x$ is divided by the series for $\cos x$. Here is a systematic way to carry out the long division. Set

$$\tan x = a_1 x + a_3 x^3 + a_5 x^5 + a_7 x^7 + \cdots,$$

valid because $\tan x$ is an odd function. Write the identity $\tan x \cos x = \sin x$ in terms of power series:

$$(a_1 x + a_3 x^3 + a_5 x^5 + a_7 x^7 + \cdots)\left(1 - \frac{x^2}{2} + \frac{x^4}{24} - \frac{x^6}{720} \cdots\right)$$
$$= x - \frac{x^3}{6} + \frac{x^5}{120} - \frac{x^7}{5040} + \cdots.$$

Multiply the two series on the left, then equate coefficients.

$$x: \qquad a_1 = 1;$$

$$x^3: \qquad a_3 - \frac{1}{2} a_1 = -\frac{1}{6};$$

$$x^5: \qquad a_5 - \frac{1}{2} a_3 + \frac{1}{24} a_1 = \frac{1}{120};$$

$$x^7: \qquad a_7 - \frac{1}{2} a_5 + \frac{1}{24} a_3 - \frac{1}{720} a_1 = -\frac{1}{5040}.$$

Solve these equations successively for a_1, a_3, a_5, a_7:

$$a_1 = 1, \qquad a_3 = \frac{1}{3}, \qquad a_5 = \frac{2}{15}, \qquad a_7 = \frac{17}{315}.$$

Alternate Solution: Write

$$\tan x = \sin x \cdot \frac{1}{\cos x}.$$

Now

$$\cos x = 1 - \left(\frac{x^2}{2} - \frac{x^4}{4!} + \frac{x^6}{6!} - + \cdots\right) = 1 - z,$$

$$\frac{1}{\cos x} = \frac{1}{1 - z} = 1 + z + z^2 + z^3 + \cdots.$$

Compute up to x^6 only:

$$z = \frac{x^2}{2} - \frac{x^4}{24} + \frac{x^6}{720} - \cdots,$$

$$z^2 = \left(\frac{x^2}{2} - \frac{x^4}{24} + \cdots\right)^2 = \left(\frac{x^2}{2}\right)^2\left(1 - \frac{x^2}{12} + \cdots\right)^2 = \frac{x^4}{4} - \frac{x^6}{24} + \cdots,$$

$$z^3 = \left(\frac{x^2}{2} - \cdots\right)^3 = \frac{x^6}{8} - \cdots.$$

The quantities z^4, z^5, $\cdots$ can be ignored; they do not contain x^6 or lower powers of x. Collect terms:

$$\begin{aligned}\frac{1}{\cos x} &= 1 + \frac{1}{2}x^2 + \left(-\frac{1}{24} + \frac{1}{4}\right)x^4 + \left(\frac{1}{720} - \frac{1}{24} + \frac{1}{8}\right)x^6 + \cdots \\ &= 1 + \frac{1}{2}x^2 + \frac{5}{24}x^4 + \frac{61}{720}x^6 + \cdots.\end{aligned}$$

Therefore

$$\begin{aligned}\tan x &= \sin x \cdot \frac{1}{\cos x} \\ &= \left(x - \frac{x^3}{6} + \frac{x^5}{120} - \frac{x^7}{5040} + \cdots\right)\left(1 + \frac{x^2}{2} + \frac{5}{24}x^4 + \frac{61}{720}x^6 + \cdots\right) \\ &= x + \left(-\frac{1}{6} + \frac{1}{2}\right)x^3 + \left(\frac{1}{120} - \frac{1}{12} + \frac{5}{24}\right)x^5 \\ &\qquad + \left(-\frac{1}{5040} + \frac{1}{240} - \frac{5}{144} + \frac{61}{720}\right)x^7 + \cdots \\ &= x + \frac{1}{3}x^3 + \frac{2}{15}x^5 + \frac{17}{315}x^7 + \cdots.\end{aligned}$$

Answer:

$$\tan x = x + \frac{1}{3}x^3 + \frac{2}{15}x^5 + \frac{17}{315}x^7 + \cdots.$$

REMARK: The power series for $\tan x$ converges to $\tan x$ for $|x| < \pi/2$. This is the largest interval about $x = 0$ in which the denominator $\cos x$ is non-zero.

EXAMPLE 3.7

What is the value of the seventh derivative of $\tan x$ at $x = 0$?

Solution: Denote $\tan x$ by $f(x)$. The coefficient of x^7 in its power series is $f^{(7)}(0)/7!$ But that coefficient is $17/315$ from the answer to the preceding problem. Hence

$$\frac{f^{(7)}(0)}{7!} = \frac{17}{315}, \qquad f^{(7)}(0) = \frac{(17)(5040)}{315}.$$

Answer: 272.

EXERCISES

Find the power series at $x = 0$:

1. $\dfrac{1}{1 - 5x}$
2. $\dfrac{x^2}{1 + x^3}$
3. e^{x^3}
4. $\cos 2x$
5. $\cosh \sqrt{x}$
6. $x(\sin x + \sin 3x)$
7. $(x - 1)e^x$
8. $\dfrac{1 - \cos x}{x^2}$
9. $\dfrac{x^2}{1 - x}$
10. $\dfrac{\sin x - x}{x^3}$
11. $\sin^2 x$ (*Hint:* Use a trigonometric identity.)
12. $\sin(x^2)$.

Compute the terms up to and including x^6 in the power series at $x = 0$:

13. $\dfrac{1}{(1 - 2x)(1 - 3x^2)}$
14. $e^x \sin(x^2)$
15. $\dfrac{1}{1 - x^2 e^x}$
16. $\dfrac{1}{1 + x^2 + x^4}$
17. $\sin^3 x$
18. $\ln \cos x$.

Find $f^{(8)}(0)$:

19. $f(x) = \dfrac{1}{1 - x^4}$
20. $f(x) = x \cos x$
21. $f(x) = e^{-x^2}$
22. $f(x) = \arctan(x^3)$.
23. In Example 3.2, the series $1 - \dfrac{x}{2!} + \dfrac{x^2}{4!} - \dfrac{x^6}{6!} + \cdots$ was shown to converge to $\cos \sqrt{x}$ for $x \geq 0$. Show that it converges to $\cosh \sqrt{-x}$ for $x \leq 0$.

4. FURTHER TECHNIQUES

The methods of the last section enable us to derive new power series from known ones. Here (without proof) is another important method for doing so.

A power series may be differentiated or integrated term-by-term within its interval of convergence.

EXAMPLE 4.1

Find power series at $x = 0$ for $(1 - x)^{-2}$ and $(1 - x)^{-3}$.

Solution: For $|x| < 1$,

$$\begin{aligned}(1 - x)^{-2} &= \frac{d}{dx}(1 - x)^{-1} \\ &= \frac{d}{dx}(1 + x + x^2 + \cdots + x^n + \cdots) \\ &= \frac{d}{dx}(1) + \frac{d}{dx}(x) + \frac{d}{dx}(x^2) + \cdots + \frac{d}{dx}(x^n) + \cdots \\ &= 1 + 2x + 3x^2 + \cdots + nx^{n-1} + \cdots .\end{aligned}$$

Differentiate again:

$$\begin{aligned}(1 - x)^{-3} &= \frac{1}{2}\frac{d}{dx}(1 - x)^{-2} = \frac{1}{2}\frac{d}{dx}(1 + 2x + 3x^2 + \cdots + nx^{n-1} + \cdots) \\ &= \frac{1}{2}[2 + 6x^2 + 12x^3 + \cdots + n(n - 1)x^{n-2} + \cdots].\end{aligned}$$

Answer: For $|x| < 1$,

$$(1 - x)^{-2} = \sum_{n=1}^{\infty} nx^{n-1}, \quad (1 - x)^{-3} = \sum_{n=2}^{\infty} \frac{n(n - 1)}{2} x^{n-2}.$$

EXAMPLE 4.2

Find the power series at $x = 0$ for $\ln(1 + x)$.

Solution: Notice that $\ln(1 + x)$ is an antiderivative of $(1 + x)^{-1} = 1 - x + x^2 - x^3 + \cdots$. Therefore

$$\begin{aligned}\ln(1 + x) &= \int_0^x \frac{dt}{1 + t} \\ &= \int_0^x [1 - t + t^2 - \cdots + (-1)^n t^n + \cdots]\, dt \\ &= \int_0^x dt - \int_0^x t\, dt + \int_0^x t^2\, dt - \cdots + (-1)^n \int_0^x t^n\, dt + \cdots \\ &= x - \frac{x^2}{2} + \frac{x^3}{3} - \cdots + (-1)^n \frac{x^{n+1}}{n + 1} + \cdots .\end{aligned}$$

Answer: For $|x| < 1$,

$$\ln(1 + x) = x - \frac{x^2}{2} + \frac{x^3}{3} - \frac{x^4}{4} + \cdots + (-1)^n \frac{x^{n+1}}{n+1} + \cdots.$$

EXAMPLE 4.3

Sum the series $x + \dfrac{x^3}{3} + \dfrac{x^5}{5} + \cdots + \dfrac{x^{2n-1}}{2n-1} + \cdots$.

Solution: By the ratio test, the series converges for $|x| < 1$ to some function $f(x)$. Write

$$f(x) = x + \frac{x^3}{3} + \frac{x^5}{5} + \cdots + \frac{x^{2n-1}}{2n-1} + \cdots.$$

Each term is the integral of a power of x. This suggests that $f(x)$ is the integral of some simple function. Differentiate term-by-term:

$$f'(x) = 1 + x^2 + x^4 + \cdots + x^{2n-2} + \cdots = \frac{1}{1 - x^2}.$$

Therefore, $f(x)$ is an antiderivative of $1/(1 - x^2)$. Since $f(0) = 0$, it follows that

$$f(x) = \int_0^x \frac{dt}{1 - t^2} = \frac{1}{2} \ln \frac{1 + t}{1 - t}\bigg|_0^x = \frac{1}{2} \ln \frac{1 + x}{1 - x}.$$

Answer: $\dfrac{1}{2} \ln \dfrac{1 + x}{1 - x}$.

EXAMPLE 4.4

Sum the series $x + 4x^2 + 9x^3 + \cdots + n^2x^n + \cdots$.

Solution: Write

$$f(x) = x + 4x^2 + 9x^3 + \cdots + n^2x^n + \cdots = xg(x),$$

where

$$g(x) = 1 + 2^2x + 3^2x^2 + \cdots + n^2x^{n-1} + \cdots.$$

Now

$$g(x) = \frac{d}{dx}(x + 2x^2 + 3x^3 + 4x^4 + \cdots + nx^n + \cdots)$$

$$= \frac{d}{dx}[x(1 + 2x + 3x^2 + 4x^3 + \cdots + nx^{n-1} + \cdots)].$$

By Example 4.1,

$$1 + 2x + 3x^2 + \cdots + nx^{n-1} + \cdots = (1 - x)^{-2}.$$

Hence

$$g(x) = \frac{d}{dx}\left[\frac{x}{(1 - x)^2}\right] = \frac{1 + x}{(1 - x)^3},$$

so

$$f(x) = xg(x) = \frac{x + x^2}{(1 - x)^3}.$$

Answer: $\dfrac{x + x^2}{(1 - x)^3}.$

CHECK: Evaluate the series at $x = 0.1$:

$$f(0.1) = \frac{1}{10} + \frac{4}{10^2} + \frac{9}{10^3} + \frac{16}{10^4} + \frac{25}{10^5} + \cdots.$$

According to the answer, the sum is

$$\frac{(0.1) + (0.1)^2}{(1 - 0.1)^3} = \frac{0.11}{(0.9)^3} \approx 0.15089\,16324.$$

It is easy to make a convincing numeral check. Start with the first term and add successive terms: 0.1, 0.14, 0.149, 0.1506, 0.15085, 0.15088 6, 0.15089 09, 0.15089 154, 0.15089 1621, 0.15089 16310.

Partial Fractions

The power series for $1/(x - 5)$ can be obtained from the geometric series. Just write

$$\frac{1}{x - 5} = -\frac{1}{5 - x} = -\frac{1}{5}\,\frac{1}{1 - \dfrac{x}{5}},$$

and expand in a geometric series:

$$\frac{1}{x - 5} = -\frac{1}{5}\left[1 + \left(\frac{x}{5}\right) + \left(\frac{x}{5}\right)^2 + \cdots\right],$$

which converges if $|x/5| < 1$, that is, $|x| < 5$.

Combined with this trick, the method of partial fractions is useful in finding power series for rational functions (quotients of polynomials).

EXAMPLE 4.5

Find a power series at $x = 0$ for $f(x) = \dfrac{1}{(x - 2)(x - 5)}.$

Solution: By partial fractions

$$\frac{1}{(x-2)(x-5)} = \frac{1}{3}\left(\frac{1}{x-5} - \frac{1}{x-2}\right).$$

Expand each fraction by the preceding trick:

$$\frac{1}{x-2} = -\frac{1}{2}\sum_{0}^{\infty}\left(\frac{x}{2}\right)^n = -\sum_{0}^{\infty}\frac{x^n}{2^{n+1}} \qquad \text{for} \quad |x| < 2,$$

$$\frac{1}{x-5} = -\frac{1}{5}\sum_{0}^{\infty}\left(\frac{x}{5}\right)^n = -\sum_{0}^{\infty}\frac{x^n}{5^{n+1}} \qquad \text{for} \quad |x| < 5.$$

Therefore if $|x| < 2$, both series converge and

$$\frac{1}{3}\left(\frac{1}{x-5} - \frac{1}{x-2}\right) = \frac{1}{3}\sum_{0}^{\infty}\left(\frac{1}{2^{n+1}} - \frac{1}{5^{n+1}}\right)x^n.$$

Answer: For $|x| < 2$,

$$\frac{1}{(x-2)(x-5)} = \frac{1}{3}\sum_{n=0}^{\infty}\left(\frac{1}{2^{n+1}} - \frac{1}{5^{n+1}}\right)x^n.$$

EXAMPLE 4.6

Find the power series at $x = 0$ for $\dfrac{-1}{x^2 + x - 1}$.

Solution: The denominator can be factored:

$$x^2 + x - 1 = (x-a)(x-b),$$

where a and b are the roots of the equation

$$x^2 + x - 1 = 0.$$

By the quadratic formula,

$$a = \frac{-1+\sqrt{5}}{2}, \qquad b = \frac{-1-\sqrt{5}}{2}.$$

Notice that $ab = -1$ because the product of the roots of $x^2 + px + q = 0$ is q. (Why?)

By partial fractions,

$$\frac{-1}{x^2+x-1} = \frac{-1}{(x-a)(x-b)} = \frac{1}{a-b}\left(\frac{1}{x-b} - \frac{1}{x-a}\right).$$

Suppose $|x| < \frac{1}{2}(-1+\sqrt{5})$, the smaller of the numbers $|a|$ and $|b|$. Now expand:

$$\frac{-1}{x^2+x-1} = \frac{1}{a-b}\left(\sum_{n=0}^{\infty} \frac{x^n}{a^{n+1}} - \sum_{n=0}^{\infty} \frac{x^n}{b^{n+1}}\right) = \sum_{n=0}^{\infty} \frac{1}{a-b}\left(\frac{1}{a^{n+1}} - \frac{1}{b^{n+1}}\right) x^n.$$

Note that

$$a - b = \frac{-1+\sqrt{5}}{2} - \frac{-1-\sqrt{5}}{2} = \sqrt{5},$$

and (since $ab = -1$)

$$\frac{1}{a} = -b = \frac{1+\sqrt{5}}{2}, \qquad \frac{1}{b} = -a = \frac{1-\sqrt{5}}{2}.$$

Answer: For $|x| < \frac{1}{2}(\sqrt{5}-1)$,

$$\frac{-1}{x^2+x-1} = \sum_{n=0}^{\infty} c_n x^n, \text{ where}$$

$$c_n = \frac{1}{\sqrt{5}}\left[\left(\frac{1+\sqrt{5}}{2}\right)^{n+1} - \left(\frac{1-\sqrt{5}}{2}\right)^{n+1}\right].$$

REMARK: It may not seem so, but the numbers c_n are actually integers! The first few are 1, 1, 2, 3, 5, 8, 13, $\cdots$; each is the sum of the previous two. See Ex. 17 below.

EXAMPLE 4.7

Find a power series for $\dfrac{1}{(1-x)(1+x^2)}$.

Solution: By partial fractions,

$$\frac{1}{(1-x)(1+x^2)} = \frac{1}{2}\left(\frac{1}{1-x} + \frac{1+x}{1+x^2}\right) = \frac{1}{2}\left(\frac{1}{1-x} + \frac{1}{1+x^2} + \frac{x}{1+x^2}\right)$$

$$= \frac{1}{2}[(1 + x + x^2 + x^3 + \cdots) + (1 - x^2 + x^4 - x^6 + - \cdots) + (x - x^3 + x^5 - x^7 + - \cdots)]$$

$$= 1 + x + x^4 + x^5 + x^8 + x^9 + \cdots.$$

Answer: For $|x| < 1$,

$$\frac{1}{(1-x)(1+x^2)} = \sum_{n=0}^{\infty} (x^{4n} + x^{4n+1}).$$

Remark: This example also can be done by multiplying together the series for $(1 - x)^{-1}$ and $(1 + x^2)^{-1}$, also by multiplying the series for $(1 - x^4)^{-1}$ by $1 + x$.

EXERCISES

Obtain the power series at $x = 0$:

1. $\dfrac{x+1}{x^2 - 5x + 6}$
2. $\dfrac{1}{(2x+1)(3x+4)}$
3. $\dfrac{1}{(x-1)(x-2)(x-3)}$
4. $\dfrac{1}{(x-3)(x^2+1)}$
5. $\dfrac{1}{1 - x + x^2}$
6. $\dfrac{1 + x^2}{1 - x^5}$.

Find the sum of the series:

7. $4 + 5x + 6x^2 + \cdots + (n+4)x^n + \cdots$
8. $1 + 4x + 9x^2 + \cdots + (n+1)^2x^n + \cdots$
9. $\dfrac{x^4}{4} + \dfrac{x^8}{8} + \dfrac{x^{12}}{12} + \cdots + \dfrac{x^{4n}}{4n} + \cdots$
   (*Hint:* Differentiate)
10. $\dfrac{x^2}{1 \cdot 2} - \dfrac{x^3}{2 \cdot 3} + \dfrac{x^4}{3 \cdot 4} - \cdots + \dfrac{(-1)^n x^n}{(n-1)n} + \cdots$
11. $2 + 3x + 4x^2 + \cdots + (n+2)x^n + \cdots$
   (*Hint:* Multiply by x and integrate).
12. Evaluate $1 + \dfrac{2}{5} + \dfrac{3}{25} + \dfrac{4}{125} + \cdots + \dfrac{n}{5^{n-1}} + \cdots$
   (*Hint:* Consider $\displaystyle\sum_{n=1}^{\infty} nx^{n-1}$).

Verify by expressing both sides in power series:

13. $\dfrac{d}{dx}(\sin x) = \cos x$
14. $\dfrac{d}{dx}(e^{-x^2}) = -2xe^{-x^2}$
15. $\displaystyle\int x^2 \cos x\, dx = 2x \cos x + (x^2 - 2) \sin x + C$
16. $\displaystyle\int \text{arc tan}\, x\, dx = x\, \text{arc tan}\, x - \frac{1}{2}\ln(1 + x^2) + C.$
17. Consider Example 4.6 from a different viewpoint. Suppose $a_0, a_1, a_2, \cdots$ is a sequence of integers satisfying $a_0 = a_1 = 1$ and $a_{n+2} = a_n + a_{n+1}$ for $n \geq 1$. Set $f(x) = a_0 + a_1x + a_2x^2 + \cdots$. Show that $xf(x) + x^2f(x) = f(x) - 1$, and conclude that $f(x) = (1 - x - x^2)^{-1}$. Now obtain a formula for a_n. (This is the method of **generating functions.**)

18. Which power series is more efficient for computing $\ln(\frac{3}{2})$, the one for $\ln(1 + x)$, or the one for $\ln[(1 + x)/(1 - x)]$? See Examples 4.2 and 4.3. Compute $\ln(\frac{3}{2})$ to 5 places.

19. Expand $\arctan x$ in a Taylor series at $x = 0$. (*Hint:* Integrate its derivative.) Conclude that

$$\arctan x = x - \frac{x^3}{3} + \frac{x^5}{5} - \frac{x^7}{7} + \cdots \quad \text{for } |x| < 1.$$

It is known that the series in Ex. 19 is valid for $x = 1$, hence that

$$1 - \frac{1}{3} + \frac{1}{5} - \frac{1}{7} + \cdots = \frac{\pi}{4}.$$

This is an interesting formula but a very poor one for computing π since its terms decrease slowly. Exercises 20–23 develop an efficient way to compute π.

20. Prove the formula $\arctan x + \arctan y = \arctan\left(\dfrac{x + y}{1 - xy}\right)$. Conclude that $\dfrac{\pi}{4} = \arctan\dfrac{1}{2} + \arctan\dfrac{1}{3}$.

21. (cont.) Use this expression for $\pi/4$ and the power series in Ex. 19 to compute π to 4 places.

22. (cont.) Show that the expression in Ex. 20 for $\pi/4$ can be modified to $\dfrac{\pi}{4} = 2\arctan\dfrac{1}{3} + \arctan\dfrac{1}{7}$. Why is this even better for computing π ?

23. (cont.) Finally modify the expression in Ex. 22 to $\dfrac{\pi}{4} = 2\arctan\dfrac{1}{5} + \arctan\dfrac{1}{7} + 2\arctan\dfrac{1}{8}$. Now compute π to 7 places.

5. BINOMIAL SERIES

The Binomial Theorem asserts that for each positive integer p,

$$(1 + x)^p = 1 + px + \frac{p(p-1)}{2!}x^2 + \frac{p(p-1)(p-2)}{3!}x^3 + \cdots + \frac{p(p-1)(p-2)\cdots(p-n+1)}{n!}x^n + \cdots + \frac{p!}{p!}x^p.$$

Standard notation for the coefficients in this identity is

$$\binom{p}{0} = 1, \qquad \binom{p}{n} = \frac{p(p-1)(p-2)\cdots(p-n+1)}{n!}, \qquad 1 \leq n \leq p.$$

With this notation the expansion of $(1 + x)^p$ can be abbreviated:

$$(1 + x)^p = \sum_{n=0}^{p} \binom{p}{n} x^n.$$

A generalization of the Binomial Theorem is the **binomial series** for $(1 + x)^p$, where p is not necessarily a positive integer.

Binomial Series Suppose p is any number. Then

$$(1+x)^p = \sum_{n=0}^{\infty} \binom{p}{n} x^n, \qquad -1 < x < 1,$$

where the coefficients in this series are

$$\binom{p}{0} = 1, \qquad \binom{p}{n} = \frac{p(p-1)(p-2)\cdots(p-n+1)}{n!}, \qquad n \geq 1.$$

REMARK: In case p happens to be a positive integer and $n > p$, then the coefficient

$$\binom{p}{n}$$

equals 0 because it has a factor $(p - p)$. In this case, the series breaks off after the term in x^p. The resulting formula,

$$(1+x)^p = \sum_{n=0}^{p} \binom{p}{n} x^n,$$

is the old Binomial Theorem again. But if p is not a positive integer or zero, then each coefficient is non-zero, so the series has infinitely many terms.

The binomial series is just the Taylor series for $y(x) = (1 + x)^p$ at $x = 0$. Notice that

$$\begin{aligned} y'(x) &= p(1+x)^{p-1}, \\ y''(x) &= p(p-1)(1+x)^{p-2}, \\ &\cdots\cdots\cdots\cdots\cdots \\ y^{(n)}(x) &= p(p-1)(p-2)\cdots(p-n+1)(1+x)^{p-n}. \end{aligned}$$

Therefore the coefficient of x^n in the Taylor series is

$$\frac{y^{(n)}(0)}{n!} = \frac{p(p-1)(p-2)\cdots(p-n+1)}{n!} = \binom{p}{n}.$$

It is not obvious, however, that the binomial series converges for $|x| < 1$ and actually equals $(1 + x)^p$. We shall omit the proof.

EXAMPLE 5.1

Find a power series for $\dfrac{1}{(1+x)^2}$.

Solution: Use the binomial series with $p = -2$. The coefficient of x^n is

$$\binom{-2}{n} = \frac{(-2)(-3)(-4)\cdots(-2-n+1)}{n!}$$

$$= (-1)^n \frac{2\cdot 3\cdot 4\cdots n\cdot(n+1)}{n!} = (-1)^n(n+1).$$

Therefore

$$(1+x)^{-2} = \sum_{n=0}^{\infty}\binom{-2}{n}x^n = \sum_{n=0}^{\infty}(-1)^n(n+1)x^n$$

$$= 1 - 2x + 3x^2 - 4x^3 + \cdots.$$

Answer: For $|x| < 1$,

$$\frac{1}{(1+x)^2} = \sum_{n=0}^{\infty}(-1)^n(n+1)x^n.$$

CHECK:

$$\frac{1}{(1+x)^2} = -\frac{d}{dx}\left(\frac{1}{1+x}\right) = -\frac{d}{dx}(1 - x + x^2 - x^3 + - \cdots)$$

$$= -(-1 + 2x - 3x^2 + - \cdots).$$

EXAMPLE 5.2

Express $\dfrac{1}{\sqrt{1+x}}$ as a power series.

Solution: Use the binomial series with $p = -\frac{1}{2}$. The coefficient of x^n is

$$\binom{-\frac{1}{2}}{n} = \frac{\left(-\frac{1}{2}\right)\left(-\frac{3}{2}\right)\left(-\frac{5}{2}\right)\cdots\left(-\frac{2n-1}{2}\right)}{n!}$$

$$= (-1)^n\frac{1\cdot 3\cdot 5\cdots(2n-1)}{2^n\cdot n!}.$$

But

$$1\cdot 3\cdot 5\cdots(2n-1) = \frac{1\cdot 2\cdot 3\cdot 4\cdots(2n)}{2\cdot 4\cdot 6\cdot 8\cdots(2n)} = \frac{(2n)!}{2^n\cdot n!}.$$

Hence the coefficient of x^n is

$$(-1)^n \frac{(2n)!}{2^n \cdot n!} \cdot \frac{1}{2^n \cdot n!} = (-1)^n \frac{(2n)!}{(2^n \cdot n!)^2}.$$

Answer: For $|x| < 1$,

$$\frac{1}{\sqrt{1+x}} = \sum_{n=0}^{\infty} (-1)^n \frac{(2n)!}{(2^n \cdot n!)^2} x^n.$$

The following example will be used in Section 7.

EXAMPLE 5.3

Express $\sqrt{1-x}$ as a power series.

Solution: Use the binomial series with $p = \frac{1}{2}$ and replace x by $-x$. The constant term is

$$\binom{\frac{1}{2}}{0} = 1.$$

The term in x^n is

$$\binom{\frac{1}{2}}{n}(-x)^n = \frac{\left(\frac{1}{2}\right)\left(-\frac{1}{2}\right)\left(-\frac{3}{2}\right)\cdots\left(-\frac{2n-3}{2}\right)}{n!}(-x)^n$$

$$= (-1)^{n-1}\frac{1 \cdot 3 \cdot 5 \cdots (2n-3)}{2^n \cdot n!}(-1)^n x^n.$$

But

$$1 \cdot 3 \cdot 5 \cdots (2n-3) = \frac{1 \cdot 2 \cdot 3 \cdot 4 \cdots (2n-2)}{2 \cdot 4 \cdot 6 \cdot 8 \cdots (2n-2)}$$

$$= \frac{(2n-2)!}{2^{n-1}(n-1)!}$$

$$= \frac{(2n)!}{(2^n \cdot n!)(2n-1)}.$$

Therefore the term in x^n is

$$-\frac{(2n)!}{(2^n \cdot n!)(2n-1)} \cdot \frac{1}{2^n \cdot n!} x^n = -\frac{(2n)!}{(2^n \cdot n!)^2(2n-1)} x^n.$$

Answer: For $|x| < 1$,

$$\sqrt{1-x} = 1 - \sum_{n=1}^{\infty} \frac{(2n)!}{(2^n \cdot n!)^2(2n-1)} x^n.$$

EXAMPLE 5.4

Estimate $\sqrt[3]{1001}$ to 7 places.

Solution: Write

$$\sqrt[3]{1001} = [1000(1 + 10^{-3})]^{1/3} = 10(1 + 10^{-3})^{1/3}.$$

Use the binomial series with $x = 10^{-3}$ and $p = \frac{1}{3}$:

$$10(1 + 10^{-3})^{1/3} = 10\left[1 + \binom{\frac{1}{3}}{1}(10^{-3}) + \binom{\frac{1}{3}}{2}(10^{-3})^2 + \cdots\right]$$

$$= 10\left[1 + \frac{1}{3}\cdot 10^{-3} - \frac{1}{9}\cdot 10^{-6} + \cdots\right].$$

The first three terms yield the estimate

$$\sqrt[3]{1001} \approx 10.0033322222 \cdots.$$

The error in this estimate is precisely the remainder

$$10\left[\binom{\frac{1}{3}}{3}(10^{-3})^3 + \binom{\frac{1}{3}}{4}(10^{-3})^4 + \cdots\right].$$

Now each binomial coefficient above is less than 1 in absolute value:

$$\left|\binom{\frac{1}{3}}{n}\right| = \left|\frac{\left(\frac{1}{3}\right)\left(-\frac{2}{3}\right)\left(-\frac{5}{3}\right)\cdots\left(-\frac{3n-4}{3}\right)}{1\cdot 2\cdot 3\cdots n}\right|$$

$$= \frac{1}{3}\cdot\frac{2}{6}\cdot\frac{5}{9}\cdots\frac{3n-4}{3n} < 1.$$

Therefore, the error (in absolute value) is less than

$$10[(10^{-3})^3 + (10^{-3})^4 + \cdots] = 10^{-8}[1 + 10^{-3} + 10^{-6} + 10^{-9} + \cdots]$$

$$= 10^{-8}\frac{1}{1 - 10^{-3}} \approx 10^{-8}.$$

Answer: To 7 places, $\sqrt[3]{1001} \approx 10.0033322.$

REMARK: The estimate is actually accurate to at least 8 places. When we estimated each binomial coefficient by 1, we were too generous since

$$\binom{\frac{1}{3}}{n} < \frac{1\cdot 2\cdot 5}{3\cdot 6\cdot 9} = \frac{10}{162}, \qquad n \geq 3.$$

(Why?) Therefore, the error estimate can be reduced by a factor of $\frac{10}{162}$, which guarantees at least 1 more place accuracy.

EXERCISES

Expand in power series at $x = 0$:

1. $\dfrac{1}{(1+x)^3}$
2. $\dfrac{1}{(1-x)^{1/3}}$
3. $\dfrac{1}{(1-4x^2)^2}$
4. $\sqrt{2-x}$

   [*Hint:* $2 - x = 2\left(1 - \dfrac{x}{2}\right)$]
5. $(1+2x^3)^{1/4}$
6. $\dfrac{1}{(3-4x^2)^2}$.

Expand in power series at $x = 1$:

7. $\sqrt{1+x}$ [*Hint:* Write $1 + x = 2 + (x-1) = 2\left(1 + \dfrac{x-1}{2}\right)$]
8. $\dfrac{1}{(3+x)^2}$.

Compute the power series at $x = 0$ up to and including the term in x^4:

9. $\sqrt{1+x^2e^x}$
10. $\dfrac{1}{(1+\sin x)^2}$
11. $(\sin 2x)\sqrt{3+x}$
12. $\dfrac{1}{\sqrt{1+x+x^2}}$

    (*Hint:* Write $x + x^2 = u$)
13. $\dfrac{\cos 2x}{\left(1+\dfrac{x}{3}\right)^4}$.

Compute to 4-place accuracy using the Binomial Theorem:

14. $\sqrt{16.1}$
15. $\sqrt[4]{82}$
16. $\dfrac{1}{(1.03)^5}$.
17. Expand arc sin x in a Taylor series at $x = 0$.

    (*Hint:* Integrate its derivative.)

6. ALTERNATING SERIES

An **alternating series** is one whose terms are alternately positive and negative. Examples:

$$1 - \frac{1}{2} + \frac{1}{3} - \frac{1}{4} + - + - \cdots,$$

$$1 - x^2 + x^4 - x^6 + - + - \cdots \qquad \text{(alternating for all } x\text{)},$$

$$x - \frac{x^2}{4} + \frac{x^3}{9} - \frac{x^4}{16} + - + - \cdots \qquad \text{(alternating only for } x > 0\text{)}.$$

Such series have some extremely useful properties, two of which we now state.

> (1) If the terms of an alternating series decrease in absolute value to zero, then the series converges.
>
> (2) If such a series is broken off at the n-th term, then the remainder (in absolute value) is less than the absolute value of the $(n + 1)$-th term.

These assertions provide a very simple convergence criterion and an immediate remainder estimate for *alternating* series. We shall not give a formal proof; rather we shall show geometrically that they make good sense.

Suppose Σa_n is an alternating series whose terms decrease in absolute value to zero. (To be definite, assume $a_1 > 0$.) Let $s_n = a_1 + a_2 + \cdots + a_n$. Plot these partial sums (Fig. 6.1). The partial sums oscillate back

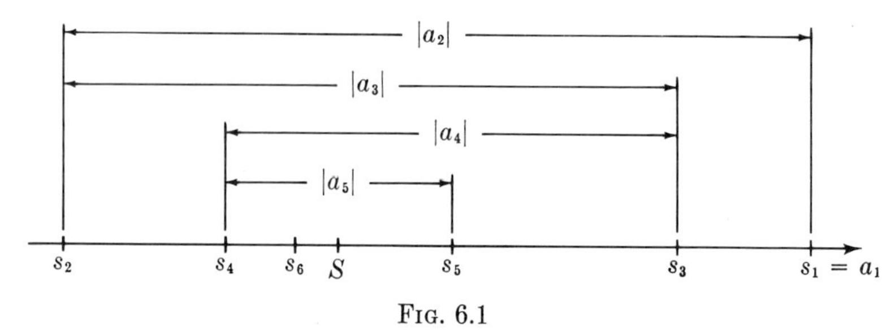

FIG. 6.1

and forth as shown. But since the terms decrease to zero, the oscillations become shorter and shorter. The odd partial sums decrease and the even ones increase, squeezing down on some number S. Thus, the series converges to S.

If the series is broken off after n terms, the remainder is $|S - s_n|$. But from Fig. 6.1,

$$|S - s_n| < |s_{n+1} - s_n| = |a_{n+1}|.$$

Thus, the remainder is less than the absolute value of the $(n + 1)$-th term.

EXAMPLE 6.1

Find all values of x for which the series

$$1 - x^2 + \frac{x^4}{2} - \frac{x^6}{3} + \frac{x^8}{4} - + \cdots$$

converges.

Solution: From the ratio test, it is easily seen that the series converges for $|x| < 1$ and diverges for $|x| > 1$. But what happens if $|x| = 1$?

At $x = \pm 1$, the series is

$$1 - 1 + \frac{1}{2} - \frac{1}{3} + \frac{1}{4} - + \cdots,$$

an alternating series whose terms decrease in absolute value to zero. Such a series is guaranteed to converge by the statement above.

Answer: Converges for $|x| \leq 1$.

EXAMPLE 6.2

The power series at $x = 0$ for $\ln(1 + x^2)$ is broken off after n terms. Estimate the remainder.

Solution: The series is obtained by integrating from 0 to x the series for the derivative of $\ln(1 + t^2)$.

$$\frac{d}{dt}[\ln(1 + t^2)] = \frac{2t}{1 + t^2} = 2t(1 - t^2 + t^4 - t^6 + - \cdots)$$

$$= 2(t - t^3 + t^5 - t^7 + - \cdots), \qquad |t| < 1.$$

It follows that

$$\ln(1 + x^2) = \int_0^x \frac{2t\,dt}{1 + t^2} = 2\left(\frac{x^2}{2} - \frac{x^4}{4} + \frac{x^6}{6} - \frac{x^8}{8} + - \cdots\right)$$

$$= x^2 - \frac{x^4}{2} + \frac{x^6}{3} - \frac{x^8}{4} + - \cdots, \qquad |x| < 1.$$

This series alternates; since $|x| < 1$, its terms decrease in absolute value to zero. Therefore, the remainder after n terms is less than the absolute value of the $(n + 1)$-th term.

Answer: $|\text{remainder}| < \dfrac{x^{2n+2}}{n + 1}$.

Occasionally we encounter an alternating series whose terms ultimately decrease in magnitude, but whose first few terms do not. If the successive terms decrease starting at the k-th term, the series will still converge, and the remainder estimate is still valid—beyond the k-th term. The front end of the series, up to the $(k - 1)$-th term, is a finite sum; it causes no trouble. The important part is the tail end, the power series starting with the k-th term. It is this series that we test for convergence.

EXAMPLE 6.3

The power series for e^{-x} at $x = 0$ is broken off after n terms. Estimate the remainder for positive values of x.

Solution: If $x > 0$ the power series

$$e^{-x} = 1 - x + \frac{x^2}{2!} - \frac{x^3}{3!} + - \cdots$$

alternates. If $0 < x \leq 1$, the terms decrease to 0, and the above remainder estimate for alternating series applies. If $x > 1$, however, the first few terms may not decrease. (Take $x = 6$:

$$e^{-6} = 1 - 6 + \frac{6^2}{2!} - \frac{6^3}{3!} + - \cdots = 1 - 6 + 18 - 36 + - \cdots .)$$

Nevertheless, for any fixed x, the terms do decrease ultimately.

To see why, note that the ratio of successive terms is

$$\frac{x^{n+1}}{(n+1)!} \Big/ \frac{x^n}{n!} = \frac{x}{n+1}.$$

Take a fixed value of x. Then

$$\frac{x}{n+1} < 1$$

as soon as $n + 1 > x$; from then on the terms decrease. Furthermore

$$\frac{x}{n+1} < \frac{1}{2}$$

as soon as $n + 1 > 2x$. From then on each term is less than one-half the preceding term. Hence, the terms decrease to 0.

Result: if the series is broken off at the n-th term, the remainder estimate for an alternating series applies, provided $n + 1 > x$.

Answer: For $n > x - 1$,

$$|\text{remainder}| < \frac{x^{n+1}}{(n+1)!}.$$

As another application of this method, recall Example 5.4. There it was shown that three terms of the binomial series for

$$10(1 + 10^{-3})^{1/3}$$

provide 7-place accuracy in computing $\sqrt[3]{1001}$. But this series alternates and its terms decrease towards 0. Therefore, one deduces that the remainder after three terms is less than the fourth term, which is

$$10\binom{\frac{1}{3}}{3}(10^{-3})^3 = \frac{\left(\frac{1}{3}\right)\left(-\frac{2}{3}\right)\left(-\frac{5}{3}\right)}{3!}10^{-8} = \frac{10}{162} \cdot 10^{-8} < 10^{-9},$$

so 8-place accuracy is assured. This estimate is both easier and more precise than the one in Example 5.4.

EXERCISES

1. The power series $\sum \frac{x^n}{n}$ converges *inside* the interval $-1 < x < 1$. Show that it converges also at $x = -1$.
2. Give an example of an alternating series that does not converge.
3. Compute $e^{-1/5}$ to 5-place accuracy.
4. Find values of x for which the formula $\cos x \approx 1 - \frac{x^2}{2!} + \frac{x^4}{4!}$ yields 5-place accuracy.
5. How many terms of the power series for $\ln x$ at $x = 10$ are needed to compute $\ln(10.5)$ with 5-place accuracy? (Assume the value of $\ln 10$ is known.)
6. How many terms of the binomial series for $\sqrt{1 + x}$ will yield 5-place accuracy for $0 \leq x \leq 0.1$?

7. APPLICATIONS TO DEFINITE INTEGRALS [optional]

Power series are used in approximating definite integrals which cannot be computed exactly.

EXAMPLE 7.1

Estimate $\int_0^x e^{-t^2}\,dt$ to 6 places for $|x| \leq \frac{1}{2}$.

Solution: A numerical integration formula such as Simpson's Rule can be used, but a power series method is simpler. Expand the integrand in a power series:

$$e^{-t^2} = 1 + (-t^2) + \frac{(-t^2)^2}{2!} + \frac{(-t^2)^3}{3!} + \cdots = 1 - t^2 + \frac{t^4}{2!} - \frac{t^6}{3!} + \cdots.$$

Since this series converges for all x, it can be integrated term-by-term:

$$\int_0^x e^{-t^2}\,dt = \int_0^x \left(1 - t^2 + \frac{t^4}{2!} - \frac{t^6}{3!} + \cdots\right) dt$$

$$= x - \frac{x^3}{3} + \frac{x^5}{5 \cdot 2!} - \frac{x^7}{7 \cdot 3!} + \frac{x^9}{9 \cdot 4!} - \cdots.$$

Because of the large denominators, this series converges rapidly if x is fairly small. For $|x| \leq \frac{1}{2}$, the sixth term is at most

$$\left(\frac{1}{2}\right)^{11} \frac{1}{11 \cdot 5!} < 4 \times 10^{-7}.$$

Since the series alternates, it follows that five terms provide 6-place accuracy.

Answer: For all x,

$$\int_0^x e^{-t^2}\,dt = x - \frac{x^3}{3} + \frac{x^5}{5\cdot 2!} - \frac{x^7}{7\cdot 3!} + \cdots.$$

For $|x| \le \frac{1}{2}$, five terms yield 6-place accuracy.

REMARK: The series converges for all values of x, but for large x it converges slowly. For example, it would be ridiculous to compute

$$\int_0^{10} e^{-t^2}\,dt$$

by this method, since more than 100 terms at the beginning are greater than 1.

Elliptic Integrals

EXAMPLE 7.2

Express the integral $\displaystyle\int_0^{\pi/2} \sqrt{1 - k^2\sin^2 t}\,dt,$ where $k^2 < 1$, as a power series in k.

Solution: By Example 5.3,

$$\sqrt{1-x} = (1-x)^{1/2} = 1 - \sum_{n=1}^{\infty} a_n x^n, \qquad |x| < 1,$$

where

$$a_n = \frac{(2n)!}{(2^n\cdot n!)^2(2n-1)}.$$

Substitute $k^2\sin^2 t$ for x. This is permissible because $k^2\sin^2 t \le k^2 < 1$.

$$(1 - k^2\sin^2 t)^{1/2} = 1 - \sum_{n=1}^{\infty} a_n k^{2n}\sin^{2n} t.$$

Now integrate term-by-term:

$$\int_0^{\pi/2} (1 - k^2\sin^2 t)^{1/2}\,dt = \frac{\pi}{2} - \sum_{n=1}^{\infty} a_n k^{2n}\int_0^{\pi/2} \sin^{2n} t\,dt.$$

The integrals on the right are evaluated by a reduction formula. Their precise values are listed in the C.R.C. Tables:

$$\int_0^{\pi/2} (\sin x)^{2n}\,dx = \frac{1\cdot 3\cdot 5 \cdots (2n-1)}{2\cdot 4\cdot 6\cdots (2n)}\cdot\frac{\pi}{2}$$

$$= \frac{1\cdot 2\cdot 3\cdot 4\cdot 5\cdots (2n)}{[2\cdot 4\cdot 6\cdots (2n)]^2}\cdot\frac{\pi}{2} = \frac{(2n)!}{(2^n\cdot n!)^2}\cdot\frac{\pi}{2}.$$

Therefore

$$a_n \int_0^{\pi/2} \sin^{2n} t\, dt = \left[\frac{(2n)!}{(2^n \cdot n!)^2(2n-1)}\right]\left[\frac{(2n)!}{(2^n \cdot n!)^2} \cdot \frac{\pi}{2}\right]$$

$$= \left[\frac{(2n)!}{(2^n \cdot n!)^2}\right]^2 \frac{1}{(2n-1)} \cdot \frac{\pi}{2}.$$

Substitute this expression to obtain the answer.

Answer: For $k^2 < 1$,

$$\int_0^{\pi/2} \sqrt{1 - k^2 \sin^2 t}\, dt$$

$$= \frac{\pi}{2}\left\{1 - \sum_{n=1}^{\infty}\left[\frac{(2n)!}{(2^n \cdot n!)^2}\right]^2 \frac{k^{2n}}{2n-1}\right\}.$$

REMARK: This integral arises in computing the arc length of an ellipse. Suppose an ellipse is given in the parametric form

$$x = a\cos\theta, \qquad y = b\sin\theta,$$

where $b > a$. If s denotes arc length,

$$\left(\frac{ds}{d\theta}\right)^2 = \left(\frac{dx}{d\theta}\right)^2 + \left(\frac{dy}{d\theta}\right)^2 = (-a\sin\theta)^2 + (b\cos\theta)^2$$

$$= b^2(\sin^2\theta + \cos^2\theta) - (b^2 - a^2)\sin^2\theta$$

$$= b^2(1 - k^2\sin^2\theta),$$

where $k^2 = (b^2 - a^2)/b^2$. The length of the ellipse is

$$4\int_0^{\pi/2} \left(\frac{ds}{d\theta}\right) d\theta = 4b\int_0^{\pi/2} \sqrt{1 - k^2\sin^2\theta}\, d\theta.$$

EXAMPLE 7.3

Estimate $\displaystyle \frac{2}{\pi}\int_0^{\pi/2} \sqrt{1 - \frac{1}{5}\sin^2 x}\, dx$ to 4-place accuracy.

Solution: By the last example with $k^2 = \frac{1}{5}$,

$$\frac{2}{\pi}\int_0^{\pi/2} \sqrt{1 - \frac{1}{5}\sin^2 x}\, dx = 1 - \sum_{n=1}^{\infty} b_n \left(\frac{1}{2n-1}\right)\left(\frac{1}{5}\right)^n,$$

where

$$b_n = \left[\frac{(2n)!}{2^{2n}(n!)^2}\right]^2 = \left[\frac{1 \cdot 3 \cdot 5 \cdots (2n-1)}{2 \cdot 4 \cdot 6 \cdots (2n)}\right]^2.$$

Break off the series after the p-th term. The remainder (error) is

$$\epsilon_p = -\sum_{n=p+1}^{\infty} b_n\left(\frac{1}{2n-1}\right)\left(\frac{1}{5}\right)^n.$$

The problem: choose p large enough that $|\epsilon_p| < 5 \times 10^{-5}$.

The error is a complicated expression. It can be enormously simplified in two steps, each of which causes a certain amount of overestimation.

First, each coefficient b_n is less than 1:

$$b_n = \left[\frac{1\cdot 3\cdot 5 \cdots (2n-1)}{2\cdot 4\cdot 6\cdots (2n)}\right]^2 = \left[\frac{1}{2}\cdot\frac{3}{4}\cdot\frac{5}{6}\cdots\frac{2n-1}{2n}\right]^2 < 1;$$

hence

$$|\epsilon_p| = \sum_{n=p+1}^{\infty} b_n\left(\frac{1}{2n-1}\right)\left(\frac{1}{5}\right)^n < \sum_{n=p+1}^{\infty}\left(\frac{1}{2n-1}\right)\left(\frac{1}{5}\right)^n.$$

Second, for $n \geq p+1$,

$$\frac{1}{2n-1} \leq \frac{1}{2(p+1)-1} = \frac{1}{2p+1};$$

hence, replacing $1/(2n-1)$ by $1/(2p+1)$,

$$|\epsilon_p| < \sum_{n=p+1}^{\infty}\left(\frac{1}{2p+1}\right)\left(\frac{1}{5}\right)^n = \frac{1}{2p+1}\sum_{n=p+1}^{\infty}\left(\frac{1}{5}\right)^n.$$

On the right is a geometric series; consequently

$$|\epsilon_p| < \frac{1}{2p+1}\left(\frac{1}{5}\right)^{p+1}\left[1+\left(\frac{1}{5}\right)+\left(\frac{1}{5}\right)^2+\cdots\right]$$

$$= \frac{1}{2p+1}\left(\frac{1}{5}\right)^{p+1}\frac{1}{1-\dfrac{1}{5}} = \frac{1}{4(2p+1)5^p}.$$

To obtain 4-place accuracy, choose p so that

$$\frac{1}{4(2p+1)5^p} < 5 \times 10^{-5}.$$

Take reciprocals:

$$4(2p+1)5^p > 20000, \qquad (2p+1)5^p > 5000.$$

Trial and error shows

$$(2\cdot 3+1)5^3 = 875, \qquad (2\cdot 4+1)5^4 = 5625.$$

Accordingly, choose $p = 4$ and estimate the integral by

$$1 - \left(\frac{1}{2}\right)^2\left(\frac{1}{5}\right) - \left(\frac{1\cdot 3}{2\cdot 4}\right)^2\left(\frac{1}{5}\right)^2\left(\frac{1}{3}\right) - \left(\frac{1\cdot 3\cdot 5}{2\cdot 4\cdot 6}\right)^2\left(\frac{1}{5}\right)^3\left(\frac{1}{5}\right) - \left(\frac{1\cdot 3\cdot 5\cdot 7}{2\cdot 4\cdot 6\cdot 8}\right)^2\left(\frac{1}{5}\right)^4\left(\frac{1}{7}\right)$$

$$= 1 - \frac{1}{20} - \frac{3}{1600} - \frac{1}{6400} - \frac{7}{409600} \approx 0.9480.$$

Answer: 0.9480.

EXERCISES

Express as a power series in x:

1. $\displaystyle\int_0^x \frac{\sin t}{t}\,dt$
2. $\displaystyle\int_0^x \frac{1-\cos t}{t}\,dt$
3. $\displaystyle\int_0^x \frac{t}{1+t^4}\,dt$
4. $\displaystyle\int_0^x \sin(t^2)\,dt.$

Compute to 5-place accuracy:

5. $\displaystyle\int_0^{0.1} e^{x^3}\,dx$
6. $\displaystyle\int_0^{0.2} e^{-x^2}\,dx$
7. $\displaystyle\int_0^{1/4} \sqrt{1+x^3}\,dx$
8. $\displaystyle\int_{3.00}^{3.01} \frac{e^x}{1+x}\,dx$ (expand at $x = 3$).
9. Compute to 4-place accuracy the arc length of an ellipse with semi-axes 40 and 41.
10. Estimate the value of x for which
$$\int_0^x \frac{t}{1+t^4}\,dt = 0.1.$$
(*Hint:* approximate the integral by the first significant term of its power series; use Ex. 3.)
11. (cont.) Refine your estimate of x to 4-place accuracy by taking the first two significant terms of the power series. Use Newton's Method to solve Approximately the resulting equation.

8. APPLICATIONS TO DIFFERENTIAL EQUATIONS [optional]

Certain differential equations can be solved in terms of power series. To illustrate the technique, we start with an easy example.

EXAMPLE 8.1

Obtain a power series solution to the differential equation $y' = xy$.

Solution: Try a solution of the form

$$y(x) = a_0 + a_1 x + a_2 x^2 + \cdots,$$

where the coefficients must be determined. Substitute this power series into the differential equation:

$$\frac{d}{dx}(a_0 + a_1 x + a_2 x^2 + \cdots) = x(a_0 + a_1 x + a_2 x^2 + \cdots),$$

$$a_1 + 2a_2 x + 3a_3 x^2 + \cdots + (k+1)a_{k+1}x^k + \cdots$$
$$= a_0 x + a_1 x^2 + \cdots + a_{k-1}x^k + \cdots.$$

Equate coefficients:

$$a_1 = 0, \quad 2a_2 = a_0, \quad 3a_3 = a_1, \quad \cdots, \quad (k+1)a_{k+1} = a_{k-1}, \quad \cdots.$$

Hence

$$a_{k+1} = \frac{a_{k-1}}{k+1}.$$

This **recurrence relation** expresses each coefficient in terms of the next to last coefficient. From it, all coefficients can be computed successively. For instance, all odd coefficients are zero: since $a_1 = 0$, it follows that $a_3 = 0$, from which it follows that $a_5 = 0$, and so on. The even coefficients can be expressed in terms of a_0. Apply the recurrence relation with $k = 1$, 3, 5:

$$a_2 = \frac{1}{2}a_0, \qquad a_4 = \frac{1}{4}a_2 = \frac{1}{4}\cdot\frac{1}{2}a_0, \qquad a_6 = \frac{1}{6}a_4 = \frac{1}{6}\cdot\frac{1}{4}\cdot\frac{1}{2}a_0.$$

The pattern is clear:

$$a_{2n} = \frac{1}{2n}\cdot\frac{1}{2n-2}\cdot\frac{1}{2n-4}\cdots\frac{1}{6}\cdot\frac{1}{4}\cdot\frac{1}{2}a_0 = \frac{1}{2^n\cdot n!}a_0.$$

Therefore the desired power series is

$$y(x) = a_0\left[1 + \frac{x^2}{2} + \frac{x^4}{2^2\cdot 2!} + \frac{x^6}{2^3\cdot 3!} + \cdots + \frac{x^{2n}}{2^n\cdot n!} + \cdots\right]$$
$$= a_0\sum_{n=0}^{\infty}\frac{x^{2n}}{2^n\cdot n!}.$$

When written in a slightly different form, this series is a familiar one:

$$y(x) = a_0\sum_{n=0}^{\infty}\frac{1}{n!}\left(\frac{x^2}{2}\right)^n = a_0\, e^{x^2/2}.$$

Answer: $a_0\, e^{x^2/2}$, a_0 any constant.

Remark: This differential equation can be solved also by separation of variables. The next example, however, is not as simple.

EXAMPLE 8.2

Obtain a power series solution of the initial-value problem

$$\begin{cases} y'' + xy' + y = 0 \\ y(0) = 0, \qquad y'(0) = 3. \end{cases}$$

Solution: Try a power series

$$y(x) = a_0 + a_1 x + a_2 x^2 + \cdots + a_n x^n + \cdots.$$

From the initial conditions, $a_0 = 0$ and $a_1 = 3$. Now substitute the series into the differential equation:

$$\begin{aligned} y'' &+ xy' + y \\ &= (2a_2 + 3 \cdot 2a_3 x + 4 \cdot 3a_4 x^2 + \cdots + n(n-1)a_n x^{n-2} + \cdots) \\ &\quad + x(a_1 + 2a_2 x + 3a_3 x^2 + \cdots + na_n x^{n-1} + \cdots) \\ &\quad + (a_0 + a_1 x + a_2 x^2 + a_3 x^3 + \cdots + a_n x^n + \cdots) = 0. \end{aligned}$$

Collect powers of x:

$$\begin{aligned} &(a_0 + 2a_2) + (2a_1 + 3 \cdot 2a_3)x + (3a_2 + 4 \cdot 3a_4)x^2 + \cdots \\ &\quad + [a_n + na_n + (n+2)(n+1)a_{n+2}]x^n + \cdots = 0. \end{aligned}$$

All coefficients on the left-hand side are zero, in particular the coefficient of x^n. It follows that

$$a_{n+2} = \left(-\frac{1}{n+2}\right) a_n.$$

From this recurrence relation, the even coefficients can be expressed as multiples of a_0 and the odd coefficients as multiples of a_1. Since $a_0 = 0$, all even coefficients equal 0. Apply the recurrence relation with $n = 1$, then $n = 3$, then $n = 5$, and so on:

$$a_3 = -\frac{1}{3}a_1, \qquad a_5 = -\frac{1}{5}a_3 = \left(-\frac{1}{5}\right)\left(-\frac{1}{3}\right)a_1 = \frac{1}{3 \cdot 5}a_1,$$

$$a_7 = -\frac{1}{7}a_5 = -\frac{1}{3 \cdot 5 \cdot 7}a_1.$$

In general,

$$\begin{aligned} a_{2n+1} &= (-1)^n \frac{1}{1 \cdot 3 \cdot 5 \cdot 7 \cdots (2n+1)} a_1 \\ &= (-1)^n \frac{2 \cdot 4 \cdot 6 \cdots (2n)}{1 \cdot 2 \cdot 3 \cdots (2n+1)} a_1 = (-1)^n \frac{2^n \cdot n!}{(2n+1)!} a_1. \end{aligned}$$

Since $a_1 = 3$, all coefficients are now determined.

Answer:

$$3 \sum_{n=0}^{\infty} (-1)^n \frac{2^n \cdot n!}{(2n+1)!} x^{2n+1}.$$

EXERCISES

Solve by the method of power series; check your answer by solving the differential equation exactly:

1. $\dfrac{dy}{dx} = x + y$

2. $\dfrac{dy}{dx} + 3y = e^x$
(*Hint:* Expand e^x in power series)

3. $\dfrac{d^2y}{dx^2} + 4y = 0$

4. $\dfrac{dy}{dx} - y = x^2$.

Obtain a power series solution:

5. $x \dfrac{d^2y}{dx^2} = y$

6. $x \dfrac{d^2y}{dx^2} + \dfrac{dy}{dx} + xy = 0$

(Ex. 6 is a special case of Bessel's Equation, which has important applications in fluid flow, electric fields, aerodynamics, and other physical problems.)

7. $x^2 \dfrac{d^2y}{dx^2} + (x^2 + x) \dfrac{dy}{dx} - y = 0$

8. $\dfrac{d^2y}{dx^2} + x \dfrac{dy}{dx} - y = 0$.

For the initial-value problem, find a power series solution at $x = 0$ up to and including the term in x^4:

9. $\dfrac{dy}{dx} = 1 - x^2 - y^2; \quad y(0) = 2$

10. $\dfrac{dy}{dx} = \dfrac{x}{x + y + 1}; \quad y(0) = 0$

11. $\dfrac{dy}{dx} = 1 - y + x^3y^2; \quad y(0) = -1$

12. $\dfrac{d^2y}{dx^2} + y = \dfrac{e^x}{1 - x}; \quad y(0) = 1$.

Find a power series solution at $x = 2$ up to and including the term in $(x - 2)^4$:

13. $x \dfrac{dy}{dx} = y^2; \quad y(2) = 1$

14. $\dfrac{d^2y}{dx^2} + (x - 2) \dfrac{dy}{dx} + y = x; \quad y(2) = 1$.

[*Hint:* Write $x = (x - 2) + 2$]

9. COMPUTER EVALUATION OF SERIES [optional]

Accurate evaluation of power series is an absolutely essential process in applied mathematics. Fortunately, computers are well-suited for this tedious job. In fact, many standard computer subroutines, such as those for e^x and $\sin x$, are simply power series evaluations.

A program for power series computation requires two basic ingredients: a method for computing the terms and a criterion for stopping.

An efficient way to compute the terms of a power series is by a recurrence relation, if such a relation can be found. Here is an example. Suppose we want to evaluate the binomial series

$$\sum_{n=0}^{\infty} \binom{p}{n} x^n.$$

Rather than compute each term independently, we note how successive terms are related. Let t_n denote the n-th term. Then

$$\begin{aligned} t_{n+1} = \binom{p}{n+1} x^{n+1} &= \frac{p(p-1) \cdots (p-n+1)(p-n)}{1 \cdot 2 \cdot 3 \cdots n(n+1)} x^{n+1} \\ &= \left[\frac{p(p-1) \cdots (p-n+1)}{1 \cdot 2 \cdot 3 \cdots n} x^n \right] \frac{p-n}{n+1} x \\ &= \binom{p}{n} x^n \cdot \frac{p-n}{n+1} x = t_n \cdot \frac{p-n}{n+1} x. \end{aligned}$$

This is the desired recurrence: each new term can be obtained from the preceding term by multiplying it by a simple factor. This a computer can do easily. Starting with the initial term $t_0 = 1$, it can rapidly grind out successive terms and add them.

Now comes the next question: when to stop. In general, we cannot tell beforehand exactly how many terms should be computed, so we let the computer decide for itself. As part of the input, we must give the computer a small number ϵ, the maximum error we shall allow in the answer (e.g., $\epsilon = 5 \times 10^{-6}$). Then we must include in the program some estimate of the remainder, with instructions to test this after each new term is added, and to halt when the remainder is less than ϵ.

Here is an example which illustrates the various aspects of power series evaluation.

EXAMPLE 9.1

Write a flowchart for computing the elliptic integral series

$$E(k) = \frac{\pi}{2} \left\{ 1 - \sum_{n=1}^{\infty} \left[\frac{(2n)!}{(2^n \cdot n!)^2} \right]^2 \frac{k^{2n}}{2n-1} \right\}, \qquad k^2 < 1.$$

Solution: Compute the expression inside the braces and, as the last step, multiply by $\pi/2$. First examine the coefficients

$$a_n = \left[\frac{(2n)!}{(2^n \cdot n!)^2} \right]^2 \cdot \frac{1}{2n-1} = \left[\frac{1 \cdot 3 \cdot 5 \cdots (2n-1)}{2 \cdot 4 \cdot 6 \cdots (2n)} \right]^2 \cdot \frac{1}{2n-1}.$$

From this expression for a_n, derive a recurrence relation:

$$a_{n+1} = \left[\frac{1 \cdot 3 \cdot 5 \cdots (2n-1)\,(2n+1)}{2 \cdot 4 \cdot 6 \cdots (2n)\,(2n+2)}\right]^2 \cdot \frac{1}{2n+1}$$

$$= \left[\frac{1 \cdot 3 \cdot 5 \cdots (2n-1)}{2 \cdot 4 \cdot 6 \cdots (2n)}\right]^2 \left[\frac{2n+1}{2n+2}\right]^2 \cdot \frac{1}{2n-1} \cdot \frac{2n-1}{2n+1}$$

$$= \left[\frac{1 \cdot 3 \cdot 5 \cdots (2n-1)}{2 \cdot 4 \cdot 6 \cdots (2n)}\right]^2 \cdot \frac{1}{2n-1} \left[\frac{2n+1}{2n+2}\right]^2 \frac{2n-1}{2n+1}$$

$$= a_n \cdot \frac{4n^2 - 1}{4(n+1)^2}.$$

A recurrence relation for the terms of this series follows easily. Let t_n denote the n-th term. Since $t_n = a_n k^{2n}$,

$$t_{n+1} = a_{n+1} k^{2n+2} = a_n \cdot \frac{4n^2 - 1}{4(n+1)^2} k^{2n} k^2 = t_n \cdot k^2 \frac{4n^2 - 1}{4(n+1)^2}.$$

From this formula, together with the initial term

$$t_1 = a_1 k^2 = \frac{1}{4} k^2,$$

the computer can rapidly calculate the terms.

Next, to decide when to halt the program, obtain a remainder estimate. Let R_N denote the remainder after N terms of the series,

$$|R_N| = \frac{\pi}{2} \sum_{n=N+1}^{\infty} a_n k^{2n}.$$

From the expression for a_n, it follows that

$$a_n \leq \frac{1}{4} \cdot \frac{1}{2n-1} \qquad \text{(equality only for } n = 1\text{)}.$$

Therefore,

$$|R_N| < \frac{\pi}{2} \sum_{n=N+1}^{\infty} \frac{1}{4} \cdot \frac{1}{2n-1} k^{2n} < \frac{\pi}{2} \cdot \frac{1}{4} \cdot \frac{1}{2N+1} \sum_{n=N+1}^{\infty} (k^2)^n.$$

Sum the geometric series on the right to obtain

$$|R_N| < \frac{\pi}{8(2N+1)} \cdot \frac{(k^2)^{N+1}}{1 - k^2} = B_N.$$

The result is a usable remainder estimate which can be tested by the computer after it adds each new term.

It is not hard to derive a recurrence relation for the remainder estimates: $B_{N+1} < k^2 B_N$. Hence the remainders decrease geometrically. If $k^2 < 0.1$, for example, then the addition of each new term improves accuracy by at least one place.

Now for the flowchart. As input, use k^2 and ϵ, the allowable error. Call the developing series SUM, and the remainder estimate TEST. The flowchart is shown in Fig. 9.1.

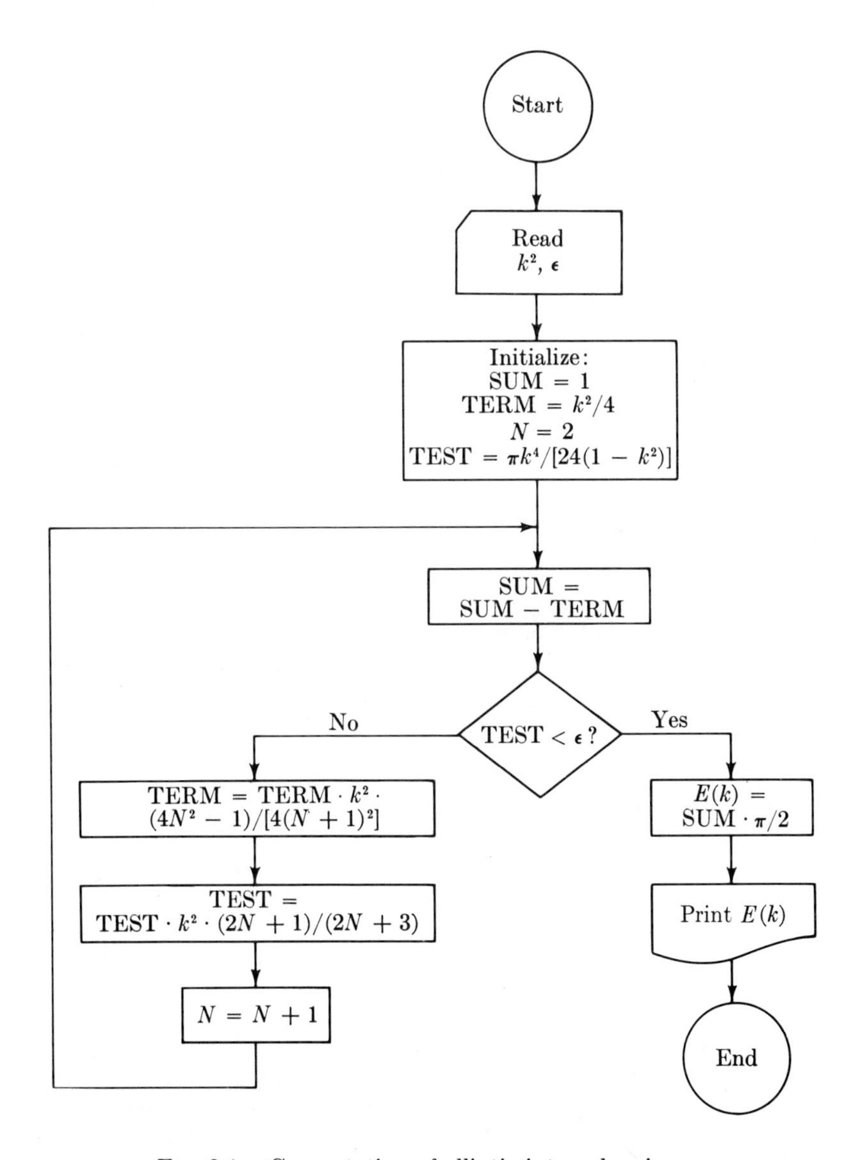

FIG. 9.1. Computation of elliptic integral series.

Truncation Errors

Whenever we use an approximation formula in solving a problem, the answer involves some error. This happens whether we are fitting a curve to data, doing a numerical integration, solving a differential equation numerically, or evaluating a power series. This error, called **truncation error,** arises because the formula used is not exact, and will occur regardless of how many digits we carry in the computation.

It is often possible to estimate the size of truncation errors. We have just done this for the elliptic integral series. For Taylor series, a standard estimate of the remainder (truncation error) is

$$|R_n| < M\left|\frac{(x-a)^{n+1}}{(n+1)!}\right|,$$

where M is the maximum of $|f^{(n+1)}(x)|$. For other approximation formulas, there are similar estimates. For example, the truncation error in Simpson's Rule (Chapter 20, Section 7, p. 408) is at most

$$\frac{h^5}{90}M, \quad \text{where } M \text{ is the maximum of } |f^{(4)}(x)| \text{ on } a \le x \le b.$$

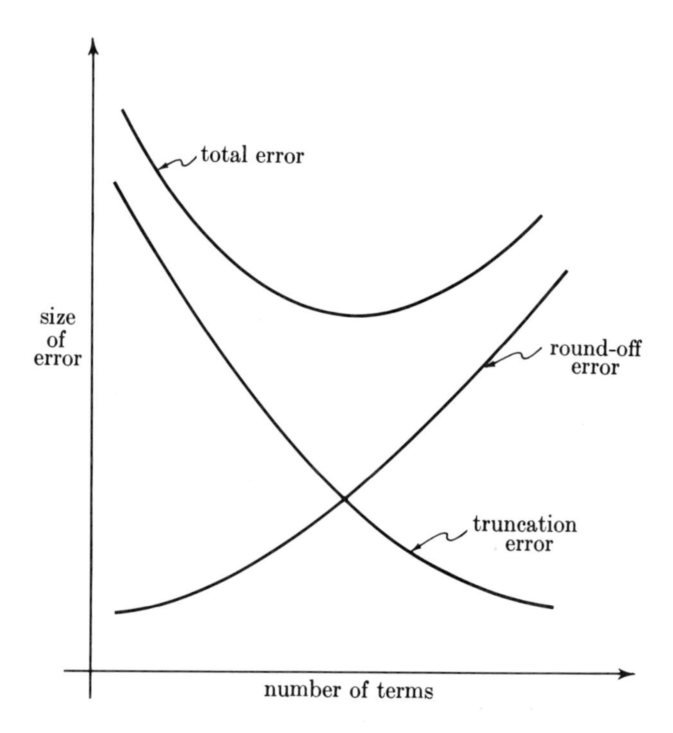

FIG. 9.2

By using higher order approximations we can make the truncation error arbitrarily small. Unfortunately, higher order approximations involve many more computation steps, each of which introduces round-off error. At some stage the accumulated round-off error becomes the major factor, so "better" approximating formulas may actually be less accurate in practice (Fig. 9.2).

Balancing round-off and truncation errors is an art studied in advanced numerical analysis.

EXERCISES

Write a computer program to evaluate the power series at 0 for the given function. Use the suggested values of x to test your program.

1. $y = e^x; \qquad x = 0, 1, 2$

2. $y = \sin x; \qquad x = 0, \frac{\pi}{6}, \frac{\pi}{4}, \frac{\pi}{2}$

3. $y = \ln(1 + x); \qquad x = -\frac{1}{2}, 0, \frac{1}{2}$

4. $y = \arcsin x$

$$= x + \frac{1}{2}\frac{x^3}{3} + \frac{1 \cdot 3}{2 \cdot 4}\frac{x^5}{5} + \frac{1 \cdot 3 \cdot 5}{2 \cdot 4 \cdot 6}\frac{x^7}{7} + \cdots + \frac{1 \cdot 3 \cdots (2n-1)}{2 \cdot 4 \cdots (2n)}\frac{x^{2n+1}}{2n+1} + \cdots; \qquad x = -\frac{1}{2}, 0, \frac{1}{2}$$

5. $y = (1 - x)^{-4} = \sum_{n=0}^{\infty} \binom{-4}{n} x^n, \qquad x = -\frac{1}{2}, 0, \frac{1}{2}.$

6. Derive the recurrence relation $B_{N+1} < k^2 B_N$ stated on p. 663.

29. Improper Integrals

1. INFINITE INTEGRALS

In scientific problems, one frequently meets definite integrals in which one (or both) of the limits is infinite. Here is an example.

Imagine a particle P of mass m at the origin. Consider the **gravitational potential** at a point $x = a$ due to P. This potential is the work required to move a unit mass from the point $x = a$ to infinity, against the force exerted by P. According to Newton's Law of Gravitation, the force is km/d^2, where d is the distance between the two masses and k is a proportionality constant. As shown in Chapter 9, the work done in moving the unit mass from $x = a$ to $x = b$ is

$$\int_a^b (\text{force})\,dx = \int_a^b \frac{km}{x^2}\,dx$$

$$= km\left(-\frac{1}{x}\right)\Bigg|_a^b = km\left(\frac{1}{a} - \frac{1}{b}\right).$$

Let $b \longrightarrow \infty$. Then $1/b \longrightarrow 0$, hence

$$\int_a^b \frac{km}{x^2}\,dx \longrightarrow km\left(\frac{1}{a} - 0\right) = \frac{km}{a}.$$

Thus km/a is the work required to move the mass from a to ∞. It is convenient to set

$$\int_a^\infty \frac{km}{x^2}\,dx = \lim_{b\to\infty} \int_a^b \frac{km}{x^2}\,dx$$

$$= \frac{km}{a}.$$

A definite integral whose upper limit is ∞, whose lower limit is $-\infty$, or both, is called an **improper integral.**

Define

$$\int_a^\infty f(x)\,dx = \lim_{b\to\infty} \int_a^b f(x)\,dx,$$

provided the limit exists. If it does, the integral is said to **converge,** otherwise to **diverge.**

Similarly, define

$$\int_{-\infty}^b f(x)\,dx = \lim_{a\to-\infty} \int_a^b f(x)\,dx,$$

provided the limit exists.

Finally, define

$$\int_{-\infty}^\infty f(x)\,dx = \int_{-\infty}^0 f(x)\,dx + \int_0^\infty f(x)\,dx,$$

provided both integrals on the right converge.

REMARK: An integral from $-\infty$ to ∞ may be split at any convenient finite point just as well as at 0.

An improper integral need not converge. As an example, take

$$\int_1^\infty \frac{dx}{x}.$$

Since

$$\int_1^b \frac{dx}{x} = \ln b,$$

and $\ln b \longrightarrow \infty$ as $b \longrightarrow \infty$, the limit

$$\lim_{b\to\infty} \int_1^b \frac{dx}{x} = \lim_{b\to\infty} \ln b$$

does not exist; the integral diverges.

Remember that a definite integral of a positive function represents the area under a curve. We interpret the improper integral

$$\int_a^\infty f(x)\,dx, \qquad f(x) \geq 0,$$

as the area of the infinite region in Fig. 1.1. If the integral converges, the area is finite; if the integral diverges, the area is infinite.

At first it may seem unbelievable that a region of infinite extent can have finite area. But it can, and here is an example. Take the region under the curve $y = 2^{-x}$ to the right of the y-axis (Fig. 1.2). The rectangles shown

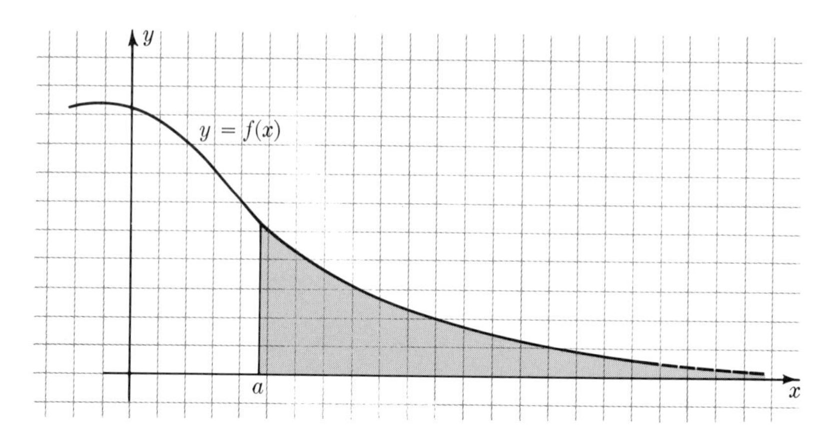

FIG. 1.1

in Fig. 1.2 have base 1 and heights 1, $\frac{1}{2}$, $\frac{1}{4}$, $\frac{1}{8}$, $\cdots$. Their total area is

$$1 + \frac{1}{2} + \frac{1}{4} + \frac{1}{8} + \cdots = 2.$$

Therefore, the shaded infinite region has finite area less than 2.

EXAMPLE 1.1

Compute the exact area of the shaded region in Fig. 1.2.

Solution. The area is given by the improper integral

$$\int_0^\infty 2^{-x}\,dx = \lim_{b\to\infty}\int_0^b 2^{-x}\,dx.$$

An antiderivative of 2^{-x} is $-2^{-x}/\ln 2$. (Use $2^{-x} = e^{-x\ln 2}$.) Hence

$$\int_0^b 2^{-x}\,dx = \left(-\frac{1}{\ln 2}\,2^{-x}\right)\Bigg|_0^b = \frac{1}{\ln 2}\,(1 - 2^{-b}),$$

$$\int_0^\infty 2^{-x}\,dx = \lim_{b\to\infty}\frac{1}{\ln 2}\,(1 - 2^{-b}) = \frac{1}{\ln 2}\cdot$$

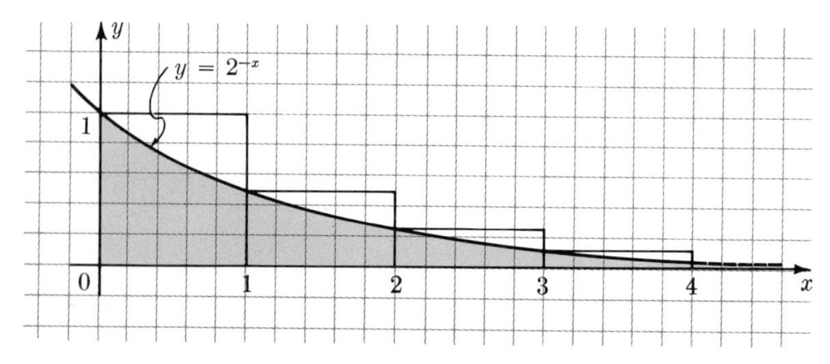

FIG. 1.2

Answer: $\dfrac{1}{\ln 2} \approx 1.4427.$

REMARK: This answer is reasonable. A look at Fig. 1.2 shows that the area of the shaded region is between 1 and 2. A closer look shows that the area is slightly less than 1.5. Why?

EXAMPLE 1.2

Evaluate $\displaystyle\int_0^\infty \frac{dx}{1 + x^2}.$

Solution:

$$\int_0^b \frac{dx}{1 + x^2} = \arctan x \Big|_0^b = \arctan b.$$

Let $b \longrightarrow \infty$. Then $\arctan b \longrightarrow \pi/2$. Hence

$$\int_0^\infty \frac{dx}{1 + x^2} = \lim_{b\to\infty} \int_0^b \frac{dx}{1 + x^2} = \lim_{b\to\infty} \arctan b = \frac{\pi}{2}.$$

Answer: $\dfrac{\pi}{2}.$

EXAMPLE 1.3

Evaluate $\displaystyle\int_{-\infty}^3 e^x\,dx.$

Solution:

$$\int_a^3 e^x\,dx = e^x \Big|_a^3 = e^3 - e^a.$$

Let $a \longrightarrow -\infty$. Then $e^a \longrightarrow 0$. Hence

$$\int_{-\infty}^3 e^x\,dx = \lim_{a\to-\infty} (e^3 - e^a) = e^3.$$

Answer: $e^3.$

The following integral arises in various applications such as electrical circuits, heat conduction, and vibrating membranes:

$$\int_0^\infty e^{-sx} f(x)\,dx.$$

It is called the **Laplace Transform** of $f(x)$.

EXAMPLE 1.4

Evaluate $\displaystyle\int_0^\infty e^{-sx}\cos x\,dx, \qquad s > 0.$

Solution: From integral tables or integration by parts,

$$\int_0^b e^{-sx}\cos x\,dx = \frac{e^{-sx}}{s^2+1}(-s\cos x+\sin x)\Big|_0^b$$

$$= \frac{e^{-sb}}{s^2+1}(-s\cos b+\sin b)+\frac{s}{s^2+1}.$$

Now let $b \longrightarrow \infty$.

$$\int_0^\infty e^{-sx}\cos x\,dx = \lim_{b\to\infty}\int_0^b e^{-sx}\cos x\,dx = 0+\frac{s}{s^2+1}.$$

Answer: $\dfrac{s}{s^2+1}$.

EXAMPLE 1.5

Evaluate $\displaystyle\int_{-\infty}^{\infty}\frac{dx}{3e^{2x}+e^{-2x}}$.

Solution: By definition, the value of this integral is

$$\int_0^\infty \frac{dx}{3e^{2x}+e^{-2x}}+\int_{-\infty}^0\frac{dx}{3e^{2x}+e^{-2x}},$$

provided *both* improper integrals converge. From integral tables,

$$\int_0^b \frac{dx}{3e^{2x}+e^{-2x}} = \frac{1}{2\sqrt{3}}\arctan(e^{2x}\sqrt{3})\Big|_0^b$$

$$= \frac{1}{2\sqrt{3}}[\arctan(e^{2b}\sqrt{3})-\arctan\sqrt{3}].$$

Now $\arctan(e^{2b}\sqrt{3}) \longrightarrow \pi/2$ as $b \longrightarrow \infty$. Note that $\arctan\sqrt{3}=\pi/3$. Hence

$$\int_0^\infty \frac{dx}{3e^{2x}+e^{-2x}} = \lim_{b\to\infty}\int_0^b\frac{dx}{3e^{2x}+e^{-2x}} = \frac{1}{2\sqrt{3}}\left[\frac{\pi}{2}-\frac{\pi}{3}\right].$$

Similarly,

$$\int_{-\infty}^0 \frac{dx}{3e^{2x}+e^{-2x}} = \lim_{a\to-\infty}\int_a^0\frac{dx}{3e^{2x}+e^{-2x}}$$

$$= \frac{1}{2\sqrt{3}}[\arctan\sqrt{3}-\arctan 0] = \frac{1}{2\sqrt{3}}\left[\frac{\pi}{3}-0\right].$$

Thus both improper integrals converge. The answer is the sum of their values.

Answer: $\dfrac{\pi}{4\sqrt{3}}$.

REMARK: Do you prefer this snappy calculation?

$$\int_{-\infty}^{\infty} \frac{dx}{3e^{2x} + e^{-2x}} = \frac{1}{2\sqrt{3}} \arctan(\sqrt{3}\, e^{2x}) \Big|_{-\infty}^{\infty}$$

$$= \frac{1}{2\sqrt{3}} (\arctan \infty - \arctan 0) = \frac{1}{2\sqrt{3}} \frac{\pi}{2} = \frac{\pi}{4\sqrt{3}}.$$

Warning. Try the same slick method on

$$\int_{-\infty}^{\infty} \frac{dx}{x^2}.$$

It fails. Why?

EXERCISES

Evaluate:

1. $\displaystyle\int_2^{\infty} \frac{dx}{x^3}$
2. $\displaystyle\int_5^{\infty} e^{-x}\, dx$
3. $\displaystyle\int_0^{\infty} xe^{-x}\, dx$
4. $\displaystyle\int_{-\infty}^{-1} \frac{dx}{x^2}$
5. $\displaystyle\int_{-\infty}^{-1} \frac{dx}{1 + x^2}$
6. $\displaystyle\int_4^{\infty} \frac{dx}{x\sqrt{x}}$
7. $\displaystyle\int_{-\infty}^{\infty} e^{-|x|}\, dx$
8. $\displaystyle\int_{-\infty}^{\infty} xe^{-x^2}\, dx$
9. $\displaystyle\int_1^{\infty} \frac{dx}{x\sqrt{9 + x^2}}$
10. $\displaystyle\int_0^{\infty} e^{-sx} \sin x\, dx \quad (s > 0)$
11. $\displaystyle\int_0^{\infty} \frac{x\, dx}{x^4 + 1} \quad (\text{let } u = x^2)$
12. $\displaystyle\int_0^{\infty} xe^{-sx}\, dx \quad (s > 0)$
13. $\displaystyle\int_0^{\infty} x^2 e^{-sx}\, dx \quad (s > 0)$
14. $\displaystyle\int_0^{\infty} x^n e^{-sx}\, dx \quad (s > 0)$
15. $\displaystyle\int_0^{\infty} e^{ax} e^{-sx}\, dx \quad (s > a)$
16. $\displaystyle\int_0^{\infty} xe^{ax} e^{-sx}\, dx \quad (s > a)$
17. $\displaystyle\int_0^{\infty} e^{-sx} \cosh x\, dx \quad (s > 1)$
18. $\displaystyle\int_0^{\infty} xe^{-sx} \sin x\, dx \quad (s > 0).$

Is the area under the curve finite or infinite?

19. $y = 1/x$; from $x = 5$ to $x = \infty$
20. $y = 1/x^2$; from $x = 1$ to $x = \infty$
21. $y = \sin^2 x$; from $x = 0$ to $x = \infty$
22. $y = (1.001)^{-x}$; from $x = 0$ to $x = \infty$.

Solve for b:

23. $\displaystyle\int_0^b e^{-x}\,dx = \int_b^\infty e^{-x}\,dx$

24. $\displaystyle\int_0^b \frac{dx}{1+x^2} = \int_b^\infty \frac{dx}{1+x^2}$.

25. Find b such that 99% of the area under $y = e^{-x}$ between $x = 0$ and $x = \infty$ is contained between $x = 0$ and $x = b$.

2. CONVERGENCE AND DIVERGENCE

Whether an improper integral converges or diverges may be a subtle matter. The following example illustrates this.

EXAMPLE 2.1

For which positive numbers p does the integral $\displaystyle\int_1^\infty \frac{dx}{x^p}$ converge? diverge?

Solution: Suppose $p \neq 1$. Then

$$\int_1^b \frac{dx}{x^p} = -\frac{1}{p-1}\,\frac{1}{x^{p-1}}\bigg|_1^b = \frac{1}{p-1}\left(1 - \frac{1}{b^{p-1}}\right).$$

As $b \longrightarrow \infty$,

$$\frac{1}{b^{p-1}} \longrightarrow 0 \qquad \text{if} \quad p - 1 > 0,$$

and

$$\frac{1}{b^{p-1}} \longrightarrow \infty \qquad \text{if} \quad p - 1 < 0$$

Hence

$$\lim_{b\to\infty} \int_1^b \frac{dx}{x^p}$$

exists if $p > 1$, does not exist if $p < 1$. That means the given integral converges if $p > 1$, diverges if $p < 1$.

If $p = 1$,

$$\int_1^b \frac{dx}{x} = \ln b \longrightarrow \infty \qquad \text{as} \quad b \longrightarrow \infty;$$

the integral diverges.

Answer: Converges if $p > 1$, diverges if $p \le 1$.

REMARK 1: Obviously the same is true of the integral

$$\int_a^\infty \frac{dx}{x^p}$$

for any positive number a.

REMARK 2: Now, a subtle question. Why should this integral converge if $p > 1$ but diverge if $p \le 1$? (See Fig. 2.1.) The curves $y = 1/x^p$

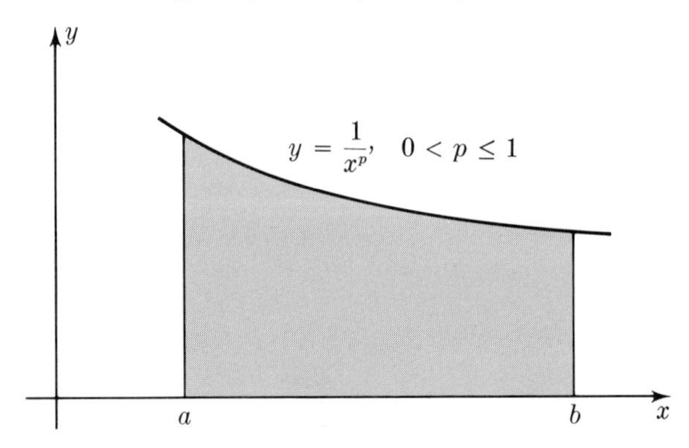

FIG. 2.1a

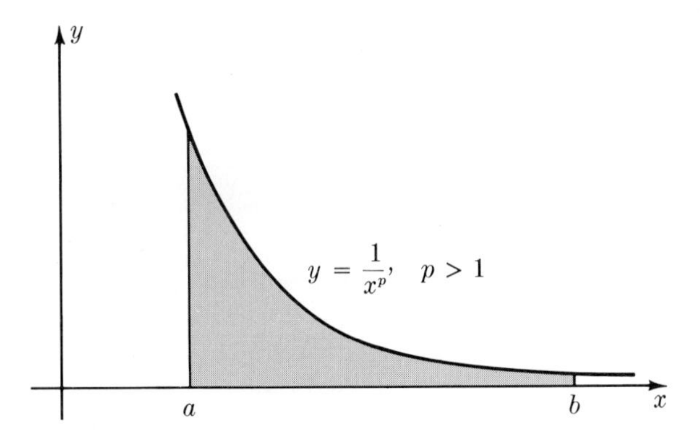

FIG. 2.1b

all decrease as x increases. The key is in their *rate* of decrease. If $p \le 1$, the curve decreases slowly enough that the shaded area (Fig. 2.1a) increases without bound as $b \longrightarrow \infty$. If $p > 1$, the curve decreases fast enough that the shaded area (Fig. 2.1b) is bounded by a fixed number, no matter how large b is.

Convergence Criteria

We know that the integral

$$\int_a^\infty \frac{dx}{x^2} \qquad (a > 0)$$

converges. Suppose that $0 \le g(x) \le 1/x^2$. Then

$$\int_a^\infty g(x)\,dx$$

also converges because the area under the curve $y = g(x)$ is even smaller than the area under $y = 1/x^2$. This illustrates a general principle:

> If $f(x) \ge 0$ and if $0 \le g(x) \le f(x)$ for $a \le x < \infty$, then the convergence of
>
> $$\int_a^\infty f(x)\,dx \quad \text{implies the convergence of} \quad \int_a^\infty g(x)\,dx.$$

(We skip the rather technical proof.)

EXAMPLE 2.2

Show that the integrals converge:

$$\text{(a)} \int_1^\infty \frac{dx}{x^2 + \sqrt{x}}; \qquad \text{(b)} \int_0^\infty \frac{e^{-x}}{x+1}\,dx; \qquad \text{(c)} \int_3^\infty \frac{\sin^2 x}{x^3}\,dx.$$

Solution: Note that

$$\frac{1}{x^2 + \sqrt{x}} < \frac{1}{x^2}, \qquad 1 \le x < \infty,$$

$$\frac{e^{-x}}{x+1} \le e^{-x}, \qquad 0 \le x < \infty,$$

$$\frac{\sin^2 x}{x^3} \le \frac{1}{x^3}, \qquad 3 \le x < \infty.$$

Since the integrals

$$\int_1^\infty \frac{dx}{x^2}, \qquad \int_0^\infty e^{-x}\,dx, \qquad \int_3^\infty \frac{dx}{x^3}$$

all converge, the given integrals converge by the preceding test.

A second important convergence criterion (also stated without proof) is this:

> Suppose $f(x) \ge 0$ and $g(x)$ is bounded, i.e., $|g(x)| \le M$ for some constant M. Then the convergence of
>
> $$\int_a^\infty f(x)\,dx \quad \text{implies the convergence of} \quad \int_a^\infty f(x)g(x)\,dx.$$

EXAMPLE 2.3

Show that the integrals converge:

(a) $\displaystyle\int_0^\infty e^{-x}\sin^3 x\,dx$, (b) $\displaystyle\int_1^\infty \frac{\ln x}{x^3}\,dx$.

Solution: Apply the above criterion.

(a) Since

$$\int_0^\infty e^{-x}\,dx$$

converges and $|\sin^3 x| \leq 1$, the given integral converges.

(b) Write

$$\frac{\ln x}{x^3} = \frac{1}{x^2}\cdot\frac{\ln x}{x}.$$

The integral

$$\int_1^\infty \frac{dx}{x^2}$$

converges and $(\ln x)/x$ is bounded. [The maximum of $(\ln x)/x$ is $1/e$ as was shown on p. 278]. Hence the given integral converges.

REMARK: Both convergence criteria apply also to improper integrals of the forms

$$\int_{-\infty}^b f(x)\,dx \qquad\text{and}\qquad \int_{-\infty}^\infty f(x)\,dx.$$

A Divergence Criterion

Here is a simple criterion for divergence of an improper integral.

If $f(x) \geq 0$ and if $g(x) \geq f(x)$ for $a \leq x < \infty$, then the divergence of

$$\int_a^\infty f(x)\,dx$$

implies the divergence of

$$\int_a^\infty g(x)\,dx.$$

This is obvious geometrically; we shall not give a formal proof. Since $g(x) \geq f(x)$, the region under $y = g(x)$ contains the region under $y = f(x)$. If the second region has infinite area, so does the first.

EXAMPLE 2.4

Show that the integrals diverge:

(a) $\displaystyle\int_1^\infty \frac{\sqrt{x}}{1+x}\,dx$ (b) $\displaystyle\int_2^\infty \frac{dx}{\sqrt{x}-\sqrt[3]{x}}$

(c) $\displaystyle\int_3^\infty \frac{\ln x}{x}\,dx.$

Solution: Note that

$$\frac{\sqrt{x}}{1+x} \ge \frac{1}{1+x}, \qquad 1 \le x < \infty,$$

$$\frac{1}{\sqrt{x}-\sqrt[3]{x}} > \frac{1}{\sqrt{x}}, \qquad 2 \le x < \infty,$$

$$\frac{\ln x}{x} \ge \frac{\ln 3}{x} > \frac{1}{x}, \qquad 3 \le x < \infty.$$

Since the integrals

$$\int_1^\infty \frac{dx}{1+x}, \quad \int_2^\infty \frac{dx}{\sqrt{x}}, \quad \int_3^\infty \frac{dx}{x}$$

all diverge, the given integrals diverge by the preceding criterion.

EXERCISES

Does the integral converge or diverge?

1. $\displaystyle\int_0^\infty \frac{dx}{x+1}$

2. $\displaystyle\int_1^\infty \frac{dx}{x^2+x}$

3. $\displaystyle\int_0^\infty \frac{x^2e^{-x}}{1+x^2}\,dx$

4. $\displaystyle\int_2^\infty e^{-x^3}\,dx$

5. $\displaystyle\int_0^\infty \cosh x\,dx$

6. $\displaystyle\int_0^\infty \frac{x\,dx}{\sqrt{x^2+3}}$

7. $\displaystyle\int_{-\infty}^\infty \frac{\sin x}{1+x^2}\,dx$

8. $\displaystyle\int_0^\infty \sin x\,dx$

9. $\displaystyle\int_1^\infty \frac{\cos x}{\sqrt{x}\,(x+4)}\,dx$

10. $\displaystyle\int_2^\infty \frac{dx}{\ln x}$

11. $\displaystyle\int_3^\infty \frac{x^3}{x^4-1}\,dx$

12. $\displaystyle\int_0^\infty \frac{dx}{1+x+e^x}.$

13. Show that $\int_2^\infty \frac{dx}{x(\ln x)^p}$ converges if $p > 1$, diverges if $p \le 1$. (*Hint:* Use the substitution $u = \ln x$.)

14. Show that $\int_3^\infty \frac{dx}{x \ln x[\ln(\ln x)]^p}$ converges if $p > 1$, diverges if $p \le 1$.

15. Denote by R the infinite region under $y = 1/x$ to the right of $x = 1$. Suppose R is rotated around the x-axis, forming an infinitely long horn. Show that the volume of this horn is finite. Its surface area, however, is infinite (the surface area is certainly larger than the area of R). Here is an apparent paradox: You can fill the horn with paint, but you cannot paint it. Where is the fallacy?

Find all values of s for which the integral converges:

16. $\int_0^\infty e^{-sx}e^x \, dx$

17. $\int_0^\infty \frac{e^{-sx}}{1 + x^2} \, dx$

18. $\int_0^\infty e^{-sx}e^{-x^2} \, dx$

19. $\int_1^\infty \frac{x^s}{(1 + x^3)^s} \, dx.$

3. RELATION TO INFINITE SERIES

There is a relation between the series

$$\frac{1}{2^2} + \frac{1}{3^2} + \cdots + \frac{1}{n^2} + \cdots$$

and the convergent integral

$$\int_1^\infty \frac{dx}{x^2}.$$

(See Fig. 3.1.) The rectangles shown in Fig. 3.1 have areas $1/2^2, 1/3^2, \cdots$. Obviously the sum of these areas is finite, being less than the finite area

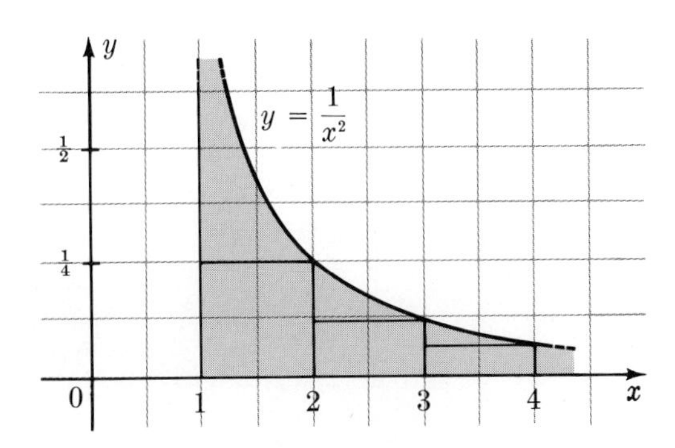

FIG. 3.1 Note: x scale is $\frac{1}{4}$ of y scale.

under the curve. Hence, the series converges. This illustrates a general principle:

> Suppose $f(x)$ is a positive decreasing function. Then the series
>
> $$f(1) + f(2) + \cdots + f(n) + \cdots$$
>
> converges if the integral
>
> $$\int_1^\infty f(x)\,dx$$
>
> converges, and diverges if the integral diverges.

The argument given above in the case $f(x) = 1/x^2$ is quite general; it shows that the series converges if $f(x)$ is a positive decreasing function and if the integral

$$\int_1^\infty f(x)\,dx$$

converges. (Actually, we have adjoined an extra term, $f(1)$, but one more term does not affect convergence or divergence.)

Suppose the integral diverges. This time the rectangles are drawn above the curve (Fig. 3.2). Their areas are $f(1)$, $f(2)$ $\cdots$. Since they contain the infinite area under the curve, the series

$$f(1) + f(2) + \cdots + f(n) + \cdots$$

diverges.

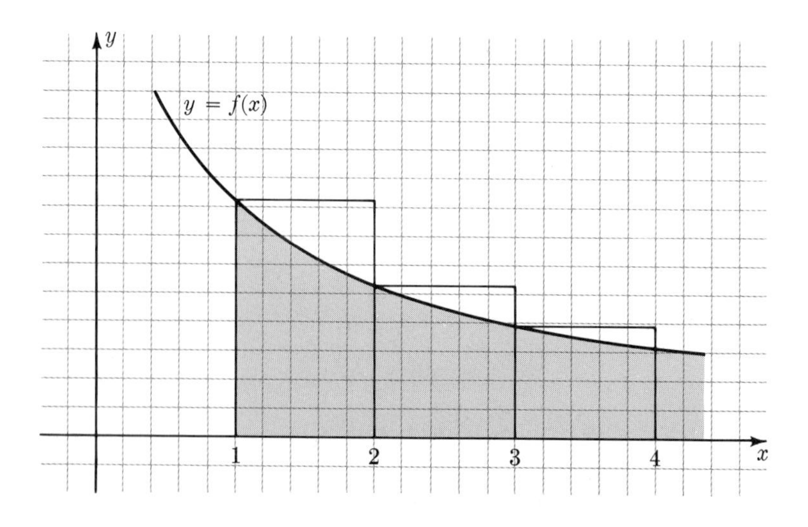

FIG. 3.2

EXAMPLE 3.1

For which positive numbers p does the series

$$1 + \frac{1}{2^p} + \frac{1}{3^p} + \cdots + \frac{1}{n^p} + \cdots$$

converge? diverge?

Solution: The series can be written

$$f(1) + f(2) + \cdots + f(n) + \cdots,$$

where $f(x) = 1/x^p$, a positive decreasing function. By the preceding principle, the given series converges or diverges as

$$\int_1^\infty \frac{dx}{x^p}$$

converges or diverges.

Answer: Converges if $p > 1$, diverges if $p \leq 1$.

EXERCISES

Does the series converge or diverge?

1. $1 + \frac{1}{\sqrt{2}} + \frac{1}{\sqrt{3}} + \frac{1}{\sqrt{4}} + \cdots$
2. $\frac{1}{e} + \frac{2}{e^2} + \frac{3}{e^3} + \frac{4}{e^4} + \cdots$
3. $1 + \frac{1}{2^3} + \frac{1}{3^3} + \frac{1}{4^3} + \cdots$
4. $\frac{1}{1+\sqrt{1}} + \frac{1}{2+\sqrt{2}} + \frac{1}{3+\sqrt{3}} + \frac{1}{4+\sqrt{4}} + \cdots$
5. $\frac{1}{2 \ln 2} + \frac{1}{3 \ln 3} + \frac{1}{4 \ln 4} + \cdots$
6. $\frac{1}{2(\ln 2)^p} + \frac{1}{3(\ln 3)^p} + \frac{1}{4(\ln 4)^p} + \cdots$
7. $\frac{1}{1+1^2} + \frac{1}{1+2^2} + \frac{1}{1+3^2} + \cdots$
8. $\frac{1}{1^2} + \frac{3}{2^2} + \frac{5}{3^2} + \frac{7}{4^2} + \cdots.$
9. Show geometrically that the sum of $1 + \frac{1}{2^2} + \frac{1}{3^2} + \frac{1}{4^2} + \cdots$ is less than 2. See Fig. 3.1. (It is known that the exact sum is $\pi^2/6$, a startling fact.)

10. Use the method of inscribing and circumscribing rectangles to show that

$$\ln(n+1) < 1 + \frac{1}{2} + \frac{1}{3} + \cdots + \frac{1}{n} < 1 + \ln n.$$

Is $1 + \frac{1}{2} + \frac{1}{3} + \cdots + \frac{1}{1000}$ more or less than 10 ?

11. Estimate how many terms of the series $1 + \frac{1}{2} + \frac{1}{3} + \frac{1}{4} + \cdots$ must be added before the sum exceeds 1000.

Does the series converge or diverge?

12. $\displaystyle\sum_{n=1}^{\infty} \frac{n}{4+n^3}$

13. $\displaystyle\sum_{n=2}^{\infty} \frac{\ln n}{1+n^2}$

14. $\displaystyle\sum_{n=1}^{\infty} \frac{n^2 e^{-n}}{1+n^2}$

15. $\displaystyle\sum_{n=2}^{\infty} \frac{1}{\sqrt{n}\ln n}\cdot$

4. OTHER IMPROPER INTEGRALS

A definite integral

$$\int_a^b f(x)\,dx, \qquad a \text{ and } b \text{ finite,}$$

is called **improper** if $f(x)$ "blows up" at one or more points in the interval $a \le x \le b$. Examples are

$$\int_0^3 \frac{dx}{x}, \qquad \int_1^5 \frac{dx}{x^2-4}, \qquad \int_6^{10} \frac{dx}{\ln(x-5)}\cdot$$

The first integrand "blows up" at $x = 0$, the second at $x = 2$, the third at $x = 6$. Such bad points are called **singularities** of the integrand.

We shall discuss integrals

$$\int_a^b f(x)\,dx$$

where $f(x)$ has exactly one singularity which occurs either at $x = a$ or at $x = b$. This is the most common case.

Consider the integral

$$\int_0^3 \frac{dx}{\sqrt{x}}$$

whose integrand has a singularity at $x = 0$. What meaning can we give to this integral?

Except at $x = 0$, the integrand is well-behaved. Hence if h is any positive number, no matter how small, the integral

$$\int_h^3 \frac{dx}{\sqrt{x}}$$

makes sense. Its value is easily computed:

$$\int_h^3 \frac{dx}{\sqrt{x}} = 2\sqrt{x}\Big|_h^3 = 2(\sqrt{3} - \sqrt{h}).$$

It is reasonable to *define*

$$\int_0^3 \frac{dx}{\sqrt{x}} = \lim_{h\to 0} \int_h^3 \frac{dx}{\sqrt{x}} = 2\sqrt{3}.$$

Next, consider the integral

$$\int_0^3 \frac{dx}{x}.$$

We try to "sneak up" on the integral as before by computing

$$\int_h^3 \frac{dx}{x} = \ln 3 - \ln h,$$

then letting $h \longrightarrow 0$. But $\ln h \longrightarrow -\infty$ as $h \longrightarrow 0$. Hence

$$\int_h^3 \frac{dx}{x} \longrightarrow \infty \qquad \text{as} \quad h \longrightarrow 0.$$

There is no reasonable value for this integral.

Motivated by these examples, we make the following definitions:

Suppose $f(x)$ has one singularity, at $x = a$, and that $a < b$. Define

$$\int_a^b f(x)\,dx = \lim_{h\to 0} \int_{a+h}^b f(x)\,dx, \qquad (h > 0)$$

provided the limit exists. If it does, the improper integral **converges,** otherwise, it **diverges.**

Similarly, if $f(x)$ has one singularity, at $x = b$, define

$$\int_a^b f(x)\,dx = \lim_{h\to 0} \int_a^{b-h} f(x)\,dx, \qquad (h > 0)$$

provided the limit exists.

EXAMPLE 4.1

For which positive numbers p does the improper integral $\displaystyle\int_0^3 \frac{dx}{x^p}$ converge? diverge?

Solution: The case $p = 1$ was discussed above; the integral diverges. Now assume $p \neq 1$. By definition, the value of the integral is

$$\lim_{h\to 0} \int_h^3 \frac{dx}{x^p},$$

provided the limit exists. Now

$$\int_h^3 \frac{dx}{x^p} = -\frac{1}{p-1}\frac{1}{x^{p-1}}\bigg|_h^3 = \frac{1}{p-1}\left(\frac{1}{h^{p-1}} - \frac{1}{3^{p-1}}\right).$$

But, as $h \longrightarrow 0$,

$$\frac{1}{h^{p-1}} \longrightarrow 0 \quad \text{if} \quad p - 1 < 0,$$

and

$$\frac{1}{h^{p-1}} \longrightarrow \infty \quad \text{if} \quad p - 1 > 0.$$

Hence the limit exists only if $p < 1$. In that case

$$\int_0^3 \frac{dx}{x^p} = \lim_{h\to 0} \int_h^3 \frac{dx}{x^p} = -\frac{1}{(p-1)3^{p-1}}.$$

Answer: Converges if $p < 1$, diverges if $p \geq 1$.

REMARK 1: The answer applies as well to

$$\int_0^b \frac{dx}{x^p}$$

for each positive number b since the upper limit plays no essential part in the discussion. Only the behavior of $1/x^p$ in the immediate neighborhood of $x = 0$ counts.

REMARK 2: If $p \geq 1$, the curve $y = 1/x^p$ increases so fast as $x \longrightarrow 0$ that the area of the shaded region (Fig. 4.1a) tends to infinity. If $p < 1$, the curve rises so slowly that the area of the shaded region (Fig. 4.1b) is bounded.

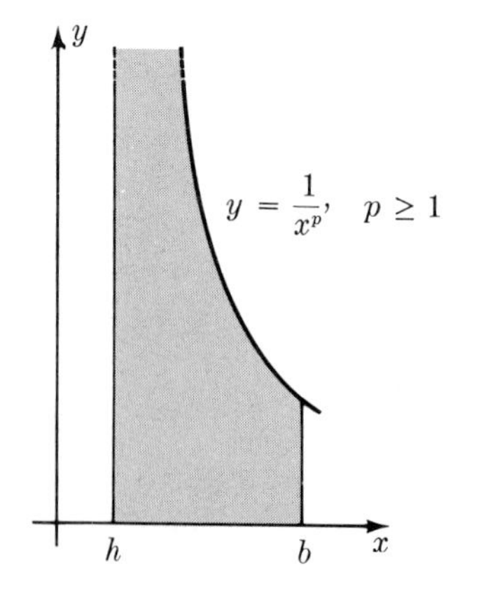

FIG. 4.1a

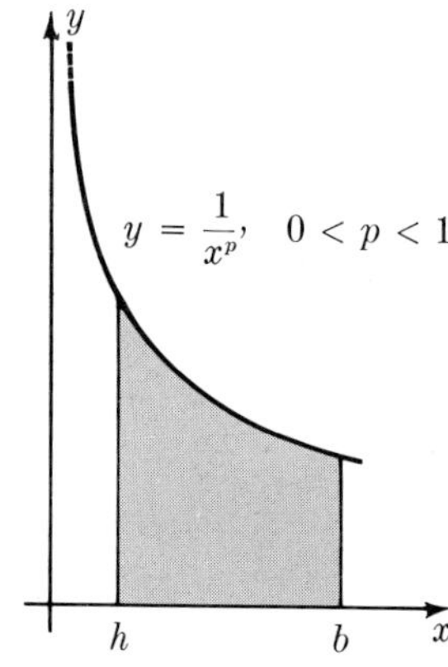

FIG. 4.1b

CAUTION: Do not confuse these results with those of Section 2 concerning

$$\int_1^\infty \frac{dx}{x^p}.$$

In fact,

$$\int_0^1 \frac{dx}{x^p} \begin{cases} \text{converges} & \text{if } p < 1, \\ \text{diverges} & \text{if } p \geq 1, \end{cases} \qquad \int_1^\infty \frac{dx}{x^p} \begin{cases} \text{diverges} & \text{if } p \leq 1, \\ \text{converges} & \text{if } p > 1. \end{cases}$$

EXAMPLE 4.2

For which positive numbers p does the improper integral

$$\int_3^4 \frac{dx}{(x-3)^p}$$

converge? diverge?

Solution: Change variable. Let $u = x - 3$. Then

$$\int_3^4 \frac{dx}{(x-3)^p} = \int_0^1 \frac{du}{u^p}.$$

But this integral was discussed above.

Answer: Converges if $p < 1$, diverges if $p \geq 1$.

The techniques of the two preceding examples yield a general fact (for a and b finite):

The integrals

$$\int_a^b \frac{dx}{(x-a)^p}, \qquad \int_a^b \frac{dx}{(b-x)^p}$$

converge if $p < 1$, diverge if $p \geq 1$.

A Convergence Criterion

The convergence criteria given in Section 2 have analogues for improper integrals with finite limits. We state (without proof) just one of these.

Suppose $f(x) \geq 0$, and that $f(x)$ has a singularity at $x = a$ or at $x = b$. Suppose $g(x)$ is a bounded function. Then the convergence of

$$\int_a^b f(x)\, dx \quad \text{implies the convergence of} \quad \int_a^b f(x)g(x)\, dx.$$

EXAMPLE 4.3

Show that the integrals converge:

$$\text{(a)} \quad \int_0^1 \frac{\cos x}{\sqrt{x}}\,dx, \qquad \text{(b)} \quad \int_1^2 \frac{dx}{\sqrt{4 - x^2}}.$$

Solution: Use the preceding criterion.

(a) Let $f(x) = 1/\sqrt{x}$ and $g(x) = \cos x$.

(b) Let $f(x) = 1/\sqrt{2 - x}$ and $g(x) = 1/\sqrt{2 + x}$.

EXERCISES

Does the integral converge or diverge?

1. $\displaystyle\int_0^1 \frac{dx}{\sqrt[3]{x}}$

2. $\displaystyle\int_0^{\pi/4} \cot x\,dx$

3. $\displaystyle\int_0^2 \frac{e^x}{x}\,dx$

4. $\displaystyle\int_0^1 \frac{dx}{x^3 - 1}$

5. $\displaystyle\int_0^1 \frac{dx}{(1 - x^2)^2}$

6. $\displaystyle\int_1^3 \frac{\sin x}{\sqrt{3 - x}}\,dx$

7. $\displaystyle\int_3^5 \frac{dx}{\sqrt{x^2 - 9}}$

8. $\displaystyle\int_0^5 \frac{\sin^3 2x}{\sqrt{x}}\,dx$

9. $\displaystyle\int_0^1 \ln x\,dx$

10. $\displaystyle\int_0^3 \frac{dx}{\sqrt{x}\,(3 + x)}$

11. $\displaystyle\int_0^{1/2} \frac{dx}{x \ln x}$

12. $\displaystyle\int_2^4 \frac{dx}{\sqrt{-x^2 + 6x - 8}}$

13. $\displaystyle\int_1^2 \frac{dx}{\sqrt[3]{x^3 - 4x^2 + 4x}}$

14. $\displaystyle\int_0^{\pi/2} \sec x\,dx$

15. $\displaystyle\int_0^1 \sqrt{\frac{1 + x}{1 - x}}\,dx$

16. $\displaystyle\int_0^{\pi/2} \frac{\cos x}{\sqrt[3]{x}}\,dx.$

30. Approximate Solutions of Differential Equations

1. INTRODUCTION

It is an unfortunate fact of life that the differential equation

$$\frac{dy}{dx} = f(x, y)$$

generally cannot be solved exactly. Even though solutions exist, usually there are no exact formulas for them. In a few special cases there are explicit solutions, but more often there are none.

These remarks apply also to the initial-value problem:

Initial-Value Problem Find a function $y = y(x)$ satisfying

$$\frac{dy}{dx} = f[x, y(x)]$$

and the initial condition

$$y(x_0) = y_0.$$

Various examples of direction fields suggest that this problem has precisely one solution; there is a unique curve $y = y(x)$ through the point (x_0, y_0) whose direction at each point is that of the field. Yet we are generally unable to write down the solution explicitly.

We shall present several methods of approximating the solution of such an initial-value problem. To compare these methods, we shall apply each of them to one particular test problem:

$$\begin{cases} \dfrac{dy}{dx} = x^2 y - 1 \\ y(0) = 2. \end{cases}$$

Each approximation method leads to a function $y(x)$ which satisfies the initial condition $y(0) = 2$ and approximately satisfies the differential equation

$$\frac{dy}{dx} = x^2y - 1.$$

For each such approximation we shall compute $y(1)$; later we shall compare the test values.

Part of the direction field is shown in Fig. 1.1. A solution curve $y(x)$ through (0, 2) has been sketched roughly. From the sketch we can make a crude guess: the value of $y(1)$ is near 1.5.

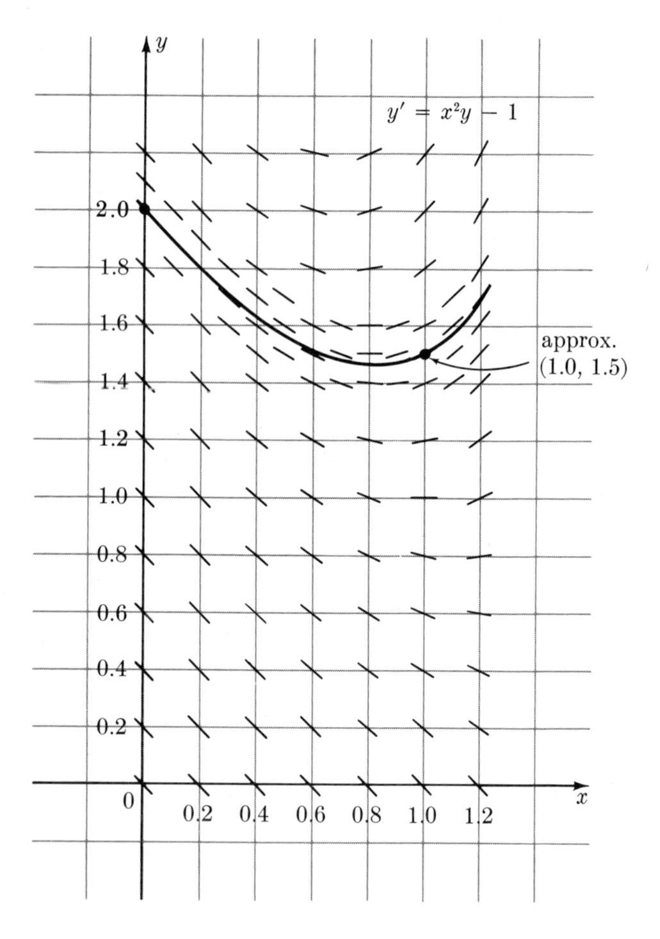

FIG. 1.1

REMARK: Most initial-value problems that arise in practice have unique solutions. However, the general question of existence and uniqueness of solutions is a very technical subject which we do not intend to discuss.

2. TAYLOR POLYNOMIAL METHOD

Consider the initial-value problem: find a function $y(x)$ satisfying

$$\begin{cases} \dfrac{dy}{dx} = f(x, y) \\ y(x_0) = y_0 . \end{cases}$$

Assume that the solution $y(x)$ has an n-th degree Taylor polynomial approximation at x_0:

$$y(x) \approx y(x_0) + \frac{y'(x_0)}{1!}(x - x_0) + \frac{y''(x_0)}{2!}(x - x_0)^2 + \cdots + \frac{y^{(n)}(x_0)}{n!}(x - x_0)^n.$$

The problem now is to determine the coefficients.

The constant term is determined by the initial condition, $y(x_0) = y_0$. The other coefficients are found by computing successive derivatives.

The coefficient $y'(x_0)/1!$ of $(x - x_0)$ can be computed directly from the differential equation:

$$y'(x_0) = f[x_0, y(x_0)] = f(x_0, y_0).$$

Now differentiate the differential equation by the Chain Rule:

$$y'' = f_x(x, y) + f_y(x, y)y'.$$

Substitute $x = x_0$:

$$y''(x_0) = f_x(x_0, y_0) + f_y(x_0, y_0)y'(x_0).$$

The coefficient $y''(x_0)/2!$ of $(x - x_0)^2$ can be computed from this formula since $y'(x_0)$ is known.

Continue this "bootstrap" operation; differentiate again:

$$\begin{aligned} y''' &= [f_{xx}(x, y) + f_{xy}(x, y)y'] + [f_{yx}(x, y) + f_{yy}(x, y)y']y' + f_y(x, y)y'' \\ &= f_{xx}(x, y) + 2f_{xy}(x, y)y' + f_{yy}(x, y)y'^2 + f_y(x, y)y''. \end{aligned}$$

Substitute $x = x_0$:

$$y'''(x_0) = f_{xx}(x_0, y_0) + 2f_{xy}(x_0, y_0)y'(x_0) + f_{yy}(x_0, y_0)y'(x_0)^2 + f_y(x_0, y_0)y''(x_0).$$

The coefficient $y'''(x_0)/3!$ of $(x - x_0)^3$ can be computed from this formula since $y'(x_0)$ and $y''(x_0)$ are known.

Continue in this way, calculating the further derivatives $y^{(4)}(x_0)$, $y^{(5)}(x_0)$, $\cdots$. The formulas for these derivatives are complicated and should not be memorized. It is best to work out each particular example directly.

EXAMPLE 2.1

Compute the 5-th degree Taylor polynomial at $x = 0$ of the function $y(x)$ that satisfies the initial-value problem

$$\frac{dy}{dx} = y \qquad y(0) = 1.$$

Solution:

$$\begin{aligned} y' &= y & y'(0) &= 1 \\ y'' &= y' = y & y''(0) &= 1 \\ y''' &= y' = y & y'''(0) &= 1 \\ y^{(4)} &= y & y^{(4)}(0) &= 1 \\ y^{(5)} &= y & y^{(5)}(0) &= 1. \end{aligned}$$

Thus

$$y(x) \approx 1 + x + \frac{x^2}{2!} + \frac{x^3}{3!} + \frac{x^4}{4!} + \frac{x^5}{5!}.$$

Answer: $1 + x + \dfrac{x^2}{2!} + \dfrac{x^3}{3!} + \dfrac{x^4}{4!} + \dfrac{x^5}{5!}.$

REMARK: This simple problem has an explicit solution, $y = e^x$. The answer to the example is the 5-th degree Taylor polynomial of e^x at $x = 0$. Denote this polynomial by $p_5(x)$. Observe that

$$p_5'(x) = 1 + x + \frac{x^2}{2!} + \frac{x^3}{3!} + \frac{x^4}{4!} = p_5(x) - \frac{x^5}{5!}.$$

Thus $p_5(x)$ satisfies the initial condition and, near $x = 0$, nearly satisfies the differential equation, $y' = y$.

EXAMPLE 2.2

Approximate the solution $y(x)$ of the initial-value problem

$$\frac{dy}{dx} = x^2y - 1 \qquad y(0) = 2$$

by its 9-th degree Taylor polynomial at $x = 0$. Estimate $y(1)$.

Solution: The value $y(0) = 2$ is given; the values $y'(0), y''(0), \cdots, y^{(9)}(0)$ must be computed. Now

$$y' = x^2y - 1, \qquad y'(0) = -1.$$

Differentiate the expression for y':

$$y'' = 2xy + x^2y', \qquad y''(0) = 0.$$

Differentiate again:

$$y''' = 2y + 4xy' + x^2y''.$$

Substitute $x = 0$ and $y(0) = 2$ to obtain

$$y'''(0) = 4.$$

Continue in this way:

$$\begin{aligned}
y^{(4)} &= 6y' + 6xy'' + x^2y''' & y^{(4)}(0) &= 6y'(0) = -6\\
y^{(5)} &= 12y'' + 8xy''' + x^2y^{(4)} & y^{(5)}(0) &= 12y''(0) = 0\\
y^{(6)} &= 20y''' + 10xy^{(4)} + x^2y^{(5)} & y^{(6)}(0) &= 80\\
y^{(7)} &= 30y^{(4)} + 12xy^{(5)} + x^2y^{(6)} & y^{(7)}(0) &= -180\\
y^{(8)} &= 42y^{(5)} + 14xy^{(6)} + x^2y^{(7)} & y^{(8)}(0) &= 0\\
y^{(9)} &= 56y^{(6)} + 16xy^{(7)} + x^2y^{(8)} & y^{(9)}(0) &= 4480.
\end{aligned}$$

The coefficients of the Taylor approximation are

$$2, \quad -1, \quad 0, \quad \frac{4}{3!} = \frac{2}{3}, \quad \frac{-6}{4!} = \frac{-1}{4}, \quad 0,$$

$$\frac{80}{6!} = \frac{1}{9}, \quad \frac{-180}{7!} = \frac{-1}{28}, \quad 0, \quad \frac{4480}{9!} = \frac{1}{81}.$$

Answer:

$$y(x) \approx 2 - x + \frac{2}{3}x^3 - \frac{1}{4}x^4 + \frac{1}{9}x^6 - \frac{1}{28}x^7 + \frac{1}{81}x^9,$$

$$y(1) \approx 1.50441.$$

This procedure illustrates a general iterative scheme common to all of the methods discussed in this chapter. It is this: select an initial approximation and a procedure for obtaining a next approximation. Follow the procedure repeatedly, at each stage applying it to the latest approximation until you are satisfied with the result. The flowchart in Fig. 2.1 (next page) illustrates the iteration in the method of Taylor coefficients.

EXERCISES

Without solving, compute the 5-th degree Taylor polynomial at $x = 0$ of the function satisfying the initial-value problem:

1. $y' = xy + x;\quad y(0) = 0$
2. $y' = xy + y;\quad y(0) = 1$
3. $y' = xy + 1;\quad y(0) = 2$
4. $y' = y + \sin x;\quad y(0) = 1$
5. $y' = x^3y - x;\quad y(0) = 1$
6. $y' = 1 + xy^2;\quad y(0) = 0$
7. $y' = x + y^3;\quad y(0) = 0$
8. $y' = -x^2 + y^2;\quad y(0) = 0.$

Compute the 4-th degree Taylor polynomial at $x = 0$ for the initial-value problem:

9. $y' = y\sin(x^2)$; $y(0) = 1$
10. $y' = 1 + x^2e^{-y}$; $y(0) = 1$
11. $y' = e^x + xy^2$; $y(0) = -1$
12. $y' = \cos(xy) - 1$; $y(0) = -1$
13. $y' = x\sinh y$; $y(0) = 0$
14. $y' = (1 + x^2)(1 + y^2)$; $y(0) = 2$.
15. The height $x(t)$ of a certain ballon satisfies $t\dot{x} = 0.5\,(t - x)^2$. When $t = 10$ sec, $x = 10$ ft. Estimate x at $t = 12$ sec to 0.1 ft accuracy.

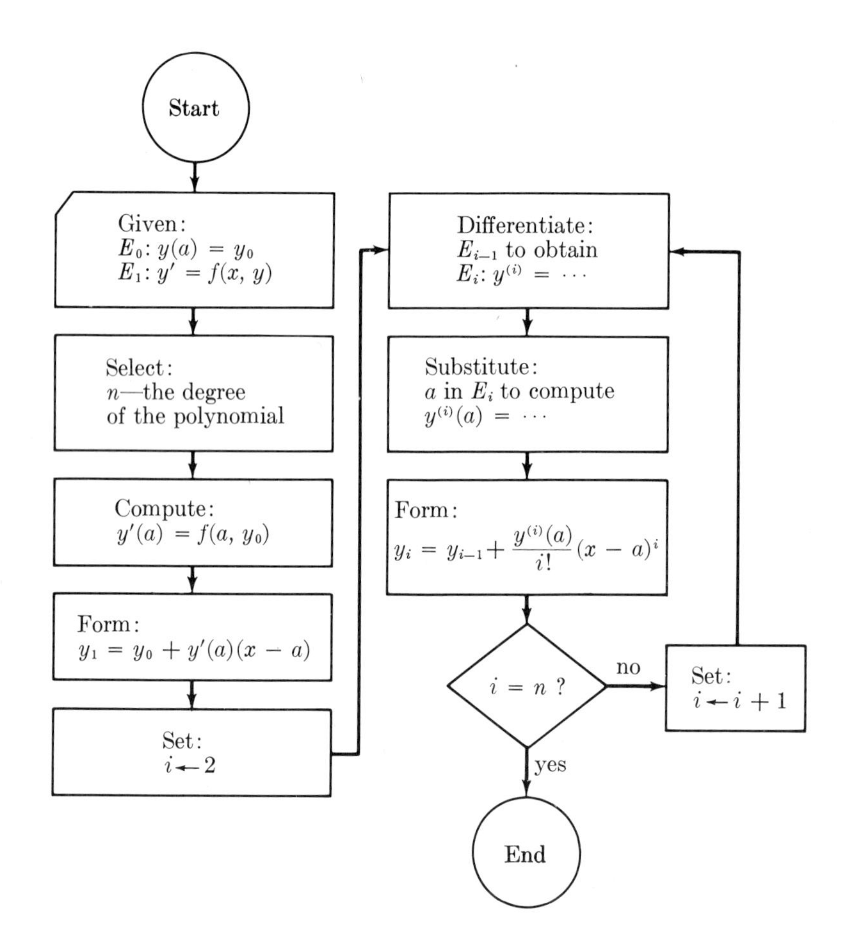

FIG. 2.1 Method of Taylor polynomials.

3. SUCCESSIVE APPROXIMATIONS

An initial-value problem may be converted into an **integral equation,** an equation in which the unknown function occurs under the integral sign.

> Solving the initial-value problem
>
> $$\frac{dy}{dx} = f[x, y(x)], \qquad y(x_0) = y_0,$$
>
> is equivalent to solving the integral equation
>
> $$y(x) = y_0 + \int_{x_0}^{x} f[t, y(t)]\, dt.$$

It is easy to verify this equivalence. On the one hand, if $y(x)$ satisfies the integral equation, then $y(x_0) = y_0$ and

$$\frac{dy}{dx} = \frac{d}{dx}\left[\int_{x_0}^{x} f[t, y(t)]\, dt\right] = f[x, y(x)].$$

On the other hand, if $y(x)$ satisfies the initial-value problem, then

$$y(x) - y_0 = y(x) - y(x_0) = \int_{x_0}^{x} \frac{dy}{dt}\, dt = \int_{x_0}^{x} f[t, y(t)]\, dt.$$

To solve the integral equation, compute a sequence

$$y_0(x), \quad y_1(x), \quad y_2(x), \quad \cdots$$

of approximate solutions by an iteration procedure.

STEP (0): Choose an approximate solution $y_0(x)$. (The constant function $y_0(x) = y_0$ is a reasonable choice.)

STEP (1): Set

$$y_1(x) = y_0 + \int_{x_0}^{x} f[t, y_0(t)]\, dt.$$

Usually, $y_1(x)$ is a better approximation than $y_0(x)$.

STEP (2): Set

$$y_2(x) = y_0 + \int_{x_0}^{x} f[t, y_1(t)]dt,$$

a better approximation still.

. . . .

STEP (n): Set

$$y_n(x) = y_0 + \int_{x_0}^{x} f[t, y_{n-1}(t)]\, dt.$$

EXAMPLE 3.1

Construct $y_4(x)$ for the initial-value problem

$$\frac{dy}{dx} = y \qquad y(0) = 1.$$

Start with $y_0(x) = 1$.

Solution: The integral equation is

$$y(x) = 1 + \int_0^x y(t)\,dt.$$

Apply the iteration procedure starting with $y_0(x) = 1$:

$$y_1(x) = 1 + \int_0^x y_0(t)\,dt = 1 + \int_0^x 1\cdot dt = 1 + x,$$

$$y_2(x) = 1 + \int_0^x (1+t)\,dt = 1 + x + \frac{1}{2}x^2,$$

$$y_3(x) = 1 + \int_0^x \left(1 + t + \frac{1}{2}t^2\right) dt = 1 + x + \frac{1}{2}x^2 + \frac{1}{6}x^3,$$

$$y_4(x) = 1 + \int_0^x \left(1 + t + \frac{1}{2}t^2 + \frac{1}{6}t^3\right) dt = 1 + x + \frac{1}{2}x^2 + \frac{1}{6}x^3 + \frac{1}{24}x^4.$$

Answer: $1 + x + \frac{1}{2}x^2 + \frac{1}{6}x^3 + \frac{1}{24}x^4.$

REMARK: The answer is the 4-th degree Taylor polynomial of e^x.

EXAMPLE 3.2

Start with the approximation $y_0(x) = 2$ and compute $y_3(x)$ for the initial-value problem

$$y' = x^2y - 1 \qquad y(0) = 2.$$

Compute $y_3(1)$ to 5 places.

Solution: The corresponding integral equation is

$$y(x) = 2 + \int_0^x [t^2y(t) - 1]\,dt.$$

The successive approximations are

$$\begin{aligned} y_1(x) &= 2 + \int_0^x [t^2y_0(t) - 1]\,dt \\ &= 2 + \int_0^x (-1 + 2t^2)\,dt = 2 - x + \frac{2}{3}x^3. \end{aligned}$$

$$\begin{aligned} y_2(x) &= 2 + \int_0^x [t^2y_1(t) - 1]\,dt \\ &= 2 + \int_0^x \left[-1 + t^2\left(2 - t + \frac{2}{3}t^3\right)\right] dt = 2 - x + \frac{2}{3}x^3 - \frac{1}{4}x^4 + \frac{1}{9}x^6. \end{aligned}$$

$$y_3(x) = 2 + \int_0^x \left[-1 + t^2\left(2 - t + \frac{2}{3}t^3 - \frac{1}{4}t^4 + \frac{1}{9}t^6\right)\right] dt.$$

Answer:

$$y_3(x) = 2 - x + \frac{2}{3}x^3 - \frac{1}{4}x^4 + \frac{1}{9}x^6 - \frac{1}{28}x^7 + \frac{1}{81}x^9;$$

$$y_3(1) \approx 1.50441.$$

REMARK 1: The answer agrees with the 9-th degree Taylor approximation to $y(x)$ as found in the previous section.

REMARK 2: The pattern of the iteration should be clear. If we want $y_4(x)$, we repeat each term in $y_3(x)$ and add the two new terms:

$$\int_0^x t^2\left(-\frac{1}{28}t^7 + \frac{1}{81}t^9\right)dt = \frac{-1}{280}x^{10} + \frac{1}{972}x^{12}.$$

This gives us $y(1) \approx y_4(1) \approx 1.50187$. Similarly the next two terms are

$$\int_0^x t^2\left(\frac{-1}{280}t^{10} + \frac{1}{972}t^{12}\right)dt = \frac{-1}{13\cdot 280}x^{13} + \frac{1}{15\cdot 972}x^{15},$$

which gives us $y(1) \approx y_5(1) \approx 1.50166$.

This **method of successive approximations,** also called the **Picard Method,** is often useful in finding approximate solutions. For the amount of computation involved, however, it is not as accurate as the methods discussed below. Still it is of great theoretical importance because it can be used to prove what we have taken for granted, that each initial-value problem for a sufficiently smooth direction field *has* a solution. The flowchart for this method appears in Fig. 3.1.

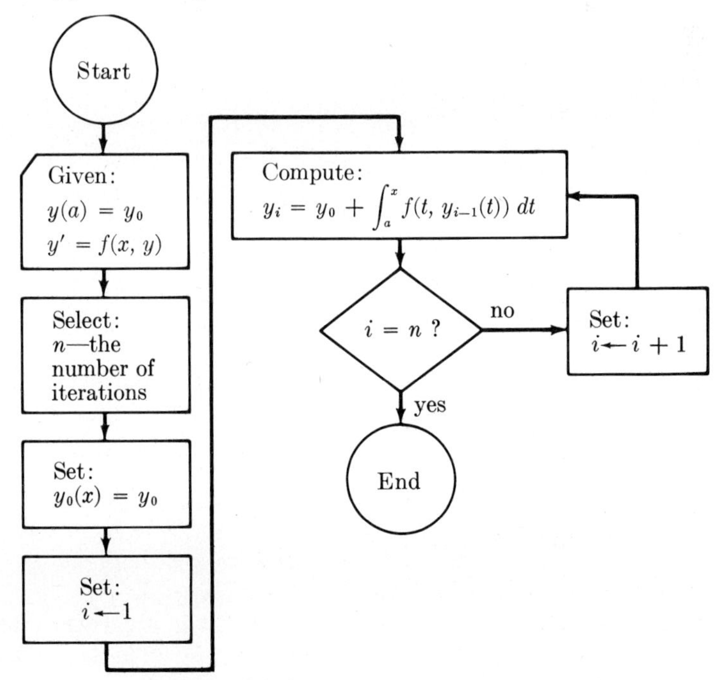

FIG. 3.1 Method of successive approximations.

EXERCISES

Compute $y_3(x)$ for the initial-value problem; choose $y_0(x) = y(0)$:

1. $y' = xy + x;\quad y(0) = 0$
2. $y' = xy + y;\quad y(0) = 1$
3. $y' = xy + 1;\quad y(0) = 2$
4. $y' = y + \sin x;\quad y(0) = 1$
5. $y' = x^3y - x;\quad y(0) = 1$
6. $y' = 1 + xy^2;\quad y(0) = 0$
7. $y' = x + y^3;\quad y(0) = 0$
8. $y' = -x^2 + y^2;\quad y(0) = 0$
9. $y' = x^2 + y^2;\quad y(0) = 0$
10. $y' = x^3;\quad y(0) = 1.$

4. THE EULER METHOD [optional]

The solution of the initial-value problem

$$\frac{dy}{dx} = f(x, y), \qquad y(x_0) = y_0$$

is a curve through (x_0, y_0) whose slope at each point is that of the direction field. In the **Euler Method,** this curve is approximated by a polygonal line (Fig. 4.1). Each of the small segments in the polygonal curve $y = p(x)$

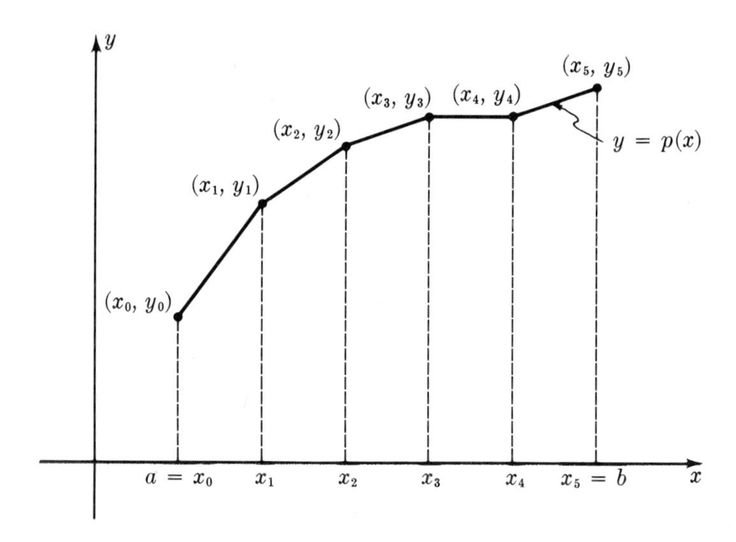

FIG. 4.1

is constructed so that its slope is the direction of the field $y' = f(x, y)$ at the left end point of the segment. If the segments are short enough, the approximation is good.

To approximate $y(b)$, where b is to the right of $x_0 = a$, divide the interval $a \le x \le b$ into n parts, each of length $h = (b - a)/n$. (The larger n, the better the approximation.) Denote the points of division by

$$a = x_0 < x_1 < x_2 \cdots < x_n = b.$$

Construct a polygonal approximation to $y(x)$ by the following steps (Fig. 4.1)

STEP (1): Draw the line through (x_0, y_0) with slope $y_0' = f(x_0, y_0)$. It meets the line $x = x_1$ at a point (x_1, y_1).

STEP (2): Draw the line through (x_1, y_1) with slope $y_1' = f(x_1, y_1)$. It meets $x = x_2$ at a point (x_2, y_2).

. . . .

STEP (n): Finally draw the line through (x_{n-1}, y_{n-1}) with slope $y'_{n-1} = f(x_{n-1}, y_{n-1})$. It meets $x = x_n = b$ at a point (x_n, y_n). Then $y(b) \approx y_n$.

Formulas for the quantities $y_0', y_1', \cdots, y'_{n-1}$ and $y_0, y_1, y_2, \cdots, y_n$ are easily found. The line drawn in Step (1) has equation

$$y = y_0 + y_0'(x - x_0).$$

Hence

$$y_1 = y_0 + y_0'(x_1 - x_0) = y_0 + y_0'h.$$

Now y_1' can be computed from the differential equation:

$$y_1' = f(x_1, y_1) = f(x_1, y_0 + y_0'h).$$

The line drawn in Step (2) has equation

$$y = y_1 + y_1'(x - x_1),$$

hence

$$y_2 = y_1 + y_1'h,$$

$$y_2' = f(x_2, y_2).$$

Continue step by step:

$$\begin{aligned} y_1 &= y_0 + y_0'h & y_1' &= f(x_1, y_1) \\ y_2 &= y_1 + y_1'h & y_2' &= f(x_2, y_2) \\ y_3 &= y_2 + y_2'h & y_3' &= f(x_3, y_3) \\ &\cdots & & \\ y_{n-1} &= y_{n-2} + y'_{n-2}h & y'_{n-1} &= f(x_{n-1}, y_{n-1}) \\ y_n &= y_{n-1} + y'_{n-1}h. & & \end{aligned}$$

EXAMPLE 4.1

Use the Euler Method to approximate $y(1) = e$ for the initial-value problem

$$y'(x) = y \qquad y(0) = 1.$$

Choose $n = 5$, and compute each value to 2 decimal places.

Solution: We have $h = \frac{1}{5} = 0.20$. Thus

$$
\begin{array}{lll}
x_0 = 0 & y_0 = 1 & y_0' = 1 \\
x_1 = 0.2 & y_1 = 1 + (1)(0.20) = 1.20 & y_1' = 1.20 \\
x_2 = 0.4 & y_2 = 1.20 + (1.20)(0.20) = 1.44 & y_2' = 1.44 \\
x_3 = 0.6 & y_3 = 1.44 + (1.44)(0.20) \approx 1.73 & y_3' \approx 1.73 \\
x_4 = 0.8 & y_4 = 1.73 + (1.73)(0.20) \approx 2.08 & y_4' \approx 2.08 \\
x_5 = 1.0 & y_5 = 2.08 + (2.08)(0.20) \approx 2.50 &
\end{array}
$$

Answer: $e \approx 2.5$.

Since $e \approx 2.718$, we see that $n = 5$ is too small for any real accuracy. In our test problem we shall use $n = 10$ and $n = 100$, and buy some computer time for the calculation.

EXAMPLE 4.2

Approximate $y(1)$ to 5-places for the initial-value problem

$$\frac{dy}{dx} = x^2y - 1 \qquad y(0) = 2.$$

Choose $n = 10$ and $n = 100$.

Solution: If $n = 10$, then $h = 0.1$; if $n = 100$, then $h = 0.01$. We tabulate the results in both cases at intervals of 0.1. See Table 4.1.

TABLE 4.1. EULER METHOD

	$h = 0.1$	$h = 0.01$
x	$y(x)$	
0.0	2.00000	2.00000
0.1	1.90000	1.90055
0.2	1.80190	1.80458
0.3	1.70911	1.71528
0.4	1.62449	1.63535
0.5	1.55048	1.56719
0.6	1.48924	1.51305
0.7	1.44286	1.47532
0.8	1.41356	1.45674
0.9	1.40402	1.46086
1.0	1.41775	1.49254

Answer: $n = 10$: $y(1) \approx 1.41775$,
$n = 100$: $y(1) \approx 1.49254$.

REMARK: The last column of Table 4.1 is the more accurate, although how accurate we cannot yet say. Later calculations will indicate that to 5 places, $y(1) = 1.50165$.

The Euler Method is far from efficient. It requires considerable calculation (large n) to insure accuracy. Nevertheless, the Euler Method is important because the later methods we study are all refinements of it.

Note that the Euler Method is a procedure for approximating the solution *at a given point only;* it does not provide an approximate formula for the solution. The same is true of the methods discussed below.

EXERCISES

0. Draw a flowchart for the Euler Method.

Use the Euler Method to approximate $y(1)$ for the initial-value problem. By hand, compute the answer to 2 places using $n = 5$; or by computer, to 5 places using $n = 10$ and $n = 100$:

1. $y' = xy + x;\quad y(0) = 0$
2. $y' = xy + y;\quad y(0) = 1$
3. $y' = xy + 1;\quad y(0) = 2$
4. $y' = y + \sin x;\quad y(0) = 1$
5. $y' = x^3 y - x;\quad y(0) = 1$
6. $y' = 1 + x^2 y;\quad y(0) = 1$
7. $y' = (1 + y)\cos x;\quad y(0) = 0$
8. $y' = (1 - x)(1 - y);\quad y(0) = -1.$
9. Isobars on a certain weather map satisfy the equation $y' = -x - \sin y$. Sketch the isobar through the point $(0, \pi/2)$ ending where $x = 1$. Use the Euler Method with 10 points to estimate $y(1)$ to 2 places.
10. A particle moves along a straight path with velocity $\dot{s} = s \ln t + t^2$ ft/sec and passes $s = 0$ at $t = 1$ sec. Estimate its position at $t = 2$ sec. Use the Euler Method with $n = 10$ and carry your work to 2 places.
11. Do Example 4.1 with $n = 10$.

5. THE HEUN METHOD [optional]

In the Euler Method, the solution curve of the initial-value problem

$$\frac{dy}{dx} = f(x, y) \qquad y(x_0) = y_0$$

is approximated by a certain polygonal graph (Fig. 4.1). The slope of each segment of this polygon matches the slope of the direction field $dy/dx = f(x, y)$ at the left end point of the segment. How well the short segment approximates a solution curve depends on how much the slope of the direction field drifts away from the slope of the segment as you move from its left end point to its right. Usually the worst deviation occurs at the right end, and as you move segment by segment from left to right, the error gets progressively worse. You can curtail the total error by increasing the number of segments, but then cumulative round-off error becomes significant.

Several refinements of the Euler Method which yield greater accuracy are discussed in this and the next section. In each of these methods the slope of successive segments is chosen not to match the direction field at *one* point, but rather to match the *average* direction field taken at two or more points along the segment.

The first refinement of the Euler Method is the Heun Method; the slope of each segment is the average of the slopes of the direction field at *both* ends of the segment constructed by the Euler Method.

Start at x_0 with initial value y_0 and initial slope $y_0' = f(x_0, y_0)$. See Fig. 5.1. Set

$$z_1 = y_0 + y_0'h.$$

(This is the next Euler Method value.) The slope at (x_1, z_1) is $z_1' = f(x_1, z_1)$.

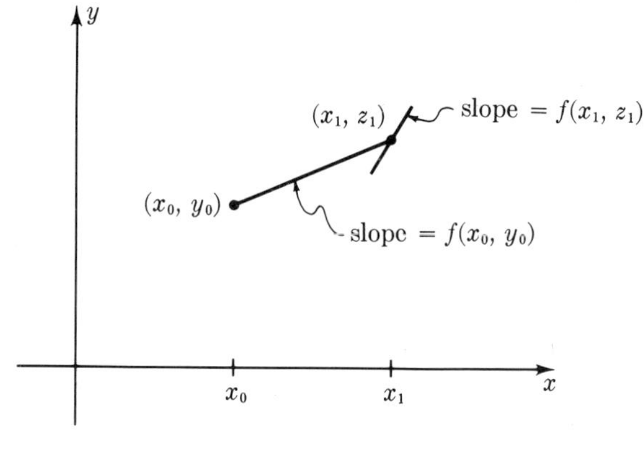

FIG. 5.1

Hence an approximation to the average slope in the interval $x_0 \le x \le x_1$ is

$$\frac{1}{2}(y_0' + z_1') = \frac{1}{2}[f(x_0, y_0) + f(x_1, z_1)].$$

Now examine Fig. 5.2.

The line through (x_0, y_0) with this average slope meets $x = x_1$ at (x_1, y_1), where

$$y_1 = y_0 + \frac{h}{2}(y_0' + z_1').$$

This locates the next point (x_1, y_1). Repeat the construction, starting at the point (x_1, y_1):

$$y_1' = f(x_1, y_1), \qquad z_2 = y_1 + y_1'h, \qquad z_2' = f(x_2, z_2),$$

$$y_2 = y_1 + \frac{h}{2}(y_1' + z_2').$$

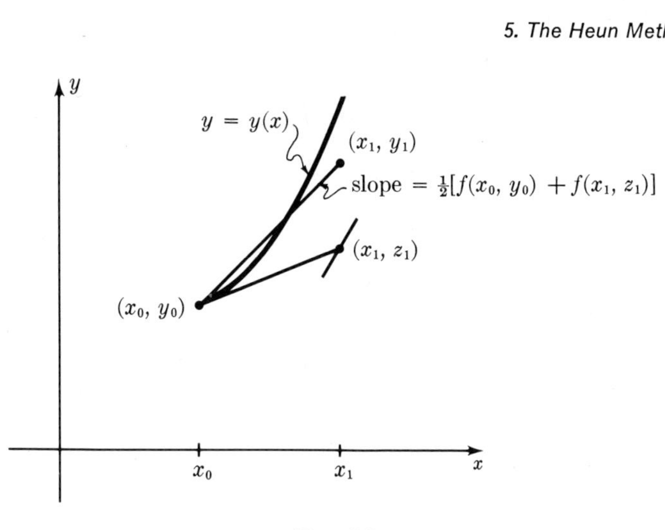

FIG. 5.2

This locates the next point (x_2, y_2). Continue the process until (x_n, y_n) is reached.

EXAMPLE 5.1

Use the Heun Method to compute $e = y(1)$ from the initial-value problem

$$\frac{dy}{dx} = y \qquad y(0) = 1.$$

Use $h = 0.20$ and 2-place accuracy.

Solution: Tabulate $x_i, z_i, z_i', y_i, y_i'$. Each row of Table 5.1 is constructed in this order, using the data in the previous row and these formulas:

$$z_{i+1} = y_i + y_i'h, \qquad z'_{i+1} = z_{i+1},$$

$$y_{i+1} = y_i + \frac{h}{2}(y_i' + z'_{i+1}), \qquad y'_{i+1} = y_{i+1}.$$

TABLE 5.1. HEUN METHOD

x_1	z_i	z_i'	y_i	y_i'
0.0			1.0	1.0
0.2	1.20	1.20	1.22	1.22
0.4	1.46	1.46	1.49	1.49
0.6	1.79	1.79	1.82	1.82
0.8	2.18	2.18	2.22	2.22
1.0	2.66	2.66	2.71	

Answer: $e \approx 2.71.$

REMARK: The same h and 2-place computation in the Euler Method led to $e \approx 2.50$. Since $e \approx 2.71828$, the improvement is striking.

We return to our test case.

EXAMPLE 5.2

Approximate $y(1)$ to 5-places for the initial-value problem

$$\frac{dy}{dx} = x^2y - 1 \qquad y(0) = 2.$$

Use the Heun Method with $n = 10$ and $n = 100$.

Solution: We used the computer on this. The print-out tabulated x_i, z_i, y_i. The results (at intervals of 0.1) are in Table 5.2, which should be compared with the corresponding Euler Method results in Table 4.1.

TABLE 5.2. HEUN METHOD

	$h = 0.1$		$h = 0.01$	
x_i	z_i	y_i	z_i	y_i
0.0		2.00000		2.00000
0.1	1.90000	1.90095	1.90063	1.90064
0.2	1.80285	1.80551	1.80491	1.80495
0.3	1.71273	1.71682	1.71601	1.71606
0.4	1.63228	1.63761	1.63662	1.63668
0.5	1.56381	1.57026	1.56913	1.56920
0.6	1.50951	1.51706	1.51582	1.51590
0.7	1.47167	1.48042	1.47910	1.47919
0.8	1.45296	1.46318	1.46180	1.46191
0.9	1.45683	1.46901	1.46759	1.46773
1.0	1.48800	1.50290	1.50149	1.50166

Answer: $n = 10$: $y(1) \approx 1.50290$,
$n = 100$: $y(1) \approx 1.50166$.

EXERCISES

0. Draw a flowchart for the Heun Method.

Use the Heun Method to approximate $y(1)$ for each initial-value problem. By hand, compute the answer to 2 places using $h = 0.20$; or by computer, to 5 places using $h = 0.10$ and $h = 0.01$:

1. $y' = xy + x$; $\quad y(0) = 0$
2. $y' = xy + y$; $\quad y(0) = 1$
3. $y' = xy + 1$; $\quad y(0) = 2$
4. $y' = y + \sin x$; $\quad y(0) = 1$

5. $y' = x^3y - x;$ $y(0) = 1$
6. $y' = 1 + x^2y;$ $y(0) = 1$
7. $y' = (1 + y) \cos x;$ $y(0) = 0$
8. $y' = (1 - x)(1 - y);$ $y(0) = -1.$
9. Solve $y' = x^2y - x + 1$ with $y(0) = 1$. Use the Heun Method with $h = 0.1$ to obtain $y(1)$ to 3 places.
10. Two particles move on the same straight path according to the equation $\dot{s} = (\cos t) - s$. Assume they are at $s = 0$ and $s = 1$, respectively, when $t = 0$. Estimate their separation to 3 places at $t = 1$. (Use the Heun Method.)
11. Do Example 5.1 with $h = 0.10$ and 4-place accuracy.

6. RUNGE-KUTTA METHODS [optional]

The methods of Euler and Heun are particular cases of a class of methods known as **Runge-Kutta Methods.** We can understand the idea behind these methods by examining a special type of initial-value problem:

$$\begin{cases} \dfrac{dy}{dx} = f(x) \\ y(a) = y_0 . \end{cases}$$

The special feature of this problem is that y does not appear in $f(x)$. The answer is found directly by integration:

$$y(b) = y_0 + \int_a^b f(x)\,dx.$$

The Euler and Heun methods provide approximate solutions to the initial-value problem. Consequently, in this case, they provide approximations to the integral. What approximations?

In the Euler Method the sequence of successive values is

$$\begin{aligned} y_1 &= y_0 + hf(x_0) \\ y_2 &= y_1 + hf(x_1) \\ &\cdots\cdots\cdots \\ y_n &= y_{n-1} + hf(x_{n-1}). \end{aligned}$$

Add these equations, then cancel $y_1 + y_2 + \cdots + y_{n-1}$. The result is

$$y_n = y_0 + h[f(x_0) + f(x_1) + \cdots + f(x_{n-1})].$$

Now y_n is the Euler approximation to $y(b)$. Compare y_n to the exact value of $y(b)$, and you see

$$\int_a^b f(x)\,dx \approx h[f_0 + f_1 + \cdots + f_{n-1}],$$

where $f_i = f(x_i)$. Thus the Euler Method approximates the definite integral of a function by rectangles with heights measured at the successive left-hand end points (Fig. 6.1).

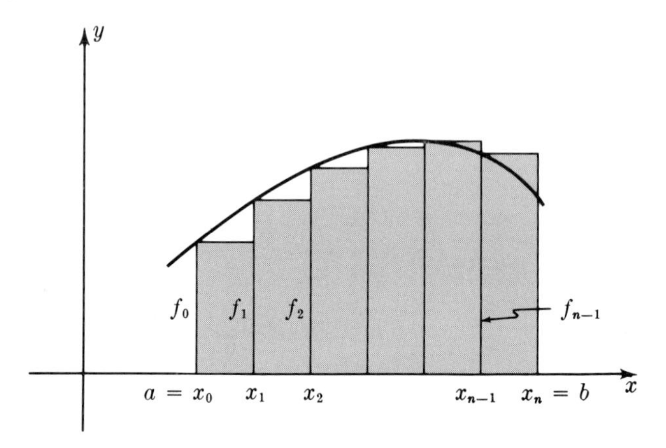

FIG. 6.1

Now consider the Heun Method in this case. Start at (x_0, y_0) with $y_0' = f_0$:

$$z_1 = z_0 + hf_0.$$

The average slope on $x_0 \leq x \leq x_1$ is approximately

$$\frac{1}{2}(f_0 + f_1),$$

since the slope at z_1 is f_1. Hence

$$y_1 = y_0 + \frac{h}{2}(f_0 + f_1).$$

Similarly

$$y_2 = y_1 + \frac{h}{2}(f_1 + f_2)$$

$$\cdots\cdots\cdots\cdots$$

$$y_n = y_{n-1} + \frac{h}{2}(f_{n-1} + f_n).$$

Add, then cancel $y_1 + \cdots + y_{n-1}$:

$$y_n = y_0 + \frac{h}{2}(f_0 + 2f_1 + 2f_2 + \cdots + 2f_{n-1} + f_n).$$

Thus, in this special case, the Heun Method also amounts to an approximate integration formula,

$$\int_a^b f(x)\,dx \approx \frac{h}{2}(f_0 + 2f_1 + \cdots + 2f_{n-1} + f_n),$$

precisely the Trapezoidal Rule.

Higher Order Runge-Kutta Methods

Runge-Kutta Methods provide piecewise polygonal approximations to the solution of initial-value problems. In each method the value y_{i+1} at x_{i+1} is computed by a formula of the type

$$y_{i+1} = y_i + m_i h,$$

where m_i is a weighted average of the values of $f(x, y)$ at (x_i, y_i) and several other points (x, y) for which $x_i < x \leq x_{i+1}$. The number of points involved in this average is the **order** of the method. For instance the Euler Method is a first-order Runge-Kutta Method with

$$m_i = f(x_i, y_i).$$

The Heun Method is a second-order Runge-Kutta Method with

$$m_i = \frac{1}{2}[f(x_i, y_i) + f(x_{i+1}, z_{i+1})], \qquad z_{i+1} = y_i + y_i'h.$$

In the special case $f(x, y) = f(x)$, the first- and second-order Runge-Kutta Method are equivalent to rectangular and trapezoidal approximations of definite integrals respectively; the third-order Runge-Kutta Method is equivalent to Simpson's Rule, and the higher order methods to various Newton-Cotes Rules.

Third Order

Let us describe a third-order Runge-Kutta Method. To do so, we must specify m_i in the formula

$$y_{i+1} = y_i + m_i h.$$

We define m_i as the weighted average of the values of $f(x, y)$ at three particular points. We set

$$p_{i+1} = f(x_i, y_i),$$

$$q_{i+1} = f\left(x_i + \frac{h}{2}, y_i + \frac{h}{2} p_{i+1}\right),$$

$$r_{i+1} = f(x_i + h, y_i + 2hq_{i+1} - hp_{i+1}).$$

Then we define the weighted average m_i by

$$m_i = \frac{1}{6}(p_{i+1} + 4q_{i+1} + r_{i+1}).$$

Thus

$$y_{i+1} = y_i + \frac{h}{6}(p_{i+1} + 4q_{i+1} + r_{i+1}).$$

In computations, starting at (x_i, y_i) we calculate in order p_{i+1}, q_{i+1}, r_{i+1}, y_{i+1}.

The choice of r_{i+1} is based on complicated error estimates which we omit. Note that in the special case $f(x, y) = f(x)$, the formula does reduce to Simpson's Rule.

EXAMPLE 6.1

Use the third-order Runge-Kutta Method to estimate $e = y(1)$ to 3 places, where $y(x)$ satisfies

$$\frac{dy}{dx} = y \qquad y(0) = 1.$$

Use $h = 0.2$.

Solution: Follow the recipe precisely, using $f(x, y) = y$. The results are shown in Table 6.1.

TABLE 6.1. THIRD-ORDER RUNGE-KUTTA METHOD

x_i	p_i	q_i	r_i	y_i
0.0				1.000
0.2	1.000	1.100	1.240	1.221
0.4	1.221	1.343	1.514	1.491
0.6	1.491	1.640	1.849	1.821
0.8	1.821	2.003	2.258	2.224
1.0	2.224	2.446	2.758	2.716

The work can be simplified in this case as follows:

$$p_{i+1} = y_i,$$

$$q_{i+1} = y_i + \frac{h}{2}p_{i+1} = \left(1 + \frac{h}{2}\right)y_i = \frac{1}{2}(2 + h)y_i,$$

$$r_{i+1} = y_i + 2hq_{i+1} - hp_{i+1} = y_i + h(2 + h)y_i - hy_i = (1 + h + h^2)y_i,$$

$$y_{i+1} = y_i + \frac{h}{6}(p_{i+1} + 4q_{i+1} + r_{i+1})$$

$$= y_i + \frac{h}{6}[y_i + (4 + 2h)y_i + (1 + h + h^2)y_i]$$

$$= y_i + \frac{h}{6}(6 + 3h + h^2)y_i = \left(1 + h + \frac{1}{2}h^2 + \frac{1}{6}h^3\right)y_i.$$

Since $h = 0.2$,

$$y_{i+1} \approx (1.2213)y_i,$$

$$y(1) = y_5 \approx (1.2213)^5 \approx 2.717.$$

(The difference between this value of y_5 and the value in Table 6.1 is due to round-off error.)

Answer: $e \approx 2.717$.

EXAMPLE 6.2

Approximate $y(1)$ to 5-places for the initial-value problem

$$\frac{dy}{dx} = x^2y - 1 \qquad y(0) = 2.$$

Use the third-order Runge-Kutta Method with $h = 0.10$ and $h = 0.01$.

Solution: The computer print-out showed x_i, hp_i, hq_i, hr_i, and y_i. We shall only tabulate y_i at intervals of 0.1. See Table 6.2.

TABLE 6.2. THIRD-ORDER RUNGE-KUTTA METHOD

	$h = 0.10$	$h = 0.01$
x_i	y_i	
0.0	2.00000	2.00000
0.1	1.90064	1.90064
0.2	1.80494	1.80494
0.3	1.71605	1.71605
0.4	1.63667	1.63667
0.5	1.56920	1.56919
0.6	1.51591	1.51589
0.7	1.47921	1.47919
0.8	1.46194	1.46191
0.9	1.46774	1.46772
1.0	1.50172	1.50165

Answer: $h = 0.10$: $y(1) \approx 1.50172$,

$h = 0.01$: $y(1) \approx 1.50165$.

Fourth Order

Next we describe a fourth-order Runge-Kutta Method. Define

$$p_{i+1} = f(x_i, y_i),$$

$$q_{i+1} = f\left(x_i + \frac{h}{2}, y_i + \frac{h}{2} p_{i+1}\right),$$

$$r_{i+1} = f\left(x_i + \frac{h}{2}, y_i + \frac{h}{2} q_{i+1}\right),$$

$$s_{i+1} = f(x_i + h, y_i + hr_{i+1}).$$

Then set

$$y_{i+1} = y_i + \frac{h}{6}(p_{i+1} + 2q_{i+1} + 2r_{i+1} + s_{i+1}).$$

EXAMPLE 6.3

Apply this method to our test problem.

Solution: We simply state the answer our computer found.

Answer: $h = 0.10$: $y(1) \approx 1.50165$,

$h = 0.01$: $y(1) \approx 1.50165$.

REMARK: The results agree to 5 places. This suggests strongly that 1.50165 is correct to 5 places.

Table 6.3 compares the results of the four numerical methods for the test problem.

TABLE 6.3. APPROXIMATIONS TO $y(1)$

Method	$h = 0.10$	$h = 0.01$
Euler	1.41775	1.49254
Heun	1.50290	1.50166
3-rd order Runge-Kutta	1.50172	1.50165
4-th order Runge-Kutta	1.50165	1.50165

EXERCISES

0. Draw flowcharts for the third- and fourth-order Runge-Kutta Methods.

Use a Runge-Kutta Method to approximate $y(1)$ for each initial-value problem. By hand, compute the answer to 2 places using the third-order method with $h = 0.20$; or by computer, to 5 places using the fourth-order method with $h = 0.10$ and $h = 0.01$.

1. $y' = xy + x;\quad y(0) = 0$
2. $y' = xy + y;\quad y(0) = 1$
3. $y' = xy + 1;\quad y(0) = 2$
4. $y' = y + \sin x;\quad y(0) = 1$
5. $y' = x^3y - x;\quad y(0) = 1$
6. $y' = 1 + x^2y;\quad y(0) = 1$
7. $y' = (1 + y)\cos x;\quad y(0) = 0$
8. $y' = (1 - x)(1 - y);\quad y(0) = -1.$
9. Under experimental treatment, the weight w in micrograms of a colony of amoebas varies with time (in hours) according to the equation $\dot{w} = we^t - 3t^2$. If the weight of the colony at $t = 0$ is 1 μgm, estimate its weight after 1 hr. Use the fourth-order Runge-Kutta Method with $h = 0.1$ to obtain a 2-place answer.
10. Work Example 6.1 with $h = 0.2$ and 5 places.

11. Work Example 6.1 with $h = 0.1$ and 5 places.

12. Use the methods of Chapter 21 to show that the test problem of this chapter has the exact solution $y(1) = e^{1/3}\left[2 - \int_0^1 e^{-t^3/3}\,dt\right].$

13. (cont.) Use Simpson's Rule to compute $y(1)$ accurately to 6 places.

14. In the special case $f(x, y) = f(x)$, the fourth-order Runge-Kutta Method yields which approximate integration formula?

31. Complex Numbers

1. INTRODUCTION

The simple equation

$$x^2 + 1 = 0$$

has no solution in terms of *real numbers*, the numbers of ordinary experience; neither do the equations

$$x^2 + 3 = 0, \qquad x^2 + x + 1 = 0, \qquad x^2 - 4x + 10 = 0.$$

Yet such equations arise in scientific computations. For this reason, the real number system is enlarged by introducing a new number i satisfying

$$\boxed{i^2 = -1.}$$

The result is the system of **complex numbers.** It consists of all expressions $a + bi$, where a and b are real numbers. These expressions are treated by the usual rules of algebra, except that i^2 is replaced by -1. Thus

$$\begin{aligned} (a + bi) + (c + di) &= (a + c) + (b + d)i, \\ (a + bi)(c + di) &= ac + bdi^2 + adi + bci \\ &= (ac - bd) + (ad + bc)i. \end{aligned}$$

Two complex numbers $a + bi$ and $c + di$ are **equal** if and only if $a = c$ and $b = d$. A real number a is considered as a special type of complex number: $a = a + 0 \cdot i$.

Remarks on Notation: Sometimes $a + ib$ is the preferred notation. For example, $-1 + i\sqrt{3}$ looks better than $-1 + \sqrt{3}\,i$ because of the possible confusion with $-1 + \sqrt{3i}$. Similarly $\cos\theta + i\sin\theta$ is better than $\cos\theta + \sin\theta\, i$. In engineering, j is often used instead of i.

In terms of complex numbers, the quadratic equation

$$x^2 + 1 = 0$$

has two roots, $\pm i$. Furthermore, every quadratic equation

$$ax^2 + bx + c = 0$$

has complex roots. By the quadratic formula, the roots are

$$x = -\frac{b}{2a} \pm \frac{\sqrt{D}}{2a},$$

where $D = b^2 - 4ac$. If D is non-negative, then $\sqrt{D}$ is a real number, so the roots are real. If D is negative, write

$$\sqrt{D} = \sqrt{(-1)(-D)} = \sqrt{-1}\sqrt{-D} = i\sqrt{-D};$$

then the equation has complex roots

$$x = \frac{-b}{2a} \pm \frac{\sqrt{-D}}{2a}\, i.$$

EXAMPLE 1.1

Find the roots of $x^2 + x + 1 = 0$ and of $x^2 - 4x + 10 = 0$.

Solution: Apply the quadratic formula. For the first equation $a = b = c = 1$, $D = -3$. For the second equation $a = 1$, $b = -4$, $c = 10$, $D = -24$.

Answer: $-\frac{1}{2} \pm \frac{\sqrt{3}}{2} i$; $2 \pm i\sqrt{6}$.

All quadratic equations with real coefficients can be solved in terms of complex numbers. What about cubic equations

$$ax^3 + bx^2 + cx + d = 0\ ?$$

If a new quantity i is needed to solve quadratics, is another new quantity needed to solve cubics, still another to solve quartics, and so on?

This question was answered in 1799 by C. F. Gauss in the Fundamental Theorem of Algebra. It asserts that each polynomial equation of any degree with real coefficients, has a complex root. Thus, the complex number system is rich enough to contain solutions for all polynomial equations. Once the number i is adjoined to the real number system, that is enough.

REMARK: Later we shall learn that each complex number has two complex square roots. Hence, the quadratic formula provides complex roots for any quadratic equation with complex coefficients. The Fundamental Theorem of Algebra also includes the statement that each polynomial equation with *complex* coefficients has complex roots.

EXERCISES

Solve and check your answers by direct substitution:

1. $x^2 - 8x + 25 = 0$
2. $x^2 + 25 = 0$
3. $x^2 + x + 2 = 0$
4. $x^2 - 6x + 9 = 0$
5. $3x^2 - 2x + 3 = 0$
6. $2x^2 + x + 2 = 0$
7. $225x^2 + 15x + 61 = 0$
8. $5x^2 - 4x + 1 = 0.$

Find all roots:

9. $x^3 - 1 = 0$
10. $x^3 + 1 = 0$
11. $x^3 - x^2 + x - 1 = 0$
12. $x^4 - 2x^2 + 1 = 0$
13. $x^4 - 1 = 0$
14. $x^4 + 5x^2 + 4 = 0.$
15. Compute $(1 + i)^2$. Use the result to solve the equation $x^2 = i$.
16. Compute $[(-1 + i\sqrt{3})/2]^3$.
17. Compute $i + i^2 + i^3 + \cdots + i^{1492}$.

2. COMPLEX ARITHMETIC

Complex numbers can be pictured as vectors in a plane. Think of the number 1 as a unit horizontal vector and the number i as a unit vertical vector (Fig. 2.1a). Then a complex number $a + bi$ is a linear combination of these two vectors (Fig. 2.1b).

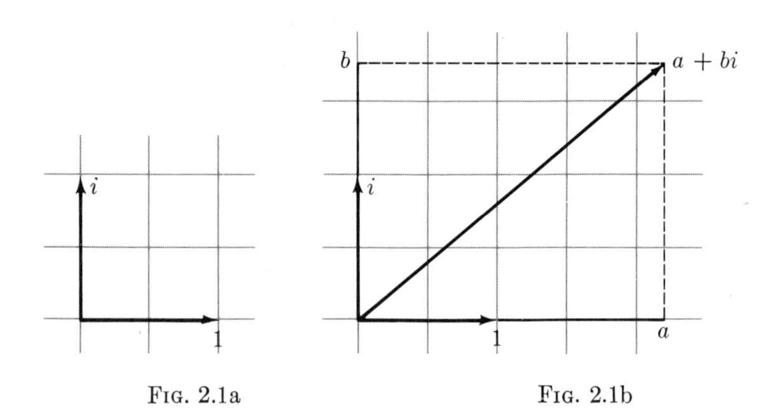

FIG. 2.1a FIG. 2.1b

The horizontal component of $a + bi$ is called its **real part,** written $\text{Re}(a + bi)$; the vertical component is called its **imaginary part,** written $\text{Im}(a + bi)$. Thus

$$\boxed{\text{Re}(a + bi) = a, \qquad \text{Im}(a + bi) = b.}$$

Note that $\text{Im}(a + bi)$ is the real number b, *not* bi.

Associated with each complex number $a + bi$ is the number $a - bi$ called its (**complex**) **conjugate** (Fig. 2.2). Conjugates are denoted by bars:

$$\boxed{\overline{a + bi} = a - bi.}$$

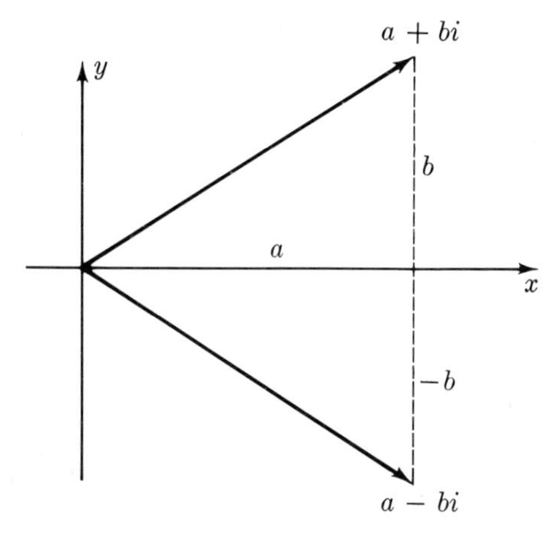

FIG. 2.2

Set $z = a + bi$; the following basic relations hold:

$$\begin{array}{|ll|} z + \bar{z} = 2a = 2\,\mathrm{Re}(z), & z - \bar{z} = 2bi = 2i\,\mathrm{Im}(z), \\ \bar{\bar{z}} = z, & z\bar{z} = a^2 + b^2. \end{array}$$

By the quadratic formula, if the roots of a quadratic equation with real coefficients are not real, then they are conjugate complex numbers.

The **modulus** or **absolute value** of a complex number $a + bi$ is

$$\boxed{|a + bi| = \sqrt{a^2 + b^2}.}$$

It is the length of the vector $a + bi$. Notice that

$$|z|^2 = z\bar{z}.$$

If $z \neq 0$, then $|z| > 0$.

The absolute value of a complex number is a measure of its size. Since complex numbers fill the plane, you cannot say that one complex number is greater than or less than another; the statements

$$i < 2, \qquad 2 + i < 4 - 3i$$

make no sense. Yet you can compare absolute values; the statements

$$|i| < |2|, \qquad |2 + i| < |4 - 3i|$$

do make sense. The latter, for example, says that the *length* of the vector $2 + i$ is less than the *length* of the vector $4 - 3i$. See Fig. 2.3.

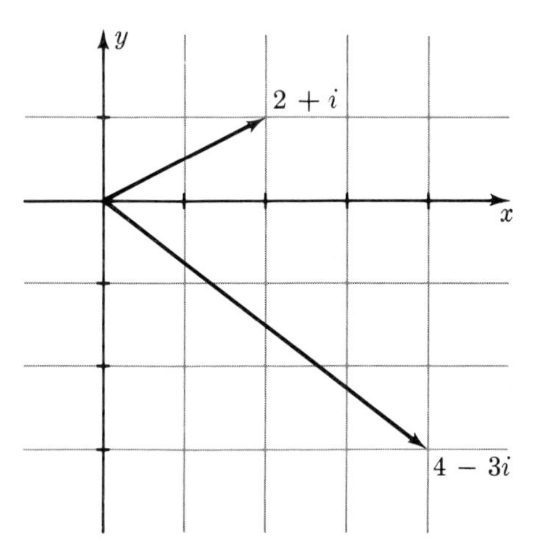

FIG. 2.3

All complex numbers of the same absolute value determine a circle (Fig. 2.4).

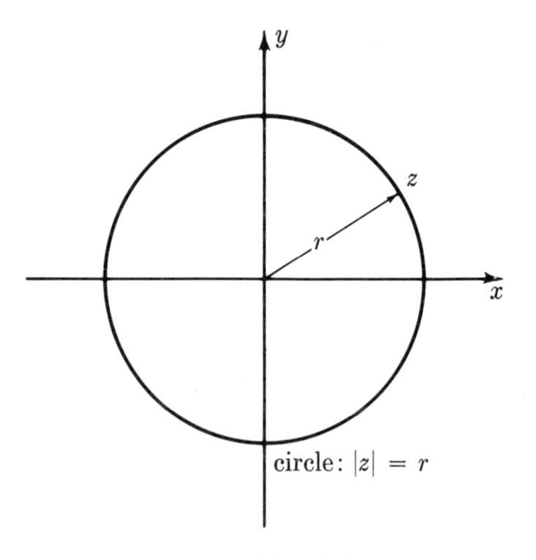

FIG. 2.4

EXAMPLE 2.1

Express $\dfrac{1 - i}{3 + 2i}$ in the form $a + bi$.

Solution: Here is a standard trick. Multiply numerator and denominator by $3 - 2i$, the conjugate of the denominator:

$$\frac{1-i}{3+2i} = \frac{1-i}{3+2i}\cdot\frac{3-2i}{3-2i} = \frac{(3-2)-(3+2)i}{3^2+2^2} = \frac{1-5i}{13}.$$

Answer: $\frac{1}{13} - \frac{5}{13}i.$

EXAMPLE 2.2

If $|z| = 1$, show that $\frac{1}{z} = \bar{z}$.

Solution: Let $z = a + bi$. Then by the trick of the last example,

$$\frac{1}{z} = \frac{1}{a+bi} = \frac{1}{a+bi}\cdot\frac{a-bi}{a-bi} = \frac{a-bi}{a^2+b^2}.$$

But $a^2 + b^2 = |z|^2 = 1$. Hence,

$$\frac{1}{z} = a - bi = \bar{z}.$$

Shorter Solution:

$$z\bar{z} = |z|^2 = 1, \quad \bar{z} = \frac{1}{z}.$$

EXERCISES

Express $\bar{z}$ in the form $a + bi$:

1. $z = \frac{1}{2} - \frac{1}{3}i$
2. $z = 0.4 + 1.7i$
3. $z = \overline{(2 - i)}$
4. $z = (2 + i)^{-1}$
5. $z = (-1 - i)(-2 + 3i)$
6. $z = (1 + i)(2 + i)(3 + i)$
7. $z = \dfrac{i}{1 - i}$
8. $z = \dfrac{1 - i}{i}$
9. $z = \dfrac{2 + i}{2 - i}$
10. $z = \dfrac{2 + 3i}{-1 + i}$.
11. Find $|z|$ in Ex. 1–10.
12. Show that $\overline{zw} = \bar{z}\bar{w}$.
13. Show that $|zw| = |z|\,|w|$.
14. Show that $|\bar{z}| = |z|$.
15. Show that the equation $\bar{z} = z$ is satisfied only by real numbers.
16. Let $f(x) = a_0x^3 + a_1x^2 + a_2x + a_3$, where the coefficients are real numbers. Let z be a complex zero of $f(x)$. Prove that $\bar{z}$ is also a zero. (*Hint:* $\bar{0} = 0$.)

17. Do the same for an n-th degree polynomial with real coefficients.
18. Show that if $(a + bi)^4 = 1$, then $a + bi$ is one of the numbers ± 1, $\pm i$.

3. POLAR FORM

When a complex number z is written $z = a + bi$, it is said to be in **rectangular form.** If $z \neq 0$, it is sometimes convenient to express z in **polar form:**

$$z = r(\cos\theta + i\sin\theta), \qquad r > 0.$$

As is seen in Fig. 3.1,

$$r = |z| = \sqrt{a^2 + b^2}; \qquad \cos\theta = \frac{a}{r}, \qquad \sin\theta = \frac{b}{r}.$$

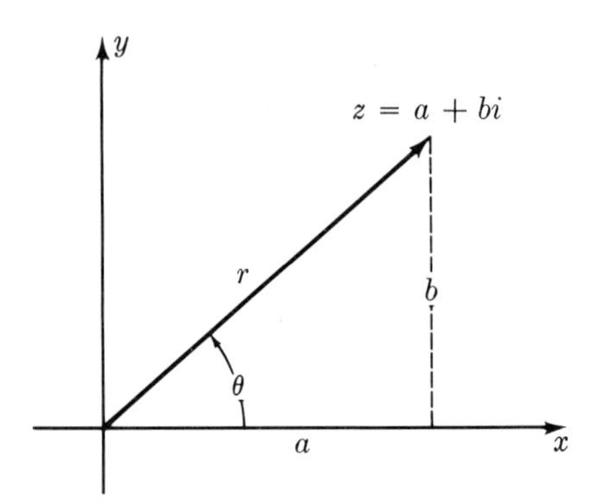

FIG. 3.1

The complex number $\cos\theta + i\sin\theta$ is a unit vector making angle θ with the positive x-axis. Hence, the polar form $z = r(\cos\theta + i\sin\theta)$ expresses the vector z as a unit vector in the same direction stretched by a factor $|z|$. The angle θ is called the **argument** of z and is written arg z. It is determined only up to a multiple of 2π. For example, $\arg(1 + i)$ can be taken to be $\pi/4$ or $9\pi/4$ or $-15\pi/4$, etc. If $z = r(\cos\theta + i\sin\theta)$, then

$$\bar{z} = r[\cos(-\theta) + i\sin(-\theta)] = r(\cos\theta - i\sin\theta).$$

See Fig. 3.2.

Polar form is particularly useful in situations involving multiplication (or division) of complex numbers; it makes multiplication easy. Suppose

$$z_1 = r_1(\cos\theta_1 + i\sin\theta_1), \qquad z_2 = r_2(\cos\theta_2 + i\sin\theta_2).$$

Then

$$\begin{aligned} z_1z_2 &= r_1r_2[(\cos\theta_1\cos\theta_2 - \sin\theta_1\sin\theta_2) + i(\sin\theta_1\cos\theta_2 + \cos\theta_1\sin\theta_2)] \\ &= r_1r_2[\cos(\theta_1 + \theta_2) + i\sin(\theta_1 + \theta_2)]. \end{aligned}$$

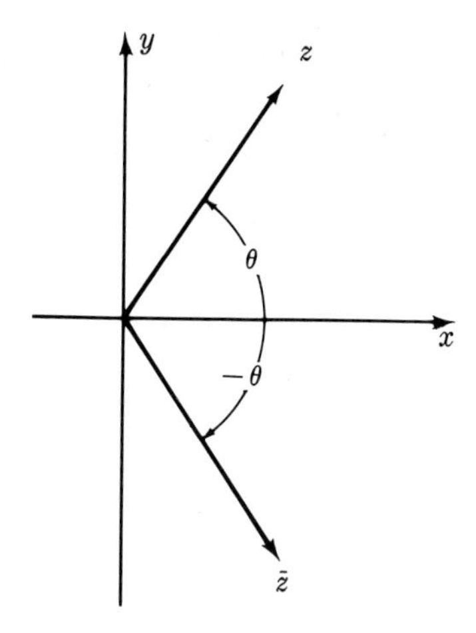

FIG. 3.2

The product is again in polar form; its modulus is r_1r_2 and its argument is $\theta_1 + \theta_2$.

> To multiply two complex numbers, multiply their absolute values and add their arguments:
>
> $$|z_1z_2| = |z_1|\,|z_2|, \qquad \arg z_1z_2 = \arg z_1 + \arg z_2 .$$
>
> Similarly, to divide two complex numbers, divide their absolute values and subtract their arguments:
>
> $$\left|\frac{z_1}{z_2}\right| = \frac{|z_1|}{|z_2|}, \qquad \arg\frac{z_1}{z_2} = \arg z_1 - \arg z_2 .$$

From this rule follows De Moivre's Theorem, a formula for powers of a complex number.

> **De Moivre's Theorem** For each positive integer n,
>
> $$[r(\cos\theta + i\sin\theta)]^n = r^n(\cos n\theta + i\sin n\theta).$$

EXAMPLE 3.1

Describe geometrically the effect of multiplying a complex number by i, by $1 - i$.

Solution: Write i and $1 - i$ in polar form:

$$i = 1\left[\cos\frac{\pi}{2} + i\sin\frac{\pi}{2}\right], \qquad 1 - i = \sqrt{2}\left[\cos\left(-\frac{\pi}{4}\right) + i\sin\left(-\frac{\pi}{4}\right)\right].$$

According to the rule for multiplication,

$$|iz| = |i|\,|z| = |z|, \qquad \arg iz = \arg z + \arg i = \arg z + \frac{\pi}{2}\cdot$$

Therefore multiplying the vector z by i simply rotates the vector 90° counterclockwise (Fig. 3.3).

According to the rule,

$$|(1 - i)z| = \sqrt{2}\,|z|, \qquad \arg(1 - i)z = \arg z - \frac{\pi}{4}\cdot$$

Therefore multiplying the vector z by $1 - i$ rotates the vector 45° clockwise and stretches it by a factor $\sqrt{2}$. See Fig. 3.3.

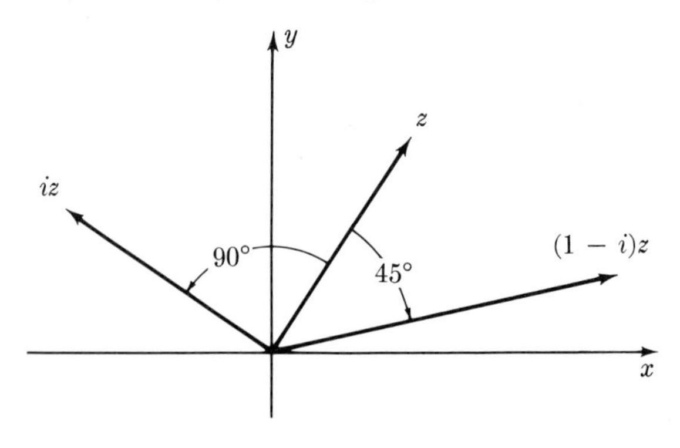

FIG. 3.3

Answer: rotation by 90°;
rotation by −45° and stretching by a factor $\sqrt{2}$.

EXAMPLE 3.2

Compute $(1 + i)^8$.

Solution: By the Binomial Theorem,

$$\begin{aligned}(1 + i)^8 &= 1 + 8i + 28i^2 + 56i^3 + 70i^4 + 56i^5 + 28i^6 + 8i^7 + i^8 \\ &= 1 + 8i - 28 - 56i + 70 + 56i - 28 - 8i + 1 = 16.\end{aligned}$$

Alternate Solution: Write $1 + i$ in polar form, then use De Moivre's Theorem:

$$1 + i = \sqrt{2}\left(\cos\frac{\pi}{4} + i\sin\frac{\pi}{4}\right),$$

$$(1 + i)^8 = (\sqrt{2})^8\left(\cos\frac{8\pi}{4} + i\sin\frac{8\pi}{4}\right) = 16(1 + 0\cdot i) = 16.$$

Answer: 16.

The following example illustrates a powerful technique for deriving trigonometric identities.

EXAMPLE 3.3

Derive the identities

$$\cos 3\theta = \cos^3\theta - 3\cos\theta\sin^2\theta,$$

$$\sin 3\theta = 3\cos^2\theta\sin\theta - \sin^3\theta.$$

Solution: Compute $(\cos\theta + i\sin\theta)^3$ two ways, by De Moivre's Theorem and by the Binomial Theorem. The results must be equal:

$$\begin{aligned}\cos 3\theta + i\sin 3\theta &= \cos^3\theta + 3\cos^2\theta\,(i\sin\theta) + 3\cos\theta\,(i\sin\theta)^2 + (i\sin\theta)^3\\ &= (\cos^3\theta - 3\cos\theta\sin^2\theta) + i(3\cos^2\theta\sin\theta - \sin^3\theta).\end{aligned}$$

Now equate real and imaginary parts on both sides of this equation.

Roots of Unity

The equation

$$z^4 = 1$$

has four complex roots, ± 1 and $\pm i$. Thus the number 1 has four 4-th roots, which are complex numbers equally spaced around the circle $|z| = 1$.

The situation for n-th roots is similar. Write the equation $z^n = 1$ in polar form, setting $z = r(\cos\theta + i\sin\theta)$:

$$[r(\cos\theta + i\sin\theta)]^n = 1(\cos 0 + i\sin 0), \qquad r > 0.$$

By De Moivre's Theorem,

$$r^n(\cos n\theta + i\sin n\theta) = 1(\cos 0 + i\sin 0).$$

Consequently,

$$r^n = 1, \qquad \cos n\theta = 1, \qquad \sin n\theta = 0.$$

It follows that

$$r = 1, \qquad n\theta = 2\pi k, \qquad \theta = \frac{2\pi k}{n}, \qquad k \text{ an integer.}$$

Thus

$$z = \cos\frac{2\pi k}{n} + i\sin\frac{2\pi k}{n},$$

where k is an integer. This formula yields exactly n distinct values: for $k = 0, 1, 2, \cdots, n - 1$. (Why?) For $k = 1$, call the root

$$\omega = \cos\frac{2\pi}{n} + i\sin\frac{2\pi}{n}.$$

Since (De Moivre again)

$$\omega^k = \cos\frac{2\pi k}{n} + i\sin\frac{2\pi k}{n},$$

the other roots are the powers of ω, namely $1 = \omega^0, \omega, \omega^2, \cdots, \omega^{n-1}$.

The equation

$$z^n = 1$$

has exactly n complex roots $1, \omega, \omega^2, \cdots, \omega^{n-1}$, where

$$\omega = \cos\frac{2\pi}{n} + i\sin\frac{2\pi}{n}.$$

These numbers are called n-th **roots of unity.** They lie equally spaced around the circle $|z| = 1$. See Fig. 3.4.

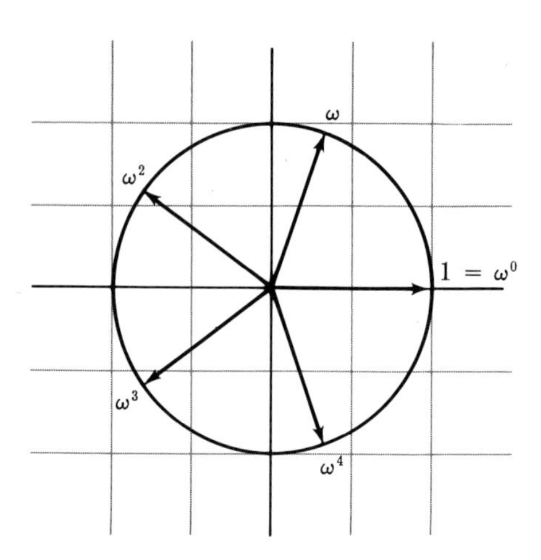

FIG. 3.4 5-th roots of unity.

EXAMPLE 3.4

Find the cube roots of 1.

Solution: Use the formulas given above with $n = 3$. The cube roots are

$$\omega^0 = 1,$$

$$\omega = \cos\frac{2\pi}{3} + i\sin\frac{2\pi}{3},$$

$$\omega^2 = \cos\frac{4\pi}{3} + i\sin\frac{4\pi}{3}.$$

Answer: $1, \ -\frac{1}{2} + \frac{\sqrt{3}}{2}\, i, \ -\frac{1}{2} - \frac{\sqrt{3}}{2}\, i.$

As a check, cube the complex numbers in the answer by the Binomial Theorem.

Not only the number 1, but any (non-zero) complex number α has complex n-th roots, exactly n of them.

Each non-zero complex number α has exactly n complex n-th roots. If

$$\alpha = r(\cos\theta + i \sin\theta),$$

set

$$\beta = r^{1/n}\left(\cos\frac{\theta}{n} + i\sin\frac{\theta}{n}\right).$$

Then the n-th roots of α are

$$\beta, \ \beta\omega, \ \beta\omega^2, \ \cdots, \ \beta\omega^{n-1},$$

where $1, \omega, \cdots, \omega^{n-1}$ are the n-th roots of unity. In polar form,

$$\beta\omega^k = r^{1/n}\left[\cos\left(\frac{\theta}{n} + \frac{2\pi k}{n}\right) + i\sin\left(\frac{\theta}{n} + \frac{2\pi k}{n}\right)\right].$$

It follows that the n-th roots of α are equally spaced points on the circle of radius $|\alpha|^{1/n}$ centered at the origin.

The assertion is easily verified. By De Moivre's Theorem

$$\beta^n = r(\cos\theta + i\sin\theta) = \alpha.$$

Hence β is an n-th root of α. Furthermore,

$$(\beta\omega^k)^n = \beta^n \cdot \omega^{kn} = \alpha \cdot 1,$$

so $\beta, \beta\omega, \beta\omega^2, \cdots, \beta\omega^{n-1}$ are n-th roots of α. There are no others; for if γ is an n-th root of α, then

$$\left(\frac{\gamma}{\beta}\right)^n = \frac{\alpha}{\alpha} = 1.$$

Hence γ/β is an n-th root of unity; $\gamma/\beta = \omega^k$ for some k. Therefore $\gamma = \beta\omega^k$.

In practice, we compute n-th roots from the above formula for $\beta\omega^k$, not by actually multiplying β and ω^k.

EXAMPLE 3.5

Find the cube roots of $\frac{27}{2}(1 + i\sqrt{3})$.

Solution: The polar form of this number is

$$27(\cos 60° + i \sin 60°).$$

Apply the above formula for n-th roots with $n = 3$ and $\theta = 60°$. The three cube roots are

$$\beta = 3(\cos 20° + i \sin 20°),$$
$$\beta\omega = 3[\cos(20 + 120)° + i \sin(20 + 120)°],$$
$$\beta\omega^2 = 3[\cos(20 + 240)° + i \sin(20 + 240)°].$$

Answer: $3(\cos 20° + i \sin 20°)$,
$3(\cos 140° + i \sin 140°)$,
$3(\cos 260° + i \sin 260°)$.
See Fig. 3.5.

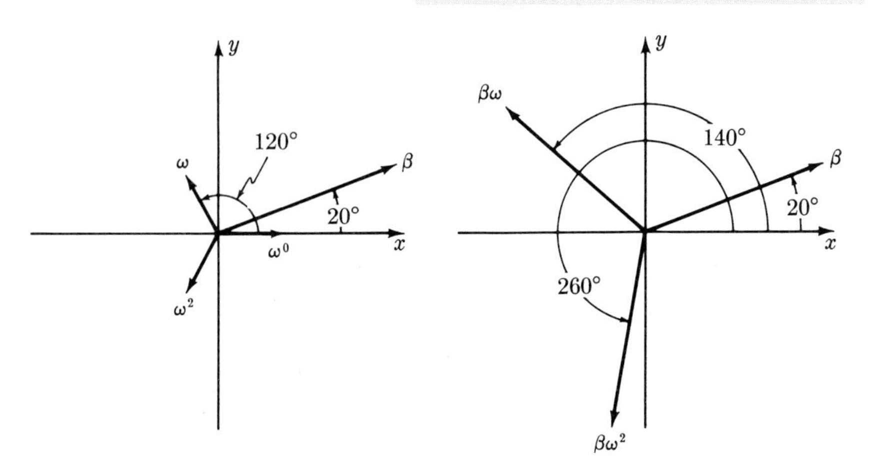

Fig. 3.5

EXERCISES

Express in polar form:

1. $-1 + i$
2. $-1 - i$
3. $1 + i\sqrt{3}$
4. $3 - 2i$
5. -4
6. $-3i$
7. $1 + 4i$
8. $4 - 3i.$

Multiply and express the answer in rectangular form:

9. $\left(\cos\frac{\pi}{6} + i \sin\frac{\pi}{6}\right)\left(\cos\frac{\pi}{3} + i \sin\frac{\pi}{3}\right)$

10. $(\cos 17° + i \sin 17°)(\cos 208° + i \sin 208°)$

11. $(\cos 135° + i \sin 135°)(1 + i)$

12. $(\cos 17° - i \sin 17°)(\cos 197° + i \sin 197°)$.

Express in polar form:

13. $\dfrac{1 + i}{1 - i}$

14. $\dfrac{1 - i}{-1 + i\sqrt{3}}$

15. $\dfrac{i\sqrt{3}}{-1 - i\sqrt{3}}$

16. $\dfrac{5}{\cos \dfrac{\pi}{5} + i \sin \dfrac{\pi}{5}}$.

Compute by De Moivre's Theorem:

17. $(1 - i)^4$

18. $(-1 + i)^6$

19. $(\sqrt{3} - i)^7$

20. $\left[\dfrac{1 + i\sqrt{3}}{2}\right]^{11}$

21. $\dfrac{1}{(1 + i)^3}$

22. $\left[\dfrac{1 + i}{\sqrt{2}}\right]^{1000}$.

23. Verify De Moivre's Theorem for $n = 2, 3, 4, 5$.

24. Find the 5-th roots of unity.

25. Find the 8-th roots of unity.

26. Find the 5-th roots of $(1 + i)$.

27. Find the 6-th roots of -1.

28. Find the 7-th roots of $-i$.

29. Find the 10-th roots of i.

30. Examine the 8-th roots of unity, $1, \omega, \omega^2, \cdots$. One root has the value 1, another (which?) is a square root of unity, and two others (which?) are 4-th roots of unity. The remaining four (which?) are called **primitive** 8-th roots of unity.

31. (cont.) Similarly analyze 6-th roots, 9-th roots, and 10-th roots of unity. Which are the primitive ones?

32. Show that each 5-th root of unity other than 1 satisfies $x^4 + x^3 + x^2 + x + 1 = 0$. (*Hint:* Factor $x^5 - 1$.)

33. Factor $x^6 - 1$. Show that each primitive 6-th root of unity satisfies $x^2 - x + 1 = 0$.

34. Set $\alpha = \dfrac{-1 \pm \sqrt{5}}{2}$ and $\omega = \dfrac{\alpha \pm \sqrt{\alpha^2 - 4}}{2}\cdot$ Show that $\omega^5 = 1$. (*Hint:* Use Ex. 32.)

35. Verify that De Moivre's Theorem holds also for exponents which are negative integers.

36. Let $1, \omega, \omega^2, \cdots, \omega^{n-1}$ be the n-th roots of unity. Show that $1 + \omega + \omega^2 + \cdots + \omega^{n-1} = 0$.

37. (cont.) Show that

$$1 + \cos\frac{2\pi}{n} + \cos\frac{4\pi}{n} + \cdots + \cos\frac{2(n-1)\pi}{n} = 0,$$

$$\sin\frac{2\pi}{n} + \sin\frac{4\pi}{n} + \cdots + \sin\frac{2(n-1)\pi}{n} = 0.$$

38. Prove that $\cos 3\theta = 4\cos^3\theta - 3\cos\theta$. (*Hint:* See Example 3.3.)

39. Express $\cos 4\theta$ in terms of $\cos\theta$.

4. COMPLEX EXPONENTIALS

Let us try to give a meaning to e^z for complex numbers z. The exponential series

$$e^x = 1 + x + \frac{x^2}{2!} + \cdots + \frac{x^n}{n!} + \cdots$$

converges for all real values of x. We now boldly substitute iy for x:

$$\begin{aligned} e^{iy} &= 1 + iy + \frac{(iy)^2}{2!} + \frac{(iy)^3}{3!} + \cdots \\ &= 1 + iy - \frac{y^2}{2!} - \frac{iy^3}{3!} + \frac{y^4}{4!} + \frac{iy^5}{5!} - \cdots \\ &= \left(1 - \frac{y^2}{2!} + \frac{y^4}{4!} - \frac{y^6}{6!} + \cdots\right) + i\left(y - \frac{y^3}{3!} + \frac{y^5}{5!} - \cdots\right). \end{aligned}$$

The quantities in parentheses are familiar; they are the series for $\cos y$ and $\sin y$. These observations suggest the following definition:

> For any complex number of the form iy with y real, define
>
> $$e^{iy} = \cos y + i\sin y.$$
>
> For a general complex number $z = x + iy$, define
>
> $$e^z = e^x \cdot e^{iy} = e^x(\cos y + i\sin y).$$

If this is to be a reasonable definition of e^z, the usual rules of exponents should hold. Let us verify that $e^{z_1} \cdot e^{z_2} = e^{z_1+z_2}$. Suppose $z_1 = x_1 + iy_1$ and $z_2 = x_2 + iy_2$. Then

$$\begin{aligned} e^{z_1} \cdot e^{z_2} &= e^{x_1}(\cos y_1 + i\sin y_1) \cdot e^{x_2}(\cos y_2 + i\sin y_2) \\ &= e^{x_1} \cdot e^{x_2}(\cos y_1 + i\sin y_1)(\cos y_2 + i\sin y_2) \\ &= e^{x_1+x_2}[\cos(y_1 + y_2) + i\sin(y_1 + y_2)] \\ &= e^{x_1+x_2+i(y_1+y_2)} = e^{(x_1+iy_1)+(x_2+iy_2)} = e^{z_1+z_2}, \end{aligned}$$

which completes the verification.

We observe that

$$e^{-iy} = \cos(-y) + i\sin(-y) = \cos y - i\sin y.$$

Hence e^{-iy} is both the reciprocal and the conjugate of e^{iy}. This is not surprising since $|e^{iy}| = 1$, and, as shown in Example 2.2, if $|z| = 1$, then $1/z = \bar{z}$.

EXAMPLE 4.1

Find the value of:

(1) e^{i}, (2) $e^{2\pi i}$, (3) $e^{\pi i}$, (4) $e^{\ln 2+(\pi i/4)}$, (5) $e^{2\pi i/n}$.

Solution: In each case, use the definition

$$e^{x+iy} = e^{x}(\cos y + i\sin y).$$

(1) $e^{i} = \cos 1 + i\sin 1 \quad (1 \text{ rad}).$

(2) $e^{2\pi i} = \cos 2\pi + i\sin 2\pi = 1.$

(3) $e^{\pi i} = \cos\pi + i\sin\pi = -1.$

(4) $e^{\ln 2+\pi i/4} = e^{\ln 2}\left(\cos\frac{\pi}{4} + i\sin\frac{\pi}{4}\right) = 2\left(\frac{\sqrt{2}}{2} + i\frac{\sqrt{2}}{2}\right) = \sqrt{2}\,(1+i).$

(5) $e^{2\pi i/n} = \cos\frac{2\pi}{n} + i\sin\frac{2\pi}{n}\cdot$

REMARK: From the answer to (5) we recognize that $e^{2\pi i/n}$ is an n-th root of unity. This agrees with the answer to (2) since

$$(e^{2\pi i/n})^{n} = e^{2\pi i} = 1.$$

Trigonometric Functions

From the basic relations

$$e^{i\theta} = \cos\theta + i\sin\theta,$$
$$e^{-i\theta} = \cos\theta - i\sin\theta,$$

follow two important formulas:

$$\cos\theta = \frac{e^{i\theta} + e^{-i\theta}}{2} = \operatorname{Re}(e^{i\theta}),$$

$$\sin\theta = \frac{e^{i\theta} - e^{-i\theta}}{2i} = \operatorname{Im}(e^{i\theta}).$$

These formulas are extremely useful and are worth memorizing. They convert problems about trigonometric functions into problems about exponentials which are often simpler to handle.

EXAMPLE 4.2

Derive the identity

$$1 + 2\cos\theta + 2\cos 2\theta + \cdots + 2\cos n\theta = \frac{\sin\left(n + \dfrac{1}{2}\right)\theta}{\sin\dfrac{1}{2}\theta}.$$

Solution: Let C_n denote the left-hand side of the identity,

$$\begin{aligned} C_n &= 1 + 2\cos\theta + 2\cos 2\theta + \cdots + 2\cos n\theta \\ &= 1 + (e^{i\theta} + e^{-i\theta}) + (e^{2i\theta} + e^{-2i\theta}) + \cdots + (e^{ni\theta} + e^{-ni\theta}). \end{aligned}$$

Rearrange:

$$\begin{aligned} C_n &= e^{-ni\theta} + e^{-(n-1)i\theta} + \cdots + e^{-i\theta} + 1 + e^{i\theta} + \cdots + e^{ni\theta} \\ &= e^{-ni\theta}(1 + e^{i\theta} + e^{2i\theta} + \cdots + e^{2ni\theta}). \end{aligned}$$

The expression in parentheses is a finite geometric series whose sum is known:

$$C_n = e^{-ni\theta} \cdot \frac{1 - e^{(2n+1)i\theta}}{1 - e^{i\theta}} = \frac{e^{-ni\theta} - e^{(n+1)i\theta}}{1 - e^{i\theta}}.$$

Here is an important trick. Remembering that $e^{i\alpha} - e^{-i\alpha} = 2i\sin\alpha$, multiply numerator and denominator by $e^{-i\theta/2}$:

$$C_n = \frac{e^{-(n+\frac{1}{2})i\theta} - e^{(n+\frac{1}{2})i\theta}}{e^{-i\theta/2} - e^{i\theta/2}} = \frac{-2i\sin\left(n + \dfrac{1}{2}\right)\theta}{-2i\sin\dfrac{\theta}{2}} = \frac{\sin\left(n + \dfrac{1}{2}\right)\theta}{\sin\dfrac{\theta}{2}}.$$

Hyperbolic Functions

There is a close connection between trigonometric and hyperbolic functions. Recall the definitions

$$\cosh x = \frac{e^x + e^{-x}}{2}, \qquad \sinh x = \frac{e^x - e^{-x}}{2}.$$

Formally replacing x by iy, we find

$$\cosh iy = \cos y, \qquad \sinh iy = i\sin y.$$

In order to give meaning to these identities, we define $\sin z$, $\cos z$, $\sinh z$, and $\cosh z$ for complex numbers z.

If z is a complex number,

$$\cos z = \frac{e^{iz} + e^{-iz}}{2}, \qquad \sin z = \frac{e^{iz} - e^{-iz}}{2i}.$$

Similarly,

$$\cosh z = \frac{e^{z} + e^{-z}}{2}, \qquad \sinh z = \frac{e^{z} - e^{-z}}{2}.$$

Thus $\cos z$ and $\cosh z$ are closely related functions, as are $\sin z$ and $\sinh z$. Directly from the definitions, one obtains the basic relations:

$$\cos z = \cosh iz, \qquad \sin z = \frac{1}{i} \sinh iz = -i \sinh iz,$$

valid for all complex numbers z. Equivalent relations are obtained on replacing z by iw:

$$\cosh w = \cos iw, \qquad i \sinh w = \sin iw,$$

valid for all complex numbers w.

It is proved in more advanced courses that each identity involving sines and cosines, valid for all real values of the variables, is valid also for all complex values of the variables. (For example, the identity

$$\cos(z + w) = \cos z \cos w - \sin z \sin w$$

holds for all complex values of z and w as well as for all real values.) In particular, in any identity involving sines and cosines of variables $z, w, \cdots$, the variables may be replaced by $iz, iw, \cdots$. It follows that each identity involving sines and cosines has a counterpart involving hyperbolic functions.

EXAMPLE 4.3

Show that $\cosh(z + w) = \cosh z \cosh w + \sinh z \sinh w$.

Solution:

$$\begin{aligned} \cosh(z + w) &= \cos i(z + w) = \cos(iz + iw) \\ &= \cos iz \cos iw - \sin iz \sin iw \\ &= \cosh z \cosh w - (i \sinh z)(i \sinh w) \\ &= \cosh z \cosh w + \sinh z \sinh w. \end{aligned}$$

EXAMPLE 4.4

Find a formula for $1 + 2\cosh z + 2\cosh 2z + \cdots + 2\cosh nz$.

Solution: Use the corresponding formula for cosines found in Example 4.2:

$$1 + 2\cosh z + \cdots + 2\cosh nz = 1 + 2\cos iz + \cdots + 2\cos niz$$
$$= \frac{\sin i\left(n + \frac{1}{2}\right)z}{\sin \frac{iz}{2}}.$$

But

$$\frac{\sin i\left(n + \frac{1}{2}\right)z}{\sin \frac{iz}{2}} = \frac{i\sinh\left(n + \frac{1}{2}\right)z}{i\sinh \frac{z}{2}}.$$

Answer:

$$1 + 2\cosh z + 2\cosh 2z + \cdots + 2\cosh nz$$
$$= \frac{\sinh\left(n + \frac{1}{2}\right)z}{\sinh \frac{z}{2}}.$$

EXERCISES

1. Evaluate
   (a) $e^{\pi i/3}$ (b) e^{1-i}
   (c) $e^{1/(1-2i)}$ (d) $\sin\left(\frac{\pi}{4} + 2i\right)$
   (e) $\cos 3i$ (f) $\cosh \pi i$.
2. Show that

$$\sin(x + iy) = \sin x \cosh y + i \sinh y \cos x,$$
$$\cos(x + iy) = \cos x \cosh y - i \sin x \sinh y.$$

3. Find all complex numbers z for which
   (a) $\sin z$ is real (b) $\cos z = 0$.
4. Show that for all complex numbers z,
   (a) $e^{z+2\pi i} = e^z$ (b) $\sin(z + 2\pi) = \sin z$
   (c) $\cos(z + 2\pi) = \cos z$ (d) $\sinh(z + 2\pi i) = \sinh z$
   (e) $\cosh(z + 2\pi i) = \cosh z$.
5. Is $|\cos z| \leq 1$ for all complex numbers z?

6. Derive the trigonometric identities:

(a) $\cos^4 x = \dfrac{1}{2^4}[\cos 4x + 4\cos 2x + 6 + 4\cos(-2x) + \cos(-4x)]$

(b) $\cos^6 x = \dfrac{1}{2^6}[\cos 6x + 6\cos 4x + 15\cos 2x + 20 + 15\cos(-2x) + 6\cos(-4x) + \cos(-6x)].$

(*Hint:* Write $\cos x = \frac{1}{2}(e^{ix} + e^{-ix})$ and use the Binomial Theorem.) Guess a general formula for $\cos^{2n} x$.

7. Derive the identity

$$1 + 2\cosh z + 2\cosh 2z + \cdots + 2\cosh nz = \frac{\sinh\left(n + \frac{1}{2}\right)z}{\sinh \frac{z}{2}}$$

by converting the left-hand side into a sum of exponentials.

8. Define $\tanh z = \dfrac{\sinh z}{\cosh z}$, $\operatorname{sech} z = \dfrac{1}{\cosh z}$. Find a relation between $\tanh^2 z$ and $\operatorname{sech}^2 z$.

9. Express $\cosh 4z$ in terms of $\cosh z$.
(*Hint:* Express $\cos 4z$ in terms of $\cos z$.)

10. Derive the hyperbolic identities:
(a) $\sinh 2z = 2\sinh z \cosh z$ (b) $\cosh 2z = 2\cosh^2 z - 1$
(c) $\sinh 3z = 4\sinh^3 z + 3\sinh z$.

5. INTEGRATION AND DIFFERENTIATION

In this section we deal with functions having complex values, for example,

$$f(x) = e^{ix} = \cos x + i \sin x.$$

Such a function can be written as

$$f(x) = u(x) + i\, v(x),$$

where $u(x)$ and $v(x)$ are real-valued functions.

> If $f(x) = u(x) + i\, v(x)$, then the derivative of $f(x)$ is defined by
>
> $$f'(x) = u'(x) + i\, v'(x).$$
>
> Similarly,
>
> $$\int f(x)\, dx = \int u(x)\, dx + i \int v(x)\, dx.$$

Many formulas for differentiation and integration extend to complex-valued functions. We shall consider just one case, complex exponentials.

If α is a complex number, then

$$\frac{d}{dx}(e^{\alpha x}) = \alpha e^{\alpha x}$$

and

$$\int e^{\alpha x}\,dx = \frac{1}{\alpha}e^{\alpha x} + C, \qquad \alpha \neq 0.$$

Let us verify the first formula; the second formula follows from it. Suppose $\alpha = a + bi$. Then

$$\frac{d}{dx}(e^{\alpha x}) = \frac{d}{dx}[e^{(a+bi)x}] = \frac{d}{dx}[e^{ax}(\cos bx + i\sin bx)]$$

$$= \frac{d}{dx}(e^{ax}\cos bx) + i\frac{d}{dx}(e^{ax}\sin bx).$$

By ordinary differentiation,

$$\frac{d}{dx}(e^{\alpha x}) = (ae^{ax}\cos bx - be^{ax}\sin bx) + i(ae^{ax}\sin bx + be^{ax}\cos bx)$$

$$= ae^{ax}(\cos bx + i\sin bx) + ibe^{ax}(\cos bx + i\sin bx)$$

$$= (a + bi)e^{ax}(\cos bx + i\sin bx) = \alpha e^{\alpha x}.$$

EXAMPLE 5.1

Evaluate $\int e^{ax}\cos bx\,dx$.

Solution: This can be done using integration by parts twice. It is easier, however, using complex exponentials. From

$$e^{(a+bi)x} = e^{ax}\cos bx + ie^{ax}\sin bx$$

follows

$$\int e^{(a+bi)x}\,dx = \int e^{ax}\cos bx\,dx + i\int e^{ax}\sin bx\,dx.$$

Therefore the desired integral is the real part of the integral on the left. By the preceding rule,

$$\int e^{(a+bi)x}\,dx = \frac{e^{(a+bi)x}}{a+bi} + C.$$

To find the real part, write

$$\frac{e^{(a+bi)x}}{a+bi} = \frac{e^{ax}(\cos bx + i\sin bx)}{a+bi}\cdot\frac{a-bi}{a-bi}$$

$$= \frac{e^{ax}[(a\cos bx + b\sin bx) + i(a\sin bx - b\cos bx)]}{a^2+b^2}.$$

Answer: $\dfrac{e^{ax}(a\cos bx + b\sin bx)}{a^2+b^2} + C.$

REMARK: By comparing imaginary parts, we get free of charge the formula

$$\int e^{ax}\sin bx\,dx = \frac{e^{ax}(a\sin bx - b\cos bx)}{a^2+b^2} + C.$$

EXAMPLE 5.2

Compute $\int_0^{2\pi} e^{inx}\,dx$, where n is a non-zero integer.

Solution:

$$\int_0^{2\pi} e^{inx}\,dx = \int_0^{2\pi}\cos nx\,dx + i\int_0^{2\pi}\sin nx\,dx = \frac{\sin nx}{n}\Big|_0^{2\pi} - i\,\frac{\cos nx}{n}\Big|_0^{2\pi} = 0.$$

Alternate Solution: An antiderivative of e^{inx} is $e^{inx}/(in)$. Hence

$$\int_0^{2\pi} e^{inx}\,dx = \frac{1}{in}\,e^{inx}\Big|_0^{2\pi} = \frac{1}{in}\,(e^{2\pi in} - e^0) = \frac{1}{in}\,(1-1) = 0.$$

(Remember, $e^{2\pi i} = 1$.)

Answer: 0.

EXAMPLE 5.3

Compute $\int_0^{2\pi}\sin kx\cos nx\,dx,\qquad k, n$ positive integers.

Solution: Write

$$\int_0^{2\pi}\sin kx\cos nx\,dx = \int_0^{2\pi}\left(\frac{e^{ikx}-e^{-ikx}}{2i}\right)\left(\frac{e^{inx}+e^{-inx}}{2}\right)dx$$

$$= \frac{1}{4i}\left[\int_0^{2\pi} e^{i(k+n)x}\,dx - \int_0^{2\pi} e^{-i(k+n)x}\,dx + \int_0^{2\pi} e^{i(k-n)x}\,dx - \int_0^{2\pi} e^{-i(k-n)x}\,dx\right].$$

Now use the result of the last example. If $k \neq n$, then all four integrals on the right are 0. If $k = n$, the first two integrals are 0, the third and fourth cancel.

Answer: 0.

EXAMPLE 5.4

Evaluate $\int x \cos^3 x \, dx$.

Solution: Write

$$x \cos^3 x = x\left(\frac{e^{ix} + e^{-ix}}{2}\right)^3 = \frac{x}{8}\left(e^{3ix} + 3e^{2ix}e^{-ix} + 3e^{ix}e^{-2ix} + e^{-3ix}\right)$$

$$= \frac{x}{8}\left(e^{3ix} + 3e^{ix} + 3e^{-ix} + e^{-3ix}\right).$$

Hence,

$$\int x \cos^3 x \, dx$$

$$= \frac{1}{8}\left(\int xe^{3ix}\, dx + 3\int xe^{ix}\, dx + 3\int xe^{-ix}\, dx + \int xe^{-3ix}\, dx\right)$$

$$= \frac{1}{8}\left(I_3 + 3I_1 + 3I_{-1} + I_{-3}\right),$$

where

$$I_n = \int xe^{inx}\, dx.$$

To evaluate I_n, integrate by parts:

$$I_n = \int xe^{inx}\, dx = \int x\, d\left(\frac{e^{inx}}{in}\right) = \frac{xe^{inx}}{in} - \int \frac{e^{inx}}{in}\, dx = \frac{xe^{inx}}{in} - \frac{e^{inx}}{(in)^2}$$

$$= e^{inx}\left(\frac{1}{n^2} + \frac{x}{in}\right) = \frac{e^{inx}}{n^2}(1 - inx).$$

Notice that I_n and I_{-n} are conjugates. Hence

$$I_n + I_{-n} = 2\,\mathrm{Re}(I_n) = 2\,\mathrm{Re}\left[\frac{(\cos nx + i \sin nx)\,(1 - inx)}{n^2}\right]$$

$$= \frac{2}{n^2}(\cos nx + nx \sin nx).$$

Now the answer follows easily:

$$\int x \cos^3 x \, dx = \frac{1}{8}(I_3 + I_{-3}) + \frac{3}{8}(I_1 + I_{-1}) + C$$

$$= \frac{1}{8}\cdot\frac{2}{3^2}(\cos 3x + 3x \sin 3x) + \frac{3}{8}\cdot 2(\cos x + x \sin x) + C.$$

Answer:

$$\frac{\cos 3x}{36} + \frac{x \sin 3x}{12} + \frac{3 \cos x}{4} + \frac{3x \sin x}{4} + C.$$

EXERCISES

1. Compute the 12-th derivative of $e^{x\sqrt{3}/2}\left(\cos\frac{x}{2} + i\sin\frac{x}{2}\right)$.

Evaluate the integral using complex exponentials:

2. $\int \cos x \cosh x \, dx$

3. $\int e^x \cos^4 x \, dx$

4. $\int_0^{\pi} \cos^{2n} x \, dx$

5. $\int_0^{2\pi} \sin kx \sin nx \, dx,$
   k, n positive integers

6. $\int_0^{2\pi} \cos kx \cos nx \, dx,$
   k, n positive integers.

7. $\int x \sin^3 x \, dx$

8. $\int x \cos^2 x \, dx$

9. Compute $\int_0^{2\pi} (a_0 + a_1 \cos x + a_2 \cos 2x + \cdots + a_n \cos nx)^2 \, dx.$
   (*Hint:* Use Ex. 6.)

6. APPLICATIONS TO DIFFERENTIAL EQUATIONS

In Chapter 22 we developed a systematic method for solving the linear differential equation

$$ay'' + by' + cy = 0.$$

Now we present another approach, via complex exponentials. The basic idea is this:

> If $y(x) = u(x) + iv(x)$ is a complex-valued function which satisfies the differential equation (with real coefficients)
>
> $$ay'' + by' + cy = 0,$$
>
> then $u(x)$ and $v(x)$ also satisfy the equation. In other words, the real and the imaginary parts of a solution are also solutions.

Let us verify this statement. We are given

$$ay'' + by' + cy = 0.$$

Hence

$$a(u'' + iv'') + b(u' + iv') + c(u + iv) = 0,$$

that is,

$$(au'' + bu' + cu) + i(av'' + bv' + cv) = 0.$$

Since the left-hand side equals 0, so do its real and imaginary parts:

$$au'' + bu' + cu = 0 \qquad \text{and} \qquad av'' + bv' + cv = 0.$$

Thus $u(x)$ and $v(x)$ are solutions.

EXAMPLE 6.1

Solve $y'' + 9y = 0$.

Solution: Try $y = e^{px}$, where p is allowed to be complex:

$$y'' + 9y = p^2e^{px} + 9e^{px} = 0, \qquad p^2 + 9 = 0, \qquad p = \pm 3i.$$

Thus e^{3ix} is a complex solution. Therefore, its real and imaginary parts, $\cos 3x$ and $\sin 3x$, are also solutions. Hence, the general solution, which involves two arbitrary constants, is $c_1 \cos 3x + c_2 \sin 3x$. Note that e^{-3ix} is another complex solution, but its real and imaginary parts are $\cos 3x$ and $-\sin 3x$, nothing new. (The general solution could be written $ae^{3ix} + be^{-3ix}$.)

Answer: $y = c_1 \cos 3x + c_2 \sin 3x.$

EXAMPLE 6.2

Solve $y'' - 5y' + 6y = 0$.

Solution: Try $y = e^{px}$:

$$y'' - 5y' + 6y = p^2e^{px} - 5pe^{px} + 6e^{px} = 0,$$

$$p^2 - 5p + 6 = 0, \qquad p = 2, 3.$$

Hence e^{2x} and e^{3x} are solutions.

Answer: $y = c_1e^{2x} + c_2e^{3x}.$

EXAMPLE 6.3

Solve $y'' + 4y' + 13y = 0$.

Solution: Try $y = e^{px}$:

$$y'' + 4y' + 13y = (p^2 + 4p + 13)e^{px} = 0,$$

$$p^2 + 4p + 13 = 0, \qquad p = -2 \pm 3i.$$

Hence $y = e^{(-2+3i)x}$ is a complex solution. Its real and imaginary parts are $e^{-2x} \cos 3x$ and $e^{-2x} \sin 3x$, also solutions.

Answer: $y = e^{-2x}(c_1 \cos 3x + c_2 \sin 3x).$

Particular Solutions

Complex exponentials can be applied also to non-homogeneous equations

$$ay'' + by' + cy = f(x),$$

where the function $f(x)$ can be expressed in terms of complex exponentials.

EXAMPLE 6.4

Find a particular solution of the differential equation $y'' + 9y = \cos 2x$.

Solution: Since $\cos 2x = \text{Re}(e^{2ix})$, find a complex solution of the differential equation

$$y'' + 9y = e^{2ix},$$

and take its real part. Try $y = ae^{2ix}$:

$$y'' + 9y = (2i)^2 ae^{2ix} + 9ae^{2ix} = -4ae^{2ix} + 9ae^{2ix} = e^{2ix},$$

$$5a = 1.$$

Thus $\frac{1}{5}e^{2ix}$ is a complex solution.

Answer: $\text{Re}\left(\frac{1}{5}e^{2ix}\right) = \frac{1}{5}\cos 2x.$

EXAMPLE 6.5

Find a particular solution of $y'' + 4y' + 13y = \cos x$.

Solution: Since $\cos x = \text{Re}(e^{ix})$, find a complex solution ot

$$y'' + 4y' + 13y = e^{ix},$$

and take its real part. Try $y = ae^{ix}$:

$$y'' + 4y' + 13y = -ae^{ix} + 4aie^{ix} + 13ae^{ix} = e^{ix}.$$

Hence

$$-a + 4ai + 13a = 1, \qquad a = \frac{1}{12 + 4i} = \frac{12 - 4i}{12^2 + 4^2} = \frac{3 - i}{40}.$$

Therefore a complex solution is

$$y = \frac{1}{40}(3 - i)e^{ix} = \frac{1}{40}(3 - i)(\cos x + i \sin x).$$

Its real part is

$$\text{Re}(y) = \frac{1}{40}(3 \cos x + \sin x).$$

Answer: $\frac{1}{40}(3 \cos x + \sin x).$

EXAMPLE 6.6

Find a particular solution of $y'' + 2y' + 3y = e^x \sin 2x$.

Solution: Since $e^x \sin 2x = \operatorname{Im}(e^{(1+2i)x})$, find a complex solution of

$$y'' + 2y' + 3y = e^{(1+2i)x},$$

and take its imaginary part. Try $y = ae^{(1+2i)x}$:

$$y'' + 2y' + 3y = a[(1+2i)^2 + 2(1+2i) + 3]e^{(1+2i)x} = e^{(1+2i)x},$$

$$a[(1 + 4i + 4i^2) + (2 + 4i) + 3] = 1,$$

$$a = \frac{1}{2+8i} = \frac{2-8i}{2^2+8^2} = \frac{1-4i}{34}.$$

Thus a complex solution is

$$y = \frac{1}{34}(1-4i)e^{(1+2i)x} = \frac{1}{34}(1-4i)e^x(\cos 2x + i \sin 2x),$$

and its imaginary part is

$$\operatorname{Im}(y) = \frac{1}{34}e^x(\sin 2x - 4\cos 2x).$$

Answer: $\dfrac{1}{34}e^x(\sin 2x - 4\cos 2x)$.

EXERCISES

Using complex exponentials, find the general real solution:

1. $y''' - y = 0$
2. $y''' + y'' + y' + y = 0$
3. $y^{(4)} - y'' - 6y = 0$
4. $y^{(4)} - y = 0$.
5. Find the general solution of $y'' - 6y' + 10y = \cos 2x$ by the method of this section.

Find a particular solution by the method of this section:

6. $y'' - a^2y = \cos bx$
7. $y'' + a^2y = x \cos bx$
8. $y'' + a^2y = e^{-bx}\cos ax$
9. $y'' + y' - 6y = 3 \sin x$
10. $L\dfrac{d^2I}{dt^2} + R\dfrac{dI}{dt} + \dfrac{1}{C}I = E\cos \omega t$
11. $\dfrac{d^2r}{d\theta^2} - \dfrac{dr}{d\theta} = a \sin \theta$.

7. APPLICATIONS TO POWER SERIES [optional]

Here is a puzzling fact about power series. We know the geometric series $1 + x + x^2 + \cdots$ diverges for $|x| \geq 1$. This is not surprising because it represents $1/(1-x)$, a function which "blows up" at $x = 1$. But the power series for $1/(1+x^2)$ also diverges for $|x| \geq 1$ even though $1/(1+x^2)$ is perfectly well-behaved for all x. What goes wrong?

The answer to this question requires a broader view of power series. When investigating a real series

$$a_0 + a_1(x - x_0) + a_2(x - x_0)^2 + \cdots + a_n(x - x_0)^n + \cdots,$$

it often helps to regard the series as a special case of the complex series

$$a_0 + a_1(z - x_0) + a_2(z - x_0)^2 + \cdots + a_n(z - x_0)^n + \cdots.$$

(The coefficients are the same, but z is allowed to be complex.) For example, the real series

$$\frac{1}{1 + x^2} = 1 - x^2 + x^4 - x^6 + \cdots$$

is a special case of the complex series

$$\frac{1}{1 + z^2} = 1 - z^2 + z^4 - z^6 + \cdots.$$

The *complex* function $1/(1 + z^2)$ is well-behaved for all *real* values of z but "blows up" at $z = i$ and $z = -i$. Hence, its power series cannot be expected to converge for all z. In fact, the series converges only in the disk $|z| < 1$. The same is true of

$$\frac{1}{1 - z} = 1 + z + z^2 + z^3 + \cdots,$$

whose convergence is limited to the disk $|z| < 1$ because of the "bad point" at $z = 1$. In both examples, the series fail to converge for $|z| > 1$ due to troublesome points on the circle $|z| = 1$.

In general, the complex power series

$$a_0 + a_1(z - z_0) + a_2(z - z_0)^2 + \cdots$$

converges to the function $f(z)$ in a region S of the complex plane if

$$|f(z) - [a_0 + a_1(z - z_0) + \cdots a_n(z - z_0)^n]| \longrightarrow 0$$

as $n \longrightarrow \infty$ for each point z of S. Formally, this definition is the same as the corresponding definition for real power series.

The basic fact (stated without proof) about convergence of complex power series is this:

> Given a power series
>
> $$a_0 + a_1(z - z_0) + a_2(z - z_0)^2 + \cdots + a_n(z - z_0)^n + \cdots,$$
>
> precisely one of three cases holds:
>
> (i) The series converges only for $z = z_0$.
>
> (ii) The series converges for all values of z.
>
> (iii) There is a positive number R such that the series converges for each z satisfying $|z - z_0| < R$ and diverges for each z satisfying $|z - z_0| > R$.

In case (iii), the series converges in the circle (Fig. 7.1) with center at z_0 and radius R. The number R is called the **radius of convergence.**

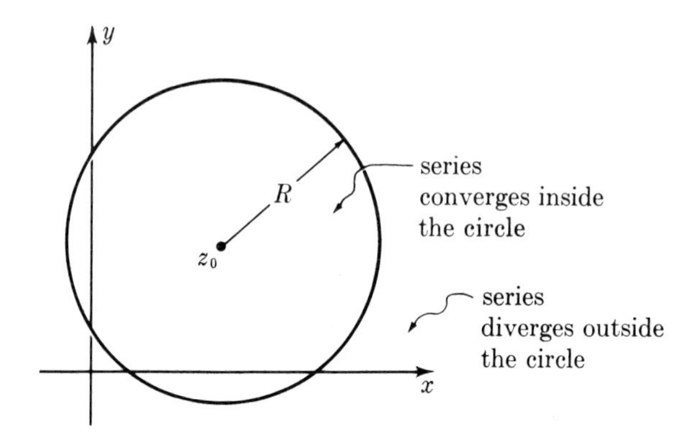

FIG. 7.1

The real power series

$$a_0 + a_1(x - x_0) + \cdots + a_n(x - x_0)^n + \cdots$$

is a special case of the complex power series

$$a_0 + a_1(z - x_0) + \cdots + a_n(z - x_0)^n + \cdots.$$

The latter converges in a circle of radius R centered at x_0 on the real axis (Fig. 7.2). The real series converges on that part of the real axis contained in this circle, namely, an interval (diameter of the circle of convergence). Half of this interval is a radius of the circle; hence the term "radius of convergence" as applied to real series.

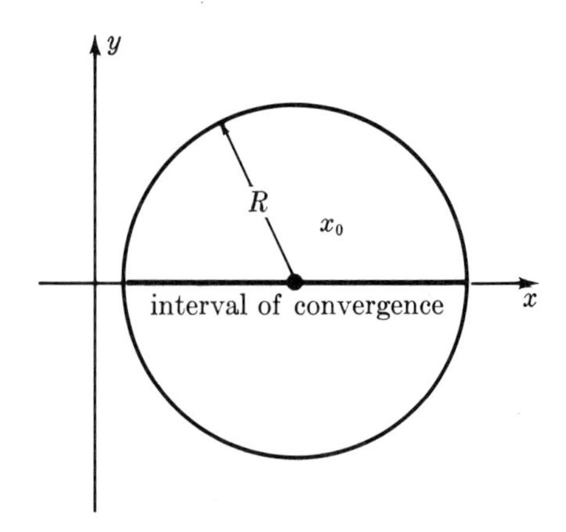

FIG. 7.2

It is shown in more advanced courses that the power series for $f(z)$ at $z = z_0$ converges inside the largest circle centered at z_0 in which $f(z)$ and all its derivatives are well-behaved. For example, take $f(z) = 1/(1 - z)$, which has one bad point, at $z = 1$. Its power series at $z = i$ converges in the circle shown in Fig. 7.3.

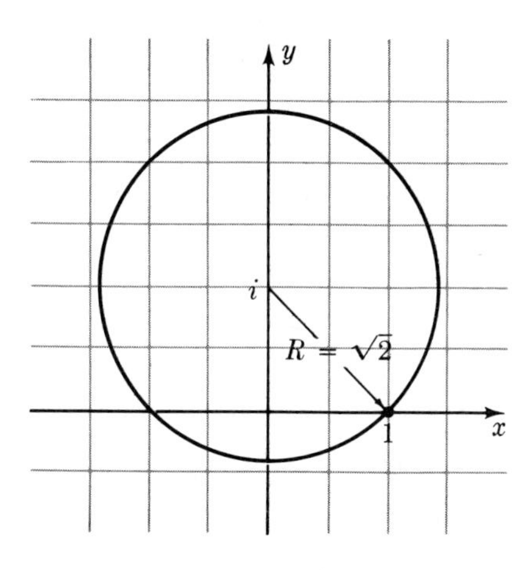

FIG. 7.3

It is known that the ratio test applies to complex power series as well as to real ones.

Partial Fractions

Complex power series have applications to problems involving real functions. An example is the power series expansion by partial fractions (p. 348), where the denominator has complex zeros.

EXAMPLE 7.1

Find the power series for $\dfrac{1}{x^2 - 6x + 10}$ at $x = 0$. What is its radius of convergence?

Solution: First, factor the denominator. Since the equation

$$x^2 - 6x + 10 = 0$$

has roots $3 + i$ and $3 - i$,

$$x^2 - 6x + 10 = [x - (3 + i)]\,[x - (3 - i)].$$

Now use partial fractions. (Use abbreviations $\alpha = 3 + i$ and $\beta = 3 - i$. Note that $\alpha\beta = 10$.)

$$\frac{1}{x^2 - 6x + 10} = \frac{1}{(x-\alpha)(x-\beta)} = \frac{1}{\alpha - \beta}\left(-\frac{1}{x-\beta} + \frac{1}{x-\alpha}\right)$$

$$= \frac{1}{\alpha-\beta}\left(\sum_{n=0}^{\infty}\frac{x^n}{\beta^{n+1}} - \sum_{n=0}^{\infty}\frac{x^n}{\alpha^{n+1}}\right) = \frac{1}{\alpha-\beta}\sum_{n=0}^{\infty}\left(\frac{1}{\beta^{n+1}} - \frac{1}{\alpha^{n+1}}\right)x^n$$

$$= \frac{1}{\alpha-\beta}\sum_{n=0}^{\infty}\left(\frac{\alpha^{n+1}-\beta^{n+1}}{(\alpha\beta)^{n+1}}\right)x^n = \frac{1}{\alpha-\beta}\sum_{n=0}^{\infty}\left(\frac{\alpha^{n+1}-\beta^{n+1}}{10^{n+1}}\right)x^n.$$

This is the answer. It can be written in a slightly different form by observing that

$$\frac{1}{\alpha-\beta} = \frac{1}{2i},$$

and that β^{n+1} is the conjugate of α^{n+1} (since β is the conjugate of α). Thus

$$\frac{\alpha^{n+1}-\beta^{n+1}}{\alpha-\beta} = \frac{1}{2i}\cdot 2i\,\mathrm{Im}(\alpha^{n+1}) = \mathrm{Im}(\alpha^{n+1}).$$

The series converges inside the largest circle centered at the origin in which $1/(x^2 - 6x + 10)$ is well-behaved. This circle passes through $3 + i$ and $3 - i$, the two points where the function "blows up." Its radius is $|3 + i| = |3 - i| = \sqrt{10}$.

Answer: $$\frac{1}{x^2 - 6x + 10} = \sum_{n=0}^{\infty}\frac{1}{10^{n+1}}\mathrm{Im}(\alpha^{n+1})x^n,$$

$\alpha = 3 + i;\quad R = \sqrt{10}.$

REMARK: Let us write out a few terms of the preceding series:

$$\alpha = 3 + i \qquad \mathrm{Im}(\alpha) = 1$$

$$\alpha^2 = (3+i)^2 = 3^2 + 6i + i^2 \qquad \mathrm{Im}(\alpha^2) = 6$$

$$\alpha^3 = (3+i)^3 = 3^3 + 3\cdot 3^2 i + 3\cdot 3i^2 + i^3 \qquad \mathrm{Im}(\alpha^3) = 3\cdot 3^2 - 1 = 26$$

$$\alpha^4 = (3+i)^4 = 3^4 + 4\cdot 3^3 i + 6\cdot 3^2 i^2 + 4\cdot 3i^3 + i^4 \qquad \mathrm{Im}(\alpha^4) = 4\cdot 3^3 - 4\cdot 3 = 96.$$

Hence

$$\frac{1}{x^2 - 6x + 10} = \frac{1}{10} + \frac{6}{10^2}x + \frac{26}{10^3}x^2 + \frac{96}{10^4}x^3 + \cdots.$$

As a quick check, let us try $x = 1$ (permissible since the series converges for $|x| < \sqrt{10}$). The left side is then $\frac{1}{5} = 0.2000$; the sum of the first four terms on the right side is

$$0.1000 + 0.0600 + 0.0260 + 0.0096 = 0.1956.$$

EXERCISES

1. Expand $1/(1 - z)$ in powers of $z - i$.
   [*Hint:* Write $(1 - z) = (1 - i) - (z - i)$.]

Find the power series at $z = 0$ and write out the first four terms explicitly. What is the radius of convergence?

2. $\dfrac{1}{z^2 + z + 1}$
3. $\dfrac{1}{z^2 + 4z + 13}$
4. $\dfrac{z + 1}{z^2 - z + 1}$
5. $\dfrac{3z + 5z^2}{1 - 2z + 5z^2}$.
6. Find the sum of the series $1 + r\cos\theta + r^2\cos 2\theta + \cdots + r^n\cos n\theta + \cdots$, $-1 < r < 1$.
   [*Hint:* $r^n \cos n\theta = \text{Re}(z^n)$, where $z = re^{i\theta}$.]
7. If $z = x + iy$, prove $\dfrac{\partial}{\partial x}(z^n) = nz^{n-1}$, $\quad \dfrac{\partial}{\partial y}(z^n) = niz^{n-1}$.
8. Write $z^n = u(x, y) + i\,v(x, y)$, in real and imaginary parts. Compute u and v for $n = \pm 1, \pm 2$, and 3.
9. (cont.) Show, using Ex. 7, that $\dfrac{\partial u}{\partial x} = \dfrac{\partial v}{\partial y}$, $\quad \dfrac{\partial u}{\partial y} = -\dfrac{\partial v}{\partial x}$.
10. (cont.) Show that $\dfrac{\partial^2 u}{\partial x^2} + \dfrac{\partial^2 u}{\partial y^2} = 0$, $\quad \dfrac{\partial^2 v}{\partial x^2} + \dfrac{\partial^2 v}{\partial y^2} = 0$.
11. (cont.) Let $f(z) = a_0 + a_1 z + a_2 z^2 + \cdots$ be the sum of a complex power series. Separate real and imaginary parts: $f(z) = u(x, y) + i\,v(x, y)$. Show that

$$\frac{\partial u}{\partial x} = \frac{\partial v}{\partial y}, \qquad \frac{\partial u}{\partial y} = -\frac{\partial v}{\partial x},$$

$$\frac{\partial^2 u}{\partial x^2} + \frac{\partial^2 u}{\partial y^2} = 0, \qquad \frac{\partial^2 v}{\partial x^2} + \frac{\partial^2 v}{\partial y^2} = 0.$$

8. MATRIX POWER SERIES [optional]

The material in this section is more advanced. It requires a working knowledge of 2 by 2 matrix algebra. Recall the formulas:

$$c\begin{pmatrix} a_{11} & a_{12} \\ a_{21} & a_{22} \end{pmatrix} = \begin{pmatrix} ca_{11} & ca_{12} \\ ca_{21} & ca_{22} \end{pmatrix},$$

$$\begin{pmatrix} a_{11} & a_{12} \\ a_{21} & a_{22} \end{pmatrix} + \begin{pmatrix} b_{11} & b_{12} \\ b_{21} & b_{22} \end{pmatrix} = \begin{pmatrix} a_{11}+b_{11} & a_{12}+b_{12} \\ a_{21}+b_{21} & a_{22}+b_{22} \end{pmatrix},$$

$$\begin{pmatrix} a_{11} & a_{12} \\ a_{21} & a_{22} \end{pmatrix}\begin{pmatrix} b_{11} & b_{12} \\ b_{21} & b_{22} \end{pmatrix} = \begin{pmatrix} a_{11}b_{11}+a_{12}b_{21} & a_{11}b_{12}+a_{12}b_{22} \\ a_{21}b_{11}+a_{22}b_{21} & a_{21}b_{12}+a_{22}b_{22} \end{pmatrix},$$

$$\operatorname{trace}\begin{pmatrix} a_{11} & a_{12} \\ a_{21} & a_{22} \end{pmatrix} = a_{11} + a_{22}, \qquad \det\begin{pmatrix} a_{11} & a_{12} \\ a_{21} & a_{22} \end{pmatrix} = a_{11}a_{22} - a_{12}a_{21},$$

$$\begin{pmatrix} a_{11} & a_{12} \\ a_{21} & a_{22} \end{pmatrix}^{-1} = \frac{1}{a}\begin{pmatrix} a_{22} & -a_{12} \\ -a_{21} & a_{11} \end{pmatrix}, \quad \text{where } a = \det\begin{pmatrix} a_{11} & a_{12} \\ a_{21} & a_{22} \end{pmatrix} \neq 0.$$

Power series whose variables take matrix values are a powerful tool in several applications of mathematics, particularly the solution of systems of differential equations. In this section we introduce the subject in its simplest form, the 2 by 2 case.

Given a power series

$$f(z) = c_0 + c_1 z + c_2 z^2 + \cdots + c_n z^n + \cdots$$

and a matrix

$$A = \begin{pmatrix} a_{11} & a_{12} \\ a_{21} & a_{22} \end{pmatrix},$$

we boldly substitute A for z:

$$f(A) = c_0 I + c_1 A + c_2 A^2 + \cdots + c_n A^n + \cdots.$$

(Note that the constant term is replaced by c_0 times the identity matrix I.)

Is this substitution meaningful? Each power A^n can be written

$$A^n = \begin{pmatrix} a_{11}^{(n)} & a_{12}^{(n)} \\ a_{21}^{(n)} & a_{22}^{(n)} \end{pmatrix},$$

where $^{(n)}$ is a superscript, not an exponent. The sum $f(A)$ is a matrix computed component-wise:

$$f(A) = c_0\begin{pmatrix} 1 & 0 \\ 0 & 1 \end{pmatrix} + \sum_{n=1}^{\infty} c_n \begin{pmatrix} a_{11}^{(n)} & a_{12}^{(n)} \\ a_{21}^{(n)} & a_{22}^{(n)} \end{pmatrix}$$

$$= \begin{pmatrix} c_0 + \sum_{n=1}^{\infty} c_n a_{11}^{(n)} & \sum_{n=1}^{\infty} c_n a_{12}^{(n)} \\ \sum_{n=1}^{\infty} c_n a_{21}^{(n)} & c_0 + \sum_{n=1}^{\infty} c_n a_{22}^{(n)} \end{pmatrix}.$$

If the series $f(z)$ has a circle of convergence $|z| < R$, it can be shown that the four series in $f(A)$ converge, provided the characteristic roots λ and μ of A satisfy $|\lambda| < R$ and $|\mu| < R$. In particular, if $R = \infty$, then $f(A)$ converges for all A.

EXAMPLE 8.1

Let $f(z) = c_0 + c_1 z + c_2 z^2 + \cdots$ and $A = \begin{pmatrix} \lambda & 0 \\ 0 & \mu \end{pmatrix}$. Find $f(A)$.

Solution:

$$A^n = \begin{pmatrix} \lambda^n & 0 \\ 0 & \mu^n \end{pmatrix},$$

hence

$$f(A) = \begin{pmatrix} c_0 + \sum_{n=1}^{\infty} c_n \lambda^n & 0 \\ 0 & c_0 + \sum_{n=1}^{\infty} c_n \mu^n \end{pmatrix} = \begin{pmatrix} f(\lambda) & 0 \\ 0 & f(\mu) \end{pmatrix}.$$

Answer: $\begin{pmatrix} f(\lambda) & 0 \\ 0 & f(\mu) \end{pmatrix}$.

EXAMPLE 8.2

Let $f(z) = c_0 + c_1 z + c_2 z^2 + \cdots$ and $A = \begin{pmatrix} \lambda & 1 \\ 0 & \lambda \end{pmatrix}$. Find $f(A)$.

Solution:

$$A = \lambda I + N,$$

where

$$N = \begin{pmatrix} 0 & 1 \\ 0 & 0 \end{pmatrix}, \qquad N^2 = 0.$$

Hence by the Binomial Theorem,

$$A^n = (\lambda I + N)^n = \lambda^n I + n\lambda^{n-1} N.$$

(Higher powers $N^2, N^3, \cdots$ are all 0.) Thus

$$f(A) = c_0 I + \sum_{n=1}^{\infty} c_n A^n = c_0 I + \sum_{n=1}^{\infty} c_n (\lambda^n I + n\lambda^{n-1} N)$$

$$= \left(c_0 + \sum_{n=1}^{\infty} c_n \lambda^n\right) I + \left(\sum_{n=1}^{\infty} n c_n \lambda^{n-1}\right) N = f(\lambda) I + f'(\lambda) N.$$

Answer: $\begin{pmatrix} f(\lambda) & f'(\lambda) \\ 0 & f(\lambda) \end{pmatrix}$.

EXAMPLE 8.3

Find e^A, where $A = \begin{pmatrix} 0 & t \\ -t & 0 \end{pmatrix}$.

Solution: Set

$$C = \begin{pmatrix} 0 & 1 \\ -1 & 0 \end{pmatrix} \quad \text{so} \quad A = tC.$$

Note that

$$C^2 = \begin{pmatrix} 0 & 1 \\ -1 & 0 \end{pmatrix}\begin{pmatrix} 0 & 1 \\ -1 & 0 \end{pmatrix} = \begin{pmatrix} -1 & 0 \\ 0 & -1 \end{pmatrix} = -I.$$

It follows readily that successive powers of C are

$$I, C, -I, -C, I, C, -I, -C, I, \cdots,$$

repeating in groups of four (compare to the powers of i). Since $A^n = t^nC^n$

$$\begin{aligned} e^A &= I + \frac{t}{1!}C - \frac{t^2}{2!}I - \frac{t^3}{3!}C + \frac{t^4}{4!}I + \frac{t^5}{5!}C - \frac{t^6}{6!}I - \frac{t^7}{7!}C + \cdots \\ &= \left(1 - \frac{t^2}{2!} + \frac{t^4}{4!} - \frac{t^6}{6!} + \cdots\right)I + \left(\frac{t}{1!} - \frac{t^3}{3!} + \frac{t^5}{5!} - \frac{t^7}{7!} + \cdots\right)C \\ &= (\cos t)I + (\sin t)C. \end{aligned}$$

Answer: $\begin{pmatrix} \cos t & \sin t \\ -\sin t & \cos t \end{pmatrix}$.

EXAMPLE 8.4

Find e^A, where $A = \begin{pmatrix} 0 & t \\ t & 0 \end{pmatrix}$.

Solution: Proceed as in the last example:

$$A = tC, \quad C = \begin{pmatrix} 0 & 1 \\ 1 & 0 \end{pmatrix}, \quad C^2 = I.$$

Successive powers of C are

$$I, C, I, C, \cdots,$$

hence

$$\begin{aligned} e^A &= I + \frac{t}{1!}C + \frac{t^2}{2!}I + \frac{t^3}{3!}C + \cdots \\ &= \left(1 + \frac{t^2}{2!} + \frac{t^4}{4!} + \cdots\right)I + \left(\frac{t}{1!} + \frac{t^3}{3!} + \cdots\right)C = (\cosh t)I + (\sinh t)C. \end{aligned}$$

Answer: $\begin{pmatrix} \cosh t & \sinh t \\ \sinh t & \cosh t \end{pmatrix}$.

The matrix exponential is a particularly useful function. It has the following properties (stated without proof):

$$\begin{aligned} e^0 &= I, \\ e^{-A} &= (e^A)^{-1} \\ e^{A+B} &= e^A e^B \quad \text{if} \quad AB = BA \quad \text{(but generally not so otherwise)}, \\ \det(e^A) &= e^{\mathrm{trace}(A)}. \end{aligned}$$

There is a further property of matrix power series worth knowing. Suppose P is non-singular and

$$B = P^{-1}AP.$$

Then

$$B^2 = (P^{-1}AP)(P^{-1}AP) = P^{-1}A^2P.$$

Similarly $B^3 = P^{-1}A^3P$, and in general $B^n = P^{-1}A^nP$. Hence

$$\begin{aligned} f(B) &= c_0 I + \sum_{n=1}^{\infty} c_n B^n = c_0 I + \sum_{n=1}^{\infty} c_n P^{-1}A^n P \\ &= P^{-1}\Big(c_0 I + \sum_{n=1}^{\infty} c_n A^n\Big)P \\ &= P^{-1}f(A)P. \end{aligned}$$

The result is

$$\boxed{f(P^{-1}AP) = P^{-1}f(A)P.}$$

This formula is useful; if A is any 2 by 2 matrix, there is a non-singular matrix P such that one alternative holds:

$$P^{-1}AP = \begin{cases} \begin{pmatrix} \lambda & 0 \\ 0 & \mu \end{pmatrix} & \text{(distinct characteristic roots)} \\ \begin{pmatrix} \lambda & 0 \\ 0 & \lambda \end{pmatrix} & \text{(equal roots,} \quad A = \lambda I) \\ \begin{pmatrix} \lambda & 1 \\ 0 & \lambda \end{pmatrix} & \text{(equal roots,} \quad A \neq \lambda I). \end{cases}$$

In each case $f(P^{-1}AP)$ is easily computed by one of the examples above, hence so is $f(A) = Pf(P^{-1}AP)P^{-1}$.

Applications to Differential Equations

Consider a system of two simultaneous differential equations:

$$\begin{cases} \dot{x} = a_{11}x + a_{12}y \\ \dot{y} = a_{21}x + a_{22}y \end{cases}$$

with constant coefficients a_{ij} and unknown functions $x(t)$ and $y(t)$. The system may be written in matrix form as

$$\dot{\mathbf{x}} = A\mathbf{x}, \qquad \text{where} \quad \mathbf{x} = \begin{pmatrix} x \\ y \end{pmatrix} \qquad \text{and} \qquad A = \begin{pmatrix} a_{11} & a_{12} \\ a_{21} & a_{22} \end{pmatrix},$$

that is

$$\begin{pmatrix} \dot{x} \\ \dot{y} \end{pmatrix} = \begin{pmatrix} a_{11} & a_{12} \\ a_{21} & a_{22} \end{pmatrix} \begin{pmatrix} x \\ y \end{pmatrix}.$$

The basic fact concerning such equations is the following.

> If $\mathbf{x}_0$ is a constant vector, then
>
> $$\mathbf{x}(t) = e^{tA}\mathbf{x}_0$$
>
> satisfies the initial-value problem
>
> $$\begin{cases} \dot{\mathbf{x}} = A\mathbf{x} \\ \mathbf{x}(0) = \mathbf{x}_0. \end{cases}$$

The second equation follows immediately from $e^0 = I$. Verification of the differential equation is routine:

$$\mathbf{x}(t) = \mathbf{x}_0 + \sum_{n=1}^{\infty} \frac{t^n}{n!} A^n \mathbf{x}_0,$$

$$\dot{\mathbf{x}} = \sum_{n=1}^{\infty} \frac{t^{n-1}}{(n-1)!} A^n \mathbf{x}_0 = A \sum_{n=1}^{\infty} \frac{t^{n-1}}{(n-1)!} A^{n-1} \mathbf{x}_0 = Ae^{tA}\mathbf{x}_0 = A\mathbf{x}.$$

REMARK: The solution is formally the same as that of the initial-value problem for one differential equation

$$\dot{\mathbf{x}} = a\mathbf{x}, \qquad \mathbf{x}(0) = \mathbf{x}_0.$$

Here the constant a can be thought of as a 1 by 1 matrix, and the solution is

$$\mathbf{x} = \mathbf{x}_0 e^{at}.$$

EXAMPLE 8.5

Solve the initial-value problem

$$\begin{cases} \dot{x} = \lambda x \\ \dot{y} = \mu y \end{cases} \qquad \begin{cases} x(0) = x_0 \\ y(0) = y_0. \end{cases}$$

Solution: In matrix form, the system is $\dot{\mathbf{x}} = A\mathbf{x}$ and $\mathbf{x}(0) = \mathbf{x}_0$, where

$$\mathbf{x} = \begin{pmatrix} x \\ y \end{pmatrix}, \qquad A = \begin{pmatrix} \lambda & 0 \\ 0 & \mu \end{pmatrix}, \qquad \mathbf{x}_0 = \begin{pmatrix} x_0 \\ y_0 \end{pmatrix}.$$

By Example 8.1,

$$e^{tA} = \begin{pmatrix} e^{\lambda t} & 0 \\ 0 & e^{\mu t} \end{pmatrix},$$

hence the solution is

$$\mathbf{x} = e^{tA}\mathbf{x}_0 = \begin{pmatrix} e^{\lambda t} & 0 \\ 0 & e^{\mu t} \end{pmatrix}\begin{pmatrix} x_0 \\ y_0 \end{pmatrix} = \begin{pmatrix} x_0 e^{\lambda t} \\ y_0 e^{\mu t} \end{pmatrix}.$$

Answer: $\begin{cases} x = x_0 e^{\lambda t} \\ y = y_0 e^{\mu t}. \end{cases}$

REMARK: The system really consists of two entirely unrelated equations. Such a system is called **uncoupled.** The next example is not so simple.

EXAMPLE 8.6

Solve the initial-value problem

$$\begin{cases} \dot{x} = \lambda x + y \\ \dot{y} = \lambda y \end{cases} \qquad \begin{cases} x(0) = x_0 \\ y(0) = y_0. \end{cases}$$

Solution: In matrix form the system is $\dot{\mathbf{x}} = A\mathbf{x}$, where

$$\mathbf{x} = \begin{pmatrix} x \\ y \end{pmatrix}, \qquad A = \begin{pmatrix} \lambda & 1 \\ 0 & \lambda \end{pmatrix}, \qquad \mathbf{x}_0 = \begin{pmatrix} x_0 \\ y_0 \end{pmatrix}.$$

Needed is e^{tA}, where

$$A = \begin{pmatrix} \lambda & 1 \\ 0 & \lambda \end{pmatrix}.$$

Write

$$tA = \begin{pmatrix} \lambda t & t \\ 0 & \lambda t \end{pmatrix} = \begin{pmatrix} \lambda t & 0 \\ 0 & \lambda t \end{pmatrix} + \begin{pmatrix} 0 & t \\ 0 & 0 \end{pmatrix} = \lambda t I + tN.$$

Since $\lambda t I$ and tN commute, that is,

$$(\lambda t I)\,(tN) = (tN)\,(\lambda t I) = \lambda t^2 N,$$

it is correct to use the matrix "law of exponents":

$$e^{(\lambda t I + tN)} = e^{\lambda t I} e^{tN} = (e^{\lambda t} I)\,(I + tN) = \begin{pmatrix} e^{\lambda t} & 0 \\ 0 & e^{\lambda t} \end{pmatrix}\begin{pmatrix} 1 & t \\ 0 & 1 \end{pmatrix} = \begin{pmatrix} e^{\lambda t} & te^{\lambda t} \\ 0 & e^{\lambda t} \end{pmatrix}.$$

Therefore the solution in matrix form is

$$\mathbf{x} = \begin{pmatrix} e^{\lambda t} & te^{\lambda t} \\ 0 & e^{\lambda t} \end{pmatrix}\begin{pmatrix} x_0 \\ y_0 \end{pmatrix}.$$

Answer: $x = (x_0 + y_0 t)e^{\lambda t}$
$y = y_0 e^{\lambda t}.$

Second-order equations can be expressed as systems of first-order equations.

EXAMPLE 8.7

Find the general solution of $\ddot{x} + k^2x = 0, \quad k > 0.$

Solution: Set

$$\dot{x} = ky.$$

Then

$$k\dot{y} + k^2x = \ddot{x} + k^2x = 0,$$

$$\dot{y} = -kx.$$

Thus the system of first order equations

$$\begin{cases} \dot{x} = ky \\ \dot{y} = -kx, \end{cases}$$

is equivalent to the given second-order equation. The system may be written

$$\dot{\mathbf{x}} = A\mathbf{x}, \qquad \text{where} \quad A = \begin{pmatrix} 0 & k \\ -k & 0 \end{pmatrix}.$$

By Example 8.3 above,

$$e^{tA} = \begin{pmatrix} \cos kt & \sin kt \\ -\sin kt & \cos kt \end{pmatrix},$$

hence the solution to the system is

$$\begin{pmatrix} x \\ y \end{pmatrix} = \begin{pmatrix} \cos kt & \sin kt \\ -\sin kt & \cos kt \end{pmatrix} \begin{pmatrix} x_0 \\ y_0 \end{pmatrix}.$$

Answer: $x = x_0 \cos kt + y_0 \sin kt$, x_0 and y_0 constants.

These examples provide only a glimpse of a vast subject, systems of differential equations. Some of the exercises hint at other possibilities.

EXERCISES

Set $A = \begin{pmatrix} -3 & 9 \\ -1 & 3 \end{pmatrix}$. Find:

1. e^A
2. $\sin A$
3. $(I - A)^{-1}$
4. $\cos A$.

Set $A = \begin{pmatrix} \lambda & 0 \\ 1 & \lambda \end{pmatrix}$. Find:

5. $f(A)$ for $f(z) = c_0 + c_1z + \cdots$
6. e^{tA}.

Compute e^A:

7. $A = \begin{pmatrix} 0 & 1 \\ 0 & 1 \end{pmatrix}$
8. $A = \begin{pmatrix} \lambda & 1 \\ 0 & 0 \end{pmatrix}$.
9. Prove the identity $e^{iA} = \cos A + i \sin A$.
10. Set $B = \ln(I - A)$. Test the formula $e^B = I - A$ in two cases (where $|\lambda| < 1$ and $|\mu| < 1$):
$$A = \begin{pmatrix} \lambda & 0 \\ 0 & \mu \end{pmatrix}, \quad A = \begin{pmatrix} \lambda & 1 \\ 0 & \lambda \end{pmatrix}.$$
11. Set
$$P = \begin{pmatrix} 5 & 2 \\ 2 & 1 \end{pmatrix}, \quad A = \begin{pmatrix} 0 & 0 \\ 1 & 0 \end{pmatrix}.$$
Verify that $e^{PAP^{-1}} = Pe^AP^{-1}$.
12. Set
$$A = \begin{pmatrix} 0 & 1 \\ 0 & 0 \end{pmatrix}, \quad B = \begin{pmatrix} 0 & 0 \\ 1 & 0 \end{pmatrix}.$$
Compute e^{A+B} and e^Ae^B. Explain the discrepancy.
13. If $A = \begin{pmatrix} a & b \\ -b & a \end{pmatrix}$, show that $e^A = e^a \begin{pmatrix} \cos b & \sin b \\ -\sin b & \cos b \end{pmatrix}$.
14. Compute e^A, where $A = \begin{pmatrix} a & b \\ b & a \end{pmatrix}$.
15. Interpret as trigonometric identities ($\exp A$ is another notation for e^A):
$$\exp\begin{pmatrix} 0 & \alpha + \beta \\ -(\alpha + \beta) & 0 \end{pmatrix} = \left[\exp\begin{pmatrix} 0 & \alpha \\ -\alpha & 0 \end{pmatrix}\right]\left[\exp\begin{pmatrix} 0 & \beta \\ -\beta & 0 \end{pmatrix}\right].$$
16. Change $\ddot{x} - 2\lambda\dot{x} + \lambda^2 x = 0$ to the system
$$\begin{cases} \dot{x} = \lambda x + y \\ \dot{y} = \lambda y \end{cases}$$
and solve.
(*Hint:* $y = \dot{x} - \lambda x$.)
17. Change $\ddot{x} - k^2 x = 0$ to the system
$$\begin{cases} \dot{x} = ky \\ \dot{y} = kx \end{cases}$$
and solve.
(*Hint:* Set $\dot{x} = ky$.)
18. Show that setting $y = \dot{x}$ changes the general second order equation
$$\ddot{x} + P(t)\dot{x} + Q(t)x = 0$$
to a linear system (with non-constant coefficients). (But Ex. 16 and Ex. 17 show this is not always the wisest choice of y.)
19. Let $\mathbf{x}$ be a solution of $\dot{\mathbf{x}} = A\mathbf{x}$. Show that $\mathbf{y} = P^{-1}\mathbf{x}$ is a solution of $\dot{\mathbf{y}} = (P^{-1}AP)\mathbf{y}$. (Here P is a non-singular constant matrix.)
20. Show that the non-homogeneous linear system $\dot{\mathbf{x}} = A\mathbf{x} + \mathbf{q}$, where
$$A = \begin{pmatrix} a_{11} & a_{12} \\ a_{21} & a_{22} \end{pmatrix}, \quad \mathbf{q} = \begin{pmatrix} q_1(t) \\ q_2(t) \end{pmatrix},$$

can be changed to $\dot{\mathbf{y}} = e^{-tA}\mathbf{q}$ by setting $\mathbf{x} = e^{tA}\mathbf{y}$. Hence the solution is

$$\mathbf{x} = e^{tA}\left(\int_0^t e^{-sA}\mathbf{q}(s)\,ds + \mathbf{x}_0\right).$$

(A matrix function is integrated component-wise.)

21. (cont.) Solve

$$\begin{cases} \dot{x} = ky + \cos kt \\ \dot{y} = -kx + \sin kt. \end{cases}$$

22. (cont.) Solve

$$\begin{cases} \dot{x} = \lambda x + y + te^{\lambda t} \\ \dot{y} = \lambda y + e^{\lambda t}. \end{cases}$$

23. Consider the most general system

$$\begin{cases} \dot{x} = f(t, x, y) \\ \dot{y} = g(t, x, y). \end{cases}$$

Interpret it as a vector equation $\dot{\mathbf{x}} = \mathbf{v}(t, \mathbf{x})$.

24. (cont.) Now place the vector $(1, f(t, x, y),\ g(t, x, y))$ at the point (t, x, y) of t, x, y-space. This defines a vector field in 3-space. Interpret a solution of the system as a curve in 3-space whose velocity vector agrees with this field. Interpret the initial-value problem.

25. (cont.) Determine the special nature of the above vector field when $\mathbf{v}(t, \mathbf{x})$ is independent of t. (A system $\dot{\mathbf{x}} = \mathbf{v}(\mathbf{x})$ is called **autonomous.**)

32. Higher Partial Derivatives

1. MIXED PARTIALS

Differentiate a function $f(x, y)$ of two variables. There are two first derivatives,

$$f_x(x, y) \qquad \text{and} \qquad f_y(x, y),$$

each itself a function of two variables. Each in turn has two first partial derivatives; these four new functions are the second derivatives of $f(x, y)$. Trace their evolution (Fig. 1.1):

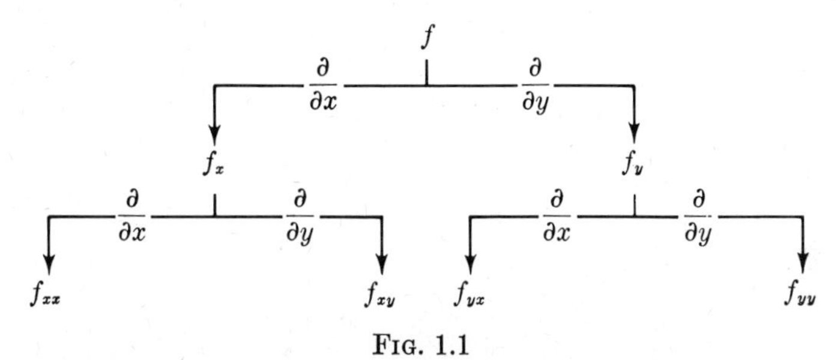

FIG. 1.1

The **pure** second partials

$$f_{xx} \qquad \text{and} \qquad f_{yy}$$

represent nothing really new. Each is found by holding one variable constant and differentiating twice with respect to the other variable.
Alternate notation:

$$f_{xx} = \frac{\partial^2 f}{\partial x^2}, \qquad f_{yy} = \frac{\partial^2 f}{\partial y^2}.$$

For example, if $f(x,y) = x^3y^4 + \cos 5y$,

$$f_x = 3x^2y^4, \qquad f_y = 4x^3y^3 - 5 \sin 5y,$$

$$f_{xx} = 6xy^4, \qquad f_{yy} = 12x^3y^2 - 25 \cos 5y.$$

The **mixed** second partials

$$f_{xy} = \frac{\partial}{\partial y}\left(\frac{\partial f}{\partial x}\right) = \frac{\partial^2 f}{\partial y\, \partial x} \qquad \text{and} \qquad f_{yx} = \frac{\partial}{\partial x}\left(\frac{\partial f}{\partial y}\right) = \frac{\partial^2 f}{\partial x\, \partial y}$$

are new. The mixed partial f_{xy} measures the rate of change in the y-direction of the rate of change of f in the x-direction. The other mixed partial f_{yx} measures the rate of change in the x-direction of the rate of change of f in the y-direction. It is not easy to see how, if at all, the two mixed partials are related to each other.

Let us compute the mixed partials of the function $f(x, y) = x^3y^4 + \cos 5y$:

$$f_x = 3x^2y^4, \qquad f_{xy} = 3 \cdot 4x^2y^3.$$

$$f_y = 4x^3y^3 - 5 \sin 5y, \qquad f_{yx} = 4 \cdot 3x^2y^3 + 0;$$

The mixed partials are equal! This is not an accident but a special case of a general phenomenon, true for functions normally encountered in applications.

Mixed second partials are equal:

$$\frac{\partial^2 f}{\partial x\, \partial y} = \frac{\partial^2 f}{\partial y\, \partial x}.$$

This fact, in no way obvious, is a deep and important property of functions. We shall not attempt a proof, but rather verify it for a useful class of functions.

If $f(x, y) = h(x)g(y)$, then

$$f_x = \frac{dh}{dx}\, g, \qquad f_y = h\, \frac{dg}{dy}.$$

Hence

$$f_{xy} = \frac{dh}{dx}\frac{dg}{dy}, \qquad f_{yx} = \frac{dh}{dx}\frac{dg}{dy} = f_{xy}.$$

Thus the mixed partials are equal if $f(x, y) = h(x)g(y)$, for example

$$x^2y^3, \quad e^x \cos y, \quad \frac{y \sin y}{1 + x^2},$$

and linear combinations of such functions.

EXERCISES

Compute $\dfrac{\partial^2 f}{\partial x\, \partial y}$ and $\dfrac{\partial^2 f}{\partial y\, \partial x}$; verify that they are equal:

1. x^4y^5
2. xy^6
3. x/y^2
4. $x + x^3y + y^4$
5. $\sin(x + y)$
6. $\cos(xy)$
7. $e^{x/y}$
8. $\arctan(x + 2y)$
9. x^my^n
10. $g(x) + h(y)$

11. x^y

12. y^x

13. $\dfrac{x+y}{x-y}$

14. $\dfrac{2xy}{x^2+y^2}$

15. $\dfrac{x-y}{1+xy}$

16. $(x-y)(x-2y)(x-3y)$

17. $\dfrac{1}{(x-y)(x-2y)}$

18. $\dfrac{x^2+y}{y^2+x}$

19. $ax^2+2bxy+cy^2$

20. $\sinh(x+y^2)$.

21. Show that each function of the form $f(x, y) = g(x+y) + h(x-y)$ satisfies the partial differential equation $\dfrac{\partial^2 f}{\partial x^2} - \dfrac{\partial^2 f}{\partial y^2} = 0$.

22. Show that each function of the form $f(x, y) = g(x) + h(y)$ satisfies the partial differential equation $\dfrac{\partial^2 f}{\partial x\,\partial y} = 0$.

23. Prove the converse of Ex. 22: each solution of $f_{xy} = 0$ has the form $f(x, y) = g(x) + h(y)$.

2. HIGHER PARTIALS

A function $z = f(x, y)$ has two distinct first partials:

$$\frac{\partial z}{\partial x}, \quad \frac{\partial z}{\partial y},$$

and three distinct second partials:

$$\frac{\partial^2 z}{\partial x^2}, \quad \frac{\partial^2 z}{\partial x\,\partial y}, \quad \frac{\partial^2 z}{\partial y^2}.$$

Because the mixed second partials are equal, so are certain mixed third partials. For example

$$\frac{\partial^2}{\partial x^2}\left(\frac{\partial z}{\partial y}\right) = \frac{\partial}{\partial y}\left(\frac{\partial^2 z}{\partial x^2}\right),$$

since

$$\frac{\partial^2}{\partial x^2}\left(\frac{\partial z}{\partial y}\right) = \frac{\partial}{\partial x}\left[\frac{\partial}{\partial x}\left(\frac{\partial z}{\partial y}\right)\right] = \frac{\partial}{\partial x}\left[\frac{\partial}{\partial y}\left(\frac{\partial z}{\partial x}\right)\right] = \frac{\partial}{\partial y}\left[\frac{\partial}{\partial x}\left(\frac{\partial z}{\partial x}\right)\right] = \frac{\partial}{\partial y}\left(\frac{\partial^2 z}{\partial x^2}\right).$$

Thus there are precisely four distinct third partials:

$$\frac{\partial^3 z}{\partial x^3}, \quad \frac{\partial^3 z}{\partial x^2\,\partial y}, \quad \frac{\partial^3 z}{\partial x\,\partial y^2}, \quad \frac{\partial^3 z}{\partial y^3}.$$

In general there are $n+1$ distinct partials of order n:

$$\frac{\partial^n z}{\partial x^k\,\partial y^{n-k}}, \qquad k = 0, 1, 2, \cdots, n.$$

EXAMPLE 2.1

Find all functions $z = f(x, y)$ which satisfy the system of partial differential equations $\dfrac{\partial^2 z}{\partial x^2} = 0, \quad \dfrac{\partial^2 z}{\partial x\, \partial y} = 0, \quad \dfrac{\partial^2 z}{\partial y^2} = 0.$

Solution: If both first partials of a function are 0 everywhere, then the function is constant. (Why?) This applies to the function $\partial z/\partial x$ since

$$\frac{\partial}{\partial x}\left(\frac{\partial z}{\partial x}\right) = \frac{\partial^2 z}{\partial x^2} = 0 \qquad \text{and} \qquad \frac{\partial}{\partial y}\left(\frac{\partial z}{\partial x}\right) = \frac{\partial^2 z}{\partial y\, \partial x} = \frac{\partial^2 z}{\partial x\, \partial y} = 0.$$

Hence

$$\frac{\partial z}{\partial x} = A.$$

Integrate with respect to x, holding y constant:

$$z = Ax + g(y),$$

where the "constant of integration" is an arbitrary function of y. Since $\partial^2 z/\partial y^2 = 0$,

$$0 = \frac{\partial^2 z}{\partial y^2} = \frac{d^2 g}{dy^2}.$$

Consequently $g(y)$ is a linear function,

$$g(y) = By + C.$$

Answer: $z = Ax + By + C.$

EXAMPLE 2.2

Find all functions $z = f(x, y)$ whose third partials are all 0.

Solution: The second partials of $\partial z/\partial x$ are all 0. By the last example,

$$\frac{\partial z}{\partial x} = Ax + By + C.$$

Integrate:

$$z = \frac{1}{2}Ax^2 + Bxy + Cx + g(y).$$

But

$$0 = \frac{\partial^3 z}{\partial y^3} = \frac{d^3 g}{dy^3}.$$

Consequently $g(y)$ is a quadratic polynomial in y.

Answer: Any quadratic polynomial in x and y.

EXAMPLE 2.3

Find all functions $z = f(x, y)$ which satisfy $\dfrac{\partial^2 z}{\partial x^2} = 0.$

Solution: Write the equation

$$\frac{\partial}{\partial x}\left(\frac{\partial z}{\partial x}\right) = 0,$$

then integrate:

$$\frac{\partial z}{\partial x} = g(y).$$

Integrate again:

$$z = g(y)x + h(y).$$

Answer: $z(x, y) = g(y)x + h(y)$, where $g(y)$ and $h(y)$ are arbitrary functions of y.

EXAMPLE 2.4

Find all functions $z = f(x, y)$ which satisfy $\dfrac{\partial^2 z}{\partial x\, \partial y} = 0.$

Solution: Write the equation

$$\frac{\partial}{\partial y}\left(\frac{\partial z}{\partial x}\right) = 0,$$

then integrate:

$$\frac{\partial z}{\partial x} = p(x).$$

Integrate again:

$$z = g(x) + h(y),$$

where $g(x)$ is an antiderivative of $p(x)$. Note that $g(x)$ is an arbitrary function of x since $p(x)$ is.

Answer: $z(x, y) = g(x) + h(y)$, where $g(x)$ and $h(y)$ are arbitrary functions of one variable.

CHECK:

$$\frac{\partial^2}{\partial x\, \partial y}[g(x) + h(y)] = \frac{\partial}{\partial x}\left(\frac{\partial}{\partial y}[g(x) + h(y)]\right) = \frac{\partial}{\partial x}[h'(y)] = 0.$$

EXAMPLE 2.5

Find all functions $z = f(x, y)$ which satisfy the system of partial differential equations $\dfrac{\partial z}{\partial x} = y, \quad \dfrac{\partial z}{\partial y} = 1.$

Solution: Integrate the first equation:

$$z = xy + g(y).$$

Substitute this into the second equation:

$$\frac{\partial}{\partial y}[xy + g(y)] = 1, \qquad x + g'(y) = 1, \qquad g'(y) = 1 - x.$$

This is impossible since the left-hand side is a function of y alone.

Answer: No solution.

This example illustrates an important point. A system of partial differential equations may have no solution at all! Could we have forseen this catastrophe for the system above? Yes; for suppose there *were* a function $f(x, y)$ satisfying

$$\frac{\partial f}{\partial x} = y \qquad \text{and} \qquad \frac{\partial f}{\partial y} = 1.$$

Then

$$\frac{\partial^2 f}{\partial y\,\partial x} = \frac{\partial}{\partial y}(y) = 1, \qquad \frac{\partial^2 f}{\partial x\,\partial y} = \frac{\partial}{\partial x}(1) = 0,$$

so the mixed partials would be unequal, a contradiction.

If the system of equations

$$\frac{\partial z}{\partial x} = p(x, y), \qquad \frac{\partial z}{\partial y} = q(x, y)$$

has a solution, then

$$\frac{\partial p}{\partial y} = \frac{\partial q}{\partial x}.$$

Indeed,

$$\frac{\partial p}{\partial y} = \frac{\partial}{\partial y}\left(\frac{\partial z}{\partial x}\right) = \frac{\partial}{\partial x}\left(\frac{\partial z}{\partial y}\right) = \frac{\partial q}{\partial x}.$$

More Variables

All that has been said applies to functions of three or more variables. For example, suppose $w = f(x, y, z)$. Then w has three first partials:

$$\frac{\partial w}{\partial x}, \quad \frac{\partial w}{\partial y}, \quad \frac{\partial w}{\partial z}.$$

The nine possible second partials may be written in matrix form:

$$\begin{bmatrix} \dfrac{\partial^2 w}{\partial x^2} & \dfrac{\partial^2 w}{\partial x\,\partial y} & \dfrac{\partial^2 w}{\partial x\,\partial z} \\[2ex] \dfrac{\partial^2 w}{\partial y\,\partial x} & \dfrac{\partial^2 w}{\partial y^2} & \dfrac{\partial^2 w}{\partial y\,\partial z} \\[2ex] \dfrac{\partial^2 w}{\partial z\,\partial x} & \dfrac{\partial^2 w}{\partial z\,\partial y} & \dfrac{\partial^2 w}{\partial z^2} \end{bmatrix}.$$

There are only 6 distinct ones since the mixed second partials are equal in pairs:

$$\frac{\partial^2 w}{\partial y\,\partial x} = \frac{\partial^2 w}{\partial x\,\partial y}, \qquad \frac{\partial^2 w}{\partial x\,\partial z} = \frac{\partial^2 w}{\partial z\,\partial x}, \qquad \frac{\partial^2 w}{\partial z\,\partial y} = \frac{\partial^2 w}{\partial y\,\partial z}.$$

EXERCISES

Compute $\dfrac{\partial^3 f}{\partial x^2\,\partial y}$ and $\dfrac{\partial^3 f}{\partial x\,\partial y^2}$:

1. x^3y^3
2. x^4y^5
3. x^2y^4
4. $x^m y^n$
5. $\cos(xy)$
6. $\sin(x^2y)$
7. $e^{xy}\sin x$
8. x^y
9. $x^{1/y}$
10. $\dfrac{x-y}{x+y}$
11. $\dfrac{1}{x^2+y^2}$
12. $\dfrac{xy}{x^2+y^2}$.
13. Find all functions $f(x, y)$ such that $\dfrac{\partial^3 f}{\partial x^2\,\partial y} = 0$.
14. Find all functions $f(x, y)$ such that $\dfrac{\partial^3 f}{\partial x^2\,\partial y} = 0$ and $\dfrac{\partial^3 f}{\partial x\,\partial y^2} = 0$.
15. Find all functions $f(x, y)$ whose 4-th partial derivatives all equal 0.
16. Find all functions $f(x, y)$ such that $\dfrac{\partial^4 f}{\partial x^2\,\partial y^2} = 0$.

Find all functions $f(x, y)$ which satisfy the system of partial differential equations:

17. $\dfrac{\partial f}{\partial x} = a,\quad \dfrac{\partial f}{\partial y} = b$
18. $\dfrac{\partial f}{\partial x} = y,\quad \dfrac{\partial f}{\partial y} = x$
19. $\dfrac{\partial f}{\partial x} = y^2,\quad \dfrac{\partial f}{\partial y} = x^2$
20. $\dfrac{\partial^2 f}{\partial x^2} = 2y^3,\quad \dfrac{\partial f}{\partial y} = 3x^2y^2$.

Write the matrix of 9 second partials:

21. $x^m y^n z^p$
22. $xy + yz + zx$
23. $\sin(x + 2y + 3z)$
24. $x^2 + yz$.

25. How many distinct third partials does $f(x, y, z)$ have?
26. (cont.) Find an explicit function for which they really are distinct.
27. Find all functions $f(x, y, z)$ satisfying $\dfrac{\partial^3 f}{\partial x\,\partial y\,\partial z} = 0$.
28. How many distinct second partials does a function $f(x, y, z, w)$ of 4 variables have? How many distinct third partials?

3. TAYLOR POLYNOMIALS

Let us recall some facts about Taylor polynomials. (See Chapter 27, Section 3, p. 597.) If $y = f(x)$, then

$$f(x) = f(a) + f'(a)(x - a) + e_1(x),$$

and

$$f(x) = f(a) + f'(a)(x - a) + \frac{1}{2}f''(a)(x - a)^2 + e_2(x),$$

where

$$|e_1(x)| \le \frac{M_2}{2!}(x - a)^2, \qquad |e_2(x)| \le \frac{M_3}{3!}|x - a|^3,$$

and where M_2 and M_3 are bounds for $|f''(x)|$ and $|f'''(x)|$ respectively.

The Taylor polynomial

$$p_1(x) = f(a) + f'(a)(x - a)$$

is constructed so that $p_1(a) = f(a)$ and $p_1'(a) = f'(a)$. The Taylor polynomial

$$p_2(x) = f(a) + f'(a)(x - a) + \frac{1}{2}f''(a)(x - a)^2$$

is constructed so that $p_2(a) = f(a)$, $p_2'(a) = f'(a)$, $p_2''(a) = f''(a)$.

In a similar way, one can construct linear and quadratic polynomials in two variables approximating a given function of two variables.

Given a function $f(x, y)$, the **first** and **second degree Taylor polynomials** of f at (x_0, y_0) are

$$p_1(x, y) = f(x_0, y_0) + \frac{\partial f}{\partial x}(x - x_0) + \frac{\partial f}{\partial y}(y - y_0),$$

$$p_2(x, y) = f(x_0, y_0) + \frac{\partial f}{\partial x}(x - x_0) + \frac{\partial f}{\partial y}(y - y_0) + \frac{1}{2}\left[\frac{\partial^2 f}{\partial x^2}(x - x_0)^2 + 2\frac{\partial^2 f}{\partial x\,\partial y}(x - x_0)(y - y_0) + \frac{\partial^2 f}{\partial y^2}(y - y_0)^2\right],$$

where all derivatives are evaluated at (x_0, y_0).

As in the case of one variable, the linear part of $p_2(x, y)$ is just $p_1(x, y)$. It is left as an exercise to prove that $p_2(x, y)$, its first, and its second partial derivatives agree at (x_0, y_0) with $f(x, y)$, its first, and its second partial derivatives.

The Taylor polynomials $p_1(x, y)$ and $p_2(x, y)$ provide approximations for $f(x, y)$ near (x_0, y_0).

> Let $p_1(x, y)$ and $p_2(x, y)$ be the Taylor polynomials of $f(x, y)$ at $\mathbf{x}_0 = (x_0, y_0)$. Then
>
> $$f(x, y) = p_1(x, y) + e_1(x, y),$$
>
> $$f(x, y) = p_2(x, y) + e_2(x, y),$$
>
> where
>
> $$|e_1(x, y)| \le C_1|\mathbf{x} - \mathbf{x}_0|^2,$$
>
> $$|e_2(x, y)| \le C_2|\mathbf{x} - \mathbf{x}_0|^3,$$
>
> and where C_1 and C_2 are constants.

This assertion is analogous to Taylor's Formula with Remainder for one variable. We shall not give a proof, but we shall illustrate its use.

Recall that in the case of one variable, the graph of $y = p_1(x)$ is the line tangent to the curve $y = f(x)$ at $(x_0, f(x_0))$. Similarly, the graph of $z = p_1(x, y)$ is the plane tangent to the surface $z = f(x, y)$ at $(x_0, y_0, f(x_0, y_0))$. Thus it is natural to approximate $f(x, y)$ by $p_1(x, y)$ near (x_0, y_0); this amounts to approximating a surface locally by its tangent plane.

EXAMPLE 3.1

Compute the Taylor polynomials $p_1(x, y)$ and $p_2(x, y)$ of the function $f(x, y) = \sqrt{x^2 + y^2}$ at (3, 4).

Solution:

$$\frac{\partial f}{\partial x} = \frac{x}{\sqrt{x^2 + y^2}}, \qquad \frac{\partial f}{\partial y} = \frac{y}{\sqrt{x^2 + y^2}},$$

$$\frac{\partial^2 f}{\partial x^2} = \frac{y^2}{(x^2 + y^2)^{3/2}}, \qquad \frac{\partial^2 f}{\partial x\,\partial y} = \frac{-xy}{(x^2 + y^2)^{3/2}}, \qquad \frac{\partial^2 f}{\partial y^2} = \frac{x^2}{(x^2 + y^2)^{3/2}}.$$

At (3, 4),

$$\frac{\partial f}{\partial x} = \frac{3}{5}, \qquad \frac{\partial f}{\partial y} = \frac{4}{5}, \qquad \frac{\partial^2 f}{\partial x^2} = \frac{16}{125}, \qquad \frac{\partial^2 f}{\partial x\,\partial y} = -\frac{12}{125}, \qquad \frac{\partial^2 f}{\partial y^2} = \frac{9}{125}.$$

Therefore

$$p_2(x, y) = 5 + \frac{3}{5}(x - 3) + \frac{4}{5}(y - 4)$$

$$+ \frac{1}{2}\left[\frac{16}{125}(x - 3)^2 - \frac{24}{125}(x - 3)(y - 4) + \frac{9}{125}(y - 4)^2\right].$$

The polynomial $p_1(x, y)$ is just the linear (first degree) part of $p_2(x, y)$.

Answer:

$$p_1(x, y) = \frac{1}{5}[25 + 3(x - 3) + 4(y - 4)];$$

$$p_2(x, y) = p_1(x, y) + \frac{1}{250}[16(x - 3)^2 - 24(x - 3)(y - 4) + 9(y - 4)^2].$$

EXAMPLE 3.2

Estimate $\sqrt{(3.1)^2 + (4.02)^2}$ by

(a) $p_1(x, y)$, (b) $p_2(x, y)$,

the Taylor polynomials in the last example.

Solution: Let $f(x, y) = \sqrt{x^2 + y^2}$.

(a) Near (3, 4),

$$f(x, y) \approx \frac{1}{5}[25 + 3(x - 3) + 4(y - 4)],$$

$$f(3.1, 4.02) \approx \frac{1}{5}[25 + 3(0.1) + 4(0.02)] = 5.076.$$

(b)

$$f(x, y) \approx p_1(x, y) + \frac{1}{250}[16(x - 3)^2 - 24(x - 3)(y - 4) + 9(y - 4)^2],$$

$$f(3.1, 4.02) \approx p_1(3.1, 4.02) + \frac{1}{250}[16(0.1)^2 - 24(0.1)(0.02) + 9(0.02)^2]$$

$$= 5.076 + \frac{0.1156}{250} = 5.0764624.$$

Answer: (a) Approximately 5.076; (b) Approximately 5.0764624. (Actual value to 7 places: 5.0764555.)

EXERCISES

Compute the Taylor polynomials $p_1(x, y)$ and $p_2(x, y)$:

1. x^2y^2; at (1, 1)
2. x^4y^3; at (2, −1)
3. $\sin(xy)$; at (0, 0)
4. e^{xy}; at (0, 0)

5. x^y; at $(1, 0)$
6. x^y; at $(1, 1)$
7. $\cos(x + y)$; at $(0, \pi/2)$
8. $1 + xy$; at $(1, 1)$
9. $\ln(x + 2y)$; at $(\frac{1}{2}, \frac{1}{4})$
10. x^2e^y; at $(1, 0)$.

Estimate, using the second-degree Taylor polynomial. Carry your work to 5 significant figures:

11. $(1.1)^{1.2}$
12. $[(1.2)^2 + 7.2)]^{1/3}$
13. $f(1.01, 2.01)$, where $f(x, y) = x^3y^2 - 2xy^4 + y^5$
14. $\sqrt{(1.99)^2 + (3.01)^2 + (6.01)^2}$.
15. Prove that $p_2(x, y)$, its first, and its second partial derivatives agree at (x_0, y_0) with $f(x, y)$, its first, and its second partial derivatives.

4. MAXIMA AND MINIMA

Consider a function of one variable, $y = f(x)$. If $f'(x_0) = 0$, then $f(x_0)$ may be a maximum or a minimum of y, or neither. To decide which, use the Taylor's approximation

$$f(x) \approx f(x_0) + \frac{1}{2}f''(x_0)\,(x - x_0)^2.$$

If $f''(x_0) > 0$, the second term on the right is positive for all $x \neq x_0$. Hence, $f(x) > f(x_0)$ for x near x_0, which means $f(x)$ has a minimum at x_0. Similarly, if $f''(x_0) < 0$, then $f(x)$ has a maximum at x_0. If $f''(x_0) = 0$, this line of reasoning is inconclusive. These remarks agree with the results of Chapter 16 concerning second derivative tests for maxima and minima.

For functions $f(x, y)$ of two variables there are similar, but more complicated results. These we take up now. For simplicity, we begin with functions $f(x, y)$ which are defined at all points (x, y).

The first basic fact is this. In order for $f(x, y)$ to have a maximum or a minimum at (x_0, y_0), both partial derivatives, f_x and f_y, must vanish at (x_0, y_0). To see why, hold y_0 fixed. If $f(x, y)$ has a maximum or a minimum at (x_0, y_0), then $f(x, y_0)$, which is a function of x alone, has a maximum or a minimum at x_0. Therefore its derivative is zero at x_0. But that says

$$\frac{\partial f}{\partial x}(x_0, y_0) = 0.$$

Similarly, hold x_0 fixed and conclude that

$$\frac{\partial f}{\partial y}(x_0, y_0) = 0.$$

If $f(x, y)$ has a maximum or a minimum at (x_0, y_0), then

$$f_x(x_0, y_0) = f_y(x_0, y_0) = 0.$$

These conditions do not guarantee a maximum or a minimum at (x_0, y_0). For example, if $f(x, y) = xy$, then $f_x(0, 0) = 0$ and $f_y(0, 0) = 0$, yet $f(x, y)$ has neither a maximum nor a minimum at $(0, 0)$. Also, these conditions do not distinguish between maxima and minima. Therefore we shall develop a second derivative criterion which furnishes additional information.

For simplicity, we shall first work at the origin. Suppose $f_x(0, 0) = 0$ and $f_y(0, 0) = 0$. Then according to Section 3

$$f(\mathbf{x}) = f(x, y) = f(0, 0) + \frac{1}{2}[Ax^2 + 2Bxy + Cy^2] + e(x, y),$$

where

$$A = f_{xx}(0, 0), \qquad B = f_{xy}(0, 0), \qquad C = f_{yy}(0, 0),$$

and

$$|e(x, y)| \leq K|\mathbf{x}|^3 = K(x^2 + y^2)^{3/2}.$$

Suppose $Ax^2 + 2Bxy + Cy^2 > 0$ for all $(x, y) \neq (0, 0)$. Then $Ax^2 + 2Bxy + Cy^2$ dominates the error $|e(x, y)|$ near $(0, 0)$, so that $[Ax^2 + 2Bxy + Cy^2] + e(x, y) > 0$ for (x, y) sufficiently near $(0, 0)$. Conclusion: $f(0, 0)$ is a minimum of $f(x, y)$. Similarly, if $Ax^2 + 2Bxy + Cy^2 < 0$ for all $(x, y) \neq (0, 0)$, then $[Ax^2 + 2Bxy + Cy^2] + e(x, y) < 0$ for all (x, y) sufficiently near $(0, 0)$. Conclusion: $f(0, 0)$ is a maximum of $f(x, y)$. However, if $Ax^2 + 2Bxy + By^2$ has both positive and negative values near $(0, 0)$, then $f(0, 0)$ is neither a maximum nor a minimum of $f(x, y)$.

The second derivative criterion is based on a careful study of the sign of the second degree polynomial

$$g(x, y) = Ax^2 + 2Bxy + Cy^2.$$

The starting point is the pair of identities

$$Ag(x, y) = (Ax + By)^2 + (AC - B^2)y^2,$$

$$Cg(x, y) = (AC - B^2)x^2 + (Bx + Cy)^2.$$

Suppose the discriminant is positive, $AC - B^2 > 0$. Then $AC > B^2 \geq 0$, so that A and C are either both positive or both negative. If both are positive, the identities above show that $g(x, y) > 0$ for each $(x, y) \neq (0, 0)$; it follows that $f(0, 0)$ is a minimum for $f(x, y)$. If A and C are both negative, then $g(x, y) < 0$ for all $(x, y) \neq (0, 0)$; hence $f(0, 0)$ is a maximum of $f(x, y)$.

Now suppose the discriminant is negative, $AC - B^2 < 0$. Three cases must be considered; surprisingly each has the same conclusion: $f(0, 0)$ is neither a maximum nor a minimum of $f(x, y)$.

CASE 1: $A = C = 0$. Then $B \neq 0$ (since $AC - B^2 < 0$) and $g(x, y) = 2Bxy$. Consequently $g(x, y)$ has one sign on the line $x = y$ and the opposite sign on the line $x = -y$.

CASE 2: $A \neq 0$. On the line $y = 0$ we have $Ag(x, y) = A^2x^2 > 0$. On the line $Ax + By = 0$ we have $Ag(x, y) = (AC - B^2)y^2 < 0$. Hence $g(x, y)$ takes both positive and negative values arbitrarily close to $(0, 0)$.

CASE 3: $C \neq 0$. This is handled like Case 2.

If the discriminant is zero, $AC - B^2 = 0$, then no conclusion about a maximum or minimum can be drawn from second derivatives alone. This is analogous to the single variable situation: $f'(0) = 0$ but $f''(0) = 0$, in which no conclusion follows.

A similar discussion applies at any point (x_0, y_0) except that x and y are replaced by $x - x_0$ and $y - y_0$.

Let $f(x, y)$ be a function of two variables and suppose

$$f_x(x_0, y_0) = 0 \quad \text{and} \quad f_y(x_0, y_0) = 0.$$

Then $f(x_0, y_0)$ is

(1) a *minimum* of $f(x, y)$ if

$$f_{xx}f_{yy} - f_{xy}^2 > 0 \quad \text{and} \quad f_{xx} > 0;$$

(2) a *maximum* of $f(x, y)$ if

$$f_{xx}f_{yy} - f_{xy}^2 > 0 \quad \text{and} \quad f_{xx} < 0;$$

(3) *neither* if

$$f_{xx}f_{yy} - f_{xy}^2 < 0.$$

In case (3) above, $f(x, y)$ is said to have a **saddle-point** at (x_0, y_0); the surface $z = f(x, y)$ is shaped like a saddle near (x_0, y_0, z_0). The tangent plane is parallel to the x, y-plane; the surface rises in some directions, falls in others; it crosses its tangent plane. For example, $f(x, y) = y^2 - x^2$ has a saddle point at $(0, 0)$. See Fig. 4.1. (The summit of a mountain pass is a saddle point.)

The case $f_{xx}f_{yy} - f_{xy}^2 = 0$ is inconclusive. Some examples will be considered in the exercises along with examples in which $f_{xx} = 0$.

There is one more point to clear up. So far, we have considered functions $f(x, y)$ defined for *all* (x, y). If $f(x, y)$ is defined only in some region R of the x, y-plane, its maximum or minimum may occur on the boundary of R. In that case the conditions $f_x = f_y = 0$ need not hold. The analogue for functions of one variable is a maximum or minimum at an end point. The conditions we have given are for maxima and minima which do not occur at a boundary point. (Of course, if $f(x, y)$ is defined for all (x, y), there are no boundary points.)

For example, consider $f(x, y) = x + y$, defined in the square $0 \leq x \leq 1$, $0 \leq y \leq 1$. Obviously $f(0, 0) = 0$ is the minimum and $f(1, 1) = 2$ is the maximum. Yet $f_x = f_y = 1$ at all points. Since the maximum and minimum occur at boundary points, f_x and f_y need not vanish.

In deriving the criterion $f_x = f_y = 0$, we tacitly assumed the maximum or minimum did not occur at a boundary point. Can you discover just where this was done?

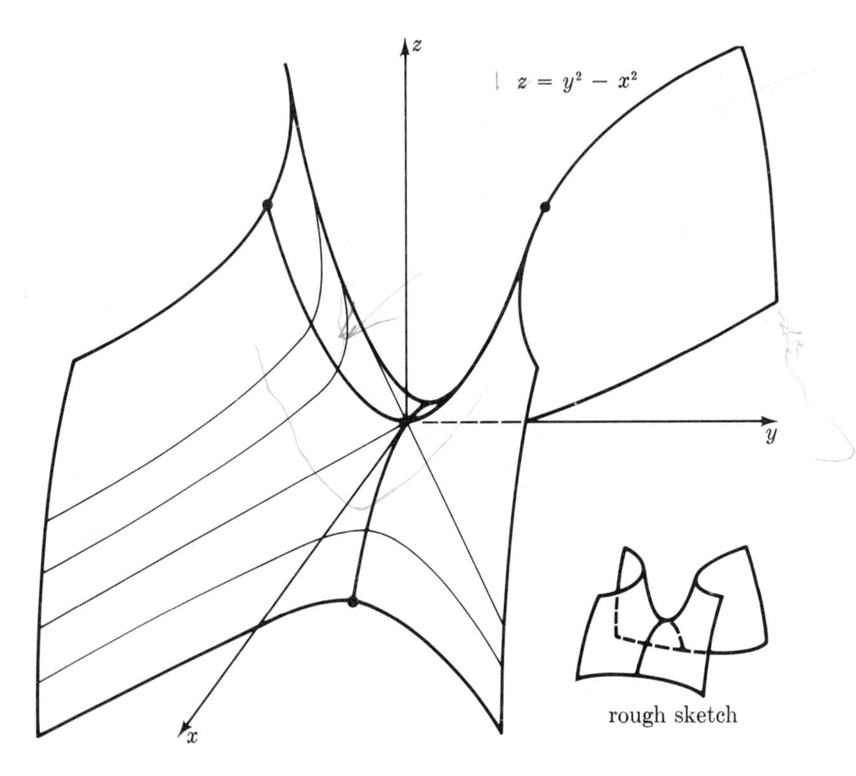

Fig. 4.1(a) Saddle point.

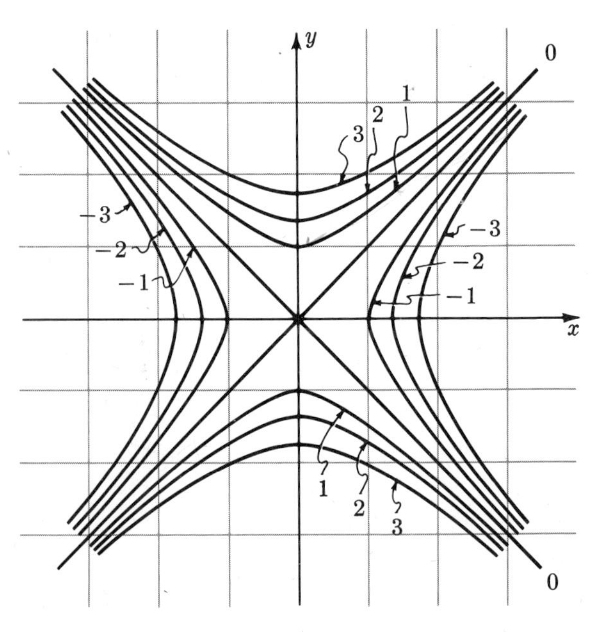

Fig. 4.1(b) Level curves of $f(x, y) = y^2 - x^2$.

EXERCISES

Does the function $z = f(x, y)$ have a possible maximum or minimum at the origin? Try the first and second derivative tests. Sketch a few level curves near the origin:

1. $z = x^2$
2. $z = xy$
3. $z = x^2 - 4y^2$
4. $z = x^2 + 2xy + y^2$
5. $z = -x^2 + 2xy - y^2$
6. $z = x^4 + y^2$
7. $z = x^4 + y^4$
8. $z = -x^2y^2$
9. $z = x^3 + y^3$
10. $z = x^2 + y^5$.
11. Suppose $f(x, y)$ is a function of two variables and $x = x(t)$ and $y = y(t)$ are functions of time. Form the composite function $g(t) = f[x(t), y(t)]$. By the Chain Rule, $\dot{g}(t) = f_x[x(t), y(t)]\,\dot{x}(t) + f_y[x(t), y(t)]\,\dot{y}(t)$. Show that
$$\ddot{g} = f_{xx}\dot{x}^2 + 2f_{xy}\dot{x}\dot{y} + f_{yy}\dot{y}^2 + f_x\ddot{x} + f_y\ddot{y}.$$
12. (cont.) Suppose also that $f_x(0, 0) = f_y(0, 0) = 0$, and that only curves $\mathbf{x}(t) = (x(t), y(t))$ are allowed which pass through $(0, 0)$ with speed 1 at $t = 0$. Suppose for each such curve $\ddot{g}(0) > 0$. Show that $\dot{g}(0) = 0$ and $f_{xx}(0, 0) > 0$ and $f_{yy}(0, 0) > 0$.
13. (cont.) Show also that $f_{xx}(0, 0)f_{yy}(0, 0) - f_{xy}(0, 0)^2 > 0$.
14. (cont.) Find a formula for d^3g/dt^3.
15. (cont.) Suppose $g(x, y) = Ax^2 + 2Bxy + Cy^2$ and $AC - B^2 = 0$. Conclude that $g(x, y) = \pm(ax + by)^2$.

5. APPLICATIONS

EXAMPLE 5.1

Show that $f(x, y) = 4x^2 - xy + y^2$ has a minimum at $(0, 0)$.

Solution:
$$f_x = 8x - y, \qquad f_y = -x + 2y.$$

Both partials are 0 at $(0, 0)$. Furthermore,
$$f_{xx} = 8, \qquad f_{xy} = -1, \qquad f_{yy} = 2.$$

Hence,
$$f_{xx} > 0, \qquad f_{xx}f_{yy} - f_{xy}^2 = 15 > 0.$$

These conditions insure a minimum at the origin.

EXAMPLE 5.2

Of all triangles of fixed perimeter, find the one with maximum area.

Solution: If the sides are a, b, c, then the area A is found by Heron's formula:
$$A^2 = s(s - a)(s - b)(s - c),$$

where $s = \frac{1}{2}(a + b + c)$ is the semiperimeter. Thus

$$2s = a + b + c, \qquad s - c = a + b - s,$$
$$A^2 = s(s - a)(s - b)(a + b - s).$$

Here s is constant and the variables are a and b. In order to maximize A, it suffices to maximize

$$f(a, b) = (s - a)(s - b)(a + b - s).$$

Now

$$f_a = (s - b)(2s - 2a - b), \qquad f_b = (s - a)(2s - a - 2b),$$
$$f_{aa} = -2(s - b), \qquad f_{ab} = -3s + 2a + 2b, \qquad f_{bb} = -2(s - a).$$

The equations $f_a = 0$ and $f_b = 0$ imply

$$2a + b = 2s, \qquad a + 2b = 2s,$$

since $s = a$ or $s = b$ is impossible in a triangle. (Why?) It follows that

$$a = b = \frac{2}{3}s.$$

For these values of a and b,

$$f_{aa} = f_{bb} = -\frac{2s}{3} < 0, \qquad f_{ab} = -\frac{s}{3},$$

$$f_{aa}f_{bb} - f_{ab}^2 = \frac{4s^2}{9} - \frac{s^2}{9} = \frac{s^2}{3} > 0.$$

These conditions insure a maximum. Now compute c:

$$c = 2s - a - b = 2s - \frac{2}{3}s - \frac{2}{3}s = \frac{2}{3}s.$$

Hence $a = b = c$.

Answer: The equilateral triangle.

EXAMPLE 5.3

Find the point on the paraboloid $z = \dfrac{x^2}{4} + \dfrac{y^2}{9}$ closest to $\mathbf{i} = (1, 0, 0)$.

Solution: Let $\mathbf{x} = (x, y, z)$ be a point on the surface. Then

$$|\mathbf{x} - \mathbf{i}|^2 = (x - 1)^2 + y^2 + z^2$$
$$= (x - 1)^2 + y^2 + \left(\frac{x^2}{4} + \frac{y^2}{9}\right)^2 = f(x, y).$$

The function $f(x, y)$ must be minimized.

$$f_x = 2(x - 1) + x\left(\frac{x^2}{4} + \frac{y^2}{9}\right),$$

$$f_y = 2y + \frac{4y}{9}\left(\frac{x^2}{4} + \frac{y^2}{9}\right) = 2y\left[1 + \frac{2}{9}\left(\frac{x^2}{4} + \frac{y^2}{9}\right)\right].$$

The condition $f_y = 0$ is satisfied if either

$$y = 0 \quad \text{or} \quad 1 + \frac{2}{9}\left(\frac{x^2}{4} + \frac{y^2}{9}\right) = 0.$$

The latter is impossible. Therefore $f_y = 0$ implies $y = 0$. But if $y = 0$, the condition $f_x = 0$ means

$$2(x - 1) + \frac{x^3}{4} = 0, \qquad \text{that is,} \qquad x^3 + 8x - 8 = 0.$$

A rough sketch shows this cubic has only one real root, and the root is near $x = 1$. By Newton's Method iterated twice, $x \approx 0.9068$.

Thus $f(x, y)$ has a possible minimum only at the point $(a, 0)$, where $a^3 + 8a - 8 = 0$. Test the second derivatives at $(a, 0)$:

$$f_{xx} = 2 + \frac{3}{4}a^2 > 0, \qquad f_{xy} = 0, \qquad f_{yy} = 2 + \frac{1}{9}a^2 > 0,$$

$$f_{xx}f_{yy} - f_{xy}^2 > 0.$$

Therefore the minimum does occur at $(a, 0)$.

Answer: $\left(a, 0, \frac{1}{4}a^2\right)$, where $a^3 + 8a - 8 = 0$, so $a \approx 0.9068$.

The following example shows that you must be careful when the variables are restricted.

EXAMPLE 5.4

Find the points on the ellipsoid $x^2 + \frac{y^2}{9} + \frac{z^2}{4} = 1$ nearest to and farthest from the origin.

Solution: The square of the distance from (x, y, z) to $(0, 0, 0)$ is

$$f(x, y) = x^2 + y^2 + z^2 = x^2 + y^2 + 4\left(1 - x^2 - \frac{y^2}{9}\right)$$

$$= 4 - 3x^2 + \frac{5}{9}y^2.$$

The first partials are

$$f_x = -6x, \qquad f_y = \frac{10}{9}\,y.$$

These vanish at (0, 0) only. However, at (0, 0),

$$f_{xx}f_{yy} - f_{xy}{}^2 = -6 \cdot \frac{10}{9} - 0 < 0.$$

Therefore, $f(x, y)$ has neither a maximum nor a minimum at (0, 0). There seems to be no possible maximum or minimum.

We have forgotten the boundary! Since

$$\frac{z}{2} = \pm\sqrt{1 - \left(x^2 + \frac{y^2}{9}\right)}$$

there is a natural restriction on x and y:

$$x^2 + \frac{y^2}{9} \leq 1.$$

Thus the function we are discussing is restricted to the region bounded by the ellipse

$$x^2 + \frac{y^2}{9} = 1.$$

Since neither maximum nor minimum occur *inside* the region, they must occur on the boundary curve (Fig. 5.1).

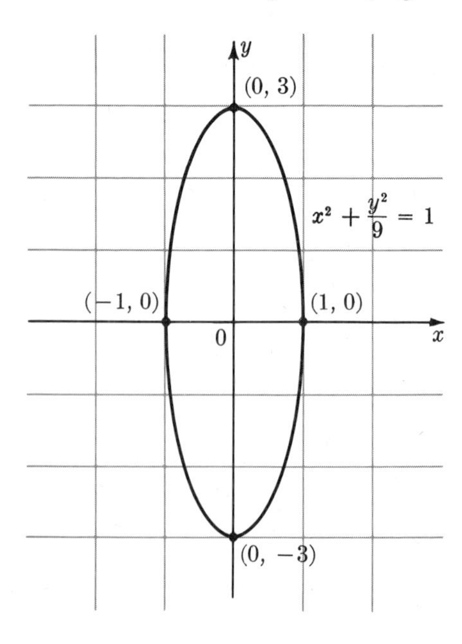

FIG. 5.1

Notice that if (x, y) is on the ellipse, then

$$x^2 + \frac{y^2}{9} = 1 \qquad \text{and} \qquad z = 0.$$

Thus the nearest and farthest points from the origin are actually points on the ellipse. By inspection (Fig. 5.1), they are the points $(\pm 1, 0, 0)$ and $(0, \pm 3, 0)$. [The points $(0, 0, \pm 2)$ are saddle points for $f(x, y)$.]

Answer: Nearest: $(\pm 1, 0, 0)$; farthest: $(0, \pm 3, 0)$.

EXERCISES

Find the point nearest the origin in the first octant and on the surface:

1. $xyz = 1$
2. $xy^2z = 1$.
3. Find the maximum and minimum of $\sin(xy)$ for $0 \leq x, y \leq \pi$.
4. Find the points on the hyperboloid $x^2 + y^2/4 - z^2/9 = 1$ nearest the origin.
5. Given n points $P_1, \cdots, P_n$ of the plane, find the point P such that $\sum_{j=1}^{n} \overline{PP_j}^2$ is least.
6. For each point (x, y, z), let $f(x, y, z)$ denote the sum of the squares of the distances of (x, y, z) to the three coordinate axes. Find the maximum of $f(x, y, z)$ on the unit sphere $x^2 + y^2 + z^2 = 1$.
7. Find the least distance between the line $\mathbf{x}(t) = (t + 1, t, -t - 3)$ and the line $\mathbf{y}(\tau) = (-2\tau + 3, \tau + 1, -\tau - 2)$.
8. Give an alternate solution to Example 5.3. By inspection, find the point closest to $\mathbf{i}$ on each cross-section $z =$ constant.
9. Approximate to 3 places the minimum first octant distance from the origin to the surface $xe^yz = 1$.

6. THREE VARIABLES [optional]

The second derivative test for a maximum or minimum of a function $f(x, y)$ was given in Section 4: There is a minimum at $\mathbf{x}_0$ if

$$f_x = f_y = 0 \qquad \text{and} \qquad f_{xx} > 0, \qquad f_{xx}f_{yy} - f_{xy}{}^2 > 0,$$

and a maximum at $\mathbf{x}_0$ if

$$f_x = f_y = 0 \qquad \text{and} \qquad f_{xx} < 0, \qquad f_{xx}f_{yy} - f_{xy}{}^2 > 0.$$

(All derivatives evaluated at $\mathbf{x}_0$.)

This test extends to functions of three or more variables. Before stating the result for three variables, it is worth making an observation. The condition $f_x = f_y = 0$ can be written $\text{grad} f = \mathbf{0}$; the condition $f_{xx}f_{yy} - f_{xy}^2 > 0$ can be written

$$\begin{vmatrix} f_{xx} & f_{xy} \\ f_{yx} & f_{yy} \end{vmatrix} > 0.$$

Here (without proof) is the test for a maximum or minimum of a function $f(x, y, z)$.

> The function $f(x, y, z)$ has a minimum at $\mathbf{x}_0$ if
>
> $$\text{grad} f = \mathbf{0}$$
>
> and if
>
> $$f_{xx} > 0, \qquad \begin{vmatrix} f_{xx} & f_{xy} \\ f_{yx} & f_{yy} \end{vmatrix} > 0, \qquad \begin{vmatrix} f_{xx} & f_{xy} & f_{xz} \\ f_{yx} & f_{yy} & f_{yz} \\ f_{zx} & f_{zy} & f_{zz} \end{vmatrix} > 0.$$
>
> It has a maximum at $\mathbf{x}_0$ if $\text{grad} f = \mathbf{0}$ and if
>
> $$f_{xx} < 0, \qquad \begin{vmatrix} f_{xx} & f_{xy} \\ f_{yx} & f_{yy} \end{vmatrix} > 0, \qquad \begin{vmatrix} f_{xx} & f_{xy} & f_{xz} \\ f_{yx} & f_{yy} & f_{yz} \\ f_{zx} & f_{zy} & f_{zz} \end{vmatrix} < 0.$$
>
> (All derivatives evaluated at $\mathbf{x}_0$.)

Notice that this test is a direct extension of the test for functions of two variables. To remember it, consider the matrix A:

$$A = \begin{pmatrix} f_{xx} & f_{xy} & f_{xz} \\ f_{yx} & f_{yy} & f_{yz} \\ f_{zx} & f_{zy} & f_{zz} \end{pmatrix}.$$

For a minimum, the test requires that three determinants be positive, the determinants of the submatrices of A indicated by dotted lines. For a maximum, these determinants must alternate: negative, positive, negative.

Let us take a very simple example. Suppose

$$f(x, y, z) = ax^2 + by^2 + cz^2,$$

where a, b, c are nonzero. Then

$$\text{grad} f = (2ax, 2by, 2cz),$$

so $\text{grad} f(\mathbf{x}) = \mathbf{0}$ only at $\mathbf{x} = \mathbf{0}$. The matrix A in this case is

$$A = \begin{pmatrix} f_{xx} & f_{xy} & f_{xz} \\ f_{yx} & f_{yy} & f_{yz} \\ f_{zx} & f_{zy} & f_{zz} \end{pmatrix} = \begin{pmatrix} 2a & 0 & 0 \\ 0 & 2b & 0 \\ 0 & 0 & 2c \end{pmatrix}.$$

According to the test, $f(\mathbf{0})$ is a minimum if the three determinants are positive:

$$2a > 0, \qquad \begin{vmatrix} 2a & 0 \\ 0 & 2b \end{vmatrix} > 0, \qquad \begin{vmatrix} 2a & 0 & 0 \\ 0 & 2b & 0 \\ 0 & 0 & 2c \end{vmatrix} > 0,$$

that is, if

$$2a > 0, \qquad 4ab > 0, \qquad 8abc > 0.$$

This is so precisely if $a > 0$, $b > 0$, and $c > 0$.

Similarly $f(\mathbf{0})$ is a maximum if

$$2a < 0, \qquad 4ab > 0, \qquad 8abc < 0,$$

which is so precisely if $a < 0$, $b < 0$, and $c < 0$.

These results agree with common sense:

$$f(\mathbf{x}) = ax^2 + by^2 + cz^2$$

so obviously $f(\mathbf{0}) = 0$. If a, b, c are all positive and $\mathbf{x} \neq \mathbf{0}$, then $f(\mathbf{x}) > 0$. If a, b, c are all negative and $\mathbf{x} \neq \mathbf{0}$, then $f(\mathbf{x}) < 0$. (If a, b, c are not all of the same sign, then $f(\mathbf{0})$ is neither a maximum nor a minimum. Why?)

EXAMPLE 6.1

Find all maxima and minima of the function

$$f(\mathbf{x}) = x^2 + 3y^2 + 4z^2 - 2xy - 2yz + 2zx.$$

Solution:

$$\operatorname{grad} f = (2x - 2y + 2z,\ -2x + 6y - 2z,\ 2x - 2y + 8z).$$

The vector equation $\operatorname{grad} f = \mathbf{0}$ amounts to the system of scalar equations (divide by 2)

$$\begin{cases} x - y + z = 0 \\ -x + 3y - z = 0 \\ x - y + 4z = 0, \end{cases}$$

whose only solution is $\mathbf{x} = \mathbf{0}$.

The matrix A is

$$\begin{pmatrix} f_{xx} & f_{xy} & f_{xz} \\ f_{yx} & f_{yy} & f_{yz} \\ f_{zx} & f_{zy} & f_{zz} \end{pmatrix} = \begin{pmatrix} 2 & -2 & 2 \\ -2 & 6 & -2 \\ 2 & -2 & 8 \end{pmatrix},$$

so the three relevant determinants are

$$2, \qquad \begin{vmatrix} 2 & -2 \\ -2 & 6 \end{vmatrix} = 8, \qquad \text{and} \qquad \begin{vmatrix} 2 & -2 & 2 \\ -2 & 6 & -2 \\ 2 & -2 & 8 \end{vmatrix} = 48.$$

All are positive. Hence by the test above, $f(\mathbf{0})$ is a minimum of f, the only one; there are no maxima.

Answer: The only extreme value of f is the minimum $f(\mathbf{0}) = 0$.

EXERCISES

The second derivative test is inconclusive at (0, 0, 0) for the given function. Determine nonetheless if the function has a maximum, a minimum, or neither at the origin:

1. $x^2 + y^2 + z^4$
2. $x^2 + y^2z^2$
3. $x^2 + y^2$
4. $x^4 + y^2 - z^6$
5. $x^2 + y^4 + z^6$
6. $x^3y^3z^3$
7. $x^4 + y^3z^3$
8. $x^4y^4 - z^5$.
9. $x^4 + y^2z^2$
10. $x^3 + y^3 + z^3$
11. $x^4y^6z^3$
12. $x^2y^2z^2$.

Find the extreme values:

13. $-2x^2 - y^2 - 3z^2 + 2xy - 2xz$
14. $x^2 + 2y^2 + z^2 + 2xy - 4yz$
15. $2x^2 + y^2 + 2z^2 + 2xy + 2yz + 2zx + x - 3z$
16. $x^2 + 3xy + y^2 - z^2 - x - 2y + z + 3$.

Determine if the function has a maximum, a minimum, or neither at the origin:

17. $x^2 + y^2 + z^2 + xy + yz + zx$
18. $x^2 + 4y^2 + 9z^2 - xy - 2yz$
19. $-x^2 - 2y^2 - z^2 + yz$
20. $x^2 + y^2 + 2z^2 - 10yz$
21. $x^2 - y^2 + 3z^2 + 12xy$
22. $3x^2 + y^2 + 4z^2 - xy - yz - zx$
23. Suppose

$$\begin{vmatrix} A & D & F \\ D & B & E \\ F & E & C \end{vmatrix} \neq 0.$$

Show that $f(x, y, z) = Ax^2 + By^2 + Cz^2 + 2Dxy + 2Eyz + 2Fzx$ can have a maximum or a minimum only at (0, 0, 0).

24. (cont.) Suppose instead that the determinant is 0. Show that there is a whole line through the origin on which $f(x, y, z) = 0$.
25. A surface $\mathbf{x} = \mathbf{x}(u, v)$ and a curve $\mathbf{p} = \mathbf{p}(t)$ are given. Suppose the distance $|\mathbf{x}(u, v) - \mathbf{p}(t)|$ of a point on the surface to a point on the curve is minimized (or maximized) for $\mathbf{x}_0$ on the surface and $\mathbf{p}_0$ on the curve. Show that the segment from $\mathbf{x}_0$ to $\mathbf{p}_0$ is normal to both the surface and the curve. Find the one exception to this assertion.
26. (cont.) Formulate the corresponding statement for two surfaces. (This is a four-variable problem!)

7. MAXIMA WITH CONSTRAINTS [optional]

Here are several problems which have a common feature.

(a) Of all rectangles with perimeter one, which has the shortest diagonal? That is, minimize $(x^2 + y^2)^{1/2}$ subject to $2x + 2y = 1$.
(b) Of all right triangles with perimeter one, which has largest area? That is, maximize $xy/2$ subject to $x + y + (x^2 + y^2)^{1/2} = 1$.
(c) Find the largest value of $x + 2y + 3z$ for points (x, y, z) on the unit sphere $x^2 + y^2 + z^2 = 1$.
(d) Of all rectangular boxes with fixed surface area, which has greatest volume? That is, maximize xyz subject to $xy + yz + zx = c$.

Each of these problems asks for the maximum (or minimum) of a function of several variables, where the variables must satisfy an equation (constraint). For example, in (a) you are asked to minimize

$$f(x, y) = x^2 + y^2,$$

where x and y must satisfy

$$g(x, y) = 2x + 2y - 1 = 0.$$

Such problems may be analyzed geometrically. Suppose you are asked to maximize a function $f(x, y)$, subject to a constraint $g(x, y) = 0$. On the same graph plot $g(x, y) = 0$ and several level curves of $f(x, y)$, noting the direction of increase of the level (Fig. 7.1). To find the largest value of

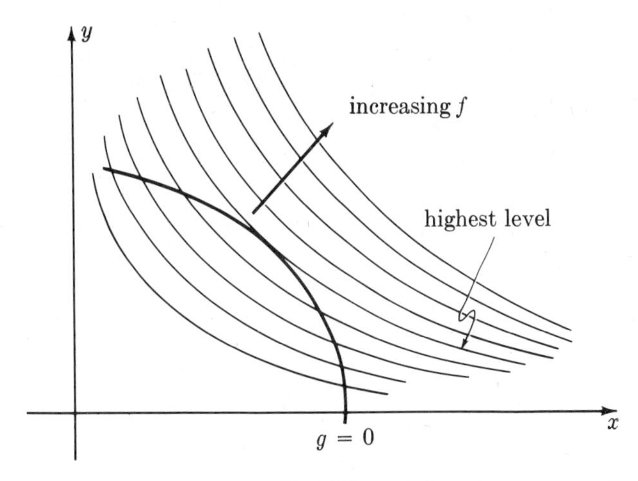

FIG. 7.1

$f(x, y)$ on the curve $g(x, y) = 0$, find the highest level curve which intersects $g = 0$. If there is a highest one and the intersection does not take place at an end point, this level curve and the graph $g = 0$ are tangent.

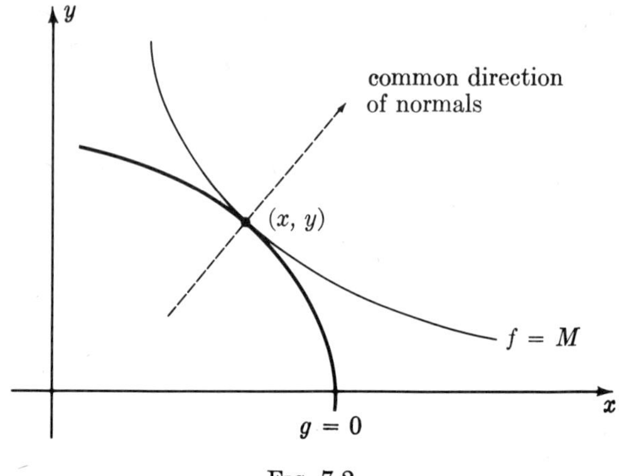

FIG. 7.2

Suppose $f(x, y) = M$ is a level curve tangent to $g(x, y) = 0$ at a point (x, y). See Fig. 7.2. Since the two graphs are tangent at (x, y), their normals at (x, y) are parallel. But the vectors

$$\operatorname{grad} f(x, y), \qquad \operatorname{grad} g(x, y)$$

point in the respective normal directions (see p. 538), hence one is a multiple of the other:

$$\operatorname{grad} f(x, y) = \lambda \operatorname{grad} g(x, y)$$

for some number λ.

This geometric argument yields a practical rule for locating points on $g(x, y) = 0$ where $f(x, y)$ may have a maximum or minimum. Note that where the condition of tangency is satisfied, there may be a maximum, a minimum, or neither (Fig. 7.3).

> To maximize or minimize a function $f(x, y)$ subject to a constraint $g(x, y) = 0$, solve the system of equations
>
> $$(f_x, f_y) = \lambda(g_x, g_y), \qquad g(x, y) = 0$$
>
> in the three unknowns x, y, λ. Each resulting point (x, y) is a candidate.
>
> The number λ is called a **Lagrange multiplier,** or simply **multiplier.**

To apply this rule, three simultaneous equations

$$\begin{cases} f_x(x, y) = \lambda g_x(x, y) \\ f_y(x, y) = \lambda g_y(x, y) \\ g(x, y) = 0, \end{cases}$$

must be solved for three unknowns x, y, λ.

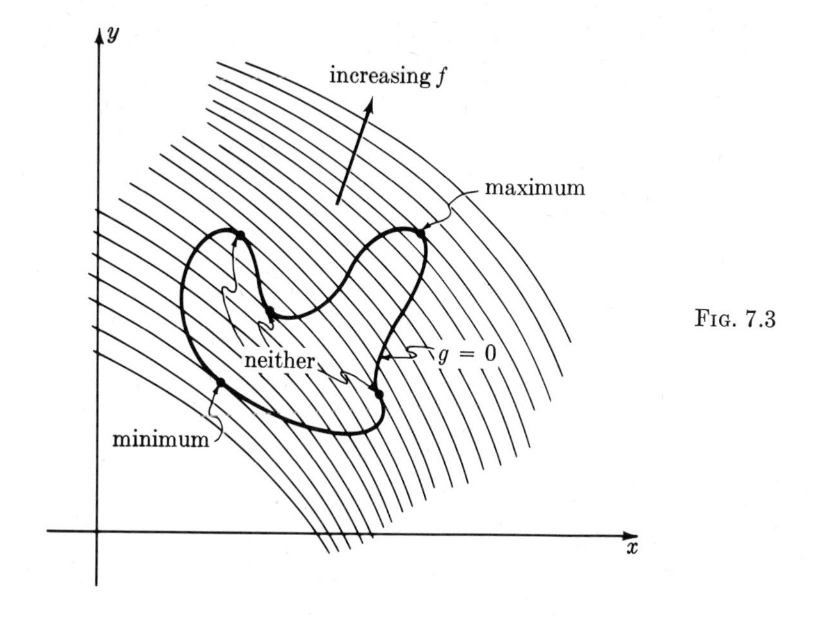

Fig. 7.3

EXAMPLE 7.1

Find the largest and smallest values of $f(x, y) = x + 2y$ on the circle $x^2 + y^2 = 1$.

Solution: Draw a figure (Fig. 7.4). As seen from the figure, f takes its maximum at a point in the first quadrant, and its minimum at a point in the

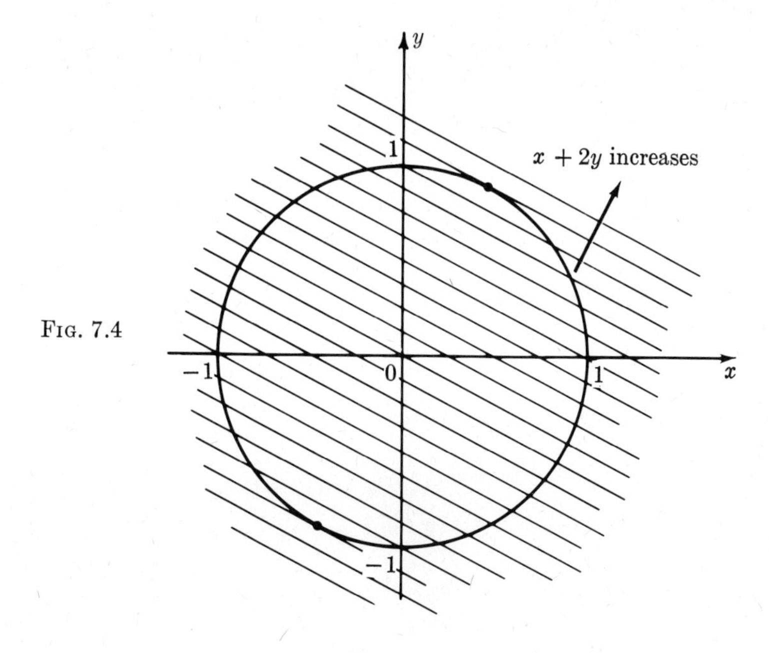

Fig. 7.4

third quadrant. Here

$$f(x, y) = x + 2y, \qquad g(x, y) = x^2 + y^2 - 1;$$

$$\operatorname{grad} f = (1, 2), \qquad \operatorname{grad} g = (2x, 2y).$$

The conditions

$$(f_x, f_y) = \lambda(g_x, g_y), \qquad g(x, y) = 0$$

become

$$(1, 2) = \lambda(2x, 2y), \qquad x^2 + y^2 = 1.$$

Thus

$$x = \frac{1}{2\lambda}, \qquad y = \frac{1}{\lambda}, \qquad \left(\frac{1}{2\lambda}\right)^2 + \left(\frac{1}{\lambda}\right)^2 = 1.$$

By the third equation,

$$\lambda^2 = \frac{5}{4}, \qquad \lambda = \pm\frac{1}{2}\sqrt{5}.$$

The value $\lambda = \frac{1}{2}\sqrt{5}$ yields

$$x = \frac{1}{\sqrt{5}}, \qquad y = \frac{2}{\sqrt{5}}, \qquad f(x, y) = \frac{5}{\sqrt{5}} = \sqrt{5};$$

the value $\lambda = -\frac{1}{2}\sqrt{5}$ yields

$$x = -\frac{1}{\sqrt{5}}, \qquad y = -\frac{2}{\sqrt{5}}, \qquad f(x, y) = -\frac{5}{\sqrt{5}} = -\sqrt{5}.$$

Answer: Largest $\sqrt{5}$; smallest $-\sqrt{5}$.

EXAMPLE 7.2

Find the largest and smallest values of xy on the segment $2x + y = 2$, $x \geq 0, y \geq 0$.

Solution: Draw a graph (Fig. 7.5). Evidently the smallest value of xy is 0, taken at either end point.

To find the largest value, use the rule with

$$f(x, y) = xy, \qquad g(x, y) = 2x + y - 2.$$

The relevant system of equations is

$$\begin{cases} (y, x) = \lambda(2, 1) \\ 2x + y - 2 = 0. \end{cases}$$

Thus $x = \lambda$, $y = 2\lambda$, and

$$2\lambda + 2\lambda - 2 = 0, \qquad \lambda = \frac{1}{2}.$$

Therefore

$$(x, y) = \left(\frac{1}{2}, 1\right), \qquad f\left(\frac{1}{2}, 1\right) = \frac{1}{2}.$$

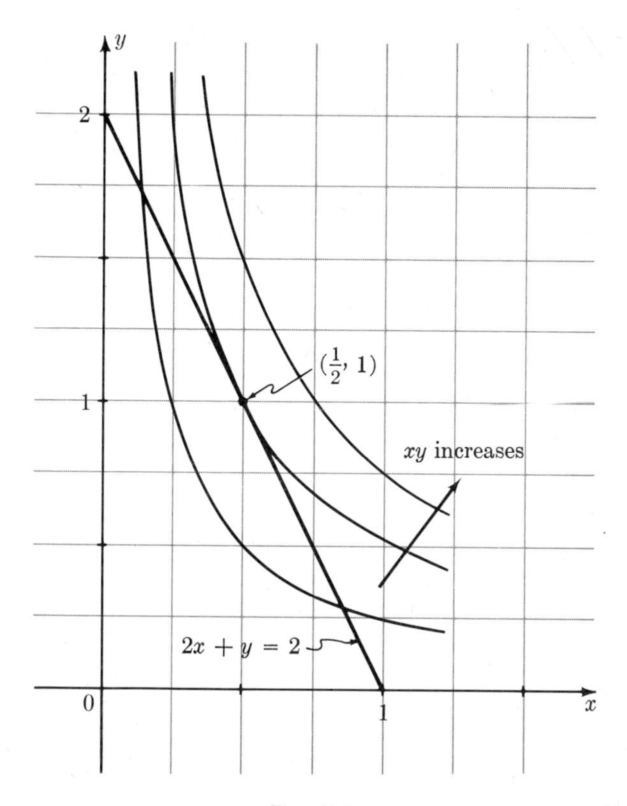

FIG. 7.5

Answer: Largest $\frac{1}{2}$; smallest 0.

EXAMPLE 7.3

Find the largest and smallest values of $x^2 + y^2$, where $x^4 + y^4 = 1$.

Solution: Graph the curve $x^4 + y^4 = 1$ and the level curves of $f(x, y) = x^2 + y^2$. See Fig. 7.6. By drawing $x^4 + y^4 = 1$ *accurately*, you see that the graph is quite flat where it crosses the axes and most sharply curved where it crosses the 45° lines $y = \pm x$. It is closest to the origin ($x^2 + y^2$ is least) at $(\pm 1, 0)$ and $(0, \pm 1)$, and farthest where $y = \pm x$.

The analysis confirms this. Set

$$f(x, y) = x^2 + y^2, \qquad g(x, y) = x^4 + y^4 - 1,$$

and solve

$$\begin{cases} (2x, 2y) = \lambda(4x^3, 4y^3) \\ x^4 + y^4 = 1. \end{cases}$$

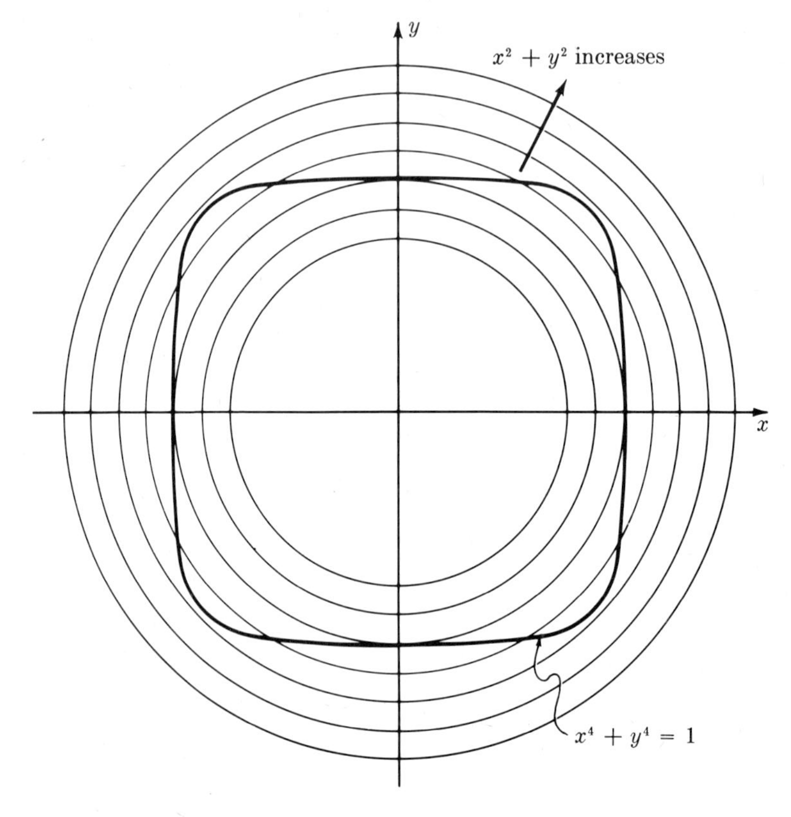

FIG. 7.6

Obvious solutions are

$$x = 0, \quad y = \pm 1, \quad \lambda = \frac{1}{2}; \qquad y = 0, \quad x = \pm 1, \quad \lambda = \frac{1}{2}.$$

Thus the points $(0, \pm 1)$ and $(\pm 1, 0)$ are candidates for the maximum or minimum. At each of these points $f(x, y) = 1$.

Suppose both $x \neq 0$ and $y \neq 0$. From

$$2x = 4\lambda x^3, \qquad 2y = 4\lambda y^3$$

follows

$$x^2 = y^2 = \frac{1}{2\lambda}.$$

Hence $\lambda = 1/(2x^2) > 0$. From $x^4 + y^4 = 1$ follows

$$\left(\frac{1}{2\lambda}\right)^2 + \left(\frac{1}{2\lambda}\right)^2 = 1, \qquad \lambda^2 = \frac{1}{2}, \qquad \lambda = \frac{1}{\sqrt{2}},$$

$$x^2 = y^2 = \frac{1}{2\lambda} = \frac{\sqrt{2}}{2} = \frac{1}{\sqrt{2}}.$$

Consequently, the four points

$$\left(\pm\frac{1}{\sqrt[4]{2}},\ \pm\frac{1}{\sqrt[4]{2}}\right)$$

are candidates for the maximum or minimum. At each of these points $f(x, y) = x^2 + y^2 = 2/\sqrt{2} = \sqrt{2}$.

Answer: Largest $\sqrt{2}$; smallest 1.

EXAMPLE 7.4

Maximize $2y - x$ on the curve $y = \sin x$ for $0 \le x \le 2\pi$.

Solution: Here $f(x, y) = 2y - x$ and $g(x, y) = y - \sin x$. The equations are

$$\begin{cases} (-1, 2) = \lambda(-\cos x, 1) \\ y = \sin x, \end{cases}$$

from which follow $\lambda = 2$, $\cos x = \frac{1}{2}$, $x = \pi/3$ or $5\pi/3$. The maximum must occur at

$$\left(\frac{\pi}{3}, \frac{\sqrt{3}}{2}\right), \quad \left(\frac{5\pi}{3}, -\frac{\sqrt{3}}{2}\right),$$

or at one of the endpoints $(0, 0)$ and $(2\pi, 0)$. But

$$f\left(\frac{\pi}{3}, \frac{\sqrt{3}}{2}\right) = \sqrt{3} - \frac{\pi}{3}, \qquad f\left(\frac{5\pi}{3}, -\frac{\sqrt{3}}{2}\right) = -\sqrt{3} - \frac{5\pi}{3},$$

$$f(0, 0) = 0, \qquad f(2\pi, 0) = -2\pi.$$

Of these numbers, $\sqrt{3} - \pi/3$ is the largest, being the only positive one.

Answer: $\sqrt{3} - \dfrac{\pi}{3}$.

REMARK: The preceding example is illustrated in Fig. 7.7. From the figure, can you tell where the maximum occurs if x is restricted to the interval $\pi/2 \le x \le 2\pi$?

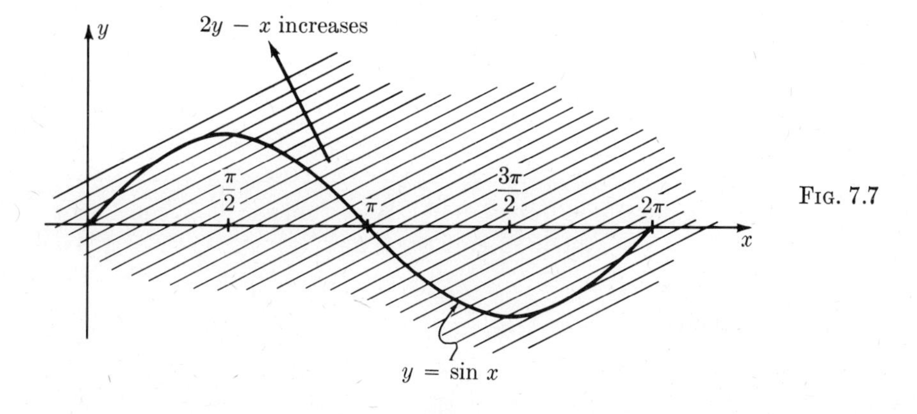

FIG. 7.7

EXERCISES

1. Find the maximum and minimum of $x + y$ on the ellipse $(x^2/4) + (y^2/9) = 1$.
2. Find the extreme values of $x - y$ on the branch $x > 0$ of the hyperbola $(x^2/9) - (y^2/4) = 1$.
3. Find the extreme values of $x - y$ on the branch $x > 0$ of the hyperbola $(x^2/4) - (y^2/9) = 1$. Explain.
4. Find the extreme values of $x - y$ on the branch $x > 0$ of the hyperbola $x^2 - y^2 = 1$. Explain.
5. Find the maximum and minimum of xy on the circle $x^2 + y^2 = 1$.
6. Find the maximum and minimum of xy on the ellipse $(x^2/4) + (y^2/9) = 1$.
7. Find the rectangle of perimeter 1 with shortest diagonal.
8. Find the right triangle of perimeter 1 with greatest area. (*Hint:* Eliminate λ.)
9. Find the right circular cone of fixed lateral area with maximum volume.
10. Find the right circular cone of fixed total surface area with maximum volume.
11. Find the right circular cylinder of fixed lateral area with maximum volume.
12. Find the right circular cylinder of fixed total surface area with maximum volume.
13. Let $0 < p < q$. Find the maximum and minimum of $x^p + y^p$ on $x^q + y^q = 1$, where $x \geq 0$ and $y \geq 0$.
14. (cont.) Let $0 < p < q$ and $x \geq 0$, $y \geq 0$. Show that
$$\left(\frac{x^p + y^p}{2}\right)^{1/p} \leq \left(\frac{x^q + y^q}{2}\right)^{1/q}.$$
15. Find the maximum and minimum of x^2y on the short arc of circle $x^2 + y^2 = 1$ between $(\frac{1}{2}\sqrt{3}, \frac{1}{2})$ and $(\frac{1}{2}\sqrt{2}, \frac{1}{2}\sqrt{2})$.

8. FURTHER CONSTRAINT PROBLEMS [optional]

In space, the problem is to maximize (minimize) a function $f(x, y, z)$ subject to a constraint $g(x, y, z) = 0$. One seeks level surfaces of $f(x, y, z)$ tangent to the surface $g(x, y, z) = 0$. See Fig. 8.1. Each point of tangency is a candidate for a maximum or minimum. At a point of tangency, the normals to the two surfaces are parallel (Fig. 8.2). But the two vectors

$$\operatorname{grad} f(\mathbf{x}), \qquad \operatorname{grad} g(\mathbf{x})$$

point in the respective normal directions, so one must be a multiple of the other. This observation leads to a practical rule (proof omitted).

> To maximize (minimize) a function $f(x, y, z)$ subject to a constraint $g(x, y, z) = 0$, solve the system of equations
>
> $$(f_x, f_y, f_z) = \lambda(g_x, g_y, g_z) \qquad g(\mathbf{x}) = 0$$
>
> in the four unknowns x, y, z, λ. Each resulting point (x, y, z) is a candidate.

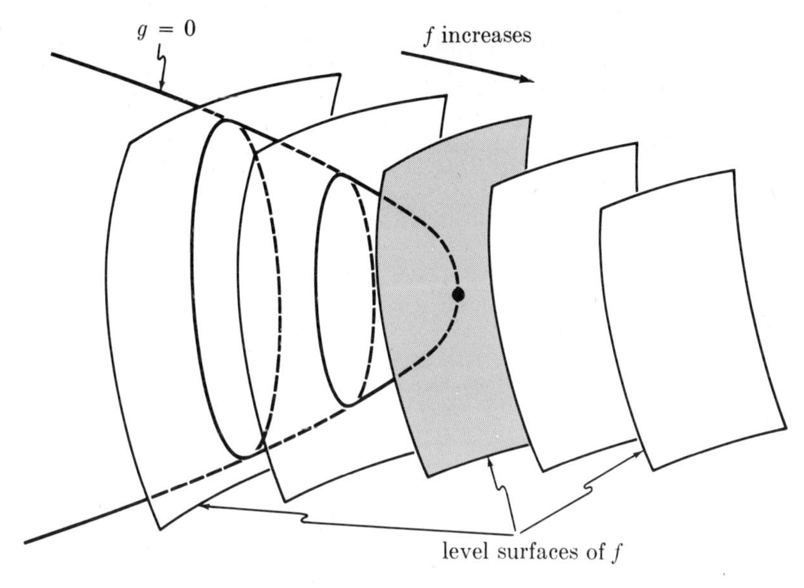

FIG. 8.1

In applications, the usual precautions concerning the boundary must be observed.

EXAMPLE 8.1

Find the longest and shortest distance from the origin to the ellipsoid $x^2 + \dfrac{y^2}{9} + \dfrac{z^2}{4} = 1$.

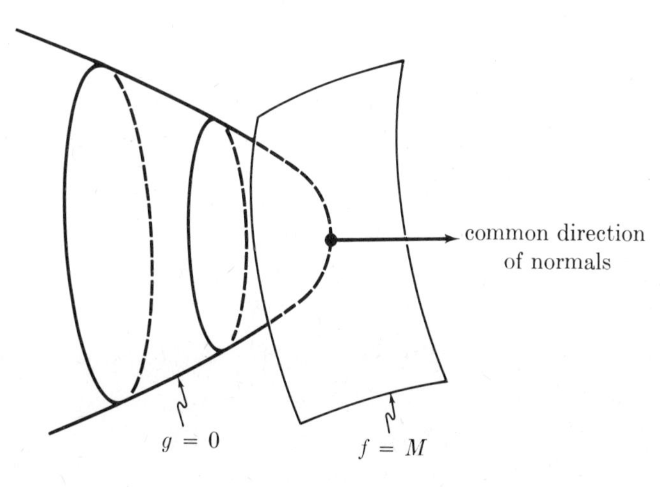

FIG. 8.2

Solution: Set $f(x, y, z)$ equal to the square of the distance,

$$f(x, y, z) = x^2 + y^2 + z^2,$$

and set

$$g(x, y, z) = x^2 + \frac{1}{9}y^2 + \frac{1}{4}z^2 - 1.$$

Then

$$\text{grad } f = (2x, 2y, 2z), \qquad \text{grad } g = \left(2x, \frac{2}{9}y, \frac{1}{2}z\right) \cdot$$

The system of equations is

$$\begin{cases} (2x, 2y, 2z) = \lambda\left(2x, \dfrac{2}{9}y, \dfrac{1}{2}z\right) \\ x^2 + \dfrac{y^2}{9} + \dfrac{z^2}{4} = 1. \end{cases}$$

If $x \neq 0$, then by the first equation $\lambda = 1$; consequently $y = z = 0$, so by the second equation $x^2 = 1$, $x = \pm 1$. Thus two candidates are the points $(\pm 1, 0, 0)$. Similarly, there are four other candidates, namely

$$\lambda = 9: \quad (0, \pm 3, 0); \qquad \lambda = 4: \quad (0, 0, \pm 2).$$

Since

$$f(\pm 1, 0, 0) = 1, \qquad f(0, \pm 3, 0) = 9, \qquad f(0, 0, \pm 2) = 4,$$

the maximum distance is $\sqrt{9}$ and the minimum distance is 1.

Answer: Longest 3; shortest 1.

REMARK: Compare this procedure with the previous solution of the same problem in Section 5, p. 765. The advantage of the present method will be crystal clear.

EXAMPLE 8.2

Find the volume of the largest rectangular solid with sides parallel to the coordinate axes that can be inscribed in the ellipsoid $x^2 + \dfrac{y^2}{9} + \dfrac{z^2}{4} = 1$.

Solution: One-eighth of the volume is

$$f(x, y, z) = xyz,$$

where $x > 0$, $y > 0$, $z > 0$. See Fig. 8.3. The constraint is $g(x, y, z) = 0$, where

$$g(x, y, z) = x^2 + \frac{y^2}{9} + \frac{z^2}{4} - 1.$$

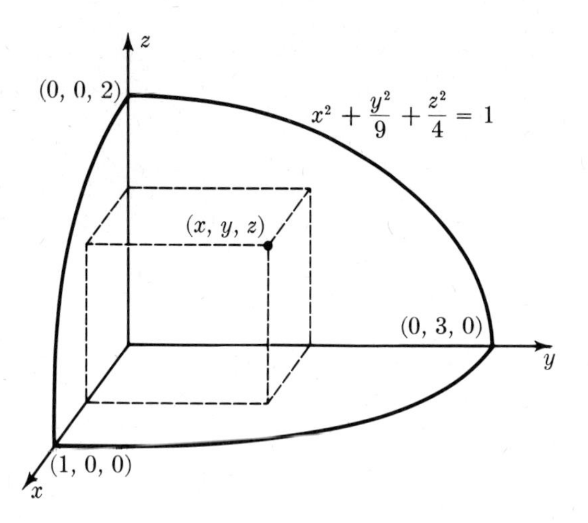

Fig. 8.3

Set grad $f = \lambda$ grad g and $g = 0$:

$$\begin{cases} (yz, zx, xy) = \lambda\left(2x, \frac{2}{9}y, \frac{1}{2}z\right) \\ x^2 + \frac{y^2}{9} + \frac{z^2}{4} = 1, \end{cases}$$

that is,

$$\begin{cases} yz = 2\lambda x \\ zx = \frac{2}{9}\lambda y \qquad x^2 + \frac{y^2}{9} + \frac{z^2}{4} = 1. \\ xy = \frac{1}{2}\lambda z \end{cases}$$

Multiply the first two equations and cancel xy:

$$z^2 = \frac{4}{9}\lambda^2.$$

Likewise

$$x^2 = \frac{1}{9}\lambda^2, \qquad y^2 = \lambda^2.$$

Substitute in the fourth equation:

$$\frac{1}{9}\lambda^2 + \frac{1}{9}\lambda^2 + \frac{1}{9}\lambda^2 = 1, \qquad \lambda^2 = 3,$$

$$x^2 = \frac{1}{3}, \qquad y^2 = 3, \qquad z^2 = \frac{4}{3}.$$

Hence

$$f(x, y, z)^2 = x^2y^2z^2 = \frac{4}{3}, \qquad f_{\max} = \frac{2\sqrt{3}}{3}.$$

Answer: $8 \cdot \frac{2\sqrt{3}}{3} = \frac{16}{3}\sqrt{3}.$

Two Constraints

Suppose the problem is to maximize (minimize) $f(x, y, z)$, where (x, y, z) is subject to *two* constraints, $g(x, y, z) = 0$ and $h(x, y, z) = 0$. Each constraint defines a surface, and these two surfaces in general have a curve of intersection (Fig. 8.4). A candidate for a maximum or minimum of $f(\mathbf{x})$ is a

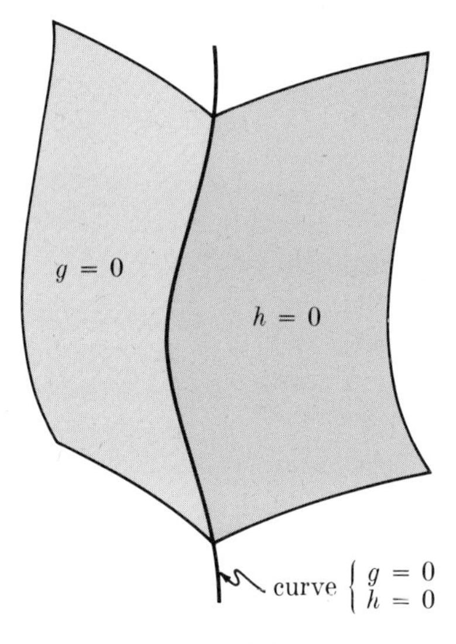

FIG. 8.4

point $\mathbf{x}$ where a level surface of f is tangent to this curve of intersection (Fig. 8.5). The vector grad $f(\mathbf{x})$ is normal to the level surface at $\mathbf{x}$, hence normal to the curve. But the vectors grad $g(\mathbf{x})$ and grad $h(\mathbf{x})$ *determine* the normal plane to the curve at $\mathbf{x}$. Hence for some constants λ and μ,

$$\text{grad} f(\mathbf{x}) = \lambda \, \text{grad}\, g(\mathbf{x}) + \mu \, \text{grad}\, h(\mathbf{x}).$$

REMARK: The existence of such **multipliers** λ and μ presupposes the linear independence of the vectors grad g and grad h at the point $\mathbf{x}$, that is, grad $g \neq \mathbf{0}$, grad $h \neq \mathbf{0}$, and neither is a multiple of the other.

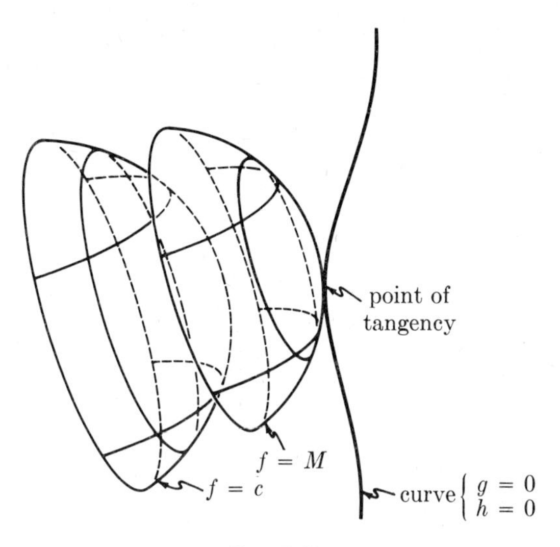

FIG. 8.5

To maximize (minimize) a function $f(x, y, z)$ subject to two constraints $g(x, y, z) = 0$ and $h(x, y, z) = 0$, solve the system of five equations

$$\begin{cases} (f_x, f_y, f_z) = \lambda(g_x, g_y, g_z) + \mu(h_x, h_y, h_z) \\ g(\mathbf{x}) = 0, \qquad h(\mathbf{x}) = 0 \end{cases}$$

in five unknowns x, y, z, λ, μ. Each resulting point (x, y, z) is a candidate.

EXAMPLE 8.3

Find the maximum and minimum of $f(x, y, z) = x + 2y + z$ on the ellipse $x^2 + y^2 = 1, \quad y + z = 1$.

Solution: Here

$$g(x, y, z) = x^2 + y^2 - 1, \qquad h(x, y, z) = y + z - 1.$$

Draw a figure showing the curve of intersection of the surfaces $g(\mathbf{x}) = 0$ and $h(\mathbf{x}) = 0$, a plane section of a right circular cylinder (Fig. 8.6). The equations to be solved are

$$\begin{cases} (1, 2, 1) = \lambda(2x, 2y, 0) + \mu(0, 1, 1) \\ x^2 + y^2 = 1, \qquad y + z = 1. \end{cases}$$

Hence

$$1 = 2\lambda x, \qquad 2 = 2\lambda y + \mu, \qquad 1 = \mu; \qquad x = \frac{1}{2\lambda}, \qquad y = \frac{1}{2\lambda};$$

$$\left(\frac{1}{2\lambda}\right)^2 + \left(\frac{1}{2\lambda}\right)^2 = 1, \qquad \lambda^2 = \frac{1}{2}, \qquad \lambda = \pm\frac{\sqrt{2}}{2}.$$

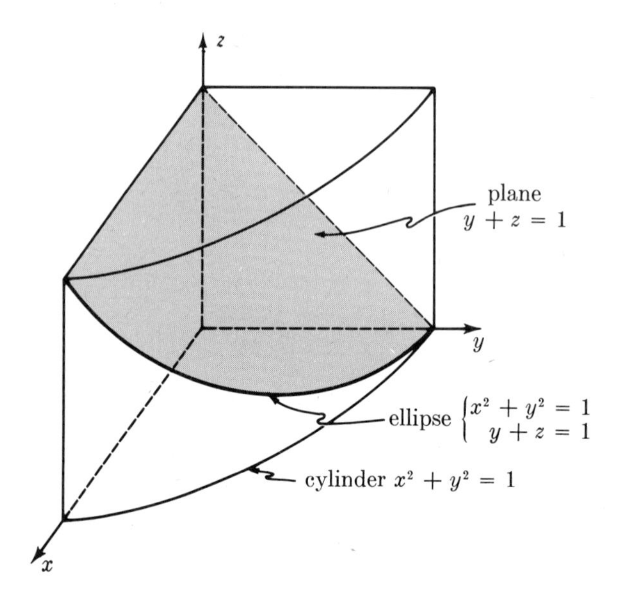

FIG. 8.6

The solution $\lambda = \sqrt{2}/2$ leads to the point

$$\mathbf{x}_1 = \left(\frac{\sqrt{2}}{2}, \frac{\sqrt{2}}{2}, 1 - \frac{\sqrt{2}}{2}\right),$$

and the solution $\lambda = -\sqrt{2}/2$ leads to the point

$$\mathbf{x}_2 = \left(-\frac{\sqrt{2}}{2}, -\frac{\sqrt{2}}{2}, 1 + \frac{\sqrt{2}}{2}\right).$$

These are the only candidates for maximum and minimum. But

$$f(\mathbf{x}_1) = 1 + \sqrt{2}, \qquad f(\mathbf{x}_2) = 1 - \sqrt{2}.$$

Answer: Maximum $1 + \sqrt{2}$; minimum $1 - \sqrt{2}$.

EXERCISES

1. Assume $a, b, c > 0$. Find the volume of the largest rectangular solid (with sides parallel to the coordinate planes) inscribed in the ellipsoid

$$\frac{x^2}{a^2} + \frac{y^2}{b^2} + \frac{z^2}{c^2} = 1.$$

2. Find the triangle of largest area with fixed perimeter.
[*Hint;* Use Heron's formula $A^2 = s(s - x)(s - y)(s - z)$, where x, y, z are the sides and s is the semiperimeter.]

3. Find the maximum and minimum of the function $x + 2y + 3z$ on the sphere $x^2 + y^2 + z^2 = 1$.

4. Find the rectangular solid of fixed surface area with maximum volume.

5. Find the rectangular solid of fixed total edge length with maximum surface area.

6. Find the rectangular solid of fixed total edge length with maximum volume.

7. Maximize xyz on $x + y + z = 1$.

8. (cont.) Conclude that $\sqrt[3]{xyz} \leq \dfrac{x + y + z}{3}$ for $x \geq 0$, $y \geq 0$, and $z \geq 0$.

9. Let $0 < p < q$. Find the maximum and minimum of $x^p + y^p + z^p$ on the surface $x^q + y^q + z^q = 1$, where $x \geq 0$, $y \geq 0$, and $z \geq 0$.

10. (cont.) Let $0 < p < q$ and $x \geq 0$, $y \geq 0$, $z \geq 0$. Show that

$$\left(\frac{x^p + y^p + z^p}{3}\right)^{1/p} \leq \left(\frac{x^q + y^q + z^q}{3}\right)^{1/q}.$$

11. Show that $xy + yz + zx \leq x^2 + y^2 + z^2$ for $x \geq 0$, $y \geq 0$, and $z \geq 0$.

12. Find the minimum of $x^2 + y^2 + z^2$ on the line $x + y + z = 1$, $x + 2y + 3z = 1$.

13. Find the maximum and minimum volumes of a rectangular solid whose total edge length is 24 ft and whose surface area is 22 ft^2.

14. Find the maximum and minimum of $x + y + z$ on the first octant portion of the curve $xyz = 1$, $x^2 + y^2 + z^2 = \frac{17}{4}$.
(*Hint:* Show that x, y, z are roots of the same quadratic equation; conclude that two of them are equal.)

33. Vector Operations

1. CROSS PRODUCT

A pair $\mathbf{v}_1$, $\mathbf{v}_2$ of non-collinear vectors determines a plane. The pair also determines a sense of turning in the plane, the rotation from $\mathbf{v}_1$ to $\mathbf{v}_2$ through the smaller of the two possible angles (Fig. 1.1).

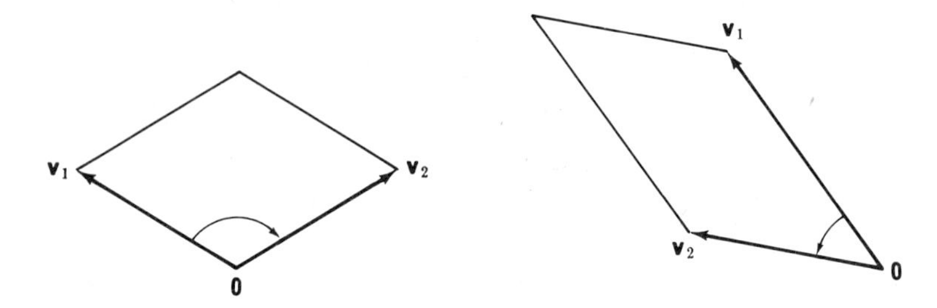

FIG. 1.1

Right-Hand Rule

Curl the fingers of your right hand in the direction of turning of the plane. Your thumb points in a direction perpendicular to the plane (Fig. 1.2).

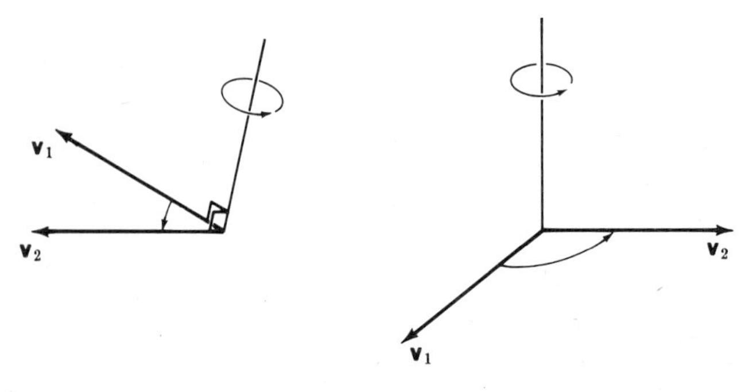

FIG. 1.2

Your thumb indicates an axis of rotation; your fingers indicate the direction of rotation. You may think of this **right-hand rule** as determining which side of the plane is the top, which way is up.

Important note: reversing the order of $\mathbf{v}_1$, $\mathbf{v}_2$, that is writing $\mathbf{v}_2$, $\mathbf{v}_1$, reverses the sense of turning.

Each pair of the three basic vectors $\mathbf{i}$, $\mathbf{j}$, $\mathbf{k}$ determines a sense of turning and a perpendicular direction (Fig. 1.3). The pair $\mathbf{i}$, $\mathbf{j}$ determines the direc-

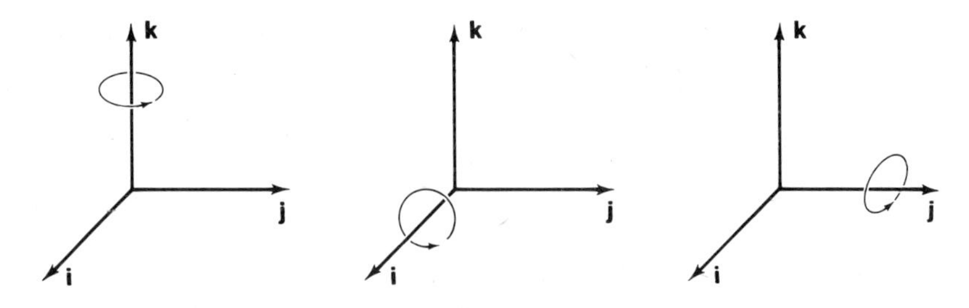

FIG. 1.3

tion of $\mathbf{k}$, the pair $\mathbf{j}$, $\mathbf{k}$ determines the direction of $\mathbf{i}$, and the pair $\mathbf{k}$, $\mathbf{i}$ determines the direction of $\mathbf{j}$. Note the cyclic order:

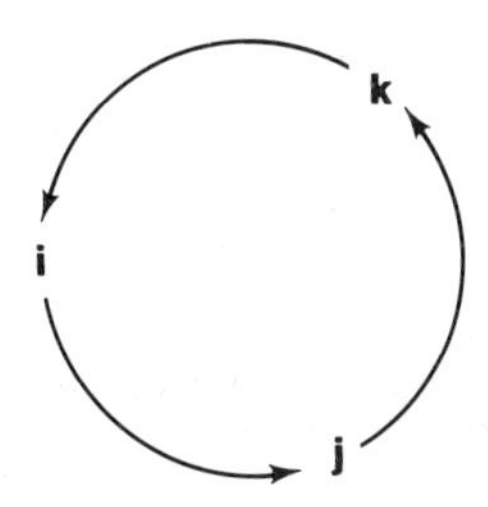

Torque

A rigid body is free to turn about the origin. A force $\mathbf{F}$ acts at a point $\mathbf{x}$ of the body. As a result the body "wants" to rotate about an axis through $\mathbf{0}$ perpendicular to the plane of $\mathbf{x}$ and $\mathbf{F}$ (unless $\mathbf{x}$ and $\mathbf{F}$ are collinear, then there is no turning). See Fig. 1.4a. As usual, the force vector $\mathbf{F}$ is *drawn* at its point of application $\mathbf{x}$. But analytically it starts at $\mathbf{0}$. See Fig. 1.4b. The positive axis of rotation is determined by the right-hand rule as applied to the pair $\mathbf{x}$, $\mathbf{F}$ in that order: $\mathbf{x}$ first, $\mathbf{F}$ second.

In physics, one speaks of the **torque** (at the origin) resulting from the force $\mathbf{F}$ applied at $\mathbf{x}$. Roughly speaking, torque is a measure of the tendency of a body to rotate under the action of forces. (Torque will be defined precisely in a moment.)

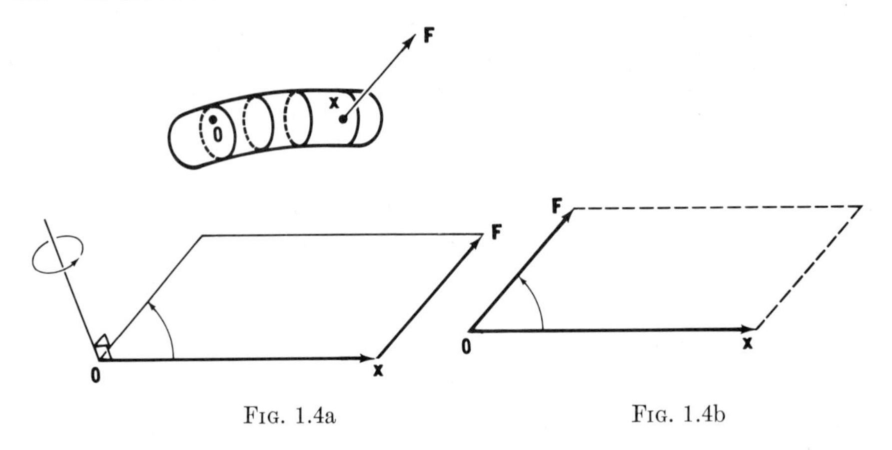

FIG. 1.4a FIG. 1.4b

By experiment, if $\mathbf{F}$ is tripled in magnitude, the torque is tripled; if $\mathbf{x}$ is moved out twice as far along the same line and the same $\mathbf{F}$ is applied there, the torque is doubled. Hence the torque is proportional to the length of $\mathbf{x}$ and to the length of $\mathbf{F}$. Therefore (Fig. 1.4b) the torque is proportional to the area of the parallelogram determined by $\mathbf{x}$ and $\mathbf{F}$.

Resolve $\mathbf{F}$ into components $\mathbf{F}^{\parallel}$ and $\mathbf{F}^{\perp}$, where $\mathbf{F}^{\parallel}$ is parallel to $\mathbf{x}$ and $\mathbf{F}^{\perp}$ is perpendicular to $\mathbf{x}$. See Fig. 1.5a. Only $\mathbf{F}^{\perp}$ produces torque; the amount of torque is the product $|\mathbf{F}^{\perp}|\,|\mathbf{x}|$ of the magnitude of $\mathbf{F}^{\perp}$ by the length $|\mathbf{x}|$ of the lever arm. But this product *is* the area of the parallelogram determined by $\mathbf{x}$ and $\mathbf{F}$ (see Fig. 1.5b).

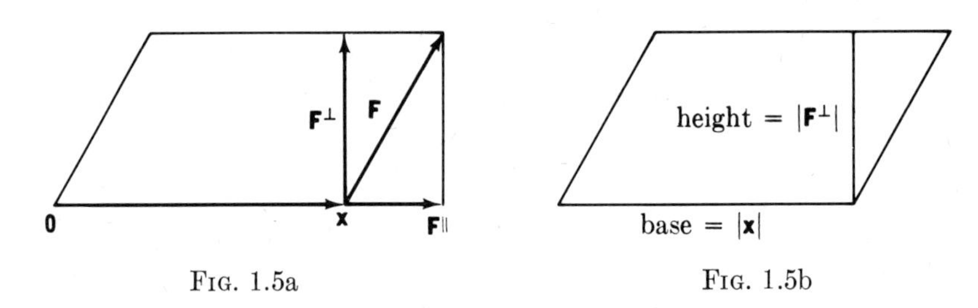

FIG. 1.5a FIG. 1.5b

The torque about the origin is completely described by one vector, written

$$\mathbf{x} \times \mathbf{F}.$$

This is a vector whose direction is determined by the right-hand rule from the pair $\mathbf{x}$, $\mathbf{F}$, and whose magnitude is the area of the parallelogram based on $\mathbf{x}$ and $\mathbf{F}$.

The length of $\mathbf{x} \times \mathbf{F}$ is the magnitude of the torque. The direction of $\mathbf{x} \times \mathbf{F}$ is the positive axis of rotation; with your right thumb along $\mathbf{x} \times \mathbf{F}$, your fingers curl in the direction of turning.

In physics, torque about the origin is *defined* to be the vector $\mathbf{x} \times \mathbf{F}$.

Cross Product

A vector $\mathbf{v} \times \mathbf{w}$ can be defined for any pair of vectors $\mathbf{v}$ and $\mathbf{w}$, not necessarily a force and a point of application.

> The **cross product** of $\mathbf{v}$ and $\mathbf{w}$, written
>
> $$\mathbf{v} \times \mathbf{w}$$
>
> is the vector whose direction is determined by the right-hand rule from the pair $\mathbf{v}$, $\mathbf{w}$, and whose magnitude is the area of the parallelogram based on $\mathbf{v}$ and $\mathbf{w}$. See Fig. 1.6.

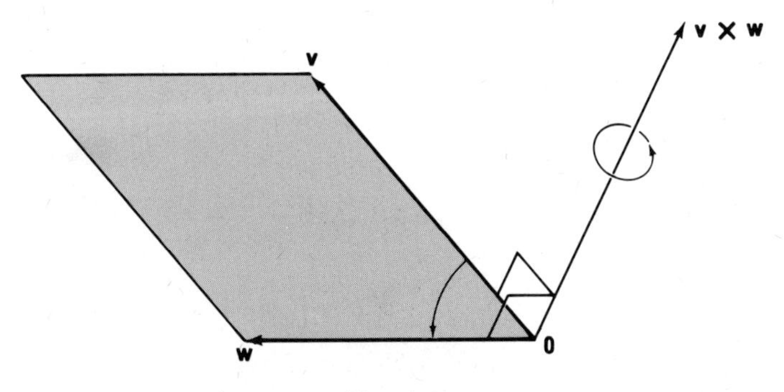

FIG. 1.6

Note that $\mathbf{v} \times \mathbf{w}$ is a vector perpendicular both to $\mathbf{v}$ and to $\mathbf{w}$. Note also that if $\mathbf{v}$ and $\mathbf{w}$ are parallel, then the parallelogram collapses, so $\mathbf{v} \times \mathbf{w} = \mathbf{0}$. In particular,

$$\mathbf{v} \times \mathbf{v} = \mathbf{0} \qquad \text{for each vector } \mathbf{v}.$$

If $\mathbf{v}$ and $\mathbf{w}$ are interchanged, the thumb reverses direction, hence

$$\mathbf{w} \times \mathbf{v} = -\mathbf{v} \times \mathbf{w}.$$

For the basic vectors $\mathbf{i}$, $\mathbf{j}$, $\mathbf{k}$, the cross products are simply

$$\mathbf{i} \times \mathbf{j} = \mathbf{k}, \qquad \mathbf{j} \times \mathbf{k} = \mathbf{i}, \qquad \mathbf{k} \times \mathbf{i} = \mathbf{j}.$$

Computation of Cross Products

A fundamental property of cross products is linearity:

$$(a_1\mathbf{v}_1 + a_2\mathbf{v}_2) \times \mathbf{w} = a_1(\mathbf{v}_1 \times \mathbf{w}) + a_2(\mathbf{v}_2 \times \mathbf{w}),$$

$$\mathbf{v} \times (b_1\mathbf{w}_1 + b_2\mathbf{w}_2) = b_1(\mathbf{v} \times \mathbf{w}_1) + b_2(\mathbf{v} \times \mathbf{w}_2).$$

The proof is rather tedious; we shall omit it.

Here is the basic rule for computation:

> If
> $$\mathbf{v} = (v_1, v_2, v_3) \quad \text{and} \quad \mathbf{w} = (w_1, w_2, w_3),$$
> then
> $$\mathbf{v} \times \mathbf{w} = (v_2w_3 - v_3w_2,\ v_3w_1 - v_1w_3,\ v_1w_2 - v_2w_1).$$

This rule is verified by a straightforward computation. Write

$$\mathbf{v} = v_1\mathbf{i} + v_2\mathbf{j} + v_3\mathbf{k}, \quad \mathbf{w} = w_1\mathbf{i} + w_2\mathbf{j} + w_3\mathbf{k}.$$

By linearity, $\mathbf{v} \times \mathbf{w}$ has nine terms, each of which involves a cross product of two of the vectors **i**, **j**, **k**. See if you can carry out the computation.

There is a simple way to remember the rule. Notice that each component of $\mathbf{v} \times \mathbf{w}$ is a two-by-two determinant:

$$\mathbf{v} \times \mathbf{w} = \left(\begin{vmatrix} v_2 & v_3 \\ w_2 & w_3 \end{vmatrix}, \begin{vmatrix} v_3 & v_1 \\ w_3 & w_1 \end{vmatrix}, \begin{vmatrix} v_1 & v_2 \\ w_1 & w_2 \end{vmatrix} \right).$$

Here are two numerical examples:

$$(4, 3, -1) \times (-2, 2, 1) = \left(\begin{vmatrix} 3 & -1 \\ 2 & 1 \end{vmatrix}, \begin{vmatrix} -1 & 4 \\ 1 & -2 \end{vmatrix}, \begin{vmatrix} 4 & 3 \\ -2 & 2 \end{vmatrix} \right)$$
$$= (3 + 2, 2 - 4, 8 + 6) = (5, -2, 14).$$

$$(1, 0, 1) \times (0, 1, 1) = \left(\begin{vmatrix} 0 & 1 \\ 1 & 1 \end{vmatrix}, \begin{vmatrix} 1 & 1 \\ 1 & 0 \end{vmatrix}, \begin{vmatrix} 1 & 0 \\ 0 & 1 \end{vmatrix} \right) = (-1, -1, 1).$$

Another device for remembering the cross product is a symbolic determinant, to be expanded by the first row:

$$(v_1, v_2, v_3) \times (w_1, w_2, w_3) = \begin{vmatrix} \mathbf{i} & \mathbf{j} & \mathbf{k} \\ v_1 & v_2 & v_3 \\ w_1 & w_2 & w_3 \end{vmatrix}$$
$$= \begin{vmatrix} v_2 & v_3 \\ w_2 & w_3 \end{vmatrix} \mathbf{i} - \begin{vmatrix} v_1 & v_3 \\ w_1 & w_3 \end{vmatrix} \mathbf{j} + \begin{vmatrix} v_1 & v_2 \\ w_1 & w_2 \end{vmatrix} \mathbf{k}.$$

Identities

We list without proofs useful identities involving cross products:

$$\mathbf{u} \cdot (\mathbf{v} \times \mathbf{w}) = \mathbf{w} \cdot (\mathbf{u} \times \mathbf{v}) = \mathbf{v} \cdot (\mathbf{w} \times \mathbf{u})$$
$$= \pm(\text{volume of the parallelepiped based on } \mathbf{u}, \mathbf{v}, \text{ and } \mathbf{w})$$
$$= \begin{vmatrix} u_1 & u_2 & u_3 \\ v_1 & v_2 & v_3 \\ w_1 & w_2 & w_3 \end{vmatrix}.$$

The product $\mathbf{u} \cdot (\mathbf{v} \times \mathbf{w})$ is sometimes written $[\mathbf{u}, \mathbf{v}, \mathbf{w}]$ and is called the **triple scalar product.**

$$\mathbf{u} \times (\mathbf{v} \times \mathbf{w}) = (\mathbf{u} \cdot \mathbf{w})\mathbf{v} - (\mathbf{u} \cdot \mathbf{v})\mathbf{w}.$$

$$|\mathbf{v} \times \mathbf{w}| = |\mathbf{v}|\,|\mathbf{w}| \sin \theta,$$

where θ is the angle from $\mathbf{v}$ to $\mathbf{w}$. See Fig. 1.7.

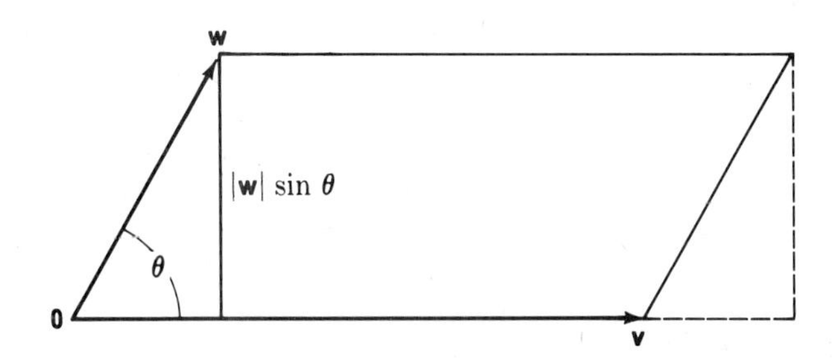

FIG. 1.7

$$|\mathbf{v} \times \mathbf{w}|^2 = |\mathbf{v}|^2|\mathbf{w}|^2 - (\mathbf{v} \cdot \mathbf{w})^2.$$

Suppose $\mathbf{v} = \mathbf{v}(t)$ and $\mathbf{w} = \mathbf{w}(t)$ are vector functions of time t. Then $\mathbf{v} \times \mathbf{w}$ is a vector function of t, and

$$\frac{d}{dt}(\mathbf{v} \times \mathbf{w}) = \dot{\mathbf{v}} \times \mathbf{w} + \mathbf{v} \times \dot{\mathbf{w}}.$$

(Note that the factors must be kept in order since $\dot{\mathbf{w}} \times \mathbf{v} = -\mathbf{v} \times \dot{\mathbf{w}}$ and $\mathbf{w} \times \dot{\mathbf{v}} = -\dot{\mathbf{v}} \times \mathbf{w}$.)

EXERCISES

Find the cross product:

1. $(-2, 2, 1) \times (4, 3, -1)$
2. $(1, 0, 1) \times (1, 1, 0)$
3. $(1, 2, 3) \times (3, 2, 1)$
4. $(3, 1, -1) \times (3, -1, -1)$
5. $(-2, -2, -2) \times (1, 1, 0)$
6. $(-1, 2, 2) \times (3, -1, 2)$
7. $(0, 0, 0) \times (1, 1, 2)$
8. $(1, -1, 1) \times (-1, 1, -1)$
9. $(2, 1, 3) \times (2, 2, -1)$
10. $(1, 2, 3) \times (4, 5, 6)$.

A force $\mathbf{F}$ is applied at point $\mathbf{x}$. Find its torque about the origin:

11. $\mathbf{F} = (-1, 1, 1), \quad \mathbf{x} = (10, 0, 0)$
12. $\mathbf{F} = (3, 0, 0) \quad \mathbf{x} = (0, 0, 1)$
13. $\mathbf{F} = (-1, 1, 1), \quad \mathbf{x} = (2, 2, -1)$
14. $\mathbf{F} = (2, -1, 5), \quad \mathbf{x} = (-7, 1, 0)$.

15. An electric charge at (0, 1, 0) exerts an attractive force of 2 dynes on a second charge at (0, 0, 1). Let the unit of length be the centimeter. Find the magnitude of the resulting torque about $\mathbf{0}$.

16. Prove the formula $|\mathbf{v} \times \mathbf{w}|^2 = |\mathbf{v}|^2|\mathbf{w}|^2 - (\mathbf{v} \cdot \mathbf{w})^2$.
(*Hint:* $\cos^2 \theta + \sin^2 \theta = 1$.)

17. (cont.) Write the above identity in components and prove it that way.

18. Check the formula for $d(\mathbf{v} \times \mathbf{w})/dt$ for $\mathbf{v} = (t, t^2, t^3)$ and $\mathbf{w} = (\cos t, \sin t, -t)$.

19. Prove that if $\mathbf{u}, \mathbf{v}, \mathbf{w}$ is a right-hand system, then $\mathbf{u} \cdot (\mathbf{v} \times \mathbf{w})$ is the volume of the parallelepiped based on $\mathbf{u}$, $\mathbf{v}$, and $\mathbf{w}$.
(*Hint:* $\mathbf{u} \cdot (\mathbf{v} \times \mathbf{w}) = |\mathbf{u}|\,|\mathbf{v} \times \mathbf{w}| \cos \theta$.)

20. Prove

$$\mathbf{u} \cdot (\mathbf{v} \times \mathbf{w}) = \begin{vmatrix} u_1 & u_2 & u_3 \\ v_1 & v_2 & v_3 \\ w_1 & w_2 & w_3 \end{vmatrix}.$$

21. Prove the formula $\mathbf{u} \times (\mathbf{v} \times \mathbf{w}) = (\mathbf{u} \cdot \mathbf{w})\mathbf{v} - (\mathbf{u} \cdot \mathbf{v})\mathbf{w}$.
(*Hint:* By the linearity of each side and by symmetry, reduce to the cases where $\mathbf{u}$, $\mathbf{v}$, and $\mathbf{w}$ are chosen from $\mathbf{i}$ and $\mathbf{j}$.)

22. (cont.) Prove the formula

$$\begin{aligned}(\mathbf{a} \times \mathbf{b}) \times (\mathbf{u} \times \mathbf{v}) &= [(\mathbf{a} \times \mathbf{b}) \cdot \mathbf{v}]\mathbf{u} - [(\mathbf{a} \times \mathbf{b}) \cdot \mathbf{u}]\mathbf{v} \\ &= [(\mathbf{u} \times \mathbf{v}) \cdot \mathbf{a}]\mathbf{b} - [(\mathbf{u} \times \mathbf{v}) \cdot \mathbf{b}]\mathbf{a}.\end{aligned}$$

Hence show that the left-hand side is a vector along the line of intersection of the plane of $\mathbf{a}$ and $\mathbf{b}$ with the plane of $\mathbf{u}$ and $\mathbf{v}$.

23. (cont.) Show that $(\mathbf{a} \times \mathbf{b}) \times (\mathbf{a} \times \mathbf{c})$ is collinear with $\mathbf{a}$.

24. Let $\mathbf{x} = \mathbf{x}(t)$ be a space curve. Show that its curvature is $k = |\mathbf{v} \times \mathbf{a}|/|\mathbf{v}|^3$.

2. APPLICATIONS

This section contains three examples of the use of cross products of vectors in physics; there are many others.

Equilibrium

Forces $\mathbf{F}_1, \cdots, \mathbf{F}_n$ are applied at points $\mathbf{x}_1, \cdots, \mathbf{x}_n$ of a rigid body (Fig. 2.1). Now a rigid body is in equilibrium when both the sum of the forces vanishes and the sum of the turning moments (torques) of the forces about $\mathbf{0}$ vanishes. Thus the conditions for equilibrium are the two vector equations:

$$\mathbf{F}_1 + \mathbf{F}_2 + \cdots + \mathbf{F}_n = \mathbf{0},$$

$$\mathbf{x}_1 \times \mathbf{F}_1 + \mathbf{x}_2 \times \mathbf{F}_2 + \cdots + \mathbf{x}_n \times \mathbf{F}_n = \mathbf{0}.$$

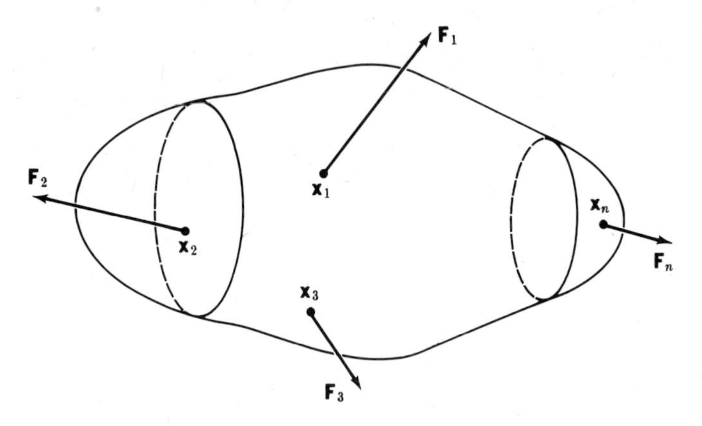

FIG. 2.1

Angular Velocity

A rigid body rotates about an axis a through $\mathbf{0}$. See Fig. 2.2a. The central angle is $\theta = \theta(t)$, so $\omega = \dot{\theta}$ is the **angular speed,** the rate of rotation in radians per second.

The **angular velocity** is defined to be the vector $\boldsymbol{\omega}$ having magnitude $\dot{\theta}$ and pointing along the (positive) axis of rotation according to the right-hand rule (Fig. 2.2b).

Suppose the actual velocity $\mathbf{v}$ of a point $\mathbf{x}$ in the rigid body is required. How can it be expressed in terms of $\mathbf{x}$ and the angular velocity $\boldsymbol{\omega}$? See Fig. 2.3.

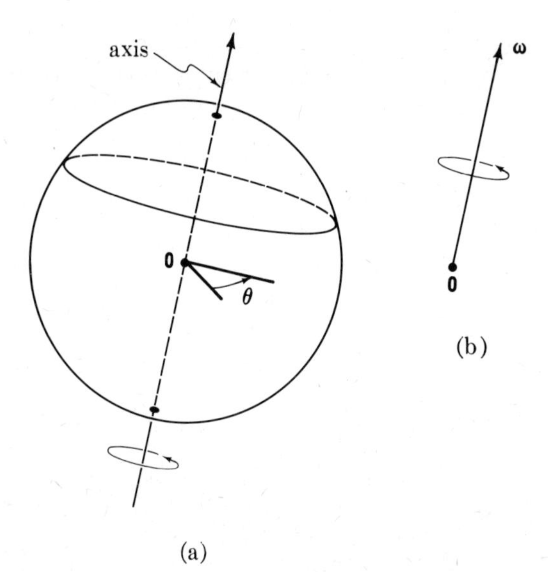

FIG. 2.2

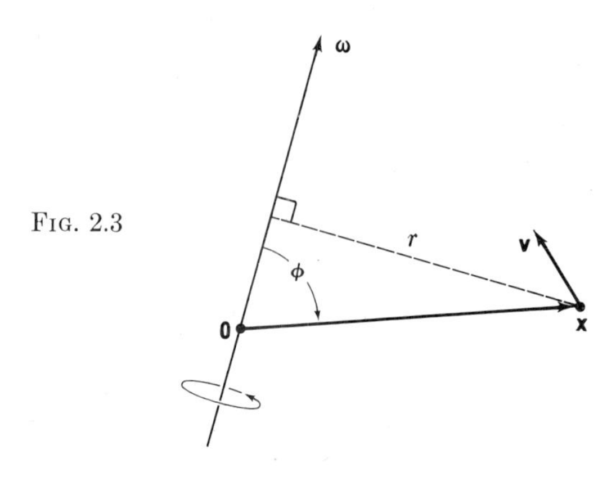

FIG. 2.3

Since the point $\mathbf{x}$ is rotating about the axis of $\boldsymbol{\omega}$, its velocity vector $\mathbf{v}$ is perpendicular to the plane of $\boldsymbol{\omega}$ and $\mathbf{x}$. By the right-hand rule, $\mathbf{v}$ points in the direction of $\boldsymbol{\omega} \times \mathbf{x}$. The speed $|\mathbf{v}|$ is the product of the angular speed $\omega = |\boldsymbol{\omega}|$ and the distance r of $\mathbf{x}$ from the axis of rotation. But $r = |\mathbf{x}| \sin \phi$, hence

$$|\mathbf{v}| = |\boldsymbol{\omega}| \cdot |\mathbf{x}| \sin \phi = |\boldsymbol{\omega} \times \mathbf{x}|.$$

Therefore:

> The velocity of a point $\mathbf{x}$ in a rigid body rotating with angular velocity $\boldsymbol{\omega}$ is
>
> $$\mathbf{v} = \boldsymbol{\omega} \times \mathbf{x}.$$

Angular Momentum

A particle of mass m has position $\mathbf{x} = \mathbf{x}(t)$ at time t and moves under the action of a (variable) force $\mathbf{F}$. The **angular momentum** of the particle with respect to the origin $\mathbf{0}$ is defined as

$$m\mathbf{x} \times \dot{\mathbf{x}}.$$

Now

$$\frac{d}{dt}[m\mathbf{x} \times \dot{\mathbf{x}}] = m\dot{\mathbf{x}} \times \dot{\mathbf{x}} + m\mathbf{x} \times \ddot{\mathbf{x}} = m\mathbf{x} \times \ddot{\mathbf{x}}$$

since $\dot{\mathbf{x}} \times \dot{\mathbf{x}} = \mathbf{0}$. But

$$m\ddot{\mathbf{x}} = \mathbf{F}$$

by Newton's Law of Motion; hence

$$m\mathbf{x} \times \ddot{\mathbf{x}} = \mathbf{x} \times m\ddot{\mathbf{x}} = \mathbf{x} \times \mathbf{F},$$

which is the torque of $\mathbf{F}$ at $\mathbf{x}$. The result is that

$$\frac{d}{dt}[m\mathbf{x} \times \dot{\mathbf{x}}] = \mathbf{x} \times \mathbf{F}.$$

Integrating,

$$m\mathbf{x} \times \dot{\mathbf{x}}\Big|_{t_0}^{t_1} = \int_{t_0}^{t_1} \mathbf{x} \times \mathbf{F}\, dt.$$

This result, called the Law of Conservation of Angular Momentum, asserts that the change in angular momentum during a motion is the time integral of the torque, called the **impulse.**

EXERCISES

1. A seesaw with unequal arms of lengths a and b is in horizontal equilibrium. Find the relations between weights A and B at the ends and the upward reaction C at the fulcrum.
2. Unit vertical forces act downward at the points $\mathbf{p}_1, \cdots, \mathbf{p}_n$ of the horizontal x, y-plane. A force $\mathbf{F}$ acts at another point $\mathbf{p}$ of the plane so that the rigid system is in equilibrium. Find $\mathbf{F}$ and $\mathbf{p}$.
3. A force $\mathbf{F}$ is applied at a point $\mathbf{x}$. Its **torque about a point p** is $(\mathbf{x} - \mathbf{p}) \times \mathbf{F}$. Suppose forces $\mathbf{F}_1, \cdots, \mathbf{F}_n$ are applied at points $\mathbf{x}_1, \cdots, \mathbf{x}_n$ of a rigid body and the body is in equilibrium. Show that the sum of the torques about $\mathbf{p}$ vanishes. (Here $\mathbf{p}$ is any point of space, not just $\mathbf{0}$.)
4. A **couple** consists of a pair of opposite forces $\mathbf{F}$ and $-\mathbf{F}$ applied at two different points $\mathbf{p}$ and $\mathbf{q}$. Show that the total torque is unchanged if $\mathbf{p}$ and $\mathbf{q}$ are displaced the same amount, i.e., replaced by $\mathbf{p} + \mathbf{c}$ and $\mathbf{q} + \mathbf{c}$. Interpret this total torque geometrically.
5. The earth turns on its axis with angular velocity 360° per day. Find the actual speed (mph) of a point on the surface (a) at the equator, (b) at the 40-th parallel, (c) at the south pole. Approximate the earth by a sphere of radius 4000 miles.
6. An electron in a uniform magnetic field follows the spiral path

$$\mathbf{x}(t) = (a \cos t, a \sin t, bt).$$

   Find its angular momentum with respect to $\mathbf{0}$.
7. A particle of unit mass moves on the unit sphere $|\mathbf{x}| = 1$ with unit speed. Show that its angular momentum with respect to $\mathbf{0}$ is a unit vector.

3. DIVERGENCE

The gradient of a function f defined on a region of space is a vector field

$$\mathbf{v} = \operatorname{grad} f$$

defined on the same region. Thus the operation grad makes vector fields out of functions. In this section and the next we introduce two further operations, div and curl. The first makes functions out of vector fields and the second makes vector fields out of vector fields. Both have important physical applications.

Consider a vector field

$$\mathbf{u} = (u, v, w),$$

where u, v, and w are functions of x, y, z defined on a region of space. The **divergence** of $\mathbf{u}$ is the function

$$\operatorname{div} \mathbf{u} = \frac{\partial u}{\partial x} + \frac{\partial v}{\partial y} + \frac{\partial w}{\partial z}.$$

EXAMPLE 3.1

Compute div $\mathbf{u}$ in each case:

(a) $\mathbf{u} = \mathbf{x} = (x, y, z)$, (b) $\mathbf{u} = (x^2y, y^2z, z^2x)$.

Solution:

$$\operatorname{div}(x, y, z) = \frac{\partial x}{\partial x} + \frac{\partial y}{\partial y} + \frac{\partial z}{\partial z} = 3,$$

$$\operatorname{div}(x^2y, y^2z, z^2x) = \frac{\partial}{\partial x}(x^2y) + \frac{\partial}{\partial y}(y^2z) + \frac{\partial}{\partial z}(z^2x) = 2xy + 2yz + 2zx.$$

Answer: (a) 3; (b) $2(xy + yz + zx)$.

The operation div, which forms div $\mathbf{u}$ out of $\mathbf{u}$, has several important properties; here are two:

$$\operatorname{div}(\mathbf{u}_1 + \mathbf{u}_2) = \operatorname{div} \mathbf{u}_1 + \operatorname{div} \mathbf{u}_2,$$

$$\operatorname{div}(f\mathbf{u}) = (\operatorname{grad} f) \cdot \mathbf{u} + f \operatorname{div} \mathbf{u},$$

where f is a function.

It is rather easy to see why the first of these is true. Here is a proof of the second:

$$\begin{aligned}
\operatorname{div}(f\mathbf{u}) = \operatorname{div}(fu, fv, fw) &= \frac{\partial(fu)}{\partial x} + \frac{\partial(fv)}{\partial y} + \frac{\partial(fw)}{\partial z} \\
&= \left(\frac{\partial f}{\partial x} u + f\frac{\partial u}{\partial x}\right) + \left(\frac{\partial f}{\partial y} v + f\frac{\partial v}{\partial y}\right) + \left(\frac{\partial f}{\partial z} w + f\frac{\partial w}{\partial z}\right) \\
&= \left(\frac{\partial f}{\partial x} u + \frac{\partial f}{\partial y} v + \frac{\partial f}{\partial z} w\right) + f\left(\frac{\partial u}{\partial x} + \frac{\partial v}{\partial y} + \frac{\partial w}{\partial z}\right) \\
&= (\operatorname{grad} f) \cdot \mathbf{u} + f \operatorname{div} \mathbf{u}.
\end{aligned}$$

Probably the most striking property of the divergence is its independence of the rectangular coordinate system. The field div $\mathbf{u}$ depends only on $\mathbf{u}$, not on the particular choice of the x, y, z-axes. This is a deep result whose proof is best postponed until an advanced calculus course.

EXAMPLE 3.2

Compute $\operatorname{div}(ye^{xyz}, ze^{xyz}, xe^{xyz})$.

Solution:

$$\begin{aligned}\operatorname{div}(ye^{xyz}, ze^{xyz}, xe^{xyz}) &= \operatorname{div}[e^{xyz}(y, z, x)]\\ &= [\operatorname{grad} e^{xyz}]\cdot(y, z, x) + e^{xyz}\operatorname{div}(y, z, x)\\ &= [e^{xyz}(yz, zx, xy)]\cdot(y, z, x) + 0.\end{aligned}$$

Answer: $(y^2z + z^2x + x^2y)e^{xyz}$.

EXAMPLE 3.3

Compute

$$\operatorname{div}\left(\frac{1}{\rho^3}\mathbf{x}\right), \qquad \text{where} \quad \rho = |\mathbf{x}|.$$

Solution: Apply the formula for $\operatorname{div}(f\mathbf{u})$ with $f(\mathbf{x}) = \rho^{-3}$ and $\mathbf{u} = \mathbf{x}$:

$$\operatorname{div}\left(\frac{1}{\rho^3}\mathbf{x}\right) = \operatorname{grad}\left(\frac{1}{\rho^3}\right)\cdot\mathbf{x} + \frac{3}{\rho^3},$$

since $\operatorname{div}\mathbf{x} = 3$ by Example 3.1. Now $\rho^2 = x^2 + y^2 + z^2$, hence $2\rho\rho_x = 2x$, so $\rho_x = x/\rho$. Likewise $\rho_y = y/\rho$ and $\rho_z = z/\rho$. Consequently,

$$\operatorname{grad}\left(\frac{1}{\rho^3}\right) = \left(\frac{-3}{\rho^4}\right)(\rho_x, \rho_y, \rho_z) = \left(\frac{-3}{\rho^4}\right)\left(\frac{x}{\rho}, \frac{y}{\rho}, \frac{z}{\rho}\right) = \frac{-3}{\rho^5}\mathbf{x}:$$

Therefore

$$\operatorname{div}\left(\frac{1}{\rho^3}\mathbf{x}\right) = \frac{-3}{\rho^5}\mathbf{x}\cdot\mathbf{x} + \frac{3}{\rho^3} = \frac{-3}{\rho^5}\rho^2 + \frac{3}{\rho^3} = 0.$$

Answer: 0.

Here are several examples showing the importance of the divergence in physics.

(1) Suppose an incompressible fluid is flowing in a region of space and $\mathbf{v} = \mathbf{v}(t, x, y, z) = (u, v, w)$ represents the velocity of the fluid at time t and position $\mathbf{x} = (x, y, z)$. Then for each t,

$$\operatorname{div}\mathbf{v} = \frac{\partial u}{\partial x} + \frac{\partial v}{\partial y} + \frac{\partial w}{\partial z} = 0.$$

This is the Equation of Continuity, or Law of Conservation of Matter. (Note t is constant in the calculation. A physicist would call this the divergence with respect to the "space variables.")

(2) Suppose a compressible fluid (gas) is flowing through a region. Let $\mathbf{v}$ denote the velocity as before and let $\rho(t, \mathbf{x})$ denote the density of the fluid

(mass per unit volume). Then the Equation of Continuity is

$$\operatorname{div}(\rho\mathbf{v}) + \frac{\partial \rho}{\partial t} = 0.$$

(3) Suppose electric charge is scattered through a region with charge density ρ esu (electrostatic units)/cm^3. Let $\mathbf{E}$ be the electric field intensity, a vector field whose magnitude is measured in dynes per unit charge. The value of the field intensity $\mathbf{E}(\mathbf{x})$ represents the force exerted by the field on a charge of 1 esu at $\mathbf{x}$. Then Coulomb's Law (also called Gauss's Law) says

$$\operatorname{div} \mathbf{E} = 4\pi\rho.$$

(4) There is a similar Gauss's Law for static magnetic fields:

$$\operatorname{div} \mathbf{H} = 4\pi\rho,$$

where $\mathbf{H}$ is the magnetic field intensity (gauss) and ρ is the pole strength per cubic centimeter.

EXERCISES

Find the divergence of the vector field:

1. $(2xy, 3, \sin z)$
2. $(x/y, y/z, z/x)$
3. (x^3, y^3, z^3)
4. (y^2z, z^2x, x^2y)
5. $(x^2, y^2, 0)$
6. (z, z, z)
7. (ρ, ρ, ρ)
8. $e^{-\rho}(1, 1, 1)$
9. $(f(\mathbf{x}), f(\mathbf{x}), f(\mathbf{x}))$
10. $g(\mathbf{x})\mathbf{x}$.

Find $\operatorname{div}(\operatorname{grad} f)$:

11. $f = \rho$
12. $f = \sin(xyz)$
13. $f = \rho^2$
14. $f = \rho^3$
15. $f = xe^{yz}$
16. $f = \rho^{-1}$
17. $f = xy/z$
18. $f = \rho^{-2}$.

Find $\operatorname{div}(\mathbf{a} \times \mathbf{u})$:

19. $\mathbf{a} = (1, 1, 1), \quad \mathbf{u} = (x^3, y^3, z^3)$
20. $\mathbf{a} = (3, 0, -1), \quad \mathbf{u} = (x^4, z^4, y^4)$.
21. Find $\operatorname{div}[(x, y, z) \times (x^2, y^2, z^2)]$.
22. Find all n satisfying $\operatorname{div}(\rho^n\mathbf{x}) = 0$.
23. For a plane field $\mathbf{u} = (u(x, y), v(x, y))$, define $\operatorname{div} \mathbf{u} = u_x + v_y$. Find all n satisfying $\operatorname{div}(r^n\mathbf{x}) = 0$, where $r^2 = x^2 + y^2$ and $\mathbf{x} = (x, y)$.
24. (This exercise shows that the divergence of a plane vector field is invariant under rotation.) Rotate coordinates in the plane through an angle α. Thus

$x = \bar{x}\cos\alpha - \bar{y}\sin\alpha$, $y = \bar{x}\sin\alpha + \bar{y}\cos\alpha$. A plane field $\mathbf{u} = (u, v)$ in x, y-coordinates has $\bar{x}, \bar{y}$-coordinates given by

$$u = \bar{u}\cos\alpha - \bar{v}\sin\alpha, \qquad v = \bar{u}\sin\alpha + \bar{v}\cos\alpha.$$

Show that

$$\frac{\partial u}{\partial x} + \frac{\partial v}{\partial y} = \frac{\partial \bar{u}}{\partial \bar{x}} + \frac{\partial \bar{v}}{\partial \bar{y}}.$$

[*Hint:* First derive

$$\left(\frac{\partial}{\partial x}, \frac{\partial}{\partial y}\right) = \left(\frac{\partial}{\partial \bar{x}}, \frac{\partial}{\partial \bar{y}}\right)\begin{pmatrix} \cos\alpha & \sin\alpha \\ -\sin\alpha & \cos\alpha \end{pmatrix}$$

and

$$\begin{pmatrix} u \\ v \end{pmatrix} = \begin{pmatrix} \cos\alpha & -\sin\alpha \\ \sin\alpha & \cos\alpha \end{pmatrix}\begin{pmatrix} \bar{u} \\ \bar{v} \end{pmatrix}.$$

Then multiply matrices.]

25. Find the conditions on the constants A, B, C forced by $\operatorname{div}[\operatorname{grad}(Ax^2 + By^2 + Cz^2)] = 0$.
26. Show that the function $f = x^3 - 3xy^2$ satisfies $\operatorname{div}(\operatorname{grad} f) = 0$.
27. (two variables again) Let $z = x + iy$. Let $u(x, y) = \operatorname{Re}(z^4)$. Show that $\operatorname{div}(\operatorname{grad} u) = 0$.
28. (cont.) Do the same for $\operatorname{Re}(z^{-1})$ and $\operatorname{Re}(z^{-2})$.

4. CURL

Consider a vector field

$$\mathbf{u} = (u, v, w),$$

where u, v, and w are functions of x, y, z on a region of space. The **curl** (or **rotation**) of $\mathbf{u}$ is the vector field

$$\operatorname{curl}\mathbf{u} = (w_y - v_z,\, u_z - w_x,\, v_x - u_y),$$

where $w_y = \partial w/\partial y$, etc.

A device for remembering the order of the various letters in curl $\mathbf{u}$ is the symbolic "determinant"

$$\operatorname{curl}\mathbf{u} = \begin{vmatrix} \mathbf{i} & \mathbf{j} & \mathbf{k} \\ \dfrac{\partial}{\partial x} & \dfrac{\partial}{\partial y} & \dfrac{\partial}{\partial z} \\ u & v & w \end{vmatrix},$$

expanded by minors of the first row:

$$\operatorname{curl}\mathbf{u} = \begin{vmatrix} \dfrac{\partial}{\partial y} & \dfrac{\partial}{\partial z} \\ v & w \end{vmatrix}\mathbf{i} - \begin{vmatrix} \dfrac{\partial}{\partial x} & \dfrac{\partial}{\partial z} \\ u & w \end{vmatrix}\mathbf{j} + \begin{vmatrix} \dfrac{\partial}{\partial x} & \dfrac{\partial}{\partial y} \\ u & v \end{vmatrix}\mathbf{k}.$$

Like the divergence, curl **u** depends only on **u**, not on the choice of the x, y, z-axes. Again this is too complicated to prove here.

Here are two identities satisfied by the curl:

$$\operatorname{curl}(\mathbf{u}_1 + \mathbf{u}_2) = \operatorname{curl} \mathbf{u}_1 + \operatorname{curl} \mathbf{u}_2,$$

$$\operatorname{curl}(f\mathbf{u}) = (\operatorname{grad} f) \times \mathbf{u} + f \operatorname{curl} \mathbf{u}.$$

The first formula is easy to verify. Here is the derivation of the second:

$$\begin{aligned}
\operatorname{curl}(f\mathbf{u}) &= \operatorname{curl}(fu, fv, fw) \\
&= ((fw)_y - (fv)_z, (fu)_z - (fw)_x, (fv)_x - (fu)_y) \\
&= ([fw_y + f_y w - fv_z - f_z v], [fu_z + f_z u - fw_x - f_x w], \\
&\qquad [fv_x + f_x v - fu_y - f_y u]) \\
&= f \cdot (w_y - v_z, u_z - w_x, v_x - u_y) + (f_y w - f_z v, f_z u - f_x w, f_x v - f_y u) \\
&= f \operatorname{curl} \mathbf{u} + (f_x, f_y, f_z) \times (u, v, w) = f \operatorname{curl} \mathbf{u} + (\operatorname{grad} f) \times \mathbf{u}.
\end{aligned}$$

EXAMPLE 4.1

Find curl **u** in each case:

(a) $\mathbf{u} = \mathbf{x} = (x, y, z)$, (b) $\mathbf{u} = (z, x, y)$,
(c) $\mathbf{u} = (yz, zx, xy)$, (d) $\mathbf{u} = (xyz, xyz, xyz)$.

Solution:

$$\operatorname{curl}(x, y, z) = \left(\frac{\partial z}{\partial y} - \frac{\partial y}{\partial z}, \frac{\partial x}{\partial z} - \frac{\partial z}{\partial x}, \frac{\partial y}{\partial x} - \frac{\partial x}{\partial y}\right) = \mathbf{0},$$

$$\operatorname{curl}(z, x, y) = \left(\frac{\partial y}{\partial y} - \frac{\partial x}{\partial z}, \frac{\partial z}{\partial z} - \frac{\partial y}{\partial x}, \frac{\partial x}{\partial x} - \frac{\partial z}{\partial y}\right) = (1, 1, 1),$$

$$\operatorname{curl}(yz, zx, xy) = \begin{vmatrix} \mathbf{i} & \mathbf{j} & \mathbf{k} \\ \dfrac{\partial}{\partial x} & \dfrac{\partial}{\partial y} & \dfrac{\partial}{\partial z} \\ yz & zx & xy \end{vmatrix} = \begin{vmatrix} \dfrac{\partial}{\partial y} & \dfrac{\partial}{\partial z} \\ zx & xy \end{vmatrix} \mathbf{i} - \begin{vmatrix} \dfrac{\partial}{\partial x} & \dfrac{\partial}{\partial z} \\ yz & xy \end{vmatrix} \mathbf{j} + \begin{vmatrix} \dfrac{\partial}{\partial x} & \dfrac{\partial}{\partial y} \\ yz & zx \end{vmatrix} \mathbf{k}$$

$$= (x - x)\mathbf{i} - (y - y)\mathbf{j} + (z - z)\mathbf{k} = \mathbf{0},$$

$$\begin{aligned}
\operatorname{curl}(xyz, xyz, xyz) &= \operatorname{curl}[xyz(1, 1, 1)] \\
&= [\operatorname{grad}(xyz)] \times (1, 1, 1) + (xyz) \operatorname{curl}(1, 1, 1) \\
&= (yz, zx, xy) \times (1, 1, 1) + \mathbf{0} \\
&= (zx - xy, xy - yz, yz - zx).
\end{aligned}$$

Answer: (a) $\mathbf{0}$; (b) $(1, 1, 1)$; (c) $\mathbf{0}$; (d) $(zx - xy, xy - yz, yz - zx)$.

A vector field $\mathbf{u}$ is called **irrotational** if curl $\mathbf{u} = \mathbf{0}$.

Each gradient field is irrotational. Briefly stated,

$$\text{curl grad } \mathbf{u} = \mathbf{0}.$$

For suppose

$$\mathbf{u} = \text{grad } f = (f_x, f_y, f_z).$$

Then

$$\text{curl } \mathbf{u} = ((f_z)_y - (f_y)_z, (f_x)_z - (f_z)_x, (f_y)_x - (f_x)_y) = (0, 0, 0) = \mathbf{0}.$$

EXAMPLE 4.2

Show that the gravitational force field due to a point mass at the origin is irrotational.

Solution: The force $\mathbf{F}$ at $\mathbf{x}$ is directed towards $\mathbf{0}$ and, by the inverse square law, has magnitude proportional to ρ^{-2}, where $\rho = |\mathbf{x}|$. Hence

$$\mathbf{F} = \frac{-k}{\rho^2}\frac{\mathbf{x}}{\rho} = \frac{-k}{\rho^3}\mathbf{x} = -k\left(\frac{x}{\rho^3}, \frac{y}{\rho^3}, \frac{z}{\rho^3}\right).$$

The first component of curl $\mathbf{F}$ is

$$-k\left[\frac{\partial}{\partial y}\left(\frac{z}{\rho^3}\right) - \frac{\partial}{\partial z}\left(\frac{y}{\rho^3}\right)\right] = \frac{3k}{\rho^4}\left(z\frac{\partial \rho}{\partial y} - y\frac{\partial \rho}{\partial z}\right).$$

This equals 0 since

$$\frac{\partial \rho}{\partial y} = \frac{y}{\rho}, \qquad \frac{\partial \rho}{\partial z} = \frac{z}{\rho}$$

(differentiate $\rho^2 = x^2 + y^2 + z^2$). Likewise the other components of curl $\mathbf{F}$ equal 0.

Alternative Solution: Observe that

$$\frac{1}{k}\mathbf{F} = -\frac{1}{\rho^3}\mathbf{x} = -\frac{1}{\rho^2}\left(\frac{x}{\rho}, \frac{y}{\rho}, \frac{z}{\rho}\right) = -\frac{1}{\rho^2}(\rho_x, \rho_y, \rho_z) = \text{grad}\left(\frac{1}{\rho}\right).$$

Hence $\mathbf{F}$ is a gradient field,

$$\mathbf{F} = \text{grad}\left(\frac{k}{\rho}\right).$$

But each gradient field is irrotational, so curl $\mathbf{F} = \mathbf{0}$.

A vector field **u** is called **solenoidal** if div **u** = 0.

> Each vector field of the form curl **u** is solenoidal. Briefly stated,
> $$\operatorname{div} \operatorname{curl} \mathbf{u} = 0.$$

For suppose $\mathbf{u} = (u, v, w)$. Then

$$\operatorname{curl} \mathbf{u} = (w_y - v_z, u_z - w_x, v_x - u_y).$$

Hence

$$\begin{aligned}\operatorname{div}(\operatorname{curl} \mathbf{u}) &= (w_y - v_z)_x + (u_z - w_x)_y + (v_x - u_y)_z \\ &= w_{yx} - v_{zx} + u_{zy} - w_{xy} + v_{xz} - u_{yz} \\ &= (w_{yx} - w_{xy}) + (u_{zy} - u_{yz}) + (v_{xz} - v_{zx}) = 0,\end{aligned}$$

since the mixed second partials are equal.

Here are several examples showing the use of the curl in physics.

(1) A vector field **v** is called **conservative** if $\mathbf{v} = -\operatorname{grad} \phi$ for some function ϕ. Each electrostatic field **E** is conservative. If $\mathbf{E} = -\operatorname{grad} \phi$, then ϕ is the **electric potential.** Since each gradient field is irrotational,

$$\operatorname{curl} \mathbf{E} = \mathbf{0}.$$

(2) Faraday's Law. Let **E** be the electric field intensity and **B** the magnetic induction in an electromagnetic field. Then

$$\operatorname{curl} \mathbf{E} = -\frac{\partial \mathbf{B}}{\partial t}\cdot$$

(3) Ampère's Law. Let **H** be the magnetic field intensity and **J** the electric current density in an electromagnetic field. Then

$$\operatorname{curl} \mathbf{H} = 4\pi \mathbf{J}.$$

Angular Velocity

We continue the discussion of angular velocity begun in Section 2, which you should review. The new relation we shall derive depends on the following result.

> Let $\mathbf{u} = (u, v, w)$ be a vector field and **a** a constant vector. Then
> $$\operatorname{curl}(\mathbf{a} \times \mathbf{u}) = (\operatorname{div} \mathbf{u})\mathbf{a} - (\mathbf{a} \cdot \operatorname{grad} u, \mathbf{a} \cdot \operatorname{grad} v, \mathbf{a} \cdot \operatorname{grad} w).$$

For let $\mathbf{a} = (a, b, c)$. Then

$$\mathbf{a} \times \mathbf{u} = (bw - cv, cu - aw, av - bu).$$

The first component of curl($\mathbf{a} \times \mathbf{u}$) is

$$\begin{aligned}(av - bu)_y - (cu - aw)_z &= (av_y - bu_y) - (cu_z - aw_z)\\ &= (av_y + aw_z) - (bu_y + cu_z)\\ &= (au_x + av_y + aw_z) - (au_x + bu_y + cu_z)\\ &= a \operatorname{div} \mathbf{u} - \mathbf{a} \cdot \operatorname{grad} u.\end{aligned}$$

Similarly the remaining components are

$$b \operatorname{div} \mathbf{u} - \mathbf{a} \cdot \operatorname{grad} v \qquad \text{and} \qquad c \operatorname{div} \mathbf{u} - \mathbf{a} \cdot \operatorname{grad} w,$$

so finally

$$\begin{aligned}\operatorname{curl}(\mathbf{a} \times \mathbf{u}) &= (a \operatorname{div} \mathbf{u}, b \operatorname{div} \mathbf{u}, c \operatorname{div} \mathbf{u}) - (\mathbf{a} \cdot \operatorname{grad} u, \mathbf{a} \cdot \operatorname{grad} v, \mathbf{a} \cdot \operatorname{grad} w)\\ &= (\operatorname{div} \mathbf{u})\mathbf{a} - (\mathbf{a} \cdot \operatorname{grad} u, \mathbf{a} \cdot \operatorname{grad} v, \mathbf{a} \cdot \operatorname{grad} w).\end{aligned}$$

Now suppose a rigid body is free to rotate about the origin $\mathbf{0}$, which is fixed. At each instant of time there is an instantaneous axis of rotation and an instantaneous angular velocity vector $\boldsymbol{\omega} = \boldsymbol{\omega}(t)$. The velocity of $\mathbf{x}$ at time t is

$$\mathbf{v} = \mathbf{v}(t, \mathbf{x}) = \boldsymbol{\omega}(t) \times \mathbf{x},$$

as derived on p. 794. Problem: compute curl $\mathbf{v}$. Note that this computation takes place at an "instant of time" because only the partials $\partial/\partial x$, $\partial/\partial y$, $\partial/\partial z$ are involved, not $\partial/\partial t$.

The formula above for curl($\mathbf{a} \times \mathbf{u}$) applies with $\mathbf{a} = \boldsymbol{\omega}(t)$ and $\mathbf{u} = \mathbf{x}$. Now

$$\operatorname{div} \mathbf{x} = \frac{\partial x}{\partial x} + \frac{\partial y}{\partial y} + \frac{\partial z}{\partial z} = 3,$$

and

$$\operatorname{grad} x = (1, 0, 0) = \mathbf{i}, \qquad \operatorname{grad} y = \mathbf{j}, \qquad \operatorname{grad} z = \mathbf{k}.$$

Hence

$$\begin{aligned}\operatorname{curl} \mathbf{v} &= \operatorname{curl}(\boldsymbol{\omega} \times \mathbf{x})\\ &= 3\boldsymbol{\omega} - (\boldsymbol{\omega} \cdot \mathbf{i}, \boldsymbol{\omega} \cdot \mathbf{j}, \boldsymbol{\omega} \cdot \mathbf{k})\\ &= 3\boldsymbol{\omega} - \boldsymbol{\omega} = 2\boldsymbol{\omega}.\end{aligned}$$

This calculation yields an important result:

> In a rotation of a rigid body about the (fixed) origin, the instantaneous angular velocity is given by
>
> $$\boldsymbol{\omega} = \frac{1}{2} \operatorname{curl} \mathbf{v},$$
>
> where $\mathbf{v}$ is the instantaneous velocity field.

This result explains the name curl or rotation.

EXERCISES

Compute the curl:

1. $(f(x), g(y), h(z))$
2. (x^2, x^2, y^2)
3. $(y^2 - z^2, z^2 - x^2, x^2 - y^2)$
4. (xe^y, ye^z, ze^x)
5. (y^4, z^4, x^4)
6. $f(x, y, z)\,\text{grad}\, g$
7. $f(x, y, z)\mathbf{c}$, $\mathbf{c}$ constant
8. (x, x^2, x^3).
9. Find $\text{curl}[(1, 1, 1) \times (x, y, z)]$.
10. Find $\text{curl}[(x, y, z) \times (x^2, y^2, z^2)]$.
11. Which fields of the form $\mathbf{u} = (f(x, y, z), 0, 0)$ are irrotational?
12. (cont.) Let $\mathbf{c}$ be a non-zero constant vector. Which fields of the form $\mathbf{u} = f(x, y, z)\mathbf{c}$ are irrotational?
13. Which fields of the form $\mathbf{u} = (f(x), 0, 0)$ are solenoidal?
14. Which fields of the form $\mathbf{u} = (f(x), g(y), h(z))$ are solenoidal?

5. THE OPERATOR ∇

We compare the formula for curl $\mathbf{u}$,

$$\text{curl}\, \mathbf{u} = \begin{vmatrix} \mathbf{i} & \mathbf{j} & \mathbf{k} \\ \dfrac{\partial}{\partial x} & \dfrac{\partial}{\partial y} & \dfrac{\partial}{\partial z} \\ u & v & w \end{vmatrix},$$

with the corresponding formula for the cross product:

$$(r, s, t) \times (u, v, w) = \begin{vmatrix} \mathbf{i} & \mathbf{j} & \mathbf{k} \\ r & s & t \\ u & v & w \end{vmatrix}.$$

The comparison suggests we consider the triple $(\partial/\partial x, \partial/\partial y, \partial/\partial z)$ as a vector in some sense. We do just this, and define an operator called **del** (also **nabla**):

$$\nabla = \left(\frac{\partial}{\partial x}, \frac{\partial}{\partial y}, \frac{\partial}{\partial z}\right) = \mathbf{i}\frac{\partial}{\partial x} + \mathbf{j}\frac{\partial}{\partial y} + \mathbf{k}\frac{\partial}{\partial z}.$$

This operator can be applied to functions and, in two ways, to vector fields. If f is a function,

$$\nabla f = \left(\mathbf{i}\frac{\partial}{\partial x} + \mathbf{j}\frac{\partial}{\partial y} + \mathbf{k}\frac{\partial}{\partial z}\right) f = \mathbf{i}\frac{\partial f}{\partial x} + \mathbf{j}\frac{\partial f}{\partial y} + \mathbf{k}\frac{\partial f}{\partial z} = \text{grad}\, f.$$

If $\mathbf{u} = (u, v, w)$ is a vector field,

$$\nabla \cdot \mathbf{u} = \left(\mathbf{i}\frac{\partial}{\partial x} + \mathbf{j}\frac{\partial}{\partial y} + \mathbf{k}\frac{\partial}{\partial z}\right) \cdot (u, v, w) = \frac{\partial u}{\partial x} + \frac{\partial v}{\partial y} + \frac{\partial w}{\partial z} = \text{div}\, \mathbf{u}.$$

Also

$$\nabla \times \mathbf{u} = \begin{vmatrix} \mathbf{i} & \mathbf{j} & \mathbf{k} \\ \dfrac{\partial}{\partial x} & \dfrac{\partial}{\partial y} & \dfrac{\partial}{\partial z} \\ u & v & w \end{vmatrix} = \text{curl } \mathbf{u}.$$

Thus gradient, divergence, and curl can be expressed in terms of ∇. Incidentally one reads ∇f as "del f," $\nabla \cdot \mathbf{u}$ as "del dot $\mathbf{u}$," and $\nabla \times \mathbf{u}$ as "del cross $\mathbf{u}$."

An important second order operator is the **Laplacian**

$$\nabla^2 = \nabla \cdot \nabla = \frac{\partial^2}{\partial x^2} + \frac{\partial^2}{\partial y^2} + \frac{\partial^2}{\partial z^2}.$$

For example, if ϕ is the potential of an electrostatic field, then the electric field intensity is $\mathbf{E} = -\nabla\phi$, and Coulomb's Law is

$$\nabla \cdot \mathbf{E} = 4\pi\rho,$$

where ρ is the charge density. Consequently

$$\nabla^2\phi = -\nabla \cdot \mathbf{E} = -4\pi\rho,$$

that is,

$$\frac{\partial^2\phi}{\partial x^2} + \frac{\partial^2\phi}{\partial y^2} + \frac{\partial^2\phi}{\partial z^2} + 4\pi\rho = 0.$$

This is an important relation called **Poisson's Equation**; it occurs in several branches of theoretical physics besides electrostatics.

EXERCISES

1. Find $\nabla^2(\rho^n)$, where $\rho^2 = x^2 + y^2 + z^2$.
2. Suppose a vector field $\mathbf{E} = (u, v, w)$ is both irrotational and solenoidal. Show that $\nabla^2 u = \nabla^2 v = \nabla^2 w = 0$.
3. Suppose a field $\mathbf{E}$ is expressible as the gradient of two different functions f and g. Find the relation between f and g.
4. Show that $\nabla \cdot (f\nabla g) = \nabla f \cdot \nabla g + f\nabla^2 g$.
5. Show that $\nabla \cdot (f\nabla f) = |\nabla f|^2 + f\nabla^2 f$.
6. Show that $\nabla \cdot (f\nabla g - g\nabla f) = f\nabla^2 g - g\nabla^2 f$.
7. Analogous to the expression for $\mathbf{u} \cdot (\mathbf{v} \times \mathbf{w})$ as a determinant, we have

$$\nabla \cdot (\mathbf{u} \times \mathbf{v}) = \begin{vmatrix} \dfrac{\partial}{\partial x} & \dfrac{\partial}{\partial y} & \dfrac{\partial}{\partial z} \\ u_1 & u_2 & u_3 \\ v_1 & v_2 & v_3 \end{vmatrix}.$$

   Does this make any sense?
8. Likewise, analogous to the formula $\mathbf{u} \times (\mathbf{v} \times \mathbf{w}) = (\mathbf{u} \cdot \mathbf{w})\mathbf{v} - (\mathbf{u} \cdot \mathbf{v})\mathbf{w}$, we have $\nabla \times (\mathbf{v} \times \mathbf{w}) = (\nabla \cdot \mathbf{w})\mathbf{v} - (\nabla \cdot \mathbf{v})\mathbf{w}$. Is this formula correct?

9. Is this reasoning acceptable?

$$\operatorname{curl}(\operatorname{grad} f) = \nabla \times (\nabla f) = (\nabla \times \nabla) f = \mathbf{0}.$$

10. Is this reasoning acceptable?

$$\operatorname{div}(\operatorname{curl} \mathbf{v}) = \nabla \cdot (\nabla \times \mathbf{v}) = \begin{vmatrix} \frac{\partial}{\partial x} & \frac{\partial}{\partial y} & \frac{\partial}{\partial z} \\ \frac{\partial}{\partial x} & \frac{\partial}{\partial y} & \frac{\partial}{\partial z} \\ v_1 & v_2 & v_3 \end{vmatrix} = 0,$$

since a determinant with two equal rows vanishes.

11. Refer to Ex. 8. Can you give some meaning to

$$\nabla \times (\nabla \times \mathbf{w}) = (\nabla \cdot \mathbf{w})\nabla - (\nabla \cdot \nabla)\mathbf{w} \,?$$

Define ∇^2 applied to a vector by $\nabla^2(u, v, w) = (\nabla^2 u, \nabla^2 v, \nabla^2 w)$.

12. Prove $\nabla^2[\mathbf{x} \times (\nabla f)] = \mathbf{x} \times [\nabla(\nabla^2 f)]$.

13. Let Ω denote the **vector angular momentum operator,**

$$\Omega = \mathbf{x} \times \nabla = \left(y \frac{\partial}{\partial z} - z \frac{\partial}{\partial y}, z \frac{\partial}{\partial x} - x \frac{\partial}{\partial z}, x \frac{\partial}{\partial y} - y \frac{\partial}{\partial x}\right).$$

Prove $\Omega \times (\Omega f) = -\Omega f$.

14. (cont.) Prove

$$\Omega \cdot (\Omega f) = -2\, \mathbf{x} \cdot (\nabla f) + |\mathbf{x}|^2 \nabla^2 f - (x, y, z) \begin{pmatrix} f_{xx} & f_{xy} & f_{xz} \\ f_{yx} & f_{yy} & f_{yz} \\ f_{zx} & f_{zy} & f_{zz} \end{pmatrix} \begin{pmatrix} x \\ y \\ z \end{pmatrix}.$$

15. (cont.) Interpret Ex. 12 as $\nabla^2(\Omega f) = \Omega(\nabla^2 f)$.

6. CYLINDRICAL COORDINATES

Cylindrical coordinates are designed to fit situations with rotational (axial) symmetry about an axis.

The **cylindrical coordinates** of a point $\mathbf{x} = (x, y, z)$ are $\{r, \theta, z\}$, where $\{r, \theta\}$ are the polar coordinates of (x, y) and z is the third rectangular coordinate (Fig. 6.1a). Each surface r = constant is a right circular cylinder, hence the name, cylindrical coordinates (Fig. 6.1b).

Through each point $\mathbf{x}$ (not on the z-axis) pass three surfaces, r = constant, θ = constant, z = constant (Fig. 6.2). Each is orthogonal (perpendicular) to the other two at their common intersection $\mathbf{x}$.

The relations between the rectangular coordinates (x, y, z) and the cylindrical coordinates $\{r, \theta, z\}$ of a point are

$$\begin{cases} x = r\cos\theta \\ y = r\sin\theta \\ z = z \end{cases} \qquad \begin{cases} r^2 = x^2 + y^2 \\ \cos\theta = \dfrac{x}{r}, \quad \sin\theta = \dfrac{y}{r} \\ z = z. \end{cases}$$

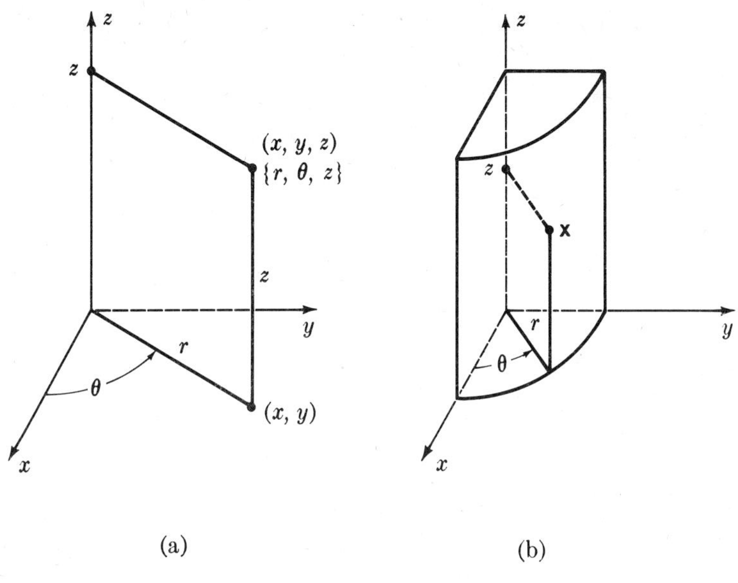

Fig. 6.1

The origin in the plane is given in polar coordinates by $r = 0$; the angle θ is undefined. Similarly, a point on the z-axis is given in cylindrical coordinates by $r = 0$, $z =$ constant; θ is undefined.

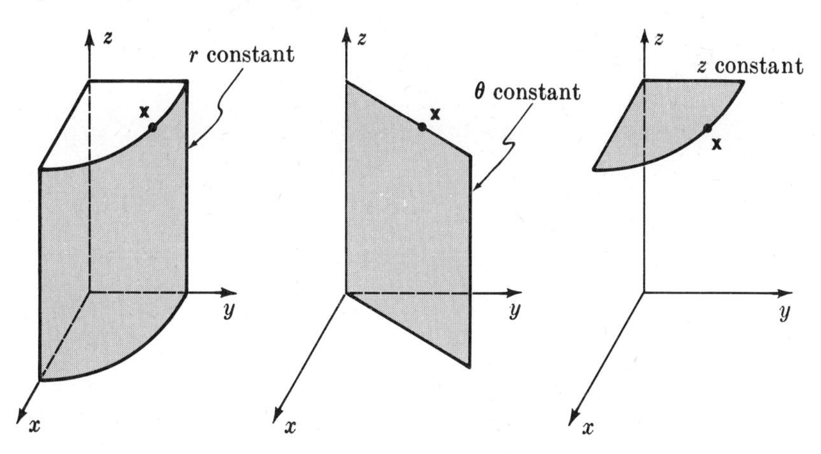

Fig. 6.2

EXAMPLE 6.1

Graph the surfaces (i) $z = 2r$, (ii) $z = r^2$.

Solution: Both are surfaces of revolution about the z-axis, as is any surface $z = f(r)$. Since z depends only on r, not on θ, the height of the surface is constant above each circle $r = c$ in the x, y-plane. Thus the level curves are circles in the x, y-plane centered at the origin.

In (i), the surface meets the first quadrant of the y, z-plane in the line $z = 2y$. (In the first quadrant of the y, z-plane, $x = 0$ and $y \geq 0$. Since $r^2 = x^2 + y^2 = y^2$, it follows that $r = y$.) Rotated about the z-axis, this line spans a cone with apex at $\mathbf{0}$. See Fig. 6.3a.

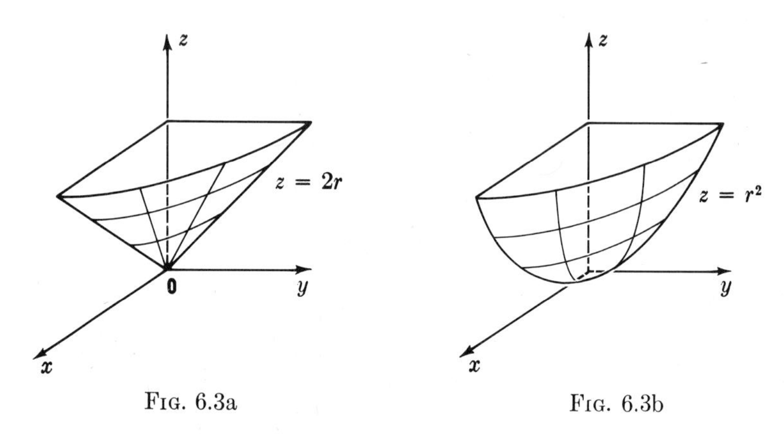

FIG. 6.3a

FIG. 6.3b

In (ii), the surface meets the y, z-plane in the parabola $z = y^2$. Rotated about the z-axis, this parabola generates a paraboloid of revolution (Fig. 6.3b).

EXAMPLE 6.2

Find the level surfaces of the function $f(r, \theta, z) = r$.

Solution: Each surface is defined by $r = c$. This is a right circular cylinder whose axis is the z-axis (Fig. 6.4).

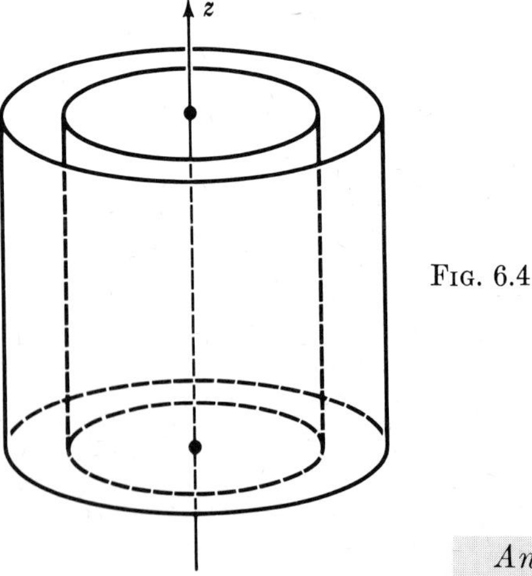

FIG. 6.4

Answer: Concentric right circular cylinders about the z-axis.

The Natural Frame

It is convenient to fit a frame of three mutually perpendicular vectors to cylindrical coordinates just as the frame **i**, **j**, **k** fits rectangular coordinates. At each point $\{r, \theta, z\}$ of space attach three mutually perpendicular unit vectors **u**, **w**, **k** chosen so

$$\begin{Bmatrix} \mathbf{u} \\ \mathbf{w} \\ \mathbf{k} \end{Bmatrix} \text{ points in the direction of increasing } \begin{Bmatrix} r \\ \theta \\ z \end{Bmatrix}.$$

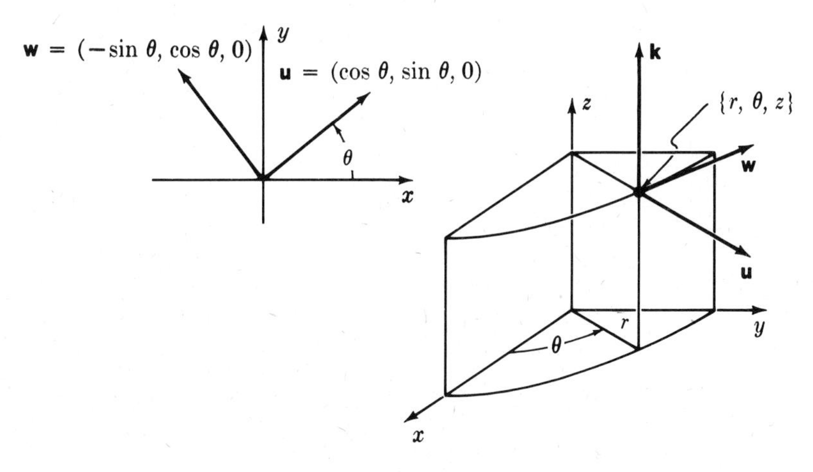

FIG. 6.5

Thus (Fig. 6.5)

$$\mathbf{u} = \frac{1}{r}(x, y, 0) = (\cos\theta, \sin\theta, 0),$$

$$\mathbf{w} = \frac{1}{r}(-y, x, 0) = (-\sin\theta, \cos\theta, 0),$$

$$\mathbf{k} = (0, 0, 1).$$

The vectors **u**, **w**, **k** form a right-hand system:

$$\mathbf{u} \times \mathbf{w} = \mathbf{k}, \qquad \mathbf{w} \times \mathbf{k} = \mathbf{u}, \qquad \mathbf{k} \times \mathbf{u} = \mathbf{w}.$$

Note that **u** and **w** depend on θ alone, while **k** is a constant vector, our old friend from the trio **i**, **j**, **k**. Note also

$$\frac{\partial \mathbf{u}}{\partial \theta} = \mathbf{w}, \qquad \frac{\partial \mathbf{w}}{\partial \theta} = -\mathbf{u}.$$

In situations with axial symmetry, it is frequently better to express vectors in terms of **u**, **w**, **k** rather than **i**, **j**, **k**.

The Operator Del

For a function $f(x, y, z)$,

$$\operatorname{grad} f = \nabla f = \mathbf{i}\frac{\partial f}{\partial x} + \mathbf{j}\frac{\partial f}{\partial y} + \mathbf{k}\frac{\partial f}{\partial z}.$$

Suppose, however, a function $f(r, \theta, z)$ is given in cylindrical coordinates. How can grad f be expressed?

Given $f(r, \theta, z)$,

$$\operatorname{grad} f = \frac{\partial f}{\partial r}\mathbf{u} + \frac{1}{r}\frac{\partial f}{\partial \theta}\mathbf{w} + \frac{\partial f}{\partial z}\mathbf{k}.$$

This formula gives the components of grad f in the three perpendicular directions **u**, **w**, **k**; it asserts that

$$(\operatorname{grad} f)\cdot\mathbf{u} = \frac{\partial f}{\partial r}, \qquad (\operatorname{grad} f)\cdot\mathbf{w} = \frac{1}{r}\frac{\partial f}{\partial \theta}, \qquad (\operatorname{grad} f)\cdot\mathbf{k} = \frac{\partial f}{\partial z}.$$

The third relation we know from the expression for grad f in rectangular coordinates. The others are applications of the Chain Rule:

$$\frac{\partial f}{\partial r} = f_x\frac{\partial x}{\partial r} + f_y\frac{\partial y}{\partial r} + f_z\frac{\partial z}{\partial r} = f_x\cos\theta + f_y\sin\theta + 0$$
$$= (f_x, f_y, f_z)\cdot(\cos\theta, \sin\theta, 0) = (\operatorname{grad} f)\cdot\mathbf{u},$$

$$\frac{1}{r}\frac{\partial f}{\partial \theta} = \frac{1}{r}\left[f_x\frac{\partial x}{\partial \theta} + f_y\frac{\partial y}{\partial \theta} + f_z\frac{\partial z}{\partial \theta}\right] = \frac{1}{r}[-f_x r\sin\theta + f_y r\cos\theta + 0]$$
$$= (f_x, f_y, f_z)\cdot(-\sin\theta, \cos\theta, 0) = (\operatorname{grad} f)\cdot\mathbf{w}.$$

The operator ∇ is given in terms of **i**, **j**, **k** by

$$\nabla = \mathbf{i}\frac{\partial}{\partial x} + \mathbf{j}\frac{\partial}{\partial y} + \mathbf{k}\frac{\partial}{\partial z}.$$

It was defined this way so that grad $f = \nabla f$. Now the above formula for grad f shows the proper expression for ∇ in terms of **u**, **w**, **k**:

In cylindrical coordinates,

$$\nabla = \mathbf{u}\frac{\partial}{\partial r} + \mathbf{w}\frac{1}{r}\frac{\partial}{\partial \theta} + \mathbf{k}\frac{\partial}{\partial z}.$$

EXAMPLE 6.3

A function $f(r)$ depends on r alone. What is its gradient?

Solution: $\partial f/\partial\theta = \partial f/\partial z = 0$. Therefore

$$\operatorname{grad} f = \nabla f = \mathbf{u}\frac{\partial f}{\partial r}.$$

Answer: $f'(r)\mathbf{u}$.

EXAMPLE 6.4

Find div $\mathbf{u}$, div $\mathbf{w}$, curl $\mathbf{u}$, curl $\mathbf{w}$.

Solution: Since $\mathbf{u}_r = \mathbf{0}$, $\mathbf{u}_\theta = \mathbf{w}$, $\mathbf{u}_z = \mathbf{0}$, it follows that

$$\operatorname{div}\mathbf{u} = \nabla\cdot\mathbf{u} = \mathbf{u}\cdot\mathbf{u}_r + \frac{1}{r}\mathbf{w}\cdot\mathbf{u}_\theta + \mathbf{k}\cdot\mathbf{u}_z = 0 + \frac{1}{r}\mathbf{w}\cdot\mathbf{w} + 0 = \frac{1}{r}.$$

Likewise,

$$\operatorname{div}\mathbf{w} = \nabla\cdot\mathbf{w} = \frac{1}{r}\mathbf{w}\cdot\mathbf{w}_\theta = -\frac{1}{r}\mathbf{w}\cdot\mathbf{u} = 0.$$

Next,

$$\operatorname{curl}\mathbf{u} = \nabla\times\mathbf{u} = \mathbf{u}\times\mathbf{u}_r + \frac{1}{r}\mathbf{w}\times\mathbf{u}_\theta + \mathbf{k}\times\mathbf{u}_z = \mathbf{0} + \frac{1}{r}\mathbf{w}\times\mathbf{w} + \mathbf{0} = \mathbf{0};$$

$$\operatorname{curl}\mathbf{w} = \nabla\times\mathbf{w} = \frac{1}{r}\mathbf{w}\times\mathbf{w}_\theta = -\frac{1}{r}\mathbf{w}\times\mathbf{u} = \frac{1}{r}\mathbf{k}.$$

Answer: $\operatorname{div}\mathbf{u} = \dfrac{1}{r}$, $\operatorname{div}\mathbf{w} = 0$,

$\operatorname{curl}\mathbf{u} = \mathbf{0}$, $\operatorname{curl}\mathbf{w} = \dfrac{1}{r}\mathbf{k}$.

EXAMPLE 6.5

Let $f(r)$ be any function of r alone. Find the curl of the force field $\mathbf{F} = f(r)\mathbf{u}$.

Solution: Use the formula for the curl of a function times a vector field, and use the results of the last two examples:

$$\nabla\times\mathbf{F} = \nabla\times(f\mathbf{u}) = (\nabla f)\times\mathbf{u} + f\,(\nabla\times\mathbf{u}) = f'(r)\mathbf{u}\times\mathbf{u} + \mathbf{0} = \mathbf{0}.$$

Alternate Solution: Choose $g(r)$ so that $g'(r) = f(r)$. Then

$$\mathbf{F} = f(r)\mathbf{u} = g'(r)\mathbf{u} = \nabla g.$$

Thus $\mathbf{F}$ is a gradient field. But a gradient field is irrotational; $\nabla\times(\nabla g) = \mathbf{0}$.

Answer: curl $\mathbf{F} = \mathbf{0}$.

EXAMPLE 6.6

Find the curl of the axially symmetric, circumferentially directed force field $\mathbf{F} = f(r)\mathbf{w}$. Apply the result to the special cases:

$$\text{(i)}\quad f(r) = \omega r, \quad \text{where } \omega \text{ is constant}, \qquad \text{(ii)}\quad f(r) = \frac{1}{r}.$$

Solution: First compute curl $\mathbf{F}$:

$$\nabla \times \mathbf{F} = (\nabla f) \times \mathbf{w} + f(\nabla \times \mathbf{w}) = f'(r)\, \mathbf{u} \times \mathbf{w} + f\frac{1}{r}\mathbf{k}$$

$$= \left(f' + \frac{1}{r}f\right)\mathbf{k} = \frac{1}{r}(rf)'\mathbf{k}.$$

Now apply this formula to the special cases:

$$\text{(i)}\quad \nabla \times (\omega r \mathbf{w}) = \frac{1}{r}(\omega r^2)'\mathbf{k} = 2\omega\mathbf{k}.$$

$$\text{(ii)}\quad \nabla \times \left(\frac{1}{r}\mathbf{w}\right) = \frac{1}{r}\left(r\frac{1}{r}\right)'\mathbf{k} = \mathbf{0}.$$

Answer: curl $\mathbf{F} = \dfrac{1}{r}(rf)'\mathbf{k}$;

(i) $\operatorname{curl}(\omega r \mathbf{w}) = 2\omega\mathbf{k}$,

(ii) $\operatorname{curl}\left(\dfrac{1}{r}\mathbf{w}\right) = \mathbf{0}$.

REMARK: The field $\mathbf{F} = \omega r \mathbf{w}$ is the velocity field of a rigid body rotating about the z-axis with angular speed ω (angular velocity $\omega\mathbf{k}$). Hence result (i) is the one for angular velocity derived on p. 803. The field $\mathbf{F} = \mathbf{w}/r$ of (ii) is the magnetic field due to a steady current flowing along the z-axis.

EXAMPLE 6.7

Find the Laplacian $\nabla^2 f$ of a function $f(r, \theta, z)$ in terms of r, θ, and z.

Solution:

$$\nabla f = f_r\mathbf{u} + \frac{1}{r}f_\theta\mathbf{w} + f_z\mathbf{k},$$

$$\nabla^2 f = \nabla \cdot (\nabla f) = \left(\mathbf{u}\frac{\partial}{\partial r} + \mathbf{w}\frac{1}{r}\frac{\partial}{\partial \theta} + \mathbf{k}\frac{\partial}{\partial z}\right) \cdot (\nabla f)$$

$$= \mathbf{u} \cdot \frac{\partial}{\partial r}\left(f_r\mathbf{u} + \frac{1}{r}f_\theta\mathbf{w} + f_z\mathbf{k}\right)$$

$$+ \frac{1}{r}\mathbf{w} \cdot \frac{\partial}{\partial \theta}\left(f_r\mathbf{u} + \frac{1}{r}f_\theta\mathbf{w} + f_z\mathbf{k}\right) + \mathbf{k} \cdot \frac{\partial}{\partial z}\left(f_r\mathbf{u} + \frac{1}{r}f_\theta\mathbf{w} + f_z\mathbf{k}\right)$$

$$= f_{rr} + \frac{1}{r}\mathbf{w} \cdot \left(f_r\mathbf{w} + \frac{1}{r}f_{\theta\theta}\mathbf{w}\right) + f_{zz} = f_{rr} + \frac{1}{r}f_r + \frac{1}{r^2}f_{\theta\theta} + f_{zz}.$$

Answer: $$\nabla^2 f = f_{rr} + \frac{1}{r} f_r + \frac{1}{r^2} f_{\theta\theta} + f_{zz}$$
$$= \frac{1}{r}(rf_r)_r + \frac{1}{r^2} f_{\theta\theta} + f_{zz}.$$

EXERCISES

1. Find the rectangular coordinates of $\{2, 5\pi/4, -3\}$, $\{1, \pi, 2\}$.
2. Find the cylindrical coordinates of $(0, -2, -2)$, $(5, -5, 0)$, $(-1, -\sqrt{3}, -1)$.
3. Interpret $\nabla f \cdot \mathbf{u}$ and $\nabla f \cdot \mathbf{w}$ as directional derivatives.
4. Show for each solenoidal field of the form $\mathbf{F} = f(r)\mathbf{u}$ that $f(r) = c/r$. (This is the electrostatic field due to a uniformly charged z-axis.)
5. Show that $\ln r$ is a solution of $\nabla^2 f = 0$.
6. Show for each integer n that $r^n \cos n\theta$ and $r^n \sin n\theta$ are solutions of $\nabla^2 f = 0$.
7. A space curve is given by $r = r(t)$, $\theta = \theta(t)$, $z = z(t)$. Show that its arc length satisfies $\dot{s} = (\dot{r}^2 + r^2\dot{\theta}^2 + \dot{z}^2)^{1/2}$.
8. (cont.) Find the length of the spiral $r = a$, $\theta = bt$, $z = ct$ for $0 \le t \le 1$.

7. SPHERICAL COORDINATES

Spherical coordinates are designed to fit situations with central symmetry. The **spherical coordinates** $[\rho, \phi, \theta]$ of a point $\mathbf{x}$ are its distance $\rho = |\mathbf{x}|$ from the origin, its elevation angle ϕ, and its azimuth angle θ. (Often θ is called the longitude and ϕ the co-latitude.) See Fig. 7.1.

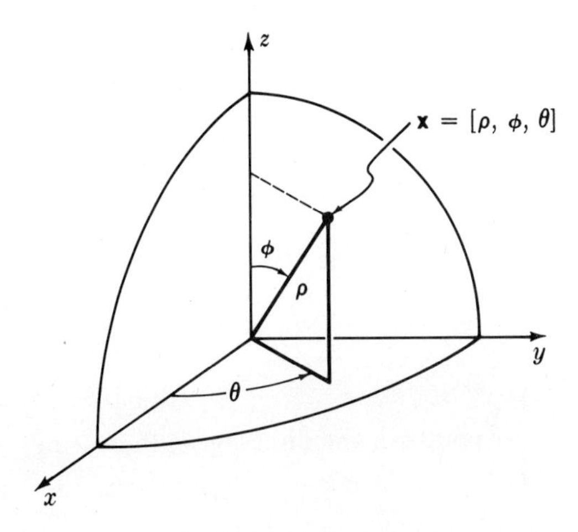

FIG. 7.1

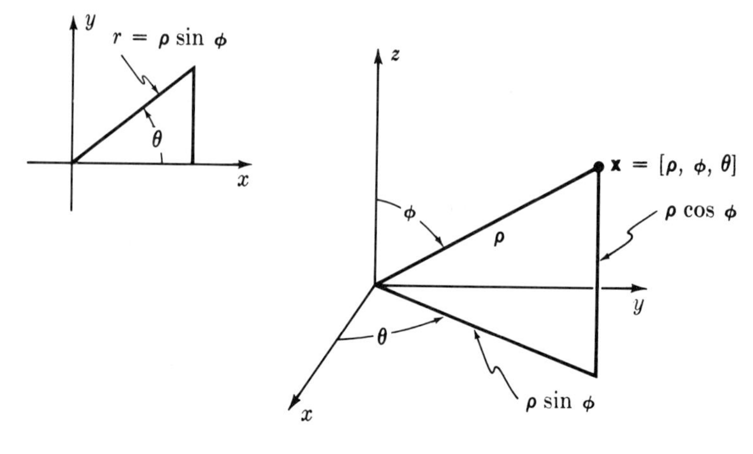

Fig. 7.2

Relations between the rectangular coordinates (x, y, z) of a point and its spherical coordinates may be read from Fig. 7.2. They are

$$\begin{cases} x = \rho \sin\phi\cos\theta \\ y = \rho\sin\phi\sin\theta \\ z = \rho\cos\phi \end{cases} \qquad \begin{cases} \rho^2 = x^2 + y^2 + z^2 \\ \cos\phi = \dfrac{z}{\rho} \\ \tan\theta = \dfrac{y}{x}. \end{cases}$$

Note that θ is not determined on the z-axis, so points of this axis are usually avoided. In general θ is determined up to a multiple of 2π, and $0 < \phi < \pi$.

The level surfaces

$$\begin{Bmatrix} \rho = \text{constant} \\ \phi = \text{constant} \\ \theta = \text{constant} \end{Bmatrix} \quad \text{are} \quad \begin{Bmatrix} \text{concentric spheres about } \mathbf{0} \\ \text{right circular cones, apex } \mathbf{0} \\ \text{planes through the } z\text{-axis} \end{Bmatrix}.$$

At each point $\mathbf{x}$ the three level surfaces intersect orthogonally (Fig. 7.3).

The Natural Frame

Select unit vectors $\boldsymbol{\lambda}, \boldsymbol{\mu}, \boldsymbol{\nu}$ at each point $\mathbf{x}$ of space (not on the z-axis) such that

$$\begin{Bmatrix} \boldsymbol{\lambda} \\ \boldsymbol{\mu} \\ \boldsymbol{\nu} \end{Bmatrix} \quad \text{points in the direction of increasing} \quad \begin{Bmatrix} \rho \\ \phi \\ \theta \end{Bmatrix}.$$

See Fig. 7.4. Then $\boldsymbol{\lambda}, \boldsymbol{\mu}, \boldsymbol{\nu}$ is a right-hand system.

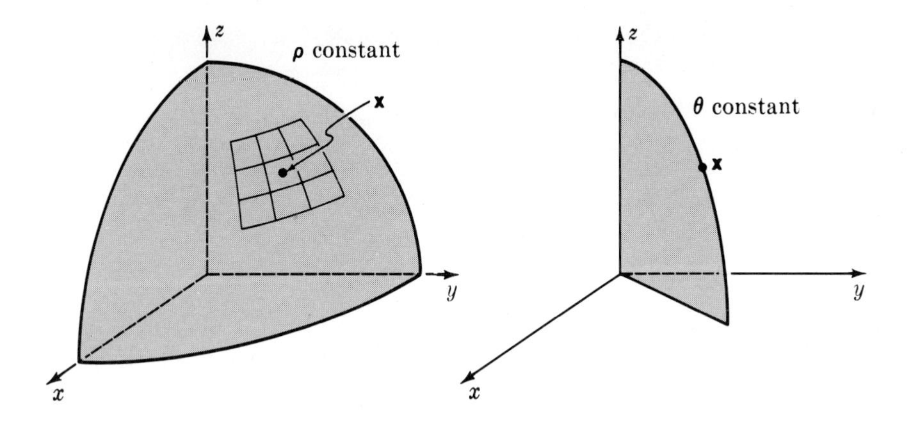

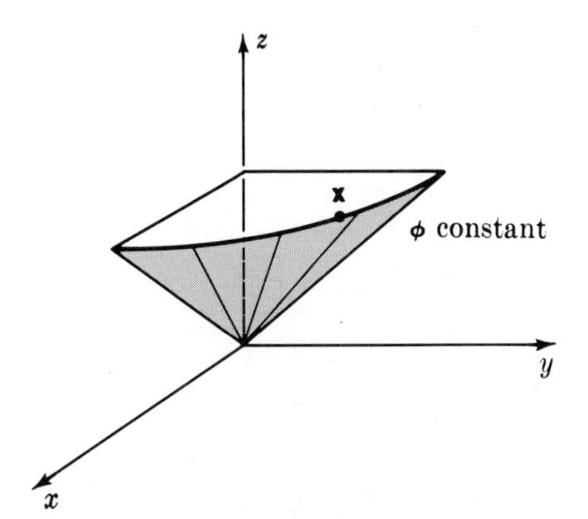

Fig. 7.3

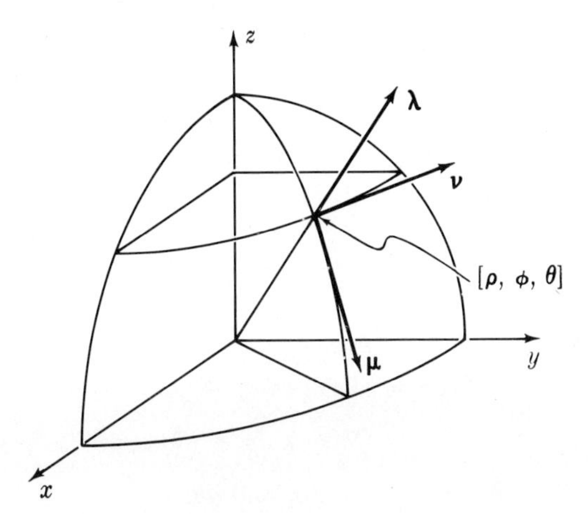

Fig. 7.4

Our immediate problem is to express $\boldsymbol{\lambda}$, $\boldsymbol{\mu}$, $\boldsymbol{\nu}$ in terms of ρ, ϕ, θ. Here is a short cut for doing so:

$$\begin{aligned}
\mathbf{x} &= \rho(\sin\phi\cos\theta,\ \sin\phi\sin\theta,\ \cos\phi),\\
d\mathbf{x} &= (\sin\phi\cos\theta,\ \sin\phi\sin\theta,\ \cos\phi)\, d\rho\\
&\quad + \rho(\cos\phi\cos\theta,\ \cos\phi\sin\theta,\ -\sin\phi)\, d\phi\\
&\quad + \rho(-\sin\phi\sin\theta,\ \sin\phi\cos\theta,\ 0)\, d\theta\\
&= \boldsymbol{\lambda}\, d\rho + \boldsymbol{\mu}\,\rho\, d\phi + \boldsymbol{\nu}\,\rho\sin\phi\, d\theta.
\end{aligned}$$

It is easy to check that $\boldsymbol{\lambda}$, $\boldsymbol{\mu}$, $\boldsymbol{\nu}$ are unit vectors. Furthermore, they point in the directions of increasing ρ, ϕ, θ respectively. Precisely, if only ρ increases, then $d\phi = d\theta = 0$, hence $d\mathbf{x} = \boldsymbol{\lambda}\, d\rho$. Similarly, if only ϕ increases, then $d\mathbf{x} = \boldsymbol{\mu}\,\rho\, d\phi$, and if only θ increases, then $d\mathbf{x} = \boldsymbol{\nu}\,\rho\sin\phi\, d\theta$. Conclusion:

$$\begin{cases}
\boldsymbol{\lambda} = (\sin\phi\cos\theta,\ \sin\phi\sin\theta,\ \cos\phi)\\
\boldsymbol{\mu} = (\cos\phi\cos\theta,\ \cos\phi\sin\theta,\ -\sin\phi)\\
\boldsymbol{\nu} = (-\sin\theta,\ \cos\theta,\ 0),
\end{cases}$$
$$d\mathbf{x} = \boldsymbol{\lambda}\, d\rho + \boldsymbol{\mu}\,\rho\, d\phi + \boldsymbol{\nu}\,\rho\sin\phi\, d\theta.$$

The Operator Del

The last formula contains a good deal of information and can be used to derive a formula for grad f in spherical coordinates.

Given $f(\rho, \phi, \theta)$,

$$\operatorname{grad} f = f_\rho\boldsymbol{\lambda} + \frac{1}{\rho} f_\phi\boldsymbol{\mu} + \frac{1}{\rho\sin\phi} f_\theta\boldsymbol{\nu}.$$

To derive this, combine

$$df = f_\rho\, d\rho + f_\phi\, d\phi + f_\theta\, d\theta$$

with

$$df = (\operatorname{grad} f)\cdot d\mathbf{x} = (\nabla f)\cdot d\mathbf{x},$$

(see p. 562) to obtain

$$\begin{aligned}
f_\rho\, d\rho + f_\phi\, d\phi + f_\theta\, d\theta &= (\nabla f)\cdot(\boldsymbol{\lambda}\, d\rho + \boldsymbol{\mu}\rho\, d\phi + \boldsymbol{\nu}\rho\sin\phi\, d\theta)\\
&= (\nabla f)\cdot\boldsymbol{\lambda}\, d\rho + (\nabla f)\cdot\boldsymbol{\mu}\rho\, d\phi + (\nabla f)\cdot\boldsymbol{\nu}\rho\sin\phi\, d\theta.
\end{aligned}$$

Now equate coefficients of $d\rho$, $d\phi$, and $d\theta$:

$$(\nabla f)\cdot\boldsymbol{\lambda} = f_\rho, \qquad (\nabla f)\cdot\boldsymbol{\mu} = \frac{1}{\rho} f_\phi, \qquad (\nabla f)\cdot\boldsymbol{\nu} = \frac{1}{\rho\sin\phi} f_\theta.$$

The vector ∇f is completely determined by these three quantities, its projections on the perpendicular directions $\boldsymbol{\lambda}$, $\boldsymbol{\mu}$, $\boldsymbol{\nu}$. Hence

$$\begin{aligned}\nabla f &= [(\nabla f)\cdot\boldsymbol{\lambda}]\,\boldsymbol{\lambda} + [(\nabla f)\cdot\boldsymbol{\mu}]\,\boldsymbol{\mu} + [(\nabla f)\cdot\boldsymbol{\nu}]\,\boldsymbol{\nu}\\ &= f_\rho\,\boldsymbol{\lambda} + \frac{1}{\rho} f_\phi\,\boldsymbol{\mu} + \frac{1}{\rho\sin\phi} f_\theta\,\boldsymbol{\nu}.\end{aligned}$$

From this formula the form of the operator ∇ can be read off:

> In spherical coordinates,
>
> $$\nabla = \boldsymbol{\lambda}\frac{\partial}{\partial\rho} + \frac{1}{\rho}\boldsymbol{\mu}\frac{\partial}{\partial\phi} + \frac{1}{\rho\sin\phi}\boldsymbol{\nu}\frac{\partial}{\partial\theta}.$$

REMARK: This procedure for finding ∇ in spherical coordinates is different from that used for cylindrical coordinates and is definitely slicker.

Applications

The spherical coordinate system is the perfect tool for handling central forces. A **central force** with center at $\mathbf{0}$ is a vector field of the form

$$\mathbf{F} = f(\rho)\boldsymbol{\lambda}.$$

EXAMPLE 7.1

Find div $\mathbf{F}$ and curl $\mathbf{F}$ for a central force $\mathbf{F} = f(\rho)\boldsymbol{\lambda}$.

Solution:

Step 1. Compute grad f. Use the formula for ∇ and the fact that $f_\theta = f_\phi = 0$:

$$\operatorname{grad} f(\rho) = \nabla f(\rho) = \left(\boldsymbol{\lambda}\frac{\partial}{\partial\rho}\right) f = f'\boldsymbol{\lambda}.$$

Step 2. Compute div $\boldsymbol{\lambda}$:

$$\operatorname{div}\boldsymbol{\lambda} = \nabla\cdot\boldsymbol{\lambda} = \boldsymbol{\lambda}\cdot\boldsymbol{\lambda}_\rho + \frac{1}{\rho}\boldsymbol{\mu}\cdot\boldsymbol{\lambda}_\theta + \frac{1}{\rho\sin\phi}\boldsymbol{\nu}\cdot\boldsymbol{\lambda}_\theta.$$

Now

$$\boldsymbol{\lambda} = (\sin\phi\cos\theta,\ \sin\phi\sin\theta,\ \cos\phi),$$

hence $\boldsymbol{\lambda}_\rho = \mathbf{0}$,

$$\boldsymbol{\lambda}_\phi = (\cos\phi\cos\theta,\ \cos\phi\sin\theta,\ -\sin\phi) = \boldsymbol{\mu},$$

$$\boldsymbol{\lambda}_\theta = (-\sin\phi\sin\theta,\ \sin\phi\cos\theta,\ 0) = \sin\phi\,\boldsymbol{\nu}.$$

Substitute in div $\boldsymbol{\lambda}$:

$$\operatorname{div}\boldsymbol{\lambda} = 0 + \frac{1}{\rho}\boldsymbol{\mu}\cdot\boldsymbol{\mu} + \frac{1}{\rho\sin\phi}\boldsymbol{\nu}\cdot(\sin\phi\,\boldsymbol{\nu}) = \frac{1}{\rho} + \frac{1}{\rho} = \frac{2}{\rho}.$$

Step 3. Compute div $\mathbf{F}$, using $\mathbf{F} = f\boldsymbol{\lambda}$:

$$\operatorname{div} \mathbf{F} = \operatorname{div}(f\boldsymbol{\lambda}) = (\operatorname{grad} f)\cdot\boldsymbol{\lambda} + f \operatorname{div}\boldsymbol{\lambda} = f'\boldsymbol{\lambda}\cdot\boldsymbol{\lambda} + f\frac{2}{\rho} = f' + \frac{2}{\rho}f.$$

Step 4. Compute curl $\boldsymbol{\lambda}$:

$$\operatorname{curl}\boldsymbol{\lambda} = \nabla\times\boldsymbol{\lambda} = \boldsymbol{\lambda}\times\boldsymbol{\lambda}_\rho + \frac{1}{\rho}\boldsymbol{\mu}\times\boldsymbol{\lambda}_\phi + \frac{1}{\rho\sin\phi}\boldsymbol{\nu}\times\boldsymbol{\lambda}_\theta$$

$$= \mathbf{0} + \frac{1}{\rho}\boldsymbol{\mu}\times\boldsymbol{\mu} + \frac{1}{\rho\sin\phi}\boldsymbol{\nu}\times(\sin\phi\,\boldsymbol{\nu}) = \mathbf{0}.$$

Step 5. Compute curl $\mathbf{F}$, using $\mathbf{F} = f\boldsymbol{\lambda}$.

$$\operatorname{curl}\mathbf{F} = \operatorname{curl}(f\boldsymbol{\lambda}) = (\operatorname{grad} f)\times\boldsymbol{\lambda} + f\operatorname{curl}\boldsymbol{\lambda} = f'\boldsymbol{\lambda}\times\boldsymbol{\lambda} + \mathbf{0} = \mathbf{0}.$$

Answer: $\operatorname{div}\mathbf{F} = f' + \frac{2}{\rho}f = \frac{1}{\rho^2}(\rho^2 f)'$,

$\operatorname{curl}\mathbf{F} = \mathbf{0}$.

REMARK 1: This example shows that each central force field $\mathbf{F}$ is irrotational. But $\mathbf{F}$ is solenoidal ($\operatorname{div}\mathbf{F} = 0$) only when $f = c/\rho^2$, i.e., only when $\mathbf{F}$ obeys the inverse square law.

REMARK 2: The computation of $\boldsymbol{\lambda}_\phi$ and $\boldsymbol{\lambda}_\theta$ in Step 2 of the solution shows the advantage of knowing all partials of the vectors $\boldsymbol{\lambda}$, $\boldsymbol{\mu}$, $\boldsymbol{\nu}$. By straightforward calculation, you find

$$\begin{cases} d\boldsymbol{\lambda} = \boldsymbol{\mu}\,d\phi + \boldsymbol{\nu}\sin\phi\,d\theta \\ d\boldsymbol{\mu} = -\boldsymbol{\lambda}\,d\phi + \boldsymbol{\nu}\cos\phi\,d\theta \\ d\boldsymbol{\nu} = -\boldsymbol{\lambda}\sin\phi\,d\theta - \boldsymbol{\mu}\cos\phi\,d\theta. \end{cases}$$

For those familiar with matrices, these formulas may be written as a single matrix equation:

$$\begin{bmatrix} d\boldsymbol{\lambda} \\ d\boldsymbol{\mu} \\ d\boldsymbol{\nu} \end{bmatrix} = \begin{bmatrix} 0 & d\phi & \sin\phi\,d\theta \\ -d\phi & 0 & \cos\phi\,d\theta \\ -\sin\phi\,d\theta & -\cos\phi\,d\theta & 0 \end{bmatrix} \begin{bmatrix} \boldsymbol{\lambda} \\ \boldsymbol{\mu} \\ \boldsymbol{\nu} \end{bmatrix},$$

which perhaps is an aid to memory. Note the skew-symmetry of the square matrix.

EXAMPLE 7.2

Find the Laplacian $\nabla^2 f$ of a function $f(\rho)$ of ρ alone.

Solution: $\operatorname{grad} f = \nabla f = f'\boldsymbol{\lambda}$, hence $\nabla^2 f = \nabla \cdot (\nabla f) = \nabla \cdot (f'\boldsymbol{\lambda})$. A result of the preceding example is

$$\nabla \cdot (f\boldsymbol{\lambda}) = \frac{1}{\rho^2}(\rho^2 f)' = f' + \frac{2}{\rho} f.$$

Replace f by f':

$$\nabla \cdot (f'\boldsymbol{\lambda}) = f'' + \frac{2}{\rho} f'.$$

To put this in a neat form, note that

$$(\rho f)'' = (f + \rho f')' = f' + (f' + \rho f'') = 2f' + \rho f'',$$

hence

$$f'' + \frac{2}{\rho} f' = \frac{1}{\rho}(\rho f)''.$$

Answer: $\nabla^2 f(\rho) = \dfrac{1}{\rho}(\rho f)''.$

EXERCISES

1. Convert to rectangular coordinates: $[1, 3\pi/4, 3\pi/4]$, $[3, \pi/2, 5\pi/4]$, $[2, 2\pi/3, \pi/3]$
2. Convert to spherical coordinates:

$$(1, 1, 1),\ (1, -1, 1),\ (-1, -1, -1),\ (-1, 1, -1).$$

3. If $\mathbf{x}$ has spherical coordinates $[\rho, \phi, \theta]$, find the spherical coordinates of $-\mathbf{x}$ and of $3\mathbf{x}$.
4. Find $\nabla \cdot \boldsymbol{\mu}$.
5. Find $\nabla \times \boldsymbol{\mu}$.
6. Find $\nabla \cdot \boldsymbol{\nu}$
7. Find $\nabla \times \boldsymbol{\nu}$.
8. Observe that the $\boldsymbol{\nu}$ of spherical coordinates equals the $\mathbf{w}$ of cylindrical coordinates. Use this information for alternate solutions of Exs. 6 and 7.
9. Show that the only solutions of Laplace's Equation $\nabla^2 f = 0$ that have the form $f = f(\rho)$ are the functions $f = a + b/\rho$, where a and b are constants.
10. The **Yukawa potential** is the function $f(\rho) = e^{-k\rho}/\rho$ with k constant. Show that $\nabla^2 f = k^2 f$.
11. Show that each vector field of the form $f(\rho)\boldsymbol{\lambda}$ is a gradient, ∇g. Hence obtain an alternate proof that $\nabla \times [f(\rho)\boldsymbol{\lambda}] = \mathbf{0}$.
12. Show for the general function $f(\rho, \phi, \theta)$ in spherical coordinates that

$$\nabla^2 f = \frac{1}{\rho}(\rho f)_{\rho\rho} + \frac{1}{\rho^2 \sin\phi}(f_\phi \sin\phi)_\phi + \frac{1}{\rho^2 \sin^2\phi} f_{\theta\theta}.$$

13. Suppose $f(\rho, \phi, \theta)$ is homogeneous of degree n. (This means $f(t\mathbf{x}) = tf(\mathbf{x})$ in rectangular coordinates. See p. 530.) Show that $f[\rho, \phi, \theta] = \rho^n g(\phi, \theta)$, where g is independent of ρ.

14. (cont.) Let f be homogeneous of degree n. Show that

$$\nabla^2 f = \rho^{n-2}[n(n+1)g \sin \phi + (g_\phi \sin \phi)_\phi + (g_\theta/\sin \phi)_\theta]/\sin \phi.$$

15. After a long calculation, it turned out that the matrix of coefficients of $\boldsymbol{\lambda}$, $\boldsymbol{\mu}$, $\boldsymbol{\nu}$ in the formulas for $d\boldsymbol{\lambda}$, $d\boldsymbol{\mu}$, $d\boldsymbol{\nu}$ is skew-symmetric. Show that this skew-symmetry is no accident, it is an immediate consequence of differentiating the relations $\boldsymbol{\lambda} \cdot \boldsymbol{\lambda} = 1$, $\boldsymbol{\lambda} \cdot \boldsymbol{\mu} = 0$, etc.

16. Find the corresponding skew-symmetric matrix M for cylindrical coordinates:

$$\begin{bmatrix} d\mathbf{u} \\ d\mathbf{w} \\ d\mathbf{k} \end{bmatrix} = M \begin{bmatrix} \mathbf{u} \\ \mathbf{w} \\ \mathbf{k} \end{bmatrix}.$$

17. (still in cylindrical coordinates) Find the formula for $d\mathbf{x}$ in terms of $\mathbf{u}$, $\mathbf{w}$, $\mathbf{k}$.

18. Suppose a space curve is given by $\rho = \rho(t)$, $\phi = \phi(t)$, $\theta = \theta(t)$. Show that its arc length satisfies $\dot{s} = (\dot{\rho}^2 + \rho^2\dot{\phi}^2 + \rho^2(\sin^2 \phi)\dot{\theta}^2)^{1/2}$.

19. Use Ex. 12 to show that if $f(\rho, \phi, \theta)$ satisfies $\nabla^2 f = 0$, then its **Kelvin Transform**

$$g(\rho, \phi, \theta) = \frac{1}{\rho} f\left(\frac{1}{\rho}, \phi, \theta\right)$$

satisfies $\nabla^2 g = 0$.

34. Multiple Integrals

1. DOUBLE INTEGRALS

Before starting this chapter, please review Chapter 26. You should understand how to set up the double integral for the volume of a solid with rectangular base, and how to evaluate the integral by iteration.

In this section we shall study volumes with non-rectangular bases. No really new idea is involved, but the iteration process is technically more difficult.

> EXAMPLE 1.1
>
> Find the volume under the surface $z = e^{-(x+y)}$, and over the region of the x, y-plane bounded by the x-axis, the line $y = x$, and the lines $x = \frac{1}{2}$ and $x = 1$.

Solution: Draw figures of the plane region (Fig. 1.1a) and the solid (Fig. 1.1b). The problem is solved by the slicing method. Slice the solid into

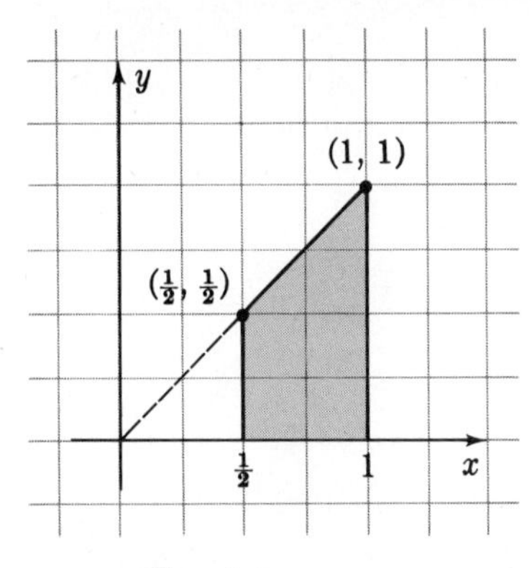

FIG. 1.1a

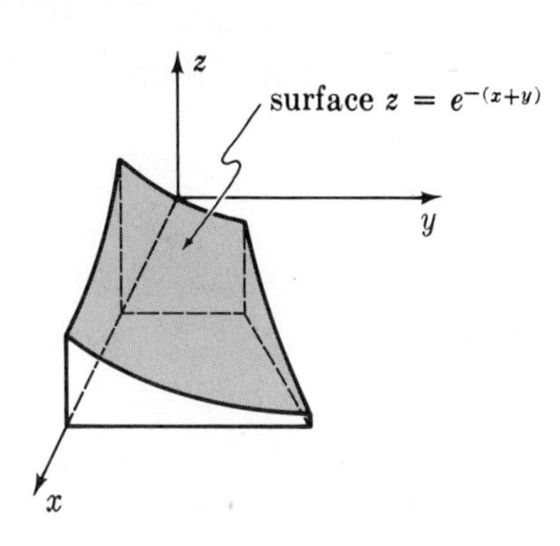

FIG. 1.1b

slabs by planes perpendicular to the x-axis. Let the area of the slab at x be denoted by $A(x)$. See Fig. 1.2a. The volume of the slab is

$$dV = A(x)\, dx,$$

so the total volume is

$$V = \int_{1/2}^{1} A(x)\, dx.$$

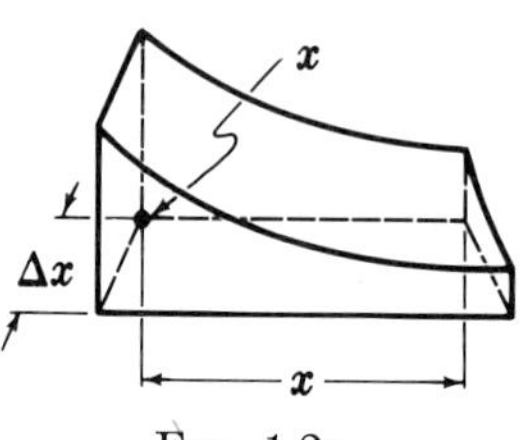

FIG. 1.2a

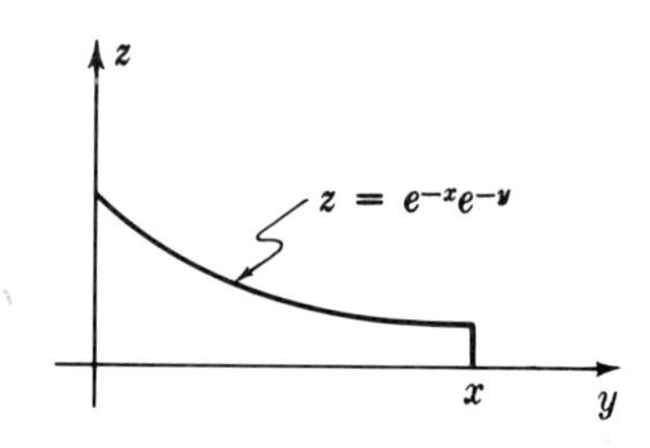

FIG. 1.2b

Now carefully draw the cross-section (Fig. 1.2b). Its base has length x, and it is bounded above by the curve $z = e^{-x}e^{-y}$. Note that x is *constant;* in this plane section z is a function of y alone. This is the absolute crux of the matter!

The area of the cross-section at x is

$$A(x) = \int_0^x e^{-x}e^{-y}\, dy = e^{-x}\int_0^x e^{-y}\, dy = -e^{-x}(e^{-y})\Big|_{y=0}^{y=x} = e^{-x} - e^{-2x}.$$

Thus

$$V = \int_{1/2}^{1} (e^{-x} - e^{-2x})\, dx = \left(\frac{1}{2}e^{-2x} - e^{-x}\right)\Big|_{1/2}^{1} = \frac{1}{2}e^{-2} - \frac{3}{2}e^{-1} + e^{-1/2}.$$

NOTE: The solution can be set up as follows:

$$V = \int_{1/2}^{1}\left(\int_0^x e^{-(x+y)}\, dy\right) dx.$$

In the inner integral the variable of integration is y, while x is treated like a constant, both in the integrand and in the upper limit. But once the inner integral is completely evaluated, x becomes the variable of integration for the outer integral.

Answer: $\frac{1}{2}e^{-2}(1 - 3e + 2e^{3/2})$.

EXAMPLE 1.2

Find the volume under the surface $z = 1 - x^2 - y^2$, lying over the square with vertices $(\pm 1, 0)$ and $(0, \pm 1)$.

Solution: First draw the square (Fig. 1.3a). Observe that by symmetry, it suffices to find the volume over the triangular portion in the first quadrant, and then to quadruple it.

Draw the corresponding portion of the solid (Fig. 1.3b). Slice by planes perpendicular to the x-axis. For each x, the plane cuts the solid in a cross-

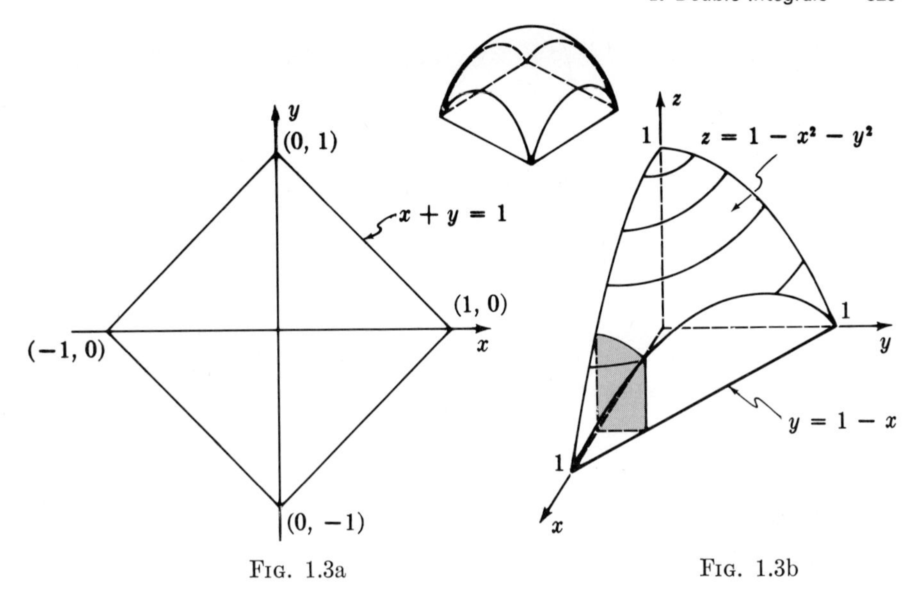

FIG. 1.3a

FIG. 1.3b

section (Fig. 1.4) whose area is

$$A(x) = \int_0^{1-x} (1 - x^2 - y^2)\, dy.$$

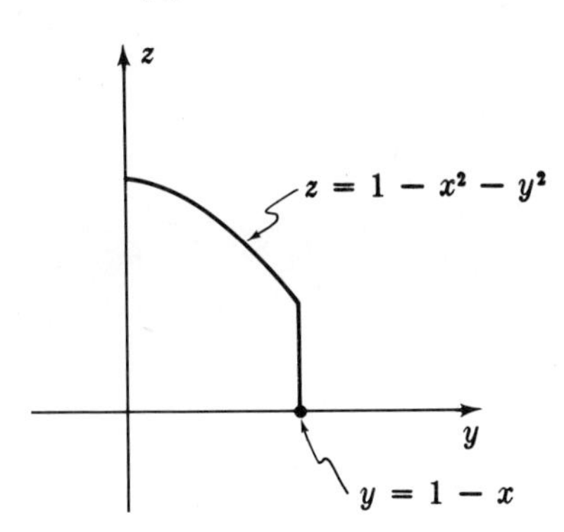

FIG. 1.4

The volume of the solid is

$$\begin{aligned} V = 4\int_0^1 A(x)\, dx &= 4\int_0^1 \left(\int_0^{1-x} (1 - x^2 - y^2)\, dy\right) dx \\ &= 4\int_0^1 \left[\left(y - x^2y - \frac{1}{3}\, y^3\right)\Bigg|_{y=0}^{y=1-x}\right] dx \\ &= 4\int_0^1 \left[(1 - x) - x^2(1 - x) - \frac{1}{3}\,(1 - x)^3\right] dx \end{aligned}$$

$$= 4\int_0^1 \left[1 - x - x^2 + x^3 - \frac{1}{3}(1 - x)^3\right] dx$$

$$= 4\left[1 - \frac{1}{2} - \frac{1}{3} + \frac{1}{4} - \frac{1}{12}\right] = 4 \cdot \frac{4}{12} = \frac{4}{3}.$$

Answer: $\frac{4}{3}$.

EXAMPLE 1.3

Find the volume under the plane $z = 1 + x + y$, and over the region bounded by the curves $x = \frac{1}{2}$, $x = 1$, $y = x^2$, $y = 2x^2$.

Solution: The plane region and solid are drawn in Fig. 1.5. The cross-section by a plane perpendicular to the x-axis at x is a trapezoid (Fig. 1.6).

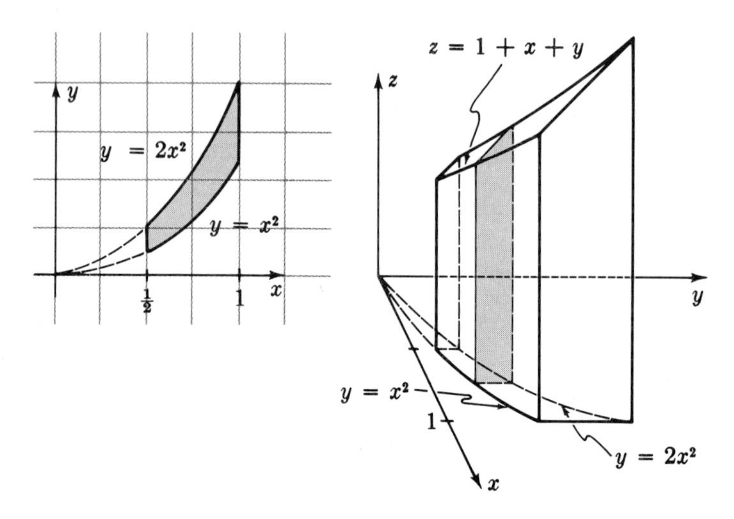

Fig. 1.5

Note the range of y, namely $x^2 \le y \le 2x^2$. Thus

$$V = \int_{1/2}^{1} A(x)\, dx,$$

where

$$A(x) = \int_{x^2}^{2x^2} (1 + x + y)\, dy = \left(y + xy + \frac{1}{2}y^2\right)\Bigg|_{y=x^2}^{y=2x^2}$$

$$= (2x^2 + 2x^3 + 2x^4) - \left(x^2 + x^3 + \frac{1}{2}x^4\right) = x^2 + x^3 + \frac{3}{2}x^4.$$

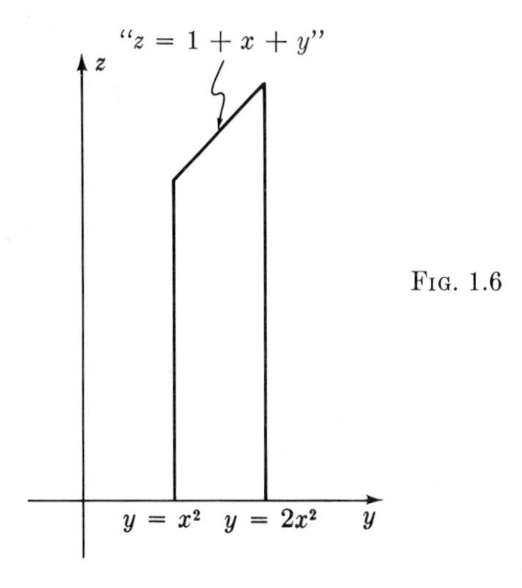

Fig. 1.6

Finally,

$$V = \int_{1/2}^{1} \left(x^2 + x^3 + \frac{3}{2}x^4\right) dx = \left(\frac{1}{3}x^3 + \frac{1}{4}x^4 + \frac{3}{10}x^5\right)\Bigg|_{1/2}^{1}$$

$$= \left(\frac{1}{3} + \frac{1}{4} + \frac{3}{10}\right) - \frac{1}{8}\left(\frac{1}{3} + \frac{1}{8} + \frac{3}{40}\right) = \frac{49}{60}.$$

Answer: $\frac{49}{60}$.

Iteration

With these examples behind us, we are prepared for the statement and solution of the double integration problem. Suppose a region R in the x, y-plane is bounded by lines $x = a$ and $x = b$, and by two curves $y = g(x)$ and $y = f(x)$. Assume $a < b$ and $g(x) < f(x)$ for each x. See Fig. 1.7. Suppose a surface $z = H(x, y)$ is given, defined over the region R. See Fig. 1.8. The

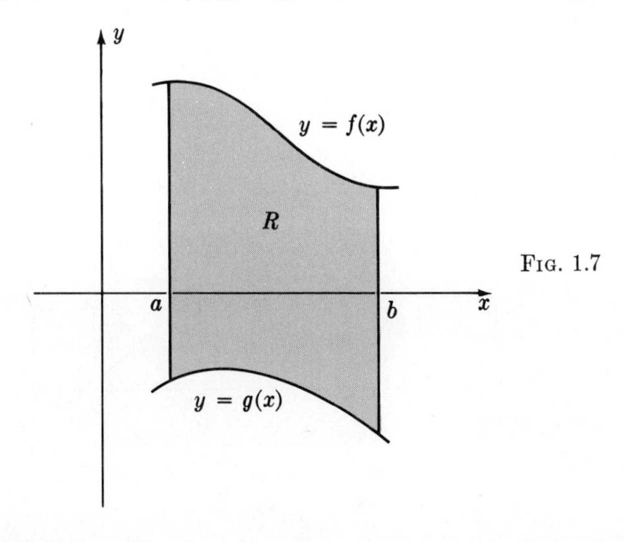

Fig. 1.7

volume of the solid column over R and under the surface is required. It is

$$V = \iint_R H(x, y)\, dx\, dy.$$

(As is usual with integrals, the portion where $H < 0$ is counted negative.)

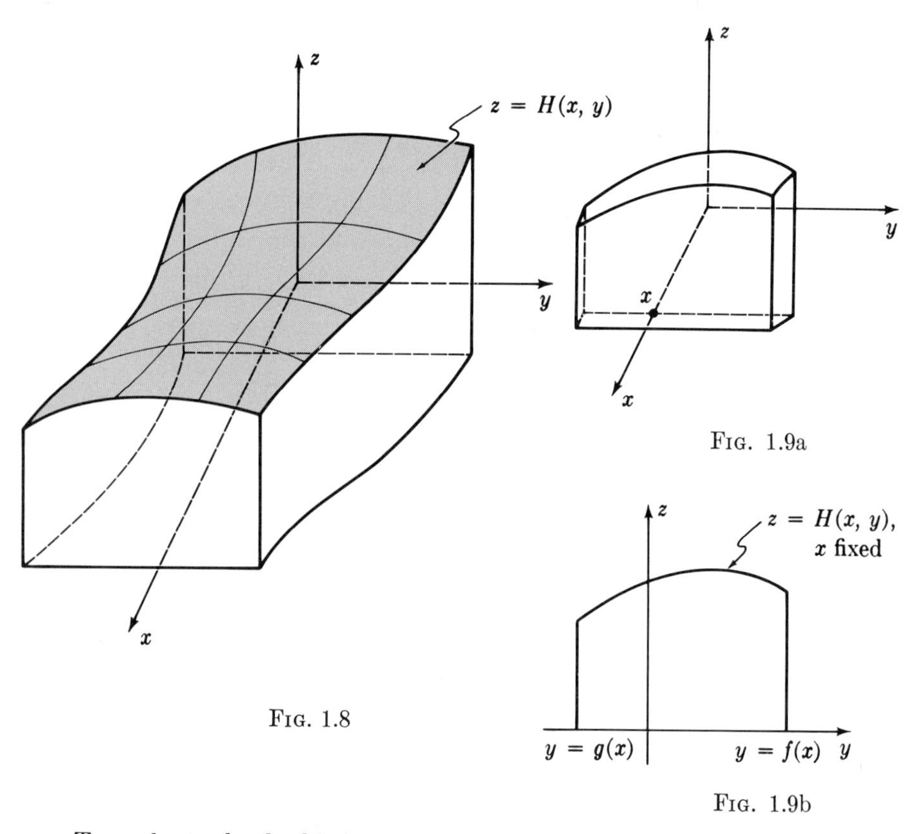

FIG. 1.8

FIG. 1.9a

FIG. 1.9b

To evaluate the double integral, consider a slab parallel to the y, z-plane at x. Its face area is

$$A(x) = \int_{g(x)}^{f(x)} H(x, y)\, dy.$$

See Fig. 1.9. In this integration, x is constant. Notice that y, the variable of integration, disappears when the definite integral $A(x)$ is evaluated.

Conclusion: the volume integral is

$$V = \iint_R H(x, y)\, dx\, dy = \int_a^b A(x)\, dx = \int_a^b \left(\int_{g(x)}^{f(x)} H(x, y)\, dy \right) dx.$$

This is called the **iteration formula.**

EXAMPLE 1.4

Find the volume under the surface $z = xy$, and over the region R bounded by $y = x$ and $y = x^2$.

Solution: The line and parabola intersect at (0, 0) and (1, 1). See Fig. 1.10. For each value of x, the range of y is

$$x^2 \le y \le x.$$

Hence

$$V = \iint_R xy\,dx\,dy = \int_0^1 \left(\int_{x^2}^x xy\,dy\right) dx = \int_0^1 \left(\frac{1}{2}xy^2\Big|_{y=x^2}^{y=x}\right) dx$$

$$= \int_0^1 \frac{1}{2}(x^3 - x^5)\,dx = \frac{1}{2}\left(\frac{1}{4} - \frac{1}{6}\right) = \frac{1}{24}.$$

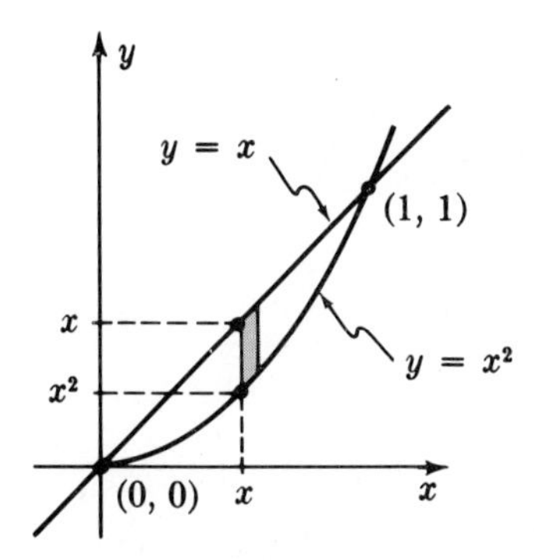

Fig. 1.10

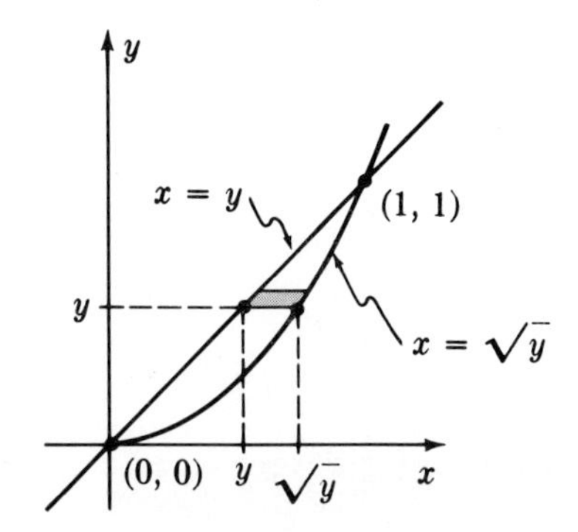

Fig. 1.11

Alternate Solution: The region R may be thought of as bounded by the curves $x = y$ (below) and $x = \sqrt{y}$ (above), where $0 \le y \le 1$. See Fig. 1.11. For each y, the range of x is $y \le x \le \sqrt{y}$. Therefore the set-up of the iteration is

$$V = \int_0^1 \left(\int_y^{\sqrt{y}} xy\,dx\right) dy = \int_0^1 \left(\frac{1}{2}x^2y\Big|_{x=y}^{x=\sqrt{y}}\right) dy$$

$$= \int_0^1 \frac{1}{2}(y^2 - y^3)\,dy = \frac{1}{2}\left(\frac{1}{3} - \frac{1}{4}\right) = \frac{1}{24}.$$

Answer: $\dfrac{1}{24}$.

The iteration method does not apply directly to every example. The boundary of R may be too complicated, in which case you must break the region into several smaller regions, and deal with each as a separate problem. In Fig. 1.12 two examples are shown. The set-up for the region in Fig. 1.12a is

$$\iint_R H(x, y)\,dx\,dy = \iint_{R_1} H(x, y)\,dx\,dy + \iint_{R_2} H(x, y)\,dx\,dy$$
$$= \int_a^c \left(\int_{g(x)}^{f_1(x)} H(x, y)\,dy \right) dx + \int_c^b \left(\int_{g(x)}^{f_2(x)} H(x, y)\,dy \right) dx.$$

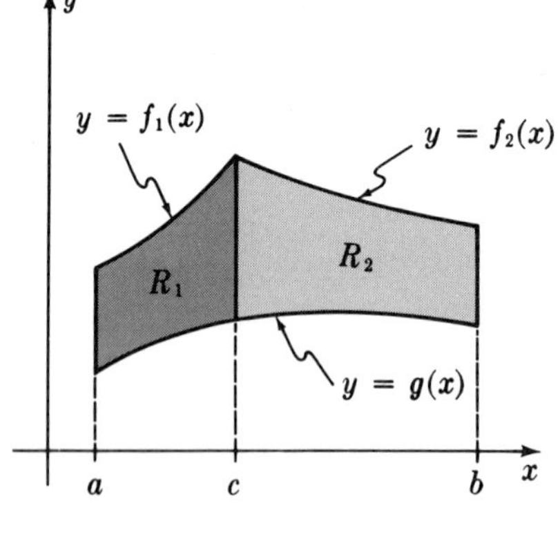

FIG. 1.12a

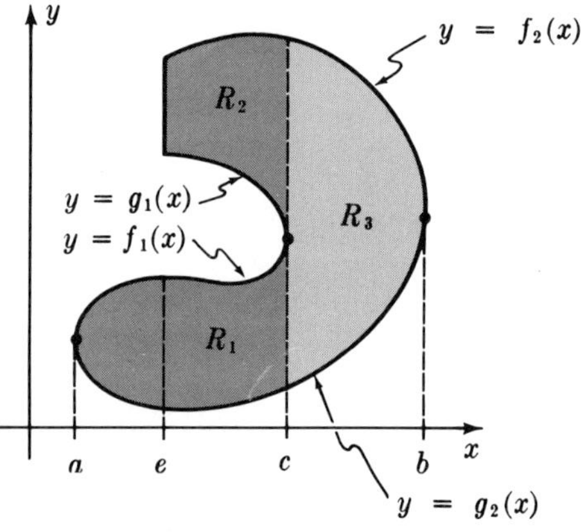

FIG. 1.12b

The set-up for the region in Fig. 1.12b is

$$\iint_R H(x, y)\,dx\,dy$$
$$= \iint_{R_1} H(x, y)\,dx\,dy + \iint_{R_2} H(x, y)\,dx\,dy + \iint_{R_3} H(x, y)\,dx\,dy$$
$$= \int_a^c \left(\int_{g_2(x)}^{f_1(x)} H(x, y)\,dy \right) dx + \int_e^c \left(\int_{g_1(x)}^{f_2(x)} H(x, y)\,dy \right) dx$$
$$+ \int_c^b \left(\int_{g_2(x)}^{f_2(x)} H(x, y)\,dy \right) dx.$$

EXAMPLE 1.5

Without evaluating, set up the calculation for

$$\iint_R \frac{dx\,dy}{x^2 + y^2}$$

over the region indicated in Fig. 1.13.

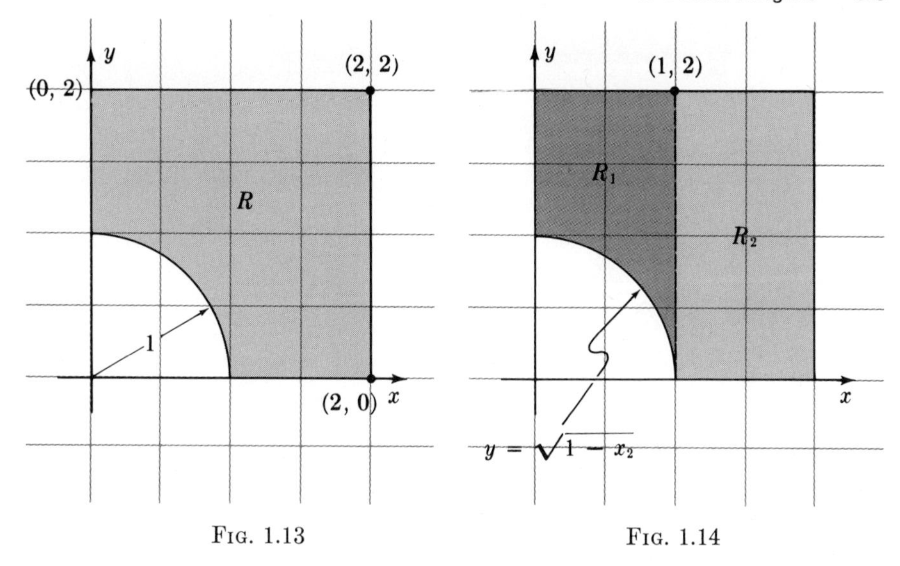

FIG. 1.13

FIG. 1.14

Solution: Draw a vertical line from (1, 0) to (1, 2). Call the left-hand part R_1 and the right-hand part R_2. See Fig. 1.14. Then

$$\iint_R \frac{dx\,dy}{x^2+y^2} = \iint_{R_1} \frac{dx\,dy}{x^2+y^2} + \iint_{R_2} \frac{dx\,dy}{x^2+y^2}$$

$$= \int_0^1 \left(\int_{\sqrt{1-x^2}}^2 \frac{dy}{x^2+y^2} \right) dx + \int_1^2 \left(\int_0^2 \frac{dy}{x^2+y^2} \right) dx.$$

Area

It is often convenient to compute areas by double integrals, since

$$\iint_R 1\,dx\,dy = \text{area}(R).$$

EXAMPLE 1.6

Find the total area A bounded by $y = x^2$ and $y = x^4$.

Solution: Draw the region (Fig. 1.15). Then

$$A = \iint_R dx\,dy = \int_{-1}^1 \left(\int_{x^4}^{x^2} dy \right) dx = \int_{-1}^1 (x^2 - x^4)\,dx = \frac{2}{3} - \frac{2}{5} = \frac{4}{15}.$$

Alternate Solution:

$$A = 2\int_0^1 \left(\int_{\sqrt{y}}^{\sqrt[4]{y}} dx \right) dy = 2\int_0^1 (\sqrt[4]{y} - \sqrt{y})\,dy = 2\left(\frac{4}{5} - \frac{2}{3}\right) = \frac{4}{15}.$$

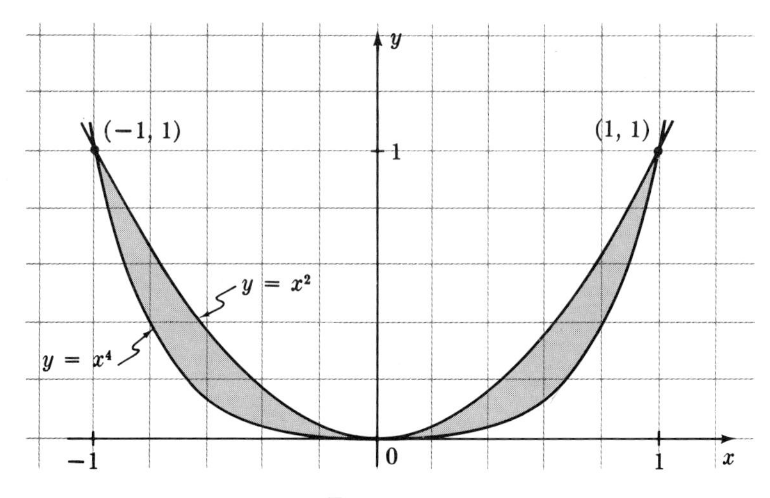

Fig. 1.15

Answer: $A = \dfrac{4}{15}$.

EXERCISES

Find the volume under the surface $z = f(x, y)$, and over the indicated region R of the x, y-plane:

1. $z = 1$; Fig. 1.16
2. $z = y$; Fig. 1.16
3. $z = x$; Fig. 1.16
4. $z = 1 + x + y$; Fig. 1.16

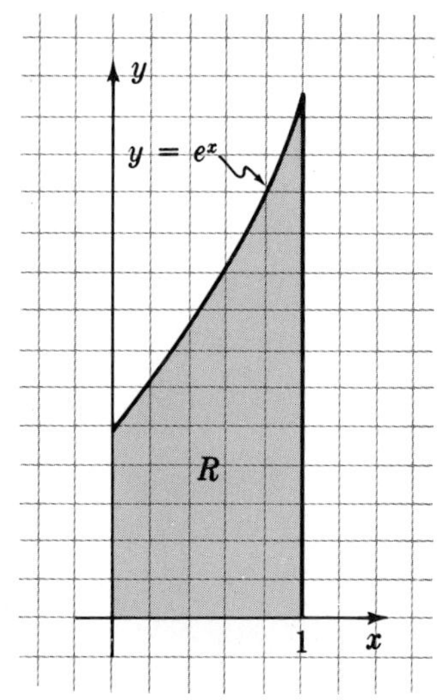

Fig. 1.16 (Ex. 1–4)

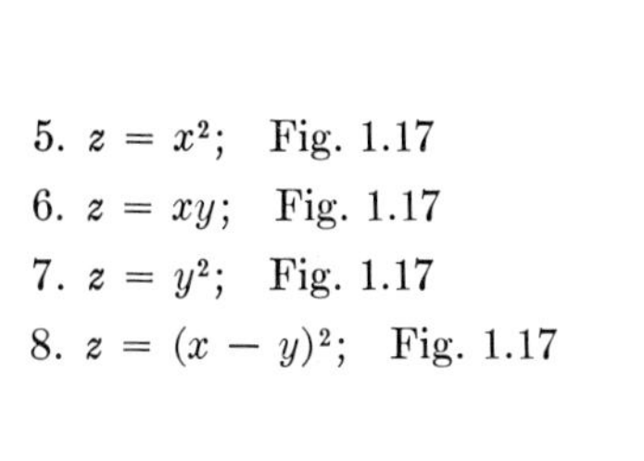

5. $z = x^2$; Fig. 1.17
6. $z = xy$; Fig. 1.17
7. $z = y^2$; Fig. 1.17
8. $z = (x - y)^2$; Fig. 1.17

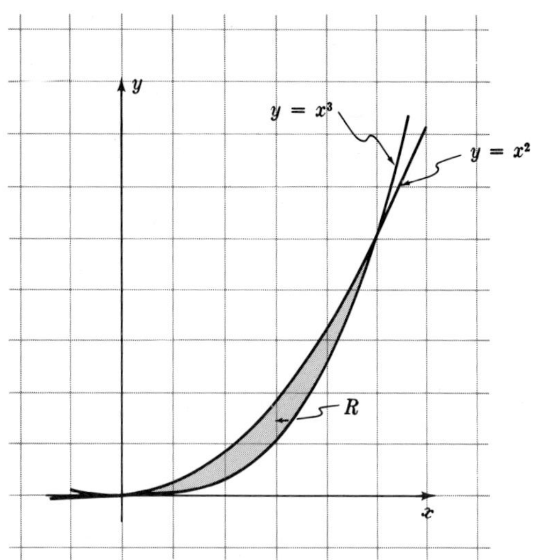

Fig. 1.17 (Ex. 5–8)

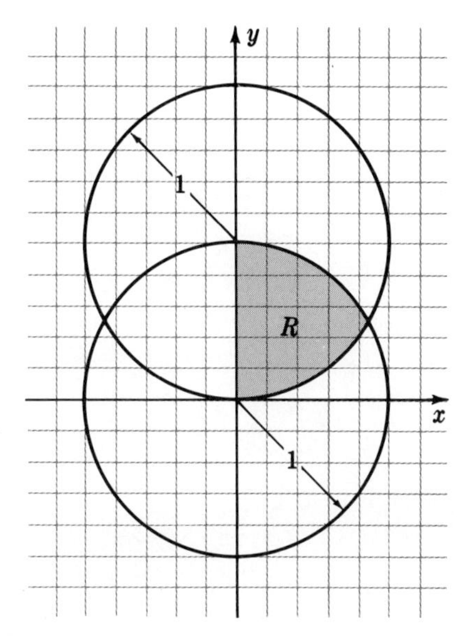

Fig. 1.18 (Ex. 9–11)

9. $z = 1$; Fig. 1.18
10. $z = 1 + x$; Fig. 1.18
11. $z = y$; Fig.18

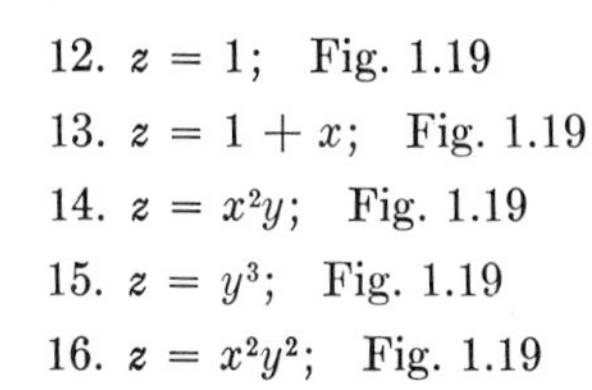

12. $z = 1$; Fig. 1.19
13. $z = 1 + x$; Fig. 1.19
14. $z = x^2y$; Fig. 1.19
15. $z = y^3$; Fig. 1.19
16. $z = x^2y^2$; Fig. 1.19

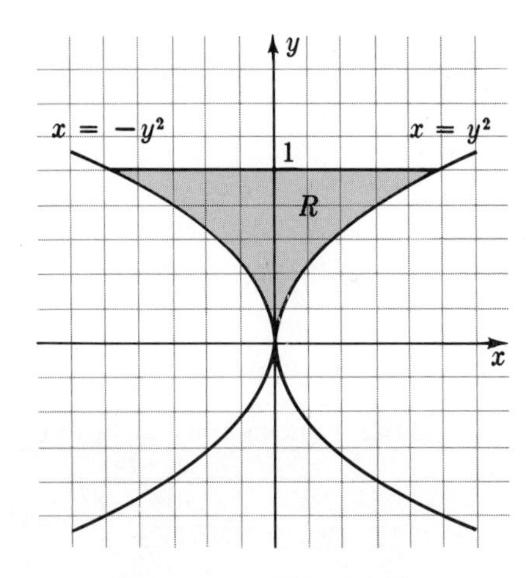

Fig. 1.19 (Ex. 12–16)

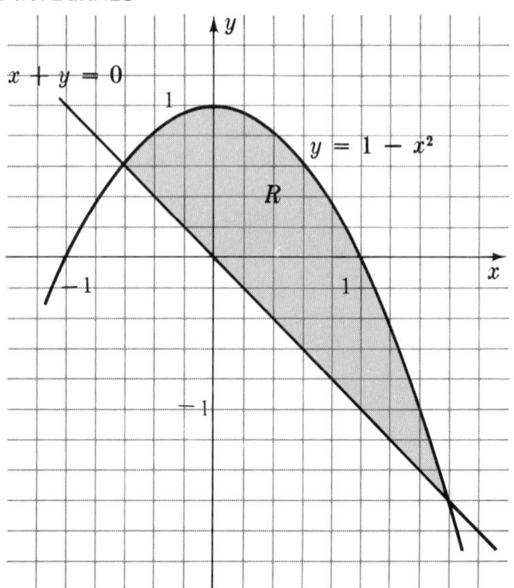

FIG. 1.20 (Ex. 17–18)

17. $z = 1$; Fig. 1.20

18. $z = 1 + x$; Fig. 1.20.

19. Describe each of the regions in Figs. 1.16–1.20 by inequalities between x and y.

Compute the double integral over the region whose boundary curves are indicated; in each case draw a figure:

20. $\iint xe^{xy}\,dx\,dy; \quad x = 1, \quad x = 3, \quad xy = 1, \quad xy = 2$

21. $\iint x^2y\,dx\,dy; \quad y = 0, \quad x = 0, \quad x = (y - 1)^2$

22. $\iint (x^3 + y^3)\,dx\,dy; \quad x^2 + y^2 = 1$

23. $\iint (x + y)^2\,dx\,dy; \quad x + y = 0, \quad y = x^2 + x$

24. $\iint (1 + xy)\,dx\,dy; \quad y = 0, \quad y = x, \quad y = 1 - x$

25. $\iint y^2\,dx\,dy; \quad y = \pm x, \quad y = \frac{1}{2}x + 3.$

26. Find $\iint\limits_R (1 + xy^2)\,dx\,dy$ over the region bounded by the parabola $x = -y^2$ and the segments from $(2, 0)$ to $(-1, \pm 1)$.

2. POLAR COORDINATES

It is often convenient to use polar coordinates if the region R has a fan shape as shown in Fig. 2.1a. You must remember that the element of area

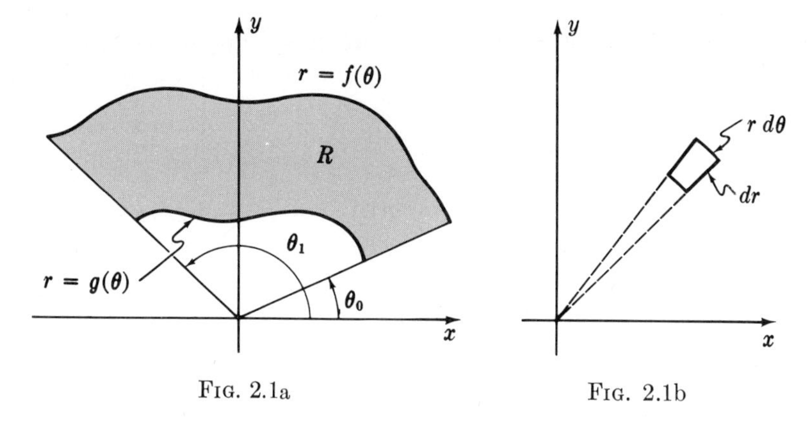

Fig. 2.1a Fig. 2.1b

(Fig. 2.1b) is $r\,dr\,d\theta$ not $dr\,d\theta$. Then there is no difficulty. The iteration formula becomes

$$\iint\limits_R H(x, y)\,dx\,dy = \iint\limits_R H(r\cos\theta, r\sin\theta)\,r\,dr\,d\theta$$

$$= \int_{\theta_0}^{\theta_1}\left(\int_{r=g(\theta)}^{r=f(\theta)} H(r\cos\theta, r\sin\theta)\,r\,dr\right)d\theta.$$

EXAMPLE 2.1

Set up in polar coordinates the integral $\displaystyle\iint\limits_R \frac{dx\,dy}{x^2+y^2}$ over the region R of Fig. 1.13. Do not evaluate.

Solution: The region R splits naturally into two regions, one the reflection of the other in the line $y = x$; the function is symmetric in this line (Fig. 2.2). In polar coordinates, the line $x = 2$ is $r\cos\theta = 2$ or $r = 2\sec\theta$.

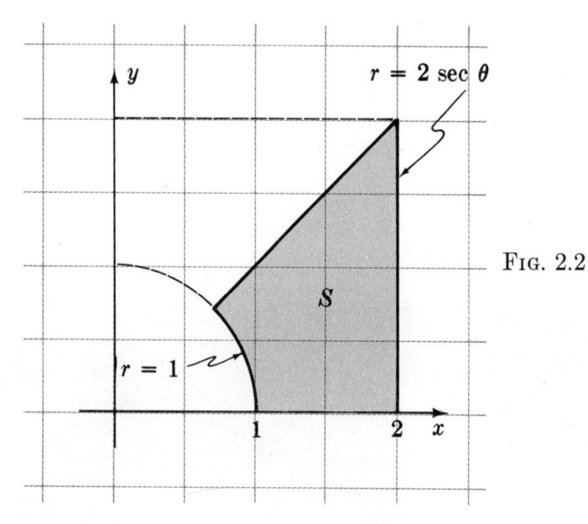

Fig. 2.2

Consequently, the lower region S is determined by $0 \leq \theta \leq \pi/4$ and $1 \leq r \leq 2 \sec \theta$. The set-up is

$$\iint_R \frac{dx\,dy}{x^2 + y^2} = 2 \iint_S \frac{r\,dr\,d\theta}{r^2}$$

$$= 2 \int_0^{\pi/4} \left(\int_1^{2 \sec \theta} \frac{dr}{r} \right) d\theta = 2 \int_0^{\pi/4} \ln(2 \sec \theta)\, d\theta.$$

EXAMPLE 2.2

Compute $\iint_R xy\,dx\,dy$ over the region R defined by $r = \sin 2\theta$, $0 \leq \theta \leq \pi/2$.

Solution: Draw a figure (Fig. 2.3):

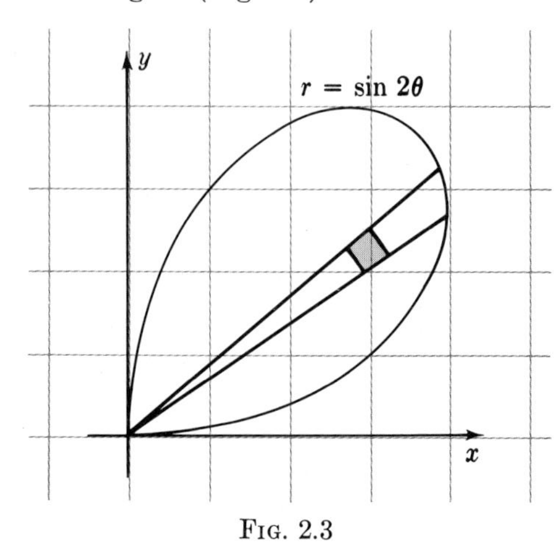

FIG. 2.3

In polar coordinates,

$$xy = (r \cos \theta)\,(r \sin \theta) = \frac{1}{2} r^2 \sin 2\theta,$$

hence

$$\iint_R xy\,dx\,dy = \iint_R \frac{1}{2} r^2 \sin 2\theta\, r\,dr\,d\theta = \frac{1}{2} \int_0^{\pi/2} (\sin 2\theta) \left(\int_0^{\sin 2\theta} r^3\,dr \right) d\theta$$

$$= \frac{1}{2} \int_0^{\pi/2} (\sin 2\theta) \left(\frac{1}{4} \sin^4 2\theta \right) d\theta = \frac{1}{8} \int_0^{\pi/2} \sin^5 2\theta\, d\theta = \frac{1}{16} \int_0^{\pi} \sin^5 \alpha\, d\alpha$$

$$= \frac{1}{8} \int_0^{\pi/2} \sin^5 \alpha\, d\alpha = \frac{1}{8} \cdot \frac{2 \cdot 4}{1 \cdot 3 \cdot 5} = \frac{1}{15} \quad \text{(by tables).}$$

Answer: $\dfrac{1}{15}$.

EXAMPLE 2.3

Find the volume common to the three solid right circular cylinders bounded by $x^2 + y^2 = 1$, $y^2 + z^2 = 1$, $z^2 + x^2 = 1$.

Solution: The first problem is drawing the solid. By symmetry, only the portion in the first octant need be shown. In Fig. 2.4a the common part

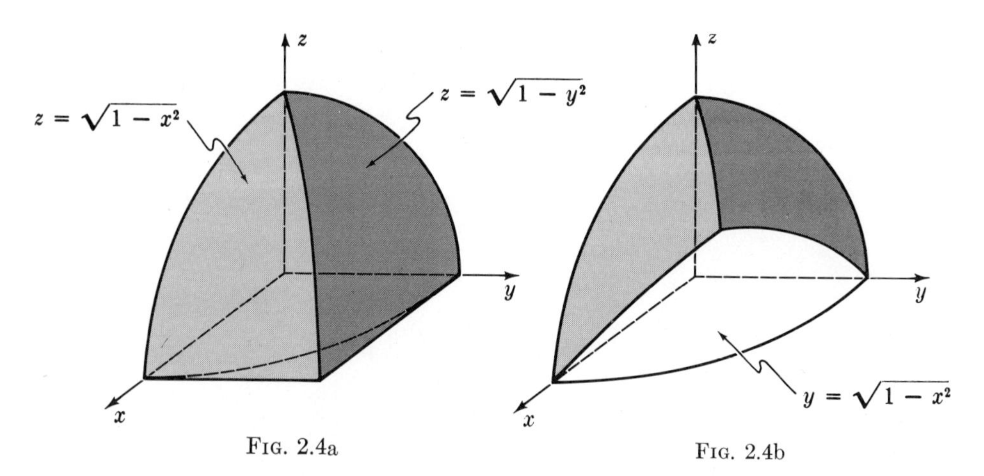

FIG. 2.4a

FIG. 2.4b

of $y^2 + z^2 = 1$ and $z^2 + x^2 = 1$ is shown. The base of the third cylinder $x^2 + y^2 = 1$ is dotted. In Fig. 2.4b this third cylinder is drawn, cutting off the desired region. One sees that the region is symmetric in the plane $x = y$, hence only the half to the left of this plane is needed (Fig. 2.5a). This is the region under the surface $z = \sqrt{1 - x^2}$ and over the wedge-shaped region R shown in Fig. 2.5b.

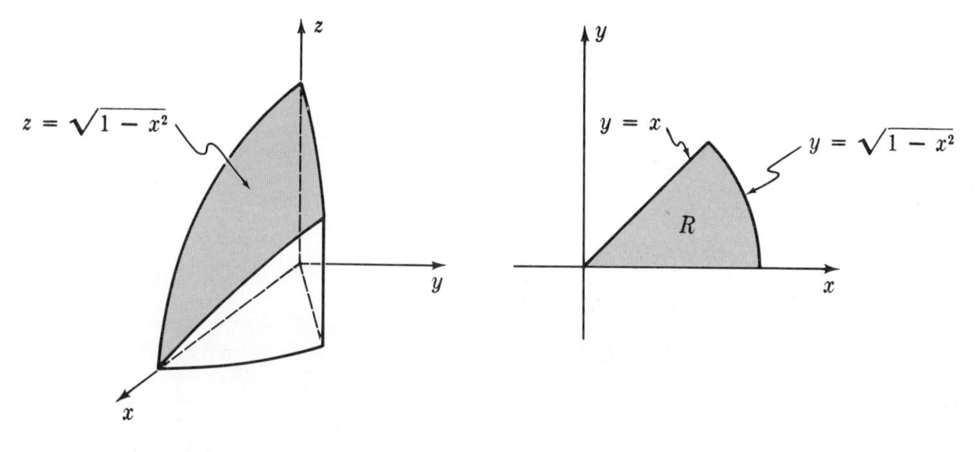

FIG. 2.5a

FIG. 2.5b

Compute the volume using polar coordinates:

$$V = 16 \iint\limits_R \sqrt{1 - x^2}\, dx\, dy = 16 \iint\limits_R \sqrt{1 - r^2 \cos^2 \theta}\; r\, dr\, d\theta$$

$$= 16 \int_0^{\pi/4} \left[\int_0^1 r \sqrt{1 - r^2 \cos^2 \theta}\, dr \right] d\theta$$

$$= 16 \int_0^{\pi/4} \left[\left(\frac{-1}{3 \cos^2 \theta} \right) (1 - r^2 \cos^2 \theta)^{3/2} \Bigg|_{r=0}^{r=1} \right] d\theta = \frac{16}{3} \int_0^{\pi/4} \frac{1 - \sin^3 \theta}{\cos^2 \theta}\, d\theta.$$

Now

$$\frac{1 - \sin^3 \theta}{\cos^2 \theta} = \frac{1 - \sin^3 \theta}{1 - \sin^2 \theta} = \frac{1 + \sin \theta + \sin^2 \theta}{1 + \sin \theta} = \sin \theta + \frac{1}{1 + \sin \theta}.$$

From tables,

$$\int \frac{d\theta}{1 + \sin \theta} = -\tan\left(\frac{\pi}{4} - \frac{\theta}{2} \right) + C,$$

hence

$$V = \frac{16}{3} \left[-\cos \theta - \tan\left(\frac{\pi}{4} - \frac{\theta}{2} \right) \right] \Bigg|_0^{\pi/4} = \frac{16}{3} \left[1 - \frac{\sqrt{2}}{2} - \tan \frac{\pi}{8} + 1 \right]$$

$$= \frac{16}{3} \left[2 - \frac{\sqrt{2}}{2} - \frac{\sqrt{2}}{2 + \sqrt{2}} \right] = 8(2 - \sqrt{2}).$$

Answer: $8(2 - \sqrt{2})$.

NOTE: The expression for $\tan(\pi/8)$ comes from the half-angle formula $\tan(\theta/2) = \sin \theta / (1 + \cos \theta)$.

A Probability Integral

The improper integral

$$I = \int_{-\infty}^{\infty} e^{-x^2}\, dx$$

is important in probability. Its exact value can be found by using polar coordinates and a clever trick. Write

$$I = \int_{-\infty}^{\infty} e^{-y^2}\, dy.$$

Then

$$I^2 = \left(\int_{-\infty}^{\infty} e^{-x^2}\, dx \right) \left(\int_{-\infty}^{\infty} e^{-y^2}\, dy \right) = \iint\limits_P e^{-(x^2 + y^2)}\, dx\, dy,$$

where P is the whole plane. Now switch from rectangular to polar coordinates. Then $x^2 + y^2 = r^2$ and $dx\,dy = r\,dr\,d\theta$. Hence

$$I^2 = \int_0^{2\pi}\left(\int_0^{\infty} e^{-r^2} r\,dr\right)d\theta = 2\pi\left(-\frac{1}{2}e^{-r^2}\right)\Big|_0^{\infty} = 2\pi\left(\frac{1}{2}\right) = \pi.$$

Since I is positive,

$$\int_{-\infty}^{\infty} e^{-x^2}\,dx = \sqrt{\pi}.$$

The function

$$\phi(x) = \frac{1}{\sqrt{2\pi}}\,e^{-x^2/2}$$

is known as the **density function** of the **normal distribution.** Its graph is the familiar bell-shaped curve (Fig. 2.6) and encloses area 1 (by a simple modification of the definite integral above).

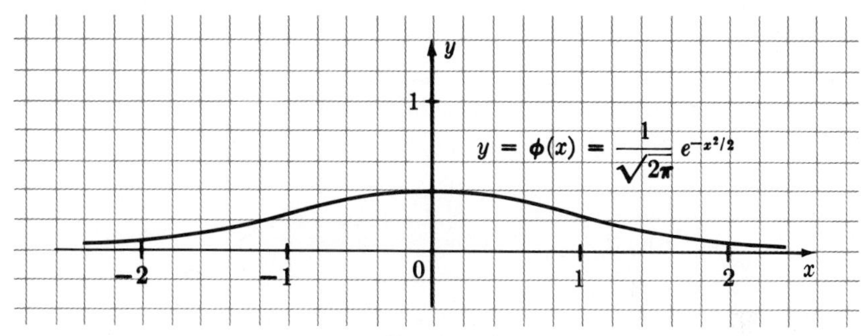

FIG. 2.6

REMARK: Expressing an antiderivative of e^{-x^2} in terms of elementary functions is known to be impossible. Thus it is quite remarkable that the integral of this function over the whole x-axis can be evaluated. The trick of passing to the plane is really ingenious.

EXERCISES

1. Find the volume under the surface $z = x^4y^4$, and over the circle $x^2 + y^2 \le 1$.
2. Find the volume under the surface $z = r^3$, and over the quarter circle $0 \le r \le 1$, $0 \le \theta \le \pi/2$.
3. Find the volume under the cone $z = 5r$, and over the rose petal with boundary $r = \sin 3\theta$, $0 \le \theta \le \pi/3$.
4. Find the area bounded by the spiral $r = \theta$, $0 \le \theta \le 2\pi$, and the x-axis from 0 to 2π.
5. Evaluate over the unit disk $0 \le r \le 1$:

$$\iint x^4\,dx\,dy, \qquad \iint x^2y^2\,dx\,dy, \qquad \iint y^4\,dx\,dy.$$

6. Evaluate over the unit disk $0 \le r \le 1$:

$$\iint x^6\,dx\,dy, \quad \iint x^4y^2\,dx\,dy, \quad \iint x^2y^4\,dx\,dy, \quad \iint y^6\,dx\,dy.$$

7. Evaluate over the unit disk $0 \le r \le 1$:

$$\iint xy^2\,dx\,dy, \quad \iint x^4y^3\,dx\,dy, \quad \iint \sin(xy)\,dx\,dy.$$

8. Evaluate over the unit disk $0 \le r \le 1$:

$$\iint \frac{dx\,dy}{1+x^2+y^2}, \quad \iint e^{x^2+y^2}\,dx\,dy.$$

9. Show that

$$\iint \frac{dx\,dy}{(x^2+y^2)^p}, \quad \text{over} \quad x^2+y^2 \ge 1$$

converges if $p > 1$ and evaluate it.
(*Hint:* Integrate over the ring-shaped region $1 \le x^2+y^2 \le a^2$. Then let $a \to \infty$.)

10. (cont.) Conclude that the double infinite series $\displaystyle\sum_{\substack{m,n=-\infty \\ (m,n)\ne(0,0)}}^{\infty} \frac{1}{(m^2+n^2)^p}$ converges if $p > 1$.

11. Find $\displaystyle\int_0^\infty e^{-tx^2}\,dx, \quad t > 0.$ (*Hint:* Set $u = x\sqrt{t}$.)

12. Find $\displaystyle\int_0^\infty x^2e^{-x^2}\,dx.$

13. Find the volume common to two right circular cylinders of radius 1 intersecting at right angles on center.

14. (cont.) Find the volume of the region in space consisting of all points (x, y, z), where one or more of the following three conditions hold:
(1) $x^2+y^2 \le 1, \; -2 \le z \le 2$
(2) $y^2+z^2 \le 1, \; -2 \le x \le 2$
(3) $z^2+x^2 \le 1, \; -2 \le y \le 2.$
(*Hint:* Use the results of Ex. 13 and Example 2.3; after that, its pure logic, not calculus.) The game of "jacks" has objects more or less in the given shape.

15. Evaluate the integral in Example 2.3 using rectangular coordinates.

3. TRIPLE INTEGRALS

Consider the following two problems: A non-homogeneous solid fills a region D of space (Fig. 3.1). At each point $\mathbf{x}$ its density is $\rho(\mathbf{x})$ gm/cm^3. What is its total mass?

Electric charge is scattered throughout a non-conductor D. At each point $\mathbf{x}$ the charge density is $\rho(\mathbf{x})$ esu/cm^3. What is the total charge on the solid?

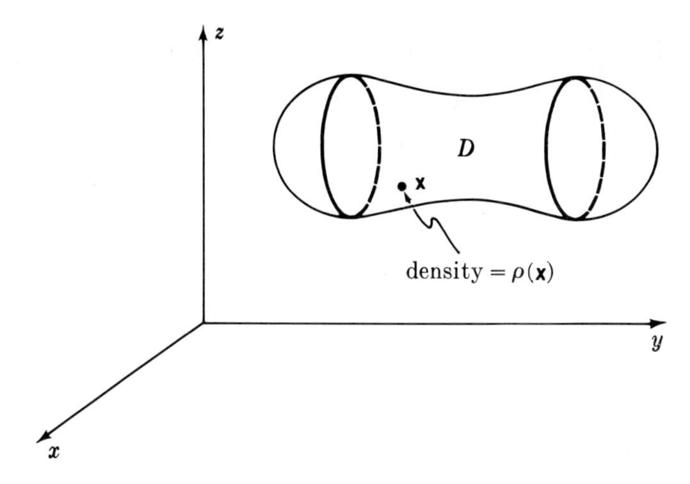

FIG. 3.1

Mathematically these problems are equivalent. In each case, first chop the region into many little boxes; second, compute the quantity (mass, charge)

$$\rho(\mathbf{x})\, dx\, dy\, dz$$

in a little box at $\mathbf{x}$; third, add all these together; last, take the limit as the boxes become smaller and smaller. You obtain the triple integral

$$\iiint_D \rho(\mathbf{x})\, dx\, dy\, dz.$$

The triple integral is usually evaluated by iteration. This means the integral is set up so that one ordinary integration reduces it to a double integral—which in turn is evaluated by iteration. Thus there are six possible orders of iteration. It is important to choose one which makes the work simple.

Iteration applies directly when the region D is the part of a cylinder bounded between two surfaces, each the graph of a function. Precisely, suppose two surfaces $z = g(x, y)$ and $z = f(x, y)$ are defined over a region R in the x, y-plane, and that $g(x, y) < f(x, y)$. See Fig. 3.2. These surfaces may be considered as the top and bottom of a region D in the cylinder over R. Thus D consists of all points (x, y, z) where (x, y) is in R and

$$g(x, y) \le z \le f(x, y).$$

In this situation the iteration formula is

$$\iiint_D \rho(\mathbf{x})\, dx\, dy\, dz = \iint_R \left(\int_{z=g(x,y)}^{z=f(x,y)} \rho(x, y, z)\, dz \right) dx\, dy.$$

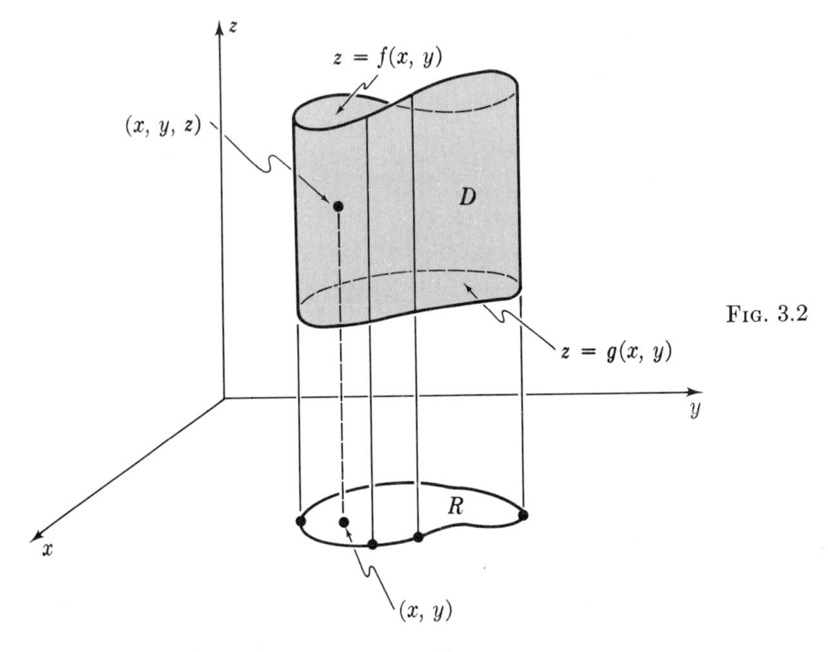

FIG. 3.2

REMARK: Some prefer the notation

$$\iint\limits_R dx\, dy \int_{g(x,y)}^{f(x,y)} \rho(x, y, z)\, dz.$$

The reason for the iteration formula is illustrated in Fig. 3.3. First, the little pieces of mass (charge) $\rho(x, y, z)\, dx\, dy\, dz$ in one column are added

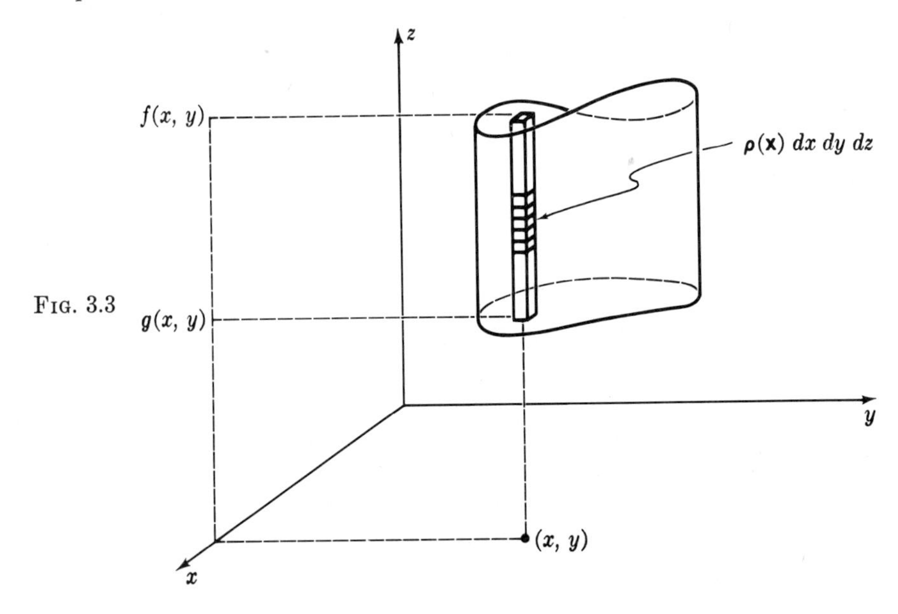

FIG. 3.3

together. The result is

$$\left(\int_{g(x,y)}^{f(x,y)} \rho(x, y, z)\, dz\right) dx\, dy.$$

Then the masses of these individual columns are totaled by a double integral over R.

EXAMPLE 3.1

Find $\iiint\limits_D (x^2 + y)\, dx\, dy\, dz$, where D is the block $1 \leq x \leq 2$, $0 \leq y \leq 1$, $2 \leq z \leq 5$.

Solution: The upper and lower boundaries of D are the planes $z = 5$ and $z = 2$.

$$\iiint\limits_D (x^2 + y)\, dx\, dy\, dz = \iint\limits_R \left(\int_2^5 (x^2 + y)\, dz\right) dx\, dy,$$

where R is the rectangle $1 \leq x \leq 2$ and $0 \leq y \leq 1$. Now x and y are constant in the inner integral:

$$\int_2^5 (x^2 + y)\, dz = 3(x^2 + y),$$

hence

$$\begin{aligned}\iiint\limits_D (x^2 + y)\, dx\, dy\, dz &= \iint\limits_R 3(x^2 + y)\, dx\, dy \\ &= \int_1^2 \left(\int_0^1 3(x^2 + y)\, dy\right) dx \\ &= \int_1^2 \left(3x^2 + \frac{3}{2}\right) dx = \frac{17}{2}.\end{aligned}$$

Answer: $\dfrac{17}{2}$.

EXAMPLE 3.2

Compute $\iiint\limits_D x^3y^2z\, dx\, dy\, dz$, where the solid D is bounded by $x = 1$, $x = 2$; $y = 0$, $y = x^2$; and $z = 0$, $z = 1/x$.

Solution: The region D is the portion between the surfaces $z = 0$ and $z = 1/x$ of a solid cylinder parallel to the z-axis. The cylinder has base R in the x, y-plane, where R is shown in Fig. 3.4a. The solid D itself is sketched in Fig. 3.4b. (A rough sketch showing the general shape is quite satisfactory.)

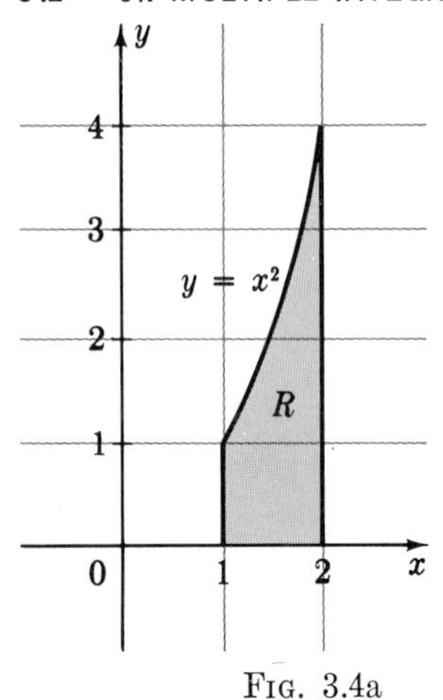

FIG. 3.4a

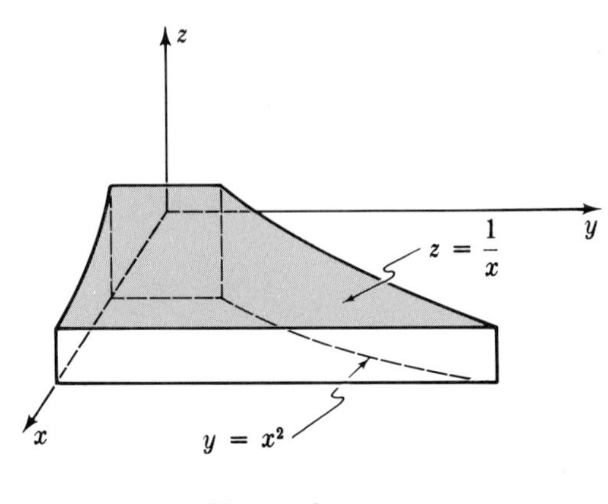

FIG. 3.4b

The iteration is

$$\iiint_D x^3y^2z\,dx\,dy\,dz = \iint_R \left(\int_0^{1/x} x^3y^2z\,dz\right) dx\,dy$$

$$= \iint_R x^3y^2 \left(\frac{1}{2}z^2 \Big|_0^{1/x}\right) dx\,dy = \frac{1}{2}\iint_R xy^2\,dx\,dy$$

$$= \frac{1}{2}\int_1^2 x \left(\int_0^{x^2} y^2\,dy\right) dx = \frac{1}{6}\int_1^2 x^7\,dx$$

$$= \frac{1}{48}(2^8 - 1) = \frac{255}{48} = \frac{85}{16}.$$

Alternate Solution: The region D may be considered as the portion between the surfaces $y = 0$ and $y = x^2$ of a solid cylinder parallel to the y-axis. The cylinder has base S in the z, x-plane (Fig. 3.5). From this view-

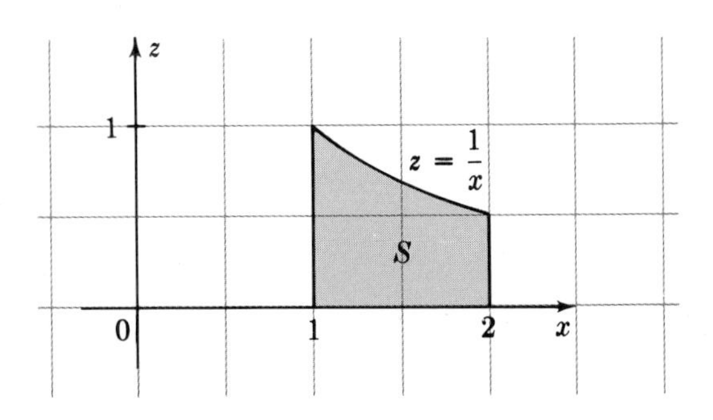

FIG. 3.5

point, the first integration is with respect to y; the iteration is

$$\iiint\limits_D x^3y^2z\,dx\,dy\,dz = \iint\limits_S x^3z\left(\int_0^{x^2} y^2\,dy\right)dx\,dz = \iint\limits_S \frac{1}{3}x^9z\,dx\,dz$$

$$= \frac{1}{3}\int_1^2 x^9\left(\int_0^{1/x} z\,dz\right)dx = \frac{1}{6}\int_1^2 x^7\,dx = \frac{1}{48}(2^8-1) = \frac{85}{16}.$$

Answer: $\dfrac{85}{16}$.

REMARK: It is bad technique to consider the region as a solid cylinder parallel to the x-axis, because the projection of the solid into the y, z-plane breaks into four parts R_1, R_2, R_3, and R_4. See Fig. 3.6. Therefore, the solid

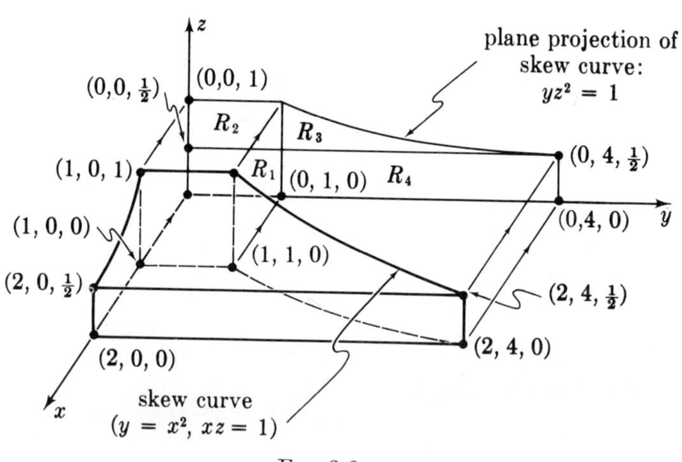

FIG. 3.6

D itself must be decomposed into four parts, and the triple integral correspondingly expressed as a sum of four triple integrals (Fig. 3.7). The resulting computation is much longer than that in either of the previous solutions.

In practice, try to pick an order of iteration which decomposes the required triple integral into as few summands as possible, hopefully only one. The typical summand has the form

$$\int_a^b\left[\int_{k(x)}^{h(x)}\left(\int_{g(x,y)}^{f(x,y)} \rho(x, y, z)\,dz\right)dy\right]dx.$$

(Possibly the variables are in some other order.) Once the integral

$$\int_{g(x,y)}^{f(x,y)} \rho(x, y, z)\,dz$$

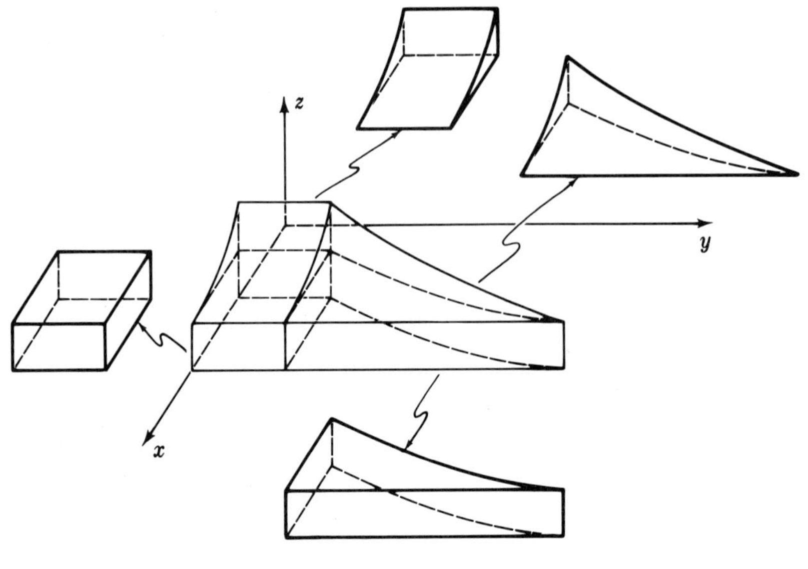

Fig. 3.7

is evaluated, the result is a function of x and y alone; z does not appear. Likewise, once the integral

$$\int_{k(x)}^{h(x)} \left(\int_{g(x,y)}^{f(x,y)} \rho(x,\ y,\ z)\ dz \right) dy$$

is evaluated, the result is a function of x alone; y does not appear.

Remember there are six possible orders of iteration for triple integrals. If you encounter an integrand you cannot find in tables, try a different order of iteration.

EXERCISES

Evaluate the triple integral over the indicated region:

1. $\iiint (2 - z)\,dx\,dy\,dz$; solid cylinder under $z = 2 - x^2$, based on the triangle with vertices (0, 0, 0), (1, 0, 0), (0, 1, 0)
2. $\iiint y\,dx\,dy\,dz$; solid cylinder under $z = x^2 + 2y^2$, based on the triangle in Ex. 1
3. $\iiint z^3\,dx\,dy\,dz$; pyramid with apex (0, 0, 1), based on the square with vertices $(\pm 1, \pm 1)$
4. The same, except the square base has vertices $(\pm 1, 0)$, $(0, \pm 1)$
5. $\iiint x^3y^2z\,dx\,dy\,dz$; the region lies in the first octant and is bounded by $x = 0$, $z = 0$, $y = x^2$, and $z = 1 - y^2$

6. $\iiint (3x^2 - z^2)\,dx\,dy\,dz$; the region in the slab $0 \le y \le 1$ bounded by $z = y^2$, $z = -y^2$, and $y = x$, $y = -x$.
7. A solid cube has side a. Its density at each point is k times the product of the 6 distances of the point to the faces of the cube, where k is constant. Find the mass.
8. Charge is distributed over the tetrahedron with vertices $\mathbf{0}$, $\mathbf{i}$, $\mathbf{j}$, $\mathbf{k}$. The charge density at each point is a constant k times the product of the 4 distances from the point to the faces of the tetrahedron. Find the total charge.
9. Find $\iiint (x + y + z)^2\,dx\,dy\,dz$ over the region in the first octant bounded by the coordinate planes, the plane $x + y + z = 2$, and the 3 planes $x = 1$, $y = 1$, $z = 1$.
10. Find $\displaystyle\iiint \frac{dx\,dy\,dz}{(x + y + z)^2}$ over the region in the first octant between the planes $x + y + z = 1$ and $x + y + z = 4$. (*Hint:* Think!)
11. Extend Simpson's Rule for double integrals (Chapter 26, Section 5) to triple integrals.
12. (cont.) For which polynomials in 3 variables is the extended Simpson's Rule exact?
13. (cont.) Estimate the integral of $\sin(xyz)$ over the cube $0 \le x, y, z \le 1$, dividing the cube into 8 equal parts. Carry your work to 3 places.

4. FURTHER EXAMPLES

A region in the plane or in space is frequently specified by a system of inequalities. A single inequality $f(x, y, z) \ge 0$ determines a region whose boundary is $f(x, y, z) = 0$. To find the region described by several such inequalities, draw the region each describes, and then take the common part of these various regions.

EXAMPLE 4.1

Draw the plane region given by $x + y \le 0$, $y \ge x^2 + 2x$.

Solution: The first inequality determines the region below (and on) the line $x + y = 0$. See Fig. 4.1a. The second inequality determines the region above (and on) the parabola $y = x^2 + 2x$. See Fig. 4.1b. The line and

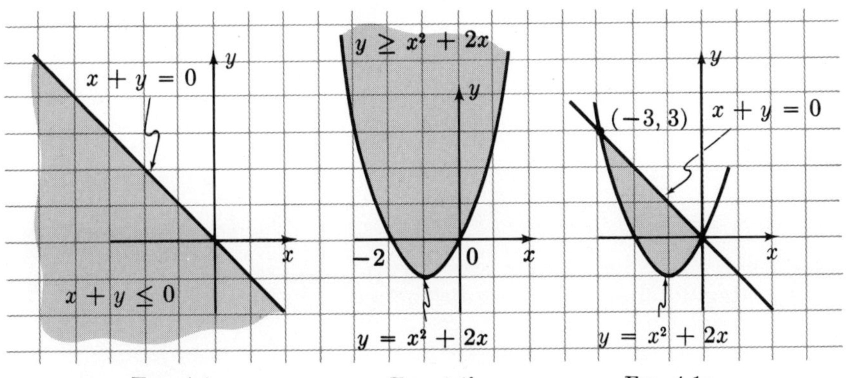

FIG. 4.1a FIG. 4.1b FIG. 4.1c

parabola intersect at $(0, 0)$ and $(-3, 3)$. The region of points (x, y) satisfying both inequalities is shown in Fig. 4.1c.

EXAMPLE 4.2

Describe the plane region R defined by $0 \leq x \leq y \leq a$. Write the integral $\iint_R f(x, y)\,dx\,dy$ as an iterated integral in both orders.

Solution: The region is described by the three inequalities

$$x \geq 0, \qquad y \geq x, \qquad y \leq a.$$

Draw the corresponding regions and take their common part, a triangle (Fig. 4.2).

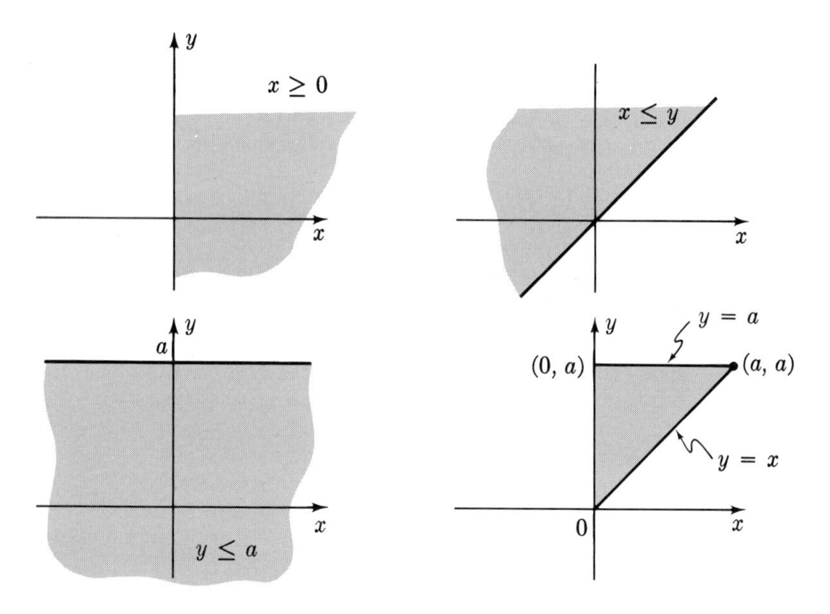

FIG. 4.2

To integrate first on x, arrange the inequalities describing the triangle in this order:

$$0 \leq y \leq a, \qquad 0 \leq x \leq y.$$

The result is

$$\iint_R f(x, y)\,dx\,dy = \int_0^a \left(\int_0^y f(x, y)\,dx \right) dy.$$

To integrate first on y, describe the region in the other order:

$$0 \leq x \leq a, \qquad x \leq y \leq a.$$

Now the result is

$$\iint_R f(x, y)\,dx\,dy = \int_0^a \left(\int_x^a f(x, y)\,dy \right) dx.$$

Answer: The triangle with vertices (0, 0), (0, a), (a, a);

$$\int_0^a \left(\int_0^y f(x, y)\,dx \right) dy = \int_0^a \left(\int_x^a f(x, y)\,dy \right) dx.$$

REMARK: A special case is interesting. Suppose $f(x, y) = g(x)$, a function of x alone. Then the right-hand side is

$$\int_0^a \left(\int_x^a g(x)\,dy \right) dx = \int_0^a g(x) \left(\int_x^a dy \right) dx = \int_0^a (a - x)g(x)\,dx.$$

The following formula is a consequence:

$$\int_0^a \left(\int_0^y g(x)\,dx \right) dy = \int_0^a (a - x)g(x)\,dx.$$

Thus the iterated integral of a function of one variable can be expressed as a simple integral.

EXAMPLE 4.3

Express the solution to the initial-value problem $\dfrac{d^2y}{dx^2} = g(x)$, $y(0) = 0$, $y'(0) = 0$ as a simple integral.

Solution: Integrate both sides of the differential equation:

$$\frac{dy}{dx} = \int_0^x g(t)\,dt.$$

Notice that $y'(0) = 0$. Now integrate again, using the initial condition $y(0) = 0$:

$$y(x) = \int_0^x y'(s)\,ds = \int_0^x \left(\int_0^s g(t)\,dt \right) ds.$$

The iterated integral can be written as a double integral, and then iterated in the other order:

$$y(x) = \iint_{0 \leq t \leq s \leq x} g(t)\,ds\,dt = \int_0^x \left(\int_t^x g(t)\,ds \right) dt = \int_0^x (x - t)g(t)\,dt.$$

Answer: $y(x) = \displaystyle\int_0^x (x - t)g(t)\,dt.$

REMARK: Since $g = y''$ this can be written

$$y(x) = \int_0^x (x - t)y''(t)\,dt.$$

Have another look at Chapter 27, Section 6, p. 614 to see how this formula is related to the first order Taylor approximation.

Tetrahedra

If a region D is specified by inequalities, it may be possible to arrange the inequalities so that limits of integration can be set up automatically. For example, suppose the inequalities can be arranged in this form:

$$a \le x \le b, \qquad h(x) \le y \le k(x), \qquad g(x, y) \le z \le f(x, y).$$

Then

$$\iiint_D \rho(x, y, z)\, dx\, dy\, dz = \int_a^b \left[\int_{h(x)}^{k(x)} \left(\int_{g(x,y)}^{f(x,y)} \rho(x, y, z)\, dz \right) dy \right] dx.$$

Tetrahedral regions can be expressed by such inequalities, and occur frequently enough that it is useful to practice setting up integrals over such regions.

EXAMPLE 4.4

A tetrahedron T has vertices at $(0, 0, 0)$, $(a, 0, 0)$, $(0, b, 0)$, $(0, 0, c)$, where $a, b, c > 0$. Set up $\iiint_T \rho(x, y, z)\, dx\, dy\, dz$ as an iterated integral.

Solution: The slanted surface (Fig. 4.3) has equation

$$\frac{x}{a} + \frac{y}{b} + \frac{z}{c} = 1.$$

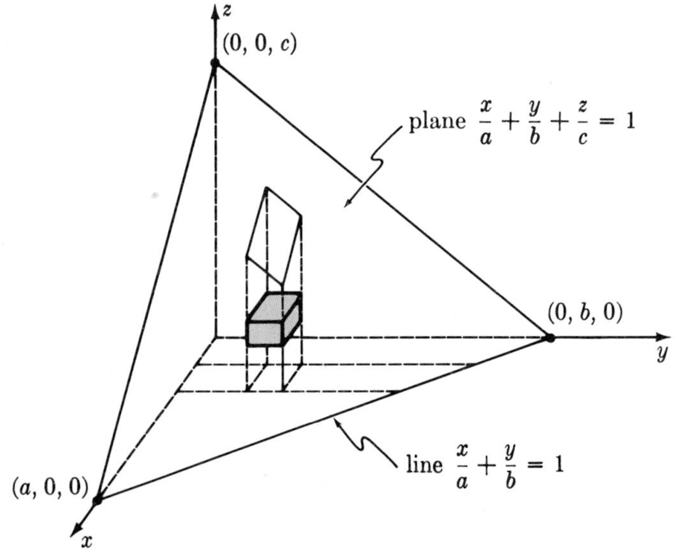

FIG. 4.3 (Iteration shown: first z, then y, finally x.)

The region is defined by the inequalities

$$0 \le x, \qquad 0 \le y, \qquad 0 \le z, \qquad \frac{x}{a} + \frac{y}{b} + \frac{z}{c} \le 1.$$

The region has enough symmetry that any order of iteration is satisfactory. For instance, choose the order of integration

$$\int \left[\int \left(\int \rho(x, y, z)\, dz \right) dy \right] dx.$$

To find the limits of integration, arrange the inequalities in this form:

$$a \le x \le b, \qquad h(x) \le y \le k(x), \qquad g(x, y) \le z \le f(x, y).$$

If x and y are fixed, then

$$0 \le z \le c\left(1 - \frac{x}{a} - \frac{y}{b}\right).$$

If x is fixed, then

$$0 \le y \le b\left(1 - \frac{x}{a} - \frac{z}{c}\right) \le b\left(1 - \frac{x}{a}\right).$$

Finally,

$$0 \le x \le a\left(1 - \frac{y}{b} - \frac{z}{c}\right) \le a.$$

The resulting system of inequalities (equivalent to the original system) is

$$0 \le x \le a, \qquad 0 \le y \le b\left(1 - \frac{x}{a}\right), \qquad 0 \le z \le c\left(1 - \frac{x}{a} - \frac{y}{b}\right).$$

The corresponding iteration is

$$\iiint_T \rho(x, y, z)\, dx\, dy\, dz$$

$$= \int_0^a \left[\int_0^{b\left(1 - \frac{x}{a}\right)} \left(\int_0^{c\left(1 - \frac{x}{a} - \frac{y}{b}\right)} \rho(x, y, z)\, dz \right) dy \right] dx.$$

EXERCISES

Sketch the region:

1. $x^2 + y^2 \le 1, \quad y + x^2 \ge 0$
2. $x^2 + y^2 \le 1, \quad -x^2 \le y \le x^2$
3. $x^2 + y^2 \ge 1, \quad (x - 2)^2 + y^2 \le 9$
4. $x \ge 3, \quad y \le -5, \quad y - x \ge -10$
5. $x + y \le 0, \quad xy \le 1, \quad (x - y)^2 \le 1$
6. $(x + y)^2 \le 1, \quad (x - y)^2 \le 1.$

Express the double integral $\iint f(x, y)\,dx\,dy$ over the region specified as the sum of one or more iterated integrals in which y is the first variable integrated:

7. $x^2 + y^2 \leq 1, \quad x^2 + (y-1)^2 \leq 1$
8. $y \geq (x+1)^2, \quad y + 2x \leq 3$
9. $x \geq 0, \quad 0 \leq y \leq \pi, \quad x \leq \sin y$
10. $x \geq 0, \quad x^2 - y^2 \geq 1, \quad x^2 + y^2 \leq 9.$
11. Describe the region $D\colon 0 \leq x \leq y \leq z \leq 1$. Iterate $\iiint_D \rho(x, y, z)\,dx\,dy\,dz$ in the orders x, y, z; z, y, x; and x, z, y.
12. Repeat Ex. 11 for $D\colon a \leq x \leq y \leq z \leq b$.
13. Repeat Ex. 11 for $D\colon 0 \leq x \leq 2y \leq 3z \leq 6$.
14. Express a triple integral over the region D determined by $(x-1)^2 + y^2 \leq 4$, $0 \leq z \leq y$, $(x+1)^2 + y^2 \leq 4$ as one iterated integral (not a sum of two or more). Sketch the region.
15. Find the integral of x over the tetrahedron with vertices $(5, 6, 3)$, $(4, 6, 3)$, $(5, 5, 3)$, $(5, 6, 2)$.
16. Set up the integral of $\rho(x, y, z)$ over the tetrahedron with vertices $(5, -5, 1)$, $(5, -5, -2)$, $(5, 1, 1)$, $(-2, -5, 1)$.
17. Express the solution to the initial-value problem $d^3y/dx^3 = g(x)$, $y(0) = y'(0) = y''(0) = 0$ as a simple integral.
18. Take four vertices of a unit cube, no two adjacent. Find the volume of the tetrahedron with these points as vertices.
19. (cont.) Now take the tetrahedron whose vertices are the remaining four vertices of the cube. The two tetrahedra intersect in a certain polyhedron. Describe it and find its volume.

5. CYLINDRICAL AND SPHERICAL COORDINATES

Cylindrical Coordinates

If a solid has axial symmetry, it is often convenient to place the z-axis on the axis of symmetry, and use cylindrical coordinates $\{r, \theta, z\}$ for the computation of integrals.

In polar coordinates $\{r, \theta\}$, the element of area is $r\,dr\,d\theta$. Hence the element of volume in cylindrical coordinates $\{r, \theta, z\}$ is

$$\boxed{dV = r\,dr\,d\theta\,dz.}$$

The expression is also a consequence of the relation

$$d\mathbf{x} = dr\,\mathbf{u} + r\,d\theta\,\mathbf{w} + dz\,\mathbf{k}$$

for the displacement of a point in terms of the natural frame $\mathbf{u}$, $\mathbf{w}$, $\mathbf{k}$ associated with the cylindrical coordinate system (Fig. 5.1). The mutually

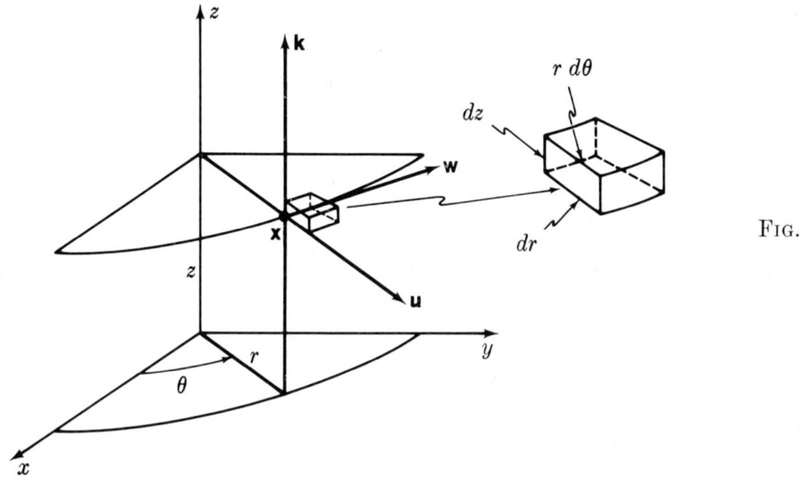

FIG. 5.1

perpendicular displacements $dr\,\mathbf{u}$, $r\,d\theta\,\mathbf{w}$, $dz\,\mathbf{k}$ span a small "rectangular" box of volume $(dr)(r\,d\theta)(dz)$.

EXAMPLE 5.1

Evaluate $\displaystyle\iiint (x^2 + y^2)^{1/2} z\,dx\,dy\,dz$, taken over the cone with apex $(0, 0, 1)$ and semicircular base bounded by the x-axis and $y = \sqrt{4 - x^2}$.

Solution: Write the integral in cylindrical coordinates:

$$I = \iiint (rz)\,r\,dr\,d\theta\,dz.$$

The lateral surface of the cone (Fig. 5.2) is given by

$$z = 1 - \frac{1}{2}r,$$

so the solid is described by

$$0 \le \theta \le \pi, \qquad 0 \le r \le 2, \qquad 0 \le z \le 1 - \frac{1}{2}r.$$

Therefore

$$I = \left(\int_0^\pi d\theta\right)\int_0^2 r^2\left(\int_0^{1-r/2} z\,dz\right)dr = \pi\int_0^2 \frac{r^2}{2}\left(1 - \frac{1}{2}r\right)^2 dr$$

$$= \frac{\pi}{2}\int_0^2 r^2\left(1 - r + \frac{1}{4}r^2\right)dr = \frac{\pi}{8}\int_0^2 (r^4 - 4r^3 + 4r^2)\,dr = \frac{2\pi}{15}.$$

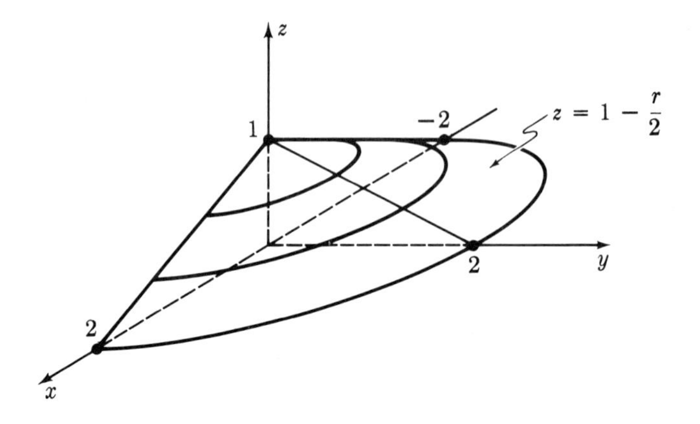

FIG. 5.2

Answer: $\dfrac{2\pi}{15}$.

EXAMPLE 5.2

A region S in space is generated by revolving the plane region bounded by $z = 2x^2$, the x-axis, and $x = 1$ about the z-axis. Mass is distributed in S so that the density at each point is proportional to the distance of the point from the plane $z = -1$, and to the square of the distance of the point from the z-axis. Compute the total mass.

Solution: The density is

$$\begin{aligned}\delta &= k(x^2 + y^2)(z + 1)\\ &= kr^2(z + 1),\end{aligned}$$

where k is a constant. The portion of the solid in the first octant is shown in Fig. 5.3. In cylindrical coordinates, the solid is described by the inequalities

$$0 \le \theta \le 2\pi, \qquad 0 \le r \le 1, \qquad 0 \le z \le 2r^2.$$

Therefore its mass is

$$\begin{aligned}M &= \iiint kr^2(z + 1)\, r\, dr\, d\theta\, dz\\ &= k\left(\int_0^{2\pi} d\theta\right) \int_0^1 r^3 \left[\int_0^{2r^2} (z + 1)\, dz\right] dr\\ &= 2\pi k \int_0^1 r^3 \left[\frac{1}{2}(2r^2)^2 + (2r^2)\right] dr = \frac{7\pi}{6}\, k.\end{aligned}$$

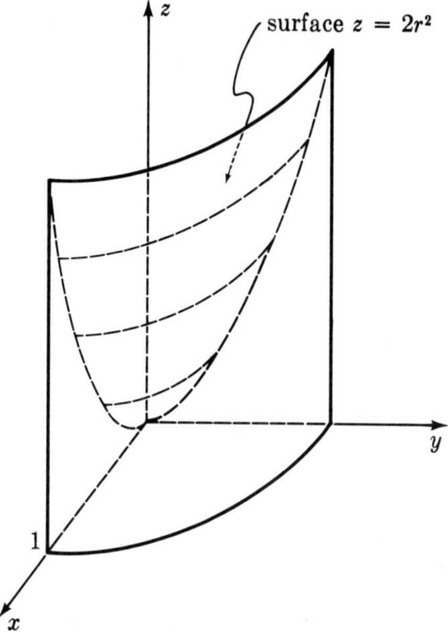

FIG. 5.3

Answer: $\dfrac{7\pi}{6}k.$

Spherical Coordinates

If a solid has central symmetry, it is often convenient to place the origin at the center of symmetry and use spherical coordinates $[\rho, \phi, \theta]$ for the computation of integrals.

The displacement of a point of space, in terms of the frame $\boldsymbol{\lambda}, \boldsymbol{\mu}, \boldsymbol{\nu}$ natural to spherical coordinates, is

$$d\mathbf{x} = d\rho\, \boldsymbol{\lambda} + \rho\, d\phi\, \boldsymbol{\mu} + \rho \sin\phi\, d\theta\, \boldsymbol{\nu}.$$

See Fig. 5.4. The element of volume of the small "rectangular" solid is the product of its sides:

$$\begin{aligned} dV &= (d\rho)\,(\rho\, d\phi)\,(\rho \sin\phi\, d\theta) \\ &= \rho^2 \sin\phi\, d\rho\, d\phi\, d\theta. \end{aligned}$$

EXAMPLE 5.3

Use spherical coordinates to find the volume of a sphere of radius a.

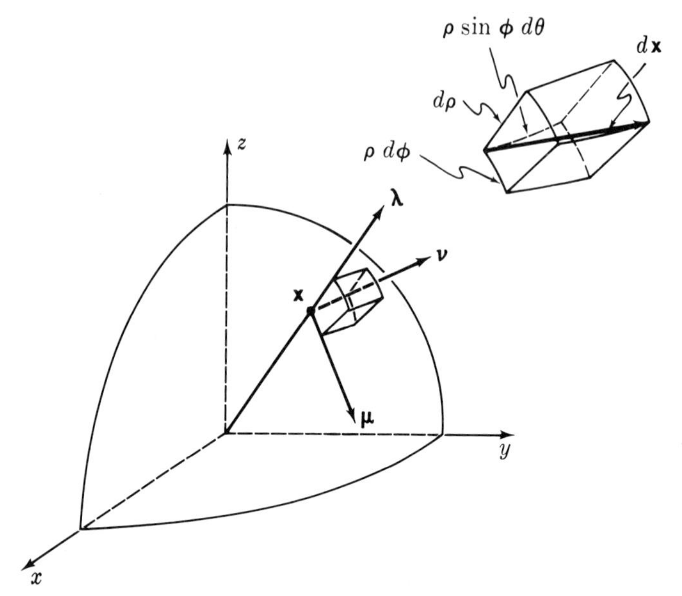

FIG. 5.4

Solution:

$$V = \iiint \rho^2 \sin\phi \, d\rho \, d\phi \, d\theta$$
$$= \left(\int_0^a \rho^2 \, d\rho\right)\left(\int_0^\pi \sin\phi \, d\phi\right)\left(\int_0^{2\pi} d\theta\right)$$
$$= \left(\frac{a^3}{3}\right)(2)\,(2\pi).$$

Answer: $\dfrac{4}{3}\pi a^3.$

EXAMPLE 5.4

Find the volume of the portion of the unit sphere which lies in the right circular cone having its apex at the origin and making angle α with the positive z-axis.

Solution: The cone is specified by $0 \le \phi \le \alpha$, so the portion of the sphere is determined by $0 \le \theta \le 2\pi$, $0 \le \phi \le \alpha$, and $0 \le \rho \le 1$. See Fig. 5.5. Hence the volume is

$$\left(\int_0^{2\pi} d\theta\right)\left(\int_0^\alpha \sin\phi \, d\phi\right)\left(\int_0^1 \rho^2 \, d\rho\right) = (2\pi)\,(1 - \cos\alpha)\left(\frac{1}{3}\right).$$

Answer: $\dfrac{2\pi}{3}(1 - \cos\alpha).$

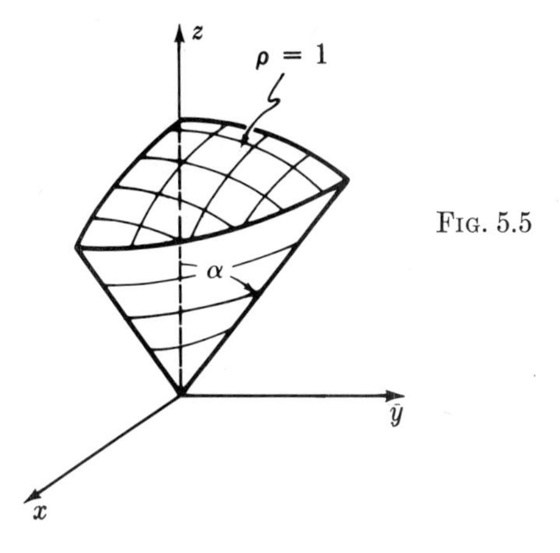

Fig. 5.5

Remark: As a check, let $\alpha \longrightarrow \pi$. Then the volume should approach the volume of a sphere of radius 1. Does it?

EXAMPLE 5.5

A solid fills the region between concentric spheres of radii a and b, where $0 < a < b$. The density at each point is inversely proportional to its distance from the center. Find the total mass.

Solution: The solid is specified by $a \le \rho \le b$; the density is $\delta = k/\rho$. Hence

$$M = \iiint \frac{k}{\rho} \rho^2 \sin\phi \, d\rho \, d\phi \, d\theta$$

$$= k\left(\int_0^{2\pi} d\theta\right)\left(\int_0^{\pi} \sin\phi \, d\phi\right)\left(\int_a^b \rho \, d\rho\right) = (2\pi k)\,(2)\left(\frac{b^2 - a^2}{2}\right).$$

Answer: $2\pi k(b^2 - a^2)$.

Remark: As $a \longrightarrow 0$, the solid tends to the whole sphere, with infinite density at the center. But $M \longrightarrow 2\pi k b^2$, which is finite.

EXAMPLE 5.6

A cylindrical hole of radius $\frac{1}{2}$ is bored through a sphere of radius 1. The surface of the hole passes through the center of the sphere. How much material is removed?

Solution: Center the sphere at **0** and let the cylinder (hole) be parallel to the z-axis, with axis through $(0, \frac{1}{2})$. See Fig. 5.6. By symmetry the volume is four times that in the first octant. The equation of the cylindrical surface is $x^2 + (y - \frac{1}{2})^2 = \frac{1}{4}$, or

$$x^2 + y^2 = y.$$

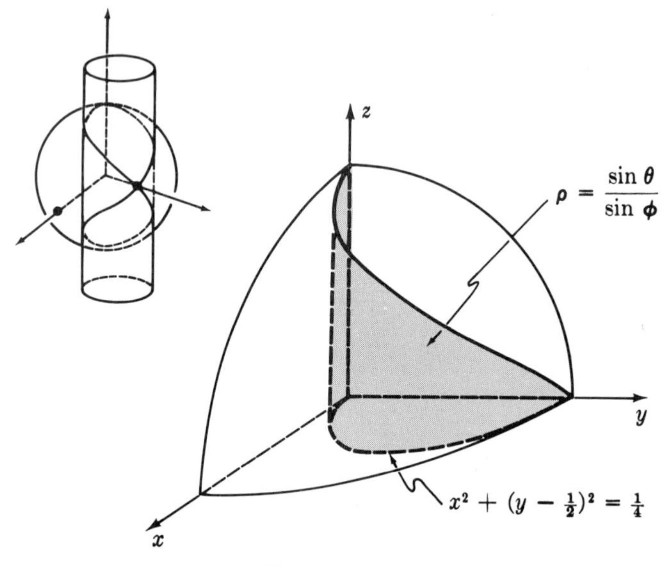

FIG. 5.6

But in spherical coordinates, $x^2 + y^2 = \rho^2 \sin^2 \phi$ and $y = \rho \sin \phi \sin \theta$. Hence the equation of the cylinder is $\rho^2 \sin^2 \phi = \rho \sin \phi \sin \theta$, or

$$\rho = \frac{\sin \theta}{\sin \phi}.$$

The cylinder intersects the sphere in the curve (Fig. 5.7)

$$\rho = 1, \qquad \frac{\sin \theta}{\sin \phi} = 1.$$

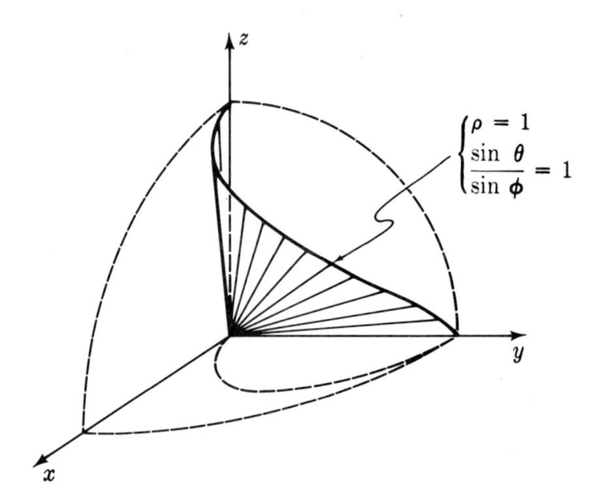

FIG. 5.7

In the first octant this is the curve

$$\rho = 1, \qquad \phi = \theta.$$

The first-octant portion of the volume naturally splits into two parts, separated by the cone $\phi = \theta$ of segments joining $\mathbf{0}$ to the curve of intersection. In the lower part, each radius from the origin ends at the cylinder, so the limits are

$$0 \le \theta \le \frac{\pi}{2}, \qquad \theta \le \phi \le \frac{\pi}{2}, \qquad 0 \le \rho \le \frac{\sin\theta}{\sin\phi}.$$

In the upper part, each radius from the origin ends on the sphere, so the limits are

$$0 \le \theta \le \frac{\pi}{2}, \qquad 0 \le \phi \le \theta, \qquad 0 \le \rho \le 1.$$

Thus $V = 4(V_1 + V_2)$, where V_1 is the volume of the lower part and V_2 that of the upper. But

$$\begin{aligned} V_1 &= \int_0^{\pi/2} d\theta \int_\theta^{\pi/2} \sin\phi\, d\phi \int_0^{\sin\theta/\sin\phi} \rho^2\, d\rho \\ &= \frac{1}{3}\int_0^{\pi/2} \sin^3\theta\, d\theta \int_\theta^{\pi/2} \frac{d\phi}{\sin^2\phi} \\ &= \frac{1}{3}\int_0^{\pi/2} \sin^3\theta \cot\theta\, d\theta = \frac{1}{3}\int_0^{\pi/2} \sin^2\theta\cos\theta\, d\theta = \frac{1}{9}, \end{aligned}$$

and

$$\begin{aligned} V_2 &= \int_0^{\pi/2} d\theta \int_0^{\theta} \sin\phi\, d\phi \int_0^1 \rho^2\, d\rho \\ &= \frac{1}{3}\int_0^{\pi/2} (1 - \cos\theta)\, d\theta = \frac{1}{3}\left(\frac{\pi}{2} - 1\right). \end{aligned}$$

Therefore

$$V = 4\left(\frac{1}{9} + \frac{\pi}{6} - \frac{1}{3}\right) = \frac{2\pi}{3} - \frac{8}{9}.$$

Answer: $\dfrac{2\pi}{3} - \dfrac{8}{9}.$

Spherical Area

Suppose a point moves on the surface of the sphere $\rho = a$. Then $d\rho = 0$, so the formula for displacement specializes to

$$d\mathbf{x} = a\, d\phi\, \boldsymbol{\mu} + a\sin\phi\, d\theta\, \boldsymbol{\nu}.$$

The area of the small "rectangular" region on the surface of the sphere corresponding to changes $d\phi$ and $d\theta$ is

$$dA = (a\,d\phi)(a \sin\phi\, d\theta) = a^2 \sin\phi\, d\phi\, d\theta.$$

See Fig. 5.8. The integral of this expression over a region is the area of that region.

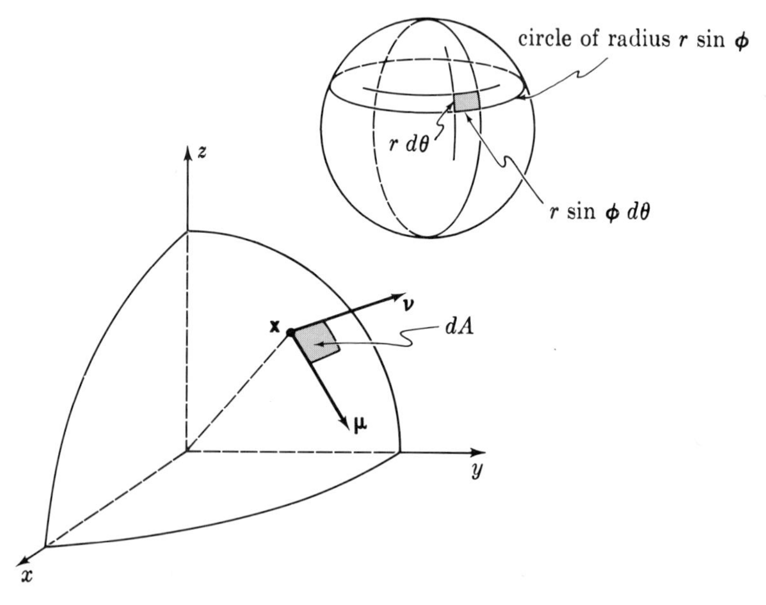

FIG. 5.8

EXAMPLE 5.7

Find the area of the polar cap, all points of co-latitude α or less on the unit sphere.

Solution: See Fig. 5.5. The region is defined on the sphere $\rho = 1$ by $0 \le \phi \le \alpha$, hence

$$A = \int_0^{2\pi} \left(\int_0^{\alpha} \sin\phi\, d\phi \right) d\theta = 2\pi(1 - \cos\alpha).$$

Answer: $2\pi(1 - \cos\alpha)$.

REMARK: Suppose R is a region on the unit sphere. The totality of infinite rays starting at **0** and passing through points of R is a cone which is called a **solid angle** (Fig. 5.9). A solid angle is measured by the area of the base region R. Since R lies on the unit sphere, its area is measured in square radians, a dimensionless unit. Thus the solid angle of the whole sphere has measure 4π rad^2, or simply 4π. The solid angle determined by the first octant has measure $4\pi/8 = \pi/2$. The solid angle of the polar cap in Example 5.7 has measure $2\pi(1 - \cos\alpha)$.

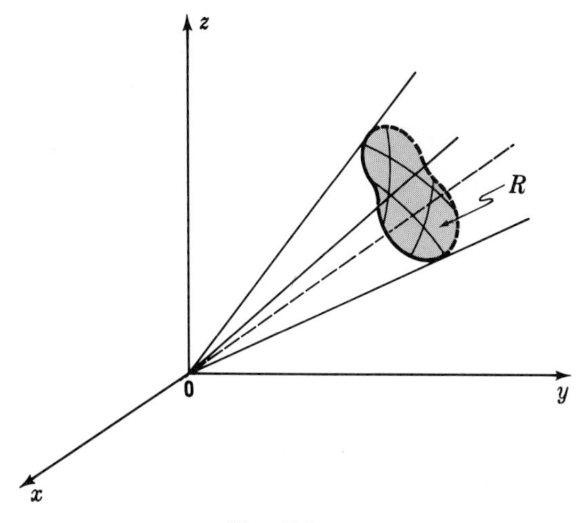

FIG. 5.9

EXERCISES

Ex. 1–13 are hard but interesting volume problems. Unless instructed otherwise, use any method you please to set up the integral. You may not be able to evaluate it, but at least try to reduce the answer to a single integral, not a double or triple integral.

1. A circular hole of radius a is bored through a solid right circular cylinder of radius b. Assume the axis of the hole intersects the axis of the solid at a right angle and that $a \le b$. Show that the volume of material removed is

$$8b^2 \int_0^\sigma (\cos^2 \theta)\ (a^2 - b^2 \sin^2 \theta)^{1/2}\, d\theta,$$

where $\sigma = \text{arc sin } a/b$.

2. Do Ex. 1 assuming this time that the axes meet obliquely at angle α. Set up, but do not evaluate the integral. Can you evaluate it when $a = b$?
3. A square hole is bored through a right circular cylinder (Fig. 5.10). How much material is removed?
4. A square hole is bored through a right circular cylinder (Fig. 5.11). How much material is removed?
5. A square hole is bored through a right circular cone (Fig. 5.12). How much material is removed? (Do not evaluate the integral.)
6. A circular hole is bored through a right circular cone, dimensions as indicated in Fig. 5.13. The axes are perpendicular and the hole just fits. How much material is removed?
7. (cont.) How much of the cone remains above the hole?
8. A sphere of radius a touches the sides of a cone of semi-apex angle α. See Fig. 5.13, *front view only*. Find the volume of the portion of the cone above the sphere.

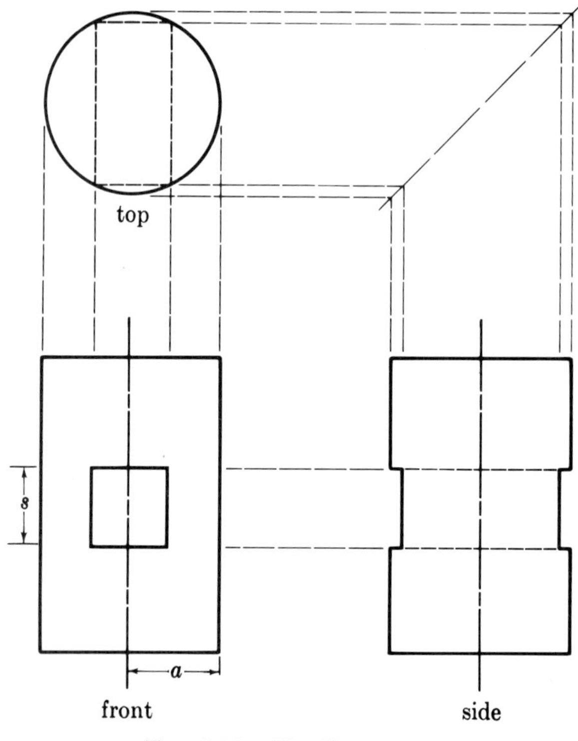

FIG. 5.10 (Ex. 3)

9. Set up Example 5.6 in cylindrical coordinates, taking the z-axis for the axis of the hole.

10. Find the volume enclosed by the ellipsoid $\dfrac{x^2}{a^2} + \dfrac{y^2}{b^2} + \dfrac{z^2}{c^2} = 1$.

    (*Hint:* See Example 3.5, Chapter 8, p. 114.)

11. A cylindrical hole of radius a is bored through a solid cylinder of radius $2a$; the hole is perpendicular to the solid cylinder and just touches a generator. Find the volume removed.

12. Find the volume of the (solid) right circular torus obtained by revolving the circle $(y - A)^2 + z^2 = a^2$, $x = 0$ about the z-axis. Assume $0 < a \leq A$.

13. Suppose in Ex. 12 that $0 < A \leq a$. Let the portion of the circle to the right of the z-axis generate volume V_1 and the portion to the left generate volume V_2. Find $V_1 - V_2$.

The **average** of a function $f(x, y, z)$ over a region R of space is

$$\bar{f} = \iiint_R f(x, y, z)\, dx\, dy\, dz \Big/ \text{volume } (R).$$

14. Find the average of r^n, $n > -2$, over the cylinder $r \leq a$, $0 \leq z \leq 1$.

15. Find the average of $\ln r$ over the cylinder $r \leq a$, $0 \leq z \leq 1$.

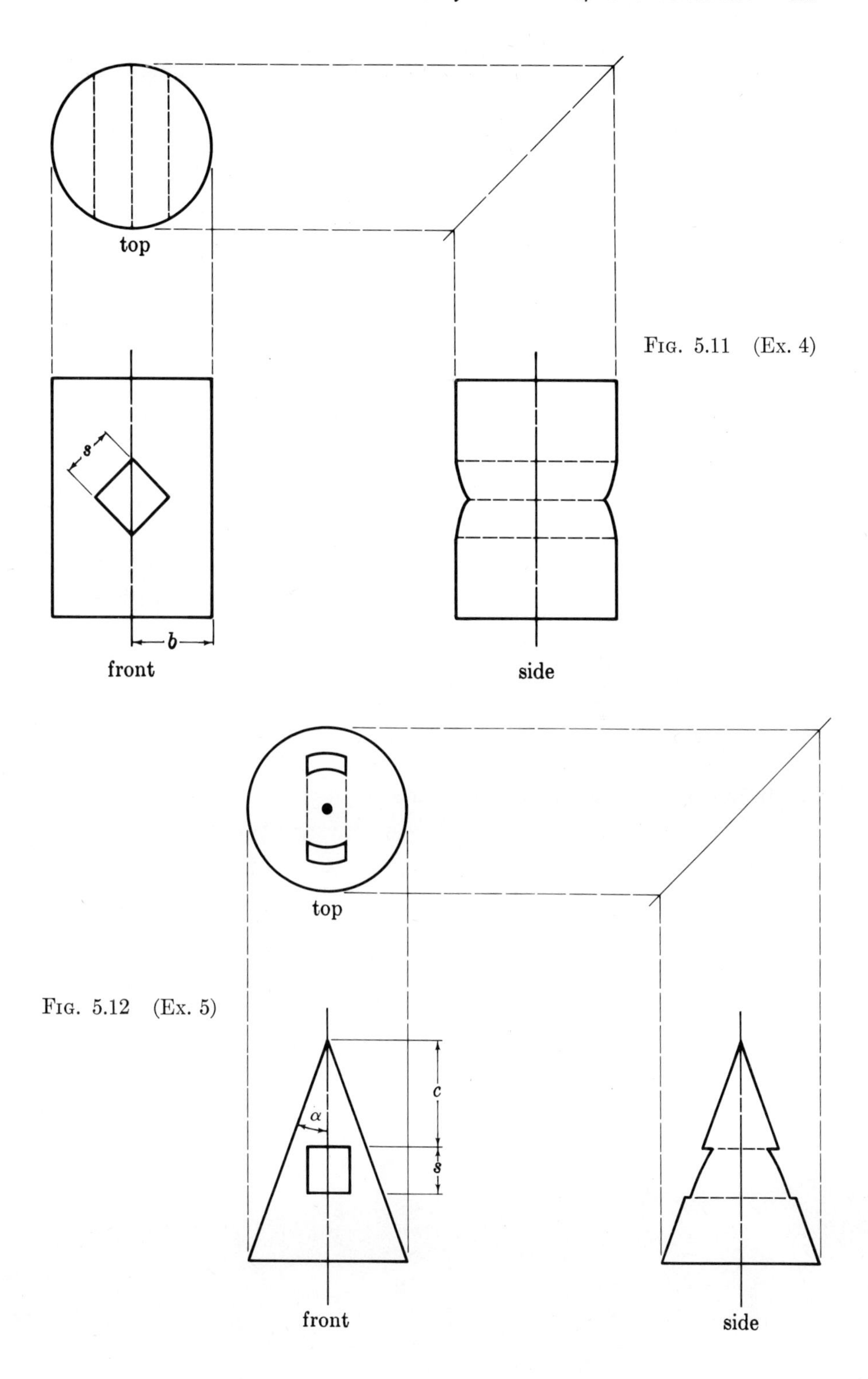

FIG. 5.11 (Ex. 4)

FIG. 5.12 (Ex. 5)

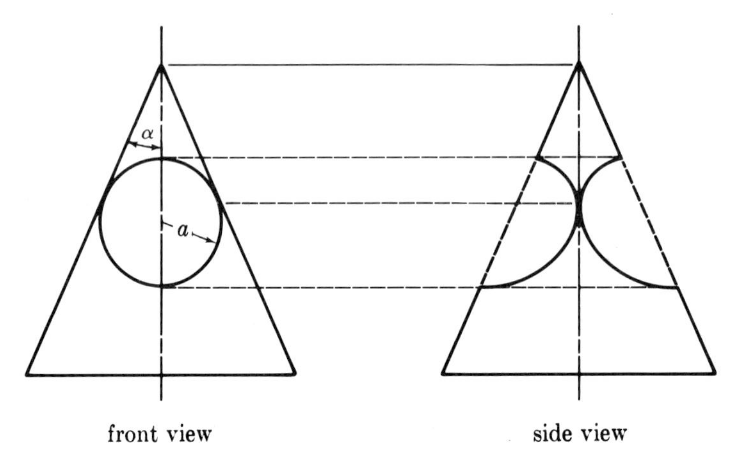

FIG. 5.13 (Ex. 6, 7, 8)

16. Find the average of ρ^n, $n > -3$, over the sphere $\rho \leq a$ of radius a.
17. Find the average of $(1 - \rho)^n$, $n > -1$, over the sphere $\rho \leq 1$.
18. Find the average of $\ln \rho$ over the sphere $\rho \leq a$.
19. Two parallel planes at distance h intersect a sphere of radius a. Find the surface area of the spherical zone between them.
20. (cont.) Two half-planes are hinged at angle α on a line through the origin. Find the solid angle determined by the wedge-shaped region between the planes.
21. (cont.) Three great circles on the unit sphere decompose the surface into 8 spherical triangles. Using the result of Ex. 20, show that the area of each is the sum of its interior angles minus π.
22. In Example 5.6, how much of the surface of the sphere is removed?

35. Applications of Multiple Integrals

1. CENTER OF GRAVITY

Centers of gravity of rectangular sheets were discussed in Chapter 26, Section 4. Review that material before proceeding to the next topic, centers of gravity of solids.

Suppose a solid R has density $\delta(\mathbf{x})$ at each point $\mathbf{x}$. Then its mass is

$$M = \iiint_R \delta(\mathbf{x})\, dx\, dy\, dz.$$

Define its **moment about the origin**

$$\mathbf{m} = \iiint_R \delta(\mathbf{x})\, \mathbf{x}\, dx\, dy\, dz,$$

and its **center of gravity**

$$\bar{\mathbf{x}} = \frac{1}{M}\mathbf{m}.$$

Note that $\mathbf{m}$ and $\bar{\mathbf{x}}$ are vectors. It is often convenient to express them in terms of components:

$$\mathbf{m} = (m_x, m_y, m_z), \qquad \bar{\mathbf{x}} = (\bar{x}, \bar{y}, \bar{z}).$$

The center of gravity may be considered as a sort of weighted average of the points of the solid. Recall in this connection that the center of gravity of a system of point-masses $M_1, \cdots, M_n$ located at $\mathbf{x}_1, \cdots, \mathbf{x}_n$ is

$$\bar{\mathbf{x}} = \frac{1}{M}(M_1\mathbf{x}_1 + M_2\mathbf{x}_2 + \cdots + M_n\mathbf{x}_n),$$

where $M = M_1 + \cdots + M_n$.

If a solid is symmetric in a coordinate plane, then the corresponding moment is zero, so the center of gravity lies on that coordinate plane. For example, suppose R is **symmetric** in the x, y-plane. This means that whenever a point (x, y, z) is in the solid, then $(x, y, -z)$ is in the solid, *and* $\delta(x, y, z) = \delta(x, y, -z)$. The contribution to m_z at (x, y, z) is

$$\delta(x, y, z)\, z\, dx\, dy\, dz;$$

it is cancelled by the contribution

$$\delta(x, y, -z)\,(-z)\, dx\, dy\, dz = -\delta(x, y, z)\, z\, dx\, dy\, dz$$

at $(x, y, -z)$. Hence $m_z = 0$, and $\bar{z} = 0$.

Similarly, if R is symmetric in a coordinate axis, then the center of gravity lies on that axis. Finally, if R is symmetric in the origin, then $\bar{\mathbf{x}} = \mathbf{0}$.

To compute the center of gravity of a solid, exploit any symmetry it has by choosing an appropriate coordinate system. Of course, express the element of volume $dV = dx\, dy\, dz$ in the coordinate system chosen.

EXAMPLE 1.1

Find the center of gravity of a uniform hemisphere of radius a and mass M.

Solution: To exploit symmetry, choose spherical coordinates, with the hemisphere defined by $\rho \leq a$ and $0 \leq \phi \leq \pi/2$. See Fig. 1.1a. The

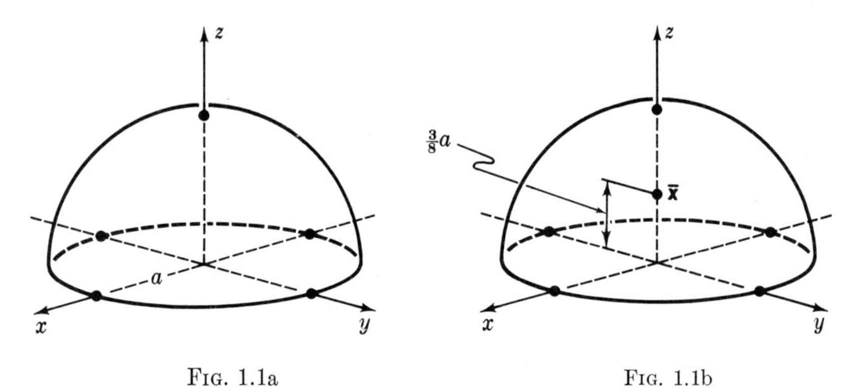

FIG. 1.1a FIG. 1.1b

density δ is constant, so the mass is δ times the volume:

$$M = \frac{1}{2}\left(\frac{4}{3}\pi a^3\right)\delta = \frac{2}{3}\pi a^3 \delta.$$

Because the hemisphere is symmetric in the y, z-plane, $m_x = 0$. Likewise $m_y = 0$; only m_z requires computation:

$$m_z = \iiint_R \delta z\, dx\, dy\, dz = \iiint_R \delta\,(\rho \cos\phi)\rho^2 \sin\phi\, d\rho\, d\phi\, d\theta$$

$$= \delta \int_0^{2\pi} d\theta \int_0^{\pi/2} \cos\phi \sin\phi\, d\phi \int_0^a \rho^3\, d\rho = \delta\,(2\pi)\left(\frac{1}{2}\right)\left(\frac{a^4}{4}\right) = \frac{\pi}{4}\,\delta\, a^4.$$

Hence

$$\bar{z} = \frac{m_z}{M} = \frac{\pi\delta a^4/4}{2\pi\delta a^3/3} = \frac{3}{8}a.$$

Answer: The center of gravity lies on the axis of the hemisphere, $\frac{3}{8}$ of the distance from the equatorial plane to the pole (Fig. 1.1b).

REMARK: The answer is independent of δ. For uniform solids in general, the constant δ cancels when you divide $\mathbf{m}$ by M, so $\bar{\mathbf{x}}$ is a purely geometric quantity. It is then called the **center of gravity** or **centroid** of the geometric region (rather than the material solid). For uniform solids, from now on take $\delta = 1$ and $M = V$, the volume.

EXAMPLE 1.2

Find the center of gravity of a uniform right circular cone.

Solution: Choose cylindrical coordinates with the apex of the cone at $\mathbf{0}$ and the base of radius a centered at $(0, 0, h)$. See Fig. 1.2a. The lateral

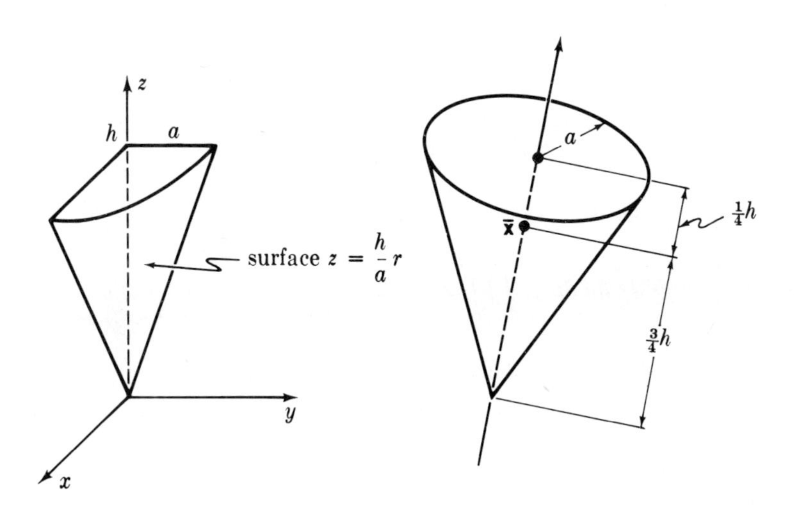

FIG. 1.2a

FIG. 1.2b

surface of the cone is

$$z = \frac{h}{a}r, \qquad \text{that is,} \qquad r = \frac{a}{h}z,$$

and the volume of the cone is

$$V = \frac{1}{3}\pi a^2 h.$$

By symmetry $m_x = m_y = 0$. Compute m_z:

$$m_z = \iiint z\,r\,dr\,d\theta\,dz = \int_0^{2\pi}\left[\int_0^h\left(\int_0^{az/h} r\,dr\right) z\,dz\right] d\theta$$

$$= \int_0^{2\pi} d\theta \int_0^h \frac{1}{2}\left(\frac{a}{h}z\right)^2 z\,dz = (2\pi)\left(\frac{a^2}{2h^2}\right)\int_0^h z^3\,dz = \frac{\pi a^2}{h^2}\cdot\frac{h^4}{4} = \frac{\pi}{4}a^2h^2.$$

Hence

$$\bar{z} = \frac{m_z}{V} = \frac{\pi a^2h^2/4}{\pi a^2h/3} = \frac{3}{4}h.$$

Answer: The center of gravity is on the cone's axis, $\frac{1}{4}$ of the distance from the base to the apex (Fig. 1.2b).

EXAMPLE 1.3

The solid $0 \le x \le 1$, $0 \le y \le 2$, $0 \le z \le 3$ has density xyz gm/cm³. Find its center of gravity.

Solution:

$$M = \iiint xyz\,dx\,dy\,dz = \int_0^1 x\,dx \int_0^2 y\,dy \int_0^3 z\,dz = \frac{1}{2}\cdot\frac{4}{2}\cdot\frac{9}{2} = \frac{9}{2}\text{ gm}.$$

$$\mathbf{m} = \iiint (x, y, z)\,xyz\,dx\,dy\,dz$$

$$= \left(\iiint x^2yz\,dx\,dy\,dz,\ \iiint xy^2z\,dx\,dy\,dz,\ \iiint xyz^2\,dx\,dy\,dz\right)$$

$$= \left(\int_0^1 x^2\,dx\int_0^2 y\,dy\int_0^3 z\,dz,\ \int_0^1 x\,dx\int_0^2 y^2\,dy\int_0^3 z\,dz,\ \int_0^1 x\,dx\int_0^2 y\,dy\int_0^3 z^2\,dz\right)$$

$$= \left(\frac{1}{3}\cdot\frac{4}{2}\cdot\frac{9}{2},\ \frac{1}{2}\cdot\frac{8}{3}\cdot\frac{9}{2},\ \frac{1}{2}\cdot\frac{4}{2}\cdot\frac{27}{3}\right)$$

$$= \frac{4\cdot 9}{3\cdot 2\cdot 2}(1, 2, 3) = 3(1, 2, 3)\text{ in gm-cm}.$$

Hence

$$\bar{\mathbf{x}} = \frac{2}{9}(3)(1, 2, 3) = \left(\frac{2}{3}, \frac{4}{3}, 2\right)\text{ in cm}.$$

Answer: $\bar{\mathbf{x}} = \left(\frac{2}{3}, \frac{4}{3}, 2\right)$ in cm.

Plane Regions; Wires

The definitions of moment and center of gravity for solids can be modified in an obvious way to fit plane regions with a mass distribution.

EXAMPLE 1.4

Find the center of gravity of a uniform semicircular disk.

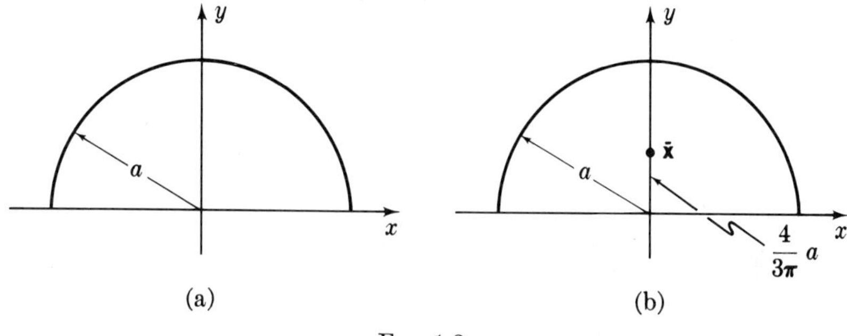

FIG. 1.3

Solution: Choose polar coordinates, taking the disk in the position $0 \le r \le a$, $0 \le \theta \le \pi$. See Fig. 1.3a. Take $\delta = 1$; the mass is the area

$$A = \frac{1}{2}(\pi a^2).$$

By symmetry $m_x = 0$. Compute m_y:

$$m_y = \iint yr\,dr\,d\theta = \iint r^2 \sin\theta\,dr\,d\theta = \int_0^\pi \sin\theta\,d\theta \int_0^a r^2\,dr = \frac{2}{3}a^3.$$

Hence

$$\bar{\mathbf{x}} = \frac{1}{A}(m_x, m_y) = \frac{2}{\pi a^2}\left(0, \frac{2}{3}a^3\right) = \left(0, \frac{4}{3\pi}a\right).$$

Answer: $\bar{\mathbf{x}} = \left(0, \frac{4}{3\pi}a\right)$. See Fig. 1.3b.

There is a useful connection between the centers of gravity of plane regions and volumes of revolution.

First Pappus Theorem Suppose a region R in the x, y-plane, to the right of the y-axis, is revolved about the y-axis. Then the volume of the resulting solid is

$$V = 2\pi\bar{x}A,$$

where A is the area of the plane region R and $\bar{x}$ is the x-coordinate of its center of gravity (Fig. 1.4).

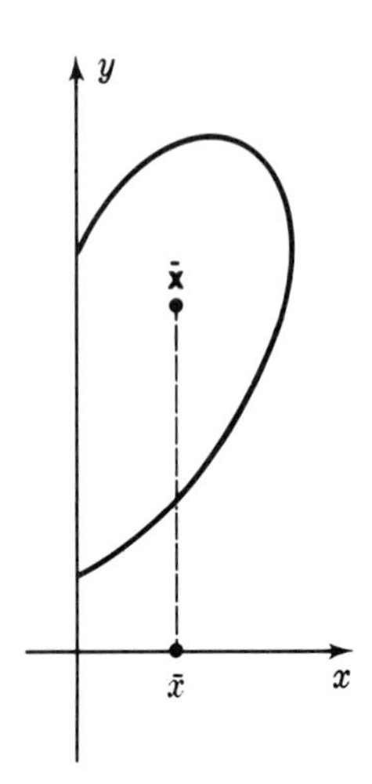

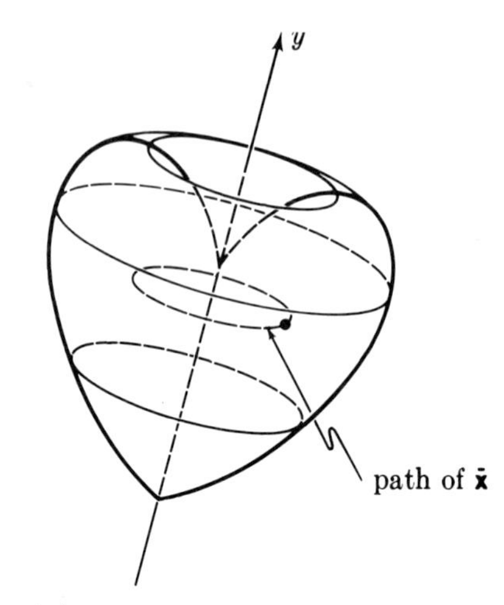

FIG. 1.4

In words, the volume is the area times the length of the circle traced by the center of gravity. Proof: A small portion $dx\,dy$ at $\mathbf{x}$ revolves into a thin ring of volume

$$dV = 2\pi x\,dx\,dy.$$

Hence

$$V = \iint_R 2\pi x\,dx\,dy = 2\pi m_x.$$

But $m_x = \bar{x}A$, hence $V = 2\pi\bar{x}A$.

A non-homogeneous wire is described by its position, a space curve $\mathbf{x} = \mathbf{x}(s)$, where $a \leq s \leq b$, and its density $\delta = \delta(s)$. (Here s denotes arc length.) Its mass is

$$M = \int_a^b \delta(s)\,ds,$$

its moment is

$$\mathbf{m} = \int_a^b \mathbf{x}(s)\delta(s)\,ds,$$

and its center of gravity is

$$\bar{\mathbf{x}} = \frac{1}{M}\mathbf{m}.$$

If the wire is uniform, then $\delta(s)$ is a constant. In this case, the center of gravity is independent of δ, hence it is a property of the curve $\mathbf{x} = \mathbf{x}(s)$ alone; you can take $\delta = 1$ and replace M by L, the length.

EXAMPLE 1.5

Find the center of gravity of the uniform semicircle $r = a$, $y \geq 0$.

Solution: The length is $L = \pi a$. The moment is

$$\mathbf{m} = \int \mathbf{x}\, ds = \int_0^{\pi} (a \cos \theta, a \sin \theta)\, a\, d\theta$$

$$= a^2 \int_0^{\pi} (\cos \theta, \sin \theta)\, d\theta = a^2(0, 2).$$

Hence

$$\bar{\mathbf{x}} = \frac{1}{L}\mathbf{m} = \frac{1}{\pi a} a^2(0, 2) = \frac{a}{\pi}(0, 2).$$

Answer: $(0, \frac{2}{\pi} a)$.

Suppose a plane curve is revolved about an axis in its plane, generating a surface of revolution. There is a useful relation between the center of gravity of the curve and the area of the surface.

Second Pappus Theorem Suppose a curve in the x, y-plane to the right of the y-axis is revolved about the y-axis. Then the area of the resulting surface is

$$A = 2\pi\bar{x}L,$$

where L is the length of the curve and $\bar{x}$ is the x-coordinate of the center of gravity (Fig. 1.5).

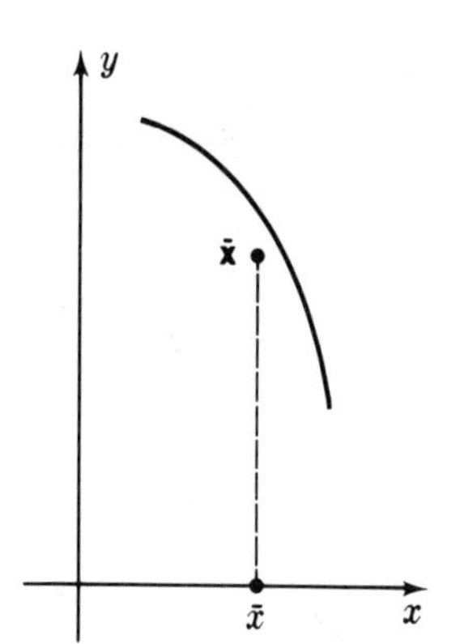

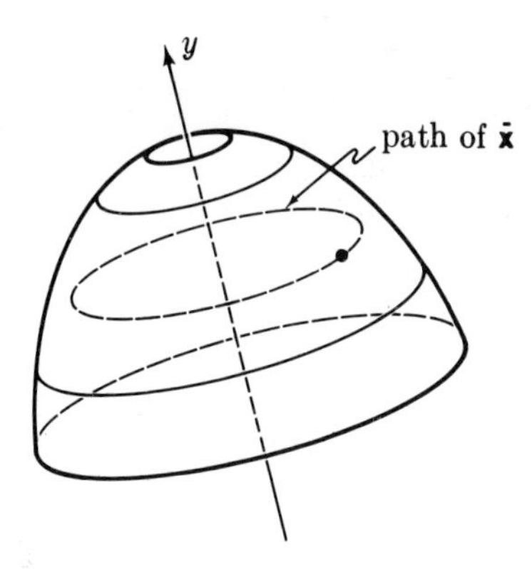

FIG. 1.5

In words, the area is the length of the curve times the length of the circle traced by the center of gravity. Proof: A short segment of length ds of the curve at the point $\mathbf{x}(s)$ revolves into the frustum of a cone with lateral area $dA = 2\pi x\, ds$. See Fig. 1.6.

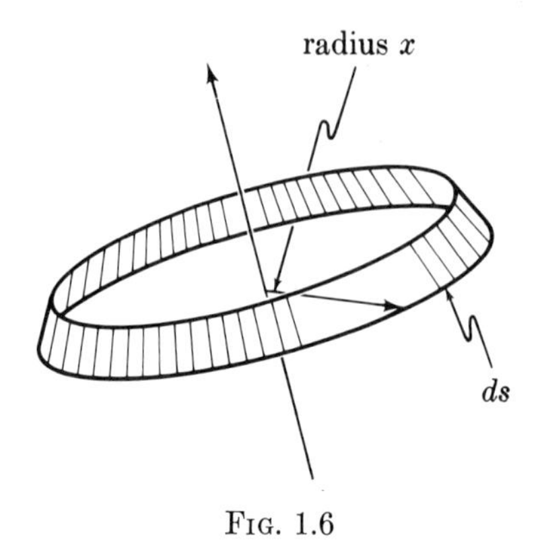

FIG. 1.6

Hence

$$A = \int_a^b 2\pi x \, ds = 2\pi \int_a^b x \, ds = 2\pi m_x = 2\pi \bar{x} L.$$

There is a useful aid to problem solving which we state for solids. With obvious modifications, it applies to wires or laminas.

Addition Law Suppose a solid R of mass M and center of gravity $\bar{\mathbf{x}}$ is cut into two pieces R_0 and R_1, of masses M_0 and M_1, and centers of gravity $\bar{\mathbf{x}}_0$ and $\bar{\mathbf{x}}_1$. Then

$$M = M_0 + M_1,$$

$$\bar{\mathbf{x}} = \frac{1}{M}(M_0 \bar{\mathbf{x}}_0 + M_1 \bar{\mathbf{x}}_1).$$

The first formula is obvious. The second is simply a decomposition of the moment integral:

$$M\bar{\mathbf{x}} = \iiint_R \delta(\mathbf{x}) \, \mathbf{x} \, dV = \iiint_{R_0} + \iiint_{R_1} = M_0 \bar{\mathbf{x}}_0 + M_1 \bar{\mathbf{x}}_1.$$

EXERCISES

1. Find the center of gravity of the first octant portion of the uniform sphere $\rho \leq a$.
2. Suppose the density of the hemisphere $\rho \leq a$, $z \geq 0$ is $\delta = a - \rho$. Find the center of gravity.
3. Find the center of gravity of the uniform spherical cone $\rho \leq a$, $0 \leq \phi \leq \alpha$.
4. Find the center of gravity of the uniform hemispherical shell $a \leq \rho \leq b$, $z \geq 0$.
5. Find the center of gravity of a uniform circular wedge (sector).

6. The plane region bounded by $z = 1$ and the parabola $z = y^2$ is revolved about the z-axis. Find the center of gravity of the resulting uniform solid.
7. Find the center of gravity of a uniform wire in the shape of a quarter circle.
8. Find the center of gravity of the uniform spiral $\mathbf{x}(t) = (a \cos t, a \sin t, bt)$, $0 \le t \le t_0$.
9. A copper wire in the shape of a semicircle of radius 100 cm is steadily tapered from 0.1 to 0.5 cm in diameter. Find its center of gravity.
10. Verify the First Pappus Theorem for a semicircle revolved about its diameter.
11. Use the First Pappus Theorem to find the volume of a right circular torus.
12. Verify the First Pappus Theorem for a right triangle revolved about a leg.
13. Use the Second Pappus Theorem to find the surface area of a right circular torus.
14. Use the Second Pappus Theorem to obtain another solution of Example 1.5.
15. Find the center of gravity of the uniform spherical cap (surface) $\rho = a, \quad 0 \le \phi \le \alpha$.
16. Find the center of gravity of the uniform spherical cap (solid) $\rho \le a, \quad a - h \le z \le a$.
17. Suppose a solid of density $\delta(\mathbf{x})$ is acted upon by a uniform (constant) gravitational field $\mathbf{f}$ so the force on a small portion at $\mathbf{x}$ is $[\delta(\mathbf{x})\, dV]\mathbf{f}$. Show that the solid is in equilibrium if a single force $-M\mathbf{f}$ is applied at $\bar{\mathbf{x}}$.
18. Find the center of gravity of the uniform triangle with vertices $\mathbf{a}$, $\mathbf{b}$, $\mathbf{c}$.
19. Find the center of gravity of the uniform tetrahedron with vertices $\mathbf{a}$, $\mathbf{b}$, $\mathbf{c}$, $\mathbf{d}$.

2. MOMENTS OF INERTIA

Recall that the kinetic energy of a moving particle of mass m and speed v is $K = \frac{1}{2}mv^2$. The kinetic energy of a system of moving particles is the sum of their individual kinetic energies. To define the kinetic energy of a moving solid body R, decompose the body into elementary masses $\delta(\mathbf{x})\, dV$, where $\delta(\mathbf{x})$ is the density at $\mathbf{x}$, and form the integral

$$K = \frac{1}{2} \iiint_R \delta(\mathbf{x})\, |\mathbf{v}(\mathbf{x}, t)|^2\, dV.$$

Here $\mathbf{v}(\mathbf{x}, t)$ is the velocity of the point $\mathbf{x}$ of the body at time t. Thus K varies with time.

We shall compute the kinetic energy of a rigid body rotating with angular velocity $\boldsymbol{\omega}$ about an axis through the origin.

First consider rotation about the z-axis with angular speed ω, so the angular velocity is $\boldsymbol{\omega} = \omega\mathbf{k}$. See Fig. 2.1. An elementary volume $dV = dx\, dy\, dz$ at $\mathbf{x}$ has mass $dM = \delta(\mathbf{x})\, dV$ and speed $\omega\sqrt{x^2 + y^2}$, since $\sqrt{x^2 + y^2}$ is the distance from $\mathbf{x}$ to the z-axis. Hence its kinetic energy is

$$dK = \frac{1}{2}(\omega\sqrt{x^2 + y^2})^2\, \delta(\mathbf{x})\, dV = \frac{1}{2}\omega^2\, \delta(\mathbf{x})\, (x^2 + y^2)\, dV.$$

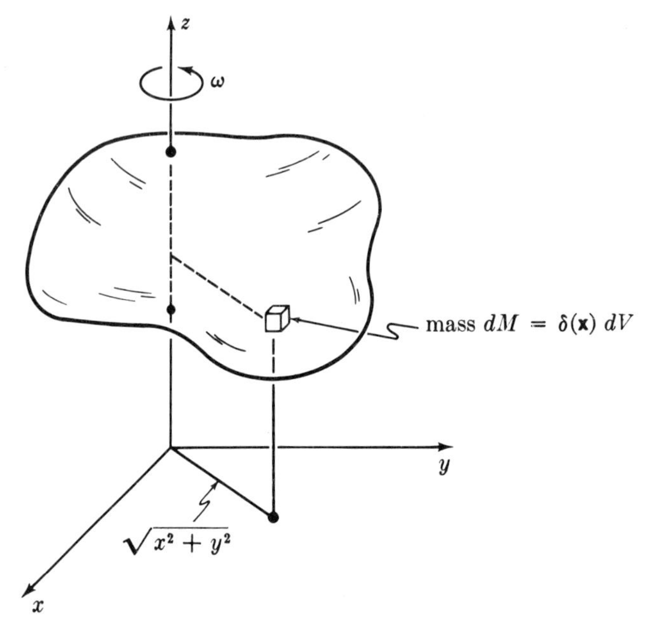

FIG. 2.1

To find the kinetic energy of the entire solid R integrate, obtaining

$$K = \frac{1}{2} I_{zz}\omega^2,$$

where

$$I_{zz} = \iiint\limits_R \delta(\mathbf{x})\,(x^2 + y^2)\,dV$$

is the **moment of inertia** of R about the z-axis.

Similarly define

$$I_{xx} = \iiint\limits_R \delta(\mathbf{x})\,(y^2 + z^2)\,dV \qquad \text{and} \qquad I_{yy} = \iiint\limits_R \delta(\mathbf{x})\,(z^2 + x^2)\,dV.$$

From these moments of inertia one can compute the kinetic energy, provided the body rotates about one of the coordinate axes:

$$K = \frac{1}{2} I_{xx}\omega^2 \qquad \text{(rotation about the } x\text{-axis)},$$

$$K = \frac{1}{2} I_{yy}\omega^2 \qquad \text{(rotation about the } y\text{-axis)}.$$

For rotation about more general axes, however, three other quantities called

products of inertia (or **mixed moments of inertia**) are needed. Define

$$I_{yz} = -\iiint\limits_R \delta(\mathbf{x})\, yz\, dV, \qquad I_{zx} = -\iiint\limits_R \delta(\mathbf{x})\, zx\, dV,$$

$$I_{xy} = -\iiint\limits_R \delta(\mathbf{x})\, xy\, dV.$$

> Suppose a solid R rotates about an axis through $\mathbf{0}$ with angular velocity
>
> $$\boldsymbol{\omega} = (\omega_x, \omega_y, \omega_z).$$
>
> Then its kinetic energy is
>
> $$K = \frac{1}{2}[I_{xx}\omega_x^2 + I_{yy}\omega_y^2 + I_{zz}\omega_z^2 + 2I_{yz}\omega_y\omega_z + 2I_{zx}\omega_z\omega_x + 2I_{xy}\omega_x\omega_y].$$

The formula is derived by direct calculation. Write

$$\boldsymbol{\omega} = \mathbf{u}\,\omega, \qquad |\mathbf{u}| = 1.$$

If $\mathbf{x}$ is any point in R, the projection of $\mathbf{x}$ on the axis of rotation is $\mathbf{x}\cdot\mathbf{u}$. Therefore by the Pythagorean Theorem, the distance of $\mathbf{x}$ from the axis is $[\,|\mathbf{x}|^2 - (\mathbf{x}\cdot\mathbf{u})^2]^{1/2}$. See Fig. 2.2. Hence the speed at $\mathbf{x}$ is $\omega[\,|\mathbf{x}|^2 - (\mathbf{x}\cdot\mathbf{u})^2]^{1/2}$,

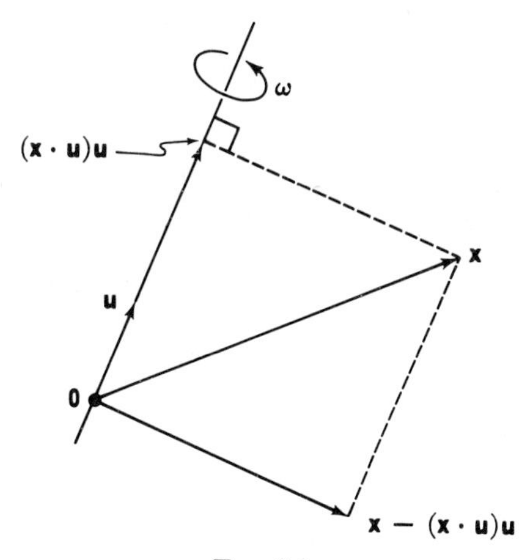

Fig. 2.2

and the kinetic energy of an elementary mass at $\mathbf{x}$ is

$$dK = \frac{1}{2}\omega^2\delta(\mathbf{x})[\,|\mathbf{x}|^2 - (\mathbf{x}\cdot\mathbf{u})^2]\,dV = \frac{1}{2}\delta(\mathbf{x})[\,\omega^2|\mathbf{x}|^2 - (\mathbf{x}\cdot\boldsymbol{\omega})^2]\,dV.$$

Into this formula substitute

$$\omega^2 = \omega_x^2 + \omega_y^2 + \omega_z^2 \qquad \text{and} \qquad \mathbf{x}\cdot\boldsymbol{\omega} = x\,\omega_x + y\,\omega_y + z\,\omega_z.$$

The result:

$$dK = \frac{1}{2}\,\delta(\mathbf{x})[\,(y^2 + z^2)\omega_x{}^2 + (z^2 + x^2)\omega_y{}^2 + (x^2 + y^2)\omega_z{}^2$$

$$- 2yz\,\omega_y\,\omega_z - 2zx\,\omega_z\,\omega_x - 2xy\,\omega_x\,\omega_y]\,dV.$$

Now integrate; the asserted formula for K follows.

Because of the formula, knowledge of the 6 moments and products

$$I_{xx},\quad I_{yy},\quad I_{zz},\quad I_{xy},\quad I_{yz},\quad I_{zx}$$

allows us to compute the kinetic energy for rotation with arbitrary angular velocity $\boldsymbol{\omega}$, regardless of the direction of its axis. The formula can be summarized in a neat matrix form:

$$K = \frac{1}{2}\,(\omega_x, \omega_y, \omega_z)\begin{bmatrix} I_{xx} & I_{xy} & I_{xz} \\ I_{yx} & I_{yy} & I_{yz} \\ I_{zx} & I_{zy} & I_{zz} \end{bmatrix}\begin{bmatrix} \omega_x \\ \omega_y \\ \omega_z \end{bmatrix}.$$

The square matrix is made symmetric by defining

$$I_{yx} = I_{xy}, \qquad I_{zy} = I_{yz}, \qquad I_{zx} = I_{xz}.$$

Geometric symmetries simplify calculation of the moments and products of inertia. For example, suppose that a solid is symmetric in the x, y-plane. Recall this means that whenever a point (x, y, z) is in the solid, then so is $(x, y, -z)$, and $\delta(x, y, -z) = \delta(x, y, z)$. Then $I_{zx} = 0$, because the contribution

$$\delta(x, y, z)\,zx\,dV$$

at each point (x, y, z) above the x, y-plane is cancelled by the contribution

$$\delta(x, y, -z)\,(-z)\,(x)\,dV = -\delta(x, y, z)\,zx\,dV$$

at the symmetric point $(x, y, -z)$ below the x, y-plane. Likewise $I_{yz} = 0$ under the same symmetry condition. For example, a uniform right circular cone with axis the z-axis has all products of inertia 0, since it is symmetric in two coordinate planes.

Applications

EXAMPLE 2.1

Compute the moments and products of inertia for a uniform sphere of radius a and mass M with center at the origin. Measure length in cm and mass in gm.

Solution: Let δ denote the constant density; then $M = 4\pi a^3\delta/3$ gm. Because the sphere is symmetric in each coordinate plane, all products of inertia are 0.

By symmetry, $I_{xx} = I_{yy} = I_{zz}$. It appears most natural to use spherical coordinates and compute I_{zz}:

$$I_{zz} = \delta \iiint (x^2 + y^2)\, dV = \delta \iiint (\rho^2 \sin^2 \phi \cos^2 \theta + \rho^2 \sin^2 \phi \sin^2 \theta)\, dV$$

$$= \delta \iiint (\rho^2 \sin^2 \phi)\rho^2 \sin \phi \, d\rho\, d\phi\, d\theta$$

$$= \delta \int_0^a \rho^4\, d\rho \int_0^\pi \sin^3 \phi\, d\phi \int_0^{2\pi} d\theta = \delta \left(\frac{a^5}{5}\right)\left(\frac{4}{3}\right)(2\pi) = \frac{8\pi a^5 \delta}{15} = \frac{2}{5} Ma^2.$$

Answer: $I_{xy} = I_{yz} = I_{zx} = 0,$

$$I_{xx} = I_{yy} = I_{zz} = \frac{2}{5} Ma^2 \text{ gm-cm}^2.$$

REMARK: Here is an alternate calculation of I_{zz} which fully exploits the symmetry of the sphere:

$$I_{zz} = I_{xx} = I_{yy} = \frac{1}{3}(I_{xx} + I_{yy} + I_{zz})$$

$$= \frac{1}{3}\delta \iiint [(y^2 + z^2) + (z^2 + x^2) + (x^2 + y^2)]\, dV$$

$$= \frac{2\delta}{3} \iiint (x^2 + y^2 + z^2)\, dx\, dy\, dz = \frac{2\delta}{3} \iiint \rho^2\, dV$$

$$= \frac{2\delta}{3} \iiint \rho^2 \rho^2 \sin \phi\, d\rho\, d\phi\, d\theta$$

$$= \frac{2\delta}{3} \int_0^a \rho^4\, d\rho \int_0^\pi \sin \phi\, d\phi \int_0^{2\pi} d\theta$$

$$= \frac{2\delta}{3}\left(\frac{a^5}{5}\right)(2)(2\pi) = \frac{8\pi}{15}\delta a^5.$$

EXAMPLE 2.2

A uniform cylinder of mass M occupies the region $-h \le z \le h$, $x^2 + y^2 \le a^2$. Find its moments and products of inertia. (The units are gm and cm.)

Solution: Let δ denote the constant density; then $M = 2\pi a^2 h\delta$. Because the cylinder is symmetric in each coordinate plane, its three products of inertia are zero.

It is clear by symmetry that $I_{xx} = I_{yy}$. However I_{zz} is probably different.

Use cylindrical coordinates to compute first I_{xx}, then I_{zz}:

$$I_{xx} = \delta \iiint (y^2 + z^2)\, dx\, dy\, dz = \delta \iiint (r^2 \sin^2 \theta + z^2)\, r\, dr\, d\theta\, dz$$

$$= \delta \int_0^a r^3\, dr \int_0^{2\pi} \sin^2 \theta\, d\theta \int_{-h}^{h} dz + \delta \int_0^a r\, dr \int_0^{2\pi} d\theta \int_{-h}^{h} z^2\, dz$$

$$= \delta \left(\frac{a^4}{4}\right) (\pi)\, (2h) + \delta \left(\frac{a^2}{2}\right) (2\pi)\, \frac{2h^3}{3}$$

$$= (2\pi a^2 h \delta) \left(\frac{a^2}{4} + \frac{h^2}{3}\right) = \frac{1}{12} M(3a^2 + 4h^2).$$

Next,

$$I_{zz} = \delta \iiint (x^2 + y^2)\, dx\, dy\, dz = \delta \iiint r^2 r\, dr\, d\theta\, dz$$

$$= \delta \int_0^a r^3\, dr \int_0^{2\pi} d\theta \int_{-h}^{h} dz = \delta \left(\frac{a^4}{4}\right) (2\pi)\, (2h) = \frac{1}{2} Ma^2.$$

Answer: $I_{xy} = I_{yz} = I_{zx} = 0,$

$$I_{xx} = I_{yy} = \frac{1}{12} M(3a^2 + 4h^2) \text{ gm-cm}^2,$$

$$I_{zz} = \frac{1}{2} Ma^2 \text{ gm-cm}^2.$$

EXAMPLE 2.3

The cylinder of Example 2.2 rotates with angular speed ω rad/sec about an axis passing through the origin and (1, 1, 1). Find its kinetic energy.

Solution: The angular velocity is

$$\boldsymbol{\omega} = (\omega_x, \omega_y, \omega_z) = \frac{\omega}{\sqrt{3}} (1, 1, 1).$$

By the formula on p. 873 and the above answer,

$$K = \frac{1}{2} (I_{xx}\omega_x{}^2 + I_{yy}\omega_y{}^2 + I_{zz}\omega_z{}^2) = \frac{1}{2} \left(\frac{\omega}{\sqrt{3}}\right)^2 [I_{xx} + I_{yy} + I_{zz}]$$

$$= \left(\frac{\omega^2}{6}\right) \left[2 \frac{M}{12} (3a^2 + 4h^2) + \frac{M}{2} a^2\right] = \frac{M\omega^2}{18} (3a^2 + 2h^2).$$

Answer: $K = \dfrac{1}{18} M\omega^2(3a^2 + 2h^2)$ erg.

Note on units: The unit of work, or energy, in the CGS system is 1 erg = 1 dyne-cm. Remember 1 dyne = 1 gm-cm/sec^2.

The following example shows that the products of inertia are not always zero.

EXAMPLE 2.4

Find the products of inertia of the first octant portion of the sphere of Example 2.1.

Solution: By symmetry $I_{xy} = I_{yz} = I_{zx}$. Choose spherical coordinates and compute that product of inertia whose formula seems the most symmetric; this is I_{xy}, since the z-axis is special in spherical coordinates:

$$\begin{aligned} I_{xy} &= -\delta \iiint xy\,dx\,dy\,dz \\ &= -\delta \iiint (\rho^2 \sin^2\phi \cos\theta \sin\theta)\rho^2 \sin\phi\,d\rho\,d\phi\,d\theta \\ &= -\delta \int_0^a \rho^4\,d\rho \int_0^{\pi/2} \sin^3\phi\,d\phi \int_0^{\pi/2} \cos\theta \sin\theta\,d\theta \\ &= -\delta\left(\frac{a^5}{5}\right)\left(\frac{2}{3}\right)\left(\frac{1}{2}\right) = -\frac{a^5}{15}\delta. \end{aligned}$$

But

$$M = \frac{1}{8}\left(\frac{4}{3}\pi a^3 \delta\right) = \frac{1}{6}\pi a^3 \delta.$$

Hence $I_{xy} = -2Ma^2/5\pi$.

Answer:

$$I_{xy} = I_{yz} = I_{zx} = -\frac{2}{5\pi} Ma^2 \text{ gm-cm}^2.$$

EXAMPLE 2.5

The solid of Example 2.4 rotates with angular velocity $\boldsymbol{\omega} = (\omega/\sqrt{3})\,(1, 1, 1)$. Find its kinetic energy.

Solution: By symmetry, the moments of inertia are $\frac{1}{8}$ those of the full sphere, and the mass M is $\frac{1}{8}$ that of the full sphere. By Example 2.1

$$I_{xx} = I_{yy} = I_{zz} = \frac{2}{5} Ma^2,$$

where M denotes the mass of the *first octant portion* of the sphere. Since

$\omega_x = \omega_y = \omega_z = \omega/\sqrt{3}$, the formula for kinetic energy yields

$$K = \frac{1}{2}\left(\frac{\omega^2}{3}\right)(3I_{xx} + 6I_{xy})$$

$$= \frac{1}{2}\left(\frac{\omega^2}{3}\right)\left(\frac{6}{5}Ma^2 - \frac{12}{5\pi}Ma^2\right) = \frac{1}{5}\left(1 - \frac{2}{\pi}\right)Ma^2\omega^2.$$

Answer: $K = \frac{1}{5}\left(1 - \frac{2}{\pi}\right)Ma^2\omega^2$ erg.

Moments and products of inertia can be defined for wires and laminas analogously. They can be computed directly or, in some cases, indirectly by limiting arguments. Examples 2.6 and 2.7 illustrate the latter.

EXAMPLE 2.6

A uniform circular sheet of mass M and radius a is centered at $\mathbf{0}$ in the x, y-plane. Find its moments of inertia. (The units are gm and cm.)

Solution: Let $h \longrightarrow 0$ in Example 2.2. The cylinder becomes a disk in the limit. Hence

$$I_{xx} = \frac{1}{12}M(3a^2 + 4h^2) \longrightarrow \frac{1}{4}Ma^2,$$

$$I_{zz} = \frac{1}{2}Ma^2.$$

Answer: $I_{xx} = I_{yy} = \frac{1}{4}Ma^2$ gm-cm^2,

$I_{zz} = \frac{1}{2}Ma^2$ gm-cm^2.

EXAMPLE 2.7

A uniform rod of mass M and length $2h$ lies along the z-axis from $z = -h$ to $z = h$. Find its moments of inertia. (The units are gm and cm.)

Solution: Again use the results of Example 2.2, only let $a \longrightarrow 0$.

Answer: $I_{xx} = I_{yy} = \frac{1}{3}Mh^2$ gm-cm^2,

$I_{zz} = 0.$

Parallel Axis Theorem

Take an axis β *anywhere* in space (not necessarily through the origin). Suppose a rigid body R rotates about this axis with angular speed ω. See Fig. 2.3. The speed at each point $\mathbf{x}$ is $\omega B_{\mathbf{x}}$, where $B_{\mathbf{x}}$ is the distance from $\mathbf{x}$

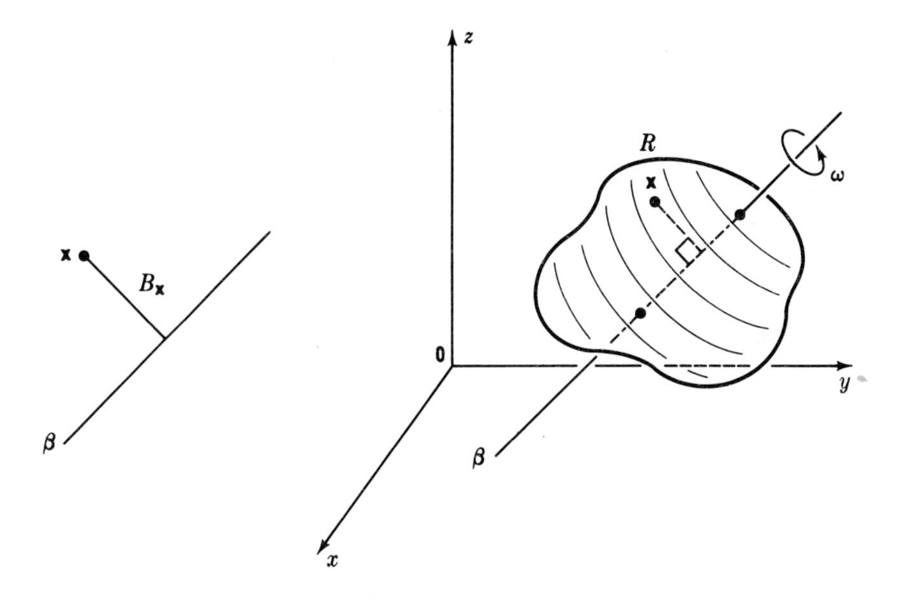

FIG. 2.3

to the axis β. Hence the kinetic energy is

$$K = \frac{1}{2} I_\beta \omega^2, \qquad \text{where} \quad I_\beta = \iiint_R B_{\mathbf{x}}^2 \, \delta(\mathbf{x}) \, dV.$$

This defines I_β, the **moment of inertia of R about the axis β**. The Parallel Axis Theorem allows us to compute I_β in terms of the moment of inertia about a parallel axis through the center of gravity.

Parallel Axis Theorem If α is an axis through the center of gravity of R, and β is an axis parallel to α, then

$$I_\beta = I_\alpha + Md^2,$$

where d is the distance between the axes and M is the mass of R.

The proof of this result in the general case is not hard, but involves some technicalities. So let us content ourselves with the special case in which $\bar{\mathbf{x}} = \mathbf{0}$ and α is the z-axis. Suppose β passes through $(a, b, 0)$. See Fig. 2.4.

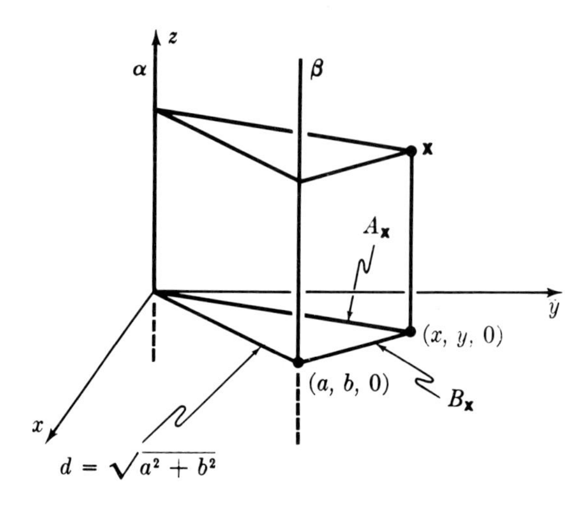

FIG. 2.4

For each point $\mathbf{x}$ of R, the distance $A_{\mathbf{x}}$ of $\mathbf{x}$ from α is given by

$$A_{\mathbf{x}}^2 = x^2 + y^2 = |(x, y)|^2,$$

and the distance $B_{\mathbf{x}}$ of $\mathbf{x}$ from β is given by

$$\begin{aligned} B_{\mathbf{x}}^2 = (x - a)^2 + (y - b)^2 &= |(x, y) - (a, b)|^2 \\ &= |(x, y)|^2 - 2(a, b) \cdot (x, y) + |(a, b)|^2 \\ &= A_{\mathbf{x}}^2 - 2(a, b) \cdot (x, y) + d^2. \end{aligned}$$

Multiply by the element of mass, $\delta(\mathbf{x})\, dV$, and integrate. Result:

$$I_\beta = I_\alpha - 2(a, b) \cdot (m_x, m_y) + Md^2.$$

But $m_x = m_y = 0$ since $\bar{\mathbf{x}} = \mathbf{0}$. Hence $I_\beta = I_\alpha + Md^2$.

EXAMPLE 2.8

Find the moment of inertia about a tangent axis of a uniform sphere of radius a and mass M.

Solution: From Example 2.1, the moment of inertia about any axis through the center (c.g.) is $2Ma^2/5$. The distance from a tangent axis β to the center is a, so the Parallel Axis Theorem implies

$$I_\beta = \frac{2}{5} Ma^2 + Ma^2 = \frac{7}{5} Ma^2.$$

Answer: $\frac{7}{5} Ma^2$.

EXERCISES

1. Find the moments of inertia of a uniform spherical shell (surface) of radius a and mass M about an axis through its center.
2. A uniform circular wire hoop $x^2 + y^2 = a^2, \quad z = 0$ has mass M. Find its moments of inertia.
3. Find the moments and products of inertia of the uniform hemisphere $\rho \leq a$, $z \geq 0$ of mass M.
4. The axis of a uniform right circular cone of mass M is the z-axis; its apex is at $\mathbf{0}$ and its base of radius a is at $z = h > 0$. Find its moments and products of inertia.
5. Find the moments and products of inertia of the uniform rectangular solid of mass M bounded by the planes $x = \pm a, \quad y = \pm b, \quad z = \pm c$.
6. Find the moments and products of inertia of the uniform rectangular solid of mass M bounded by the coordinate planes and the planes $x = a$, $y = b$, $z = c$.
7. A uniform rod of length L and mass M lies along the positive x-axis on the interval $a \leq x \leq a + L$. Find I_{yy}.
8. The circle $y = 0, \quad (x - A)^2 + z^2 = a^2, \quad 0 < a \leq A$ is revolved about the z-axis. Suppose the resulting anchor ring (solid torus) is a uniform solid of mass M. Find I_{zz}.
9. Find I_{zz} for the toroidal shell, the surface of the solid torus in Ex. 8.
10. Find the moments of inertia of the uniform cylindrical shell $x^2 + y^2 = a^2$, $-h \leq z \leq h$.
11. Find I_{zz} for the uniform solid paraboloid of revolution of mass M bounded by $az = x^2 + y^2$ and $z = h$.
12. Find the moments of inertia of the uniform solid ellipsoid of mass M,

$$\frac{x^2}{a^2} + \frac{y^2}{b^2} + \frac{z^2}{c^2} = 1.$$

(*Hint:* Stretch a suitable amount in each direction until the solid becomes a sphere; set $x = ua$, $y = vb$, and $z = wc$.)

13. Find I_{zz} for the uniform solid elliptic paraboloid of mass M bounded by

$$z = \frac{x^2}{a} + \frac{y^2}{b}, \qquad z = h.$$

(*Hint:* Use the result of Ex. 11 and the hint of Ex. 12.)

14. Find I_{zz} for the uniform solid of mass M bounded by the hyperboloid of revolution $\dfrac{z^2}{c^2} = \dfrac{1}{a^2}(x^2 + y^2) - 1$ and the planes $z = \pm h$.
15. Find the moments of inertia of the uniform solid of mass M in the region $-a \leq x, y, z \leq a, \quad x^2 + y^2 + z^2 \geq a^2$.
16. Find the moment of inertia of a uniform solid right circular cylinder about a generator of its lateral surface. (This is important in problems concerning rolling.)

17. Suppose a rigid body R is rotating about an axis through $\mathbf{0}$ with angular velocity $\boldsymbol{\omega}$. Show that the angular momentum, $\mathbf{J} = \iiint [\delta(\mathbf{x})\,\mathbf{x} \times \mathbf{v}]\,dV$ ($\mathbf{v}$ is velocity) is given by

$$\mathbf{J} = (\omega_x, \omega_y, \omega_z)\begin{bmatrix} I_{xx} & I_{xy} & I_{xz} \\ I_{yx} & I_{yy} & I_{yz} \\ I_{zx} & I_{zy} & I_{zz} \end{bmatrix}.$$

18. Show that $I_{xx} + I_{yy} + I_{zz}$ is **isotropic,** i.e., independent of the choice of coordinate axes.

19. (cont.) Show that the sum of three determinants

$$\begin{vmatrix} I_{xx} & I_{xy} \\ I_{yx} & I_{yy} \end{vmatrix} + \begin{vmatrix} I_{yy} & I_{yz} \\ I_{zy} & I_{zz} \end{vmatrix} + \begin{vmatrix} I_{zz} & I_{zx} \\ I_{xz} & I_{xx} \end{vmatrix}$$

is also isotropic.

20. (cont.) Show that the determinant

$$\begin{vmatrix} I_{xx} & I_{xy} & I_{xz} \\ I_{yx} & I_{yy} & I_{yz} \\ I_{zx} & I_{zy} & I_{zz} \end{vmatrix}$$

is isotropic.

3. SURFACE AREA

Just as a curve in space is described by a vector function $\mathbf{x} = \mathbf{x}(t)$ of one variable t, so a surface in space is described by a vector function $\mathbf{x} = \mathbf{x}(u, v)$ of two variables u and v. As (u, v) ranges over a certain region D of the u, v-plane, the point $\mathbf{x} = \mathbf{x}(u, v)$ runs over a portion of the surface in space (Fig. 3.1). For reasons that will be clear shortly, assume that at each point the vectors $\partial\mathbf{x}/\partial u$ and $\partial\mathbf{x}/\partial v$ are linearly independent, i.e., are not collinear.

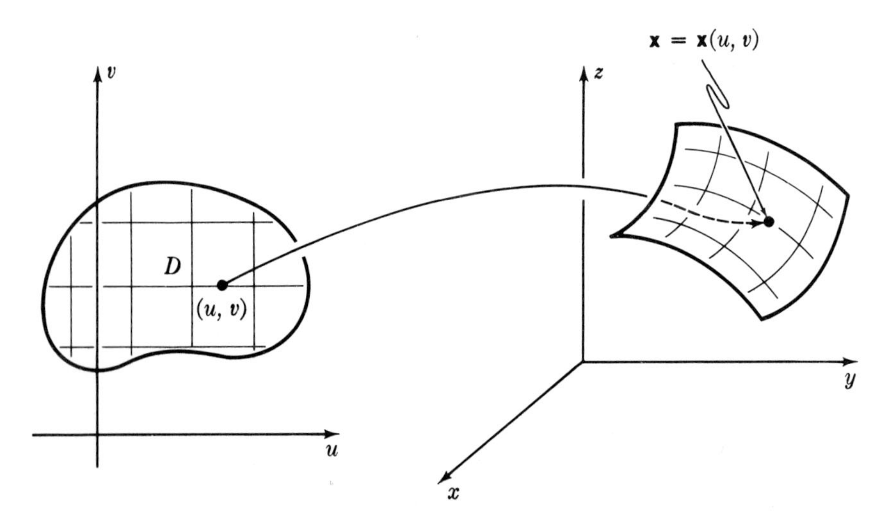

Fig. 3.1

Give small displacements du and dv to u and v. The corresponding change in $\mathbf{x}$ is

$$d\mathbf{x} = \mathbf{x}_u\,du + \mathbf{x}_v\,dv.$$

The vectors $\mathbf{x}_u$ and $\mathbf{x}_v$ are tangent vectors to the surface (Fig. 3.2). The small rectangle with sides du and dv corresponds to a small region of the surface

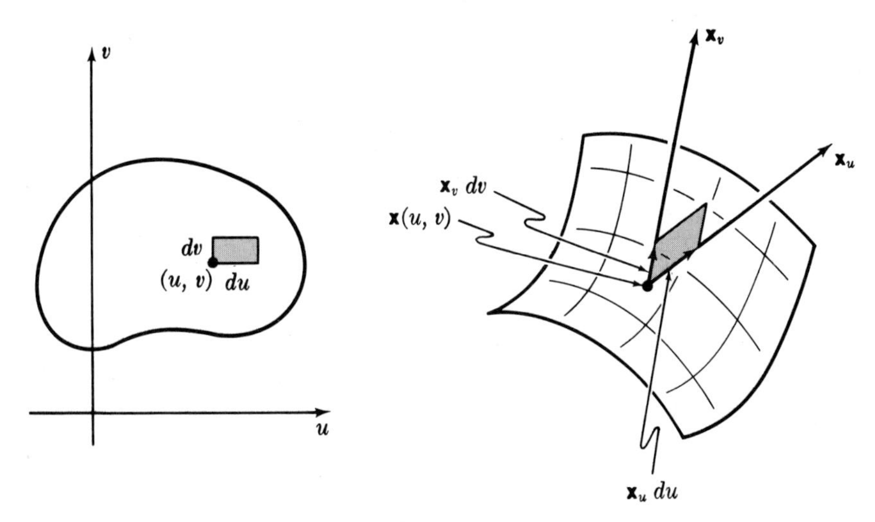

FIG. 3.2

very closely approximated by a parallelogram in the tangent plane at $\mathbf{x} = \mathbf{x}(u, v)$. The area of the parallelogram is

$$dA = |(\mathbf{x}_u\,du) \times (\mathbf{x}_v\,dv)| = |\mathbf{x}_u \times \mathbf{x}_v|\,du\,dv.$$

Here the cross product $\mathbf{x}_u \times \mathbf{x}_v$ is not zero precisely because the vectors $\mathbf{x}_u$ and $\mathbf{x}_v$ are not collinear. Now

$$\mathbf{x}_u \times \mathbf{x}_v = \begin{vmatrix} \mathbf{i} & \mathbf{j} & \mathbf{k} \\ x_u & y_u & z_u \\ x_v & y_v & z_v \end{vmatrix} = \begin{vmatrix} y_u & z_u \\ y_v & z_v \end{vmatrix}\mathbf{i} + \begin{vmatrix} z_u & x_u \\ z_v & x_v \end{vmatrix}\mathbf{j} + \begin{vmatrix} x_u & y_u \\ x_v & y_v \end{vmatrix}\mathbf{k}.$$

Consequently,

$$|\mathbf{x}_u \times \mathbf{x}_v|^2 = \begin{vmatrix} y_u & z_u \\ y_v & z_v \end{vmatrix}^2 + \begin{vmatrix} z_u & x_u \\ z_v & x_v \end{vmatrix}^2 + \begin{vmatrix} x_u & y_u \\ x_v & y_v \end{vmatrix}^2.$$

The final result of this calculation is a practical formula for surface area:

$$\boxed{dA = \sqrt{\begin{vmatrix} y_u & z_u \\ y_v & z_v \end{vmatrix}^2 + \begin{vmatrix} z_u & x_u \\ z_v & x_v \end{vmatrix}^2 + \begin{vmatrix} x_u & y_u \\ x_v & y_v \end{vmatrix}^2}\,du\,dv.}$$

Integrated over the domain D in the u, v-plane, this yields the area of the corresponding portion of surface.

It is worth emphasizing that the integration takes place on a domain of the u, v-plane, not in a region of the x, y, z-space.

EXAMPLE 3.1

Find the area of the spiral ramp $\mathbf{x} = (u\cos v, u\sin v, bv)$ corresponding to the domain $D: 0 \le u \le a,\ 0 \le v \le c$.

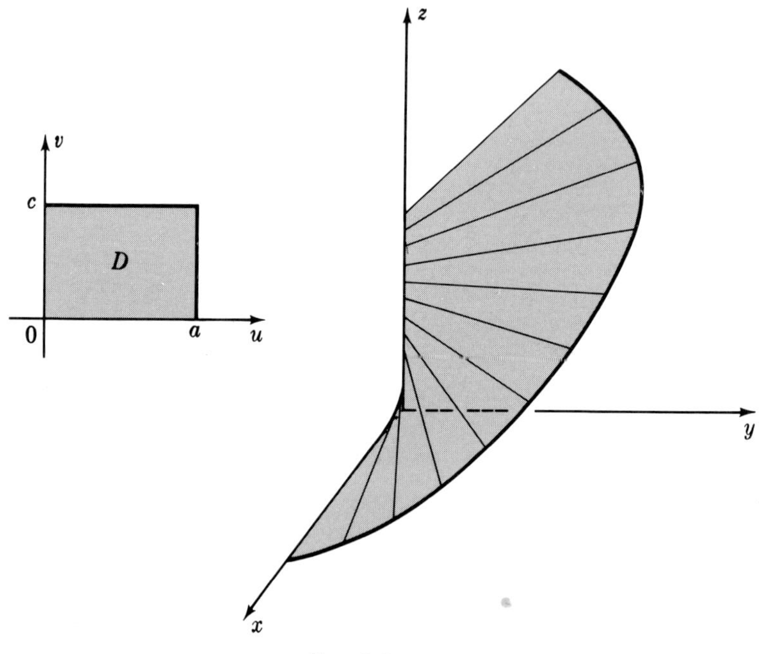

FIG. 3.3

Solution: Although not necessary, it is nice to sketch the surface (Fig. 3.3). Since

$$\mathbf{x}_u = (\cos v, \sin v, 0) \qquad \text{and} \qquad \mathbf{x}_v = (-u\sin v, u\cos v, b),$$

the element of area is

$$\begin{aligned} dA &= \sqrt{\begin{vmatrix} \sin v & 0 \\ u\cos v & b \end{vmatrix}^2 + \begin{vmatrix} 0 & \cos v \\ b & -u\sin v \end{vmatrix}^2 + \begin{vmatrix} \cos v & \sin v \\ -u\sin v & u\cos v \end{vmatrix}^2}\, du\, dv \\ &= \sqrt{b^2\sin^2 v + b^2\cos^2 v + u^2}\, du\, dv = \sqrt{b^2 + u^2}\, du\, dv. \end{aligned}$$

As (u, v) ranges over the rectangle D, the point $\mathbf{x}(u, v)$ runs over the spiral ramp. Hence

$$A = \iint_D \sqrt{b^2 + u^2}\, du\, dv = \left(\int_0^a \sqrt{b^2 + u^2}\, du\right)\left(\int_0^c dv\right).$$

Answer:

$$\frac{c}{2}\left[a\sqrt{a^2 + b^2} + b^2 \ln\left(\frac{a + \sqrt{a^2 + b^2}}{b}\right)\right].$$

Non-parametric Surfaces

A surface in the form $\mathbf{x} = \mathbf{x}(u, v)$ is sometimes called a **parametric surface,** and u and v are called **parameters.** Equally important are **non-parametric surfaces,** surfaces given as graphs of functions $z = f(x, y)$. See Fig. 3.4.

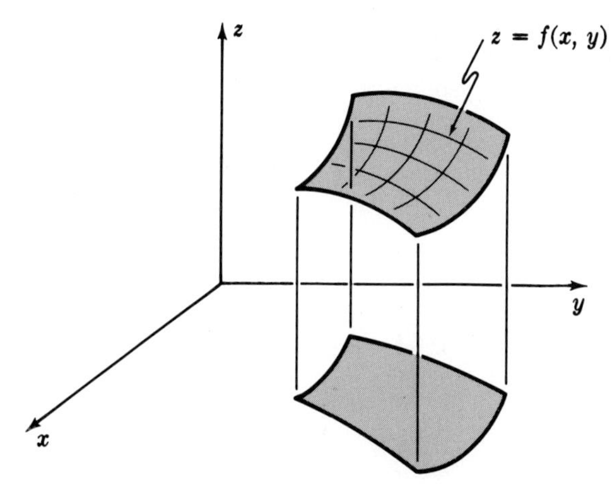

FIG. 3.4

Such a surface $z = f(x, y)$ may be treated like a parametric surface by the simple device of naming x and y the parameters. The variable point on the surface is

$$\mathbf{x} = (x, y, f(x, y)).$$

Then

$$\frac{\partial \mathbf{x}}{\partial x} = (1, 0, f_x), \qquad \frac{\partial \mathbf{x}}{\partial y} = (0, 1, f_y).$$

Consequently

$$\frac{\partial \mathbf{x}}{\partial x} \times \frac{\partial \mathbf{x}}{\partial y} = (1, 0, f_x) \times (0, 1, f_y) = (-f_x, -f_y, 1),$$

and the resulting formula for the element of area is

$$\boxed{dA = \sqrt{1 + f_x^2 + f_y^2}\, dx\, dy,}$$

a most useful expression.

The formula has a geometric interpretation. Recall that the unit normal (p. 534) to the surface is

$$\mathbf{N} = \frac{1}{\sqrt{1 + f_x^2 + f_y^2}} (-f_x, -f_y, 1).$$

Its third component (direction cosine) is

$$\cos\gamma = \frac{1}{\sqrt{1 + f_x^2 + f_y^2}}.$$

Here γ is the angle between the normal and the z-axis. Thus

$$(\cos\gamma)\, dA = dx\, dy,$$

which means that the small piece of surface of area dA projects onto a small portion of the x, y-plane of area $dx\, dy$.

EXAMPLE 3.2

Find the area of the portion of the hyperbolic paraboloid (saddle surface) $z = -x^2 + y^2$ defined on the domain $x^2 + y^2 \leq a^2$.

Solution: First sketch the surface, then the portion corresponding to the range $x^2 + y^2 \leq a^2$. See Fig. 3.5. By the formula,

$$dA = \sqrt{1 + f_x^2 + f_y^2}\, dx\, dy = \sqrt{1 + (-2x)^2 + (2y)^2}\, dx\, dy$$
$$= \sqrt{1 + 4x^2 + 4y^2}\, dx\, dy.$$

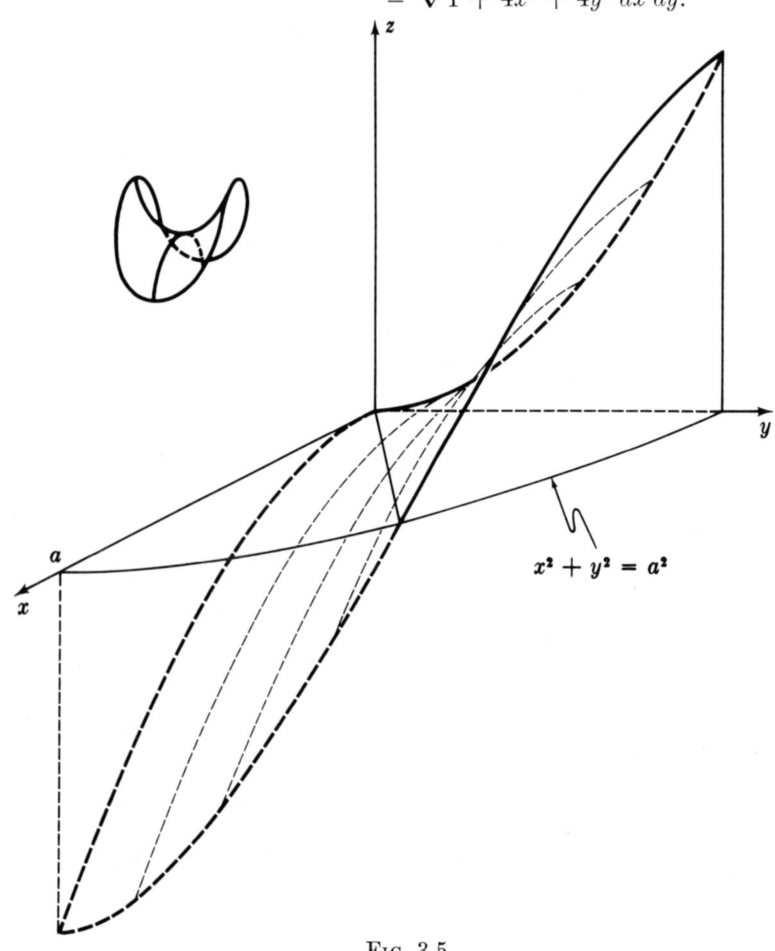

FIG. 3.5

Use polar coordinates:

$$A = \iint \sqrt{1 + 4r^2}\, r\, dr\, d\theta$$

$$= \int_0^a r\sqrt{1 + 4r^2}\, dr \int_0^{2\pi} d\theta = \frac{1}{12}[(1 + 4a^2)^{3/2} - 1] \cdot 2\pi.$$

Answer: $\frac{\pi}{6}[(1 + 4a^2)^{3/2} - 1]$.

EXERCISES

Use the first method of this section to set up the surface area as an integral. Evaluate when you can:

1. sphere of radius a
2. lateral surface of a right circular cylinder of radius a and height h
3. lateral surface of a right circular cone of radius a and lateral height L
4. right circular torus obtained by revolving a circle of radius a about an axis in its plane at distance A from its center (where $a < A$)
5. ellipsoid $\dfrac{x^2}{a^2} + \dfrac{y^2}{b^2} + \dfrac{z^2}{c^2} = 1$

   [*Hint:* Use spherical coordinates and $\mathbf{x} = (a \sin\phi \cos\theta, b \sin\phi \sin\theta, c \cos\phi)$.]
6. (cont.) Reduce the double integral to a simple integral in the special case $a = b$.

Set up the area for the given non-parametric surface. Evaluate when you can:

7. $z = xy$; $-1 \le x \le 1$, $-1 \le y \le 1$ (do not evaluate)
8. $z = ax + by$; (x, y) in a domain D
9. $z = x^2 + y^2$; $x^2 + y^2 \le 1$
10. $z = \sqrt{1 - x^2 - y^2}$; $x^2 + y^2 \le 1$.
11. Find the area of the triangle with vertices $(a, 0, 0)$, $(0, b, 0)$, and $(0, 0, c)$.
12. Prove that $\iint \mathbf{N}\, dA = \mathbf{0}$ for a closed surface.

    (*Hint:* Show that $\mathbf{i} \cdot \iint \mathbf{N}\, dA = 0$, etc.)
13. From each point of the space curve $\mathbf{x} = \mathbf{x}(s)$ draw a segment of length 1 in the direction of the unit tangent. These segments sweep out a surface. Show that its area is $\frac{1}{2}\int k(s)\, ds$, where $k(s)$ is the curvature and the integral is taken over the length of the curve.
14. (cont.) Interpret for a circle. Can you show that $\int k(s)\, ds = 2\pi$ for any closed oval (convex curve) in the plane? If so, interpret Ex. 13 for ovals.
15. Let $\mathbf{x} = \mathbf{x}(s)$ be a curve of length L on the unit sphere $\mathbf{x} = 1$. Connect each point of the curve to the origin, forming a surface. Show that the area of this surface is $\frac{1}{2}L$.

36. Calculus Theory

1. INTRODUCTION

In this book we presented calculus as a working tool. Our presentation was intuitive rather than rigorous, because we were more concerned with the use of calculus than with its foundations. Now, in this final chapter, we shall sketch some of the main theoretical considerations behind calculus. This theory is covered in detail in courses on Functions of Real Variables or Mathematical Analysis.

The two basic processes of calculus are differentiation and integration. Both are applied to functions. Calculus really begins with a study of basic properties of functions. This study requires a thorough knowledge of the real number system and the notion of a limit.

2. REAL NUMBERS AND LIMITS

The basic number system of science is the real number system. It has two operations, addition and multiplication (and their inverses, subtraction and division). The system is a **field,** meaning roughly, that all the usual rules of arithmetic hold.

Among the real numbers are the **rational numbers,** those real numbers expressible as quotients of integers, for example, 6/1, 2/3, $-9/5$, 37/2965, $3.1416 = 31{,}416/10{,}000$.

Each real number can be written as an unending decimal. The rational numbers are precisely the decimals which repeat from some point on. For example

$$0.125000 \cdots = \frac{1}{8}, \qquad 0.257000 \cdots = \frac{257}{1000},$$

$$1.333 \cdots = \frac{4}{3}, \qquad 0.09090909 \cdots = \frac{1}{11},$$

$$2.53697697697697 \cdots = \frac{253{,}444}{99{,}900} = \frac{63{,}361}{24{,}975}.$$

Not all decimals repeat, for instance

$$0.101001000100001000001 \cdots,$$

where each string of zeros is one longer than the preceding string. Thus not all real numbers are rational.

Many numbers are known to be irrational, for example, $\sqrt{2}$. Similarly, $\sqrt{3}$, $\sqrt{5}$, and $\sqrt[3]{2}$ are irrational, as are π, e, and $\ln 17$.

The rational number system and the real number system are similar in many ways. The rational numbers form a field; all the usual rules of arithmetic apply to them. In fact most computations in real life are carried out entirely within the rational field.

But there is a very important distinction between the rational and real number systems: the rational number system is full of "gaps" or "holes," while the real system is not; it is "complete."

To understand this, consider the sequence of rational numbers

$$1, \quad 1.4, \quad 1.41, \quad 1.414, \quad 1.4142, \quad 1.41421, \quad 1.414213, \quad \cdots.$$

Their squares approach 2 as closely as we wish, so the sequence seems to be approaching a number whose square is 2. But there is no rational number whose square is exactly 2; there is a "hole" in the rational numbers where $\sqrt{2}$ should be. In the real number system, there *is* a number $\sqrt{2}$.

Similarly

$$3, \quad 3.1, \quad 3.14, \quad 3.141, \quad 3.1415, \quad 3.14159, \quad 3.141592, \quad \cdots$$

is a sequence of rational numbers approximating the circumference of a circle of unit diameter. But there is no rational number exactly equal to the circumference; there is a "hole" in the rational numbers where π should be.

The real numbers can be constructed rigorously by "filling in the holes" or "completing the rational numbers." Or the real numbers can be described by a set of axioms, the critical one being an axiom of **completeness.** There are many equivalent forms of this axiom. Here is one of the simplest:

> Suppose $a_1 < a_2 < a_3 < \cdots \leq B$. This means $\{a_n\}$ is a strictly increasing sequence of real numbers, all bounded by a fixed real number B. Completeness asserts that the sequence $\{a_n\}$ has a limit, that is, there exists a real number A such that
>
> $$\lim_{n \to \infty} a_n = A.$$

The limit concept is precisely this:

$$|a_n - A| \longrightarrow 0 \qquad \text{as} \quad n \longrightarrow \infty.$$

In words, $|a_n - A|$ can be made smaller than any prescribed number (however small) by making n sufficiently large. More precisely still, if $\epsilon > 0$, there is a whole number $N = N(\epsilon)$ so that: if $n > N$, then $|a_n - A| < \epsilon$.

Here is how completeness fills the holes in the rational numbers. Reconsider the sequence

$$1, \quad 1.4, \quad 1.41, \quad 1.414, \quad 1.4142, \quad 1.41421, \quad 1.414213, \quad \cdots .$$

It is an increasing sequence of real numbers bounded by 1.5. Therefore, completeness guarantees that the sequence has a limit A in the real number system, and it can be shown that $A^2 = 2$. Thus $\sqrt{2}$ is included in the real numbers; the gap in the rationals is filled.

Notice that completeness merely asserts the existence of a limit for a bounded, increasing sequence; it does not tell what the limit is. Here are some examples of increasing sequences $a_1, a_2, a_3, \cdots$.

(1)
$$a_n = 1 - \frac{1}{n}.$$

In this case $a_1 < a_2 < a_3 \cdots < 1$ and obviously $a_n \longrightarrow 1$.

(2)
$$a_n = 10^{-1!} + 10^{-2!} + \cdots + 10^{-n!}.$$

It is an easy matter to check that

$$a_1 < a_2 < a_3 < \cdots < 1$$

(look at the decimal expansion of a_n), so completeness asserts that $\{a_n\}$ has a limit. But this limit is not rational (the decimal expansion does not repeat) nor does it seem to be familiar; it is hard to say much about this limit except that it is there.

(3)
$$a_n = \left(1 + \frac{1}{n}\right)^n.$$

We shall see later that

$$a_1 < a_2 < \cdots < 3$$

and indeed that $a_n \longrightarrow e \approx 2.718$.

(4)
$$a_n = 1 + \frac{1}{2^2} + \frac{1}{3^2} + \cdots + \frac{1}{n^2}.$$

This is an increasing sequence, bounded by 2 because

$$a_n < 1 + \int_1^\infty \frac{dx}{x^2} = 2.$$

By completeness, $\lim a_n$ exists. We can estimate the limit, but at this time, we cannot find it exactly. In Advanced Calculus it is shown that the limit is $\pi^2/6$. Therefore

$$\sum_{n=1}^{\infty} \frac{1}{n^2} = \frac{\pi^2}{6},$$

a remarkable fact.

(5) $$a_n = 1 + \frac{1}{2^3} + \frac{1}{3^3} + \cdots + \frac{1}{n^3}.$$

This increasing sequence is bounded by $\frac{3}{2}$ because

$$a_n < 1 + \int_1^{\infty} \frac{dx}{x^3} = \frac{3}{2}.$$

We conclude that $a_n \longrightarrow A$, that is,

$$\sum_{n=1}^{\infty} \frac{1}{n^3} = A.$$

At the time of writing, little is known about this number A. It is not even known if A is rational.

Limits

Review the definition of limit given above. Notice that general sequences, not only increasing ones, can have limits. For instance, the sequence

$$1 - \tfrac{1}{2}, \quad 1 + \tfrac{1}{3}, \quad 1 - \tfrac{1}{4}, \quad 1 + \tfrac{1}{5}, \quad \cdots$$

is neither increasing nor decreasing, yet it has a perfectly good limit, 1.

Let us list some theorems concerning limits of sequences.

S_1. If

$$\lim_{n\to\infty} a_n = A \qquad \text{and} \qquad \lim_{n\to\infty} b_n = B,$$

then

$$\lim_{n\to\infty}(a_n + b_n) = A + B.$$

S_2. Same hypotheses; then

$$\lim_{n\to\infty} a_n b_n = AB.$$

S_3. If the hypotheses of S_1 are satisfied, if each $b_n \neq 0$, and if $B \neq 0$, then

$$\lim_{n\to\infty} \frac{a_n}{b_n} = \frac{A}{B}.$$

Each of these statements can be proved formally as a consequence of the definition of limit and the properties of the real system. Let us sketch the ideas for S_1 and S_2. The proof of S_1 is based on the relations

$$|(a_n + b_n) - (A + B)| = |(a_n - A) + (b_n - B)| \leq |a_n - A| + |b_n - B|.$$

The problem is to show that

$$|(a_n + b_n) - (A + B)|$$

can be made arbitrarily small by taking n sufficiently large. But by hypothesis, $|a_n - A|$ can be made small and so can $|b_n - B|$, hence their sum can be made small. But this sum exceeds the quantity $|(a_n + b_n) - (A + B)|$ which we want to control, so S_1 follows.

The proof of S_2 in turn is based on the estimate

$$\begin{aligned} |a_nb_n - AB| &= |(a_nb_n - a_nB) + (a_nB - AB)| \\ &\leq |a_n(b_n - B)| + |B(a_n - A)| \\ &= |a_n| \cdot |b_n - B| + |B| \cdot |a_n - A|. \end{aligned}$$

Since $|a_n - A| \longrightarrow 0$ as $n \longrightarrow \infty$ and $|B|$ is constant, it is fairly clear that the term $|B| \cdot |a_n - A| \longrightarrow 0$. To say the same of the other term $|a_n| \cdot |b_n - B|$ requires a trick, because $|a_n|$ is not constant. Now $|a_n - A| \longrightarrow 0$, so a_n sticks close to A. After n exceeds some N_1, then $|a_n| < |A| + 1$. Thus for $n > N_1$,

$$|a_n| \cdot |b_n - B| \leq (|A| + 1)|b_n - B| \longrightarrow 0.$$

These are the main ideas for proving S_2, the rest of the proof is the technical drudgery of writing it down properly.

The statements S_1, S_2, and S_3 can be summarized in three brief sentences: (1) Limit of sum equals sum of limits. (2) Limit of product equals product of limits. (3) If defined, limit of quotient equals quotient of limits.

3. CONTINUOUS FUNCTIONS

In Calculus we are more interested in limits of real functions than of sequences. A real function $f(x)$ first of all is defined on an interval (its **domain**), perhaps an **open interval** $a < x < b$, or a **closed interval** $a \leq x \leq b$, or an open half line $a < x < \infty$, $\cdots$, or the whole line $-\infty < x < \infty$.

> The assertion
>
> $$\lim_{x \to x_0} f(x) = A$$
>
> means that if $\epsilon > 0$, however small, then there is a number $\delta > 0$ (δ depends on ϵ) so that if $|x - x_0| < \delta$, then $|f(x) - A| < \epsilon$.

It can be proved that this definition of $\lim f(x) = A$ as $x \longrightarrow x_0$ is equivalent to the following one:

If $x_1, x_2, x_3, \cdots$ is any sequence satisfying $\lim x_n = x_0$, then $\lim f(x_n) = A$. Briefly, if $x_n \longrightarrow x_0$, then $f(x_n) \longrightarrow A$.

> A function $f(x)$ defined on an interval is said to be **continuous** if for each point x_0 in the interval
>
> $$\lim_{x \to x_0} f(x) = f(x_0),$$
>
> where values of x on the left-hand side are restricted to the domain of f. In other words, if $x_n \longrightarrow x_0$, then $f(x_n) \longrightarrow f(x_0)$.

Thus a continuous function is predictable; you can predict its value at x_0 from its values near (but not at) x_0.

Examples. The functions in Fig. 3.1 are continuous wherever defined; those in Fig. 3.2 are not continuous at $x = 0$, but are continuous everywhere else.

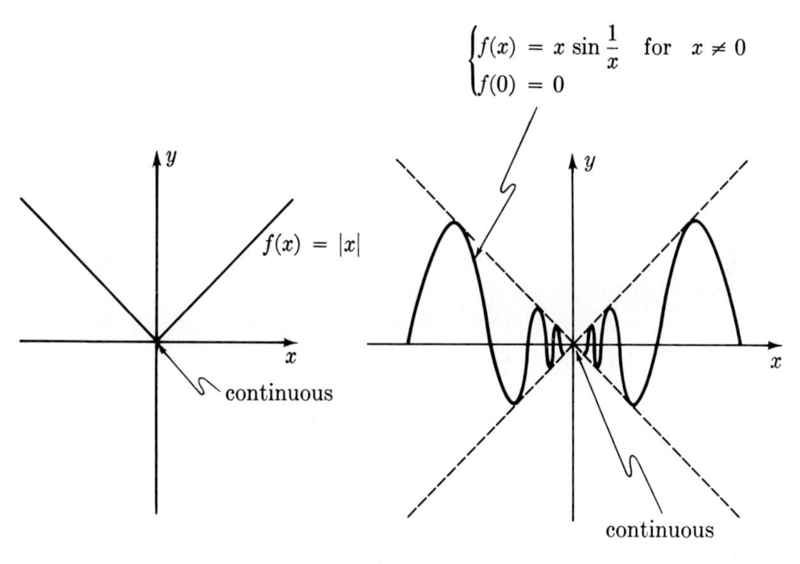

FIG. 3.1a FIG. 3.1b

From the theorems S_1, S_2, S_3 concerning limits of sequences, there follow easily corresponding theorems for limits of functions.

F_1. If

$$\lim_{x \to x_0} f(x) = A \quad \text{and} \quad \lim_{x \to x_0} g(x) = B,$$

then

$$\lim_{x \to x_0} [f(x) + g(x)] = A + B.$$

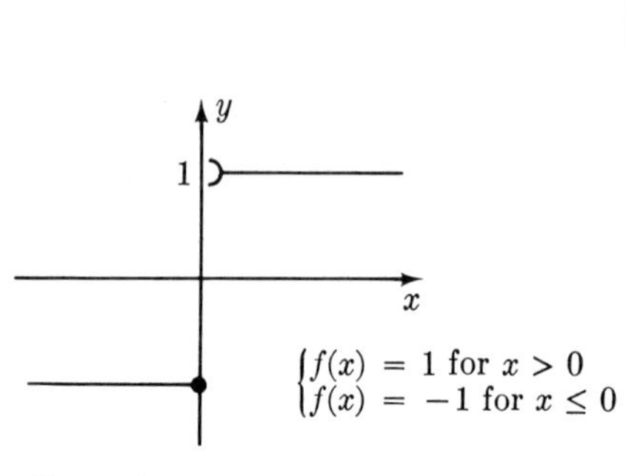

discontinuous at $x = 0$

FIG. 3.2a

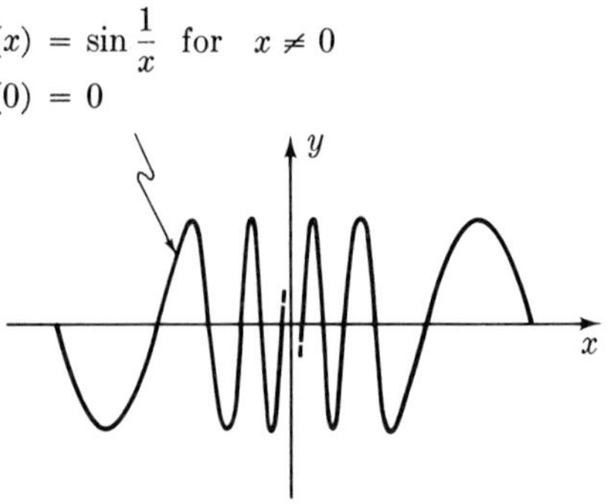

$\begin{cases} f(x) = \sin \frac{1}{x} \quad \text{for} \quad x \neq 0 \\ f(0) = 0 \end{cases}$

y

x

discontinuous at $x = 0$

FIG. 3.2b

F_2. Same hypotheses; then

$$\lim_{x \to x_0} [f(x)g(x)] = AB.$$

F_3. Same hypotheses and in addition, $g(x)$ never 0 and $B \neq 0$; then

$$\lim_{x \to x_0} \frac{f(x)}{g(x)} = \frac{A}{B}.$$

These three theorems in turn can be translated into three properties of continuous functions.

C_1. If $f(x)$ and $g(x)$ are continuous on the same domain, so is $f(x) + g(x)$.
C_2. Same hypotheses; then $f(x)g(x)$ is continuous.
C_3. Same hypotheses and in addition, $g(x)$ never 0 on the domain; then $f(x)/g(x)$ is continuous.

Two Properties of Continuous Functions

We are going to state two of the most important properties of continuous functions. The proofs are deep, going right back to the completeness of the real number system.

Here is the first.

> **Intermediate Value Theorem** Suppose $f(x)$ is continuous on the closed interval $a \leq x \leq b$. Suppose $f(a)$ and $f(b)$ have opposite signs. Then there is an x_0 in the interval such that $f(x_0) = 0$.

This theorem is the theoretical basis for finding solutions to equations. If $f(a)$ and $f(b)$ have opposite signs, then the equation $f(x) = 0$ has a solution between a and b. The Bisection Method of Chapter 14, Section 3, p. 206 is directly based on this Intermediate Value Theorem.

Here is a curious example. Suppose heat is continuously distributed on a piece of wire in the shape of a circle. Then there must be at least one pair of diametrically opposite points at which the temperatures are equal. Can you see how this follows from the Intermediate Value Theorem?

Here is the second property of continuous functions.

Maximum Value Theorem Suppose $f(x)$ is continuous on the closed interval $a \leq x \leq b$. Then there is an x_0 on this interval such that

$$f(x_0) \geq f(x)$$

for each x on the interval.

There is also a corresponding Minimum Value Theorem; it follows at once by applying the Maximum Value Theorem to $-f(x)$.

These theorems are the theoretical basis for the study of maxima and minima. They assure the existence of a maximum and a minimum of each continuous function on a closed interval.

Note carefully the words *closed interval.* The Maximum Value Theorem is simply false without them. For example, the function $f(x) = x$ has neither a maximum nor a minimum on any open interval $a < x < b$; the function $f(x) = x^2$ has a minimum but no maximum on the whole x-axis.

Several Variables

The concept of continuity can be extended to functions of several real variables, that is, vector variables. Briefly $f(\mathbf{x})$ is continuous at $\mathbf{x}_0$ if

$$f(\mathbf{x}) \longrightarrow f(\mathbf{x}_0) \qquad \text{as} \quad \mathbf{x} \longrightarrow \mathbf{x}_0.$$

In other words, as $|\mathbf{x} - \mathbf{x}_0| \longrightarrow 0$, then $|f(\mathbf{x}) - f(\mathbf{x}_0)| \longrightarrow 0$.

The properties C_1, C_2, and C_3 are valid for vector variables; so are the Intermediate Value Theorem and the Maximum Value Theorem for each function continuous on a rectangular region $a_1 \leq x \leq b_1$, $a_2 \leq y \leq b_2$, $a_3 \leq z \leq b_3$.

4. DERIVATIVES

The formal definition of derivative is based on the limit concept.

Let $f(x)$ be continuous on an interval. Suppose

$$f'(x_0) = \lim_{x \to x_0} \frac{f(x) - f(x_0)}{x - x_0}$$

exists for a point x_0 of the interval. Then $f(x)$ is **differentiable** at x_0 and the value of the limit is the **derivative** of $f(x)$ at x_0.

Of course the definition of the derivative is motivated by the thought that the tangent line to a curve is the limit of secants. If a function has a derivative at each point of its domain, then its graph is a smooth curve, without corners.

A function may be continuous without being differentiable. For example $f(x) = \sqrt[3]{x}$ is continuous, but not differentiable at $x = 0$. Another example is $f(x) = x^{2/3}$. See Fig. 4.1. Still another is the function $y = |x|$ of Fig. 3.1a.

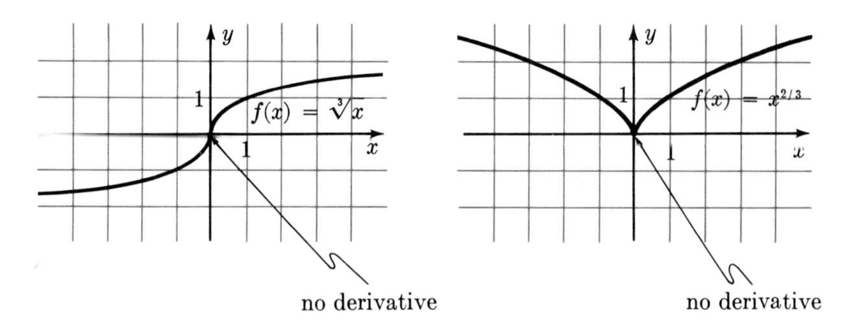

FIG. 4.1

Our discussions of maxima and minima (Chapters 3 and 16) assume the given function is differentiable at all points. However, if a function is non-differentiable at some point, its maximum or minimum could occur right there. For example, the minimum of $f(x) = |x|$ on $-1 \leq x \leq 1$ occurs at $x = 0$, the one point where $f(x)$ is non-differentiable.

Therefore our discussion should be amended as follows. Candidates for the point(s) where a continuous function $f(x)$ takes its maximum in a closed interval are

(i) the end points,

(ii) points where $f'(x) = 0$,

(iii) points where $f(x)$ has no derivative.

Fortunately, most functions that arise in practice are differentiable so case (iii) is rare.

The formal properties of limits, F_1, F_2, and F_3, can be used to prove easily the formal properties of derivatives of sums, products, and quotients. All derivatives of combinations of simple functions are derived from these properties and one other, the Chain Rule. Let us formulate it precisely and examine its proof.

Chain Rule Suppose $g(x)$ is differentiable on an interval $a < x < b$ and that its values lie in an interval $c < y < d$. Suppose $f(y)$ is differentiable on the interval $c < y < d$. Then the composite function $f[g(x)]$ is differentiable on the interval $a < x < b$ and

$$\frac{d}{dx} f[g(x)] = f'[g(x)]g'(x).$$

This is usually proved by starting with a trick:

$$\frac{f[g(x)] - f[g(x_0)]}{x - x_0} = \left(\frac{f[g(x)] - f[g(x_0)]}{g(x) - g(x_0)}\right)\left(\frac{g(x) - g(x_0)}{x - x_0}\right).$$

Let $x \longrightarrow x_0$. Then

$$\frac{g(x) - g(x_0)}{x - x_0} \longrightarrow g'(x_0).$$

Also $g(x) \longrightarrow g(x_0)$, so

$$\frac{f[g(x)] - f[g(x_0)]}{g(x) - g(x_0)} \longrightarrow f'[g(x_0)].$$

Since the limit of a product is the product of the limits, we conclude that

$$\frac{f[g(x)] - f[g(x_0)]}{x - x_0} \longrightarrow f'[g(x_0)]g'(x_0).$$

There is, however, an error in this reasoning. For in introducing the denominator $g(x) - g(x_0)$, maybe we are dividing by 0; horror! Thus there are two cases. Either (1) $g(x) \neq g(x_0)$ when x is sufficiently near x_0 and the proof works, or (2) there are numbers x arbitrarily close to x_0 for which $g(x) = g(x_0)$.

In this second case, the formula

$$\{f[g(x_0)]\}' = f'[g(x_0)]g'(x_0)$$

is still true, but for a different reason: both sides are zero. In fact $g'(x_0) = 0$ since $[g(x) - g(x_0)]/(x - x_0) = 0$ for numbers x arbitrarily close to x_0, and $\{f[g(x_0)]\}' = 0$ for exactly the same reason: $\{f[g(x)] - f[g(x_0)]\}/(x - x_0)$ is 0 for numbers x arbitrarily near x_0, those for which $g(x) = g(x_0)$.

Mean Value Theorem

One of the central theoretical results in calculus is the Mean Value Theorem. From it stem many results, for example Taylor's Formula, the error in interpolation, Lhospital's Rule (below), and the Fundamental Theorem (below).

Mean Value Theorem If $f(x)$ is continuous on the interval $a \le x \le b$, and if $f(x)$ is differentiable on the interval $a < x < b$, then there is a point x_0 in the interval $a < x < b$ such that

$$\frac{f(b) - f(a)}{b - a} = f'(x_0).$$

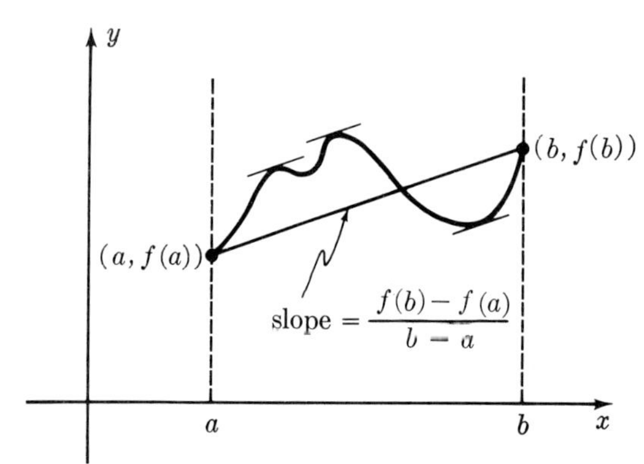

FIG. 4.2

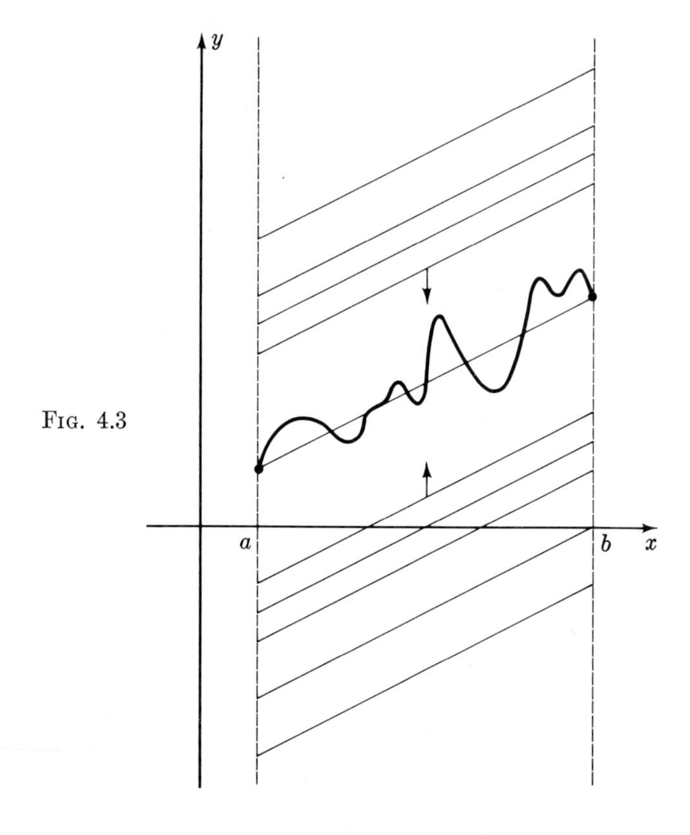

FIG. 4.3

Geometrically this means that at some point the tangent is parallel to the chord joining the end points (Fig. 4.2). (There may be several x_0 that work.)

Here is an intuitive proof of the Mean Value Theorem. By the Maximum Value Theorem, $f(x)$ is bounded. Take parallels to the chord, above and below, so far away that they miss the graph (Fig. 4.3). Now let these parallels move in together from opposite sides. The first to hit the graph is a tangent. The four possible cases are shown in Fig. 4.4. Note that in the first case the function is linear, so each tangent is parallel to the chord.

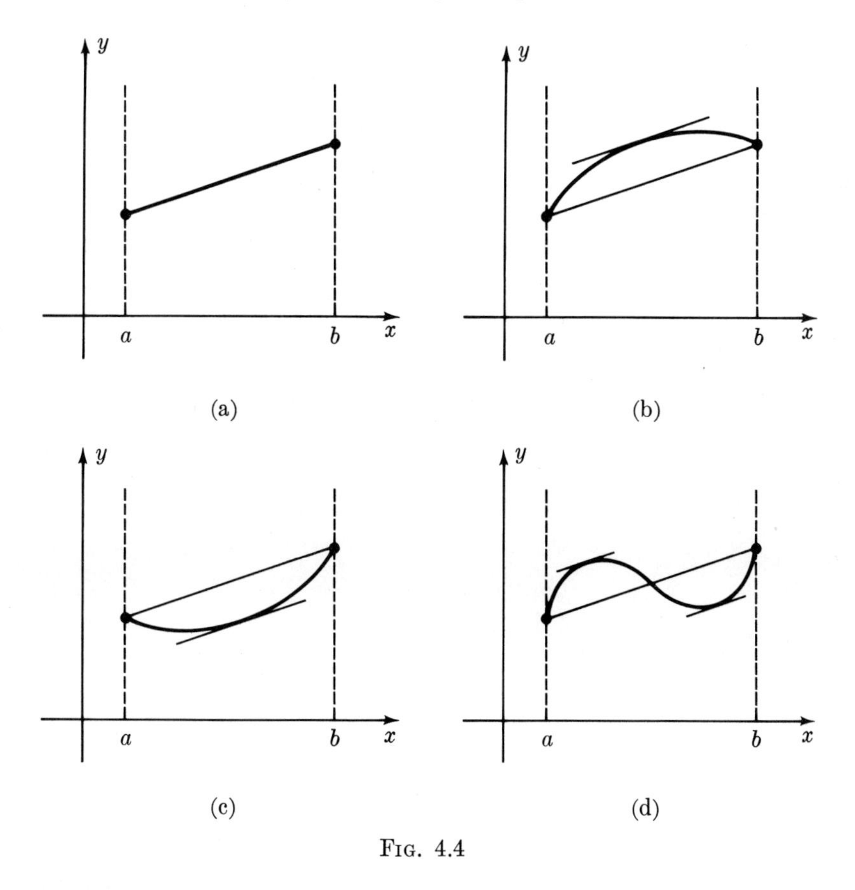

FIG. 4.4

An immediate application of the Mean Value Theorem is the result we took for granted in our study of direction fields (Chapter 4).

If $f(x)$ is continuous on $a \le x \le b$, differentiable on $a < x < b$, and $f'(x) = 0$ for all x on this interval, then $f(x)$ is constant.

For suppose $a < x \le b$. Then the slope $[f(x) - f(a)]/(x - a)$ of the chord joining $(a, f(a))$ to $(x, f(x))$ is equal to the slope $f'(x_0)$ of the tangent for some x_0 between a and x. But $f'(x_0) = 0$, so $f(x) = f(a)$. This is true for all x satisfying $a < x \leq b$.

Lhospital's Rule

There is a rule for finding limits such as

$$\lim_{x\to 0} x \ln x, \qquad \lim_{x\to 0} \frac{1 - \cos x}{x^2}$$

$$\lim_{x\to\infty} xe^{-x}, \qquad \lim_{x\to 0} x^x.$$

Each of these can be put in the form

$$\lim_{x\to a} \frac{f(x)}{g(x)},$$

where a is finite or ∞, and where both

$$f(x) \longrightarrow 0 \qquad \text{and} \qquad g(x) \longrightarrow 0,$$

or both

$$f(x) \longrightarrow \pm\infty \qquad \text{and} \qquad g(x) \longrightarrow \pm\infty.$$

Lhospital's Rule Suppose $f(x)$ and $g(x)$ have continuous derivatives near $x = a$, that $g(x) \neq 0$ and $g'(x) \neq 0$, and that

$$\lim_{x\to a} f(x) = \lim_{x\to a} g(x) = 0$$

or

$$\lim_{x\to a} f(x) = \lim_{x\to a} g(x) = \pm\infty.$$

Suppose

$$\lim_{x\to a} \frac{f'(x)}{g'(x)} = L.$$

Then

$$\lim_{x\to a} \frac{f(x)}{g(x)} = L.$$

A complete proof of Lhospital's Rule is difficult, and depends on a complicated form of the Mean Value Theorem. Here is an intuitive justification for the case a finite and $g'(a) \neq 0$: For x near a, the approximations $f(x) \approx (x - a)f'(a)$ and $g(x) \approx (x - a)g'(a)$ hold, hence

$$\frac{f(x)}{g(x)} \approx \frac{(x - a)f'(a)}{(x - a)g'(a)} = \frac{f'(a)}{g'(a)}.$$

We consider some applications of Lhospital's Rule.

EXAMPLE 4.1

Use Lhospital's Rule to show that

(1) $\lim_{x \to 0} \frac{\sin x}{x} = 1$ (2) $\lim_{x \to 0} \frac{1 - \cos x}{x^2} = \frac{1}{2}$

(3) $\lim_{x \to 0} x \ln x = 0$ (4) $\lim_{x \to \infty} xe^{-x} = 0$

(5) $\lim_{x \to 0} x^x = 1.$

Solution:

(1) $\sin x \longrightarrow 0$ as $x \longrightarrow 0$,

$$\frac{(\sin x)'}{(x)'} = \frac{\cos x}{1} \longrightarrow \cos 0 = 1,$$

hence

$$\lim_{x \to 0} \frac{\sin x}{x} = 1.$$

(2) $1 - \cos x \longrightarrow 0$ and $x^2 \longrightarrow 0$ as $x \longrightarrow 0$,

$$\frac{(1 - \cos x)'}{(x^2)'} = \frac{\sin x}{2x} = \frac{1}{2} \frac{\sin x}{x} \longrightarrow \frac{1}{2},$$

hence

$$\lim_{x \to 0} \frac{1 - \cos x}{x^2} = \frac{1}{2}.$$

(3) $\ln x \longrightarrow -\infty$ and $1/x \longrightarrow \infty$ as $x \longrightarrow 0$,

$$\frac{(\ln x)'}{(1/x)'} = \frac{1/x}{-1/x^2} = -x \longrightarrow 0,$$

hence

$$\lim_{x \to 0} x \ln x = \lim_{x \to 0} \frac{\ln x}{1/x} = 0.$$

(4) $e^x \longrightarrow \infty$ as $x \longrightarrow \infty$,

$$\frac{(x)'}{(e^x)'} = \frac{1}{e^x} = e^{-x} \longrightarrow 0,$$

hence

$$\lim_{x \to \infty} xe^{-x} = \lim_{x \to \infty} \frac{x}{e^x} = 0.$$

(5) $\lim_{x \to 0} x^x = \lim_{x \to 0} e^{\ln x^x} = \lim_{x \to 0} e^{x \ln x}.$

Since the exponential is a continuous function,

$$\lim_{x\to 0} e^{h(x)} = e^L, \qquad \text{where} \quad L = \lim_{x\to 0} h(x).$$

Hence

$$\lim_{x\to 0} x^x = e^L, \qquad \text{where} \quad L = \lim_{x\to 0} x \ln x = 0,$$

$$\lim_{x\to 0} x^x = e^0 = 1.$$

We close this section with several examples which show the power of the Mean Value Theorem.

EXAMPLE 4.2

Show that $\left(1 + \frac{1}{x}\right)^x$ increases for $x > 0$ and approaches e as $x \longrightarrow \infty$. Briefly,

$$\left(1 + \frac{1}{x}\right)^x \uparrow e.$$

Solution: Set $f(x) = (1 + 1/x)^x$ and

$$g(x) = \ln f(x) = x \ln\left(1 + \frac{1}{x}\right) = \frac{\ln\left(1 + \frac{1}{x}\right)}{\frac{1}{x}} = \frac{\ln(1 + t)}{t},$$

where $t = 1/x$. Let $x \longrightarrow \infty$ so $t \longrightarrow 0$. By Lhospital's Rule,

$$\lim_{x\to\infty} g(x) = \lim_{t\to 0} \frac{\ln(1 + t)}{t} = \lim_{t\to 0} \frac{1}{1 + t} = 1.$$

Hence $f(x) \longrightarrow e$ as $x \longrightarrow \infty$.

This much is relatively easy. To prove that $f(x)$ increases is quite a bit harder. Since the logarithm is an increasing function, it suffices to show that $g(x)$ increases. We must show that $g'(x) > 0$ for $x > 0$. Now

$$g'(x) = \ln\left(1 + \frac{1}{x}\right) - \frac{1}{x + 1} = [\ln(x + 1) - \ln x] - \frac{1}{x + 1}.$$

By the Mean Value Theorem,

$$\ln(x + 1) - \ln x = \frac{1}{x_0},$$

where $x < x_0 < x + 1$. Thus

$$g'(x) = \frac{1}{x_0} - \frac{1}{x+1} > 0$$

as desired.

EXAMPLE 4.3

Suppose $p(x) = (x - a_0)(x - a_1) \cdots (x - a_n)$ where $a_0 < a_1 < \cdots < a_n$. Show that $p'(x)$ has n distinct real zeros.

Solution: By the Mean Value Theorem, $[p(a_1) - p(a_0)]/(a_1 - a_0) = p'(b_1)$ for some b_1 satisfying $a_0 < b_1 < a_1$. Hence $p'(b_1) = 0$. Similarly $p'(b_2) = 0$ where $a_1 < b_2 < a_2, \cdots, p'(b_n) = 0$ where $a_{n-1} < b_n < a_n$. Thus $p'(x)$ has zeros $b_1 < b_2 < \cdots < b_n$. These are all of the zeros of $p'(x)$ since $p'(x)$ is a polynomial of degree n and hence has at most n distinct zeros.

REMARK: By counting multiplicities carefully, you can show similarly that if a polynomial has all real zeros, then so does its derivative.

In Chapter 27, Section 7, p. 615, we postponed a derivation of the error estimate in Lagrange interpolation until now. The next example covers a special case. Once understood, it can be extended easily to the general case.

EXAMPLE 4.4

Suppose $|f'''(x)| \leq M$ for $a < x < b$ and that

$$f(x_0) = f(x_1) = f(x_2) = 0,$$

where $a < x_0 < x_1 < x_2 < b$. Prove that for each x on the interval $a < x < b$,

$$|f(x)| \leq \frac{M}{6}|(x - x_0)(x - x_1)(x - x_2)|.$$

Solution: If $x = x_0, x_1$, or x_2, the result is certainly true. Suppose $x \neq x_0, x_1$, or x_2. Set

$$A = \frac{f(x)}{(x - x_0)(x - x_1)(x - x_2)}$$

and

$$g(t) = f(t) - A(t - x_0)(t - x_1)(t - x_2).$$

Then $g(x_0) = g(x_1) = g(x_2) = 0$ *and* $g(x) = 0$. What is more, x_0, x_1, x_2, x are 4 *distinct* points of the interval $a < t < b$. By the Mean Value Theorem, $g'(t) = 0$ at 3 *distinct* points of the interval. By the same theorem again, $g''(t) = 0$ at 2 *distinct* points of the interval. Once again, $g'''(t) = 0$ at a point t_0 of the interval. But $g'''(t) = f'''(t) - 6A$, hence $f'''(t_0) = 6A$ for some t_0 satisfying $a < t_0 < b$. Consequently

$$\left| \frac{f(x)}{(x - x_0)(x - x_1)(x - x_2)} \right| = |A| = \frac{1}{6}|f'''(t_0)| \le \frac{1}{6}M.$$

Our final example contains the interesting result that the second derivative can be obtained directly as the limit of a second difference quotient, without passing through the first derivative.

EXAMPLE 4.5

Suppose $f(x)$ is twice differentiable near $x = a$ and that $f''(x)$ is continuous at $x = a$. Prove that

$$\lim_{h \to 0} \left(\frac{f(a + h) - 2f(a) + f(a - h)}{h^2} \right) = f''(a).$$

Solution: First note that the result is plausible. For suppose $f(x)$ has a Taylor expansion at a,

$$f(a + x) \approx f(a) + f'(a)x + \frac{1}{2}f''(a)x^2.$$

Then

$$f(a + h) - 2f(a) + f(a - h) \approx f''(a)h^2.$$

The proof requires two applications of the Mean Value Theorem. Set

$$g(x) = f(x) - f(x - h).$$

Then

$$\begin{aligned} g(a + h) - g(a) &= [f(a + h) - f(a)] - [f(a) - f(a - h)] \\ &= f(a + h) - 2f(a) + f(a - h). \end{aligned}$$

By the Mean Value Theorem

$$g(a + h) - g(a) = hg'(x_0),$$

where x_0 is between a and $a + h$. But $g'(x) = f'(x) - f'(x - h)$, hence

$$f(a + h) - 2f(a) + f(a - h) = h[f'(x_0) - f'(x_0 - h)].$$

By the Mean Value Theorem again,

$$f'(x_0) - f'(x_0 - h) = hf''(x_1),$$

where x_1 is between $x_0 - h$ and x_0, and hence is between $a - h$ and $a + h$. Consequently

$$\frac{f(a + h) - 2f(a) + f(a - h)}{h^2} = f''(x_1),$$

where $a - h < x_1 < a + h$. Let $h \longrightarrow 0$. Then $x_1 \longrightarrow a$ so $f''(x_1) \longrightarrow f''(a)$ since $f''(x)$ is continuous at $x = a$. The result follows.

5. DEFINITE INTEGRAL

There are several approaches to a rigorous theory of the definite integral. We shall discuss one based on squeezing the function between rectangles (Fig. 5.1).

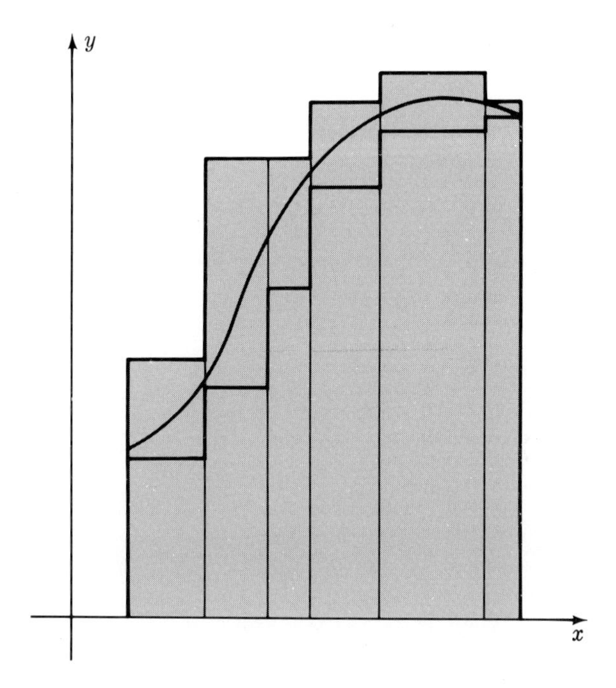

FIG. 5.1

Step Functions

A **step function** $f(x)$ on an interval $a \leq x \leq b$ is a function which is piecewise constant (Fig. 5.2). The interval $a \leq x \leq b$ is divided into subintervals

$$a = x_0 < x_1 < \cdots < x_n = b,$$

and

$$f(x) = c_i \qquad \text{for} \quad x_{i-1} < x < x_i.$$

The values of $f(x)$ at the partition points $x_0, x_1, \cdots$ can be anything.

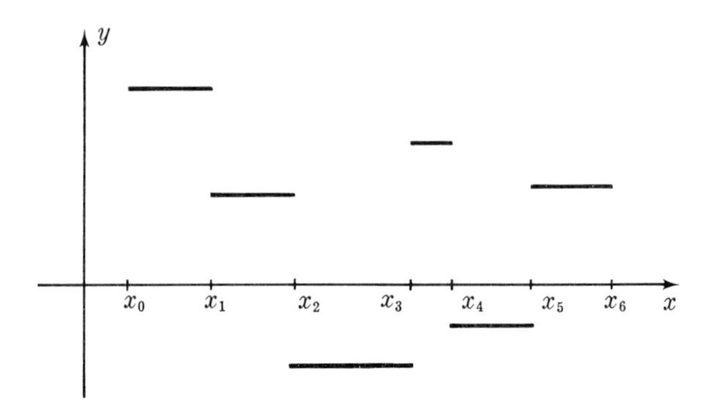

FIG. 5.2

The **definite integral** of a step function $f(x)$ is defined in the obvious way:

$$\int_a^b f(x)\,dx = \sum_{i=1}^{n} c_i(x_i - x_{i-1}).$$

For a positive step function, this is just the total area of the rectangles defined by the function.

Several properties follow easily from the definition:

P_1. $\displaystyle\int_a^b [f(x) + g(x)]\,dx = \int_a^b f(x)\,dx + \int_a^b g(x)\,dx;$

P_2. $\displaystyle\int_a^b [cf(x)]\,dx = c\int_a^b f(x)\,dx;$

P_3. if $f(x) \geq 0$, then $\displaystyle\int_a^b f(x)\,dx \geq 0;$

P_4. if $f(x) \geq g(x)$, then $\displaystyle\int_a^b f(x)\,dx \geq \int_a^b g(x)\,dx.$

Step functions are used to approximate other functions. We have already seen a practical application in rectangular approximation. More important is their use in theory.

Definite Integrals

Let $f(x)$ be any function defined for $a \leq x \leq b$ that is bounded:

$$m \leq f(x) \leq M.$$

We wish to define

$$\int_a^b f(x)\,dx$$

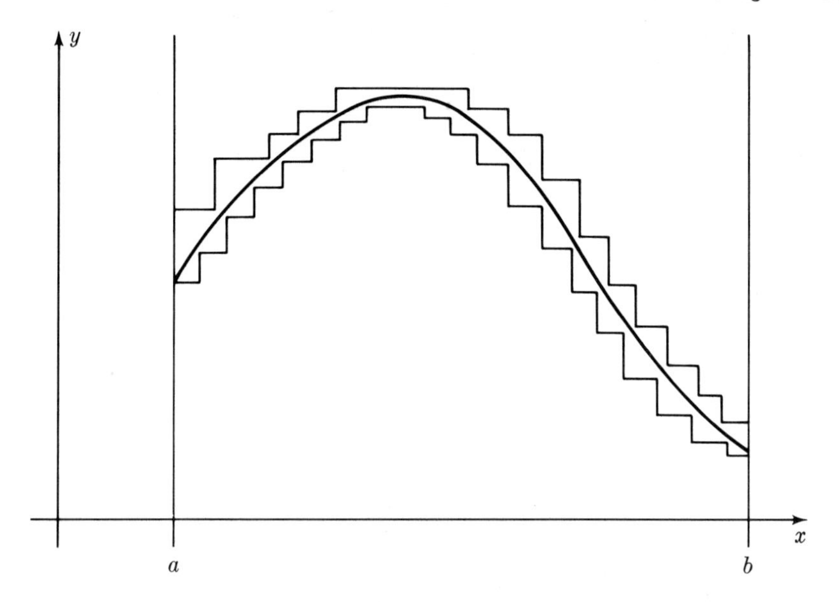

FIG. 5.3

based, if possible, on the integrals of step functions. The idea is to squeeze $f(x)$ tightly between step functions (Fig. 5.3). We hope that the integrals of the approximating step functions approach some number I. If so we shall define

$$\int_a^b f(x)\,dx = I.$$

Here is the precise mathematical formulation.

> Suppose there is a number I with the following property. If $\epsilon > 0$, then there are step functions $g(x)$ and $h(x)$ such that
>
> (1) $$g(x) \le f(x) \le h(x);$$
>
> (2) $$\int_a^b h(x)\,dx - \int_a^b g(x)\,dx < \epsilon;$$
>
> (3) $$\int_a^b g(x)\,dx < I < \int_a^b h(x)\,dx.$$
>
> Then the function $f(x)$ is called **integrable** and its **integral** is
>
> $$I = \int_a^b f(x)\,dx.$$

This is the formal definition of integrable function. Each step function is integrable—you can prove that easily—what is hard to prove is that there are any other ntegrable functions. One of the main achievements of this theory is the proof of the following basic result.

> Each continuous function $f(x)$ on the interval $a \leq x \leq b$ is integrable.

The properties $P_1, \cdots, P_4$ of the integral of step functions are valid for any integrable functions. In addition, one has

P_5. $\displaystyle\int_a^b f(x)\,dx + \int_b^c f(x)\,dx = \int_a^c f(x)\,dx;$

P_6. if $m \leq f(x) \leq M$, then $\displaystyle m(b-a) \leq \int_a^b f(x)\,dx \leq M(b-a).$

These formal properties are routine, although rather tedious to prove.

Fundamental Theorem of Calculus

The most important property of the integral for us is the justly famous Fundamental Theorem of Calculus, the connecting link between Differential and Integral Calculus.

> **Fundamental Theorem of Calculus** Suppose $F(x)$ is continuous on the interval $a \leq x \leq b$, and suppose that $F(x)$ is differentiable on the interval $a < x < b$. Then
>
> $$\int_a^b F'(x)\,dx = F(b) - F(a).$$

A complete proof of the Fundamental Theorem is not easy, but the idea behind it is. Divide the interval $a \leq x \leq b$ into many small pieces:

$$a = x_0 < x_1 < \cdots < x_n = b.$$

On the typical small division, $x_{i-1} \leq x \leq x_i$, the difference $F(x_i) - F(x_{i-1})$ is a good approximation to the integral of $F'(x)$. Why? Because by the Mean Value Theorem,

$$F(x_i) - F(x_{i-1}) = F'(t_i)\,(x_i - x_{i-1}),$$

where t_i is some number satisfying $x_{i-1} < t_i < x_i$. This is the idea; the rest of the proof is difficult analysis.

It is worth recalling the great practical value of the Fundamental Theorem. Very briefly it is this: the integral is a powerful tool in applications, but extremely hard to compute directly from its definition. Antiderivatives, which are not very interesting in themselves, are fairly easy to compute. According to the Fundamental Theorem, computation of a definite integral can be reduced to computation of an antiderivative.

Multiple Integrals

The technique of squeezing between step functions works for multiple integrals also. To simplify matters consider only functions defined on rectangles $a \le x \le b$, $c \le y \le d$. Such a rectangle is divided into small rectangles by lines parallel to the axes. A **step function** is a function constant on the interior of each rectangle. (It does not matter what its values are on the boundaries.) As before, a function is integrable on the rectangle if it can be squeezed between step functions whose integrals are arbitrarily close together. Again one can *prove* that each continuous function is integrable.

The main result is the Iteration Formula. It reduces the evaluation of multiple integrals to the evaluation of single integrals.

Iteration Formula If $f(x, y)$ is continuous on the rectangle R: $a \le x \le b$, $c \le y \le d$, then

$$\iint_R f(x, y)\,dx\,dy = \int_a^b \left(\int_c^d f(x, y)\,dy \right) dx = \int_c^d \left(\int_a^b f(x, y)\,dx \right) dy.$$

This is another case of a result whose proof requires great care.

The extension of these matters to non-rectangular regions requires technique, but presents no real difficulties.

6. PARTIAL DERIVATIVES

After the derivative of a function of one variable is defined, there is nothing really new in the definition of the partial derivative. Remember that all but one variable are held fixed, so that the several variable situation may be considered like the one variable case. There are several basic theoretical results which are central to the subject, and we now discuss them.

Chain Rule Suppose $f(x, y)$, $f_x(x, y)$, and $f_y(x, y)$ are continuous functions near a point (a, b). Suppose $x(t)$ and $y(t)$ are differentiable at t_0 and $(x(t_0), y(t_0)) = (a, b)$. Then the composite function

$$G(t) = f[x(t), y(t)]$$

is differentiable at t_0 and

$$\left.\frac{dG}{dt}\right|_{t_0} = f_x(a, b)\,\dot{x}(t_0) + f_y(a, b)\,\dot{y}(t_0).$$

The proof of this result starts with the subtraction-and-addition trick:

$$\begin{aligned} G(t_0 + h) - G(t_0) &= f[x(t_0 + h), y(t_0 + h)] - f(a, b) \\ &= \Big\{f[x(t_0 + h), y(t_0 + h)] - f[a, y(t_0 + h)]\Big\} \\ &\qquad + \Big\{f[a, y(t_0 + h)] - f(a, b)\Big\}. \end{aligned}$$

Each of the two quantities in braces is the difference of two values of $f(x, y)$ in which one variable is fixed, so it can be attacked by the Mean Value Theorem:

$$f[x(t_0 + h), y(t_0 + h)] - f[a, y(t_0 + h)] = [x(t_0 + h) - a]f_x[x_1, y(t_0 + h)],$$

where x_1 is between a and $x(t_0 + h)$; similarly

$$f[a, y(t_0 + h)] - f(a, b) = [y(t_0 + h) - b]f_y(a, y_1),$$

where y_1 is between b and $y(t_0 + h)$.

Substitute these expressions into the first relation and divide by h:

$$\begin{aligned} \frac{G(t_0 + h) - G(t_0)}{h} &= \left(\frac{x(t_0 + h) - a}{h}\right) f_x[x_1, y(t_0 + h)] \\ &\qquad + \left(\frac{y(t_0 + h) - b}{h}\right) f_y(a, y_1). \end{aligned}$$

Now let $h \longrightarrow 0$. Then $(x_1, y(t_0 + h)) \longrightarrow (a, b)$ and $(a, y_1) \longrightarrow (a, b)$, so by the assumed continuity,

$$f_x[x_1, y(t_0 + h)] \longrightarrow f_x(a, b) \qquad \text{and} \qquad f_y(a, y_1) \longrightarrow f_y(a, b).$$

Also

$$\frac{x(t_0 + h) - a}{h} \longrightarrow \dot{x}(t_0) \qquad \text{and} \qquad \frac{y(t_0 + h) - b}{h} \longrightarrow \dot{y}(t_0),$$

therefore

$$\frac{G(t_0 + h) - G(t_0)}{h} \longrightarrow \dot{x}(t_0) f_x(a, b) + \dot{y}(t_0) f_y(a, b).$$

Mixed Partials

The real distinction between one and several variables occurs at the second derivative level, where mixed partials appear. Their equality is a key theoretical result.

> Suppose the four partials $f_x(x, y), f_y(x, y), f_{xy}(x, y)$, and $f_{yx}(x, y)$ are continuous near (a, b). Then
> $$f_{xy}(a, b) = f_{yx}(a, b).$$

The proof of this result begins with the **mixed second difference,** computed in two ways:

$$\begin{aligned} \Delta^{(2)} &= [f(a+h, b+k) - f(a+h, b)] - [f(a, b+k) - f(a, b)] \\ &= [f(a+h, b+k) - f(a, b+k)] - [f(a+h, b) - f(a, b)]. \end{aligned}$$

Each of these expressions for $\Delta^{(2)}$ will be altered by the Mean Value Theorem. For the first one, set

$$g(x) = f(x, b+k) - f(x, b).$$

Then

$$\Delta^{(2)} = g(a+h) - g(a) = hg'(x_0),$$

where x_0 is between a and $a + h$. But by the Mean Value Theorem again,

$$g'(x_0) = f_x(x_0, b+k) - f_x(x_0, b) = kf_{xy}(x_0, y_0),$$

where y_0 is between b and $b + k$. Thus

$$\Delta^{(2)} = hkf_{xy}(x_0, y_0),$$

where x_0 is between a and $a + h$, and y_0 is between b and $b + k$.

Proceed similarly with the second expression for $\Delta^{(2)}$. The result is

$$\Delta^{(2)} = hkf_{yx}(x_1, y_1),$$

where x_1 is between a and $a + h$, and y_1 is between b and $b + k$. Conclusion:

$$\begin{aligned} hkf_{xy}(x_0, y_0) &= hkf_{yx}(x_1, y_1), \\ f_{xy}(x_0, y_0) &= f_{yx}(x_1, y_1). \end{aligned}$$

Now let $h \longrightarrow 0$ and $k \longrightarrow 0$. Then $(x_0, y_0) \longrightarrow (a, b)$ and $(x_1, y_1) \longrightarrow (a, b)$. But f_{xy} and f_{yx} are continuous by assumption, hence $f_{xy}(x_0, y_0) \longrightarrow f_{xy}(a, b)$ and $f_{yx}(x_1, y_1) \longrightarrow f_{yx}(a, b)$. Therefore

$$f_{xy}(a, b) = f_{yx}(a, b).$$

Taylor Approximations

The Taylor approximations to a function of several variables can be derived from the one variable case by a simple device.

For example, to estimate

$$f(a+h, b+k) - f(a, b)$$

by a second degree polynomial, work instead with the function

$$g(t) = f(a + ht, b + kt).$$

Then

$$g(1) \approx g(0) + g'(0) + \tfrac{1}{2}g''(0).$$

By the Chain Rule,

$$\begin{aligned} g'(t) &= hf_x(a + ht, b + kt) + kf_y(a + ht, b + kt), \\ g''(t) &= h^2 f_{xx}(a + ht, b + kt) \\ &\qquad + 2hkf_{xy}(a + ht, b + kt) + k^2 f_{yy}(a + ht, b + kt), \end{aligned}$$

so the estimate for $g(1)$, spelled out in terms of f, is

$$\begin{aligned} f(a + h, b + k) &\approx f(a, b) + [hf_x(a, b) + kf_y(a, b)] \\ &\qquad + \tfrac{1}{2}[h^2 f_{xx}(a, b) + 2hkf_{xy}(a, b) + k^2 f_{yy}(a, b)]. \end{aligned}$$

This is the second-degree Taylor approximation.

7. REFERENCES

The material presented in this chapter is only a minimal sketch of calculus theory. To do the subject justice would require at least a semester course. For those readers interested in learning more about the subject here are three (among many) possible sources.

1. R. Bartle, *The Elements of Real Analysis*, 2d ed, Wiley, New York, 1964.
2. R. C. Buck, *Advanced Calculus*, 2d ed, McGraw-Hill, New York, 1965.
3. W. Rudin, *Principles of Mathematical Analysis*, 2d ed, McGraw-Hill, New York, 1964.

We list one more reference, a delightful book on the history and development of calculus.

4. O. Toeplitz, *The Calculus; a Genetic Approach*, Univ. of Chicago Press, Chicago, 1963.

TABLE 1 Trigonometric Functions

Degrees	sin	cos	tan	cot	
0°	.0000	1.000	.0000	—	90°
1°	.0175	.9998	.0175	57.29	89°
2°	.0349	.9994	.0349	28.64	88°
3°	.0523	.9986	.0524	19.08	87°
4°	.0698	.9976	.0699	14.30	86°
5°	.0872	.9962	.0875	11.43	85°
6°	.1045	.9945	.1051	9.514	84°
7°	.1219	.9925	.1228	8.144	83°
8°	.1392	.9903	.1405	7.115	82°
9°	.1564	.9877	.1584	6.314	81°
10°	.1736	.9848	.1763	5.671	80°
11°	.1908	.9816	.1944	5.145	79°
12°	.2079	.9781	.2126	4.705	78°
13°	.2250	.9744	.2309	4.331	77°
14°	.2419	.9703	.2493	4.011	76°
15°	.2588	.9659	.2679	3.732	75°
16°	.2756	.9613	.2867	3.487	74°
17°	.2924	.9563	.3057	3.271	73°
18°	.3090	.9511	.3249	3.078	72°
19°	.3256	.9455	.3443	2.904	71°
20°	.3420	.9397	.3640	2.747	70°
21°	.3584	.9336	.3839	2.605	69°
22°	.3746	.9272	.4040	2.475	68°
23°	.3907	.9205	.4245	2.356	67°
24°	.4067	.9135	.4452	2.246	66°
25°	.4226	.9063	.4663	2.145	65°
26°	.4384	.8988	.4877	2.050	64°
27°	.4540	.8910	.5095	1.963	63°
28°	.4695	.8829	.5317	1.881	62°
29°	.4848	.8746	.5543	1.804	61°
30°	.5000	.8660	.5774	1.732	60°
31°	.5150	.8572	.6009	1.664	59°
32°	.5299	.8480	.6249	1.600	58°
33°	.5446	.8387	.6494	1.540	57°
34°	.5592	.8290	.6745	1.483	56°
35°	.5736	.8192	.7002	1.428	55°
36°	.5878	.8090	.7265	1.376	54°
37°	.6018	.7986	.7536	1.327	53°
38°	.6157	.7880	.7813	1.280	52°
39°	.6293	.7771	.8098	1.235	51°
40°	.6428	.7660	.8391	1.192	50°
41°	.6561	.7547	.8693	1.150	49°
42°	.6691	.7431	.9004	1.111	48°
43°	.6820	.7314	.9325	1.072	47°
44°	.6947	.7193	.9657	1.036	46°
45°	.7071	.7071	1.000	1.000	45°
	cos	sin	ctn	tan	Degrees

TABLE 2 Trigonometric Functions for Angles in Radians*

Rad.	Sin	Tan	Cot	Cos	Rad.	Sin	Tan	Cot	Cos
.00	.00000	.00000	∞	1.00000	**.50**	.47943	.54630	1.8305	.87758
01	.01000	.01000	99.997	0.99995	.51	.48818	.55936	1.7878	.87274
.02	.02000	.02000	49.993	.99980	.52	.49688	.57256	1.7465	.86782
.03	.03000	.03001	33.323	.99955	.53	.50553	.58592	1.7067	.86281
.04	.03999	.04002	24.987	.99920	.54	.51414	.59943	1.6683	.85771
.05	.04998	.05004	19.983	.99875	.55	.52269	.61311	1.6310	.85252
.06	.05996	.06007	16.647	.99820	.56	.53119	.62695	1.5950	.84726
.07	.06994	.07011	14.262	.99755	.57	.53963	.64097	1.5601	.84190
.08	.07991	.08017	12.473	.99680	.58	.54802	.65517	1.5263	.83646
.09	.08988	.09024	11.081	.99595	.59	.55636	.66956	1.4935	.83094
.10	.09983	.10033	9.9666	.99500	**.60**	.56464	.68414	1.4617	.82534
.11	.10978	.11045	9.0542	.99396	.61	.57287	.69892	1.4308	.81965
.12	.11971	.12058	8.2933	.99281	.62	.58104	.71391	1.4007	.81388
.13	.12963	.13074	7.6489	.99156	.63	.58914	.72911	1.3715	.80803
.14	.13954	.14092	7.0961	.99022	.64	.59720	.74454	1.3431	.80210
.15	.14944	.15114	6.6166	.98877	.65	.60519	.76020	1.3154	.79608
.16	.15932	.16138	6.1066	.98723	.66	.61312	.77610	1.2885	.78999
.17	.16918	.17166	5.8256	.98558	.67	.62099	.79225	1.2622	.78382
.18	.17903	.18197	5.4954	.98384	.68	.62879	.80866	1.2366	.77757
.19	.18886	.19232	5.1997	.98200	.69	.63654	.82534	1.2116	.77125
.20	.19867	.20271	4.9332	.98007	**.70**	.64422	.84229	1.1872	.76484
.21	.20846	.21314	4.6917	.97803	.71	.65183	.85953	1.1634	.75836
.22	.21823	.22362	4.4719	.97590	.72	.65938	.87707	1.1402	.75181
.23	.22798	.23414	4.2709	.97367	.73	.66687	.89492	1.1174	.74517
.24	.23770	.24472	4.0864	.97134	.74	.67429	.91309	1.0952	.73847
.25	.24740	.25534	3.9163	.96891	.75	.68164	.93160	1.0734	.73169
.26	.25708	.26602	3.7591	.96639	.76	.68892	.95045	1.0521	.72484
.27	.26673	.27676	3.6133	.96377	.77	.69614	.96967	1.0313	.71791
.28	.27636	.28755	3.4776	.96106	.78	.70328	.98926	1.0109	.71091
.29	.28595	.29841	3.3511	.95824	.79	.71035	1.0092	.99084	.70385
.30	.29552	.30934	3.2327	.95534	**.80**	.71736	1.0296	.97121	.69671
.31	.30506	.32033	3.1218	.95233	.81	.72429	1.0505	.95197	.68950
.32	.31457	.33139	3.0176	.94924	.82	.73115	1.0717	.93309	.68222
.33	.32404	.34252	2.9195	.94604	.83	.73793	1.0934	.91455	.67488
.34	.33349	.35374	2.8270	.94275	.84	.74464	1.1156	.89635	.66746
.35	.34290	.36503	2.7395	.93937	.85	.75128	1.1383	.87848	.65998
.36	.35227	.37640	2.6567	.93590	.86	.75784	1.1616	.86091	.65244
.37	.36162	.38786	2.5782	.93233	.87	.76433	1.1853	.84365	.64483
.38	.37092	.39941	2.5037	.92866	.88	.77074	1.2097	.82668	.63715
.39	.38019	.41105	2.4328	.92491	.89	.77707	1.2346	.80998	.62941
.40	.38942	.42279	2.3652	.92106	**.90**	.78333	1.2602	.79355	.62161
.41	.39861	.43463	2.3008	.91712	.91	.78950	1.2864	.77738	.61375
.42	.40776	.44657	2.2393	.91309	.92	.79560	1.3133	.76146	.60582
.43	.41687	.45862	2.1804	.90897	.93	.80162	1.3409	.74578	.59783
.44	.42594	.47078	2.1241	.90475	.94	.80756	1.3692	.73034	.58979
.45	.43497	.48306	2.0702	.90045	.95	.81342	1.3984	.71511	.58168
.46	.44395	.49545	2.0184	.89605	.96	.81919	1.4284	.70010	.57352
.47	.45289	.50797	1.9686	.89157	.97	.82489	1.4592	.68531	.56530
.48	.46178	.52061	1.9208	.88699	.98	.83050	1.4910	.67071	.55702
.49	.47063	.53339	1.8748	.88233	.99	.83603	1.5237	.65631	.54869
.50	.47943	.54630	1.8305	.87758	**1.00**	.84147	1.5574	.64209	.54030
Rad.	Sin	Tan	Cot	Cos	Rad.	Sin	Tan	Cot	Cos

* Tables 2–5 are reproduced from "Handbook of Tables for Mathematics" (R. C. Weast and S. M. Selby, eds.), 3rd ed., Chemical Rubber Co., Cleveland, Ohio, and used by permission.

TABLE 2 Trigonometric Functions for Angles in Radians (Continued)

Rad.	Sin	Tan	Cot	Cos	Rad.	Sin	Tan	Cot	Cos
1.00	.84147	1.5574	.64209	.54030	**1.50**	.99749	14.101	.07091	.07074
1.01	.84683	1.5922	.62806	.53186	1.51	.99815	16.428	.06087	.06076
1.02	.85211	1.6281	.61420	.52337	1.52	.99871	19.670	.05084	.05077
1.03	.85730	1.6652	.60051	.51482	1.53	.99917	24.498	.04082	.04079
1.04	.86240	1.7036	.58699	.50622	1.54	.99953	32.461	.03081	.03079
1.05	.86742	1.7433	.57362	.49757	1.55	.99978	48.078	.02080	.02079
1.06	.87236	1.7844	.56040	.48887	1.56	.99994	92.621	.01080	.01080
1.07	.87720	1.8270	.54734	.48012	1.57	1.00000	1255.8	.00080	.00080
1.08	.88196	1.8712	.53441	.47133	1.58	.99996	−108.65	−.00920	−.00920
1.09	.88663	1.9171	.52162	.46249	1.59	.99982	−52.067	−.01921	−.01920
1.10	.89121	1.9648	.50897	.45360	**1.60**	.99957	−34.233	−.02921	−.02920
1.11	.89570	2.0143	.49644	.44466	1.61	.99923	−25.495	−.03922	−.03919
1.12	.90010	2.0660	.48404	.43568	1.62	.99879	−20.307	−.04924	−.04918
1.13	.90441	2.1198	.47175	.42666	1.63	.99825	−16.871	−.05927	−.05917
1.14	.90863	2.1759	.45959	.41759	1.64	.99761	−14.427	−.06931	−.06915
1.15	.91276	2.2345	.44753	.40849	1.65	.99687	−12.599	−.07937	−.07912
1.16	.91680	2.2958	.43558	.39934	1.66	.99602	−11.181	−.08944	−.08909
1.17	.92075	2.3600	.42373	.39015	1.67	.99508	−10.047	−.09953	−.09904
1.18	.92461	2.4273	.41199	.38092	1.68	.99404	− 9.1208	−.10964	−.10899
1.19	.92837	2.4979	.40034	.37166	1.69	.99290	− 8.3492	−.11977	−.11892
1.20	.93204	2.5722	.38878	.36236	**1.70**	99166	− 7.6966	−.12993	−.12884
1.21	.93562	2.6503	.37731	.35302	1.71	.99033	− 7.1373	−.14011	−.13875
1.22	.93910	2.7328	.36593	.34365	1.72	.98889	− 6.6524	−.15032	−.14865
1.23	.94249	2.8198	.35463	.33424	1.73	.98735	− 6.2281	−.16056	−.15853
1.24	.94578	2.9119	.34341	.32480	1.74	.98572	− 5.8535	−.17084	−.16840
1.25	.94898	3.0096	.33227	.31532	1.75	.98399	− 5.5204	−.18115	−.17825
1.26	.95209	3.1133	.32121	.30582	1.76	.98215	− 5.2221	−.19149	−.18808
1.27	.95510	3.2236	.31021	.29628	1.77	.98022	− 4.9534	−.20188	−.19789
1.28	.95802	3.3413	.29928	.28672	1.78	.97820	− 4.7101	−.21231	−.20768
1.29	.96084	3.4672	.28842	.27712	1.79	.97607	− 4.4887	−.22278	−.21745
1.30	.96356	3.6021	.27762	.26750	**1.80**	.97385	− 4.2863	−.23330	−.22720
1.31	.96618	3.7471	.26687	.25785	1.81	.97153	− 4.1005	−.24387	−.23693
1.32	.96872	3.9033	.25619	.24818	1.82	.96911	− 3.9294	−.25449	−.24663
1.33	.97115	4.0723	.24556	.23848	1.83	.96659	− 3.7712	−.26517	−.25631
1.34	.97348	4.2556	.23498	.22875	1.84	.96398	− 3.6245	−.27590	−.26596
1.35	.97572	4.4552	.22446	.21901	1.85	.96128	− 3.4881	−.28669	−.27559
1.36	.97786	4.6734	.21398	.20924	1.86	.95847	− 3.3608	−.29755	−.28519
1.37	.97991	4.9131	.20354	.19945	1.87	.95557	− 2.2419	−.30846	−.29476
1.38	.98185	5.1774	.19315	.18964	1.88	.95258	− 3.1304	−.31945	−.30430
1.39	.98370	5.4707	.18279	.17981	1.89	.94949	− 3.0257	−.33051	−.31381
1.40	.98545	5.7979	.17248	.16997	**1.90**	.94630	− 2.9271	−.34164	−.32329
1.41	.98710	6.1654	.16220	.16010	1.91	.94302	− 2.8341	−.35284	−.33274
1.42	.98865	6.5811	.15195	.15023	1.92	.93965	− 2.7463	−.36413	−.34215
1.43	.99010	7.0555	.14173	.14033	1.93	.93618	− 2.6632	−.37549	−.35153
1.44	.99146	7.6018	.13155	.13042	1.94	.93262	− 2.5843	−.38695	−.36087
1.45	.99271	8.2381	.12139	.12050	1.95	.92896	− 2.5095	−.39849	−.37018
1.46	.99387	8.9886	.11125	.11057	1.96	.92521	− 2.4383	−.41012	−.37945
1.47	.99492	9.8874	.10114	.10063	1.97	.92137	− 2.3705	−.42185	−.38868
1.48	.99588	10.983	.09105	.09067	1.98	.91744	− 2.3058	−.43368	−.39788
1.49	.99674	12.350	.08097	.08071	1.99	.91341	− 2.2441	−.44562	−.40703
1.50	.99749	14.101	.07091	.07074	**2.00**	.90930	− 2.1850	−.45766	−.41615
Rad.	Sin	Tan	Cot	Cos	Rad.	Sin	Tan	Cot	Cos

TABLE 3 Four-Place Mantissas for Common Logarithms

N	0	1	2	3	4	5	6	7	8	9	Proportional Parts								
											1	2	3	4	5	6	7	8	9
10	0000	0043	0086	0128	0170	0212	0253	0294	0334	0374	*4	8	12	17	21	25	29	33	37
11	0414	0453	0492	0531	0569	0607	0645	0682	0719	0755	4	8	11	15	19	23	26	30	34
12	0792	0828	0864	0899	0934	0969	1004	1038	1072	1106	3	7	10	14	17	21	24	28	31
13	1139	1173	1206	1239	1271	1303	1335	1367	1399	1430	3	6	10	13	16	19	23	26	29
14	1461	1492	1523	1553	1584	1614	1644	1673	1703	1732	3	6	9	12	15	18	21	24	27
15	1761	1790	1818	1847	1875	1903	1931	1959	1987	2014	*3	6	8	11	14	17	20	22	25
16	2041	2068	2095	2122	2148	2175	2201	2227	2253	2279	3	5	8	11	13	16	18	21	24
17	2304	2330	2355	2380	2405	2430	2455	2480	2504	2529	2	5	7	10	12	15	17	20	22
18	2553	2577	2601	2625	2648	2672	2695	2718	2742	2765	2	5	7	9	12	14	16	19	21
19	2788	2810	2833	2856	2878	2900	2923	2945	2967	2989	2	4	7	9	11	13	16	18	20
20	3010	3032	3054	3075	3096	3118	3139	3160	3181	3201	2	4	6	8	11	13	15	17	19
21	3222	3243	3263	3284	3304	3324	3345	3365	3385	3404	2	4	6	8	10	12	14	16	18
22	3424	3444	3464	3483	3502	3522	3541	3560	3579	3598	2	4	6	8	10	12	14	15	17
23	3617	3636	3655	3674	3692	3711	3729	3747	3766	3784	2	4	6	7	9	11	13	15	17
24	3802	3820	3838	3856	3874	3892	3909	3927	3945	3962	2	4	5	7	9	11	12	14	16
25	3979	3997	4014	4031	4048	4065	4082	4099	4116	4133	2	3	5	7	9	10	12	14	15
26	4150	4166	4183	4200	4216	4232	4249	4265	4281	4298	2	3	5	7	8	10	11	13	15
27	4314	4330	4346	4362	4378	4393	4409	4425	4440	4456	2	3	5	6	8	9	11	13	14
28	4472	4487	4502	4518	4533	4548	4564	4579	4594	4609	2	3	5	6	8	9	11	12	14
29	4624	4639	4654	4669	4683	4698	4713	4728	4742	4757	1	3	4	6	7	9	10	12	13
30	4771	4786	4800	4814	4829	4843	4857	4871	4886	4900	1	3	4	6	7	9	10	11	13
31	4914	4928	4942	4955	4969	4983	4997	5011	5024	5038	1	3	4	6	7	8	10	11	12
32	5051	5065	5079	5092	5105	5119	5132	5145	5159	5172	1	3	4	5	7	8	9	11	12
33	5185	5198	5211	5224	5237	5250	5263	5276	5289	5302	1	3	4	5	6	8	9	10	12
34	5315	5328	5340	5353	5366	5378	5391	5403	5416	5428	1	3	4	5	6	8	9	10	11
35	5441	5453	5465	5478	5490	5502	5514	5527	5539	5551	1	2	4	5	6	7	9	10	11
36	5563	5575	5587	5599	5611	5623	5635	5647	5658	5670	1	2	4	5	6	7	8	10	11
37	5682	5694	5705	5717	5729	5740	5752	5763	5775	5786	1	2	3	5	6	7	8	9	10
38	5798	5809	5821	5832	5843	5855	5866	5877	5888	5899	1	2	3	5	6	7	8	9	10
39	5911	5922	5933	5944	5955	5966	5977	5988	5999	6010	1	2	3	4	5	7	8	9	10
40	6021	6031	6042	6053	6064	6075	6085	6096	6107	6117	1	2	3	4	5	6	8	9	10
41	6128	6138	6149	6160	6170	6180	6191	6201	6212	6222	1	2	3	4	5	6	7	8	9
42	6232	6243	6253	6263	6274	6284	6294	6304	6314	6325	1	2	3	4	5	6	7	8	9
43	6335	6345	6355	6365	6375	6385	6395	6405	6415	6425	1	2	3	4	5	6	7	8	9
44	6435	6444	6454	6464	6474	6484	6493	6503	6513	6522	1	2	3	4	5	6	7	8	9
45	6532	6542	6551	6561	6571	6580	6590	6599	6609	6618	1	2	3	4	5	6	7	8	9
46	6628	6637	6646	6656	6665	6675	6684	6693	6702	6712	1	2	3	4	5	6	7	7	8
47	6721	6730	6739	6749	6758	6767	6776	6785	6794	6803	1	2	3	4	5	5	6	7	8
48	6812	6821	6830	6839	6848	6857	6866	6875	6884	6893	1	2	3	4	4	5	6	7	8
49	6902	6911	6920	6928	6937	6946	6955	6964	6972	6981	1	2	3	4	4	5	6	7	8
50	6990	6998	7007	7016	7024	7033	7042	7050	7059	7067	1	2	3	3	4	5	6	7	8
51	7076	7084	7093	7101	7110	7118	7126	7135	7143	7152	1	2	3	3	4	5	6	7	8
52	7160	7168	7177	7185	7193	7202	7210	7218	7226	7235	1	2	2	3	4	5	6	7	7
53	7243	7251	7259	7267	7275	7284	7292	7300	7308	7316	1	2	2	3	4	5	6	6	7
54	7324	7332	7340	7348	7356	7364	7372	7380	7388	7396	1	2	2	3	4	5	6	6	7
N	0	1	2	3	4	5	6	7	8	9	1	2	3	4	5	6	7	8	9

* Interpolation in this section of the table is inaccurate.

TABLE 3 Four-Place Mantissas for Common Logarithms (Continued)

N	0	1	2	3	4	5	6	7	8	9	Proportional Parts								
											1	2	3	4	5	6	7	8	9
55	7404	7412	7419	7427	7435	7443	7451	7459	7466	7474	1	2	2	3	4	5	5	6	7
56	7482	7490	7497	7505	7513	7520	7528	7536	7543	7551	1	2	2	3	4	5	5	6	7
57	7559	7566	7574	7582	7589	7597	7604	7612	7619	7627	1	2	2	3	4	5	5	6	7
58	7634	7642	7649	7657	7664	7672	7679	7686	7694	7701	1	1	2	3	4	4	5	6	7
59	7709	7716	7723	7731	7738	7745	7752	7760	7767	7774	1	1	2	3	4	4	5	6	7
60	7782	7789	7796	7803	7810	7818	7825	7832	7839	7846	1	1	2	3	4	4	5	6	6
61	7853	7860	7868	7875	7882	7889	7896	7903	7910	7917	1	1	2	3	4	4	5	6	6
62	7924	7931	7938	7945	7952	7959	7966	7973	7980	7987	1	1	2	3	3	4	5	6	6
63	7993	8000	8007	8014	8021	8028	8035	8041	8048	8055	1	1	2	3	3	4	5	5	6
64	8062	8069	8075	8082	8089	8096	8102	8109	8116	8122	1	1	2	3	3	4	5	5	6
65	8129	8136	8142	8149	8156	8162	8169	8176	8182	8189	1	1	2	3	3	4	5	5	6
66	8195	8202	8209	8215	8222	8228	8235	8241	8248	8254	1	1	2	3	3	4	5	5	6
67	8261	8267	8274	8280	8287	8293	8299	8306	8312	8319	1	1	2	3	3	4	5	5	6
68	8325	8331	8338	8344	8351	8357	8363	8370	8376	8382	1	1	2	3	3	4	4	5	6
69	8388	8395	8401	8407	8414	8420	8426	8432	8439	8445	1	1	2	2	3	4	4	5	6
70	8451	8457	8463	8470	8476	8482	8488	8494	8500	8506	1	1	2	2	3	4	4	5	6
71	8513	8519	8525	8531	8537	8543	8549	8555	8561	8567	1	1	2	2	3	4	4	5	5
72	8573	8579	8585	8591	8597	8603	8609	8615	8621	8627	1	1	2	2	3	4	4	5	5
73	8633	8639	8645	8651	8657	8663	8669	8675	8681	8686	1	1	2	2	3	4	4	5	5
74	8692	8698	8704	8710	8716	8722	8727	8733	8739	8745	1	1	2	2	3	4	4	5	5
75	8751	8756	8762	8768	8774	8779	8785	8791	8797	8802	1	1	2	2	3	3	4	5	5
76	8808	8814	8820	8825	8831	8837	8842	8848	8854	8859	1	1	2	2	3	3	4	5	5
77	8865	8871	8876	8882	8887	8893	8899	8904	8910	8915	1	1	2	2	3	3	4	4	5
78	8921	8927	8932	8938	8943	8949	8954	8960	8965	8971	1	1	2	2	3	3	4	4	5
79	8976	8982	8987	8993	8998	9004	9009	9015	9020	9025	1	1	2	2	3	3	4	4	5
80	9031	9036	9042	9047	9053	9058	9063	9069	9074	9079	1	1	2	2	3	3	4	4	5
81	9085	9090	9096	9101	9106	9112	9117	9122	9128	9133	1	1	2	2	3	3	4	4	5
82	9138	9143	9149	9154	9159	9165	9170	9175	9180	9186	1	1	2	2	3	3	4	4	5
83	9191	9196	9201	9206	9212	9217	9222	9227	9232	9238	1	1	2	2	3	3	4	4	5
84	9243	9248	9253	9258	9263	9269	9274	9279	9284	9289	1	1	2	2	3	3	4	4	5
85	9294	9299	9304	9309	9315	9320	9325	9330	9335	9340	1	1	2	2	3	3	4	4	5
86	9345	9350	9355	9360	9365	9370	9375	9380	9385	9390	1	1	2	2	3	3	4	4	5
87	9395	9400	9405	9410	9415	9420	9425	9430	9435	9440	0	1	1	2	2	3	3	4	4
88	9445	9450	9455	9460	9465	9469	9474	9479	9484	9489	0	1	1	2	2	3	3	4	4
89	9494	9499	9504	9509	9513	9518	9523	9528	9533	9538	0	1	1	2	2	3	3	4	4
90	9542	9547	9552	9557	9562	9566	9571	9576	9581	9586	0	1	1	2	2	3	3	4	4
91	9590	9595	9600	9605	9609	9614	9619	9624	9628	9633	0	1	1	2	2	3	3	4	4
92	9638	9643	9647	9652	9657	9661	9666	9671	9675	9680	0	1	1	2	2	3	3	4	4
93	9685	9689	9694	9699	9703	9708	9713	9717	9722	9727	0	1	1	2	2	3	3	4	4
94	9731	9736	9741	9745	9750	9754	9759	9763	9768	9773	0	1	1	2	2	3	3	4	4
95	9777	9782	9786	9791	9795	9800	9805	9809	9814	9818	0	1	1	2	2	3	3	4	4
96	9823	9827	9832	9836	9841	9845	9850	9854	9859	9863	0	1	1	2	2	3	3	4	4
97	9868	9872	9877	9881	9886	9890	9894	9899	9903	9908	0	1	1	2	2	3	3	4	4
98	9912	9917	9921	9926	9930	9934	9939	9943	9948	9952	0	1	1	2	2	3	3	4	4
99	9956	9961	9965	9969	9974	9978	9983	9987	9991	9996	0	1	1	2	2	3	3	3	4
N	0	1	2	3	4	5	6	7	8	9	1	2	3	4	5	6	7	8	9

TABLE 4 Antilogarithms

	0	1	2	3	4	5	6	7	8	9	Proportional Parts								
											1	2	3	4	5	6	7	8	9
.00	1000	1002	1005	1007	1009	1012	1014	1016	1019	1021	0	0	1	1	1	1	2	2	2
.01	1023	1026	1028	1030	1033	1035	1038	1040	1042	1045	0	0	1	1	1	1	2	2	2
.02	1047	1050	1052	1054	1057	1059	1062	1064	1067	1069	0	0	1	1	1	1	2	2	2
.03	1072	1074	1076	1079	1081	1084	1086	1089	1091	1094	0	0	1	1	1	1	2	2	2
.04	1096	1099	1102	1104	1107	1109	1112	1114	1117	1119	0	1	1	1	1	2	2	2	2
.05	1122	1125	1127	1130	1132	1135	1138	1140	1143	1146	0	1	1	1	1	2	2	2	2
.06	1148	1151	1153	1156	1159	1161	1164	1167	1169	1172	0	1	1	1	1	2	2	2	2
.07	1175	1178	1180	1183	1186	1189	1191	1194	1197	1199	0	1	1	1	1	2	2	2	2
.08	1202	1205	1208	1211	1213	1216	1219	1222	1225	1227	0	1	1	1	1	2	2	2	3
.09	1230	1233	1236	1239	1242	1245	1247	1250	1253	1256	0	1	1	1	1	2	2	2	3
.10	1259	1262	1265	1268	1271	1274	1276	1279	1282	1285	0	1	1	1	1	2	2	2	3
.11	1288	1291	1294	1297	1300	1303	1306	1309	1312	1315	0	1	1	1	2	2	2	3	3
.12	1318	1321	1324	1327	1330	1334	1337	1340	1343	1346	0	1	1	1	2	2	2	2	3
.13	1349	1352	1355	1358	1361	1365	1368	1371	1374	1377	0	1	1	1	2	2	2	3	3
.14	1380	1384	1387	1390	1393	1396	1400	1403	1406	1409	0	1	1	1	2	2	2	3	3
.15	1413	1416	1419	1422	1426	1429	1432	1435	1439	1442	0	1	1	1	2	2	2	3	3
.16	1445	1449	1452	1455	1459	1462	1466	1469	1472	1476	0	1	1	1	2	2	2	3	3
.17	1479	1483	1486	1489	1493	1496	1500	1503	1507	1510	0	1	1	1	2	2	2	3	3
.18	1514	1517	1521	1524	1528	1531	1535	1538	1542	1545	0	1	1	1	2	2	2	3	3
.19	1549	1552	1556	1560	1563	1567	1570	1574	1578	1581	0	1	1	1	2	2	3	3	3
.20	1585	1589	1592	1596	1600	1603	1607	1611	1614	1618	0	1	1	1	2	2	3	3	3
.21	1622	1626	1629	1633	1637	1641	1644	1648	1652	1656	0	1	1	2	2	2	3	3	3
.22	1660	1663	1667	1671	1675	1679	1683	1687	1690	1694	0	1	1	2	2	2	3	3	3
.23	1698	1702	1706	1710	1714	1718	1722	1726	1730	1734	0	1	1	2	2	2	3	3	4
.24	1738	1742	1746	1750	1754	1758	1762	1766	1770	1774	0	1	1	2	2	2	3	3	4
.25	1778	1782	1786	1791	1795	1799	1803	1807	1811	1816	0	1	1	2	2	2	3	3	4
.26	1820	1824	1828	1832	1837	1841	1845	1849	1854	1858	0	1	1	2	2	3	3	3	4
.27	1862	1866	1871	1875	1879	1884	1888	1892	1897	1901	0	1	1	2	2	3	3	3	4
.28	1905	1910	1914	1919	1923	1928	1932	1936	1941	1945	0	1	1	2	2	3	3	4	4
.29	1950	1954	1959	1963	1968	1972	1977	1982	1986	1991	0	1	1	2	2	3	3	4	4
.30	1995	2000	2004	2009	2014	2018	2023	2028	2032	2037	0	1	1	2	2	3	3	4	4
.31	2042	2046	2051	2056	2061	2065	2070	2075	2080	2084	0	1	1	2	2	3	3	4	4
.32	2089	2094	2099	2104	2109	2113	2118	2123	2128	2133	0	1	1	2	2	3	3	4	4
.33	2138	2143	2148	2153	2158	2163	2168	2173	2178	2183	0	1	1	2	2	3	3	4	4
.34	2188	2193	2198	2203	2208	2213	2218	2223	2228	2234	1	1	2	2	3	3	4	4	5
.35	2239	2244	2249	2254	2259	2265	2270	2275	2280	2286	1	1	2	2	3	3	4	4	5
.36	2291	2296	2301	2307	2312	2317	2323	2328	2333	2339	1	1	2	2	3	3	4	4	5
.37	2344	2350	2355	2360	2366	2371	2377	2382	2388	2393	1	1	2	2	3	3	4	4	5
.38	2399	2404	2410	2415	2421	2427	2432	2438	2443	2449	1	1	2	2	3	3	4	4	5
.39	2455	2460	2466	2472	2477	2483	2489	2495	2500	2506	1	1	2	2	3	3	4	5	5
.40	2512	2518	2523	2529	2535	2541	2547	2553	2559	2564	1	1	2	2	3	4	4	5	5
.41	2570	2576	2582	2588	2594	2600	2606	2612	2618	2624	1	1	2	2	3	4	4	5	5
.42	2630	2636	2642	2649	2655	2661	2667	2673	2679	2685	1	1	2	2	3	4	4	5	6
.43	2692	2698	2704	2710	2716	2723	2729	2735	2742	2748	1	1	2	3	3	4	4	5	6
.44	2754	2761	2767	2773	2780	2786	2793	2799	2805	2812	1	1	2	3	3	4	4	5	6
.45	2818	2825	2831	2838	2844	2851	2858	2864	2871	2877	1	1	2	3	3	4	5	5	6
.46	2884	2891	2897	2904	2911	2917	2924	2931	2938	2944	1	1	2	3	3	4	5	5	6
.47	2951	2958	2965	2972	2979	2985	2992	2999	3006	3013	1	1	2	3	3	4	5	5	6
.48	3020	3027	3034	3041	3048	3055	3062	3069	3076	3083	1	1	2	3	4	4	5	6	6
.49	3090	3097	3105	3112	3119	3126	3133	3141	3148	3155	1	1	2	3	4	4	5	6	6
	0	1	2	3	4	5	6	7	8	9	1	2	3	4	5	6	7	8	9

TABLE 4 Antilogarithms (Continued)

	0	1	2	3	4	5	6	7	8	9	Proportional Parts 1	2	3	4	5	6	7	8	9
.50	3162	3170	3177	3184	3192	3199	3206	3214	3221	3228	1	1	2	3	4	4	5	6	7
.51	3236	3243	3251	3258	3266	3273	3281	3289	3296	3304	1	2	2	3	4	5	5	6	7
.52	3311	3319	3327	3334	3342	3350	3357	3365	3373	3381	1	2	2	3	4	5	5	6	7
.53	3388	3396	3404	3412	3420	3428	3436	3443	3451	3459	1	2	2	3	4	5	6	6	7
.54	3467	3475	3483	3491	3499	3508	3516	3524	3532	3540	1	2	2	3	4	5	6	6	7
.55	3548	3556	3565	3573	3581	3589	3597	3606	3614	3622	1	2	2	3	4	5	6	7	7
.56	3631	3639	3648	3656	3664	3673	3681	3690	3698	3707	1	2	3	3	4	5	6	7	8
.57	3715	3724	3733	3741	3750	3758	3767	3776	3784	3793	1	2	3	3	4	5	6	7	8
.58	3802	3811	3819	3828	3837	3846	3855	3864	3873	3882	1	2	3	4	4	5	6	7	8
.59	3890	3899	3908	3917	3926	3936	3945	3954	3963	3972	1	2	3	4	5	5	6	7	8
.60	3981	3990	3999	4009	4018	4027	4036	4046	4055	4064	1	2	3	4	5	6	6	7	8
.61	4074	4083	4093	4102	4111	4121	4130	4140	4150	4159	1	2	3	4	5	6	7	8	9
.62	4169	4178	4188	4198	4207	4217	4227	4236	4246	4256	1	2	3	4	5	6	7	8	9
.63	4266	4276	4285	4295	4305	4315	4325	4335	4345	4355	1	2	3	4	5	6	7	8	9
.64	4365	4375	4385	4395	4406	4416	4426	4436	4446	4457	1	2	3	4	5	6	7	8	9
.65	4467	4477	4487	4498	4508	4519	4529	4539	4550	4560	1	2	3	4	5	6	7	8	9
.66	4571	4581	4592	4603	4613	4624	4634	4645	4656	4667	1	2	3	4	5	6	7	9	10
.67	4677	4688	4699	4710	4721	4732	4742	4753	4764	4775	1	2	3	4	5	7	8	9	10
.68	4786	4797	4808	4819	4831	4842	4853	4864	4875	4887	1	2	3	4	6	7	8	9	10
.69	4898	4909	4920	4932	4943	4955	4966	4977	4989	5000	1	2	3	5	6	7	8	9	10
.70	5012	5023	5035	5047	5058	5070	5082	5093	5105	5117	1	2	4	5	6	7	8	9	11
.71	5129	5140	5152	5164	5176	5188	5200	5212	5224	5236	1	2	4	5	6	7	8	10	11
.72	5248	5260	5272	5284	5297	5309	5321	5333	5346	5358	1	2	4	5	6	7	9	10	11
.73	5370	5383	5395	5408	5420	5433	5445	5458	5470	5483	1	3	4	5	6	8	9	10	11
.74	5495	5508	5521	5534	5546	5559	5572	5585	5598	5610	1	3	4	5	6	8	9	10	12
.75	5623	5636	5649	5662	5675	5689	5702	5715	5728	5741	1	3	4	5	7	8	9	10	12
.76	5754	5768	5781	5794	5808	5821	5834	5848	5861	5875	1	3	4	5	7	8	9	11	12
.77	5888	5902	5916	5929	5943	5957	5970	5984	5998	6012	1	3	4	5	7	8	10	11	12
.78	6026	6039	6053	6067	6081	6095	6109	6124	6138	6152	1	3	4	6	7	8	10	11	13
.79	6166	6180	6194	6209	6223	6237	6252	6266	6281	6295	1	3	4	6	7	9	10	11	13
.80	6310	6324	6339	6353	6368	6383	6397	6412	6427	6442	1	3	4	6	7	9	10	12	13
.81	6457	6471	6486	6501	6516	6531	6546	6561	6577	6592	2	3	5	6	8	9	11	12	14
.82	6607	6622	6637	6653	6668	6683	6699	6714	6730	6745	2	3	5	6	8	9	11	12	14
.83	6761	6776	6792	6808	6823	6839	6855	6871	6887	6902	2	3	5	6	8	9	11	13	14
.84	6918	6934	6950	6966	6982	6998	7015	7031	7047	7063	2	3	5	6	8	10	11	13	15
.85	7079	7096	7112	7129	7145	7161	7178	7194	7211	7228	2	3	5	7	8	10	12	13	15
.86	7244	7261	7278	7295	7311	7328	7345	7362	7379	7396	2	3	5	7	8	10	12	13	15
.87	7413	7430	7447	7464	7482	7499	7516	7534	7551	7568	2	3	5	7	9	10	12	14	16
.88	7586	7603	7621	7638	7656	7674	7691	7709	7727	7745	2	4	5	7	9	11	12	14	16
.89	7762	7780	7798	7816	7834	7852	7870	7889	7907	7925	2	4	5	7	9	11	13	14	16
.90	7943	7962	7980	7998	8017	8035	8054	8072	8091	8110	2	4	6	7	9	11	13	15	17
.91	8128	8147	8166	8185	8204	8222	8241	8260	8279	8299	2	4	6	8	9	11	13	15	17
.92	8318	8337	8356	8375	8395	8414	8433	8453	8472	8492	2	4	6	8	10	12	14	15	17
.93	8511	8531	8551	8570	8590	8610	8630	8650	8670	8690	2	4	6	8	10	12	14	16	18
.94	8710	8730	8750	8770	8790	8810	8831	8851	8872	8892	2	4	6	8	10	12	14	16	18
.95	8913	8933	8954	8974	8995	9016	9036	9057	9078	9099	2	4	6	8	10	12	15	17	19
.96	9120	9141	9162	9183	9204	9226	9247	9268	9290	9311	2	4	6	8	11	13	15	17	19
.97	9333	9354	9376	9397	9419	9441	9462	9484	9506	9528	2	4	7	9	11	13	15	17	20
.98	9550	9572	9594	9616	9638	9661	9683	9705	9727	9750	2	4	7	9	11	13	16	18	20
.99	9772	9795	9817	9840	9863	9886	9908	9931	9954	9977	2	5	7	9	11	14	16	18	20
	0	1	2	3	4	5	6	7	8	9	1	2	3	4	5	6	7	8	9

TABLE 5 Exponential Functions

x	e^x	$\text{Log}_{10}(e^x)$	e^{-x}
0.00	1.0000	0.00000	1.000000
0.01	1.0101	.00434	0.990050
0.02	1.0202	.00869	.980199
0.03	1.0305	.01303	.970446
0.04	1.0408	.01737	.960789
0.05	1.0513	0.02171	0.951229
0.06	1.0618	.02606	.941765
0.07	1.0725	.03040	.932394
0.08	1.0833	.03474	.923116
0.09	1.0942	.03909	.913931
0.10	1.1052	0.04343	0.904837
0.11	1.1163	.04777	.895834
0.12	1.1275	.05212	.886920
0.13	1.1388	.05646	.878095
0.14	1.1503	.06080	.869358
0.15	1.1618	0.06514	0.860708
0.16	1.1735	.06949	.852144
0.17	1.1853	.07383	.843665
0.18	1.1972	.07817	.835270
0.19	1.2092	.08252	.826959
0.20	1.2214	0.08686	0.818731
0.21	1.2337	.09120	.810584
0.22	1.2461	.09554	.802519
0.23	1.2586	.09989	.794534
0.24	1.2712	.10423	.786628
0.25	1.2840	0.10857	0.778801
0.26	1.2969	.11292	.771052
0.27	1.3100	.11726	.763379
0.28	1.3231	.12160	.755784
0.29	1.3364	.12595	.748264
0.30	1.3499	0.13029	0.740818
0.31	1.3634	.13463	.733447
0.32	1.3771	.13897	.726149
0.33	1.3910	.14332	.718924
0.34	1.4049	.14766	.711770
0.35	1.4191	0.15200	0.704688
0.36	1.4333	.15635	.697676
0.37	1.4477	.16069	.690734
0.38	1.4623	.16503	.683861
0.39	1.4770	.16937	.677057
0.40	1.4918	0.17372	0.670320
0.41	1.5068	.17806	.663650
0.42	1.5220	.18240	.657047
0.43	1.5373	.18675	.650509
0.44	1.5527	.19109	.644036
0.45	1.5683	0.19543	0.637628
0.46	1.5841	.19978	.631284
0.47	1.6000	.20412	.625002
0.48	1.6161	.20846	.618783
0.49	1.6323	.21280	.612626
0.50	1.6487	0.21715	0.606531

x	e^x	$\text{Log}_{10}(e^x)$	e^{-x}
0.50	1.6487	0.21715	0.606531
0.51	1.6653	.22149	.600496
0.52	1.6820	.22583	.594521
0.53	1.6989	.23018	.588605
0.54	1.7160	.23452	.582748
0.55	1.7333	0.23886	0.576950
0.56	1.7507	.24320	.571209
0.57	1.7683	.24755	.565525
0.58	1.7860	.25189	.559898
0.59	1.8040	.25623	.554327
0.60	1.8221	0.26058	0.548812
0.61	1.8404	.26492	.543351
0.62	1.8589	.26926	.537944
0.63	1.8776	.27361	.532592
0.64	1.8965	.27795	.527292
0.65	1.9155	0.28229	0.522046
0.66	1.9348	.28663	.516851
0.67	1.9542	.29098	.511709
0.68	1.9739	.29532	.506617
0.69	1.9937	.29966	.501576
0.70	2.0138	0.30401	0.496585
0.71	2.0340	.30835	.491644
0.72	2.0544	.31269	.486752
0.73	2.0751	.31703	.481909
0.74	2.0959	.32138	.477114
0.75	2.1170	0.32572	0.472367
0.76	2.1383	.33006	.467666
0.77	2.1598	.33441	.463013
0.78	2.1815	.33875	.458406
0.79	2.2034	.34309	.453845
0.80	2.2255	0.34744	0.449329
0.81	2.2479	.35178	.444858
0.82	2.2705	.35612	.440432
0.83	2.2933	.36046	.436049
0.84	2.3164	.36481	.431711
0.85	2.3396	0.36915	0.427415
0.86	2.3632	.37349	.423162
0.87	2.3869	.37784	.418952
0.88	2.4109	.38218	.414783
0.89	2.4351	.38652	.410656
0.90	2.4596	0.39087	0.406570
0.91	2.4843	.39521	.402524
0.92	2.5093	.39955	.398519
0.93	2.5345	.40389	.394554
0.94	2.5600	.40824	.390628
0.95	2.5857	0.41258	0.386741
0.96	2.6117	.41692	.382893
0.97	2.6379	.42127	.379083
0.98	2.6645	.42561	.375311
0.99	2.6912	.42995	.371577
1.00	2.7183	0.43429	0.367879

TABLE 5 Exponential Functions (Continued)

x	e^x	$\text{Log}_{10}(e^x)$	e^{-x}
1.00	2.7183	0.43429	0.367879
1.01	2.7456	.43864	.364219
1.02	2.7732	.44298	.360595
1.03	2.8011	.44732	.357007
1.04	2.8292	.45167	.353455
1.05	2.8577	0.45601	0.349938
1.06	2.8864	.46035	.346456
1.07	2.9154	.46470	.343009
1.08	2.9447	.46904	.339596
1.09	2.9743	.47338	.336216
1.10	3.0042	0.47772	0.332871
1.11	3.0344	.48207	.329559
1.12	3.0649	.48641	.326280
1.13	3.0957	.49075	.323033
1.14	3.1268	.49510	.319819
1.15	3.1582	0.49944	0.316637
1.16	3.1899	.50378	.313486
1.17	3.2220	.50812	.310367
1.18	3.2544	.51247	.307279
1.19	3.2871	.51681	.304221
1.20	3.3201	0.52115	0.301194
1.21	3.3535	.52550	.298197
1.22	3.3872	.52984	.295230
1.23	3.4212	.53418	.292293
1.24	3.4556	.53853	.289384
1.25	3.4903	0.54287	0.286505
1.26	3.5254	.54721	.283654
1.27	3.5609	.55155	.280832
1.28	3.5966	.55590	.278037
1.29	3.6328	.56024	.275271
1.30	3.6693	0.56458	0.272532
1.31	3.7062	.56893	.269820
1.32	3.7434	.57327	.267135
1.33	3.7810	.57761	.264477
1.34	3.8190	.58195	.261846
1.35	3.8574	0.58630	0.259240
1.36	3.8962	.59064	.256661
1.37	3.9354	.59498	.254107
1.38	3.9749	.59933	.251579
1.39	4.0149	.60367	.249075
1.40	4.0552	0.60801	0.246597
1.41	4.0960	.61236	.244143
1.42	4.1371	.61670	.241714
1.43	4.1787	.62104	.239309
1.44	4.2207	.62538	.236928
1.45	4.2631	0.62973	0.234570
1.46	4.3060	.63407	.232236
1.47	4.3492	.63841	.229925
1.48	4.3929	.64276	.227638
1.49	4.4371	.64710	.225373
1.50	4.4817	0.65144	0.223130

x	e^x	$\text{Log}_{10}(e^x)$	e^{-x}
1.50	4.4817	0.65144	0.223130
1.51	4.5267	.65578	.220910
1.52	4.5722	.66013	.218712
1.53	4.6182	.66447	.216536
1.54	4.6646	.66881	.214381
1.55	4.7115	0.67316	0.212248
1.56	4.7588	.67750	.210136
1.57	4.8066	.68184	.208045
1.58	4.8550	.68619	.205975
1.59	4.9037	.69053	.203926
1.60	4.9530	0.69487	0.201897
1.61	5.0028	.69921	.199888
1.62	5.0531	.70356	.197899
1.63	5.1039	.70790	.195930
1.64	5.1552	.71224	.193980
1.65	5.2070	0.71659	0.192050
1.66	5.2593	.72093	.190139
1.67	5.3122	.72527	.188247
1.68	5.3656	.72961	.186374
1.69	5.4195	.73396	.184520
1.70	5.4739	0.73830	0.182684
1.71	5.5290	.74264	.180866
1.72	5.5845	.74699	.179066
1.73	5.6407	.75133	.177284
1.74	5.6973	.75567	.175520
1.75	5.7546	0.76002	0.173774
1.76	5.8124	.76436	.172045
1.77	5.8709	.76870	.170333
1.78	5.9299	.77304	.168638
1.79	5.9895	.77739	.166960
1.80	6.0496	0.78173	0.165299
1.81	6.1104	.78607	.163654
1.82	6.1719	.79042	.162026
1.83	6.2339	.79476	.160414
1.84	6.2965	.79910	.158817
1.85	6.3598	0.80344	0.157237
1.86	6.4237	.80779	.155673
1.87	6.4883	.81213	.154124
1.88	6.5535	.81647	.152590
1.89	6.6194	.82082	.151072
1.90	6.6859	0.82516	0.149569
1.91	6.7531	.82950	.148080
1.92	6.8210	.83385	.146607
1.93	6.8895	.83819	.145148
1.94	6.9588	.84253	.143704
1.95	7.0287	0.84687	0.142274
1.96	7.0993	.85122	.140858
1.97	7.1707	.85556	.139457
1.98	7.2427	.85990	.138069
1.99	7.3155	.86425	.136695
2.00	7.3891	0.86859	0.135335

TABLE 5 Exponential Functions (Continued)

x	e^x	$\text{Log}_{10}(e^x)$	e^{-x}
2.00	7.3891	0.86859	0.135335
2.01	7.4633	.87293	.133989
2.02	7.5383	.87727	.132655
2.03	7.6141	.88162	.131336
2.04	7.6906	.88596	.130029
2.05	7.7679	0.89030	0.128735
2.06	7.8460	.89465	.127454
2.07	7.9248	.89899	.126186
2.08	8.0045	.90333	.124930
2.09	8.0849	.90768	.123687
2.10	8.1662	0.91202	0.122456
2.11	8.2482	.91636	.121238
2.12	8.3311	.92070	.120032
2.13	8.4149	.92505	.118837
2.14	8.4994	.92939	.117655
2.15	8.5849	0.93373	0.116484
2.16	8.6711	.93808	.115325
2.17	8.7583	.94242	.114178
2.18	8.8463	.94676	.113042
2.19	8.9352	.95110	.111917
2.20	9.0250	0.95545	0.110803
2.21	9.1157	.95979	.109701
2.22	9.2073	.96413	.108609
2.23	9.2999	.96848	.107528
2.24	9.3933	.97282	.106459
2.25	9.4877	0.97716	0.105399
2.26	9.5831	.98151	.104350
2.27	9.6794	.98585	.103312
2.28	9.7767	.99019	.102284
2.29	9.8749	.99453	.101266
2.30	9.9742	0.99888	0.100259
2.31	10.074	1.00322	.099261
2.32	10.176	1.00756	.098274
2.33	10.278	1.01191	.097296
2.34	10.381	1.01625	.096328
2.35	10.486	1.02059	0.095369
2.36	10.591	1.02493	.094420
2.37	10.697	1.02928	.093481
2.38	10.805	1.03362	.092551
2.39	10.913	1.03796	.091630
2.40	11.023	1.04231	0.090718
2.41	11.134	1.04665	.089815
2.42	11.246	1.05099	.088922
2.43	11.359	1.05534	.088037
2.44	11.473	1.05968	.087161
2.45	11.588	1.06402	0.086294
2.46	11.705	1.06836	.085435
2.47	11.822	1.07271	.084585
2.48	11.941	1.07705	.083743
2.49	12.061	1.08139	.082910
2.50	12.182	1.08574	0.082085

x	e^x	$\text{Log}_{10}(e^x)$	e^{-x}
2.50	12.182	1.08574	0.082085
2.51	12.305	1.09008	.081268
2.52	12.429	1.09442	.080460
2.53	12.554	1.09877	.079659
2.54	12.680	1.10311	.078866
2.55	12.807	1.10745	0.078082
2.56	12.936	1.11179	.077305
2.57	13.066	1.11614	.076536
2.58	13.197	1.12048	.075774
2.59	13.330	1.12482	.075020
2.60	13.464	1.12917	0.074274
2.61	13.599	1.13351	.073535
2.62	13.736	1.13785	.072803
2.63	13.874	1.14219	.072078
2.64	14.013	1.14654	.071361
2.65	14.154	1.15088	0.070651
2.66	14.296	1.15522	.069948
2.67	14.440	1.15957	.069252
2.68	14.585	1.16391	.068563
2.69	14.732	1.16825	.067881
2.70	14.880	1.17260	0.067206
2.71	15.029	1.17694	.066537
2.72	15.180	1.18128	.065875
2.73	15.333	1.18562	.065219
2.74	15.487	1.18997	.064570
2.75	15.643	1.19431	0.063928
2.76	15.800	1.19865	.063292
2.77	15.959	1.20300	.062662
2.78	16.119	1.20734	.062039
2.79	16.281	1.21168	.061421
2.80	16.445	1.21602	0.060810
2.81	16.610	1.22037	.060205
2.82	16.777	1.22471	.059606
2.83	16.945	1.22905	.059013
2.84	17.116	1.23340	.058426
2.85	17.288	1.23774	0.057844
2.86	17.462	1.24208	.057269
2.87	17.637	1.24643	.056699
2.88	17.814	1.25077	.056135
2.89	17.993	1.25511	.055576
2.90	18.174	1.25945	0.055023
2.91	18.357	1.26380	.054476
2.92	18.541	1.26814	.053934
2.93	18.728	1.27248	.053397
2.94	18.916	1.27683	.052866
2.95	19.106	1.28117	0.052340
2.96	19.298	1.28551	.051819
2.97	19.492	1.28985	.051303
2.98	19.688	1.29420	.050793
2.99	19.886	1.29854	.050287
3.00	20.086	1.30288	0.049787

TABLE 5 Exponential Functions (Continued)

x	e^x	$\text{Log}_{10}(e^x)$	e^{-x}
3.00	20.086	1.30288	0.049787
3.01	20.287	1.30723	.049292
3.02	20.491	1.31157	.048801
3.03	20.697	1.31591	.048316
3.04	20.905	1.32026	.047835
3.05	21.115	1.32460	0.047359
3.06	21.328	1.32894	.046888
3.07	21.542	1.33328	.046421
3.08	21.758	1.33763	.045959
3.09	21.977	1.34197	.045502
3.10	22.198	1.34631	0.045049
3.11	22.421	1.35066	.044601
3.12	22.646	1.35500	.044157
3.13	22.874	1.35934	.043718
3.14	23.104	1.36368	.043283
3.15	23.336	1.36803	0.042852
3.16	23.571	1.37237	.042426
3.17	23.807	1.37671	.042004
3.18	24.047	1.38106	.041586
3.19	24.288	1.38540	.041172
3.20	24.533	1.38974	0.040762
3.21	24.779	1.39409	.040357
3.22	25.028	1.39843	.039955
3.23	25.280	1.40277	.039557
3.24	25.534	1.40711	.039164
3.25	25.790	1.41146	0.038774
3.26	26.050	1.41580	.038388
3.27	26.311	1.42014	.038006
3.28	26.576	1.42449	.037628
3.29	26.843	1.42883	.037254
3.30	27.113	1.43317	0.036883
3.31	27.385	1.43751	.036516
3.32	27.660	1.44186	.036153
3.33	27.938	1.44620	.035793
3.34	28.219	1.45054	.035437
3.35	28.503	1.45489	0.035084
3.36	28.789	1.45923	.034735
3.37	29.079	1.46357	.034390
3.38	29.371	1.46792	.034047
3.39	29.666	1.47226	.033709
3.40	29.964	1.47660	0.033373
3.41	30.265	1.48094	.033041
3.42	30.569	1.48529	.032712
3.43	30.877	1.48963	.032387
3.44	31.187	1.49397	.032065
3.45	31.500	1.49832	0.031746
3.46	31.817	1.50266	.031430
3.47	32.137	1.50700	.031117
3.48	32.460	1.51134	.030807
3.49	32.786	1.51569	.030501
3.50	33.115	1.52003	0.030197

x	e^x	$\text{Log}_{10}(e^x)$	e^{-x}
3.50	33.115	1.52003	0.030197
3.51	33.448	1.52437	.029897
3.52	33.784	1.52872	.029599
3.53	34.124	1.53306	.029305
3.54	34.467	1.53740	.029013
3.55	34.813	1.54175	0.028725
3.56	35.163	1.54609	.028439
3.57	35.517	1.55043	.028156
3.58	35.874	1.55477	.027876
3.59	36.234	1.55912	.027598
3.60	36.598	1.56346	0.027324
3.61	36.966	1.56780	.027052
3.62	37.338	1.57215	.026783
3.63	37.713	1.57649	.026516
3.64	38.092	1.58083	.026252
3.65	38.475	1.58517	0.025991
3.66	38.861	1.58952	.025733
3.67	39.252	1.59386	.025476
3.68	39.646	1.59820	.025223
3.69	40.045	1.60255	.024972
3.70	40.447	1.60689	0.024724
3.71	40.854	1.61123	.024478
3.72	41.264	1.61558	.024234
3.73	41.679	1.61992	.023993
3.74	42.098	1.62426	.023754
3.75	42.521	1.62860	0.023518
3.76	42.948	1.63295	.023284
3.77	43.380	1.63729	.023052
3.78	43.816	1.64163	.022823
3.79	44.256	1.64598	.022596
3.80	44.701	1.65032	0.022371
3.81	45.150	1.65466	.022148
3.82	45.604	1.65900	.021928
3.83	46.063	1.66335	.021710
3.84	46.525	1.66769	.021494
3.85	46.993	1.67203	0.021280
3.86	47.465	1.67638	.021068
3.87	47.942	1.68072	.020858
3.88	48.424	1.68506	.020651
3.89	48.911	1.68941	.020445
3.90	49.402	1.69375	0.020242
3.91	49.899	1.69809	.020041
3.92	50.400	1.70243	.019841
3.93	50.907	1.70678	.019644
3.94	51.419	1.71112	.019448
3.95	51.935	1.71546	0.019255
3.96	52.457	1.71981	.019063
3.97	52.985	1.72415	.018873
3.98	53.517	1.72849	.018686
3.99	54.055	1.73283	.018500
4.00	54.598	1.73718	0.018316

TABLE 5 Exponential Functions (Continued)

x	e^x	$\text{Log}_{10}(e^x)$	e^{-x}
4.00	54.598	1.73718	0.018316
4.01	55.147	1.74152	.018133
4.02	55.701	1.74586	.017953
4.03	56.261	1.75021	.017774
4.04	56.826	1.75455	.017597
4.05	57.397	1.75889	0.017422
4.06	57.974	1.76324	.017249
4.07	58.557	1.76758	.017077
4.08	59.145	1.77192	.016907
4.09	59.740	1.77626	.016739
4.10	60.340	1.78061	0.016573
4.11	60.947	1.78495	.016408
4.12	61.559	1.78929	.016245
4.13	62.178	1.79364	.016083
4.14	62.803	1.79798	.015923
4.15	63.434	1.80232	0.015764
4.16	64.072	1.80667	.015608
4.17	64.715	1.81101	.015452
4.18	65.366	1.81535	.015299
4.19	66.023	1.81969	.015146
4.20	66.686	1.82404	0.014996
4.21	67.357	1.82838	.014846
4.22	68.033	1.83272	.014699
4.23	68.717	1.83707	.014552
4.24	69.408	1.84141	.014408
4.25	70.105	1.84575	0.014264
4.26	70.810	1.85009	.014122
4.27	71.522	1.85444	.013982
4.28	72.240	1.85878	.013843
4.29	72.966	1.86312	.013705
4.30	73.700	1.86747	0.013569
4.31	74.440	1.87181	.013434
4.32	75.189	1.87615	.013300
4.33	75.944	1.88050	.013168
4.34	76.708	1.88484	.013037
4.35	77.478	1.88918	0.012907
4.36	78.257	1.89352	.012778
4.37	79.044	1.89787	.012651
4.38	79.838	1.90221	.012525
4.39	80.640	1.90655	.012401
4.40	81.451	1.91090	0.012277
4.41	82.269	1.91524	.012155
4.42	83.096	1.91958	.012034
4.43	83.931	1.92392	.011914
4.44	84.775	1.92827	.011796
4.45	85.627	1.93261	0.011679
4.46	86.488	1.93695	.011562
4.47	87.357	1.94130	.011447
4.48	88.235	1.94564	.011333
4.49	89.121	1.94998	.011221
4.50	90.017	1.95433	0.011109

x	e^x	$\text{Log}_{10}(e^x)$	e^{-x}
4.50	90.017	1.95433	0.011109
4.51	90.922	1.95867	.010998
4.52	91.836	1.96301	.010889
4.53	92.759	1.96735	.010781
4.54	93.691	1.97170	.010673
4.55	94.632	1.97604	0.010567
4.56	95.583	1.98038	.010462
4.57	96.544	1.98473	.010358
4.58	97.514	1.98907	.010255
4.59	98.494	1.99341	.010153
4.60	99.484	1.99775	0.010052
4.61	100.48	2.00210	.009952
4.62	101.49	2.00644	.009853
4.63	102.51	2.01078	.009755
4.64	103.54	2.01513	.009658
4.65	104.58	2.01947	0.009562
4.66	105.64	2.02381	.009466
4.67	106.70	2.02816	.009372
4.68	107.77	2.03250	.009279
4.69	108.85	2.03684	.009187
4.70	109.95	2.04118	0.009095
4.71	111.05	2.04553	.009005
4.72	112.17	2.04987	.008915
4.73	113.30	2.05421	.008826
4.74	114.43	2.05856	.008739
4.75	115.58	2.06290	0.008652
4.76	116.75	2.06724	.008566
4.77	117.92	2.07158	.008480
4.78	119.10	2.07593	.008396
4.79	120.30	2.08027	.008312
4.80	121.51	2.08461	0.008230
4.81	122.73	2.08896	.008148
4.82	123.97	2.09330	.008067
4.83	125.21	2.09764	.007987
4.84	126.47	2.10199	.007907
4.85	127.74	2.10633	0.007828
4.86	129.02	2.11067	.007750
4.87	130.32	2.11501	.007673
4.88	131.63	2.11936	.007597
4.89	132.95	2.12370	.007521
4.90	134.29	2.12804	0.007447
4.91	135.64	2.13239	.007372
4.92	137.00	2.13673	.007299
4.93	138.38	2.14107	.007227
4.94	139.77	2.14541	.007155
4.95	141.17	2.14976	0.007083
4.96	142.59	2.15410	.007013
4.97	144.03	2.15844	.006943
4.98	145.47	2.16279	.006874
4.99	146.94	2.16713	.006806
5.00	148.41	2.17147	0.006738

TABLE 5 Exponential Functions (Continued)

x	e^x	$\text{Log}_{10}(e^x)$	e^{-x}
5.00	148.41	2.17147	0.006738
5.01	149.90	2.17582	.006671
5.02	151.41	2.18016	.006605
5.03	152.93	2.18450	.006539
5.04	154.47	2.18884	.006474
5.05	156.02	2.19319	0.006409
5.06	157.59	2.19753	.006346
5.07	159.17	2.20187	.006282
5.08	160.77	2.20622	.006220
5.09	162.39	2.21056	.006158
5.10	164.02	2.21490	0.006097
5.11	165.67	2.21924	.006036
5.12	167.34	2.22359	.005976
5.13	169.02	2.22793	.005917
5.14	170.72	2.23227	.005858
5.15	172.43	2.23662	0.005799
5.16	174.16	2.24096	.005742
5.17	175.91	2.24530	.005685
5.18	177.68	2.24965	.005628
5.19	179.47	2.25399	.005572
5.20	181.27	2.25833	0.005517
5.21	183.09	2.26267	.005462
5.22	184.93	2.26702	.005407
5.23	186.79	2.27136	.005354
5.24	188.67	2.27570	.005300
5.25	190.57	2.28005	0.005248
5.26	192.48	2.28439	.005195
5.27	194.42	2.28873	.005144
5.28	196.37	2.29307	.005092
5.29	198.34	2.29742	.005042
5.30	200.34	2.30176	0.004992
5.31	202.35	2.30610	.004942
5.32	204.38	2.31045	.004893
5.33	206.44	2.31479	.004844
5.34	208.51	2.31913	.004796
5.35	210.61	2.32348	0.004748
5.36	212.72	2.32782	.004701
5.37	214.86	2.33216	.004654
5.38	217.02	2.33650	.004608
5.39	219.20	2.34085	.004562
5.40	221.41	2.34519	0.004517
5.41	223.63	2.34953	.004472
5.42	225.88	2.35388	.004427
5.43	228.15	2.35822	.004383
5.44	230.44	2.36256	.004339
5.45	232.76	2.36690	0.004296
5.46	235.10	2.37125	.004254
5.47	237.46	2.37559	.004211
5.48	239.85	2.37993	.004169
5.49	242.26	2.38428	.004128
5.50	244.69	2.38862	0.004087

x	e^x	$\text{Log}_{10}(e^x)$	e^{-x}
5.50	244.69	2.38862	0.0040868
5.55	257.24	2.41033	.0038875
5.60	270.43	2.43205	.0036979
5.65	284.29	2.45376	.0035175
5.70	298.87	2.47548	.0033460
5.75	314.19	2.49719	0.0031828
5.80	330.30	2.51891	.0030276
5.85	347.23	2.54062	.0028799
5.90	365.04	2.56234	.0027394
5.95	383.75	2.58405	.0026058
6.00	403.43	2.60577	0.0024788
6.05	424.11	2.62748	.0023579
6.10	445.86	2.64920	.0022429
6.15	468.72	2.67091	.0021335
6.20	492.75	2.69263	.0020294
6.25	518.01	2.71434	0.0019305
6.30	544.57	2.73606	.0018363
6.35	572.49	2.75777	.0017467
6.40	601.85	2.77948	.0016616
6.45	632.70	2.80120	.0015805
6.50	665.14	2.82291	0.0015034
6.55	699.24	2.84463	.0014301
6.60	735.10	2.86634	.0013604
6.65	772.78	2.88806	.0012940
6.70	812.41	2.90977	.0012309
6.75	854.06	2.93149	0.0011709
6.80	897.85	2.95320	.0011138
6.85	943.88	2.97492	.0010595
6.90	992.27	2.99663	.0010078
6.95	1043.1	3.01835	.0009586
7.00	1096.6	3.04006	0.0009119
7.05	1152.9	3.06178	.0008674
7.10	1212.0	3.08349	.0008251
7.15	1274.1	3.10521	.0007849
7.20	1339.4	3.12692	.0007466
7.25	1408.1	3.14863	0.0007102
7.30	1480.3	3.17035	.0006755
7.35	1556.2	3.19206	.0006426
7.40	1636.0	3.21378	.0006113
7.45	1719.9	3.23549	.0005814
7.50	1808.0	3.25721	0.0005531
7.55	1900.7	3.27892	.0005261
7.60	1998.2	3.30064	.0005005
7.65	2100.6	3.32235	.0004760
7.70	2208.3	3.34407	.0004528
7.75	2321.6	3.36578	0.0004307
7.80	2440.6	3.38750	.0004097
7.85	2565.7	3.40921	.0003898
7.90	2697.3	3.43093	.0003707
7.95	2835.6	3.45264	.0003527
8.00	2981.0	3.47436	0.0003355

TABLE 5 Exponential Functions (Continued)

x	e^x	$\text{Log}_{10}(e^x)$	e^{-x}
8.00	2981.0	3.47436	0.0003355
8.05	3133.8	3.49607	.0003191
8.10	3294.5	3.51779	.0003035
8.15	3463.4	3.53950	.0002887
8.20	3641.0	3.56121	.0002747
8.25	3827.6	3.58293	0.0002613
8.30	4023.9	3.60464	.0002485
8.35	4230.2	3.62636	.0002364
8.40	4447.1	3.64807	.0002249
8.45	4675.1	3.66979	.0002139
8.50	4914.8	3.69150	0.0002035
8.55	5166.8	3.71322	.0001935
8.60	5431.7	3.73493	.0001841
8.65	5710.1	3.75665	.0001751
8.70	6002.9	3.77836	.0001666
8.75	6310.7	3.80008	0.0001585
8.80	6634.2	3.82179	.0001507
8.85	6974.4	3.84351	.0001434
8.90	7332.0	3.86522	.0001364
8.95	7707.9	3.88694	.0001297
9.00	8103.1	3.90865	0.0001234
9.05	8518.5	3.93037	.0001174
9.10	8955.3	3.95208	.0001117
9.15	9414.4	3.97379	.0001062
9.20	9897.1	3.99551	.0001010
9.25	10405	4.01722	0.0000961
9.30	10938	4.03894	.0000914
9.35	11499	4.06065	.0000870
9.40	12088	4.08237	.0000827
9.45	12708	4.10408	.0000787
9.50	13360	4.12580	0.0000749
9.55	14045	4.14751	.0000712
9.60	14765	4.16923	.0000677
9.65	15522	4.19094	.0000644
9.70	16318	4.21266	.0000613
9.75	17154	4.23437	0.0000583
9.80	18034	4.25609	.0000555
9.85	18958	4.27780	.0000527
9.90	19930	4.29952	.0000502
9.95	20952	4.32123	0.0000477
10.00	22026	4.34294	0.0000454

Answers to Selected Exercises

CHAPTER 1

Section 2, page 5

1. $m = \frac{1}{4}$ **3.** $m = -\frac{3}{2}$ **5.** $m = 2$ **7.** $m = \frac{1}{3}$ **9.** $y = x$
11. $y = \frac{3}{2}x - 3$ **13.** $y = x + 3$ **15.** $y = 5x - 17$
17. $m = 1/(5280 \cdot 12) = 1/63360$ **19.** like the line $y = 1$

Section 4, page 11

1. $2x$ **3.** -4 **5.** 1 **7.** 3 **9.** 8 **11.** 6, 14, 22
13. 13 **15.** 6 **17.** -32 **19.** -12, 24, 2
21. -3, -3.5, -3.9, -3.999 **23.** (3, 9) only

Section 5, page 14

1. $3x^2$ **3.** 0 **5.** 0, 27, 27 **7.** 108 **9.** 48, $3a^2$
11. 7, 4.75, 3.31, 3.003001 **13.** 19, 15.25, 12.61, 12.006001
15. $(-2, -8)$, $(2, 8)$ **17.** The line has slope -1; the curve has slope $3x^2 \geq 0$. **19.** $0 < x < \frac{2}{3}$

Section 6, page 16

1. $y' = -1/x^2$ **3.** -1, -1, $-1/a^2$, $-1/a^2$ **5.** $-1/a^2$, $-a^2$
7. $-1/b^2$ **9.** at $x = \pm\sqrt{2}$ **11.** no; $-1/x^2 < 0$, but $3x^2 \geq 0$
13. $-1/110$

Section 7, page 18

1. $y = 4x - 4$ **3.** $y = -x - 2$ **5.** $y = -2x - 1$
7. $y = -\frac{1}{9}x - \frac{2}{3}$ **9.** $y = 10x - 25$ **11.** $y = \pm 8x - 16$
13. $y = 20x - 100$, $(0, -100)$, $(5, 0)$
15. $y = -81x - 18$, $(0, -18)$, $(-\frac{2}{9}, 0)$; $y = -81x + 18$, $(0, 18)$, $(\frac{2}{9}, 0)$
17. 2 **19.** $(0, -9)$ **21.** $y = -9x \pm 6$

Section 8, page 24

1. $3x^2 - 1/x^2$ **3.** $2x + 4$ **5.** $-1/(x + 1)^2$ **7.** $2(t - 8)$
9. $6x$ **11.** $3x^2 - 4x$ **13.** $6x^2 - 10x + 2$ **15.** $-2/(2x - 1)^2$
17. $72(36x - 2)$ **19.** $4\pi r + 2\pi$ **21.** $-(2x + 1)/x^2(x + 1)^2$
23. $(2x - 1)/x^2(x - 1)^2$ **25.** $-(x^2 + 4x + 2)/x^2(x + 1)^2$
27. $y = 9x - 15$ **29.** $y = -12x + 19$ **31.** 17.60 ohms/°K

Section 10, page 31

1. $10x$, 10 **3.** $21x^2 + 8x - 3$, $42x + 8$ **5.** $3P^2$, $6P$
7. $3(x - 2)^2$, $6(x - 2)$ **9.** $18x$, 18 **11.** 5, 17, 6, 6 **13.** 8
15. $-3/(3t - 1)^2$ **17.** 23π **19.** 0 **21.** $0, \frac{2}{3}$
23. $\frac{1}{2}(1 - \sqrt[3]{2})$ **25.** -1 **27.** The tangents are $y = 2ax - a^2$ and $y = 2(a + 1)x - (a + 1)^2$; they intersect at $x = a + \frac{1}{2}$, $y = a^2 + a$. Eliminate a to find the relation $y = x^2 - \frac{1}{4}$. **29.** $(-1, 5)$

CHAPTER 2

Section 2, page 38

1. concave up
3. concave down: $x < -1$; inflection point: $(-1, 0)$; concave up: $x > -1$
5. concave down: $x < 2$; inflection point: $(2, 4)$; concave up: $x > 2$
7. concave up: $x < 0$; inflection point: $(0, -5)$; concave down: $x > 0$

CHAPTER 3

Section 1, page 44

1. 2 **3.** 4 **5.** $\frac{7}{4}$ **7.** 28 ft **9.** 12
11. 4 hr after work starts **13.** $r = \sqrt[3]{26/\pi}$ in., $h = 2r$ in. **15.** 2
17. $32/(8 + 3\pi)$ in.

Section 2, page 51

1. $r = h = \sqrt[3]{V/\pi}$ **3.** $\frac{1}{2}a \times \frac{1}{2}b$
5. radius $= 1/16\pi$ mi $= 330/\pi$ ft, straight edge $= 1/16$ mi $= 330$ ft
7. $(1/2, \sqrt{2}/2)$ **9.** 1 unit **11.** \$2725.00
13. $100\sqrt{24} \approx 490$ mph

CHAPTER 4

Section 1, page 56

1. $1/(x+1)^2$ **3.** $10/(2x+5)^2$ **5.** $-\sin 2x$

Section 2, page 62

1. $f(x) = \frac{1}{2}x + c$ **3.** $y = -\frac{1}{2}x^2 + c$ **5.** $f(x) = \frac{1}{3}x^3 + c$
7. $y = 1/x + c$

Section 3, page 65

1. $ax + c$ **3.** $-\frac{1}{2}x^2 + c$ **5.** $\frac{4}{3}x^3 + \frac{1}{2}x^2 + c$
7. $4/x + c$ **9.** $\frac{1}{2}x^2 + x + c$ **11.** $x^3 + x^2 + x + c$
13. $\frac{2}{3}x^3 + 7/x + c$ **15.** $\frac{1}{3}(x+4)^3 - 1/x + c$
17. $y = \frac{1}{2}x^2 - 5x + 13/2$ **19.** $y = x^3 - 2x^2 - 1/x + 19/2$

CHAPTER 5

Section 1, page 68

1. 900, 914.4, 916 ft/sec
3. 10,500 ft, -80 ft/sec, $-80\sqrt{105} \approx -820$ ft/sec
5. $\frac{3}{4}$ sec, 16 ft/sec up or down (speed, not velocity)
7. (*a*) increasing for $t < 2$ or $t > 4$, decreasing for $2 < t < 4$; (*b*) increasing for $t > 3$, decreasing for $t < 3$; (*c*) 20 ft forward from 0 to 2 sec, 4 ft backwards from 2 to 4 sec, and 20 ft forward from 4 to 6 sec; total 44 ft

Section 2, page 72

1. $y = -8x^2 + 12$ **3.** $y = -16t^2 + 64t$
5. $y = \frac{1}{3}t^3 - \frac{1}{2}t^2 + 0.25t + 1.75$ **7.** 40 ft **9.** 6 ft/sec^2
11. 26,500 ft **13.** $300 + 3600 + 257\frac{1}{7} = 4157\frac{1}{7}$ ft

CHAPTER 6

Section 1, page 75

1. $\pi/3$, $5\pi/6$, $-4\pi/3$, $13\pi/6$, $-5\pi/2$, 5π
3. $\pi/4$, $-\pi$, $3\pi/2$, $7\pi/2$, $-3\pi/4$, $11\pi/4$

5. $90°, -120°, 300°, 540°, -480°, 3000°$

7. $45°, -315°, 225°, -630°, 2340°, 720°$

13. $(\sqrt{2}/2, \sqrt{2}/2), (-\sqrt{2}/2, -\sqrt{2}/2), (-\sqrt{2}/2, \sqrt{2}/2)$

15. $(-1, 0), (-1/2, -\sqrt{3}/2), (\sqrt{3}/2, 1/2)$ **17.** $\pi/4, 3\pi/4$

19. $5\pi/6, 7\pi/6$ **21.** $3\pi/2$ **23.** $\pi/4, 5\pi/4$

Section 3, page 83

1. $\pi \cos \pi t$ **3.** $-ab \sin ax$ **5.** $\frac{1}{2}\pi \cos(\frac{1}{2}\pi\theta)$ **7.** $1 - \sin t$

9. $\dot{x} = \cos(t + t_0)$ **11.** $2t - \sin t$ **13.** $8x + 12 \cos 4x$

15. $-1/t^2 + \sin t$ **17.** $y = \sin t - 1$ **19.** $y = \sin x + 1$

25. $\pi/2 < t < 3\pi/2$ **27.** $(\frac{1}{3}\pi + 2\pi k, \sqrt{3}/2)$
and $(-\frac{1}{3}\pi + 2\pi k, -\sqrt{3}/2)$, where $k = 0, \pm 1, \pm 2, \cdots$

Section 4, page 88

7. starts at $s = 0$ and oscillates with period (time of complete cycle) 2π between $s = -2$ and $s = 2$; also $v(t) = 2 \cos t$ and $a(t) = -2 \sin t$

9. starts at $s = 3$ and oscillates with period 2 between $s = -3$ and $s = 3$; also $v(t) = -3\pi \sin \pi t$ and $a(t) = -3\pi^2 \cos \pi t$

11. starts at $y = 7$ and oscillates with period $2\pi/\omega = 4\pi/\sqrt{g}$ sec between $y = 3$ and 7; also $v(t) = -2\omega \sin \omega t$ and $a(t) = -2\omega^2 \cos \omega t = -\frac{1}{2}g \cos \omega t$

13. The function is even.

Section 5, page 89

1. 0.472 **3.** -9.522 **5.** 0.123 **7.** 3.443

9. before: 0.735; after: 0.734 **11.** 0.855, 0.854 **13.** 1.061, 1.062

15. $-0.006, +0.001$ **17.** before: 20; after: 16 **19.** 16, 15

21. 1, 2

CHAPTER 7

Section 1, page 93

1. $1, 2, \frac{1}{4}, 8$ **3.** $27, \frac{1}{9}, 81, 3$ **5.** $3, \frac{1}{3}, 9, 1/81$ **7.** 37.83, 16.12

9. 3.758, 0.08486 **11.** $1, 125, \frac{1}{8}$ **13.** $4, 1/49, 4$ **17.** $x \geq 1$

19. $-\frac{3}{2} < x < 12$

Section 3, page 103

1. only if $c = 0$ **5.** $3e^{3x}$ **7.** $2e^{2x+1}$ **9.** $\frac{1}{2}(e^x - e^{-x})$
11. $3e^{3t} - 8e^{2t}$ **13.** $12e^{4x+1}$ **15.** $e^x + 2e^{-2x}$
17. $(\log_e 10)10^x$ **19.** $(\log_e 5)5^{x-1}$ **21.** $(4 \log_e 10)10^{4x-1}$
23. $(2 \log_e 10)(10^{2x} - 10^{-2x})$ **35.** $y = \frac{1}{2}e^{2x} + c$ **37.** $y = e^x + 1$
43. $y'' = y$, hence $y^{(58)} = y$

Section 4, page 106

1. $M = M_0 e^{-\lambda t}$, where $\lambda = \log 2/\log e^{3.64} \approx 0.1904$; about 5.77 days
3. $\log 2/\log e \approx 69.315\%$
5. $M = (3 \times 10^6)3^{t/2}$, $t = \log 4/\log 3 \approx 1.26$ hr **7.** $30(\frac{5}{6})^5 \approx 12.1$ in.
9. Let $a^{138.3} = 2$ and $b^{3.64} = 2$. Then $t = [\log(3/2)]/[\log(b/a)] \approx 2.2$ days.

CHAPTER 8

Section 3, page 115

1. 2 **3.** 45/2 **5.** 9 **7.** 72 **9.** 2 **11.** 5 **13.** $e - 1$
15. $e^4 - 5$ **17.** $\frac{2}{3}$ **19.** 1

Section 4, page 120

1. 3/2 **3.** 0 **5.** $\frac{1}{2}$ **7.** 2 **9.** 20/3 **11.** $\frac{7}{2} - e^{-1}$
13. $\frac{1}{6}(b - a)^3$ **15.** 3/2 **17.** 29/6

Section 5, page 127

1. 1 **3.** $\frac{1}{3}$ **5.** 0 **7.** 16/3 **9.** approx: 21/64, exact: $\frac{1}{3}$
11. approx. ≈ 0.64395, exact: $2/\pi \approx 0.63662$ **13.** $39
15. 0.326 in. **17.** 1, 0 **19.** $e - 1, \frac{1}{2}(e + 1)$ **21.** 11/2, 11/2

CHAPTER 9

Section 2, page 134

1. 36 **3.** 4 **5.** $\frac{36}{25}$ **7.** $2\sqrt{2}$ **9.** $2(e^2 + 1)$ **11.** $\frac{3}{2}$
13. $\frac{3}{2}$

Section 3, page 138

1. 6 **3.** $6/\pi$ **5.** $14/3$ **7.** 48 **9.** $\frac{1}{2}$ **11.** $8/\pi$

13. distance $= e - 2 + e^{-1}$, displacement $= e - 2 - e^{-1}$

15. distance $= 4\sqrt{2}$, displacement $= 0$

Section 4, page 140

1. 84 dyne-cm **3.** 2500 ft-lb **5.** 1, 4, 16 ft-lb

7. $(\frac{12}{43})(10^6)(5280)$, $(\frac{12}{23})(10^6)(5280)$ ft-lb

CHAPTER 10

Section 1, page 143

1. $n = 4$: 3.8281; $n = 10$: 3.8325; exact: $\frac{23}{6} = 3.8333 \cdots$

3. $n = 4$: 6.3230; $n = 10$: 6.3784; exact: $e^2 - 1 \approx 6.3891$

5. $n = 10$: 0.49927; $n = 100$: 0.49999; exact: 0.5

7. $n = 20$: 3.41674, $n = 200$: 3.41615; exact: $3 - \cos 2 \approx 3.41615$

9. $\Delta x = 0.5$: 2.0817; $\Delta x = 0.2$: 2.0816, exact: $12 \log_{10} 12 - 10 - 2M \approx$ 2.0816

11. $\Delta x = 0.2$: 1.09714; $\Delta x = 0.01$: 1.09861; exact: $\ln 3 \approx 1.09861$

13. $\Delta x = 0.2$: 1.74146; $\Delta x = 0.01$: 1.74159; exact: $\sin 2 - 2 \cos 2 \approx$ 1.74160

15. $\Delta x = 0.2$: 6.69284; $\Delta x = 0.01$: 6.69315; exact: $6 + \ln 2 \approx 6.69315$

Section 2, page 147

1. $n = 4$: 10.000; $n = 10$: 9.440; exact: $9\frac{1}{3} = 9.333 \cdots$

3. $n = 10$: 0.45931; $n = 50$: 0.45968; exact: $1 - \cos 1 \approx 0.45970$

5. $\Delta x = \pi/12$: 0.78540; $\Delta x = \pi/180$: 0.78540

7. $\Delta x = 0.1$: 0.53333; $\Delta x = 0.01$: 0.53333

9. $\Delta x = 0.1$: 1.86326; $\Delta x = 0.01$: 1.86275

11. $\Delta x = 0.1$: 2.67663; $\Delta x = 0.005$: 2.53357

13. rect: 2.3325; trap: 2.3350; exact: $2\frac{1}{3} = 2.3333 \cdots$

15. rect: 0.95685; trap: 0.95565; exact: $\cos 1 - \cos 2 \approx 0.95645$

17. 592.8 ft

CHAPTER 11

Section 2, page 161

5. $\sqrt{26}$ **7.** $\sqrt{11}$ **9.** $5, \sqrt{13}, 2\sqrt{5}$ **13.** no

15. right angle at $(-1, 4, -7)$, area $6\sqrt{41}$ **17.** no **19.** yes

Section 3, page 164

13. $y = -3$ **15.** $x + y - 2z = 0$

Section 4, page 167

11. $(x - 1)^2 + (y - 2)^2 + (z - 3)^2 = 16$

13. $(x - 2)^2 + (z - 4)^2 = 9$ **15.** $(x - 1)^2 + (z - 1)^2 = 2(y - 1)^2$

CHAPTER 12

Section 2, page 178

1. 117π **3.** $4\pi/5$ **5.** 72π **7.** 102π **9.** $\frac{1}{4}\pi(e^8 + 4e^4 + 3)$

11. $\frac{1}{3}\pi h(R_0^2 + R_0R_1 + R_1^2)$ **13.** $\frac{1}{3}\pi(a - h)^2(2a + h)$ **15.** $\pi h^3/6$

17. 8

Section 3, page 184

1. $\frac{1}{3}\pi$ **3.** 3π **5.** $176\pi/3$ **7.** does not apply **9.** $15\pi/2$

11. $\frac{1}{3}(112)\ (\pi + 8)$ **13.** $(2/a)\ (e^{2ab} - 1)$

CHAPTER 13

Section 1, page 186

1. $b = a \sin\beta/\sin\alpha$

3. $r = A/s$, where $s = \frac{1}{2}(a + b + c)$ and the area A satisfies $A^2 = s(s - a)\ (s - b)\ (s - c)$

Section 3, page 192

1. 1, 2 **3.** $3y, 3x$ **5.** $4x/(y + 1), -2x^2/(y + 1)^2$

7. $\sin y, x \cos y$ **9.** $2 \cos 2x, -3 \sin 3y$

11. $2y \cos 2xy, 2x \cos 2xy$ **13.** $1/y - y/x^2, -x/y^2 + 1/x$

15. e^y, xe^y **17.** ye^{xy}, xe^{xy} **19.** $2e^{2x} \sin y, e^{2x} \cos y$

21. $4x, 1$ **23.** $32, 0, 3x^3z^2$

Section 4, page 198

1. max = 4 at (0, 0) **3.** min = 0 at (2, −3)

5. min = 4 at (0, 0)

7. max = $2\sqrt{3}/9$ at $(\sqrt{3}/3, 0)$, min = $-2\sqrt{3}/9$ at $(-\sqrt{3}/3, 0)$

9. cube of side 2 ft **11.** $x = y = (2V/3)^{1/3}, z = (9V/4)^{1/3}$

Section 5, page 200

1. $z = f(x)$ **3.** $z = x^2 + 3y + c$ **5.** $z = -1/x + g(y)$

7. $z = -\cos x + \sin y + c$ **13.** $w = ax + by + cz + k$

15. $z = e^x \sin y$ **17.** $z = \frac{1}{3}x^3y + x \cos y + \frac{1}{2}y^2 + y$

CHAPTER 14

Section 2, page 206

1. $y = 1, -x^2$ **3.** $y = 27(x - 2), (x - 3)^2(x + 6)$

5. $y = -3x - 2, 9(x + 1)^2/(3x + 4)$ **7.** $y = 1, \sin x - 1$

9. $(x - 3)(x + 6)$

Section 3, page 215

1. 1.71 **3.** −1.84 **5.** 3.162 **7.** −1.148 **9.** 0.540

11. 1.895

Section 4, page 221

7. $1 - x + \frac{1}{2}x^2 - \frac{1}{6}x^3$ **9.** $2 + 2x$

11. $-20(x - \pi/40) + \frac{1}{3}(4000)(x - \pi/40)^3$ **13.** $1 + 2x + 2x^2 + \frac{4}{3}x^3$

CHAPTER 15

Section 1, page 227

1. $6x^2 - 2x$ **3.** $6e^{3x}$ **5.** $-3/(x + 2)^2$ **7.** $-5e^{-x} - 4 \sin 4x$

9. $7e^{x+4} + 5e^{5x}$ **11.** 0

Section 2, page 231

1. $(x^2 + 2x)e^x$ **3.** $(-3x + 7)(x - 1)^3 e^{-3x}$
5. $\cos^2 x - \sin^2 x = \cos 2x$ **7.** $6(\sin 2x + 1)^2 \cos 2x$
9. $e^x(\sin x + x \sin x + x \cos x)$ **11.** $2^n n x^{n-1}$
13. $x \sin^2 x\,(2 \sin x + 3x \cos x)$ **15.** $\frac{1}{2}e^{x/2}$ **17.** $-\frac{3}{2}x^{-5/2}$
19. $25(5x - 1)(3x + 1)$ **21.** $3(3x - 2)^2(x - 2)(15x - 22)$
23. $(x \cos x - \sin x)/x^2$

Section 3, page 235

1. $(1 - x)/e^x$ **3.** $-x(x + 2)/(x^2 + x + 1)^2$ **5.** $-12x/(x^2 + 1)^7$
7. $-2/(\sin x - \cos x)^2$ **9.** $-(\cos 2x)^3(8x \sin 2x + 3 \cos 2x)/x^4$
11. $-1/[(x + 1)(x - 1)^3]^{1/2}$
13. $[(\sin x)^{m-1}/(\cos x)^{n+1}](m \cos^2 x + n \sin^2 x)$
15. $-(\sin x + x \cos x)/x^2 \sin^2 x$

Section 4, page 237

1. $3x^2 \cos x^3$ **3.** $2e^{\sin 2x} \cos 2x$ **5.** $5(x^3 + x - 3)^4 (3x^2 + 1)$
7. $(2ax + b)e^{ax^2+bx+c}$ **9.** $e^x e^{e^x}$
11. $2[(x^3 + 2x)^5 + x^2][5(x^3 + 2x)^4 (3x^2 + 2) + 2x]$ **13.** $2x(e^{x^2} - e^{-x^2})$

Section 5, page 238

1. 12 **3.** $4x^3$ **5.** 4 **7.** $\frac{1}{3}$ **9.** -1 **11.** $1/2\sqrt{2}$
13. $\sqrt[4]{2}/8$

Section 7, page 243

1. $-4/(2x - 1)^2$ **3.** $-e^{1/x}/x^2$ **5.** $e^{ax}(a \cos bx - b \sin bx)$
7. $(1 - x^2)^{-3/2}$ **9.** $(\cos\sqrt{x/6})/(2\sqrt{6x})$
11. $(3x^3 - 2a^2x)/(x^2 - a^2)^{1/2}$ **13.** $3k \sin^2 kx \cos kx$
15. $\sin(1/x) - [\cos(1/x)]/x$ **17.** $-16/(8x - x^2)^{3/2}$
19. $(2x \cos 2x - 3 \sin 2x)/x^4$ **21.** $(2x + 3)e^{-3/x}$
23. $-4 \sin x \cos x = -2 \sin 2x$ (note $y = \cos^2 x - \sin^2 x = \cos 2x$)
25. $3x^2(1 + e^{2x} - 2xe^{2x})/(e^{2x} + 1)^4$
27. $e^{-ax}(\sin bx - ax \sin bx + bx \cos bx)$
29. $-3 \sin u \cos u\,(1 + \cos^2 u)$, where $u = \frac{1}{2}x$ **31.** $4/(e^x + e^{-x})^2$

CHAPTER 16

Section 1, page 248

1. 24π ft/sec **3.** 4 ft/sec **5.** 192 cm^3/sec **7.** 8 ft^2/min
9. $640/\sqrt{65}$ mph **11.** $\frac{3}{2}$ ft/sec **13.** $10\sqrt{15/\pi}$ ft
15. $+10\pi$ ft^2/sec **17.** 80 ft/min (don't use calculus)

Section 2, page 255

1. $1, -1$ **3.** $1, 1/e$ **5.** $1, -\frac{1}{3}$ **7.** $4/3, 2\sqrt{3}/3$
9. $4\sqrt{3}/9, 0$ **11.** $5/27$, no min **13.** $4\pi, 0$

Section 3, page 262

1. $160 \times 53\frac{1}{3}$ **3.** width $= \sqrt[3]{10}$, height $= \frac{1}{2}\sqrt[3]{10}$
5. $\frac{1}{2}a, \frac{1}{2}b$ **7.** $x/2a + y/2b = 1$ **9.** 2048π in^3
11. $y = -4x/\sqrt{3} + \frac{16}{3}$ **13.** 500 **15.** 6 ft from short post
17. Row to the point $\sqrt{3}$ miles upshore from the nearest point, then walk.
19. $3r$

CHAPTER 17

Section 1, page 267

1. $x = \frac{1}{3}(y + 7)$ **3.** $x = \frac{1}{4}(y + 1)$
5. $x = (4y + 7)/(-y + 2), y \neq 2$ **7.** $x = (3y - 2)/(-y + 1), y \neq 1$
9. $x = -1/y, y \neq 0$ **11.** $x = \frac{1}{2}y^2 + 4, y > 0$
19. $f[g(x)] = x$; but $f[g(x)] = g(x)^3 + g(x)$
21. $f[f^{-1}(x)] = x$, so the answer is x

Section 2, page 274

1. $\frac{3}{5}x^{-4/5}$ **3.** $\frac{1}{4}(10^{1/4})x^{-3/4}$ **5.** $\frac{7}{6}(x^2 - x + 4)^{1/6}(2x - 1)$
7. $e^x(e^x - 2)(e^x + 1)^{1/2}(e^x - 1)^{-3/2}$ **9.** $e^{2\sqrt{x}}/\sqrt{x}$
11. $(p/8)(\sin x)^{(p/8)-1}(\cos x)$ **13.** $(1/7a)(x/a)^{-6/7}$ **15.** $-\frac{1}{5}x^{-6/5}$

Section 3, page 277

1. $a + 2$ **3.** $1/x$ **5.** $\frac{1}{2}$ **7.** 1 **9.** 2.99573
11. 0.67853 **13.** 2.48, −1.51, 5.52, 0.90 **15.** 6

Section 4, page 281

1. $1/x$ **3.** $4/x$ **5.** $-1/x$ **7.** $\cot x$ **9.** 1

11. $1/x \ln x$ **13.** $[2/(x^2+1)] - [1/x^2]\ln(x^2+1)$

15. $2(\ln x)/x$ **17.** $[x(\ln x)(\ln\ln x)]^{-1}$ **19.** $-[x(\ln x)^2]^{-1}$

25. $-1/x^2 < 0$; concave downwards.

Section 5, page 283

1. $(3x^2/2)(x^3+2)^{-1/2}$ **3.** $6\left(\dfrac{x^2+4}{x+7}\right)^6\left(\dfrac{2x}{x^2+4} - \dfrac{1}{x+7}\right)$

5. $\frac{2}{3}(2x+3)^{-2/3}(-6x^2-9x+1)e^{-x^2}$ **7.** $x^{x-2}(x-1+x\ln x)$

9. $x^{1/x}(1-\ln x)/x^2$ **11.** $(\ln 3)3^{\ln x}/x = (\ln 3)x^{\ln 3 - 1}$

13. $10^x \ln 10$ **15.** $\dfrac{e^x(x^3-1)}{\sqrt{2x+1}}\left[1 + \dfrac{3x^2}{x^3-1} - \dfrac{1}{2x+1}\right]$

Section 6, page 284

1. $\ln 5$ **3.** $\ln \frac{7}{4}$ **5.** $\frac{1}{2}\ln 7$ **7.** $\ln 2$ **9.** $\frac{1}{3}\ln 4$

11. $\ln(b/a) = \ln(a^{-1}/b^{-1})$ **13.** $\frac{1}{4}\ln 13$ **15.** $\frac{1}{4}\ln\frac{5}{3}$ **17.** $\frac{52}{3} + \ln 5$

19. $\displaystyle\int_a^{2a} dx/x = \ln 2a - \ln a = \ln 2$, for any $a > 0$

Section 7, page 288

1. ∞ **3.** 0 **5.** ∞ **7.** 0 **9.** 0 **13.** e^a, $a = 10^{10}$

15. In common logs, the inequality is $(\log x)^2/x < M^2 \cdot 10^{-10}$.
For $x > 10^{13}$, $(\log x)^2/x < 13^2 \times 10^{-13} = 1.69 \times 10^{-11} < M^2 \cdot 10^{-10}$.
Hence $x > 10^{13} \approx e^{29.9}$ works. **17.** $\max = e^{1/e} \approx 1.445$

Section 8, page 290

1. 0 **3.** $e - 1$ **5.** Set $x = u^3$; therefore, $u \to \infty$ as $x \to \infty$. Then the function is $e^u/u^{30} \to \infty$. **9.** no; try $x = 10^{10}$

11. (smallest) x^3, e^x, 10^x, e^{10x}, e^{x^2}, e^{e^x} (largest)

CHAPTER 18

Section 1, page 295

1. π **3.** 1 **5.** 2 **7.** π **9.** π **11.** 2π **13.** odd

15. even **17.** odd **19.** none

21. no: for functions, even $\times$ odd = odd, odd $\times$ odd = even

Section 2, page 298

1. $\sin x = (1 + \cot^2 x)^{-1/2} = (\sec^2 x - 1)^{1/2}/\sec x$

3. $\cot^2 x = (\cos^2 x)/(1 - \cos^2 x)$

5. $\cos^4 x = \frac{1}{4}(\cos^2 2x + 2\cos 2x + 1)$

7. $\sin 15° = \sin(60° - 45°) = \sin 60° \cos 45° - \cos 60° \sin 45°$;
$\sin^2 15° = \frac{1}{2}(1 - \cos 30°)$

9. $1 - \cos x = 2\sin^2 \frac{1}{2}x$, $1 + \cos x = 2\cos^2 \frac{1}{2}x$; divide.

11. $\sin x \cos y = \frac{1}{2}[\sin(x + y) + \sin(x - y)]$;
$\sin x \sin y = \frac{1}{2}[\cos(x - y) - \cos(x + y)]$

Section 3, page 300

1. $\frac{1}{3}\cos(x/3)$ **3.** $-3\cos^2 x \sin x$ **5.** $\cot 2x - 2x\csc^2 2x$

7. $\csc^2 x \cot x\,(2 - \csc^2 x)^{-1/2}$

9. $(-\frac{10}{3})\,(\cos^4 \frac{1}{3}x)\,(\sin \frac{1}{3}x)\,(1 + \cos^5 \frac{1}{3}x)$ **11.** $2(\cos x)/(1 - \sin x)^2$

13. $12x \sec^3(2x^2)\tan(2x^2)$

15. $-\sin x$; successive derivatives repeat in cycles of four

17. $dy/dx = -y\sqrt{y^2 - 1}$

19. $d(\ \ )/dx = 4\sec^4 x \tan x - 4\tan^3 x \sec^2 x$
$= 4\sec^2 x \tan x\,(\sec^2 x - \tan^2 x) = 4\sec^2 x \tan x$

21. $d(\ \ )/dx = 1 + \sec x \tan x - \sec^2 x = \sec x \tan x - \tan^2 x =$
$\tan x\,(\sec x - \tan x) = (\sin x/\cos x)\,(1 - \sin x)/\cos x$. But
$(1 - \sin x)/\cos^2 x = 1/(1 + \sin x)$

Section 4, page 306

3. $5\pi\sqrt{2}/36$ in^2/sec; when the vertex angle is 90°

5. $(6 - 2\sqrt{6})\pi/3$ **7.** $(6 - \sqrt{3})$ ft; the angle at C is 120°

9. 4π in./min to the left **10.** $(a^{2/3} + b^{2/3})^{3/2}$; this occurs when
$\tan^3 \theta = b/a$, found by minimizing the length $L = a\sec\theta + b\csc\theta$

11. In this case, $L^2 = (a\sec\theta + b\csc\theta)^2 + c^2$ so it is essentially the same problem. Answer: $[(a^{2/3} + b^{2/3})^3 + c^2]^{1/2}$

Section 5, page 315

1. $\pi/4$ **3.** $\pi/3$ **5.** $\pi/2$ **7.** $4/5$ **9.** $\pi/2$

11. Let $\theta = \arccos x$. Then $\cos\theta = x$ and $\cos 2\theta = 2\cos^2\theta - 1 = 2x^2 - 1$;
hence, $2\theta = \arccos(2x^2 - 1)$. **13.** Use Ex. 12

Section 6, page 317

1. $1/(9 - x^2)^{1/2}$ **3.** $2x/(1 + x^4)$ **5.** $6(\text{arc sin } 3x)\ (1 - 9x^2)^{-1/2}$
7. $1/(1 + x^2)$ **9.** $1/(1 + x^2)$ **11.** $2 \text{ arc tan } 2x$

Section 7, page 321

1. $(TB)^{1/2}$ ft **3.** $\pi/3$ **5.** $\sqrt{7}/210$ rad/sec

Section 8, page 326

1. $11 \cosh 2x + 3 \sinh 2x$ **3.** $\cosh 2x = 2\cosh^2 x - 1 = 2 \sinh^2 x + 1$
5. Set $x = \tanh y = (e^y - e^{-y})/(e^y + e^{-y})$. Then $(1 + x)/(1 - x) = e^{2y}$; hence, $\frac{1}{2} \ln[(1 + x)/(1 - x)] = y = \tanh^{-1} x$.
7. $5 \cosh 5x$ **9.** 0 **11.** $e^{2x} \cosh x$

CHAPTER 19*

Section 2, page 332

1. $\sin 2x$ **3.** e^{x^2} **5.** $2(1 - x^2)^{1/2}$ **7.** $\frac{1}{4} \sin^4 x$
9. $-\frac{1}{2}(1 + x^2)^{-2}$ **11.** $-\sec x$ **13.** $\tan 5x$
15. $\ln(e^x + x^2 + 1)$ **17.** $\frac{2}{3}(4x^2 + 1)^{3/2}$ **19.** $e^{\sqrt{x}}$

Section 3, page 336

1. $(3x + 1)^6/18$ **3.** $(x^2 + 1)^4/8$ **5.** $\frac{1}{3} \sin 3x$
7. $(-\cos ax + \sin ax)/a$ **9.** $\frac{1}{3}e^{3x}$ **11.** $e^{2\sqrt{x}}$ **13.** $\frac{2}{5}(1 + 5x)^{1/2}$
15. $\frac{1}{4} \ln(1 + x^4)$ **17.** $\frac{1}{6} \ln^2|3x + 1|$ **19.** $\frac{1}{3}(5 - 3x)^{-1}$
21. $\frac{1}{2} \tan^4(x/2)$

Section 4, page 338

1. $\frac{2}{5}(x + 3)^{3/2}(x - 2)$ **3.** $\frac{1}{3}(x - 5)\ (2x + 5)^{1/2}$
5. $\ln|x - 1| - 2(x - 1)^{-1} - \frac{1}{2}(x - 1)^{-2}$ **7.** $2\sqrt{x} - 2 \ln(1 + \sqrt{x})$
9. $\frac{4}{3} \ln(1 + x^{3/4})$ **11.** $\frac{1}{2} \text{ arc tan } e^{2x}$ **13.** $\frac{1}{5} \text{ arc tan}(5x + 2)$
15. $\text{arc sin}(x/a)$ **17.** $\frac{1}{3}[x - (1/\sqrt{3}) \text{ arc tan}(x\sqrt{3})]$
19. $\frac{1}{3}x^3 - 2x + 2\sqrt{2} \text{ arc tan}(x/\sqrt{2})$ **21.** $\frac{1}{2}[x^2 - \ln(x^2 + 1)]$

Section 5, page 342

1. $\frac{49}{3}$ **3.** $\frac{1}{2}(e^9 - e)$ **5.** $\frac{11}{36}$ **7.** $2 - \sqrt{2}$ **9.** $2 - 2 \ln \frac{3}{2}$
11. $\frac{1}{3}(16 - 9\sqrt{3})$

---

* The constant of integration is always omitted.

Section 6, page 347

1. x **3.** $\frac{1}{5}\sin^5 x - \frac{1}{7}\sin^7 x$ **5.** $\frac{1}{5}\cos^5 x - \frac{1}{3}\cos^3 x$

7. $\frac{1}{3}\sin 3x - \frac{2}{9}\sin^3 3x + \frac{1}{15}\sin^5 3x$

9. $\frac{1}{32}(\sin 4x + 8\sin 2x + 12x)$ **11.** $\frac{1}{3}\tan^3 x - \tan x + x$

13. $(\sec^n kx)/kn$ **15.** $\tan x + \sec x$ **17.** 1 **19.** 0

21. $\frac{1}{2}\arctan\frac{1}{2}(x+1)$ **23.** $\arcsin\frac{1}{3}(x-3)$

25. $-(4x - x^2)^{1/2} + 2\arcsin\frac{1}{2}(x-2)$

27. $(1/2\sqrt{3})\ln|x^2 - \frac{2}{3} + [(3x^4 - 4x^2 + 1)/3]^{1/2}|$

29. $(1/b)\ln|ax/(b - ax)|$

Section 7, page 353

1. $\frac{1}{2}\left(\frac{1}{x-1} - \frac{1}{x+1}\right)$ **3.** $1 + \frac{1}{3}\left(\frac{4}{x-2} - \frac{1}{x+1}\right)$

5. $\frac{1}{2}\left(\frac{-1}{x+1} + \frac{4}{x+2} - \frac{3}{x+3}\right)$ **7.** $1 - \frac{2}{x^2+1} + \frac{1}{(x^2+1)^2}$

9. $\frac{1}{5}\left(\frac{2}{x-1} + \frac{-2x+3}{x^2+4}\right)$ **11.** $\ln|(x-2)/(x-1)|$

13. $\ln|x^3/(x+1)^2|$ **15.** $3\ln|x/(x^2+1)^{1/2}| + 2\arctan x$

17. $\frac{1}{169}\{\ln[(x^2+9)^2/(x-2)^4] - 13/(x-2) - \frac{5}{3}\arctan(x/3)\}$

19. $\frac{1}{2}x^2 + \frac{1}{3}\ln|(x-1)/(x^2+x+1)^{1/2}| + (1/\sqrt{3})\arctan(2x+1)/\sqrt{3}$

21. $\frac{1}{2}x^2 - 3x + 8\ln|x+2| - \ln|x+1|$

23. $\frac{1}{17}[13\ln|x-3| + 2\ln(x^2+2x+2) - \arctan(x+1)]$

25. $\frac{1}{3}\ln|(2+\sin\theta)/(5+\sin\theta)|$

Section 8, page 357

1. $\ln[x + (1+x^2)^{1/2}]$ **3.** $-\frac{1}{3}(8+x^2)(4-x^2)^{1/2}$

5. $-(16-x^2)^{1/2}/(16x)$ **7.** $\ln|x + (x^2+a^2)^{1/2}| - (x^2+a^2)^{1/2}/x$

9. $\frac{1}{2}\arctan x - \frac{1}{2}x/(1+x^2)$

11. $\operatorname{arg\,sinh}(x/a) = \ln[x + (a^2+x^2)^{1/2}] - \ln a$

13. $\frac{1}{2}x(a^2+x^2)^{1/2} - \frac{1}{2}a^2 \operatorname{arg\,sinh}(x/a)$

Section 9, page 363

1. $\sin x - x\cos x$ **3.** $\frac{1}{4}e^{2x}(2x-1)$ **5.** $\frac{2}{9}x^{3/2}(3\ln x - 2)$

7. $x\arctan x - \frac{1}{2}\ln(1+x^2)$ **9.** $\frac{1}{2}(1+x^2)\arctan x - \frac{1}{2}x$

11. $e^{2x}(2\sin 3x - 3\cos 3x)/13$ **13.** $x\sinh x - \cosh x$

15. $(x^2/a)\sin ax + (2x/a^2)\cos ax - (2/a^3)\sin ax$ **17.** 2

19. $(24\ln 2 - 7)/9$ **21.** $10e - 5$

Section 10, page 367

1. Set $u = \ln^n x$, $v = x$, and integrate by parts.

3. Set $u = x^n$, $v = -\cos x$, and integrate by parts.

7. $(2 \times 4 \times 6)/(3 \times 5 \times 7) = \frac{16}{35}$ **9.** $\frac{1}{9} - \frac{1}{7} + \frac{1}{5} - \frac{1}{3} + 1 - \frac{1}{4}\pi$

Section 12, page 372

1. $(e^{-2x}/29)(-2\sin 5x - 5\cos 5x)$

3. $\frac{1}{64}[-4x(1 - 4x^2)^{3/2} + 2x(1 - 4x^2)^{1/2} + \text{arc sin } 2x]$

5. $\frac{1}{25}[5x - \sqrt{10}\text{ arc tan}(\frac{1}{2}x\sqrt{10})]$

7. $12(x^2 - 8)\sin\frac{1}{2}x - 2x(x^2 - 24)\cos\frac{1}{2}x$

9. $$\frac{1}{10}\sqrt{10x^2+7} - \frac{6}{\sqrt{10}}\ln|x\sqrt{10} + \sqrt{10x^2+7}| + \frac{2}{\sqrt{7}}\ln\left|\frac{\sqrt{7} + \sqrt{10x^2+7}}{x}\right|$$

11. $-\frac{1}{2}\pi$ **13.** $\frac{77}{325}(e^{3\pi} - 1)$ **15.** $\frac{3}{4} + 5\ln\frac{7}{8}$

CHAPTER 20

Section 1, page 376

1. $\frac{1}{3}(2x^2 - 7x + 6)$ **3.** $\frac{1}{84}(-5x^2 + 63x - 106)$ **5.** $\frac{1}{3}(x + 7)$

7. $\frac{1}{2}(x^2 - 11x + 26)$ **9.** $\frac{1}{2}(-x + 4)$

11. $\frac{1}{6}(-x^3 + 3x^2 + 10x + 12)$ **13.** $\frac{1}{6}(-x^3 + 10x)$ **15.** $2x - 4$

17. $\frac{1}{6}(7x^3 + 27x^2 + 14x - 6)$ **19.** $\frac{1}{10}(-x^2 - 3x + 40)$

21. $\frac{1}{2}[(e - 1)^2x^2 - (e - 1)(e - 3)x + 2]$ **23.** $-\frac{1}{6}(x^2 - 4x)$

25. $6x^3 - 11x^2 + 5x$ **27.** $\frac{1}{24}(-x^3 + 7x^2 + 6x)$ **29.** Estimate by $s = \frac{1}{6}(t^3 + 21t^2 - 82t + 120)$. The minimum is $s \approx 7.70$ ft at $t \approx 1.74$ hr.

31. $\int_0^3 (-\frac{1}{2}x^3 + 2x^2 - x)\,dx = 3.375$

Section 2, page 379

1. $\frac{1}{12}(x^3 + 17x^2 + 18x - 24)$

3. $\frac{1}{48}(-x^3 + 18x^2 - 128x + 384)$

5. $\frac{1}{192}(-x^4 + 40x^3 - 212x^2 + 80x + 384)$

7. $\frac{1}{3}(x^3 - 6x^2 + 8x)$ **9.** $\frac{1}{2}(8x^2 - 18x + 9)$

11. $\int_{-1/2}^{1} (-\frac{1}{3})(4x^3 - x)\, dx = -\frac{3}{16} = -0.1875$

13. $\frac{2}{3}(1 + 4e^{-4} + e^{-16}) \approx 0.72$

Section 3, page 385*

1. 136 **3.** 8.55 **5.** $-\frac{4}{45} \approx -0.09$ **7.** 1.22

9. 42.3266, 42.4301, 42.4387 **11.** 22.8784, 22.8664, 22.8654

13. -4.1641, -2.5272, -2.3123 **15.** 0.886227 **17.** 0.902570

Section 4, page 389

1. Simpson: 1.65, Newton–Cotes: 1.64 **3.** 14.12, 14.10

5. -1.84, -1.81 **7.** 0.8517, 0.8456, 0.8490

9. 0.4930, 0.4931, 0.4931 **11.** 0.2040, 0.2141, 0.2133

13. Three-Eighths: $\frac{39}{8} = 4.875$ **15.** 5-pt. N–C: $\frac{116}{9} \approx 12.89$

17. Weddle: $\frac{63}{10} = 6.3$, 7-pt. N–C: $\frac{881}{140} \approx 6.293$

	Simp.	$\frac{3}{8}$	5-pt. N–C	7-pt. N–C	9-pt. N–C	Weddle
19.	0.684199	0.684038	0.684527	0.684286	0.683844	0.684327
21.	849.902	849.906	849.900	849.900	849.899	849.900
23.	238.000	230.250	237.422	246.114	218.351	244.200

Section 5, page 397

1. even **3.** odd **5.** even **7.** neither **9.** $x = -1$

11. $x = -1$ **13.** $x = \pi/4$ **15.** $x = 0$ **17.** $\int_1^2 (x^3 - 5x)\, dx$

19. 0 **21.** $\int_0^{\pi} \sin x\, dx = 2\int_0^{\pi/2} \sin x\, dx$ **23.** $4\int_0^{\pi/2} \sin^2 x\, dx$

25. $-2 \sin 4 \int_0^{4\pi} \cos \frac{x}{12}\, dx$

---

*Answers from a computer depend on the computer, the compiler, and the algorithm used. Hence numerical answers may differ slightly from those given.

Section 6, page 404

1. hint: $1 - k^2 < 1 - k^2 \sin^2 \theta < 1$ **3.** $x^3 + 3x + 1 > 5$

5. $\sin^2 x < 1$ **7.** $3 + 2x > 2x$ **9.** $x^2 < x^2 + x + 1 < (x+1)^2$

11. $9 < 3 + 2x < 13$ **13.** $1 < 1 + \sin^2 x < 2$

15. $\ln^2 2 < \ln^2 x < \ln^2 5$

17. $\displaystyle\int_0^{100} - \int_0^{10} = \int_{10}^{100} < e^{-10} \int_{10}^{100} dx = 90e^{-10} \approx .0041$

Section 7, page 410

1. Trapezoidal: $(\sin 1)/(12 \times 6^2) < 2 \times 10^{-3}$;
Simpson: $(\sin 1)/(180 \times 6^4) < 4 \times 10^{-6}$;
Three-Eighths: $(\sin 1)/(80 \times 6^4) < 9 \times 10^{-6}$

3. $\frac{1}{54} < 2 \times 10^{-2}$, $1/(45 \times 3^3) < 9 \times 10^{-4}$, $(20 \times 3^3)^{-1} < 2 \times 10^{-3}$

5. $(3^7 \times 1400)^{-1} < 4 \times 10^{-7}$, $11/(3^7 \times 1400) < 4 \times 10^{-6}$

7. $(9 \times 73e)/(2^9 \times 1400) < 3 \times 10^{-3}$,
$(10 \times 43e + 9 \times 73e/4)/(2^7 \times 1400) < 10^{-2}$

9. 7-point N–C (4 times) works. For $0 \le x \le 6$,
$|(x^2e^{-x})^{(8)}| = |(x^2 - 16x + 56)e^{-x}| \le 6^2 + 16 \cdot 6 + 56 = 188 = M$,
$|\epsilon| \le 4(9/1400)\,(188)\,(1/4)^9 < 2 \times 10^{-5}$

11. $\displaystyle\frac{M}{12}\left(\frac{h}{10}\right)^2 (b - a) = \frac{1}{100}\left[\frac{M}{12} h^2(b - a)\right]$

17. for $2n + 1$ points: $h = (b - a)/2n$,

$$|\epsilon| \le n\left(\frac{Mh^5}{90}\right) = \frac{Mh^4}{90}(nh) = \frac{Mh^4}{180}(b - a)$$

19. $\displaystyle h = (b - a)/4n,\ |\epsilon| \le n\left(\frac{8Mh^7}{945}\right) = \frac{8Mh^6}{945}(nh) = \frac{2Mh^6}{945}(b - a)$

CHAPTER 21

Section 1, page 415

1. $y = \frac{1}{4}x^4$ **3.** $y = 1 + 3 \ln x$ **5.** $y = 6 - e^{-x}$ **7.** $y = 10e^{x-1}$

9. $y = \frac{1}{2}x^2 - \cos x + 1$ **11.** $y = (x^2 + 1)^{1/2} - 1$

13. $y = x + \frac{1}{3}x^3 + \frac{1}{4}x^4 + \frac{1}{6}x^6$ **15.** $y = x + c/x$

17. $\ln(x^2 + y^2) = 2x + c$

Section 2, page 421

1. $y^2 = \frac{2}{3}x^3 + c$ **3.** $y = (\sqrt{x} + c)^2$ **5.** $y = \ln[2/(c - x^2)]$

7. $y = x[1 + 2/(c - \ln x)]$ **9.** $y = 3/\sqrt{x}$

11. $y = (4 - x)/(2 - x)$ **13.** (Slope at P) $= -$(slope $\overline{OP}$); $xy = c$; rectangular hyperbolas with axes $y = \pm x$

15. (Slope at P) $=$ (slope $\overline{OP}$); $y = cx$; straight lines through (0, 0)

Section 3, page 425

1. linear **3.** linear **5.** non-linear

7. $y = \frac{1}{3}x^3 + c$; yes, $y = \frac{1}{3}x^3$ is a particular solution and $y = c$ is the general solution of $y' = 0$.

Section 4, page 427

1. $y = c/x^3$ **3.** $y = c\cos\theta$ **5.** $y = ce^{-x}/x$

7. $y = 3(1 - x^2)^{1/2}$ **9.** $y = \frac{4}{5}x^{-2}$

Section 5, page 434

1. $y = \frac{1}{2}x - \frac{1}{4}$ **3.** $y = \frac{1}{5}e^x$ **5.** $y = \frac{1}{4}e^x(2x^2 - 2x + 1)$

7. $\frac{1}{29}(15\sin x - 6\cos x)$ **9.** $y = \frac{1}{2}e^{-x}(\sin x + \cos x)$

11. $i = E(R\cos t + L\sin t)/(R^2 + L^2)$

13. $y = -(x^5 + 5x^4 + 20x^3 + 60x^2 + 120x + 120)$

15. $y = xe^x$ **17.** $y = x^3 - x$ **19.** $y = \frac{1}{2}(x^2 + 1)\ln(x^2 + 1)$

21. $y = ce^{-3x} + \frac{1}{3}$ **23.** $y = ce^{-x} + \frac{1}{3}xe^{2x} - \frac{1}{9}e^{2x} + 1$

25. $i = ce^{-Rt/L} + E(R\cos\omega t + L\omega\sin\omega t)/(R^2 + L^2\omega^2)$

27. $y = ce^{-x^2} + 1$ **29.** $y = ce^{-3x} + xe^{-3x}$

33. The slope is constant c along each line $x + y = c$. But each such line has slope -1, so $c = -1$ gives the solution $x + y = -1$.

Section 6, page 443

1. $y = cx^3$ **3.** $100(\log 10)/(\log 2) \approx 332$ years

5. $11^3 \times 10^2 \approx 133000$ **7.** about 32.1 sec

9. $100(2 + \sqrt{3}) \approx 373.2°\text{C}$ **13.** $8000/\sqrt{2g} \approx 997$ sec

CHAPTER 22

Section 2, page 450

1. $x = ae^t + be^{5t}$ **3.** $r = a \cos 2\theta + b \sin 2\theta$

5. $y = e^{x/4}[a \cos(x\sqrt{7}/4) + b \sin(x\sqrt{7}/4)]$ **7.** $x = a + be^{-6t}$

9. $y = ae^{-t} + be^{-4t}$ **11.** $x = e^{2a^2t}(bt + c)$ **13.** $y = a \sin 3x$

15. $x = e^{2t}(at + 1)$ **17.** $r = e^{\theta}(-e^{-\pi} \cos \theta + b \sin \theta)$

19. $r = a(\cos \theta - \sqrt{3} \sin \theta)$

Section 3, page 453

1. $x = \frac{1}{3}t$ **3.** $x = -\frac{1}{2}t - \frac{1}{4}$ **5.** $x = -t^2 - 10t - 38$

7. $x = \frac{1}{7}e^{3t}$ **9.** $y = \frac{1}{9}e^{-2x} - x - 1$

11. $y = \frac{4}{53}(-2 \cos 2x - 7 \sin 2x)$ **13.** $y = \frac{1}{6}e^x(\cos x + \sin x)$

15. $y = \dfrac{e^x}{1^2 + 1} + \dfrac{e^{2x}}{2^2 + 1} + \dfrac{e^{3x}}{3^2 + 1} + \dfrac{e^{4x}}{4^2 + 1} + \dfrac{e^{5x}}{5^2 + 1}$

17. $x = a \cos t + b \sin t + t^2 - 2$

19. $x = ae^{2t} + be^{-3t} + \frac{1}{36}e^{-t}(1 - 6t)$

21. $x = a + be^{-3t} + \frac{1}{10}(-2 \cosh 2t + 3 \sinh 2t)$

23. $i = e^{-2t/3}(a \cos t\sqrt{2}/3 + b \sin t\sqrt{2}/3) + \frac{10}{17}(4 \sin t - \cos t)$

25. $x = 5 \cos t - \sin t + 2t - 5$ **27.** $x = \frac{1}{16}(9e^{2t} + 7e^{-2t} - 2 \sin 2t)$

29. $x = 10e^t - \frac{56}{5}e^{3t/4} + \frac{1}{5}e^{2t}$

31. Because e^{2t} is a solution of the homogeneous equation. $x = \frac{1}{4}te^{2t}$

33. $z\dot{w} + (2\dot{z} + pz)w = r$

Section 5, page 465

1. $\sqrt{2} \cos 3(t - \frac{1}{12}\pi) = \sqrt{2} \sin 3(t + \frac{1}{12}\pi)$

3. $2 \cos 2(t + \frac{1}{12}\pi) = 2 \sin 2(t + \frac{1}{3}\pi)$

5. $5 \cos(t - \alpha) = 5 \sin(t - \beta)$, where $\alpha = \text{arc} \cos(-\frac{3}{5}) \approx 126° 52'$ and $\beta = \text{arc} \sin \frac{3}{5} \approx 36° 52'$

7. $x = 6 \cos \pi(t \pm \frac{1}{3})$ **9.** $90/\pi$ **11.** 9

13. π, assuming the equilibrium position is 8 ft from the ceiling; $2\pi\sqrt{L/g}$ if it is L ft from the ceiling

15. $2\pi/[(62.4)\pi g/100]^{1/2} \approx 0.791$ sec

CHAPTER 23

Section 2, page 476

1. $(5, 2, 4)$ **3.** 29 **5.** $(1, -16, 9)$ **7.** $(\frac{5}{3}, 0, 1)$ **9.** $\sqrt{69}$

11. $\sqrt{11}$ **13.** $\arccos(-1/\sqrt{15})$ **15.** $\sqrt{2}/2, 0, \sqrt{2}/2$

17. $2/\sqrt{14}, 1/\sqrt{14}, -3/\sqrt{14}$ **19.** $\pi/2$

Section 3, page 480

1. $\dot{\mathbf{x}} = (e^t, 2e^{2t}, 3e^{3t})$ **3.** $(1, 3, 4)$

5. $\mathbf{v} = (2t, 3t^2 + 4t^3, 0)$, $|\mathbf{v}| = (4t^2 + 9t^4 + 24t^5 + 16t^6)^{1/2}$

7. $\mathbf{v} = (-A\omega \sin \omega t, A\omega \cos \omega t, B)$, $|\mathbf{v}| = (A^2\omega^2 + B^2)^{1/2}$

9. $d|\mathbf{x}|^2/dt = d(\mathbf{x} \cdot \mathbf{x})/dt = 2\mathbf{x} \cdot \dot{\mathbf{x}} = 0$, hence $|\mathbf{x}|^2 = \text{const}$

Section 4, page 487

1. $(a_1^2 + a_2^2 + a_3^2)^{1/2}$ **3.** $2\pi\sqrt{2}$ **5.** $\displaystyle\int_{x_0}^{x_1} (1 + a^2n^2x^{2n-2})^{1/2}\, dx$

7. $\sqrt{17} - \frac{1}{4}\ln(\sqrt{17} - 4)$ **9.** $(\sqrt{2}/2)\,(1, -\sin t, \cos t)$

11. $(a_1^2 + a_2^2 + a_3^2)^{-1/2}(a_1, a_2, a_3)$

Section 5, page 493

1. $2\sqrt{5}/25$ **3.** $\sqrt{19}/7\sqrt{14}$

5. $\dot{\mathbf{N}} \cdot \mathbf{N} = \frac{1}{2}\, d(\mathbf{N} \cdot \mathbf{N})/dt = 0$; $\dot{\mathbf{T}} \cdot \mathbf{N} + \mathbf{T} \cdot \dot{\mathbf{N}} = 0$, $\dot{\mathbf{T}} = k\mathbf{N}$, hence
$\dot{\mathbf{N}} \cdot \mathbf{T} = -k$. Thus $\dot{\mathbf{N}} = -k\mathbf{T} + 0\mathbf{N} = -k\mathbf{T}$.

7. $(\pi/2, 1)$ **9.** $k = |y''|/(\cdot\ \cdot\ \cdot)$. But $y'' = 0$ at an inflection.

11. $d\alpha/dt = (d\alpha/ds)\,(ds/dt) = 3e^2/(1 + e^4)^{3/2}$ rad/sec

13. $k = (2/x^3)/[1 + (-1/x^2)^2]^{3/2} \longrightarrow 0$ as $x \longrightarrow 0$ or $x \longrightarrow \infty$

CHAPTER 24

Section 1, page 499

1. Use the solution of Example 1.3. The shell strikes the hill when $y/x = \tan\beta$, that is, when $t = (2v_0/g)\,(\sin\alpha - \tan\beta\cos\alpha)$

3. $\mathbf{v} = (1, 2t)$, $\mathbf{a} = (0, 2)$

5. $\mathbf{v} = \mathbf{T}$ and $\mathbf{a} = d\mathbf{T}/ds = k\mathbf{N}$ since $ds/dt = 1$. Hence, the tangential component of $\mathbf{a}$ is always $\mathbf{0}$ and the normal component is $\mathbf{a}$ itself. Since $\mathbf{x} =$

$(x, \sin x)$, $\mathbf{T}(ds/dx) = (1, \cos x)$ and $\mathbf{a}(ds/dx) + \mathbf{T}(d^2s/dx^2) = (0, -\sin x)$. Hence, $(ds/dx)^2 = 1 + \cos^2 x$, $(ds/dx)(d^2s/dx^2) = -\sin x \cos x$. Evaluate at $x = 0$: $\mathbf{a} = \mathbf{0}$; at $x = \pi/2$: $\mathbf{a} = (0, -\sqrt{2}/2)$.

7. $\mathbf{v} = (-a\omega \sin \omega t, a\omega \cos \omega t, b)$, $\mathbf{a} = -a\omega^2(\cos \omega t, \sin \omega t, 0)$. Since these are perpendicular, the tangential component of $\mathbf{a}$ is $\mathbf{0}$; the normal component is $\mathbf{a}$.

Section 2, page 503

1. 3 **3.** $6\pi^2$ **5.** $\frac{2}{3}$ **7.** 1 **9.** g **11.** $(\frac{3}{2}, 2, \frac{5}{2})$
13. $(0, \frac{2}{5}, 0)$

Section 3, page 510

3. θ is the angle at the center of the rolling circle, measured clockwise from the downward vertical to the moving point. The center is $(a\theta, a)$ and the vector is (moving point) $-$ (center) $= -a(\sin \theta, \cos \theta)$; hence, the moving point is $\mathbf{x} = (a\theta, a) - a(\sin \theta, \cos \theta)$.

5. $8a$ **7.** $y = x^2$, $z = x^3$

9. $y = \pm\cosh t$, $z = \pm\sinh t$, $x = \cosh 2t$; alternate: $y = (1 + t^2)/(1 - t^2)$, $z = 2t/(1 - t^2)$, $x = (t^4 + 4t^2 + 1)/(1 - t^2)^2$

Section 4, page 516

1. $(0, 1)$ **3.** $(\sqrt{2}, \sqrt{2})$ **5.** $(\frac{1}{2}\sqrt{3}, \frac{1}{2})$ **7.** $\{\sqrt{2}, 7\pi/4\}$
9. $\{2, 5\pi/6\}$ **11.** $\frac{1}{2}a[\sqrt{2} + \ln(1 + \sqrt{2})]$

13. $6a \int_0^{\pi/6} (8 \sin^2 3\theta + 1)^{1/2}\, d\theta$ **15.** $\pi a^2/4$ **17.** $\pi a^2/2$

19. $3\pi a^2/2$ **21.** a^2

23. $(2b^2 + a^2)(\pi/2 - \operatorname{arc}\cos b/a) + 3b\sqrt{a^2 - b^2}$

Section 6, page 521

9. Rotate $\pi/4$ rad: $-\frac{1}{2}\bar{x}^2 + (\frac{1}{2})5\bar{y}^2 = 1$. Hence a hyperbola with $a = \sqrt{\frac{2}{5}}$, $b = \sqrt{2}$. **11.** $\alpha = \pi/8$, $\cos 2\alpha = \sin 2\alpha = \sqrt{2}/2$.
$\bar{A} = 2 - \sqrt{2} > 0$, $\bar{C} = 2 + \sqrt{2} > 0$; hence an ellipse with $a = (2 - \sqrt{2})^{-1/2}$ and $b = (2 + \sqrt{2})^{-1/2}$.

CHAPTER 25

Section 1, page 525

1. $y = 0$ **3.** $x - y + 2z = 1$ **5.** $(1, -1, -1)$
7. $(1, 0, 2)$ **9.** $x - y + 3z = 0$ **11.** $3/\sqrt{26}$
13. $x + y + z = -1$

Section 2, page 529

1. $(9t^2 + 2t)e^{t^2(3t+1)}$ **3.** $4t^3 \cos(1/t) + t^2 \sin(1/t) - 2t$ **5.** $9t^8$
7. $4te^{-t}[(2 - t) \sin 4t + 4t \cos 4t]$
9. $(s^2 - t)^2(2s^2 + t)/2s^3t^2, -(s^2 - t)^2(2s^2 + t)/4s^2t^3$
11. $(st^4 + st^2 + t)/z$, $(s + s^2t + 2s^2t^3)/z$, where $z^2 = 1 + s^2t^4 + (1 + st)^2$
13. 3.5π ft^3/hr **17.** 1 **19.** 2 **21.** -1
27. $y^2 = x^2(cx - 1)$

Section 3, page 536

1. $z = 0$ **3.** $4x + 8y - z = 8$ **5.** $4x - 13y + z = -20$
7. $(-1, -1, 1)/\sqrt{3}$ **9.** $(0, -3, 1)/\sqrt{10}$ **11.** $y = \pm x$, $z = 0$

Section 4, page 538

1. lines; grad $z = (1, -2)$ **3.** parabolas; grad $z = (2x, 1)$
5. planes normal to $(1, 1, 1)$ **9.** $(1, 1, 1)$, $(2x, 8y, 18z)$, (yz, zx, xy), $(2x, 2y, -2z)$ **11.** $-2r^{-3}(\cos 3\theta, \sin 3\theta)$

Section 5, page 543

1. 1, 1, 1 **3.** 0, -3, 1 **5.** 1 **7.** 6

Section 6, page 547

1. $-2, -\frac{4}{3}$
3. $F = \text{grad}[\frac{1}{3}(x^3 + y^3 + z^3) + xyz]$, $\frac{1}{3}(a^3 + b^3 + c^3) + abc$.
5. Let θ be the polar angle. Then grad $\theta = (-y/(x^2 + y^2), x/(x^2 + y^2))$; θ is only defined in a region which excludes a curve starting at **0** and continuing out indefinitely. $\int = 2\pi$. **7.** $(n - 2)^{-1}a^{2-n}$

Section 7, page 552

1. $(1 - \sin y)/(x \cos y - 1)$ **3.** $y(e^{xy} - 3y)/x(6y - e^{xy})$

5. $(e^x \sin y + e^y \sin x)/(e^y \cos x - e^x \cos y)$ **7.** $-2x^3/9y^5$

Section 8, page 557

1. $y = \frac{1}{6}(9t - 1)$ **3.** $y = 2t - \frac{1}{2}$ **5.** $y = 0.19t + 4.85$

7. $y = -\frac{4}{5}$ **9.** $y = -0.35t + 5.7$ **11.** $y = t - \frac{1}{6}$

13. $y = 6(3 - e)t + (4e - 10) \approx 1.69t + 0.87$

15. $y = \int_a^b y(t)\, dt/(b - a)$

Section 9, page 562

1. $2, -1$ **3.** $1, -1$ **5.** $e, -e$ **7.** $e \ln 4, e^{-1} \ln 2$

9. $\sqrt{3}, -1$ **11.** $\frac{5}{4}, -1$ **13.** $\frac{1}{2}(2 + \sqrt{3}), -\frac{1}{4}(2 + \sqrt{3})$

15. max ≈ 2.81971 at $(2.029, 1.571)$, min ≈ -5.81447 at $(4.913, 4.712)$

17. max ≈ 97.3729 at $(5.000, 3.651)$, min ≈ -125.000 at $(0.00, 5.00)$

CHAPTER 26

Section 2, page 569

1. $\frac{15}{2}$ **3.** $\frac{1}{9}$ **5.** $\frac{1}{16}$ **7.** 0 **9.** 0 **11.** $\frac{8}{3}$ **13.** 0

15. 0 **17.** $\pi/2$ **19.** $\frac{9}{2} \ln 3 + 3 \ln 2 - \frac{15}{4}$

Section 3, page 574

1. $\ln \frac{4}{3}$ **3.** $\frac{4}{9}$ **5.** $\frac{1}{2}$ **7.** 0 **9.** 18

11. $(3^{n+2} - 2^{n+3} + 1)/(n + 1)(n + 2)$ **15.** $A = \iint f(x, y)\, dx\, dy$

Section 4, page 583

1. 16/3 **3.** 16/3 **5.** 46/3 **7.** 27/4 gm **9.** 6.5 gm

11. (7/15, 7/15) **13.** $(\pi - 1, (\frac{3}{8}\pi^2 - \pi + 1)/(\frac{1}{2}\pi - 1))$

15. (0, 14/5) **17.** $15\pi/8$ **19.** 15/16 **21.** 2π

23. $(4\pi/3)[a^3 - (a^2 - b^2)^{3/2}]$ **27.** $(\frac{2}{3}a, 0)$

Section 5, page 589

1. 1.1946 **3.** 4.5368; exact: $\frac{127}{28} = 4.53\dot{5}7142\dot{8}$

7. Both expressions are $A + \frac{1}{2}B + \frac{1}{2}C + \frac{1}{4}D$.

CHAPTER 27

Section 2, page 597

1. $(x-1)^2+7(x-1)+8$ **3.** $2(x+1)^3-(x+1)^2+9(x+1)$

5. $2(x+2)^4-11(x+2)^3+18(x+2)^2$

7. $5(x+1)^5-21(x+1)^4+31(x+1)^3-19(x+1)^2+5(x+1)$

9. $x^4-7x^3+5x^2+3x-6$ **11.** 62.23 **13.** -157.2

Section 3, page 605

1. $2x-\dfrac{8}{3!}x^3+\dfrac{2^5}{5!}x^5-\cdots+(-1)^{n-1}\dfrac{2^{2n-1}}{(2n-1)!}x^{2n-1}$,

$|r_{2n-1}(x)|\le\dfrac{2^{2n+1}}{(2n+1)!}|x|^{2n+1}$

3. $x+x^2+\dfrac{x^3}{2!}+\dfrac{x^4}{3!}+\cdots+\dfrac{x^n}{(n-1)!}$,

$|r_n(x)|\le\dfrac{(x+n+1)e^x}{(n+1)!}x^{n+1}$ if $x\ge0$, $|r_n(x)|\le|x|^{n+1}/n!$ if $x<0$

5. $(x-1)+3(x-1)^2/2!+2(x-1)^3/3!-2(x-1)^4/4!+$
$4(x-1)^5/5!+\cdots+(-1)^n2(x-1)^n/n(n-1)(n-2)$;
if $n\ge4$: $|r_n(x)|\le2(n-2)!(x-1)^{n+1}/(n+1)!$ for $x\ge1$,
$|r_n(x)|\le2(n-2)!(1-x)^{n+1}/(n+1)!x^{n-1}$ for $0<x\le1$

7. $x^2-x^3+\frac12x^4-\frac16x^5+\cdots+(-1)^nx^n/(n-2)!$;
$|r_n(x)|\le x^{n+1}/(n-1)!$ for $x\ge0$,
$|r_n(x)|\le[x^2-2(n+1)x+n(n+1)]|x|^{n+1}/e^x(n+1)!$ for $x\le0$

9. $x^2-x^4/3!+x^6/5!+\cdots+(-1)^{n-1}x^{2n}/(2n-1)!$;
$|r_{2n}(x)|\le[|x|+2n+2]|x|^{2n+2}/(2n+2)!$

11. $1+x-\frac12x^2-\frac16x^3+x^4/4!+x^5/5!+\cdots+\sigma_nx^n/n!$, where
$\sigma_{4k}=\sigma_{4k+1}=1$, $\sigma_{4k+2}=\sigma_{4k+3}=-1$; $|r_n(x)|\le|x|^{n+1}/(n+1)!$

13. $2x+2x^5/5!+2x^9/9!+\cdots+2x^{4n+1}/(4n+1)!$;
$|r_{4n+1}(x)|\le(1+\cosh x)|x|^{4n+3}/(4n+3)!$ **15.** $|\text{error}|\le\frac{3}{128}<0.024$

Section 4, page 611

1. $1-x^2+\frac12x^4-\frac16x^6+x^8/4!-x^{10}/5!+x^{12}/6!$

3. $1/(11\times2^{11})<5\times10^{-5}$

5. $(2/3\pi^3)[16(1-\sqrt2)x^3+\pi^2(4\sqrt2-1)x]$

7. $|\sin x - p_9(x)| < (\pi/2)^{11}/11! < 3.6 \times 10^{-6}$

9. $\sin(5\pi/8) \approx 1 - (\frac{1}{8}\pi)^2/2! + (\frac{1}{8}\pi)^4/4! - (\frac{1}{8}\pi)^6/6! \approx 0.92388$, $|\text{error}| < 2 \times 10^{-8}$

Section 5, page 613

1. $\Sigma_1^\infty (-1)^{i-1} 3^{2i-1} x^{2i-1}/(2i-1)!$ 3. $\Sigma_1^\infty (-1)^{i-1} 2^{2i-1} x^{2i}/(2i)!$

5. $\Sigma_0^\infty (-1)^i 2^i x^i/i!$ 7. $\Sigma_0^\infty 3^i (x + \frac{2}{3})^i/i!$ 9. $\Sigma_0^\infty 3^i x^i$

11. $\Sigma_0^\infty (-1)^i (x-2)^i/2^{i+1}$

Section 7, page 616

1. $|\text{error}| < \frac{2}{3}(\ln^3 2)|x(x-1)(x-2)| < 9 \times 10^{-2}$

3. $|\text{error}| \le e^2 2^{n+1}/(n+1)n^{n+1} < 6 \times 10^{-4}$ for $n = 6$

Section 8, page 621

1. $3 + 2(x-1) + \frac{3}{2}(x-1)(x-2) = \frac{1}{2}(3x^2 - 5x + 8)$

3. $-5 + 3x + \frac{3}{2}x(x-1) - \frac{4}{3}x(x-1)(x-2) = -\frac{4}{3}x^3 + \frac{11}{2}x^2 - \frac{7}{6}x - 5$

5. $4 - 5(x+1) + 4x(x+1) - \frac{7}{3}x(x+1)(x-1) = -\frac{7}{3}x^3 + 4x^2 + \frac{4}{3}x - 1$

7. $3 - 3(x+2) + \frac{3}{2}(x+2)(x+1) - \frac{5}{6}x(x+2)(x+1) + \frac{1}{2}x(x+2)(x+1)(x-1) = \frac{1}{2}x^4 + \frac{1}{6}x^3 - \frac{3}{2}x^2 - \frac{7}{6}x$

9. $1 + \frac{1}{2}(x+2) - \frac{3}{8}x(x+2) + \frac{5}{48}x(x+2)(x-2) - \frac{3}{128}x(x+2)(x-2)(x-4)$

11. $2 + x + \frac{1}{2}x(x-1) - \frac{1}{6}x(x-1)(x-2) + \frac{1}{8}x(x-1)(x-2)(x-3) - \frac{3}{40}x(x-1)(x-2)(x-3)(x-4) = -\frac{3}{40}x^5 + \frac{7}{8}x^4 - \frac{85}{24}x^3 + \frac{49}{8}x^2 - \frac{143}{60}x + 2$

CHAPTER 28

Section 1, page 626

1. $1/(4-x)$, 1 3. $1/(1+x^2)$, 1 5. e^{x+1}, ∞

7. $1/(1-125x^3)$, $\frac{1}{5}$ 9. $e^{-(x-1)^4}$, ∞ 11. $1/(1-e^x)$, $x < 0$

13. $e^{-\sin x}$, all x 15. $1/(1-2\sqrt{x})$, $0 \le x < \frac{1}{4}$

17. $1/(1-a^2-x^2)$, $|x| < (1-a^2)^{1/2}$ if $|a| < 1$, no x if $|a| \ge 1$

19. $\cos x^{1/3}$, all x

Section 2, page 630

1. 1 **3.** 1 **5.** 2 **7.** 1 **9.** $1/e$ **11.** $\frac{4}{3}$ **13.** 1
15. 5 **17.** 1 **19.** ∞

Section 3, page 637

1. $\Sigma_0^\infty 5^n x^n$ **3.** $\Sigma_0^\infty x^{3n}/n!$ **5.** $\Sigma_0^\infty x^n/(2n)!$

7. $-1 + \Sigma_2^\infty (n-1)x^n/n!$ **9.** $\Sigma_0^\infty x^{n+2}$

11. $\Sigma_1^\infty (-1)^{n+1}2^{2n-1}x^{2n}/(2n)!$

13. $1 + 2x + 7x^2 + 14x^3 + 37x^4 + 74x^5 + 175x^6$

15. $1 + x^2 + x^3 + \frac{3}{2}x^4 + \frac{13}{6}x^5 + \frac{73}{24}x^6$ **17.** $x^3 - \frac{1}{2}x^5$

19. $8!$ **21.** $8!/4! = 1680$

Section 4, page 643

1. $\Sigma_0^\infty (3/2^{n+1} - 4/3^{n+1})x^n$, $|x| < 2$

3. $\frac{1}{2}\Sigma_0^\infty (1/2^n - 1 - 1/3^{n+1})x^n$, $|x| < 1$

5. $\Sigma_0^\infty (-1)^n(x^{3n} + x^{3n+1})$, $|x| < 1$ **7.** $(4 - 3x)/(1 - x)^2$, $|x| < 1$

9. $-\frac{1}{4}\ln(1 - x^4)$, $|x| < 1$ **11.** $(2 - x)/(1 - x)^2$, $|x| < 1$

Section 5, page 649

1. $\Sigma_0^\infty \frac{1}{2}(-1)^n(n+1)(n+2)x^n$ **3.** $1 + \Sigma_1^\infty (n+1)4^n x^{2n}$

5. $1 + \frac{1}{2}x^3 + \Sigma_2^\infty (-1)^{n-1}[3 \cdot 7 \cdot 11 \cdots (4n-5)]x^{3n}/2^n \cdot n!$

7. $\sqrt{2}\,\{1 + \frac{1}{4}(x-1) + \Sigma_2^\infty (-1)^n[3 \cdot 5 \cdot 7 \cdots (2n-3)] \times$
$(x-1)^n/4^n \cdot n!\}$

9. $1 + \frac{1}{2}x^2 + \frac{1}{2}x^3 + \frac{1}{8}x^4$ **11.** $\sqrt{3}(2x + \frac{1}{3}x^2 - \frac{49}{36}x^3 - \frac{47}{216}x^4)$

13. $1 - \frac{4}{3}x - \frac{8}{9}x^2 + \frac{52}{27}x^3 - \frac{91}{81}x^4$

15. $3(1 + \frac{1}{81})^{1/4} \approx 3 + \frac{1}{108} - 2^{-5} \times 3^{-6} \approx$
$3.00000 + 0.00926 - 0.00004 \approx 3.0092$

17. $$\sum_0^\infty \frac{(2n)!}{2^{2n}(n!)^2}\frac{x^{2n+1}}{2n+1}$$

Section 6, page 653

1. The alternating series $1 - \frac{1}{2} + \frac{1}{3} - \cdots$ has $1/n \to 0$.

3. $(\frac{1}{5})^5/5! < 4 \times 10^{-6}$, hence $e^{-1/5} \approx 1 - \frac{1}{5} + \frac{1}{2}(\frac{1}{5})^2 - \frac{1}{6}(\frac{1}{5})^3 + \frac{1}{24}(\frac{1}{5})^4 \approx 0.81874$ **5.** 4, including the constant

Section 7, page 657

1. $\Sigma_0^\infty(-1)^n x^{2n+1}/(2n+1)(2n+1)!$ **3.** $\Sigma_0^\infty(-1)^n x^{4n+2}/(4n+2)$

5. 0.10003 **7.** 0.25049 **9.** $k = \frac{9}{41}$, length $= (2\pi)(41)(1 - S)$, $S \approx (\frac{1}{2})^2(\frac{9}{41})^2 + (\frac{1}{2}\cdot\frac{3}{4})^2(\frac{1}{3})(\frac{9}{41})^4 \approx 0.01216$; length ≈ 254.5

Section 8, page 660

1. $y = ce^x - x - 1$ **3.** $y = a\cos 2x + b\sin 2x$

5. $y = a_1\Sigma_1^\infty nx^n/(n!)^2$ **7.** $(2a_1/x)(e^{-x} + x - 1)$

9. $y = 2 - 3x + 6x^2 - \frac{34}{3}x^3 + \frac{61}{3}x^4 + \cdots$

11. $y = -1 + 2x - x^2 + \frac{1}{3}x^3 + \frac{1}{6}x^4 + \cdots$

13. $y = 1 + \frac{1}{2}(x-2) + \frac{1}{8}(x-2)^2 + \frac{1}{24}(x-2)^3 + \frac{1}{96}(x-2)^4 + \cdots$

CHAPTER 29

Section 1, page 671

1. $\frac{1}{8}$ **3.** 1 **5.** $\pi/4$ **7.** 2 **9.** $\frac{1}{3}\ln(3 + \sqrt{10})$

11. $\pi/4$ **13.** $2/s^3$ **15.** $1/(s - a)$ **17.** $s/(s^2 - 1)$

19. infinite **21.** infinite **23.** $\ln 2 \approx 0.69315$

25. $\ln 100 \approx 4.60517$

Section 2, page 676

1. D **3.** C **5.** D **7.** C **9.** C **11.** D

15. $V = \pi\int_1^\infty dx/x^2 = \pi$, $A \geq \int_1^\infty dx/x = \infty$. The word "paint" implies a layer of uniform thickness. **17.** $s \geq 0$ **19.** $s > \frac{1}{2}$

Section 3, page 679

1. D **3.** C **5.** D **7.** C **9.** $\Sigma_1^\infty 1/n^2 < 1 + \int_1^\infty dx/x^2 = 2$

11. about $e^{1000} \approx 1.97 \times 10^{434}$ **13.** C **15.** D

Section 4, page 684

1. C **3.** D **5.** D **7.** C **9.** C **11.** D **13.** C

15. C

CHAPTER 30

Section 2, page 689

1. $\frac{1}{2}x^2 + \frac{1}{8}x^4$ **3.** $2 + x + x^2 + \frac{1}{3}x^3 + \frac{1}{4}x^4 + \frac{1}{15}x^5$

5. $1 - \frac{1}{2}x^2 + \frac{1}{4}x^4$ **7.** $\frac{1}{2}x^2$ **9.** $1 + \frac{1}{3}x^3$

11. $-1 + x + x^2 - \frac{1}{2}x^3 - \frac{5}{24}x^4$ **13.** 0

15. $10 + \frac{1}{60}(t - 10)^3 - \frac{1}{800}(t - 10)^4$, $x(12) \approx 10.1$ ft

Section 3, page 694

1. $\frac{1}{2}x^2 + \frac{1}{8}x^4 + \frac{1}{48}x^6$ **3.** $2 + x + x^2 + \frac{1}{3}x^3 + \frac{1}{4}x^4 + \frac{1}{15}x^5 + \frac{1}{24}x^6$

5. $1 - \frac{1}{2}x^2 + \frac{1}{4}x^4 - \frac{1}{12}x^6 + \frac{1}{32}x^8 - \frac{1}{120}x^{10} + \frac{1}{384}x^{12}$

7. $\frac{1}{2}x^2 + \frac{1}{56}x^7 + \frac{1}{16 \cdot 56}x^{12} + \frac{3}{34 \cdot (56)^2}x^{17} + \frac{1}{22 \cdot (56)^3}x^{22}$

9. $\frac{1}{3}x^3 + \frac{1}{63}x^7 + \frac{2}{33 \cdot 63}x^{11} + \frac{1}{15 \cdot (63)^2}x^{15}$

Section 4, page 697

1. $n = 5$: 0.46, $n = 10$: 0.54711, $n = 100$: 0.63782

3. $n = 5$: 4.18, $n = 10$: 4.42448, $n = 100$: 4.67751

5. $n = 5$: 0.74, $n = 10$: 0.71312, $n = 100$: 0.69382

7. $n = 5$: 1.26, $n = 10$: 1.28826, $n = 100$: 1.31667

9. 0.32 **11.** 2.59374

Section 5, page 700

1. $h = 0.2$: 0.64, $h = 0.1$: 0.64788, $h = 0.01$: 0.64871

3. $h = 0.2$: 4.70, $h = 0.1$: 4.70753, $h = 0.01$: 4.70813

5. $h = 0.2$: 0.70, $h = 0.1$: 0.69332, $h = 0.01$: 0.69175

7. $h = 0.2$: 1.30, $h = 0.1$: 1.31576, $h = 0.01$: 1.31974

9. 2.075 **11.** 2.7141

Section 6, page 706

1. $h = 0.2$: 0.65, $h = 0.1$: 0.64872, $h = 0.01$: 0.64872

3. $h = 0.2$: 4.71, $h = 0.1$: 4.70813, $h = 0.01$: 4.70813

5. $h = 0.2$: 0.69, $h = 0.1$: 0.69173, $h = 0.01$: 0.69173

7. $h = 0.2$: 1.32, $h = 0.1$: 1.31978, $h = 0.01$: 1.31978

9. 2.10 **11.** 3-rd: 2.71818, 4-th: 2.71828 **13.** 1.501646

CHAPTER 31

Section 1, page 710

1. $4 \pm 3i$ **3.** $\frac{1}{2}(-1 \pm i\sqrt{7})$ **5.** $\frac{1}{3}(1 \pm 2i\sqrt{2})$

7. $\frac{1}{30}(-1 \pm 9i\sqrt{3})$ **9.** $1, \frac{1}{2}(-1 \pm i\sqrt{3})$ **11.** $1, \pm i$

13. $\pm 1, \pm i$ **15.** $(1 + i)^2 = 2i;\ \pm(1 + i)/\sqrt{2}$ **17.** 0

Section 2, page 713

1. $\frac{1}{2} + \frac{1}{3}i$ **3.** $2 - i$ **5.** $5 + i$ **7.** $-\frac{1}{2} - \frac{1}{2}i$ **9.** $\frac{3}{5} - \frac{4}{5}i$

11. [1] $\frac{1}{6}\sqrt{13}$, [3] $\sqrt{5}$, [5] $\sqrt{26}$, [7] $\frac{1}{2}\sqrt{2}$, [9] 1

Section 3, page 720

1. $\sqrt{2}(\cos \frac{3}{4}\pi + i \sin \frac{3}{4}\pi)$ **3.** $2(\cos \frac{1}{3}\pi + i \sin \frac{1}{3}\pi)$

5. $4(\cos \pi + i \sin \pi)$

7. $\sqrt{17}(\cos \theta + i \sin \theta)$, where $\tan \theta = 4,\ 0 < \theta < \frac{1}{2}\pi$ **9.** i

11. $-\sqrt{2}$ **13.** $\cos \frac{1}{2}\pi + i \sin \frac{1}{2}\pi$

15. $\frac{1}{2}\sqrt{3}(\cos 7\pi/6 + i \sin 7\pi/6)$ **17.** -4 **19.** $-64\sqrt{3} + 64i$

21. $-\frac{1}{4} - \frac{1}{4}i$ **25.** $\pm 1, \pm i, (\pm 1 \pm i)/\sqrt{2}$, all signs

27. $\pm i, \frac{1}{2}(\pm\sqrt{3} \pm i)$, all signs

29. $\cos \theta + i \sin \theta,\ \theta = \frac{1}{20}\pi \pm \frac{1}{5}\pi k,\ 0 \leq k \leq 9$

39. $\cos 4\theta = 8 \cos^4 \theta - 8 \cos^2 \theta + 1$

Section 4, page 726

1. $\frac{1}{2}(1 + i\sqrt{3}),\ e(\cos 1 - i \sin 1),\ e^{1/5}(\cos 2\pi/5 + i \sin 2\pi/5),$
$(\cosh 2 + i \sinh 2)/\sqrt{2},\ \cosh 3,\ -1$

3. Set $z = x + iy$. By **2,** $\sin z$ is real only if $\sinh y \cos x = 0$, i.e., $y = 0$ or $x = \frac{1}{2}\pi \pm k\pi$, $k = 0, 1, 2, \cdots$. Likewise $\cos z$ is real only if
$\cos x \cosh y = 0$ and $\sin x \sinh y = 0$, that is,
$\cos x = 0$ and $y = 0$, $z = \frac{1}{2}\pi \pm k\pi$.

5. no; for example, $\cos iy = \cosh y \to \infty$ as $y \to \infty$

9. $\cosh 4x = 8 \cosh^4 x - 8 \cosh^2 x + 1$

Section 5, page 731

1. $e^{x\sqrt{3}/2}(\cos \frac{1}{2}x + i \sin \frac{1}{2}x)$

3. $\frac{1}{8}e^x[\frac{1}{17}(\cos 4x + 4 \sin 4x) + \frac{4}{5}(\cos 2x + 2 \sin 2x) + 3] + c$

5. 0, if $k \neq n$; π, if $k = n$

7. $\frac{1}{36}(-\sin 3x + 3x \cos 3x + 27 \sin x - 27x \cos x) + c$

9. $\pi(2a_0^2 + a_1^2 + a_2^2 + \cdots + a_n^2)$

Section 6, page 734

1. $y = ae^x + e^{-x/2}(b \cos \frac{1}{2}x\sqrt{3} + c \sin \frac{1}{2}x\sqrt{3})$

3. $y = ae^{x\sqrt{3}} + be^{-x\sqrt{3}} + c \cos x\sqrt{2} + d \sin x\sqrt{2}$

5. $y = e^{3x}(a \cos x + b \sin x) + \frac{1}{30}(\cos 2x - 2 \sin 2x)$

7. $y = (a^2 - b^2)^{-2}[(a^2 - b^2)x \cos bx + 2b \sin bx]$ if $a \neq b$;
$y = \frac{1}{4}a^{-2}(ax^2 \sin ax + x \cos ax + a \sin ax)$ if $a = b$

9. $y = \frac{3}{50}(-\cos x - 7 \sin x)$ **11.** $\frac{1}{2}a(\cos \theta - \sin \theta)$

Section 7, page 739

1. $\Sigma_0^\infty (\frac{1}{2})^{n+1}(1 + i)^{n+1}(z - i)^n$

3. $\frac{1}{3}\Sigma_0^\infty (\frac{1}{13})^{n+1} \operatorname{Im} (-2 + 3i)^{n+1}z^n$, $R = \sqrt{13}$

5. $1 - \Sigma_0^\infty \operatorname{Re} (1 + 2i)^{n+1}z^n$, $R = \sqrt{5}/5$

Section 8, page 746

1. $e^A = I + A = \begin{pmatrix} -2 & 9 \\ -1 & 4 \end{pmatrix}$ **3.** $(I - A)^{-1} = I + A$

5. $f(A) = \begin{pmatrix} f(\lambda) & 0 \\ f'(\lambda) & f(\lambda) \end{pmatrix}$ **7.** $I + (e - 1)A = \begin{pmatrix} 1 & e - 1 \\ 0 & e \end{pmatrix}$

11. $P^{-1} = \begin{pmatrix} 1 & -2 \\ -2 & 5 \end{pmatrix}$, $e^A = \begin{pmatrix} 1 & 0 \\ 1 & 1 \end{pmatrix}$, $e^{PAP^{-1}} = \begin{pmatrix} 3 & -4 \\ 1 & -1 \end{pmatrix}$

13. $A = aI + \begin{pmatrix} 0 & b \\ -b & 0 \end{pmatrix}$, so $e^A = e^{aI}e^{\begin{pmatrix} 0 & b \\ -b & 0 \end{pmatrix}}$, etc.

15. $\cos(\alpha + \beta) = \cos \alpha \cos \beta - \sin \alpha \sin \beta$,
$\sin(\alpha + \beta) = \sin \alpha \cos \beta + \sin \beta \cos \alpha$

17. $x = x_0 \cosh kt + y_0 \sinh kt$ **19.** $\dot{\mathbf{y}} = P^{-1}\dot{\mathbf{x}} = P^{-1}A\mathbf{x} = P^{-1}AP\mathbf{y}$

CHAPTER 32

Section 1, page 750

1. $20x^3y^4$ **3.** $-2/y^3$ **5.** $-\sin(x + y)$ **7.** $-e^{x/y}(x + y)/y^3$

9. $mnx^{m-1}y^{n-1}$ **11.** $x^{y-1}(1 + y \ln x)$ **13.** $-2(x + y)/(x - y)^3$

15. $2(y - x)/(1 + xy)^3$ **17.** $(-9x^2 + 25xy - 18y^2)/(x - y)^3(x - 2y)^3$

19. $2b$ **21.** $f_{xx} = f_{yy} = g''(x + y) + h''(x - y)$

Section 2, page 755

1. $18xy^2$, $18x^2y$ **3.** $8y^3$, $24xy^2$

5. $-2y\cos(xy) + xy^2\sin(xy)$, $-2x\cos(xy) + x^2y\sin(xy)$

7. $e^{xy}(\sin x)(2y + xy^2 - x) + 2e^{xy}(\cos x)(1 + xy)$,
$x^2e^{xy}\cos x + (x^2y + 2x)e^{xy}\sin x$

9. $(x^{1/y}/x^2y^4)[(y-1)(\ln x) + y^2 - 2y]$, $(x^{1/y}/xy^5)[2y^2 + 4y(\ln x) + \ln^2 x]$

11. $8y(y^2 - 5x^2)/(x^2+y^2)^4$, $8x(x^2 - 5y^2)/(x^2+y^2)^4$

13. $f(x, y) = g(x) + xh(y) + k(y)$ **15.** cubic polynomials

17. $ax + by + c$ **19.** none

21. $$\begin{bmatrix} m(m-1)x^{m-2}y^nz^p & mnx^{m-1}y^{n-1}z^p & mpx^{m-1}y^nz^{p-1} \\ mnx^{m-1}y^{n-1}z^p & n(n-1)x^my^{n-2}z^p & npx^my^{n-1}z^{p-1} \\ mpx^{m-1}y^nz^{p-1} & npx^my^{n-1}z^{p-1} & p(p-1)x^my^nz^{p-2} \end{bmatrix}$$

23. $$\begin{bmatrix} -\sin w & -2\sin w & -3\sin w \\ -2\sin w & -4\sin w & -6\sin w \\ -3\sin w & -6\sin w & -9\sin w \end{bmatrix}, \qquad w = x + 2y + 3z$$

25. 10 **27.** $g(x, y) + h(y, z) + k(z, x)$

Section 3, page 758

1. $p_2(x, y) = 1 + 2(x-1) + 2(y-1) + (x-1)^2 + 4(x-1)(y-1) + (y-1)^2$, $p_1(x, y) = 1 + 2(x-1) + 2(y-1)$, the linear part of $p_2(x, y)$

3. $p_1(x, y) = 0$, $p_2(x, y) = xy$ **5.** $p_1(x, y) = 1$, $p_2(x, y) = 1 + (x-1)y$

7. $p_1(x, y) = p_2(x, y) = -x - (y - \frac{1}{2}\pi)$

9. $p_1(x, y) = (x - \frac{1}{2}) + 2(y - \frac{1}{4})$,
$p_2(x, y) = p_1(x, y) - \frac{1}{2}[(x - \frac{1}{2})^2 + 4(x - \frac{1}{2})(y - \frac{1}{4}) + 4(y - \frac{1}{4})^2]$

11. 1.1200; exact: $(1.1)^{1.2} \approx 1.12117$

13. 3.9993; exact: 3.99929499

Section 4, page 763

1. min **3.** neither **5.** max **7.** min **9.** neither

Section 5, page 767

1. (1, 1, 1) **3.** 1, −1

5. $P = (\Sigma_1^n x_k/n, \Sigma_1^n y_k/n)$, where $P_k = (x_k, y_k)$ **7.** $\sqrt{2}$

9. distance ≈ 1.207 at $x \approx 0.7533$, $y \approx 0.5674$, $z \approx 0.7527$

Section 6, page 770

1. min **3.** min **5.** min **7.** neither **9.** min
11. neither **13.** max 0 **15.** min $-\frac{5}{2}$ at $(-\frac{1}{2}, -1, \frac{3}{2})$
17. min **19.** max **21.** neither
23. The only solution of the system of homogeneous linear equations $f_x = 0, f_y = 0, f_z = 0$ in x, y, z is $(0, 0, 0)$.

Section 7, page 778

1. $\sqrt{13}, -\sqrt{13}$
3. None; no level curve $x + y = c$ is tangent to the hyperbola.
5. $\frac{1}{2}, -\frac{1}{2}$ **7.** square **9.** height $= \sqrt{2}$ (radius) **11.** none
13. max $2^{1-p/q}$, min 1 **15.** $\frac{2}{9}\sqrt{3}, \frac{1}{4}\sqrt{2}$

Section 8, page 784

1. $8abc/3\sqrt{3}$ **3.** $\sqrt{14}, -\sqrt{14}$ **5.** cube
7. $\frac{1}{27}$, at $(\frac{1}{3}, \frac{1}{3}, \frac{1}{3})$ **9.** max $3^{1-p/q}$, min 1
11. Maximize $xy + yz + zx$ subject to $x^2 + y^2 + z^2 = 1$.
13. max: $\frac{2}{9}(27 + \sqrt{3})$ ft^3; min: $\frac{2}{9}(27 - \sqrt{3})$ ft^3; sides: $\mu, \mu, 6 - 2\mu$, where $\mu = \frac{1}{3}(6 \mp \sqrt{3})$ ft

CHAPTER 33

Section 1, page 791

1. $(-5, 2, -14)$ **3.** $(-4, 8, -4)$ **5.** $(2, -2, 0)$ **7.** $\mathbf{0}$
9. $(-7, 8, 2)$ **11.** $(0, -10, 10)$ **13.** $(3, -1, 4)$ **15.** $\sqrt{2}$
17. $(v_2w_3 - v_3w_2)^2 + (v_3w_1 - v_1w_3)^2 + (v_1w_2 - v_2w_1)^2 = (v_1^2 + v_2^2 + v_3^2)(w_1^2 + w_2^2 + w_3^2) - (v_1w_1 + v_2w_2 + v_3w_3)^2$

Section 2, page 795

1. $A + B = C, aA = bB$
3. $\Sigma(\mathbf{x}_j - \mathbf{p}) \times \mathbf{F}_j = \Sigma\mathbf{x}_j \times \mathbf{F}_j - \mathbf{p} \times \Sigma\mathbf{F}_j = \mathbf{0} - \mathbf{0} = \mathbf{0}$
5. $8000\pi/24 \approx 1047$ mph, $8000\pi(\cos 40°)/24 \approx 802$ mph, 0

Section 3, page 798

1. $2y + \cos z$ **3.** $3(x^2 + y^2 + z^2)$ **5.** $2(x + y)$
7. $(x + y + z)/\rho$ **9.** $f_x + f_y + f_z$ **11.** $2/\rho$ **13.** 6
15. $xe^{yz}(y^2 + z^2)$ **17.** $2xy/z^3$ **19.** 0 **21.** 0 **23.** $n = -2$
25. $A + B + C = 0$ **27.** $u = x^4 - 6x^2y^2 + y^4$, etc.

Section 4, page 804

1. $\mathbf{0}$ **3.** $-2(y + z, z + x, x + y)$ **5.** $-4(z^3, x^3, y^3)$
7. $(\text{grad } f) \times \mathbf{c}$ **9.** $2(1, 1, 1)$ **11.** those for which $f = g(x)$
13. those for which $f = g(y, z)$

Section 5, page 805

1. $n(n + 1)\rho^{n-2}$ **3.** $f - g = \text{const}$
7. Yes; expand by the first row.

Section 6, page 813

1. $(-\sqrt{2}, -\sqrt{2}, -3)$, $(-1, 0, 2)$
3. Recall (p. 540) that $D_{\mathbf{u}}f = (\nabla f) \cdot \mathbf{u}$

Section 7, page 819

1. $(-\frac{1}{2}, \frac{1}{2}, -\frac{1}{2}\sqrt{2})$, $(-\frac{3}{2}\sqrt{2}, -\frac{3}{2}\sqrt{2}, 0)$, $(\frac{1}{2}\sqrt{3}, \frac{3}{2}, -1)$
3. $[\rho, \pi - \phi, \theta + \pi]$, $[3\rho, \phi, \theta]$ **5.** $\rho^{-1}\boldsymbol{\nu}$ **7.** $\rho^{-1}(\cot\phi)\boldsymbol{\lambda} - \rho^{-1}\boldsymbol{\mu}$
9. By Example 7.2, $(\rho f)'' = 0$, $\rho f = a\rho + b$.
11. $f(\rho)\boldsymbol{\lambda} = \nabla F(\rho)$, where $F' = f$
13. $f[\rho, \phi, \theta] = f(\rho\sin\phi\cos\theta, \rho\sin\phi\sin\theta, \rho\cos\phi)$

$$= \rho^n f(\sin\phi\cos\theta, \sin\phi\sin\theta, \cos\phi)$$

15. $\boldsymbol{\lambda} \cdot d\boldsymbol{\lambda} = 0$, $(d\boldsymbol{\lambda}) \cdot \boldsymbol{\mu} + \boldsymbol{\lambda} \cdot d\boldsymbol{\mu} = 0$, etc.
17. $d\mathbf{x} = \mathbf{u}\, dr + \mathbf{w}\, r\, d\theta + \mathbf{k}\, dz$

CHAPTER 34

Section 1, page 830

1. $e - 1$ **3.** 1 **5.** $\frac{1}{30}$ **7.** $\frac{1}{70}$ **9.** $\frac{2}{3}\pi - \frac{1}{2}\sqrt{3}$
11. $\frac{1}{3}\pi - \frac{1}{4}\sqrt{3}$ **13.** $\frac{2}{3}$ **15.** $\frac{1}{3}$ **17.** $\frac{5}{6}\sqrt{5}$
19. Fig. 1.16: $0 \le x \le 1$, $0 \le y \le e^x$; Fig. 1.17: $0 \le x \le 1$, $x^3 \le y \le x^2$;

Fig. 1.18: $-\frac{1}{2}\sqrt{3} \le x \le \frac{1}{2}\sqrt{3}$, $1-(1-x^2)^{1/2} \le y \le (1-x^2)^{1/2}$; Fig. 1.19: $0 \le y \le 1$, $-y^2 \le x \le y^2$; Fig. 1.20: $\frac{1}{2}(1-\sqrt{5}) \le x \le \frac{1}{2}(1+\sqrt{5})$, $-x \le y \le 1-x^2$ **21.** $\frac{1}{168}$ **23.** $\frac{32}{105}$ **25.** 104

Section 2, page 837

1. $3\pi/640$ **3.** $\frac{20}{27}$ **5.** $\pi/8, \pi/24, \pi/8$
7. 0, 0, 0 by odd symmetry **9.** $\pi/(p-1)$ **11.** $\sqrt{\pi}/2\sqrt{t}$
13. $\frac{16}{3}$

Section 3, page 844

1. $\frac{59}{60}$ **3.** $\frac{1}{15}$ **5.** $\frac{1}{315}$ **7.** $ka^9/216$ **9.** $\frac{16}{5}$

Section 4, page 849

7. $\displaystyle\int_{-\sqrt{3}/2}^{\sqrt{3}/2}\left(\int_{1-\sqrt{1-x^2}}^{\sqrt{1-x^2}} f\,dy\right)dx$ **9.** $\displaystyle\int_0^1\left(\int_{\arcsin x}^{\pi-\arcsin x} f\,dy\right)dx$

11. tetrahedron with vertices at (0, 0, 0), (0, 0, 1), (0, 1, 1), (1, 1, 1);

$$\int_0^1\left[\int_0^z\left(\int_0^y \rho\,dx\right)dy\right]dz,\quad \int_0^1\left[\int_x^1\left(\int_y^1 \rho\,dz\right)dy\right]dx,\quad \int_0^1\left[\int_y^1\left(\int_0^y \rho\,dx\right)dz\right]dy$$

13. tetrahedron with vertices (0, 0, 0), (0, 0, 2), (0, 3, 2), (6, 3, 2);

$$\int_0^2\left[\int_0^{3z/2}\left(\int_0^{2y} \rho\,dx\right)dy\right]dz,\quad \int_0^6\left[\int_{x/2}^3\left(\int_{2y/3}^2 \rho\,dz\right)dy\right]dx,\quad \int_0^3\left[\int_{2y/3}^2\left(\int_0^{2y} \rho\,dx\right)dz\right]dy$$

15. $\displaystyle\int_4^5\left[\int_{10-x}^6\left(\int_{13-x-y}^3 x\,dz\right)dy\right]dx = \frac{19}{24}$

17. $\displaystyle y(x) = \frac{1}{2}\int_0^x (x-t)^2 g(t)\,dt$

Section 5, page 859

1. Section by planes parallel to the plane of the two axes; θ is the central angle measured from the hole axis in a plane perpendicular to the cylinder axis. **3.** $2a^2 s \arcsin(s/2a) + \frac{1}{2}s^2\sqrt{4a^2-s^2}$

5. $\displaystyle V = \frac{s^2}{2b}\int_c^{c+s}\left[\sqrt{x^2-b^2} + \frac{x^2}{b}\,\text{arc csc}\left(\frac{x}{b}\right)\right]dx,$ where $b = s/2\tan\alpha$

9. $2\int_0^{2\pi}\left(\int_0^{1/2} r(\frac{3}{4} - r\cos\theta - r^2)^{1/2}\,dr\right)d\theta$

11. V is the volume common to $x^2 + y^2 \le 4a^2$ and $(x - a)^2 + z^2 \le a^2$, that is, $z^2 \le 2ax - x^2$. Cut into rectangular slabs perpendicular to the x-axis:

$$V = 4\int_0^{2a} (\sqrt{4a^2 - x^2})\,(\sqrt{2ax - x^2})\,dx$$

$$= 4\int_0^{2a} (2a - x)\sqrt{x(2a + x)}\,dx = [\tfrac{44}{3}\sqrt{2} - 6\ln(3 + \sqrt{8})]a^3$$

13. $2\pi^2Aa^2$. Slice into washers by planes z = constant. Note that $y_1 + y_0 = 2A$.

15. $(\ln a) - \frac{1}{2}$ **17.** $6/[(n + 1)(n + 2)(n + 3)]$ **19.** $2\pi ah$

CHAPTER 35

Section 1, page 870

1. $(\frac{3}{8}a, \frac{3}{8}a, \frac{3}{8}a)$ **3.** $(0, 0, \frac{3}{8}a(1 + \cos\alpha))$

5. For the wedge $0 \le r \le a$ and $-\alpha \le \theta \le \alpha$, $\bar{\mathbf{x}} = (\frac{2}{3}(\sin\alpha)\alpha^{-1}a, 0)$.

7. For the wire $r = a$ and $0 \le \theta \le \pi/2$, $\bar{\mathbf{x}} = (2a/\pi, 2a/\pi)$.

9. Let the wire be $\mathbf{x}(\theta) = 100(\cos\theta, \sin\theta)$, where $0 \le \theta \le \pi$ and $\delta(\theta) = 0.01 + 0.24\theta/\pi$. Then $\bar{x} = -4800/13\pi^2 \approx -37.4$ cm, $\bar{y} = 200/\pi \approx 63.7$ cm

11. $V = 2\pi^2Aa^2$ for a circle of radius a revolved about an axis in its plane at distance A from the center of the circle. (See Ex. **12**, p. 860)

13. Data is in Ex. **11**, area $= 4\pi^2Aa$. **15.** $(0, 0, \frac{1}{2}a(1 + \cos\alpha))$

19. $\frac{1}{4}(\mathbf{a} + \mathbf{b} + \mathbf{c} + \mathbf{d})$

Section 2, page 881

1. $I_{xx} = I_{yy} = I_{zz} = \frac{2}{3}Ma^2$ **3.** products: 0, $I_{xx} = I_{yy} = I_{zz} = \frac{2}{5}Ma^2$

5. products: 0, $I_{xx} = \frac{1}{3}M(b^2 + c^2)$, etc. **7.** $\frac{1}{3}M(3a^2 + 3aL + L^2)$

9. $\frac{1}{2}M(2A^2 + 3a^2)$ **11.** $I_{zz} = \frac{1}{3}Mha$ **13.** $\frac{1}{6}Mh(a + b)$

15. $\frac{2}{5}Ma^2(10 - \pi)/(6 - \pi)$

Section 3, page 887

1. $\int_0^{2\pi}\left(\int_0^{\pi} a^2\sin\phi\,d\phi\right)d\theta = 4\pi a^2$ **3.** $\int_0^{2\pi}\left(\int_0^{a}\frac{L}{a}r\,dr\right)d\theta = \pi aL$

5. $\displaystyle\int_0^{2\pi}\left(\int_0^{\pi}[c^2\sin^2\phi\,(b^2\cos^2\theta + a^2\sin^2\theta) + a^2b^2\cos^2\phi]^{1/2}\sin\phi\,d\phi\right)d\theta$

7. $\displaystyle 4\int_0^1\left(\int_0^1(1 + x^2 + y^2)^{1/2}\,dx\right)dy$

9. $\displaystyle\int_0^{2\pi}\left(\int_0^1 r\sqrt{1 + 4r^2}\,dr\right)d\theta = \tfrac{1}{6}\pi(5\sqrt{5} - 1)$

11. $\tfrac{1}{2}(a^2b^2 + b^2c^2 + c^2a^2)^{1/2}$

13. $\mathbf{x}(s, u) = \mathbf{x}(s) + u\mathbf{T}(s)$, $0 \le u \le 1$, $0 \le s \le L$; $\mathbf{x}_s = \mathbf{T} + uk\mathbf{N}$,
$\mathbf{x}_u = \mathbf{T}$, $dA = |\mathbf{x}_s \times \mathbf{x}_u|\,ds\,du = uk\,ds\,du$,

$$A = \int_0^1 u\left(\int_0^L k\,ds\right)du = \tfrac{1}{2}\int_0^L k\,ds$$

Index

N

O

P

Q

R

S

10. $\int \sin^2 ax\, dx = \frac{x}{2} - \frac{\sin 2ax}{4a}$

11. $\int \cos^2 ax\, dx = \frac{x}{2} + \frac{\sin 2ax}{4a}$

12. $\int \sin^n ax\, dx = -\frac{\sin^{n-1} ax \cos ax}{na} + \frac{n-1}{n} \int \sin^{n-2} ax\, dx$

13. $\int \cos^n ax\, dx = \frac{\cos^{n-1} ax \sin ax}{na} + \frac{n-1}{n} \int \cos^{n-2} ax\, dx$

14. $\int \sin ax \cos bx\, dx = -\frac{\cos(a+b)x}{2(a+b)} - \frac{\cos(a-b)x}{2(a-b)} \qquad (a \neq \pm b)$

15. $\int \sin ax \sin bx\, dx = \frac{\sin(a-b)x}{2(a-b)} - \frac{\sin(a+b)x}{2(a+b)} \qquad (a \neq \pm b)$

16. $\int \cos ax \cos bx\, dx = \frac{\sin(a-b)x}{2(a-b)} + \frac{\sin(a+b)x}{2(a+b)} \qquad (a \neq \pm b)$

17. $\int x \sin ax\, dx = \frac{\sin ax}{a^2} - \frac{x \cos ax}{a}$

18. $\int x \cos ax\, dx = \frac{\cos ax}{a^2} + \frac{x \sin ax}{a}$

19. $\int x^n \sin ax\, dx = -\frac{x^n \cos ax}{a} + \frac{n}{a} \int x^{n-1} \cos ax\, dx$

20. $\int x^n \cos ax\, dx = \frac{x^n \sin ax}{a} - \frac{n}{a} \int x^{n-1} \sin ax\, dx$

21. $\int \tan ax\, dx = -\frac{1}{a} \ln |\cos ax|$

22. $\int \cot ax\, dx = \frac{1}{a} \ln |\sin ax|$

23. $\int \tan^2 ax\, dx = \frac{1}{a} \tan ax - x$

24. $\int \cot^2 ax\, dx = -\frac{1}{a} \cot ax - x$

25. $\int \tan^n ax\, dx = \frac{\tan^{n-1} ax}{(n-1)a} - \int \tan^{n-2} ax\, dx \qquad (n > 1)$

26. $\int \cot^n ax\, dx = -\frac{\cot^{n-1} ax}{(n-1)a} - \int \cot^{n-2} ax\, dx \qquad (n > 1)$